W9-DFP-379

# Trissel's™ Stability of Compounded Formulations

**5th edition**

# Trissel's™ Stability of Compounded Formulations

5th edition

## Lawrence A. Trissel, FASHP
Research Consultant
TriPharma Research
St. Augustine, Florida

**American Pharmacists Association®**
Improving medication use. Advancing patient care.
APhA
WASHINGTON, D.C.

Managing Editor: Vicki Meade, Meade Communications
Acquiring Editor: Julian Graubart
Copy Editors: Amy Morgante, Vicki Meade
Proofreader: Betty Bruner
Indexer: Mary Coe
Cover Designer: Mariam Safi, APhA Creative Services
Layout: Circle Graphics

© 2012 by the American Pharmacists Association
APhA was founded in 1852 as the American Pharmaceutical Association.

Published by the American Pharmacists Association
2215 Constitution Avenue, N.W.
Washington, DC 20037-2985
www.pharmacist.com
www.pharmacylibrary.com

To comment on this book via email, send your message to the publisher at aphabooks@aphanet.org

**LIBRARY OF CONGRESS CATALOGING-IN-PUBLICATION DATA**

Trissel, Lawrence A.
  Trissel's stability of compounded formulations / Lawrence A. Trissel. — 5th ed.
    p. ; cm.
  Stability of compounded formulations
  Includes bibliographical references and index.
  ISBN 978-1-58212-167-3 (alk. paper)
  I. American Pharmacists Association. II. Title. III. Title: Stability of compounded formulations.
  [DNLM: 1. Drug Stability—Handbooks. 2. Drug Compounding—Handbooks. QV 735]

  615.1'9—dc23
                    2012019045

**HOW TO ORDER THIS BOOK**
Online: www.pharmacist.com/shop_apha
By phone: 800-878-0729 (from the United States and Canada)
VISA®, MasterCard®, and American Express® cards accepted

*To those who recognize the value of clinical pharmaceutics research*

*And to Pam, for everything you do.*

# Contents

Preface          ix

Monographs          1

Appendix I          521

Appendix II          523

References          528

Index          571

# Preface

Compounding medicines is a burgeoning health care activity that involves much of the health care community, including physicians, pharmacists, nurses, and many others in one way or another. Furthermore, compounded medications are used by millions of patients each year.

Compounding medications for individual patients is an essential aspect of modern health care that has roots stretching back to antiquity. Indeed, the traditional mortar and pestle, tools going back many centuries, remain symbols of the profession of pharmacy to this day. Only in the 20th century did the industrial revolution and the large-scale commercial manufacturing of medicines replace the compounder's expertise for most patients' medication needs. While today most medications are manufactured by large-scale commercial concerns, the patient's need for specialized and individualized medications remains, and the public and other professions continue to expect pharmacists to be able to compound medicines. Furthermore, the need is sufficiently great that health care personnel beyond pharmacists are involved in compounding medications, including physicians, physician's assistants, nurses, physical therapists, inhalation therapists, and others.

The clinical need for individual patient prescription compounding continues and is growing. Many patients have medication needs that simply are not met by commercially manufactured products. Pediatric and geriatric patients, for whom mass-produced medicines often are inappropriate, unusable, or unavailable, particularly need these services. Patients with allergies to dyes, preservatives, or other components of manufactured drugs have a persistent need for medication compounding. Furthermore, compounded parenteral medications are the norm for patients in hospitals, ambulatory patients, and even home care patients. In addition, removing therapeutically useful commercially manufactured drug products from the market because they generate insufficient financial revenue for large-scale drug manufacturers is a persistent and all too common occurrence. When a useful drug product is removed from the market for financial reasons, it can leave patients' clinical needs unmet. The ability to compound custom-tailored medications remains an essential component in delivering an appropriate level of pharmaceutical care. And, of course, the repeated and continuing shortages of commercially manufactured drugs add to the ongoing critical need for medication compounding.

A whole new generation of cutting-edge research medications, innovative biologicals, and genetically custom-tailored medications may place new demands on the compounding expertise of health care providers to deliver treatments to patients. The value of this function, although frequently overlooked or discounted by some, continues to grow in importance to patient care.

Adequate reference materials to aid in the compounding process are important for serving patient needs and fulfilling our professional obligations. Resource materials for the stability and compatibility of compounded sterile preparations are much more abundant than they are for nonparenteral medications. Over the years, a few reference works aimed at nonparenteral compounded medications, largely on the order of recipe books, have been produced. Such works can be useful and certainly have their place. Other works have accumulated information on a limited set of preparations used at single institutions. These too can be useful but are not comprehensive. My intention in *Trissel's™ Stability of Compounded Formulations* is to provide a more comprehensive resource of compounded drug stability that summarizes the body of published information.

This fifth edition of *Trissel's™ Stability of Compounded Formulations* provides summaries of 1714 published articles, including 481 new to this edition, on the stability of various compounded preparations, including orals, enterals, ophthalmics, topicals, and other specialized preparations. The book comprises 510 monographs on various drug entities, including 77 new monographs, arranged alphabetically by nonproprietary name.

All information in *Trissel's™ Stability of Compounded Formulations* is fully referenced to the original published sources. This helps the reader obtain and review the original work, if desired. The information in this book consists of summaries of each original work, focusing on those pieces of practical information essential to the stability of compounded medications. Each monograph is written in a standardized framework of subheadings so the presentation is consistent. The content is focused principally, but not exclusively, on nonparenteral dosage forms. Compounding personnel are encouraged also to utilize the resource books and electronic databases that summarize the more voluminous information on sterile drug stability and compatibility.

The first section, Properties, is devoted to the pharmaceutical characteristics of the drug, including physical characteristics, solubilities, pH, pK$_a$, and osmolality information, where available. Following this section is a discussion of General Stability Considerations of the drug and its formulations, including recommendations for packaging and storage conditions. Stability Reports of Compounded Preparations summarizes information from published articles on the subject. The summaries are organized by categories of preparations such as orals, ophthalmics, topicals, and compatibility with enteral feeds, and they include the formulas and compounding procedures used in the studies. If information on the compatibility of compounded preparations with other drugs is available, it is included in the next section, Compatibility with Other Drugs. Unfortunately, this kind of information is too often unavailable for nonparenteral compounded preparations.

Given the volume and diversity of therapeutic entities that are compounded, occasionally stability information from a published research report is unavailable for a specific compounded medication. For these cases, the *United States Pharmacopeia* (USP) provides general guidance on assigning beyond-use dating when no specific stability information is available. The USP guidance is summarized in Appendices I and II. The guidance, of necessity, is very general, and should be used only when relevant specific stability information is unavailable. The guidance should also be used in conjunction with professional judgment and common sense. There are examples of medications that are less stable than the USP guidance permits. To protect the patient, it is incumbent on the compounding professional to interpret the available information in a conservative manner to ensure the pharmaceutical integrity of compounded dosage forms.

Readers of *Trissel's™ Stability of Compounded Formulations* should recognize that discussing the proper therapeutic use of the preparations described in this book is beyond the scope of this work. Appropriate clinical use, including indication, adverse effects, dosage, etc., must be considered for each patient's clinical situation individually. Furthermore, the art and science of pharmaceutical compounding, for both sterile and nonsterile preparations, requires highly specialized knowledge and a set of skills too often neglected in pharmacy education. Consequently, pharmacists called upon to compound medications for their patients must ensure that they have adequate specialized knowledge, training, and experience to safely prepare the prescribed medication. Such knowledge and training may be obtained from postgraduate courses,

from specialized journals such as the *International Journal of Pharmaceutical Compounding*, and from other sources, such as *The Art, Science, and Technology of Pharmaceutical Compounding* by Loyd V. Allen, Jr., PhD, which is a highly recommended reference book for compounding pharmacists.

I would like to extend my appreciation to some individuals whose help and encouragement have made five editions of *Trissel's™ Stability of Compounded Formulations* a reality. First, my very great thanks go to the American Pharmacists Association, Julian Graubart, Senior Director, Books and Electronic Products, L. Luan Corrigan, managing editor for four editions, and Vicki Meade, managing editor for this fifth edition, for turning this ever-expanding manuscript into a book. Their assistance and effort in bringing this work into being are appreciated, and they are a pleasure to work with. I would also like to acknowledge Dr. David Newton and Dr. Loyd Allen, Jr., who have helped me expand my understanding of this complex topic area.

And finally, my deepest thanks again go to my wife, Pam, who has had to put up with me spending a seemingly endless procession of evenings and nights, early mornings, weekends, holidays, and vacations with papers, proofs, and publishing deadlines for nearly 40 years. Thank you, Pam, for your forbearance, tolerance, and loving support that made those efforts possible.

## Final Words

I have spent many decades compiling, writing, revising, and proofing various research articles, books, electronic databases, and other works. It has been a calling and a true labor of love for me to provide informational resources that benefit the members of our profession and the patients they serve. I think of it as the contribution I am here to perform. In *Ecclesiastes 9:10* it is written, "Whatsoever thy hand findeth to do, do it with thy might." I have tried to conduct all my professional undertakings in a manner consistent with this statement.

Though much sacrifice of my time and other life goals was required, I have been willing and eager to continue all these years in the knowledge that my efforts were providing a useful and valuable informational resource in patient care. For the many colleagues, friends, and others who have expressed their gratitude for my efforts, it is I who am grateful. I am grateful for those kind words that have encouraged me throughout the late nights and early mornings, weekends, holidays, and vacations, all seemingly countless in number. And to the members of the profession of pharmacy, especially those in the trenches of patient care who have found my work useful, I am glad I could help. Thank you for this opportunity to serve.

However, the end of manuscripts, papers, and proofs is now at hand. The time has come to pass on this cherished work to others. While I will not be a part of the future, I will always take pride in my contributions of the past. So thank you again my colleagues and friends. It has been a pleasure.

*LAT*
December 2011

# Aceclofenac

## Properties

Aceclofenac occurs as a white crystalline material.[1]

### Solubility

Aceclofenac is poorly water soluble, having an aqueous solubility of 0.18 mg/mL.[1653]

### pH

An extemporaneously prepared injection was found to have a pH of 8 to 8.5 after reconstitution due to the presence of excipients in the formulation.[1653]

## General Stability Considerations

Aceclofenac oral tablets are to be stored at controlled room temperature and kept in the original packages.[7]

## Stability Reports of Compounded Preparations

### Injection

Injections, like other sterile drugs, should be prepared in a suitable clean air environment using appropriate aseptic procedures. When prepared from nonsterile components, an appropriate and effective sterilization method must be employed.

Maheshwari and Indurkhya[1653] developed and evaluated an extemporaneously prepared lyophilized injection of aceclofenac from powder. The authors used mixed hydrotropic agents (urea and sodium citrate) as excipients in the formulation to overcome the poor aqueous solubility of aceclofenac. The presence of the two hydrotropic agents increased the aqueous solubility of aceclofenac by about 250-fold.

The aceclofenac injection was prepared by dissolving the proper amount of drug powder in an aqueous solution containing urea 20% and sodium citrate 10% in an amber volumetric flask to yield an aceclofenac concentration of 100 mg/2.5 mL. The flask head space was then flushed with nitrogen for 15 minutes. This initial solution was then diluted with an equal volume of water for injection. The diluted solution was filtered through 0.22-µm filters (Millipore) into a sterilized glass bottle and again flushed with nitrogen. The diluted and sterile filtered solution was then filled as 5-mL of solution containing 100 mg of aceclofenac into 15-mL washed and sterilized amber glass vials fitted with slotted lyophilization rubber vial stoppers. The vials were then lyophilized, resulting in vials containing a lyophilized cake of aceclofenac 100 mg for injection.

The finished vials were reconstituted with 2.5 mL of sterile water for injection, resulting in a clear, colorless, liquid injection containing aceclofenac 40 mg/mL having a pH of 8 to 8.5. No change in color, clarity, or pH occurred during 30 days of storage under refrigeration, at room temperature, and at 40°C and 75% relative humidity. Stability-indicating HPLC analysis found little loss of aceclofenac occurred during 30 days of storage at these same temperature and humidity conditions.

The authors also evaluated dilution of the reconstituted aceclofenac injection formulation in dextrose 5% and sodium chloride 0.9%, infusion solutions that are likely to be considered for intravenous administration of the drug. Dilutions ranged from 1:1 to 1:500, drug to solution. In dextrose 5% no precipitation occurred at any of the dilution concentrations for eight hours at room temperature. However, after 24 hours, precipitation was observed in the 1:20 to 1:500 dilutions; in the 1:500 dilution the precipitate redissolved. In sodium chloride 0.9% no precipitation occurred at any of the dilution concentrations for eight hours at room temperature as well. However, after 24 hours, precipitation was observed in the 1:30 to 1:500 dilutions; in the 1:500 dilution the precipitate again redissolved.

The authors concluded that the compounded injection was sufficiently compatible and stable that it could be administered by intravenous infusion. However, it should be noted that no toxicity or safety information on the proposed injection was presented. Use of an injection without such toxicity and safety information cannot be recommended. ▪

# Acetaminophen
## (Paracetamol)

## Properties

Acetaminophen is a white crystalline powder with a bitter taste.[1–3,6]

### Solubility

Acetaminophen has an aqueous solubility of about 14 mg/mL, increasing to about 50 mg/mL in boiling water and to about 100 to 142 mg/mL in ethanol.[1–3,6]

### pH

Acetaminophen oral solution has a pH between 3.8 and 6.1; the oral suspension has a pH between 4 and 6.9.[4] Acetaminophen liquid (Tylenol, McNeil) is reported to have a pH of 4.7.[19] A saturated solution of acetaminophen has a pH of about 6.[6]

### $pK_a$

Acetaminophen has a $pK_a$ of 9.51.[6]

### Osmolality

Acetaminophen oral solution 65 mg/mL (Roxane) had an osmolality of 5400 mOsm/kg. Acetaminophen with codeine elixir (Wyeth) had an osmolality of 4700 mOsm/kg.[233,905]

## General Stability Considerations

Acetaminophen oral dosage forms should be stored in tight containers, and suppositories should be stored in well-closed containers at controlled room temperature.[4,6] Acetaminophen is very stable in aqueous solution, exhibiting maximum stability at pH 5 to 7 with a half-life at pH 6 of almost 22 years.[6]

## Stability Reports of Compounded Preparations
### Enteral

#### ACETAMINOPHEN COMPATIBILITY SUMMARY

*Compatible with:* Enrich • Ensure • Ensure HN • Ensure Plus • Ensure Plus HN • Osmolite • Osmolite HN • Precitene • TwoCal HN • Vital • Vivonex T.E.N.

**STUDY 1:** Cutie et al.[19] added 10 mL of acetaminophen elixir (Tylenol, McNeil) to varying amounts (15 to 240 mL) of Ensure, Ensure Plus, and Osmolite (Ross Laboratories) with vigorous agitation to ensure thorough mixing. The acetaminophen elixir was physically compatible, distributing uniformly in all three enteral products with no phase separation or granulation.

**STUDY 2:** Burns et al.[739] reported the physical compatibility of acetaminophen elixir (Children's Tylenol) 15 mL with 10 mL of three enteral formulas, including Enrich, TwoCal HN, and Vivonex T.E.N. Visual inspection found no physical incompatibility with any of the enteral formulas.

**STUDY 3:** Altman and Cutie[850] reported the physical compatibility of acetaminophen elixir (Tylenol, McNeil) 10 mL with varying amounts (15 to 240 mL) of Ensure HN, Ensure Plus HN, Osmolite HN, and Vital after vigorous agitation to ensure thorough mixing. The acetaminophen elixir was physically compatible, distributing uniformly in all four enteral products with no phase separation or granulation.

**STUDY 4:** Ortega de la Cruz et al.[1101] reported the physical compatibility of an unspecified amount of oral liquid acetaminophen (Apiretal, ERN) with 200 mL of Precitene (Novartis) enteral nutrition diet for the 24-hour observation period. No particle growth or phase separation was observed.

### Oral

**STUDY 1 (ORAL LIQUID):** Johnson et al.[867] evaluated the stability of two oral suspension formulations providing tramadol hydrochloride 7.5 mg/mL and acetaminophen 65 mg/mL prepared from commercial Ultracet tablets (Ortho-McNeil). Ultracet tablets were crushed using a mortar and pestle. The tablet powder was suspended in an equal parts mixture of Ora-Plus and strawberry syrup and also an equal parts mixture of Ora-Plus and Ora-Sweet SF for a sugar-free preparation. The suspensions were packaged in amber plastic prescription bottles and stored refrigerated at 3 to 5°C and at room temperature of 23 to 25°C for 90 days.

No change in color, odor, or taste occurred. Stability-indicating HPLC analysis found no loss of tramadol hydrochloride or acetaminophen occurred in either formulation in 90 days at either room temperature or under refrigeration.

**STUDY 2 (TABLET REPACKAGING):** Haywood et al.[1595] evaluated the stability of commercial acetaminophen tablets (Panamax, Sanofi-Synthelabo) repackaged in dose administration aids (Webster-pak) over eight weeks stored at 24 to 26°C and 58.5 to 61.5% relative humidity both protected from light and also exposed to tungsten, fluorescent, and indirect daylight. The tablets were also tested in simulated home use. In all cases, the tablets were physically stable; weight uniformity, friability, hardness, disintegration, and dissolution were all satisfactory. In tablets under all test conditions, the acetaminophen content at all time points remained within the British Pharmacopeial range of 95 to 105%. The authors concluded that acetaminophen tablets repackaged in dose administration aids were chemically and physically stable for at least eight weeks.

### Rectal

Allen[1274] reported on a compounded formulation of acetaminophen 10 mg/mL for use as a rectal solution. The solution had the following formula:

| | | |
|---|---|---|
| Acetaminophen, micronized | | 1 g |
| Pluronic P-105 | | 44 g |
| Propylene glycol | | 52 mL |
| Purified water | qs | 100 mL |

The recommended method of preparation was to thoroughly mix the acetaminophen powder and propylene glycol and then add the Pluronic P-105. Finally, purified water sufficient to bring the volume to 100 mL was to be added. The solution was to be packaged in tight, light-resistant containers. The author recommended a beyond-use date of six months at room temperature because this formula is a commercial medication in some countries with an expiration date of two years or more. ▣

# ▣ Acetazolamide
# Acetazolamide Sodium

## Properties

Acetazolamide is a white to yellowish white odorless crystalline powder. Acetazolamide sodium is available as a white lyophilized powder. Acetazolamide sodium 275 mg is equivalent to about 250 mg of acetazolamide.[1–3]

### Solubility

The solubility of acetazolamide in water is 0.7 mg/mL and in buffered aqueous solution is around 0.8 to 1.2 mg/mL over the range of pH 1.7 and 6.9. At alkaline pH the solubility increases, but the rate of decomposition by hydrolysis also increases. The solubility of acetazolamide in organic solvents is greatest in polyethylene glycol 400 at 87.8 mg/mL. Solubilities in propylene glycol, ethanol, and glycerin are 7.4, 3.9, and 3.7 mg/mL, respectively.[180,181]

Acetazolamide sodium is freely soluble in water.[2]

### pH

A 10% aqueous solution of acetazolamide sodium in water has a pH between 9 and 10. Reconstituted acetazolamide sodium injection has a pH around 9.2.[2,4,7]

### pK$_a$

Acetazolamide has pK$_a$ values of 7.2 and 9.1.[1,2]

## General Stability Considerations

Acetazolamide tablets should be stored in well-closed containers at controlled room temperature.[4] Reconstituted acetazolamide sodium injection is stated to be chemically stable for one week under refrigeration. Because the injection contains no preservative, use within 24 hours is recommended. In older package inserts, the manufacturer stated that the reconstituted injection was chemically stable for two weeks at room temperature and four weeks refrigerated.[2,8]

The aqueous stability of acetazolamide sodium decreases at pH values above 9. A 0.25-mg/mL solution at pH 8.8 at room temperature lost 4% in three days, but at pH 10.8 and 12.7 it lost 12 and 17%, respectively, after four days. The optimum pH for stability has been cited as 4 to 5.[181,272,543,1126]

Freezing diluted solutions of acetazolamide sodium 0.375 mg/mL in 5% dextrose injection or 0.9% sodium chloride injection at −10°C resulted in a loss of less than 3% in 44 days.[273]

## Stability Reports of Compounded Preparations
### Oral
**USP OFFICIAL FORMULATION:**

Acetazolamide 2.5 g
Vehicle for Oral Solution, NF (sugar-containing or sugar-free)
Vehicle for Oral Suspension, NF (1:1) qs 100 mL

(See the vehicle monographs for information on the individual vehicles.)

Use acetazolamide powder or commercial tablets. If using tablets, crush or grind to fine powder. Add 20 mL of the vehicle mixture, and mix to make a uniform paste. Add additional vehicle almost to volume in increments with thorough mixing. Quantitatively transfer to a suitable calibrated tight and light-resistant bottle, bring to final volume with the vehicle mixture, and thoroughly mix, yielding acetazolamide 25 mg/mL oral suspension. The final liquid preparation should have a pH between 4 and 5. Store the preparation at controlled room temperature or under refrigeration between 2 and 8°C. The beyond-use date is 60 days from the date of compounding at either storage temperature.[4,5]

**STUDY 1:** Alexander et al.[108] reported on the stability of an extemporaneous acetazolamide suspension. Acetazolamide suspension was prepared by triturating 300 acetazolamide 250-mg tablets (Diamox, Lederle) and levigating with about 100 mL of sorbitol 10%. An additional 200 mL of sorbitol 70% was then added. Carboxymethylcellulose 15 g was mixed in 500 mL of purified water. Aluminum magnesium silicate (Veegum, Vanderbilt Chemical) also was added to a separate 500 mL of purified water. All of these were then combined in a container and mixed until a homogeneous mixture was obtained. Syrup, USP, 600 mL was added and mixed thoroughly with a Dispersator. Finally, FD&C red no. 40, 3 mL of strawberry flavor, 75 mL of glycerin, and 60 mL of methylparaben 2.5% with propylparaben 1% solution were added. A sufficient quantity of purified water was then incorporated to bring the mixture to 3000 mL. The suspension had a theoretical acetazolamide concentration of 25 mg/mL. The pH was adjusted to 5 with 36% hydrochloric acid.

The finished suspension was packaged in eight-ounce amber glass bottles with child-resistant closures. The bottles were stored at 5, 22, and 30°C protected from light, as well as at elevated temperatures, for 79 days. The bottles were shaken for 10 minutes prior to sampling to ensure complete dispersion. The acetazolamide content was assessed using a stability-indicating HPLC assay. Losses of 2, 3, and 5% occurred at 5, 22, and 30°C, respectively, after 79 days of storage. Losses were greater at elevated temperatures. All samples were homogeneous, showed no evidence of caking, and had very good redispersibility. The authors calculated the shelf life of the acetazolamide 25-mg/mL suspension to be 371 days when stored at 22°C and approximately four years when stored at 4°C.

**STUDY 2:** Allen and Erickson[543] evaluated the stability of three acetazolamide 25-mg/mL oral suspensions extemporaneously compounded from tablets. Vehicles used in this study were (1) an equal parts mixture of Ora-Sweet and Ora-Plus (Paddock), (2) an equal parts mixture of Ora-Sweet SF and Ora-Plus (Paddock), and (3) cherry syrup (Robinson Laboratories) mixed 1:4 with simple syrup. Twelve acetazolamide 250-mg tablets (Mutual Pharmaceutical) were crushed and comminuted to fine powder using a mortar and pestle. About 20 mL of the test vehicle was added to the powder and mixed to yield a uniform paste. Additional vehicle was added geometrically and brought to the final volume of 120 mL, mixing thoroughly after each addition. The process was repeated for each of the three test suspension vehicles. Samples of each of the finished suspensions were packaged in 120-mL amber polyethylene terephthalate plastic prescription bottles and stored at 5 and 25°C in the dark.

No visual changes or changes in odor were detected during the study. Stability-indicating HPLC analysis found less than 6% acetazolamide loss in any of the suspensions stored at either temperature after 60 days of storage.

**STUDY 3:** Parasrampuria and Gupta[180] evaluated the stability of acetazolamide in various liquid formulations for oral administration. The final formulations (Table 1) were prepared by dissolving acetazolamide in polyethylene glycol 400 and adding propylene glycol. Menthol was dissolved in the ethanol and added to the acetazolamide mixture. The artificial sweeteners and flavors were mixed with simple syrup; sodium benzoate and sorbitol then were added. The aqueous solutions of the dyes were added to the buffer solution and mixed with flavored syrup. Finally, the aqueous solution and the glycol mixture were combined to form the final formulations, which were then stored at 25 and 37°C. A stability-indicating HPLC assay was used to assess the concentration of acetazolamide. Little or no loss of acetazolamide occurred during 178 days of storage at 25°C. Furthermore, losses at elevated temperature were about 10% in each formulation when stored at 37°C for 179 days. From these data an expiration date of two years can be extrapolated. ■

**TABLE 1.** Composition of Acetazolamide Oral Liquids[180]

| Component | Concentration |
| --- | --- |
| Acetazolamide | 5 mg/mL |
| Polyethylene glycol 400 | 7% (vol/vol) |
| Propylene glycol | 53% (vol/vol) |
| Sorbitol solution 70% (wt/wt) | 15% (vol/vol) |
| Simple syrup [sucrose 85% (wt/vol)] | 15% (vol/vol) |
| Saccharin sodium | 0.18% (wt/vol) |
| Aspartame | 0.18% (wt/vol) |
| Sodium benzoate | 0.2% (wt/vol) |
| FD&C red no. 40 | 1.2 ppm |
| FD&C blue no. 1 | 0.05% (vol/vol) |
| Raspberry flavor | 0.05% (vol/vol) |
| Natural Pharma Sweet Flavor | 0.3% (vol/vol) |
| Menthol | 0.002% (wt/vol) |
| Ethanol | 0.5% (vol/vol) |
| Phosphate or citrate buffer solution | 0.1 M (pH 4) |

# Acetylcholine Chloride

## Properties
Acetylcholine chloride occurs as a very hygroscopic white or nearly white crystalline powder or as odorless crystals.[1–3]

### Solubility
Acetylcholine chloride is very soluble in water and ethanol.

### pH
The reconstituted acetylcholine chloride commercial ophthalmic solution has a pH in the range of 5 to 8.2.[2]

### Osmolality
The reconstituted acetylcholine chloride commercial ophthalmic solution is isotonic.[2]

## General Stability Considerations
Acetylcholine chloride is very hygroscopic and must be stored in tightly closed containers and protected from exposure to light.[1,3]

Prior to reconstitution, the commercial ophthalmic solution in intact containers should be stored at controlled room temperature and protected from freezing. After reconstitution, the commercial ophthalmic solution is very unstable and should be used immediately; any unused portion should be discarded.[2]

## Stability Reports of Compounded Preparations
### Topical
Sletten et al.[982] reported the stability of 10% acetylcholine chloride in water solutions intended for administration by iontophoresis for the quantitative sudomotor axon reflex test. The test solutions were stored at −20°C, 4°C, ambient room temperature, and elevated temperature of 50°C with all samples protected from exposure to light. HPLC analysis found rapid loss at the highest temperature but very little loss and similar stability in 84 days in the dark at −20 and 4°C. Frozen storage is unnecessary for the solution. At ambient room temperature, adequate stability was maintained for up to 28 days in the dark. However, the authors still recommended low temperature storage for acetylcholine chloride solutions, although they noted that if the solution was left at room temperature for the day of use, the integrity of the drug would not be adversely affected.

# Acetylcysteine

## Properties
Acetylcysteine is a white crystalline powder having a slight acetic odor.[2,3] Commercial products are available as the sodium salt.[2]

### Solubility
Acetylcysteine is freely soluble in water, having an aqueous solubility of about 200 mg/mL. In ethanol, its solubility is about 250 mg/mL.[2,3]

### pH
A 1% aqueous solution of acetylcysteine has a pH between 2 and 2.8.[4] The commercial solution is made with the aid of sodium hydroxide and has a pH between 6 and 7.5.[2,4]

### Osmolality
Acetylcysteine inhalation solution (Senju Pharmaceutical Co.) is extremely hypertonic, having an osmolality of approximately 2259 mOsm/kg as measured by freezing point depression. Dilution with an equal quantity of sodium chloride 0.45% resulted in a decrease to about 876 mOsm/kg.[928]

## General Stability Considerations
Acetylcysteine solution is a sterile solution and should be packaged in tight containers stored at controlled room temperature between 15 and 30°C and protected from light.[2–4,7] Acetylcysteine is unstable in air, oxidizing to the disulfide form.[409] Acetylcysteine 20% placed in cups and continually exposed to air for seven days congealed.[1088] Following exposure to air after opening, the vial should be stored under refrigeration. Opened vials should be discarded after 96 hours.[2,7]

Solutions become discolored and liberate hydrogen sulfide on contact with rubber and some metals, especially iron and copper, and when autoclaved. The drug is compatible with glass, plastics, aluminum, silver, chromed metal, or stainless steel. An acetylcysteine solution that is slightly purple does not significantly impair safety or efficacy.[2,3]

Acetylcysteine is incompatible with amphotericin B, ampicillin, erythromycin lactobionate, and tetracyclines, as well as hydrogen peroxide and iodized oil.[2,3]

Kerc et al.[1435] evaluated the compatibility of acetylcysteine with a wide variety of common tablet excipients using differential scanning calorimetry, Fourier transform infrared spectroscopy, HPLC, and thin-layer chromatography (TLC). Acetylcysteine was found to exhibit no interaction with microcrystalline cellulose (Avicel PH101), carboxymethylcellulose, amorphous silicon dioxide (Aerosil 200), polyvinylpyrrolidone, cross-linked polyvinylpyrrolidone (Polyplasdone XL), corn starch, saccharose, and magnesium stearate. However, possible decreased acetylcysteine thermal stability occurred with corn starch, saccharose, and magnesium stearate. Possible interactions of acetylcysteine were found with lactose, polyethylene glycol 4000 and 6000, glycine, adipic acid, and saccharin sodium.

## Stability Reports of Compounded Preparations
### Oral
**STUDY 1:** Siden and Johnson[987] evaluated the stability of an oral liquid formulation for acetylcysteine 86.5 mg/mL prepared from acetylcysteine sodium inhalation solution (Abbott). The intent was to prepare a stable formulation with the sulfurous smell and taste of the acetylcysteine masked by the flavoring. The oral liquid was prepared by adding 7 mL of sweetener (FLAVORx) and 7 mL of strawberry creamsicle flavoring (FLAVORx) to 90 mL of commercial 10% acetylcysteine sodium solution for inhalation. The oral liquid was packaged in two-ounce amber plastic prescription bottles. Samples were stored at room temperature of 23 to 25°C exposed to normal fluorescent light and under refrigeration at 3 to 5°C in a non-transparent refrigerator for 35 days.

Visual examination found no detectable change in color. Three individuals compared the smell and taste and found no changes after 35 days. The taste and smell were well masked by the flavoring. Stability-indicating HPLC analysis found acetylcysteine losses after 35 days of storage of about 8% at room temperature and about 3% refrigerated.

**STUDY 2:** Kiser et al.[1088] evaluated the stability of acetylcysteine 600 mg/3 mL (20%, Hospira) packaged in oral syringes (Exacta-Med Oral Dispenser, Baxa Corporation) and stored for six months at room temperature of 23 to 25°C exposed to normal fluorescent light and also under refrigeration at 3 to 5°C protected from exposure to light except for sampling time points. Visual examination found the solution remained clear with no precipitate formation. The room temperature samples underwent about 5% loss in six months, and the refrigerated samples incurred only about 2% loss in six months.

**STUDY 3:** To et al.[1398] evaluated the stability of an oral capsule formulation of acetylcysteine for use in the prevention of contrast-dye-induced nephropathy. Acetylcysteine 600-mg capsules were prepared using lactose as filler. Acetylcysteine crystalline powder and lactose were ground separately using a mortar and pestle and sifted through a 250-μm sieve. Size 00 gelatin capsules were filled with the mixture, resulting in each capsule containing acetylcysteine 600 mg and lactose 110 mg. The capsules were packaged in amber polystyrene containers with and also without a 1-g sachet of desiccant (Dri-Pax). The sample capsules were stored for up to 420 days refrigerated at 2 to 8°C, at room temperature of 18 to 25°C, and at elevated temperature of 40°C. The samples were split between some stored

at ambient humidity and the rest stored at controlled high humidity of about 75%.

Most of the capsules remained stable throughout the study. All refrigerated and room temperature samples remained stable with no physical changes and HPLC analyses showing little or no loss; the maximum change in drug concentration after 420 days of storage was only 2.7%. Even the capsules stored at 40°C and ambient humidity demonstrated little degradation. Only the samples stored at 40°C at 75% humidity demonstrated substantial instability. The capsules developed a hydrogen sulfide odor after 28 days. HPLC analysis found acetylcysteine losses were about 15% in 84 days and up to 40% at 421 days of such storage. The presence of the desiccant made only a minor difference in the drug's stability.

**STUDY 4:** Fohl et al.[1608] evaluated the stability of extemporaneously compounded acetylcysteine 1 and 10% solutions for treatment of meconium ileus. Acetylcysteine injection (Hospira) was diluted to concentrations of 1% (10 mg/mL) and 10% (100 mg/mL) with bacteriostatic sodium chloride 0.9% injection (Hospira) preserved with benzyl alcohol. The solutions were packaged in amber 2-ounce prescription bottles and stored at room temperature of 20 to 25°C for 90 days. The solutions did not change in color but developed a pungent odor of hydrogen sulfide after 90 days of storage. Stability-indicating HPLC analysis found an acetylcysteine loss of 9% in 60 days in both concentrations. Drug losses after 90 days of storage were 15% in the 10-mg/mL solution and 11% in the 100-mg/mL solution. The authors stated that these oral solutions in amber plastic bottles were stable for at least 60 days at room temperature.

**STUDY 5:** Thielens[1669] evaluated the effects of pH, temperature, and the presence of edetate disodium 0.1% on the stability of acetylcysteine in aqueous solutions. The greatest acetylcysteine stability was found in the solutions with edetate disodium present, the pH adjusted to 6.5, and when stored under refrigeration. The author noted that acetylcysteine could be incorporated into any cough syrup formulation that met these conditions.

### Ophthalmic
Ophthalmic preparations, like other sterile drugs, should be prepared in a suitable clean air environment using appropriate aseptic procedures. When prepared from nonsterile components, an appropriate and effective sterilization method must be employed.

**STUDY 1:** Fawcett et al.[409] reported the stability of an extemporaneously prepared acetylcysteine 10% eyedrop formulation. The eyedrops were prepared from 7.5 mL of acetylcysteine 20% in ampules (Parvolex, Glaxo) and 0.5 mL of benzalkonium chloride 0.16%, diluted with 7 mL of hydroxypropyl methylcellulose (hypromellose) 0.5% eye drops (Isopto Tears, Alcon). The solution was drawn into a plastic syringe and filtered through a 0.2-μm filter (Millex GV) into a sterile dropper bottle (Isopto Tears) (low-density polyethylene). Samples were stored under refrigeration at 2 to 8°C, at room temperature of 25°C, and at an elevated temperature of 37°C for 90 days. There was no decomposition of acetylcysteine over 90 days when stored under refrigeration. The values of $t_{90}$ for the 25 and 37°C samples were 48 and 23 days, respectively. Samples stored under refrigeration for 90 days and then evaluated under simulated use conditions at 25°C had a 10% loss of acetylcysteine in about 26 days. Samples stored under refrigeration during simulated use (after the original 90-day storage under refrigeration) exhibited no decomposition over 30 days.

**STUDY 2:** Anaizi et al.[571] studied the stability of 10% acetylcysteine ophthalmic solution compounded from the commercial solution. The ophthalmic diluent was prepared by triturating chlorobutanol in a mortar, weighing 500 mg of the chlorobutanol triturate, and transferring it to a beaker. About 80 mL of artificial tears (Liquifilm Tears, Allergan) composed of 1.4% polyvinyl alcohol and 0.5% chlorobutanol was added to the beaker and stirred until the powdered chlorobutanol dissolved. The solution was transferred to a graduated cylinder and was brought to 100 mL with the addition of artificial tears. The ophthalmic diluent was passed through a 0.22-μm filter into sterile vials. To prepare the 10% acetylcysteine ophthalmic solution, 7.5 mL of the ophthalmic diluent and 7.5 mL of sterile 20% acetylcysteine solution (Mucosil-20, Dey) were aseptically transferred to a 15-mL sterile low-density polyethylene ophthalmic dropper bottle by syringe and mixed. The final solution contained 0.025% disodium edetate and 0.5% chlorobutanol including the Mucosil-20 contribution. Samples of this solution were stored under refrigeration at about 4°C. Stability-indicating HPLC analysis of the samples found 8 to 10% acetylcysteine loss after 60 days of storage under refrigeration.

**STUDY 3:** Anaizi et al.[571] also evaluated the stability of similarly prepared 10% acetylcysteine ophthalmic solutions containing varying disodium edetate concentrations of 0.025, 0.05, 0.075, and 0.1% at 24°C. Stability-indicating HPLC analysis found greater than 10% acetylcysteine loss in as little as seven days at room temperature. The 0.075% disodium edetate samples exhibited 6 to 8% loss in 30 days but as much as 11% loss in 40 days. The 0.1% disodium edetate samples were stable for 50 days at room temperatures, exhibiting 7 to 9% acetylcysteine loss.

Anaizi et al.[571] combined equal quantities of 20% acetylcysteine solution with Isopto Tears (Alcon) having 0.05% disodium edetate and 0.01% benzalkonium chloride. However, a fine suspended precipitate was observed after 12 days of storage at both 4 and 24°C. The precipitated material redissolved if the pH was raised from the original 7.4 to 9.

### Injection

Injections, like other sterile drugs, should be prepared in a suitable clean air environment using appropriate aseptic procedures. When prepared from nonsterile components, an appropriate and effective sterilization method must be employed.

**STUDY 1:** Dribben et al.[889] reported the stability of acetylcysteine inhalation solution compounded as an admixture in 5% dextrose injection for use in treating acetaminophen poisoning. Acetylcysteine 20% inhalation (Mucomyst, Abbott) 30 g was passed through a 0.22-μm filter as it was introduced into 1-L polyvinyl chloride bags of 5% dextrose injection, yielding a concentration of 30 mg/mL. The solution was stored at room temperature of 25°C and 65% relative humidity for 72 hours. No particulate matter, discoloration, or other visible phenomena were observed. HPLC analysis found less than 10% acetylcysteine loss occurred in 60 hours but 10 to 15% loss occurred in 72 hours.

**STUDY 2:** Van Loenen et al.[1670] evaluated the stability of an acetylcysteine injection compounded from powder. The injection was prepared by adding sodium bicarbonate to neutralize the acetylcysteine in solution and adding edetate disodium 0.05% with the absence of oxygen in the 2-mL ampules. Iodometric assay and several thin layer chromatography assays were performed to determine acetylcysteine stability. Based on drug decomposition at elevated temperatures up to 120°C, the time to 10% drug loss ($t_{90}$) was calculated to be about 30 years in this formulation stored at 25°C. The authors noted that, in the absence of oxygen in the sealed ampules, acetylcysteine can be autoclaved at 120°C, and is stable for an extended shelf life.

## Compatibility with Other Drugs
### ACETYLCYSTEINE COMPATIBILITY SUMMARY
*Compatible with:* Arformoterol tartrate • Betamethasone sodium phosphate • Budesonide • Cromolyn sodium • Fenoterol hydrobromide • Formoterol fumarate • Ipratropium bromide • Netilmicin sulfate • Sodium chloride 7% (hypertonic)
*Incompatible with:* Albuterol sulfate • Isoetharine • Metaproterenol sulfate • Terbutaline sulfate

**STUDY 1 (CROMOLYN):** Lesko and Miller[659] evaluated the chemical and physical stability of 1% cromolyn sodium nebulizer solution (Fisons) 2 mL admixed with 2 mL of acetylcysteine 20% inhalation solution. The admixture was visually clear and colorless and remained chemically stable by HPLC analysis for 60 minutes after mixing at 22°C. About 4% cromolyn sodium loss and 6% acetylcysteine loss occurred.

**STUDY 2 (ARFORMOTEROL):** Bonasia et al.[1151] evaluated the physical compatibility and chemical stability of arformoterol 15 mcg/2 mL (as the tartrate) inhalation solution (Brovana, Sepracor) with acetylcysteine 800 mg/4 mL (Mucomyst, Hospira), ipratropium bromide 0.2 mg/2 mL (Atrovent, Astra-Zeneca), and budesonide 0.25 mg/2 mL and 0.125 mg/2 mL (Pulmicort Respules, AstraZeneca). The admixtures were prepared and mixed for homogeneity. Visual inspection found no evidence of precipitation or other physical incompatibility. HPLC analysis found less than 2% change in any drug concentration over the 30-minute test period.

**STUDY 3 (MULTIPLE DRUGS):** Owsley and Rusho[517] evaluated the compatibility of several respiratory therapy drug combinations. Acetylcysteine 10% solution (Mucosil-10, Dey) was combined with the following drug solutions: albuterol sulfate 5-mg/mL inhalation solution (Proventil, Schering), isoetharine 1% solution (Bronkosol, Sanofi Winthrop), metaproterenol sulfate 5% solution (Alupent, Boehringer Ingelheim), and terbutaline sulfate 1-mg/mL solution (Brethine, Geigy). The test solutions were filtered through 0.22-μm filters into clean vials. The combinations were evaluated over 24 hours (temperature unspecified) using the USP particulate matter test. None of the combinations was found to be compatible. Acetylcysteine combined with the albuterol sulfate, isoetharine, metaproterenol sulfate, and terbutaline sulfate all formed unacceptable levels of larger particulates (≥10 μm).

**STUDY 4 (MULTIPLE DRUGS):** Rieutord et al.[841] evaluated the compatibility and stability of an aerosol inhalation solution composed of acetylcysteine 1 g/5 mL, netilmicin (as the sulfate) 100 mg/1 mL, and betamethasone (as the sodium phosphate) 4 mg/1 mL. HPLC analysis found little or no change in concentration of any of the three drugs in one hour at 23 to 27°C. The authors indicated that this brief period should provide a sufficient time frame for a nurse or patient to prepare the mixture and administer it.

**STUDY 5 (MULTIPLE DRUGS):** Lee et al.[928] evaluated the stability of acetylcysteine 20% mixed in equal volumes with ipratropium bromide 250 mcg/mL and separately with fenoterol hydrobromide 0.625 mcg/mL in combination inhalation solutions. HPLC analysis of the solutions found acetylcysteine and fenoterol hydrobromide were stable for seven hours at room temperature of 25°C, exhibiting losses of about 7 and 6%, respectively. The acetylcysteine mixture with ipratropium bromide was less stable. Ipratropium bromide losses in the inhalation mixture were about 7% in one hour and 11% in two hours at room temperature; the acetylcysteine remained stable for two hours with a loss of about 2%.

**STUDY 6 (MULTIPLE DRUGS):** Gronberg et al.[1027] evaluated the stability and compatibility of budesonide (Pulmicort, concentration unspecified) with several inhalation solutions, including (1) acetylcysteine (Lysomucil) 100 mg/mL 2 parts to 3 parts budesonide, (2) fenoterol hydrobromide (Berotec) 5 mg/mL 8 parts to 1 part budesonide, (3) ipratropium bromide (Atrovent) 0.25 mg/mL 2 parts to 1 part budesonide, and (4) ipratropium bromide 0.125 mg/mL plus fenoterol hydrobromide 0.31 mg/mL (Duovent) in equal parts with budesonide. Samples were stored at room temperature of 22 to 25°C protected from exposure to light. HPLC analysis found all mixtures to be compatible, maintaining greater than 90% of the initial concentrations for at least 18 hours.

**STUDY 7 (HYPERTONIC SODIUM CHLORIDE):** Fox et al.[1239] evaluated the physical compatibility of hypertonic sodium chloride 7% with several inhalation medications used in treating cystic fibrosis. Acetylcysteine injection (American Regent) 200 mg/mL mixed in equal quantities with extemporaneously compounded hypertonic sodium chloride 7% did not exhibit any visible evidence of physical incompatibility, and the measured turbidity did not increase over the one-hour observation period.

**STUDY 8 (LEVALBUTEROL):** Bonasia et al.[1329] evaluated the compatibility and stability of levalbuterol hydrochloride inhalation solution (Sepracor) with several drugs including acetylcysteine (Bristol-Meyers Squibb). The drugs were mixed with sodium chloride 0.9% and then with each other to prepare the following solution:

> Levalbuterol 1.25 mg/0.5 mL diluted with 5 mL of sodium chloride 0.9%
> Acetylcysteine 800 mg/4 mL diluted with 0.5 mL of sodium chloride 0.9%

The drug mixture was prepared in triplicate and evaluated initially and after 30 minutes at an ambient temperature of 21 to 25°C. The authors indicated that a 30-minute study duration was chosen because it represents the approximate time it takes for a patient to administer drugs by nebulization at home. No visible evidence of physical incompatibility was observed with the samples. HPLC analysis found no loss of any of either of the drugs in the samples.

**STUDY 9 (FORMOTEROL):** Akapo et al.[1330] evaluated the compatibility and stability of formoterol fumarate 20 mcg/2 mL inhalation solution (Performist, Dey) mixed with acetylcysteine 200 mg/2 mL (American Regent) for administration by inhalation and evaluated over 60 minutes at room temperature of 23 to 27°C. The test samples were physically compatible by visual examination and measurement of osmolality, pH, and turbidity. HPLC analysis of drug concentrations found little or no loss of either of the drugs within the study period. ▪

# Acyclovir
# Acyclovir Sodium
## (Aciclovir; Aciclovir sodium)

## Properties

Acyclovir occurs as a white to off-white crystalline powder.[1,3,4]

### Solubility

Acyclovir is soluble in diluted mineral acids (such as hydrochloric acid) and alkali hydroxides (such as sodium hydroxide).[3,4] Acyclovir is slightly soluble in water having a solubility of about 2.5 mg/mL at 37°C.[3,4,7] It is very slightly soluble[3] or insoluble[4] in ethanol. Acyclovir is freely soluble in dimethylsulfoxide.[3]

Acyclovir sodium has an aqueous solubility at 25°C exceeding 100 mg/mL.[7]

### pH

Acyclovir (as the sodium salt) reconstituted with sterile water for injection to an acyclovir concentration of 50 mg/mL has a pH of approximately 11.[7]

### pK$_a$

Acyclovir has pK$_a$ values of 2.27 and 9.25.[7]

### Osmolality

Acyclovir sodium reconstituted and diluted for injection to a concentration of 7 mg/mL in dextrose 5% or in sodium chloride 0.9% is nearly isotonic.[8]

## General Stability Considerations

Acyclovir powder, oral capsules, and oral tablets should be packaged in tight containers and stored at controlled room temperature. They should be protected from exposure to light and moisture.[4] GlaxoSmithKline states that acyclovir dispersible tablets should not be removed from the original packaging and placed in dosing compliance aids.[1622] Acyclovir oral suspension should be packaged in tight containers and stored at controlled room temperature protected from exposure to light. Acyclovir for injection should also be stored at controlled room temperature and protected from exposure to light for long-term storage. Acyclovir ointment should be packaged in tight containers and stored at controlled room temperature in a dry place.[4]

Reconstituted acyclovir sodium injection should be stored at controlled room temperature and used within 12 hours. Refrigeration may result in precipitation. After dilution for administration in a compatible infusion solution, the solution should be stored at controlled room temperature and used within 24 hours.[2,8]

Barboza et al.[1446] evaluated the compatibility of acyclovir with a number of potential excipients for extended release oral dosage forms. Using differential scanning calorimetry, thermogravimetry, and x-ray powder diffraction, the authors found that acyclovir was incompatible with magnesium stearate. However, acyclovir was found to be compatible with ethylcellulose (Ethocel), hydroxypropylmethylcellulose (HPMC KM15, Methocel), polyethylene oxide (Polyox), lactose, microcrystalline cellulose (Microcel), colloidal silicon dioxide (Aerosil), sodium starch glycolate (Primojel), sodium croscamellose (AcDiSol), and stearic acid.

Monajjemzadeh et al.[1450] evaluated the incompatibility of acyclovir with lactose in physical mixtures and commercial tablets using a variety of analytical techniques, including differential scanning calorimetry, thin-layer chromatography, tandem mass spectrometry, and stability-indicating HPLC analysis. An acyclovir-lactose condensation product was detected in only one brand of commercial tablets. The authors concluded that the incompatibility is less likely to form in the dry state compared to aqueous liquids.

## Stability Reports of Compounded Preparations
### Injection

Injections, like other sterile drugs, should be prepared in a suitable clean air environment using appropriate aseptic procedures. When prepared from nonsterile components, an appropriate and effective sterilization method must be employed.

Ling and Gupta[1259] evaluated the stability of acyclovir sodium 10 mg/mL in sodium chloride 0.9% packaged in polypropylene syringes for pediatric administration. The drug was found to be stable for 30 days stored at controlled room temperature, but storage under refrigeration caused the drug to precipitate.

### Oral

Desai et al.[1258] evaluated the stability of acyclovir at a low concentration of 0.2 mg/mL in 50% sucrose and maltitol solutions and 26% fructose and glucose solutions stored at 50°C. The sucrose, fructose, and glucose mixtures turned yellow then light orange over 13 weeks of storage. HPLC analysis found hydrolysis of sucrose to fructose and glucose that interacted with the acyclovir resulting in extensive loss of the drug, especially at low pH. With the pH of the sucrose solution adjusted to pH values of 4, 6, and 7, acyclovir losses were about 60%, 22%, and 12%, respectively. Similar losses were also found in fructose and glucose solutions, leading the authors to believe the hydrolysis of sucrose led to the instability of the drug.

However, in 50% maltitol, no color change was observed in 13 weeks. In addition, HPLC analysis found less than 5% loss of acyclovir after 13 weeks at 50°C. The authors concluded that acyclovir was unstable in a sucrose solution and that maltitol was the preferred sweetener for an oral liquid preparation. ■

# Adapalene

## Properties
Adapalene is a white to off-white crystalline retinoic acid analog that is similar to tretinoin.[2,4,7]

### Solubility
Adapalene is sparingly soluble in ethanol and practically insoluble in water.[7]

### $pK_a$
Adapalene has a $pK_a$ of 4.2.[1]

## General Stability Considerations
Adapalene is stated to be stable to light exposure.[1] The commercial topical adapalene cream (Differin) should be stored at controlled room temperature and protected from freezing.[7]

## Compatibility with Other Drugs
**STUDY 1:** Martin et al.[636] evaluated the stability of retinoid preparations, including adapalene 0.1% gel, combined with an equal quantity of benzoyl peroxide 10% lotion. The mixture was packaged in 10-mL plastic syringes and stored exposed to normal room fluorescent light over 24 hours. Retinoid content was assessed by HPLC analysis. The adapalene combination remained stable with no loss of the retinoid.

**STUDY 2:** Ohtani et al.[1522] evaluated the stability and compatibility of adapalene gel (Differin) mixed with topical clindamycin and also nadifloxacin for use in treating acne. The authors reported that the mixtures were stable. HPLC analysis found the concentrations of adapalene, clindamycin, and nadifloxacin remained acceptable. ■

# Adderall
## (Obetrol)

## Properties
Adderall (Shire Richwood) is a 3:1 mixture of two dextroamphetamine salts (sulfate and saccharate) and two levoamphetamine salts (sulfate and aspartate).[3] Dextroamphetamine sulfate and levoamphetamine sulfate are bitter-tasting white crystalline powders.[1,3]

### Solubility
Dextroamphetamine sulfate and levoamphetamine sulfate are freely soluble in water to about 100 and 111 mg/mL, respectively, and are slightly soluble in ethanol to about 1.25 and 2 mg/mL, respectively.[1]

### pH
The pH of a 5% solution is between 5 and 6.[1]

## General Stability Considerations
Adderall should be packaged in tight, light-resistant containers and should be stored at controlled room temperature.[2,4,7]

## Stability Reports of Compounded Preparations
### Oral
Justice et al.[708] reported the stability of three oral suspension formulations of Adderall (Shire Richwood), a 3:1 mixture of dextroamphetamine and levoamphetamine salts. The Adderall 1-mg/mL suspensions were prepared by crushing commercial 10-mg tablets using a mortar and pestle and gradually adding Ora-Sweet, Ora-Plus, or a 1:1 mixture of Ora-Sweet and Ora-Plus with mixing. The suspensions were packaged in glass bottles and stored at 25°C in a controlled temperature chamber at 60% relative humidity protected from exposure to light for 30 days.

No visible changes to the three suspensions occurred, and no microbial growth was observed. Stability-indicating gas chromatography–mass spectroscopy found less than 10% loss of either the dextroamphetamine or levoamphetamine components, with the ratio remaining at 3:1 throughout the 30-day study period. ■

# Adenosine

## Properties

Adenosine is an endogenous nucleoside that occurs as a white crystalline powder.[1,7]

### Solubility

Adenosine is soluble in water and practically insoluble in ethanol.[1,7]

### pH

The commercial injections of adenosine (Adenocard and Adenoscan) have a pH between 4.5 and 7.5.[7]

## General Stability Considerations

Adenosine powder should be packaged in well-closed light-resistant containers and stored at controlled room temperature.[3,4] Commercial injections of adenosine should be stored at controlled room temperature. Refrigeration may result in crystallization; crystals may be redissolved by warming to room temperature.[7]

## Stability Reports of Compounded Preparations

### Injection

Injections, like other sterile drugs, should be prepared in a suitable clean air environment using appropriate aseptic procedures. When prepared from nonsterile components, an appropriate and effective sterilization method must be employed.

**STUDY 1:** Ketkar et al.[884] evaluated the stability of adenosine 3-mg/mL injection repackaged as 25 mL in 60-mL polypropylene syringes sealed with tip caps and in glass vials stored frozen at −15°C, refrigerated at 5°C, and at room temperature of 25°C. The injection remained visually clear throughout the study at all three of the temperatures. HPLC analysis found that no adenosine loss occurred in 28 days at −15°C, 14 days at 5°C, and seven days at 25°C.

**STUDY 2:** Proot et al.[699] reported that adenosine 2 mg/mL extemporaneously prepared in sodium chloride 0.9% and packaged in glass vials was stable for at least six months stored refrigerated at 4°C, at room temperature near 22°C, and at body temperature of 37°C. Stability-indicating HPLC analysis found that no loss of adenosine occurred. At 60 and 72°C, extremes of temperature designed to decompose the drug, 10% adenosine loss occurred in 250 and 91 days, respectively. Based on these data, the authors indicated a shelf life of five years at room temperature could be expected.

**STUDY 3:** Naud et al.[885] reported the stability of adenosine 6 mcg/mL in 0.9% sodium chloride packaged in glass ampules. High-temperature accelerated drug decomposition studies projected the drug would be stable for at least five years at room temperature of 25°C and refrigerated at 5°C.

**STUDY 4:** Meijer and van Loenen[1000] evaluated the stability of an extemporaneously compounded adenosine 3-mg/mL injection in sodium chloride 0.9% and having a pH in the range of 5.5 to 7.5. The injection was stable to autoclaving at 120°C for 20 minutes and was calculated to have a shelf life of at least two years based on the small amount of loss by HPLC analysis that occurred in 12 months stored at 20°C. Even at an elevated storage temperature of 40°C, less than 8% loss occurred in 12 months.

### Inhalation

Inhalations, like other sterile drugs, should be prepared in a suitable clean air environment using appropriate aseptic procedures. When prepared from nonsterile components, an appropriate and effective sterilization method must be employed.

Martinez-Garcia et al.[1457] evaluated the stability of adenosine 5'-phosphate solutions of 30 mcg/mL and 400 mg/mL in sterile sodium chloride 0.9% for use as bronchoprovocation solutions. Adenosine 5'-phosphate powder was dissolved in sodium chloride 0.9% and passed through a 0.22-µm filter into storage flasks. Samples were stored at room temperature of 20 to 25°C and refrigerated at 4°C for up to 25 weeks and analyzed by HPLC. Room temperature samples exhibited more than 10% drug loss in nine and 15 days for the 30-mcg/mL and 400-mg/mL solutions, respectively. However, both concentrations stored at 4°C exhibited no loss over 25 weeks of storage. ▪

# Albendazole
# Albendazole Sulfoxide

## Properties

Albendazole occurs as a white to faintly yellowish or off-white powder[4,7] or as colorless crystals.[1] Albendazole sulfoxide (also named ricobendazole) is a major metabolite of albendazole used in some countries as a veterinary anthelmintic agent by subcutaneous injection.[1397]

### Solubility

Albendazole is soluble in strong acids but practically insoluble in water and ethanol.[4,7]

Albendazole sulfoxide has an aqueous solubility of 62 mcg/mL. Its solubility was low over the pH range of 4 to 9 but increased below pH 2 and above pH 11. The solubility of albendazole in ethanol and propylene glycol is 1.2 and 2.6 mg/mL, respectively. In dimethyl sulfoxide, its solubility is 16.5 mg/mL.[1396]

### pH

Albendazole oral suspension has a pH in the range of 4.5 to 5.5.[4]

### $pK_a$

Albendazole sulfoxide has $pK_a$ values near 3.5 to 3.7 and 9.5 to 9.8.[1396]

## General Stability Considerations

Albendazole tablets and oral suspension should be packaged in tight containers and stored at controlled room temperature.[4,7]

Ragno et al.[1020] stated that benzimidazole class drugs, including albendazole, are known to be sensitive to light. The European Pharmacopoeia requires protection from light during storage.[3] However, the USP does not include light protection as a requirement.[4]

Albendazole and related benzimidazole carbamate drugs (including albendazole sulfoxide) exhibit pH-dependent hydrolysis in aqueous media. The minimum hydrolysis rate occurs near pH 5. Higher rates of hydrolysis occur at acidic and especially at alkaline pH values.[1395, 1397]

## Stability Reports of Compounded Preparations
### Oral

Ragno et al.[1020] evaluated the stability of albendazole in the solid state as well as in liquid form in an ethanol and water medium (concentrations not provided). Exposure of the solid drug to high temperatures up to 50°C and intense light from a Xenon arc lamp did not result in loss of the drug within 10 hours. Similarly, no loss due to thermal exposure resulted in the liquid form as well. However, exposure of the liquid form to intense light resulted in rapid and extensive loss with more than 80% of the albendazole decomposing in 10 hours. This loss was much more extensive than with other benzimidazoles tested by the authors. The authors indicated that they believed water must be present for photodegradation to occur because no loss occurred in the solid form. ■

# Albuterol Sulfate
## (Salbutamol Sulphate)

## Properties

Albuterol sulfate is a white crystalline powder. Aqueous solutions are colorless to slightly yellow. Albuterol sulfate 1.2 mg is approximately equivalent to albuterol 1 mg.[1–3]

### Solubility

Albuterol sulfate has an aqueous solubility of 250 mg/mL. It is slightly soluble in ethanol.[1–3]

### pH

Oral solutions have a pH of 3.5 to 4.5. The inhalation solution has a pH of 3 to 5.[2,7]

### $pK_a$

The $pK_a$ values of albuterol are 9.3 and 10.3.[2]

## General Stability Considerations

Albuterol sulfate dosage forms should be stored in well-closed, light-resistant containers.[2–4,7] The manufacturer has stated that albuterol sulfate oral dosage forms should not be removed from the original packaging and placed in dosing compliance aids because of hygroscopicity of the drug.[1622]

The metered inhalers should be stored and used at controlled room temperature. The nebulization solutions should be stored at controlled room temperature or between 2 and 25°C, depending on the specific product. Oral dosage forms should be stored between 2 and 30°C. Discolored products should be discarded.[2–4,7]

Albuterol aqueous solutions were found to be very stable in the acidic pH range of 2.2 to 5; heating to 85°C resulted in less than 10% loss in 48 hours. However, at mildly alkaline pH 8, albuterol decomposed much more rapidly at 85°C, with only 12.5% remaining after 48 hours.[587]

Urmi et al.[1285] evaluated the effects of several common excipients on albuterol sulfate stability at normal storage and stress conditions (50°C for 45 days). Calcium salts (carbonate, sulfate, and stearate) adversely affected the stability of albuterol sulfate. Calcium lactate caused discoloration. Ethyl cellulose delayed dissolution of tablets. Dextrose was found to show the least effects on dissolution, diffusion, and stability. Infrared spectrophotometry studies on an albuterol sulfate-dextrose mixture showed evidence of reversible binding.

Malkki-Laine et al.[1541] evaluated the effect on albuterol sulfate stability of various buffers and antioxidants. Albuterol sulfate decomposition is somewhat faster in citrate buffer compared to acetate and phosphate buffers. The antioxidants sodium bisulfite and sodium metabisulfite exhibited a slight stabilizing effect on albuterol sulfate. No decomposition of albuterol sulfate solutions at pH 3 was found when autoclaved for 20 minutes at 120°C.

## Stability Reports of Compounded Preparations
### Inhalation
Inhalations, like other sterile drugs, should be prepared in a suitable clean air environment using appropriate aseptic procedures. When prepared from nonsterile components, an appropriate and effective sterilization method must be employed.

**STUDY 1:** Allen[151] reported the stability of an extemporaneously compounded preservative-free albuterol sulfate inhalation solution. The solution was made to the prescribed concentration by weighing the appropriate amount of albuterol sulfate powder and dissolving it in 0.9% sodium chloride injection. The solution was filtered through a sterilizing filter (0.22 µm) and packaged in sterile containers. Refrigerated storage was recommended; room temperature storage was limited to two weeks. Although the solution is chemically stable, the microbiological stability of this preservative-free product is dependent on the ability of the compounder to ensure the sterility of the product initially and during storage.

**STUDY 2:** Hunter et al.[587] evaluated the stability of albuterol inhalation solution (Allen & Hanburys) diluted to 200 mcg/mL in 0.9% sodium chloride injection for continuous nebulization. The samples were packaged in 250-mL polyvinyl chloride (PVC) bags (Kendall Canada), 50-mL polyolefin bags (Abbott), 1-mL polypropylene syringes with needle (Terumo), 20-mL borosilicate glass tubes (Wheaton), and 1.5-mL polypropylene microcentrifuge tubes (Baxter). The samples were stored under refrigeration at 4°C and at room temperature of about 24°C for seven days. Stability-indicating HPLC analysis found little or no loss of albuterol in any of the samples at either temperature.

However, Hunter et al.[587] also noted that freezing samples at −20°C in the plastic syringes until analyzed resulted in unacceptably wide variation in the measured drug concentrations. This variation did not occur in samples frozen in other containers or in samples from the syringes transferred to glass vials before freezing. Although the cause of the variability in measured drug concentration was not identified, the authors concluded that albuterol solutions should not be frozen in plastic syringes.

**STUDY 3:** Blondino and Baker[930] evaluated the physical and chemical stability of four common drugs administered by inhalation: albuterol sulfate 2.5 mg/3 mL, budesonide 0.25 mg/3 mL, cromolyn sodium 20 mg/3 mL, and ipratropium bromide 0.5 mg/3 mL, individually in an acidified (pH 4) 15% ethanol in water solution. The budesonide was dissolved in the ethanol, which was then added to a solution of the other drugs dissolved in the acidified sterile water for irrigation. No visible changes in color or clarity were observed in any sample. Albuterol sulfate by itself underwent substantial losses over eight weeks. Losses were 11, about 13, and 24% at refrigerated, room, and elevated 40°C temperature respectively. However, albuterol sulfate was more stable in the four-drug combination. See *Compatibility with Other Drugs* below.

**STUDY 4:** The amount of foam generated during nebulization of commercial 0.083% albuterol solution (Proventil, Schering) was noted by Marino et al.[161] to be considerably more than an equivalent dose of the concentrate diluted with 0.9% sodium chloride. The authors speculated that the more pronounced foaming was due to the higher concentration of the cationic surfactant preservative benzalkonium chloride. Although foaming is more prominent with the commercial product compared to the extemporaneously diluted concentrate, the droplet size remains unchanged and the amount of drug residue is minimal, indicating that the foam does not cause an inadequate dose to be delivered.[162]

**STUDY 5:** Gammon et al.[1303] reported on the stability of a number of drugs used by paramedics when exposed to the temperature range that is found in ambulances as documented by Brown et al.[1304] and Allegra et al.[1318] Undiluted albuterol sulfate 0.083% solution was stored at temperatures that cycled every 24 hours from −6 to 54°C (2.12 to 129.2°F) for 28 days. The drug was exposed to a total of 336 hours at each of the temperature extremes. The mean kinetic temperature was 33°C. HPLC and ultraviolet spectrophotometry found drugs losses of about 11% occurred over the 28-day test period.

### Enteral
Ortega de la Cruz et al.[1101] reported the physical compatibility of an unspecified amount of oral liquid albuterol sulfate (Ventolin, Glaxo) with 200 mL of Precitene (Novartis) enteral

nutrition diet for a 24-hour observation period. No particle growth or phase separation was observed.

## Rectal

Taha et al.[725] reported the stability of albuterol sulfate in four rectal suppository formulations present in a concentration of 2 mg per suppository. The formulas were selected from a variety of other formulas because of their acceptable physical properties and highest release of drug. The formulas used are listed here from best to worst stability:

1. Witepsol W25 plus methylcellulose 3%
2. Witepsol H15 plus methylcellulose 3%
3. Suppocire NA plus Eudispert hv gel 6%
4. PEG 4000 plus PEG 1000 (1:1)

The bases were melted and the albuterol sulfate was incorporated with thorough mixing to yield a uniform dispersion. The finished albuterol sulfate 2-mg suppositories were packaged in glass containers and stored at room temperature protected from exposure to light for a year.

The suppositories demonstrated acceptable physical characteristics, including hardness, melting time, and drug uniformity. Stability-indicating HPLC analysis found that formulas 1, 2, and 3 maintained adequate stability, with 94, 93, and 91% of the albuterol sulfate remaining after 12 months of storage. However, the PEG formula stability was unacceptable, with only 62% of the albuterol sulfate remaining after 12 months.

## Compatibility with Other Drugs

### ALBUTEROL SULFATE COMPATIBILITY SUMMARY

*Compatible with:* Budesonide • Dexamethasone • Fluticasone propionate • Ipratropium bromide • Sodium chloride 7% (hypertonic) • Tobramycin sulfate

*Incompatible with:* Acetylcysteine • Atropine sulfate • Dornase alfa • Gentamicin sulfate • Sodium bicarbonate

*Uncertain or variable compatibility with:* Colistin • Cromolyn sodium

**STUDY 1 (MULTIPLE DRUGS):** Owsley and Rusho[517] evaluated the compatibility of several respiratory therapy drug combinations. Albuterol sulfate 5-mg/mL inhalation solution (Proventil, Schering) was combined with the following drug solutions: acetylcysteine 10% solution (Mucosil-10, Dey), atropine sulfate 1 mg/mL (American Regent), cromolyn sodium 10-mg/mL nebulizer solution (Intal, Fisons), dexamethasone (form unspecified) 4 mg/mL (American Regent), and sodium bicarbonate 8.4% injection (Abbott). The test solutions were filtered through 0.22-μm filters into clean vials. The combinations were evaluated over 24 hours (temperature unspecified) using the USP particulate matter test. Only the albuterol sulfate–dexamethasone combination

was found to be compatible. Albuterol sulfate combined with acetylcysteine, atropine sulfate, cromolyn sodium, and sodium bicarbonate all formed unacceptable levels of larger particulates (≥10 μm).

**STUDY 2 (IPRATROPIUM):** Jacobson and Peterson[652] evaluated the stability of admixtures of ipratropium bromide and albuterol sulfate nebulizer solutions mixed in equal quantities. The drugs were found to retain more than 90% of the initial concentrations after five days stored at 4 and 22°C protected from light or at 22°C exposed to fluorescent light.

**STUDY 3 (CROMOLYN):** Emm et al.[628] evaluated the stability and compatibility of cromolyn sodium combinations during short-term preparation and use. Cromolyn sodium 1% nebulizer solution (Fisons) was combined with albuterol sulfate 0.5% (Glaxo) and studied over 90 minutes. No visible signs of precipitation or measured changes in pH were found, and HPLC analysis found less than 10% loss of either drug within 90 minutes. The authors did note that unpublished reports indicate that a turbid hazy mixture may result from admixing cromolyn sodium solution with albuterol sulfate 0.083%.

**STUDIES 4 AND 5 (BUDESONIDE):** Smaldone et al.[1026] and McKenzie and Cruz-Rivera[892] evaluated the compatibility and stability of two concentrations of budesonide inhalation suspension with other inhalation medications, including albuterol sulfate. Budesonide (Pulmicort Respules, AstraZeneca) 0.25 mg/2 mL and also 0.5 mg/2 mL was mixed with albuterol sulfate 2.5 mg/0.5 mL. The inhalations were added to new Pari LC Plus nebulizer cups and mixed using a vortex mixer. The samples were examined initially after mixing and over 30 minutes.

Visual inspection found no change in color, formation of a precipitate, or nonresuspendability of any of the samples. Stability-indicating HPLC analysis found that the drugs remained adequately stable throughout the study period, with little or no loss of budesonide or albuterol sulfate in most of the samples over 30 minutes.

**STUDY 6 (MULTIPLE DRUGS):** Roberts and Rossi[897] reported the compatibility of albuterol sulfate (Ventolin) as both single-use and multiple-use forms with budesonide suspension (Pulmicort Respules), cromolyn sodium, and ipratropium bromide inhalation as both single-use (Atrovent UDV) and multiple-use (Atrovent) forms. Visual inspection found that the budesonide and ipratropium bromide combinations were compatible, with no visible cloudiness. The combinations with cromolyn sodium (Intal) were more problematic. The unpreserved single-use form of albuterol sulfate was compatible with cromolyn sodium while the multiple-use form preserved with benzalkonium chloride was not. Marked and immediate cloudiness appeared due to the formation of an oily complex of cromolyn sodium with benzalkonium chloride.

**STUDY 7 (MULTIPLE DRUGS):** Blondino and Baker[930] evaluated the physical and chemical stability of four common drugs administered by inhalation; albuterol sulfate 2.5 mg/3 mL, budesonide 0.25 mg/3 mL, cromolyn sodium 20 mg/3 mL, and ipratropium bromide 0.5 mg/3 mL, individually in an acidified (pH 4) 15% ethanol in water solution and in a four-drug combination. Budesonide was dissolved in ethanol, which was then added to a solution of the other drugs dissolved in the acidified sterile water for irrigation. No visible changes in color or clarity were observed in any sample. Albuterol sulfate by itself underwent substantial losses over eight weeks. Losses were 11, about 13, and 24% at refrigerated, room temperature, and elevated temperature of 40°C. However, albuterol sulfate was more stable in the four-drug combination, exhibiting little or no loss under refrigeration, 3% loss at room temperature, and about 10% loss at 40°C in eight weeks. The other three drugs remained stable under refrigeration and at room temperature with no more than 6% loss in eight weeks, but budesonide and ipratropium bromide underwent substantial losses of 20 to 25% at 40°C in that time frame.

**STUDY 8 (COLISTIN):** Roberts et al.[1022] evaluated the compatibility and stability of albuterol sulfate (Delta West) 5 mg/mL and colistimethate sodium (Coly-Mycin M Parenteral) 33.3 mg/mL of colistin base in equal volumes for use as an inhalation solution in the treatment of cystic fibrosis. Albuterol sulfate inhalation solution that contained benzalkonium chloride as a preservative caused the rapid formation of a cloudy precipitate of the colistin. Precipitation persisted for at least 10 hours before finally resolving. Use of preservative-free albuterol sulfate inhalation solution did not result in a cloudy precipitate.

HPLC analysis found albuterol losses of 7 and 14% in 1 and 24 hours, respectively. Evaluation of colistin by bioassay found 2% loss in one hour, but no additional testing was performed because the 14% loss of albuterol was considered unacceptable.

**STUDY 9 (TOBRAMYCIN):** Gooch[1023] evaluated the compatibility and stability of albuterol sulfate (Ventolin) 1.4 mg/mL and tobramycin sulfate (Nebcin) 11.4 mg/mL in sodium chloride 0.9% for inhalation in treating infections in cystic fibrosis patients. Samples were prepared in screw-cap glass containers and stored refrigerated at 4°C for seven days. No precipitation or color change was observed. HPLC analysis found no loss of albuterol in seven days. Fluorescence immunoassay of tobramycin found no loss in seven days as well.

**STUDY 10 (MULTIPLE DRUGS):** Hood and White[1025] reported that albuterol sulfate inhalation solution was compatible with tobramycin sulfate, although no drug concentrations were stated. Adding ipratropium bromide to the inhalation admixture was also stated to be compatible, again without citing drug concentrations.

**STUDY 11 (IPRATROPIUM AND CROMOLYN):** Nagtegaal et al.[1046] evaluated the long-term stability of two inhalation solution formulations containing albuterol sulfate 1.5 mg/mL with ipratropium bromide 62.5 mcg/mL, both with and without cromolyn sodium 5 mg/mL. The drugs were compounded in a solution composed of sodium chloride 8.8 mg/mL in water for injection. The pH was adjusted to 5.6 for the formulation with cromolyn sodium and to 5.0 without the cromolyn. The inhalation solutions were filtered through 0.2-μm filters, packaged in glass ampules, and autoclaved at 121°C for 15 minutes for sterilization. Samples of the sterilized solutions were stored at room temperature and also at elevated temperatures of 45, 65, and 85°C for accelerated degradation studies.

Ipratropium was the least stable component of the inhalation solutions. Stability-indicating HPLC analysis found the albuterol sulfate with ipratropium bromide inhalation solution (no cromolyn sodium) to be stable, with less than 10% ipratropium loss occurring in 18 months at room temperature and a calculated stability of 3.8 years refrigerated. With cromolyn sodium present in the three-drug inhalation solution, the ipratropium was less stable, exhibiting less than 10% loss for 10 months at room temperature and a calculated stability of 1.9 years refrigerated.

**STUDY 12 (DORNASE ALFA):** Kramer et al.[1216] evaluated the compatibility and stability of mixtures of inhalation solution admixtures with dornase alfa (Pulmozyme, Genentech). The contents of one ampule of dornase alfa (Pulmozyme Respule) were mixed with ipratropium bromide (Atrovent and Atrovent LS, Boehringer Ingelheim) and albuterol sulfate (Sultanol and Sultanol forte FI, GlaxoSmithKline) in the following combinations, and the mixtures were stored at room temperature exposed to room light:

*Mixture 1*
Pulmozyme 2.5 mg/2.5 mL
Atrovent 0.5 mg/2 mL

*Mixture 2*
Pulmozyme 2.5 mg/2.5 mL
Sultanol forte FI 0.5 mL

*Mixture 3*
Pulmozyme 2.5 mg/2.5 mL
Atrovent LS 2 mL
Sultanol 0.5 mL

*Mixture 4*
Pulmozyme 2.5 mL
Atrovent 0.5 mg/2 mL
Sultanol forte FI 0.5 mL

Dornase alfa was physically incompatible with the unpreserved unit-dose forms of Atrovent and Sultanol forte FI, with visible formation of particulates upon mixing. Dornase alfa activity measured using a kinetic colorimetric DNase activity assay remained unchanged over the five-hour test period. Stability-indicating HPLC analysis found that the ipratropium bromide and albuterol sulfate concentrations also remained unchanged.

However, dornase alfa mixed with Atrovent LS and Sultanol inhalation solution, which both contain the preservative benzalkonium chloride, resulted in rapid loss of dornase alfa activity, with losses of 20 to 50% within one to two hours. The ipratropium bromide and albuterol sulfate concentrations remained unchanged. Dornase alfa was found to lose activity due to the preservative benzalkonium chloride and also edetate disodium found in the Atrovent LS formulation. These combinations also resulted in visible particulate formation upon mixing.

Because the unpreserved formulations of ipratropium bromide and albuterol sulfate caused particulate formation when mixed with dornase alfa and because the preserved formulations of these drugs caused extensive and rapid loss of dornase alfa activity along with particulate formation, all of these tested mixtures of inhalation solutions were determined to be incompatible.

**STUDY 13 (FLUTICASONE AND IPRATROPIUM):** Kamin et al.[1231] evaluated the physicochemical compatibility and stability of fluticasone-17-propionate nebulizer suspension (Flutide forte "ready-to-use," GlaxoSmithKline) mixed with ipratropium bromide (Atrovent LS, Boehringer Ingelheim) and albuterol sulfate (Sultanol, GlaxoSmithKline) for combined use in a nebulizer. The test mixtures were prepared in 10-mL glass containers. Fluticasone-17-propionate 2 mg/2 mL was mixed with ipratropium bromide 0.25 mg/2 mL and albuterol 2.5 mg/0.5 mL (as sulfate). No physical changes were observed during the five-hour study period. Stability-indicating HPLC analysis of the drug concentrations found the drug concentrations remained near 100% over five hours as well.

**STUDY 14 (HYPERTONIC SODIUM CHLORIDE):** Fox et al.[1239] evaluated the physical compatibility of hypertonic sodium chloride 7% with several inhalation medications used in treating cystic fibrosis. Albuterol (as sulfate) respiratory solution (Dey) 0.083% mixed in equal quantities with extemporaneously compounded hypertonic sodium chloride 7% did not exhibit any visible evidence of physical incompatibility, and the measured turbidity did not increase over the one-hour observation period.

**STUDY 15 (DORNASE ALFA, IPRATROPIUM, TOBRAMYCIN):** Kramer et al.[1383] evaluated the compatibility and stability of a variety of two- and three-drug solution admixtures of inhalation drugs, including albuterol sulfate. The mixtures were evaluated using chemical assays including HPLC, DNase activity assay, and fluorescence immunoassay as well as visual inspection, pH measurement, and osmolality determination. The drug combinations tested and the compatibility results that were reported are shown below.

*Mixture 1*
Albuterol sulfate (Sultanol) 5 mg/1 mL
Dornase alfa (Pulmozyme) 2500 units/2.5 mL
Ipratropium bromide (Atrovent LS) 0.25 mg/1 mL
    Result: Physically and chemically incompatible

*Mixture 2*
Albuterol sulfate (Sultanol Forte) 2.5 mg/2.5 mL
Dornase alfa (Pulmozyme) 2500 units/2.5 mL
    Result: Physically and chemically incompatible

*Mixture 3*
Albuterol sulfate (Sultanol) 5 mg/1 mL
Ipratropium bromide (Atrovent LS) 0.25 mg/1 mL
Tobramycin (Tobi) 300 mg/5 mL
    Result: Physically and chemically compatible

*Mixture 4*
Albuterol sulfate (Sultanol) 5 mg/1 mL
Ipratropium bromide (Atrovent LS) 0.25 mg/1 mL
Tobramycin sulfate (Gernebcin) 80 mg/2 mL
    Result: Physically and chemically compatible

**STUDY 16 (GENTAMICIN SULFATE):** Tomonaga et al.[1605] evaluated the compatibility and stability of gentamicin sulfate injection mixed with six drugs for inhalation including Ventolin (albuterol sulfate). Using a microbiological activity assay, it was found that in all of these inhalation solutions the gentamicin sulfate content deteriorated substantially in most combinations within one day and within all combinations within three days stored both under refrigeration at 4°C and at room temperature of 20 and 30°C. ■

# Alendronate Sodium

## Properties
Alendronate sodium is a white free-flowing crystalline non-hygroscopic powder.[1,3,4,7]

### Solubility
Alendronate sodium is soluble in water; very slightly soluble in methanol, dimethyl sulfoxide, and propylene glycol; and practically insoluble in acetone, ethanol, isopropanol, and chloroform.[3,4,7]

### pH
Alendronate sodium 1% solution in water has a pH of 4 to 5.[3]

### pK$_a$
Alendronic acid has pK$_a$ values of 2.72, 8.73, 10.5, and 11.6.[1]

## General Stability Considerations
Alendronate sodium powder is packaged in well-closed containers and stored at controlled room temperature. Alendronate sodium commercial oral tablets are packaged in tight containers and stored at controlled room temperature.[4] Alendronate sodium oral solution is stored at 25°C and protected from freezing.[7]

## Stability Reports of Compounded Preparations
### Dental
Reddy et al.[1168] reported on an extemporaneously prepared topical alendronate sodium 1% dental gel (formula follows) for the treatment of bone resorptive lesions in periodontitis.

**TOPICAL DENTAL GEL FORMULA**

| | |
|---|---|
| Alendronate sodium | 1% |
| Carbopol 934P | 1, 1.5, and 2% |
| Trolamine (Triethanolamine) | 1% |
| Methylparaben | 0.1% |
| Propylparaben | 0.05% |
| Distilled water | qs |

Alendronate sodium powder was dissolved in the required amount of distilled water. Weighed amounts of Carbopol 934P to yield 1, 1.5, and 2% concentrations were added to the alendronate sodium solution and were allowed to soak for two hours. Trolamine (triethanolamine) was added to form the gel. The pH was adjusted to 6.8. (The authors did not indicate how the pH was adjusted.) Methylparaben and propylparaben were dissolved in ethanol (ethanol amount not specified) and added to the gel. The gel was sterilized by autoclaving for 30 minutes at 121°C.

Drug release studies performed at 37°C found that the amount of alendronate sodium released varied by Carbopol 934P concentration and, therefore, viscosity. The formulas containing Carbopol 934P 1, 1.5, and 2% released about 50, 40, and 30%, respectively, in seven hours. A clinical study of patients receiving local delivery of the alendronate sodium gel for six months found potent inhibition of bone resorption, increased bone formation, and statistically significant improvement in clinical parameters of periodontitis. ∎

# Allergen Extracts

## Properties
Allergen extracts are a category of injectable sterile products used in the diagnosis and immunotherapeutic treatment of allergies. There are hundreds of allergens that are available as sterile aqueous extracts from allergenic source material in extracting solution, singly and in various mixtures, for intradermal or subcutaneous use after dilution with a sterile diluent to a concentration appropriate for the sensitivity of the individual. Highly sensitive individuals may require dilution to as much as 1:10,000,000 (vol/vol). Allergen extracts must never be administered intravenously.

Allergen extract injections often contain additional substances such as sodium chloride, sodium bicarbonate, glycerin up to 50%, and antimicrobial preservatives such as phenol.[7,1571]

## General Stability Considerations
The manufacturers state that allergen extracts must be stored under refrigeration at 2 to 8°C and protected from freezing.[7]

Allergen extract concentrates are diluted for use; sterile albumin and normal saline with phenol 0.4% or sterile glycerin 50% may be used for dilution. Like the allergen extract concentrates, the dilutions should also be stored refrigerated at 2 to 8°C.[7,1571]

Glycerin down to 10% and human serum albumin have been shown to preserve the activity of allergens compared with simple dilution in sodium chloride 0.9%.[1572] Others have shown the superiority of glycerin 50% and high concentrations of sugars, which probably prevents denaturation of the allergen proteins and inhibits enzymatic activity that modifies the allergen.[1573–1575] Phenol 0.4%, an antimicrobial preservative, has been shown not to have a significant detrimental effect on the stability of allergen extracts.[1576–1578]

The FDA defines expiration dates for allergen extracts in glycerin 50% to be three years after shipping and allows the manufacturer three years of refrigerated storage before shipping for a total possible period of six years. Aqueous saline-phenol extracts are given 1.5 years after shipping and another 1.5 years of refrigerated storage before shipping for a total possible period of three years. However, precipitation can dramatically reduce the shelf life of an allergen extract. A number of commercial allergen extracts have been found to have activity reductions of 0 to 17% within these time periods.[1571]

All allergen extracts lose activity over time, even under ideal storage conditions; a fresh extract could have a much higher activity than an older batch even though it is the same formula and concentration. Consequently, when beginning a new lot of allergen extract, a substantially reduced initial dose (such as 50% reduction) is required. Similarly, an allergen extract from a different manufacturer will have differences in the manufacturing methods that may result in different product potencies necessitating a substantially reduced initial dose. In addition, if changing from an alum-absorbed or alum-precipitated allergen extract to an aqueous or glycerinated one, the safest course is to restart the treatment as if the patient had never had an allergen extract previously.[7]

## Stability Reports of Compounded Preparations
### Injection

Injections, like other sterile drugs, should be prepared in a suitable clean air environment using appropriate aseptic procedures. When prepared from nonsterile components, an appropriate and effective sterilization method must be employed.

Plunkett[1571] evaluated the stability of several allergen extracts diluted with human serum albumin (HSA), glycerin 50% (GLY50), and sodium chloride 0.9% (NS). The commercial allergen extracts were diluted 1:125 vol/vol and also 1:625 vol/vol. The dilutions all contained phenol 0.4% as an antimicrobial preservative. The diluted allergen extracts were stored under refrigeration for 12 months. Using enzyme-linked immunoabsorbent assays (ELISAs) to determine the percentage of allergen remaining, Plunkett reported the results shown below. Lowering the allergen concentration can cause a detrimental effect on the stability of an allergen. Significant loss was found with the allergens tested in periods of three to six months. However, adding human serum albumin to the diluent can slow the loss of the allergens. ■

**TABLE 2.** Major Allergen Content Remaining (%) of Dilutions Stored Refrigerated[1571]

| Allergen Extract | Diluent | 3 months | 6 months | 12 months |
|---|---|---|---|---|
| *Dilution 1:125 vol/vol* | | | | |
| Grass | HSA | 60 | 76 | 51 |
| | GLY50 | 84 | 57 | 30 |
| | NS | 54 | 60 | 16 |
| Bermuda Grass | HSA | 73 | 67 | 43 |
| | GLY50 | 55 | 43 | 15 |
| | NS | 62 | 53 | 9 |
| English Plantain | HSA | 63 | 60 | 54 |
| | GLY50 | 45 | 19 | 19 |
| | NS | 50 | 32 | 22 |
| Red/White Birch | HSA | 93 | 79 | 75 |
| | GLY50 | 66 | 35 | 31 |
| | NS | 82 | 41 | 34 |
| *Dilution 1:625 vol/vol* | | | | |
| Grass | HSA | 79 | 74 | 54 |
| | GLY50 | 45 | 12 | 7 |
| | NS | 29 | 24 | 6 |
| Bermuda Grass | HSA | 68 | 51 | 34 |
| | GLY50 | 32 | 16 | 4 |
| | NS | 26 | 16 | 0 |
| English Plantain | HSA | 53 | - | 32 |
| | GLY50 | 28 | - | 0 |
| | NS | 29 | - | - |
| Red/White Birch | HSA | 76 | 63 | 49 |
| | GLY50 | 71 | 24 | 0 |
| | NS | 41 | 21 | 13 |

# Allopurinol

## Properties

Allopurinol is a fluffy white to off-white almost odorless powder.[1–3,7]

### Solubility

Allopurinol has an aqueous solubility of 0.48 and 0.8 mg/mL at 25 and 37°C, respectively. Solubilities in ethanol and dimethyl sulfoxide (DMSO) are 0.3 and 4.6 mg/mL, respectively. It is soluble in aqueous solutions of alkali hydroxides.[1–3,7]

### pH

The injection of allopurinol sodium has a pH of 10.8.[274]

### $pK_a$

Allopurinol has a $pK_a$ variously reported as 9.4[2] and 10.2.[1]

## General Stability Considerations

Allopurinol tablets should be stored in well-closed, light-resistant containers at controlled room temperature.[2,4,7] Maximum stability of allopurinol in solution is reported to occur at pH 3.1 to 3.4.[551]

## Stability Reports of Compounded Preparations
### Oral

**USP OFFICIAL FORMULATION (ORAL SUSPENSION):**

> Allopurinol 2 g
> Glycerin 5 mL
> Vehicle for Oral Suspension, NF 45 mL
> Vehicle for Oral Solution, NF qs 100 mL

(See the vehicle monographs for information on the individual vehicles.)

Using commercial allopurinol tablets, crush or grind to fine powder. Mix the powdered tablets with the glycerin to form a smooth paste. Add the Vehicle for Oral Suspension; then add the Vehicle for Oral Solution to bring to volume, and mix well yielding allopurinol 20 mg/mL oral suspension. The final liquid preparation should have a pH between 6.5 and 7.5. Adjust the pH if necessary. Transfer to a suitable tight and light-resistant container. Store the preparation at controlled room temperature. The beyond-use date is 60 days from the date of compounding.[4,5]

**STUDY 1 (ORAL SUSPENSION):** The stability of allopurinol 20 mg/mL in an extemporaneous oral suspension was studied by Dressman and Poust.[87] Allopurinol 100-mg tablets (Burroughs-Wellcome) were crushed and mixed with Cologel (Lilly) as one-third of the total volume, shaken, and brought to final volume with a 2:1 mixture of simple syrup and cherry syrup, followed by vigorous shaking for 30 seconds and ultrasonication for at least two minutes. The allopurinol 20-mg/mL suspension was packaged in amber glass bottles and stored at 5°C and ambient room temperature. An HPLC assay was used to determine the allopurinol content. Less than 3% loss occurred in 56 days at either temperature.

**STUDY 2 (ORAL SUSPENSION):** Allen and Erickson[543] evaluated the stability of three allopurinol 20-mg/mL oral suspensions extemporaneously compounded from tablets. Vehicles used in this study were (1) an equal parts mixture of Ora-Sweet and Ora-Plus (Paddock), (2) an equal parts mixture of Ora-Sweet SF and Ora-Plus (Paddock), and (3) cherry syrup (Robinson Laboratories) mixed 1:4 with simple syrup. Eight allopurinol 300-mg tablets (Mylan) were crushed and comminuted to fine powder using a mortar and pestle. About 20 mL of the test vehicle was added to the powder and mixed to yield a uniform paste. Additional vehicle was added geometrically and brought to the final volume of 120 mL, mixing thoroughly after each addition. The process was repeated for each of the three test suspension vehicles. Samples of each of the finished suspensions were packaged in 120-mL amber polyethylene terephthalate plastic prescription bottles and stored at 5 and 25°C in the dark.

No visual changes or changes in odor were detected during the study. Stability-indicating HPLC analysis found less than 5% allopurinol loss in any of the suspensions stored at either temperature after 60 days of storage.

**STUDY 3 (ORAL SUSPENSION):** Alexander et al.[551] evaluated the physical and chemical stability of an extemporaneously compounded allopurinol 20-mg/mL oral suspension prepared from tablets. The formula for 3000 mL of the suspension is shown in Table 3. The suspending agents Veegum 22.5 g and sodium carboxymethylcellulose medium viscosity 22.5 g were each hydrated in 500 mL of water. A total of 200 allopurinol 300-mg tablets (Mylan) were crushed and triturated to fine powder in a glass mortar. The powder was incorporated in portions with the hydrated Veegum and sodium carboxy-methylcellulose. Lycasin (Roquette), a syrup vehicle composed of 75% (wt/wt) maltitol, was added slowly with continuous mixing until a homogeneous mixture was obtained. The parabens concentrate and the sodium bisulfite and saccharin sodium dissolved in 75 mL of water were added to the mixture and homogenized. Wild cherry and vanilla flavorings were incorporated. The mixture was transferred to a graduate and brought to a final volume of 3000 mL with distilled water. The suspension was packaged as 120 mL in six-ounce amber glass prescription bottles with

**TABLE 3.** Allopurinol 20-mg/mL Oral Suspension Formula[551]

| Component | Amount |
|---|---|
| Allopurinol 300-mg tablets | 200 |
| Veegum | 22.5 g |
| Sodium carboxymethylcellulose | 22.5 g |
| Lycasin syrup | 1500 g |
| Sodium bisulfite | 1.5 g |
| Saccharin sodium | 1.5 g |
| Parabens stock solution[a] | 60 mL |
| Wild cherry flavor | 2 mL |
| Vanilla flavor | 2 mL |
| Distilled water | qs  3000 mL |

[a]Methylparaben 10% and propylparaben 2% in neat propylene glycol.

child-resistant caps and stored at elevated temperatures (50 to 80°C).

All of the samples remained homogeneous, with no signs of caking, settling, or changes in viscosity. Redispersibility and pourability remained unchanged. However, a color change from off-white to yellow occurred in these elevated temperature samples. Based on stability-indicating HPLC analysis of the samples at elevated temperatures, the time to 10% decomposition ($t_{90}$) at 25°C was calculated to be about 8.3 years.

**STUDY 4 (MOUTHWASH):** Stiles and Allen[146] described the extemporaneous formulation of an allopurinol mouthwash. Two 250-mL portions of distilled water were prepared; one was chilled and the other was heated to boiling. Methylcellulose 2.5 g was sprinkled on the surface of the hot water as it was stirred; then the chilled water was used to bring the volume to 500 mL. The colloidal vehicle was stirred occasionally until clear. Allopurinol tablets 600 mg were crushed to fine powder and then added to the colloidal methylcellulose vehicle. (Artificial sweeteners and flavoring agents may be added if desired.) The product was packaged in amber bottles and stored under refrigeration. This allopurinol mouthwash suspension is similar to the one described by Dressman and Poust,[87] and the authors indicated that the products should have a similar stability of eight weeks at room or refrigerator temperature.[146]

**STUDY 5 (MOUTHWASH):** Loprinzi et al.[417] reported on the stability of an allopurinol 1-mg/mL mouthwash prepared from tablets. The mouthwash was prepared by crushing one 300-mg tablet in a glass mortar and pestle and mixing the powder with 100 mL of Cologel. The mixture was both shaken and mixed in a blender to create a suspension. This suspension was brought to 300 mL with a 2:1 mixture of simple syrup and cherry syrup. The mouthwash was stored under refrigeration and at room temperature. The allopurinol content was assessed using a stability-indicating HPLC assay. Little or no loss of allopurinol occurred during eight months of storage at either temperature.

Clark and Sievin[148] reported that allopurinol 1 mg/mL in 3% methylcellulose was used as a mouthwash in fluorouracil-induced oral mucositis.

## Injection

Injections, like other sterile drugs, should be prepared in a suitable clean air environment using appropriate aseptic procedures. When prepared from nonsterile components, an appropriate and effective sterilization method must be employed.

**STUDY 1:** Schuster et al.[496] reported the stability of allopurinol sodium as a 10-mg/mL injection in 0.45% sodium chloride injection with the pH adjusted to 10.5 to 11.5 using sodium hydroxide solution. The injection was filtered using a 0.2-μm filter and was filled into vials. The product was sterilized by autoclaving at 121°C for 20 minutes. HPLC analysis of the product found it to be stable for three months under refrigeration and nine months frozen.

**STUDY 2:** Lee and Wang[700] reported the stability of an allopurinol 2.5-mg/mL injection formulated in an equal parts mixture of dimethyl sulfoxide (DMSO) and propylene glycol and packaged in amber glass ampules. The ampules were stored at elevated temperatures of 50 to 70°C to determine the allopurinol decomposition rate. The estimated shelf life was determined to be more than 39 years at room temperature.

NOTE: No biological or toxicological testing to determine the safety of the drug in the unusual cosolvent system was performed. Whether this dosage form would be suitable for actual use is not known.

## Rectal

Lee and Wang[700] evaluated the release of allopurinol from 6-mg suppositories composed of varying ratios of polyethylene glycol (PEG) 4000 and PEG 1500 prepared by the hot melt method. The drug was formulated in 1.21-g suppositories having PEG 4000/PEG 1500 ratios of 10/2.5 to 2.5/10. HPLC analysis of in vitro dissolution found similar results for all of the suppository formulas, which yielded 100% release of the drug. Incorporation of sodium lauryl sulfate in the suppositories decreased the allopurinol release rate substantially and was not recommended. ■

# Aloe

## Properties

Aloe is a genus of succulent plants in the family *Liliaceae* that has triangular spearlike leaves with thorny ridges. The plant is native to parts of Africa, southern Arabia, and Madagascar. *Aloe barbadensis* (also known as Curacao aloe), *Aloe ferox*, *Aloe africana*, and *Aloe spicata* have been used medicinally.[1,3,4]

Aloe vera (*Aloe barbadensis*) gel is the fresh mucilaginous gel from the leaf center used as an emollient and for wound healing. Aloe vera should not be confused with the dried latex of the plant called aloes or drug aloe. Aloe vera does not include the plant's sap, which contains the anthroquinones.[1,3,4]

The dried latex of the plant, also called aloes or drug aloe, has been used as a cathartic.[1,4] The dried juice of *Aloe barbadensis* contains not less than 28% of hydroxyanthracene derivatives, while the dried juice of *Aloe capensis* contains not less than 18% of hydroxyanthracene derivatives. The hydroxyanthracene derivatives occur as dark brown masses tinged with green or as a greenish-brown powder and have a sour, disagreeable odor.[3,4]

### Solubility

Aloe yields not less than 50% water-soluble extractives.[4] The hydroxyanthracene derivatives are partially soluble in boiling water and soluble in hot ethanol.[3]

## General Stability Considerations

Aloe extracts should be stored in airtight containers and protected from exposure to light.[3]

## Stability Reports of Compounded Preparations
### Topical

Kodym and Bujak[820] evaluated the stability of several formulations of topical ointments of 3% aloe extract with and without added neomycin sulfate. The concentration of the two principal aloe constituents, aloenin and aloin, were determined by thin-layer chromatography and ultraviolet spectroscopy. The antimicrobial activity of neomycin was determined by microbiological assay.

The most successful ointment was prepared in white Vaseline, liquid paraffin, solid paraffin, and cholesterol (amounts not specified). No loss of either principal aloe component or of neomycin activity occurred during two years of storage at 20°C. However, ointment formulations that contained water or propylene glycol underwent about 34 or 42% loss of aloe components and 40 or 19% loss of neomycin activity, respectively. ∎

# Alprazolam

## Properties

Alprazolam is a white or off-white crystalline powder.[2,3] The USP cautions that care should be taken to prevent inhalation of and exposure of the skin to alprazolam particles.[4]

### Solubility

Alprazolam is insoluble in water but soluble in ethanol.[1–3]

## General Stability Considerations

Alprazolam products should be stored in tight, light-resistant containers at controlled room temperatures of 20 to 25°C.[4,7]

## Stability Reports of Compounded Preparations
### Oral

**USP OFFICIAL FORMULATION (ORAL SUSPENSION):**

Alprazolam 100 mg
Vehicle for Oral Solution, NF (sugar-containing or sugar-free)
Vehicle for Oral Suspension, NF (1:1) qs 100 mL

(See the vehicle monographs for information on the individual vehicles.)

Use alprazolam powder or commercial tablets. If using tablets, crush or grind to fine powder. Add about 20 mL of the vehicle mixture and mix to make a uniform paste. Add additional vehicle almost to volume in increments with thorough mixing. Quantitatively transfer to a suitable calibrated tight and light-resistant bottle, bring to final volume with the vehicle mixture, and thoroughly mix yielding alprazolam 1-mg/mL oral suspension. The final liquid preparation should have a pH between pH 4 and 5. Store the preparation at controlled room temperature or under refrigeration between 2 and 8°C. The beyond-use date is 60 days from the date of compounding at either storage temperature.[4,5]

**STUDY 1:** Allen and Erickson[594] evaluated the stability of three alprazolam 1-mg/mL oral suspensions extemporaneously compounded from tablets. Vehicles used in this study were (1) an equal parts mixture of Ora-Sweet and Ora-Plus (Paddock), (2) an equal parts mixture of Ora-Sweet SF and Ora-Plus (Paddock), and (3) cherry syrup (Robinson Laboratories) mixed 1:4 with simple syrup. Sixty alprazolam 2-mg tablets (Geneva) were crushed and comminuted to fine powder using a mortar and pestle. About 40 mL of the test vehicle was added to the powder and mixed to yield a uniform paste. Additional vehicle was added geometrically and brought to the final volume of 120 mL, mixing thoroughly after each addition. The process was repeated for each of the three test suspension vehicles. Samples of each of the finished suspensions were packaged in 120-mL amber polyethylene terephthalate plastic prescription bottles and stored at 5 and 25°C.

No visual changes or changes in odor were detected during the study. Stability-indicating HPLC analysis found 7 to 9% alprazolam loss in the cherry syrup and less than 5% loss in either of the Ora-Plus-containing suspensions stored at either temperature after 60 days of storage.

**STUDY 2:** de la Paz et al.[1437] evaluated the stability of an alprazolam 1-mg/mL oral solution formulation. Other components in the oral solution were not specified. Three batches of alprazolam 1-mg/mL oral liquid were packaged in amber glass containers with polypropylene caps and were stored at room temperature. HPLC analysis found alprazolam losses of 4 to 6% occurred over six months of storage. ■

# ■ Alprostadil

## Properties

Alprostadil is a white to off-white, odorless hygroscopic powder.[3,7]

### Solubility

Alprostadil is essentially insoluble in water, having a solubility of 0.08 mg/mL at 35°C,[7] but is freely soluble in ethanol.[3] Alprostadil complexed to α-cyclodextrin (Edex, Schwartz Pharma) is freely soluble in water but practically insoluble in ethanol.[7]

### pH

Alprostadil complexed to α-cyclodextrin has a pH from 4 to 8 after reconstitution.[7]

## General Stability Considerations

Alprostadil products should be stored under refrigeration at 2 to 8°C.[4] Products may be stored at room temperatures up to 25°C for three months after dispensing.[7] Alprostadil complexed to α-cyclodextrin may be stored at room temperatures up to 25°C; brief temperature excursions up to 30°C are permitted.[7] Intraurethral inserts should be stored at 2 to 8°C and should be protected from temperatures above 30°C. Room temperature storage up to 30°C is permitted for 14 days after dispensing.[7]

Alprostadil was shown to exhibit loss due to sorption to glass vial and plastic syringe surfaces.[2]

## Stability Reports of Compounded Preparations
### Injection

Injections, like other sterile drugs, should be prepared in a suitable clean air environment using appropriate aseptic technique. When prepared from nonsterile components, an appropriate and effective sterilization method must be employed.

**STUDY 1:** Gatti et al.[579] studied the stability of alprostadil (prostaglandin E1) prepared in a physiological solution. Prostin VR injection (Upjohn) is a solution of alprostadil 0.5 mg/mL in ethanol. Prostin VR injection was aseptically diluted in an unspecified physiological solution, yielding a concentration of 10 mcg/mL. The dilution was stored at about 4°C in the dark, at ambient temperature in the dark, and at ambient temperature exposed to sunlight. HPLC analysis found that the samples stored under refrigeration exhibited about 10% loss in 30 days. The samples at ambient temperature with or without exposure to light were less stable, exhibiting 15 to 18% loss in 15 days. The following studies found much longer stability periods under refrigeration.

**STUDY 2:** Fraccaro et al.[924] reported the stability of alprostadil (Prostin VR, Upjohn) diluted to a concentration of 10 mcg/mL in sodium chloride 0.9% or in sterile water for injection, packaged in glass vials, and stored under refrigeration. HPLC analysis found less than 10% alprostadil loss occurred over 90 days at 2 to 8°C. The drug loss reached 13% after 120 days.

**STUDY 3:** Shulman and Fyfe[925] evaluated the shelf life of alprostadil (Prostin VR, Upjohn) diluted to 20 mcg/mL in sodium chloride 0.9% and packaged in glass ampules. Based on decomposition rates at elevated temperatures, shelf lives for 5% drug loss were calculated to be 4.8 days at 25°C and 52 days at 4°C. Shelf lives for 10% drug loss were calculated to be 9.8 days at 25°C and 106 days at 4°C.

**STUDY 4:** Uebel et al.[919] evaluated the stability of alprostadil (prostaglandin E1) diluted in normal saline. Prostin VR injection (Pharmacia and Upjohn) was diluted to a concentration of 40 mcg/mL in sodium chloride 0.9% and packaged as 0.5 mL of the dilution in polypropylene plastic syringes. HPLC analysis found less than 10% drug loss occurred in six weeks stored at room temperature near 21°C and in 24 weeks refrigerated at 5°C.

**STUDY 5 (THREE-DRUG MIXTURE):** Trissel and Zhang[876] evaluated the stability of a three-drug mixture of alprostadil (prostaglandin E1) 12.5 mcg/mL, papaverine hydrochloride

4.5 mg/mL, and phentolamine mesylate 0.125 mg/mL (commonly called the "Knoxville Formula") prepared from commercial injections diluted in bacteriostatic sodium chloride 0.9%. The injection mixture was packaged in commercial empty sterile vials and stored at room temperature near 23°C, refrigerated at 4°C, and frozen at −20 and −70°C.

All the samples remained clear and colorless throughout the study. Stability-indicating HPLC analysis found that alprostadil was the least stable of the three drug components and was the limiting factor in the combination injection. At room temperature, alprostadil losses of 8 and 13% occurred in five and seven days, respectively. Under refrigeration, losses of about 6% in one month and 11% in two months occurred. Frozen at −20 and −70°C, alprostadil losses did not exceed 5% in six months. Subjecting vials frozen at −20°C to four freeze–thaw cycles with warming the vials to room temperature each time resulted in no loss of any drug. The authors recommended a beyond-use date of six months or one month

stored frozen at −20°C or under refrigeration, respectively, for prepared batches that have passed a sterility test. The authors also recommended that room temperature exposure be limited and that vials should be returned to refrigeration as soon as possible after use.

**STUDY 6: (TWO- AND THREE-DRUG MIXTURES):** Soli et al.[1531] evaluated the stability of alprostadil (Upjohn) 0.004 mg/mL in two- and three-drug combinations with papaverine hydrochloride (Fluka) 6 mg/mL and phentolamine mesylate (Ciba-Geigy) 0.4 mg/mL diluted in sodium chloride 0.9%. The injection mixtures were packaged in sterile glass vials and stored refrigerated at 2 to 8°C for 60 days.

No visible precipitation or color change was reported. HPLC analysis found alprostadil losses of about 10 to 13% occurred in five days. No loss of papaverine hydrochloride and about 3 to 4% loss of phentolamine mesylate occurred in 60 days. ■

# Alteplase
## (t-PA)

## Properties
Alteplase, a glycoprotein consisting of 527 amino acids, is a white to off-white powder that is a recombinantly produced plasminogen activator. Alteplase has a specific activity of 580,000 IU/mg.[2,7]

### Solubility
Alteplase has an aqueous solubility of less than 0.4 mcg/mL. Consequently, arginine is incorporated into the formulation, which substantially increases the aqueous solubility.[2,7]

### pH
The reconstituted injection has a pH of approximately 7.3[7] with a range of 7.1 to 7.5.[4]

### Osmolality
The reconstituted injection has an osmolality of 215 mOsm/kg.[7]

## General Stability Considerations
Alteplase products should be stored between 2 and 30°C and protected from excessive light.[4,7] A vacuum should be present in the 50-mg vials, but no vacuum is present in the 100-mg vials. Alteplase is stable in the range of pH 5 to 7.5. The manufacturer states that the reconstituted solution should be stored at 2 to 30°C and should be used within eight hours, because of the absence of antibacterial preservatives. Exposure of the reconstituted solution to light does

not adversely affect solution stability within its use period.[2,7] Alteplase exposure to ultrasound for 60 minutes in the presence or absence of Optison had no adverse effects on the stability of the drug.[1268]

## Stability Reports of Compounded Preparations
### Injection
Injections, like other sterile drugs, should be prepared in a suitable clean air environment using appropriate aseptic procedures. When prepared from nonsterile components, an appropriate and effective sterilization method must be employed.

**STUDY 1:** Jaffe et al.[622] reported on repackaging tissue plasminogen activator in syringes for frozen storage. A 50-mg vial of alteplase (Genentech) reconstituted with 50 mL of sterile water for injection (1 mg/mL) was diluted 1:4 in balanced saline solution to yield 250 mcg/mL. Aliquots of 0.3 mL (75 mcg) of the diluted alteplase were drawn into 1-mL tuberculin syringes, capped with a 30-gauge needle, and stored frozen at −70°C. (The needle was replaced with a fresh, sterile needle immediately before administration to patients.) Activity was assessed using a solid-phase fibrin assay technique. Activity was found to be retained under these conditions for at least one year. Furthermore, culturing for bacterial and fungal contamination was negative.

**STUDY 2:** Ward and Weck[623] objected to this dilution and frozen storage technique. They noted that the alteplase formulation is optimized for stability, and dilution to a concentration lower than 500 mcg/mL could compromise the solubility of the protein by diluting the solubilizing effect of the excipients. Additionally, only 0.9% sodium chloride or 5% dextrose solution should be used; using a diluent containing calcium or magnesium salts may create a reaction with the phosphate present in the alteplase formulation, resulting in precipitation. In trying the dilution reported by Jaffe et al.,[622] Ward and Weck found precipitated protein forming after 24 hours at room temperature. Storage frozen at −20°C also resulted in changes in light scattering upon thawing. The authors recommended not diluting alteplase with balanced saline solution and not storing solutions for any length of time at room temperature or at −20°C.

**STUDY 3:** Grewing et al.[624] reported on a modified formulation of alteplase to permit −20°C storage. Reconstituted alteplase injection 1 mg/mL was diluted with additional polysorbate 80, L-arginine, and phosphoric acid solution to a concentration of 50 mcg/mL. The exact composition was not identified but it likely duplicated the vehicle of the alteplase injection. The dilution with this excipient vehicle prevented protein precipitation when stored frozen at −20°C. Clinical activity of doses of 10 to 15 mcg (0.2 to 0.3 mL) used in ophthalmic procedures

was found to be unchanged after frozen storage for up to six months.

**STUDY 4:** Wiernikowski et al.[774] reported the stability of an extemporaneously prepared dilution of alteplase (Genentech) 1 mg/mL packaged in 2 mL plastic syringes and stored frozen at −30°C. A commercial activity assay (Spectrolyse/fibrin t-PA) was used to assess the drug's activity. Frozen at −30°C, at least 95% of the alteplase activity was retained for 22 weeks.

**STUDY 5:** Calis et al.[886] reported the stability of alteplase (Genentech) 1 mg/mL in sterile water for injection packaged in polypropylene syringes frozen at −20°C for six months. They also evaluated the stability of similar solutions in glass vials frozen at −70°C for two weeks, thawed and kept at 23°C for 24 hours, and then refrozen at −70°C for 19 days. In all samples little or no loss of alteplase bioactivity occurred.

Allen[129] commented that an alteplase dilution of 250 mcg/mL in 0.9% sodium chloride could be prepared and repackaged in syringes that could then be placed in individual plastic bags prior to freezing. These individual bags could then serve as the dispensing package. ▪

# ▪ Alum
## (Potassium Alum; Aluminum Potassium Sulfate)

## Properties

Alum is the dodecahydrate form of aluminum potassium sulfate and occurs as colorless, transparent, odorless, hard, large crystals or as a white crystalline powder. It has a sweetish astringent taste.[1,3,4] If subjected to a temperature around 200°C, it will become anhydrous aluminum potassium sulfate also known as burnt alum.[1]

### Solubility

Alum is soluble in water to about 138.9 mg/mL and to about 3.3 gm/mL in boiling water. Alum is freely but slowly soluble in glycerol but insoluble in ethanol.[1,3,4]

### pH

Aqueous solutions of alum are acidic; a 0.2-M solution has a pH near 3.3.[1,3]

## General Stability Considerations

Alum should be packaged in tight containers and should be stored at controlled room temperature.[4]

## Stability Reports of Compounded Preparations
### Topical

Allen[1487] reported on a compounded formulation of alum 6.5% topical cream for use in responsive dermatological conditions. The cream had the following formula:

| | |
|---|---|
| Alum | 6.5 g |
| Cetostearyl alcohol | 4 g |
| Octyldodecanol | 5 g |
| Lanolin alcohol | 4 g |
| Ethoxylated castor oil | 2 g |
| White petrolatum | 2 g |
| Cetylpyridium ammonium chloride | 2.5 g |
| Purified water | 74 g |

The recommended method of preparation was to melt the cetostearyl alcohol, ethoxylated castor oil, lanolin alcohol, octyldodecanol, and white petrolatum at 60°C. Dissolve the alum and cetylpyridium ammonium chloride in the purified water previously heated to 62°C. The two phases should then be combined and mixed well while cooling to form a uniform mixture. The cream should be packaged in a convenient container for topical use. The author recommended a beyond-use date of six months at room temperature because this formula is a commercial medication in some countries with an expiration date of two years or more. ▪

# ▪ Ambroxol Hydrochloride

## Properties
Ambroxol hydrochloride is a white or yellowish crystalline powder.[2,3]

### Solubility
Ambroxol hydrochloride is sparingly soluble in water and soluble in methanol.[3]

### pH
An aqueous 1% solution of ambroxol hydrochloride has a pH of 4.5 to 6.[3]

## General Stability Considerations
Ambroxol hydrochloride powder should be packaged in well-closed containers and stored at controlled room temperature and protected from exposure to light.[3]

## Stability Reports of Compounded Preparations
### Enteral
Ortega de la Cruz et al.[1101] reported the physical compatibility of an unspecified amount of oral liquid ambroxol hydrochloride (Mucosan, Boehringer Ingelheim) with 200 mL of Precitene (Novartis) enteral nutrition diet for a 24-hour observation period. No particle growth or phase separation was observed. ▪

# ▪ Amikacin Sulfate

## Properties
Amikacin sulfate occurs as a white or almost white crystalline powder.[1,3,4] Amikacin sulfate injection is colorless to pale yellow or light straw in color.[7,8]

### Solubility
Amikacin sulfate is freely soluble in water, but practically insoluble in ethanol and acetone.[3,4]

### pH
A 1% aqueous solution of amikacin sulfate has a pH of 2 to 4.[3,4] Amikacin sulfate injection has a pH in the range of 3.5 to 5.5.[7]

### Osmolality
Amikacin sulfate 500 mg in 100 mL of dextrose 5% and sodium chloride 0.9% has osmolalities of 319 mOsm/kg and 349 mOsm/kg, respectively.[8]

## General Stability Considerations
Amikacin powder should be packaged in airtight containers and stored at controlled room temperature.[3,4] Amikacin sulfate injection is stored at controlled room temperature.[7] Amikacin sulfate and its solutions are subject to air oxidation, resulting in color darkening. The color change is not an indication of drug concentration. Autoclaving commercial amikacin sulfate vials resulted in no loss of drug.[1137]

Amikacin sulfate in concentrations of 0.25 to 5 mg/mL is stable (1) for at least 60 days refrigerated followed by 24 hours at room temperature; (2) frozen at $-15°C$ for 30 days, thawed and stored at room temperature for 24 hours; and (3) frozen at $-15°C$ for 30 days, thawed and stored refrigerated for 24 hours, then stored at room temperature for 24 hours.[1138] Amikacin sulfate 20 mg/mL in dextrose 5% stored frozen at $-20°C$ for 30 days exhibited less than 6% drug loss.[1139]

Amikacin sulfate, like other aminoglycoside antibiotics, is incompatible with penicillins and other beta-lactam antibiotics.[8]

## Stability Reports of Compounded Preparations
### Injection
Injections, like other sterile drugs, should be prepared in a suitable clean air environment using appropriate aseptic procedures. When prepared from nonsterile components, an appropriate and effective sterilization method must be employed.

Zbrozek et al.[1140] evaluated the stability of amikacin sulfate 750 mg with 1 mL of added sodium chloride 0.9% (750 mg/4 mL) packaged in Becton Dickinson polypropylene syringes. About 2% drug loss occurred in 48 hours when stored at 23°C exposed to fluorescent light.

## Ophthalmic

Ophthalmic preparations, like other sterile drugs, should be prepared in a suitable clean air environment using appropriate aseptic procedures. When prepared from nonsterile components, an appropriate and effective sterilization method must be employed.

**STUDY 1:** Chedru-Legros et al.[1219] evaluated the physical and chemical stability of amikacin sulfate fortified ophthalmic solution prepared from the commercial injection. Amikacin sulfate (Bristol-Myers Squibb) was diluted in sodium chloride 0.9%, yielding a 50-mg/mL concentration. The solution was passed through a 0.22-μm Millipore filter, packaged in clear glass containers, and stored frozen at −20°C for the 75-day test period. The amikacin sulfate solutions had a pH of 6.5 and an osmolality of 367 mOsm/kg, neither of which changed substantially throughout the study. HPLC analysis found little or no change in drug concentration over the 75-day study period. The authors recommended that the amikacin sulfate fortified ophthalmic solution could be stored frozen for 75 days. However, after thawing, refrigerated storage and discarding after three days was recommended.

**STUDY 2:** Chedru-Legros et al.[1591] also evaluated the physical and chemical stability of several fortified ophthalmic solutions. Amikacin sulfate injection powder (Bristol-Myers Squibb) 50 mg/mL in balanced salt solution (Alcon), dextrose 5% (B. Braun), and sodium chloride 0.9% (B. Braun) packaged in 20-mL glass bottles fitted with pipettes mounted on screw caps were stored frozen at −20 °C protected from exposure to light for six months. The solution was filtered through 0.22-μm Millipore filters as it was added to each bottle. No visible instability was observed, and osmolality and pH were acceptable. HPLC analysis of amikacin sulfate in sodium chloride 0.9% (but not the other diluents) found little or no change in amikacin sulfate concentration at any time point during six months of frozen storage. The authors concluded such fortified antibiotic solutions could be stored frozen at −20 °C for six months.

**STUDY 3:** Demir-bas et al.[1459] evaluated the stability of amikacin sulfate 20-mg/mL eyedrops in sodium chloride 0.9% compounded aseptically. The eyedrops were packaged in opaque glass vials and were stored at 2 to 8°C. During three months of refrigerated storage, no precipitation or discoloration was observed, and the pH of 5.8 and the osmolality of 323 mOsm/kg did not change. HPLC analysis found that the amikacin concentration did not change by more than 5% over the three-month test period.

**STUDY 4:** Sikora et al.[1528] evaluated the stability of two amikacin sulfate 3-mg/mL formulations with increased viscosities.

In addition to amikacin sulfate, the formulations contained polyvinyl alcohol 16.7 and 22.2 mg/mL to increase viscosity to about 3.6 and 7.1 cP, respectively, along with phosphate buffer, benzalkonium chloride 0.1% preservative, and sodium chloride 8 mg/mL for isotonicity. The solution was sterilized by filtration through a 0.3-μm prefilter followed by a 0.22-μm sterilizing filter. The eyedrops were packaged in polyethylene 5-mL dropper bottles and sealed with caps. Samples were stored at room temperatures of 25 and 30°C at 60% relative humidity for 12 months.

The ophthalmic drops remained clear and colorless and exhibited little or no change in pH, osmolality, and viscosity. Microbiological assays found little or no loss of amikacin sulfate activity occurred at either temperature.

**STUDY 5:** Lin et al.[1620] evaluated the antibiotic activity of amikacin sulfate 1% and 10% in dextrose 5% extemporaneously prepared for use as eyedrops by microbiological assay. Amikacin injection was diluted in dextrose 5% to concentrations of 10 and 100 mg/mL. Samples were stored refrigerated at 4°C and stored frozen at −18°C for 28 days. The 1% concentration was for clinical use while the 10% concentration was for use as a stock solution. The MICs indicated that the differences between time zero and 28 days were statistically insignificant.

## Compatibility with Other Drugs

**STUDY 1 (VANCOMYCIN):** Lin et al.[955] reported the activity retention and physiological characteristics of a mixed ophthalmic solution of amikacin (Bristol-Myers Squibb) 20 mg/mL and vancomycin (Lilly) 50 mg/mL prepared from reconstituted injections in sterile water for injection. The mixed antibiotic solution packaged in standard ophthalmic dispensing bottles remained clear and colorless throughout 14 days of storage refrigerated at 4°C. The admixed ophthalmic solution had a pH of 5 to 5.2 and an osmolarity of about 200 mOsm/L. Antimicrobial activity evaluated by the disk diffusion method found no significant differences in the zones of inhibition.

**STUDY 2 (VANCOMYCIN):** Hui et al.[1153] reported the compatibility of amikacin sulfate 0.4 mg/0.1 mL and vancomycin hydrochloride 1 mg/0.1 mL for ophthalmic use prepared in sodium chloride 0.9% and in balanced salt solution plus (BSS Plus), both with and without dexamethasone sodium phosphate 0.4 mg/0.1 mL. The drugs were mixed together in 4 mL of sodium chloride 0.9%, BSS Plus, and vitreous obtained from cadaver eyes. The samples were incubated at 37°C. Amikacin sulfate and vancomycin hydrochloride concentrations were evaluated using TDx analysis, while dexamethasone sodium phosphate was assayed using HPLC.

No precipitation was observed for amikacin sulfate mixed with vancomycin hydrochloride in all three media and in the human vitreous either with or without dexamethasone sodium phosphate present. In addition, no substantial loss of either antibiotic occurred within 48 hours. However, about 13% dexamethasone sodium phosphate loss occurred. The authors indicated that the amikacin–vancomycin two-drug combination was preferred for use in the treatment of infective endophthalmitis. ∎

# Amiloride Hydrochloride

## Properties

Amiloride hydrochloride as the dihydrate is a yellow to greenish-yellow crystalline powder.[2,3]

### Solubility

Amiloride hydrochloride has an aqueous solubility of 5.2 mg/mL[2,753] and a solubility in ethanol of 19.6 mg/mL at 25°C.[2] It is freely soluble in dimethyl sulfoxide (DMSO) but is practically insoluble in acetone and ethyl acetate.[1,3]

### $pK_a$

Amiloride hydrochloride has a $pK_a$ of 8.7.[2,7]

## General Stability Considerations

Amiloride hydrochloride commercial tablets should be stored in well-closed containers at controlled room temperature and should be protected from exposure to moisture, freezing, and excessive heat.[2,7]

Amiloride hydrochloride stability in aqueous media was found to be substantially better in the pH range of 4 to 5 compared to the stability near pH 6 to 7.[753] Amiloride in solutions at alkaline pH undergoes photodegradation at a threefold faster rate than at acidic pH.[824]

## Stability Reports of Compounded Preparations
### Oral

Fawcett et al.[753] reported the stability of amiloride hydrochloride 1 mg/mL oral liquid preparations. Amiloride hydrochloride tablets (Midamor, Merck) and powder (Sigma) were dissolved in glycerin 40% in sterile water with and without hydroxybenzoate 0.1%. The oral liquids were packaged in amber high-density polyethylene prescription bottles with polypropylene screw caps. The oral liquids prepared from tablets were stored refrigerated at 2 to 8°C and at room temperature of 23 to 27°C protected from exposure to light. The oral liquids prepared from powder were stored at room temperature of 23 to 27°C only.

The visual appearance of all samples remained unchanged over 90 days. Stability-indicating HPLC analysis found the stabilities of the various formulations varied by source of drug and whether the preservative was present; the formulation prepared from bulk powder was much more stable. The times to 10% drug loss ($t_{90}$) at room temperature were 61 days and more than 90 days for the powder-prepared formulation (about pH 4 to 5) without and with hydroxybenzoate, respectively. For preparations from tablets (about pH 6 to 7), the $t_{90}$ at room temperature was 5 and 7.5 days without and with hydroxybenzoate, respectively. Refrigeration of the preparations from tablets extended the $t_{90}$ to 30 days and to more than 90 days without and with hydroxybenzoate, respectively. The difference in stability periods was attributed to the lower pH of the preparations made from powder. ∎

# Aminacrine
# Aminacrine hydrochloride
## (Aminoacridine hydrochloride)

## Properties

Aminacrine occurs as yellow needlelike crystals.[1] Aminacrine hydrochloride occurs as a pale yellow crystalline material.[1]

### Solubility

Aminacrine is freely soluble in ethanol and is soluble in acetone.[1] Aminacrine hydrochloride has an aqueous solubility of about 3.3 mg/mL yielding a faint yellow solution with a bluish-violet fluorescence. It has a solubility in ethanol of about 6.7 mg/mL.[1]

### pH

Aminacrine is a moderately strong base.[1] Aminacrine hydrochloride is near neutrality in aqueous solution.[1]

### $pK_a$
Aminacrine has a $pK_a$ of 4.53.[1]

## Stability Reports of Compounded Preparations
### Vaginal Cream

Allen[1707] reported on a cream formulation of aminacrine hydrochloride 0.1% for infections of the vagina and exocervix, including moniliasis and trichomonal vaginitis. The cream formulation was prepared using the following formula:

| | | |
|---|---|---|
| Aminacrine hydrochloride | 100 mg | |
| Thymol | 5 mg | |
| Glyceryl monostearate | 9.5 g | |
| Cetostearyl alcohol | 3.2 g | |
| Polyoxyl 40 stearate | 1.9 g | |
| Liquid paraffin | 10 g | |
| Cetrimide | 450 mg | |
| Isopropyl alcohol | 1 mL | |
| Perfume | qs | |
| Purified water | qs | 100 g |

To prepare the cream, the glyceryl monostearate, cetostearyl alcohol, polyoxyl 40 stearate, and 5 g of liquid paraffin were heated to 60°C. The aminacrine hydrochloride was mixed with the remainder of the liquid paraffin and slowly heated to 60°C. The thymol and perfume were dissolved in the isopropyl alcohol. The cetrimide was dissolved in 74 mL of purified water and heated to 60°C. The aminacrine hydrochloride mixture, cetrimide mixture, and the first multicomponent mixture were combined with stirring to form an emulsion and allowed to cool to about 45°C. Then the thymol and perfume mixture was added and mixed well while the completed mixture cooled. Additional purified water (if needed) was then added to bring to 100 g and mixed well. The cream was packaged in a tight, light-resistant container.

The author stated that the cream could receive a beyond-use date of six months according to USP standards. ■

# ▧ Aminohippurate Sodium
## (PAH)

## Properties
para-Aminohippuric acid occurs as a white crystalline powder that discolors upon exposure to light.[3,4]

### Solubility
para-Aminohippuric acid is sparingly soluble in water, but it is water soluble as the sodium salt.[1]

### pH
The commercial aminohippurate sodium 200-mg/mL injection has a pH of 6.7 to 7.6.[7]

### pKₐ
Aminohippurate sodium has a $pK_a$ of 3.83.[7]

## General Stability Considerations
para-Aminohippuric acid should be stored protected from light[3,4] to avoid excessive oxidation and color darkening.[991] The commercial injection is stored at controlled room temperature. This injection varies in color from colorless to yellowish brown. The efficacy of the drug is not affected by color variations within this color range.[7]

## Stability Reports of Compounded Preparations
### Injection
Injections, like other sterile drugs, should be prepared in a suitable clean air environment using appropriate aseptic procedures. When prepared from nonsterile components, an appropriate and effective sterilization method must be employed.

Boersma et al.[772] reported the stability of extemporaneously compounded aminohippurate sodium 250-mg/mL injection and projected the shelf life. The formulation that was evaluated is shown in Table 4. The sodium salt of para-aminohippuric acid was formed in situ during preparation. Accelerated decomposition of the drug at 80°C over one week was determined using HPLC analysis. Based on the absence of decomposition, the stability of the injection was projected to be 24 months at room temperature. ■

**TABLE 4.** Aminohippurate Sodium 250-mg/mL Injection[772]

| Component | | Amount |
|---|---|---|
| para-Aminohippuric acid | | 1000 g |
| Sodium metabisulfite | | 4 g |
| Sodium hydroxide | | 210 g |
| Sodium hydroxide solution 4 moles/L | qs | pH 7 |
| Water for injection | qs | 4000 mL |

# Aminolevulinic Acid
## (ALA)

## Properties

Aminolevulinic acid as the hydrochloride is a white to off-white odorless crystalline solid material used as a photosensitizer.[7]

### Solubility

Aminolevulinic acid as the hydrochloride is very soluble in water and slightly soluble in ethanol and methanol. It is practically insoluble in mineral oil.[7]

## General Stability Considerations

Aminolevulinic acid in solution was found to be stable at acidic pH values below 6[809,966-969] but is unstable at physiologic or alkaline pH values.[807,809,966-969] In aqueous solution at pH 2.35, aminolevulinic acid exhibited no decomposition in 37 days at 50°C. However, at pH 4.8, the half-life ($t_{50}$) was only 257 hours. At pH 7.4, the half-life was three hours.[809] Aminolevulinic acid stability in solutions was also found to be greater at lower concentrations compared to higher ones.[809] In addition to pH and concentration, the drug's stability is also dependent on the degree of oxygenation of the solution.[966,969]

The commercial topical product (Levulan Kerastick, Berlex) should be stored at controlled room temperature. After dissolution of the drug in the Levulan Kerastick, the manufacturer indicates that it should be used within two hours and discarded after that time.[7]

## Stability Reports of Compounded Preparations
### Injection

Injections, like other sterile drugs, should be prepared in a suitable clean air environment using appropriate aseptic procedures. When prepared from nonsterile components, an appropriate and effective sterilization method must be employed.

de Blois et al.[807] evaluated the stability of 5-aminolevulinic acid formulated for intracutaneous injection as a photosensitizer for photodynamic therapy of basal cell carcinomas. The drug was formulated at concentrations ranging from 0.1 to 5%, with the 2% concentration being isotonic. A 1% solution was found to have an osmolarity of 142 mOsm/L. The addition of sodium chloride to reach isotonicity was required for all concentrations below 2%. Autoclaving was found to result in essentially total loss of the drug and cannot be used. The aminolevulinic acid injections developed a yellow discoloration that progressed to orange and deep red with decomposition. HPLC analysis found faster decomposition at higher drug concentrations. The authors recommended a shelf life of 90 days to ensure less than 10% drug loss for formulations in the 0.1 to 5% range.

### Topical

**STUDY 1:** Valenta et al.[943] reported the stability and skin penetration of aminolevulinic acid using a new gel formulation prepared in a manner similar to a previous study.[944] The gel was prepared by heating the emulgins, polyoxyethylene-5-cetyloleylether 0.75 g and polyoxyethylene-10-cetyloleylether 0.75 g, to 70 to 80°C with 0.8 g of cetylsteryl-2-ethylhexanote. The 2.75 g of hot water was poured into the mixture and stirred vigorously. A clear semisolid very elastic gel was formed. About 1.5% aminolevulinic acid was then incorporated into the gel. The gel was stored at 4°C for 90 days. The gel remained transparent throughout the study, and HPLC analysis found that no loss of aminolevulinic acid occurred. Porcine skin penetration of the drug from the gel was about 66% compared to 80% for a patch. However, the aminolevulinic acid was less stable in the patch with about 60% loss in 14 days.

**STUDY 2:** McCarron et al.[966] evaluated the stability of aminolevulinic acid in two nonaqueous topical organogel formulations. The gels consisted of 5% (wt/wt) Carbopol ETD 2050 in anhydrous glycerol or in anhydrous polyethylene glycol 400. After overnight refrigerated storage to allow for removal of air bubbles, the aminolevulinic acid was added with the aid of gentle stirring to a final concentration of 20% (wt/wt). Test samples were stored for up to six months refrigerated at 5°C, at room temperature, and at elevated temperatures of 37 and 60°C. HPLC analysis found extensive and rapid drug loss in the elevated temperature samples and to a somewhat lesser extent in the room temperature samples. The drug loss was accompanied by orange or brown discoloration, which is characteristic of the decomposition products of aminolevulinic acid. The refrigerated samples lost approximately 10 to 15% in about two to three weeks and more than 50% in 204 days. The polyethylene glycol gel samples began to develop a yellow discoloration as the end of the study approached, but the glycerol gel samples remained colorless. Neither of the gels demonstrated stability that was as good as bioadhesive patches, which lost about 14% in 196 days at 5°C. ■

# Aminophylline

## Properties

Aminophylline, a combination of theophylline (84 to 87%) with ethylenediamine (13.5 to 15%), is a white or slightly yellow powder or granules with a slightly ammoniacal odor and bitter taste.[1-3]

### Solubility

Aminophylline has an aqueous solubility of 200 mg/mL but may require the addition of ethylenediamine to ensure complete dissolution. It is insoluble in ethanol.[1-3]

### pH

Aminophylline oral solution has a pH of 8.5 to 9.7. The enema has a pH of 9 to 9.5. The injection pH is 8.6 to 9.[2-4]

### pKa

Aminophylline has a pKa of 5.[2]

### Osmolality

The osmolality of the injection is 170 mOsm/kg.[8]

Aminophylline liquid 21 mg/mL (Fisons) had an osmolality of 450 mOsm/kg.[233,905]

## General Stability Considerations

Aminophylline exposed to the air may gradually absorb carbon dioxide and free theophylline, becoming turbid or developing crystals. Consequently, aminophylline oral products should be stored in tight containers and should be protected from light. The suppositories should be stored under refrigeration, while the injection may be stored at controlled room temperature. Refrigeration encourages crystal formation in some injections; the injections should be inspected carefully for particulate matter or discoloration prior to use.[1-4] Although light protection is recommended, no loss of theophylline 50-mg/mL solution was found after eight weeks of storage during which the solution was exposed to fluorescent light.[278]

Aminophylline 40 mg/mL or less is reported to be stable over pH 3.5 to 8.6 for at least 48 hours at 25°C.[275] Edward[276] noted the potential for formation of theophylline crystals at pH values below 8. At 4 mg/mL, aminophylline was stable for 86 days.[277]

## Stability Reports of Compounded Preparations

### Injection

Injections, like other sterile drugs, should be prepared in a suitable clean air environment using appropriate aseptic procedures. When prepared from nonsterile components, an appropriate and effective sterilization method must be employed.

Nahata et al.[279] evaluated the stability of aminophylline (Abbott) 5 mg/mL in bacteriostatic 0.9% sodium chloride injection with 0.9% benzyl alcohol in plastic syringes (Becton Dickinson). The aminophylline content was determined using HPLC analysis. After 91 days of storage, aminophylline losses were 2 and 3% at 4 and 22°C, respectively.

### Oral

**STUDY 1:** Swerling[77] evaluated the stability of an aminophylline 5-mg/mL dilution prepared from Somophylline (Fisons) with sterile water. The preparation was packaged in one-ounce amber bottles and stored for 86 days (temperature unspecified). The theophylline content was determined using an enzyme-multiplied immunoassay (EMIT) technique. No change in the theophylline concentration occurred during the study.

**STUDY 2:** Chong et al.[938] reported the stability of aminophylline 3 and 22 mg/mL in oral suspension formulations. Commercial aminophylline injection 25 mg/mL (Abbott) was diluted in an equal parts (1:1) mixture of Ora-Sweet and Ora-Plus to yield the 3- and 22-mg/mL concentrations. The oral suspensions were packaged in 50-mL amber glass prescription bottles and stored for 91 days at refrigerated temperature (4°C) and at room temperature (25°C) exposed to normal laboratory fluorescent light. The 3-mg/mL concentration was physically stable with no change in appearance, odor, or pH (pH range 5.5 to 6). HPLC analysis found no loss of aminophylline during 91 days of storage. Similar results were found for the 22-mg/mL concentration stored at room temperature. No change in appearance, odor, or pH (pH range 8.5 to 9.2) and no loss of aminophylline by HPLC analysis occurred in 91 days. However, refrigerated storage of the 22-mg/mL concentration at 4°C was unacceptable due to development of crystalline precipitation within seven days and an accompanying 20 to 25% loss of drug upon analysis.

### Rectal

**STUDY 1:** Brower et al.[270] reported the decomposition of aminophylline incorporated into suppository bases. The ethylenediamine component of the aminophylline reacted with fatty acids in cocoa butter and in coconut and palm kernel oils, resulting in a progressive loss in the ethylenediamine content. Total loss of ethylenediamine occurred in about 3.5 months at 40°C. The theophylline content was much less affected, with about a 7% loss in four months at 40°C. The suppositories became hard and the melting point increased until it exceeded the temperature of a steam bath. A large amount of a white waxy material was found in the hardened suppositories. The ethylenediamine component is believed to react with glyceryl esters in suppository bases to form

**TABLE 5.** Aminophylline Suppository Formulas Tested for Stability[424]

| Formula | Aminophylline | Witepsol | | | Polyethylene Glycol | | Water | Number Made |
| | | H-15 | E-75 | W-35 | 6000 | 1540 | | |
|---|---|---|---|---|---|---|---|---|
| 1 | 74 g | 1 kg | | | | | | 740 |
| 2 | 74 g | | 1 kg | | | | | 740 |
| 3 | 74 g | | | 1 kg | | | | 740 |
| 4 | 74 g | 500 g | 500 g | | | | | 740 |
| 5 | 100 g | | | | 850 g | 510 g | 260 mL | 1000 |

insoluble high-melting-point amide decomposition products. The reaction may occur during the preparation process and during storage at room temperature.

**STUDY 2:** Fujii et al.[424] evaluated five suppository formulations containing aminophylline 100 mg per suppository. The suppository formulas are shown in Table 5. In formulas 1 through 4, the bases were melted at 70°C and then cooled to 45°C, at which point sieved (100 mesh) aminophylline powder was added and mixed. The mixture was transferred into a suppository plugger at 32 to 36°C and squeezed into 1.5-mL containers. For formula 5, the polyethylene glycol 6000 and polyethylene glycol 1540 were melted at 80°C. Aminophylline was dissolved in distilled water heated to 80°C. The polyethylene glycols and aminophylline solution were mixed at 70°C, and the mixture was cooled while being stirred. At 55°C, the mixture was transferred to a suppository plugger and squeezed into containers with stirring at 47°C. Suppositories were stored at 5 and 24°C for 12 months. The contents of ethylenediamine and theophylline were estimated using wet chemistry reactions.

The appearance of decomposition products was observed by thin-layer chromatography.

The physical properties of all suppository formulas met acceptance criteria. However, progressive discoloration occurred during storage in all formulas, with 3 being the most discolored (dark yellow) and 5 the least discolored (light yellow). Both theophylline and ethylenediamine levels were initially uniform among all formulas, but formulas 2 and 3 had ethylenediamine contents 10% less than the theophylline level, an indication of ethylenediamine decomposition. After 12 months of storage, theophylline levels exhibited no loss in any formulation at either temperature. However, ethylenediamine content decreased in formulas 1 through 4, with nearly total losses in 12 months at room temperature and losses of 13 to 34% under refrigeration. The polyethylene glycol formula, 5, retained 90 and 83% of the ethylenediamine after 6 and 12 months of storage at room temperature, respectively. No loss occurred in the refrigerated samples. Thin-layer chromatography identified two decomposition products from aminophylline suppositories made with the Witepsol bases.[424] ■

# ■ 4-Aminopyridine
(4-AP)

## Properties

4-Aminopyridine is a white crystalline odorless solid material.[883]

### Solubility

4-Aminopyridine is moderately soluble in water to about 80 mg/mL. The hydrochloride salt has an aqueous solubility of 500 mg/mL.[883]

## General Stability Considerations

4-Aminopyridine as the pure solid material is stated to be stable at normal storage temperatures as long as it is kept dry.[883]

## Stability Reports of Compounded Preparations
### Oral

Trissel et al.[721] reported the stability of 4-aminopyridine compounded in 5-mg oral hard-gelatin capsules. In addition to the drug, each capsule also contained silica gel micronized and lactose hydrous, NF. The capsules were packaged in amber polypropylene prescription vials and stored for one month at 37°C and for six months at room temperature near 23°C and refrigerated at 4°C.

No visible changes or weight changes occurred in the capsules during storage. Stability-indicating HPLC analysis found the drug remained stable under all of the storage conditions with little loss in six months at room temperature or refrigerated and about 4% loss in one month at 37°C. ■

# Aminosalicylate Sodium
## (PAS Sodium)

## Properties

Aminosalicylate sodium (para-aminosalicylate dihydrate sodium) occurs as a practically odorless white to cream crystalline powder.[1,4] It has a sweet saline taste.[4]

### Solubility

Aminosalicylate sodium is freely soluble in water having an aqueous solubility of about 500 mg/mL. It is sparingly soluble in ethanol and acetone and only very slightly soluble in ether and chloroform.[1,4]

### pH

A aminosalicylate dihydrate sodium 1% aqueous solution has a pH of 7.[1] A 2% aqueous solutions has a pH in the range of 6.5 to 8.5.[4]

## General Stability Considerations

Aminosalicylate sodium powder and oral tablets should be packaged in tight, light-resistant containers and stored at controlled room temperature protected from excessive heat. Aminosalicylate sodium solutions decompose slowly and become darker in color. Solutions should be used within 24 hours after preparation, and no solution should be used if it is darker in color than freshly prepared solutions.[4]

## Stability Reports of Compounded Preparations
### Oral

Hergert et al.[1713] evaluated the stability of a compounded formulation of aminosalicylate sodium for oral administration. The formulation consisted of a powder blend prepared using a double cone mixer and containing desiccated aminosalicylate sodium mixed with aspartame and vanillin to mask the taste and colloidal silicon dioxide as an adsorbent and stabilizer. The final powder mixture contained aminosalicylate sodium 5.6 g or 92.87% (wt/wt), aspartame 1.66% (wt/wt), vanillin 0.83% (wt/wt), and colloidal silicon dioxide 4.64% (wt/wt) and was packaged as 6.03 g of powder blend in 60-mL tight, light-resistant glass bottles. The drug was prepared for use as an oral solution by adding 60 mL of water to the bottles to dissolve the powder blend just before administration.

Samples of the packaged powder blend were stored at 23 to 27°C and 60% relative humidity for three months. The powder blend remained easily dissolved over the storage period. Stability-indicating HPLC analysis found little loss of intact aminosalicylate sodium with the formation of only 0.18% decomposition product. ∎

# Amiodarone Hydrochloride

## Properties

Amiodarone hydrochloride is a white to cream crystalline powder.[2,3]

### Solubility

At 25°C, amiodarone hydrochloride has a solubility in water of about 0.72 mg/mL and a solubility in ethanol of 12.8 mg/mL.[2,3]

### pH

The pH of the injection is reported to be 4.08.[8]

### pKa

Amiodarone has a pKa of about 6.6.[2]

## General Stability Considerations

Amiodarone hydrochloride products should be stored in tight containers at controlled room temperature and protected from light.[2]

## Stability Reports of Compounded Preparations
### Oral

**STUDY 1:** The extemporaneous formulation of an amiodarone hydrochloride 5-mg/mL oral suspension was described by Nahata and Hipple.[160] Three amiodarone hydrochloride 200-mg tablets were crushed and mixed thoroughly with 90 mL of 1% methylcellulose and 10 mL of syrup. The suspension was brought to 120 mL with purified water. A stability period of seven days under refrigeration was used; however, chemical stability testing was not performed.

Nahata[606] subsequently evaluated the stability of an amiodarone 5-mg/mL oral suspension compounded from tablets. Amiodarone 200-mg tablets (Wyeth-Ayerst) were crushed to powder using a mortar and pestle. An equal parts mixture of simple syrup and 1% methylcellulose was added to the tablet powder with mixing, yielding a 5-mg/mL concentration. The suspension was packaged in two-ounce plastic and glass prescription bottles and stored under refrigeration at 4°C and in a temperature-controlled water bath at

25°C. Stability-indicating HPLC analysis found that about 8 to 10% loss occurred in 42 days at 25°C and in 91 days at 4°C in both types of containers.

**STUDY 2:** Nahata et al.[857] reported the stability of amiodarone 5-mg/mL oral suspensions compounded from tablets and using commercially available vehicles. Amiodarone 200-mg tablets (Cordarone, Wyeth-Ayerst) were crushed to powder using a mortar and pestle. The powder was incorporated geometrically into equal quantity (1:1) mixtures of Ora-Sweet and Ora-Plus and also Ora-Sweet SF (sugar free) and Ora-Plus. The pH was adjusted to 6 to 7 with sodium bicarbonate 5% solution. The amiodarone 5-mg/mL suspension was packaged in plastic prescription bottles and stored at room temperature of 25°C and refrigerated at 4°C for 91 days.

No change in physical appearance or odor occurred. HPLC analysis found the room temperature samples were stable, with about 7 and 10% losses in 42 and 56 days, respectively. The refrigerated samples were stable for 91 days, with about 8% loss. The authors recommended refrigerated storage because of the potential for microbiological growth. However, short-term room temperature storage (as during travel) was satisfactory.

**STUDY 3:** Alexander and Thyagarajapuram[864] reported the shelf life of an extemporaneously compounded oral suspension of amiodarone hydrochloride 20 mg/mL prepared from commercial tablets (Paceron, Upsher Smith). The proper number of commercial tablets was triturated using a mortar and pestle. Tween 80 1.25 mL was added to the tablet powder with trituration. The vehicle that was used consisted of carboxymethylcellulose sodium (viscosity 7MF) 0.75%, Veegum 0.75%, sucrose 25% (from simple syrup), and strawberry flavor concentrate. The suspension was packaged in amber glass bottles and stored under refrigeration at 4 to 6°C, at room temperature of 30°C, and at elevated temperatures of 40, 50, and 60°C.

At room temperature and under refrigeration, no visible changes were observed. Stability-indicating HPLC–MS analysis was used to determine the decomposition rate of amiodarone at each of the temperatures, and the data were used to project stability periods. Based on these results, the shelf life was calculated to be 193 days at room temperature of 25°C and 677 days refrigerated at 4°C. ■

# ■ Amitriptyline Hydrochloride

## Properties
Amitriptyline hydrochloride occurs as white or almost white, odorless or nearly odorless crystals or a crystalline powder with a bitter burning taste.[1,2,3,4]

### Solubility
Amitriptyline hydrochloride is freely soluble in water, ethanol, and chloroform.[1,3,4] The drug is insoluble in ether.[4]

### pH
Amitriptyline hydrochloride 1% in water has a natural pH between 5 and 6. Amitriptyline hydrochloride injection has a pH range of 4 to 6.[4]

### $pK_a$
Amitriptyline hydrochloride has a $pK_a$ of 9.4.[1]

## General Stability Considerations
Amitriptyline hydrochloride bulk material and oral tablets should be packaged in well-closed, light-resistant containers and stored at controlled room temperature. Amitriptyline hydrochloride injection is to be packaged in single- or multiple-dose Type I glass containers and stored at controlled room temperature protected from light and excessive heat or freezing.[4,7]

Amitriptyline hydrochloride in solution exposed to light forms a ketone and, in three or four days, a precipitate. In the absence of light, these effects were not observed.[1410] Autoclaving amitriptyline hydrochloride dissolved in water or pH 6.8 phosphate buffer for 30 minutes at 115°C in the presence of oxygen resulted in decomposition of the drug. The presence of sodium metabisulfite or ferric and cupric ions increases the rate of amitriptyline hydrochloride decomposition.[1411]

## Stability Reports of Compounded Preparations
### Oral
Gupta[1409] evaluated the stability of amitriptyline hydrochloride 1 mg/mL prepared in two oral liquid formulations. The oral liquids were prepared by mixing amitriptyline hydrochloride powder with flavoring and 2 mL of glycerin. Sufficient simple syrup was added to bring to volume, and the preparation was mixed well. A second version of the formulation was also prepared incorporating edetate disodium 0.1% into the formulation. The completed liquid preparations were packaged in amber glass prescription bottles and stored at 25°C for 37 days.

Both formulas did not change in physical appearance throughout the study. The formula with edetate disodium 0.1% exhibited a reduction in pH from about pH 5 to 4. The formula without the edetate disodium maintained the pH near 5 throughout the study. Stability-indicating HPLC analysis found that the formula with edetate disodium 0.1% was more stable, exhibiting no loss of drug in 37 days. The formula without the edetate disodium exhibited more rapid drug loss with 9% loss occurring in 21 days and 17% loss occurring in 37 days. The author stated the edetate disodium stabilizes the drug by chelating metal ions that leach from the glass containers. ■

# Amlodipine Besylate

## Properties

Amlodipine besylate is a white or almost white crystalline powder.[1,3,7] Amlodipine besylate 6.9 mg is approximately equivalent to amlodipine 5 mg.[3]

### Solubility

Amlodipine besylate is slightly soluble in water and isopropanol and sparingly soluble in ethanol.[1,3,7]

## General Stability Considerations

Amlodipine besylate powder should be stored in airtight containers and protected from exposure to light.[3] The commercial tablets should be packaged in tight, light-resistant containers and stored at controlled room temperature.[7]

## Stability Reports of Compounded Preparations
### Oral

Nahata et al.[718] reported the stability of amlodipine besylate in two extemporaneously compounded oral suspensions.

Amlodipine besylate 5-mg tablets (Norvasc, Pfizer) were pulverized using a mortar and pestle. The powder was mixed with (1) an equal parts mixture of simple syrup and 1% methylcellulose, and (2) an equal parts mixture of Ora-Plus suspending medium and Ora-Sweet syrup (Paddock), resulting in a 1-mg/mL concentration. The suspensions were packaged in amber plastic prescription bottles and stored at room temperature of 25°C in a water bath and refrigerated at 4°C for 91 days.

No change in physical appearance or odor of the suspensions occurred. Stability-indicating HPLC analysis found that amlodipine besylate was more stable in the Ora-Plus–Ora-Sweet mixture. Drug losses of about 4 and 9% occurred in 91 days at 4 and 25°C, respectively. In the simple syrup–1% methylcellulose mixture, a drug loss of 6% occurred in 91 days at 4°C. However, at room temperature 8% loss occurred in 56 days and 11% loss occurred in 70 days. ■

# Ammonium Lactate

## Properties

Ammonium lactate is prepared by neutralizing DL-lactic acid with ammonium hydroxide.[1,7]

### Solubility

Ammonium lactate is soluble in water, glycerin, and 95% ethanol but practically insoluble in ethanol or isopropanol.[1]

### pH

The 12% cream and lotion have pH values of 4.5 to 5.5.[7]

## General Stability Considerations

The commercial cream and lotion should be stored at controlled room temperature between 15 and 30°C.[7]

## Compatibility with Other Drugs

Patel et al.[635] evaluated the compatibility of calcipotriene 0.005% ointment mixed by levigation in equal quantities with other topical medications. The mixtures were stored at 5 and 25°C. Drug concentration was assessed by HPLC analysis. The combination with ammonium lactate 12% lotion was found to be incompatible. Liquid separation appeared in 24 hours, and there was more than 10% loss of calcipotriene content. ■

# Amoxicillin

## Properties

Amoxicillin (as the trihydrate) is a white or almost white practically odorless crystalline powder aminopenicillin having a bitter taste. Approximately 1.15 g of the trihydrate is equivalent to 1 g of amoxicillin.[1–3,6]

### Solubility

Amoxicillin solubility in water has been listed as 4 mg/mL[1] and 2.7 mg/mL.[6] In absolute ethanol, the solubility has been cited as 3.4 mg/mL.[1]

### pH

A 0.2% aqueous amoxicillin solution has a pH of 3.5 to 6. The oral suspensions have pH values of 5 to 7.5.[2–4] The injectable suspension has a pH of 5 to 7.[4] Amoxicillin suspension (Larotid, Roche) is reported to have a pH of 5.6.[19]

### pKₐ

Amoxicillin has three dissociable protons with $pK_a$ values of 2.63, 7.55, and 9.64 at 23°C.[6,1379]

### Osmolality

Amoxicillin suspension 250 mg/5 mL (Utimox, Parke-Davis) had an osmolality of 1775 mOsm/kg,[232] and the Squibb product had an osmolality of 2250 mOsm/kg.[233,905]

## General Stability Considerations

Amoxicillin oral products should be stored in tight containers at controlled room temperature.[4] Elevated temperature and humidity may sufficiently free the intrinsic water molecules of crystallization to participate in amoxicillin hydrolysis.[6] Following reconstitution, the oral suspensions are stable for 14 days at room or refrigeration temperature. However, refrigeration is the recommended storage condition.[2,3] Freezing aqueous solutions of amoxicillin results in a significantly increased degradation rate.[1263]

Amoxicillin is subject to hydrolytic cleavage of the β-lactam ring, especially at alkaline pH. It also may self-catalyze hydrolysis. Concurrently, the drug also undergoes dimerization, the predominant route of degradation. Dimerization assumes greater importance as the concentration of amoxicillin increases.[6]

The pH of maximum stability is 5.77. The optimum stability range in aqueous solutions is pH 5.8 to 6.5.[464]

## Stability Reports of Compounded Preparations

### Oral

Allen and Lo[65] tested the stability of oral amoxicillin trihydrate (Larotid, Roche) repackaged into unit-dose containers at several temperatures, including frozen storage. The amoxicillin trihydrate was reconstituted to 250 mg/5 mL. Samples of 5 mL were placed in 26-mL amber glass vials with screw-cap closures and stored at −20, −10, 5, and 25°C. Penicillin content was determined spectrophotometrically. Samples frozen at −20°C lost 10% in 60 days; at −10°C, drug losses were 6% in 30 days and 12% in 60 days. However, the effect of freezing on product bioavailability, if any, is unknown.[68,69] Samples under refrigeration lost 7.5% in 30 days and 13% in 60 days. Samples at room temperature lost 7.5% in 10 days.[65]

### Enteral

#### AMOXICILLIN COMPATIBILITY SUMMARY

*Compatible with:* Enrich • Ensure • Ensure Plus • Osmolite • Precitene • TwoCal HN • Vivonex T.E.N.

**STUDY 1:** Cutie et al.[19] added 5 mL of amoxicillin suspension (Larotid, Roche) to varying amounts (15 to 240 mL) of Ensure, Ensure Plus, and Osmolite (Ross Laboratories) with vigorous agitation to ensure thorough mixing. The amoxicillin suspension was physically compatible, distributing uniformly in all three enteral products with no phase separation or granulation.

**STUDY 2:** Burns et al.[739] reported the physical compatibility of amoxicillin suspension (Amoxil) 250 mg/5 mL with 10 mL of three enteral formulas, including Enrich, TwoCal HN, and Vivonex T.E.N. Visual inspection found no physical incompatibility with any of the enteral formulas.

**STUDY 3:** Ortega de la Cruz et al.[1101] reported the physical compatibility of an unspecified amount of oral liquid amoxicillin (Clamoxyl, Glaxo) with 200 mL of Precitene (Novartis) enteral nutrition diet for a 24-hour observation period. No particle growth or phase separation was observed. ■

# Amoxicillin Trihydrate–Clavulanate Potassium
## (Co-Amoxiclav)

## Properties

Amoxicillin (as the trihydrate) and clavulanate potassium are provided as fixed combination products. (See also *Amoxicillin* monograph.) Clavulanate potassium is a white to off-white crystalline powder that is a β-lactamase inhibitor. Approximately 1.19 g of the potassium salt is equivalent to 1 g of clavulanic acid.[2,3]

### Solubility

Amoxicillin solubility in water has been listed as 4 mg/mL[1] and as 2.7 mg/mL.[6] In absolute ethanol, the solubility has been cited as 3.4 mg/mL.[1] Clavulanate potassium is very soluble in water and slightly soluble in ethanol.[2,3]

### pH

Reconstituted oral suspensions have pH values of 4.2 to 6.6.[4]

### pKₐ

Amoxicillin has three dissociable protons with $pK_a$ values of 2.63, 7.55, and 9.64 at 23°C.[6,1379] The $pK_a$ of clavulanic acid is 2.7.[2]

## General Stability Considerations

Amoxicillin and clavulanate potassium products should be stored in tight containers at below 24°C.[2–4] (See *Amoxicillin* monograph.) Exposure to humidity should be avoided due to the moisture sensitivity of clavulanate potassium.[1157] The optimum stability range of amoxicillin in aqueous solutions has been reported as pH 5.8 to 6.5.[464,1379] Clavulanate potassium stability in aqueous solution is maximal between pH 6 and 6.3[1157] or near pH 6.4[1401] with the stability more sensitive to alkaline pH than to acid pH.[1400] Clavulanate potassium is the least stable component of the drug combination.[1157] Clavulanic acid stability is also influenced by ionic strength of the solution with increased degradation of about 4% by sodium sulfate, magnesium sulfate, calcium chloride, and sodium chloride in decreasing order of effect.[1400] The manufacturer indicates that reconstituted amoxicillin and clavulanate potassium suspensions stored under refrigeration at 2 to 8°C are stable for 10 days.[2,3,7] However, freezing aqueous solutions of amoxicillin and clavulanate potassium results in significantly increased degradation rates for both components.[1263]

## Stability Reports of Compounded Preparations
### Oral

**STUDY 1:** Tu et al.[36] evaluated the stability of reconstituted amoxicillin (as the trihydrate) and clavulanic acid (as the potassium salt) oral suspension (Augmentin, Beecham) 25 mg/mL + 6.25 mg/mL and also of amoxicillin (as the trihydrate) and clavulanic acid (as the potassium salt) 50 mg/mL + 12.5 mg/mL, respectively, both in the original containers and in five different kinds of oral syringes. The suspension was reconstituted with purified water, and 5 mL of the suspension was drawn into the oral syringes or retained in the original containers. The samples were stored at −10, 5, and 25°C for 14 days, and the drug content was assessed using a stability-indicating HPLC technique. Amoxicillin in both concentrations was much more stable than was clavulanate potassium. The calculated times to 10% decomposition ($t_{90}$) for each drug in each container at 5 and 25°C are given in Table 6. The results confirm the manufacturer's recommended expiration dating for refrigerated storage of the suspension in the original containers. None of the oral syringes stored under refrigeration met the expiration date requirement. There was little or no degradation of amoxicillin trihydrate or clavulanate potassium in any container over 14 days when stored frozen at −10°C.

**STUDY 2:** Mehta et al.[1157] evaluated the stability of co-amoxiclav oral suspension in the original containers stored at room temperature of 20°C and under refrigeration at 8°C. HPLC stability results found the drug combination was stable for two days at room temperature and seven days refrigerated. This is a shorter time period than the 11 days that was found by Tu et al.[36] for the drug in the original containers. The manufacturer stated that co-amoxiclav oral suspension is stable under refrigeration for 10 days in the original container.

### Enteral

Ortega de la Cruz et al.[1101] reported the physical compatibility of an unspecified amount of oral liquid co-amoxiclav (Augmentin, Glaxo) with 200 mL of Precitene (Novartis) enteral nutrition diet for a 24-hour observation period. No particle growth or phase separation was observed. ∎

**TABLE 6.** Stability of Amoxicillin Trihydrate and Clavulanate Potassium Oral Suspension in Various Containers[36]

| Container | Time (Days) to 10% Decomposition (t90) | | |
|---|---|---|---|
| | Temperature (°C) Amoxicillin[a] 25 mg/mL + | Amoxicillin Clavulanic Acid[b] | Clavulanate 6.25 mg/mL |
| Original bottles | 25 | 18 | 1.3 |
| | 5 | —[c] | 11.1 |
| Exacta-Med (Baxa) | 25 | 9.3 | 0.5 |
| | 5 | — | 4.5 |
| B-D Plastic Oral | 25 | 8.1 | 0.5 |
| Dispensing Syringe | | | |
| (Becton Dickinson) | 5 | — | 1.7 |
| Apex Oral Syringe | 25 | 3.4 | 0.7 |
| (Apex) | 5 | — | 1.2 |
| Perfectip Oral | 25 | 7.5 | 0.6 |
| Syringe (Burron) | 5 | — | 1.4 |
| B-D Glaspak | 25 | 10.3 | 0.6 |
| Pediatric Oral Liquid | | | |
| Administration Device | | | |
| (Becton Dickinson) | 5 | — | 1.5 |
| | Amoxicillin[a] 50 mg/mL + | Clavulanic Acid[b] | 12.5 mg/mL |
| Original bottles | 25 | 14 | 0.5 |
| | 5 | — | 11.1 |
| Exacta-Med | 25 | 8.7 | 0.4 |
| | 5 | — | 3.2 |
| B-D Plastic Oral | 25 | 8.1 | 2.4 |
| Dispensing Syringe | 5 | — | 4.1 |
| Apex Oral Syringe | 25 | 4.7 | 0.6 |
| | 5 | — | 2.4 |
| Perfectip Oral | 25 | 8.7 | 0.5 |
| Syringe | 5 | — | 2.5 |
| B-D Glaspak | 25 | 10.6 | 0.5 |
| Pediatric Oral | | | |
| Liquid Administration | | | |
| Device | 5 | — | 1.5 |

[a]Present as the trihydrate.
[b]Present as the potassium salt.
[c]No loss of amoxicillin occurred.

# Amphotericin B

## Properties

Amphotericin B is a mixture of antifungal polyenes that is a yellow to orange crystalline odorless or practically odorless powder.[1–3,6]

### Solubility

Amphotericin B is insoluble in water at pH 6 to 7 and soluble at pH 2 or 11 to about 0.1 mg/mL. It is also insoluble in ethanol. Solubility in dimethyl sulfoxide (DMSO) is 30 to 40 mg/mL. In the injection, amphotericin B is dispersed by micelle formation; sodium desoxycholate is incorporated into the parenteral form of amphotericin B as a solubilizing agent, which creates the colloidal dispersion of the drug.[1–3,6]

### pH

The commercial injection reconstituted as directed to a concentration of 4 mg/mL has a pH in the range of 5 to 6.[7] A 10-mg/mL aqueous dispersion of amphotericin B has a pH between 7.2 and 8.[4] A 5-mg/mL dispersion in sterile water for injection has a pH of 5.6.[1165] A 3% aqueous dispersion has a pH of 6 to 8. For parenteral administration, amphotericin B must be admixed in 5% dextrose injection having a pH of at least 4.2.[2,3] Amphotericin B lotion has a pH of 5 to 7.[4]

## General Stability Considerations

The parenteral formulation should be stored under refrigeration and protected from light.[4] However, the vials are stable for at least one month at room temperature. Topical formulations should be stored in well-closed containers at room temperature and protected from freezing.[2–4,280]

The reconstituted vials are stable for one week under refrigeration and for one day at room temperature, according to the manufacturer. Other information indicates that aqueous solutions may be stable for more than a week even at 28°C. Amphotericin B is subject to photolytic degradation; greater degradation may occur with longer exposure and/or higher light intensity. Although light protection is needed for long-term storage, little or no concentration difference is noted after light exposure of 8 to 24 hours.[2,6,281–283]

Optimum clarity and stability of amphotericin B injection occur at pH 6 to 7. If the pH drops below 6, the colloidal dispersion develops turbidity; below pH 5 the particles coagulate.[2,6,284,285] Amphotericin B is incompatible with chloride-containing solutions.[8]

Because amphotericin B injection is a colloidal dispersion, filtration with filters less than 1 μm may remove the drug.[2,3,285]

## Stability Reports of Compounded Preparations
### Oral

**STUDY 1:** Kumar et al.[618] evaluated the stability of an amphotericin B 50-mg/mL oral rinse product designed for use in patients with xerostomia. Amphotericin B oral suspension (Fungizone, Apothecon) 20 mL was mixed with 20 mL of Optimoist (Colgate-Palmolive) saliva substitute using a magnetic stirrer. Samples were packaged in screw-cap sample vials and stored at 4 and 25°C as well as at elevated temperatures.

Stability-indicating HPLC analysis found that amphotericin B activity was retained for three weeks at both temperatures with a loss of about 10%. At 40°C with and without 75% relative humidity, amphotericin B stability was retained for only one week, with 15 to 24% loss occurring in two weeks. Phase separation occurred in all samples during storage, but redispersion was easily accomplished. The sedimentation rate was slightly faster in the Fungizone–Optimoist mixture compared to Fungizone alone. Elevated temperature samples exhibited a color change from yellow to orange. Although the presence of chloride and phosphate ions in Optimoist may contribute to a reduced shelf life of amphotericin B, the authors stated that amphotericin B oral suspension combined with Optimoist could be stored for up to three weeks after mixing.

**STUDY 2:** Groeschke et al.[1050] reported the stability of amphotericin B for oral suspension (Fungizone) prepared as an antifungal mouth rinse containing amphotericin B 7.4 mg in 1.4% sodium bicarbonate. The preparations were packaged in glass and polypropylene containers and stored over 15 days at temperatures ranging from 4 to 37°C exposed to and protected from light. Inspection found no change in color, odor, or apparent microbial contamination, and the suspensions were easily resuspended when stored refrigerated, at room temperature, or at 37°C whether protected from light or exposed to artificial light. However, exposure to natural sunlight resulted in the suspension turning brownish and developing flocculation from seven days on. HPLC analysis of refrigerated samples found about 4 to 7% loss after 15 days. Room temperature and 37°C samples exhibited 5 to 10% loss in three days and 10 to 12% loss in four days.

**STUDY 3:** Dentinger et al.[1082] evaluated the stability of an extemporaneously compounded amphotericin B 100-mg/mL oral suspension compounded from amphotericin B powder (Medisca). The suspending medium was prepared by first dissolving sodium phosphate, dibasic (anhydrous) 4.77 g, sodium phosphate, monobasic 2.88 g, sodium benzoate 300 mg, sodium metabisulfite 450 mg, and citric acid 2.25 g in 90 mL of sterile water for irrigation. With low heat, 37.5 mL of carboxymethylcellulose 1% solution was added with mixing. To this mixture, 3 mL of methylparaben 10% and propylparaben 2% in propylene glycol was added a drop at a time, while stirring. Cherry flavor 1.2 mL was added to 45 mL of glycerin and added to the suspending medium. Amphotericin B 30 g was added to a porcelain mortar and wetted with sufficient suspending medium to create a paste. The paste was transferred to a calibrated beaker, rinsing the mortar with suspending medium. After adding the remaining suspending medium to the beaker, the mixture was brought to about 290 mL with sterile water for irrigation. The pH was then adjusted to 5.3 with citric acid 200-mg/mL solution, and the suspension was brought to a final volume of 300 mL with sterile water for injection and mixed well. The suspension was packaged in amber polyethylene terephthalate prescription bottles with child-resistant caps and stored at room temperature of 22 to 25°C for 93 days.

There was no change in pH, color, or odor of the suspension throughout storage. No visible microbial growth was observed. The samples were resuspended with ease at all time points. Stability-indicating HPLC analysis found that no loss of amphotericin B occurred over the 93-day storage period.

### Ophthalmic

Ophthalmic preparations, like other sterile drugs, should be prepared in a suitable clean air environment using appropriate aseptic procedures. When prepared from nonsterile components, an appropriate and effective sterilization method must be employed.

**STUDY 1:** Charlton et al.[1666] evaluated the stability of extemporaneously compounded fortified amphotericin B ophthalmic solution by measuring zones of inhibition as an assessment of the antibiotic activity. pH and osmolality were determined as well. The amphotericin B concentration and the other component(s) in the fortified ophthalmic solution were not cited. The authors reported that no loss of amphotericin B activity was found after 28 days of storage at 4 and 25°C. Similarly, no substantive change in pH or osmolality occurred.

**STUDY 2:** Allen[604] reported on an amphotericin B 2-mg/mL ophthalmic product. A 50-mg vial of the commercial amphotericin B injection was reconstituted with 10 mL of sterile water for injection. The vial was shaken and allowed to stand until the colloidal dispersion was complete, resulting in a 5-mg/mL concentration. Then 4 mL was withdrawn and transferred to a sterile ophthalmic container. Six milliliters of sterile water for injection was added and thoroughly mixed, yielding the 2-mg/mL ophthalmic product. This product was stable for seven days when stored under refrigeration. Amphotericin B is incompatible with electrolytes. In aqueous mixtures it is a colloidal dispersion that should not be filtered through a filter with a pore size of less than 1 μm. This ophthalmic product may produce pain upon instillation.

**STUDY 3:** Peyron et al.[852,870] reported the stability of amphotericin B 5-mg/mL ophthalmic solution prepared in dextrose 5% injection. Vials of amphotericin B injection 50 mg (Fungizone, Bristol-Myers Squibb) were reconstituted with 10 mL

of dextrose 5% injection, resulting in a 5-mg/mL concentration. The reconstituted solution was packaged in low-density polyethylene dropper bottles and stored refrigerated at 2 to 6°C and at room temperature of 20 to 25°C exposed to and protected from light. The refrigerated samples did not exhibit a physical instability, and stability-indicating HPLC analysis found no loss of amphotericin B during 120 days of storage. However, the samples stored at room temperature precipitated in 14 days when exposed to light and in 17 days when protected from light, and HPLC analyses found amphotericin B losses to be around 15%.

**STUDY 4:** Achach and Peroux[1058] evaluated the stability of several anti-infective drugs in ophthalmic solutions including amphotericin B. Compounded amphotericin B 5 mg/mL in dextrose 5% was stored under refrigeration at 4°C for 12 days. No changes in color or turbidity were observed, and no change in osmolality was found. The analytical method was not specified, but the authors reported that little or no amphotericin B loss occurred in 12 days and recommended a 12-day use period.

**STUDY 5:** Morand et al.[1165] evaluated the stability of extemporaneous amphotericin B liposome eyedrops. Commercial amphotericin B liposomal injection (AmBisome) was reconstituted to a concentration of 5 mg/mL using sterile water for injection, was passed through a 0.22-μm filter, and packaged in glass vials with polyvinyl chloride stoppers.

NOTE: The manufacturer indicates that this liposomal product should not be filtered through a filter with a pore size of less than 1 μm.

The eyedrops were found to have a pH of 5.6 and an osmolality of 350 mOsm/kg. Samples were stored refrigerated at 2 to 8°C and at room temperature for 180 days. No precipitation or color change was observed throughout the study. The mean liposome diameter remained unchanged, and HPLC analysis found no substantial reduction in the amphotericin B concentration at either storage temperature over 180 days.

### Otic

Kinzel et al.[268] reported the use of an extemporaneous amphotericin B 0.25% otic solution. The solution was prepared by diluting the contents of an amphotericin B 50-mg vial with 20 mL of sterile water for injection. The solution was transferred to an amber glass dropper bottle for use. The solution was given a seven-day expiration period under refrigeration, based on the manufacturer's stability information for the reconstituted amphotericin B 5-mg/mL injection.

### Inhalation and Nasal Spray

Inhalation preparations, like other sterile drugs, should be prepared in a suitable clean air environment using appropriate aseptic procedures. When prepared from nonsterile components, an appropriate and effective sterilization method must be employed.

**STUDY 1:** Pesko[420] prepared a 0.5% amphotericin B nasal spray from a 50-mg vial of the injection. Sterile water for irrigation 10 mL was added and mixed well until all solid matter was dispersed completely, resulting in the appearance of a clear yellow solution. This solution was placed in a nasal-spray bottle or reusable atomizer delivering about 0.3 mL/spray. The preparation's stability was stated to be seven days (as for the reconstituted injection) when stored under refrigeration and protected from light.

**STUDY 2:** Juarez et al.[1110] evaluated the stability of amphotericin B (Squibb) 1 mg/mL that had been reconstituted and diluted with sterile water for injection for use as an inhalation solution and packaged in sterile glass vacuum containers. Samples were stored at room temperature of 20 to 25°C exposed to fluorescent light and also protected from exposure to light, and refrigerated at 4 to 8°C protected from exposure to light for 30 days.

No visible changes in color, precipitation, or phase separation were observed during the study. Stability-indicating HPLC analysis found about 4% amphotericin B loss after 30 days in the room temperature samples exposed to fluorescent light. Little or no loss was found in 30 days in the room temperature and refrigerated samples protected from exposure to light.

**STUDY 3:** Fittler et al.[1214] evaluated the stability of amphotericin B (Bristol-Myers Squibb) 5 mg/mL in sterile water and in dextrose 5% for use as a nasal spray. Light exposure or protection was not addressed. Bioassay found that the drug was stable for 30 days refrigerated at 4°C. The presence or absence of dextrose 5% did not affect drug stability.

## Compatibility with Other Drugs

Feron et al.[467] evaluated the interaction of sucralfate with several antimicrobial drugs, including amphotericin B. Sucralfate 500 mg was added to 40 mL of HPLC-grade water and adjusted to pH 3.5 with hydrochloric acid. The amphotericin B was added to yield a final concentration of 0.025 mg/mL (25 mcg/mL). Drug concentration was determined initially and numerous times over 90 minutes of storage by spectroscopy. Amphotericin B losses occurred very rapidly and extensively (95% after five minutes). Furthermore, the loss was not reversible by adjusting pH to near neutrality. The authors noted that the mechanism of interaction is unclear. However, concurrent enteral administration of sucralfate and amphotericin B would result in a substantially lower concentration than would be achieved in the absence of sucralfate. ■

# ■ Ampicillin

## Properties

Ampicillin (as the trihydrate or in the anhydrous form) is an aminopenicillin that occurs as a white practically odorless crystalline powder.[1–3]

### Solubility

Ampicillin is slightly soluble in water and practically insoluble in ethanol (about 4 mg/mL). The aqueous solubilities of the anhydrous and trihydrate forms at 20°C are 13 and 6 mg/mL, respectively. The drug also dissolves in dilute solutions of acids and alkali hydroxides.[1–3,6]

### pH

A 1% aqueous ampicillin solution has a pH of 3.5 to 6. Ampicillin oral suspension has a pH value between 5 and 7.5, while the injectable suspension has a pH of 5 to 7. Ampicillin trihydrate suspension (Penbritin, Ayerst) has a measured pH of 5.7. Ampicillin sodium injection is much more alkaline, having a pH of 8 to 10.[2–4,19]

### Osmolality

Ampicillin suspension 250 mg/5 mL (Principen, Squibb) was reported to have an osmolality of 3070 mOsm/kg by Niemiec et al.,[232] while Dickerson and Melnik[233] reported the osmolality to be 2250 mOsm/kg. Ampicillin suspension 500 mg/5 mL (Bristol) had an osmolality of 1850 mOsm/kg. The powder for injection reconstituted with sterile water for injection has an osmolality of 602 mOsm/kg at 100 mg/mL and around 675 to 700 mOsm/kg at 125 mg/mL.[232,233,286,287,905]

## General Stability Considerations

Ampicillin oral products should be stored in tight containers at controlled room temperature.[4] Elevated temperature and humidity free the intrinsic molecules of crystallization water sufficiently to participate in solid-state ampicillin hydrolysis.[6] When reconstituted, the oral ampicillin suspensions are stable for seven days at room temperature and 14 days under refrigeration.[2,3]

Ampicillin is subject to hydrolytic cleavage of the β-lactam ring, both at acidic and alkaline conditions. Maximum stability occurs at pH 5.8. Increasing concentrations of dextrose, sucrose, or other carbohydrates enhance decomposition of ampicillin.[6]

Jaffe et al.[297] reported the stability of five manufacturers' ampicillin (anhydrous or trihydrate) 250-mg/5-mL oral liquids following reconstitution (Penbritin, Ayerst; Omnipen,

Wyeth; Principen, Squibb; Polycillin, Bristol; SK-Ampicillin, Smith Kline & French). A spectrophotometric assay specific for the intact drug in the presence of its degradation products was used to assess the ampicillin content of products stored at 5, 25, and 35°C. All tested products retained at least 90% of the labeled amount after 27 days of refrigerated storage and seven days of room temperature storage. However, no product retained the required concentration after seven days at 35°C.

The stability of the reconstituted injection is variable, depending on the concentration, pH, and temperature. The parenteral form (sodium salt) is most stable at lower concentrations, at around pH 5 to 6, and at refrigeration temperatures. The decomposition rate increases as concentration increases, as pH diverges higher or lower than 5 to 6, and with storage at room temperature or freezing.[6,8,264,288–294]

## Stability Reports of Compounded Preparations

### Oral

**STUDY 1 (REPACKAGED):** Allen and Lo[65] tested the stability of oral ampicillin suspension (Omnipen, Wyeth) repackaged into unit-dose containers at several temperatures, including frozen storage. The ampicillin suspension was reconstituted to 250 mg/5 mL. Samples of 5 mL were placed in 26-mL amber glass vials with screw-cap closures and stored at −20, −10, 5, and 25°C. Penicillin content was analyzed spectrophotometrically. Samples frozen at either −20 or −10°C lost less than 10% in 60 days. However, the effect, if any, of freezing on bioavailability of the product is unknown.[68,69] Samples lost about 11% in 10 days under refrigeration and at room temperature.[65]

**STUDY 2 (REPACKAGED):** Sylvestri and Makoid[33] tested the stability of ampicillin trihydrate oral suspension (Principen 125, Squibb) 25 mg/mL repackaged into 5-mL amber polypropylene plastic oral syringes (Exacta-Med Liquid Dispensers). After reconstitution with distilled water, 5 mL of the oral suspension was drawn into the syringes, which were then capped and stored at −20, 4, and 25°C. A spectrophotometric analysis stated to be specific for the β-lactam ring was used to assess intact ampicillin over 47 days of storage. Although there was considerable variation in the results, the authors concluded that less than 10% of the drug was lost in 47 days at −20 and at 4°C; at 25°C losses of 10% were estimated to occur at about 30 days. The expiration dates recommended by the manufacturer for the suspension in the original container (seven days at room temperature and 14 days under refrigeration) can be extended to include the suspension packaged in these oral syringes.

**STUDY 3 (EXTEMPORANEOUS):** Brown and Kayes[264] described the formulation of a sugar-free ampicillin suspension extemporaneously prepared from Penbritin (Beecham) capsules. The suspension was prepared using tragacanth as the suspending agent in the hydroalcoholic vehicle described in Table 7. The pH was buffered to 4.5 to 6. The suspension was packaged in brown glass screw-cap bottles. The authors calculated the shelf life from spectrophotometric analysis and iodometric titration. Acceptable concentration was maintained for samples stored for 24 days at 5°C and for 4.5 days at 20°C.

### Enteral

**AMPICILLIN COMPATIBILITY SUMMARY**

*Compatible with*: Ensure • Ensure Plus • Osmolite

Cutie et al.[19] added 5 mL of ampicillin trihydrate suspension (Penbritin, Ayerst) to varying amounts (15 to 240 mL) of Ensure, Ensure Plus, and Osmolite (Ross Laboratories) with vigorous agitation to ensure thorough mixing. The ampicillin suspension was physically compatible, distributing uniformly in all three enteral products with no phase separation or granulation. ∎

**TABLE 7.** Suspension Formulation for Ampicillin[264]

| Component | Percent (wt/vol) | |
|---|---|---|
| Tragacanth | | —[a] |
| Alcohol 90% | | 22.5 |
| Compound orange spirit | | 0.2 |
| Compound tartrazine solution | | 0.2 |
| Amaranth solution | | 0.05 |
| Cetomacrogol 1000 | qs | |
| Syrup | | 12.5 |
| Sodium citrate | qs pH | 6.15 |
| Water | qs | 100 |

[a]Concentration unspecified.

# ■ Antacids

## Properties
This category of oral medications is composed primarily of the inorganic alkaline salts of metals. Examples include aluminum, calcium, and magnesium carbonates; sodium bicarbonate; and aluminum hydroxide and magnesium hydroxide.

## pH
Aqueous solutions or dispersions have pH values ranging from neutral to alkaline, depending on the salt form.[3]

## General Stability Considerations
Antacid products should be stored in tight containers at controlled room temperature.[3,4]

## Stability Reports of Compounded Preparations
*Enteral*
### ANTACIDS COMPATIBILITY SUMMARY
*Compatible with:* Vital
*Uncertain or variable compatibility with:* Osmolite • Osmolite HN

Fagerman and Ballou[52] reported on the compatibility of several antacid products mixed in equal quantities with three enteral feeding formulas. Only Mylanta was compatible visually with Vital, Osmolite, and Osmolite HN (Ross), with no obvious thickening, granulation, or precipitation. Mylanta II was compatible visually with Vital, but it underwent unacceptable thickening, forming an increasingly viscous mixture upon standing when combined with Osmolite and Osmolite HN. Both Amphojel and Riopan were compatible visually with Vital and Osmolite HN but underwent unacceptable thickening with Osmolite. ■

# ■ Anthralin
## (Dithranol)

## Properties
Anthralin is a yellow to yellowish-brown odorless, tasteless crystalline powder.[3,4]

## Solubility
Anthralin is insoluble in water, slightly soluble in ethanol, soluble in acetone, and soluble in dilute solutions of alkaline hydroxides.[1,3,4]

## pH
Anthralin filtrate from a suspension in water has a neutral pH.[1,4]

## General Stability Considerations
Anthralin should be stored in tight containers in a cool place at 8 to 15°C protected from exposure to light.[3,4] The USP states that anthralin topical cream and ointment should also be packaged in tight containers and stored in a cool place protected from exposure to light.[4] However, commercial anthralin topical cream (Psoriatec) is labeled for room temperature storage. Excessive heat should be avoided.[7]

Anthralin stability is concentration dependent with better stability at higher concentrations.[1182] Anthralin is unstable in aqueous media and is degraded not only by water but also by high temperature, oxygen, and metal contact.[847–849] Because anthralin is poorly stable in aqueous environments,[848,849] oil-in-water emulsions are preferred and salicylic acid and ascorbic acid are usually incorporated into the topical preparations as stabilizers,[847] although salicylic acid has had mixed results.[1182] Other potential antioxidants including citric acid, benzoic acid, tartaric acid, mannitol, methionine, cysteine, and malic acid were found to be ineffective.[1182]

## Stability Reports of Compounded Preparations
*Topical*
**STUDY 1:** Anthralin in concentrations from 0.1 to 1% was prepared in Nutra-D cream base. Samples were stored under normal room temperature and protected from light. The anthralin cream samples were evaluated for drug strength and for color changes over seven weeks. The cream underwent a distinct color change after four weeks. The anthralin concentration decreased rapidly after four weeks irrespective of the initial anthralin concentration.[1682]

**STUDY 2:** Allen[704] reported the stability of anthralin 1% in a lipid crystals cream containing glyceryl laurate, glyceryl myristate, citric acid, sodium hydroxide, and purified water. The anthralin was incorporated into the glyceryl laurate and glyceryl myristate heated to 70°C and then added to the aqueous phase (water with citric acid and sodium hydroxide), with subsequent controlled cooling. The anthralin was demonstrated to be stable for at least six months at room temperature in this formulation.[705]

**TABLE 8.** Formulations of Anthralin Tested for Stability[847]

| Component | Anthralin Concentration 0.1% | | 0.3% | 0.5% |
|---|---|---|---|---|
| Anthralin | | 3 g | 9 g | 15 g |
| Cetiol V | | 603 g | 609 g | 615 g |
| Cetomacrogol wax | | 450 g | 450 g | 450 g |
| Liquid paraffin | | 450 g | 450 g | 450 g |
| Salicylic acid powder | | 30 g | 30 g | 30 g |
| Sorbic acid | | 4.5 g | 4.5 g | 4.5 g |
| Ascorbic acid | | 1.5 g | 1.5 g | 1.5 g |
| Distilled water | qs | 3000 g | 3000 g | 3000 g |

**STUDY 3:** Wuis et al.[847] reported the stability of anthralin at concentrations ranging from 0.1 to 0.5%. The formulations prepared and evaluated are cited in Table 8. The oil-in-water creams were prepared by heating cetiol V, cetomacrogol wax, and liquid paraffin to 70°C. Sorbic acid was dissolved in boiling distilled water followed by ascorbic acid; the solution was allowed to cool to 70°C. The salicylic acid was added using a heated mortar. Anthralin was added to the lipid phase and stirred with a high-speed mixer until dissolved. The aqueous phase was then added and mixed with a high-speed mixer for a short period of time to reduce the inclusion of air. The cream was mixed manually until it cooled, and it was packaged in aluminum tubes. The 0.3% concentration of the preparation was also packaged in polypropylene tubes.

Stability-indicating HPLC analysis found that anthralin concentrations for all three concentrations in aluminum tubes remained within the British Pharmacopoeia limits of 85% or greater for 12 months refrigerated at 4°C. At room temperature of 20°C, the anthralin content remained above 95% for one month, but unacceptable losses occurred in three months. In polypropylene tubes, decomposition proceeded more rapidly; 30% loss occurred in three months at room temperature and in 12 months refrigerated.

**STUDY 4:** Green et al.[1182] evaluated the stability of anthralin in concentrations ranging from 2 to 0.1% in three topical bases packaged in amber ointment jars and stored at room temperature protected from exposure to light. The best stability occurred when white soft paraffin was used as the base. The time to 10% anthralin loss was more than 12 months at concentrations greater than 1% and in three months at a concentration of 0.2%. In Lassar's Paste, anthralin is also relatively stable, with 10% loss occurring in about 13, 10, and three months at concentrations of 2, 1, and 0.25%, respectively. In Unguentum Merck, anthralin underwent relatively rapid decomposition, with 10% loss occurring in about 10 days at 2% concentration and in as little as one day at 0.1% concentration.

**STUDY 5:** Allen[1404] reported on a compounded formulation of anthralin scalp lotion, also named African Scalp Lotion. The lotion had the following formula:

| | | |
|---|---|---|
| Anthralin | | 250 mg |
| Salicylic acid | | 2.4 g |
| Peanut oil | | 50 mL |
| Coal tar solution | qs | 100 g |

The recommended method of preparation was to dissolve the anthralin and salicylic acid in the coal tar solution. Then the peanut oil was added and the mixture was stirred, making certain the lotion was thoroughly mixed. The lotion was to be packaged in tight, light-resistant containers. The author recommended a beyond-use date of six months at room temperature because this formula is a commercial medication in some countries with an expiration date of two years or more.

**STUDY 6:** Braun and Wiegrebe[1405] reported that anthralin in petrolatum was stable for at least 29 days at room temperature of 19 to 25°C. Anthralin was also found to be similarly stable in the presence of zinc oxide and salicylic acid in ointments. However, about 20% anthralin loss was reported after six weeks of storage. The authors recommended limiting the use period of anthralin ointments to less than one month.

**STUDY 7:** Lee[1474] studied the stability of anthralin 1 and 4% (wt/wt) in yellow soft paraffin with and without salicyclic acid 1% (wt/wt). The topical preparations were packaged in clear soda glass jars with samples stored protected from exposure to light at ambient room temperature and refrigerated at about 6°C. Anthralin 0.5% (wt/wt) was also prepared in zinc and salicylic acid paste (Lassar's Paste), packaged in soda glass jars, and stored protected from light at ambient room temperature.

HPLC analysis found that anthralin concentrations in the yellow soft paraffin formulations with and without salicyclic acid exhibited good stability. The time to 10% anthralin loss was calculated to be 397 days at ambient temperature. However, some discoloration occurred at the anthralin-soda glass interface; this discoloration did not occur if the preparation was refrigerated. The authors stated that plastic containers might be better for packaging anthralin topical preparations because of this discoloration. The authors also indicated that the presence of salicyclic acid and higher anthralin concentrations in the formulations seemed to improve anthralin stability. In Lassar's Paste, the time to 10% anthralin loss was calculated to be 251 days at ambient temperature.

**STUDY 8:** Weller et al.[1475] evaluated the stability of anthralin in concentrations from 0.05% up to 1% (wt/wt) in emulsifying ointment, BP, [cetostearyl alcohol 45% (wt/wt), sodium lauryl sulfate 5% (wt/wt), soft paraffin 50% (wt/wt)] with salicyclic acid 0.5%. In addition, the stability of anthralin 0.05% (wt/wt) in emulsifying ointment, BP, with ascorbyl palmitate 0.1% (wt/wt) as an antioxidant and anthralin 0.05, 0.1, and 0.5% (wt/wt) in Lassar's Paste was tested. The topical preparations were packaged in amber glass jars and were stored at room temperature.

HPLC analysis found that in the formulations containing salicyclic acid with anthralin concentrations of 0.5% (wt/wt) or greater, 10% anthralin loss occurred in about 16 weeks. At lower anthralin concentrations, 39% loss occurred within just four weeks. Similarly, the anthralin 0.05% (wt/wt) formulations in Lassar's Paste exhibited 10% anthralin loss in just four weeks. However, in the formulation containing ascorbyl palmitate, little or no anthralin loss occurred in 52 weeks.

**STUDY 9:** Thoma and Holzmann[1667] reported the stabilities of anthralin in three topical formulations, a stick, a cream, and a gel. Anthralin was found to be stable for at least six months in all three formulations when stored refrigerated. At room temperature, however, unacceptable losses occurred in less than six months with the stick formulation exhibiting the greatest decomposition.

**STUDY 10:** Hager and Kaestner[1710] evaluated the stability of anthralin when mixed in an oily formulation with salicyclic acid. The formulations also contained castor oil, peanut oil, and Emulgator MF. Samples were stored at room temperature of 20 to 25°C and refrigerated at 2 to 4°C for six months. The stability of anthralin was found to be dependent on storage temperature and exposed to atmospheric oxygen. The authors recommended that the preparations be stored in completely filled containers for up to two months at room temperature or under refrigeration for up to six months. ▪

# ▪ Antipyrine
## (Phenazone)

## Properties
Antipyrine is a bitter-tasting white powder or crystals.[1,3]

### Solubility
Antipyrine is very soluble in water, having an aqueous solubility of 1 g in less than 1 mL. In ethanol the solubility is 1 g in 1.3 mL.[1,3]

### pH
Antipyrine aqueous solution is pH neutral.[1]

## General Stability Considerations
Antipyrine products should be stored in tight containers.[4]

## Stability Reports of Compounded Preparations
### Injection
Injections, like other sterile drugs, should be prepared in a suitable clean air environment using appropriate aseptic procedures. When prepared from nonsterile components, an appropriate and effective sterilization method must be employed.

Mutch and Hutson[166] reported the stability of an extemporaneously prepared injection of antipyrine and caffeine 100 and 20 mg/mL, respectively. Antipyrine powder (Sigma Chemical) and caffeine powder (Spectrum Chemical) were weighed and dissolved in warm sterile water for injection to yield the appropriate concentrations. The solution was filled into 10-mL clear glass vials with rubber stoppers and autoclaved at 121°C for 30 minutes. The finished vials were stored upright and inverted at 5, 20, 40, and 60°C to evaluate shelf life. In addition, 10 mL of the finished injection was added to 100- and 250-mL polyvinyl chloride (PVC) bags of 0.9% sodium chloride injection and stored at 5 and 20°C. Both vials and admixtures at 20°C were exposed to normal room light and to intense fluorescent light. The content of antipyrine and caffeine was then evaluated by HPLC analysis.

No changes in drug content were found in the intact vials at any storage temperature after storage for up to six months; contact of the solution with the rubber stopper had no effect on the content of either drug. Similarly, the drugs were stable in the admixtures in PVC bags for up to 48 hours. Furthermore, no turbidity, flocculation, precipitation, or pH change was noted for any sample. However, the samples exposed to intense fluorescent light developed a yellow discoloration within one month that continued to intensify with time. Exposure of an antipyrine solution to the intense light showed that the discoloration was associated with that component. Even though no detectable changes in drug content occurred during the study, protection from light was recommended. ▪

# Apomorphine Hydrochloride

## Properties
Apomorphine hydrochloride is a white to slightly yellow to green-tinged grayish crystalline powder or crystals.[3]

### Solubility
Apomorphine hydrochloride is soluble in water to about 20 mg/mL, increasing to 50 to 58 mg/mL at 80°C. It is also soluble in ethanol to about 20 mg/mL.[1,3]

### pH
A 1% aqueous solution has a reported pH of 4 to 5[3] while a 3.33-mg/mL aqueous solution is stated to have a pH of 4.8.[1]

## General Stability Considerations
Apomorphine hydrochloride crystals and powder exhibit increased green discoloration upon exposure to air and light. Aqueous solutions decompose upon storage; solutions should not be used if they turn green or brown and/or contain a precipitate.[1,3] Apomorphine hydrochloride products should be stored in tight, light-resistant containers.[4] The USP states that containers of the drug for compounding should contain not more than 350 mg and should be taken only for immediate use in compounding.[4]

## Stability Reports of Compounded Preparations
### Injection
Injections, like other sterile drugs, should be prepared in a suitable clean air environment using appropriate aseptic procedures. When prepared from nonsterile components, an appropriate and effective sterilization method must be employed.

**STUDY 1:** Maloney[492] reported on the formulation of an apomorphine hydrochloride 1-mg/mL solution for subcutaneous injection. The formulation also contained 0.1% sodium metabisulfite. It was sterile filtered (filter pore size unspecified) and filled into 5-mL vials. The air was purged from the vials with nitrogen, and the vials were sealed. The solution was stored under refrigeration protected from light and at room temperature both protected from and exposed to light. Apomorphine oxidizes to green quinolinedione derivatives that are inactive. The refrigerated samples remained clear and colorless after storage for one year. The room temperature samples protected from light developed a green tinge after one year. The room temperature samples exposed to light developed a green tinge after six months. Sterile filtration was used because sterilization in an autoclave at 116°C for 30 minutes caused the development of a green color. Definitive analytical testing for stability was not performed.

**STUDY 2:** de Castro et al.[629] reported the results of a clinical trial using a 2.5-mg/mL apomorphine hydrochloride injection in sterile water prepared with 0.1% sodium metabisulfite as an antioxidant. Ultraviolet spectrophotometric analysis of the apomorphine hydrochloride preparation over periods of up to two months stored at room temperature found no change in absorbance. The formulation was found to be clinically effective in 19 of 20 patients who received the apomorphine hydrochloride injection subcutaneously.

**STUDY 3:** Kin et al.[1509] evaluated the stability of apomorphine hydrochloride 1 mg/mL in 0.125% sodium metabisulfite aqueous solution. Chlorobutanol 0.25% was also present as an antimicrobial preservative. Samples were stored refrigerated at 4°C protected from exposure to light. HPLC analysis found less than 5% loss of apomorphine hydrochloride occurred during six months of storage. However, if the solution was diluted 10-fold to a concentration of 0.1 mg/mL and stored under the same conditions, substantial drug decomposition occurred within as little as three weeks. Even so the characteristic blue-green discoloration associated with apomorphine hydrochloride decomposition did not become visible until after six weeks of storage.

**STUDY 4:** Decker et al.[1406] reported on apomorphine hydrochloride formulated as an injection in 1% reduced l-ascorbic acid aqueous solution. The solution was sterilized by filtration through a Millipore 0.22-μm filter. The apomorphine hydrochloride injection was packaged in vials, and the air was replaced with nitrogen gas. The vials were stored refrigerated at 5°C and protected from exposure to light. The authors stated the apomorphine hydrochloride injection remained clear and retained drug concentration for at least one year.

### Nasal Solution
Fuentes de Frutos et al.[1190] evaluated the stability of apomorphine hydrochloride injection packaged in dosifiers (dosificador) for intranasal administration in Parkinson's disease. Apomorphine hydrochloride injection (Laboratorio Aguettant) 10 mg/mL was placed in ethylene oxide sterilized dosifiers. Ultraviolet spectrophotometric analysis found less than 3% drug loss in 84 days in ambient room temperature and light. It should be noted that the ultraviolet spectrophotometric analytical method was not demonstrated to be stability indicating.

# Aprepitant

## Properties

Aprepitant is a white to off-white crystalline material.[7]

### Solubility

Aprepitant is practically insoluble in water and sparingly soluble in ethanol.[1,7]

### pH

Dupuis et al.[1365] reported that the pH of extemporaneously compounded aprepitant 20-mg/mL oral liquid in a 1:1 mixture of Ora-Sweet and Ora-Plus has a pH near 4.3.

### $pK_a$

Aprepitant has a $pK_a$ of 4.38.[1366]

## General Stability Considerations

Aprepitant commercial capsules are packaged in tight containers and stored at controlled room temperature.[7]

## Stability Reports of Compounded Preparations

### Oral

Dupuis et al.[1365] evaluated the stability of an aprepitant 20-mg/mL oral liquid formulation compounded from commercial oral capsules. The granular contents of the capsules containing aprepitant nanoparticles were emptied into a mortar and ground to fine powder using a pestle. Grinding the granules to fine powder prior to mixing with the vehicle resulted in a smoother, less grainy suspension. A small amount of a 1:1 mixture of Ora-Sweet and Ora-Plus was added to the powder and triturated to a fine paste ensuring that there were no lumps. Additional vehicle mixture was added to form a liquid that was transferred to a graduate with rinsing of the mortar with the vehicle blend. The oral liquid was then brought to volume using additional vehicle blend. Samples were packaged in amber glass and amber polyethylene terephthalate plastic containers and were stored at room temperature of 23°C and refrigerated at 4°C for 15 weeks.

All samples remained creamy white with no color change and no change in consistency throughout the study period. There was no detectable settling or caking with the oral liquid remaining extremely well suspended even without shaking. The pH remained unchanged at about 4.3 throughout the study. Stability-indicating HPLC analysis of the aprepitant content found less than 10% loss in 100 days when refrigerated and in 66 to 85 days at room temperature. The authors recommended that this compounded preparation be stored refrigerated and that a beyond-use date of 90 days be used. ■

# Arformoterol Tartrate

## Properties

Arformoterol tartrate is white to off-white. Arformoterol 7.5 mcg/mL (as the tartrate) inhalation solution is colorless.[7]

### Solubility

Arformoterol tartrate is slightly soluble in water.[7]

### pH

Arformoterol 7.5 mcg/mL (as the tartrate) inhalation solution is adjusted to pH 5 with citric acid and sodium citrate.[7]

### Osmolality

Arformoterol 7.5 mcg/mL (as the tartrate) inhalation solution is isotonic.[7]

## General Stability Considerations

Arformoterol 7.5 mcg/mL (as the tartrate) inhalation solution is packaged in 2-mL low-density polyethylene vials in a foil pouch. The vials should be stored refrigerated at 2 to 8°C and protected from exposure to light and excessive heat. The drug is stable for up to six weeks at controlled room temperature. Once the foil pouch is opened, the inhalation solution should be used immediately.[7]

## Compatibility with Other Drugs

Bonasia et al.[1151] evaluated the physical compatibility and chemical stability of arformoterol 15 mcg/2 mL (as the tartrate) inhalation solution (Brovana, Sepracor) with acetylcysteine 800 mg/4 mL (Mucomyst, Hospira), ipratropium bromide 0.2 mg/2 mL (Atrovent, AstraZeneca), and budesonide 0.25 mg/2 mL and 0.125 mg/2 mL (Pulmicort Respules, AstraZeneca). The admixtures were prepared and mixed for homogeneity. Visual inspection found no evidence of precipitation or other physical incompatibility. HPLC analysis found less than 2% change in any drug concentration over the 30-minute test period. ■

# Arsenic Trioxide

## Properties
Arsenic trioxide is a white solid crystalline material that is subject to sublimation.[1,7]

### Solubility
Arsenic trioxide is sparingly and very slowly soluble in cold water but soluble 1 in 15 (approximately 67 mg/mL) in boiling water. It is also soluble in dilute hydrochloric acid and alkaline hydroxide or carbonate solutions. It is practically insoluble in ethanol.[1,7]

### pH
Arsenic trioxide 1 mg/mL injection (Trisenox) is adjusted to pH 7.5 to 8.5 with sodium hydroxide.[7,934]

## General Stability Considerations
Arsenic trioxide injection is stored at controlled room temperature and protected from freezing. The manufacturer indicates that arsenic trioxide injection after dilution for administration is stable for 24 hours at room temperature and 48 hours under refrigeration. However, other information from the manufacturer states that the drug mixed in 5% dextrose injection or in 0.9% sodium chloride injection is stable for up to 14 days at both room temperature and when refrigerated.[7,942]

## Stability Reports of Compounded Preparations
### Oral
Kumana et al.[934] reported the stability of arsenic trioxide 1 mg/mL oral solution used to treat hematological malignancies. Arsenic trioxide is slowly and sparingly soluble in water. To prepare this oral solution, 500 mg of arsenic trioxide powder was added to a beaker containing 150 mL of sterile water resulting in a suspension. To dissolve the suspended powder, 3 M sodium hydroxide was added dropwise. When the suspended powder had completely dissolved, an additional 250 mL of sterile water was added. With the use of a pH meter, the solution pH was slowly titrated back to pH 8 using 6 M hydrochloric acid. Dilute hydrochloric acid was then used to reach a final pH of 7.2, and sterile water was added to reach a final volume of 500 mL and a 1-mg/mL concentration. The shelf life of the solution proved to be at least three months with no appearance of fungal growth, a known problem of arsenic trioxide solutions. ■

#  Artesunate
## (Artemether)

## Properties
Artesunate occurs as a white crystalline powder.[1]

### Solubility
Artesunate is a water soluble semi-synthetic derivative of artemisinin.[7]

### pH
Artesunate injection is reconstituted with sodium bicarbonate 5% injection and should have a pH near 7 to 8.[7]

### pKₐ
Artesunate has a $pK_a$ of 4.6.[1264]

## General Stability Considerations
Artesunate oral tablets and injection are to be stored at controlled room temperature not exceeding 30°C and protected from exposure to light.[7]

## Stability Reports of Compounded Preparations
### Rectal
Gaudin et al.[1265] evaluated the drug release and stability of an extemporaneously prepared artesunate rectal gel used for the treatment of malaria in children unable to take solid oral dosage forms. Because of the relative aqueous instability of artesunate, a two-part rectal gel formulation was developed that kept the artesunate as part of a dry blend until the two parts are mixed together. For a sufficient amount to make 100 g of finished gel, the dry blend was composed of artesunate 5 g (having a mean particle size of less than 150 μm) combined with Carbopol 974P 800 mg and blended together. The liquid part was composed of absolute ethanol 10 g, sodium hydroxide 5 M 1.64 g, and deionized water 86.36 g. The extemporaneous preparation of the gel was performed by mixing 9.42 g of the liquid phase into a vial containing 0.58 g of the dry blend and mixing vigorously using a vortex mixer for 90 seconds. It was then permitted to stand at room temperature and became homogeneous and ready-to-use after one hour. The gel was packaged in glass vials with rubber stoppers for stability testing.

Stability-indicating HPLC analysis of the premixed dry blend component found that the artesunate was stable at 45°C and 60% relative humidity; nearly 100% of the drug was present after 175 days of storage. After extemporaneously combining the dry and liquid parts to prepare the rectal gel, artesunate was found to retain 94% of the drug content after 72 hours at room temperature of 22°C. ■

# Ascorbic Acid

## Properties

Ascorbic acid occurs as white or slightly yellow crystals or powder that is nearly odorless but has a sharp acidic taste. The injection contains sodium ascorbate or ascorbic acid with sodium hydroxide, sodium carbonate, or sodium bicarbonate.[1–3,602]

### Solubility

Ascorbic acid is freely soluble in water, having a solubility of about 333 mg/mL. Other solubilities include 33 mg/mL in 95% ethanol, 20 mg/mL in absolute ethanol, 10 mg/mL in glycerol, and 50 mg/mL in propylene glycol.[1,602] One USP or international unit (IU) of activity is equivalent to 0.05 mg of the USP reference standard.[1]

### pH

Ascorbic acid injection has a pH of 5.5 to 7.[4] A 5% solution of ascorbic acid has a pH of 2.1 to 2.6.[3]

### $pK_a$

Ascorbic acid has $pK_a$ values of 4.2 and 11.6.[2,6,602]

## General Stability Considerations

Ascorbic acid is light sensitive and will darken upon exposure to light; slight discoloration does not impair therapeutic activity. It is also easily oxidized to dehydroascorbic acid by exposure to air and in an alkaline environment as well as in the presence of iron or other heavy metals and copper ions.[1,2,6,602,856] The drug also degrades in anaerobic conditions, yielding furfural and carbon dioxide.[6]

Ascorbic acid products should be packaged in tight, light-resistant containers.[2,4] Upon long-term storage, ampules containing solutions of greater than 100 mg/mL may produce carbon dioxide buildup due to decomposition. The increased pressure necessitates careful opening of the ampules.[2]

The pH of maximum stability in both aerobic and anaerobic conditions is near pH 3 and also near pH 6. The highest rates of decomposition occur near pH 4 and below pH 1.5.[6] Ascorbic acid 1% aqueous solutions at various pH values from 3 to 7 stored at 45°C demonstrated losses in 24 hours of 14% at pH 3, 20% at pH 4 and 5, and 24% at pH 7.[812]

Yillar et al.[1381] reported that choline ascorbate, an alternative form of vitamin C, was found to be more stable than ascorbic acid in solution. Lower rates and extents of decomposition were found due to increased temperature and solution pH in the range of pH 4 to 8 compared to ascorbic acid.

## Stability Reports of Compounded Preparations

### Oral

The stability of ascorbic acid in a number of vehicles for oral use has been summarized by Connors et al.[6] Best stability was observed in syrup, USP, glycerin, propylene glycol, sorbitol, and equal parts mixtures of glycerin with propylene glycol or sorbitol. At least 90% of the ascorbic acid was retained for 240 days at room temperature in these vehicles. The worst stability at room temperature was in 4% carboxymethylcellulose (16% loss in 30 days) and in distilled water (about 10% loss in 30 days). The addition of B-complex vitamins enhances ascorbic acid stability.

### Enteral

Ortega de la Cruz et al.[1101] reported the physical compatibility of an unspecified amount of oral liquid ascorbic acid (Redoxon, Roche) with 200 mL of Precitene (Novartis) enteral nutrition diet for a 24-hour observation period. No particle growth or phase separation was observed.

### Ophthalmic

Ophthalmic preparations, like other sterile drugs, should be prepared in a suitable clean air environment using appropriate aseptic procedures. When prepared from nonsterile components, an appropriate and effective sterilization method must be employed.

**STUDY 1:** Allen[602] reported on a 10% ascorbic acid ophthalmic solution used to treat alkali burns of the cornea. Ascorbic acid 1 g was dissolved in about 9 mL of sterile water for injection. Dilute sodium hydroxide was added dropwise to bring the pH in the range of 5.5 to 7. Additional sterile water for injection was added to bring the volume to 10 mL, and the solution was mixed well. Alternatively, the proper amount of ascorbic acid injection can be used as the source of the drug. The solution was filtered through a 0.2-µm filter into sterile ophthalmic containers, leaving little or no headspace. The product was stable for 96 hours stored under refrigeration and protected from light.

**STUDY 2:** Barnes[1635] evaluated the stability of ascorbic acid 10% ophthalmic solution packaged in amber glass eyedrop bottles and low-density polyethylene Steri-Dropper bottles (Helapet). The ascorbic acid eyedrops were compounded using the following formula:

| | | |
|---|---|---|
| Ascorbic acid | | 10 g |
| Potassium bicarbonate | | 5.7 g |
| EDTA | | 200 mg |
| Sodium metabisulfite | | 300 mg |
| Water | qs | 100 mL |

The solution was sterilized by passing it through a 0.22-µm filter as it was packaged in the glass and plastic containers. The containers were stored under refrigeration at 6 to 8°C for 84 days. The ascorbic acid content was determined using a titration assay. In the glass containers, little loss of ascorbic acid was found over 84 days. However, in the polyethylene containers, losses of 5 and 8% occurred in 57 and 84 days respectively. The authors attributed the losses in polyethylene containers to the oxygen-sensitive nature of ascorbic acid and concluded that the glass containers were preferable.

## Topical

**STUDY 1:** Spiclin et al.[842] reported the stability of lipophilic ascorbyl palmitate and hydrophilic sodium ascorbyl phosphate at concentrations ranging from 0.25 to 2% in the oil-in-water (o/w) and water-in-oil (w/o) microemulsions cited in Table 9. To formulate the microemulsions, ascorbyl palmitate was dissolved in the PEG-8 caprylic/capric triglycerides (Labrasol) surfactant while sodium ascorbyl phosphate was dissolved in purified water and then the other three components were added. The microemulsions formed spontaneously with gentle hand mixing. The various preparations were stored in glass containers at 21 to 23°C.

HPLC analysis found ascorbyl palmitate decomposition was catalyzed by oxygen, metal ions, and/or light exposure. Ascorbyl palmitate was less stable than sodium ascorbyl phosphate. However, less degradation occurred at higher concentrations and in w/o microemulsions. In w/o microemulsions, concentrations of ascorbyl palmitate remaining after 28 days ranged from 36% in the 2% concentration to none in the 0.25% concentration. In contrast, sodium ascorbyl phosphate was stable in both o/w and w/o microemulsions, retaining 95% of the initial concentration after 60 days.

**TABLE 9.** Composition of Microemulsions[842]

| Component | Oil in Water | Water in Oil |
|---|---|---|
| Caprylic/capric triglycerides | 7.43% | 24.75% |
| PEG-8 caprylic/capric glycerides | 38.02% | 47.53% |
| Polyglyceryl-6 dioleate | 9.50% | 11.88% |
| Purified water | 45.05% | 15.84% |

**STUDY 2:** Austria et al.[856] compared the stability of two ascorbic acid derivatives, ascorbyl palmitate and magnesium ascorbyl phosphate, in topical oil-in-water emulsion cosmetic products (formulation not described). HPLC analysis found the magnesium ascorbyl phosphate to be very stable, retaining up to 95% after 60 days of storage at 42°C protected from light, which is the equivalent of one year at room temperature. However, ascorbyl palmitate was unstable, exhibiting 27% loss in two months at room temperature.

**STUDY 3:** Maia et al.[1087] evaluated ascorbic acid stability in oil-in-water emulsion and aqueous gel topical bases with and without the antioxidant compounds sodium metabisulfite or glutathione. The ascorbic acid 10% topical preparations were stored at 5, 24, and 40°C for 90 days and were evaluated by HPLC analysis for drug concentrations. The formulations that contained sodium metabisulfite or glutathione consistently demonstrated slightly better stability with less ascorbic acid loss compared to the formulations without either antioxidant. Under refrigeration and at room temperature, ascorbic acid losses were 7 and 13%, respectively, compared to 5 and 8 to 10%, respectively, with either of the antioxidants present.

**STUDY 4:** Rozman and Gasperlin[1267] evaluated the stability of ascorbic acid 0.4% (wt/wt) and vitamin E 1% (wt/wt) in several topical microemulsion formulations. Oil/water, water/oil, and gel-like microemulsions were evaluated. The microemulsions had the components shown in Table 10. The surfactant and cosurfactant were blended in a 1:1 mass ratio. The isopropyl myristate and distilled water were then added and mixed using a magnetic stirrer for 30 minutes.

The finished topical preparations were packaged in glass containers and stored at controlled room temperature of 21 to 23°C. Stability-indicating HPLC analysis found that the vitamin E was stable with no difference between any of the formulations, with light exposure, and with the presence of ascorbic acid. Indeed, the presence of ascorbic acid protected the vitamin E from decomposition over 56 days of storage. Vitamin C decomposed during storage about 30 to 40% over 28 days. ■

**TABLE 10.** Percentage (wt/wt) Composition of Microemulsion Formulations[1267]

| Component | Water/Oil | Oil/Water | Gel-like |
|---|---|---|---|
| Distilled water | 10 | 45 | 60 |
| Isopropyl myristate | 60 | 25 | 10 |
| Tween 40[a] | 15 | 15 | 15 |
| Imwitor 308[b] | 15 | 15 | 15 |

[a]Polyoxyethylene (20) sorbitan monopalmitate
[b]Glyceryl caprylate

# Aspirin
## (Acetylsalicylic Acid)

## Properties

Aspirin occurs as colorless or white crystals or crystalline powder that is odorless or has a slight odor of acetic acid.[1-3]

### Solubility

Aspirin is slightly soluble in water, having an aqueous solubility of 3.3 mg/mL at 25°C and 10 mg/mL at 37°C. In ethanol, the drug is freely soluble, having a solubility of 200 mg/mL.[2,3,6]

### $pK_a$

Aspirin has a $pK_a$ of 3.5[1,2] or 3.6.[6]

## General Stability Considerations

Oral aspirin products should be stored in tight containers.[3,4] The drug is stable in dry air but gradually hydrolyzes in contact with moisture from air, forming acetic acid and salicylic acid. The only effective method of inhibiting aspirin hydrolysis is to prevent contact with water.[2,3,6] Most manufacturers state that aspirin tablets should not be removed from the original packaging and placed in dosing compliance aids because of the drug's hygroscopicity and sensitivity to atmospheric moisture.[1622] The rate of hydrolysis is increased by heat and is pH dependent.[2,3,6] The drug is most stable around pH 2 to 3, somewhat less stable at pH 5 to 9, and the least stable at pH values below 1 and above 9. At pH 7 and 25°C, aspirin has a half-life of about 52 hours and a time to 10% loss of about eight hours.[6] Aspirin stability was found to be adversely affected by contact with antacids.[6]

Aspirin suppositories should be packaged in well-closed containers and stored at 2 to 15°C.[4]

Aspirin forms a damp pasty mass if triturated with acetanilide, antipyrine, aminopyrine, phenacetin (acetophenetidin), phenol, or phenyl salicylate.[6]

## Stability Reports of Compounded Preparations
### Oral

**STUDY 1:** Al-Achi et al.[569] evaluated the weight variation and content uniformity of aspirin 325-mg chewable lozenges prepared from aspirin powder in a gelatin base. The density factor for aspirin (the weight of drug that displaces 1 g of lozenge base) was determined to be 1.09. The gelatin base and three batches of 24 of the chewable lozenges were prepared according to the method of Allen.[570] Although the weight of the lozenges was reasonably close (1.083 ± 0.026 g), content uniformity of the aspirin determined by the USP assay was not. Aspirin content ranged from a mean of 0.369 mg to 0.204 mg between batches. The authors indicated that quality control testing may be needed to ensure appropriate aspirin content between batches of the chewable lozenges.

**STUDY 2:** A study of aspirin stability in oral liquid formulations reported the best stability to occur in a vehicle composed of 35% ethanol, 40% propylene glycol, and 25% glycerin. Approximately 8.7% aspirin hydrolysis occurred in eight weeks (temperature unspecified).[6]

**STUDY 3:** One aspirin effervescent buffered tablet (Alka-Seltzer) dissolved in 90 mL of water (resulting in pH 6 to 7) retained at least 90% concentration through 10 hours at room temperature and through 90 hours under refrigeration.[2]

### Rectal

Regdon et al.[1668] reported on the stability of aspirin formulated into two suppository bases, Witepsol H 32 and Witepsol H 35. Aspirin 15% was incorporated as a finely divided powder into the bases that had been melted at 45°C. The mixtures were held at 45°C for 24 hours before molding the suppositories and storing them at room temperature for eight months. Approximately 0.6% aspirin loss occurred over eight months of room temperature storage in both bases. ■

# Atenolol

## Properties

Atenolol is a white or almost white crystalline powder.[1-3]

### Solubility

Atenolol has an aqueous solubility of 26.5 mg/mL at 37°C. It is also soluble in ethanol.[2,3]

### pH

The injection has a pH of 5.5 to 6.5.[7]

### $pK_a$

The $pK_a$ of atenolol is 9.6.[1]

## General Stability Considerations

Atenolol oral tablets should be stored in well-closed, light-resistant containers at controlled room temperature. The injection also should be stored at room temperature and should be protected from light during long-term storage.[2,7] Atenolol exhibits maximum stability at pH 4.[773] Exposure of atenolol solutions to ultraviolet light resulted in drug decomposition at both physiologic and acid pH.[792]

Donyai[1625] evaluated the stability of atenolol 100-mg oral tablets made by CP Pharmaceuticals and by Alpharma when transferred from the original packaging and placed in multi-compartment dose compliance aids both with and without aspirin 75-mg tablets stored in the same compartments. The physical and chemical stability of the tablets was assessed during storage at room temperature of 25°C and at stress conditions of 40°C and 75% relative humidity for 28 days. Differences in physical parameters such as changes in appearance and reduction in tablet hardness were found in some of the samples. However, stability-indicating HPLC analysis found no loss of drug, and there was no evidence of changes in bioavailability. The author concluded that it was unlikely that any of the minor changes would have a great impact on the quality of patient care.

## Stability Reports of Compounded Preparations
### Oral
**USP OFFICIAL FORMULATION (ORAL LIQUID):**

Atenolol 200 mg
Glycerin 5 mL
Vehicle for Oral Suspension, NF 45 mL
Vehicle for Oral Solution, Sugar Free, NF qs 100 mL

NOTE: Do not use sucrose-containing vehicle. (See the vehicle monographs for information on the individual vehicles.)

Using commercial atenolol tablets, crush or grind to fine powder. Mix the powdered tablets with the glycerin to form a smooth paste. Add the Vehicle for Oral Suspension or an equivalent volume of the Vehicle for Oral Solution, Sugar Free. The Vehicle for Oral Suspension may be omitted if desired. Add additional Vehicle for Oral Solution, Sugar Free in increments to bring to volume, and mix well yielding atenolol 2-mg/mL oral liquid. Transfer to a suitable tight and light-resistant container. Store the preparation at controlled room temperature. The beyond-use date is 60 days from the date of compounding.[4,5]

**STUDY 1:** The stability of atenolol compounded in a commercial oral diluent was determined by Garner et al.[127] Six 50-mg atenolol tablets (Tenormin, ICI Pharma) were crushed and triturated with 50 mL of Diluent (flavored) for Oral Use (Roxane) containing 1% ethanol and 0.05% saccharin in a cherry-flavored 33% polyethylene glycol 8000 base. The mixture was brought to 150 mL with two 50-mL portions of oral diluent to yield an atenolol concentration of 2 mg/mL. The mixture was shaken mechanically for 15 minutes and then packaged in two-ounce amber glass vials and stored under refrigeration (5°C) and at room temperature (25°C) for 40 days. The atenolol content was assessed using a stability-indicating HPLC assay. The atenolol concentration remained above 90% in all samples whether the samples were shaken before testing or not. Sediment formed in samples but was due to insoluble tablet excipients. Also, microbial growth occurred in some samples.

**STUDY 2:** Patel et al.[556] evaluated the stability of several oral liquid formulations of atenolol 2 mg/mL prepared from powder and also from tablets crushed and ground to powder using a mortar and pestle. The vehicles used for this testing included simple syrup, Ora-Sweet, Ora-Sweet SF, Ora-Plus, and a 1% methylcellulose 2000 dispersion. To compound the various formulations, the atenolol powder from either source was levigated with about 2 mL of glycerin, forming a paste. Approximately 20 mL of the vehicle was added to the paste, mixed thoroughly, and transferred to a graduated cylinder. The mortar and pestle were rinsed with remaining vehicle and brought to their final volume to yield an atenolol 2-mg/mL concentration. The oral liquids were packaged in amber prescription bottles and stored (temperature unspecified) over 90 days.

The formulations prepared with atenolol powder in simple syrup, Ora-Sweet, and Ora-Sweet SF were clear solutions. The formulations prepared using Ora-Plus were cloudy because the vehicle is not clear. Formulations compounded from tablets had insoluble tablet components that exhibited sedimentation. No changes in appearance or consistency were observed in the samples throughout the study. HPLC analysis found atenolol in the Ora-Sweet SF samples remained stable; less than 10% loss occurred in 90 days using either atenolol source. Stability in 1% methylcellulose was stated to be good, but no data were presented. Syrup and Ora-Sweet were less acceptable vehicles for atenolol; about 10% drug loss appeared in one to two weeks in products prepared from either atenolol source. Ora-Plus samples were difficult to sample because the product's viscosity prevented adequate analysis.

**STUDY 3:** Modamio et al.[745] reported the stability of atenolol (ICI) 0.25 mg/mL in phosphate buffer (pH 7.4) at elevated temperatures. Stability-indicating HPLC analysis found that little or no loss of atenolol occurred in 84 hours, even at 90°C.

**STUDY 4:** Foppa et al.[1152] evaluated the stability of atenolol 10-mg/mL oral liquid prepared by adding atenolol to simple syrup. The oral liquid was packaged in both clear and amber glass containers. Samples were stored refrigerated at 4°C and at room temperature of 25°C. HPLC analysis found about 10% loss in seven days at room temperature and in nine days refrigerated in both clear and amber containers.

**STUDY 5 (REPACKAGED TABLETS):** Chan et al.[1586] reported on the stability of atenolol 100-mg immediate-release tablets (Wockhardt) repackaged from the original containers into 28-compartment plastic compliance aids with transparent lids. Samples were stored for 28 days at ambient room temperature near 25°C and at elevated conditions of 40°C and 75% relative humidity. At room temperature, little or no change in appearance, weight uniformity, disintegration, and dissolution occurred; the tablets passed all of the relevant compendial tests. However, the tablets were softer than those in the original packaging. At elevated temperature and humidity, the tablets became paler and moist in appearance, significantly softer, more friable, and failed the tests for disintegration and dissolution. HPLC analysis found little or no loss of drug content occurred at either storage condition. The authors stated that the repackaging of atenolol tablets into plastic compliance aids might be possible for storage at 25°C but should be avoided in locations with hotter and more humid weather conditions.

## Compatibility with Other Drugs

Kumar et al.[1419] evaluated the compatibility of atenolol, with 29 drug product excipients. Atenolol was mixed 1:1 with each of the excipients and stored for one month at 40°C and 75% relative humidity. The samples were observed for physical changes and analyzed by stability-indicating HPLC for

**TABLE 11.** Excipients Compatible with Atenolol[1419]

| |
| --- |
| Brilliant scarlet 3R (Cochineal Red A; Ponceau 4R) |
| Calcium carbonate |
| Calcium phosphate dibasic |
| FD&C Red 40 (Allura red; Food Red 17) |
| Ferric oxide |
| Hydroxypropyl cellulose |
| Hydroxypropyl methylcellulose |
| Lactose |
| Magnesium oxide |
| Mannitol |
| Microcrystalline cellulose |
| Povidone |
| Silicon dioxide |
| Sodium starch glycolate |
| Starch |
| Stearic acid |
| Talc |
| Titanium dioxide |
| Zinc stearate |

chemical changes. Little or no physical or chemical change was observed with most of the excipients tested. See Table 11. However, atenolol was incompatible with ascorbic acid, citric acid, and butylated hydroxyanisole. The mixtures became wet and/or discolored. Analysis found atenolol losses of 12 to 21%. ▪

## ▪ Atropine
## Atropine Sulfate

### Properties

Atropine is a racemic mixture of *d*- and *l*-hyoscyamine obtained by synthesis or isolated from plants of the *Solanaceae* genus, including *Atropa belladonna* (deadly nightshade) and *Datura stramonium* (Jimson weed), among others. Atropine occurs as white crystals or crystalline powder.[1–3]

Atropine sulfate occurs as odorless, colorless granules or as a white crystalline powder having a very bitter taste.[1–3]

### Solubility

Atropine has an aqueous solubility of 2.2 mg/mL and a solubility in ethanol of 500 mg/mL.[1,6] Atropine sulfate is very soluble in water, having a solubility of approximately 2 to 2.5 g/mL. It has solubilities of 200 mg/mL in ethanol at 25°C and 400 mg/mL in glycerol.[1–3]

### pH

A 2% atropine sulfate solution in water has a pH of 4.5 to 6.2.[3] Atropine sulfate ophthalmic solution has a pH of 3.5 to 6, while atropine sulfate injection has a pH of 3 to 6.5.[4]

### pKa

Atropine has a pKa of 9.8.[6]

### General Stability Considerations

Atropine sulfate effloresces on exposure to air and changes slowly upon exposure to light. It should be stored in tight containers and protected from light.[2–4] The ophthalmic ointment and solution should be stored at controlled room temperature and protected from heat and from freezing.[2]

The major route of atropine decomposition is hydrolysis. The drug also is readily oxidizable, and stability may be improved by the presence of antioxidants. Atropine is most stable in the pH range of 2 to 4, with the minimum rate of decomposition occurring at pH 3.5. However, the drug undergoes increasing hydrolysis as the pH increases or decreases from this range. Sterilization of atropine sulfate solutions for 20 minutes at 120°C resulted in no appreciable loss in the pH range of 2.8 to 6, but hydrolysis became rapid at pH above 6. Fungal contamination of atropine sulfate solutions may also result in loss of drug activity.[6]

## Stability Reports of Compounded Preparations
### Ophthalmic

Ophthalmic preparations, like other sterile drugs, should be prepared in a suitable clean air environment using appropriate aseptic procedures. When prepared from nonsterile components, an appropriate and effective sterilization method must be employed.

**STUDY 1 (WITH COCAINE):** Miethke[491] reported the stability of a cocaine hydrochloride 10-mg/mL–atropine sulfate 2-mg/mL mixed solution for ophthalmic use. The solution was prepared by dissolving 2 g of cocaine hydrochloride and 400 mg of atropine sulfate in 200 mL of water for injection. The solution was filtered through a 0.22-µm cellulose ester filter (Optex) and filled as 2 mL in 10-mL containers. The solution was sterilized in an autoclave at 121°C for 15 minutes. HPLC analysis determined that the solution was stable for 15 months, presumably at room temperature.

**STUDY 2:** Garcia-Valldecabres et al.[880] reported the change in pH of several commercial ophthalmic solutions, including atropine sulfate 5 mg/mL (Colircusi Atropina), over 30 days after opening. Slight variation in solution pH was found, but no consistent trend was noted and no substantial change occurred.

### Injection

Injections, like other sterile drugs, should be prepared in a suitable clean air environment using appropriate aseptic procedures. When prepared from nonsterile components, an appropriate and effective sterilization method must be employed.

**STUDY 1:** Dix et al.[874] reported the stability of atropine sulfate 1 mg/mL prepared from bulk powder in sodium chloride 0.9%. Large amounts were prepared to simulate the need that might arise from mass chemical terrorism. The powder was dissolved in sodium chloride 0.9% and diluted to 5 mg/mL, and the solution was passed through a sterilizing filter. This stock solution was diluted further to 1 mg/mL for use. Samples were stored refrigerated at 4 to 8°C, at room temperature of 20 to 25°C, and at an elevated temperature of 32 to 36°C for 72 hours. All samples were protected from exposure to light using an amber occlusive cover. All of the samples remained clear and colorless throughout the study. HPLC analysis found that little or no loss of atropine occurred in 72 hours.

**STUDY 2:** Casasin Edo et al.[887] evaluated the stability of atropine 1-mg/mL injection (form unspecified) packaged in polypropylene syringes. Spectrophotometric and potentiometric analyses found that little or no loss occurred in four weeks at room temperature with no exposure to direct light.

**STUDY 3:** Donnelly and Corman[1294] reported the stability of extemporaneously compounded atropine sulfate 2-mg/mL injection for use in the event of a terrorist attack using nerve gas. The injection was prepared by dissolving atropine sulfate powder in sodium chloride 0.9% injection to yield a 2-mg/mL solution. The atropine sulfate injection was adjusted to pH 3.5 with 0.1 N sulfuric acid and was sterilized by filtration through a 0.2-µm mixed cellulose ester filter. The atropine sulfate 2-mg/mL injection was packaged as 3 mL of injection in Becton Dickinson 5-mL polypropylene plastic syringes with sealed tips. Sample syringes were stored for 364 days refrigerated at 5°C protected from exposure to light, for 364 days at room temperature of 23°C exposed to light, and for 28 days at elevated temperature of 35°C exposed to light.

No visible haze, precipitation, or color change was observed in any of the samples. Stability-indicating HPLC analysis found no loss of atropine sulfate occurred within 364 days during refrigerated and room temperature storage and for 28 days at elevated temperature.

**STUDY 4:** Geller et al.[1319] evaluated the stability of a fortified atropine sulfate 2-mg/mL injection. The fortified injection with its high concentration was intended for the possibility of treating mass casualties of a nerve agent attack using intramuscular injection. The test injection was compounded by emptying commercial atropine sulfate 0.1-mg/mL prefilled syringes (Amphastar) into a sterile evacuated glass container, removing the rubber stopper, and adding the necessary amount of atropine sulfate powder to bring the atropine sulfate concentration to 2 mg/mL. The rubber stopper was then replaced in the bottle, and the solution was shaken until it appeared clear. The solution was passed through a 0.22-µm filter and was packaged as 3 mL of solution in 3-mL syringes. Sample syringes were stored refrigerated at 5°C, at room temperature of 24°C, and at elevated temperatures of 35 and 45°C for up to four weeks.

HPLC analysis of atropine sulfate concentrations found that excessive drug losses exceeding 10% occurred within two weeks at 35 and 45°C. Samples at room temperature lost 11% within six weeks. Refrigerated samples fared the best, exhibiting little or no loss over the entire eight-week study period.

**STUDY 5:** Schier et al.[1485] reported the stability of commercial atropine sulfate injections past their expiration periods for possible use in the event of chemical terrorism. Samples of injections that were aged by 3, 5, 16, and about 60 years from production were tested using a gas chromatography/mass spectrometry (GC/MS) analytical technique. All samples were found to still have substantial amounts of atropine sulfate with very little decomposition product present. The authors pointed to a previous study by Kondrizter and Zvirblis[1486] that estimated the half-life of atropine sulfate in aqueous solution at pH 4 and a temperature of 20°C to be about 1800 years. Of course, a 10°C higher temperature will reduce the half-life to an estimated 473 years. Schier et al. concluded that in the event of an emergency situation such as chemical terrorism, atropine sulfate injection past its expiration date may still be useful and efficacious because treating cholinergic crisis is based exclusively on the clinical response to treatment.

## Compatibility with Other Drugs

### ATROPINE COMPATIBILITY SUMMARY

*Compatible with:* Sodium chloride 7% (hypertonic)
*Incompatible with:* Albuterol sulfate • Hydroxybenzoate
  preservatives • Isoetharine • Metaproterenol sulfate •
  Sodium bicarbonate • Terbutaline sulfate
*Uncertain or variable compatibility with:* Cromolyn sodium

**STUDY 1:** Hydroxybenzoate preservatives are incompatible with atropine sulfate. In one report, total loss of atropine sulfate occurred in an oral liquid preparation in about two to three weeks.[637]

**STUDY 2:** Owsley and Rusho[517] evaluated the compatibility of several respiratory therapy drug combinations. Atropine sulfate 1 mg/mL (American Regent) was combined with the following drug solutions: albuterol sulfate 5-mg/mL inhalation solution (Proventil, Schering), cromolyn sodium 10-mg/mL nebulizer solution (Intal, Fisons), isoetharine 1% solution (Bronkosol, Sanofi Winthrop), metaproterenol sulfate 5% solution (Alupent, Boehringer Ingelheim), sodium bicarbonate 8.4% injection (Abbott), and terbutaline sulfate 1-mg/mL solution (Brethine, Geigy). The test solutions were filtered through 0.22-µm filters into clean vials. The combinations were evaluated over 24 hours (temperature unspecified) using the USP particulate matter test. None of the combinations was found to be compatible. Atropine sulfate combined with the albuterol sulfate, cromolyn sodium, isoetharine, metaproterenol sulfate, sodium bicarbonate, and terbutaline sulfate all formed unacceptable levels of larger particulates (≥10 µm).

**STUDY 3:** Emm et al.[628] evaluated the stability and compatibility of cromolyn sodium combinations during short-term preparation and use. Cromolyn sodium 1% nebulizer solution (Fisons) was combined with atropine sulfate 0.2 and 0.5% (Dey) and studied over 90 minutes. No visible signs of precipitation or measured changes in pH were found, and HPLC analysis of the prepared combinations found less than 10% loss of any of the drugs within 90 minutes.

**STUDY 4 (HYPERTONIC SODIUM CHLORIDE):** Fox et al.[1239] evaluated the physical compatibility of hypertonic sodium chloride 7% with several inhalation medications used in treating cystic fibrosis. Atropine sulfate (American Regent) 1 mg/mL mixed in equal quantities with extemporaneously compounded hypertonic sodium chloride 7% did not exhibit any visible evidence of physical incompatibility, and the measured turbidity did not increase over the one-hour observation period. ▪

## ▪ Azathioprine

### Properties

Azathioprine is a yellow odorless crystalline powder that is a purine antagonist antimetabolite. Azathioprine has been cited as a known carcinogen.[1-3]

### Solubility

Azathioprine is slightly soluble in water, about 0.13 mg/mL at 25°C. It also is slightly soluble in ethanol. It dissolves in solutions of dilute alkali hydroxides, although it undergoes slow decomposition to mercaptopurine.[1-3,6]

### pH

The reconstituted injection (of the sodium salt) has a pH range of 9.8 to 11.[4]

### $pK_a$

The $pK_a$ has been cited as 8.2[1] and 7.87.[6]

### General Stability Considerations

Azathioprine tablets should be stored in well-closed containers at controlled room temperature and protected from light.[2,4] The injection also should be stored at controlled room temperature and protected from light. The drug is stable in neutral or acid solutions, but in alkaline solutions hydrolysis to mercaptopurine occurs. The pH of maximum stability is around 5.5 to 6.5. The presence of sulfhydryl compounds such as cysteine may also result in hydrolysis.[2,6,7,295]

Azathioprine decomposition increases as the concentration of alkali increases. An aqueous solution in 0.02 M sodium hydroxide was chromatographically stable at ambient temperature for eight days in the dark and for four days with exposure to light. In 0.04 M and greater sodium hydroxide, drug decomposition was evident after two days.[6]

The reconstituted injection (as the sodium salt) is chemically stable for about two weeks at room temperature. However, it contains no antibacterial preservative and use within 24 hours is recommended.[2,7]

## Stability Reports of Compounded Preparations
### Injection

Injections, like other sterile drugs, should be prepared in a suitable clean air environment using appropriate aseptic procedures. When prepared from nonsterile components, an appropriate and effective sterilization method must be employed.

Johnson and Porter[296] compared the stability of azathioprine reconstituted and stored in the original vial to the solution repackaged in plastic syringes (Jelco). The containers were stored at 20 to 25°C under fluorescent light. No loss of drug by HPLC analysis or precipitation occurred during 16 days of storage in either container. However, storage at 4°C resulted in a visible precipitate in four days.

### Oral

#### USP OFFICIAL FORMULATION (ORAL SUSPENSION):

    Azathioprine 5 g
    Vehicle for Oral Solution, NF (sugar-containing or
       sugar-free)
    Vehicle for Oral Suspension, NF (1:1) qs 100 mL

(See the vehicle monographs for information on the individual vehicles.)

Use azathioprine powder or commercial tablets. If using tablets, crush or grind to fine powder. Add about 10 mL of the vehicle mixture and mix to make a uniform paste. Add additional vehicle almost to volume in increments with thorough mixing. Quantitatively transfer to a suitable calibrated tight and light-resistant bottle, bring to final volume with the vehicle mixture, and thoroughly mix yielding azathioprine 50 mg/mL oral suspension. The final liquid preparation should have a pH between pH 3.8 and 4.8. Store the preparation at controlled room temperature or under refrigeration between 2 and 8°C. The beyond-use date is 60 days from the date of compounding at either storage temperature.[4,5]

**STUDY 1:** The stability of azathioprine 50 mg/mL in an extemporaneous oral suspension was studied by Dressman and Poust.[87] Azathioprine 50-mg tablets (Burroughs-Wellcome) were crushed and mixed with Cologel (Lilly) as one-third of the total volume, shaken, and brought to final volume with a 2:1 mixture of simple syrup and cherry syrup, followed by vigorous shaking for 30 seconds and ultrasonication for at least two minutes. The azathioprine 50-mg/mL suspension was packaged in amber glass bottles and stored at 5°C and ambient room temperature. Semiquantitative ultraviolet spectrophotometry was used to determine the azathioprine content of the suspensions. Little or no loss occurred in 56 days at room temperature or in 84 days at 5°C.

**STUDY 2:** Allen and Erickson[543] evaluated the stability of three azathioprine 50-mg/mL oral suspensions extemporaneously compounded from tablets. Vehicles used in this study were (1) an equal parts mixture of Ora-Sweet and Ora-Plus (Paddock), (2) an equal parts mixture of Ora-Sweet SF and Ora-Plus (Paddock), and (3) cherry syrup (Robinson Laboratories) mixed 1:4 with simple syrup. A total of 120 azathioprine 50-mg tablets (Burroughs Wellcome) were crushed and comminuted to fine powder using a mortar and pestle. About 40 mL of the test vehicle was added to the powder and mixed to yield a uniform paste. Additional vehicle was added geometrically and brought to the final volume of 120 mL, mixing thoroughly after each addition. The process was repeated for each of the three test suspension vehicles. Samples of each of the finished suspensions were packaged in 120-mL amber polyethylene terephthalate plastic prescription bottles and stored at 5 and 25°C in the dark.

No visual changes or changes in odor were detected during the study. Stability-indicating HPLC analysis found less than 4% azathioprine loss in any of the suspensions stored at either temperature after 60 days of storage.

**STUDY 3:** Le Quan et al.[1453] evaluated the stability of azathioprine 5-mg/mL pediatric oral liquid prepared from commercial tablets in a mixture of Ora-Sweet and Ora-Plus. The oral liquid was packaged in light-resistant containers and was stored at 4°C for 30 days. The oral liquid exhibited no change in color or pH. HPLC analysis of drug concentrations found less than 5% change over 30 days of storage. ■

# Azithromycin

## Properties
Azithromycin occurs as a white or almost white crystalline material.[1,3,7]

### Solubility
Azithromycin is practically insoluble in water but freely soluble in dehydrated ethanol and dichloromethane.[3]

### pH
Azithromycin oral suspension has a pH between 8.5 and 11.[4] Reconstituted azithromycin injection at 100 mg/mL is buffered with citric acid to a pH of 6.4 to 6.6.[7]

### pKa
Azithromycin has apparent $pK_a$ values of 9.16 and 9.37.[1426]

## General Stability Considerations
Azithromycin bulk powder and oral suspension powder should be packaged in tight containers and stored at controlled room temperature. Azithromycin oral capsules and tablets should be packaged in well-closed containers and stored at controlled room temperature. Single-dose packets of oral suspension powder should be stored between 5 and 30°C.[4,7] Azithromycin for injection vials should be stored at controlled room temperature.[7]

Reconstituted azithromycin oral suspension should be stored between 5 and 30°C. After reconstitution, the oral suspension should be used according to the manufacturer's labeling for the specific product.[7]

Reconstituted azithromycin injection at 100 mg/mL is stable for 24 hours at controlled room temperature. Diluted for use to 1 to 2 mg/mL in a compatible infusion solution, the drug is stable for 24 hours at controlled room temperature and for seven days refrigerated.[7]

## Stability Reports of Compounded Preparations
### Ophthalmic
Ophthalmic preparations, like other sterile drugs, should be prepared in a suitable clean air environment using appropriate aseptic procedures. When prepared from nonsterile components, an appropriate and effective sterilization method must be employed.

Moreno et al.[1425] evaluated factors that affect the stability of azithromycin ophthalmic solution. The ophthalmic solution had an azithromycin concentration of 1.667 mg/mL in an unspecified physiological solution. The ophthalmic solution was subjected to a variety of stresses to observe the effects on the drug's stability. A microbiological assay technique was used to assess stability. Extremes of pH were evaluated using hydrochloric acid and sodium hydroxide 0.1 mol/L along with heat of 70°C; these resulted in nearly total loss of the drug in six hours. Exposure to hydrogen peroxide 0.3% yielded a similar result. Exposure of the azithromycin ophthalmic solution to sunlight and ultraviolet light (at 254 and 284 nm) resulted in losses of 11, 38, and 20%, respectively, in 48 hours. The authors concluded that azithromycin ophthalmic solution pH and exposure to light should be controlled for stability.

# Bacitracin

## Properties
Bacitracin is a polypeptide antibiotic that has been variously described as a grayish-white[1] or white to pale buff [2,3] hygroscopic powder having little or no odor and a very bitter taste.[1,3] In the United States, it has an activity of not less than 40 bacitracin activity units per milligram[4] and not less than 50 units per milligram for parenteral use.[2] In Europe, it has not less than 60 bacitracin activity units per milligram.[3]

### Solubility
Bacitracin is freely soluble in water and ethanol.[1–3]

### pH
The pH of a 10,000-unit/mL aqueous solution is 5.5 to 7.5.[4] A 1% solution has a pH of 6 to 7.[3]

## General Stability Considerations
Bacitracin products should be stored in tight containers.[4] The injection should be stored at 2 to 15°C protected from sunlight, while the ophthalmic and topical ointments should be stored at controlled room temperature.[2]

## Stability Reports of Compounded Preparations
### Ophthalmic
Ophthalmic preparations, like other sterile drugs, should be prepared in a suitable clean air environment using appropriate aseptic procedures. When prepared from nonsterile components, an appropriate and effective sterilization method must be employed.

Osborn et al.[169] evaluated the stability of several antibiotics in three artificial tears solutions composed of 0.5% hydroxypropyl methylcellulose. The three artificial tears solutions (Lacril, Tearisol, Isopto Tears) were used to reconstitute and dilute bacitracin initially to 9600 units/mL. The product was packaged in plastic squeeze bottles and stored at 25°C. A serial dilution bioactivity test was used to estimate antibiotic activity remaining during seven days of storage at room temperature: 83% in Lacril, 80% in Tearisol, and 84% in Isopto Tears.

### Irrigation
Irrigations, like other sterile drugs, should be prepared in a suitable clean air environment using appropriate aseptic procedures. When prepared from nonsterile components, an appropriate and effective sterilization method must be employed.

The stability of bacitracin solution frozen in glass vials and plastic syringes was evaluated by Souney et al.[94] Bacitracin powder 50,000 units per vial (Pfizer) was reconstituted with 9.8 mL of sterile water to yield a 5000-unit/mL solution. Five milliliters of the reconstituted solution was packaged in 10-mL plastic syringes (Becton Dickinson) and 10-mL glass vials (Abbott). The syringes and vials were stored frozen at −15°C for 20 weeks. Bacitracin activity was assessed microbiologically. No significant difference in bacitracin activity was found at any time in either container.

# ■ Baclofen

## Properties
Baclofen is a white to off-white nearly odorless crystalline material.[2,3]

### Solubility
Baclofen is slightly soluble in water and in alcohol.[2,3] Maximum aqueous solubility is stated by the drug manufacturer to be about 4.5 mg/mL, although this may take a long time to achieve. While aqueous solubilization at higher concentrations has been attempted, simply adding the drug to aqueous vehicles at higher concentrations does not result in complete dissolution. Solubility of 0.024 mg/mL in ethanol has been reported.[586]

### pH
The injection has a pH of 5 to 7.[2]

### pK$_a$
Baclofen has pK$_a$ values of 5.4 and 9.5.[249]

## General Stability Considerations
Baclofen tablets should be stored in well-closed containers[4] at controlled room temperature.[2] The injection for intrathecal use should be stored at controlled room temperature and protected from freezing; it should not be autoclaved. The drug is stable at 37°C, the temperature in implantable pumps. Dilution must be performed only with preservative-free 0.9% sodium chloride injection.[2] The manufacturer states that baclofen tablets should not be removed from the original packaging and placed in dosing compliance aids.[1622]

## Stability Reports of Compounded Preparations
### Oral
#### USP OFFICIAL FORMULATION:
  Baclofen 500 mg
  Vehicle for Oral Solution, NF (sugar-containing or
    sugar-free)
  Vehicle for Oral Suspension, NF (1:1) qs 100 mL

(See the vehicle monographs for information on the individual vehicles.)

Use baclofen powder or commercial tablets. If using tablets, crush or grind to fine powder. Add about 5 mL of the vehicle mixture, and mix to make a uniform paste. Add additional vehicle almost to volume in small increments with thorough mixing. Quantitatively transfer to a suitable calibrated tight and light-resistant bottle, bring to final volume with the vehicle mixture, and thoroughly mix yielding baclofen 5 mg/mL oral suspension. The final liquid preparation should have a pH between pH 4.2 and 5.2. Store the preparation under refrigeration between 2 and 8°C. The beyond-use date is 35 days from the date of compounding.[4,5]

**STUDY 1:** Johnson and Hart[124] reported the stability of extemporaneously prepared baclofen oral liquid formulations. Fifteen baclofen 20-mg tablets (Lioresal, Ciba-Geigy) were crushed to fine powder. The powder was wetted with glycerin, USP, and triturated to a paste. Simple syrup, NF, was added in three increments, bringing the total volume to 60 mL and yielding a theoretical baclofen concentration of 5 mg/mL. Although baclofen is soluble in water, a suspension is formed due to the tablet excipients. The suspension was packaged in two-ounce type III amber glass bottles with child-resistant caps and stored under refrigeration at 4°C.

In addition, baclofen analytical-grade powder was dissolved in simple syrup, NF, to yield a 5-mg/mL syrup that was packaged and stored similarly to the suspension. The baclofen content was assessed by a stability-indicating HPLC assay during 35 days of storage. Less than 5% loss occurred in either the suspension or the syrup stored at 5°C for 35 days. No changes in color or odor occurred in either preparation; additionally, no precipitate, turbidity, or evidence of microbial growth was observed in any of the preparations tested.

**STUDY 2:** Allen and Erickson[544] evaluated the stability of three baclofen 10-mg/mL oral suspensions extemporaneously compounded from tablets. Vehicles used in this study were (1) an equal parts mixture of Ora-Sweet and Ora-Plus (Paddock), (2) an equal parts mixture of Ora-Sweet SF and Ora-Plus (Paddock), and (3) cherry syrup (Robinson Laboratories) mixed 1:4 with simple syrup. A total of 120 baclofen 10-mg tablets (Zenith) were crushed and comminuted to fine powder using a mortar and pestle. About 40 mL of the test vehicle was added to the powder and mixed to yield a uniform paste. Additional vehicle was added geometrically and brought to the final volume of 120 mL, mixing thoroughly after each addition. The process was repeated for each of the three test suspension vehicles. Samples of each of the finished suspensions were packaged in 120-mL amber polyethylene terephthalate plastic prescription bottles and stored at 5 and 25°C in the dark.

No visual changes or changes in odor were detected during the study. Stability-indicating HPLC analysis found less than 4% baclofen loss in any of the suspensions stored at either temperature after 60 days of storage.

**STUDY 3:** Woods[586] reported on an extemporaneous oral liquid formulation of baclofen recommended by the manufacturer

**TABLE 12.** Formula of Baclofen 2-mg/5-mL Oral Liquid Prepared from Tablets[586]

| Component | | Amount |
|---|---|---|
| Baclofen 10-mg tablets | | 4 |
| Sodium carboxymethylcellulose | | 1 g |
| Ethanol 95% | | 25 mL |
| Methylparaben | | 140 mg |
| Propylparaben | | 28 mg |
| Water | qs | 100 mL |

of Lioresal tablets. The formula is shown in Table 12. An expiration period of two weeks when stored at room temperature protected from light was suggested, although no stability data were presented.

## *Injection*

Injections, like other sterile drugs, should be prepared in a suitable clean air environment using appropriate aseptic procedures. When prepared from nonsterile components, an appropriate and effective sterilization method must be employed.

Moberg-Wolff[1448] evaluated 29 samples of compounded baclofen intrathecal injection obtained from six compounding pharmacies. The intended concentrations included 2, 3, 4, 5, and 6 mg/mL. However, analysis found 41% of the samples were 5% or more above or below the intended concentration, and 22% were more than 10% above or below the intended concentration. The 6-mg/mL samples had visible precipitation. The author indicated that adverse clinical results could result from the variability among the pharmacy compounded baclofen intrathecal injections. ■

# ■ Beclomethasone Dipropionate
## (Beclometasone Dipropionate)

## Properties

Beclomethasone dipropionate, a synthetic halogenated corticosteroid, is a white to creamy white odorless crystalline powder.[3,4,7]

## *Solubility*

Beclomethasone dipropionate monohydrate is very slightly soluble in water and freely soluble in ethanol and acetone.[4,7,1633] Beclomethasone dipropionate anhydrous has an aqueous solubility of less than 5 mcg/mL and a solubility in ethanol of 22 mg/mL.[2]

## *pH*

Beclomethasone dipropionate nasal spray has a pH in the range of 4.5 to 7 for Beconase and 5.5 to 6.8 (target pH 6.4) for Vancenase.[7]

## General Stability Considerations

Beclomethasone dipropionate bulk powder should be stored at room temperature protected from light.[3,4] The oral inhalation form of beclomethasone dipropionate (QVAR) is stored at controlled room temperature near 25°C. The nasal inhalation aerosol is stored at 2 to 30°C for Beconase or at 15 to 30°C for Vancenase. The nasal sprays are stored at 15 to 30°C for Beconase AQ or at 2 to 25°C for Vancenase AQ.[2,7]

## Stability Reports of Compounded Preparations
### *Topical*
**STUDY 1:** Cornarakis-Lentzos and Cowin[750] reported the stability of beclomethasone dipropionate (Propaderm, Allen

& Hanburys) cream diluted 1 in 2 and also 1 in 10 with cetomacrogol (polyethylene glycol) cream, BP, and the ointment diluted in the same proportions with white soft paraffin ointment, BP (petroleum jelly). The cream and ointment dilutions were stored at ambient temperature for 12 months. The appearance of the dilutions was satisfactory throughout the study. Stability-indicating HPLC analysis indicated that no substantial drug loss occurred over 12 months. The pattern of drug concentrations over time was the same for the original concentration and for the dilutions.

**STUDY 2:** Ray-Johnson[775] reported the stability of beclomethasone dipropionate ointment (Propaderm, Glaxo) diluted to a concentration of 0.013% (1:1 mixture) using Unguentum Merck, an ambiphilic ointment that combines the properties of both an oil-in-water and a water-in-oil emulsion. The diluted ointment was packaged in screw-cap jars and stored refrigerated at 4°C, at room temperature of 18 to 23°C, and at elevated temperature of 32°C. After storage for 32 weeks, the authors reported that HPLC analysis did not find unacceptable drug loss under refrigeration or at room temperature. However, about 25% loss occurred at 32°C.

### *Rectal*
Stolk et al.[1633] evaluated the stability of beclomethasone dipropionate 2 mg/40 mL rectal enema with and without mesalazine 1g/40 mL. The suspending medium was composed of Carbopol 934 P 3.5 g, methyloxybenzoate 15%, propylene glycol 5 mL in distilled water 465 mL with the pH adjusted to pH 7 and resulted in 500 g of carbomer-water 0.7% gel.

Concentrated beclomethasone dipropionate in ethanol was added to the gel to prepare the 2 mg/4mL suspension. For the mesalazine-containing suspension, mesalazine 1 g/40 mL was incorporated. The enema suspensions were packaged in 50-mL brown glass bottles and were stored refrigerated at 4°C, room temperature of 20°C, and elevated temperature of 37°C for the enemas with mesalazine but only 20°C for the enema without mesalazine.

HPLC analysis of drug concentrations found no loss of beclomethasone dipropionate in four weeks at 20°C. Similarly, no loss of beclomethasone dipropionate occurred in the mesalazine-containing enemas in 44 days at 4, 20, and 37°C. The enemas exhibited only small amounts of visible beclomethasone precipitation; the precipitation could be resuspended easily. However, the mesalazine content led to a progressive darkening of color.

## Compatibility with Other Drugs

Ohishi et al.[1714] evaluated the stability of Propaderm ointment mixed in equal quantity (1:1) with zinc oxide ointment. Samples were stored refrigerated at 5°C. HPLC analysis of the ointment mixtures found that the drugs were stable for 32 weeks at 5°C. ■

# Belladonna with Phenobarbital

## Properties

Belladonna with phenobarbital is a combination product containing the alkaloids of belladonna (atropine sulfate, hyoscyamine sulfate, scopolamine hydrobromide) along with phenobarbital.[2]

## pH

Belladonna with phenobarbital elixir (Donnatal Elixir, Robins) has a pH of 4.8.[19]

## Osmolality

Belladonna with phenobarbital elixir (Robins) had an osmolality of 1050 mOsm/kg.[233]

## Stability Reports of Compounded Preparations
### Enteral
**BELLADONNA COMPATIBILITY SUMMARY**

*Compatible with:* Enrich • Ensure • Ensure HN • Ensure Plus • Ensure Plus HN • Osmolite • Osmolite HN • TwoCal HN • Vital • Vivonex T.E.N.

**STUDY 1:** Cutie et al.[19] added 10 mL of belladonna with phenobarbital elixir (Donnatal, Robins) to varying amounts (15 to 240 mL) of Ensure, Ensure Plus, and Osmolite (Ross Laboratories) with vigorous agitation to ensure thorough mixing. The elixir was physically compatible, distributing uniformly in all three enteral products with no phase separation or granulation.

**STUDY 2:** Burns et al.[739] reported the physical compatibility of belladonna with phenobarbital elixir (Donnatal, Robins) 5 mL with 10 mL of three enteral formulas, including Enrich, TwoCal HN, and Vivonex T.E.N. Visual inspection found no physical incompatibility with any of the enteral formulas.

**STUDY 3:** Altman and Cutie[850] reported the physical compatibility of belladonna with phenobarbital elixir (Donnatal, Robins) 10 mL with varying amounts (15 to 240 mL) of Ensure HN, Ensure Plus HN, Osmolite HN, and Vital that had been agitated vigorously to ensure thorough mixing. The elixir was physically compatible, distributing uniformly in all four enteral products with no phase separation or granulation. ■

# Benazepril Hydrochloride

## Properties

Benazepril hydrochloride occurs as a white to off-white crystalline powder.[1,4]

## Solubility

Benazepril hydrochloride is soluble in water and ethanol.[1,4]

## General Stability Considerations

The USP states that benazepril hydrochloride powder and tablets should be packaged in well-closed containers and stored at controlled room temperature below 30°C.[4] Novartis states that benazepril hydrochloride tablets should be dispensed in tight containers, stored below 30°C, and protected from moisture.[7]

## Stability Reports of Compounded Preparations
### Oral
**STUDY 1:** The Novartis package insert[7] for Lotensin describes the following compounded formulation of benazepril hydrochloride 2-mg/mL oral liquid. The oral liquid had the following formula:

| | | |
|---|---|---|
| Benazepril hydrochloride 20-mg tablets | | 15 tablets |
| Ora-Plus | | 75 mL |
| Ora-Sweet | qs | 150 mL |

The recommended preparation procedure was to add 75 mL of Ora-Plus suspending vehicle to an amber polyethylene terephthalate (PET) bottle containing 15 benazepril hydrochloride (Lotensin) tablets and immediately shake for at least two minutes, allow to stand for one hour, and then shake for at least an additional minute. Add 75 mL of Ora-Sweet syrup and shake to disperse the ingredients. The oral liquid should be stored refrigerated at 2 to 8°C. The oral liquid should be shaken before each use. Novartis states that the oral liquid can be stored for up to 30 days.

**STUDY 2:** Allen[1371] reported on a similar compounded formulation of benazepril hydrochloride 2-mg/mL oral liquid. The oral liquid had the following formula:

| | | |
|---|---|---|
| Benazepril hydrochloride 20-mg tablets | | 10 tablets |
| Ora-Plus | | 50 mL |
| Ora-Sweet | qs | 100 ml |

The recommended method of preparation was to place 10 benazepril hydrochloride (Lotensin) tablets in a calibrated container and add 50 mL of Ora-Plus. Shake the mixture for at least two minutes, allow to stand for one hour, and then shake for at least an additional minute. The mixture should then be brought to volume with Ora-Sweet and shaken well. The oral liquid should be packaged in tight, light-resistant containers. The author referred to the Novartis labeling in recommending a beyond-use date of up to 30 days stored under refrigeration.

# Bendroflumethiazide
## (Bendrofluazide)

## Properties

Bendroflumethiazide is a white to cream almost odorless crystalline powder.[2,3]

### Solubility

Bendroflumethiazide is practically insoluble in water but has a solubility in ethanol of 43 to 59 mg/mL.[2,3]

### $pK_a$

The drug has a $pK_a$ of 8.5.[2]

## General Stability Considerations

Bendroflumethiazide tablets should be packaged in tight containers and stored at room temperature protected from excessive heat.[2-4]

## Stability Reports of Compounded Preparations
### Oral

Barnes and Nash[425] reported the stability of extemporaneously prepared bendroflumethiazide 1.25-mg capsules for infant administration. The dose was administered by opening the hard gelatin capsules and placing the contents in the infant's food. The capsules were prepared by crushing bendroflumethiazide 5-mg tablets with a mortar and pestle and diluting with sufficient lactose to yield an individual capsule weight of 100 mg. Two batches were prepared. For the first batch, 25 tablets were crushed and diluted with lactose to 10 g. For the second batch, 15 tablets were crushed and diluted to 6 g with lactose. The capsules were stored at room temperature exposed to daylight, at ambient temperature in the dark with 75% relative humidity, and at elevated temperatures of 45 and 60°C. The bendroflumethiazide content was assessed using a stability-indicating HPLC analysis. Decomposition product formation also was observed using thin-layer chromatography.

The elevated temperature capsules were difficult to disperse after three months of storage and had a yellow discoloration. Both ambient temperature conditions retained acceptable dispersion characteristics with no discoloration throughout the study. Losses in the ambient temperature capsules ranged from 4 to 7% after a year. The authors calculated the shelf life from the lower 95% confidence limit of pooled data as seven months at room temperature both exposed to light and under 75% relative humidity.

# Benzocaine

## Properties

Benzocaine is a local anesthetic that occurs as odorless color-less or white crystals or crystalline powder.[2,3]

### Solubility

Benzocaine has an aqueous solubility of 0.4 mg/mL and a solubility in ethanol of 200 mg/mL. It is soluble in olive and almond oils to about 20 to 33 mg/mL and is also soluble in dilute acids.[1,3,6]

### $pK_a$

Benzocaine has a pK$_a$ of 2.5.[1,6]

## General Stability Considerations

Most benzocaine products should be packaged in tight containers protected from light and stored at controlled room temperature. The products should be protected from heating and from freezing.[2,4] The lozenges and gel should be packaged in well-closed containers.[4]

Benzocaine is subject to hydrolysis of its ester linkage, yielding ethanol and *p*-aminobenzoic acid. Appreciable hydrolysis will not occur in the absence of water. Specific base catalysis occurs at increasing rates as the pH increases from pH 8. Specific acid catalysis is also expected to occur. Benzocaine degradation is greatest at pH 11, is intermediate at pH 2, and is slowest at pH 7.[1004] Minimizing contact with water, bases, and acids will enhance stability.[6]

Triturating benzocaine with menthol, camphor, resorcinol, or phenol results in the formation of a soft or liquid mass.[6]

## Stability Reports of Compounded Preparations
### Topical

Allen[600] described a topical gel formulation of 2% benzocaine for use as an anesthetic. Benzocaine 2 g was dissolved in 90 mL of ethanol. Carbomer 934 2 g was dispersed in this solution and purified water was added with thorough mixing to bring the volume to 100 mL. A few drops of triethanolamine (trolamine) were added to thicken the preparation to the desired consistency. The gel should be packaged in a tight, light-resistant container. The author indicated that a use period of 30 days is appropriate for this product, although no stability data were provided. ■

# Benzoyl Peroxide

## Properties

Benzoyl peroxide hydrous is a white amorphous or granular powder having a characteristic odor.[3] The USP specifies that the bulk substance contain not less than 65% and not more than 82% of anhydrous benzoyl peroxide. It contains about 26% water for the purpose of reducing flammability and explosiveness.[3,4]

### Solubility

Benzoyl peroxide is sparingly soluble in water and ethanol.[1,3] It has a solubility of about 33 mg/mL in olive oil.[1]

The solubility of benzoyl peroxide is inversely related to solvent polarity, with greater solubility in semipolar solvents. The solubility of benzoyl peroxide in mixtures of polyethylene glycol 400 (PEG 400) and water decreased as the ratio of water to PEG 400 increased. A 67,000-fold increase in solubility occurred between 10% PEG 400 (0.6 mcg/g) and 100% PEG 400 (40 mg/g).[1189]

### pH

Benzoyl peroxide gel and lotion have a pH in the range of 2.8 to 6.6.[4]

## General Stability Considerations

Benzoyl peroxide exhibits maximum stability in the pH range of 4.5 to 5.[1006] Benzoyl peroxide bulk should be stored in the original container. It loses water on exposure to air and may explode if the water content becomes too low.[3,4] Storage recommendations for the bulk substance have varied from room temperature[4] to refrigeration[3] in containers designed to reduce static charges. The bulk substance should not be transferred to other containers, and unused material should not be returned to the original container. Unused material should be destroyed by mixing with 10% sodium hydroxide solution. Destruction is complete if the addition of a crystal of potassium iodide results in no formation of free iodine.[3,4]

Benzoyl peroxide may explode if subjected to grinding, percussion, or heat. Hydrous benzoyl peroxide may explode at temperatures exceeding 60°C and may cause fires in the presence of reducing substances.[3,4]

Benzoyl peroxide products should be packaged in tight containers and stored at controlled room temperatures.[4,7]

In various solvents, benzoyl peroxide stability was found to be dependent on the specific solvent used. At room temperature of 25°C, benzoyl peroxide in acetone, acetonitrile, and 50% acetonitrile in water was relatively stable with little loss in eight days. However, in ethanol and methanol, losses were much greater and more rapid. In eight days, benzoyl peroxide losses were 40% and nearly 100% in ethanol and methanol, respectively.[1096]

In addition to the solvent system used, the stability of benzoyl peroxide has been reported to be dependent on the concentration and storage temperature of benzoyl peroxide, with more rapid decomposition at higher concentration and higher temperature. Benzoyl peroxide stability is also reported to be dependent on drug solubility, with improved stability occurring when the drug is least soluble. For example, the drug was least stable in 100% PEG 400, but stability improved as water was added.[1189]

## Stability Reports of Compounded Preparations
### Topical

**STUDY 1:** Gupta[1180] evaluated the effects of several common topical formulation components on the stability of benzoyl peroxide. Benzoyl peroxide 2% (wt/wt) packaged in glass ointment jars decomposed very rapidly in polyethylene glycol ointment base, USP. Benzoyl peroxide was more stable in acetone and especially mixtures of acetone and ethanol. Benzoyl peroxide in mixtures of acetone and propylene glycol was less stable, indicating the propylene glycol increased the rate of benzoyl peroxide decomposition. Addition of acetanilide, benzoic acid, or 8-hydroxyquinoline sulfate (intended as stabilizers) had no influence on the rate of loss. However, incorporation of 0.1% 5-chloro-8-hydroxyquinoline did improve the stability of benzoyl peroxide modestly. The author indicated that benzoyl peroxide decomposition is complex and susceptible to many factors.

**STUDY 2:** Majekodunmi et al.[1261] evaluated the stability of benzoyl peroxide 1% (wt/vol) at 37°C in a variety of single solvents as well as binary and ternary mixtures both with and without antioxidants. Stability-indicating HPLC analysis found that, in general, the stability decreased from ternary to binary to single solvents. The presence of antioxidants had no significant effect on stability. The best stability in a single solvent was found with ethyl benzoate and in C12-15 alkyl benzoate, each with a half-life of about 7.5 weeks. The best benzoyl peroxide stability overall was found in a ternary mixture of ethanol-ethyl benzoate-C12-15 alkyl benzoate (60:20:20) with a half-life of 18 weeks. When converted to a gel by the addition of Cab-O-Sil, the half-life was extended to about 23 weeks at 37°C, which was substantially better than two commercial benzoyl peroxide products that were also evaluated.

## Compatibility with Other Drugs

Martin et al.[636] evaluated the stability of two retinoid preparations, adapalene 0.1% gel and tretinoin 0.025% gel, combined with an equal quantity of benzoyl peroxide 10% lotion. The mixtures were packaged in 10-mL plastic syringes and stored exposed to normal room fluorescent light over 24 hours. Retinoid content was assessed by HPLC analysis. The adapalene combination remained stable, with no loss of the retinoid. However, the tretinoin mixture underwent substantial and rapid decomposition of about 50% in two hours and 95% in 24 hours. Even without exposure to fluorescent light, the strongly oxidizing benzoyl peroxide reduced the tretinoin concentration to about 80% of the initial amount in 24 hours. ■

# ■ Benzyl Benzoate

## Properties

Benzyl benzoate occurs as colorless crystals or leaflets or as a colorless oily liquid. It has an aromatic odor but a sharp burning taste.[1,3,4]

CAUTIONS: Benzoyl benzoate in direct contact with skin may result in irritation. Use of benzyl benzoate is contraindicated in cats.[1]

### Solubility

Benzyl benzoate is insoluble in water and glycerol. It is miscible with ethanol, ether, and oils.[1,3,4]

### pH

Benzyl benzoate lotion, USP, has a pH between 8.5 and 9.2.[4]

## General Stability Considerations

Benzyl benzoate should be packaged in tight, well-filled, light-resistant containers. It should be stored at controlled room temperature and be protected from exposure to excessive heat.[3,4]

Benzyl benzoate lotion, USP, should be packaged in tight containers and stored at controlled room temperature.[4]

## Stability Reports of Compounded Preparations
### Topical

**STUDY 1:** Benzyl benzoate lotion, USP,[4] is stated to contain between 26 and 30% (wt/wt) of benzyl benzoate. The USP formula is shown below:

| | | |
|---|---|---|
| Benzyl benzoate | | 250 mL |
| Triethanolamine | | 5 g |
| Oleic acid | | 20 g |
| Purified water | qs | 1000 mL |

To prepare the USP benzyl benzoate lotion, mix the triethanolamine with the oleic acid. Benzyl benzoate is then added to this mixture and mixed well using a suitable size container. Purified water 250 mL is then added and the mixture is shaken well. The lotion is brought to volume with additional purified water and again shaken thoroughly. The lotion is to have a pH between 8.5 and 9.2. The lotion is to be packaged in suitable tight containers and labeled for storage at room temperature.

The USP does not cite a specific beyond-use date for Benzyl Benzoate Lotion, but USP Chapter <795> states that such compounded preparations should be labeled with a beyond-use date that is not later than the intended duration of therapy or 30 days, whichever is earlier.[4]

**STUDY 2:** Allen[1403] reported on a compounded formulation of benzyl benzoate 100-mg/mL topical lotion for use as a scabicide and pediculoside. The lotion had the following formula:

| | |
|---|---|
| Benzyl benzoate | 10 g |
| Cremophor RH 40 | 22 g |
| Ethanol | 41 g |
| Purified water | qs 100 g |

The recommended method of preparation was to heat the benzyl benzoate and Cremophor RH 40 to about 60°C and then add purified water solely with rapid stirring. The mixture then should be removed from the heat source and the ethanol added and mixed well. The lotion was to be packaged in tight, light-resistant containers. The author recommended a beyond-use date of six months at room temperature because this formula is a commercial medication in some countries with an expiration date of two years or more. ■

# ■ Betamethasone

## Properties

Betamethasone is a white to almost white odorless crystalline powder. Betamethasone is available in a number of forms having the same general description, including the acetate, benzoate, dipropionate, sodium phosphate, and valerate.[3,4]

### Solubility

Betamethasone is practically insoluble in water and sparingly soluble in ethanol and acetone.[3,4]

Betamethasone acetate is practically insoluble in water and freely soluble in ethanol and acetone.[3,4]

Betamethasone benzoate is insoluble in water and soluble in ethanol.[3,4]

Betamethasone dipropionate is insoluble in water, sparingly soluble in ethanol, and freely soluble in acetone.[3,4]

Betamethasone sodium phosphate is freely soluble in water and insoluble in acetone.[3,4]

Betamethasone valerate is practically insoluble in water, soluble in ethanol, and freely soluble in acetone.[3,4]

### pH

Betamethasone sodium phosphate injection has a pH of 8 to 9 (target 8.5). Betamethasone sodium phosphate and betamethasone acetate injectable suspension has a pH of 6.8 to 7.2.[2,4]

## General Stability Considerations

Betamethasone tablets, oral solution, and syrup are stored at controlled room temperature and should be protected from exposure to light. Betamethasone sodium phosphate injection and betamethasone sodium phosphate–betamethasone acetate injectable suspension should be stored at controlled room temperature and protected from freezing.[4]

The numerous official betamethasone topical products, including aerosol, gel, ointment, cream, and lotion preparations, are typically packaged in tight containers and stored at controlled room temperature.[4] The manufacturer states that betamethasone soluble tablets should not be removed from the original packaging and placed in dosing compliance aids.[1622]

## Stability Reports of Compounded Preparations
### Injection

Injections, like other sterile drugs, should be prepared in a suitable clean air environment using appropriate aseptic procedures. When prepared from nonsterile components, an appropriate and effective sterilization method must be employed.

Betamethasone sodium phosphate injection has been commercially available in the formulation shown in Table 13. A similar commercial injection is stated to have an expiration date of 24 months stored at controlled room temperature.[888]

### Topical
**STUDY 1:** Cornarakis-Lentzos and Cowin[750] reported the stability of betamethasone valerate (Betnovate, Glaxo) cream

**TABLE 13.** Betamethasone Sodium Phosphate Injection[7]

| Component | Amount |
|---|---|
| Betamethasone (present as 4 mg of the sodium phosphate) | 3 mg |
| Dibasic sodium phosphate | 10 mg |
| Edetate disodium | 0.1 mg |
| Phenol | 5 mg |
| Sodium bisulfite | 3.2 mg |
| Sodium hydroxide to adjust to | pH 8.5 |
| Water for injection | qs 1 mL |

diluted one in two and also one in 10 with cetomacrogol (polyethylene glycol) cream, BP, and the ointment diluted in the same proportions with white soft paraffin ointment, BP (petroleum jelly). The cream and ointment dilutions were stored at ambient temperature for 12 months. The appearance of the dilutions was satisfactory throughout the study. Stability-indicating HPLC analysis found that no substantial drug loss occurred over 12 months. The pattern of drug concentrations over time was the same for the original concentration and in the dilutions.

**STUDY 2:** Not all topical diluents result in stability. Mehta et al.[746] and Ryatt et al.[1602] reported the stability of betamethasone valerate 0.1% (Betnovate, Glaxo) diluted one part in four parts (final concentration 0.025% wt/wt) in emulsifying ointment (Evans). HPLC analysis found that loss of the betamethasone occurred. The half-life (time to 50% loss) was about four hours, with the formation of betamethasone 21-valerate (which is 15 times less potent) and free betamethasone.

**STUDY 3:** Barnes et al.[766] reported the instability of betamethasone dipropionate 0.05% (Diprosone, Kirby-Warrick Pharmaceuticals) and betamethasone valerate 0.1% (Harris Pharmaceuticals) ointments diluted one in 10 with compound zinc paste, BP (25% zinc oxide) (Thornton & Ross) and stored at controlled room temperature of 24 to 26°C. HPLC analysis found that the betamethasone dipropionate dilution lost about 7% in seven days and about 13% in 14 days. The betamethasone valerate dilution was worse, degrading 40% in seven days.

**STUDY 4:** Boonsaner et al.[770] reported the stability of betamethasone valerate cream (Betnelan-V) diluted one part in two and one part in three with cold cream with 5% urea and with Beeler's basis cream at room temperature and at 40°C. Unlike some other topical dilutions, these two diluents did not result in loss of the drug. HPLC analysis found no evidence of loss of betamethasone valerate in 30 days at room temperature. At 40°C, no loss occurred in Beeler's basis cream dilution in 30 days, but about 10% loss occurred in the cold cream dilution.

**STUDY 5:** Yip and Po[771] reported the stability of betamethasone valerate 0.1% ointment (Betnovate, Glaxo) diluted 3:1, 2:1, and 1:1 with emulsifying ointment, BP; Plastibase (Squibb); and white soft paraffin (petroleum jelly) at room temperature about 25°C using thin-layer chromatographic analysis. Dilution with emulsifying ointment resulted in rapid decomposition of the drug, with 10% loss occurring in less than an hour. Plastibase dilutions resulted in variable stability, depending on the amount of the Plastibase used. Ten percent loss of betamethasone valerate occurred in 35 hours in a 3:1 dilution but in less than an hour in a 1:1 dilution. The drug was stable in white soft paraffin, with 10% loss occurring in about 1700 hours.

**STUDY 6:** Ray-Johnson[775] reported the stability of betamethasone valerate ointment (Betnovate, Glaxo) diluted to a concentration of 0.05% (1:1 mixture) using Unguentum Merck, an ambiphilic ointment that combines the properties of both an oil-in-water and a water-in-oil emulsion. The diluted ointment was packaged in screw-cap jars and stored refrigerated at 4°C, at room temperature of 18 to 23°C, and at elevated temperature of 32°C. HPLC analysis found no evidence of drug loss after storage for 32 weeks at any of the temperatures.

**STUDY 7:** Ohtani et al.[831] found that dilution of a commercial betamethasone butyrate propionate topical ointment (Antebate) with four different moisturizing creams did not result in decomposition of methylparaben and propylparaben preservatives during storage at room temperature over three months. No microbial contamination occurred, even when preparations were touched two times daily with a finger. However, if the aqueous phase separated from the mixture, microbial contamination occurred within one week after the mixture was touched.

**STUDY 8:** Alberg et al.[1060] evaluated the use of a modified hydrophilic ointment base containing ethanol for suitability as a topical vehicle for betamethasone valerate 0.1%. The base was composed of 9% (wt/wt) cetosteryl alcohol, 10.5% (wt/wt) liquid paraffin, 10.5% (wt/wt) white petrolatum, 10% (wt/wt) ethanol 96%, and 60% (wt/wt) water. After incorporation of the betamethasone valerate, samples were stored at 21°C for four months and 31°C for 3.5 months. Scanning electron microscopy showed the formation of needlelike crystals greater than 100 µm that were not agglomerated. In addition, a visible layering occurred. Differential scanning calorimetry found that the drug had changed to a different crystalline structure that was solvated and crystallized from a supersaturated solution, making the topical product unsuitable for use. Although when stored at 4°C no particle growth was detected in the cream, the authors indicated that this modified base was not an appropriate vehicle for betamethasone valerate.

## Compatibility with Other Drugs
### BETAMETHASONE COMPATIBILITY SUMMARY
*Compatible with:* Acetylcysteine • Netilmicin • Tazarotene • Zinc Oxide
*Uncertain or variable compatibility with:* Mupirocin • Urea

**STUDY 1 (TAZAROTENE):** Hecker et al.[815] evaluated the compatibility of tazarotene 0.05% gel combined in equal quantity with other topical preparations, including betamethasone dipropionate 0.05% cream, gel, ointment, and lotion. The mixtures were sealed and stored at 30°C for two weeks. HPLC analysis found that less than 10% tazarotene loss occurred in all combinations except betamethasone dipropionate gel, with a loss of 13.1%. Betamethasone dipropionate exhibited less than 10% loss in this time frame.

**STUDY 2 (MUPIROCIN):** Mourya et al.[859] reported the stability of mupirocin and betamethasone in a combination ointment that provided mupirocin 2% (wt/wt) (Glenmark Pharmaceuticals) and betamethasone dipropionate 0.05% (wt/wt) in a base composed of polyethylene glycol 400 and 4000 packaged in aluminum collapsible tubes. Samples were stored at room temperature of 25°C as well as at elevated temperatures of 37 and 40°C, with 75% relative humidity. No physical changes in color or ointment consistency were observed. HPLC analysis found little or no loss of either mupirocin or betamethasone in 90 days at any of the storage temperatures.

**STUDY 3 (MUPIROCIN):** Jagota et al.[920] evaluated the compatibility and stability of mupirocin 2% ointment (Bactroban, Beecham) with betamethasone 0.1% lotion, cream, and ointment (Valisone, Schering) mixed in a 1:1 proportion over periods of up to 60 days at 37°C. The physical compatibility was assessed by visual inspection, while the chemical stability of mupirocin was evaluated by stability-indicating HPLC analysis. The study found that betamethasone 0.1% lotion was physically incompatible immediately upon mixing with mupirocin 2%. Betamethasone 0.1% cream and ointment mixed with mupirocin 2% also resulted in separation and layering and may be unacceptable for patient use. However, they could be remixed back to homogeneity, and the mupirocin loss was less than 10% for 15 days (cream) and 60 days (ointment).

**STUDY 4 (ACETYLCYSTEINE, NETILMICIN):** Rieutord et al.[841] evaluated the compatibility and stability of an aerosol inhalation solution composed of acetylcysteine 1 g/5 mL, netilmicin (as the sulfate) 100 mg/1 mL, and betamethasone (as the sodium phosphate) 4 mg/1 mL. HPLC analysis found little or no change in concentration of any of the three drugs in one hour at 23 to 27°C. The authors indicated that this brief period should provide a sufficient time frame for a nurse or patient to prepare the mixture and administer it.

**STUDY 5 (UREA):** Tezuka[1708] evaluated the stability of betamethasone in a series of compounded topical mixtures with urea-containing products. The betamethasone-17-valerate topical products evaluated in the testing included Rinderon V ointment, cream, and lotion and Rinderon VG lotion. The urea-containing products included Urepeal, Urepeal L, Keratinamin Kowa ointment, Pastaron, Pastaron 20, Pastaron soft, and Pastaron 10 lotion. The products were mixed in equal quantity by weight, and samples were stored at room temperature and elevated temperature of 40°C and 75% relative humidity for four weeks. There were no visible changes to the samples. All samples except the Urepeal mixtures resulted in substantial decomposition of the betamethasone when stored at room temperature and were much worse at elevated temperature with losses up to 84%. The authors recommended a short expiration period and storage at room temperature.

**STUDY 6 (ZINC OXIDE):** Nagatani et al.[1526] evaluated the stability of several steroids in steroid ointments when mixed with zinc oxide ointment. The steroid ointments included betamethasone valerate (Rinderon VG). Samples of the mixed ointments were stored at 5 and 30°C until analyzed. Analysis found no loss of betamethasone valerate when stored at 5 and 30°C.

**STUDY 7 (ZINC OXIDE):** Ohishi et al.[1714] evaluated the stability of Rinderon ointment mixed in equal quantity (1:1) with zinc oxide ointment. Samples were stored refrigerated at 5°C. HPLC analysis of the ointment mixtures found that the drugs were stable for 32 weeks at 5°C. ◼

# ◼ Bethanechol Chloride

## Properties

Bethanechol chloride occurs as hygroscopic colorless or white crystals or white crystalline powder. It has a characteristic aminelike or fishy odor.[2,3]

### Solubility

Bethanechol is reported to be very soluble in water,[586] having an aqueous solubility of about 1.67 g/mL.[1] It has a solubility in 95% ethanol of 80 mg/mL.[1]

### pH

A 1% aqueous solution has a pH of 5.5 to 6.5, while a 0.5% solution has a pH of 5.5 to 6.0.[1,3,4] The injection has a pH of 5.5 to 7.5.[4,586]

## General Stability Considerations

Bethanechol chloride products should be packaged in tight containers, stored at controlled room temperature, and protected from temperatures above 40°C. The injection should be protected from freezing.[4,7]

Aqueous solutions of bethanechol chloride may be sterilized by autoclaving at 120°C for 20 minutes without discoloring or loss of concentration.[1,7]

## Stability Reports of Compounded Preparations
### Oral
**USP OFFICIAL FORMULATION (ORAL SOLUTION):**

    Bethanechol chloride 500 mg

    Vehicle for Oral Solution, NF (sugar-containing or sugar-free) qs 100 mL

(See the vehicle monograph for information on the Vehicle for Oral Solution, NF.)

Use bethanechol chloride powder. Add about 20 mL of the vehicle and mix well. Add additional vehicle almost to volume in increments with thorough mixing. Quantitatively transfer to a suitable calibrated tight and light-resistant bottle, bring to final volume with the vehicle mixture, and thoroughly mix yielding bethanechol chloride 5-mg/mL oral solution. The final liquid preparation should have a pH between pH 3.9 and 4.9. Store the preparation at controlled room temperature or under refrigeration between 2 and 8°C. The beyond-use date is 60 days from the date of compounding at either storage temperature.[4,5]

### USP OFFICIAL FORMULATION (ORAL SUSPENSION):

Bethanechol chloride 500 mg
Vehicle for Oral Solution, NF (sugar-containing or sugar-free)
Vehicle for Oral Suspension, NF (1:1) qs 100 mL

(See the vehicle monographs for information on the individual vehicles.)

Use bethanechol chloride powder or commercial tablets. If using tablets, crush or grind to fine powder. Add about 20 mL of the vehicle mixture and mix to make a uniform paste. Add additional vehicle almost to volume in increments with thorough mixing. Quantitatively transfer to a suitable calibrated tight and light-resistant bottle, bring to final volume with the vehicle mixture, and thoroughly mix yielding bethanechol chloride 5-mg/mL oral suspension. The final liquid preparation should have a pH between pH 3.9 and 4.9. Store the preparation at controlled room temperature or under refrigeration between 2 and 8°C. The beyond-use date is 60 days from the date of compounding at either storage temperature.[4,5]

**STUDY 1:** Schlatter and Saulnier[498] evaluated the stability of bethanechol chloride 1-mg/mL oral solutions prepared from tablets and bulk powder. To make each of the solutions from tablets, four bethanechol chloride 25-mg tablets were crushed to fine powder in a glass mortar and transferred to a graduated cylinder. One solution was prepared by adding sterile water for injection (pH 5.4) to bring the volume to 100 mL. The other solution utilized sterile water for irrigation (pH 6.5) to bring the solution to volume. The mixtures were sonicated for 15 minutes and were filtered through a 0.22-µm filter to remove insoluble excipients. A solution from bethanechol chloride bulk powder (Sigma) was prepared by dissolving the drug in sterile water for irrigation. All of the solutions yielded a nominal bethanechol chloride concentration of 1 mg/mL. The solutions were filled into amber glass bottles and stored under refrigeration at 4°C protected from light.

The drug content of the solution was assessed using ultraviolet–visible spectrophotometry. The solution prepared from bulk powder was stable for at least 90 days, exhibiting 5% loss. In contrast, the solutions prepared from tablets were less stable. In sterile water for irrigation, the drug was stable through 30 days but lost 16% after 60 days. Stability was somewhat worse in sterile water for injection, presumably because of its slightly lower pH. After 15 days, 8% of the drug had been lost, becoming a 13% loss in 30 days. It appears that tablet component(s) may play a role in degradation of the drug.

**STUDY 2:** Gupta and Maswoswe[554] studied the stability of several oral liquid preparations of bethanechol chloride 1 mg/mL prepared from tablets, powder, and injection. Bethanechol chloride 5-mg/mL injection (Merck Sharp & Dohme) was mixed with water to yield a 1-mg/mL concentration. Bethanechol chloride powder (Merck) was formulated with phosphate buffers to yield 1-mg/mL solutions having pH values ranging from 3.0 to 6.8. The mixtures prepared from injection or powder were clear solutions. For the tablet-derived formulation, the 10-mg tablets (Sidmak) were ground to fine powder using a mortar and pestle and mixed for five minutes with the appropriate amount of water and also syrup. Another portion of tablets ground to fine powder was triturated with 4 mL of 0.05 N hydrochloric acid and brought to 100 mL with distilled water. The mixtures prepared from tablets were not clear; the mixtures were allowed to stand, and the decanted clear solution was used for analysis. The various mixtures were packaged in amber bottles and stored at 25°C. The USP–NF stability-indicating colorimetric assay method was used to evaluate bethanechol chloride stability.

Analysis of the samples found that the dilution of the injection in water, the various powder formulations in phosphate buffers, and the formulation prepared from tablets using acidified water all provided nearly 100% of the expected concentration of bethanechol chloride and retained concentration, exhibiting no loss of drug after storage for 40 days at 25°C. However, the mixtures prepared from tablets in plain water and syrup were not acceptable. In plain water, the decanted liquid initially had a bethanechol chloride concentration of only 86.5%; the concentration did not reach 100% until it had stood for 12 days. In syrup, no bethanechol chloride was detectable using the colorimetric analysis.

**STUDY 3:** Allen and Erickson[595] evaluated the stability of three bethanechol chloride 5-mg/mL oral suspensions extemporaneously compounded from tablets. Vehicles used in this study were (1) an equal parts mixture of Ora-Sweet and Ora-Plus (Paddock), (2) an equal parts mixture of Ora-Sweet SF and Ora-Plus (Paddock), and (3) cherry syrup (Robinson Laboratories) mixed 1:4 with simple syrup. Twelve bethanechol chloride 50-mg tablets (Sidmak) were crushed and comminuted to fine powder using a mortar and pestle. About

20 mL of the test vehicle was added to the powder and mixed to yield a uniform paste. Additional vehicle was added geometrically and brought to the final volume of 120 mL, mixing thoroughly after each addition. The process was repeated for each of the three test suspension vehicles. Samples of each of the finished suspensions were packaged in 120-mL amber polyethylene terephthalate plastic prescription bottles and stored at 5 and 25°C in the dark.

No visual changes or changes in odor were detected during the study. Stability-indicating HPLC analysis found less than 8% bethanechol chloride loss in any of the suspensions stored at either temperature after 60 days of storage. ▪

# Bevacizumab

## Properties

Bevacizumab is recombinant humanized monoclonal IgG1 antibody. Bevacizumab injection is a clear to slightly opalescent and colorless to pale brown sterile liquid injection intended for intravenous infusion after dilution in 100 mL of sodium chloride 0.9%.[7]

## pH

Bevacizumab injection has a pH of 6.2.[7]

## General Stability Considerations

Bevacizumab injection in intact single-use vials should be stored refrigerated at 2 to 8°C and protected from exposure to light. The drug should be protected from freezing and shaking. The manufacturer states that bevacizumab may be used for up to eight hours after dilution of the dose for infusion in sodium chloride 0.9% when stored under refrigeration. Bevacizumab should not be mixed with dextrose-containing solutions.[7]

## Stability Reports of Compounded Preparations
### Ophthalmic

Ophthalmic preparations, like other sterile drugs, should be prepared in a suitable clean air environment using appropriate aseptic procedures. When prepared from nonsterile components, an appropriate and effective sterilization method must be employed.

NOTE: Genentech no longer supplies bevacizumab for the purpose of compounding into ophthalmic preparations.

Bakri et al.[1053] evaluated the stability of bevacizumab (Avastin, Genentech) 25-mg/mL injection repackaged for ophthalmic use into 1-mL polypropylene tuberculin syringes (Becton Dickinson) in quantities of 1.25 mg in 0.05 mL and 2.5 mg in 0.1 mL. The syringes were stored refrigerated at 4°C for six months. In addition, a single syringe was also stored frozen at −10°C. An immunoassay measuring bevacizumab binding to VEGF-165 was used to determine the drug concentration throughout the study. About 9 to 10% loss of bevacizumab activity in the refrigerated samples occurred within three months and increased to 13 to 16% in six months. The single frozen syringe also lost about 12% in six months, which was not an improvement over refrigerated storage. ▪

# Bimatoprost

## Properties

Bimatoprost is a synthetic prostamide analog used in ophthalmic treatment that occurs as a powder.[1,7]

## Solubility

Bimatoprost is slightly soluble in water and very soluble in ethanol and methanol.[1,7]

## pH

Bimatoprost 0.03% ophthalmic solution is in the range of pH 6.8 to 7.8.[7]

## Osmolality

Bimatoprost 0.03% ophthalmic solution is isotonic having an osmolality of 290 mOsm/kg.[7]

## General Stability Considerations

Commercial bimatoprost 0.03% ophthalmic solution should be stored in the original container at 2 to 25°C.[7]

## Stability Reports of Compounded Preparations
### Ophthalmic

Ophthalmic preparations, like other sterile drugs, should be prepared in a suitable clean air environment using appropriate

aseptic procedures. When prepared from nonsterile components, an appropriate and effective sterilization method must be employed.

Paolera et al.[1272] evaluated and compared the stability of bimatoprost 0.03% and latanoprost 0.005% ophthalmic solutions during in-use conditions with patients in Sao Paulo, Brazil. Patients were instructed to store and use their ophthalmic solutions as usual and return them between days 28 and 34. The sample solutions were transferred to an analytical laboratory arriving on days 35 through 42. HPLC analysis of the active drugs found no loss of bimatoprost after initiating use of the ophthalmic solution. However, the latanoprost samples averaged only 88% of the active drug remaining. ◼

# Bisoprolol Fumarate

## Properties
Bisoprolol fumarate occurs as a white crystalline powder.[7]

### Solubility
Bisoprolol fumarate is soluble in water and ethanol.[1,7]

## General Stability Considerations
Bisoprolol fumarate commercial tablets (Zebeta, Wyeth-Ayerst) should be stored at controlled room temperature and protected from exposure to moisture.[7]

## Stability Reports of Compounded Preparations
### Solution
Modamio et al.[745] reported the stability of bisoprolol fumarate (Merck) 0.25 mg/mL in phosphate buffer (pH 7.4) at elevated temperatures. Stability-indicating HPLC analysis found that little or no loss of bisoprolol occurred in 84 hours, even at 90°C. ◼

# Bitolterol Mesylate

## Properties
Bitolterol mesylate is a white crystalline powder.[1] The commercial inhalation solution (Tornalate, Dura Pharmaceuticals) contains bitolterol mesylate 0.2% (2 mg/mL) in an aqueous vehicle also containing ethanol 25% (vol/vol), citric acid, propylene glycol, and sodium hydroxide.[7]

### Solubility
Bitolterol mesylate is soluble in dimethyl sulfoxide (DMSO).[7]

### pH
Bitolterol mesylate inhalation solution 0.2% has a pH of 3 to 3.4.[7]

## General Stability Considerations
Bitolterol mesylate inhalation solution should be stored at controlled room temperature and should not be used if it is discolored or contains a precipitate.[7]

## Compatibility with Other Drugs
Joseph[1024] reported that 2.5 mL of preservative-free ipratropium bromide 0.025% inhalation solution is physically and chemically compatible with 1.25 mL of bitolterol mesylate 0.2% solution for at least one hour. ◼

# Boronophenylalanine–Fructose Complex

## Properties
Boronophenylalanine [4-(borono 10B)-L-phenylalanine] is a white to slightly yellow powder.[1074] Fructose is a sugar that is a white crystalline powder or colorless crystals.[4]

### Solubility
Boronophenylalanine is soluble at least to 50 mg/mL in 1 M hydrochloric acid.[1074] However, it is insufficiently soluble in water at neutral pH for intravenous infusion.[1075] To improve solubility it is complexed with fructose,[1075, 1076] which is freely soluble in water and ethanol.[4] Boronophenylalanine solubility in 0.3 M fructose aqueous solution is about 33 mg/mL at pH 7.98.[1076]

### pH
Fructose injections have a pH in the range of 3 to 6.[4] The investigational boronophenylalanine intravenous infusion solution that has been reported has a pH of 7.4.[1075]

## pKₐ

Boronophenylalanine has $pK_a$ values of 2.46, 8.46, and 9.72.[1077, 1078]

## General Stability Considerations

Boronophenylalanine powder should be stored at controlled room temperature and is stated to have a shelf life of 18 months.[1074]

Fructose should be packaged in well-closed containers and stored at controlled room temperature. Fructose injections should be packaged in single-dose containers of Type I or Type II glass and stored at controlled room temperature.[4]

## Stability Reports of Compounded Preparations

### Injection

Injections, like other sterile drugs, should be prepared in a suitable clean air environment using appropriate aseptic procedures. When prepared from nonsterile components, an appropriate and effective sterilization method must be employed.

Van Rij et al.[1075] evaluated the stability of an extemporaneously compounded boronophenylalanine 30-mg/mL injection intended for intravenous infusion for use in boron neutron capture therapy. The infusion solution was prepared by dissolving 30 g of boronophenylalanine in a solution consisting of 10.8 g of sodium hydroxide in 500 mL of water for injection. The pH was adjusted with 2 M sodium hydroxide to pH 10.5. The solution became clear at that time. Fructose 28.5 g was then added. The pH was then adjusted to 7.4 using 4 M hydrochloric acid, and additional water for injection was added, bringing the total volume to 1000 mL. This clear solution was passed through a 0.22-μm Millex GS filter into a sterile Type II soda-lime infusion bottle and capped with an Omniflex FM257 bromobutyl cap. This solution was stored overnight under refrigeration. The next day the pH was checked and adjusted to pH 7.4 if necessary. This final solution was again filtered into a new sterile Type II glass infusion bottle, capped with an Omniflex FM257 bromobutyl cap, and wrapped in aluminum foil for light protection. Samples were stored under refrigeration at 2 to 8°C and at room temperature of 15 to 25°C.

Stability-indicating HPLC analysis found a slow decline in the drug concentration at both storage temperatures. Drug loss near 10% was found after about 20 days of storage. The shelf life was calculated using the lower 95% confidence limit for 10% loss. The authors indicated the shelf life should be limited to 12 days. ■

# ■ Brimonidine Tartrate

## Properties

Brimonidine tartrate occurs as a white to off-white, pale yellow, or even pale pink crystalline material.[1,7]

### Solubility

Brimonidine tartrate has been stated to have an aqueous solubility of 0.6 mg/mL. In the commercial ophthalmic solution formulation (Alphagan P, Allergan) the manufacturer states that the drug has a solubility of 1.4 mg/mL.[7] However, other manufacturers and published articles indicate the aqueous solubility of brimonidine tartrate is 34 mg/mL.[7, 1418] There is no readily apparent reason for the divergent solubility statements.

### pH

Commercial Alphagan P ophthalmic solution has a pH of 7.4 to 8 for the 0.1% concentration and a pH of 6.6 to 7.4 for the 0.15% concentration. Other commercial brimonidine tartrate ophthalmic solutions with benzalkonium chloride preservative have a pH in the range of 5.6 to 6.6.[7]

### pKₐ

Brimonidine tartrate has a $pK_a$ of 7.4.[1417,1418]

### Osmolality

Commercial Alphagan P ophthalmic solution has an osmolality in the range of 250 to 350 mOsm/kg.[7]

## General Stability Considerations

Brimonidine tartrate as the commercial Alphagan ophthalmic solution products are stored at controlled room temperature.[7] Commercial brimonidine tartrate ophthalmic solutions with benzalkonium chloride bear expiration dates of 24 to 36 months depending on the product. Many of the ophthalmic solution products state that they should be discarded 28 days after opening.[7] Exposure of brimonidine tartrate 0.2% solution to eight hours of direct sunlight resulted in an 18% loss of the drug.[1418]

## Stability Reports of Compounded Preparations

### Ophthalmic

Ophthalmic solutions, like other sterile drugs, should be prepared in a suitable clean air environment using appropriate aseptic procedures. When prepared from nonsterile components, an appropriate and effective sterilization method must be employed.

Ali et al.[1418] evaluated the stability of an ophthalmic formulation of brimonidine tartrate 2 mg/mL. The ophthalmic formulation also contained citric acid monohydrate, sodium citrate, polyvinyl alcohol, sodium chloride, and benzalkonium chloride 0.05 mg/mL preservative in water. Samples were stored at 40°C and 25% relative humidity and at ambient room temperature protected from exposure to light. Stability-indicating hydrophilic interaction liquid chromatography (HILIC) analysis found no loss of brimonidine tartrate after 48 hours of storage. ■

# ■ Bromhexine Hydrochloride

## Properties
Bromhexine hydrochloride is a white or nearly white crystalline powder.[3]

### Solubility
Bromhexine hydrochloride is very slightly soluble in water and it is slightly soluble in ethanol.[3]

## General Stability Considerations
Bromhexine hydrochloride oral liquid products should be stored at controlled room temperature and protected from exposure to light.[3]

## Stability Reports of Compounded Preparations
### Inhalation
Inhalation preparations, like other sterile drugs, should be prepared in a suitable clean air environment using appropriate aseptic procedures. When prepared from nonsterile components, an appropriate and effective sterilization method must be employed.

Tomonaga et al.[1605] evaluated the compatibility and stability of gentamicin sulfate injection mixed with six drugs for inhalation including Bisolvon (bromhexine hydrochloride). Using a microbiological activity assay, it was found that in all of these inhalation solutions the gentamicin sulfate content deteriorated substantially in most combinations within one day and within all combinations within three days stored both under refrigeration at 4°C and at room temperature of 20 and 30°C.

### Enteral
Ortega de la Cruz et al.[1101] reported the physical compatibility of an unspecified amount of oral liquid bromhexine (Bisolvon, Boehringer Ingelheim) and with bromhexine hydrochloride compound with diphenhydramine hydrochloride, ephedrine hydrochloride, and codeine hydrochloride (Bisolvon Compositum) with 200 mL of Precitene (Novartis) enteral nutrition diet for a 24-hour observation period. No particle growth or phase separation was observed. ■

# ■ Brompheniramine Maleate

## Properties
Brompheniramine maleate is a white odorless crystalline powder.[2,3]

### Solubility
Brompheniramine maleate has an aqueous solubility of about 200 mg/mL at 25°C; in ethanol its solubility is 66.7 mg/mL at 25°C.[2]

### pH
A 1 to 2% aqueous solution has a pH near 4 to 5.[1,3,4] Brompheniramine maleate elixir has a pH range of 2.5 to 3.5.[4] Brompheniramine maleate elixir (Dimetane, Robins) had a measured pH of 2.7.[19] Brompheniramine maleate with phenylpropanolamine hydrochloride elixir (Dimetapp, Robins) had a measured pH of 2.6.[19] The injection has a pH of 6.3 to 7.3.[4]

### $pK_a$
Brompheniramine has $pK_a$ values of 3.59 and 9.12.[2]

## General Stability Considerations
Brompheniramine maleate oral products should be stored in tight (tablets) or well-closed (elixir) containers at controlled room temperature. The elixir should be protected from light during storage. The injection should be stored at controlled room temperature and protected from freezing and exposure to light during storage. At temperatures below 0°C, crystals may form in the injection; the crystals can be redissolved by warming to 30°C.[2]

## Stability Reports of Compounded Preparations
### Oral
Gupta and Gupta[1641] evaluated the stability of a compounded oral liquid formulation of brompheniramine maleate

0.4 mg/mL prepared from powder. In addition to the drug, the oral liquid formulation included in each 100 mL ethanol 3.5 mL, citric acid anhydrous 250 mg, water 20 mL, and raspberry flavor. The mixture was brought to volume with simple syrup containing sodium benzoate 0.1%. The oral liquid was packaged in amber bottles and stored at 25°C for 202 days. The physical appearance did not change, and the pH remained constant over the observation period. Stability-indicating HPLC analysis found little or no loss of brompheniramine maleate occurred in 202 days indicating this formulation was stable for the entire test period.

### Enteral

**BROMPHENIRAMINE MALEATE COMPATIBILITY SUMMARY**
*Compatible with:* Vivonex T.E.N.
*Incompatible with:* Enrich • Ensure • Ensure HN • Ensure
    Plus • Ensure Plus HN • Osmolite • Osmolite HN • Vital
*Uncertain or variable compatibility with:* TwoCal HN

**STUDY 1:** Cutie et al.[19] added 10 mL of brompheniramine maleate elixir (Dimetane, Robins) and separately 10 mL of brompheniramine maleate with phenylpropanolamine hydrochloride elixir (Dimetapp, Robins) to varying amounts (15 to 240 mL) of Ensure, Ensure Plus, and Osmolite (Ross Laboratories) with vigorous agitation. Both elixirs were physically incompatible in all three enteral products; brompheniramine maleate formed an adhesive gelatinous material that clogged feeding tubes. The authors stated that the enteral foods show some breakdown, although the problem could be minimized by adding the brompheniramine maleate with phenylpropanolamine hydrochloride elixir slowly, with stirring.

**STUDY 2:** Burns et al.[739] reported the physical compatibility of (1) brompheniramine maleate elixir (Dimetane, Robins) 5 mL and separately (2) brompheniramine maleate with phenylpropanolamine hydrochloride elixir (Dimetapp, Robins) 5 mL with 10 mL of three enteral formulas, including Enrich, TwoCal HN, and Vivonex T.E.N. For the brompheniramine maleate elixir (1), visual inspection found no physical incompatibility with TwoCal HN and Vivonex T.E.N. However, a thin granular material formed with Enrich. For brompheniramine maleate with phenylpropanolamine hydrochloride elixir (2), visual inspection found no physical incompatibility with Vivonex T.E.N. However, mixed with Enrich and TwoCal HN, thickening with particulates occurred.

**STUDY 3:** Altman and Cutie[850] reported the physical incompatibility of 10 mL of brompheniramine maleate elixir (Dimetane, Robins) and separately 10 mL of brompheniramine maleate with phenylpropanolamine hydrochloride elixir (Dimetapp, Robins) with varying amounts (15 to 240 mL) of Ensure HN, Ensure Plus HN, Osmolite HN, and Vital after vigorous agitation to ensure thorough mixing. Both elixirs were physically incompatible in Ensure HN, Ensure Plus HN, and Osmolite HN enteral products; brompheniramine maleate formed an adhesive gelatinous material that clogged feeding tubes. The authors stated that the Ensure HN, Ensure Plus HN, and Osmolite HN enteral foods show some breakdown with Dimetapp elixir, although the problem could be minimized by adding the elixir slowly with stirring. Both elixirs were compatible with Vital. ■

# ■ Budesonide

## Properties
Budesonide is a white to off-white tasteless odorless crystalline powder.[3,7]

## Solubility
Budesonide is practically insoluble in water and sparingly soluble in ethanol.[7]

## pH
Budesonide nasal spray has a target pH of 4.5.[7]

## General Stability Considerations
Budesonide capsules (Entocort EC) should be packaged in tight containers and stored at controlled room temperature.[7]

Budesonide inhalation suspension (Pulmicort Respules) should be stored at controlled room temperature and protected from exposure to light. After opening of the inhalation suspension foil envelope, the shelf life of the unused ampules is two weeks when protected from light by the foil envelope. Opened ampules of budesonide should be used promptly.[7]

Budesonide nasal spray (Rhinocort Aqua) should be stored at controlled room temperature and protected from exposure to light and from freezing.[7]

## Stability Reports of Compounded Preparations
### Inhalation
Inhalations, like other sterile drugs, should be prepared in a suitable clean air environment using appropriate aseptic procedures. When prepared from nonsterile components, an appropriate and effective sterilization method must be employed.

Blondino and Baker[930] evaluated the physical and chemical stability of four common drugs administered by inhalation, albuterol sulfate 2.5 mg/3 mL, budesonide 0.25 mg/ 3 mL, cromolyn sodium 20 mg/3 mL, and ipratropium bromide 0.5 mg/3 mL, individually in an acidified (pH 4) 15% ethanol in water solution. The budesonide was dissolved in the ethanol, which was then added to a solution of the other drugs dissolved in the acidified sterile water for irrigation. No visible changes in color or clarity of any of the samples were observed. Budesonide by itself underwent losses at some temperatures over eight weeks. Losses were 0, about 8 to 10, and 49% at refrigerated, room, and elevated temperature of 40°C, respectively. However, budesonide was more stable in the four-drug combination. See *Compatibility with Other Drugs* below.

### Topical

Munoz et al.[1177] evaluated the stability of two topical oil-in-water cream formulations of budesonide 0.025%. The complete formulas were not presented, but the emulsifying agent in one cream formula was polysorbate 80 (pH range 5.5 to 6.5) and the other used sodium lauryl sulfate (pH range 4.0 to 5.0). HPLC analysis found little or no loss of budesonide in either formulation over 12 months at room temperature and under refrigeration. The authors concluded that pH did not have a substantial influence on the chemical stability of budesonide in the cream formulations tested.

## Compatibility with Other Drugs

### BUDESONIDE COMPATIBILITY SUMMARY

*Compatible with:* Acetylcysteine • Albuterol sulfate • Arformoterol tartrate • Cromolyn sodium • Dornase alfa • Fenoterol hydrobromide • Formoterol fumarate • Ipratropium bromide • Levalbuterol hydrochloride • Salbutamol sodium • Terbutaline sulfate • Tobramycin • Tobramycin sulfate

**STUDY 1 (ARFORMOTEROL):** Bonasia et al.[1151] evaluated the physical compatibility and chemical stability of arformoterol 15 mcg/2 mL (as the tartrate) inhalation solution (Brovana, Sepracor) with budesonide 0.25 mg/2 mL and 0.125 mg/2 mL (Pulmicort Respules, AstraZeneca). The admixtures were prepared and mixed for homogeneity. Visual inspection found no evidence of precipitation or other physical incompatibility. HPLC analysis found less than 2% change in either drug concentration over the 30-minute test period.

**STUDIES 2 AND 3 (MULTIPLE DRUGS):** Smaldone et al.[1026] and McKenzie and Cruz-Rivera[892] evaluated the compatibility and stability of two concentrations of budesonide inhalation suspension with four other inhalation medications. Budesonide (Pulmicort Respules, AstraZeneca) 0.25 mg/2 mL and also 0.5 mg/2 mL was mixed with the inhalation medications in the concentrations noted in Table 14.

The inhalations were added to new Pari LC Plus nebulizer cups and mixed using a vortex mixer. The samples were examined initially after mixing and over 30 minutes. Visual inspection found no change in color, formation of a precipitate, or nonresuspendibility of any of the samples.

**TABLE 14.** Inhalation Medications Mixed with Budesonide Inhalation Suspension[892]

| Component | Amount |
| --- | --- |
| Albuterol sulfate | 2.5 mg/0.5 mL |
| Cromolyn sodium | 20 mg/2 mL |
| Ipratropium bromide | 0.5 mg/2.5 mL |
| Levalbuterol hydrochloride | 0.63 mg/3 mL and 1.25 mg/3 mL |

Stability-indicating HPLC analysis found that all of the drugs remained adequately stable throughout the study period with little or no loss of budesonide in most of the samples over 30 minutes. Budesonide concentrations ranged from about 97 to 105%, reflecting normal analytical variation for all samples except budesonide 0.25 mg/2 mL with ipratropium bromide, which was determined to be about 93%. Whether this is simply an analytical anomaly is uncertain. All of the other drugs were very near the starting concentrations, ranging from 98 to 103%.

**STUDY 4 (MULTIPLE DRUGS):** Roberts and Rossi[897] reported the compatibility of budesonide inhalation suspension (Pulmicort Respules) with ipratropium (Atrovent) and salbutamol sodium (Ventolin) as both single and multiple-use forms as well as with cromolyn sodium (Intal) and terbutaline multiple-use (Bricanyl). Visual inspection found all of the combinations were compatible, with no visible cloudiness.

**STUDY 5 (MULTIPLE DRUGS):** Blondino and Baker[930] evaluated the physical and chemical stability of four common drugs administered by inhalation: albuterol sulfate 2.5 mg/3 mL, budesonide 0.25 mg/3 mL, cromolyn sodium 20 mg/3 mL, and ipratropium bromide 0.5 mg/3 mL, individually in an acidified (pH 4) 15% ethanol in water solution and in a four-drug combination. The budesonide was dissolved in the ethanol, which was then added to a solution of the other drugs dissolved in the acidified sterile water for irrigation. No visible changes in color or clarity in any of the samples were observed. Budesonide by itself underwent losses at some temperatures over eight weeks. Losses were 0, about 8 to 10, and 49% at refrigerated, room, and elevated temperature of 40°C, respectively. However, budesonide was more stable in the four-drug combination, exhibiting no loss under refrigeration, 6% loss at room temperature, and about 27% loss at 40°C in eight weeks. The other three drugs remained stable under refrigeration and at room temperature with no more than 6% loss in eight weeks, but albuterol sulfate and ipratropium bromide underwent substantial losses of 10 and 19%, respectively, at 40°C in that time frame.

**STUDY 6 (MULTIPLE DRUGS):** Gronberg et al.[1027] evaluated the stability and compatibility of budesonide (Pulmicort, concentration unspecified) with several inhalation solutions, including (1) acetylcysteine (Lysomucil) 100 mg/mL 2 parts to 3 parts budesonide, (2) fenoterol hydrobromide (Berotec) 5 mg/mL 8 parts to 1 part budesonide, (3) ipratropium bromide (Atrovent) 0.25 mg/mL 2 parts to 1 part budesonide, and (4) ipratropium bromide 0.125 mg/mL plus fenoterol hydrobromide 0.31 mg/mL (Duovent) in equal parts with budesonide. Samples were stored at room temperature of 22 to 25°C protected from exposure to light. HPLC analysis found all mixtures to be compatible; tested samples maintained greater than 90% of the initial concentrations for at least 18 hours.

**STUDY 7 (LEVALBUTEROL):** Bonasia et al.[1329] evaluated the compatibility and stability of levalbuterol hydrochloride inhalation solution (Sepracor) with several drugs including budesonide (AstraZeneca). The drugs were mixed with sodium chloride 0.9% and then with each other to prepare the following solution:

> Levalbuterol 1.25 mg/0.5 mL diluted with 2 mL of
>     sodium chloride 0.9%
> Budesonide 0.5 mg/2 mL diluted with 2 mL of
>     sodium chloride 0.9%

The drug mixture was prepared in triplicate and evaluated initially and after 30 minutes at an ambient temperature of 21 to 25°C. The authors indicated that a 30-minute study duration was chosen because it represents the approximate time it takes for a patient to administer drugs by nebulization at home. No visible evidence of physical incompatibility was observed with the samples. HPLC analysis found no loss of any of either of the drugs in the samples.

**STUDY 8 (FORMOTEROL):** Akapo et al.[1330] evaluated the compatibility and stability of formoterol fumarate 20 mcg/2 mL inhalation solution (Performist, Dey) mixed with budesonide 0.5 mg/2 mL (AstraZeneca) for administration by inhalation and evaluated over 60 minutes at room temperature of 23 to 27°C. The test samples were physically compatible by visual examination and measurement of osmolality, pH, and turbidity. HPLC analysis of drug concentrations found little or no loss of either of the drugs within the study period.

**STUDY 9 (DORNASE ALFA, TOBRAMYCIN):** Kramer et al.[1383] evaluated the compatibility and stability of a variety of drug solution admixtures of inhalation drugs, including budesonide. The mixtures were evaluated using chemical assays including HPLC, DNase activity assay, and fluorescence immunoassay as well as visual inspection, pH measurement, and osmolality determination. The drug combinations tested and the compatibility results that were reported are shown below.

*Mixture 1*
Budesonide (Pulmicort) 1 mg/2 mL
Dornase alfa (Pulmozyme) 2500 units/2.5 mL
    Result: Physically and chemically compatible

*Mixture 2*
Budesonide (Pulmocort) 1 mg/2 mL
Tobramycin (Tobi) 300 mg/5 mL
    Result: Physically and chemically compatible

*Mixture 3*
Budesonide (Pulmocort) 1 mg/2 mL
Tobramycin sulfate (Gernebcin) 80 mg/2 mL
    Result: Physically and chemically compatible ▪

# ▪ Bupivacaine Hydrochloride

## Properties
Bupivacaine hydrochloride is a white odorless crystalline powder or colorless crystals.[2,3]

### Solubility
Bupivacaine hydrochloride has an aqueous solubility of 40 mg/mL. In ethanol its solubility is 125 mg/mL.[1]

### pH
A 1% bupivacaine hydrochloride aqueous solution has a pH of 4.5 to 6.[3,4] The commercial injections have a pH of 4 to 6.5.[4] Injections that also contain epinephrine bitartrate have a pH of 3.3 to 5.5.[2]

### $pK_a$
Bupivacaine has a $pK_a$ of 8.1.[7]

## General Stability Considerations
Bupivacaine hydrochloride products should be stored at controlled room temperature and protected from freezing; products that contain epinephrine bitartrate should be protected from light. Bupivacaine hydrochloride products that contain a precipitate or are darker in color than slightly yellow should not be used.[2,7] Bupivacaine hydrochloride injections that do not contain epinephrine in the formulation may be autoclaved at 121°C and 15 psi for 15 minutes.[2]

## Stability Reports of Compounded Preparations
### Injection
Injections, like other sterile drugs, should be prepared in a suitable clean air environment using appropriate aseptic procedures. When prepared from nonsterile components, an appropriate and effective sterilization method must be employed.

**TABLE 15.** Epidural Analgesic Formula[128]

| Component | Amount |
|---|---|
| Fentanyl citrate 50 mcg/mL | 2.5 mL |
| Bupivacaine hydrochloride 0.5% | 8.75 mL |
| Epinephrine hydrochloride 1:100,000 | 6.9 mL |
| 0.9% Sodium chloride injection, USP | 81.85 mL |

**STUDY 1:** Allen[128] reported the compounding of an epidural analgesic formulation. The product consisted of fentanyl citrate 1.25 mcg/mL, bupivacaine hydrochloride 0.4375 mg/mL, and epinephrine hydrochloride 0.6875 mcg/mL in 0.9% sodium chloride injection, USP, and was compounded from the formula cited in Table 15. Packaged in light-resistant sterile containers, the preparation was physically compatible and chemically stable for 20 days stored under refrigeration at 3°C.

**STUDY 2:** Kjonniksen et al.[1226] reported the long-term stability of a similar three-drug analgesic mixture for epidural administration. The preparation contained fentanyl citrate 1.93 mcg/mL, bupivacaine hydrochloride 0.935 mg/mL, and epinephrine bitartrate 2.07 mcg/mL with sodium metabisulfite approximately 2 mcg/mL and disodium edetate approximately 0.2 mcg/mL in sodium chloride 0.9%. The formulation was prepared by first compounding a concentrate that was packaged in vials and autoclaved. This concentrate was then diluted 1:11 in 500 mL of sodium chloride 0.9% in polyolefin-lined Excel multilayer bags. The bags were stored refrigerated at 2 to 8°C for 180 days, followed by four days at room temperature of 22°C exposed to ambient laboratory illumination. No visible changes were reported. Stability-indicating HPLC analysis found that none of the drugs exhibited changes in concentration greater than 4%. The authors concluded that this drug combination was stable and could be given a shelf life of six months under refrigeration based on the chemical stability.

## Compatibility with Other Drugs

Injections, like other sterile products, should be prepared in a suitable clean air environment using appropriate aseptic procedures. When prepared from nonsterile components, an appropriate and effective sterilization method must be employed.

Hudson et al.[1192] reported the stability of diamorphine hydrochloride and hyperbaric bupivacaine hydrochloride in 8% dextrose at concentrations of 0.1 and 4 mg/mL, respectively. The drug mixture was packaged in Becton Dickinson 5-mL polypropylene plastic syringes sealed with tip caps. The samples were stored refrigerated at 7°C, room temperature of 25°C and 65% relative humidity, and elevated temperature of 40°C and 75% relative humidity. Stability-indicating HPLC analysis found no bupivacaine hydrochloride loss in 91 days under any of these storage conditions. However, 10% diamorphine hydrochloride losses were calculated to occur in 26 days refrigerated, seven days at room temperature, and four days at 40°C. ∎

# ▉ Busulfan

## Properties

Busulfan is a white or almost white crystalline powder.[2,3] This substance is a known carcinogen.[1] It also is an irritant, and contact with skin and mucous membranes should be avoided.[3]

### Solubility

Busulfan is practically insoluble in water,[1] having a solubility of about 0.1 mg/mL[1115]; however, it slowly dissolves as hydrolysis occurs. It is soluble in ethanol to about 1 mg/mL.[1]

## General Stability Considerations

Busulfan tablets should be stored in well-closed containers at controlled room temperature.[2,4]

## Stability Reports of Compounded Preparations
### Oral

**STUDY 1:** Allen[133] reported on the stability of a busulfan oral suspension. Busulfan tablets (Myleran, Burroughs-Wellcome) were crushed to fine powder and mixed thoroughly in increments with simple syrup to form a suspension. The stability of busulfan as a suspension is greatly temperature-dependent. A 10-mg/5-mL suspension was stable for at least 30 days under refrigeration. However, at room temperature (25°C) losses of 4% in two days and 9% in three days occurred. The suspension should be stored under refrigeration to maintain a reasonable stability period.

**STUDY 2:** Partin et al.[1122] evaluated the stability of an extemporaneously compounded busulfan 2-mg/mL oral suspension formulation. Commercial busulfan tablets were crushed and mixed with simple syrup. Incorporation of a suspending agent was found to be unnecessary. Samples were stored at room temperature of 25°C and refrigerated at 4°C for 30 days. Fourier-transform IR analysis of the busulfan concentration found that about 10% decomposition occurred within three days at room temperature but the samples remained stable throughout the 30-day study period when stored under refrigeration. ∎

# Caffeine

## Properties

Caffeine, as both the monohydrate and anhydrous form, occurs as a white crystalline powder or matted white crystalline needles that are odorless but have a bitter taste.[1-3] Caffeine may be combined with an equal amount of citric acid to yield citrated caffeine, which is a white bitter-tasting powder.[2]

### Solubility

Caffeine is soluble in water, about 16.7[3] to 21.7 mg/mL,[1] and ethanol, about 15.2 mg/mL.[1] Mixed with citric acid (citrated caffeine), the aqueous solubility increases to about 250 mg/mL in warm water.[1]

### pH

Aqueous solutions of caffeine are neutral.[3] Caffeine and sodium benzoate injection has a pH of 6.5 to 8.5.[4]

## General Stability Considerations

Caffeine (anhydrous) should be stored in well-closed containers. The monohydrate should be stored in tight containers to prevent efflorescence.[1,3,4]

## Stability Reports of Compounded Preparations

### Injection

Injections, like other sterile drugs, should be prepared in a suitable clean air environment using appropriate aseptic procedures. When prepared from nonsterile components, an appropriate and effective sterilization method must be employed.

**STUDY 1 (REPACKAGED):** Nahata et al.[484] evaluated the stability of caffeine injection 10 mg/mL (formulated as the base) packaged in syringes and stored for 60 days at room temperature and 4°C. The caffeine injection was drawn into 10-mL plastic syringes and 10-mL glass syringes (both Becton Dickinson). HPLC analysis found little or no change in caffeine concentration in either type of syringe throughout the study period; the largest change in concentration was 3.8%.

**STUDY 2 (REPACKAGED):** Fraser[769] reported the stability of commercial caffeine citrate 10-mg/mL injection (Sabex) packaged in polypropylene syringes (Becton Dickinson) sealed with friction tip caps (Monoject). The syringes were stored at room temperature of about 20 to 23°C exposed to fluorescent light for 180 days. No visible changes in color or clarity were observed. Stability-indicating HPLC analysis found that no loss of caffeine citrate occurred over the 180-day storage period.

**STUDY 3 (EXTEMPORANEOUSLY PREPARED):** Donnelly and Tirona[202] evaluated the stability of an extemporaneously prepared injection of citrated caffeine. Ten grams of drug was dissolved in sterile water for injection to produce a solution containing caffeine 10 mg/mL (as the citrate). The solution was sterilized by filtration through a 0.22-μm filter and was packaged in 2-mL clear sterile pyrogen-free glass vials. The samples were stored at 22 and 4°C for 342 days. The caffeine content was assessed using a stability-indicating HPLC assay. No loss of caffeine occurred in any sample over 342 days; similarly, no change in color or clarity was observed.

**STUDY 4 (EXTEMPORANEOUSLY PREPARED):** Nahata et al.[221] evaluated the stability of caffeine base at 10 mg/mL, extemporaneously compounded and mixed in various parenteral solutions. The injection was compounded with benzyl alcohol 9 mg/mL in sterile water for injection and added to the parenteral solutions in Table 16 to yield a final caffeine concentration of 5 mg/mL. The solutions were stored at 21°C under fluorescent light for 24 hours and analyzed for caffeine content using a stability-indicating HPLC assay. Little change in caffeine concentration occurred (maximum change 4%). Furthermore, unlike citrated caffeine injection, the caffeine base injection did not interfere with the limulus lysate test for pyrogens.

**STUDY 5 (EXTEMPORANEOUSLY PREPARED):** Nahata et al.[494] also reported on formulating an extemporaneously prepared caffeine citrate injection. Caffeine base 10 g and citric acid, USP, 10.94 g were dissolved in 1 liter of bacteriostatic water for injection, USP,

**TABLE 16.** Parenteral Solutions Tested for Compatibility with Caffeine Injection[221]

Dextrose 5% in water

Dextrose 5% in 0.2% sodium chloride injection

Dextrose 5% in 0.2% sodium chloride injection with potassium chloride 20 mEq/L

Dextrose 10% in water

Dextrose 10% in 0.2% sodium chloride injection with potassium chloride 5 mEq/L

Parenteral nutrition solutions composed of Travasol 1.1, 2.2, or 4.25% with dextrose 10% in water and various electrolytes

to yield a caffeine concentration of 10 mg/mL. The injection was sterilized by filtration and filled into 30-mL sterile pyrogen-free glass vials. The USP Bacterial Endotoxins Test using Limulus Amebocyte Lysate (LAL) reagent to test for pyrogen content could not be performed. The preparation appeared to interfere with the test, yielding a false-negative result. Consequently, the USP Pyrogen Test (rabbit test) had to be performed instead.

**STUDY 6 (EXTEMPORANEOUSLY PREPARED):** Nahata et al.[483] reported the stability of the extemporaneously prepared caffeine citrate injection at a concentration of 5 mg/mL in the solutions cited in Table 16 except that the 10% dextrose in 0.2% sodium chloride injection had 20 mEq/L of potassium chloride. Once again, the caffeine citrate concentrations remained stable in these solutions over 24 hours at room temperature analyzed using an HPLC assay. The largest loss, 6.4%, was found in the 10% dextrose in 0.2% sodium chloride injection with 20 mEq/L of potassium chloride admixture.

**STUDY 7 (EXTEMPORANEOUSLY PREPARED):** Loeb[1510] reported on the extemporaneous compounding of caffeine and sodium benzoate injection from information provided by Eli Lilly and company. The following formula was used:

| | | |
|---|---|---|
| Caffeine anhydrous | | 123.675 mg |
| Sodium benzoate | | 131.325 mg |
| Water for injection | qs | 1 mL |

The solution was filtered through a 0.22-μm filter into glass vials and assigned a 24-hour expiration period.

**STUDY 8 (WITH ANTIPYRINE):** Mutch and Hutson[166] reported the stability of an extemporaneously prepared injection of antipyrine and caffeine 100 and 20 mg/mL, respectively. Antipyrine powder (Sigma Chemical) and caffeine powder (Spectrum Chemical) were weighed and dissolved in warm sterile water for injection to yield the appropriate concentrations. The solution was filled into 10-mL clear glass vials with rubber stoppers and autoclaved at 121°C for 30 minutes. The finished vials were stored upright and inverted at 5, 20, 40, and 60°C to evaluate shelf life. In addition, 10 mL of the finished injection was added to 100- and 250-mL polyvinyl chloride (PVC) bags of 0.9% sodium chloride injection and stored at 5 and 20°C. Both vials and admixtures at 20°C were exposed to normal room light and to intense fluorescent light. The content of the two drugs was evaluated by HPLC.

No changes in drug content were found in the intact vials at any storage temperature after storage for up to six months; contact of the solution with the rubber stopper had no effect on the content of either drug. Similarly, the drugs were stable in the admixtures in PVC bags for up to 48 hours. Furthermore, no turbidity, flocculation, precipitation, or pH change was noted for any sample. However, the samples exposed to intense fluorescent light developed a yellow discoloration within one month that continued to intensify with time. Exposure of an antipyrine solution to the intense light showed that the discoloration was associated with that component. Even though no detectable changes in drug content occurred during the study, protection from light was recommended.

## Oral

Barnes et al.[669,690] evaluated the stability of oral caffeine citrate liquid formulations for neonatal use. One formulation tested contained caffeine citrate 1% (wt/vol), potassium sorbate 0.1% (wt/vol), and sodium citrate 1% (wt/vol) in water. The other formulation was identical except for the inclusion of sorbitol 20% (wt/vol). Sodium citrate was included to buffer the mixture to pH 4.5. This addition ensures sufficient concentrations of unionized sorbate, which is the form that has antimicrobial activity. The oral liquids were packaged as 100 mL in 200-mL amber glass bottles and stored at 5, 25, 32, and 45°C. The refrigerated and room temperature samples had no visual changes. The samples at elevated temperatures developed a yellow color after 12 months of storage. Stability-indicating HPLC analysis found no loss of caffeine citrate in either formulation at any of the temperatures. However, potassium sorbate concentrations did decline in some samples. Based on these results, a shelf life of one year was calculated for storage at 25°C. The authors indicated that the formulation without sorbitol was preferable because of the potential for sorbitol to cause gastrointestinal effects.

## Enteral

Eisenberg and Kang[24] evaluated the stability of a citrated caffeine preparation for enteral use. The product was compounded from citrated caffeine powder 10 g, which was dissolved in 250 mL of sterile water for irrigation, USP. A combination of simple syrup and cherry syrup (2:1) was used to bring the product to 500 mL with a nominal initial caffeine concentration of 10 mg/mL. The caffeine concentration of three separate lots was monitored over 90 days of storage (two lots) and 120 days (one lot) (temperature unspecified) by HPLC analysis. Little or no loss of caffeine was found. ■

# Calcipotriene

## Properties

Calcipotriene, a synthetic vitamin D₃ derivative, is a white or off-white crystalline substance.[7]

### Solubility

The calcipotriene solution is formulated in a vehicle that is composed of 51% (vol/vol) isopropanol, propylene glycol, hydroxypropyl cellulose, sodium citrate, menthol, and water.[7]

## General Stability Considerations

Calcipotriene is a relatively unstable molecule and is inactivated by acidic pH.[635] Calcipotriene products should be stored at controlled room temperature between 15 and 25°C and protected from freezing. The lotion should also be protected from exposure to light.[7]

## Compatibility with Other Drugs

### CALCIPOTRIENE COMPATIBILITY SUMMARY

*Compatible with:* Halobetasol propionate 0.05% ointment
  • Halobetasol propionate 0.05% cream • Tazarotene 0.05% gel

*Incompatible with:* Ammonium lactate 12% lotion • Hydrocortisone valerate 0.2% ointment • Salicylic acid 6% • Tar gel 5%

**STUDY 1:** Patel et al.[635] evaluated the compatibility of calcipotriene 0.005% ointment mixed by levigation with equal quantities of other topical medications. The mixtures were stored at 5 and 25°C. Drug concentration was assessed by HPLC analysis. The combinations are shown in Table 17.

Mixed with halobetasol propionate 0.05% ointment, no changes in physical appearance and little or no change in either drug concentration occurred in 13 days of storage at either temperature. Similarly, mixed with halobetasol propionate 0.05% cream, no changes in physical appearance occurred in 13 days. Both drugs remained stable for 48 hours, but analysis was suspended at that point.

**TABLE 17.** Compatibility of Calcipotriene 0.005% Ointment Mixed in Equal Quantities with Other Medications[635]

| Drug Mixed with Calcipotriene Ointment | Result |
| --- | --- |
| Ammonium lactate 12% lotion | Incompatible |
| Halobetasol propionate 0.05% ointment | Compatible |
| Halobetasol propionate 0.05% cream | Compatible |
| Hydrocortisone valerate 0.2% ointment | Incompatible |
| Salicylic acid 6%[a] | Incompatible |
| Tar gel 5% | Incompatible |

[a]Calcipotriene 0.005% ointment mixed with salicylic acid to yield a final salicylic acid concentration of 6%.

However, the other combinations did not perform as well. Mixed with hydrocortisone valerate 0.2% ointment, no physical changes were observed, but over 10% calcipotriene loss occurred within 72 hours at both temperatures. Mixed with salicylic acid to yield a 6% concentration, physical separation occurred in 24 hours, and calcipotriene content dropped precipitously, losing almost all of the drug in about five hours. Mixed with ammonium lactate 12% lotion, liquid separation appeared in 24 hours, as did a more than 10% loss of calcipotriene. Mixed with 5% tar gel (composition unspecified), the color became darker after six days at both temperatures. A 20% loss in calcipotriene content in the 25°C sample also occurred in six days, but only about a 4% loss occurred at 4°C in this time frame.

**STUDY 2:** Hecker et al.[815] evaluated the compatibility of tazarotene 0.05% gel combined in equal quantity with other topical products, including calcipotriene 0.005% cream and ointment. The mixtures were sealed and stored at 30°C for two weeks. HPLC analysis found that less than 10% tazarotene loss occurred in both. Calcipotriene in the combinations could not be assayed due to interference with tazarotene. ∎

# Calcitonin-Salmon

## Properties

Calcitonin is a calcium-regulating hormone secreted in mammals by the thyroid gland and the ultrabrachial gland in birds and fish.[1] Calcitonin-human is a synthetic 32-amino acid polypeptide in the same sequence as naturally occurring calcitonin in humans. Calcitonin-salmon is a polypeptide that has the structure of naturally occurring salmon calcitonin I.[3]

Commercial calcitonin products may be synthetically prepared.[1] Calcitonin-salmon is a white or nearly white powder.[3]

### Solubility

Calcitonin-salmon is freely soluble in water.[3]

## General Stability Considerations

Calcitonin-salmon powder should be stored under refrigeration at 2 to 8°C and protected from exposure to light. If the calcitonin substance is sterile, it must be packaged in sterile, airtight, tamper-proof containers.[3] Calcitonin-salmon injection and nasal spray (Miacalcin, Novartis) are also stored under refrigeration and protected from freezing. The manufacturer states that calcitonin-salmon nasal spray is stable for up to 35 days (Miacalcin) or 30 days (Fortical) if stored in the upright position at controlled room temperature.[7]

Lee et al.[1211] reported that the maximum stability of calcitonin-salmon occurred at pH 3.3.

## Stability Reports of Compounded Preparations
### Nasal Spray
Cook and Shenoy[720] reported the stability of calcitonin-salmon nasal spray (Miacalcin Nasal Spray, Novartis) at storage temperatures of 25, 40, and 60°C, elevated temperatures that might occur during shipping under less than ideal conditions. The nasal spray was stored for three days at each of the elevated temperatures and at 60% relative humidity followed by two weeks at 5°C and two more weeks at 25°C. HPLC analysis found calcitonin-salmon losses of 3 and 9% after storage at 25 and 40°C, respectively, followed by an additional four weeks of storage as noted previously. When stored at 60°C for three days followed by the four weeks of storage, an unacceptable loss of 19% occurred. ■

# ■ Calcitriol

## Properties
Calcitriol occurs as white to almost white crystals or crystalline powder.[1,3,4]

### Solubility
Calcitriol is practically insoluble in water but is freely soluble in ethanol and soluble in ether and fatty oils. It is slightly soluble in methanol and ethyl acetate.[1,3,4]

### pH
Calcitriol injection may range from pH 5.9 to 7.7.[7]

### Osmolality
Calcitriol injection is an isotonic solution.[7]

## General Stability Considerations
Calcitriol is sensitive to air, heat, and light.[1,3,4] Calcitriol should be packaged in tight, light-resistant containers under at atmosphere of nitrogen and stored refrigerated at 2 to 8°C. Once opened, containers of calcitriol should be used immediately.[3]

Calcitriol injection in intact containers should be stored at controlled room temperature and protected from exposure to light. The drug should also be protected from freezing and exposure to excessive heat, although brief exposures to temperatures of 40°C do not adversely impact the stability. Unused injection should be discarded because of the absence of an antimicrobial preservative.[7]

Calcitriol injection undergoes sorption to polyvinyl chloride (PVC) plastic containers. Studies have shown losses can vary but may range up to 50% in two hours and 75% in 20 hours. The drug does not sorb as extensively to glass and polypropylene containers and syringes; losses have ranged from 10 to 20% in 24 hours.[1533,1535,1536]

## Stability Reports of Compounded Preparations
### Topical
Lebwohl et al.[1532] evaluated the effect of ultraviolet light on the stability of calcitriol ointment. Two and four grams of calcitriol ointment 3 mcg/g was applied to a 40-square-centimeter area of clear plastic film that permits light transmission to create thin and thick ointment layers. The samples were irradiated with ultraviolet A and both broad-band and narrow-band ultraviolet B light as well as ambient room light and fluorescent light.

HPLC analysis of drug concentrations found 93 to 100% degradation of the calcitriol occurred with all three ultraviolet light exposures. There was no significant loss of calcitriol from fluorescent and ambient room light exposure. In addition, the ointment substantially reduced transmission of ultraviolet light through the ointment by up to 41% in the thin layer and 87% in the thick layer. The authors recommended that when calcitriol ointment is to be used in conjunction with phototherapy, the ointment should be applied after the ultraviolet exposure rather than before.

### Injection
Injections, like other sterile drugs, should be prepared in a suitable clean air environment using appropriate aseptic procedures. When prepared from nonsterile components, an appropriate and effective sterilization method must be employed.

**STUDY 1:** Pecosky et al.[1533] reported that calcitriol 0.5 mcg/mL diluted in dextrose 5%, sodium chloride 0.9%, or sterile water for injection and calcitriol 2 mcg/mL and 1 mcg/mL undiluted was stable by HPLC analysis for eight hours at room temperature exposed to light when packaged in polypropylene syringes from Becton Dickinson.

**STUDY 2:** Vargas-Ruiz et al.[1534] reported that undiluted calcitriol 0.8, 1, and 2 mcg/mL packaged in tuberculin plastic syringes was stable for at least seven days stored under refrigeration. ■

# Calcium Glubionate

## Properties
Each gram of calcium glubionate monohydrate (calcium gluconate and calcium lactobionate monohydrate) provides approximately 1.6 mmol of calcium. One gram of calcium is provided by 15.2 g of calcium glubionate monohydrate.[3]

### pH
Calcium glubionate syrup has a pH between 3.4 and 4.5.[4] Calcium glubionate syrup (Neo-Calglucon, Dorsey) had a measured pH of 4.[19]

### Osmolality
Calcium glubionate syrup 0.36 mg/mL (Sandoz) had an osmolality of 2550 mOsm/kg.[233]

## General Stability Considerations
Calcium glubionate syrup should be stored in tight containers at or below 30°C but protected from freezing.[4]

## Stability Reports of Compounded Preparations
### Enteral
**CALCIUM GLUBIONATE COMPATIBILITY SUMMARY**
*Compatible with:* Vital • Vivonex T.E.N.
*Incompatible with:* Enrich • Ensure • Ensure HN • Ensure Plus • Ensure Plus HN • Osmolite • Osmolite HN • TwoCal HN

**STUDY 1:** Cutie et al.[19] added 5 mL of calcium glubionate syrup (Neo-Calglucon, Dorsey) to varying amounts (15 to 240 mL) of Ensure, Ensure Plus, and Osmolite (Ross Laboratories), with vigorous agitation. The calcium glubionate syrup was physically incompatible with all three enteral products, immediately becoming an adhesive gelatinous mass that clogged feeding tubes.

Cutie et al.[19] also noted that calcium ions in concentrations exceeding 1.15% coagulated all three enteral products studied.

**STUDY 2:** Burns et al.[739] reported the physical compatibility of calcium glubionate syrup (Neo-Calglucon, Dorsey) 5 mL with 10 mL of three enteral formulas: Enrich, TwoCal HN, and Vivonex T.E.N. Visual inspection found no physical incompatibility with the Vivonex T.E.N. enteral formula. However, when mixed with Enrich and TwoCal HN, thickening occurred and particulates formed.

**STUDY 3:** Altman and Cutie[850] reported the physical compatibility of calcium glubionate syrup (Neo-Calglucon, Dorsey) 5 mL with varying amounts (15 to 240 mL) of Ensure HN, Ensure Plus HN, Osmolite HN, and Vital after vigorous agitation to ensure thorough mixing. The calcium glubionate syrup was physically incompatible with Ensure HN, Ensure Plus HN, and Osmolite HN, immediately becoming an adhesive gelatinous mass that clogged feeding tubes. However, it was physically compatible with Vital, distributing uniformly with no phase separation or granulation. ■

# Calcium Gluconate

## Properties
Calcium gluconate is an odorless tasteless crystalline or granular powder in anhydrous or monohydrate form.[1,3,4] Each gram of calcium gluconate monohydrate represents 2.2 mmol of calcium; 11.2 g of calcium gluconate monohydrate provides about 1 g of calcium.[4]

### Solubility
Calcium gluconate is sparingly and slowly soluble in about 30 parts of cold water and about five parts of boiling water. It is insoluble in ethanol and other organic solvents. Supersaturated aqueous solutions are prepared using calcium D-saccharate.[1,3,4]

### pH
Aqueous solutions of calcium gluconate are neutral to litmus, with the pH of an aqueous solution of calcium gluconate being about pH 6 to 7.[1,3,4] Calcium gluconate injection has a pH between 6 and 8.2.[4,7]

### Osmolality
Calcium gluconate 10% injection is very hypertonic, having an osmolality about 0.68 mOsm/mL.[7]

## General Stability Considerations
Calcium gluconate and calcium gluconate tablets should be packaged in well-closed containers and stored at controlled room temperature. Calcium gluconate injection is packaged in single-dose containers and stored at controlled room temperature. The injection should be protected from freezing. The injection should not be used if it contains a precipitate.[4,7]

## Stability Reports of Compounded Preparations
### Topical
Rotheli-Simmen et al.[1113] evaluated the physical stability of several topical formulations of calcium gluconate 2.5% (wt/wt) for use in treating hydrofluoric acid burns. Several topical gel formulations were attempted using methylcellulose, KY-Jelly,

and polyacrylate (Carbopol), but all proved to be incompatible and resulted in immediate calcium gluconate precipitation. A hydroxyethylcellulose gel formulation proved to be marginally compatible with the high concentration calcium gluconate, resulting in just a slight trace of precipitation.

The hydroxyethylcellulose gel was prepared by dissolving 2.5 g of calcium gluconate in 87 mL of purified water with heating. While still warm, 3 g of hydroxyethylcellulose was added with continuous stirring until gel formation occurred. Propylene glycol 5 g was then incorporated. The gel was heat treated using steam at 100°C for 30 minutes to reduce the microbial load. The finished gel was packaged in clear glass containers with screw caps. The gel was evaluated for physical stability (homogeneity, precipitation, discoloration, viscosity change) and stored at room temperature for six months, with no changes occurring.

The addition of the antimicrobial preservative chlorhexidine to the formulation resulted in discoloration that was deemed an incompatibility. However, the absence of a preservative in the formulation resulted in positive microorganism growth, including fungi. The authors noted that aseptic preparation of the gel is mandatory. ▪

# Captopril

## Properties

Captopril is a white to off-white crystalline powder with a slight sulfide odor.[2,3] Even the commercial tablets may exhibit a slight odor.[2]

### Solubility

Captopril is freely soluble in water and ethanol.[1–3] The aqueous solubility is 160 mg/mL.[6,7]

### pH

A 2% aqueous solution of captopril has a pH of 2 to 2.6.[3]

### $pK_a$

The $pK_a$ values of captopril are 3.7 and 9.8.[1]

## General Stability Considerations

Captopril products should be stored in tight containers protected from moisture at temperatures not exceeding 30°C.[2–4,7] In the dry state, captopril is relatively stable up to 50°C.[298] However, adequate control of humidity and temperature is necessary for drug stability.[6]

Captopril, a thiol, undergoes oxidative degradation in aqueous solutions. The rate is influenced by the solution pH and oxygen content and can be catalyzed by metal ions. The drug is more stable at acidic pH than in basic solutions. The oxidative rate is also dependent on captopril concentration; more concentrated solutions oxidize more slowly than lower concentrations.[6,544]

Timmins et al.[215] found the stability of captopril in aqueous solution to be pH dependent. Above pH 4, the decomposition rate increases with increasing pH. At pH 4 and below, captopril decomposition is pH independent, with little difference in degradation rate at least down to pH 2.[6,817] Lee and Notari[263] reported that the oxidation product captopril disulfide is the predominant or only decomposition product in the pH range of 6.6 to 8.

Kristensen et al.[1331] evaluated the stability of captopril in aqueous solution at pH 3. The drug exhibited both concentration-dependent and temperature-dependent decomposition rates. At 1 mg/mL, the decomposition followed zero order degradation, while at higher concentrations the decomposition kinetics were complex and mixed. The time to 10% loss ($t_{90}$) at 5, 25, and 36°C was found to be 26 days, 7 days, and 4 days, respectively.

Timmins et al.[215] reported that the addition of copper 5 ppm or iron 5 ppm to captopril aqueous solutions markedly increased the degradation rate. Unfortunately, copper and iron are the most likely contaminants in formulation additives, containers, closures, etc. It was suggested that edetate sodium, which chelates the catalytic metal ions, could improve stability. Furthermore, the antioxidant propyl gallate and the reducing agent sodium metabisulfite markedly increase the degradation rate of captopril.

Lee and Notari[263] found that the formation of captopril disulfide decomposition product occurred with or without copper ions, although copper ions increased the rate. Furthermore, the drug was stabilized by the addition of the chelating agent edetate (EDTA) or 8-hydroxyquinoline, indicating that inactivation of trace metal ions was the mechanism of stabilization.

Stulzer et al.[1256] evaluated the compatibility of captopril with a selection of common tablet excipients including microcrystalline cellulose, ethylcellulose, methylcellulose, monohydrated lactose, lactose Supertab, polyvinylpyrrolidone, colloidal silicon dioxide, stearic acid, and magnesium stearate. The authors evaluated the compatibility of captopril with each of these potential excipients using differential scanning calorimetry (DSC), thermogravimetric analysis, x-ray powder diffraction, and Fourier transform infrared spectroscopy. All of the potential excipients tested were found to be compatible with captopril except magnesium stearate. DSC indicated a possible drug-excipient interaction. However, the interaction was not confirmed by the other techniques.

## Stability Reports of Compounded Preparations
### *Oral*

#### SUMMARY

The varying results on captopril's solution stability are difficult to resolve. Many factors accelerate the oxidation of captopril to its disulfide decomposition product, such as available oxygen in the container headspace, pH values exceeding 4, captopril concentration, and variations in the metal-ion content that catalyze oxidation. Some metals appear in excipients and other inactive tablet components; presumably, variations in the content of such trace metals could cause variability in the stability results. Certainly, the absence of oxidation in captopril solutions prepared from pure drug rather than tablets would indicate that the metals from the tablets themselves tend to catalyze the oxidation. The addition of an antioxidant such as ascorbic acid may extend captopril stability, as does the addition of edetate disodium that chelates metal ions. (See Studies 4, 13, and 14 below.) However, not all studies arrived at similar conclusions. (See Study 18 below.) Prudence would indicate that a relatively short expiration period be used because of these variables.

#### USP OFFICIAL FORMULATION (ORAL SOLUTION):

> Captopril 75 mg
> Vehicle for Oral Solution, NF (sugar-containing or
>    sugar-free) qs 100 mL

(See the vehicle monograph for information on the Vehicle for Oral Solution, NF.)

Use captopril powder. Add about 10 mL of the vehicle and mix well. Add additional vehicle almost to volume in increments with thorough mixing. Quantitatively transfer to a suitable calibrated tight and light-resistant bottle, bring to final volume with the vehicle mixture, and thoroughly mix, yielding captopril 0.75-mg/mL oral solution. The final liquid preparation should have a pH between pH 3.8 and 4.3. Store the preparation under refrigeration between 2 and 8°C. The beyond-use date is seven days from the date of compounding.[4,5]

#### USP OFFICIAL FORMULATION (ORAL SUSPENSION):

> Captopril 75 mg
> Vehicle for Oral Solution, NF (sugar-containing or
>    sugar-free)
> Vehicle for Oral Suspension, NF (1:1) qs 100 mL

(See the vehicle monographs for information on the individual vehicles.)

Use captopril powder or commercial tablets. If using tablets, crush or grind to fine powder. Add about 10 mL of the vehicle mixture and mix to make a uniform paste. Add additional vehicle almost to volume in increments with thorough mixing. Quantitatively transfer to a suitable calibrated tight and light-resistant bottle, bring to final volume with the vehicle mixture, and thoroughly mix, yielding captopril 0.75-mg/mL oral suspension. The final liquid preparation should have a pH between pH 3.8 and 4.3. Store the preparation under refrigeration between 2 and 8°C. The beyond-use date is seven days from the date of compounding at either storage temperature.[4,5]

**STUDY 1:** Taketomo et al.[105] evaluated the stability of captopril in powder papers. Captopril powder was prepared by triturating 12.5-mg tablets (Capoten, Squibb) and adding sufficient lactose, USP (Humco), to yield 2 mg/100 mg of powder. The powder was mixed further in a blender (Waring) to ensure a homogeneous mixture. The powder papers were stored in prescription vials, zipper-lock bags, and Moisture Proof Barrier Bags (Baxa) at 25°C and protected from light for 24 weeks. An HPLC assay was used to assess captopril content. There was little difference in the stability of captopril in the three containers during the first 12 weeks. However, at 24 weeks, the disulfide decomposition product of captopril was found in the samples in zipper-lock bags. Samples in the prescription vials and Moisture Proof Barrier Bags were stable through 24 weeks.

**STUDY 2:** Nahata et al.[14] studied captopril stability in three oral liquid dosage forms utilizing syrup, NF, distilled water, and distilled water with sodium ascorbate as vehicles. Captopril in syrup was prepared by triturating three 50-mg tablets with a mortar and pestle with 70 mL of syrup, NF, and 70 mL of methylcellulose solution. The mortar and pestle were rinsed with additional syrup, and the mixture was brought to a 150-mL total volume with syrup, yielding a nominal 1-mg/mL captopril concentration. To prepare the distilled water solution, three 50-mg tablets were dissolved in 150 mL; the mixture was shaken to effect disintegration and dissolution. A nominal 1-mg/mL concentration was prepared. Captopril in distilled water with sodium ascorbate was prepared similarly. Two 50-mg captopril tablets were dissolved, and 500 mg of sodium ascorbate injection was added. Sufficient distilled water was then added to bring the total volume to 100 mL, yielding nominal concentrations for the captopril and sodium ascorbate of 1 and 5 mg/mL, respectively. The solutions were stored in glass prescription bottles at 4 and 22°C.

The solutions were analyzed using a stability-indicating HPLC technique. Captopril in syrup was the least stable; acceptable drug concentrations were maintained for only seven days, both at 4 and 22°C. In distilled water, captopril also was stable for seven days at 22°C but lost 10% in 14 days at 4°C. Captopril was most stable in distilled water with sodium ascorbate. The drug exhibited 10% decomposition in 14 days at 22°C; under refrigeration no loss occurred within 56 days. The presence of the antioxidant sodium ascorbate decreases the oxidation of captopril in aqueous solutions.

**STUDY 3:** Chan et al.[16] evaluated the effects of varying concentrations and sources of captopril on solution stability. Captopril 1- and 10-mg/mL solutions were prepared from 25-mg tablets (Bristol-Myers Squibb), and a 10-mg/mL solution was prepared from the pure reference standard powder. Sterile water for irrigation was the diluent. The solutions were filtered through a prefilter and a 0.2-µm filter, resulting in clear, colorless solutions. The solutions were stored in sterile evacuated glass containers at 23°C protected from light for 28 days. Analysis was performed using a stability-indicating HPLC technique initially and after 14 and 28 days. The solutions also were evaluated for microbial growth after 28 days. The 10-mg/mL solution underwent significant decomposition, exhibiting 10% loss in 14 days and 19% loss in 28 days. A crystalline precipitate formed in some samples. The 10-mg/mL solution prepared from pure reference standard did not undergo decomposition in 28 days. Similarly, the 1-mg/mL solution exhibited little decomposition in 28 days. The authors attributed this difference to the higher content of metals inherent in the tablet formulation. No bacterial or fungal growth occurred in any filtered solution.

**STUDY 4:** Nahata et al.[200] reported the stability of two extemporaneous formulations of captopril oral liquid. The products were prepared by slowly disintegrating and dissolving two captopril 50-mg tablets (Squibb) by shaking with distilled water. One ascorbic acid 500-mg tablet or 500 mg of sodium ascorbate injection then was added. Distilled water was used to bring to 100 mL, yielding a captopril 1-mg/mL concentration. The preparations were packaged in glass prescription bottles and stored at 4 and 22°C for up to 56 days. A stability-indicating HPLC assay was used to determine captopril levels. The formulation made with the ascorbic acid tablet was somewhat more stable than the formulation containing sodium ascorbate injection. At 22°C, the ascorbic acid formulation lost 8% in 28 days; the sodium ascorbate formulation lost 10% in 14 days. At 4°C, both preparations were stable for up to 56 days.

**STUDY 5:** Andrews and Essex[245] evaluated the stability of captopril aqueous solution 1 mg/mL prepared from crushed tablets (Capoten) and stored at room temperature in tightly closed glass containers for five days. Analysis found 96.6% of the initial captopril concentration after five days; the inactive disulfide increased during storage from 0.5 to 5.3%.

**STUDY 6:** Pramar et al.[462] also studied the stability of captopril solutions prepared from both powder and crushed tablets. A captopril concentration of 5 mg/mL was prepared from the powder or tablets (Squibb) ground to fine powder. Vehicles included water (source unspecified), 25% syrup in water also containing citric acid and sodium benzoate, and 25% Syrpalta in water, the last for samples prepared from tablets only. Unfortunately, the source and nature of the water used

in all of these vehicles were unspecified. Captopril powder was found to be the most stable when mixed in water alone, exhibiting 10% decomposition in 10 days at 25°C and 9% loss in 27 days at 4°C. Prepared from crushed tablets, captopril exhibited greater decomposition, with 11% loss in six days at 25°C and 10% loss in 20 days at 4°C. Decomposition in either syrup vehicle was substantially faster. The authors postulated that the losses were due to oxidation and were accelerated by other tablet components. The source(s) of the higher rates of loss in the syrups was not identified.

**STUDY 7:** Using stability-indicating HPLC analysis, Mathew and Gupta[536] evaluated the influence of several factors on the stability of captopril in aqueous systems. Ionic strength had little effect, but increased captopril decomposition resulted as phosphate or citrate buffer concentrations increased from 0.05 to 0.2 M. Increased degradation of captopril occurred at lower captopril concentrations. Both 10- and 5-mg/mL solutions exhibited similar lower decomposition rates than 2-mg/mL and particularly 1-mg/mL solutions, the least stable concentration tested. Along with a nitrogen purge, inclusion of cosolvents in the formulation improved captopril stability. Formulations including 10% ethanol with 30% polyethylene glycol 400, 30% glycerin, or 30% propylene glycol all improved captopril stability. The addition of edetate (EDTA) sodium 0.1% enhanced solution stability but incorporation of ascorbic acid 0.2% along with edetate did not. All of the nitrogen-purged cosolvent formulations retained at least 95% of the captopril concentration through 622 days stored at 5°C. At room temperature, captopril in nitrogen-purged cosolvent formulations including edetate sodium 0.1% exhibited 4 to 8% loss in 111 days at 25°C. The authors concluded that the right combination of cosolvents, chelating agent, and a minimized oxygen level decreases thiol oxidation and enhances captopril stability in aqueous solution.

**STUDY 8:** Allen and Erickson[544] evaluated the stability of three captopril 0.75-mg/mL oral suspensions extemporaneously compounded from tablets. Vehicles used in this study were (1) an equal parts mixture of Ora-Sweet and Ora-Plus (Paddock), (2) an equal parts mixture of Ora-Sweet SF and Ora-Plus (Paddock), and (3) cherry syrup (Robinson Laboratories) mixed 1:4 with simple syrup. One captopril 100-mg tablet (Squibb) was crushed and comminuted to fine powder using a mortar and pestle. About 10 mL of the test vehicle was added to the powder and mixed to yield a uniform paste. Additional vehicle was added geometrically and brought to the final volume of 134 mL, mixing thoroughly after each addition. The process was repeated for each of the three test suspension vehicles. Samples of each of the finished suspensions were packaged in 120-mL amber polyethylene terephthalate plastic prescription bottles and stored at 5 and 25°C in the dark.

No visual changes or changes in odor were detected during the study. Stability-indicating HPLC analysis found substantial decomposition in all samples tested. In the Ora-Plus-containing suspensions, 10% or less captopril loss occurred in 14 days stored at 5°C and in about seven days at 25°C. However, in the cherry syrup, stability was only retained for two days at either temperature. More than 10% loss occurred after that time.

**STUDY 9:** Lye et al.[549] evaluated a number of factors influencing captopril stability in aqueous solutions and determined the stability of a captopril 1-mg/mL oral liquid prepared from tablets and powder. Two captopril 12.5-mg tablets (Bristol-Myers Squibb) ground to fine powder and 25 mg of captopril powder (Sigma Chemical) were incorporated into the vehicles listed in Table 18 and then packaged in amber glass containers. Samples were stored at 5°C for 30 days.

The stability of captopril was evaluated using a stability-indicating HPLC assay. Unlike previous investigators, Lye et al. reported slightly better stability with the formulations prepared from tablets compared to those prepared from powder. The authors speculated that the differing result may be related to the lower concentration (1 mg/mL) compared to the 10-mg/mL solution tested by Chan et al.[16] They also reported[549] longer stability periods, at least for some vehicles, compared to other reports. A difference was observed in captopril stability in sterile water for irrigation compared to highly purified analytical-grade water presumably having a lower metal-ion content. Captopril was unacceptably

**TABLE 18.** Stability of Various Captopril 1-mg/mL Formulations Reported by Lye et al.[549]

| Vehicle | Less Than 10% Loss | |
|---|---|---|
| | Tablets | Powder |
| ***Vehicles Providing Best Captopril Stability*** | | |
| Syrup (undiluted) | 30 days | 30 days |
| Sterile water for irrigation–syrup (50:50) with edetate disodium 0.1% | — | 30 days |
| Methylcellulose 2% with edetate disodium 0.1% | — | 30 days |
| Analytical-grade water | 15 days | 30 days |
| ***Vehicles Providing Less Captopril Stability*** | | |
| Sterile water for irrigation | 7 days | 10 days |
| Sterile water for irrigation–syrup (50:50) | 3 days | 4 days |
| Methylcellulose 2% | 1 day | 3 days |
| Methylcellulose 4% | <1 day | 1 day |

unstable in an equal parts mixture of syrup and sterile water for irrigation and in methylcellulose 2 or 4% in sterile water for irrigation, which has a higher metal-ion content. The best stability was found with undiluted syrup and in the two vehicles incorporating 0.1% edetate disodium, indicating that a principal decomposition factor could be the available metal-ion content. Captopril losses of 10% or less were found in the time periods noted in Table 18.

**STUDY 10:** Liu et al.[836] reported the effect of 0.03 M citric acid buffer on the stability of captopril 1 mg/mL in a 10% sucrose solution. In this low-concentration sucrose solution, captopril stability was substantially enhanced by the citric acid. About 7% captopril loss occurred in 30 days. Without the citric acid present, 16% loss occurred in 30 days. This stabilizing effect was not observed at higher sucrose concentrations.

**STUDY 11:** Fajolle et al.[952] prepared an oral suspension of captopril 0.75 mg/mL in Ora-Sweet and Ora-Plus. The oral suspension was deemed palatable by patients and easy to administer by nurses. Microbiological assessment met the requirements of the European Pharmacopoeia. Stability of the drug was not evaluated.

**STUDY 12:** Escribano Garcia et al.[954] reported that captopril 1 mg/mL in distilled chemically pure water exhibited substantially better stability than if mixed in tap water or mineral water. See *Use of Tap Water* below. HPLC analysis of the samples in distilled water packaged in polyvinyl chloride (PVC) plastic containers found that about 2 or 3% captopril loss occurred in 30 days at 4°C. Refrigeration was found to improve the microbiological quality compared to room temperature storage.

**STUDY 13:** Berger-Gryllaki et al.[1205] evaluated the long-term stability of extemporaneously compounded oral liquids of captopril 1 mg/mL for pediatric use. Ten formulations were evaluated. The most stable formulation was captopril powder dissolved in purified water obtained by reverse osmosis with edetate disodium 1 mg/mL. The oral solution was packaged in well-closed glass containers and stored under refrigeration at 2 to 8°C, room temperature of 20 to 24°C, and elevated temperature of 38 to 42°C for two years. The pH remained constant at about pH 3.4, and no microbial growth was observed throughout the study. Stability-indicating HPLC analysis found no captopril loss in the solutions stored refrigerated or at room temperature in two years. The solution stored at elevated temperature exhibited about 4% loss in 12 months.

**STUDY 14 (ORAL LIQUID):** Brustugun et al.[1427] evaluated the long-term stability of captopril 1 and 5 mg/mL in an extemporaneously compounded oral liquid formulation. Captopril powder was used to prepare the oral liquids in sterile water for irrigation with sorbitol (as a 70% solution) 287 mg/mL, edetate disodium 0.1 mg/mL, and sodium benzoate 1 mg/mL. The pH of the solution was adjusted to pH 3.9 using 1 M hydrochloric acid if necessary. The oral liquids were packaged in amber glass bottles with an air-filled headspace, and stored at room temperature of 22°C for 12 months. Subsequently, the 1-mg/mL solution was subjected to additional study simulating an in-use situation. For this testing the captopril 1-mg/mL solution was stored refrigerated at 2 to 8°C and opened every weekday for sampling over one month.

All samples were clear and colorless initially and throughout the studies. Captopril 1 mg/mL was reported to have no taste or smell associated with it. However, the captopril 5-mg/mL solution had a slightly sulfurous taste and odor. The authors stated that mixing the solution with a palatable liquid like lemonade could ameliorate the taste and smell. The pH values of all of the test solutions varied only slightly throughout the studies. Microbiological testing found no bacterial or fungal growth. Stability-indicating HPLC analysis found no captopril loss occurred in either the 1- or 5-mg/mL concentrations during 12 months of room temperature storage. In addition, the 1-mg/mL solution stored refrigerated and subjected to repeated opening and sampling to simulate in-use conditions also demonstrated little or no change in captopril concentration over one month.

### Use of Tap Water

Mirkin and Newman[10] and O'Dea et al.[11] noted that captopril has been administered to pediatric patients as a solution prepared in water. However, Lee and Notari[13] and Timmins et al.[215] noted that captopril stability is affected by metals such as copper and iron. Tap water from different supplies can vary greatly, containing differing amounts of minerals; these minerals affect drug stability in solution. The use of tap water for preparing captopril oral liquid dosage forms is not recommended, because of the variability of water supplies in differing locales and the lack of information on inherent impurities. Crawford[120] pointed to the compendial requirement to use purified water in compendial dosage forms; tap water may not be used. However, patients or their parents may use tap water for captopril dilution anyway, not having ready access to purified water. The tap water stability information may be useful in these cases.[121,122]

**STUDY 15 (TAP WATER):** Pereira and Tam[12] studied the stability of a captopril 1-mg/mL solution prepared in common tap water from their community supply (Edmonton, Alberta). The solutions were prepared using 25-mg tablets (Capoten, Squibb) that were crushed separately and added to 25 mL of tap water in glass flasks. The solutions were shaken vigorously for two minutes at an ambient temperature of around 20°C to effect dissolution. Initial analysis showed that complete dissolution had occurred with an initial captopril concentration of approximately 1 mg/mL. Samples were stored at 5, 25,

50, and 75°C for up to 28 days. Captopril content was determined using a stability-indicating HPLC technique. Captopril losses exceeding 10% occurred within 1 and 3.6 days at 75 and 50°C, respectively. The shelf life at 25 and 5°C was much longer; the calculated times to reach 10% decomposition were 11.8 and 28 days, respectively. Captopril solutions in tap water should be stored under refrigeration to maximize stability, and patients should be advised not to expose the solution to high temperatures.

**STUDY 16 (TAP WATER):** Anaizi and Swenson[15] compared the stability of captopril in tap water from their community supply (Rochester, N.Y.) with its stability in sterile water for irrigation and in sterile water for irrigation containing citric acid 40 mg/mL as a potential chelator of trace metals. Captopril 1-mg/mL solutions were prepared by triturating 25-mg tablets (Squibb) and bringing to volume with the appropriate diluent. The solutions were then filtered through 0.2-μm sterile filters. Samples were transferred to polypropylene tubes for storage under refrigeration protected from light. The solutions were analyzed immediately and after 7, 14, 21, and 28 days of storage using a stability-indicating HPLC technique.

Rapid loss of captopril content occurred in the tap water samples; the rate was more than twice the decomposition rate of captopril in sterile water for irrigation. Approximately 70% was lost after seven days at 5°C, largely forming the inactive disulfide form. In sterile water for irrigation, the drug was stable for three days at 5°C, with 22% loss occurring in seven days. The addition of citric acid had no effect on captopril's stability in sterile water for irrigation.

**STUDY 17 (TAP WATER):** Sabol et al.[535] evaluated the stability of captopril 1-mg/mL oral solutions prepared from tablets in distilled water and tap water. Ten captopril 25-mg tablets (Capoten, Bristol-Myers Squibb) were crushed to fine powder in a glass mortar. The powder was triturated with about 10 to 15 mL of either distilled water or tap water, and the mixtures were transferred to graduated cylinders. The mortar and pestle were rinsed several times with the respective diluents, and the rinses were added to the mixtures in the graduates. Sufficient distilled or tap water was added to bring the final volumes to 250 mL and was followed by thorough mixing. The products yielded a final captopril concentration of 1 mg/mL. Undissolved tablet components were allowed to settle, and 5-mL aliquots were transferred to amber glass vials that were then capped. These sample vials were stored at room temperature near 23°C and under refrigeration at 2°C. HPLC analysis found captopril stabilities that paralleled the results of Nahata et al.[14] In distilled water, 10% drug loss occurred in seven days at 23°C, and 8 to 9% loss occurred in 14 days at 2°C. In tap water, 10% captopril loss occurred in as little as three days under refrigeration and in less than one day at room temperature.

**STUDY 18 (TAP, MINERAL, AND DISTILLED WATER):** Garcia et al.[1626] reported the stability of captopril 1 mg/mL in tap water, mineral water, and distilled water stored at 4 and 25°C in PVC plastic containers and protected from exposure to light. In distilled water alone, HPLC analysis found that captopril was stable for 30 days stored at 4°C but was much less stable at 25°C. Captopril was also much less stable in tap and mineral water at either temperature. Incorporation of edetate disodium, ascorbic acid, sodium ascorbate, or sodium metabisulfite all reduced captopril stability in aqueous solutions compared to the stability in distilled water alone.

### Topical

Huang et al.[817] evaluated the stability of captopril incorporated into 5% topical gel formulations for transdermal delivery.

Several formulations were used with 5% carboxymethylcellulose and hydroxymethylcellulose gels having shelf lives of 29 and 27 days, respectively, at 30°C and 70% relative humidity. These were substantially better than absorption cream base (seven days) or topical solution (5.6 days). The addition of 1% EDTA to the carboxymethylcellulose gel formulation extended the shelf life to 320 days; if 1% acetic acid was also incorporated, the shelf life became 418 days.

## Compatibility with Other Drugs

Huang et al.[817] evaluated the stability of captopril incorporated into 5% topical gel formulations with the addition of 0.2% clobetasol and 0.05% diphenhydramine. No differences in oxidative rate constants were found with either agent, indicating they are compatible with captopril. ∎

# ∎ Carbamazepine

## Properties

Carbamazepine is a white to off-white crystalline powder.[2,3] It is reported to be bitter, with an unpleasant aftertaste.[586]

### Solubility

Carbamazepine is insoluble in water but soluble in ethanol.[2,3,7]

## General Stability Considerations

Carbamazepine tablets should be stored in tight containers (glass is recommended), preferably at controlled room temperature; they should be protected from temperatures above 30°C. The tablet bioavailability may be reduced one-third by moisture due to formation of a dihydrate that causes tablet hardening. Consequently, containers should be kept tightly closed and stored away from areas subjected to high humidity (i.e., bathrooms). Use of silica gel sachets also has been suggested.[2–4,299,300]

Carbamazepine suspension should be stored in tight, light-resistant containers at temperatures not exceeding 30°C. The suspension also should be protected from freezing.[2,4]

### Sorption

The loss of carbamazepine from the oral suspension to nasogastric feeding tubes was evaluated by Clark-Schmidt et al.[59] Samples of 10 mL of carbamazepine suspension (Tegretol, Ciba-Geigy) 100 mg/5 mL were drawn into syringes and administered undiluted or diluted with an equal volume of flush solution (sterile water, 0.9% sodium chloride, 5% dextrose) through adult (12 French, Seamless Division of Professional Medical Services) or pediatric (5 French, American Pharmaseal Co.) nasogastric tubes. After the suspension was administered, the nasogastric tubes were flushed twice with 50 mL of the flush solution. Effluent samples were analyzed for carbamazepine by HPLC. Carbamazepine losses were generally greater in the samples administered undiluted, the worst providing only 76% of the intended dose. Losses in the diluted sample were usually less than 10%, with about 90 to 98% of the dose being delivered. Tube size made little difference. The undiluted suspension appears to stick to the tube walls to a greater extent than does the diluted suspension. The authors recommended diluting the suspension with an equal volume of diluent before administering, followed by adequate flushing.

## Stability Reports of Compounded Preparations
### Oral

**STUDY 1 (REPACKAGED):** Lowe et al.[60] evaluated the stability of commercial carbamazepine suspension (Tegretol, Ciba-Geigy) 20 mg/mL repackaged in several types of single-dose containers in 2- and 8-mL portions. Containers evaluated included glass oral syringes (2 mL in 5-mL amber Solopak syringe only), amber glass vials (Solopak), amber polypropylene oral syringes (Baxa), and polypropylene vials (Solopak). The sample containers were capped, stored at 24°C, and exposed to fluorescent light for 12 weeks. Carbamazepine content was assessed using a stability-indicating HPLC assay. No changes in odor, color, suspension consistency, or flow were observed. However, significant loss of drug content (about 20 to 30%) occurred in all containers within 12 weeks. Although the results were somewhat erratic, the authors concluded that the repackaged suspension was stable for eight weeks at room temperature.

**STUDY 2 (EXTEMPORANEOUS):** Burckart et al.[158] reported on the stability of an extemporaneously prepared carbamazepine

**TABLE 19.** Custom Vehicles for Carbamazepine Suspensions[158]

| Component | | Vehicle 1 | Vehicle 2 |
|---|---|---|---|
| Sucrose | | 125 g | 95 g |
| Sorbitol 70% | | 50 mL | 40 mL |
| Glycerin | | 12.5 mL | 8.5 mL |
| Saccharin sodium | | 250 mg | 170 mg |
| Methylparaben | | 500 mg | 340 mg |
| Methylcellulose 400 | | 6.9 g | 4.7 g |
| Methylcellulose 4000 | | 3.1 g | 2.1 g |
| FD&C yellow | | 750 mg | 510 mg |
| Lemon-lime flavor | | 1.5 mL | 1 mL |
| Purified water | qs | 500 mL | 500 mL |

suspension. Carbamazepine 200-mg tablets (Tegretol, Ciba-Geigy) were crushed to fine powder and incorporated into four suspending vehicles, sorbitol 70%, simple syrup, and two custom vehicles, at a concentration of 200 mg/5 mL (Table 19). The suspensions were packaged in amber bottles and stored at 4, 25, and 37°C. In addition, the suspensions were packaged in unit-dose syringes and stored at 4°C only. Carbamazepine was assessed using a stability-indicating enzyme-multiplied immunoassay (EMIT, Syva) over 90 days.

The refrigerated sorbitol suspension froze but resuspended when thawed and shaken vigorously. The suspension in simple syrup separated over 90 days, but it resuspended when shaken vigorously. Particle size in these two suspensions increased during storage from two- to fourfold. However, carbamazepine concentrations remained above 90% of initial concentration in both suspensions throughout the study. In custom vehicle 1, the suspension became viscid and difficult to pour; foam formed when the suspension was shaken. Furthermore, the carbamazepine concentrations were somewhat erratic. In custom vehicle 2, the suspension was homogeneous, easy to pour, and produced less foam upon shaking. Also, carbamazepine concentrations were above 90% of the initial concentration in all samples throughout the study. The authors concluded that either simple syrup or custom vehicle 2 was an acceptable vehicle for carbamazepine suspensions.

### Enteral

**STUDY 1:** Bass et al.[214] found a reduction in the bioavailability of carbamazepine suspension when given with an enteral

product (Osmolite, Ross). A 10 French 104-cm nasogastric feeding tube (Entriflex, Biosearch) was used to administer carbamazepine suspension 500 mg/25 mL (Tegretol, Ciba-Geigy) both in a fasting state and with enteral feeding. After the carbamazepine dosing, the tube was flushed with 60 mL of water. The relative bioavailability of carbamazepine given with the enteral feeding was 90.1% of that given in a fasting state.

**STUDY 2:** Williams[1290] noted decreased absorption from carbamazepine suspension delivered through enteral feeding tubes and cited the need to dilute carbamazepine suspension with sterile water or sodium chloride 0.9% to reduce adherence to feeding tubes. In addition, it was noted that close monitoring of carbamazepine serum concentrations is warranted.

### Rectal

Ammar and Marzouk[1059] evaluated the stability of carbamazepine fast-release polyethylene glycol (PEG) rectal suppositories. The formulas tested varied by the molecular weight of the PEG. All of the formulas exhibited acceptable hardness, solubility time, and content uniformity. Accelerated stability testing at 30 and 40°C demonstrated that about 2 to 5% of the drug was lost during the test period.

### Injection

Injections, like other sterile drugs, should be prepared in a suitable clean air environment using appropriate aseptic procedures. When prepared from nonsterile components, an appropriate and effective sterilization method must be employed.

Jain et al.[411] reported the stability of an extemporaneously prepared carbamazepine 3.6-mg/mL injection formulated in 50% (wt/vol) sodium salicylate solution. The injection was sterilized by filtration through a 0.2-μm filter; autoclaving resulted in loss of about 16% of the carbamazepine. Samples of the injection were stored at 8 and 23°C as well as at elevated temperatures. About 2 to 3% carbamazepine loss occurred after 49 days stored at 8 or 23°C. At elevated temperatures of 50 and 60°C, 32 to 36% loss occurred in seven days, accompanied by the formation of a deep yellow color. The safety of using this formulation was not reported.

## Compatibility with Other Drugs

Mixing chlorpromazine hydrochloride oral solution with carbamazepine oral solution is reported to result in the formation of an orange rubbery precipitate.[2] ■

# Carbenicillin Disodium
# Carbenicillin Indanyl Sodium

## Properties

Carbenicillin disodium is a white to almost white hygroscopic powder. Each 1.1 g of the disodium salt is equivalent to 1 g of carbenicillin. Each gram provides 4.7 to 6.5 mEq of sodium.[3,6] Carbenicillin indanyl sodium also is a white to nearly white powder and has a bitter taste.[2]

## Solubility

Carbenicillin disodium has an aqueous solubility of 1 g in 1.2 mL; its solubility in ethanol is 40 mg/mL.[3,6] The indanyl sodium form is also soluble in water.[6]

## pH

A 1% solution of carbenicillin disodium in water has a pH of 6.5 to 8, while a 5% solution has a pH of 5.5 to 7.5.[3]

## General Stability Considerations

Carbenicillin products should be stored in tight containers at controlled room temperature.[3,4]

In aqueous solution, carbenicillin is subject to hydrolysis of the β-lactam ring. Maximum stability occurs at pH 6.5; the degradation rate increases rapidly as the pH increases or decreases from 6.5.[6,301] At pH 2, the half-life of carbenicillin disodium at 21°C is 140 minutes.[302]

## Stability Reports of Compounded Preparations
### Ophthalmic

Ophthalmic preparations, like other sterile drugs, should be prepared in a suitable clean air environment using appropriate aseptic procedures. When prepared from nonsterile components, an appropriate and effective sterilization method must be employed.

El-Shattawy[218] evaluated the stability of 0.3% carbenicillin disodium in 36 formulations of ophthalmic ointment bases. Each of 18 different formulations was duplicated with 0.1% parabens (methylparaben–propylparaben, 2:1) as preservatives and with 0.1% tocopheryl acetate as an antioxidant. The ointments were stored at 5°C for eight months. Carbenicillin activity was assessed microbiologically. Carbenicillin was unstable in all formulations containing Aerosil, cetyl alcohol, cetosteryl alcohol, Span 40, Span 80, white wax, and wool alcohols. The activity was well retained in formulations composed primarily of 10 to 20% liquid paraffin, 75 to 90% yellow soft paraffin, and 4 to 20% wool fat. Parabens and tocopheryl acetate had no effect on the antibiotic activity.

# Carbidopa

## Properties

Carbidopa is a white to creamy or yellowish white odorless or practically odorless crystalline powder.[3,4] Carbidopa is available as the monohydrate, but the drug potency is expressed in terms of the anhydrous form.[2]

## Solubility

Carbidopa is slightly soluble in water and practically insoluble in ethanol and acetone. It is freely soluble in 3 N hydrochloric acid.[4]

## General Stability Considerations

Carbidopa should be packaged in well-closed light-resistant containers and stored at controlled room temperature.[4] Pappert et al.[1159] reported that carbidopa is very unstable in water with or without ascorbate present. Losses of about 30 and 60% occurred in eight and 24 hours, respectively, whether stored at 25°C, 12°C, or frozen. Exposure to light and the presence or absence of ascorbate did not change the rapid loss of carbidopa in aqueous solution.

## Stability Reports of Compounded Preparations
### Oral

**STUDY 1:** Nahata et al.[829] evaluated the stability of oral suspension formulations of levodopa 5 mg/mL and carbidopa 1.25 mg/mL. Tablets of levodopa and carbidopa were ground to fine powder using a mortar and pestle. A small amount of an equal parts mixture of Ora-Plus and Ora-Sweet was incorporated into the powder to make a paste. Additional vehicle mixture was incorporated geometrically to reach the final volume. The suspensions were packaged in amber plastic prescription bottles and stored at room temperature of 25°C and refrigerated at 4°C.

Stability-indicating HPLC analysis found that the levodopa and carbidopa concentrations were 96 and 92%, respectively, after 28 days of storage at room temperature. After 42 days of storage refrigerated, levodopa and carbidopa concentrations were 94 and 93%, respectively. The physical appearance of the suspensions did not change during these time periods, but the suspensions became a darker yellow during longer storage as more extensive decomposition occurred. Addition of ascorbic acid 2 mg/mL to the formulation resulted in more rapid loss of both drugs.

**STUDY 2:** Lopez Lozano and Moreno Cano[1481] evaluated the stability of an oral liquid formulation of levodopa/carbidopa prepared in an ascorbic acid-containing aqueous solution. The authors intended this to be preparable by patients at home when needed. The patient would crush the daily number of levodopa/carbidopa tablets that they take and mix them into 500 mL of cold water containing a crushed 1-g tablet of ascorbic acid. If the patient was taking five tablets daily, as an example, that would result in L-dopa/carbidopa/ascorbic acid concentrations of 1/0.25/2 mg/mL, respectively.

Any remaining undissolved material was considered unimportant. The test solution was packaged in a glass bottle and stored refrigerated at 0 to 4°C wrapped in aluminum-foil for protection from exposure to light. The patient would shake the liquid and withdraw the correct volume for the dose (100 mL in the example) and drink it. Notably, the dose could be changed while keeping the volume of liquid to drink the same.

HPLC analysis found about 3% levodopa loss occurred in 24 hours in the example solution stored under refrigeration and protected from light. Unfortunately, the carbidopa concentration was not tested although it may be similar to the levodopa stability. ■

# ■ Carboplatin

## Properties
Carboplatin is a white crystalline powder.[1,7]

### Solubility
Carboplatin is soluble in water to approximately 14 mg/mL and is virtually insoluble in ethanol, acetone, and dimethylacetamide.[1,3,7]

### pH
A 10-mg/mL aqueous solution of carboplatin has a pH of 5 to 7.[4,7]

### Osmolality
Carboplatin 10 mg/mL in sterile water for injection is hypotonic, having an osmolality of 94 mOsm/kg.[8]

## General Stability Considerations
Carboplatin for injection should be stored at controlled room temperature and protected from exposure to light.[7] Carboplatin is stable in the pH range of 4 to 6.5, but it is unstable at higher pH values.[9]

The manufacturer of the commercial drug states that reconstituted carboplatin is stable for eight hours at room temperature.[7] However, other reports indicate the drug is stable for longer periods. Trissel et al.[893] reported that carboplatin 15 mg/mL in water and 0.5 to 2 mg/mL in dextrose 5% injection exhibited no decomposition in 24 hours at room temperature. Williams[894] reported that carboplatin 1 mg/mL in water exhibited no decomposition in five days under refrigeration and only 3% loss in 24 hours at body temperature. A 7-mg/mL solution was also stable, exhibiting 4% loss in seven days at room temperature.

Contact of carboplatin with aluminum equipment displaces the platinum molecule, resulting in formation of a black precipitation and loss of drug concentration. Consequently, carboplatin should not be prepared using equipment that contains aluminum parts.[2]

## Stability Reports of Compounded Preparations
### Injection
Injections, like other sterile drugs, should be prepared in a suitable clean air environment using appropriate aseptic procedures. When prepared from nonsterile components, an appropriate and effective sterilization method must be employed.

Sewell et al.[895] reported that reconstituted carboplatin injection 10 mg/mL repackaged into plastic syringes was stable, exhibiting no loss of drug in five days refrigerated and about 3% loss in 24 hours at 37°C.

### Rectal
Chen et al.[806] reported the stability of carboplatin 1.13 mg formulated in 100-mg lipid suppositories. No color changes occurred at room temperature, but gray discoloration formed at an elevated temperature of 40°C. HPLC analysis found that less than 4% carboplatin loss occurred in three months at 25°C and 75% relative humidity. ■

# Carboprost Tromethamine

## Properties

Carboprost tromethamine occurs as a white to off-white crystalline powder.[4,7] Carboprost and tromethamine are in a ratio of 1:1.[3]

### Solubility

Carboprost tromethamine is soluble in water, having an aqueous solubility of 75 mg/mL at room temperature.[4,7]

### pH

Carboprost tromethamine injection has a pH between 7 and 8.[4]

## General Stability Considerations

Carboprost tromethamine powder should be packaged in well-closed containers and stored in a freezer. Great care should be taken to avoid inhaling particles of carboprost tromethamine as well as avoiding skin contact.[4]

Carboprost tromethamine injection is packaged in single-dose or multiple-dose vials, preferably made of Type I glass. The injection is stored under refrigeration.[4,7]

## Stability Reports of Compounded Preparations
### Injection

Injections, like other sterile drugs, should be prepared in a suitable clean air environment using appropriate aseptic procedures. When prepared from nonsterile components, an appropriate and effective sterilization method must be employed.

Chu et al.[1235] evaluated the preparation and stability of an extemporaneously prepared injection of carboprost tromethamine prepared as a dilution of the commercial injection. One 250 mcg in 1 mL portion of carboprost tromethamine injection (Hemabate, Pharmacia) was mixed with 1 mL of water or sodium chloride 0.9% and was transferred to amber glass vials. The vials were stored at room temperature and refrigerated at 4°C for six days.

Using a stability-indicating LC-MS analytical method, carboprost tromethamine was stable at both room temperature and at 4°C for six days with little or no loss of drug occurring. ■

# Carmustine

## Properties

Carmustine is a light yellow powder that melts in the range of 27°C[9] up to 32°C.[1,3] The injectable formulation consists of a white or yellow vacuum-dried flaky powder.[2,8,9] Upon reconstitution, a colorless to pale yellow solution is formed.[2,7]

### Solubility

Carmustine is soluble in water, 4 mg/mL, and in 50 to 95% ethanol, 150 mg/mL.[1,9] The drug also is highly soluble in lipids.[2]

### pH

The reconstituted and diluted injection has a pH of 5.6 to 6.[7]

## General Stability Considerations

Vials of carmustine should be stored under refrigeration at 2 to 8°C and protected from light.[7,9] However, according to the manufacturer, the drug is stable in intact vials for seven days at room temperatures that do not exceed 25°C.[280,303,304] Temperatures above 27°C result in drug liquefaction to an oily film. The manufacturer recommends discarding vials containing melted drug.[7] Chan and Zackheim[305] reported that storage at 37°C resulted in about 10% loss in seven days.

The reconstituted solution is stable for about eight hours at room temperature protected from light.[2,7] Losses at room temperature of 6, 8, and 20%, in three, six, and 21 hours, respectively, were reported.[306,307] Refrigeration improves the stability of the reconstituted solution; the solution is stable for 24 hours protected from light,[2,7] with losses of about 4%.[306]

Maximum stability of carmustine has been reported variously to occur at pH 3.5 to 5,[9] 5.2 to 5.5,[308] and 3.3 to 4.8.[309] A pH value greater than 6 results in a greatly increased degradation rate; losses of 10% occur in two hours at pH 6.5 but not until 5.5 hours at optimum pH.[308,309] Phosphate buffers are reported to greatly accelerate decomposition.[9]

The use of glass containers is recommended for carmustine packaging.[2,7] Loss of carmustine has occurred due to sorption to various plastic materials, including polyvinyl chloride, ethylene vinyl acetate, and polyurethane. Polyolefins such as polyethylene have not been found to sorb the drug.[309–311]

## Stability Reports of Compounded Preparations
### Topical

Carmustine has been used topically in several forms and concentrations. It has been dissolved in 95% ethanol as a concentrate for dilution with water up to a 10-mg/60-mL concentration (10% ethanol) for total body application. It also has been used as a 20-mg/mL solution in ethanol for topical application.[17]

**STUDY 1:** Chan and Zackheim[18] reported that carmustine 0.5 to 0.6 mg/mL diluted in absolute alcohol or 95% ethanol and evaluated by ultraviolet spectroscopy was stable for up to three months stored under refrigeration at 0 to 5°C. However, aqueous solutions were much less stable; carmustine losses of 50% occurred in 19 days in distilled water. The authors recommended preparing aqueous solutions immediately before use.

**STUDY 2:** Levin et al.[17] evaluated the stability of carmustine for topical application using HPLC and mass spectroscopy. Commercial vials of carmustine injection (Bristol-Myers), both intact and diluted to 2 mg/mL with 95% ethanol, were subjected to various storage temperatures. The intact vials were stable for at least eight days at 4 and 23°C with no decomposition. However, exposure to the elevated temperature of 37°C resulted in 10% loss in seven days. The 2-mg/mL solution in 95% ethanol exhibited no decomposition within 24 hours stored under refrigeration, at room temperature, or even at 37°C. When a carmustine 2-mg/mL solution was prepared in 91% isopropanol, less than 10% decomposition occurred in three months under refrigeration. ■

# Carnitine

## Properties

Carnitine occurs as hygroscopic white crystals or crystalline powder.[1,3,4] Also see the Levocarnitine monograph.

### Solubility

Carnitine is freely soluble in water and hot ethanol but practically insoluble in acetone, ether, and benzene.[1,4]

### pH

The pH range of a 5% aqueous solution has been cited as pH 5.5 to 9.5.[4,1281] The European Pharmacopeia indicates a 5% aqueous solution has a pH of 6.5 to 8.5.[3]

Levocarnitine injection has a pH range of 6 to 6.5.[4,7] Commercial levocarnitine oral solution has a pH between 4 and 6.[4]

## General Stability Considerations

Levocarnitine is commercially available. Levocarnitine powder, levocarnitine tablets, and levocarnitine oral solution should be packaged in tight containers, and stored at controlled room temperature. Levocarnitine injection in single-dose containers should be stored below 25°C and protected from exposure to light leaving the vials in their cartons until time of use.[4,7] The injection should also be protected from freezing.[4]

## Stability Reports of Compounded Preparations
### Oral

Allen[1281] reported on a compounded formulation of carnitine 400-mg/mL with ubidecarenone (coenzyme $Q_{10}$) 10-mg/mL oral solution. The solution had the following formula:

| | | |
|---|---|---|
| Carnitine | | 40 g |
| Ubidecarenone (Coenzyme $Q_{10}$) | | 1 g |
| Polyethylene glycol 400 | | 1 g |
| Cremophor RH40 | | 4 g |
| Methylparaben | | 200 mg |
| Purified water | qs | 100 mL |

The recommended method of preparation was to place the ubidecarenone (coenzyme $Q_{10}$), polyethylene glycol 400, Cremophor RH40, methylparaben, and 50 mL of purified water powder in an appropriate container, heat the mixture to about 60°C, and stir thoroughly. The mixture was then cooled, the carnitine was added along with sufficient purified water to bring the volume to 100 mL, and the preparation was thoroughly mixed. The solution was to be packaged in tight, light-resistant containers. The author recommended a beyond-use date of six months at room temperature because this formula is a commercial medication in some countries with an expiration date of two years or more. ■

# Carprofen

## Properties

Carprofen is a white crystalline nonsteroidal anti-inflammatory for veterinary use.[3,7]

### Solubility

Carprofen is practically insoluble in water but freely soluble in ethanol.[7]

### pH

The manufacturer notes that the pH of the injection for subcutaneous administration is adjusted with sodium hydroxide and/or hydrochloric acid, but the actual pH is not provided.[7]

## General Stability Considerations

Carprofen oral tablets are packaged in bottles and blister packs and should be stored at controlled room temperature. Carprofen chewable tablets are packaged in bottles and should also be stored at controlled room temperature. Carprofen injection for subcutaneous use only is packaged in amber glass multiple-dose vials and should be stored under refrigeration.[7]

## Stability Reports of Compounded Preparations
### Oral

Hawkins et al.[1052] evaluated the stability of carprofen 1.25, 2.5, and 5 mg/mL in extemporaneously compounded oral liquid formulations for use in small exotic pets. Carprofen (Rimadyl, Pfizer) oral tablets were compounded into oral liquid formulations using two diluents: (1) a 1:1 mixture of methylcellulose 1% gel and simple syrup and (2) a 1:1 mixture of Ora-Plus and Ora-Sweet. The oral liquids were packaged in amber glass prescription bottles and stored at room temperature of 20 to 24°C and refrigerated at 3 to 5°C for 28 days.

Carprofen at all three concentrations in the 1:1 Ora-Plus and Ora-Sweet suspending vehicle yielded suspensions acceptable in appearance. During storage, the suspensions were easily redispersed with shaking to apparent uniformity. HPLC analysis yielded somewhat erratic and variable results. From the data, it can be estimated that the 2.5- and 5-mg/mL concentrations lost about 10% in 21 days at both storage temperatures. The 1.25-mg/mL concentration retained adequate drug concentration for seven days under refrigeration and 14 days at room temperature. Losses beyond those times ranged from 13 to 30%.

Carprofen in the 1:1 mixture of methylcellulose 1% gel and simple syrup mixture yielded an unacceptable mixture with formation of solid crusty material and poor drug content and uniformity. This vehicle appears to be unacceptable for use. ■

# ■ Carteolol Hydrochloride

## Properties

Carteolol hydrochloride exists as white crystals or as a crystalline powder.[3]

### Solubility

Carteolol hydrochloride is soluble in water, slightly soluble in ethanol, and sparingly soluble in methanol.[3]

### pH

Carteolol hydrochloride 1% aqueous solution has a pH of 5 to 6.[3,4]

The official pH range of the ophthalmic solution is 6 to 8.[4] One manufacturer of carteolol hydrochloride ophthalmic solution (Occupress, Novartis) indicates that the product is in the pH range of 6.2 to 7.2, while Falcon Pharmaceuticals cites the official pH range of 6 to 8.[7] One carteolol hydrochloride 20-mg/mL ophthalmic solution (Elebloc, Cusi) was measured to have a pH of about 6.7.[880]

### pK$_a$

Carteolol hydrochloride has a pK$_a$ of 9.7.[1080]

## General Stability Considerations

Carteolol hydrochloride powder should be packaged in well-closed containers and stored at controlled room temperature. Carteolol hydrochloride ophthalmic solution and oral tablets should be packaged in tight containers and stored at controlled room temperature.[4] The manufacturers indicate that the ophthalmic solution should be stored protected from exposure to light.[7]

## Stability Reports of Compounded Preparations
### Ophthalmic

Ophthalmic preparations, like other sterile drugs, should be prepared in a suitable clean air environment using appropriate aseptic procedures. When prepared from nonsterile components, an appropriate and effective sterilization method must be employed.

Garcia-Valldecabres et al.[880] reported the change in pH of several commercial ophthalmic solutions, including carteolol hydrochloride 20 mg/mL (Elebloc, Cusi), over 30 days after opening. Slight variation in solution pH was found, but no consistent trend and no substantial change occurred. ■

# ◾ **Carvedilol**

## Properties

Carvedilol is a white to off-white crystalline powder.[1,3,7]

### Solubility

Carvedilol is practically insoluble in water and dilute acids, but it is sparingly soluble in ethanol and freely soluble in dimethyl sulfoxide (DMSO).[1,3,7]

## General Stability Considerations

Carvedilol tablets should be packaged in tight, light-resistant containers and stored at controlled room temperature protected from moisture.[7]

Lanzanova et al.[1428] reported stress-testing carvedilol. The study found the drug was relatively stable exposed to extreme acid (1 N hydrochloric acid) and neutral pH in water, but exhibits an increased rate of decomposition at alkaline pH (1 N sodium hydroxide). In addition, carvedilol was subject to oxidation, exhibiting rapid degradation when exposed to 7.5% hydrogen peroxide. The drug was also stable exposed to light for seven days in a photostability chamber and exposed to daylight for 100 days at room temperature. These results were also confirmed by Rizwan et al.[1445] Similarly, increased decomposition was reported by Buontempo et al.[1675] in an alkaline aqueous solution at pH 8.2.

## Stability Reports of Compounded Preparations
### Oral

**STUDY 1:** Yamreudeewong et al.[978] evaluated the stability of carvedilol oral liquid compounded from oral tablets. Three carvedilol 25-mg tablets (Coreg, GlaxoSmithKline) were triturated to fine powder using a mortar and pestle, and 20 mL of deionized water was added and mixed to form a paste. Sorbitol 70% 40 mL was added to the paste and additional deionized water was added to bring the mixture to a final volume of 120 mL. The nominal carvedilol concentration was 0.625 mg/mL. The oral liquid was packaged in amber polyethylene terephthalate (PET) prescription bottles and stored at ambient room temperature between 22 and 26°C and under refrigeration at 4 to 8°C.

HPLC analysis of the drug concentrations gave variable results, indicating the analysis and/or sampling was apparently not well controlled. Drug concentrations at room temperature approached 97 to 98% of the initial concentration throughout the study. However, refrigerated samples were reported to be less stable. The anomalous results indicated losses of 10 and 8% in four and eight weeks, respectively. Although the authors reported carvedilol to be stable in the oral liquid, the results of this study do not definitively support that conclusion.

**STUDY 2:** Buontempo et al.[1675] evaluated the stability of three oral liquid formulations of carvedilol 1 mg/mL, an easier concentration for dosing, as noted by Allen.[1674] Two oral solutions, one alkaline at pH 8.2 and one acidic adjusted to pH 4.2 with citric acid, were prepared. In addition, an oral suspension was prepared. The oral solutions were compounded by solubilizing the drug with 0.5% (wt/vol) of polyvinylpyrrolidone (PVP-K30) in 40 mL of propylene glycol at 50°C. After dissolution of the PVP, 0.1% (wt/vol) of carvedilol was added. The solution was cooled and 10 mL of glycerin with 0.1% raspberry essence was added. Then sorbitol 40 mL was added and homogenized. Then 0.1% (wt/vol) of sodium saccharin was added and the volume was brought to 100 mL, yielding carvedilol 1 mg/mL in an alkaline solution having a pH of 8.2. To create a similar oral solution with an acidic pH, citric acid 0.1% (wt/vol) was added, resulting in a pH of 4.2. Finally, an oral suspension was prepared using a suspending vehicle similar to Ora-Plus. The suspending vehicle included in each 100 mL xanthan gum 200 mg, carboxymethylcellulose sodium 25 mg, glycerin 2 mL, sorbitol 70% solution 5 mL, sodium saccharin 200 mg, citric acid 100 mg, dibasic sodium phosphate 60 mg, methylparaben 100 mg, potassium sorbate 150 mg, simethicone 100 mg, and purified water to bring to volume.[1674] The three formulations were packaged in amber glass vials and stored refrigerated at 4°C, at room temperature of 25°C, and at elevated temperature of 40°C for 56 days.

Stability-indicating HPLC analysis found no carvedilol loss in 56 days in the acidic solution and the oral suspension when stored at 4 and 25°C. At 40°C, carvedilol in the acidic solution exhibited no loss, but in the oral suspension about 15% loss occurred in 56 days. In the alkaline solution, carvedilol losses ranged from 8 to 12% in 56 days at all three storage temperatures, with refrigerated storage exhibiting the greatest loss and room temperature storage exhibiting the least loss at 8%. The authors concluded that carvedilol in any of the formulations was stable for 56 days if stored at room temperature. ◾

# Cefaclor

## Properties
Cefaclor is a white to off-white or slightly yellow crystalline powder.[1,3,4]

### Solubility
Cefaclor is slightly soluble in water and practically insoluble in methanol.[1,3,4]

### pH
A 2.5% aqueous suspension of cefaclor has a pH between 3 and 4.5.[3,4] Cefaclor oral suspension has a pH in the range of 2.5 to 5.[4]

## General Stability Considerations
Cefaclor powder, oral capsules, and oral suspension should be packaged in tight containers and stored at controlled room temperature.[3,4] Cefaclor extended-release tablets should be packaged in tight, light-resistant containers and stored at controlled room temperature.[3]

## Stability Reports of Compounded Preparations
### Enteral
Ortega de la Cruz et al.[1101] reported the physical compatibility of an unspecified amount of cefaclor oral suspension (Ceclor, Lilly) with 200 mL of Precitene (Novartis) enteral nutrition diet for a 24-hour observation period. No particle growth or phase separation was observed. ■

# Cefadroxil

## Properties
Cefadroxil is a white to off-white or yellowish-white crystalline powder provided as the monohydrate.[2,3,7]

### Solubility
Cefadroxil is soluble in water and slightly soluble in ethanol.[2,3]

### pH
A 5% cefadroxil suspension in water has a pH of 4 to 6.[3] When reconstituted, the commercial suspensions have pH values of 4.5 to 6.[4]

### pKa
Cefadroxil has $pK_a$ values of 2.64, 7.30, and 9.69 at 35°C.[6]

## General Stability Considerations
Cefadroxil products should be stored in tight containers, preferably at controlled room temperature, and protected from temperatures exceeding 40°C. Hydrolytic decomposition may occur in the solid state from moisture in the air.[4,6,7] The estimated time to 10% loss in the solid state was 6.6 years at 20°C and 5.4 years at 22°C.[1007] The reconstituted suspension has an expiration of 14 days when stored at 2 to 8°C.[4,6,7]

Cefadroxil degradation in aqueous solutions is the result of the cleaving of the β-lactam moiety by any of several methods, including intramolecular aminolysis, water-catalyzed hydrolysis, and hydroxide-ion nucleophilic attack. Below pH 5, the cefadroxil degradation rate is pH independent.[6] In pH 3 and 5 buffers, less than 2% loss occurred in 24 hours at 3°C and at room temperature.[1007] However, as the pH increases above 6, the degradation rate increases rapidly.[6] At pH 7 and 8, 23 to 27% decomposition was found in 24 hours at room temperature.[1007] General acid and base catalysis occurs in citrate and phosphate buffer systems.[6]

## Stability Reports of Compounded Preparations
### Oral
STUDY 1: Nahata and Jackson[107] evaluated the stability of cefadroxil (Duricef, Mead Johnson) oral suspension. The bottles of powder were reconstituted with 67 mL of deionized water to yield 100 mg/mL. The reconstituted suspension was stored at 4 and 22°C for six weeks. Samples were taken after the bottles had been shaken mechanically for 1.5 hours to ensure an even dispersion. The cefadroxil content was assessed using a stability-indicating HPLC assay. Little or no loss of cefadroxil in suspension (<4%) occurred in six weeks when stored under refrigeration. At room temperature, there was little or no loss of cefadroxil in suspension (<5%) in 10 days.

STUDY 2: Vidal et al.[1343] evaluated the pharmaceutical parameters for acceptability of four (unnamed) brands of cefadroxil oral suspension. Samples were stored at 25°C and 60% relative humidity for 12 months and accelerated conditions of 40°C 75% relative humidity for six months. Samples were also evaluated after reconstitution when stored at room temperature and under refrigeration.

Requirements for deliverable volume were fulfilled and no important changes in pH were found in any of the samples. None of the samples exhibited visual signs of caking, and all samples were easily redispersed throughout the study. All suspensions were found by HPLC analysis to be chemically stable and meet USP content uniformity requirements throughout the storage periods at all conditions.

However, one brand exhibited significant decreases in dissolution percentages over time, becoming as little as 59% in one case. The other three brands were found to have dissolution profiles that were unchanged under the test conditions. A decreased dissolution percentage could adversely affect in vivo bioavailability. No reason for the difference in dissolution of this brand was identified. ∎

# Cefazolin Sodium

## Properties

Cefazolin sodium is a white to off-white or yellowish-white hygroscopic crystalline powder with a bitter salty taste. The dry powder may darken during storage, but this change does not reflect degradation. Cefazolin sodium 1.05 g is equivalent to 1 g of cefazolin; each gram of cefazolin sodium provides approximately 2 mEq (48.3 mg) of sodium.[1–3,7]

### Solubility

Cefazolin sodium is freely soluble in water and common aqueous solutions and slightly soluble in ethanol.[1–3]

### pH

A 10% cefazolin sodium solution in water has a pH of 4 to 6.[3,4] The reconstituted cefazolin sodium injection has a pH of 4 to 6.[4]

### Osmolality

The osmolality of reconstituted cefazolin sodium 225 mg/mL in sterile water for injection has been measured by freezing-point depression to be 636 mOsm/kg.[287] Several extemporaneously compounded cefazolin sodium ophthalmic solutions in citrate buffers resulted in osmolarities from 342 to 387 mOsm/L.[1065]

## General Stability Considerations

Cefazolin sodium vials should be stored at controlled room temperature and protected from temperatures above 40°C.[4,7] Both the intact vials and the reconstituted solution should be protected from light. The manufacturers state that the reconstituted solution is stable for 24 hours at room temperature and for 10 days under refrigeration at 2 to 8°C.[7]

Other information indicates that the drug may be stable longer. Bornstein et al.[312] reported that cefazolin sodium 250 mg/mL in water for injection exhibited an 8 to 10% loss in four days at 25°C and a 3% loss in 14 days at 5°C. Using ultraviolet spectroscopy, Borst et al.[313] found a 10% loss in 13 days when cefazolin sodium 100 and 200 mg/mL in sterile water for injection was stored in plastic syringes at 24°C. Under refrigeration at 4°C, the solutions sustained less than 10% loss in 28 days.

Cefazolin sodium aqueous solutions in original containers at concentrations between 125 and 330 mg/mL were reported to be stable when stored frozen at −20°C for 12 weeks.[2,7] Borst et al.[313] reported a similar result for reconstituted cefazolin sodium 100 and 200 mg/mL in sterile water for injection stored in plastic syringes. Less than 10% loss by ultraviolet spectroscopy occurs in three months of frozen storage at −15°C. Thawed solutions are stable for 24 hours at room temperature or for 10 days at 2 to 8°C.[2,7]

Under refrigeration, reconstituted solutions have been reported to form crystals.[314] Similarly, reconstitution of cefazolin sodium with 0.9% sodium chloride injection may result in crystal formation even at room temperature.[315]

Cefazolin sodium in aqueous solutions is relatively stable over pH 4.5 to 8.5. However, rapid hydrolysis occurs above pH 8.5, and below pH 4.5 precipitation of the insoluble free acid of cefazolin may occur.[2,316]

## Stability Reports of Compounded Preparations
### Ophthalmic

Ophthalmic preparations, like other sterile drugs, should be prepared in a suitable clean air environment using appropriate aseptic procedures. When prepared from nonsterile components, an appropriate and effective sterilization method must be employed.

**USP OFFICIAL FORMULATION (OPHTHALMIC SOLUTION FROM POWDER):**

| | | |
|---|---|---|
| Cefazolin sodium (powder) | | 35 mg |
| Thimerosal | | 0.2 mg |
| Sodium chloride injection 0.9% | qs | 10 mL |

Dissolve cefazolin powder and thimerosal powder diluting quantitatively, and stepwise if necessary, in sodium chloride 0.9% to yield a solution containing cefazolin sodium 3.5 mg/mL. Filter a 10-mL portion through a 0.2-μm filter producing a clear sterile ophthalmic solution. Package the ophthalmic solution in a tight, sterile ophthalmic solution container. The final ophthalmic solution should have a pH between pH 4.5 and 6. Store the preparation under refrigeration between 2 and 8°C. The beyond-use date is five days from the date of compounding.[4,5]

**USP OFFICIAL FORMULATION (OPHTHALMIC SOLUTION FROM INJECTION):**

| | | |
|---|---|---|
| Cefazolin sodium (injection) | | 35 mg |
| Thimerosal | | 0.2 mg |
| Sodium chloride injection 0.9% | qs | 10 mL |

Dissolve thimerosal powder diluting quantitatively, and step-wise if necessary, in sodium chloride 0.9% to yield a solution containing thimerosal 0.3 mg/mL. Add 9.8 mL of the resulting solution to a 500-mg vial of cefazolin sodium injection and mix well to yield a stock solution. Transfer 3.3 mL of this stock solution to a 50-mL volumetric flask and dilute to volume with sodium chloride 0.9% injection. Filter a 10-mL portion of this solution through a 0.2-μm filter producing a clear sterile ophthalmic solution. Package the ophthalmic solution in a tight, sterile ophthalmic solution container. The final ophthalmic solution should have a pH between pH 4.5 and 6. Store the preparation under refrigeration between 2 and 8°C. The beyond-use date is five days from the date of compounding.[4,5]

**STUDY 1:** Ahmed and Day[40] tested the stability of cefazolin sodium in several ophthalmic formulations. Cefazolin sodium injection (Kefzol, Lilly) was reconstituted with 0.9% sodium chloride injection to yield a stock solution. A 0.17-mL aliquot of the stock solution was added to 4.83 mL of each of three buffer solutions (acetate pH 4.5, acetate pH 5.7, phosphate pH 7.5) and four commercial artificial tears products (Liquifilm Tears, Allergan; Liquifilm Forte, Allergan; Isopto Tears, Alcon; Tearisol, Cooper Vision) as well as more 0.9% sodium chloride injection to yield test formulations having nominal concentrations of 3.33 mg/mL. A stability-indicating HPLC assay was used to determine the cefazolin content of the samples during seven days of storage at 4, 25, and 35°C.

In solutions with alkaline pH values, including the phosphate buffer, Tearisol, and Isopto Tears, a faint yellow color and particulates formed when solutions were stored at 35°C for three or four days. Particulates also formed in the 25°C samples of these solutions, but a color change was not seen. Refrigerated samples did not exhibit a color change or particulates. All other test solutions (acetate buffers, Liquifilm products, 0.9% sodium chloride) that had acidic pH values did not exhibit color changes or particulates at any temperature over seven days. All artificial tears preparations developed an objectionable odor when stored at 35°C; even at 25°C the odor was noticeable. Cefazolin content remained above 90% in all samples for three and seven days at 25 and 4°C, respectively. Only the 0.9% sodium chloride solution retained more than 90% concentration after seven days at 25°C. At 35°C, cefazolin losses were variable among the preparations, depending on the pH of the solutions; higher losses were noted in the more neutral to alkaline solutions compared to mildly acidic solutions (pH 4.5 to 5.9). The authors recommended storage under refrigeration for the cefazolin-containing ophthalmic

preparations. A seven-day expiration for refrigerated preparations would be appropriate based on this information. The authors also recommended an expiration period of three days if stored at room temperature.

**STUDY 2:** Charlton et al.[582] studied the stability of cefazolin 33 mg/mL as the sodium salt in an ophthalmic solution. The solution was prepared by aseptically adding 15 mL of artificial tears (Liquifilm Tears, Allergan) to a 500-mg vial of cefazolin sodium (SmithKline Beecham) for injection. The ophthalmic solution samples were stored in the original artificial tears bottles at 4 and 25°C for 28 days. The osmolality remained unchanged. The pH changed over time and was different by nearly a pH unit between the two storage temperatures, although it remained within the acceptable range. Antimicrobial activity, estimated by Kirby-Bauer microbial growth inhibition, exceeded the minimum but was somewhat better in the refrigerated samples. The authors recommended use for 28 days but only with storage under refrigeration.

**STUDY 3:** Bowe et al.[627] evaluated the stability of cefazolin 50 mg/mL as the sodium salt in a fortified ophthalmic preparation. A vial of the lyophilized injection was reconstituted with methylcellulose artificial tears. Samples were stored at 24 and 4°C in the dark for four weeks. Antibiotic potency was determined by measuring minimum inhibitory concentration against common ophthalmic pathogens. No loss of antibiotic activity was found at either storage temperature over four weeks, but changes in pH and ultraviolet absorbance in the room temperature samples after seven days led the authors to recommend refrigerated storage.

**STUDY 4:** How et al.[670] reported on the stability of two formulations of cefazolin sodium 50-mg/mL ophthalmic solutions prepared from the injection. For each of the products, three 1-g vials of cefazolin sodium were reconstituted with sterile water for injection. The unpreserved formula for single use was prepared by diluting the reconstituted drug with an equal quantity of 0.9% sodium chloride to yield a 50-mg/mL concentration in 0.45% sodium chloride. The preserved preparation for multiple use was made by combining 30 mL of the reconstituted cefazolin sodium injection with 3 mL of a sterile-filtered (0.22-μm filter) concentrate containing thimerosal 0.1% (wt/vol) and glycerol 20% (wt/vol) in water for injection. This mixture was then brought to the final volume of 60 mL with water for injection. The final preserved ophthalmic solution had a 50-mg/mL antibiotic concentration in thimerosal 0.005% and glycerol 1%. The ophthalmic solutions were packaged in 10-mL aliquots in amber glass eye-drop bottles. Stability-indicating HPLC analysis found about 6% cefazolin loss in 42 days at 4°C but 10% loss in five (preserved) to seven (unpreserved) days at 25°C. Osmolality and pH changes were minimal at 4°C, but a much greater increase occurred at 25°C during the 42 days of storage.

**STUDY 5:** Kommanaboyina et al.[731] reported the stability of cefazolin sodium 3.5 mg/mL ophthalmic solution prepared from Ancef injection with thimerosal in sodium chloride 0.9%. The solution was packaged in sterile ophthalmic dropper bottles and was stored at 7, 17, 25, and 40°C. Only the elevated 40°C temperature resulted in substantial loss of about 25% in 12 days. Little loss occurred at the other temperatures in 300 hours, especially under refrigeration. The authors recommended very conservative beyond-use dates of two weeks under refrigeration and six days at room temperature.

**STUDY 6:** Arici et al.[796] reported the stability of cefazolin sodium 33.3-mg/mL stock ophthalmic solutions prepared from injection in sodium chloride 0.9% and in artificial tears (Liquifilm Tears) and packaged in ophthalmic squeeze bottles at room temperature of 24°C and refrigerated at 4°C. No loss in microbiological activity against *Staphylococcus aureus* and *Pseudomonas aeruginosa* was found over 28 days of storage at both temperatures in both vehicles. Small pH increases of 0.5 pH unit or less occurred.

**STUDY 7:** Kodym et al.[1065] evaluated the effects of various antimicrobial preservatives on the stability of cefazolin sodium in ophthalmic formulas. Chlorhexidine 0.01%, benzalkonium chloride 0.005%, and thimerosal at concentrations over 0.003% resulted in opalescence and precipitation leading to sedimentation that renders these preservatives unacceptable. Thimerosol at concentrations lower than 0.003% was ineffective as a preservative. Cefazolin 1% ophthalmic solutions in citrate buffer (pH 6.15–6.2) with or without polyvinyl alcohol for viscosity and with or without phenylmercuric borate 0.001% and β-phenylethyl alcohol 0.4% were delivered through a 0.22-μm filter and stored for 30 days at 4 and 20°C. The refrigerated ophthalmic solution remained clear and colorless while the solution stored at room temperature developed a slight yellow color. Antimicrobial activity remained near 100% at both storage temperatures.

**STUDY 8:** Rojanarata et al.[1619] evaluated the stability of two fortified cefazolin sodium 50-mg/mL ophthalmic solutions compounded using commercial artificial tears. Cefazolin sodium injection 1 g was reconstituted with 4.5 mL of sterile water for injection. One mL of the reconstituted solution was mixed with 3 mL of Tears Naturale (Alcon) or Natear (Silom Medical). The Tears Naturale contained 0.001% polyquaternium-1, while the Natear contained 0.028% sodium perborate as preservatives. The fortified cefazolin sodium ophthalmic solutions were stored at room temperature of 28°C protected from exposure to light and refrigerated at 4°C for 28 days. The solutions were clear and colorless and had an osmolarity of 420 mOsm/L, which can be tolerated by the human eye.

The refrigerated samples exhibited no changes in color and clarity throughout the study period. Stability-indicating HPLC analysis found little or no loss of cefazolin sodium over 28 days of refrigerated storage. However, the samples stored at 28°C developed a yellowish discoloration and an unpleasant odor after three days of storage. Drug decomposition exceeded 10% within three days as well. The authors concluded that the fortified cefazolin sodium ophthalmic solutions prepared in these two commercial artificial tears could be stored under refrigeration for up to 28 days.

*Topical*
Wong et al.[470] studied the stability of cefazolin sodium in three concentrations of Pluronic F-127, a polyethylene-polyoxypropylene-polyoxyethylene block copolymer. Pluronic F-127 gels of 20, 25, and 30% were prepared by slowly adding the copolymer to cold water and stirring with a magnetic stir bar for 30 minutes; gels were then stored overnight under refrigeration to ensure complete dissolution. Cefazolin sodium was dissolved in each clear, viscous solution to yield a concentration of 100 mg/mL. Samples were then stored at elevated temperatures of 35, 45, and 55°C and were analyzed for cefazolin content using an HPLC analytical technique. The degradation rate of cefazolin was found to decrease with increasing Pluronic F-127 concentration, presumably due to the nonionic micellar nature of the gels. From the data developed, the time to 10% decomposition was determined to be 130.6 hours when stored at 4°C. ■

# ■ Cefepime Hydrochloride

## Properties
Cefepime hydrochloride occurs as a white to off-white crystalline powder.[3,4] It is not hygroscopic.[4]

### Solubility
Cefepime hydrochloride is freely soluble in water and methanol.[3,4]

### pH
Aqueous reconstituted solutions of cefepime hydrochloride injection have a pH in the range of 4 to 6.[4,7]

## General Stability Considerations
Cefepime hydrochloride bulk powder should be packaged in tight, light-resistant containers and stored at controlled

room temperature protected from light.[3,4] Cefepime hydrochloride for injection should also be packaged in tight, light-resistant containers suitable for sterile solids and stored under refrigeration or at controlled room temperature. As the aqueous reconstituted solution, cefepime hydrochloride is stated to be stable for 24 hours at room temperature[7] and no more than seven days under refrigeration.[4,7]

## Stability Reports of Compounded Preparations
### Ophthalmic
Ophthalmic preparations, like other sterile drugs, should be prepared in a suitable clean air environment using appropriate aseptic procedures. When prepared from nonsterile components, an appropriate and effective sterilization method must be employed.

Kodym et al[1698] evaluated the stability of cefepime hydrochloride 1 and 5% eyedrop formulations prepared in citrate buffer (pH 6.24, 298 mOsm/L). The eyedrops were formulated with and without polyvinyl alcohol and β-phenylethyl alcohol 0.4% and phenylmercuric borate 0.001% preservatives. The preservatives thimerosol, benzalkonium chloride, and chlorhexidine diacetate were found to result in precipitation and turbidity in cefepime hydrochloride solutions. Samples were stored refrigerated at 4°C and at room temperature of 20°C.

Stability-indicating HPLC analysis found that 10% cefepime hydrochloride loss occurred in 27 days in the 1% formulation and in 21 days in the 5% formulation. Storage at 20°C resulted in more than 10% loss in three days in both 1 and 5% formulations.

### Injection
Injections, like other sterile drugs, should be prepared in a suitable clean air environment using appropriate aseptic procedures.

When prepared from nonsterile components, an appropriate and effective sterilization method must be employed.

**STUDY 1:** Stewart et al.[1699,1700] reported that cefepime hydrochloride solutions of 100 and 200 mg/mL in sterile water for injection, dextrose 5%, and sodium chloride 0.9% packaged in polypropylene syringes from Becton Dickinson with tip caps were stable (less than 10% loss) for 24 hours at room temperature, 14 days refrigerated, and 90 days frozen at −20°C. After 48 hours at room temperature, some syringes had lost up to 13% of the drug concentration.

Storage of thawed solutions for three to five days refrigerated and for an additional day at room temperature resulted in less than 10% total loss, but losses increased to 11 to 19% after two days of room temperature storage. If thawed solutions were stored refrigerated for seven days (incurring about 7% loss), the loss after one additional day stored at room temperature increased up to 12 to 15%. Restricting storage of thawed solutions to three to five days refrigerated allowed for adequate drug concentration for one additional day at room temperature.

**STUDY 2:** Ling and Gupta[1701] reported that cefepime hydrochloride 20 mg/mL in sodium chloride 0.9% packaged in polypropylene syringes from Becton Dickinson was stable at room temperature for at least two days with 5% loss but not four days with 11% loss. About 3% loss occurred in 21 days when refrigerated. The solution's initial yellow color became darker over time. ▇

# ▇ Ceftazidime

## Properties
Ceftazidime is a white to cream-colored crystalline powder.[3,4,7] Ceftazidime pentahydrate 1.16 g is equivalent to 1 g of ceftazidime anhydrous.[3]

## Solubility
Ceftazidime is slightly soluble in water and practically insoluble in ethanol (<1 mg/mL) and acetone.[2,4]

## pH
A 5-mg/mL aqueous solution of ceftazidime has a pH of 3 to 4.[4] The commercial injection has a pH in the range of 5 to 8.[7]

## pKa
Ceftazidime has pKa values of 1.9, 2.7, and 4.1.[2]

## Osmolality
The commercial frozen solutions are adjusted to an osmolality of 300 mOsm/kg.[2] Reconstituted ceftazidime added to dextrose 5% in water and to sodium chloride 0.9% to yield a 50-mg/mL concentration has osmolalities of 321 and 330 mOsm/kg, respectively.[8] Several extemporaneously compounded ophthalmic solutions in citrate buffers resulted in osmolarities from 331 to 389 mOsm/L.[1121]

## General Stability Considerations
The commercial vials should be stored at controlled room temperature and protected from exposure to light. The frozen infusions should be stored at −20°C.[7]

The manufacturer indicates that the reconstituted solution is stable for 24 hours at room temperature, seven days refrigerated, and at least three months frozen.[7]

## Stability Reports of Compounded Preparations
### Ophthalmic

Ophthalmic preparations, like other sterile drugs, should be prepared in a suitable clean air environment using appropriate aseptic procedures. When prepared from nonsterile components, an appropriate and effective sterilization method must be employed.

**STUDY 1:** Barnes[675] evaluated the stability of ceftazidime (Glaxo) 50 mg/mL prepared from injection in Sno Tears (Smith & Nephew). Stability-indicating HPLC analysis found that about 7% ceftazidime loss occurred in seven days stored refrigerated at 7°C. The solution developed an unacceptable yellow discoloration after that time.

**STUDY 2:** Barnes and Nash[823] also reported the stability of ceftazidime (Glaxo-Wellcome) 50 mg/mL prepared from injection in Sno Tears (Smith & Nephew) packaged in amber glass eyedrop bottles and stored at room temperature of 24 to 26°C and refrigerated at 6 to 8°C for 14 days.

At room temperature, the ophthalmic solution developed a yellow discoloration within 24 hours. HPLC analysis found that 35% loss occurred in seven days. When refrigerated, the ophthalmic solution developed a yellow discoloration after seven days of storage. HPLC analysis found that less than 10% loss occurred in seven days, but the statistical confidence limit indicated that restriction of the refrigerated storage period to five days was appropriate.

**STUDY 3:** Peyron et al.[870] reported the stability of ceftazidime 20 mg/mL intended for ophthalmic use prepared from injection in sodium chloride 0.9%. HPLC analysis found that about 9% loss occurred in 21 days stored under refrigeration. However, at room temperature of 25°C, 10% or more loss occurred at the first assay point of four days.

**STUDY 4:** Achach and Peroux[1058] evaluated the stability of several anti-infective drugs in ophthalmic solutions including ceftazidime. Compounded ceftazidime 50 mg/mL in sodium chloride 0.9% was stored under refrigeration at 4°C for 12 days. No changes in color or turbidity were observed, and no change in osmolality was found. The analytical method was not specified, but the authors reported that ceftazidime retained 96% through four days but retained only 88% after eight days. The authors recommended a four-day use period.

**STUDY 5:** Kodym et al.[1121] evaluated the effects of various antimicrobial preservatives on the stability of ceftazidime in ophthalmic formulas. Chlorhexidine 0.01%, benzalkonium chloride 0.005%, and thimerosal at concentrations over 0.003% resulted in opalescence and precipitation, leading to sedimentation that renders these preservatives unacceptable. Thimerosal at concentrations lower than 0.003% was

ineffective as a preservative. Ceftazidime 1% ophthalmic solutions in citrate buffer (pH 6.18 to 6.3) with or without polyvinyl alcohol for viscosity and with or without phenylmercuric borate 0.001% and β-phenylethyl alcohol 0.4% were delivered through a 0.22-μm filter and stored for 30 days at 4 and 20°C. The refrigerated ophthalmic solution remained clear and colorless, while the solution stored at room temperature developed a slight yellow color. Antimicrobial activity remained near 100% under refrigeration for 30 days, but 10 to 13% loss of ceftazidime activity occurred after 18 days of storage at room temperature.

**STUDY 6:** Dobrinas et al.[1195] evaluated the stability of ophthalmic injections, including ceftazidime (GlaxoSmithKline) 22.5 mg/mL, in balanced salts solution simulating aqueous humor solution. The solutions were passed through 0.22-μm filters for sterilization. The samples were packaged as 1 mL in syringes (syringe composition not cited) and were stored frozen at less than 18°C for six months. HPLC analysis found less than 5% drug concentration change and the formation of no decomposition products. In addition, the drug remained stable after thawing and storing at room temperature for six hours.

**STUDY 7:** Chedru-Legros et al.[1219] evaluated the physical and chemical stability of ceftazidime-fortified ophthalmic solution prepared from the commercial injection. Ceftazidime (GlaxoSmithKline) was diluted in sodium chloride 0.9%, yielding a 50-mg/mL concentration. The solution was passed through a Millipore 0.22-μm filter, packaged in clear glass containers, and stored frozen at −20°C over a 75-day test period. The ceftazidime solutions had a pH of 6.5 and an osmolality of 488 mOsm/kg, neither of which changed substantially throughout the study. HPLC analysis found little or no change in drug concentration over the 75-day study period. The authors recommended that the ceftazidime-fortified ophthalmic solution could be stored frozen for 75 days. However, after thawing, refrigerated storage and discarding after three days was recommended.

**STUDY 8:** Chedru-Legros et al.[1591] also evaluated the physical and chemical stability of several fortified ophthalmic solutions. Ceftazidime injection powder (GlaxoSmithKline) 50 mg/mL in balanced salt solution (Alcon), dextrose 5% (B. Braun), and sodium chloride 0.9% (B. Braun) packaged in 20-mL glass bottles fitted with pipettes mounted on screw caps were stored frozen at −20°C protected from exposure to light for six months. The solution was filtered through 0.22-μm Millipore filters as it was added to each bottle. No visible instability was observed, and osmolality and pH were acceptable. HPLC analysis of ceftazidime in sodium chloride 0.9% (but not the other diluents) found little or no change in ceftazidime concentration at any time point during six months of frozen storage. The authors concluded such fortified antibiotic solutions could be stored frozen at −20°C for six months.

**STUDY 9:** Xu et al.[1347] evaluated the stability of ceftazidime (Hospira) 1 mg/mL in sodium chloride 0.9% and in balanced salt solution packaged in plastic syringes and stored at room temperature, refrigerated, and frozen at −20°C. Stability-indicating HPLC analysis found no difference in the drug's stability in the two diluents with about 9 to 10% ceftazidime loss in three days at room temperature and in 17 days refrigerated. No drug loss occurred in the frozen samples during 60 days of storage.

**STUDY 10:** Karampatakis et al.[1597] evaluated the stability of ceftazidime 50-mg/mL ophthalmic drops in balanced salt solution (BSS). Ceftazidime injection (GlaxoSmithKline) 1-g vials were reconstituted using BSS (Alcon) resulting in a 50-mg/mL solution. The solution was packaged in polypropylene containers and stored at room temperature of 24°C and under refrigeration at 4°C both exposed to and protected from exposure to light. Organoleptic evaluations were performed, and microbiological assays against *Pseudomonas aeruginosa* and *Staphylococcus aureus* were used to determine antibiotic activity.

At room temperature the solution developed a yellow discoloration within 24 hours that darkened upon further storage. An offensive odor developed within 72 hours, and the solution pH increased by one pH unit. Unacceptable antibiotic activity loss was evident within seven days. The solution stored under refrigeration exhibited only a pale yellow discoloration, and unacceptable loss of antibiotic activity became evident after 14 days of refrigerated storage. Drug stability was unaffected by exposure to or protection from exposure to light.

**STUDY 11:** Hill and Barnes[1639] evaluated the stability of phenylmercuric acetate 0.002% wt/vol antimicrobial preservative in compounded ceftazidime 5% wt/vol (50 mg/mL) eyedrops compounded from the injection. A vial of ceftazidime injection (Glaxo) was reconstituted using phenylmercuric acetate 0.002% solution yielding the 5% ceftazidime concentration. The solution was packaged in amber glass eyedrop bottles and samples were stored refrigerated at 6 to 8°C and at room temperature of 24 to 26°C for up to 35 days. HPLC analysis of the phenylmercuric acetate concentration found extensive loss occurred stored under refrigeration. Stability was even worse at room temperature where complete loss of the phenylmercuric acetate occurred within 14 days.

*Injection*
Injections, like other sterile drugs, should be prepared in a suitable clean air environment using appropriate aseptic procedures. When prepared from nonsterile components,

an appropriate and effective sterilization method must be employed.

Stewart et al.[896] reported the stability of ceftazidime (Fortaz, Glaxo) 100 and 200 mg/mL reconstituted with sterile water for injection and repackaged in Becton Dickinson polypropylene syringes and glass vials. HPLC analysis found not more than 5% loss in eight hours at room temperature and 96 hours refrigerated.

## Compatibility with Other Drugs

**STUDY 1 (VANCOMYCIN):** Lifshitz et al.[915] reported the incompatibility of vancomycin hydrochloride with ceftazidime in intravitreal and subconjunctival injections. Two cases of immediate precipitation, described as yellowish-white in color, occurred upon sequential administration of vancomycin hydrochloride and ceftazidime. The intravitreal doses of vancomycin hydrochloride and ceftazidime were 1 mg/0.1 mL and 2.2 mg/0.1 mL, respectively. The subconjunctival doses of vancomycin hydrochloride and ceftazidime were 25 mg/0.25 mL and 100 mg/0.5 mL, respectively. The intravitreal opacities created by the precipitation cleared gradually over two months.

**STUDY 2 (VANCOMYCIN):** McLellan and Papadopoulos[916] reported another case of immediate precipitation upon sequential subconjunctival administration of ceftazidime 100 mg/0.5 mL and vancomycin hydrochloride 25 mg/0.5 mL. The white precipitation appeared immediately and dissipated over time. Intravitreal administration of ceftazidime 2 mg/0.1 mL and vancomycin hydrochloride 1 mg/0.1 mL were reported to result in no precipitate formation in this one case. However, the reports of precipitation with the combined ophthalmic use of these drugs and the reports of possible concentration-dependent precipitation in injections indicate the need for caution.

**STUDY 3 (VANCOMYCIN):** Kwok et al.[1154] reported the precipitation of ceftazidime 2.2 mg/0.1 mL and vancomycin hydrochloride 1 mg/0.1 mL for ophthalmic use in sodium chloride 0.9% and balanced salts solution plus glutathione (BSS Plus). The drugs were mixed together in 4 mL of sodium chloride 0.9%, BSS Plus, and vitreous obtained from cadaver eyes. The samples were incubated at 37°C. Ceftazidime and vancomycin hydrochloride were analyzed by HPLC.

Ceftazidime precipitated by itself and in combination with vancomycin hydrochloride in all three media. The extent of precipitation was about 54% if prepared in sodium chloride 0.9% and about 88% if prepared in BSS Plus. Vancomycin precipitation was negligible. After precipitation, the concentration of ceftazidime may not have been sufficiently high for antibacterial activity against common microorganisms. ■

# Ceftizoxime Sodium

## Properties

Ceftizoxime sodium occurs as a white to pale yellow crystalline powder.[4]

### Solubility

Ceftizoxime sodium is freely soluble in water.[4]

### pH

Ceftizoxime sodium 1 g in 10 mL of water has a pH between 6 and 8. Reconstituted ceftizoxime sodium injection also has a pH of 6 to 8. Ceftizoxime sodium frozen premixed infusion solution has a pH between 5.5 and 8.[4,7]

### Osmolality

Ceftizoxime 1 g as the sodium salt dissolved in 13 mL of sterile water for injection is isotonic.[7]

## General Stability Considerations

Ceftizoxime sodium powder should be packaged in tight containers and stored at controlled room temperature. Ceftizoxime (as sodium) for injection is also stored at controlled room temperature. Ceftizoxime (as sodium) injection is available as a frozen premixed infusion solution that is to be stored frozen at −20°C.[4,7]

## Stability Reports of Compounded Preparations
### Topical

Wong et al.[1688] evaluated the stability of ceftizoxime in aqueous Pluronic F-127 gels. The gel samples were stored at temperatures of 35, 45, and 55°C, and the drug decomposition rates were determined. Based on the rates of decomposition, 10% decomposition in 30% Pluronic F-127 gel stored refrigerated at 4°C was predicted to occur in 314 hours or about 13 days. ■

# Cefuroxime Axetil

## Properties

Cefuroxime axetil is a white to cream amorphous powder. Cefuroxime axetil is the 1-(acetyloxy)ethyl ester of cefuroxime; 1.2 g is approximately equal to 1 g of cefuroxime.[2,3]

## General Stability Considerations

Cefuroxime axetil tablets should be stored in well-closed containers at controlled room temperature[4] and protected from excessive moisture.[2]

The powder for suspension should be stored between 2 and 30°C. Following reconstitution it is stored between 2 and 25°C and used within 10 days.[2,7,1249]

## Stability Reports of Compounded Preparations
### Oral

**STUDY 1:** Pramar et al.[726] evaluated the stability of cefuroxime axetil oral suspensions prepared from crushed tablets to yield a 10-mg/mL concentration suspended in freshly prepared simple syrup with a pH of 6.8, commercial simple syrup with citric acid 0.1% and sodium benzoate (C&D Flavor Company) with a pH of 3.3, and the commercial simple syrup with added citrate buffer to adjust to pH 5.5. The suspensions were packaged in amber bottles and stored refrigerated at 4 to 6°C for 28 days. The suspensions retained their original appearances and pH values throughout the study. Stability-indicating HPLC analysis found little or no cefuroxime axetil loss over the 28-day study period when refrigerated.

**STUDY 2:** Farrington et al.[997] evaluated the stability of cefuroxime axetil 125-mg/5-mL and 250-mg/5-mL oral suspensions compounded from commercial tablets. The tablets were ground to fine powder and triturated with Ora-Plus suspending vehicle to form a paste. Ora-Sweet syrup vehicle and peppermint oil were incorporated to help mask the bitter taste of the drug. The oral suspensions were packaged in amber plastic prescription bottles and polypropylene oral syringes and stored at room temperature of 25°C and under refrigeration at 5°C. Stability-indicating HPLC analysis found less than 5% drug loss in seven days at room temperature and in 17 days refrigerated.

## Compatibility with Common Beverages and Foods
### CEFUROXIME AXETIL COMPATIBILITY SUMMARY

*Compatible with:* Apple juice • Chocolate milk • Grape juice • Orange juice

**STUDY 1:** St. Claire et al.[159] reported the stability of cefuroxime axetil (Ceftin) in apple juice. Cefuroxime axetil 125- and 250-mg tablets were dispersed in three brands of apple juice (Gerber, Mott's Natural, Tropicana). Each tablet was added to 40 mL of apple juice at 5°C and vortexed briefly to ensure complete dispersion. The samples were stored at room temperature for 24 hours and assayed using an HPLC technique for cefuroxime axetil content. Little or no loss of drug occurred during the 24-hour period at either concentration;

96 to 100% of the initial concentration remained at the end of the study.

**STUDY 2:** The stability of cefuroxime axetil (Ceftin, Glaxo) as crushed 125- and 250-mg tablets in orange juice, grape juice, and chocolate milk also was evaluated by St. Claire and Caudill.[97] The tablets were added to 40 mL of the beverages that had been stored at 5°C, and the mixtures were shaken using a Vortex mixer and then left standing for two hours at room temperature. The concentration of cefuroxime axetil was then determined initially and at two hours using an HPLC assay. Cefuroxime axetil losses of 5% or less in two hours occurred in each sample. These beverages are suitable for use as aids in administering the drug. ■

# ■ Cefuroxime Sodium

## Properties

Cefuroxime sodium is a white to off-white or faintly yellow powder. Solutions may range from light yellow to amber. Both the powder and solution may darken during storage without affecting concentration. Cefuroxime sodium 1.05 g is approximately equivalent to cefuroxime 1 g. Each gram of cefuroxime sodium contains 2.4 mEq (54.2 mg) of sodium.[1-3,7]

### Solubility

Cefuroxime sodium has an aqueous solubility of about 200 mg/mL; in ethanol its solubility is about 1 mg/mL.[1,2] Reconstitution of the sterile powder to 220 mg/mL for intramuscular use results in the formation of a suspension; dilution to 100 mg/mL or less ensures complete dissolution.[2,7]

### pH

Reconstituted vials of cefuroxime sodium have a pH of 6 to 8.5. The frozen premixed solutions of the drug have pH values ranging from 5 to 7.5.[2,4,7]

### $pK_a$

Cefuroxime has a pK$_a$ of 2.45.[2]

## General Stability Considerations

Cefuroxime sodium sterile powder should be stored at controlled room temperature and protected from light. The reconstituted solution should be stored under refrigeration. It is stable for 24 hours at room temperature and 48 hours under refrigeration. More dilute solutions of 1 to 30 mg/mL are stable for 24 hours at room temperature and seven days under refrigeration.[2,7]

The frozen premixed cefuroxime sodium injection should be stored at −20°C. Thawed commercially frozen solutions are stable for 24 hours at room temperature and for 28 days under refrigeration. Extemporaneously prepared frozen solutions from the sterile powder at 7.5 to 30 mg/mL are stable for six months at −20°C. Solutions should be thawed at room temperature and should not be refrozen. Thawed extemporaneously prepared solutions are stable for 24 hours at room temperature and for seven days under refrigeration.[2,7]

## Stability Reports of Compounded Preparations
### Ophthalmic

Ophthalmic preparations, like other sterile drugs, should be prepared in a suitable clean air environment using appropriate aseptic procedures. When prepared from nonsterile components, an appropriate and effective sterilization method must be employed.

**STUDY 1:** Hebron and Scott[260] reported the stability of cefuroxime sodium 5% in extemporaneously prepared eyedrop formulations. Cefuroxime sodium (Zinacef, Glaxo) was dissolved in the artificial tears solutions cited in Table 20 to yield a nominal concentration of 50 mg/mL. The ophthalmic solutions were stored in clear glass or plastic bottles at 4 and 25°C in the dark for 28 days. The cefuroxime sodium content was assessed after 14 and 28 days of storage using HPLC. The samples also were visually inspected for turbidity, precipitate, or color change. As with other cephalosporins, a yellow color developed in the solutions; all solutions became yellow within seven days when stored at 25°C, while only the alkaline solutions became yellow in seven days at 4°C.

Artificial tears products that contained benzalkonium chloride as a preservative developed turbidity upon addition of cefuroxime sodium, except for Sno Tears, which contains only 0.004% of the preservative; Sno Tears developed turbidity after seven days of storage at 25°C but remained clear through 28 days stored at 4°C. Many formulations went on to develop a precipitate (Table 20). Only solutions that were preservative-free (distilled water, preservative-free sodium chloride eyedrops, Hypromellose PF) or were preserved with chlorhexidine acetate (BJ6 eyedrops, Thornton & Ross) remained free of turbidity and precipitation at both temperatures throughout the 28-day study. The authors found turbidity and/or precipitation when cefuroxime sodium was combined with solutions having benzalkonium chloride concentrations above 0.002%. They speculated that a relatively insoluble compound was formed due to ion pairing between the acid group of the cefuroxime and the amide group of the benzalkonium.

Analysis of these pharmaceutically elegant formulations showed that acceptable cefuroxime content was retained at

**TABLE 20.** Stability of Cefuroxime Sodium in Eyedrop Formulations[260]

| Solution | Results |
|---|---|
| BJ6 eyedrops (Thornton & Ross) | Physically stable, with a calculated shelf life of 28 days at 4°C and one day at 25°C. |
| BJ6 eyedrops (Daniels) | Turbidity forms immediately, with precipitation in seven days at 25°C. |
| Hypotears | Turbidity forms immediately, with precipitation in 14 and seven days at 4 and 25°C, respectively. |
| Hypromellose BP | Turbidity forms immediately, with precipitation in 14 and seven days at 4 and 25°C, respectively. |
| Hypromellose PF | Physically stable, with a calculated shelf life of 31 days at 4°C and one day at 25°C. |
| Isopto Alkaline | Turbidity forms immediately and persists at both 4 and 25°C. |
| Isopto Plain | Turbidity forms immediately, with precipitation in 21 and seven days at 4 and 25°C, respectively. |
| Liquifilm Tears | Turbidity forms immediately, with precipitation in 21 and seven days at 4 and 25°C, respectively. |
| Sno Tears | Physically stable, with a calculated shelf life of 40 days at 4°C. Turbidity forms in seven days and precipitation occurs in 21 days at 25°C. |
| Sodium chloride eyedrops BP | Turbidity forms immediately, with precipitation in seven days at both 4 and 25°C. |
| Sodium chloride eyedrops PF | Physically stable, with 3% loss of cefuroxime in 14 days and 11% loss in 28 days at 4°C. |
| Tears Naturale | Turbidity forms immediately, with precipitation in 14 and seven days at 4 and 25°C, respectively. |
| Water, distilled | Physically stable, with a calculated shelf life of 39 days at 4°C and one day at 25°C. |

4°C storage for 14 days in preservative-free sodium chloride eyedrops and Hypromellose PF; acceptable cefuroxime content was retained for 28 days in distilled water, BJ6 with chlorhexidine (Thornton & Ross), and Sno Tears.

**STUDY 2:** Oldham[761] reported the stability of cefuroxime sodium 5.5% in several ophthalmic formulations. The ophthalmic solutions were prepared from commercial injection (Zinacef, Glaxo) and utilized phosphate or citrate buffers along with phenylmercuric nitrate 0.002% preservative. The ophthalmic solutions were packaged in amber glass and low-density polyethylene ophthalmic containers.

Based on the results of the study, the authors recommended that simple dilution of the commercial injection to a 5.5% concentration with sterile water for injection was the preferred ophthalmic formulation. Stability-indicating HPLC analysis found that the buffered solutions did not exhibit improved stability compared to a simple aqueous solution. Furthermore, the compatibility of cefuroxime with phenylmercuric nitrate was uncertain. The authors recommended beyond-use periods of 24 hours at room temperature, 21 days refrigerated, and up to 12 months frozen at −30°C.

**STUDY 3:** Kodym et al.[1215] evaluated the effects of various antimicrobial preservatives on the stability of cefuroxime sodium in ophthalmic formulas. Chlorhexidine 0.01%, benzalkonium chloride 0.005%, and thimerosal at concentrations of more than 0.003% resulted in opalescence and precipitation, leading to sedimentation that renders these preservatives unacceptable. Thimerosal at concentrations lower than 0.003% was ineffective as a preservative. Cefuroxime sodium 1% ophthalmic solutions in citrate buffer (pH 6.15 to 6.20) with or without polyvinyl alcohol for viscosity and with or without phenylmercuric borate 0.001% and β-phenylethyl alcohol 0.4% were delivered through a 0.22-μm filter and stored for 30 days at 4 and 20°C. The refrigerated ophthalmic solution remained clear and colorless, while the solution stored at room temperature developed a yellow color. Antimicrobial activity remained near 100% under refrigeration for 30 days, but 10 to 11% loss of cefuroxime activity occurred after 18 days of storage at room temperature.

**STUDY 4:** Hill and Barnes[1639] evaluated the stability of phenylmercuric acetate 0.002% wt/vol antimicrobial preservative in compounded cefuroxime 5% wt/vol (50 mg/mL) eyedrops compounded from the injection. A vial of cefuroxime injection (Glaxo) was reconstituted using phenylmercuric acetate 0.002% solution yielding the 5% cefuroxime concentration. The solution was packaged in amber glass eyedrop bottles, and samples were stored refrigerated at 6 to 8°C and at room temperature of 24 to 26°C for up to 35 days. HPLC analysis of the phenylmercuric acetate concentration found extensive loss occurred stored under refrigeration. Stability was even worse at room temperature where complete loss of the phenylmercuric acetate occurred within 14 days. ■

# Celecoxib

## Properties
Celecoxib occurs as a pale yellow solid material.[1]

### Solubility
Celecoxib is nearly insoluble in water, having an aqueous solubility of about 6 or 7 mcg/mL. The drug is slightly more soluble at alkaline pH; at pH 11 the solubility is about 48 mcg/mL. It is more soluble in some organic solvents: in ethanol about 69 mg/mL, in propylene glycol about 30 mg/mL, and in polyethylene glycol 400 about 414 mg/mL.[1504]

### pH
Celecoxib oral liquid compounded from capsules exhibited a pH of 4.4 to 4.5.[1503]

### $pK_a$
Celecoxib has a $pK_a$ of 11.1.[1505]

## General Stability Considerations
Celecoxib commercial oral capsules are to be stored at controlled room temperature.[7]

## Stability Reports of Compounded Preparations
### Oral
Donnelly et al.[1503] evaluated the stability of extemporaneously compounded celecoxib 10-mg/mL oral suspension prepared from commercial capsules (Pfizer). The contents of the commercial capsules were ground to powder using a mortar and pestle, and the powder was combined with Ora-Blend (Paddock) suspending/flavoring vehicle resulting in a 10-mg/mL concentration. The oral liquid was packaged in amber polyvinyl chloride (PVC) plastic bottles. Samples were stored at both room temperature of 23°C and refrigerated at 5°C for 93 days. The samples exhibited no changes in color, odor, or ease of resuspension. Stability-indicating HPLC analysis of drug concentrations found little or no loss of drug occurred over the storage period. ■

# Cephalexin

## Properties
Cephalexin (as the monohydrate) is a white to off-white crystalline powder with a bitter taste. Cephalexin monohydrate 1.05 g is approximately equivalent to 1 g of the anhydrous form.[2–4]

### Solubility
Cephalexin is slightly soluble in water,[3,4,7] variously cited as 1 or 2 mg/mL[7] and 1 in 100,[3] and insoluble in ethanol.[2,3,7]

### pH
A 50-mg/mL suspension of cephalexin has a pH of 3 to 5.5.[4] A 0.5% aqueous solution of cephalexin has a pH between 4 and 5.5.[3] Cephalexin suspension (Keflex, Dista) has a measured pH of 4.1,[19] with a range of pH 3 to 6.[4] Cephalexin hydrochloride 10 mg/mL has a pH of 1.5 to 3.[4]

### $pK_a$
Cephalexin has $pK_a$ values of 5.2 and 7.3.[1]

### Osmolality
Cephalexin suspension 250 mg/5 mL (Keflex, Dista) was found by Niemiec et al.[232] to have an osmolality of 2445 mOsm/kg, while Dickerson and Melnik[233] reported the osmolality to be 1950 mOsm/kg.

## General Stability Considerations
Cephalexin products should be stored in tight containers at controlled room temperature. The reconstituted suspension should be stored under refrigeration at 2 to 8°C and discarded 14 days after preparation.[2,4,7]

The calculated time to 10% loss of cephalexin in the dry state is 4.7 years at 20°C and 3.8 years at 22°C. In pH 3 and 5 buffer solutions, less than 2% cephalexin loss occurred in 24 hours at 3°C and at room temperature. However, at pH 7 and 8, 23 to 27% loss occurred in 24 hours at room temperature.[1007]

## Stability Reports of Compounded Preparations
### Oral
The stability of cephalexin oral suspension (Keflex, Dista) 25 mg/mL stored in polypropylene oral syringes (Exacta-Med, Baxa) was evaluated by Sylvestri et al.[37] The suspension was reconstituted with distilled water and 5-mL portions were drawn into the oral syringes, which were capped and stored at −20, 4, and 25°C along with samples of the drug in the original containers. The cephalexin content was assessed using the USP iodometric assay specific for the intact β-lactam ring. No difference was found in the stability of cephalexin oral suspension in the original containers compared to the oral syringes. Losses of 5, 7, and 9% occurred in both containers at −20, 5, and 25°C, respectively, in 90 days.

## Enteral

### CEPHALEXIN COMPATIBILITY SUMMARY

*Compatible with:* Enrich • Ensure • Ensure HN • Ensure Plus • Ensure Plus HN • Isocal • Osmolite • Osmolite HN • Precitene • Sustacal • Sustacal HC • TwoCal HN • Vital • Vivonex T.E.N.

**STUDY 1:** Cutie et al.[19] added 5 mL of cephalexin suspension (Keflex, Dista) to varying amounts (15 to 240 mL) of Ensure, Ensure Plus, and Osmolite (Ross Laboratories) with vigorous agitation to ensure thorough mixing. The cephalexin suspension was physically compatible, distributing uniformly in all three enteral products with no phase separation or granulation.

**STUDY 2:** Strom and Miller[44] evaluated the stability of cephalexin (Keflex, Dista) from 250-mg capsules and oral suspension 250 mg/5 mL when mixed with three enteral formulas at full- and half-strength. The capsule contents and the suspension were dispersed in approximately 10 mL of deionized water and added to 240 mL of Isocal, Sustacal, and Sustacal HC (Mead Johnson) full- or half-strength, mixed with deionized water. The mixtures were stirred with a magnetic stirrer for five minutes; duplicate samples of each mixture were stored at 24°C.

The mixtures were evaluated visually for precipitation and phase separation, but no physical incompatibilities were observed. A stability-indicating HPLC assay was employed to assess the cephalexin content at various time points over 24 hours. The cephalexin content was between 92 and 100% after 24 hours in each of the enteral mixtures prepared from the capsules. A cephalexin content between 90 and 97% was found for each mixture prepared from the oral suspension except for the full-strength Sustacal HC. In this mixture, cephalexin from the oral suspension retained 92% of initial concentration after 10 hours but only 86% after 24 hours. Studies to determine the amount of bound and free cephalexin found about 85 to 87% free drug in all half-strength enteral mixtures. The stability of the enteral formula components was not evaluated.

**STUDY 3:** Burns et al.[739] reported the physical compatibility of cephalexin oral suspension (Keflex) 500 mg/10 mL with 10 mL of three enteral formulas, including Enrich, TwoCal HN, and Vivonex T.E.N. Visual inspection found no physical incompatibility with any of the enteral formulas.

**STUDY 4:** Altman and Cutie[850] reported the physical compatibility of cephalexin suspension (Keflex, Dista) 5 mL with varying amounts (15 to 240 mL) of Ensure HN, Ensure Plus HN, Osmolite HN, and Vital after vigorous agitation to ensure thorough mixing. The cephalexin suspension was physically compatible, distributing uniformly in all four enteral products, with no phase separation or granulation.

**STUDY 5:** Ortega de la Cruz et al.[1101] reported the physical compatibility of an unspecified amount of oral liquid cephalexin (Cefalexgobens, Norman) with 200 mL of Precitene (Novartis) enteral nutrition diet for a 24-hour observation period. No particle growth or phase separation was observed. ▪

# ▪ Cephalothin Sodium

## Properties

Cephalothin sodium is a white to off-white crystalline powder that is almost odorless. Cephalothin sodium 1.06 g is approximately equivalent to 1 g of cephalothin. Sodium bicarbonate has been used to adjust the pH of the commercial products.[3]

### Solubility

Cephalothin sodium is freely soluble in water and aqueous solutions; a solubility of about 285 mg/mL has been cited. The drug is only slightly soluble in ethanol.[3,6]

### pH

A 10% aqueous solution has a pH of 4.5 to 7.[3] The reconstituted vials of powder for injection and the frozen premixed solution have pH values of 6 to 8.5.[2,4]

### pKa

The $pK_a$ of the conjugate acid is 2.22 at 35°C.[6]

## General Stability Considerations

Cephalothin sodium products should be stored in airtight containers at controlled room temperature not exceeding 30°C and protected from light.[3]

Cephalothin powder may gradually darken during storage, but this change does not adversely affect the drug if only slight discoloration occurs. Aqueous solutions were reported to be stable for 96 hours under refrigeration at 2 to 8°C and for 12 hours at room temperature. The drug may precipitate if stored under refrigeration, but the precipitate redissolves by warming to room temperature with continuous agitation. Precipitation also may occur if the solution has a pH less than 5.

Cephalothin is subject to hydrolysis of the β-lactam ring and the 3-acetoxy group. Its decomposition in solution is independent of pH in the range of 3 to 8. However, decomposition accelerates at pH below 2 or above 8.[6,316,317]

Cephalothin sodium solutions are stable in the frozen state. Extemporaneously prepared frozen solutions retained stability for six weeks in the original containers,[318] for up to 30 days in polyvinyl chloride bags,[319,320] and for as long as nine months in glass syringes.[321]

## Stability Reports of Compounded Preparations
### Ophthalmic
Ophthalmic preparations, like other sterile drugs, should be prepared in a suitable clean air environment using appropriate aseptic procedures. When prepared from nonsterile components, an appropriate and effective sterilization method must be employed.

Osborn et al.[169] evaluated the stability of several antibiotics in three artificial tears solutions composed of 0.5% hydroxypropyl methylcellulose. The three artificial tears (Lacril, Tearisol, Isopto Tears) were used to reconstitute and dilute cephalothin sodium initially to 65 mg/mL. The preparation was packaged in plastic squeeze bottles and stored at 25°C. A serial dilution bioactivity test was used to estimate antibiotic activity. The amount of antibiotic remaining after seven days at room temperature was 73% in Lacril, 80% in Tearisol, and 73% in Isopto Tears. ▪

# ▪ Charcoal

## Properties
Charcoal is a fine black odorless tasteless powder free from gritty matter. Charcoal is obtained from various organic and vegetable matter and is formed by destructive distillation and carbonization processes.[3,4]

## Solubility
Charcoal is practically insoluble in all common solvents.[3]

## General Stability Considerations
Charcoal should be packaged in well-closed containers and stored at controlled room temperature.[4]

## Stability Reports of Compounded Preparations
### Injection
Injections, like other sterile drugs, should be prepared in a suitable clean air environment using appropriate aseptic

procedures. When prepared from nonsterile components, an appropriate and effective sterilization method must be employed.

Bonhomme-Faivre et al.[798–800] evaluated the use of several excipients in 10% charcoal injection used for tattooing of breast tumors. Peat charcoal SX4 was crushed in a stainless steel micronizer and dispersed in water for injection using a turbine mixer at 200 rpm for 10 minutes to yield a 10% concentration. On average, 50% of the particles were in the size range of 2 to 5 µm. Various excipients were added in an attempt to modify the viscosity or particle charge to improve injectibility. The various formulations were sterilized by autoclaving at 120°C for 20 minutes. None of the excipients provided improvements that were not accompanied by unacceptable toxicity. The authors concluded that the best formulation for intratumoral injections was simply the sterilized suspension of the peat charcoal 10% in water for injection. ▪

#  Chloral Hydrate

## Properties
Chloral hydrate occurs as transparent colorless or white crystals and has an aromatic penetrating, slightly acrid odor and a caustic bitter taste.[1,3,4]

## Solubility
Chloral hydrate is very soluble in water[3,4] to about 4 g/mL[3] or 8 g/mL[1] at room temperature and is also soluble in olive oil to about 0.71 g/mL.[1] It is freely soluble in ethanol to about 0.77g/mL.[1,3,4] The solubility in glycerin is about 2 g/mL.[1]

## pH
A 10% chloral hydrate solution in water has a pH of 3.5 to 5.5.[3]

## General Stability Considerations
Chloral hydrate volatilizes slowly when exposed to air. It should be stored in airtight containers. It melts at about 55°C and is light sensitive, decomposing upon exposure.[2–4] Aqueous solutions of chloral hydrate are incompatible with alkaline materials.[2] The half-life (50% loss) of chloral hydrate solution at 20°C is stated to be 28 months,[1002] but at pH 12 it is only 25 minutes at room temperature. At pH 9 and 8, the half-lives are four and 17.5 days, respectively.[993] The presence of sucrose does not influence chloral hydrate decomposition.[1002]

Chloral hydrate forms a liquid mixture when triturated with a number of organic compounds, including camphor, menthol, antipyrine (phenazone), phenol, quinine salts, and thymol.[1,3]

Chloral hydrate capsules should be stored in tight containers at controlled room temperature. Chloral hydrate oral solution (formerly syrup) should be packaged in tight, light-resistant containers and stored at controlled room temperature.[4]

## Stability Reports of Compounded Preparations
### Oral
**STUDY 1:** Kakehi et al.[818] reported the stability of chloral hydrate in syrup at room temperature of 30°C and at elevated temperature of 60°C. Evaluation of pH and decomposition products by capillary electrophoresis demonstrated no substantial change in chloral hydrate concentration over three months at room or elevated temperature.

**STUDY 2:** Taguchi et al.[1003] reported the stability of chloral hydrate syrup formulated at a chloral hydrate concentration of 40 mg/mL (4%) in 50% simple syrup with pineapple flavor. The chloral hydrate content was found to be stable for at least two weeks when stored in cold and dark conditions. However, a loss of color was observed during the storage period.

### Rectal Solution
Yoshida et al.[877] evaluated the stability of chloral hydrate 50 mg/mL (5%) solution packaged in both glass and polypropylene containers stored refrigerated at 4°C in the dark and in glass containers at room temperature of 25°C exposed to light and light-protected over 90 days. Analysis found that chloral hydrate stability was nearly identical under all of the conditions, with no loss in 30 days and about 3 to 5% loss occurring in 90 days. ◼

# ◼ Chlorambucil

## Properties
Chlorambucil is a white to off-white crystalline or granular powder.[2,3] The USP recommends that great care be taken to prevent inhalation of or skin exposure to chlorambucil.[4]

### Solubility
Chlorambucil is nearly insoluble in water; however, in ethanol its solubility is about 1 g in 1.5 mL. It is also soluble in dilute alkali hydroxides.[2,3,6]

### $pK_a$
Chlorambucil has apparent $pK_a$ values of 1.3 and 5.8.[2,7]

## General Stability Considerations
Chlorambucil tablets should be stored in well-closed, light-resistant containers at controlled room temperature.[2–4]

Chlorambucil is subject to rapid hydrolysis in the presence of water. The hydrolysis is independent of pH between 4.5 and 10. The minimum rate of hydrolysis occurs at about pH 2. Hydrolysis also is slowed by the presence of alcohol or propylene glycol in the solution and by refrigeration.[6]

Ehrsson et al.[211] reported on the degradation of chlorambucil in aqueous solutions at pH 1.5 to 10. Chlorambucil was dissolved in 0.1 mL of methanol and mixed with the appropriate buffer solutions. The solution was stored at 25°C and analyzed using a reversed-phase HPLC assay. The degradation rate was highest above pH 5. At pH 5 to 10, the degradation rate was static and did not increase. Below pH 5 the degradation rate decreased; the lowest degradation rate found occurred at pH 1.5.

An aqueous injection prepared at pH 6.5 to achieve adequate solubility undergoes 5% decomposition in nine minutes at 25°C. A solution containing 40% (vol/vol) ethanol or 45% (wt/vol) propylene glycol increases the time about fivefold. A chlorambucil solution in dehydrated ethanol stored in the freezer at −10°C has a shelf life of 31 days. It has been suggested that extemporaneous injections be prepared fresh from a stock solution in dehydrated ethanol diluted with 40% (vol/vol) ethanol or 45% (wt/vol) propylene glycol and refrigerated until use.[6]

Using a potentiometric assay, Stewart and Owen[486] evaluated factors that affect the stability of chlorambucil in solutions. The pH, solvent polarity, and temperature all influenced the drug's stability. The degradation rate was rapid and was independent of pH in the pH range of 4.5 to 10. Below pH 4.5 the rate of hydrolysis decreased. Unfortunately, a pH of 6.5 or above is required for adequate solubility, but the time to 5% loss is only nine minutes. In mixed solvent systems incorporating 45% propylene glycol or 40% alcohol in water, the times to 5% loss were about 45 minutes for propylene glycol and 49 minutes for alcohol at 25°C. Chlorambucil dissolved in absolute alcohol exhibited a shelf life of 6.3 days at 4°C and 31 days at −10°C, but to prevent water entry during storage, care must be taken to ensure the preparation is anhydrous and well stoppered. Higher concentrations of chloride decreased the rate of hydrolysis. However, the presence of salts (other than chlorides) and buffer species had little effect on hydrolysis. In addition, due to drug loss, the use of heat sterilization was shown to be impossible.

Bosanquet and Clarke[485] also evaluated factors affecting the stability of chlorambucil dissolved in ethanol and diluted with normal saline to a concentration of 20 mcg/mL. Rapid dilution to this low concentration was necessary to avoid precipitation, because the drug passes through a supersaturated phase during which precipitation can readily occur.

Stability-indicating HPLC analysis found several factors that affected the drug's stability. The chlorambucil solution was stable (not more than 5% loss) for three and eight months frozen at −20 and −70°C, respectively. Storage at higher temperatures resulted in more rapid loss of the drug. Under refrigeration and at room temperature, loss of 5% occurred in 166 and 22 minutes, respectively. Compared to storage in the dark, exposure to sunlight and to ambient room light reduced the drug's half-life by 41.5 and 11%, respectively. Chlorambucil loss also occurred, presumably due to sorption to polyvinyl chloride (PVC) containers; loss occurred at twice the rate in PVC as in glass, polystyrene, polypropylene, and polyethylene containers. Filtration through cellulose acetate, polysulfone, and polytetrafluoroethylene filter units substantially reduced the drug concentration. Sterilization by filtration was not recommended. Chlorambucil was found to have up to 30% greater stability when normal saline rather than phosphate-buffered saline was used for dilution, presumably due to the higher chloride-ion concentration in the normal saline.

The inability to use heat sterilization due to drug loss and filtration due to sorption makes the preparation of a sterile form of chlorambucil very problematic.

## Stability Reports of Compounded Preparations
### Oral
The stability of chlorambucil 2 mg/mL in an extemporaneous oral suspension was studied by Dressman and Poust.[87] Chlorambucil 2-mg tablets (Burroughs-Wellcome) were crushed and mixed with Cologel (Lilly) as one-third of the total volume, shaken, and brought to final volume with a 2:1 mixture of simple syrup and cherry syrup, followed by vigorous shaking for 30 seconds and ultrasonication for at least two minutes. The chlorambucil 2-mg/mL suspension was packaged in amber glass bottles and stored at 5°C and at ambient room temperature. An HPLC assay was used to determine the chlorambucil content of the suspensions. Chlorambucil decomposed rapidly in suspension, with losses of 15% in one day at room temperature and 10% in seven days at 5°C. The authors recommended storing the suspension under refrigeration and discarding it after seven days. ■

# ■ Chloramphenicol
# Chloramphenicol Palmitate
# Chloramphenicol Sodium Succinate

## Properties
Chloramphenicol is a white, grayish- or yellowish-white fine crystalline powder, fine needlelike crystals, or elongated plates.[2,3]

Chloramphenicol palmitate is a fine white or almost white unctuous crystalline powder with a faint odor and a mild, bland taste. Chloramphenicol palmitate 1.7 g is approximately equivalent to 1 g of chloramphenicol.[2,3]

Chloramphenicol sodium succinate is a white to yellowish-white hygroscopic powder. Approximately 1.4 g of the sodium succinate form is equivalent to 1 g of chloramphenicol. Each gram of chloramphenicol as the sodium succinate contains approximately 2.3 mEq of sodium.[2,3]

### Solubility
Chloramphenicol has an aqueous solubility of 2.5 mg/mL (1:400) at 25°C. It is freely soluble in ethanol, having a solubility of about 400 mg/mL.[2,3,6]

Chloramphenicol palmitate is practically insoluble in water, with a solubility of 1.05 mg/mL, but has a solubility of about 22 mg/mL in ethanol.[1–3]

Chloramphenicol sodium succinate is freely soluble in water (1:<1) and ethanol (1:1).[2,3]

### pH
As an aqueous suspension of 25 mg/mL, chloramphenicol has a pH between 4.5 and 7.5. Chloramphenicol injection has a pH of 5 to 8, while the ophthalmic solutions have a pH near 7 to 7.5 (unless unbuffered). The oral solution has a pH of 5 to 8.5, and the otic solution has a pH of 4 to 8.[4]

Reconstituted chloramphenicol sodium succinate injection has a pH value of 6.4 to 7.[4]

### pKa
Chloramphenicol has a $pK_a$ of 5.5.[2]

### Osmolality
Reconstituted chloramphenicol sodium succinate injection has an osmolality of 533 mOsm/kg.[287]

## General Stability Considerations
Chloramphenicol products should be stored in tight containers at controlled room temperature.[2,4] It has been recommended that oral suspensions of the palmitate be stored in tight, light-resistant containers at controlled room temperature.[4] Aboshiha et al.[1322] reported that in tropical countries, refrigeration of chloramphenicol eyedrops may be

required to prevent the formation of excessive decomposition products.

Chloramphenicol is a relatively stable antibiotic. The pH of maximum stability is 6, but the drug is stable from pH 2 to 7. Aqueous degradation is primarily hydrolytic cleavage of the amide linkage. Chloramphenicol is unstable at alkaline pH. Decomposition may be accelerated by heat and light. Protection from light has been recommended for storage of solutions. The use of amber glass bottles for ophthalmic solutions affords the best light protection. Exposure to light may result in solutions slowly turning yellow and developing an orangish-yellow precipitate.[3,6] The reconstituted injection (as the sodium succinate) is stable for up to 30 days at controlled room temperature and as much as six months frozen. Cloudy solutions should not be used.[2,275,325]

## Stability Reports of Compounded Preparations
### Ophthalmic

Ophthalmic preparations, like other sterile drugs, should be prepared in a suitable clean air environment using appropriate aseptic procedures. When prepared from nonsterile components, an appropriate and effective sterilization method must be employed.

**STUDY 1:** de Vries et al.[21] found substantial formation of toxic decomposition products in chloramphenicol eyedrop formulations exposed to sunlight. The ophthalmic drops were prepared from chloramphenicol 250 mg, boric acid 1.5 g, and borax 300 mg, with and without phenylmercuric borate 4 mg in 100 mL. The solutions were stored in glass vials exposed to sunlight (1 mW/cm² at 360 nm). HPLC analysis found the formation of toxic *p*-nitrobenzaldehyde within minutes of exposure to sunlight in the ophthalmic drops both with and without preservative. Only wavelengths of 285 nm and longer induce this photodecomposition. The authors recommended preparing chloramphenicol eyedrops in diffuse light, with the exclusion of excess light as much as possible to avoid photodecomposition. Furthermore, the preparation should be stored in amber glass vials. Patients should be advised to store chloramphenicol eyedrops in the dark. The authors also found some photodecomposition in a concentration simulating that found intraocularly (0.01 mg/mL).

**STUDY 2:** Attia et al.[46] evaluated the stability of chloramphenicol 1% in eight ophthalmic ointments of varying composition. See Table 21 for the formulas of the bases yielding the greatest chloramphenicol stability. The ointments were prepared by melting white soft paraffin in an oven at 100°C for one hour, with activated charcoal to remove irritating materials present, and then filtering through filter paper. The oleaginous components were sterilized at 160°C for two hours prior to preparing the ointments. For the absorption bases,

**TABLE 21.** Ointment Bases Yielding Greatest Chloramphenicol Stability[46]

***Absorption Base Formula 1:*** White paraffin 70%, liquid paraffin 15%, anhydrous lanolin 15%, benzalkonium chloride 0.01%, sodium pyrophosphate 0.5%, edetate sodium 0.3%

***Absorption Base Formula 2:*** White paraffin 70%, liquid paraffin 20%, cetyl alcohol 10%, benzalkonium chloride 0.01%, sodium pyrophosphate 0.5%, edetate sodium 0.3%

***Emulsion Base (o/w) Formula 3:*** White paraffin 25%, liquid paraffin 12%, cetyl alcohol 25%, distilled water 33%, Tween 40 5%, benzalkonium chloride 0.01%, sodium pyrophosphate 0.5%, edetate sodium 0.3%

***Emulsion Base (w/o) Formula 4:*** White paraffin 30%, cetyl alcohol 15%, propylene glycol 20%, distilled water 30%, Span 40 5%, benzalkonium chloride 0.01%, sodium pyrophosphate 0.5%, edetate sodium 0.3%

chloramphenicol was incorporated into the melted base at low temperatures, with continuous stirring for homogeneity until the ointments became cold. The emulsion formulations were prepared by suspending the chloramphenicol in water and incorporating into the aqueous phase and then stirring into the warmed and melted oleaginous phase.

During 12 months of storage at 25 and 35°C, chloramphenicol was most stable in ointment formulas 1 and 2, the absorption base with antioxidant and chelating agents. In emulsion bases, chloramphenicol was most stable in formula 3, an oil-in-water emulsion, followed by formula 4, a water-in-oil emulsion, both with antioxidant and chelating agents. The other formulations were less stable than these four.

**STUDY 3:** Lv et al.[1037] reported the stability of chloramphenicol 0.27% in two different microemulsion formulations composed of (1) sorbitan monooleate (Span 20) and polyethylene glycol sorbitan monooleate (Tween 20) in equal proportions and (2) sorbitan monooleate (Span 80) and polyethylene glycol sorbitan monooleate (Tween 80) in equal proportions along with n-butanol and isopropyl palmitate 10% in water. HPLC analysis found that the quantity of decomposition products formed during accelerated stability testing at 40 and 75°C over three months was approximately half the amount of decomposition products formed in conventional commercial chloramphenicol eyedrops.

**STUDY 4:** Lv et al.[1016] also reported a substantial improvement in chloramphenicol stability in an ophthalmic microemulsion formulation without an alcohol component compared to a conventional ophthalmic solution. Using an oil-in-water microemulsion prepared from sorbitan monooleate (Span 20) and polyethylene glycol sorbitan monooleate (Tween 20) 4.26%, isopropyl myristate 0.47%, and water qs, chloramphenicol 0.27% was found to be in the hydrophilic shells of the microemulsion drops. HPLC analysis found that the quantity of decomposition products formed during accelerated stability testing at 40 and 75°C over three months was approximately half the amount of decomposition products formed in conventional commercial chloramphenicol eyedrops.

*Enteral*

Ortega de la Cruz et al.[1101] reported the physical compatibility of an unspecified amount of oral liquid chloramphenicol (Chloromycetin, Parke Davis) with 200 mL of Precitene (Novartis) enteral nutrition diet for a 24-hour observation period. No particle growth or phase separation was observed. ■

# ■ Chlorhexidine Diacetate

## Properties

Chlorhexidine diacetate occurs as a white or almost white crystalline powder.[1,3,1413,1414]

### Solubility

Chlorhexidine diacetate is sparingly soluble in water, having an aqueous solubility of 19 mg/mL. It is soluble in ethanol to about 67 mg/mL and slightly soluble in glycerol, propylene glycol, and polyethylene glycol.[1,3,1413,1414]

### pH

Chlorhexidine gluconate 1% aqueous solution has a natural pH of 5.5 to 7. Chlorhexidine gluconate oral rinse has a pH of 5 to 7.[4]

## General Stability Considerations

Chlorhexidine diacetate bulk material should be packaged in well-closed containers and be protected from exposure to light. It is stable at controlled room temperature but is hygroscopic and should be kept in a dry place. Chlorhexidine salts in aqueous solution at acidic or neutral pH may be autoclaved. Chlorhexidine gluconate was found to undergo only a small amount of decomposition upon autoclaving.

However, at alkaline pH a larger amount of decomposition occurs.[1413,1414]

## Stability Reports of Compounded Preparations
### Topical

Allen[1413,1414] reported on a compounded formulation of chlorhexidine diacetate 2% for use as a topical gel. The gel had the following formula:

| | |
|---|---|
| Chlorhexidine diacetate | 2 g |
| Propylene glycol | 30 g |
| Poloxamer F-127 | 22 g |
| Purified water | 46 g |

The recommended method of preparation was to dissolve the chlorhexidine diacetate in the propylene glycol heated to about 70°C and mix thoroughly. The poloxamer-F127 is then added slowly with stirring. Finally, the purified water was to be added and mixed well. The gel was then to be cooled slowly to room temperature with continued stirring. The gel was to be packaged in tight, light-resistant containers. The author recommended a beyond-use date of six months at room temperature because this formula is a commercial medication in some countries with an expiration date of two years or more. ■

# ■ Chloroquine Hydrochloride
# Chloroquine Phosphate

## Properties

Chloroquine is a white to slightly yellow odorless crystalline powder with a bitter taste. The hydrochloride, a white crystalline substance, is prepared using hydrochloric acid. Chloroquine hydrochloride 123 mg is approximately equivalent to 100 mg of chloroquine. The phosphate is a white or almost white odorless hygroscopic crystalline powder with a bitter taste. Chloroquine phosphate 161 mg is approximately equivalent to 100 mg of chloroquine.[1–3,7]

### Solubility

Chloroquine is only slightly soluble in water.[2,3] The hydrochloride and the phosphate are freely soluble in water. The phosphate has an aqueous solubility of about 250 mg/mL but is almost insoluble in ethanol.[3,7]

### pH

Chloroquine hydrochloride injection has a pH between 5.5 and 6.5.[4] A 1% chloroquine phosphate solution has a pH of about 4.5.[1]

## General Stability Considerations

Chloroquine products should be stored in well-closed containers. The injection (as the hydrochloride) should be stored at controlled room temperature and protected from freezing and temperatures over 40°C.[2-4] The phosphate is light sensitive, discoloring upon light exposure.[2,3]

## Stability Reports of Compounded Preparations
### Oral

**STUDY 1:** Closson[41] reported that chloroquine hydrochloride injection (Aralen HCl, Sanofi Winthrop) was added to simple syrup to make a 20-mg/mL oral pediatric dosage form. The product was incubated at 49°C for 63 hours; no visible changes in physical appearance or consistency occurred. The product then was frozen at −6°C for eight hours; it became a white frozen solid and reliquefied upon warming to its original colorless, slightly hazy appearance. No chemical analysis was performed.

**STUDY 2:** Odusote and Nasipuri[422] evaluated the stability of three syrup formulations (Table 22) containing chloroquine phosphate 16 mg/mL prepared from bulk powder. The sucrose syrup formulation was prepared by adding the chloroquine phosphate powder and sodium benzoate to heated syrup and stirring until complete dissolution occurred. The flavor and color were then added. The other two formulations were prepared by dispersing the methylcellulose in some hot water followed by ice-cold water and subsequently keeping the mixture in a freezer. The other materials then were added and mixed well. After preparation, the syrups were packaged in amber bottles and stored at 5, 25, and 40°C for 12 weeks. Chloroquine content was assessed spectrophotometrically. No change in chloroquine concentration and no observable physical change were found in samples from any storage temperature during storage. If clear bottles were used instead of amber, exposure to light resulted in about 6% drug loss in eight weeks. If the syrup pH was adjusted from the original pH 4.5 to 4.9 down to pH 3.5 with citric acid (used to help mask the bitter taste), the chloroquine concentration remained constant over 12 weeks at all temperatures.

However, if 1% sodium carboxymethylcellulose was substituted as the viscosity agent, a white turbidity or precipitate appeared (depending on concentration), along with a sudden drop in viscosity, indicating an interaction with the chloroquine phosphate. Consequently, the authors recommended using only methylcellulose as the viscosity-imparting agent.[422]

**STUDY 3:** Mirochnick et al.[625] attempted to determine the stability of a chloroquine phosphate suspension prepared from tablets. Although no loss of drug was found by HPLC analysis, substantial increases in drug concentration at various time points indicated that a nonuniform dispersion of the drug might have existed.

**STUDY 4:** Allen and Erickson[594] evaluated the stability of three chloroquine phosphate 15-mg/mL oral liquids extemporaneously compounded from tablets. Vehicles used in this study were (1) an equal parts mixture of Ora-Sweet and Ora-Plus (Paddock), (2) an equal parts mixture of Ora-Sweet SF and Ora-Plus (Paddock), and (3) cherry syrup (Robinson Laboratories) mixed 1:4 with simple syrup. Three chloroquine phosphate 500-mg tablets (Sanofi Winthrop) were crushed and comminuted to fine powder using a mortar and pestle. About 15 mL of the test vehicle was added to the powder and mixed to yield a uniform paste. Additional vehicle was added geometrically and brought to the final volume of 100 mL, mixing thoroughly after each addition. The process was repeated for each of the three test suspension vehicles. Samples of each of the finished suspensions were packaged in 120-mL amber polyethylene terephthalate plastic prescription bottles and stored at 5 and 25°C. Because the phosphate salt is freely soluble in water, the drug is in solution in these products.

No visual changes or changes in odor were detected during the study. Stability-indicating HPLC analysis found little

**TABLE 22.** Chloroquine Phosphate Formulations Tested for Stability by Odusote and Nasipuri[422]

| Component | 1 | 2 | 3 |
|---|---|---|---|
| Chloroquine phosphate | 1.6 g | 1.6 g | 1.6 g |
| Sodium benzoate | 0.2 g | 0.2 g | 0.2 g |
| Saccharin sodium | — | 0.05 g | — |
| Talin[a] | — | — | 0.05 g |
| Essence of lemon grass | 0.5 mL | 0.5 mL | 0.5 mL |
| Yellow food color | 0.2 mL | 0.2 mL | 0.2 mL |
| Sucrose syrup 84% (wt/vol) | qs 100 mL | — | — |
| Methylcellulose 1.12% solution | — | qs 100 mL | qs 100 mL |

[a]Sweetening agent derived from the fruit of *Thaumatococcus danielii*.

or no drug loss in any of the liquid products stored at either temperature after 60 days of storage.

**STUDY 5:** Van Doorne et al.[423] evaluated the suitability of several antimicrobial preservatives for use in chloroquine phosphate 16-mg/mL syrup containing sucrose 66%. Chloroform had been used previously, but it is carcinogenic, potentially toxic to liver and kidneys, and is volatile, resulting in loss of protection over time. Benzalkonium chloride is unsuitable because of its taste and incompatibilities. The best result was obtained using sorbic acid 1.5 g/L along with citric acid 2 g/L to reduce the pH to 4. Methylparaben 1.8 g/L with propylparaben 0.2 g/L also was acceptable, although the latter system was not as effective against *Aspergillus niger*.

**STUDY 6:** Chandibhamar et al.[962] evaluated the stability of a taste-masked suspension of chloroquine. Chloroquine phosphate 1.7 g was dissolved in 40 mL of simple syrup containing glycerin 5% (vol/vol). Then 30 mL of hot syrup containing pamoate sodium 1.3 g and sodium bicarbonate 0.56 g was slowly mixed in at a rate of 5 mL per minute with constant stirring of 30 to 80 rpm. A precipitate of chloroquine pamoate was produced. The suspension was adjusted to pH 6 and sodium benzoate 100 mg, amaranth, and raspberry flavor were added. The suspension was brought to 100 mL with additional syrup. The suspension exhibited a very slow rate of sedimentation. Spectroscopic analysis found the suspension remained stable over at least 42 days at 25°C with more than 98% of the ion pair remaining and only 1.75% of free chloroquine present. Elevated temperature of 45°C resulted in an approximate doubling of the rate of free chloroquine formation. Bioavailability of this suspension was comparable with chloroquine phosphate liquid.

### Topical

Brouwers et al.[1333] developed a topical gel containing chloroquine phosphate for use as a microbicide against HIV-1 infection. The gel was prepared by adding a mixture of hydroxyethyl cellulose 1.6% wt/wt and glycerol 2.5% to a solution of methyl- and propylparabens 0.18% and 0.02%, respectively. When mixed, a clear and homogeneous gel formed. The pH of the gel was decreased by adding lactic acid 0.05% and adjusting to pH 4.5 by adding sodium hydroxide 1 M. Chloroquine phosphate powder was added to the gel in varying amounts of 0.3, 1.3, 3, 10, and 30 mg/g of gel and

mixed thoroughly to ensure complete dissolution and uniformity. Entrapped air was removed by using reduced pressure. The completed gels were packaged in capped syringes. The gels were clear and homogeneous with an osmolality of 300 mOsm/kg, a pH of 4.6, and a viscosity of 1.4 Pa s.

Samples containing chloroquine phosphate 3 mg/g were stored at 40°C and 75% relative humidity for three months. Little or no change in gel mass, pH, and osmolality occurred. Viscosity decreased by 14%, which is similar to changes observed in previously reported observations for hydroxyethyl cellulose gels. HPLC analysis found that little or no change in chloroquine concentrations occurred over the three-month test period.

### Injection

Injections, like other sterile drugs, should be prepared in a suitable clean air environment using appropriate aseptic procedures. When prepared from nonsterile components, an appropriate and effective sterilization method must be employed.

Allen[1357] reported on a compounded formulation of chloroquine phosphate 64.5-mg/mL injection. The injection had the following formula:

| | | |
|---|---|---|
| Chloroquine phosphate | | 6.45 g |
| Benzyl alcohol | | 2 g |
| Sterile water for injection | qs | 100 mL |

The recommended method of preparation is to dissolve the chloroquine phosphate powder in about 90 mL of sterile water for injection. The benzyl alcohol is then added and stirred until dissolved. Sterile water for injection sufficient to bring the volume to 100 mL is added and the solution is mixed well. The solution is to be filtered through a suitable 0.2-µm sterilizing filter and packaged in sterile containers. If no sterility test is performed, the USP specifies a beyond-use date of 24 hours at room temperature or three days stored under refrigeration because of concern for inadvertent microbiological contamination during preparation. However, if an official USP sterility test for each batch of drug is performed, the author recommended a beyond-use date of six months at room temperature because this formula is similar or the same as a commercial medication in some countries with an expiration date of two years or more. ■

# Chlorothiazide
# Chlorothiazide Sodium

## Properties

Chlorothiazide is a white or nearly white odorless crystalline powder having a bitter taste.[2,3] Chlorothiazide sodium 537 mg is equivalent to chlorothiazide 500 mg.[3]

Chlorothiazide sodium for injection is prepared by neutralizing chlorothiazide solution with sodium hydroxide. It is a white lyophilized powder containing mannitol.[2]

### Solubility

Chlorothiazide is very slightly soluble in water, with a solubility at pH 7 of 0.65 mg/mL and 0.5 mg/mL at pH 4.1.[6] It is soluble in alkaline solutions,[1] having a solubility of 100 mg/mL in 1 N sodium hydroxide.[6] It is slightly soluble in alcohol, with a solubility of 2 mg/mL.[6,586]

### pH

Chlorothiazide oral suspension has a pH of 3.2 to 4. Chlorothiazide sodium for injection has a pH of 9.2 to 10.[4]

### pK$_a$

Chlorothiazide has pK$_a$ values of 6.7 or 6.85 and 9.5.[1,2]

## General Stability Considerations

Chlorothiazide tablets are packaged in well-closed containers, and the oral suspension is packaged in tight containers.[4] The products should be stored at controlled room temperature and protected from freezing.[7] The injection should be stored at controlled room temperature. After reconstitution, the solution should be stored between 2 and 25°C.[7] The reconstituted solution is stable for 24 hours, but precipitation may occur if the pH is less than 7.4.[2,7]

Chlorothiazide is subject to hydrolysis in both acidic and basic media.[6] The maximum stability of chlorothiazide has been variably reported to occur in the range of pH 3.5 to 4.[586]

## Stability Reports of Compounded Preparations
### Oral

**STUDY 1:** Woods[586] reported on an extemporaneous oral suspension formulation of chlorothiazide 50 mg/mL prepared

**TABLE 23.** Formula of Chlorothiazide 50-mg/mL Oral Suspension Prepared from Tablets[586]

| Component | Amount | |
|---|---|---|
| Chlorothiazide 500-mg tablets | | 10 |
| Sodium carboxymethylcellulose | | 2 g |
| Glycerol | | 10 mL |
| Parabens | | 0.1% |
| Citric acid monohydrate (approximately 800 mg is required to adjust the pH) | qs | pH 4 |
| Water | qs | 100 mL |

from tablets. The formula is shown in Table 23. An expiration period of 30 days when stored under refrigeration protected from light was suggested, although no stability data were presented. If a preservative-free formulation omitting the parabens is prepared, the author recommended restricting the use period to seven days.

**STUDY 2:** Tucker et al.[758] reported the effects of two preparation methods on physical stability and dosing variations in two extemporaneously prepared formulations of chlorothiazide 40-mg/mL suspensions prepared from 500-mg tablets (Frosst). The tablets were triturated by hand or comminuted in a Vitamizer blender and then mixed with lemon syrup, concentrated chloroform water, distilled water, and then tragacanth powder to form a 2% tragacanth concentration. The suspension prepared with hand-triturated tablets was found to have 30% of the particles larger than 220 μm, whereas the suspension prepared using the blender had all of the particles less than 220 μm. Spectrophotometric analysis found variations ranging from one-half to four times the required dose in the first and last dose samples with the hand-triturated suspension. A second suspension prepared using the blender and increasing the tragacanth to 4% resulted in much greater consistency in dosing, but also had some variability in viscosity over the six-day study period. ■

# Chlorpheniramine Maleate

## Properties

Chlorpheniramine maleate is a white odorless crystalline powder.[2,3]

## Solubility

Chlorpheniramine maleate has solubilities of 160[1] to 250[2,3] mg/mL in water and about 100 mg/mL in ethanol.[2,3]

## pH

Chlorpheniramine maleate injection has a pH of 4 to 5.2.[4] Chlorpheniramine maleate with phenylpropanolamine hydrochloride syrup (Triaminic, Dorsey) had a measured pH of 4.6.[19]

## pKa

Chlorpheniramine has a $pK_a$ of about 9.2.[2]

## General Stability Considerations

Chlorpheniramine maleate products should be stored in tight containers at controlled room temperature and protected from temperatures exceeding 40°C. The liquid forms should be protected from light (to prevent discoloration) and from freezing.[2,4]

## Stability Reports of Compounded Preparations

### Nasal

Soliman et al.[1601] evaluated numerous possible formulations for a chlorpheniramine maleate 1% (wt/wt) nasal gel with 0.1% (wt/wt) methylparaben and 0.01% (wt/wt) propylparaben preservatives. Gelling agents included sodium carboxymethyl cellulose 3, 4, 5, and 6%, hydroxypropyl methylcellulose 4, 5, and 6%, Carbopol 934 0.5, 0.75, 1, 1.25, and 1.5%, and Pluronic F127 20, 25, and 30%. The formulations were evaluated for physical characteristics such as color, odor, and spreadability, sensitization or irritation potential,

pH, rheological behavior, drug release and diffusion, and drug stability during freeze thaw cycle (50°C and −5°C each for 48 hours) and at 25°C for one year.

All gels exhibited good spreadability and none generated any sensitization reactions. The gels all had pH values in the range of pH 5.5 to 6.5, similar to nasal fluids pH. None of the gels resulted in any loss of drug content over 12 months at room temperature. Freeze-thaw, however, resulted in rheological changes to some of the gels. Drug release and diffusion was found to vary substantially among the various gel formulations. The authors stated that two gel formulations yielded the best overall results: sodium carboxymethyl cellulose 3% and Carbopol 934 0.5%.

### Enteral

**CHLORPHENIRAMINE MALEATE COMPATIBILITY SUMMARY**
*Compatible with:* Ensure • Ensure HN • Ensure Plus • Ensure Plus HN • Osmolite • Osmolite HN • Vital

**STUDY 1:** Cutie et al.[19] added 5 mL of chlorpheniramine maleate with phenylpropanolamine hydrochloride syrup (Triaminic, Dorsey) to varying amounts (15 to 240 mL) of Ensure, Ensure Plus, and Osmolite (Ross Laboratories) with vigorous agitation to ensure thorough mixing. The syrup was physically compatible, distributing uniformly in all three enteral products, with no phase separation or granulation.

**STUDY 2:** Altman and Cutie[850] reported the physical compatibility of chlorpheniramine maleate with phenylpropanolamine hydrochloride syrup (Triaminic, Dorsey) 5 mL with varying amounts (15 to 240 mL) of Ensure HN, Ensure Plus HN, Osmolite HN, and Vital that had been vigorously agitated to ensure thorough mixing. The Triaminic syrup was physically compatible, distributing uniformly in all four enteral products, with no phase separation or granulation. ■

# Chlorpromazine Hydrochloride

## Properties

Chlorpromazine hydrochloride is a white to creamy-white crystalline powder having a very bitter taste. Approximately 111 mg of the hydrochloride is equivalent to 100 mg of chlorpromazine.[2,3]

## Solubility

The aqueous solubility of chlorpromazine hydrochloride has been variously cited as 1 g/2.5 mL[1] and about 1 g/mL.[2,3] Solubility in ethanol has been cited as 667 mg/mL.[2,3]

## pH

A freshly prepared 10% chlorpromazine hydrochloride aqueous solution has a pH of 3.5 to 4.5.[3] Chlorpromazine hydrochloride injection has a pH of 3.4 to 5.4.[4] The oral concentrate solution has a pH of 2.3 to 4.1.[4] Thorazine brand of oral chlorpromazine hydrochloride concentrate had a pH of 3.[19]

## Osmolality

The osmolality of chlorpromazine hydrochloride injection 25 mg/mL is 262 mOsm/kg.[326]

## General Stability Considerations

Chlorpromazine hydrochloride may exhibit color changes upon prolonged light exposure. Yellow, pink, and violet colors may form. A slight yellow discoloration does not indicate significant concentration loss, but a markedly discolored solution should be discarded. Consequently, chlorpromazine hydrochloride tablets are packaged in well-closed light-resistant containers, while liquid products should be stored in tight containers protected from light at controlled room temperature. They also should be protected from temperatures above 40°C, and the liquid products should be protected from freezing.[2–4]

Chlorpromazine hydrochloride exhibits maximum stability in aqueous solution at pH 6.[327] In alkaline media, the drug is subject to oxidation.[2] Also, titration with an alkaline solution results in precipitation of chlorpromazine base at about pH 6.7.[328]

Chlorpromazine hydrochloride undergoes sorptive loss to polyvinyl chloride containers and administration sets. Sorption also occurs to cellulose propionate burette chambers and Silastic tubing. However, no significant loss due to sorption seems to occur with polyethylene or polypropylene materials.[330–332]

## Stability Reports of Compounded Preparations
### Oral

Dugas[76] reported the stability of chlorpromazine hydrochloride oral concentrate (Thorazine, Smith Kline & French) diluted with distilled water to 50, 200, and 500 mg/15 mL. The dilutions were repackaged (15 mL in 30-mL amber glass unit-dose vials sealed with aluminum closures or in clear glass screw-cap vials) and were stored at 27°C exposed to light or in the dark. Chlorpromazine content was determined using gas–liquid chromatography. Chlorpromazine hydrochloride loss was 8% or less after 20 weeks of storage in the dark. For samples exposed to light in amber containers, losses of 5 to 14% occurred; in clear glass containers losses of up to 40% were found.

### Injection

Injections, like other sterile drugs, should be prepared in a suitable clean air environment using appropriate aseptic procedures. When prepared from nonsterile components, an appropriate and effective sterilization method must be employed.

DeVane and Wailand[329] found by HPLC analysis that a 1-mg/mL dilution of chlorpromazine hydrochloride injection in 0.9% sodium chloride injection stored in 5-mL vials at 18 to 23°C in the dark remained stable for up to 30 days.

### Enteral
#### CHLORPROMAZINE HYDROCHLORIDE COMPATIBILITY SUMMARY
*Compatible with:* Vital • Vivonex T.E.N.
*Incompatible with:* Enrich • Ensure • Ensure HN • Ensure Plus
• Ensure Plus HN • Osmolite • Osmolite HN • TwoCal HN

**STUDY 1:** Cutie et al.[19] added 1 mL of chlorpromazine hydrochloride solution concentrate (Thorazine, SmithKline Beecham) to varying amounts (15 to 240 mL) of Ensure, Ensure Plus, and Osmolite (Ross Laboratories) with vigorous agitation to ensure thorough mixing. The chlorpromazine hydrochloride concentrate was physically incompatible with the enteral products, forming particles and granules at the point of mixing. The particles and granules may clog feeding tubes.

**STUDY 2:** Burns et al.[739] reported the physical compatibility of chlorpromazine hydrochloride solution concentrate 100 mg/1 mL with 10 mL of three enteral formulas, including Enrich, TwoCal HN, and Vivonex T.E.N. Visual inspection found no physical incompatibility with Vivonex T.E.N. However, when mixed with Enrich, a soft thickened consistency resulted. When mixed with TwoCal HN, a thin consistency and hard particulates developed.

**STUDY 3:** Altman and Cutie[850] reported the physical compatibility of 1 mL of chlorpromazine hydrochloride solution concentrate (Thorazine, SmithKline Beecham) with varying amounts (15 to 240 mL) of Ensure HN, Ensure Plus HN, Osmolite HN, and Vital after vigorous agitation to ensure thorough mixing. The chlorpromazine hydrochloride concentrate was physically incompatible with Ensure HN, Ensure Plus HN, and Osmolite HN enteral products, forming particles and granules at the point of mixing. The particles and granules may clog feeding tubes. Chlorpromazine hydrochloride concentrate was physically compatible with Vital.

## Compatibility with Other Drugs
### CHLORPROMAZINE HYDROCHLORIDE COMPATIBILITY SUMMARY
*Incompatible with:* Carbamazepine • Lithium citrate
• Thioridazine

**STUDY 1:** Wilson[72] reported precipitate formation when lithium citrate syrup (Lithionate-S, Rowell) and chlorpromazine hydrochloride oral concentrate (Thorazine, Smith Kline & French) were mixed together directly.

**STUDY 2:** Theesen et al.[78] evaluated the compatibility of lithium citrate syrup with oral neuroleptic solutions including chlorpromazine hydrochloride concentrate. Lithium citrate syrup (Lithionate-S, Rowell) and lithium citrate syrup (Roxane) at a lithium-ion concentration of 1.6 mEq/mL were combined in volumes of 5 and 10 mL with 0.1, 1, 5, and 10 mL of chlorpromazine hydrochloride oral concentrate 100 mg/mL (both Lederle and Smith Kline & French) in duplicate, reversing the order of mixing between samples. Samples were stored at 4 and 25°C for six hours and evaluated visually for physical compatibility. In all samples, an opaque turbidity formed immediately, eventually separating into two layers. Thin-layer chromatography of the sediment layers of incompatible combinations showed high concentrations of the neuroleptic drugs. The authors attributed the incompatibilities to the excessive ionic strength of the lithium citrate syrup. Salting out of the neuroleptic drug salts results in a hydrophobic, viscid layer that adheres tenaciously to container surfaces. The problem persists even when moderate to high doses are diluted 10-fold in water.

**STUDY 3:** Raleigh[454] evaluated the compatibility of lithium citrate syrup (Philips Roxane) 1.6 mEq lithium/mL with chlorpromazine hydrochloride concentrate (concentration unspecified). The chlorpromazine hydrochloride concentrate was combined with the lithium citrate syrup in a manner described as "simple random dosage combinations"; the exact amounts tested and test conditions were not specified. A milky precipitate formed upon mixing.

Raleigh[454] also evaluated the compatibility of thioridazine suspension (Sandoz) (concentration unspecified) with chlorpromazine hydrochloride concentrate (concentration unspecified). The chlorpromazine hydrochloride concentrate was combined with the thioridazine suspension in the manner described as "simple random dosage combinations." The exact amounts tested and test conditions were not specified. A dense white curdy precipitate formed.

**STUDY 4:** Mixing chlorpromazine hydrochloride oral solution with carbamazepine oral solution is reported to result in the formation of an orange rubbery precipitate.[2]

## Compatibility with Common Beverages and Foods

### CHLORPROMAZINE HYDROCHLORIDE
### COMPATIBILITY SUMMARY

*Compatible with:* Grapefruit juice • Lemon-lime soda • Puddings • Soups • Water

*Incompatible with:* Apple juice • Cranberry juice • Ginger ale • Grape juice • Pineapple juice • Root beer • Tang

*Uncertain or variable compatibility with:* Apricot juice • Coffee • Cola sodas • Milk • Orange juice • Orange soda • Prune juice • Tea • Tomato juice • V8 vegetable juice

**STUDY 1:** Theesen et al.[78] also noted that fruit and vegetable beverages, which are often used to administer the neuroleptics, also are effective salting-out agents. Unfortunately, the color and opacity of the beverages impair observing the phenomenon. However, chlorpromazine hydrochloride concentrate was observed to be incompatible with orange, prune, and cranberry juices as well as Tang.

**STUDY 2:** Geller et al.[242] reported that chlorpromazine liquid was compatible with water, milk, apricot juice, grapefruit juice, orange juice, tomato juice, vegetable juice (V8), and lemon-lime soda. Chlorpromazine was stated to be incompatible with coffee, tea, apple juice, cola soda, root beer, and ginger ale.

**STUDY 3:** Kerr[256] also cited the compatibility of chlorpromazine with coffee, tea, cola sodas, orange juice, orange sodas, and soups and puddings. However, cranberry juice, prune juice, and Tang were stated to be incompatible.

**STUDY 4:** Lever and Hague[1011] reported the precipitation and/or color change of chlorpromazine hydrochloride oral concentrate 50 mg mixed with 15 mL of a number of taste-masking liquids, including Coca-cola, carbonated orange soda, milk, coffee, tea, pineapple juice, orange juice, V8 vegetable juice, grape juice, apple juice, and tomato juice. Apricot juice and prune juice were difficult to assess and may have masked turbidity and/or precipitation. The precipitation occurred within a few minutes of mixing and clung to the sides of the container, indicating unacceptable variable dosage may occur. The authors recommended mixing chlorpromazine hydrochloride oral concentrate freshly with distilled water only.

The problematic nature of visual observation of incompatibilities in opaque liquids may be the reason Geller et al.[242] reported both orange juice (from frozen concentrate) and prune juice as compatible with chlorpromazine, whereas Theesen et al.[78] and Lever and Hague[1011] noted an incompatibility. To complicate matters, Kerr[256] reported chlorpromazine to be compatible with orange juice but incompatible with prune juice. Because the compatibility determinations require making judgments, it is also possible that different investigators come to differing compatibility conclusions. ■

# Chlorprothixene

## Properties

Chlorprothixene is a yellow crystalline powder having a slight amine odor.[1,3]

### Solubility

Chlorprothixene is essentially insoluble in water, having an aqueous solubility of about 0.6 mg/mL. Its solubility in ethanol is about 34 mg/mL.[1,3] The injection is prepared with hydrochloric acid to form the hydrochloride.[4]

### pH

Chlorprothixene injection has a pH between 3 and 4, while the pH of the oral suspension is between 3.5 and 4.5.[4]

## General Stability Considerations

Chlorprothixene products should be stored in tight, light-resistant containers protected from light.[3,4]

## Compatibility with Common Beverages and Foods

Geller et al.[242] reported that chlorprothixene liquid was visually compatible with water, milk, coffee, fruit juices, and carbonated beverages, although visual observation in opaque liquids is problematic. ▪

# Chlortetracycline Hydrochloride

## Properties

Chlortetracycline hydrochloride is a yellow odorless crystalline powder having a bitter taste.[1,3]

### Solubility

Chlortetracycline hydrochloride has an aqueous solubility of about 13 mg/mL.[3] The solubility in ethanol is reported to be 1.7 mg/mL.[1,3]

### pH

A 1% chlortetracycline hydrochloride aqueous solution has a pH of 2.3 to 3.3.[3] A saturated aqueous solution has a pH of 2.8 to 2.9.[1]

## General Stability Considerations

Chlortetracycline hydrochloride products should be stored in tight, light-resistant containers. The drug may darken upon exposure to light and moisture.[3,4]

## Stability Reports of Compounded Preparations

### Ophthalmic

Ophthalmic preparations, like other sterile drugs, should be prepared in a suitable clean air environment using appropriate aseptic procedures. When prepared from nonsterile components, an appropriate and effective sterilization method must be employed.

Attia et al.[46] evaluated the stability of chlortetracycline hydrochloride 0.5% in 10 ophthalmic ointments of varying composition (Table 24). The ointments were prepared by melting white soft paraffin in an oven at 100°C for one hour with activated charcoal to remove irritating materials present and filtering through filter paper. The oleaginous components were sterilized at 160°C for two hours prior to preparing the ointments. For the absorption bases, chlortetracycline hydrochloride was incorporated into the melted base at low temperatures with continuous stirring for homogeneity until the ointments became cold. The emulsion formulations were prepared by incorporating the drug into the aqueous phase and then stirring into the warmed and melted oleaginous phase. Drug concentration was determined spectrophotometrically. Chlortetracycline hydrochloride was most stable during 12 months of storage at 25 and 35°C in ointment formulas 1 and 2, the absorption bases with antioxidant and chelating agents. In emulsion bases, chlortetracycline hydrochloride was most stable in 4, a water-in-oil emulsion, followed by 3, an oil-in-water emulsion. All other formulations were less stable than these four. ▪

**TABLE 24.** Ointment Bases Yielding Greatest Chlortetracycline Hydrochloride Stability[46]

*Absorption Base Formula 1:* White paraffin 70%, liquid paraffin 15%, anhydrous lanolin 15%, benzalkonium chloride 0.01%, sodium pyrophosphate 0.5%, edetate sodium 0.3%

*Absorption Base Formula 2:* White paraffin 70%, liquid paraffin 20%, cetyl alcohol 10%, benzalkonium chloride 0.01%, sodium pyrophosphate 0.5%, edetate sodium 0.3%

*Emulsion Base (o/w) Formula 3:* White paraffin 25%, liquid paraffin 12%, cetyl alcohol 25%, distilled water 33%, Tween 40 5%, benzalkonium chloride 0.01%, sodium pyrophosphate 0.5%, edetate sodium 0.3%

*Emulsion Base (w/o) Formula 4:* White paraffin 30%, cetyl alcohol 15%, propylene glycol 20%, phosphate buffer (pH 6.5) 30%, Span 40 5%, benzalkonium chloride 0.01%, sodium pyrophosphate 0.5%, edetate sodium 0.3%

# Cholecalciferol
## (Vitamin D₃)

## Properties

Cholecalciferol, a naturally occurring form of vitamin D, is white or almost white odorless crystalline needles.[3,6]

### Solubility

Cholecalciferol is insoluble in water but soluble in ethanol and fatty oils.[3,6]

## General Stability Considerations

Cholecalciferol is sensitive to heat, light, oxygen, and humidity.[3,6] It should be stored in hermetically sealed containers under nitrogen in a cool place protected from light.[4] Above 40°C and greater than 45% humidity it is quite unstable, exhibiting about 35% loss in seven days. Commercially formulated products are stabilized with antioxidants and solubilizers.[6] Cholecalciferol solution should be packaged in tight, light-resistant containers and stored at controlled room temperature.[4]

Huyghebaert et al.[1233] evaluated the stability of the bulk powder of the cold-water-soluble form of cholecalciferol after opening the original container. The cold-water-soluble form of vitamin D₃ is embedded in a coarse gelatin matrix. After opening the original container, packages were stored without applying an inert gas layer for eight months at both room temperature and under refrigeration at 8°C. Stability-indicating HPLC analysis found the bulk substance remained within the specifications of 95 to 105% for drug content for the entire eight-month period.

## Stability Reports of Compounded Preparations

### Injection

Injections, like other sterile drugs, should be prepared in a suitable clean air environment using appropriate aseptic procedures. When prepared from nonsterile components, an appropriate and effective sterilization method must be employed.

Bertino et al.[79] evaluated the stability of an extemporaneous formulation of cholecalciferol injection 40,000 IU/mL. The crystalline powder of cholecalciferol was dissolved in a solution composed of propylene glycol with ethanol 10%, filtered through a compatible 0.22-μm filter, and packaged in 2-mL vials. The vials were stored at 4°C and protected from light. The cholecalciferol content was assessed by HPLC. Approximately 7.3% loss occurred after 199 days of storage.

Cholecalciferol prepared as an aqueous injection of 300,000 IU/mL with 10% Tween 80 and 10% ethanol and as an oil injection of 600,000 IU/mL was found to undergo losses of 15 and 25%, respectively, after autoclaving.[6]

### Oral

**STUDY 1:** Cholecalciferol was prepared as a syrup with 0.15% Tween 80 and various stabilizers, including the following:

- ethyl gallate 0.01%
- butylated hydroxytoluene (BHT) 0.01%
- citraconic acid 0.01% and BHT 0.01%
- citric acid 0.1% and BHT 0.01%
- BHT 0.01% and ascorbic acid 0.01%

When stored at 17°C, the syrups were stable for six months. At 37°C, the syrups were stable for two months.[6]

**STUDY 2:** Huyghebaert et al.[1233] evaluated the stability of the cold-water-soluble form of cholecalciferol prepared as oral capsules containing 1000 units of vitamin D₃. The cold-water-soluble form of cholecalciferol is embedded in a coarse gelatin matrix. The vitamin powder was mixed with lactose 80 mesh by gently shaking the two powders together in a glass bottle. More vigorous mixing using a mortar and pestle was not employed in order to avoid destruction of the vitamin's gelatin matrix. The powder mixture was filled into size 2 hard-gelatin capsules that were then stored at room temperature and refrigerated at 8°C. Stability-indicating HPLC analysis found that the cholecalciferol content in the capsules remained within the specifications of 90 to 110% for drug content for the entire two-month period at both storage temperatures. ■

# Cidofovir

## Properties

Cidofovir occurs as a fluffy white crystalline powder.[1,7]

### Solubility

Cidofovir has a solubility greater than 170 mg/mL in aqueous solutions at pH 6 to 8.[7]

### pH

Cidofovir injection 75 mg/mL (Gilead) is adjusted to pH 7.4 with sodium hydroxide and/or hydrochloric acid during manufacturing.[7]

### Osmolality

Cidofovir injection is a hypertonic concentrate that must be diluted for use.[7,1615]

In dextrose 5%, cidofovir 0.21 and 8.12 mg/mL has osmolalities 241 and 286 mOsm/kg, respectively.[1615]

In sodium chloride 0.9%, cidofovir 0.21 and 8.12 mg/mL has osmolalities 241 and 286 mOsm/kg, respectively.[1615]

## General Stability Considerations

Cidofovir injection is stored at controlled room temperature between 20 and 25°C. Upon dilution in 100 mL of sodium chloride 0.9% for clinical use, the manufacturer indicates that use within 24 hours is required. If not used immediately, storage under refrigeration at 2 to 8°C is necessary, but use within 24 hours of preparation is still required.[2,7]

Cidofovir 0.2 and 8.1 mg/mL in sodium chloride 0.9% in polyvinyl chloride and polyethylene-polypropylene containers stored frozen at −20°C was physically and chemically stable for five days.[1616]

## Stability Reports of Compounded Preparations

### Ophthalmic

Ophthalmic preparations, like other sterile drugs, should be prepared in a suitable clean air environment using appropriate aseptic procedures. When prepared from nonsterile components, an appropriate and effective sterilization method must be employed.

Stiles et al.[1617] evaluated in vitro the antiviral activity of cidofovir 0.5% in sodium chloride 0.9% packaged in glass and plastic vials for use as a veterinary ophthalmic solution. Samples were stored at 4, −20, and −80°C for 180 days. The antiviral activity against feline herpes virus (FHV-1) was tested after storage for 30, 60, 120, and 180 days. The antiviral activity was found to be unchanged at any point during the six-month test period.

### Injection

Injections, like other sterile drugs, should be prepared in a suitable clean air environment using appropriate aseptic procedures. When prepared from nonsterile components, an appropriate and effective sterilization method must be employed.

Hennere et al.[1618] evaluated the stability of cidofovir 6.25 mg/mL in sodium chloride 0.9% packaged in polypropylene syringes. No visible particulate matter or color changes occurred. Stability-indicating HPLC analysis found little or no loss of cidofovir in 150 days in samples stored refrigerated at 4 to 8°C, at room temperature of 25°C exposed to or protected from light, and at an elevated temperature of 32°C. ■

# Cimetidine
# Cimetidine Hydrochloride

## Properties

Cimetidine is a white to off-white crystalline powder having a bitter taste and a characteristic odor. Cimetidine hydrochloride also is a white crystalline powder.[2,3,7]

### Solubility

Cimetidine is slightly soluble in water;[2,3] the solubility at 37°C has been cited as 1.14%.[1] Cimetidine is soluble in ethanol.[2,3]

Cimetidine hydrochloride is very soluble in water and soluble in alcohol.[2]

### pH

Cimetidine hydrochloride injection has a pH of 3.8 to 6. The premixed infusion has a pH of 5 to 7.[4,7]

### pK$_a$

Cimetidine has a pK$_a$ of 6.8, and cimetidine hydrochloride has a pK$_a$ of 7.11.[2]

### Osmolality

Cimetidine hydrochloride liquid 300 mg/5 mL (Tagamet, Smith Kline & French) was found by Niemiec et al.[232] to have an osmolality of 4035 mOsm/kg, while Dickerson and Melnik[233] reported the osmolality to be 5550 mOsm/kg.

## General Stability Considerations

Cimetidine products should be stored in tight, light-resistant containers at controlled room temperature.[2–4] Cimetidine hydrochloride injection should not be refrigerated because of the potential for precipitation. The precipitate can be redissolved without drug degradation by rewarming the solution.[333]

## Stability Reports of Compounded Preparations
### Oral

**STUDY 1 (REPACKAGED):** The stability of commercial cimetidine hydrochloride oral liquid (Tagamet, Smith Kline & French) 60 mg/mL repackaged as 2 mL in 5-mL amber polypropylene oral syringes (Baxa) and 15-mL amber glass vials (Wheaton Scientific) with suitable closures was investigated by Christensen et al.[86] Samples were stored at 4 and 25°C as well as at several elevated temperatures, protected from light and in constant moisture conditions for 180 days. An HPLC assay was used to assess the drug content. Losses were less than 7% at 4 and 25°C in both containers in 180 days. Substantially greater losses were incurred at higher temperatures. The authors calculated the estimated shelf life at 25°C, based on a cimetidine hydrochloride loss of 10%. In polypropylene oral syringes the calculated room temperature shelf life was 317 days; in the glass vials it was 332 days. After repackaging, no loss of cimetidine hydrochloride occurred in the oral liquid remaining in the manufacturer's original container when it was stored at room temperature for 180 days.

**STUDY 2 (EXTEMPORANEOUS):** Tortorici[63] reported the stability of an extemporaneous oral suspension of cimetidine 300 mg/5 mL. The suspension was prepared by triturating 24 cimetidine 300-mg tablets (Tagamet, Smith Kline & French), levigating with 10 mL of glycerin, and bringing to 120 mL with simple syrup. The mixture was placed in a blender and blended until smooth. Cimetidine was stable, retaining 98% concentration after 17 days of storage at 4°C. Furthermore, no obvious fungal or bacterial growth occurred during storage, although true microbiological stability was not tested.[67] Sparkman[66] noted that soaking the tablets in sterile water for three to five minutes before triturating dissolved the film coating, facilitating a more elegant preparation.

### Injection

Injections, like other sterile drugs, should be prepared in a suitable clean air environment using appropriate aseptic procedures. When prepared from nonsterile components, an appropriate and effective sterilization method must be employed.

Nahata et al.[493] evaluated the stability of extemporaneously compounded cimetidine hydrochloride 15-mg/mL injection diluted for pediatric use. Commercial cimetidine hydrochloride injection (SmithKline Beecham) containing 150 mg/mL of drug was diluted with sterile water for injection to yield a 15-mg/mL solution. The diluted solution

was transferred to sterile glass vials and stored at 4 and 22°C for up to 91 days. Visual inspection found no changes in the appearance of the solutions. Stability-indicating HPLC analysis determined that at room temperature stability was retained through 14 days, but 10% loss of cimetidine occurred after 28 days. Under refrigeration, stability was better. Stability was maintained through 42 days, but 13% loss of cimetidine occurred after 56 days.

### Enteral
#### CIMETIDINE COMPATIBILITY SUMMARY
*Compatible with:* Enrich • Isocal • Sondalis Iso • TwoCal HN • Vital • Vivonex T.E.N.
*Incompatible with:* Osmolite • Osmolite HN
*Uncertain or variable compatibility with:* Sustacal • Sustacal HC

**STUDY 1:** Strom and Miller[44] evaluated the stability of cimetidine (Tagamet, Smith Kline & French) from 300-mg tablets and oral liquid 300 mg/5 mL when mixed with three enteral formulas at full- and half-strength. The ground tablets and the oral liquid were dispersed in approximately 10 mL of deionized water and added to 240 mL of Isocal, Sustacal, and Sustacal HC (Mead Johnson) full- or half-strength mixed with deionized water. The mixtures were stirred with a magnetic stirrer for five minutes; duplicate samples of each mixture were stored at 24°C. The mixtures were evaluated visually for precipitation and phase separation, but no physical incompatibilities were observed. A stability-indicating HPLC assay was employed to assess the cimetidine content at various times over 24 hours. The cimetidine content was between 97 and 100% after 24 hours in each enteral mixture prepared from either drug source. Studies to determine the amount of bound and free cimetidine found about 84% free drug in the half-strength Isocal but about 69% free in the two Sustacal products that have higher protein contents. The stability of the enteral formula components was not evaluated.

**STUDY 2:** Fagerman and Ballou[52] reported on the compatibility of cimetidine hydrochloride oral liquid (Tagamet, Smith-Kline Beecham) 300 mg/5 mL mixed in equal quantities with three enteral feeding formulas. The cimetidine hydrochloride was compatible visually with Vital (Ross) but underwent unacceptable thickening on contact with Osmolite and Osmolite HN (Ross). A precipitate formed in the mixtures upon standing.

**STUDY 3:** Rochard and Rogeron[53] tested the stability of cimetidine hydrochloride injection (Tagamet, Smith Kline & French) when mixed with Sondalis Iso (Sopharga Laboratory) enteral feed at a 2-mg/mL concentration. The mixture was stored in glass containers at 4 and 21°C for seven days. No visible color change, precipitation, or solution separation occurred. Cimetidine content (determined by a stability-indicating HPLC assay) was retained above 90% through 48 hours. This result is sufficient to permit drug administration of the enteral product to hospitalized patients.

## Cinnarizine

### Properties
Cinnarizine is a white or nearly white powder.[3]

#### Solubility
Cinnarizine is practically insoluble in water, having an aqueous solubility of 20 mcg/mL. Cinnarizine is slightly soluble in ethanol and methanol, soluble in acetone, and freely soluble in dichloromethane.[3]

#### pH
Aqueous solutions having a pH less than pH 4 are needed to increase cinnarizine solubility,[1429] but the drug is labile in acidic solutions.[1430] As a consequence, Shi et al.[1429] prepared an injectable emulsion formulation having a pH pf 8.5.

### General Stability Considerations
Cinnarizine should be stored at room temperature and protected from exposure to light.[3] Cinnarizine in aqueous solutions is unstable, making a suitable aqueous injection formulation difficult to obtain. Solubility of the drug can be increased by lowering the solution pH below 4.[1429] However, cinnarizine is labile in acidic solutions with a pH below 4.[1430] Shi et al.[1429] tested cinnarizine stability over a pH range of 3.12 to 9.44. Cinnarizine exhibited rapid decomposition in the pH range of 3.12 to 7.32. Drug decomposition rates declined in the pH range of 8.5 to 9.44. The authors also reported that the injectable emulsion formulation they developed (see *Injection,* below) was stable during sterilization at 121°C over 15 minutes, while aqueous solutions of cinnarizine were not.

### Stability Reports of Compounded Preparations
#### Injection
Injections, like other sterile drugs, should be prepared in a suitable clean air environment using appropriate aseptic procedures. When prepared from nonsterile components, an appropriate and effective sterilization method must be employed.

Shi et al.[1429] developed an injectable emulsion formulation of cinnarizine 1 mg/mL to overcome the solubility and stability problems of the drug. A variety of oil phase components were evaluated along with variations in surfactants,

**STUDY 4:** Burns et al.[739] reported the physical compatibility of cimetidine hydrochloride elixir 300 mg/5 mL with 10 mL of three enteral formulas, including Enrich, TwoCal HN, and Vivonex T.E.N. Visual inspection found no physical incompatibility with any of the enteral formulas.

formulation procedure, and sterilization periods. The authors found that the optimal lipid injection formulation, considering both chemical stability of the drug and emulsion stability, used an equal parts mixture of long-chain triglycerides (LCT) and medium-chain triglycerides (MCT). The authors reported that the optimal formulation of the injectable emulsion had the following composition:

Cinnarizine 1 mg/mL
LCT 5% (wt/vol)
MCT 5% (wt/vol)
Egg lecithin 3% (wt/vol)
Tween 80 0.2% (wt/vol)
Pluronic F68 surfactant 0.4% (wt/vol)
Glycerol 2.5% (wt/vol)
Sodium oleate 0.03% (wt/vol)
EDTA 0.02% (wt/vol)

The optimal preparation method was to dissolve egg lecithin in a mixture of LCT and MCT heated to 80°C. The aqueous phase consisted of glycerin and sodium oleate along with the auxiliary surfactants Tween 80 and Pluronic F68 surfactant also heated to 80°C and agitated until uniformly dissolved. The water phase was added slowly to the oil phase with high-speed shear mixing using an Ultra Turrax IKA T18 basic mixer at 10,000 rpm for five minutes to create a coarse emulsion. The pH was adjusted to pH 8.5 with 0.1 mol/L sodium hydroxide or hydrochloric acid as needed. This coarse emulsion was then subjected to high-pressure homogenization using a Niro Soavi NS10012k homogenizer at 700 bar for eight cycles at 40°C to create the final emulsion. The emulsion was packaged in vials and sealed and sterilized at 121°C for 15 minutes.

The stability of the finished injectable emulsion was evaluated at 4 and 25°C for 12 months. The physical appearance, pH, particle size distribution, entrapment efficiency, and zeta potential remained unchanged throughout the study period. HPLC analysis found little or no cinnarizine loss occurred over 12 months at either temperature. The authors calculated that the shelf life of the product under refrigeration was about four years.

# Ciprofloxacin
# Ciprofloxacin Hydrochloride

## Properties

Ciprofloxacin hydrochloride is a faintly yellowish to yellow crystalline powder. Approximately 291.1 mg of the hydrochloride is equivalent to ciprofloxacin 250 mg.[2–4,7]

Ciprofloxacin also is a faint to light yellow crystalline powder. Ciprofloxacin for injection is made with the aid of lactic acid to prepare the lactate salt in situ. The injection is colorless to slightly yellow. Ciprofloxacin lactate 127 mg is approximately equivalent to 100 mg of ciprofloxacin.[2–4,7]

### Solubility

Ciprofloxacin is practically insoluble in water and ethanol.[7] Ciprofloxacin hydrochloride has an aqueous solubility of 36 mg/mL at 25°C[2] and is very slightly soluble in ethanol.[3]

### pH

Ciprofloxacin injection (as the lactate) in vials has a pH of 3.3 to 3.9. In polyvinyl chloride bags, the pH range is 3.5 to 4.6.[4,7] The ophthalmic solution has a pH of 3.5 to 5.5.[4]

A 2.5% ciprofloxacin hydrochloride solution in water has a pH of 3 to 4.5.[3,4]

### $pK_a$

The $pK_a$ values of ciprofloxacin hydrochloride are 6 and 8.8.[2]

## General Stability Considerations

Ciprofloxacin hydrochloride products should be stored in well-closed (tablets) or tight, light-resistant containers at controlled room temperature and protected from intense ultraviolet light.[2,4] The manufacturer stated that ciprofloxacin capsules should not be removed from the original packaging and placed in dosing compliance aids because of the potential for water absorption and sensitivity to light.[1622] Aqueous solutions of the hydrochloride in the pH range of 1.5 to 7.5 are stable for at least two weeks at room temperature.[2] However, Teraoka et al.[648] reported that substantial losses of ciprofloxacin occur in solutions having a pH over 6.

Ciprofloxacin is light sensitive. Tiefenbacher et al.[692] reported that ciprofloxacin in an aqueous solution degrades slowly when exposed to natural daylight. A 2% loss occurred in 12 hours, and a 9% loss occurred after 96 hours when exposed to mixed natural daylight and fluorescent light.

Ciprofloxacin microcapsules for oral suspension should be stored at less than 25°C and protected from freezing. After addition of the diluent, the suspension is stable for 14 days stored at room temperatures less than 30°C and under refrigeration protected from freezing.[2,7]

Ciprofloxacin (as the lactate) injections should be stored between 5 and 30°C for the vials and between 5 and 25°C for infusion bags protected from light, excessive heat, and freezing.[2,7]

Ciprofloxacin forms complexes with magnesium 2+ ions in a 1:1 ratio. The absorption and anti-infective activity of the complex is reduced compared to ciprofloxacin alone.[7,1327]

Clark et al.[1506] evaluated the in vitro antimicrobial efficacy of ciprofloxacin with dexamethasone otic solution (Ciprodex, Alcon) after opening. Antimicrobial susceptibility testing against *Pseudomonas aeruginosa* and *Staphylococcus aureus* on agar plates was evaluated over four months of storage. The zones of microbial inhibition remained constant throughout the study period.

## Stability Reports of Compounded Preparations
### Oral

**STUDY 1:** Johnson et al.[616] evaluated the stability of a ciprofloxacin 50-mg/mL oral suspension compounded from tablets. Four ciprofloxacin 750-mg tablets were crushed to fine powder in a glass mortar. Separately, 30 mL of Ora-Plus suspending medium (Paddock) and 30 mL of simple syrup were mixed thoroughly. A small amount of the vehicle mixture was added to wet the tablet powder and was mixed to a fine paste. About 15 mL of the vehicle mixture was added, triturated well, and transferred to a two-ounce calibrated amber plastic prescription bottle. The mortar was rinsed with another 15 mL of vehicle and the rinse was added to the bottle. The mortar was rinsed again and the volume brought to a final 60 mL with additional vehicle mixture, yielding the ciprofloxacin 50-mg/mL oral suspension. Samples were stored at 4 and 24°C for 56 days. No visible changes in color, odor, or microbial growth occurred. Because of the bitter taste and prolonged bitter aftertaste, the authors recommended mixing this suspension with an equal amount of chocolate syrup just prior to administration.

**STUDY 2:** Nahata et al.[1112] evaluated the stability of two ciprofloxacin 50-mg/mL oral suspension formulations prepared from ciprofloxacin 500-mg commercial tablets (Miles). The tablets were ground to powder using a mortar and pestle. The tablet powder was incorporated into an equal parts mixture of (1) simple syrup and 1% methylcellulose and (2) Ora-Plus and Ora-Sweet to yield a ciprofloxacin concentration of 50 mg/mL. The suspensions were packaged in amber plastic prescription bottles (plastic composition not cited) and stored at room temperature of 25°C and refrigerated at 4°C for 91 days.

No visible changes in the physical appearance of either oral suspension formulation were observed during the study. Stability-indicating HPLC analysis found that ciprofloxacin losses were about 9% in 91 days at room temperature and 4% in 91 days refrigerated in both formulations.

## Enteral

**STUDY 1:** Piccolo et al.[198] evaluated the effect of an enteral nutritional supplement on ciprofloxacin bioavailability. Ciprofloxacin 750-mg tablets were taken by volunteers who then took 240 mL of Resource (Sandoz Nutrition). There was a 25% reduction in ciprofloxacin absorption.

**STUDY 2:** Mueller et al.[1090] evaluated the effect of enteral feeding with Ensure (Ross Laboratories) on the bioavailability of orally administered ciprofloxacin (Miles). The test subjects consumed 120 mL of Ensure every 30 minutes for a total of five doses. With the second administration of Ensure, the subjects took a ciprofloxacin 750-mg tablet that had been uniformly crushed and mixed into the Ensure. The cup with the crushed ciprofloxacin was rinsed with 60 mL of Ensure, which the test subjects ingested. The total amount of Ensure was 660 mL. No other food or drink was ingested. Ciprofloxacin bioavailability was evaluated over a 24-hour period. The Ensure reduced the percentage of relative ciprofloxacin bioavailability to 72% compared to taking the drug with water.

## Ophthalmic

Ophthalmic preparations, like other sterile drugs, should be prepared in a suitable clean air environment using appropriate aseptic procedures. When prepared from nonsterile components, an appropriate and effective sterilization method must be employed.

Jain et al.[1283] reported on the development and stability of ciprofloxacin 0.3% ophthalmic sustained release gel. Because ciprofloxacin hydrochloride proved to be incompatible with Carbopol 980 NF (polyacrylic acid) developing a lumpy precipitate, the drug was complexed in a 1:2 ratio to an ion exchange resin, Inion 254F (polystyrene cross-linked with divinylbenzene). The ion exchange resin was dispersed in water with constant stirring for 15 minutes. Then the drug was added with stirring until complexation was complete, forming ciprofloxacin hydrochloride resinate. The ophthalmic gel was prepared by dissolving disodium edetate 0.01% wt/vol, mannitol 5% wt/vol, and benzalkonium chloride 0.02% wt/vol in 75 mL of water. Methocel K100LV (hydroxypropyl methylcellulose) 2% was added to this solution and allowed to hydrate for 20 minutes. Then Carbopol 980F 0.5% was dispersed and allowed to hydrate with stirring. This solution was filtered through a 0.45-µm cellulose acetate filter. The ciprofloxacin hydrochloride resinate equivalent to ciprofloxacin 0.3% was then added with stirring. The mixture was adjusted to pH 3.8 to 4.2 using 0.5 N sodium hydroxide

solution. The mixture was brought to 100 mL with additional water. The mixture was packaged in amber glass vials with rubber closures and sterilized by autoclaving at 121°C and 15 psi for 20 minutes. The final ophthalmic gel was clear and colorless.

The stability of the ciprofloxacin ophthalmic gel was evaluated using a stability-indicating HPLC analytical method. Samples of the gel were stored at 4, 25, 37, and 45°C for three months. There was no change in gel clarity, pH, viscosity, and in vitro drug release. HPLC analysis found no loss of ciprofloxacin with the drug content in the samples remaining near 100%. The authors estimated that the ciprofloxacin 0.3% ophthalmic gel would have a shelf life of two years.

## Compatibility with Other Drugs

Hui et al.[873] reported the compatibility of ciprofloxacin 0.2 mg/0.1 mL and vancomycin hydrochloride 1 mg/0.1 mL for ophthalmic use in sodium chloride 0.9%, balanced salts solution plus glutathione (BSS Plus), and human vitreous obtained from cadaver eyes. The drugs were mixed together in 4 mL of sodium chloride 0.9%, BSS Plus, and human vitreous. The samples were incubated at 37°C. Ciprofloxacin was analyzed by HPLC while vancomycin was evaluated using TDx analysis.

Ciprofloxacin precipitated by itself and in combination with vancomycin hydrochloride in all three media. The extent of precipitation (up to 45%) was similar in both sodium chloride 0.9% and BSS Plus. Vancomycin precipitation was negligible. Even after precipitation, the ciprofloxacin concentration was greater than the minimum inhibitory concentration. The authors indicated that the two drugs could be used together in the treatment of infective endophthalmitis.

## Compatibility with Common Beverages and Foods

Sadrieh et al.[956] evaluated the stability and palatability of crushed ciprofloxacin tablets in common foods and beverages in this FDA-sponsored study of compounding for pediatric patients using counterterrorism drugs. Ciprofloxacin proved to be most stable in water and chocolate syrup, with 2 and 8% drug losses, respectively, after seven days refrigerated. However, water was the least palatable of the media tested and was rated as the least acceptable based on taste. Ciprofloxacin was found to be stable for only 24 hours in the other media that were evaluated. Palatability rankings from best to worst were chocolate syrup, Log Cabin syrup, strawberry jelly, apple juice, low-fat chocolate milk, and water. ◼

# Cisapride

## Properties

Cisapride, as the monohydrate, is a white to slightly beige odorless powder.[3,7] Approximately 1.04 mg of the monohydrate is equivalent to 1 mg of the anhydrous form.[7]

### Solubility

Cisapride is practically insoluble in water.[7]

## General Stability Considerations

Cisapride products should be stored at controlled room temperature and protected from moisture and light.[7]

## Stability Reports of Compounded Preparations

### Oral

**STUDY 1:** Horn and Anderson[178] evaluated the stability of cisapride in three extemporaneous oral suspensions. Twelve cisapride 10-mg tablets (Propulsid, Janssen) were crushed to fine powder for each formulation. For the first formulation, the powder was triturated with 20 mL of cherry syrup and brought to 120 mL with additional cherry syrup, forming a cisapride 1-mg/mL suspension. In the second formulation, the powder was triturated with 12 mL of propylene glycol, followed by trituration with 20 mL of cherry syrup; the mixture was brought to 120 mL with additional cherry syrup. For the third formulation, the powder was triturated with 12 mL of propylene glycol, followed by trituration with 18 mL of sodium bicarbonate injection 1 mEq/mL. The mixture, brought to 120 mL with cherry syrup, had a pH of 6.5 to 7.5. The suspensions were packaged in four-ounce amber bottles and stored at 22°C in ambient room light and at 5°C in the dark. An HPLC analytical method was used to assess the cisapride content of the suspensions. Extensive decomposition losses of 27 and 44% occurred in formulation 1 in seven days at 5 and 22°C, respectively; similarly, in formulation 2, losses were 16 and 45% in seven days at 5 and 22°C, respectively. However, in formulation 3, with the pH buffered to near neutrality, cisapride was stable, exhibiting little or no loss in 21 days at either storage temperature.[179]

**STUDY 2:** Nahata et al.[475] evaluated the stability of an extemporaneously compounded cisapride 1-mg/mL suspension prepared from tablets. Twenty cisapride (Propulsid, Janssen) 10-mg tablets were crushed using a mortar and pestle, and small volumes of simple syrup were slowly incorporated to form a paste. The paste was transferred to a graduated cylinder, and the mortar and pestle were rinsed with additional syrup to bring the volume to 100 mL. Methylcellulose 1% was then added to bring the final volume to 200 mL and yield a nominal cisapride concentration of 1 mg/mL. The suspension was packaged in amber plastic prescription bottles and stored at 4 and 25°C. Stability-indicating HPLC analysis determined that little or no cisapride loss occurred in 91 days at 4°C. However, at 25°C, cisapride content was retained through 28 days with about 5% loss, although 14% loss occurred after 42 days. No change in the physical appearance of the suspension was observed.

**STUDY 3:** Allen and Erickson[594] evaluated the stability of three cisapride 1-mg/mL oral suspensions extemporaneously compounded from tablets. Vehicles used in this study were (1) an equal parts mixture of Ora-Sweet and Ora-Plus (Paddock), (2) an equal parts mixture of Ora-Sweet SF and Ora-Plus (Paddock), and (3) cherry syrup (Robinson Laboratories) mixed 1:4 with simple syrup. Twelve cisapride 10-mg tablets (Janssen) were crushed and comminuted to fine powder using a mortar and pestle. About 20 mL of the test vehicle was added to the powder and mixed to yield a uniform paste. Additional vehicle was added geometrically almost to volume, mixing thoroughly after each addition. The pH was adjusted to 7 with sodium bicarbonate. The mixture was brought to the final volume of 120 mL and mixed. The process was repeated for each of the three test suspension vehicles. Samples of each of the finished suspensions were packaged in 120-mL amber polyethylene terephthalate plastic prescription bottles and stored at 5 and 25°C.

No visual changes or changes in odor were detected during the study. Stability-indicating HPLC analysis found less than 9% cisapride loss in any of the suspensions stored at either temperature after 60 days of storage. ■

# Clindamycin Hydrochloride
# Clindamycin Palmitate Hydrochloride
# Clindamycin Phosphate

## Properties

Clindamycin is a yellow amorphous solid.[1]

Clindamycin hydrochloride is a white to almost white odorless or almost odorless crystalline powder. It may contain a variable amount of water. Approximately 1.13 g of the hydrochloride is equivalent to 1 g of clindamycin.[1–3]

Clindamycin palmitate hydrochloride is a white to off-white amorphous powder with a characteristic odor. Approximately 1.6 g of the palmitate hydrochloride is equivalent to 1 g of clindamycin.[2,3]

Clindamycin phosphate is a white to off-white odorless or almost odorless hygroscopic crystalline powder. Approximately 1.2 g of the phosphate is equivalent to 1 g of clindamycin.[2,3]

### Solubility

The hydrochloride, palmitate hydrochloride, and phosphate forms of clindamycin are freely soluble in water.[2,6] One gram of the palmitate hydrochloride is soluble in about 5 mL of water.[6] Clindamycin phosphate has an aqueous solubility of about 400 mg/mL.[2] In ethanol, the hydrochloride and phosphate are slightly soluble and the palmitate hydrochloride has a solubility of 250[6] to 333 mg/mL.[3]

### pH

A clindamycin hydrochloride 100-mg/mL aqueous solution has a pH of 3 to 5.5. The oral solution also has a pH of 3 to 5.5.[4] Clindamycin palmitate hydrochloride for oral solution at 10 mg/mL has a pH of 2.5 and 5.[2,4] Clindamycin phosphate injection has a pH of 5.5 to 7,[2,4] but the pH is usually near 6 to 6.3.[334] Clindamycin phosphate vaginal cream has a pH of 3 to 6,[7] while the topical solution has a pH of 4 to 7.[4]

### pKa

The pKa of clindamycin has been variously cited as 7.45[2] and 7.72.[6] Clindamycin hydrochloride has a pKa of 7.6.[1]

### Osmolality

Clindamycin phosphate injection 150 mg/mL has an osmolality that has been measured by freezing-point depression to be about 795 mOsm/kg[286] to 835 mOsm/kg;[287] however, the manufacturer indicates that it is usually around 825 to 880 mOsm/kg.[8]

## General Stability Considerations

Clindamycin products should be stored in tight containers at controlled room temperature. The liquids, gel, and cream should be protected from freezing. Crystals may form if the injection is refrigerated or frozen. The crystals may be resolubilized by rewarming to room temperature. To avoid thickening, the reconstituted granules for oral solution should not be refrigerated; the reconstituted solution is stable for two weeks at room temperature.[2,7,334]

Clindamycin may undergo several hydrolytic decomposition reactions, including hydrolysis of the thioglycoside group at low pH and hydrolysis of the 7-chloro above pH 5. Amide hydrolysis is another important reaction. Maximum stability occurs at pH 4, but solutions are reasonably stable over the pH range of 1 to 6.5.[6]

## Stability Reports of Compounded Preparations
### Injection

Injections, like other sterile drugs, should be prepared in a suitable clean air environment using appropriate aseptic procedures. When prepared from nonsterile components, an appropriate and effective sterilization method must be employed.

**STUDY 1 (DILUTED):** Nahata et al.[493] evaluated the stability of extemporaneously compounded clindamycin phosphate 15-mg/mL injection diluted for pediatric use. Commercial clindamycin phosphate injection (Abbott) containing 150 mg/mL of drug was diluted with sterile water for injection to yield a 15-mg/mL solution. The diluted solution was transferred to sterile glass vials and stored at 4 and 22°C for 91 days. Visual inspection found no changes in the appearance of the solutions. Stability-indicating HPLC analysis determined that no loss of clindamycin occurred at either temperature throughout the study period.

**STUDY 2 (REPACKAGED):** Repackaging clindamycin phosphate injection 150 mg/mL into plastic syringes (Becton Dickinson and 3M) resulted in retention of clindamycin activity for at least 48 hours at room temperature under fluorescent light.[335,336]

### Topical

**STUDY 1:** Migton et al.[42] evaluated the stability of clindamycin as the hydrochloride and phosphate salts in four topical vehicles as 1% solutions. Vehicle A was composed of isopropanol 70%, propylene glycol 10%, and water 20% (pH 5.5). Vehicle B (Ionax, Owen) was composed of isopropanol 48%, polyoxyethylene ethers, acetone, salicylic acid, and allantoin (pH 2.9). Vehicle C (Sebanil, Texas Pharmaceutical) was composed of ethanol 40%, acetone, polysorbate 20, fragrance, and water (pH 5). Vehicle D (Cleocin vehicle, Upjohn) was composed of isopropanol 50%, propylene glycol, and water (pH 5).

**TABLE 25.** Times to 10% Decomposition of Topical Clindamycin Formulations[42]

| | $t_{90}$ (Months) | | | |
|---|---|---|---|---|
| | Glass | | Plastic | |
| Vehicle | 25°C | 40°C | 25°C | 40°C |
| Clindamycin Hydrochloride | | | | |
| A | 10 | 4 | 10 | 4 |
| B | 8 | 4 | 8 | 4 |
| C | 12 | 4 | 10 | 4 |
| Clindamycin Phosphate | | | | |
| A | 12 | 4 | 10 | 2 |
| B | 10 | 4 | 6 | 2 |
| C | 12 | 4 | 12 | 2 |
| D | 13.6 | 2 | — | — |

The 1% solutions were prepared using the appropriate amounts of clindamycin hydrochloride (1 mg = 858 mcg clindamycin base) or clindamycin phosphate (1 mg = 785 mcg clindamycin base) as bulk powders. To achieve complete dissolution, the formulations were mixed for 24 hours using magnetic stirring bars but no heat, which could affect stability. The finished formulations were packaged in glass screw-cap bottles or polypropylene plastic bottles and stored at 25 and 40°C for the stability study. A stability-indicating HPLC assay was used to determine remaining clindamycin content.

The times to 10% decomposition ($t_{90}$) are shown in Table 25. Clindamycin in vehicle B exhibited the worst stability, probably related to its low pH. The other vehicles exhibited approximately comparable stabilities.

**STUDY 2:** Lacina et al.[175] reported crystal formation in a clindamycin topical formulation. Clindamycin (as the phosphate) 1% was prepared in an equal parts mixture of 99% isopropyl alcohol and propylene glycol. The crystals appeared after one to two weeks, and their formation was accelerated by refrigeration. Similarly, clindamycin (as the hydrochloride) prepared in an isopropyl alcohol and propylene glycol solution developed crystals. In a series of 17 prescriptions for topical clindamycin, the three that developed precipitates had a water content of only 7.1%.

**STUDY 3:** Orr et al.[176] evaluated a variety of formulation approaches to prepare topical 1% clindamycin from capsules

(containing the hydrochloride) or injection (containing the phosphate). The capsule contents include a substantial amount of lactose that must be considered in anticipating adequate solubility. The authors recommended a mixture of 70% isopropyl or ethyl alcohol, 20% water, and 10% propylene glycol as the ideal vehicle to solubilize 1% clindamycin from either capsule or injection sources. The topical preparation made from the injection must contain at least 20% water to solubilize the phosphate salt. Lower water concentrations have resulted in precipitate formation. To prepare a topical 1% clindamycin solution from the capsules, 30 mL of the recommended solvent system was added to the contents of four capsules and shaken for 30 seconds and repeated three or four times over 10 to 15 minutes. The clear liquid was decanted through a filter. Another 30 mL of the vehicle was added to the solids and shaken for 30 seconds, and the liquid was once again poured through the filter. Enough vehicle was passed through the filter to bring the volume to 60 mL. Clindamycin retained activity in the 70% ethanol–20% water–10% propylene glycol vehicle for at least six months.[177] However, a six- to eight-week expiration period has been recommended.[176]

**STUDY 4:** Hirota et al.[415] evaluated the stability of an extemporaneously prepared clindamycin phosphate 1% lotion prepared from the injection. Clindamycin phosphate injection 600 mg (4 mL) was mixed with 48 mL of isopropyl alcohol 70% and 6 mL of propylene glycol and brought to 60 mL with distilled water. The clindamycin content was determined using an HPLC assay. The lotion retained 92% concentration after 180 days at 5°C and 88% concentration after 180 days at 25°C.

**STUDY 5:** Algra et al.[631] evaluated the stability of an extemporaneously compounded 1% clindamycin topical solution. The contents of a suitable number of clindamycin hydrochloride capsules were added to a mixture composed of 70% ethanol and 10% propylene glycol (vol/vol). The resulting mixture was tested both unfiltered and after filtration to remove undissolved capsule components. Both forms demonstrated continued antibacterial activity against *Staphylococcus aureus* for at least six months after preparation when stored at room temperature.

## Compatibility with Other Drugs

Ohtani et al.[1522] evaluated the stability and compatibility of adapalene gel (Differin) mixed with topical clindamycin for use in treating acne. The authors reported that the mixtures were stable. HPLC analysis found the concentrations of adapalene and clindamycin remained acceptable. ■

# Clobetasol Propionate

## Properties

Clobetasol propionate is a white to almost white or cream-colored crystalline powder.[3,4]

### Solubility

Clobetasol propionate is practically insoluble in water at about 2 mcg/mL, sparingly soluble in ethanol at about 10 mg/mL, and soluble in acetone.[2,3,4]

### pH

Clobetasol propionate cream has a pH of 4.5 to 7. Clobetasol propionate topical solution has a pH of 4.5 to 6.[4] Clobetasol 1 mg/mL as butyrate ophthalmic solution was found to have a pH of about 5.1 to 5.3.[880]

## General Stability Considerations

Clobetasol propionate should be packaged in tight containers and protected from exposure to light.[3,4]

Clobetasol propionate cream should be stored at controlled room temperature and not refrigerated. Clobetasol propionate gel should be stored between 2 and 30°C. The topical solution should be stored between 4 and 25°C and not exposed to an open flame. The topical foam should be stored at controlled room temperature between 20 and 25°C.[2]

Clobetasol propionate in aqueous solution exhibits pH-dependent stability. HPLC analysis found that a minimum rate of decomposition occurs at pH near 3.2, with 10% decomposition calculated to occur in 761 days at room temperature. Decomposition accelerates as the pH deviates higher or lower.[1038]

## Stability Reports of Compounded Preparations

### Topical

Ray-Johnson[775] reported the stability of clobetasol propionate ointment (Dermovate, Glaxo) diluted to a concentration of 0.025% (1:1 mixture) using Unguentum Merck, an ambiphilic ointment that combines the properties of both an oil-in-water and a water-in-oil emulsion. The diluted ointment was packaged in screw-cap jars and stored refrigerated at 4°C, at room temperature of 18 to 23°C, and at elevated temperature of 32°C. After storage for 32 weeks, the authors reported that HPLC analysis did not find unacceptable drug loss at any of the temperatures.

### Ophthalmic

Ophthalmic preparations, like other sterile drugs, should be prepared in a suitable clean air environment using appropriate aseptic procedures. When prepared from nonsterile components, an appropriate and effective sterilization method must be employed.

Garcia-Valldecabres et al.[880] reported the change in pH of several commercial ophthalmic solutions, including clobetasone 1 mg/mL (Cortoftal) as the butyrate over 30 days after opening. Slight variation in solution pH was found, but no consistent trend and no substantial change occurred.

## Compatibility with Other Drugs

**CLOBETASOL PROPIONATE COMPATIBILITY SUMMARY**

*Compatible with:* Captopril • Tacrolimus • Tazarotene • Zinc oxide

**STUDY 1:** Hecker et al.[815] evaluated the compatibility of tazarotene 0.05% gel combined in equal quantity with other topical products, including clobetasol propionate 0.05% cream, gel, ointment, and scalp solution. The mixtures were sealed and stored at 30°C for two weeks. HPLC analysis found that less than 10% tazarotene loss occurred in all combinations except clobetasol propionate 0.05% gel, which had a loss of 18.8%. Clobetasol propionate exhibited less than 10% loss in this time frame.

**STUDY 2:** Huang et al.[817] evaluated the stability of captopril incorporated into 5% topical gel formulations with the addition of 0.2% clobetasol. No difference in oxidative rate constant was found, indicating clobetasol is compatible with captopril.

**STUDY 3 (TACROLIMUS):** Pappas et al.[1444] evaluated the compatibility and stability of tacrolimus 0.1% ointment (Protopic, Fujisawa) mixed in a 1:1 ratio with clobetasol propionate 0.05% (Dermovate, TaroPharma). Samples were packaged in sealed glass vials and stored at 25°C (60% relative humidity), 30°C (60% relative humidity), and 40°C (75% relative humidity) for 28 days. HPLC analysis found no substantial loss of tacrolimus and clobetasol propionate during the study period.

**STUDY 4 (ZINC OXIDE):** Ohishi et al.[1714] evaluated the stability of Dermovate and Kindavate (clobetasol propionate) ointments each mixed in equal quantity (1:1) with zinc oxide ointment. Samples were stored refrigerated at 5°C. HPLC analysis of the ointment mixtures found that the drugs were stable for 32 weeks at 5°C. ■

# Clomipramine Hydrochloride

## Properties
Clomipramine hydrochloride occurs as a white to off-white or faintly yellow, slightly hygroscopic crystalline powder.[1,3,4,7]

### Solubility
Clomipramine hydrochloride is freely soluble in water,[1,4,7] soluble in ethanol,[3] and practically insoluble in ethyl ether and methylene chloride.[1,3,7]

### pH
Clomipramine hydrochloride 100 mg/mL aqueous solution has a pH between 3.5 and 5.[3,4]

## General Stability Considerations
Clomipramine hydrochloride powder and oral capsules should be packaged in well-closed containers and be stored at controlled room temperature protected from exposure to light and moisture.[3,4,7]

## Stability Reports of Compounded Preparations
### Oral
Blanchard and Donnelly[1630] evaluated the stability of clomipramine hydrochloride 5 mg/mL compounded as an oral suspension in simple syrup and stored at 10°C for 30 days. Stability-indicating gas chromatographic analysis of drug concentrations found little or no loss of clomipramine occurred during the study period.

# Clonazepam

## Properties
Clonazepam is an off-white to light yellow crystalline powder having a faint odor.[2,3]

### Solubility
Clonazepam is essentially insoluble in water,[2,3] having a solubility of less than 0.1 mg/mL at 25°C.[1] In ethanol, the drug is slightly soluble.[2,3]

### $pK_a$
Clonazepam has $pK_a$ values of 1.5 and 10.5.[1,2]

## General Stability Considerations
Clonazepam products should be stored in tight, light-resistant containers at controlled room temperature.[2-4] Clonazepam solutions exhibit loss due to sorption to polyvinyl chloride (PVC) containers.[557-560] However, sorption did not appear to be a problem with a suspension prepared from tablets packaged in a PVC bottle.[557] (See *Oral* section that follows.)

## Stability Reports of Compounded Preparations
### Oral
#### USP OFFICIAL FORMULATION (ORAL SUSPENSION):
> Clonazepam 10 mg
> Vehicle for Oral Solution, NF (sugar-containing or sugar-free)
> Vehicle for Oral Suspension, NF (1:1) qs 100 mL

(See the vehicle monographs for information on the individual vehicles.)

Use clonazepam powder or commercial tablets. If using tablets, crush or grind to fine powder. Add about 10 mL of the vehicle mixture and mix to make a uniform paste. Add additional vehicle almost to volume in increments with thorough mixing. Quantitatively transfer to a suitable calibrated tight and light-resistant bottle, bring to final volume with the vehicle mixture, and thoroughly mix yielding clonazepam 0.1 mg/mL oral suspension. The final liquid preparation should have a pH between 3.6 and 4.6. Store the preparation at controlled room temperature or under refrigeration between 2 and 8°C. The beyond-use date is 60 days from the date of compounding at either storage temperature.[4,5]

**STUDY 1:** The extemporaneous formulation of a clonazepam 0.1-mg/mL oral suspension was described by Nahata and Hipple.[160] Five clonazepam 2-mg tablets were crushed and mixed thoroughly with 10 mL of purified water. The suspension was brought to 100 mL with 1% methylcellulose. A stability period of 14 days under refrigeration was used, although chemical stability testing was not performed.

**STUDY 2:** Allen[456] reported on extemporaneously prepared clonazepam suspensions. The suspensions were prepared by crushing the required number of tablets using a mortar and pestle and wetting the powder with a small amount of flavored vehicle to form a paste. The remaining vehicle was added and thoroughly mixed. Recommended vehicles included cherry syrup or equal quantities of Ora-Sweet/Ora-Plus or Ora-Sweet SF/Ora-Plus. These preparations are stated to retain at least 90% concentration for at least 60 days in these vehicles.

**STUDY 3:** Allen and Erickson[543] evaluated the stability of three clonazepam 0.1-mg/mL oral suspensions extemporaneously compounded from tablets. Vehicles used in this study were (1) an equal parts mixture of Ora-Sweet and Ora-Plus (Paddock), (2) an equal parts mixture of Ora-Sweet SF and Ora-Plus (Paddock), and (3) cherry syrup (Robinson Laboratories) mixed 1:4 with simple syrup. Six clonazepam 2-mg tablets (Roche) were crushed and comminuted to fine powder using a mortar and pestle. About 10 mL of the test vehicle was added to the powder and mixed to yield a uniform paste. Additional vehicle was added geometrically and brought to the final volume of 120 mL, mixing thoroughly after each addition. The process was repeated for each of the three test suspension vehicles. Samples of each finished suspension were packaged in 120-mL amber polyethylene terephthalate plastic prescription bottles and stored at 5 and 25°C in the dark.

No visual changes or changes in odor were detected during the study. Stability-indicating HPLC analysis found less than 5% clonazepam loss in any of the suspensions stored at either temperature after 60 days of storage.

**STUDY 4:** Roy and Besner[557] studied the stability of a clonazepam 0.1-mg/mL oral suspension prepared from tablets. The vehicle for the suspension was prepared in two steps. A 1% methylcellulose suspension was prepared by dissolving 2 g of sodium benzoate in boiling water and adding 10 g of methylcellulose 1500 mPa (cps), with stirring for several minutes. Ice cold water was then incorporated quickly, bringing the volume to 1000 mL. The thick, creamy, white mixture was stirred for 10 minutes and transferred to a liter bottle. The bottle was stored on its side under refrigeration overnight (minimum four hours), which changed the preparation to a clear gel. Then 700 mL of the 1% methylcellulose was added to 300 mL of simple syrup containing methylparaben 1.1 mg/mL and mixed well. The final vehicle was transferred to a liter bottle, mixed by rolling, and allowed to stand for four hours. To compound the clonazepam suspension, 20 clonazepam 0.5-mg tablets were crushed and ground to fine powder using a mortar and pestle. A small amount of sterile water was incorporated into the powder to wet it and form a paste. The vehicle was then incorporated geometrically, bringing the suspension to a final volume of 100 mL. It was packaged in an amber polyvinyl chloride bottle. HPLC analysis found no clonazepam loss after 60 days stored at 4°C.

# Clonidine Hydrochloride

## Properties
Clonidine hydrochloride is a white or almost white odorless crystalline powder with a bitter taste.[2,3,7]

### Solubility
Clonidine hydrochloride has an aqueous solubility of about 77 mg/mL[1] and is also soluble in ethanol.[1–3]

### pH
A 1 in 20 aqueous solution of clonidine hydrochloride has a pH of 3.5 to 5.5.[4] The natural pH of clonidine hydrochloride 0.15 to 1.5 mg/mL in 0.9% sodium chloride injection has been reported to be about 6 to 6.5.[553]

## General Stability Considerations
Clonidine hydrochloride tablets should be stored in well-closed, light-resistant containers at less than 30°C.[4] The transdermal patches also should be stored below 30°C.[2,7]

## Stability Reports of Compounded Preparations
### Injection
Injections, like other sterile drugs, should be prepared in a suitable clean air environment using appropriate aseptic procedures. When prepared from nonsterile components, an appropriate and effective sterilization method must be employed.

**STUDY 1:** Trissel et al.[553] evaluated the stability of clonidine hydrochloride 0.15-, 0.5-, and 1.5-mg/mL intrathecal injections prepared from powder. The calculated amount of bulk clonidine hydrochloride powder (Ash-Stevens) was weighed and transferred to a sterile plastic screw-cap bottle. The powder was dissolved in about 75% of the total amount of sterile 0.9% sodium chloride injection with shaking and swirling for several minutes until the powder completely dissolved. The solution was brought to its final volume with additional sterile 0.9% sodium chloride injection. The pH was adjusted to the target of 6.5 with sodium hydroxide when necessary. The measured osmolalities of the three injection concentrations, all around 285 mOsm/kg, were nearly isotonic. The bulk solution was passed through a 0.22-μm filter and filled into 20-mL Type I flint glass vials. The vials were sterilized by autoclaving at 121°C and 15 psi for 30 minutes. Samples of the vials were then stored at 4, 23, and 37°C for shelf-life determinations. Stability-indicating HPLC analysis found no loss of clonidine hydrochloride as a result of the autoclave sterilizing. Furthermore, no loss occurred after three months of storage at 37°C and for up to two years at 4 and 23°C.

**STUDY 2:** Trissel and Xu[918] evaluated the stability of clonidine hydrochloride 1.5 mg/mL packaged in Becton Dickinson polypropylene syringes sealed with tip caps compared to the stability of the drug in original glass vials. The syringes and vials were stored at room temperature of 23°C and elevated temperature of 37°C. Stability-indicating HPLC analysis found no loss of clonidine hydrochloride in three months in either the glass vials or polypropylene syringes.

**STUDY 3:** Hollenwaeger et al.[1335] developed and evaluated extemporaneously compounded high-concentration clonidine hydrochloride 2-mg/mL injection. Clonidine hydrochloride powder was dissolved in sodium chloride 0.9%. In one sample solution, the pH was adjusted to 4.1 with hydrochloric acid. In a second sample solution, the pH was adjusted to 7.3 with sodium hydroxide. The solutions were passed through 0.22-μm Minisart filters and filled into 5-mL glass vials sealed with brombutyl Teflon rubber stoppers. The vials of injection were sterilized by autoclaving at 121°C for 25 minutes. Samples were stored at 40°C for six months. Several vials were also exposed to daylight for a month.

The clear, colorless injection solution was unchanged visually throughout the study period; pH remained acceptable as well. HPLC analysis found little or no change in drug concentration in any of the samples, including those exposed to daylight for a month. The authors stated that the clonidine hydrochloride 2-mg/mL injection was physically and chemically stable for six months.

*Oral*

The stability of clonidine hydrochloride extemporaneously compounded as a 0.1-mg/mL oral liquid was assessed by Levinson and Johnson.[47] Oral liquids were prepared by crushing 0.2-mg tablets (Catapres, Boehringer Ingelheim), adding 2 mL of purified water to form a paste, and then mixing with simple syrup, NF (Humco), to yield 0.1 mg/mL. The final product was a suspension because of insoluble tablet excipients. An oral solution also was prepared from analytical-grade clonidine hydrochloride powder dissolved in 2 mL of purified water and added to simple syrup, NF. The oral suspension and solution were stored in Type III amber glass bottles with child-resistant caps and refrigerated at 4°C in the dark for 28 days. No visually observable change in color, odor, or pH occurred. A stability-indicating HPLC assay was used to determine the clonidine hydrochloride content. Both the suspension and the solution remained stable, with about 6 to 8% drug loss during the study. ∎

# ∎ Clopidogrel Bisulfate

## Properties
Clopidogrel bisulfate (clopidogrel hydrogen sulfate) occurs as a white to off-white crystalline powder.[1,3,4]

## Solubility
Several common references (Merck Index, Martindale, USP) state that clopidogrel bisulfate is soluble in water and methanol but practically insoluble in ether.[1,3,4] However, the official labeling for clopidogrel bisulfate tablets states that the drug is insoluble in water at neutral pH but freely soluble at pH 1.[7]

## pH
Clopidogrel bisulfate oral liquid compounded from tablets exhibited pH 2.65.[1508]

## pKa
Clopidogrel bisulfate has a pKa of 4.55.[1507]

## General Stability Considerations
Clopidogrel bisulfate powder and commercial oral tablets should be packaged in well-closed containers and stored at controlled room temperature.[4,7]

## Stability Reports of Compounded Preparations
*Oral*

Skillman et al.[1508] evaluated the stability of extemporaneously compounded clopidogrel bisulfate 5-mg/mL oral suspension prepared from commercial tablets (Sanofi-Aventis). The commercial tablets were ground to powder using a mortar and pestle, and the powder was combined with an equal parts mixture of Ora-Plus and Ora-Sweet (Paddock) suspending/flavoring vehicle resulting in a 5-mg/mL concentration. The oral liquid was packaged in amber plastic bottles with child-resistant caps. Samples were stored at both room temperature of 23 to 25°C and refrigerated at 2 to 8°C for 60 days. The samples exhibited no changes in visual appearance, odor, or taste. Stability-indicating HPLC analysis of drug concentrations found little or no loss of drug occurred over the storage period. ∎

# Clotrimazole

## Properties
Clotrimazole is an odorless white to pale yellow crystalline powder.[1,2,4]

### Solubility
Clotrimazole is practically insoluble in water, but it is freely soluble in acetone, ethanol, and methanol.[4]

### pH
Clotrimazole is a weak base.[1] Commercial topical cream and lotion formulations of clotrimazole have pH values in the range of 5 to 7.[2,4] The drug in commercial topical solution formulations has a pH in the range of 4.5 to 8. Dissolution of the 500-mg clotrimazole vaginal tablets, which have lactic acid present, results in a pH in the range of 3.5 to 3.8.[2]

## General Stability Considerations
Clotrimazole hydrolyzes upon heating in aqueous acids.[1] Topical cream, lotion, and solutions of clotrimazole should be packaged in tight containers and stored between 2 and 30°C. The lozenges, vaginal inserts, and vaginal tablets should be packaged in well-closed containers and stored at controlled room temperature.[2,4]

## Stability Reports of Compounded Preparations
### Nasal
Mathews et al.[1416] evaluated a series of compounded clotrimazole 1% gel formulations and compared them to commercial clotrimazole 1% creams for use in treating nasal aspergillosis in dogs. The commercial lotion product was unacceptable for application because it was insufficiently viscous to be retained in the frontal sinuses. The commercial cream formulation (Lotrimin, Teva) was insufficiently viscous as well. Several hydroxypropyl cellulose, poloxamer, and sodium carboxymethylcellulose gels were then evaluated in vitro at 37°C and in the sinuses of dogs. Of these, the hydroxypropyl cellulose 10 and 20% and sodium carboxymethylcellulose 3 and 4% gels induced unacceptable inflammation. The poloxamer (Pluronic 127) 25 and 30% gels were better. The compounded formulations were liquids at room temperature to aid in dissolving the drug but gelled in situ upon being warmed to 37°C, which aided in retaining the clotrimazole gel in the sinuses. HPLC analysis found clotrimazole concentrations of 32 to 62% remaining even after four weeks at 37°C.

## Compatibility with Other Drugs
Jagota et al.[920] evaluated the compatibility and stability over periods of up to 60 days at 37°C of mupirocin 2% ointment (Bactroban, Beecham) with clotrimazole 1% lotion, cream, and solution (Lotrimin, Schering) mixed in a 1:1 proportion. The physical compatibility was assessed by visual inspection, and the chemical stability of mupirocin was evaluated by stability-indicating HPLC analysis. The study found that mupirocin 2% ointment was physically compatible and chemically stable for 60 days with clotrimazole 1% solution.

However, clotrimazole 1% lotion was found to be physically incompatible due to separation upon mixing with mupirocin ointment. Mixing clotrimazole 1% cream with mupirocin ointment also resulted in separation, but analysis of the mupirocin content found that less than 10% loss occurred over 60 days. ■

# Cloxacillin Sodium

## Properties
Cloxacillin sodium is a white or almost white hygroscopic odorless crystalline powder.[3,4]

### Solubility
Cloxacillin sodium is freely soluble in water and methanol and soluble in ethanol.[3,4]

### pH
The European Pharmacopoeia states that the pH of a 10% aqueous solution of cloxacillin sodium is between 5 and 7.[3] The United States Pharmacopeia (USP) states that the pH of a 1% aqueous solution of cloxacillin sodium is between 4.5 and 7.5. The USP also states that cloxacillin sodium oral solution has a pH of 5 to 7.5.[4]

## General Stability Considerations
Cloxacillin sodium powder, capsules, injection, and powder for oral solution should be packaged in tight containers and stored at controlled room temperature not exceeding 25°C.[3,4,7] Reconstituted cloxacillin sodium injection is stated by the manufacturer to be stable for 24 hours at room temperature and 48 hours refrigerated.[7] Lynn reported that reconstituted cloxacillin sodium injection

degraded about 5% in seven days at 5°C and about 15% in four days at 23°C.[1105]

Cloxacillin sodium exhibits the greatest stability in the pH range of 5.5 to 7, with the minimum rate of drug degradation occurring at pH 6.3.[1103,1104]

## Stability Reports of Compounded Preparations
*Enteral*

Ortega de la Cruz et al.[1101] reported the physical compatibility of an unspecified amount of cloxacillin sodium oral suspension (Orbenin, Beecham) with 200 mL of Precitene (Novartis) enteral nutrition diet for a 24-hour observation period. No particle growth or phase separation was observed. ■

# ■ Clozapine

## Properties

Clozapine exists as a yellow crystalline powder.[1,3,4]

### Solubility

Clozapine is essentially insoluble in water having an aqueous solubility less than 0.01%. Clozapine is more soluble in acetone and ethyl acetate with solubilities of more than 5% and in dehydrated ethanol with a solubility of 4%.[1,3,4] Clozapine dissolves in dilute acetic acid.[3]

### $pK_a$

Clozapine has $pK_a$ values of 3.7 and 7.6.[1]

## General Stability Considerations

Clozapine powder and commercial oral tablets are to be packaged in well-closed containers and stored at controlled room temperature.[4,7]

## Stability Reports of Compounded Preparations
*Oral*

**STUDY 1:** Ramuth et al.[1555] evaluated the stability of clozapine 20-mg/mL oral suspension prepared from crushed commercial oral tablets in a compounded vehicle called "Guy's Hospital Paediatric Base". The vehicle contained sucrose, carboxymethylcellulose BP, methyl hydroxybenzoate BP, and propyl hydroxybenzoate BP in water. The oral suspension was packaged in amber glass bottles stored at room temperature for 18 days. The suspension settled rapidly but resuspended readily by vigorous shaking even after 18 days of storage. HPLC analysis found the oral suspension contained only about 80 to 90% of the intended clozapine concentration immediately after compounding. However, the suspension demonstrated no consistent change in drug concentration throughout the rest of the study. Because no microbiological testing was performed, the authors recommended limiting the use of the product to seven days stored under refrigeration.

**STUDY 2:** Walker et al.[1556] evaluated the stability of clozapine 20-mg/mL oral liquid preparations in six different liquid vehicles. The six vehicles were Ora-Sweet, Ora-Plus, Ora-Sweet/Ora-Plus (1:1), simple syrup, methylcellulose 1500 CPS, and "Guy's Hospital Paediatric Base" (see Study 1 above). The oral liquid preparations were packaged in 60-mL amber low-density polyethylene containers and were stored at room temperature of 23°C without protection from fluorescent room light for 63 days. The samples were visually inspected and analyzed for clozapine concentrations using a stability-indicating HPLC analytical method.

No visually apparent changes occurred in color or consistence, and caking did not occur in any of the formulations. HPLC analysis found that clozapine was stable for 63 days at room temperature in all six formulations with all concentrations remaining at or above 96% throughout the study. ■

# ■ Coal Tar
# Coal Tar Solution
## (Liquor Carbonis Detergens)

## Properties

Coal tar is a black or nearly black viscous liquid produced from the destructive distillation of bituminous coal at 900 to 1100°C. It has a penetrating naphthalenelike odor and produces a sharp burning sensation on the tongue. It is composed of tar acids, phenols, cresols, and other phenolic derivatives, as well as hydrocarbons polymerized to thousands of differing but related compounds.[2,3] It has a weight of about 1.15 g/mL.[3]

Coal tar solution, liquor carbonis detergens, is a preparation of coal tar, alcohol, and polysorbate 80. Mixing coal tar solution with water results in a fine dispersion of coal tar.[2]

### Solubility
Coal tar is slightly soluble in water and partially soluble in ethanol[2,3] and acetone.[597]

### pH
A saturated aqueous solution of coal tar is alkaline to litmus.[3]

## General Stability Considerations
Coal tar products should be stored in tight or well-closed containers. Coal tar becomes more viscous with exposure to air. It should be protected from freezing.[2,3]

## Stability Reports of Compounded Preparations
### Topical
**STUDY 1:** Pesko[255] reported on an extemporaneous topical formulation for scalp psoriasis. Liquor carbonis detergens (LCD) 20 g was allowed to sit exposed to the environment at room temperature for several hours to permit evaporation of the alcohol. Salicylic acid 20 g was mixed with a small amount of castor oil, and the evaporated LCD was added and mixed thoroughly. A sufficient amount of Unibase then was added to the mixture to make 200 g of a liquidlike cream. The author recommended a 30-day expiration period, although no stability study was reported.

**STUDY 2:** Allen[597] described a topical ointment for psoriasis. Each component in the formula shown in Table 26 was accurately weighed. The sulfur and salicylic acid powders were mixed and reduced to fine powder, and a few drops of glycerin were incorporated to make a paste. The paste was

**TABLE 26.** Psoriasis Topical Ointment Formula[597]

| Component | Amount |
| --- | --- |
| Precipitated sulfur | 3 g |
| Salicylic acid | 1 g |
| Glycerin | Several drops |
| Fluocinonide (Lidex) 0.05% ointment | 24 g |
| Aquaphor or Aquabase | 70 g |
| Coal tar | 2 g |

mixed with the fluocinonide ointment (Lidex, Medicis) and the mixture was incorporated into the Aquaphor or Aquabase geometrically with thorough mixing. The crude coal tar was added last and mixed well. The final ointment was packaged in tight containers and protected from light. The author indicated that a use period of six months is appropriate for this ointment, although no stability data were provided.

**STUDY 3:** Allen[1391] reported on a compounded formulation of coal tar 2% and allantoin 2.5% formulated in a topical cream. The cream had the following formula:

| | | |
| --- | --- | --- |
| Coal tar | | 2 g |
| Allantoin | | 2.5 g |
| Stearic acid | | 16 g |
| Oleyl alcohol | | 6 g |
| Lanolin | | 2 g |
| Triethanolamine | | 600 mg |
| Methylparaben | | 100 mg |
| Purified water | qs | 100 g |

The recommended method of preparation was to heat about 70 mL of purified water to 80°C and add the methylparaben and triethanolamine. Then the stearic acid, oleyl alcohol, lanolin, coal tar, and allantoin were also heated to 80°C. The two heated mixtures were then mixed together and mixed thoroughly. Additional purified water to bring the mixture to 100 g was then added, and the mixture was allowed to cool while stirring. The cream was to be packaged in tight, light-resistant containers. The author recommended a beyond-use date of six months at room temperature because this formula is a commercial medication in some countries with an expiration date of two years or more.

## Compatibility with Other Drugs
Patel et al.[635] evaluated the compatibility of calcipotriene 0.005% ointment mixed by levigation in equal quantities with other topical medications. The mixtures were stored at 5 and 25°C. Drug concentration was assessed by HPLC analysis. The combination was found to be incompatible. Mixed with 5% tar gel (composition unspecified), the color became darker after six days at both temperatures. A 20% loss in calcipotriene content in the 25°C sample also occurred in six days, but only about 4% loss occurred at 4°C in this time frame. ∎

# Cocaine Hydrochloride

## Properties
Cocaine hydrochloride is a hygroscopic white crystalline powder or colorless crystals. It has a characteristic salty, slightly bitter taste.[1-3] Approximately 1.12 g of the hydrochloride is equivalent to 1 g of cocaine.[3]

### Solubility
Cocaine hydrochloride is very soluble in water, with an aqueous solubility of about 1 g in 0.4 or 0.5 mL. In ethanol it is freely soluble, with a solubility of about 285 to 312 mg/mL.[1-3] It is also soluble in glycerol.[1]

### pKa
The $pK_a$ of cocaine is 8.81 at 25°C and 8.7 at 30°C.[6]

## General Stability Considerations
Cocaine hydrochloride crystals or powder should be protected from moisture and light by storage in well-closed, light-resistant containers. Topical solutions should be stored at controlled room temperature.[2,3]

Cocaine is subject to ester hydrolysis in aqueous media.[6] The pH of maximum stability in aqueous solution was stated to be 1.95,[257] around 3,[6] and 2.5 to 3[405] with pH playing a prominent role above 5.5.[685] The drug is unstable in neutral or alkaline solutions but is relatively stable in acidic media,[6] although the aqueous stability at pH 7.5 was extended 23-fold when refrigerated, compared to room temperature storage.[685]

Solutions are sterilized by autoclaving.[3] However, the commercial topical solutions should not be autoclaved or sterilized with ethylene oxide due to potential damage to the container and possible concentration loss.[2]

## Stability Reports of Compounded Preparations
### Oral
Poochikian and Cradock[405] evaluated the stability of the active components in several formulas of Brompton mixtures. The stabilities of heroin 1 mg/mL and cocaine hydrochloride 0.5 mg/mL were assessed using a stability-indicating HPLC technique. The influence of alcohol and syrup on the stability of these drugs was tested by preparing formulations with varying compositions of the two vehicle components. The samples were stored at 5 and 25°C for 65 days. Cocaine content was stable in all mixtures at both temperatures throughout the study and did not approach its $t_{90}$ (time to 10% decomposition). Varying the syrup concentration had no effect on cocaine stability. The hydrolysis of heroin occurs to a greater extent and is the primary determinant of the useful shelf life of the product. Heroin stability increased as the alcohol content increased (Table 27). Syrup had a much less

**TABLE 27.** Stability of Heroin at Various Alcohol Concentrations with Syrup 25%[405]

| Alcohol (%) | $t_{90}$ 5°C (days) | $t_{90}$ 25°C (days) |
|---|---|---|
| 0 | 16 | 6 |
| 12.5 | 25 | 10 |
| 25 | 40 | 15 |
| 40 | 75 | 30 |

pronounced effect on heroin stability than alcohol did. When morphine rather than heroin was incorporated in the 12.5% alcohol–25% syrup vehicle, the stability of the cocaine was adversely affected, although it still resulted in a useful shelf life ($t_{90}$ of the cocaine = 10 days at 25°C).[405]

### Ophthalmic
Ophthalmic preparations, like other sterile drugs, should be prepared in a suitable clean air environment using appropriate aseptic procedures. When prepared from nonsterile components, an appropriate and effective sterilization method must be employed.

**STUDY 1:** The modification of cocaine hydrochloride topical solution for use as an ophthalmic solution was described by Fiscella et al.[123] Cocaine hydrochloride 4% topical solution (Roxane and Astra) has a pH around 2, which is too low for ophthalmic use. Adding 0.3 mL of 8.4% sodium bicarbonate injection to 4 mL of the cocaine hydrochloride solution raised the final pH to 3.9, which should be more comfortable. The final cocaine hydrochloride concentration was approximately 3.7%. The modified cocaine hydrochloride solution was aseptically filtered through a 0.22-μm filter and packaged in sterile ophthalmic dropper bottles. Cocaine hydrochloride solutions at pH 4 were stable for at least 45 days, and at a pH of 4 the sodium benzoate preservative in the cocaine hydrochloride topical solution was effective. However, the authors recommended limiting shelf life to one month because of both sterility and stability concerns.

**STUDY 2:** Murray and Al-Shora[257] reported, using ultraviolet spectroscopy, that an unbuffered autoclaved 1% cocaine hydrochloride ophthalmic solution (Cocaine Eye Drops BPC) exhibited 10% loss in three months at 25°C and six months at 5°C. A 1% cocaine hydrochloride solution buffered to pH 2.2 (near the pH of maximum stability) increased the stability period but was expected to cause eye irritation. Buffering a 1% cocaine hydrochloride ophthalmic solution to pH 6.8 increased the decomposition rate, with one-week losses of 54 and 12% at 25 and 5°C, respectively.

**STUDY 3 (WITH ATROPINE):** Miethke[491] reported the stability of a cocaine hydrochloride 10-mg/mL–atropine sulfate 2-mg/mL mixed solution for ophthalmic use. The solution was prepared by dissolving 2 g of cocaine hydrochloride and 400 mg of atropine sulfate in 200 mL of water for injection. The solution was filtered through a 0.22-μm cellulose ester filter (Optex) and filled as 2 mL in 10-mL containers. The solution was sterilized in an autoclave at 121°C for 15 minutes. HPLC analysis determined that the solution was stable for 15 months, presumably at room temperature.

## Topical

**USP OFFICIAL FORMULATION (T-A-C TOPICAL SOLUTION):**

| | | |
|---|---|---|
| Cocaine hydrochloride | | 4 g |
| Tetracaine hydrochloride | | 1 g |
| Epinephrine hydrochloride injection (1:1000) | | 25 mL |
| Benzalkonium chloride | | 10 mg |
| Edetate disodium | | 6.4 mg |
| Sodium chloride injection 0.9% | | 35 mL |
| Purified water | qs | 100 mL |

The cocaine hydrochloride and tetracaine hydrochloride powders should be dissolved in 25 mL of purified water. Add the epinephrine hydrochloride (1:1000) injection and mix. Dissolve the edetate disodium separately in sodium chloride 0.9% and bring to a final volume of 35 mL with additional sodium chloride 0.9%. Dissolve the benzalkonium chloride separately in a quantity of purified water and bring to a final volume of 10 mL with additional purified water. Combine the three solutions, add additional purified water to bring the solution to volume, and mix well. Package the solution in a suitable sterile, tight, and light-resistant container. The final liquid preparation should have a pH between pH 4 and 6. Store the preparation under refrigeration between 2 and 8°C. The beyond-use date is 30 days from the date of compounding.[4,5]

**STUDY 1:** The stability of aqueous solutions of cocaine hydrochloride was reported by Gupta.[167] Cocaine hydrochloride solutions of 1, 2, and 10% were prepared and stored at 24°C. In addition, cocaine hydrochloride solutions were prepared in various phosphate and carbonate buffers with pH values ranging from 1.6 to 8.1. A stability-indicating HPLC technique was used to assess the drug content of the solutions during the study. The 1, 2, and 10% aqueous solutions of cocaine hydrochloride had pH values of 3.6, 3.4, and 2.8, respectively. All three concentrations were chemically stable, with no loss of drug due to hydrolysis over 45 days at 24°C. In the buffer solutions at pH 1.7 to 4.5, no hydrolysis of cocaine hydrochloride occurred. At pH values of 5.8 up to 8, increasing amounts of hydrolysis occurred. Furthermore, increasing temperature played a role in increasing the hydrolysis rate at the higher pH values.

**TABLE 28.** T-A-C Topical Solution Formulations[150]

| Component | Formula 1 | Formula 2 |
|---|---|---|
| Tetracaine hydrochloride 2% | 2 mL | 1 mL |
| Epinephrine hydrochloride 0.1% | 1 mL | 2 mL |
| Cocaine hydrochloride 4% solution | 1 mL | |
| Cocaine hydrochloride powder | | 0.472 g[a] |
| 0.9% Sodium chloride injection | qs | 4 mL |

[a]Cocaine hydrochloride 11.8%.

**STUDY 2:** Stiles and Allen[150] reported on two formulations for topical anesthesia. The T-A-C solutions (Table 28) consisted of tetracaine hydrochloride (Pontocaine, Winthrop), epinephrine hydrochloride (Adrenalin, Parke-Davis), and cocaine hydrochloride. The first formula is a simple combination of three commercial products. The second formula is a cocaine-fortified solution prepared from cocaine hydrochloride powder. The T-A-C solutions should be terminally filtered to remove any contaminants. The solutions are unstable in light and air and should be freshly prepared for use.

**STUDY 3:** Pesko[253] also reported on a T-A-C topical formulation (Table 29). The author recommended mixing the components aseptically with filtration through a 0.22-μm filter; an expiration period of 14 days for samples protected from light was recommended, although stability studies were not conducted.

**STUDY 4:** Jeter and Mueller[404] described the preparation of a sterile T-A-C gel formulation. The product was prepared from T-A-C sterile solution composed of tetracaine hydrochloride 0.5%, epinephrine hydrochloride 1:2000, and cocaine hydrochloride 11.8% that had been sterilized using a 0.2-μm filter (Millex-GS). T-A-C solution 12 mL was added to a 1-g jar of sterile Gelfoam powder and thoroughly mixed. The mixture was divided into six 2.5-g portions using a gas-sterilized measuring spoon, packaged into heat-sterilized and depyrogenated amber open-mouth vials, and sealed with gas-sterilized tamper-evident aluminum lids. The preparation

**TABLE 29.** T-A-C Topical Anesthetic Solution Formula[253]

| Component | Amount |
|---|---|
| Tetracaine hydrochloride 2% | 50 mL |
| Epinephrine hydrochloride 0.1% | 2.5 mL |
| Cocaine hydrochloride 10% | 40 mL |
| Sterile water for irrigation | 7.5 mL |

should be protected from light to prevent epinephrine oxidation, which turns the mixture brown. The authors used a three-month date on the preparation, although specific stability was not assessed.

**STUDY 5:** Accordino et al.[1057] evaluated the stability of a cocaine hydrochloride 50-mg/mL (5% wt/vol) solution for topical use. The solution also contained chlorobutanol 5 mg/mL dissolved in ethanol, resulting in about 1% ethanol in the final product; the solution was brought to volume with purified water. It was then packaged in amber glass vials with polypropylene screw caps and Teflon/silicone seals. The vials were stored refrigerated at 2 to 8°C, at room temperature of 25°C, and at elevated temperature of 37°C for 27 weeks.

No color change or turbidity was observed in any of the samples. Stability-indicating HPLC analysis found that little or no cocaine loss occurred in 27 weeks in the refrigerated and room temperature samples. The 37°C elevated temperature samples underwent about 5 to 6% loss of cocaine in 27 weeks. ■

# ■ Codeine
# Codeine Phosphate
# Codeine Sulfate

## Properties

Codeine phosphate and sulfate are small colorless or white crystals or a white crystalline powder.[2,3]

### Solubility

Codeine is slightly soluble in water to about 1 g in 120 mL; 1 g can be dissolved in 60 mL of water heated to 80°C. Codeine is freely soluble in ethanol to about 1 g in 2 mL and about 1.2 mL of hot ethanol. Codeine is almost insoluble in solutions of alkali hydroxides.[1,4] When heated in an amount of water that is insufficient for complete dissolution, it melts into oily drops that crystallize upon cooling.[4]

Codeine phosphate is freely soluble in water, having an aqueous solubility of about 400 to 435 mg/mL. It is slightly soluble in ethanol, having a solubility around 3 mg/mL.[1–3]

Codeine sulfate has a solubility of about 33 mg/mL in water. In ethanol, its solubility is about 0.77 mg/mL.[1,3]

### pH

A saturated aqueous solution of codeine has a pH of 9.8.[1] A 2 or 4% codeine phosphate aqueous solution has a pH of 4 to 5.[1,3] Codeine phosphate injection has a pH of 3 to 6.[2–4]

## General Stability Considerations

Codeine phosphate and sulfate products should be stored in tight or well-closed, light-resistant containers at controlled room temperature. The injection should be stored at controlled room temperature and protected from freezing, temperatures exceeding 40°C, and light.[2,4]

## Stability Reports of Compounded Preparations
### Oral

**STUDY 1:** Dentinger and Swenson[1179] reported the stability of an oral liquid formulation of codeine phosphate 3 mg/mL. Codeine phosphate powder 600 mg was dissolved in 2.5 mL of sterile water for irrigation. Ora-Sweet syrup was added to bring the volume to 200 mL, and the mixture was stirred for 10 minutes using a stir bar. The oral liquid was packaged in Kerr amber polyethylene terephthalate bottles with child-resistant caps and in Baxa 5-mL polyethylene oral syringes with silicon elastomer plunger tips and sealed with Baxa high-density polyethylene syringe tip caps. The samples were stored at room temperature of 22 to 25°C for 98 days.

No visible changes in color or clarity and no change in odor were observed. No visible evidence of microbial growth occurred. The pH remained at 4.2 throughout the study. Stability-indicating HPLC analysis found that no loss of codeine phosphate occurred in 98 days in either the prescription bottles or the oral syringes.

**STUDY 2:** Feyel-Dobrokhotov et al.[1312] evaluated the stability of Saint-Lazare potion, an antidiarrheal oral liquid containing codeine and ethylmorphine hydrochloride (codethyline hydrochloride) 6.52 mg/mL in a vehicle containing unspecified amounts of absolute alcohol, simple syrup, and purified water. The preparation was packaged in clear and amber glass bottles and stored at controlled room temperature of 22 to 28°C for 26 weeks. HPLC analysis found little change in drug concentrations in either container during the study period.

### Enteral
**CODEINE PHOSPHATE COMPATIBILITY SUMMARY**
*Compatible with:* Osmolite • Osmolite HN • Vital

Fagerman and Ballou[52] reported on the compatibility of codeine (salt form and concentration unspecified) mixed in equal quantities with three enteral feeding formulas. The codeine was visually compatible with Vital, Osmolite, and Osmolite HN (Ross), with no obvious thickening, granulation, or precipitation. ■

# Colistin Sulfate
# Colistimethate Sodium

## Properties
Colistin sulfate is a mixture of polypeptides. It is a white to slightly yellow odorless hygroscopic powder.[2,3] The United States Pharmacopeia (USP) specifies not less than 500 mcg of colistin per milligram of bulk substance.[4]

### Solubility
Colistin sulfate is freely soluble in water and slightly soluble in ethanol.[3]

### pH
A 1% colistin sulfate solution has a pH between 4 and 7.[3,4]

## General Stability Considerations
Colistin sulfate products should be stored in tight containers and protected from light.[3,4] Colistin base may precipitate from solutions having a pH above 7.5.[3] Colistin in aqueous solutions is stable at acidic pH from 2 to 6, exhibiting maximum stability at pH 3.4.[832] Colistimethate sodium conversion to the active but more toxic colistin in aqueous solution is reported to be temperature and concentration dependent. Greater conversion occurs at room temperature than under refrigeration; greater conversion occurs at low concentrations such as 4 mg/mL than at higher concentrations such as 77.5 and 200 mg/mL.[1334]

Li et al.[1203] evaluated the stability of colistin sodium and its prodrug, colistimethate sodium, in aqueous media. Colistin sodium evaluated by HPLC analysis was found to be stable in water at both 4 and 37°C for up to 60 days and 120 hours, respectively. Colistimethate sodium in water, phosphate buffer, and plasma underwent rapid conversion to colistin sodium at 37°C. Approximately 60% of the colistimethate sodium prodrug converted to the active form within 48 hours at 37°C. If stored refrigerated at 4°C, the prodrug appeared to remain intact for up to two days. However, about 30% of the colistimethate sodium had converted after 60 days of refrigerated storage. The authors indicated that colistimethate sodium should be freshly prepared before use to avoid excessive formation of the more toxic colistin.

## Stability Reports of Compounded Preparations
### Inhalation
Inhalations, like other sterile drugs, should be prepared in a suitable clean air environment using appropriate aseptic procedures. When prepared from nonsterile components, an appropriate and effective sterilization method must be employed.

**STUDY 1:** McCoy[1202] reported the death of a patient from acute respiratory distress syndrome associated with colistimethate sodium prepared as an inhalation solution. The likely cause of the fatality was the conversion of the colistimethate sodium prodrug to the biologically active form, colistin sodium, which is less likely to be tolerated. McCoy pointed to the study of Li et al.[1203] (see *General Stability Considerations*) that demonstrated substantial conversion to colistin in a relatively short period. McCoy recommended that if nebulization of colistimethate sodium is required, the drug should be reconstituted just before administration to avoid excessive conversion to the active colistin, which may cause airway and alveolar injury.

**STUDY 2:** Wallace et al.[1334] evaluated the conversion of colistimethate sodium to the active but more toxic colistin in an inhalation solution. The extemporaneously compounded inhalation solution was prepared from colistimethate sodium powder dissolved in water, and the solution was adjusted to isotonicity with sodium chloride. The final colistimethate sodium concentration in the inhalation solution was 77.5 mg/mL. The inhalation solution was sterilized by filtration and was packaged in glass bottles with rubber stoppers. Samples were stored refrigerated at 2 to 6°C and at room temperature of 23 to 27°C for 52 weeks. Osmolality and pH remained unchanged throughout the study. Unlike more dilute solutions, conversion of colistimethate sodium to colistin proceeded very slowly; less than 0.1% of the original colistimethate sodium content was found to have converted to the more toxic colistin after 52 weeks of storage at either temperature.

## Compatibility with Other Drugs
### COLISTIN SULFATE COMPATIBILITY SUMMARY
*Incompatible with:* Sucralfate
*Uncertain or variable compatibility with:* Albuterol

**STUDY 1:** Feron et al.[467] evaluated the interaction of sucralfate with the antimicrobial drugs colistin sulfate, tobramycin sulfate, and amphotericin B resulting from enteral coadministration. Sucralfate 500 mg was added to 40 mL of HPLC-grade water and the solution was adjusted to pH 3.5 with hydrochloric acid. The antimicrobial drugs were added to separate samples to yield final concentrations of colistin 0.05 mg/mL, tobramycin 0.05 mg/mL, and amphotericin B 0.025 mg/mL. The concentration of each drug was analyzed initially and numerous times over 90 minutes of storage. The colistin concentration was assessed by HPLC, tobramycin was assessed by enzyme immunoassay, and amphotericin B was assessed by spectroscopy. Losses of all three antimicrobials occurred very rapidly and extensively. Colistin, tobramycin, and amphotericin B losses were 88, 99, and 95%, respectively, after five minutes. Furthermore, the loss was not reversible by adjusting pH to near neutrality. The authors noted that the mechanism of interaction is unclear. However, concurrent enteral administration of sucralfate and these antimicrobials would result in substantially lower concentrations than would be achieved in the absence of sucralfate.

**STUDY 2:** Roberts et al.[1022] evaluated the compatibility and stability of albuterol sulfate (Delta West) 5 mg/mL and colistimethate sodium (Coly-Mycin M Parenteral) 33.3 mg/mL of colistin base in equal volumes for use as an inhalation solution in the treatment of cystic fibrosis. Albuterol sulfate inhalation solution that contained benzalkonium chloride as a preservative caused the rapid formation of a cloudy precipitate of colistin. The precipitation persisted for at least 10 hours before finally resolving. Preservative-free albuterol sulfate inhalation solution did not cause cloudy precipitation to occur.

HPLC analysis found albuterol losses of 7 and 14% in one and 24 hours, respectively. Evaluation of colistin by bioassay found 2% loss in one hour, but no additional testing was performed because the 14% loss of albuterol was considered unacceptable. ■

# ■ Co-trimoxazole
## (Trimethoprim–Sulfamethoxazole)

## Properties

Co-trimoxazole is a fixed combination of trimethoprim and sulfamethoxazole in a ratio of 1:5.[2,7] Trimethoprim is a white to cream or light yellow odorless crystalline powder with a bitter taste.[1-3,7] Sulfamethoxazole is a white to off-white nearly odorless crystalline powder.[1-3]

### Solubility

Trimethoprim is very slightly soluble in water, having an aqueous solubility of about 0.4 mg/mL. It is slightly soluble in ethanol.[1,3] At 25°C, sulfamethoxazole aqueous solubility is 0.29 mg/mL; in ethanol, its solubility is 20 mg/mL.[2,3] It also is soluble in dilute alkali hydroxide solutions.[3]

### pH

Co-trimoxazole suspension has a pH of 5 to 6.5.[2] Bactrim suspension (Roche) had a pH of 5.6.[19] Co-trimoxazole injection has a pH of approximately 10.[2]

### $pK_a$

Trimethoprim has a $pK_a$ of 6.6.[1]

## General Stability Considerations

Co-trimoxazole oral products should be stored in tight or well-closed, light-resistant containers at controlled room temperature. The injection concentrate should be stored at controlled room temperature and not refrigerated.[2]

The solubility of trimethoprim is partially dependent on the solution pH. Its solubility is lower in alkaline solutions.[337]

## Stability Reports of Compounded Preparations
### Injection

Injections, like other sterile drugs, should be prepared in a suitable clean air environment using appropriate aseptic procedures. When prepared from nonsterile components, an appropriate and effective sterilization method must be employed.

Kaufman et al.[338] reported on the stability of co-trimoxazole injection concentrate repackaged in polypropylene syringes and stored for 2.5 days at room temperature exposed to fluorescent lights. HPLC analysis found no loss of either drug.

### Oral

**STUDY 1:** Wu et al.[846] reported the stability of two extemporaneously compounded formulations of sulfamethoxazole with trimethoprim (co-trimoxazole). The suspensions were prepared by grinding commercial tablets to fine powder using a mortar and pestle and incorporating the powder into (1) simple syrup and (2) a 1% methylcellulose–simple syrup 7:3 mixture. The suspensions were packaged in amber glass containers and stored at ambient room temperature or refrigerated at 4 to 8°C. Analysis of each drug component after extraction was performed using the British Pharmacopoeia spectrophotometric methods.

The suspensions did not exhibit caking or flocculation and were easily redispersed. Analytical results were variable, possibly due to difficulties in consistent extraction. However, sulfamethoxazole concentration determinations were all within 10% of the initial concentration in the simple syrup formulation throughout the 94-day study period. The trimethoprim concentration determinations showed variability but no trend toward drug loss. The methylcellulose–simple syrup mixture formulation showed similar results throughout the 28-day period it was studied.

**STUDY 2:** Midhat et al.[827] reported the stability of an extemporaneously compounded oral suspension of sulfamethoxazole 40 mg/mL and trimethoprim 8 mg/mL in a suspension vehicle composed of carboxymethylcellulose sodium, glycerin, polysorbate 80, saccharin sodium, methyl hydroxybenzoate, ethanol, flavor, and purified water. HPLC analysis found 8% loss of sulfamethoxazole and 10% loss of trimethoprim after 21 months of room temperature storage.

### Enteral
#### CO-TRIMOXAZOLE COMPATIBILITY SUMMARY
*Compatible with:* Enrich • Ensure • Ensure HN • Ensure Plus • Ensure Plus HN • Osmolite • Osmolite HN • Precitene • TwoCal HN • Vital • Vivonex T.E.N.

**STUDY 1:** Cutie et al.[19] added 10 mL of co-trimoxazole suspension (Bactrim, Roche) to varying amounts (15 to 240 mL) of Ensure, Ensure Plus, and Osmolite (Ross Laboratories) with vigorous agitation to ensure thorough mixing. The co-trimoxazole elixir was physically compatible, distributing uniformly in all three enteral products with no phase separation or granulation. A slight increase in viscosity was measured when the suspension was mixed with the smaller volumes of enteral products.

**STUDY 2:** Burns et al.[739] reported the physical compatibility of co-trimoxazole suspension (Septra) 20 mL with 10 mL of three enteral formulas, including Enrich, TwoCal HN, and Vivonex T.E.N. Visual inspection found no physical incompatibility with any of the enteral formulas.

**STUDY 3:** Altman and Cutie[850] reported the physical compatibility of co-trimoxazole suspension (Bactrim, Roche) 10 mL with varying amounts (15 to 240 mL) of Ensure HN, Ensure Plus HN, Osmolite HN, and Vital after vigorous agitation to ensure thorough mixing. The co-trimoxazole suspension was physically compatible, distributing uniformly in all four enteral products with no phase separation or granulation. A slight increase in viscosity was measured when the suspension was mixed with the smaller volumes of enteral products.

**STUDY 4:** Ortega de la Cruz et al.[1101] reported the physical compatibility of an unspecified amount of oral liquid co-trimoxazole (Abactrim, Andreu) with 200 mL of Precitene (Novartis) enteral nutrition diet for a 24-hour observation period. No particle growth or phase separation was observed. ■

# ■ Cromolyn Sodium
## (Sodium Chromoglycate)

## Properties
Cromolyn sodium is a white or almost white odorless hygroscopic crystalline powder having a slightly bitter aftertaste.[1–3]

### Solubility
Cromolyn sodium is freely soluble in water, having a solubility of 100 mg/mL at 20°C.[1] The drug is practically insoluble in ethanol.[3]

### pH
Cromolyn sodium inhalation solution has a pH target of 5.5 and a pH range of 4 to 7.[2,7] The nasal solution and the 4% ophthalmic solution also have a pH range of 4 to 7.[4,7]

### $pK_a$
Cromolyn sodium is dibasic, having $pK_{a1}$ and $pK_{a2}$ of 2.[2]

## General Stability Considerations
Cromolyn sodium products should be packaged in tight, light-resistant containers. They should be stored at controlled room temperature and protected from temperatures exceeding 37°C and from direct light. The oral concentrate should not be used if it is discolored or contains a precipitate.[2–4] The oral and nebulizer solutions should remain stored in the foil pouches until ready for use.[7]

Cromolyn sodium is stable in solution over a pH range of 2 to 7. However, the drug is unstable in neutral and alkaline solutions, undergoing hydrolysis. Cromolyn sodium may precipitate in the presence of magnesium or calcium salts[2,659] and benzalkonium-chloride-preserved solutions.[768,1023]

## Stability Reports of Compounded Preparations
### Inhalation
Inhalations, like other sterile drugs, should be prepared in a suitable clean air environment using appropriate aseptic procedures. When prepared from nonsterile components, an appropriate and effective sterilization method must be employed.

Blondino and Baker[930] evaluated the physical and chemical stability of four common drugs administered by inhalation, albuterol sulfate 2.5 mg/3 mL, budesonide 0.25 mg/3 mL, cromolyn sodium 20 mg/3 mL, and ipratropium bromide 0.5 mg/3 mL, individually in an acidified (pH 4) 15% ethanol in water solution. The budesonide was dissolved in the ethanol, which was then added to a solution of the other drugs dissolved in the acidified sterile water for irrigation. No visible changes in color or clarity of any of the samples were observed. Cromolyn sodium by itself underwent little loss over eight weeks. Losses were negligible at refrigerated and room temperatures, and there was about a 4% loss at elevated temperature of 40°C. Cromolyn sodium was also stable in the four-drug combination. See *Compatibility with Other Drugs* below.

## Compatibility with Other Drugs
### CROMOLYN SODIUM COMPATIBILITY SUMMARY
*Compatible with:* Acetylcysteine • Budesonide • Epinephrine hydrochloride • Formoterol fumarate • Isoproterenol hydrochloride • Levalbuterol hydrochloride
*Uncertain or variable compatibility with:* Albuterol sulfate • Atropine sulfate • Ipratropium bromide • Isoetharine hydrochloride • Metaproterenol sulfate • Sodium chloride 7% (hypertonic) • Terbutaline sulfate

**STUDY 1:** Lesko and Miller[659] evaluated the chemical and physical stability of 1% cromolyn sodium nebulizer solution (Fisons) admixed with several bronchodilator inhalation solutions. The admixtures were prepared by mixing 2 mL of 1% cromolyn sodium inhalation solution with the amounts of the bronchodilators shown in Table 30. All admixtures were visually clear and colorless and remained chemically stable by HPLC analysis for 60 minutes after mixing at 22°C. Cromolyn sodium was stable, exhibiting less than 4% loss in any sample. The greatest bronchodilator losses were acetylcysteine with 6% loss and isoetharine hydrochloride with 4% loss.

**STUDY 2:** Owsley and Rusho[517] evaluated the compatibility of several respiratory therapy drug combinations. Cromolyn sodium 10-mg/mL nebulizer solution (Intal, Fisons) was combined with the following drug solutions: albuterol sulfate 5-mg/mL inhalation solution (Proventil, Schering), atropine sulfate 1 mg/mL (American Regent), isoetharine 1% solution (Bronkosol, Sanofi Winthrop), metaproterenol sulfate 5% solution (Alupent, Boehringer Ingelheim), and terbutaline sulfate 1-mg/mL solution (Brethine, Geigy). The test solutions were filtered through 0.22-µm filters into

**TABLE 30.** Cromolyn Sodium 1% Compatibility for 60 Minutes with Bronchodilator Inhalation Solutions[659]

| Bronchodilator Solutions | Amount[a] |
| --- | --- |
| Acetylcysteine 20% | 2.0 mL |
| Epinephrine hydrochloride 2.25% | 1.0 mL |
| Isoetharine hydrochloride 1% | 0.5 mL |
| Isoproterenol hydrochloride 0.5% | 0.5 mL |
| Metaproterenol sulfate 5% | 0.3 mL |
| Metaproterenol sulfate | [b] |
| Terbutaline sulfate 0.1% | 0.5 mL |

[a]Cromolyn sodium 1% 2 mL combined with the amounts of bronchodilator indicated.
[b]Metaproterenol sulfate 5% 0.3 mL diluted with 2.5 mL of 0.9% sodium chloride prior to mixing with the cromolyn sodium 1%.

clean vials. The combinations were evaluated over 24 hours (temperature unspecified) using the USP particulate matter test. None of the combinations was compatible during this longer term storage. Cromolyn sodium combined with the albuterol sulfate, atropine sulfate, isoetharine, metaproterenol sulfate, and terbutaline sulfate all formed unacceptable levels of larger particulates (≥10 µm).

**STUDY 3:** Emm et al.[628] evaluated the stability and compatibility of cromolyn sodium combinations during short-term preparation and use. Cromolyn sodium 1% nebulizer solution (Fisons) was combined with the following broncho-active drug solutions and studied over 90 minutes: albuterol sulfate 0.5% (Glaxo), atropine sulfate 0.2 and 0.5% (Dey), and metaproterenol sulfate 5 and 0.6% (Alupent, Boehringer Ingelheim). No visible signs of precipitation or measured changes in pH were found, and HPLC analysis found less than 10% loss of any of the drugs within 90 minutes. Most combinations showed no loss at all. The authors did note that longer term (24-hour) storage may result in discoloration of metaproterenol sulfate solutions; such discolored solutions should not be used. Furthermore, unpublished reports indicate that a turbid, hazy mixture may result from admixing cromolyn sodium solution with albuterol sulfate 0.083%.

**STUDY 4:** Iacono et al.[767] reported the compatibility and stability of 1:1 inhalation solution mixtures of ipratropium bromide 0.25 mg/mL (Atrovent, Boehringer Ingelheim) and cromolyn sodium 10 mg/mL (Intal, Fisons) for use as nebulizer solution combinations. Cloudy turbidity developed immediately upon mixing and persisted for 24 hours. The mixture cleared by 48 hours. Filtration did not isolate a precipitate. HPLC analysis found that no loss of either drug occurred within time periods up to 111 minutes (the study period limit). Addition of albuterol sulfate (Respolin, Riker) yielded the same cloudiness in the mixture and similar analytical results. Albuterol was not analyzed. If Intal Spincaps were used as the source of cromolyn sodium, mixing with ipratropium bromide did not result in cloudiness, suggesting an excipient in the Intal nebulizer solution might be responsible.

**STUDY 5:** Turner[768] reported that the cloudiness of the ipratropium bromide and cromolyn sodium inhalation solution mixture was caused by the formation of a complex of benzalkonium chloride preservative in the ipratropium bromide solution with cromolyn sodium. In an equal parts mixture of the two inhalation solutions, only about 0.8% of the cromolyn sodium was complexed. The complex is an oil and not a crystalline precipitate. Turner indicated that the oily complex would nebulize readily. If unpreserved ipratropium bromide inhalation solution is combined with cromolyn sodium, no cloudiness occurs.

**STUDIES 6 AND 7:** Smaldone et al.[1026] and McKenzie and Cruz-Rivera[892] evaluated the compatibility and stability of two concentrations of budesonide inhalation suspension with other inhalation medications, including cromolyn sodium. Budesonide (Pulmicort Respules, AstraZeneca) 0.25 mg/ 2 mL and also 0.5 mg/2 mL was mixed with cromolyn sodium 20 mg/2 mL. The inhalations were added to new Pari LC Plus nebulizer cups and mixed using a vortex mixer. The samples were examined initially after mixing and over 30 minutes.

Visual inspection found no change in color, formation of a precipitate, or nonresuspendability of any of the samples. Stability-indicating HPLC analysis found that all of the drugs remained adequately stable throughout the study period with little or no loss of budesonide and cromolyn sodium in most of the samples over 30 minutes.

**STUDY 8:** Roberts and Rossi[897] reported the compatibility of cromolyn sodium inhalation (Intal) with budesonide suspension (Pulmicort Respules), ipratropium bromide as both single (Atrovent UDV) and multiple-use (Atrovent) forms, albuterol sulfate (Ventolin) single-dose and multiple-dose forms, and terbutaline multiple-use (Bricanyl). Visual inspection found that the combinations with budesonide and terbutaline sulfate were compatible, with no visible cloudiness. The combinations with cromolyn sodium (Intal) were more problematic. The unpreserved single-use forms of ipratropium bromide and of albuterol sodium were compatible with cromolyn sodium, but the multiple-use forms preserved with benzalkonium chloride were not. Marked and immediate cloudiness appeared due to the formation of an oily complex of cromolyn sodium with benzalkonium chloride.

**STUDY 9:** Blondino and Baker[930] evaluated the physical and chemical stability of four common drugs administered by inhalation, albuterol sulfate 2.5 mg/3 mL, budesonide 0.25 mg/3 mL, cromolyn sodium 20 mg/3 mL, and ipratropium bromide 0.5 mg/3 mL, individually in an acidified (pH 4) 15% ethanol in water solution and in a four-drug combination. The budesonide was dissolved in the ethanol, which was then added to a solution of the other drugs dissolved in the acidified sterile water for irrigation. No visible changes in color or clarity of any of the samples were observed. Cromolyn sodium by itself underwent little loss over eight weeks. Losses were negligible at refrigerated and room temperatures, and about a 4% loss occurred at elevated temperature of 40°C. Cromolyn sodium was also stable in the four-drug combination. The other three drugs remained stable under refrigeration and at room temperature, with no more than 6% loss in eight weeks, but budesonide and ipratropium bromide underwent substantial losses of 20 to 25% at 40°C in that time frame.

**STUDY 10 (ALBUTEROL AND IPRATROPIUM):** Nagtegaal et al.[1046] evaluated the long-term stability of two inhalation solution formulations containing albuterol sulfate 1.5 mg/mL with ipratropium bromide 62.5 mcg/mL, both with and without cromolyn sodium 5 mg/mL. The drugs were compounded in a solution composed of sodium chloride 8.8 mg/mL in water for injection. The pH was adjusted to 5.6 for the formulation with cromolyn sodium and to 5 for the formulation without the cromolyn. The inhalation solutions were packaged in glass ampules, filtered through a 0.2-μm filter, and autoclaved at 121°C for 15 minutes for sterilization. Samples of the sterilized solutions were stored at room temperature and also at elevated temperatures of 45, 65, and 85°C for accelerated degradation studies.

Ipratropium was the least stable component of the inhalation solutions. Stability-indicating HPLC analysis found the albuterol sulfate with ipratropium bromide inhalation solution (no cromolyn sodium) to be stable, with less than 10% ipratropium loss occurring in 18 months at room temperature and a calculated stability of 3.8 years refrigerated. With cromolyn sodium present in the three-drug inhalation solution, the ipratropium was less stable, exhibiting less than 10% loss for 10 months at room temperature and a calculated stability of 1.9 years refrigerated.

**STUDY 11 (LEVALBUTEROL):** Bonasia et al.[1329] evaluated the compatibility and stability of levalbuterol hydrochloride inhalation solution (Sepracor) with several drugs including cromolyn sodium (Alpharma). The drugs were mixed with sodium chloride 0.9% and then with each other to prepare the following solution:

Levalbuterol 1.25 mg/0.5 mL diluted with 2 mL of sodium chloride 0.9%
Cromolyn sodium 20 mg/2 mL diluted with 0.5 mL of sodium chloride 0.9%

The drug mixture was prepared in triplicate and evaluated initially and after 30 minutes at an ambient temperature of 21 to 25°C. The authors indicated that 30-minutes study duration was chosen because it represents the approximate time it takes for a patient to administer drugs by nebulization at home. No visible evidence of physical incompatibility was observed with the samples. HPLC analysis found no loss of any of either of the drugs in the samples.

**STUDY 12 (FORMOTEROL):** Akapo et al.[1330] evaluated the compatibility and stability of formoterol fumarate 20 mcg/2 mL inhalation solution (Perforomist, Dey) mixed with cromolyn sodium 20 mg/2 mL (Dey), for administration by inhalation and evaluated over 60 minutes at room temperature of 23 to 27°C. The test samples were physically compatible by visual examination and measurement of osmolality, pH, and turbidity. HPLC analysis of drug concentrations found little or no loss of either of the drugs within the study period.

**STUDY 13 (HYPERTONIC SODIUM CHLORIDE):** Fox et al.[1239] evaluated the physical compatibility of hypertonic sodium chloride 7% with several inhalation medications used in treating cystic fibrosis. Cromolyn sodium inhalation solution (Dey) 10 mg/mL mixed in equal quantities with extemporaneously compounded hypertonic sodium chloride 7% did not exhibit any visible evidence of physical incompatibility. However, the measured turbidity underwent a small increase over the one-hour observation period. ▪

# ▪ Curcumin

## Properties

Curcumin occurs as an orange-yellow crystalline powder. It is odorless or has a mild odor. It is the main coloring component of tumeric.[1,3,4]

### Solubility

Curcumin is insoluble in water and ether but soluble in ethanol, methanol. ethyl acetate, and glacial acetic acid, and slightly soluble in acetone.[1,4]

## General Stability Considerations

Exposure to light and heat should be avoided for curcumin storage.[4] Curcumin gives a brownish-red color with alkali and a light yellow color with acids.[1]

Song et al.[1452] evaluated curcumin photostability at concentrations of 5, 50, and 500 mcg/mL in absolute alcohol. Samples were exposed to ultraviolet light at 254 nm and protected from exposure to light. HPLC analysis found curcumin was stable for 24 hours in the dark but was unstable exposed to ultraviolet light. Curcumin degraded more rapidly at the lower concentrations with total loss in 12 and 24 hours at 5 and 50 mcg/mL and 50% loss in 24 hours at 500 mcg/mL.

## Stability Reports of Compounded Preparations

### Topical

**STUDY 1:** Patel et al.[1287] reported on the formulation and stability of curcumin in a transdermal film for use as an anti-inflammatory drug. A number of transdermal film formulations were evaluated with the one that performed best being composed of hydroxypropyl methylcellulose K4M 1% (wt/vol) polymer, oleic acid 20% (vol/vol) penetration enhancer, and di-*n*-butyl phthalate 2% (vol/vol) plasticizer. The hydroxypropyl methylcellulose K4M was dissolved in a solvent system composed of chloroform:methanol:dichloromethane (4:4:2). The di-*n*-butyl phthalate, oleic acid, and a weighed amount of curcumin were added with stirring.

The polymeric drug solution was poured onto a mercury surface and dried for 24 hours at room temperature in a dust-free environment. The films were cut into standardized pieces and a backing membrane of polypropylene film was glued on.

Differential scanning calorimetry and infrared spectroscopy suggested that curcumin was compatible with the polymer mixture. Drug content was found to be between 97.3 and 100.2% by ultraviolet spectrophotometry. Samples of the curcumin polymer film were stored at ambient laboratory conditions for 365 days and at 40°C and 75% relative humidity for 90 days. Drug content losses found were about 2 to 3% in 365 days at ambient conditions and 3 to 5% in 90 days at accelerated stress conditions. Skin permeation and anti-inflammatory activity were found to be superior to other film formulations in test animals.

**STUDY 2:** Patel et al.[1342] also reported on the formulation and stability of curcumin 2% in a topical gel for use as an anti-inflammatory drug. A number of topical formulations based on Carbopol 934P and hydroxypropylcellulose with menthol incorporated as a skin penetration enhancer were evaluated. The two formulations that performed best were composed of curcumin 2% in (1) Carbopol 934P 2%, triethanolamine 1%, menthol 12.5%, ethanol 30%, and distilled water to make 100%; (2) hydroxypropylcellulose 2%, menthol 12.5%, ethanol 30%, and distilled water to make 100%. Menthol concentrations from 2.5% up to 12.5% were tried with the highest concentration resulting in the greatest penetration of the curcumin. The high concentration of menthol also resulted in decreased viscosity in both gel formulations. Both formulations were free of skin irritation and performed well in anti-inflammatory testing in animals. Drug content was found to be between 99.2 and 100.1% by ultraviolet spectrophotometry. Differential scanning calorimetry and infrared spectroscopy suggested that curcumin was compatible with the topical gel mixtures. ▪

# Cyclopentolate Hydrochloride

## Properties
Cyclopentolate hydrochloride is a white crystalline powder with a characteristic odor that develops upon standing.[3,4]

### Solubility
Cyclopentolate hydrochloride is very soluble in water and freely soluble in ethanol.[4]

### pH
Aqueous solutions of cyclopentolate hydrochloride are acidic, with a 1% solution having a pH range of 4.5 to 5.5. Cyclopentolate hydrochloride ophthalmic solution has a pH of 3 to 5.5.[3,4]

### pK$_a$
Cyclopentolate has a pK$_a$ of 8.4.[1068]

## General Stability Considerations
Cyclopentolate hydrochloride should be packaged in tight containers and stored under refrigeration. Cyclopentolate hydrochloride ophthalmic solution should be packaged in tight containers and stored at controlled room temperature.[4]

## Stability Reports of Compounded Preparations
### Ophthalmic
Ophthalmic preparations, like other sterile drugs, should be prepared in a suitable clean air environment using appropriate aseptic procedures. When prepared from nonsterile components, an appropriate and effective sterilization method must be employed.

**STUDY 1:** Garcia-Valldecabres et al.[880] reported the change in pH of several commercial ophthalmic solutions, including cyclopentolate hydrochloride 10 mg/mL (Colircusi Ciclople-jico) over 30 days after opening. Slight variation in solution pH was found, but no consistent trend and no substantial change occurred.

**STUDY 2:** Aye and Koch[1010] reported that cyclopentolate 0.5 and 1% ophthalmic solutions with benzalkonium chloride and polyvinylpyrrolidinone were stable for more than one year at ambient temperature.

**STUDY 3 (GEL WITH PHENYLEPHRINE, TROPICAMIDE, LIDO-CAINE):** Bailey et al.[1525] evaluated the stability of a four-drug combination ophthalmic gel for use as a preoperative medication for cataract surgery. The final drug concentrations in the preoperative gel were cyclopentolate hydrochloride 0.51 mg/mL, phenylephrine hydrochloride 5.1 mg/mL, and tropicamide 0.51 mg/mL in the lidocaine hydrochloride jelly 2%. The gel was prepared by combining 0.3 mL each of cyclopentolate hydrochloride ophthalmic solution 1%, USP, phenylephrine hydrochloride ophthalmic solution 10%, USP, and tropicamide ophthalmic solution 1%, USP in preservative-free lidocaine hydrochloride jelly 2%, USP. The gel was packaged as 0.5 mL in 1-mL polycarbonate plastic syringes with sealed tips. The samples were stored protected from light in an environmental chamber at room temperature of 23 to 27°C and 60% relative humidity and also at refrigerated temperature of 2 to 4°C for 60 days.

The ophthalmic gel exhibited little change in pH and remained sterile and endotoxin-free throughout the 60-day study. Stability-indicating HPLC analysis of drug concentrations over 60 days of storage found no loss of any of the drugs when stored under refrigeration and not more than 4% loss of any drug when stored at room temperature. ■

# Cyclophosphamide

## Properties
Cyclophosphamide (as the monohydrate) is a white or almost white crystalline powder.[2,3] This substance is a known carcinogen.[1]

### Solubility
Cyclophosphamide has an aqueous solubility of about 40 mg/mL[1–3] and a solubility in ethanol of about 1 g/mL.[6]

### pH
A 1% cyclophosphamide aqueous solution has a pH between 3.9 and 7.1.[4] A 22-mg/mL solution had a pH of 6.87.[339]

### Osmolarity
The reconstituted dry-filled injection at 20 mg/mL has an osmolarity of 352 mOsm/L. The reconstituted lyophilized injection at 20 or 25 mg/mL has an osmolarity of 172 or 219 mOsm/L, respectively.[2,7]

## General Stability Considerations
Cyclophosphamide products should be stored at or below 25°C and protected from temperatures exceeding 30°C. The tablets should be stored in tight containers.[2–4]

Cyclophosphamide is susceptible to hydrolysis in aqueous solutions. The hydrolysis occurs at a constant rate over a pH range of 2 to 10.[6] This range approximates the quite broad pH range of maximum stability reported as pH 2 to 11. At pH less than 2 and above 11, increased decomposition occurs.[649]

The reconstituted injection should be used within 24 hours if stored at room temperature or within six days if stored under refrigeration.[2,7] Stability periods of up to two months have been cited for refrigerated solutions.[6]

The use of heat to speed dissolution is not recommended due to the potential for decomposition if the temperature reaches the range of 70 to 80°C for as little as 15 minutes. If the temperature can be controlled not to exceed the range of 50 to 60°C for 15 minutes, decomposition is minimized.[339]

## Stability Reports of Compounded Preparations
### Injection

Injections, like other sterile drugs, should be prepared in a suitable clean air environment using appropriate aseptic procedures. When prepared from nonsterile components, an appropriate and effective sterilization method must be employed.

Kirk et al.[340] studied the stability of reconstituted cyclophosphamide injection 20 mg/mL repackaged into polypropylene syringes (Plastipak syringes) with blind Luer locking hubs stored under refrigeration and in the freezer. Analysis showed about 3% loss in four weeks and 10% loss in 11 to 14 weeks at 4°C. Although frozen storage at −20°C slowed the decomposition rate even further (to about 4% loss in 19 weeks), seepage of the solution past the plunger occurred due to contraction of the plunger from contact with the syringe wall. This effect increases the risk of bacterial contamination of the product. Furthermore, cyclophosphamide precipitated during thawing, and redissolving it required vigorous efforts. This precipitation of concentrated solutions during thawing does not seem to occur in solutions below 8 mg/mL.

### Oral

**STUDY 1 (LIQUID):** Brooke et al.[223] evaluated the stability of cyclophosphamide dissolved in aromatic elixir, USP, for oral administration. Cyclophosphamide monohydrate bulk powder and also cyclophosphamide for injection (Cytoxan, Bristol) were dissolved in aromatic elixir, USP, to yield a cyclophosphamide concentration of 2 mg/mL. The solutions were packaged in two-ounce clear glass bottles with plastic caps and stored in the refrigerator at 5°C, at room temperature of 25°C, and at elevated temperatures. The cyclophosphamide content was assessed using vapor phase chromatography. All solutions remained clear and colorless. The solutions were calculated to retain at least 98.6% of the initial concentration for 14 days at 5°C and 96.4% of the initial concentration for one day at 25°C.

**STUDY 2 (LIQUID):** Kennedy et al.[1455] evaluated the stability of cyclophosphamide 10 mg/mL in two oral liquid formulations. Cyclophosphamide injection (Baxter) was reconstituted with sodium chloride 0.9% injection to yield a concentration of 20 mg/mL. The injection was diluted with an equal quantity of syrup, USP, (simple syrup) and also in Ora-Plus (Paddock) resulting in the 10-mg/mL drug concentration. The oral liquids were packaged in 3-mL amber polypropylene oral syringes (Becton Dickinson). Samples were stored at room temperature near 22°C and refrigerated at 4°C for up to 56 days. Visual observation found no changes in odor or color in any of the samples. Similarly, microscopic examination found no evidence of microbiological growth at either storage temperature. Stability-indicating HPLC analysis found little or no loss of cyclophosphamide in 56 days stored under refrigeration. At room temperature, 10% drug loss was calculated to occur in about 10 days in syrup and about six days in Ora-Plus. However, the authors recommended more conservative room temperature beyond-use periods of eight days and three days, respectively, in the two vehicles based on the lower 95% confidence limit.

**STUDY 3 (CAPSULES):** Bouligand et al.[949] evaluated the long-term stability of lower dose cyclophosphamide capsules containing 10 and 25 mg of drug extemporaneously compounded for use in pediatric dosing. The capsules were prepared taking into account that 500 mg of cyclophosphamide base is equivalent to 534.5 mg of the monohydrate that was used in the compounding operation. The formulation included lactose and carmine and was filled into gelatin capsules, providing doses of 10 and 25 mg of cyclophosphamide. Storage conditions were not specified but were probably controlled room temperature. High-performance thin-layer chromatography analysis of the finished capsules found that little or no loss of drug occurred within the 70-day study period. ■

# Cycloserine

## Properties

Cycloserine occurs as a white to pale yellow crystalline powder. It is odorless or nearly so.[1,4,7]

### Solubility

Cycloserine is freely soluble in water and slightly soluble in propylene glycol.[1,4]

### pH

Aqueous solutions of 10% cycloserine have a pH between 5.5 and 6.5.[1,4]

## General Stability Considerations

Cycloserine powder and oral capsules should be packaged in tight containers and stored at controlled room temperature.[4,7] Cycloserine is hygroscopic and deteriorates upon absorbing water.[4]

Aqueous solutions of cycloserine buffered to pH 10 with sodium carbonate are stated to be stable without loss for one week under refrigeration.[1] Neutral and acidic solutions of cycloserine are unstable. The drug forms salts with acids and bases.[1298]

## Compatibility with Common Beverages and Foods

Peloquin et al.[1135] evaluated the stability of several oral anti-tuberculosis drugs, including cycloserine, for compatibility with common foods to facilitate administration to children. Cycloserine capsules were opened and the contents mixed with 30 g of Hunt's Snackpack no-sugar-added chocolate pudding, Safeway grape jelly, and Safeway creamy peanut butter. The drug–food mixtures were thoroughly mixed and analyzed (no method indicated) for drug content initially and after one, two, and four hours had elapsed. Drug recovery at all time points was ≥93% from the pudding, ≥89% from the jelly, and ≥80% from the peanut butter. The authors recommended the use of the no-sugar-added chocolate pudding or the jelly to aid in administering the drugs to children. ∎

# Cyclosporine

## Properties

Cyclosporine is a white or nearly white crystalline powder.[2,3] The oral solution, prepared in olive oil with a surfactant and 12.5% alcohol, is a clear yellow oily liquid. The concentrate for injection prepared in a mixture of Cremophor EL and alcohol is a clear brownish-yellow liquid.[2]

### Solubility

Cyclosporine is relatively insoluble in water, having an aqueous solubility of 0.04 mg/mL. In ethanol, it has a solubility of more than 80 mg/mL at 25°C. It is also soluble in other organic solvents and lipids.[1,2]

## General Stability Considerations

Cyclosporine oral products should be stored in tight, light-resistant containers.[2–4] The oral solution should be stored at less than 30°C, but refrigeration should be avoided since coalescence and separation of the components occurs. The capsules should be stored at less than 25°C.[2] The manufacturer stated that cyclosporine capsules should not be removed from the original packaging and placed in dosing compliance aids because of loss of ethanol due to vaporization.[1622] The injection concentrate should be stored at less than 30°C and protected from light. However, light protection is not required for intravenous admixtures.[2,7,341]

Cyclosporine capsules for microemulsion and oral solution for microemulsion should be stored in the original containers at 25°C. At temperatures below 20°C, the emulsion may gel or develop a flocculation or sedimentation. The effects can be reversed by warming the products to 25°C. The oral solution for microemulsion should be used within two months after opening.[2]

Cyclosporine may be lost due to sorption to administration sets. Ptachcinski et al.[341] found significant loss of cyclosporine from infusion solutions delivered through Abbott administration sets. However, Parr et al.[342] found no loss due to sorption to polyvinyl chloride (PVC) containers or after six-hour delivery using a pump. A loss upon filtration through a 0.22- or 0.45-μm filter was found based on the initial amount of drug present. However, the loss from complete infusion was not quantified.

The Cremophor EL surfactant leaches phthalate plasticizer from PVC containers and delivery equipment. The use of non-PVC containers and administration sets is recommended.[2,343,344]

## Stability Reports of Compounded Preparations
### Oral

Ptachcinski et al.[32] tested the stability of cyclosporine oral solution (Sandimmune, Sandoz) 100 mg/mL when 5 mL was repackaged into 10-mL plastic oral syringes (Baxa). The filled syringes were stored at 25°C for 28 days both exposed to room light and protected from light in light-protection plastic bags. Cyclosporine content was evaluated using a stability-indicating HPLC technique. No loss of cyclosporine occurred in any sample during the 28-day study; the concentrations of the oral solution in the syringes and in the controls in the original amber bottles were no different. Furthermore, there was no difference in cyclosporine content in the light-exposed and light-protected samples.

### Enteral

Ortega de la Cruz et al.[1101] reported the physical incompatibility of an unspecified amount of oral liquid cyclosporine (Sandimmune, Sandoz) with 200 mL of Precitene (Novartis) enteral nutrition diet. Phase separation was observed.

### Ophthalmic

Ophthalmic preparations, like other sterile drugs, should be prepared in a suitable clean air environment using appropriate aseptic procedures. When prepared from nonsterile components, an appropriate and effective sterilization method must be employed.

**STUDY 1:** Allen[130] reported on the extemporaneous compounding of cyclosporine ophthalmic solution. The exterior of the bottle of 10% oral solution was cleaned with alcohol and allowed to stand opened for 24 hours to permit evaporation of the alcohol (12.5%) from the oral solution. Four volumes of corn, olive, arachis, or castor oil was added to the oral solution and thoroughly mixed. The diluted solution then was filtered through a 0.2-μm filter into sterile ophthalmic containers. If filled full with no headspace, the drug was stable at room temperature or under refrigeration for six months or to the expiration date on the original product bottle, whichever was shorter. The preparation, however, does not contain a preservative and the potential for microbiological growth exists. Limiting the shelf life to one month or less was recommended.

**STUDY 2:** Mueller[196] described a similar 2% cyclosporine ophthalmic preparation extemporaneously made using cyclosporine capsules. The fluid in cyclosporine capsules has a corn oil base and contains 100 mg/mL of drug. An 18-gauge needle was used to puncture one end of the capsule to provide a vent. Another 18-gauge needle attached to a syringe was used to puncture the other end of the capsule and draw out the contents. Approximately 12 to 15 capsules yield 12 mL of fluid. The fluid was transferred to a sterile beaker in a laminar airflow hood and allowed to sit for 24 hours to permit evaporation of the alcohol content. The residual cyclosporine fluid was brought to 60 mL with corn oil, USP, and mixed well, yielding a cyclosporine concentration of 2%. The mixture was sterilized by filtration through a 0.2-μm filter (Millex

FG) into sterile 2-mL empty glass vials. To ensure maximum stability the vials were filled completely, leaving no air space (requires about 3.4 mL). The filled vials were stored under refrigeration and assigned a six-month expiration date based on information from Sandoz Pharmaceuticals. The oil and aqueous phases may separate during refrigerated storage. To redisperse, vials should be allowed to warm to room temperature and then shaken well. For dispensing, the fully dispersed contents of the vials were transferred to sterile dropper bottles in a laminar airflow hood, given a one-month expiration date, and labeled with a statement not to refrigerate the preparation.

**STUDY 3:** Fiscella et al.[751] reported the stability of cyclosporine 1% ophthalmic solution prepared from the injection diluted in artificial tears (Tears Plus) containing polyvinyl alcohol 1.4% and povidone 0.6%. One milliliter of cyclosporine 50-mg/mL injection was added to 4 mL of artificial tears, yielding a 10-mg/mL concentration. The ophthalmic solution was packaged in 2.5-mL sterile ophthalmic bottles and stored frozen at −20°C for one month. Samples were thawed and stored for an additional 28 days under refrigeration or for seven days at room temperature. HPLC analysis found that no loss of cyclosporine occurred in any of the samples under any of the storage conditions. Patient preference was for this preparation in artificial tears rather than for the oil-based cyclosporine product.

**STUDY 4:** Frøyland et al.[811] reported the stability of cyclosporine 10 mg/mL in peanut oil for use as eyedrops. Commercial Sandimmune infusion concentrate (Novartis) 98 g was transferred to a glass flask and sterilized peanut oil (150°C for four hours) was added to bring the final weight to 551 g. Medical-grade nitrogen gas filtered through a 0.22-μm filter was passed through the solution to evaporate the ethanol present in the injection. Such nitrogen gas purging was found to reduce the ethanol content to 2.5 or 1% for 30 or 60 minutes of purging, respectively. The ophthalmic solution was packaged in glass dropper bottles and stored at 22°C protected from exposure to light. HPLC analysis found no loss of cyclosporine over 20 weeks of storage. However, the peroxide value increased to unacceptable levels, limiting the use period to four weeks at 22°C.

**STUDY 5:** Pasto et al.[833] evaluated the microbial contamination of extemporaneously prepared cyclosporine 2% ophthalmic solution during patient use. The oily ophthalmic solution was prepared from commercial oral solution (Sandimmune, Sandoz) mixed with sterilized olive oil. Microbial contamination was found in five of 19 samples returned by patients when a 20-day use period was utilized. Reducing the time period of use to 10 days resulted in no detected microbial contamination among 115 samples. The authors concluded that the cyclosporine 2% ophthalmic solution should be restricted to a 10-day use period because of the potential for microbial growth.

**STUDY 6:** Hernando and Alvarez[1521] evaluated a number of oily and aqueous formulations of cyclosporine 2% ophthalmic drops. They reported that their best formulation was cyclosporine 2% in isopropyl myristate. The cyclosporine readily dissolved in the isopropyl myristate. The solution was filtered through 0.2-μm nylon sterile filters or autoclaved at 121°C for 20 minutes for sterilization. The eyedrops were packaged in transparent glass containers and stored at 40°C and 75% relative humidity for four weeks. The authors reported this formulation remained stable throughout storage at the elevated temperature and humidity conditions. The authors proposed that this formulation be included in the Spanish National Formulary.

## Topical

**STUDY 1:** Ghnassia et al.[516] evaluated the stability of a topical paste of cyclosporine 9.6 mg/g incorporated into Orabase adhesive base. A total of 191.2 g of Orabase was weighed and placed on an ointment slab. A 20.2-mL volume of cyclosporine 100-mg/mL oral solution (Sandoz) was measured and incorporated into an equal quantity of the Orabase using spatulation. The remainder of the Orabase was incorporated geometrically with thorough blending after each addition. The paste was packaged in aluminum ointment tubes, filling 7 g of paste into 10-g tubes. The tube end was folded three times and crimped, and the tubes were stored at 37, 21, and 2°C for 31 days. Stability-indicating HPLC analysis found 3% or less cyclosporine loss at any point during the 31 days in samples stored at all three temperatures. Fluorescence polarization immunoassay results were more variable but also showed little tendency to drug loss.

**STUDY 2:** Prins et al.[1213] evaluated a topical emulsion formulation of cyclosporine 70 mg/mL for use in treating nail psoriasis. The emulsion was prepared by mixing oral Neoral liquid with pharmaceutical-grade maize oil to yield the 70-mg/mL concentration. The topical emulsion was packaged in a brown glass bottle. Immediately upon mixing, a white and hazy or a turbid mixture was obtained. However, upon storage (presumably at room temperature) for as little as four

hours, the emulsion separated into layers. Using an immunoassay for cyclosporine, it was found that there was an unequal distribution of the cyclosporine between the layers. The drug concentration in the upper layer (mostly corn oil) was 47 mg/mL, while in the lower layer (mostly ethanol and propylene glycol from Neoral) it was 102 mg/mL. The authors indicated that this emulsion instability could have led to the failure of the topical formulation to successfully treat the patient.

**STUDY 3:** Tamura et al.[1227] evaluated the stability of a topical gel formulation of cyclosporine prepared from the commercial oral liquid (Sandoz Pharma) using the following formula:

| | | |
|---|---|---|
| Cyclosporin A oral solution | | 20 mL |
| Carboxyvinyl polymer | | 1 g |
| Propylene glycol | | 12 g |
| Diisopropanolamine | | 1.2 g |
| Polyoxyethylene glyceryl monostearate | | 5 g |
| Purified water | qs ad | 100 g |

Preliminary testing found that a concentration of 3 to 7% of the polyoxyethylene glyceryl monostearate was needed to prepare gels with fine uniform particles. Concentrations lower than 3% and at 10% resulted in larger particle size. Consequently, the 5% concentration was selected for stability testing.

Carboxyvinyl polymer was dissolved in a mixture of propylene glycol in purified water in a water bath at 50°C. The cyclosporine oral liquid and polyoxyethylene glyceryl monostearate were mixed separately at 50°C and then mixed with the first mixture, creating a viscous liquid that was emulsified using a homogenizer. Finally, the diisopropanolamine (a neutralizing agent to be near pH 7) in the remainder of the purified water was added and thoroughly mixed in a mortar.

Using this formulation, a gel was prepared that had good spreadability and consistency. It was not sticky and washed off easily. The particle size did not change substantially over 28 days of storage. HPLC analysis found that the cyclosporine A concentration was unchanged, remaining at 100% over 28 days stored at 20°C. This hydrogel topical preparation also was found to have superior permeability compared to some alternative formulations, including white petrolatum and hydrophilic petrolatum. ■

# Cyproheptadine Hydrochloride

## Properties

Cyproheptadine hydrochloride is a white to slightly yellow odorless or nearly odorless crystalline powder. Anhydrous cyproheptadine hydrochloride 10 mg is approximately equivalent to 11 mg of the sesquihydrate.[2,3,7]

### Solubility

Cyproheptadine hydrochloride has an aqueous solubility variously cited as 3.64 mg/mL[2,3] and approximately 5 mg/mL.[1] In ethanol the solubility is 28.6 mg/mL.[2,3]

### pH

Cyproheptadine hydrochloride syrup has a pH of 3.5 to 4.5.[4,7]

### pKₐ

Cyproheptadine has a $pK_a$ of 9.3.[2]

## General Stability Considerations

Cyproheptadine hydrochloride oral tablets should be stored in well-closed containers, while the syrup should be stored in tight containers. The products should be kept at controlled room temperature and protected from temperatures exceeding 40°C. The syrup also should be protected from freezing.[2,4,7]

Abounassif et al.[939] reported that cyproheptadine did not demonstrate photo-instability when subjected to ultraviolet radiation at 254 nm.

## Stability Reports of Compounded Preparations

### Oral

Gupta[1124] reported the stability of a compounded cyproheptadine 0.4-mg/mL oral liquid. Cyproheptadine hydrochloride sesquihydrate 46.3 mg, equivalent to 40 mg of free base, was dissolved in a mixture of ethanol 5 mL and glycerin 5 mL. To this mixture purified water 40 mL, sodium saccharin 100 mg, sucrose 30 g, citric acid anhydrous 100 mg, sorbic acid 100 mg, and raspberry flavor 0.1 mL were added. The mixture was brought to 100 mL with additional purified water. The oral liquid was packaged in amber glass bottles and stored at room temperature of 25°C. The oral liquid did not change in physical appearance or pH throughout the study. Stability-indicating HPLC analysis found no loss of cyproheptadine hydrochloride within 180 days at room temperature.

### Enteral

**CYPROHEPTADINE HYDROCHLORIDE**
**COMPATIBILITY SUMMARY**
*Compatible with:* Osmolite • Osmolite HN • Vital

Fagerman and Ballou[52] reported on the compatibility of cyproheptadine hydrochloride (Periactin, Merck) 2 mg/5 mL mixed in equal quantities with three enteral feeding formulas. The cyproheptadine hydrochloride syrup was visually compatible with Vital, Osmolite, and Osmolite HN (Ross), with no obvious thickening, granulation, or precipitation. ■

# Cyproterone Acetate

## Properties

Cyproterone acetate is a white or almost white crystalline powder.[3]

### Solubility

Cyproterone acetate is practically insoluble in water, sparingly soluble in dehydrated ethanol, and freely soluble in acetone.[3]

## General Stability Considerations

Cyproterone acetate should be stored at room temperature protected from exposure to light.[3] Commercial oral tablets are stored at controlled room temperature and have a shelf life of five years.[7]

## Stability Reports of Compounded Preparations

### Topical

Valenta[844] reported the stability of the antiandrogen cyproterone acetate in a variety of extemporaneously compounded topical cream and lotion formulations. HPLC analysis found no loss of cyproterone acetate and no formation of decomposition products in six weeks at room temperature in any of the topical formulations. ■

# Cysteamine Bitartrate
# Cysteamine Hydrochloride
## (Mercaptamine Bitartrate; Mercaptamine Hydrochloride)

## Properties

Cysteamine hydrochloride occurs as a white to gray hygroscopic crystalline material.[1,1171] Cysteamine bitartrate occurs as a white powder.[7]

### Solubility

Cysteamine hydrochloride is soluble in water and ethanol.[1] Aqueous solutions of cysteamine hydrochloride are clear and colorless.[1170] Cysteamine bitartrate is highly soluble in water.[7]

### pH

Cysteamine hydrochloride as a 1 M solution in water at 25°C has a pH of 3.5 to 5.[1170]

## General Stability Considerations

Cysteamine hydrochloride powder should be stored under refrigeration and is subject to oxidation upon standing.[1170,1171] Cysteamine bitartrate commercial capsules should be packaged in tight, light-resistant containers and stored at controlled room temperature protected from exposure to light and to moisture.[7]

## Stability Reports of Compounded Preparations
### Oral

Brodrick et al.[1171] evaluated the stability of cysteamine hydrochloride extemporaneously compounded as a syrup composed as follows:

| | |
|---|---|
| Cysteamine hydrochloride | 10 g |
| Water-soluble peppermint essence | 0.6 mL |
| Saccharin solution 1:15 | 10 mL |
| Syrup to | 100 mL |

Cysteamine hydrochloride has an unpleasant odor and an extremely bitter taste that is difficult to mask. This very sweet formulation proved to be a tolerable formulation. HPLC analysis found the drug to be stable for one month in the syrup stored under refrigeration and protected from exposure to light but less stable as a simple aqueous solution. ■

# Danazol

## Properties

Danazol is a synthetic derivative of ethisterone and occurs as a white to pale yellow crystalline powder.[2–4]

### Solubility

Danazol is practically insoluble in water, sparingly soluble in ethanol, and soluble in acetone.[2–4]

## General Stability Considerations

Danazol should be stored in airtight containers at controlled room temperature and protected from exposure to light. The commercial capsules should be packaged in well-closed containers and stored at controlled room temperature.[2–4,7]

## Stability Reports of Compounded Preparations

### Oral

Gad Kariem et al.[822] evaluated the photolability of a danazol 1.6-mg/mL solution in glass vials at room temperature exposed to daylight and in quartz cells irradiated in an ultraviolet light chamber for up to 120 minutes. Stability-indicating HPLC analysis found a slight decrease beginning within three days of exposure to daylight and a drop to about 93% after seven days. A photodegradation product had clearly formed. Photosensitivity to ultraviolet light was confirmed in the light chamber. ■

# Dantrolene Sodium

## Properties

Dantrolene sodium is an orange powder. It is available as the hydrate containing about 15% water.[1,2,7]

### Solubility

Dantrolene sodium is slightly soluble in water, with an aqueous solubility of about 0.015 mg/mL; its aqueous solubility is increased somewhat in alkaline solutions.[1,2,7,426] The free dantrolene acid is virtually insoluble in water, with an aqueous solubility of less than 0.001 mg/mL.[426]

### pH

The reconstituted injection has a pH of approximately 9.5.[2,7]

### $pK_a$

Dantrolene is a weak acid, having a $pK_a$ of about 7.5.[2]

## General Stability Considerations

Dantrolene sodium capsules should be stored in well-closed containers at controlled room temperature and protected from temperatures exceeding 40°C. The powder for injection should be stored at controlled room temperature of 15 to 30°C and protected from extended exposure to light.[2,7]

The powder for injection should be reconstituted with sterile water for injection; solutions containing a bacteriostatic agent should not be used. The reconstituted injection is stable for six hours stored at controlled room temperature and protected from direct light. The reconstituted dantrolene sodium injection is incompatible with 5% dextrose injection, 0.9% sodium chloride injection, and other acidic solutions.[2,7]

## Stability Reports of Compounded Preparations
### Oral

**STUDY 1:** The extemporaneous formulation of a dantrolene sodium 25-mg/5-mL oral suspension was described by Nahata and Hipple.[160] The contents of five dantrolene sodium 100-mg capsules were mixed with a small amount of simple syrup; citric acid 150 mg dissolved in 10 mL of purified water was added and mixed. The suspension was brought to 100 mL with simple syrup. A stability period of two days under refrigeration was used, although chemical stability testing was not performed.

**STUDY 2:** Fawcett et al.[426] evaluated the stability of an oral suspension of dantrolene sodium 5 mg/mL prepared from 50-mg capsules. The suspension was prepared by emptying the contents of 10 capsules in a mortar and triturating with a solution of citric acid 150 mg in 10 mL of water. Syrup containing 0.15% methyl hydroxybenzoate was used to bring the volume to 100 mL. The presence of citric acid results in a suspension with a pH of 4.9 and ensures the conversion of dantrolene sodium to the insoluble free acid form. The suspension was packaged in amber high-density polyethylene bottles with polypropylene caps and stored in the dark at 5, 25, and 40°C for 150 days. Dantrolene content was assessed using a stability-indicating HPLC assay. Losses of dantrolene from the suspension were about 2 and 4% at 5 and 25°C, respectively, after 150 days. Even at 40°C, losses did not exceed 6%. Furthermore, there was no change in suspension appearance. A representative sample in a clear glass container also showed no significant decomposition at room temperature.

Alternatively, it has been suggested that a single oral dose can be given by emptying the contents of the correct number of capsules into fruit juice to serve as a vehicle for administration.[2]

**STUDY 3 (WITH TOLPERISONE HCL):** Hirakawa et al.[1689] evaluated the stability and compatibility of dantrolene sodium and tolperisone hydrochloride in a dry ground mixture. The mixture was prepared for use in a 4-year-old patient with cerebral palsy. After storage for seven days, the powder color changed from orange to yellow. In a follow-up stability study, the authors stored the powder mixture for 30 days at both 37 and 25°C and at 23 and 75% relative humidity protected from exposure to light. The color of the powder mixture changed after seven days under all storage conditions except 23% relative humidity at 25°C. At 37°C and 75% relative humidity, HPLC analysis found a tolperisone hydrochloride loss of 38% and a dantrolene sodium loss of 12%. Addition of lactose to tolperisone hydrochloride also caused more than 10% loss of tolperisone hydrochloride. The authors concluded that the color of dantrolene sodium is affected by tolperisone hydrochloride, and the tolperisone hydrochloride content is decreased by contact with both dantrolene sodium and lactose. ∎

# ▉ Dapsone

## Properties
Dapsone is a white, creamy white, or yellowish-white odorless crystalline powder with a bitter taste.[2,3,7]

### Solubility
Dapsone is very slightly soluble in water but soluble in ethanol (about 33.3 mg/mL).[1-3] Aqueous solubility was found to be about 10 to 12 mg/mL in water at 35°C.[780] It is insoluble in fixed and vegetable oils.[7]

### pK$_b$
Dapsone has a pK$_b$ of 13.0.[1]

## General Stability Considerations
Dapsone tablets should be stored in well-closed, light-resistant containers at controlled room temperature and protected from temperatures exceeding 40°C. Dapsone may discolor if exposed to light; no chemical change associated with the discoloration has been detected.[2-4,7]

## Stability Reports of Compounded Preparations
### Oral

**STUDY 1:** Nahata et al.[716] reported the longer-term stability of two dapsone 2-mg/mL oral suspensions extemporaneously compounded from 25-mg tablets (Jacobus). The tablets were crushed using a mortar and pestle and mixed with an equal parts mixture of (1) Ora-Plus suspending vehicle and Ora-Sweet syrup (Paddock) and (2) citric acid 20-mg/mL aqueous solution and simple syrup (1:3). The suspensions were packaged in amber polyethylene terephthalate prescription bottles and stored at room temperature of 25°C (in a water bath) and refrigerated at 4°C for 91 days.

The refrigerated samples exhibited no change in pH, color, odor, or taste throughout the study, but the room temperature samples developed a yellow discoloration and a slight change in pH after 28 days. Stability-indicating HPLC analysis found 3 to 4% loss at room temperature and 2 to 3% loss refrigerated in 91 days.

**STUDY 2:** Kaila et al.[780] reported the calculated shelf life of a dapsone 15-mg/mL oral suspension extemporaneously compounded from tablets. Dapsone 100-mg tablets (Jacobus) were triturated using a mortar and pestle, and orange oil 1 mL/L of suspension was added to the tablet powder with additional trituration. Hydrated carboxymethylcellulose 0.75% wt/vol was added slowly with trituration. Hydrated Veegum HV 0.75% wt/vol and then sorbitol solution to yield 60% final concentration were added individually with trituration. Finally, methylparaben and propylparaben concentrate was added to yield concentrations of 1 and 0.2%, respectively. The suspension was packaged in amber glass prescription bottles and stored at various temperatures ranging from 4 to 70°C and analyzed using stability-indicating HPLC to calculate shelf life. The calculated shelf life periods were 31 days at 25°C and 230 days refrigerated at 4°C. ■

# ■ Daunorubicin Hydrochloride

## Properties
Daunorubicin hydrochloride is an anthracycline glycoside antineoplastic antibiotic. It is a red or red-orange hygroscopic crystalline powder.[2,3] The bulk substance contains 842 to 1030 mcg of the base per milligram.[4] Daunorubicin hydrochloride is an irritant; contact with skin and mucous membranes should be avoided.[3,4]

### Solubility
Daunorubicin hydrochloride is soluble in aqueous solutions and slightly soluble in ethanol.[2,3]

### pH
A 0.5% solution in water has a pH of 4.5 to 6.5.[3] The reconstituted injection also has a pH of 4.5 to 6.5.[4]

### pKa
Daunorubicin hydrochloride has a $pK_a$ of 10.3.[2]

### Osmolality
Daunorubicin hydrochloride 5 mg/mL reconstituted with sterile water for injection has an osmolality of 141 mOsm/kg.[8]

## General Stability Considerations
Daunorubicin hydrochloride for injection should be stored at controlled room temperature protected from light. The reconstituted solution is stable for 24 hours at room temperature and 48 hours under refrigeration protected from light.[2,4]

Daunorubicin hydrochloride becomes more stable as the pH becomes more acidic in the range of pH 7.4 to 4.5.[9] The drug is unstable in alkaline solutions above pH 8. Decomposition in strongly alkaline solutions is indicated by a color change from red to blue-purple.[1,2,8,9]

Although daunorubicin hydrochloride solutions are degraded upon exposure to light, the extent of the loss is reported not to be significant at drug concentrations of 0.5 mg/mL or above. However, lower concentrations, especially below 0.1 mg/mL, require light protection.[3,8,9]

Admixed in aqueous infusion solutions such as 5% dextrose injection and 0.9% sodium chloride injection, daunorubicin hydrochloride is stable for periods up to 43 days at −20, 4, and 25°C protected from light.[638]

## Stability Reports of Compounded Preparations
### Injection
Injections, like other sterile drugs, should be prepared in a suitable clean air environment using appropriate aseptic procedures. When prepared from nonsterile components, an appropriate and effective sterilization method must be employed.

**STUDY 1:** Wood et al.[638] reported that reconstituted daunorubicin hydrochloride 2 mg/mL repackaged into polypropylene syringes exhibited little or no loss by HPLC analysis when stored at 4°C.

**STUDY 2:** Islam and Asker[573] studied the photoprotective effect of the common antioxidant sodium sulfite on daunorubicin hydrochloride (approximately 50 mcg/mL) stability in aqueous solutions. The addition of 0.048% sodium sulfite was found to markedly reduce the photodegradation of daunorubicin hydrochloride in solutions exposed to fluorescent light. The photoprotection was most effective within the pH range of 3.7 to 8.2; this range includes the common pH range for intravenous infusions of the drug. The authors postulated that the addition of sodium sulfite would markedly improve daunorubicin stability during intravenous administration to patients. ■

# Deferoxamine Mesylate
(Desferrioxamine Mesylate)

## Properties
Deferoxamine mesylate is a white to off-white, nearly odorless powder.[2,3]

### Solubility
Deferoxamine mesylate is freely soluble in water, with a solubility at 20°C of about 200 mg/mL. It is slightly soluble in ethanol.[1–3]

### pH
A 1% solution has a pH of 4 to 6, as does the reconstituted injection.[4] A 10% solution has a pH of 3.7 to 5.5.[3]

## General Stability Considerations
Deferoxamine mesylate for injection should be stored at 25°C or below. Solutions reconstituted with sterile water for injection and stored at room temperature protected from light are stable for up to one week. The reconstituted solutions should not be refrigerated. Reconstitution with other solutions may result in precipitation. Solutions that become turbid should not be used.[7]

## Stability Reports of Compounded Preparations
### Injection
Injections, like other sterile drugs, should be prepared in a suitable clean air environment using appropriate aseptic procedures. When prepared from nonsterile components, an appropriate and effective sterilization method must be employed.

Stiles et al.[529] evaluated the stability of deferoxamine mesylate (Ciba) 250 mg/mL in sterile water for injection. The solution was packaged as 3 mL in polypropylene infusion pump 10-mL syringes for the CADD Micro and CADD-LD pumps (Pharmacia Deltec) sealed with Luer tip caps (Becton Dickinson) and stored at 30°C. HPLC analysis found that at least 95% of the drug was retained after 14 days of storage. ■

# Desonide

## Properties
Desonide is a synthetic corticosteroid that occurs as small plates or as a white to off-white odorless powder.[1,2,1069]

### Solubility
Desonide is insoluble in water.[2,1069]

## General Stability Considerations
Desonide otic liquid should be packaged in tight, light-resistant containers and can be stored at room temperature.[1069]

## Stability Reports of Compounded Preparations
### Otic
Gupta[1070] reported the stability of an extemporaneously compounded desonide 0.5-mg/mL otic liquid that was similar to the commercial product Tridesilon (Miles) that is no longer available. The compounding procedure of Allen[1069] was used to prepare the otic liquid. Desonide 50 mg and glacial acetic acid 2 mL were added to 50 mL of glycerin and mixed thoroughly. Propylene glycol was added to bring the volume to 100 mL with thorough mixing. The desonide 0.5 mg/mL otic liquid was packaged in two-ounce amber glass bottles and stored at room temperature near 25°C for 180 days.

The desonide 0.5 mg/mL otic liquid was initially clear and remained clear throughout the study. Stability-indicating HPLC analysis found a slow loss of desonide. Desonide losses were about 4% in 91 days and 8% in 180 days. ■

# Desoximetasone

## Properties
Desoximetasone, an anti-inflammatory corticosteroid, is a white to practically white odorless crystalline powder.[3,4]

### Solubility
Desoximetasone is insoluble in water and freely soluble in ethanol and acetone.[1,3,4]

## General Stability Considerations
Desoximetasone should be preserved in well-closed containers and stored at controlled room temperature. Desoximetasone topical cream, gel, and ointment should be packaged in collapsible tubes and stored at controlled room temperature.[4]

## Compatibility with Other Drugs

*Compatible with:* Camphor • Menthol • Phenol • Tacrolimus
*Incompatible with:* Liquor carbonis detergens • Salicylic acid
*Conditionally Compatible with:* Urea

**STUDY 1:** Levitt et al.[727] evaluated the stability of desoximetasone and tacrolimus in an equal parts mixture of their commercial ointments. A 1:1 (wt/wt) mixture of Topicort (Taro Pharmaceuticals) containing desoximetasone 0.25% and Protopic (Fujisawa Healthcare) containing tacrolimus 0.1% was prepared and packaged in glass vials. The samples were stored at the following conditions of temperature and relative humidity: 25°C/60% RH, 30°C/60% RH, 40°C/75% RH for 28 days.

No change in appearance occurred with the ointment, which remained white until a slight yellow discoloration was observed after 28 days at all storage conditions. HPLC analysis found that desoximetasone remained stable throughout the study, with little or no loss occurring. Tacrolimus also was stable. A wider range in concentration occurred (90 to 107%), but this appears to have been more likely due to analytical variability. After 28 days, all samples were near their starting concentrations.

**STUDY 2:** Krochmal et al.[776] reported the stability of desoximetasone 0.25% cream (Topicort) mixed individually with (1) salicylic acid 2%, (2) liquor carbonis detergens 5%, (3) urea 10%, and (4) a combination of camphor 0.25%, menthol 0.25%, and phenol 0.25%. The topical mixtures were packaged in amber glass jars with polyethylene-lined screw caps and were stored for two months at unspecified ambient temperature.

The salicylic acid mixture underwent physical separation, but uniformity could be obtained with remixing. The liquor carbonis detergens mixture underwent severe separation; uniformity could not be reestablished with remixing. No other physical incompatibilities were observed. HPLC analysis found about 20% loss of desoximetasone in two months in the urea mixture. Acceptable desoximetasone concentrations were found in the other mixtures over two months of storage.

**STUDY 3:** Pappas et al.[1444] evaluated the compatibility and stability of tacrolimus 0.1% ointment (Protopic, Fujisawa) mixed in a 1:1 ratio with desoximetasone 0.25% (TopiCort, TaroPharma). Samples were packaged in sealed glass vials and stored at 25°C (60% relative humidity), 30°C (60% relative humidity), and 40°C (75% relative humidity) for 28 days. HPLC analysis found no substantial loss of tacrolimus and desoximetasone during the study period. ∎

# Dexamethasone
# Dexamethasone Acetate
# Dexamethasone Sodium Phosphate

## Properties

Dexamethasone and dexamethasone acetate are white or nearly white odorless crystalline powders. Dexamethasone acetate 1.1 mg is approximately equal to 1 mg of dexamethasone. Dexamethasone sodium phosphate is a white or slightly yellow nearly odorless and very hygroscopic crystalline powder. Dexamethasone sodium phosphate 1.3 mg is approximately equivalent to dexamethasone 1 mg.[2,3]

### Solubility

Dexamethasone is practically insoluble in water, with a solubility of 0.1 mg/mL at 25°C. The drug is sparingly soluble in ethanol, with a solubility of about 24 mg/mL.[1–3,407]

Dexamethasone acetate is practically insoluble in water and freely soluble in ethanol.[2,3]

Dexamethasone sodium phosphate is freely soluble in water, with a solubility of 500 mg/mL. It is slightly soluble in ethanol.[2,3]

### pH

A 1% dexamethasone sodium phosphate aqueous solution has a pH between 7.5 and 10.5. Dexamethasone sodium phosphate injection has a pH of 7 to 8.5. Dexamethasone ophthalmic suspension has a pH of 5 to 6; the dexamethasone sodium phosphate ophthalmic solution has a pH of 6.6 to 7.8. Dexamethasone acetate suspension has a pH of 5 to 7.5.[2,4] Dexamethasone oral solution has a pH of 3 to 5, while the oral elixir (Merck Sharp & Dohme) had a pH of 3.3 when measured.[19]

### Osmolality

Dexamethasone elixir 0.1 mg/mL (Organon) had an osmolality of 3350 mOsm/kg, while dexamethasone solution 1 mg/mL (Roxane) had an osmolality of 3100 mOsm/kg.[233] The osmolality of dexamethasone sodium phosphate 4-mg/mL injection is about 356 mOsm/kg.[8]

## General Stability Considerations

Solutions and suspensions of dexamethasone and its salts are heat labile; the formulated products should not be autoclaved for sterilization. In general, dexamethasone products should be packaged in tight or well-closed containers stored at controlled room temperature and protected from heat. The injections should be protected from light and freezing.[2,4]

The use of the sodium phosphate salt to prepare oral products results in reduced bioavailability compared to oral administration of dexamethasone alcohol.[639]

## Stability Reports of Compounded Preparations
### Injection

Injections, like other sterile drugs, should be prepared in a suitable clean air environment using appropriate aseptic procedures. When prepared from nonsterile components, an appropriate and effective sterilization method must be employed.

**STUDY 1:** Lugo and Nahata[500] reported the stability of dexamethasone sodium phosphate diluted with bacteriostatic 0.9% sodium chloride injection to 1 mg/mL. The solution was packaged in 10-mL sterile vials and stored at 4 and 22°C under fluorescent light. HPLC analysis found little or no loss of dexamethasone in 28 days at either storage condition. Furthermore, no discoloration or precipitation was observed visually.

**STUDY 2:** Lau et al.[640] reported on the stability of dexamethasone sodium phosphate 10 mg/mL repackaged into 1- and 2.5-mL glass syringes and 1- and 3-mL plastic syringes and stored at 4 and 23°C. In the glass syringes, less than 5% loss occurred in 91 days. In the plastic syringes, less than 7% loss occurred in the 3-mL syringes in 55 days and less than 3% loss occurred in the 1-mL syringes in 35 days, the maximum time the samples were evaluated.

### Oral

**STUDY 1:** Accordino et al.[407] reported the stability of an extemporaneously prepared hydroalcoholic oral solution of dexamethasone. The oral solution was prepared from dexamethasone powder using the formula in Table 31. The finished solution was packaged in 50-mL amber glass bottles with polyethylene closures. Samples were stored for 26 weeks under refrigeration at 2 to 8°C, at 25°C both in the dark and exposed to sunlight, and at 37°C. A stability-indicating HPLC assay was used to determine dexamethasone content. All samples under refrigeration and at room temperature (in the dark and exposed to sunlight) were stable, retaining concentration throughout the study. However, at 37°C, dexamethasone losses of 5 to 9% were found after 26 weeks.

**TABLE 31.** Dexamethasone Oral Solution Evaluated for Stability[407]

| Component | | Quantity |
|---|---|---|
| Dexamethasone, BP | | 100 mg |
| Ethanol 95% | | 15 mL |
| Propylene glycol | | 20 mL |
| Glycerol | | 50 mL |
| Raspberry flavor | | 0.5 mL |
| Saccharin sodium | | 300 mg |
| Purified water | qs | 100 mL |

**STUDY 2:** Wen-Lin Chou et al.[1073] evaluated the stability of dexamethasone 0.5- and 1-mg/mL oral suspensions prepared from dexamethasone sodium phosphate injection. The dexamethasone suspensions were prepared by adding the correct amount of dexamethasone sodium phosphate injection (Sabex Inc.) to a 1:1 mixture of Ora-Sweet and Ora-Plus (Paddock Laboratories) to yield dexamethasone concentrations of 0.5 and 1 mg/mL. The suspensions were packaged in 40-mL amber plastic bottles and stored for 91 days at room temperature of 25°C and under refrigeration at 4°C. The bottles were exposed only to normal laboratory fluorescent light.

All suspension samples remained unchanged in physical appearance and odor, and the suspensions were easily resuspended throughout the 91-day test period. In addition, no pH fluctuations were found. Stability-indicating HPLC analysis found little or no loss of dexamethasone content within 91 days. Although some assay variability occurred, the concentrations were generally found to be above 95% of the starting concentration and no sample was less than 90%.

**STUDY 3:** Hames et al.[1251] evaluated the palatability of a compounded dexamethasone 1-mg/mL oral liquid suspension in an equal parts mixture of Ora-Sweet and Ora-Plus because palatability is an important factor in medication compliance for children. The palatability of this compounded dexamethasone oral liquid was compared to a commercial prednisolone 1-mg/mL oral liquid (Pediapred). There was a statistically significant preference for the compounded dexamethasone oral liquid over the prednisolone oral liquid. The preference was especially strong in males.

**STUDY 4:** Tiphine et al.[1554] reported the instability of dexamethasone acetate 1 mg/mL formulated in an oral suspension. The oral suspension was formulated by suspending the insoluble dexamethasone acetate 100 mg in 100 mL of an equal parts mixture of Ora-Sweet and Ora-Plus. The oral suspension was packaged in bottles (composition not noted), and samples were stored at room temperature of 20 to 25°C and refrigerated at 2 to 8°C for up to 210 days. Stability of both sealed bottles and bottles that were repeatedly opened from day 10 through day 50 to simulate in-use conditions was evaluated. HPLC analysis found that the dexamethasone concentration was not stable, apparently increasing substantially over time. The authors postulated an unknown decomposition product was interfering with the assays, or the drug was not remaining suspended, resulting in nonuniform concentrations that could have led to the increases. Although not mentioned in the article, the possibility of inadequate analysis cannot be ruled out given the high degree of variability of the concentration results. In any event, this study did not support the use of this simple oral formulation for the administration of dexamethasone acetate.

## Topical

Vonbach et al.[1336] evaluated the long-term stability of a three-drug solution to be used for bladder iontophoresis. The solution contained lidocaine hydrochloride 20 mg/mL, dexamethasone sodium phosphate 0.2 mg/mL, and epinephrine hydrochloride 0.01 mg/mL in water. The solution was packaged in amber glass vials and clear glass vials. The samples in amber glass vials were stored frozen at −20°C, refrigerated at 4°C, at room temperature of 20°C, and at elevated stress temperature of 40°C. Samples in clear vials were stored at 20°C exposed to daylight.

HPLC analysis found that epinephrine hydrochloride was the least stable component. At 40°C and exposed to daylight extensive and rapid loss of epinephrine hydrochloride occurred. At 20°C in amber vials, about 10% epinephrine hydrochloride loss occurred in 28 days. Under refrigeration, the loss was about 6% in six months; when frozen, no loss occurred in six months. For lidocaine hydrochloride little or no loss occurred under any of the test conditions. For dexamethasone sodium phosphate, losses never exceeded 5% in six months at −20, 4, and 20°C. The authors stated that this three-drug iontophoresis solution was stable for three months refrigerated and for six months frozen at −20°C.

## Ophthalmic

Ophthalmic preparations, like other sterile drugs, should be prepared in a suitable clean air environment using appropriate aseptic procedures. When prepared from nonsterile components, an appropriate and effective sterilization method must be employed.

Dobrinas et al.[1195] evaluated the stability of ophthalmic injections including dexamethasone sodium phosphate (Mepha Pharma) 4 mg/mL in balanced salts solution simulating aqueous humor solution. The solutions were passed through 0.22-μm filters for sterilization. The samples were packaged as 1 mL in syringes (syringe composition not cited) and were stored frozen at less than 18°C for six months. HPLC analysis found less than a 5% drug concentration change and no decomposition products. In addition, the drug remained stable after thawing and storing at room temperature for six hours.

## Enteral

**DEXAMETHASONE COMPATIBILITY SUMMARY**
*Compatible with:* Enrich • Ensure • Ensure HN • Ensure Plus • Ensure Plus HN • Osmolite • Osmolite HN • TwoCal HN • Vital • Vivonex T.E.N.

**STUDY 1:** Cutie et al.[19] added 5 mL of dexamethasone elixir (Decadron, Merck Sharp & Dohme) to varying amounts (15 to 240 mL) of Ensure, Ensure Plus, and Osmolite (Ross Laboratories) after vigorous agitation to ensure thorough mixing. The dexamethasone elixir was physically compatible, distributing uniformly in all three enteral products with no phase separation or granulation.

**STUDY 2:** Burns et al.[739] reported the physical compatibility of dexamethasone elixir (Decadron) 2.5 mL with 10 mL of three enteral formulas, including Enrich, TwoCal HN, and Vivonex T.E.N. Visual inspection found no physical incompatibility with any of the enteral formulas.

**STUDY 3:** Altman and Cutie[850] reported the physical compatibility of dexamethasone elixir (Decadron, Merck Sharp & Dohme) 5 mL with varying amounts (15 to 240 mL) of Ensure HN, Ensure Plus HN, Osmolite HN, and Vital after vigorous agitation to ensure thorough mixing. The elixir was physically compatible, distributing uniformly in all four enteral products with no phase separation or granulation.

## Compatibility with Other Drugs
### DEXAMETHASONE COMPATIBILITY SUMMARY
*Compatible with:* Albuterol sulfate • Metaproterenol sulfate • Sodium chloride 7% (hypertonic) • Zinc oxide
*Incompatible with:* Isoetharine • Terbutaline sulfate

**STUDY 1 (MULTIPLE DRUGS):** Owsley and Rusho[517] evaluated the compatibility of several respiratory therapy drug combinations. Dexamethasone (form unspecified) 4 mg/mL (American Regent) was combined with the following drug solutions: albuterol sulfate 5-mg/mL inhalation solution (Proventil, Schering), isoetharine 1% solution (Bronkosol, Sanofi Winthrop), metaproterenol sulfate 5% solution (Alupent, Boehringer Ingelheim), and terbutaline sulfate 1-mg/mL solution (Brethine, Geigy). The test solutions were filtered through 0.22-μm filters into clean vials. The combinations were evaluated over 24 hours (temperature unspecified) using the USP particulate matter test. The dexamethasone–albuterol sulfate and dexamethasone–metaproterenol sulfate combinations were compatible. Dexamethasone combined with isoetharine and terbutaline sulfate formed unacceptable levels of larger particulates (≥10 μm).

**STUDY 2 (HYPERTONIC SODIUM CHLORIDE):** Fox et al.[1239] evaluated the physical compatibility of hypertonic sodium chloride 7% with several inhalation medications used in treating cystic fibrosis. Dexamethasone 4 mg/mL (as sodium phosphate) injection (American Regent) mixed in equal quantities with extemporaneously compounded hypertonic sodium chloride 7% did not exhibit any visible evidence of physical incompatibility, and the measured turbidity did not increase over the one-hour observation period.

**STUDY 3 (ZINC OXIDE):** Ohishi et al.[1714] evaluated the stability of Decaderm (dexamethasone) and Methaderm (dexamethasone propionate) ointments each mixed in equal quantity (1:1) with zinc oxide ointment. Samples were stored refrigerated at 5°C. HPLC analysis of the ointment mixtures found that the drugs were stable for 32 weeks at 5°C. ■

# Dexchlorpheniramine Maleate

## Properties
Dexchlorpheniramine maleate is a white odorless crystalline powder.[3,4]

### Solubility
Dexchlorpheniramine maleate is very soluble in water, with a solubility of 1 part drug in 1.1 parts of water. The drug is also freely soluble in ethanol to about 1 g in 2 mL.[3,4]

### pH
The United States Pharmacopeia states that 1% dexchlorpheniramine maleate aqueous solution has a pH of 4 to 5, and the European Pharmacopoeia cites a pH of 4.5 to 5.5.[3,4]

## General Stability Considerations
Dexchlorpheniramine maleate powder, tablets, and oral solution should be packaged in tight containers and stored at controlled room temperature.[3,4]

## Stability Reports of Compounded Preparations
### Enteral
Ortega de la Cruz et al.[1101] reported the physical compatibility of an unspecified amount of oral liquid dexchlorpheniramine maleate (Polaramine, Schering) with 200 mL of Precitene (Novartis) enteral nutrition diet for a 24-hour observation period. No particle growth or phase separation was observed. ■

# Dexpanthenol

## Properties
Dexpanthenol is a clear, viscous, somewhat hygroscopic liquid having a slight characteristic odor and a bitter taste.[1,3,4]

### Solubility
Dexpanthenol is freely soluble in water, propylene glycol, ethanol, and methanol, and slightly soluble in ether and glycerin.[1,3,4]

### pH
Dexpanthenol has a natural pH of 9.[1] Dexpanthenol 5% in water has a pH not greater than 10.5.[3]

## General Stability Considerations
Dexpanthenol should be packaged in tight containers and stored at controlled room temperature. Dexpanthenol is subject to some crystallization upon standing.[3,4]

## Stability Reports of Compounded Preparations
### Topical
Allen[1277] reported on a compounded formulation of piroxicam 5% and dexpanthenol 5% topical gel. The gel had the following formula:

| | | |
|---|---|---|
| Piroxicam | | 5 g |
| Dexpanthenol | | 5 g |
| Propylene glycol | | 25 g |
| Ethanol | | 5 g |
| Triethanolamine | | 0.4 g |
| Pluronic F-127 | | 23 g |
| Purified water | qs | 100 g |

The recommended method of preparation was to thoroughly dissolve the piroxicam powder in propylene glycol and then add the dexpanthenol and triethanolamine. The Pluronic F-127 was added to about half of the water, cooled to 5°C, and then added to the piroxicam/dexpanthenol solution. Finally, the ethanol was added, followed by adding sufficient purified water to bring the volume to 100 mL. The topical gel was then to be stored at 5°C until the air bubbles escaped. The gel was to be packaged in tight, light-resistant containers. The author recommended a beyond-use date of six months at room temperature because this formula is a commercial medication in some countries, with an expiration date of two years or more. ■

# Dextromethorphan Hydrobromide

## Properties

Dextromethorphan hydrobromide occurs as almost white crystals or crystalline powder with a faint odor.[3,4]

### Solubility

Dextromethorphan hydrobromide is sparingly soluble in water, with a solubility of about 1 g in 65 mL. The drug is freely soluble in ethanol.[3,4]

### pH

A 1% aqueous solution of dextromethorphan hydrobromide has a pH in the range of 5.2 to 6.5.[3,4]

## General Stability Considerations

Dextromethorphan hydrobromide powder should be packaged in tight containers and stored at controlled room temperature.

Dextromethorphan hydrobromide oral solution should be packaged in tight light-resistant containers and stored at controlled room temperature.[4]

## Stability Reports of Compounded Preparations

### Enteral

Ortega de la Cruz et al.[1101] reported the physical compatibility of an unspecified amount of oral liquid dextromethorphan hydrobromide (Romilar, Roche) with 200 mL of Precitene (Novartis) enteral nutrition diet for a 24-hour observation period. No particle growth or phase separation was observed. ■

# 3,4-Diaminopyridine
## (3,4-DAP)

## Properties

3,4-Diaminopyridine occurs as a white to beige crystalline powder or needles.[1]

### Solubility

3,4-Diaminopyridine is readily soluble in water with a solubility of about 24 mg/mL.[1,882] It is also readily soluble in ethanol.[1]

## General Stability Considerations

3,4-Diaminopyridine was reported to be relatively stable in the solid state, with a stability period not exceeding 12 months. Degradation experiments found that oxidation was the principal degradation route, while irradiation resulted in relatively little decomposition.[1095]

## Stability Reports of Compounded Preparations

### Oral

**STUDY 1:** Trissel et al.[721] reported the stability of 3,4-diaminopyridine compounded in 5-mg oral hard-gelatin capsules. In addition to the drug, each capsule also contained silica gel micronized and lactose hydrous, NF. The capsules were packaged in amber polypropylene prescription vials and stored for one month at 37°C and for six months at room temperature near 23°C and refrigerated at 4°C.

No visible changes or weight changes occurred in the capsules during storage. Stability-indicating HPLC analysis found that the drug remained stable under all of the storage conditions, with 4 to 5% loss in six months at room temperature and refrigeration and about 4% loss in one month at 37°C.

**STUDY 2:** Benziane et al.[879] evaluated 3,4-diaminopyridine 2.5-mg capsule stability. Using ultraviolet spectroscopic analysis, the authors found that the drug was stable at temperatures between 4 and 60°C for up to two months. However, at 60°C, the capsule powder turned brown after three months. It should be noted that ultraviolet analysis may not be stability indicating. ■

# Diamorphine Hydrochloride
## (Diacetylmorphine Hydrochloride; Heroin)

## Properties

Diamorphine hydrochloride occurs as a white or nearly white crystalline powder. The powder is nearly odorless initially but develops an odor characteristic of acetic acid upon storage.[3]

### Solubility

Diamorphine hydrochloride is freely soluble in water, soluble in ethanol, and practically insoluble in ether.[3]

## General Stability Considerations

Diamorphine hydrochloride is incompatible with mineral acids and alkalis as well as chlorocresol.[3]

Commercial diamorphine hydrochloride injection vials should be stored at controlled room temperature and protected from exposure to light.[3]

Diamorphine hydrochloride in a 1-mg/mL aqueous solution in glass ampules exhibited 10% loss in about 50 days stored at room temperature.[1141]

Diamorphine hydrochloride in aqueous solution exhibits pH-dependent stability. The drug is stated to be most stable in the pH range of 3.8 to 4.5[1141] or pH 4 to 5.[1192] The degradation rate greatly increases at neutral or basic pH.[1142] In addition, the drug must remain below pH 6 to stay in solution.[1143]

## Stability Reports of Compounded Preparations
### Injections

Injections, like other sterile products, should be prepared in a suitable clean air environment using appropriate aseptic procedures. When prepared from nonsterile components, an appropriate and effective sterilization method must be employed.

**STUDY 1 (COMPOUNDED INJECTION):** Klous et al.[931] reported the stability of an extemporaneously compounded lyophilized bulk injection of diamorphine hydrochloride (heroin). Diamorphine hydrochloride, BP, 3 g in 10 mL of sterile water for injection was filtered through a 0.22-μm Millipak 40 filter into sterile glass 30-mL vials. The vials were partially closed with washed and sterilized siliconized gray bromobutyl rubber stoppers and lyophilized. Other excipients for use as bulking agents were not needed. The vials were stored at 25°C and 60% relative humidity. The stability was evaluated at this storage condition for 24 months and at an accelerated decomposition condition of 40°C and 75% relative humidity for three months.

Stability-indicating HPLC analysis found that not more than 5 or 6% loss of drug occurred in two years at room temperature as the lyophilized dry powder. Even at elevated temperature of 40°C, less than 2% loss was found in 14 weeks. The vials were reconstituted with 18 mL of sterile water for injection to yield a drug solution of 150 mg/mL. This reconstituted solution was found to exert an antimicrobial effect on a number of commonly encountered microorganisms, including *E. scherichia coli*, *E. faecalis*, *Pseudomonas aeruginosa*, *Staphylococcus* species, *Bacillus subtilus*, and *Candida albicans*. The authors recommended a beyond-use period after reconstitution of 12 hours, although they provided no data to support this recommendation.

**STUDY 2 (REPACKAGED):** Allwood[1144] reported the stability of diamorphine hydrochloride 2 and 20 mg/mL diluted in water for injection and packaged in Becton Dickinson polypropylene syringes sealed with blind hubs. When stored at 20°C, the diamorphine hydrochloride underwent about 5% loss over 18 days by HPLC analysis.

**STUDY 3 (REPACKAGED):** Gove et al.[391] reported that diamorphine hydrochloride injection repackaged into plastic syringes with syringe caps (Braun) was stable and within acceptable degradation limits for at least 40 days stored at room temperature.

**STUDY 4 (REPACKAGED):** Kleinberg et al.[1145] evaluated the stability of diamorphine hydrochloride 1 and 20 mg/mL in sodium chloride 0.9% packaged in glass syringes. At ambient room temperature, the 1-mg/mL concentration was stable for seven days, and the 20-mg/mL concentration was stable for 12 days.

### Topical

Zeppetella et al.[959] reported the stability of morphine sulfate and diamorphine hydrochloride topical gels prepared in Intrasite gel (Smith & Nephew) for use in treating decubitis ulcers. Morphine sulfate (Arum Pharmaceutical) and diamorphine hydrochloride (Arum Pharmaceutical) were incorporated into separate aliquots of the Intrasite gel to form 1.25-mg/mL concentrations. The samples were stored refrigerated at 4°C in the dark, at room temperature in the dark, and at room temperature exposed to normal mixed fluorescent light and daylight. HPLC analysis of the morphine sulfate samples found no loss of morphine and no evidence of common decomposition products after 28 days of storage under any of the storage conditions.

However, the diamorphine hydrochloride samples were much less stable. The analysis demonstrated rapid loss of intact diamorphine exceeding 10%, along with the formation of 6-monoacetylmorphine within a few days at all temperatures. At room temperature, morphine also began to appear after about two weeks. The authors indicated this may be acceptable for pain control if the diamorphine hydrochloride is viewed as a prodrug for morphine along with its own analgesic action.

## Compatibility with Other Drugs

Injections, like other sterile products, should be prepared in a suitable clean air environment using appropriate aseptic procedures. When prepared from nonsterile components, an appropriate and effective sterilization method must be employed.

Hudson et al.[1192] reported the stability of diamorphine hydrochloride and hyperbaric bupivacaine hydrochloride in 8% dextrose at concentrations of 0.1 and 4 mg/mL, respectively. The drug mixture was packaged in Becton Dickinson 5-mL polypropylene plastic syringes sealed with tip caps. The samples were stored refrigerated at 7°C, room temperature of 25°C and 65% relative humidity, and elevated temperature of 40°C and 75% relative humidity. Stability-indicating HPLC analysis found no bupivacaine hydrochloride loss in 91 days at any storage condition. However, 10% diamorphine hydrochloride losses were calculated to occur in 26 days refrigerated, seven days at room temperature, and four days at 40°C. ■

# ■ Diazepam

## Properties

Diazepam is an odorless off-white to yellow crystalline powder.[2,3]

### Solubility

Diazepam is practically insoluble in water, having an aqueous solubility of about 0.05 mg/mL.[6,8] The solubility in ethanol is 62.5 mg/mL;[2,3] in 95% ethanol the solubility is 41 mg/mL.[6] Diazepam is sparingly soluble in propylene glycol, having a solubility of 17 mg/mL.[2,6]

### pH

Diazepam injection has a pH of 6.2 to 6.9.[4]

### pK$_a$

Diazepam has a pK$_a$ of 3.3[6] or 3.4.[1,2,139]

### Osmolality

The osmolality of diazepam injection was 7775 mOsm/kg.[345]

## General Stability Considerations

Diazepam products should be stored at controlled room temperature and protected from light. The tablets and capsules should be stored in tight, light-resistant containers.[2–4]

Diazepam undergoes hydrolysis in aqueous solution, particularly below pH 3.[6,348] Maximum stability occurs at about pH 5,[6,139] with good stability over pH 4 to 8.[139,348] Solubility and stability are enhanced in mixed aqueous solvent systems incorporating ethanol and propylene glycol or polyethylene glycol.[6]

Diazepam injection in aqueous solutions is subject to loss due to sorption to filters[352] and to some plastic materials such as polyvinyl chloride containers and administration sets, cellulose propionate burettes, and Silastic tubing. However, little or no loss due to sorption has been reported with plastics such as polyethylene, polypropylene, polybutadiene, and ethylene vinyl acetate and to glass containers.[8]

## Stability Reports of Compounded Preparations
### Injection

Injections, like other sterile drugs, should be prepared in a suitable clean air environment using appropriate aseptic procedures. When prepared from nonsterile components, an appropriate and effective sterilization method must be employed.

**STUDY 1:** Levin et al.[347] reported that diazepam injection repackaged into parenteral cartridges (Tubex) with a 2-mL fill was stable for three months at room temperature.

**STUDY 2:** Smith and Nuessle[350] reported the stability of diazepam injection repackaged into glass unit-dose syringes with slit rubber plunger-stoppers packaged in light-resistant bags and stored at 4 and 30°C. Under refrigeration, about 5% was lost in 90 days; at 30°C about 9 to 10% was lost in 90 days. Sorption to the rubber plunger-stopper was believed to be the cause.

**STUDY 3:** Using ultraviolet spectroscopy, Wisnes et al.[349] found that diazepam injection repackaged into plastic syringes composed of polypropylene and polyethylene had no loss in four hours. However, Speaker et al.[351] reported diazepam loss when the injection was repackaged and stored in plastic syringes for a longer time. Diazepam injection was repackaged into 3-mL plastic syringes made by Becton Dickinson, Sherwood Monoject, and Terumo. The syringes were stored in the dark at −20, 4, and 25°C. Approximately 6% loss occurred in 24 hours at 25°C. At 4°C, about 4 to 8% loss occurred in seven days, while 5 to 13% loss occurred in 30 days at −20°C. The losses were attributed to sorption to the elastomeric plunger seal and/or plastic surfaces.

## Oral

**STUDY 1:** Newton et al.[172] evaluated the stability of two extemporaneous formulations of diazepam 1-mg/mL oral suspensions. The suspensions were prepared from diazepam 10-mg tablets (Valium, Roche). For the first suspension, four tablets were triturated and incorporated into 40 mL of simple syrup. For the second suspension, eight tablets were triturated and levigated with 4 mL of ethanol and 24 mL of propylene glycol. Simple syrup 20 mL and chocolate syrup 32 mL then were added to make 80 mL of suspension. The suspensions were stored in amber glass screw-cap bottles at room temperature exposed to ambient light for 14 days. Diazepam content in the suspensions was assessed using a spectrophotometric analysis. The first suspension, in simple syrup, yielded a subpotent concentration averaging 0.88 mg/mL with only 25% of the diazepam dissolved. The variations were attributed to the small fraction of drug in solution and to nonuniform dispersion of the suspended tablet fragments. The second suspension, in the mixed solvents, yielded concentrations of around 1.03 mg/mL throughout the 14-day study, with 90% of the diazepam dissolved.

**STUDY 2:** Allen[139] reported the extemporaneous formulation of a diazepam oral suspension. The suspension was prepared by crushing 12 oral diazepam 10-mg tablets to obtain fine powder. Ethanol 6 mL and propylene glycol 36 mL were added with thorough mixing. The product was brought to 120 mL with a good tasting syrup having a pH between 4 and 8 (such as Syrpalta or chocolate syrup) to cover the bad taste of diazepam. The product was a suspension because of insoluble tablet excipients. An oral solution could be prepared from 24 mL of diazepam injection combined with 3.6 mL of ethanol and 26.4 mL of propylene glycol and a suitable good tasting vehicle to bring the volume to 120 mL. The stability of the diazepam suspension or solution is at least two weeks at room temperature.

**STUDY 3:** Strom and Kalu[265] reported the stability of a diazepam 1-mg/mL suspension prepared from tablets according to the formula in Table 32. Diazepam 10-mg tablets were triturated to fine powder and levigated with ethanol and propylene glycol. The mixture was incorporated into the suspension vehicle and the pH was adjusted to 4.2 with 6 N hydrochloric acid. The product was packaged in 60-mL amber glass bottles with child-resistant closures and stored at 5, 22, and 40°C for 60 days. A stability-indicating HPLC assay was used to determine diazepam content throughout the study. No loss of diazepam was found at room temperature or under refrigeration. Furthermore, the samples remained homogeneous, with no caking or settling. However, approximately 10% diazepam loss occurred at 60 days at 40°C. Also, there was a slight color change after 24 days. No visible signs of microbial growth were noted in any sample. The authors concluded that this formulation was stable for 60 days stored at room or refrigeration temperature.

**TABLE 32.** Diazepam 1-mg/mL Suspension Formulation[265]

| Component | Quantity |
| --- | --- |
| Diazepam 10-mg tablets | 10 |
| Ethanol 95% | 3.6 mL |
| Propylene glycol | 5.0 mL |
| Sucrose | 55 g |
| Magnesium aluminum silicate | 2.0 g |
| Carboxymethylcellulose sodium (medium viscosity) | 1.0 g |
| Raspberry flavor | qs |
| Red color | qs |
| Purified water | qs 100 mL |

## Enteral

### DIAZEPAM COMPATIBILITY SUMMARY

*Compatible with:* Isocal • Sustacal • Sustacal HC

Strom and Miller[44] evaluated the stability of diazepam (Valium, Roche) from 10-mg tablets and diazepam oral liquid (Roxane) 5 mg/5 mL when mixed with three enteral formulas at full- and half-strength. The ground tablets and the oral liquid were dispersed in approximately 10 mL of deionized water and added to 240 mL of Isocal, Sustacal, and Sustacal HC (Mead Johnson) full- or half-strength mixed with deionized water. The mixtures were stirred with a magnetic stirrer for five minutes; duplicate samples of each mixture were stored at 24°C. The mixtures were visually evaluated for precipitation and phase separation, but no physical incompatibilities were observed. A stability-indicating HPLC assay was employed to assess the diazepam content at various times over 24 hours. The diazepam content was between 97 and 103% after 24 hours in each enteral mixture prepared from either drug source. The stability of the enteral formula components was not evaluated.

## Rectal

**STUDY 1:** Mayer et al.[346] studied the stability of water-free diazepam suppositories. Storage for 12 weeks at 35°C resulted in 1.3% decomposition.

**STUDY 2:** Alldredge et al.[795] evaluated the stability of diazepam 5 mg/mL in a rectal gel formulation composed of propylene glycol, ethyl alcohol (10%), hydroxypropyl methylcellulose, methylcellulose, sodium benzoate, benzyl alcohol (1.5%), benzoic acid, and water (Diastat). Little or no loss of diazepam occurred during three freeze–thaw cycles of −30°C for 24 hours followed by ambient temperature for 24 hours, and separately during intense light exposure at 25°C for one month. Long-term storage at 30°C for 24 months and 40°C for eight months also resulted in little diazepam loss and less than 1% degradation products. ■

# Dichlorobenzyl Alcohol

## Properties
Dichlorobenzyl alcohol occurs as a white to slightly yellow crystalline material.[1,1280]

### Solubility
Dichlorobenzyl alcohol has an aqueous solubility of 1 mg/mL at 20°C. It is also soluble to 45, 95, and 103 mg/mL in propylene glycol, acetone, and N-methyl pyrrolidone, respectively, at 20°C.[1280]

## General Stability Considerations
Dichlorobenzyl alcohol should be packaged in tight, light-resistant containers and stored at controlled room temperature.[1279]

## Stability Reports of Compounded Preparations
### Dental
Allen[1279] reported on a compounded formulation of dichlorobenzyl alcohol 1% (10-mg/mL) gel for use as a dental antiseptic. The gel had the following formula:

| | | |
|---|---|---|
| 2,4-Dichlorobenzyl alcohol | | 1 g |
| Sodium carboxymethylcellulose | | 2 g |
| Purified water | qs | 100 g |

The recommended method of preparation was to disperse the sodium carboxymethylcellulose in nearly all of the purified water heated to 70°C. This solution was allowed to cool and the 2,4-dichlorobenzyl alcohol was added and mixed well. Additional purified water was then added to bring the preparation to 100 g and the formulation was mixed well. The gel was to be packaged in tight, light-resistant containers. The author recommended a beyond-use date of six months at room temperature because this formula is a commercial medication in some countries with an expiration date of two years or more. ■

# Diclofenac Sodium

## Properties
Diclofenac sodium is a white to off-white hygroscopic crystalline powder.[4]

### Solubility
Diclofenac sodium is soluble in ethanol and sparingly soluble in water.[4] Reported solubility in water (pH 5.2) was greater than 9 mg/mL, and in phosphate buffer (pH 7.2) it was 6 mg/mL.[1] Sodium bicarbonate injection (1 mL of 4.2%) is added to 100 to 500 mL of dextrose 5% or sodium chloride 0.9% to buffer the solutions and keep the drug in solution.[7]

### $pK_a$
Diclofenac sodium has a $pK_a$ of 4.[1]

## General Stability Considerations
The commercial diclofenac sodium injection is stored at controlled room temperature and protected from exposure to heat and light and from freezing. Buffering dextrose 5% or sodium chloride 0.9% with sodium bicarbonate injection is required to keep the drug in solution for administration. Admixtures should not be stored; any admixture containing crystals or precipitation should be discarded.[7]

## Stability Reports of Compounded Preparations
### Oral
Donnelly et al.[1501] evaluated the stability of extemporaneously compounded diclofenac sodium 10-mg/mL oral suspension prepared from commercial enteric-coated tablets (Novopharm). The tablets were ground to powder using a mortar and pestle, and the powder was combined with Ora-Blend (Paddock) suspending/flavoring vehicle resulting in a 10-mg/mL concentration. The oral liquid was packaged in amber polyvinyl chloride (PVC) plastic bottles. Samples were stored at both room temperature of 23°C and refrigerated at 5°C for 93 days. The samples exhibited no changes in color, odor, or ease of resuspension. Stability-indicating HPLC analysis of drug concentrations found little or no loss of drug occurred over the storage period.

### Injection
Injections, like other sterile drugs, should be prepared in a suitable clean air environment using appropriate aseptic procedures. When prepared from nonsterile components, an appropriate and effective sterilization method must be employed.

**STUDY 1:** Gupta[976] evaluated the stability of a compounded diclofenac sodium 25-mg/mL injection in a vehicle containing polysorbate 80 1.25 mL, ethanol 12.5 mL, and bacteriostatic water for injection qs 100 mL. The injection was packaged in 20-mL clear glass vials and stored at room temperature protected from exposure to light and under refrigeration. At room temperature, the appearance of the injection remained unchanged and stability-indicating HPLC analysis found little or no loss of diclofenac sodium in 23 days. Under refrigeration, the drug precipitated and required more than two hours to dissolve completely. The author indicated refrigeration was inappropriate for this formulation.

**STUDY 2:** Allen[1346] reported on a compounded formulation of diclofenac sodium 75-mg/mL injection. The injection had the following formula:

| | | |
|---|---|---|
| Diclofenac sodium | | 7.5 g |
| Benzyl alcohol | | 12 g |
| Propylene glycol | | 63 g |
| Sodium metabisulfite | | 300 mg |
| Sodium hydroxide 0.1 N | | qs |
| Sterile water for injection | qs | 100 mL |

The recommended method of preparation was to dissolve the diclofenac sodium in a mixture containing the propylene glycol, benzyl alcohol, and about 20 mL of sterile water for injection. The sodium metabisulfite should then be added and mixed until dissolved. The pH of the mixture should be measured; adjust to pH 8 to 9 with freshly prepared sodium hydroxide 0.1 N solution as necessary. Sufficient sterile water for injection should then be added to bring to volume and the solution should be mixed well. The solution should be sterilized by passing the solution through a compatible 0.2-μm filter. The sterilized injection should be filled to capacity (no headspace) in appropriate sterile light-resistant containers. The author recommended a beyond-use date of 14 days stored at room temperature.

## Ophthalmic

Ophthalmic solutions, like other sterile drugs, should be prepared in a suitable clean air environment using appropriate aseptic procedures. When prepared from nonsterile components, an appropriate and effective sterilization method must be employed.

**STUDY 1 (SOLUTION):** Hirowatari et al.[929] reported the stability of a three-component ophthalmic solution for preoperative use. The ophthalmic solution was prepared by mixing commercially available ophthalmic solutions to yield final concentrations of phenylephrine hydrochloride 1.83%, tropicamide 0.17%, and diclofenac sodium 0.03%. The ophthalmic solution admixture was dispensed as 2-mL aliquots packaged in light-resistant bottles. The physical and chemical stability of the mixture was evaluated over one month stored at 10°C protected from exposure to light. Visual inspection found the solution remained unchanged, being clear and colorless; measured pH remained unchanged as well. HPLC analysis found no loss of diclofenac and tropicamide and about 2% loss of phenylephrine in one month.

**STUDY 2 (SOLUTION):** Ahuja et al.[1399] evaluated the stability of dozens of potential formulations for aqueous solutions of

diclofenac sodium and oily ophthalmic solutions of diclofenac. The best aqueous formulations of diclofenac sodium 0.1% were prepared in Sorenson's phosphate buffer 0.667 M (pH 7.4) with a tonicity modifier incorporated to make the solution isotonic with or without an antimicrobial preservative. The authors found that formulations that incorporated sodium chloride 0.9% as the tonicity modifier maintained the best stability. Ahuja et al.[1399] reported that all of the formulations at pH 7.4 with sodium chloride 0.9% retained at least 90% by HPLC analysis of the diclofenac sodium content for 12 months stored at room temperature. The aqueous formulations with the best combination of stability and corneal permeation incorporated phenylmercuric nitrate, methyl- and propylparaben, or sodium metabisulfite as preservatives. Lower pH solutions not prepared in phosphate buffer resulted in diclofenac precipitation. Formulations that contained benzalkonium chloride and EDTA resulted in unacceptable opalescence.

Ahuja et al.[1399] also evaluated the stability of diclofenac base 0.2 to 1% in oily formulations for ophthalmic use. Several vegetable oils were used as vehicles and tested with and without benzyl alcohol 0.5% (vol/vol) as a preservative. In mustard and olive oils, greater than 10% diclofenac loss occurred over six months of room temperature storage, but no visible changes were observed. Diclofenac dissolved in arachis, castor, sunflower, safflower, soybean, and sesame oils both with and without benzyl alcohol exhibited losses of 8% or less by HPLC analysis after 12 months of room temperature storage. The best stability was found with diclofenac 1% in castor oil with benzyl alcohol 0.5% (vol/vol); only 5% diclofenac loss occurred after 12 months of room temperature storage.

**STUDY 3 (GEL):** Sankar et al.[1538] evaluated the stability of diclofenac sodium 0.1% in three ophthalmic gel formulations using hydroxypropylcellulose, sodium carboxymethylcellulose, and methylcellulose as gelling agents. Of these, the gel formulated using hydroxypropylmethylcellulose 4% was the best. The gel also contained propylene glycol 10%, benzalkonium chloride 0.01%, and was buffered to pH 7.2 with potassium dihydrogen orthophosphate 0.908 g and disodium hydrogen orthophosphate 2.38 g with purified water to bring the mass to 100 g. The gel was sterilized by autoclaving at 15 lbs and 121°C for 20 minutes. This gel was white, translucent, and homogeneous and remained with that appearance throughout the study period of six weeks. Ultraviolet spectrophotometry found the diclofenac sodium content remained above 95% of the initial concentration for at least six weeks refrigerated at 7 to 9°C, at ambient room temperature of 25 to 28°C, and at elevated temperature of 35 to 39°C. In vitro drug release was found to be about 96% in nine hours.

## Topical

**STUDY 1:** Csoka et al.[1206] evaluated the stability of diclofenac sodium formulated as a transdermal "soft-patch" type gel system. The soft-patch is a semisolid form-retaining gel system consisting of flexible plates containing the drug. The two gels evaluated were a stearate gel and a lipophilic gel. The gels were packaged in sealed plastic containers and stored refrigerated at 4°C, room temperature of 25°C, and elevated temperature of 40°C. The gels had the following compositions:

Stearate Gel (1000 g)
   Propylene glycol 500 g
   Glycerol 330 g
   Stearic acid 60 g
   Aqueous sodium hydroxide 10.7% (wt/vol) 100 mL
   Diclofenac sodium 10 g
Lipophilic Gel (1000 g)
   Adeps solidus 600 g
   Glycerol 390 g
   Diclofenac sodium 10 g

Diclofenac sodium was found to decompose faster in the lipophilic gel. Based on stability-indicating HPLC analysis of drug decomposition, the expiration dates at room temperature were calculated to be 29 months in the stearate gel and 11 months in the lipophilic gel. Under refrigeration, the calculated expiration dates were 62 months in the stearate gel and 32 months in the lipophilic gel. No change in color was observed when the gels were stored at room and refrigeration temperatures. At 40°C, yellow to brownish discoloration occurred within 20 months.

**STUDY 2:** Shivare et al.[1449] evaluated nine acrylamide polymer gel formulations for the topical delivery of diclofenac sodium. The topical formulations were prepared using acrylamide 5 g, diclofenac sodium 3 g, isopropanol 5 g, sodium metabisulfite 100 mg, and distilled water with various other components. The authors found the best of the nine formulations also included cocodiethanolamide 1 g. The topical water-washable gel was prepared by dissolving 3 g of diclofenac sodium in 5 g of isopropanol. Cocodiethanolamide 1 g was added and dissolved, forming solution A. Polyacrylamide 5 g was added to 75 g of distilled water containing 100 mg of sodium metabisulfite and stirred until dissolution occurred, forming Solution B. Solutions A and B were then mixed thoroughly, and the final gel weight was brought to 100 g with additional distilled water.

The preferred formulation was clear, homogeneous, had a pH near 6.8, and exhibited better speadability than a current commercial topical form of diclofenac sodium. The preferred formulation exhibited no skin irritation in human volunteers, and exhibited skin permeability comparable with the commercial form. Accelerated stability studies (analytical method not cited) found little or no loss of diclofenac content after three months stored at 40°C. ▪

# ■ Dicloxacillin Sodium

## Properties
Dicloxacillin sodium as the monohydrate is a white to off-white crystalline powder; each 1.09 g is equivalent to 1 g of dicloxacillin. Each milligram of dicloxacillin sodium contains 850 mcg of cloxacillin.[2,3]

### Solubility
Dicloxacillin sodium is freely soluble in water and soluble in ethanol.[1–3]

### pH
A 10-mg/mL aqueous solution has a pH between 4.5 and 7.5, as does the reconstituted oral suspension.[4]

### pK$_a$
Dicloxacillin sodium has a pK$_a$ of 2.7 to 2.8.[2]

### Osmolality
As measured by a vapor pressure osmometer, a 62.5-mg/5-mL suspension (Pathocil, Wyeth-Ayerst) had an osmolality of 2980 mOsm/kg.[232]

## General Stability Considerations
Dicloxacillin sodium products should be stored in tight containers at controlled room temperature and protected from temperatures exceeding 40°C.[2,4] Aqueous solutions of dicloxacillin sodium are most stable in the pH range of 5.5 to 6.5.[992] The reconstituted oral suspension is stable for seven days at controlled room temperature and for 14 days stored under refrigeration at 2 to 8°C.[2]

## Stability Reports of Compounded Preparations
### Oral
Sylvestri et al.[34] compared the stability of dicloxacillin sodium oral suspension (Dynapen, Bristol) packaged in 5-mL clear polypropylene oral syringes (Exacta-Med, Baxa) to the suspension in the original containers. The suspension was prepared with distilled water and 5 mL was drawn into the syringes, which were then capped. The remaining suspension was left in the original containers. The samples were stored at −20, 4, and 25°C and were assayed at various times using a modified USP iodometric antibiotic assay. The time to 10% decomposition (t$_{90}$) of the oral suspension

in the original containers was 79, 44, and 37 days at −20, 4, and 25°C, respectively. However, the $t_{90}$ for the suspension packaged in the oral syringes was substantially reduced at each temperature: 25, 12.5, and 9.3 days at −20, 4, and 25°C, respectively. The authors concluded that the manufacturer's recommended stability period for the reconstituted suspension in the original containers (14 days under refrigeration) cannot be extended to the suspension in the oral syringes. ▪

## ■ Dicyclomine Hydrochloride

### Properties
Dicyclomine hydrochloride is a white or almost white odorless crystalline powder having a bitter taste.[2,3]

#### Solubility
Dicyclomine hydrochloride has an aqueous solubility of 77 mg/mL and a solubility in ethanol of 200 mg/mL at 25°C.[2,3]

#### pH
A 1% aqueous solution has a pH between 5 and 5.5.[4] The injection has a pH of 3 to 6. The oral solution has a pH of 3 to 5.[2]

#### pK$_a$
Dicyclomine hydrochloride has a $pK_a$ of 9.[2]

### General Stability Considerations
Dicyclomine hydrochloride products should be stored in tight or well-closed containers protected from light, moisture, and excessive heat, preferably below 30°C. The liquid products should be protected from freezing.[2,4,7]

Dicyclomine hydrochloride is stable over pH 2 to 6.5. It is unstable in alkaline solutions, with the possible precipitation of the free base.[2]

### Stability Reports of Compounded Preparations
#### Enteral
**DICYCLOMINE HYDROCHLORIDE COMPATIBILITY SUMMARY**
*Compatible with:* Vital
*Incompatible with:* Osmolite • Osmolite HN

Fagerman and Ballou[52] reported on the compatibility of dicyclomine syrup (Bentyl, Lakeside) 10 mg/5 mL mixed in equal quantities with three enteral feeding formulas. The syrup was visually compatible with Vital (Ross) but underwent unacceptable thickening that could not be improved with agitation when combined with Osmolite and Osmolite HN (Ross). ▪

## ■ Diflorasone Diacetate

### Properties
Diflorasone diacetate is a white to pale yellow or buff-colored crystalline powder.[2–4]

#### Solubility
Diflorasone diacetate is insoluble in water, soluble in methanol and acetone, and sparingly soluble in ethyl acetate.[3,4]

### General Stability Considerations
Diflorasone diacetate should be stored in tight containers at controlled room temperature. Commercial diflorasone diacetate topical cream and ointment should be packaged in collapsible tubes and stored at controlled temperature.[3,4]

### Stability Reports of Compounded Preparations
#### Topical
Hecker et al.[815] evaluated the compatibility of tazarotene 0.05% gel combined in equal quantity with other topical products, including diflorasone diacetate 0.05% cream and ointment. The mixtures were sealed and stored at 30°C for two weeks. HPLC analysis found that the tazarotene loss was less than 10% in all samples. Diflorasone diacetate also exhibited less than 10% loss in this time frame. ▪

# Diflucortolone Valerate

## Properties
Diflucortolone valerate is a white to creamy white crystalline powder.[3]

### Solubility
Diflucortolone valerate is practically insoluble in water and slightly soluble in methanol.[3]

## General Stability Considerations
Diflucortolone valerate should be stored at controlled room temperature protected from light. Commercial diflucortolone valerate topical cream and ointment should be stored at controlled room temperature below 25°C for the cream and below 30°C for the ointment.[3]

## Stability Reports of Compounded Preparations
### Topical
Ray-Johnson[775] reported the stability of diflucortolone valerate ointment (Nerisone, Schering) diluted to a concentration of 0.05% (1:1 mixture) using Unguentum Merck, an ambiphilic ointment that combines the properties of both an oil-in-water emulsion and a water-in-oil emulsion. The diluted ointment was packaged in screw-cap jars and stored refrigerated at 4°C, at room temperature of 18 to 23°C, and at elevated temperature of 32°C. After storage for 32 weeks, the authors reported that HPLC analysis found that about 20 to 25% drug loss occurred at all storage temperatures.

## Compatibility with Other Drugs
**STUDY 1:** Nagatani et al.[1526] evaluated the stability of several steroids in steroid ointments when mixed with zinc oxide ointment. The steroid ointments included diflucortolone valerate (Nerisona). Samples of the mixed ointments were stored at 5 and 30°C until analyzed. Analysis found no loss of diflucortolone valerate when stored at 5 and 30°C.

**STUDY 2:** Ohishi et al.[1714] evaluated the stability of diflucortolone valerate (Nerisona) ointment mixed in equal quantity (1:1) with zinc oxide ointment. Samples were stored refrigerated at 5°C. HPLC analysis of the ointment mixtures found that the drugs were stable for 32 weeks at 5°C. ■

# Diflunisal
## (Difluorophenylsalicylic Acid)

## Properties
Diflunisal is a white to off-white practically odorless powder.[3,4]

### Solubility
Diflunisal is insoluble in water at neutral or acidic pH, freely soluble in ethanol, and soluble in acetone and ethyl acetate.[3,4] It is readily soluble in aqueous media at alkaline pH.[7]

## General Stability Considerations
Diflunisal powder and commercial tablets should be stored in well-closed containers at controlled room temperature.[2,4] Diflunisal dissolved in dilute alkaline solution is described as moderately stable.[7]

## Stability Reports of Compounded Preparations
### Oral
Kizu et al.[819] evaluated the stability of diflunisal 2% ointment for oral application. The ointment was prepared by grinding 2 g of diflunisal crystals (Sigma) using a mortar and pestle and incorporating 2 g of liquid paraffin to form a suspension. Plastibase (Taisho) 20 g was gradually incorporated, followed by 20 g of carmellose (carboxymethylcellulose) sodium (Wako Pure Chemicals) as an adhesive. Additional Plastibase was incorporated to make 100 g. The ointment was stored at 5, 20, and 30°C for 100 days.

The ointment demonstrated adequate spreadability. HPLC analysis of diflunisal concentrations found little or no change in drug concentration at any of the temperatures throughout 100 days of storage. Application to human volunteers demonstrated absorption of the diflunisal, and application to eight patients resulted in pain relief. ■

# ■ Digoxin

## Properties

Digoxin is a cardiac glycoside that occurs as odorless clear to white crystals or as a white crystalline powder having a bitter taste.[2,3,6]

### Solubility

Digoxin is practically insoluble in water, having an aqueous solubility of 0.08 mg/mL. It is slightly soluble (1 in 122) in diluted (80%) alcohol and very slightly soluble in propylene glycol.[2,3,6]

### pH

Digoxin injection has a pH of 6.8 to 7.2.[7] Digoxin elixir (Lanoxin, Burroughs-Wellcome) had a pH of 7.[19]

### Osmolality

Digoxin elixir 50 mcg/mL (Burroughs-Wellcome) had an osmolality of 1350 mOsm/kg.[233] The pediatric injection (Burroughs-Wellcome) had an osmolality of 9105 mOsm/kg by freezing-point depression and 5885 mOsm/kg by vapor pressure.[287]

## General Stability Considerations

Digoxin products should be stored in tight containers at controlled room temperature and protected from excessive temperatures and exposure to light.[2,4]

Digoxin is subject to acid-catalyzed hydrolysis at pH less than 3 and especially at very acidic pH of 1 to 2. It is stable at higher pH and is not hydrolyzed in aqueous solution at pH 5 to 8.[6,353–356]

## Stability Reports of Compounded Preparations

### Injection

Injections, like other sterile drugs, should be prepared in a suitable clean air environment using appropriate aseptic procedures. When prepared from nonsterile components, an appropriate and effective sterilization method must be employed.

Digoxin injection (Burroughs-Wellcome) 0.25 mg/mL repackaged into parenteral cartridges (Tubex) retained concentration for three months at room temperature.[347]

### Enteral

#### DIGOXIN COMPATIBILITY SUMMARY

*Compatible with:* Enrich • Ensure • Ensure HN • Ensure Plus • Ensure Plus HN • Osmolite • Osmolite HN • TwoCal HN • Vital • Vivonex T.E.N.

*Incompatible with:* Precitene

**STUDY 1:** Cutie et al.[19] added 1 mL of digoxin elixir (Lanoxin, Burroughs-Wellcome) to varying amounts (15 to 240 mL) of Ensure, Ensure Plus, and Osmolite (Ross Laboratories) with vigorous agitation to ensure thorough mixing. The digoxin elixir was physically compatible, distributing uniformly in all three enteral products with no phase separation or granulation. However, after standing for 24 to 48 hours, the mixture became a rubbery mass.

**STUDY 2:** Holtz et al.[54] evaluated the stability of digoxin oral liquid (Lanoxin, Burroughs-Wellcome) at 0.125 mg/L and 0.25 mg/L in full-strength Ensure, Ensure Plus, and Osmolite (Ross) over 12 hours at room temperature. Visual inspection revealed no changes such as clumping, gelling, separation, precipitation, or increased viscosity in the mixtures. Osmolalities of the Ensure mixtures were largely unaffected by drug addition. Digoxin concentrations in Ensure were assessed using a fluorescence polarization immunoassay (TDX, Abbott). About 8 to 9% of the digoxin was lost in 12 hours. The other two enteral products could not be analyzed due to assay interference.

**STUDY 3:** Burns et al.[739] reported the physical compatibility of digoxin oral liquid (Lanoxin) 0.25 mg/5 mL with 10 mL of three enteral formulas, including Enrich, TwoCal HN, and Vivonex T.E.N. Visual inspection found no physical incompatibility with any of the enteral formulas.

**STUDY 4:** Altman and Cutie[850] reported the physical compatibility of digoxin elixir (Lanoxin, Burroughs-Wellcome) 2 mL with varying amounts (15 to 240 mL) of Ensure HN, Ensure Plus HN, Osmolite HN, and Vital after vigorous agitation to ensure thorough mixing. The elixir was physically compatible, distributing uniformly in all four enteral products with no phase separation or granulation. However, after standing for 24 to 48 hours, the mixture became a rubbery mass.

**STUDY 5:** Ortega de la Cruz et al.[1101] reported the physical incompatibility of an unspecified amount of oral liquid digoxin (Lanacordin, Kern) with 200 mL of Precitene (Novartis) enteral nutrition diet. Particle growth was observed. ■

# Diltiazem Hydrochloride

## Properties

Diltiazem hydrochloride is an odorless white to off-white crystalline powder or small crystals having a bitter taste.[1–3]

### Solubility

Diltiazem hydrochloride is freely soluble in water and slightly soluble in absolute ethanol.[1–3]

### pH

Diltiazem hydrochloride injection has a pH of 3.7 to 4.1.[7]

## General Stability Considerations

Diltiazem hydrochloride oral products should be stored in tight, light-resistant containers at controlled room temperature.[4,7] Diltiazem hydrochloride injection should be stored under refrigeration at 2 to 8°C and protected from freezing. It may be stored for up to one month at controlled room temperature but then should be destroyed.[7] Diltiazem hydrochloride for injection should be stored at controlled room temperature; freezing should be avoided. Reconstituted solutions are stable for 24 hours at controlled room temperature.[7]

Andrisano et al.[791] reported on diltiazem hydrochloride 10 mg/mL in water exposed to UVA–UVB radiation for 28 hours with a solar simulator. HPLC analysis found only 5.6% degradation with this intense light exposure. The authors concluded that diltiazem hydrochloride is photostable and does not require light protection.

Diltiazem hydrochloride is most stable in aqueous solutions in the pH range of 5 to 6,[544,650] but decomposition increases substantially at pH 7 to 8.[650] A diltiazem hydrochloride 0.1-mg/mL aqueous solution at pH 7 exhibited losses of 3 to 4% in 24 hours.[650] Diltiazem hydrochloride degradation in aqueous solution is enhanced at elevated temperature.[357]

## Stability Reports of Compounded Preparations
### Oral
#### USP OFFICIAL FORMULATION (ORAL SOLUTION):

Diltiazem hydrochloride 1.2 g
Vehicle for Oral Solution, NF (sugar-containing or
  sugar-free) qs 100 mL

(See the vehicle monograph for information on the Vehicle for Oral Solution, NF.)

Use diltiazem hydrochloride powder. Add about 10 mL of the vehicle and mix well. Add additional vehicle almost to volume in increments with thorough mixing. Quantitatively transfer to a suitable calibrated tight and light-resistant bottle, bring to final volume with the vehicle mixture, and thoroughly mix yielding diltiazem hydrochloride 12 mg/mL

oral solution. The final liquid preparation should have a pH between pH 3.7 and 4.7. Store the preparation at controlled room temperature or under refrigeration between 2 and 8°C. The beyond-use date is 60 days from the date of compounding at either storage temperature.[4,5]

#### USP OFFICIAL FORMULATION (ORAL SUSPENSION):

Diltiazem hydrochloride 1.2 g
Vehicle for Oral Solution, NF (sugar-containing or
  sugar-free)
Vehicle for Oral Suspension, NF (1:1) qs 100 mL

(See the vehicle monographs for information on the individual vehicles.)

Use diltiazem hydrochloride powder or commercial tablets. If using tablets, crush or grind to fine powder. Add about 10 mL of the vehicle mixture and mix to make a uniform paste. Add additional vehicle almost to volume in increments with thorough mixing. Quantitatively transfer to a suitable calibrated tight and light-resistant bottle, bring to final volume with the vehicle mixture, and thoroughly mix yielding diltiazem hydrochloride 12 mg/mL oral suspension. The final liquid preparation should have a pH between pH 3.7 and 4.7. Store the preparation at controlled room temperature or under refrigeration between 2 and 8°C. The beyond-use date is 60 days from the date of compounding at either storage temperature.[4,5]

**STUDY 1:** Suleiman et al.[261] reported the stability of diltiazem hydrochloride 1 mg/mL in various sugar solutions. The solutions were stored in amber bottles at elevated temperatures of 40 to 60°C. An HPLC method was used to assess the diltiazem content. The shelf life of diltiazem hydrochloride (calculated time to 10% decomposition) in the stable sugar solutions is presented in Table 33. Lactose 0.28 M yielded a reduced shelf life compared to the other sugars. The addition of 0.1 or 0.2 M sodium dihydrogen phosphate reduced the stability of

**TABLE 33.** Stability of Diltiazem Hydrochloride 1 mg/mL in 0.28 M Sugar Solutions Stored at 25°C[261]

| Sugar Solution (0.28 M) | $t_{90}$ (days) |
| --- | --- |
| Dextrose | 77.5 |
| Fructose | 103 |
| Mannitol | 57.5 |
| Lactose | 26.8 |
| Sorbitol | 57.4 |
| Sucrose | 57.5 |

diltiazem hydrochloride compared to the plain sugar solutions. The authors concluded that diltiazem hydrochloride oral liquid preparations could be compounded extemporaneously with a reasonable shelf life of 50 days in most cases.

**STUDY 2:** Allen and Erickson[544] evaluated the stability of three diltiazem hydrochloride 12-mg/mL oral suspensions extemporaneously compounded from tablets. Vehicles used in this study were (1) an equal parts mixture of Ora-Sweet and Ora-Plus (Paddock), (2) an equal parts mixture of Ora-Sweet SF and Ora-Plus (Paddock), and (3) cherry syrup (Robinson Laboratories) mixed 1:4 with simple syrup. Sixteen diltiazem hydrochloride 90-mg tablets (Lederle) were crushed and comminuted to fine powder using a mortar and pestle. About 10 mL of the test vehicle was added to the powder and mixed to yield a uniform paste. Additional vehicle was added geometrically and brought to the final volume of 120 mL, mixing thoroughly after each addition. The process was repeated for each of the three test suspension vehicles. Samples of each of the finished suspensions were packaged in 120-mL amber polyethylene terephthalate plastic prescription bottles and stored at 5 and 25°C in the dark.

No visual changes or changes in odor were detected during the study. Stability-indicating HPLC analysis found less than 7% diltiazem hydrochloride loss in any of the suspensions stored at either temperature after 60 days of storage.

## Topical

Buur et al.[1125] evaluated the stability of diltiazem hydrochloride 50, 100, and 250 mg/mL in a transdermal gel formulation for veterinary use. The formulation was prepared by weighing the correct amount of drug and then adding 1.5 mL of ethoxydiglycol in a plastic syringe; the mixture was swirled to cause dissolution of the drug. Lipoderm gel was then added to bring the volume to 5 mL. Aliquots of 1 mL were transferred to 1-mL tuberculin syringes and tip caps were used to seal the syringes. The syringes were stored protected from exposure to light in brown opaque zipper-lock bags for 60 days. Samples were stored at room temperature of 25°C, refrigerated at 4°C, and stored frozen at −20°C.

The room temperature samples separated in 14 days into a lipid and an organic layer. Samples stored in the refrigerator developed a crystalline appearance, while frozen samples separated into a lipid and an organic layer upon thawing. HPLC analysis found a high degree of nonuniformity in the initial drug concentration among the tested samples, which was attributed to variability in the compounding process. Although the analytical method was not validated to be stability indicating, the authors reported significant drug loss in 30 days in the 250-mg/mL concentration and in 60 days in the 100-mg/mL concentration. However, because of the absence of stability-indicating validation, it is not possible to know if this represents the true drug stability in the formulation.

## Rectal

Dhawan et al.[1402] reported the stability of diltiazem hydrochloride 2% gel formulations for use in treating anal fissures. Diltiazem hydrochloride 2 g was dissolved in 10 mL of water. Methylcellulose 1, 2, and 3% and hydroxypropyl methylcellulose 1, 2, 3, and 4% were prepared by dissolving in warm water and stirring for five minutes. Methylparaben was dissolved separately in warm water and then added to the gel. The diltiazem hydrochloride solution was then added to the preserved gel and was brought to a final amount of 100 g with additional water. The gels were then stirred using a mechanical stirrer for 15 minutes. The gels were filled into collapsible aluminum tubes and were stored at 8 to 15°C and at accelerated conditions of 40°C and 75% relative humidity for up to six months. Various polyethylene oxide gels with diltiazem hydrochloride were also prepared in a similar manner but were found to be unacceptable. The gels were visually inspected for color, clarity, and transparency. The color and texture were also evaluated. Ultraviolet spectrophotometric estimation of drug content was performed as well.

The methylcellulose and hydroxypropyl methylcellulose gels were all found to be acceptable, but the methylcellulose 3% and hydroxypropyl methylcellulose 4% gel were deemed optimal. Clinical use of the gels resulted in patient improvement after eight weeks of treatment. No difference in clinical results was found among the gels tested. ▪

# Dimenhydrinate

## Properties
Dimenhydrinate is a white odorless crystalline powder or colorless crystals having a bitter numbing taste. The pure material contains 53 to 55.5% diphenhydramine and 44 to 47% 8-chlorotheophylline.[2,3,4]

### Solubility
Dimenhydrinate is slightly soluble in water and freely soluble in ethanol and propylene glycol.[2,3,4]

### pH
A saturated solution of dimenhydrinate in water has a natural pH of 7.1 to 7.6.[3] Dimenhydrinate injection has a pH of 6.4 to 7.2.[4]

## General Stability Considerations
Dimenhydrinate powder and oral tablets should be packaged in well-closed containers and stored at controlled room temperature. Dimenhydrinate injection should be stored at controlled room temperature and protected from freezing. Dimenhydrinate oral solution should be packaged in tight containers and stored at controlled room temperature and protected from freezing.[4]

## Stability Reports of Compounded Formulations
### Nasal
Belgamwar et al.[1441] evaluated 12 variations of nasal gel formulas containing dimenhydrinate 2.5% wt/vol for the treatment of motion sickness. The optimal formulation for gel strength, viscosity, muco-adhesiveness, and drug stability had the following composition:

> Dimenhydrinate 2.5% (wt/vol) in 15% PEG 400
> Gellan gum 0.3% (wt/vol)
> Carbopol 934P 0.15% (wt/vol)
> Mannitol 4% (wt/vol)
> Benzalkonium chloride 0.001% (wt/vol)
> Hydrochloric acid 0.1 N qs
> Ultra pure water qs

The gellan gum was dispersed in ultra pure water and stirred for 20 minutes in a water bath at 100°C and then cooled to room temperature. Dimenhydrinate 2.5% solubilized in 15% polyethylene glycol 400 was added during the cooling. The solution was allowed to equilibrate for 16 hours under refrigeration. Carbopol 934P was added slowly with stirring. Mannitol and benzalkonium chloride were added simultaneously. The pH was adjusted with 0.1N hydrochloric acid to a pH range of 4.5 to 5.5. The final formulation was packaged in amber glass vials with rubber closures and autoclaved at 121°C and 15 Pa for 20 minutes. The finished sterilized gel was stored for 90 days at 32°C and 60% relative humidity. No changes in drug content or other attributes of the nasal gel occurred. ■

# Dinoprostone
## (Prostaglandin E₂)

## Properties
Dinoprostone is a white or nearly white crystalline powder or colorless crystals.[2,3]

### Solubility
Dinoprostone is practically insoluble in water with a solubility of 1.3 mg/mL. However, it is freely soluble in ethanol and soluble in 25% ethanol in water.[2,3,7]

### pKₐ
Dinoprostone has a pKₐ of 4.6.[2]

## General Stability Considerations
Dinoprostone suppositories should be stored frozen at –20°C but warmed to room temperature for use. The vaginal inserts should be stored frozen as well. The drug release characteristics of the vaginal inserts may be altered by exposure to moisture in the air. The gel should be stored under refrigeration.[7]

## Stability Reports of Compounded Preparations
### Vaginal
**STUDY 1 (SUPPOSITORY):** Gannon et al.[519] studied the stability of an extemporaneous vaginal suppository formulation of dinoprostone for cervical ripening. The suppositories, which contained 3 mg of dinoprostone, were compounded from commercial dinoprostone 20-mg suppositories (Prostin E2, Upjohn) diluted with fatty acid blend suppository base (Gallipot).

To prepare 24 of the suppositories, each having a total weight of 520 mg, four Prostin E2 suppositories were removed from frozen storage, unwrapped, and placed on a pill tile.

Each suppository was cut into about eight small sections with a razor blade and then crushed, using a firm metal spatula. The crushed suppositories were levigated to create a clump of material. Fatty acid blend suppository base 10.585 g was slowly mixed with the crushed suppositories geometrically until all of the base was incorporated as homogeneously as possible. The mixture was transferred to a glass beaker, placed in a shallow water bath (0.5 to 1 inch of water) and gently heated at low temperature on a hot plate. As melting began, a glass stirring rod was used to mix the melting mass; complete melting occurred in about five minutes. The melted mixture was removed from the water bath and poured into a 24-well urethral suppository mold until each well was completely filled. An excess of the melted mixture should remain in the beaker. The suppositories were allowed to cool for 25 minutes at room temperature. Any overflow mixture was scraped off with a spatula. The mold was opened and each suppository removed, wrapped in foil, and placed in a vial.

Samples of the suppositories were stored at room temperature, under refrigeration at 4°C, and frozen at −20°C. The stability of dinoprostone in the suppositories was assessed using an HPLC analytical technique. Less than 10% loss occurred in 24 hours at room temperature, 72 hours under refrigeration, and up to six months frozen at −20°C.

**STUDY 2 (SUPPOSITORY):** Goodwin et al.[1184] evaluated the stability of dinoprostone 2.5-mg vaginal suppositories used for cervical ripening and induction. Dinoprostone in ethanol 1% (Prostin E2, Upjohn) was incorporated into Witepsol S55 base melted at 37°C at a concentration of 2.5 mg of dinoprostone per 4-g suppository. The melt was poured into molds that were then put under refrigeration at 4°C. HPLC analysis was conducted, and the time to 10% drug loss ($t_{90}$) was calculated to be 50 days under refrigeration.

**STUDY 3 (GEL):** Gauger[1185] described the preparation of dinoprostone gel for cervical ripening prepared from commercial Prostin E2 20-mg suppositories. The hydroxyethylcellulose gel base was prepared by heating 97 mL of sterile water for irrigation to just below boiling and then adding 3 g of Tylose MH 300 (Hoechst) with constant stirring until the gel thickened. The gel was poured into 60-mL glass vials, sealed, and autoclaved at 120°C and 15 psi for 20 minutes. The sterilized gel was stored under refrigeration for 24 hours to allow for complete hydration. The 3% gel density at room temperature was 1 g/1 mL.

A 20-mg dinoprostone suppository was then weighed, the amount of sterile 3% hydroxyethylcellulose gel necessary to bring the total mass to 66.67 g was drawn into a syringe, and the syringe was allowed to stand upright for several minutes to permit bubbles to escape. In a laminar airflow workbench on a sterile tray, a Prostin E2 suppository was sliced into thin sections and allowed to soften at room temperature for 15 minutes. Using a syringe and needle, one drop of 1% methylene blue injection was placed onto each slice. The methylene blue injection served as a marker for homogeneity. Using a sterile spatula, the 3% sterile hydroxyethylcellulose gel and the softened suppository slices were thoroughly mixed to yield a uniform blue colored gel. The final gel has a density of 0.95 g/1 mL at room temperature and contains 3-mg of dinoprostone per 10.5 mL. In the laminar airflow workbench, 10.5-mL portions of the completed gel were packaged in sterile plastic syringes and tested for sterility, density, and viscosity. The gel in syringes was stored at −20°C and labeled with a beyond-use date of 35 days, although stability testing of the gel was not performed. See *Study 4* below.

Gauger[1186] also noted that commercial K-Y jelly is sterile hydroxyethylcellulose and has also been used to deliver diluted dinoprostone. Essentially the same procedure was used, but the author did not believe that sterilization of the gel was necessary.

**STUDY 4 (GEL):** Lee[1232] evaluated the stability of dinoprostone 1 mg in 20 mL of methylhydroxyethylcellulose 5% gel. A 0.1-mL portion of commercial Prostin E2 injection 10 mg/mL was added to 20 mL of a 5% gel of methylhydroxyethylcellulose and thoroughly mixed. Samples were stored at temperatures ranging from refrigeration at 4°C, room temperature of 25°C, and up to an elevated temperature of 32°C. Aliquots were removed for analysis by gas chromatography after derivatization. From the reaction constants that were determined, the time to 5% loss of dinoprostone was calculated to be eight days under refrigeration and 32 hours at room temperature. ■

# ◼ Diphenhydramine Hydrochloride

## Properties

Diphenhydramine hydrochloride is a white or almost white, odorless crystalline powder having a bitter taste. Exposure of the powder to light results in darkening.[1-3]

### Solubility

Diphenhydramine hydrochloride has an aqueous solubility of about 1 g/mL; in ethanol the solubility is about 500 mg/mL.[1-3]

### pH

Diphenhydramine hydrochloride 1% aqueous solution has a pH of about 5.5.[1] The injection has a pH of 4 to 6.5.[4] Diphenhydramine hydrochloride elixir (Benadryl, Parke-Davis) had a measured pH of 5.2.[19] Diphenhydramine hydrochloride with dextromethorphan syrup (Benylin DM syrup, Parke-Davis) had a measured pH of 4.8.[19]

### $pK_a$

Diphenhydramine has a $pK_a$ of approximately 9.[2]

### Osmolality

Diphenhydramine hydrochloride elixir 2.5 mg/mL (Roxane) had an osmolality of 850 mOsm/kg.[233] The 50-mg/mL injection had a measured osmolality of 240 mOsm/kg; the 10-mg/mL injection had a measured osmolality of 65 mOsm/kg.[326]

## General Stability Considerations

Diphenhydramine hydrochloride products generally should be stored in tight containers at controlled room temperature. The injection and elixir should be stored in light-resistant containers. Liquid products should be protected from freezing.[2]

## Stability Reports of Compounded Preparations

### Enteral

#### DIPHENHYDRAMINE HYDROCHLORIDE

#### COMPATIBILITY SUMMARY

*Compatible with:* Enrich • Ensure • Ensure HN • Ensure Plus • Ensure Plus HN • Osmolite • Osmolite HN • TwoCal HN • Vital • Vivonex T.E.N.

**STUDY 1:** Cutie et al.[19] added 10 mL of diphenhydramine hydrochloride elixir (Benadryl, Parke-Davis) and also 10 mL of diphenhydramine hydrochloride with dextromethorphan syrup (Benylin DM, Parke-Davis) to varying amounts (15 to 240 mL) of Ensure, Ensure Plus, and Osmolite (Ross Laboratories), with vigorous agitation to ensure thorough mixing. Both products were physically compatible, distributing uniformly in all three enteral products with no phase separation or granulation.

**STUDY 2:** Burns et al.[739] reported the physical compatibility of diphenhydramine hydrochloride elixir (Benadryl, Parke-Davis) 10 mL with 10 mL of three enteral formulas, including Enrich, TwoCal HN, and Vivonex T.E.N. Visual inspection found no physical incompatibility with any of the enteral formulas.

**STUDY 3:** Altman and Cutie[850] reported the physical compatibility of 10 mL of diphenhydramine hydrochloride elixir (Benadryl, Parke-Davis) and also 10 mL of diphenhydramine hydrochloride with dextromethorphan syrup (Benylin DM, Parke-Davis) with varying amounts (15 to 240 mL) of Ensure HN, Ensure Plus HN, Osmolite HN, and Vital after vigorous agitation to ensure thorough mixing. Both products were physically compatible, distributing uniformly in all four enteral products with no phase separation or granulation.

## Compatibility with Other Drugs

**STUDY 1 (CAPTOPRIL):** Huang et al.[817] evaluated the stability of captopril incorporated into 5% topical gel formulations with the addition of 0.05% diphenhydramine. No difference in oxidative rate constant was found, indicating that diphenhydramine is compatible with captopril.

**STUDY 2 (LIDOCAINE):** Gupta[990] evaluated the stability and compatibility of diphenhydramine hydrochloride elixir 2.5 mg/mL (Alpharma) combined in equal volume with lidocaine hydrochloride 2% viscous solution (Roxane) for use in treating stomatitis. This concentrate would be diluted to lower concentrations as needed for the individual patient. After mixing, the concentrated drug combination was packaged in amber bottles and stored at room temperature of 24 to 26°C for 21 days. The mixture did not undergo any changes in physical appearance during the study. Stability-indicating HPLC analysis found no loss of either drug in 21 days at room temperature. ◼

# Diphenoxylate Hydrochloride with Atropine Sulfate

## Properties

Diphenoxylate hydrochloride is a white or almost white odorless crystalline powder.[2,3]

Atropine sulfate is a white crystalline powder or colorless crystals having a bitter taste.[1–3]

### Solubility

Diphenoxylate hydrochloride is slightly soluble in water, having an aqueous solubility of 0.8 mg/mL. In ethanol, it has a solubility of 3 mg/mL.[1–3]

Atropine sulfate has an aqueous solubility of 1 g/0.4 to 0.5 mL and a solubility in ethanol of 0.2 g/mL at 25°C.[1–3]

### pH

A saturated solution of diphenoxylate hydrochloride in water has a pH of about 3.3.[3] An aqueous solution of atropine sulfate has a pH of about 5.4.[1]

Diphenoxylate hydrochloride with atropine sulfate oral solution has a pH of 3 to 4.3 when diluted with an equal quantity of water.[4] Diphenoxylate hydrochloride 0.5 mg/mL with atropine sulfate 0.005 mg/mL liquid (Lomotil, Searle) had a measured pH of 3.3.[19]

### Osmolality

Diphenoxylate hydrochloride–atropine sulfate suspension (Roxane) had an osmolality of 8800 mOsm/kg.[233]

## General Stability Considerations

Diphenoxylate hydrochloride with atropine sulfate products should be stored in tight (solution) or well-closed (tablets), light-resistant containers at controlled room temperature and protected from temperatures exceeding 40°C. The solution should also be protected from freezing.[2,4]

## Stability Reports of Compounded Preparations
### Enteral

**DIPHENOXYLATE HYDROCHLORIDE WITH ATROPINE SULFATE COMPATIBILITY SUMMARY**

*Compatible with:* Enrich • Ensure • Ensure HN • Ensure Plus • Ensure Plus HN • Osmolite • Osmolite HN • TwoCal HN • Vital • Vivonex T.E.N.

**STUDY 1:** Cutie et al.[19] added 5 mL of diphenoxylate hydrochloride with atropine sulfate liquid (Lomotil Liquid, Searle) to varying amounts (15 to 240 mL) of Ensure, Ensure Plus, and Osmolite (Ross Laboratories) with vigorous agitation to ensure thorough mixing. The product was physically compatible, distributing uniformly in all three enteral products with no phase separation or granulation. The authors recommended adding the drug slowly with agitation.

**STUDY 2:** Burns et al.[739] reported the physical compatibility of diphenoxylate hydrochloride with atropine sulfate liquid (Lomotil Liquid, Searle) 5 mL with 10 mL of three enteral formulas, including Enrich, TwoCal HN, and Vivonex T.E.N. Visual inspection found no physical incompatibility with any of the enteral formulas.

**STUDY 3:** Altman and Cutie[850] reported the physical compatibility of diphenoxylate hydrochloride with atropine sulfate liquid (Lomotil Liquid, Searle) 5 mL with varying amounts (15 to 240 mL) of Ensure HN, Ensure Plus HN, Osmolite HN, and Vital with vigorous agitation to ensure thorough mixing. The diphenoxylate hydrochloride with atropine sulfate liquid was physically compatible, distributing uniformly in all four enteral products with no phase separation or granulation. The authors recommended adding the drug slowly with agitation. ■

# Dipyridamole

## Properties

Dipyridamole is an odorless yellow crystalline powder or needles having a bitter taste.[2,3,7]

### Solubility

Dipyridamole is slightly soluble in water but very soluble in ethanol. It is soluble in dilute acid solutions having a pH below 3.3.[2,3,7]

## General Stability Considerations

Dipyridamole is subject to photodecomposition and should be protected from exposure to light.[1055] Dipyridamole products should be packaged in tight, light-resistant containers and stored at controlled room temperature.[2,4]

## Stability Reports of Compounded Preparations
### Oral

**USP OFFICIAL FORMULATION:**

Dipyridamole 1 g
Vehicle for Oral Solution, NF (sugar-containing or sugar-free)
Vehicle for Oral Suspension, NF (1:1) qs 100 mL

(See the vehicle monographs for information on the individual vehicles.)

Use dipyridamole powder or commercial tablets. If using tablets, crush or grind to fine powder. Add 20 mL of the vehicle mixture, and mix to make a uniform paste. Add additional vehicle almost to volume in increments with thorough mixing. Quantitatively transfer to a suitable calibrated tight and light-resistant bottle, bring to final volume with the vehicle mixture, and thoroughly mix yielding dipyridamole 10 mg/mL oral suspension. The final liquid preparation should have a pH between 3.8 and 4.8. Store the preparation at controlled room temperature or under refrigeration between 2 and 8°C. The beyond-use date is 60 days from the date of compounding at either storage temperature.[4,5]

**STUDY 1:** The extemporaneous formulation of a dipyridamole 10-mg/mL oral suspension was described by Nahata and Hipple.[160] Four dipyridamole 25-mg tablets were crushed, brought to 10 mL with purified water, and mixed thoroughly. A stability period of three days was used, although chemical stability testing was not done.

**STUDY 2:** Allen and Erickson[544] evaluated the stability of three dipyridamole 10-mg/mL oral suspensions extemporaneously compounded from tablets. Vehicles used in this study were (1) an equal parts mixture of Ora-Sweet and Ora-Plus (Paddock), (2) an equal parts mixture of Ora-Sweet SF and Ora-Plus (Paddock), and (3) cherry syrup (Robinson Laboratories) mixed 1:4 with simple syrup. Twenty-four dipyridamole 10-mg tablets (Geneva) were crushed and comminuted to fine powder using a mortar and pestle. About 20 mL of the test vehicle was added to the powder and mixed to yield a uniform paste. Additional vehicle was added geometrically and brought to the final volume of 120 mL, mixing thoroughly after each addition. The process was repeated for each of the three test suspension vehicles. Samples of each of the finished suspensions were packaged in 120-mL amber polyethylene terephthalate plastic prescription bottles and stored at 5 and 25°C in the dark.

No visual changes or changes in odor were detected during the study. Stability-indicating HPLC analysis found less

**TABLE 34.** Sugar-Free Suspending Solution Used for Dipyridamole Suspension[1055]

| Component | | Amount |
| --- | --- | --- |
| Sodium saccharin | | 1 g |
| Potassium sorbate | | 1 g |
| Citric acid | | 1 g |
| Propylene glycol | | 100 mL |
| Methylcellulose 4000 cps | | 5 g |
| Berry citrus blend flavor | | 2.5 mL |
| Distilled water | qs | 1000 mL |

than 8% dipyridamole loss in any of the suspensions stored at either temperature after 60 days of storage.

**STUDY 3:** Ameer et al.[1055] evaluated the stability of an oral suspension of dipyridamole 10 mg/mL compounded from commercial dipyridamole tablets (Persantine, Boehringer Ingelheim). The tablets were ground to fine powder using a mortar and pestle. A sugar-free vehicle having the composition shown in Table 34 was used as the suspending vehicle. A small amount of the vehicle was added to the ground tablet powder to form a slurry. The slurry was transferred to a graduated cylinder and the mortar was repeatedly rinsed with the vehicle, with the rinsings added to the graduated cylinder to ensure complete transfer of the drug. The suspension was brought to volume with additional vehicle to yield a 10-mg/mL concentration. The dipyridamole suspension was packaged in amber glass prescription bottles and stored at room temperature of 25°C and refrigerated at 4°C.

Stability-indicating HPLC analysis found the dipyridamole to be stable at both storage conditions for the 30-day study period. Little or no loss occurred in the samples stored under refrigeration; room temperature samples lost about 4%. The authors did not report any changes in appearance of the suspension, but they did note that the refrigerated samples were more difficult to resuspend. ■

# ■ Disopyramide Phosphate

## Properties

Disopyramide phosphate is a white or almost white odorless crystalline powder composed of a racemic mixture of two isomers.[2,3] Disopyramide phosphate 1.3 g is approximately equivalent to 1 g of disopyramide.[3]

### Solubility

Disopyramide phosphate is freely soluble in water, having an aqueous solubility of 50 mg/mL. In ethanol, it has a solubility of 20 mg/mL.[2,3]

### pH

Disopyramide phosphate as a saturated (5%) aqueous solution has a pH of 4 to 5.[3,4]

### pK$_a$

Disopyramide has a pK$_a$ variously reported as 10.2 and 10.4.[1,2]

## General Stability Considerations

Disopyramide phosphate capsules should be stored in well-closed containers at controlled room temperature or lower.[2,4]

## Stability Reports of Compounded Preparations
### Oral

The stability of disopyramide phosphate in cherry syrup was evaluated by Mathur et al.[82] The powder contents of disopyramide phosphate 100-mg conventional capsules (Norpace, Searle) were removed and triturated in a glass mortar with cherry syrup and brought to volume. (The extended-release formulation cannot be used.) Two concentrations, 1 and 100 mg/mL, were prepared and packaged in 60-mL amber glass prescription bottles with child-resistant closures. Samples were stored at 5°C, 30°C, and ambient room temperature under intense fluorescent light for four weeks. A stability-indicating HPLC assay was used to determine the disopyramide phosphate content. No loss of disopyramide phosphate occurred in any sample during the four weeks. Furthermore, no change in pH, odor, or color was observed. Microbiological evaluation found levels of bacteria and fungi well below maximum acceptable levels. ■

# ■ Disulfiram

## Properties

Disulfiram is a white or almost white odorless crystalline powder having a bitter taste.[2,3]

### Solubility

Disulfiram is very slightly soluble in water, with a solubility of 0.2 mg/mL[1] and a solubility of 33 to 38 mg/mL in ethanol.[1–3]

## General Stability Considerations

Disulfiram tablets should be packaged in tight, light-resistant containers and stored at controlled room temperature protected from elevated temperatures above 40°C.[2–4]

## Stability Reports of Compounded Preparations
### Oral

The stability of disulfiram extemporaneously prepared as a suspension was reported by Gupta.[74] The suspension was prepared by mixing 2 g of acacia, USP, with 100 mg of sodium benzoate in a mortar. Water 40 mL was added, and the suspension was stirred to form a smooth mixture. Five disulfiram 500-mg tablets (Antabuse, Ayerst) crushed to powder or 2.5 g of pure disulfiram powder was added and stirred to form a smooth consistency. More water was added to bring the suspensions to 100 mL each. The suspensions were packaged in amber glass bottles and stored at 24°C under fluorescent lights. Disulfiram content was assessed by a stability-indicating HPLC assay. The suspensions were stable, exhibiting no disulfiram loss in 178 days (from tablets) and 295 days (from powder).

### Injection

Injections, like other sterile drugs, should be prepared in a suitable clean air environment using appropriate aseptic procedures. When prepared from nonsterile components, an appropriate and effective sterilization method must be employed.

Phillips et al.[489] evaluated the gamma irradiation sterilization of a compounded disulfiram suspension formulation intended for subcutaneous injection. Because disulfiram has a low melting point (70°C), heat sterilization is not possible. Because of its poor water solubility, filtration is not feasible. Disulfiram bulk powder (Antabuse, Ayerst) that was manufactured aseptically was used in this study. In a laminar airflow hood, 1 g of powder was weighed and poured into the barrel of a plastic 20-mL syringe from which the plunger had been removed and that had been fitted with a plastic tip seal. The plunger was then replaced in the barrel and depressed to the 10-mL mark. The syringe was placed in a zipper-lock plastic bag and sealed. Sterilization was performed by exposure to a Cs-137 irradiator emitting 660-keV gamma rays. Total exposure was 50,000 rads. HPLC analysis found no decomposition of the disulfiram to its decomposition product diethyldithiocarbamate from sterilization. Sterility testing determined that no microorganisms were present. To prepare this formulation for administration by subcutaneous injection, it was intended that about 10 mL of 0.9% sodium chloride injection be drawn into the syringe and shaken to suspend the disulfiram. It should be emphasized that the clinical suitability of this formulation for subcutaneous injection was not determined; the volume described exceeds maximum subcutaneous volumes. ■

# Docusate Sodium
## (Dioctyl Sodium Sulfosuccinate)

## Properties

Docusate sodium is a white or almost white hygroscopic waxy mass or flakes having a characteristic odor of octyl alcohol.[2,3]

### Solubility

Docusate sodium is sparingly soluble in water, having an aqueous solubility of about 14 to 15 mg/mL. Higher concentrations form a thick gel. In ethanol, it has a solubility of about 333 mg/mL.

### pH

Docusate sodium solution has a pH between 4.5 and 6.9. Docusate sodium syrup has a pH of 5.5 to 6.5.[4]

## General Stability Considerations

Docusate sodium should be stored in tight containers at controlled room temperature. The syrup should be in tight, light-resistant containers.[4]

## Stability Reports of Compounded Preparations
### Enteral
**DOCUSATE SODIUM COMPATIBILITY SUMMARY**

*Compatible with:* Enrich • Ensure • Ensure HN • Ensure Plus • Ensure Plus HN • Osmolite • Osmolite HN • TwoCal HN • Vital • Vivonex T.E.N.

**STUDY 1:** Cutie et al.[19] added 15 mL of docusate sodium syrup (Lederle) to varying amounts (15 to 240 mL) of Ensure, Ensure Plus, and Osmolite (Ross Laboratories) with vigorous agitation to ensure thorough mixing. The docusate sodium syrup was physically compatible, distributing uniformly in all three enteral products with no phase separation or granulation.

**STUDY 2:** Burns et al.[739] reported the physical compatibility of docusate sodium liquid (concentration and amount unspecified) of three enteral formulas, including Enrich, TwoCal HN, and Vivonex T.E.N. Visual inspection found no physical incompatibility with any of the enteral formulas.

**STUDY 3:** Altman and Cutie[850] reported the physical compatibility of docusate sodium syrup (Lederle) 15 mL with varying amounts (15 to 240 mL) of Ensure HN, Ensure Plus HN, Osmolite HN, and Vital after vigorous agitation to ensure thorough mixing. The docusate sodium syrup was physically compatible, distributing uniformly in all four enteral products with no phase separation or granulation. ■

# Dolasetron Mesylate

## Properties

Dolasetron mesylate is a white to off-white powder.[4,7]

### Solubility

Dolasetron mesylate is freely soluble in water and propylene glycol and slightly soluble in ethanol and normal saline.[4,7]

### pH

Dolasetron mesylate injection has a pH of 3.2 to 3.8.[7]

## General Stability Considerations

Dolasetron mesylate should be stored in a well-closed container at controlled room temperature and protected from exposure to light.[4] Commercial dolasetron mesylate tablets and injection also should be stored at controlled room temperature and protected from exposure to light.[7]

## Stability Reports of Compounded Preparations
### Oral
**USP OFFICIAL FORMULATION (ORAL SOLUTION):**
> Dolasetron mesylate 1 g
> Vehicle for Oral Solution, NF (sugar-containing or sugar-free) qs 100 mL

(See the vehicle monograph for information on the Vehicle for Oral Solution, NF.)

Use dolasetron mesylate powder. Add about 15 mL of the vehicle and mix well. Add additional vehicle almost to volume in increments with thorough mixing. Quantitatively transfer to a suitable calibrated tight and light-resistant bottle, bring to final volume with the vehicle, and thoroughly mix yielding dolasetron mesylate 10 mg/mL oral solution. The final liquid preparation should have a pH between 3.6 and 4.6. Store the preparation under refrigeration between 2 and 8°C. The beyond-use date is 90 days from the date of compounding.[4,5]

**USP OFFICIAL FORMULATION (ORAL SUSPENSION):**

   Dolasetron mesylate 1 g
   Vehicle for Oral Solution, NF (sugar-containing or
      sugar-free)
   Vehicle for Oral Suspension, NF (1:1) qs 100 mL

(See the vehicle monographs for information on the individual vehicles.)

Use dolasetron mesylate powder or commercial tablets. If using tablets, crush or grind to fine powder. Add about 20 mL of the vehicle mixture and mix to make a uniform paste. Add additional vehicle almost to volume in increments with thorough mixing. Quantitatively transfer to a suitable calibrated tight and light-resistant bottle, bring to final volume with the vehicle mixture, and thoroughly mix

yielding dolasetron mesylate 10 mg/mL oral suspension. The final liquid preparation should have a pH between 3.6 and 4.6. Store the preparation under refrigeration between 2 and 8°C. The beyond-use date is 90 days from the date of compounding.[4,5]

Johnson et al.[865] evaluated the stability of dolasetron mesylate 10-mg/mL oral suspensions prepared from commercial 50-mg tablets (Anzemet, Aventis) in two similar vehicles. The tablets were crushed thoroughly using a mortar and pestle. The powder was suspended in an equal parts mixture of Ora-Plus and strawberry syrup and separately in an equal parts mixture of Ora-Plus and Ora-Sweet SF (sugar free). The suspensions were packaged in amber plastic prescription bottles and were stored at room temperature of 23 to 25°C and refrigerated at 3 to 5°C for 90 days. No detectable changes in color, odor, or taste occurred. Stability-indicating HPLC analysis found that no dolasetron mesylate loss occurred at either temperature in 90 days. ∎

# ∎ Domperidone
# Domperidone Maleate

## Properties

Domperidone and domperidone maleate are a white or nearly white powders.[3] Domperidone maleate 12.72 mg is equivalent to domperidone base 10 mg.[7]

### Solubility

Domperidone is practically insoluble in water but slightly soluble in ethanol and soluble in dimethylformamide.[3]

Domperidone maleate is very slightly soluble in water and ethanol and sparingly soluble in dimethylformamide.[3]

### pH

Extemporaneously compounded oral suspensions prepared from commercial tablets crushed and suspended in an equal parts mixture of Ora-Sweet and Ora-Plus were found to have pH values near 4.3 to 4.6.[953]

## General Stability Considerations

Domperidone and domperidone maleate should be stored at controlled room temperature and protected from exposure

to light.[3] Commercial domperidone maleate film-coated tablets should also be stored at controlled room temperature not exceeding 25 or 30°C (depending on the specific labeling) and protected from light and moisture.[7]

## Stability Reports of Compounded Preparations
### Oral

Ensom et al.[953] evaluated the stability of extemporaneously compounded oral suspensions of domperidone maleate equivalent to domperidone 1 and 10 mg/mL. The domperidone suspensions were prepared using domperidone maleate 10-mg tablets (Motilium, Novopharm) in an equal parts mixture of Ora-Sweet syrup and Ora-Plus suspending medium. The oral suspensions were packaged in amber plastic prescription bottles and stored at room temperature of 25°C and refrigerated at 4°C for 91 days. No change in physical appearance or odor of the suspensions was noted; after settling, resuspension was accomplished easily. Stability-indicating HPLC analysis found domperidone losses did not exceed 8% in 91 days at either temperature. ∎

# Dornase Alfa

## Properties

Dornase alfa is a purified form of recombinant human deoxyribonuclease I, a glycoprotein having a 260 amino acid sequence and a molecular weight of 37,000 daltons. The drug is provided as a sterile, clear, colorless inhalation solution (Pulmozyme, Genentech) for the treatment of cystic fibrosis. Each milliliter of the inhalation solution contains dornase alfa 1 mg, calcium chloride dihydrate 0.15 mg, and sodium chloride 8.77 mg. The inhalation solution is packaged in 2.5-mL single-use ampules.[7]

## pH

Dornase alfa inhalation solution has a pH of 6.3.[7]

## Osmolality

Dornase alfa inhalation solution is near isotonicity.[7]

## General Stability Considerations

Dornase alfa single-use ampules should be stored in their protective foil pouch under refrigeration at 2 to 8°C and should be protected from exposure to strong light.[7] Dornase alfa is incompatible with solutions containing benzalkonium chloride and edetate disodium.[1216]

## Compatibility with Other Drugs

### DORNASE ALFA COMPATIBILITY SUMMARY

*Compatible with:* Budesonide
*Incompatible with:* Albuterol sulfate • Fluticasone propionate
 • Ipratropium bromide • Tobramycin • Tobramycin sulfate

**STUDY 1 (ALBUTEROL AND IPRATROPIUM):** Kramer et al.[1216] evaluated the compatibility and stability of inhalation solution admixtures with dornase alfa (Pulmozyme, Genentech). The contents of one ampule of dornase alfa (Pulmozyme Respule) was mixed with ipratropium bromide (Atrovent and Atrovent LS, Boehringer Ingelheim) and albuterol sulfate (Sultanol and Sultanol forte FI, GlaxoSmithKline) in the following combinations; these mixtures were stored at room temperature exposed to room light:

*Mixture 1*
Pulmozyme 2.5 mg/2.5 mL
Atrovent 0.5 mg/2 mL

*Mixture 2*
Pulmozyme 2.5 mg/2.5 mL
Sultanol forte FI 0.5 mL

*Mixture 3*
Pulmozyme 2.5 mg/2.5 mL
Atrovent LS 2 mL
Sultanol 0.5 mL

*Mixture 4*
Pulmozyme 2.5 mL
Atrovent 0.5 mg/2 mL
Sultanol forte FI 0.5 mL

Dornase alfa was physically incompatible with the unpreserved unit-dose forms of Atrovent and Sultanol forte FI, with visible formation of particulates upon mixing. Dornase alfa activity measured using a kinetic colorimetric DNase activity assay remained unchanged over the five-hour test period. Stability-indicating HPLC analysis found that the ipratropium bromide and albuterol sulfate concentrations also remained unchanged.

However, dornase alfa mixed with Atrovent LS and Sultanol inhalation solution, which both contain the preservative benzalkonium chloride in the formulations, resulted in dornase alfa activity losses of 20 to 50% rapidly within one to two hours. The ipratropium bromide and albuterol sulfate concentrations remained unchanged. Dornase alfa was found to lose activity due to the preservative benzalkonium chloride and also edetate disodium found in the Atrovent LS formulation. These combinations also resulted in visible particulate formation upon mixing.

Because the unpreserved formulations of ipratropium bromide and albuterol sulfate caused particulate formation when mixed with dornase alfa and because the preserved formulations of these drugs caused extensive and rapid loss of dornase alfa activity along with particulate formation, all of these mixtures of inhalation solutions tested were determined to be incompatible.

**STUDY 2 (TOBRAMYCIN):** Kramer et al.[1301] also evaluated the compatibility and stability of inhalation solution admixtures of tobramycin inhalation solutions with dornase alfa (Pulmozyme, Roche). Dornase alfa (Pulmozyme Respule) 2.5 mL (2500 units) was mixed with 300 mg/5 mL of tobramycin inhalation solution (Tobi, Chiron) or tobramycin sulfate inhalation solution (Gernebcin, Infectopharm) 80 mg/2 mL and was stored at room temperature exposed to room light.

The admixtures were incompatible. Visible brown discoloration was observed within an hour after mixing, and the odor of the mixtures worsened. Fluorescence immunoassay (TDx/TDxFLx) assay of tobramycin concentrations found no loss over 24 hours in either combination. However, kinetic colorimetric DNase activity assays of dornase alfa found decreased activity. Dornase alfa mixed with Tobi resulted in a decrease of dornase alfa of about 23% in 24 hours. With Gernebcin, dornase alfa activity decreased 24% in as little as one hour and nearly 40% loss in 24 hours after mixing. The more rapid and extensive loss with Gernebcin N was attributed to the presence of sodium metabisulfite in the Gernebcin formulation. When dornase alfa was mixed with sodium metabisulfite 0.05% solution, dornase alfa loss of 40% occurred in about one hour.

**STUDY 3 (MULTIPLE DRUGS):** Kramer et al.[1383] evaluated the compatibility and stability of a variety of two- and three-drug solution admixtures of inhalation drugs, including dornase alfa (Pulmozyme, Roche) 2500 units/2.5 mL. The mixtures were evaluated using chemical assays including HPLC, DNase activity assay, and fluorescence immunoassay as well as visual inspection, pH measurement, and osmolality determination. The drug combinations tested and the compatibility results that were reported are shown below.

*Mixture 1*
Dornase alfa (Pulmozyme) 2500 units/2.5 mL
Tobramycin (Tobi) 300 mg/5 mL
    Result: Physically and chemically incompatible

*Mixture 2*
Dornase alfa (Pulmozyme) 2500 units/2.5 mL
Tobramycin sulfate (Gernebcin) 80 mg/2 mL
    Result: Physically and chemically incompatible

*Mixture 3*
Albuterol sulfate (Sultanol) 5 mg/1 mL
Dornase alfa (Pulmozyme) 2500 units/2.5 mL
Ipratropium bromide (Atrovent LS) 0.25 mg/1 mL
    Result: Physically and chemically incompatible

*Mixture 4*
Dornase alfa (Pulmozyme) 2500 units/2.5 mL
Ipratropium bromide (Atrovent) 0.5 mg/2 mL
    Result: Physically and chemically incompatible

*Mixture 5*
Albuterol sulfate (Sultanol Forte) 2.5 mg/2.5 mL
Dornase alfa (Pulmozyme) 2500 units/2.5 mL
    Result: Physically and chemically incompatible

*Mixture 6*
Budesonide (Pulmicort) 1 mg/2 mL
Dornase alfa (Pulmozyme) 2500 units/2.5 mL
    Result: Physically and chemically compatible

*Mixture 7*
Dornase alfa (Pulmozyme) 2500 units/2.5 mL
Fluticasone propionate (Flutide forte) 2 mg/2 mL
    Result: Chemically incompatible ▪

# Doxepin Hydrochloride

## Properties
Doxepin hydrochloride is a white crystalline powder having a slightly aminelike odor.[2,3] Doxepin hydrochloride 113 mg is approximately equivalent to 100 mg of doxepin.[3]

### Solubility
Doxepin hydrochloride is freely soluble in water, having an aqueous solubility of about 200 mg/mL. In ethanol, the solubility is about 1 g/mL.[2,3]

### pH
Doxepin hydrochloride oral concentrate has a pH of 4 to 7.[4] Sinequan oral concentrate (Pfizer) had a measured pH of 5.7.[19]

### pK$_a$
Doxepin has a pK$_a$ of 8.[2]

## General Stability Considerations
Doxepin hydrochloride products should be stored in well-closed (capsules) or tight (solution), light-resistant containers at controlled room temperature. The products should be protected from exposure to direct sunlight.[2–4]

## Stability Reports of Compounded Preparations
*Enteral*
**DOXEPIN HYDROCHLORIDE COMPATIBILITY SUMMARY**
*Compatible with:* Ensure • Ensure HN • Ensure Plus • Ensure Plus HN • Osmolite • Osmolite HN • TwoCal HN • Vital • Vivonex T.E.N.
*Uncertain or variable compatibility with:* Enrich

**STUDY 1:** Cutie et al.[19] added 2 mL of doxepin hydrochloride concentrate (Sinequan, Pfizer) to varying amounts (15 to 240 mL) of Ensure, Ensure Plus, and Osmolite (Ross Laboratories) with vigorous agitation to ensure thorough mixing. The doxepin hydrochloride concentrate was physically compatible, distributing uniformly in all three enteral products with no phase separation and very little granulation.

**STUDY 2:** Burns et al.[739] reported the physical compatibility of doxepin hydrochloride concentrate (Sinequan) 100 mg/10 mL with 10 mL of three enteral formulas, including Enrich, TwoCal HN, and Vivonex T.E.N. Visual inspection found no physical incompatibility with the TwoCal HN or the Vivonex T.E.N. enteral formulas. However, when mixed with Enrich, a slight separation with small particulates formed. The authors indicated the particulates were not likely to lead to clogging.

**STUDY 3:** Altman and Cutie[850] reported the physical compatibility of 2 mL of doxepin hydrochloride concentrate (Sinequan, Pfizer) with varying amounts (15 to 240 mL) of Ensure HN, Ensure Plus HN, Osmolite HN, and Vital after vigorous agitation to ensure thorough mixing. The doxepin hydrochloride concentrate was physically compatible, distributing uniformly in all four enteral products with no phase separation and very little granulation.

## Compatibility with Other Drugs
**DOXEPIN HYDROCHLORIDE COMPATIBILITY SUMMARY**
*Compatible with:* Lithium citrate syrup
*Incompatible with:* Thioridazine suspension
*Uncertain or variable compatibility with:* Methadone
     hydrochloride

**STUDY 1:** Doxepin hydrochloride oral solution concentrate may be mixed with methadone hydrochloride in Gatorade, lemonade, orange juice, sugar water, Tang, or water. The two drugs should not be mixed in grape juice.[7]

**STUDY 2:** Raleigh[454] evaluated the compatibility of lithium citrate syrup (Philips Roxane) 1.6 mEq lithium/mL with doxepin concentrate 10 mg/mL (as the hydrochloride). The doxepin concentrate was combined with the lithium citrate syrup in a manner described as "simple random dosage combinations"; the exact amounts tested and test conditions were not specified. No precipitate was observed to form upon mixing.

**STUDY 3:** Raleigh[454] also evaluated the compatibility of thioridazine suspension (Sandoz) (concentration unspecified) with doxepin concentrate 10 mg/mL (as the hydrochloride). The doxepin concentrate was combined with the thioridazine suspension in the manner described as "simple random dosage combinations." The exact amounts tested and test conditions were not specified. A dense white curdy precipitate formed.

## Compatibility with Common Beverages and Foods
Doxepin hydrochloride oral solution concentrate can be diluted with approximately 120 mL of water, whole or skimmed milk, or orange, grapefruit, tomato, prune, or pineapple juice. The oral concentrate is incompatible with a number of carbonated beverages.[7] ■

# ■ Doxycycline
# Doxycycline Hyclate

## Properties
Doxycycline is a light yellow crystalline powder. Doxycycline hyclate is a yellow crystalline powder.[3,4,7]

### Solubility
Doxycycline is very slightly soluble in water but freely soluble in dilute acid and alkali hydroxide solutions. It is sparingly soluble in ethanol.[4]

Doxycycline hyclate is soluble in water and in solutions of alkali hydroxides and carbonates. It is slightly soluble in ethanol.[4]

### pH
Doxycycline in a 10-mg/mL aqueous suspension has a pH of 5 to 6.5.[3,4] Doxycycline calcium oral suspension has a pH in the range of 6.5 to 8.[4]

Doxycycline hyclate 10-mg/mL aqueous solution has a pH of 2 to 3.[3,4] Doxycycline hyclate injection reconstituted as directed has a pH of 1.8 to 3.3.[4]

### Osmolality
Doxycycline hyclate injection diluted to 1 mg/mL in dextrose 5% or in sodium chloride 0.9% was nearly isotonic, having measured osmolalities of 292 and 310 mOsm/kg, respectively.[1147]

## General Stability Considerations
Doxycycline and doxycycline hyclate substances should be packaged in tight containers; the drug substances should be stored at controlled room temperature and protected from exposure to light.[3,4]

Doxycycline and doxycycline hyclate oral capsules and tablets should be packaged in tight, light-resistant containers.[4,7]

Doxycycline hyclate for injection in intact vials is also stored at controlled room temperature and protected from exposure to light.[4,7] After dilution to a concentration of 0.1 to 1 mg/mL in a suitable infusion solution for administration, the drug must be protected from exposure to direct sunlight. The drug is stable in such infusion solutions for 48 hours at controlled room temperature and 72 hours stored under refrigeration and protected from exposure to both natural sunlight and artificial light.[7]

### Topical
Xiao et al.[1680] evaluated the stability of a topical gel formulation of doxycycline prepared using hydroxypropyl methylcellulose and glycerin. The gel was stored refrigerated at 2 to 8°C for up to six months. The gel remained homogeneous and transparent, and no change in pH was found. Ultraviolet spectrophotometric analysis found the doxycycline gel remained acceptable for clinical use.

## Compatibility with Common Beverages and Foods

Sadrieh et al.[956] evaluated the stability and palatability of crushed doxycycline tablets in common foods and beverages in this FDA-sponsored study of compounding for pediatric patients using counterterrorism drugs. Doxycycline proved to be most stable in low-fat chocolate milk with drug losses of less than 2% after seven days refrigerated. Doxycycline was found to be stable for only 24 hours in the other media that were evaluated. Palatability rankings from best to worst were chocolate pudding, chocolate milk, low-fat chocolate milk, simple syrup with sour apple flavor, apple juice with added sugar, low-fat milk, yogurt with cherry flavor, strawberry jelly, grape jelly, and water. ■

# ■ Droperidol

## Properties

Droperidol is a white to light tan amorphous or microcrystalline powder.[2,3]

### Solubility

Droperidol has a solubility of about 0.1 mg/mL in water and 3.4 to 7.14 mg/mL in ethanol.[1,2]

### pH

Droperidol injection has a pH adjusted to 3 to 3.8[4] with lactic acid, which facilitates solubilization.

### pK$_a$

Droperidol has a pK$_a$ of 7.64.[1]

## General Stability Considerations

Droperidol injection should be stored at controlled room temperature and protected from exposure to light.[4,7]

Droperidol in infusion solutions does not usually undergo sorption to polyvinyl chloride (PVC) containers. However, sorption to the PVC container occurred when an infusion solution was prepared in lactated Ringer's injection. A 15% loss in drug content occurred in 48 hours, and a 25% loss occurred in seven days at 27°C.[358]

### Injection

Injections, like other sterile drugs, should be prepared in a suitable clean air environment using appropriate aseptic procedures. When prepared from nonsterile components, an appropriate and effective sterilization method must be employed.

McCluskey and Lovely[1685] reported the stability of droperidol 0.625 mg/mL diluted in sodium chloride 0.9% and packaged as 1.1-mL aliquots in 12-mL Terumo polypropylene syringes sealed with Baxa syringe tip closures. The samples were stored in an environmental chamber at 23 to 27 °C and 55 to 65% relative humidity protected from exposure to light for 180 days. The samples remained visually clear, colorless, and particulate-free and pH changes were minimal. Stability-indicating HPLC analysis found no droperidol loss occurred.

## Compatibility with Common Beverages and Foods

Hirsch[244] noted that the addition of droperidol liquid products to coffee or tea resulted in the formation of an insoluble precipitate. ■

# Emetine Hydrochloride

## Properties

Emetine hydrochloride, the dihydrochloride salt of emetine alkaloid, occurs as white or very slightly yellowish, odorless, crystals or crystalline powder.[1,3,4] Emetine hydrochloride contains water of crystallization varying from three to eight $H_2O$.[1] Emetine is the principal alkaloid of ipecac.[1,3,4]

Emetine hydrochloride injection, USP, contains an amount of anhydrous emetine hydrochloride equivalent to not less than 84% and not more than 94% of the labeled amount of emetine hydrochloride.[4]

### Solubility

Emetine hydrochloride is freely soluble in water with 1 g dissolving in about 7 mL. Emetine hydrochloride is also freely soluble in ethanol.[1,3,4]

### pH

The natural pH of emetine hydrochloride 20 mg/mL in water is pH 5.6.[1] Emetine hydrochloride injection, USP, has a pH in the range of 3 to 5.[4] The European Pharmacopeia cites the pH range of emetine hydrochloride injection as 4 to 6.[3] Emetine hydrochloride injection prepared as described by Allen (See *Injection* below) has a pH in the range of 2.7 to 3.3.[1293]

### Osmolality

Emetine hydrochloride injection prepared as described by Allen (See *Injection* below) is likely to be very hypotonic.[1293]

## General Stability Considerations

Emetine hydrochloride as powder and in solutions turns yellow upon exposure to light and heat.[1] Emetine hydrochloride powder should be packaged in tight, light-resistant containers and stored at controlled room temperature.[4] Emetine hydrochloride injection should be packaged in single-dose, light-resistant containers made of Type I glass.[4]

## Stability Reports of Compounded Preparations

### Injection

Injections, like other sterile drugs, should be prepared in a suitable clean air environment using appropriate aseptic procedures. When prepared from nonsterile components, an appropriate and effective sterilization method must be employed.

Allen[1293] reported on a compounded formulation of emetine hydrochloride 30-mg/mL injection. The injection had the following formula:

| | | |
|---|---|---|
| Emetine hydrochloride | 3 g | |
| Sodium hydroxide or hydrochloric acid to adjust pH | qs | |
| Sterile water for injection | qs | 100 mL |

The recommended method of preparation was to dissolve thoroughly the emetine hydrochloride powder in about 90 mL of sterile water for injection. The pH of the solution was then adjusted to 2.7 to 3.3 with a 10% solution of sodium hydroxide or hydrochloric acid. [NOTE: This pH range does not conform to the USP pH range for emetine hydrochloride injection, USP.[4]] The solution was brought to volume with additional sterile water for injection and mixed well. The solution was sterilized by filtration through a 0.22-μm filter and packaged in suitable light-resistant sterile vials. The author recommended a beyond-use date of six months at room temperature because this formula is a commercial medication in some countries with an expiration date of two years or more. ∎

# Enalapril Maleate

## Properties

Enalapril maleate is a white to off-white crystalline powder.[1–3] Enalapril maleate is a prodrug that is converted to enalaprilat. Enalaprilat from the injection is unsuitable for oral use because of poor bioavailability.[586]

### Solubility

Enalapril maleate has an aqueous solubility of around 25 mg/mL at room temperature.[1,408] In ethanol, its solubility is about 80 mg/mL.[1]

### pH

A 1% aqueous solution of enalapril maleate has a pH of around 2.6.[1]

### pKa

The drug has apparent $pK_a$ values of 3 and 5.4.[1]

## General Stability Considerations

Enalapril maleate tablets should be packaged in well-closed containers and stored at less than 30°C. Exposure to temperatures greater than 50°C even transiently should be avoided.[2,4,7] The manufacturer has stated that enalapril maleate tablets should not be removed from the original packaging and placed in dosing compliance aids because of drug sensitivity to moisture.[1622]

Degradation of enalapril is pH dependent.[408] The pH of maximum stability is about 3.[408,1126] Above pH 5, an increased rate of decomposition occurs.[408]

## Stability Reports of Compounded Preparations
### Oral

**STUDY 1:** The manufacturer recommends in the Vasotec tablet package insert[7] preparation of an oral liquid suspension formulation that provides enalapril maleate at 1 mg/mL. The suspension is prepared as follows: Bicitra 50 mL is added to a polyethylene terephthalate (PET) bottle containing ten 20-mg tablets of Vasotec and shaken for at least two minutes. The concentrate should stand for 60 minutes. Following the 60-minute hold time, the concentrate is shaken for an additional minute. Ora-Sweet SF 150 mL is added to the concentrate in the PET bottle, and the suspension is shaken to disperse the ingredients. The suspension should be refrigerated at 2 to 8°C (36 to 46°F) and can be stored for up to 30 days. The suspension should be shaken before each use.

**STUDY 2:** Boulton et al.[408] reported the extraction efficiency and stability of oral solutions prepared from enalapril maleate tablets. The solutions containing enalapril maleate 1 and 0.1 mg/mL were prepared by grinding commercial 20-mg tablets (Merck) to fine powder with a glass mortar and pestle. One tablet was used for the 0.1-mg/mL concentration, and 10 tablets were used for the 1-mg/mL concentration. The finely ground tablet powder was triturated with isotonic citrate buffer (pH 5) used as the vehicle and then filtered through Whatman No. 1 filter paper into 200-mL volumetric flasks. Representative samples also were sonicated for 30 minutes to determine if extractability could be improved. One 1-mg/mL formulation was preserved with hydroxybenzoate solution, whereas the others were tested unpreserved. The solutions were brought to 200 mL with additional isotonic citrate buffer. Samples were packaged in 25-mL amber high-density polyethylene bottles with polypropylene screw caps and stored at 5, 25, and 40°C. The enalapril content of the solution was evaluated over 90 days using a stability-indicating HPLC assay.

The extraction efficiency of enalapril from the tablets by the isotonic citrate buffer was about 93%, which improved to about 97% with sonication. The samples stored under refrigeration at 5°C exhibited no detectable decomposition during 90 days. The samples stored at 25°C had a higher decomposition rate; the 0.1-mg/mL solution reached its $t_{90}$ in 55 days, while the 1-mg/mL concentration with and without preservative reached $t_{90}$ in about 43 days. At 40°C, 10% loss occurred in about nine days in all solutions.[408]

**STUDY 3:** Nahata et al.[581] evaluated the stability of three different enalapril maleate 1-mg/mL oral suspensions prepared from tablets. Twenty enalapril maleate 10-mg tablets (Merck) were crushed to powder using a mortar and pestle. Three vehicles were tested for use in this formulation: deionized water; citrate buffer, pH 5; and an equal parts mixture of Ora-Sweet and Ora-Plus (Paddock). The citrate buffer solution was prepared from 0.353 g of citric acid monohydrate granular, USP; 1.01 g of sodium citrate dihydrate granular, USP; and 0.54 g of sodium chloride in 100 mL of distilled water. A small amount of the test vehicle was triturated with the tablet powder to form a smooth paste. Additional amounts of the vehicle were incorporated in increments until a pourable mixture was formed. This mixture was then transferred to a graduated cylinder and brought to a volume of 200 mL with additional vehicle. The 1-mg/mL final suspensions were packaged as 20 mL of suspension in 60-mL plastic prescription bottles and stored at 4 and 25°C for up to 91 days.

The appearance of the suspensions in the three vehicles remained unchanged throughout the study period. Stability-indicating HPLC analysis found a 5% or less enalapril loss in all three vehicles stored at 4°C and a 7 to 8% loss in the citrate buffer and Ora-Plus and Ora-Sweet vehicles at 25°C after storage for 91 days. However, the suspension in deionized water stored at 25°C exhibited 10% enalapril loss after 56 days.

**STUDY 4:** Allen and Erickson[594] evaluated the stability of three enalapril maleate 1-mg/mL oral suspensions extemporaneously compounded from tablets. Vehicles used in this study were (1) an equal parts mixture of Ora-Sweet and Ora-Plus (Paddock), (2) an equal parts mixture of Ora-Sweet SF and Ora-Plus (Paddock), and (3) cherry syrup (Robinson Laboratories) mixed 1:4 with simple syrup. Six enalapril maleate 20-mg tablets (Merck Sharp & Dohme) were crushed and comminuted to fine powder using a mortar and pestle. About 15 mL of the test vehicle was added to the powder and mixed to yield a uniform paste. Additional vehicle was added geometrically and brought to the final volume of 120 mL, mixing thoroughly after each addition. The process was repeated for each of the three test suspension vehicles. Samples of each finished suspension were packaged in 120-mL amber polyethylene terephthalate plastic prescription bottles and stored at 5 and 25°C.

No visual changes or changes in odor were detected during the study. Stability-indicating HPLC analysis found less than 6% enalapril maleate loss in any of the suspensions stored at either temperature after 60 days of storage.

**STUDY 5:** Schlatter and Saulnier[1471] evaluated the stability of enalapril maleate 0.1 and 1 mg/mL oral solutions prepared from oral tablets dissolved in sterile water for irrigation. After tablet dissolution, the solutions were filtered through 0.22-μm filters. The solutions were clear and colorless after filtration. The solutions were packaged in amber glass bottles and stored at 20 to 24°C protected from exposure to light for 28 days. Crystals formed in some of the test samples after 14 days of storage; the composition of the crystals was not identified. Stability-indicating HPLC analysis found no substantive difference in stability between the two concentrations. Enalapril concentrations were about 91 and 96% for the 0.1 and 1 mg/mL concentrations after 14 days of storage. By 21 days of storage the drug concentration had declined to about 80%. The authors stated that the enalapril maleate solution prepared this way were suitable for home preparation. They stated that although the solutions were stable for 14 days, using a seven-day expiration would be wiser.

**STUDY 6:** Supattanakul[932] evaluated the stability of an enalapril maleate oral syrup prepared incorporating citric acid to adjust pH. A formulation having pH 3 was selected for stability testing by HPLC. After 159 days of storage at room temperature, about 8% loss of drug occurred; little or no loss occurred when stored under refrigeration. By 185 days (six

months) of storage, the amount of drug loss exceeded 10% at room temperature and was about 7 to 8% under refrigeration.

**STUDY 7:** Sosnowska et al.[1451] and Sosnowska et al.[1466] evaluated the stability of enalapril maleate 0.1 and 1 mg/mL oral suspensions prepared from commercial oral tablets. Enalapril maleate 20-mg tablets (Enarenal, Polpharma) were crushed to fine powder and combined with (1) hydroxymethylcellulose 0.5% and (2) in a 1:10 mixture of raspberry syrup and hydroxymethylcellulose 0.5%. The raspberry syrup was used to help mask the unpleasant taste of enalapril maleate. The pH was adjusted with citric acid to 3, and methyl hydroxybenzoate 0.2% was added as a preservative. The oral suspensions were packaged in orange glass bottles, and samples were stored refrigerated at 2 to 8°C and also at room temperature of 23 to 25°C protected from exposure to light.

The viscosities and pH values of the samples exhibited no changes over the 30-day study period. HPLC analysis found little change in enalapril maleate concentration varying from the initial concentration by no more than 2% in any of the samples over 30 days of storage for any of the samples.

*Topical*

Aqil et al.[1289] reported on the formulation and stability of enalapril maleate in a transdermal film for use as an antihypertensive drug. Three transdermal film formulations were evaluated with the one that performed best being composed of Eudragit E-100 500 mg, PVP K-30 500 mg, enalapril maleate 29% (wt/wt), dichloromethane 5 mL, isopropanol 5 mL, piperidine hydrochloride 10% (wt/wt) penetration enhancer, and dibutylphthalate 5% (wt/wt) plasticizer. The Eudragit E-100, PVP K-30, piperidine hydrochloride, dibutylphthalate, and the enalapril maleate were dissolved in the isopropanol and dichloromethane solvent system. The polymeric drug solution was poured onto a mercury surface and dried for 24 hours at about 32°C and a relative humidity of 40 to 50%. Aluminum foil was used as the backer and wax paper as the release liner. The films were cut into standardized circular pieces of 6.74 cm². Drug release studies were conducted using USP Apparatus 5. Samples for stability testing were stored at 40 and 75°C for three months, and the predicted shelf life period was calculated from the stressed storage conditions.

Ultraviolet spectrophotometry and thin layer chromatography found the absence of chemical interaction of enalapril maleate with the additives in the formulation. The in vitro drug release profile showed about 87% drug release occurred in a test run over 48 hours. HPLC analysis from the accelerated stress temperature studies found a very low drug degradation rate; the calculated shelf life at ambient conditions was at least two years. Skin permeation and antihypertensive activity were found to be superior to oral enalapril maleate tablets in test animals. ▪

# Entecavir

## Properties
Entecavir occurs as a white to off-white powder.[1,7]

### Solubility
Entecavir has an aqueous solubility of 2.4 mg/mL.[1,7]

### pH
An entecavir saturated aqueous solution at 25°C has a pH of 7.9.[1,7]

## General Stability Considerations
Entecavir oral tablets should be packaged in tight containers and stored at controlled room temperature. Entecavir oral solution containers at controlled room temperature should be stored in the outer carton and protected from exposure to light.[7]

## Stability Reports of Compounded Preparations
### Oral
Desai et al.[1258] evaluated the stability of entecavir at a low concentration of 0.2 mg/mL in 50% sucrose and maltitol solutions and 26% fructose and glucose solutions stored at 50°C. Over 13 weeks of storage, the sucrose, fructose, and glucose mixtures turned yellow, then light orange. HPLC analysis found hydrolysis of sucrose to fructose and glucose that interacted with the entecavir, resulting in extensive loss of the drug, especially at low pH. With the pH of the sucrose solution adjusted to pH values of 4, 6, and 7, entecavir losses were about 68%, 25%, and 23%, respectively. Similar losses were also found in fructose and glucose solutions, leading the authors to believe the hydrolysis of sucrose led to the instability of the drug.

However, in 50% maltitol, no color change was observed in 13 weeks. In addition, HPLC analysis found less than 5% loss of entecavir after 13 weeks at 50°C. The authors concluded that entecavir was unstable in a sucrose solution and that maltitol was the preferred sweetener for an oral liquid preparation. ■

# Ephedrine Hydrochloride
# Ephedrine Sulfate

## Properties
Ephedrine hydrochloride occurs as fine white odorless crystals or as a crystalline powder.[1,3,4]

Ephedrine sulfate occurs as fine white odorless crystals or as a powder.[1,3,4]

### Solubility
Ephedrine hydrochloride is freely soluble in water (1 in 3), soluble in ethanol (1 in 14), and practically insoluble in ether.[1,3,4]

Ephedrine sulfate is freely soluble in water (1 in 1.3) and sparingly soluble in ethanol (1 in 90).[1,3,4]

### pH
Ephedrine sulfate injection has a pH of 4.5 to 7.[4]

### Osmolality
Ephedrine sulfate 50-mg/mL injection has a calculated osmolarity of 350 mOsm/L.[7]

## General Stability Considerations
Ephedrine hydrochloride may be affected by exposure to light, and ephedrine sulfate darkens upon exposure to light.[1,4]

Ephedrine hydrochloride and ephedrine sulfate powders should be stored in well-closed and light-resistant containers at controlled room temperature. Ephedrine sulfate capsules, nasal solution, and oral solution should be packaged in tight, light-resistant containers and stored at controlled room temperature. Ephedrine sulfate injection is also available in light-resistant containers and is stored at controlled room temperature.[4]

## Stability Reports of Compounded Preparations
### Injection
Injections, like other sterile drugs, should be prepared in a suitable clean air environment using appropriate aseptic procedures. When prepared from nonsterile components, an appropriate and effective sterilization method must be employed.

**STUDY 1 (SULFATE REPACKAGED):** Casasin Edo and Roca Massa[1118] reported that ephedrine 10 mg/mL in polypropylene syringes was stable, with spectrophotometric and potentiometric analyses finding little or no loss of drug in four weeks at room temperature exposed to light.

**STUDY 2 (SULFATE REPACKAGED):** Storms et al.[1119] reported that ephedrine sulfate 5 mg/mL in sodium chloride 0.9% in

10-mL Becton Dickinson polypropylene syringes was stable, with stability-indicating HPLC analysis finding less than 3% loss of ephedrine sulfate in 60 days, both at room temperature exposed to fluorescent light and under refrigeration.

**STUDY 3 (SULFATE REPACKAGED):** Lewis et al.[1120] reported that undiluted ephedrine sulfate 50 mg/mL packaged as 1 mL of drug solution in 3-mL Becton Dickinson polypropylene syringes and Micro-Mate glass syringes and stored at room temperature of 24 to 27°C exhibited no change in appearance and no precipitate formation throughout the four-day study. Stability-indicating HPLC analysis found that little or no ephedrine sulfate loss occurred in four days as well.

**STUDY 4 (HYDROCHLORIDE COMPOUNDED):** Griffiths et al.[1117] reported the long-term stability of an extemporaneously compounded solution (from powder) of ephedrine hydrochloride 20 mg/2 mL in sodium chloride 0.6%, filtered through a 0.22-µm sterilizing filter, and packaged in 2.5-mL PlastiPak (Becton Dickinson) syringes sealed with TEC 1000 (B. Braun) tip caps. Syringes were stored at room temperature of 23 to 27°C and at an elevated temperature of 38 to 42°C for 12 months. Samples of the syringes initially passed sterility testing and electronic particle content analysis.

No visible changes were observed, and stability-indicating HPLC analysis found little or no loss of ephedrine hydrochloride over 12 months of storage at either storage temperature. ■

# Epinephrine
## (Adrenaline)

## Properties

Epinephrine is an endogenous catecholamine that is the active principle of the medulla of the adrenal gland. It may be prepared synthetically or obtained from the adrenal glands of mammals. It is a white or creamy-white odorless crystalline powder or granules.[2,3]

The endogenous substance and official preparations are levorotatory. The drug is also available as a racemic mixture of the hydrochlorides, exhibiting about half the activity of the levorotatory isomer.[2]

### Solubility

Epinephrine is slightly soluble in water and in ethanol. It dissolves in dilute solutions of mineral acids, forming water-soluble salts such as the hydrochloride.[1–3] Epinephrine bitartrate is soluble in water to about 333 mg/mL.[6]

### pH

Epinephrine solutions are alkaline to litmus.[3] Epinephrine hydrochloride injection has a pH of 2.2 to 5. Epinephrine hydrochloride ophthalmic solution has a pH of 2.2 to 4.5.[4] Racemic epinephrine oral inhalation has a pH of 2 to 3.5.[2]

### pKa

The drug has $pK_a$ values of 8.7 and 9.9 at 20°C.[6]

## General Stability Considerations

Epinephrine products should be packaged in tight light-resistant containers and stored at controlled room temperature.[2–4] Liquid products should be protected from freezing.[2]

Epinephrine darkens on exposure to light and air due to oxidation.[2,3,6] Injections are usually packaged under nitrogen.[6]

Withdrawal of doses from vials introduces air that results in oxidation, with the solution becoming pink as adrenochrome forms to brown as melanin forms. Discolored solutions or those that contain a precipitate should not be used.[2]

Epinephrine is unstable in neutral or alkaline solutions[3] and is destroyed by oxidizing agents.[2] Stable formulations should have a pH high enough so that racemization is not significant but is low enough to slow oxidation. The pH of optimum stability is 3 to 4.[6] Above pH 5.5,[8] the drug as the hydrochloride is unstable in aqueous solutions such as 5% dextrose injection.

Epinephrine is also unstable to heat. Grant et al.[1545] reported 64% loss of epinephrine in 1:1000 solutions when heated to 70°C for up to 12 weeks. Church et al.[1544] reported the total loss of epinephrine in commercial epinephrine cartridges subjected to 65°C for seven days; losses of about 31% occurred when the temperature was cycled being 65°C for eight hours in 24 hours instead of being held constant. Rudland et al.[1546] reported some losses up to about 40% from physician-returned epinephrine vials from doctor's bags subjected to summer temperatures of 40 to 65°C in vehicles.

Sterilization of the outer surface of epinephrine hydrochloride ampules by autoclaving at 121°C for 15 minutes resulted in no detectable loss of epinephrine content. However, 8% loss resulted if the ampules were subjected to autoclaving two times.[8]

## Stability Reports of Compounded Preparations
### Injection

Injections, like other sterile drugs, should be prepared in a suitable clean air environment using appropriate aseptic procedures. When prepared from nonsterile components, an appropriate and effective sterilization method must be employed.

**STUDY 1 (REPACKAGING):** Donnelly and Yen[641] evaluated the stability of epinephrine hydrochloride 1 and 7 mg/10 mL in sterile water for injection repackaged in 10-mL glass vials or plastic syringes. At room temperature, the 1-mg/mL concentration loss was about 6% in seven days and 13% in 14 days. The 7-mg/mL concentration lost no more than 5% in 56 days.

**STUDY 2 (REPACKAGING):** Rawas-Qalaji et al.[1465] evaluated the stability of epinephrine hydrochloride 1 mg/mL drawn into unsealed 1-mL Becton Dickinson polypropylene syringes for use in first aid treatment of anaphylaxis when stored at 38°C at high and low humidity and exposed to or protected from exposure to light. The intent was to simulate conditions that might occur in areas of the world that are hot and where conditions are suboptimal and epinephrine autoinjectors are unavailable. HPLC analysis of drug content found substantial loss of epinephrine occurred within two to three months under all storage conditions evaluated. The authors stated that such syringes should be replaced "every few months" because of this drug loss. However, the maintenance of sterility in these unsealed syringes was not evaluated, which is a critical issue.

**STUDY 3 (REPACKAGING):** Kerddonfak et al.[1596] evaluated the stability of epinephrine hydrochloride 1 mg/mL drawn into unsealed 1-mL plastic syringes (Nipro) with 23-gauge needles for use in first aid treatment. Samples were stored for three months at 23 to 29°C protected from exposure to light. HPLC analysis found little or no loss of epinephrine hydrochloride occurred over three months of storage. Visual examination found that the solutions remained clear and colorless. However, brown particles were found at the needle. The particles were negative for bacteria or fungus. The authors speculated that the brown particles were formed by the epinephrine reacting with the atmosphere and resulting in oxidation of the epinephrine to adrenochrome, which is a pink-brown in color. It is also possible that melanin had formed with its characteristic brown color. After four months of storage, the epinephrine hydrochloride solutions had turned pinkish-brown and were deemed unacceptable. Although the authors stated that the prefilled epinephrine syringes were stable for three months at room temperature protected from exposure to light, caution would indicate a shorter utility period may be warranted because of the oxidation that leads to the brown particulates.

**STUDY 4 (THREE DRUGS):** Allen[128] reported the compounding of an epidural analgesic formulation. The product consisted of fentanyl citrate 1.25 mcg/mL, bupivacaine hydrochloride 0.4375 mg/mL, and epinephrine hydrochloride 0.6875 mcg/mL in 0.9% sodium chloride injection, USP, and was compounded from the formula cited in Table 35. The product packaged in light-resistant sterile containers was physically compatible and chemically stable for 20 days stored under refrigeration at 3°C.

**TABLE 35.** Epidural Analgesic Formula[128]

| Component | Amount |
| --- | --- |
| Fentanyl citrate 50 mcg/mL | 2.5 mL |
| Bupivacaine hydrochloride 0.5% | 8.75 mL |
| Epinephrine hydrochloride 1:100,000 | 6.9 mL |
| 0.9% Sodium chloride injection, USP | 81.85 mL |

**STUDY 5 (THREE DRUGS):** Kjonniksen et al.[1226] reported the long-term stability of a similar three-drug analgesic mixture for epidural administration. The preparation contained fentanyl citrate 1.93 mcg/mL, bupivacaine hydrochloride 0.935 mg/mL, and epinephrine bitartrate 2.07 mcg/mL with sodium metabisulfite approximately 2 mcg/mL and disodium edetate approximately 0.2 mcg/mL in sodium chloride 0.9%. The formulation was prepared by first compounding a concentrate that was packaged in vials and autoclaved. This concentrate was then diluted 1 to 11 in 500 mL of sodium chloride 0.9% in polyolefin-lined Excel multilayer bags. The bags were stored refrigerated at 2 to 8°C for 180 days followed by four days at room temperature of 22°C exposed to ambient laboratory illumination. No visible changes were reported. Stability-indicating HPLC analysis found that none of the drugs exhibited changes in concentration greater than 4%. The authors concluded that this drug combination was stable and could be given a shelf life of six months under refrigeration based on the chemical stability.

**STUDY 6: (THREE DRUGS):** Tuleu et al.[1266] buffered lidocaine hydrochloride with epinephrine 5 mcg/mL to around pH 7.5 to 8 with sodium bicarbonate. Stability-indicating HPLC analysis found no loss of lidocaine hydrochloride but nearly 30% loss of epinephrine in six hours at 24°C exposed to light. If kept in the dark, epinephrine losses were about 9% in 20 hours at 24°C. The authors recommended not preparing the buffered injection well in advance of use.

## Inhalation

Inhalation preparations, like other sterile drugs, should be prepared in a suitable clean air environment using appropriate aseptic procedures. When prepared from nonsterile components, an appropriate and effective sterilization method must be employed.

Mizushima et al.[1677] evaluated the stability of an inhalation solution containing epinephrine and prednisolone succinate. The solution was packaged in 50-mL sterile bottles and was stored for four weeks at room temperature of 23 to 27°C exposed to room light, at room temperature protected from exposure to light using a brown light-resistant bag, and in a cool dark place at 5 to 10°C protected from exposure to light using a brown light-resistant bag.

The solutions stored at room temperature were markedly discolored whether exposed to or protected from light. HPLC analysis found loss of both drugs and formation of the decomposition products adrenochrome and free prednisolone. Stored in a cool dark place, there was no apparent discoloration and losses of epinephrine and prednisolone succinate were 14 and 8%, respectively.

## Topical
### USP OFFICIAL FORMULATION (T-A-C TOPICAL SOLUTION):

| | | |
|---|---|---|
| Cocaine hydrochloride | | 4 g |
| Tetracaine hydrochloride | | 1 g |
| Epinephrine hydrochloride injection (1:1000) | | 25 mL |
| Benzalkonium chloride | | 10 mg |
| Edetate disodium | | 6.4 mg |
| Sodium chloride injection 0.9% | | 35 mL |
| Purified water | qs | 100 mL |

The cocaine hydrochloride and tetracaine hydrochloride powders should be dissolved in 25 mL of purified water. Add the epinephrine hydrochloride (1:1000) injection and mix. Dissolve the edetate disodium separately in sodium chloride 0.9% and bring to a final volume of 35 mL with additional sodium chloride 0.9%. Dissolve the benzalkonium chloride separately in a quantity of purified water and bring to a final volume of 10 mL with additional purified water. Combine the three solutions, add additional purified water to bring the solution to volume, and mix well. Package the solution in a suitable sterile, tight, and light-resistant container. The final liquid preparation should have a pH between pH 4 and 6. Store the preparation under refrigeration between 2 and 8°C. The beyond-use date is 30 days from the date of compounding.[4,5]

**STUDY 1:** Larson et al.[539] evaluated the stability of a topical anesthetic solution described by Schilling et al.[540] The topical anesthetic was composed of lidocaine hydrochloride 40 mg/L, racemic epinephrine hydrochloride 2.25 mg/mL, tetracaine hydrochloride 5 mg/mL, and sodium metabisulfite 0.63 mg/mL as an aqueous solution. A 500-mL quantity of the solution was prepared by measuring 100 mL of 20% lidocaine hydrochloride injection (Abbott), 50 mL of racemic epinephrine 2.25% as the hydrochloride (Nephron), 125 mL of 2% tetracaine hydrochloride (Winthrop), 315 mg of sodium metabisulfite (Gallipot), and 225 mL of sterile water for irrigation. All ingredients were combined and mixed well. The finished solution had a blue tint. The solution was packaged in both clear and amber glass bottles with tight-fitting caps, and samples were stored under refrigeration at 4°C in the dark, at room temperatures near 18°C exposed to ambient room light, and at an elevated temperature of 35°C exposed to ambient room light.

The refrigerated samples remained clear with a blue tint, and HPLC analysis of the samples found little or no loss of any of the three drug components after 26 weeks of storage. Epinephrine was the least stable of the drug components at the higher temperatures. Samples in amber containers

**TABLE 36.** T-A-C Topical Solution Formulations[150]

| Component | Formula 1 | Formula 2 |
|---|---|---|
| Tetracaine hydrochloride 2% | 2 mL | 1 mL |
| Epinephrine hydrochloride 0.1% | 1 mL | 2 mL |
| Cocaine hydrochloride 4% solution | 1 mL | |
| Cocaine hydrochloride powder | | 0.472 g[a] |
| 0.9% Sodium chloride injection | | qs  4 mL |

[a]Cocaine hydrochloride 11.8%.

at room temperature exposed to light were not discolored through four weeks of storage but became discolored in eight weeks, exhibiting about 5% epinephrine loss at that time; the epinephrine loss increased to 9% in 16 weeks. Discoloration was thought to be the result of a highly colored epinephrine decomposition product that was visible when epinephrine decomposition of 2% or more occurred. The lidocaine and tetracaine components both retained at least 95% of the initial concentration throughout 26 weeks of room temperature storage. In clear containers exposed to light, discoloration appeared in as little as one to two weeks. The authors[539] recommended that the solution be packaged only in amber glass bottles and be given expiration dating of 26 weeks under refrigeration and four weeks at room temperature.

The results of a controlled clinical trial of this topical anesthetic solution by Schilling et al.[540] indicated that adequate anesthesia resulted throughout the time period evaluated for stability by Larson et al.[539]

**STUDY 2:** Stiles and Allen[150] reported on two formulations for topical anesthesia. The T-A-C solutions (Table 36) consisted of tetracaine hydrochloride (Pontocaine, Winthrop), epinephrine hydrochloride (Adrenalin, Parke-Davis), and cocaine hydrochloride. The first formula is a simple combination of three commercial products. The second formula is a cocaine-fortified solution prepared from cocaine hydrochloride powder. The T-A-C solutions should be terminally filtered to remove any contaminants. The solutions are unstable in light and air and should be freshly prepared for use.

**STUDY 3:** Pesko[253] also reported on a T-A-C topical formulation (Table 37). The author recommended mixing the components aseptically with filtration through a 0.22-μm filter; an expiration

**TABLE 37.** T-A-C Topical Anesthetic Solution Formula[253]

| Component | Amount |
|---|---|
| Tetracaine hydrochloride 2% | 50 mL |
| Epinephrine hydrochloride 0.1% | 2.5 mL |
| Cocaine hydrochloride 10% | 40 mL |
| Sterile water for irrigation | 7.5 mL |

period of 14 days for samples protected from light was recommended, although stability studies were not conducted.

**STUDY 4:** Jeter and Mueller[404] described the preparation of a sterile T-A-C gel formulation. The product was prepared from T-A-C sterile solution composed of tetracaine hydrochloride 0.5%, epinephrine hydrochloride 1:2000, and cocaine hydrochloride 11.8% that had been sterilized using a 0.2-μm filter (Millex-GS). T-A-C solution 12 mL was added to a 1-g jar of sterile Gelfoam powder and thoroughly mixed. The mixture was divided into six 2.5-g portions using a gas-sterilized measuring spoon, packaged into heat-sterilized and depyrogenated amber open-mouth vials, and sealed with gas-sterilized tamper-evident aluminum lids. The product should be protected from light to prevent epinephrine oxidation, which turns the mixture brown. The authors used a three-month date on the product, although specific stability was not assessed.

**STUDY 5:** Vonbach et al.[1336] evaluated the long-term stability of a three-drug solution to be used for bladder iontophoresis. The solution contained lidocaine hydrochloride 20 mg/mL, dexamethasone sodium phosphate 0.2 mg/mL, and epinephrine hydrochloride 0.01 mg/mL in water. The solution was packaged in amber glass vials and clear glass vials. The samples in amber glass vials were stored frozen at −20°C, refrigerated at 4°C, at room temperature of 20°C, and at elevated stress temperature of 40°C. Samples in clear vials were stored at 20°C exposed to daylight.

HPLC analysis found that epinephrine hydrochloride was the least stable component. At 40°C and exposed to daylight, extensive and rapid loss of epinephrine hydrochloride occurred. At 20°C in amber vials, about 10% epinephrine hydrochloride loss occurred in 28 days. Under refrigeration, the loss was about 6% in six months; when frozen, no loss occurred in six months. For lidocaine hydrochloride, little or no loss occurred under any of the test conditions. For dexamethasone sodium phosphate, losses never exceeded 5% in six months at −20, 4, and 20°C. The authors stated that this three-drug iontophoresis solution was stable for three months refrigerated and for six months frozen at −20°C.

## Ophthalmic

Ophthalmic preparations, like other sterile drugs, should be prepared in a suitable clean air environment using appropriate aseptic procedures. When prepared from nonsterile components, an appropriate and effective sterilization method must be employed.

**STUDY 1:** Gairard et al.[741] reported the stability of epinephrine 1 mg/L in Aqsia (laboratoire Opsia) intraocular irrigation solution in plastic bags. Aqsia contains sodium chloride, potassium chloride, calcium chloride, magnesium chloride, sodium bicarbonate, and sodium citrate in water. The test solutions were stored at room temperature of 25°C, refrigerated at 4°C, and frozen at −20°C. HPLC analysis found that the epinephrine was stable for two days with about 10% loss at room temperature and for 14 days under refrigeration. Refrigeration for 21 days resulted in about 13% loss. No loss occurred after 35 days when stored frozen. There was no difference in drug concentration between thawing the frozen solutions at ambient temperature and using microwave thawing.

**STUDY 2 (WITH SULFISOXAZOLE):** Kato et al.[1580] evaluated the stability of an equal volume mixture of epinephrine and sulfisoxazole ophthalmic solution. Samples were stored refrigerated at 5°C and at room temperature of 30°C in the dark. Under refrigeration, the epinephrine content was reduced to 50 and to 10% of the initial concentration in 5 and 30 days, respectively. The rate of loss increased at room temperature. The color of the mixed solution changed from colorless to dark brown. No change in the concentration of sulfisoxazole occurred in the samples.

## Compatibility with Other Drugs

Lesko and Miller[659] evaluated the chemical and physical stability of 1% cromolyn sodium nebulizer solution (Fisons) 2 mL admixed with 1.0 mL of epinephrine hydrochloride 2.25% inhalation solution. The admixture was visually clear and colorless and remained chemically stable by HPLC analysis for 60 minutes after mixing at 22°C.

# Epoetin Alfa

## Properties

Epoetin alfa is a 165 amino acid glycoprotein having a molecular weight of 30,400 daltons produced by recombinant DNA technology. The drug is formulated as a clear or slightly turbid colorless liquid injection.[3,7] Epoetin alfa has an activity of not less than 100,000 units per milligram of active substance.[3]

Epoetin alfa injections also contain human albumin, sodium citrate, citric acid, and sodium chloride. Sodium phosphate monobasic and dibasic may also be present as buffers in some dosage forms. The multiple-dose vials also contain benzyl alcohol as a preservative.[7]

## pH

The commercial epoetin alfa injection in single-dose containers has a pH of 6.6 to 7.2. In multiple-dose containers the pH ranges from 5.8 to 6.4.[7]

## Osmolality
Epoetin alfa injections are isotonic.[7]

## General Stability Considerations
Epoetin alfa should be stored in airtight containers below −20°C. Repeated thawing and refreezing should be avoided.[3]

Epoetin alfa injections should be stored under refrigeration at 2 to 8°C and protected from freezing and shaking to avoid foaming and inactivation of the protein. However, the manufacturer notes that a small amount of flocculated protein may be present in the injection and does not indicate substantial loss of drug activity.[7]

Epoetin alfa injections contain human albumin for stability. Substantial dilution resulting in an albumin concentration below approximately 0.05% may result in inactivation of the drug.[3]

## Stability Reports of Compounded Preparations
### Enteral
Calhoun et al.[881] evaluated the stability of filgrastim 225 ng/mL and epoetin alfa 4400 microunits/mL in simulated amniotic fluid for orogastric or nasogastric administration to preterm neonates. The fluid contained sodium chloride 115 mEq/L, sodium acetate 17 mEq/L, potassium chloride 4 mEq/L, and human serum albumin 0.05%. The samples were packaged in polyvinyl chloride (PVC) intravenous infusion bags (Viaflex, Baxter) and were stored at room temperature for 24 hours, refrigerated for 24 hours, and frozen at −80°C for three weeks. In addition, samples were passed through PVC feeding tubes, and the delivered concentrations of the two drugs were measured. Filgrastim and epoetin alfa concentrations were assessed using enzyme-linked immunosorbent assay (ELISA).

No substantial loss of filgrastim was found after 24 hours under refrigeration or when frozen for three weeks. However, room temperature storage resulted in a steady loss of filgrastim, totaling about 50% loss in 24 hours. No significant loss of filgrastim occurred as a result of passing the drug through the PVC feeding tubes.

Epoetin alfa was found to be stable for 24 hours at room temperature and frozen for three weeks. Frozen storage for four weeks resulted in a significant loss of about 18%. Approximately 20% loss of epoetin alfa also occurred during priming of the feeding tube, but after the 10-mL priming volume, no further loss occurred during delivery.

The authors recommended storing the filgrastim–epoetin alfa mixture frozen at −80°C for a maximum of three weeks or refrigerated for up to 24 hours. Aliquoted frozen solution should be thawed and then refrigerated prior to use. It may be warmed to room temperature just before administration.

### Injection
Injections, like other sterile drugs, should be prepared in a suitable clean air environment using appropriate aseptic procedures. When prepared from nonsterile components, an appropriate and effective sterilization method must be employed.

Naughton et al.[898] reported that epoetin alfa (Amgen) 20,000 units/1 mL packaged in 1-mL hubless Medsaver (Becton Dickinson) plastic syringes and stored under refrigeration for six weeks underwent no loss of biological activity. ■

# ■ Erythromycin

## Properties
Erythromycin base is white to slightly yellow odorless hygroscopic crystals or crystalline powder with a bitter taste. It is a mixture of macrolide antibiotics containing not less than 850 mcg of erythromycin per milligram calculated on the anhydrous basis.[2–4,6]

## Solubility
Erythromycin base is slightly soluble in water,[3] having an aqueous solubility of about 2 mg/mL.[1] The base has a solubility of about 200 mg/mL in ethanol.[3]

## pH
A 0.7-mg/mL aqueous solution of erythromycin has a pH between 8 and 10.5.[3,4]

## pKa
The pKa of erythromycin has been variously cited as 8.6,[6] 8.8,[1] and 8.9.[2]

## General Stability Considerations
Erythromycin tablets and capsules are stored in tight containers at controlled room temperature and protected from temperatures exceeding 40°C. The tablets prepared from enteric-coated pellets should be stored at temperatures not exceeding 30°C. Erythromycin ophthalmic ointment and topical solution products also are stored at controlled room temperature, while the gel is stored at less than 27°C.[2,4]

Erythromycin base exhibits greatest aqueous stability at pH 7 to 7.5; it is more susceptible to decomposition in acidic solutions than in alkaline solutions. The solution pH is a primary determinant of stability, but it may also be affected by buffers and metal ions. Sodium buffers are better than potassium buffers; citrate and phosphate buffers are better than acetate. Calcium and magnesium ions have no effect on erythromycin degradation, but aluminum, iron, and copper ions greatly enhance the degradation rate. The powder and the solution at pH 4 and 8 are photostable.[6,1028]

Erythromycin base is more stable in water-soluble and oleaginous ointment bases; it is less stable in emulsion bases.[6]

## Stability Reports of Compounded Preparations
### Ophthalmic

Ophthalmic preparations, like other sterile drugs, should be prepared in a suitable clean air environment using appropriate aseptic procedures. When prepared from nonsterile components, an appropriate and effective sterilization method must be employed.

Bialer et al.[227] evaluated the stability of an extemporaneous and a commercial 0.5% erythromycin ophthalmic ointment during storage at room temperature and at elevated temperatures. The extemporaneous ointment was prepared by incorporating white petrolatum and mineral oil (79.5:19.9 by weight), which simulates the Ilotycin (Lilly) ointment base. Erythromycin gluceptate powder was dissolved in ethanol and diluted in phosphate-buffered saline (pH 7.5) to 50 mg/mL. This concentrate was stored at −70°C until use. Aliquots were removed and thawed for incorporation into the ointment base. A commercial 0.5% erythromycin ophthalmic ointment (Fougera) also was tested. Samples were stored at room temperature for 15 days and then for 24 hours at 37°C. Additional samples were stored at 45°C for six hours to simulate the temperatures in neonatal radiant warmers. Erythromycin bioactivity was assessed using a microbiological agar diffusion assay. No loss of erythromycin occurred after 15 days of storage at room temperature. Exposure to 37°C for 24 hours or 45°C for six hours resulted in only 3% loss in bioactivity.

### Topical

**STUDY 1:** Vandenbossche et al.[203] reported the stability of 1.5% erythromycin in six topical formulations. The formulations included two oil-in-water (o/w) emulsion bases composed of cetyl alcohol 15%, white beeswax 1%, propylene glycol 10%, sodium lauryl sulfate 2%, methylparaben 0.08%, propylparaben 0.02%, and deionized water 72% as prepared (pH 6.3) and with adjustment to pH 8.5 with sodium hydroxide. Two water-in-oil (w/o) emulsion bases were composed of white beeswax 8%, spermaceti 10%, cetiol V 60%, sorbitan monooleate 2%, methylparaben 0.08%, propylparaben 0.02%, and deionized water 20% with and without the same amount of sodium hydroxide as was used in the o/w emulsion. An alcoholic solution was tested that was composed of ethanol 40%, propylene glycol 20%, and

deionized water 40%. Finally, an alcoholic gel was prepared from the alcoholic solution by adding 2% hydroxyethyl-cellulose derivative (Idroramosan, Arion). Erythromycin 1.5% was incorporated into each base. The four emulsions were packaged in aluminum ointment tubes and stored at 4 and 25°C. The alcoholic solution and the gel were packaged in an amber glass bottle and in an aluminum ointment tube, respectively, and were stored at 25°C with continuous exposure to light. A microbiological agar-well diffusion technique was used to assess the antimicrobiological activity of the erythromycin in the samples.

In the o/w emulsions, the erythromycin content decreased to 70% in one week at 25°C; samples with the pH adjusted to 8.5 were even worse. At 4°C, 90% of the original activity at both pH values was retained during the first month.

In the w/o emulsions, the erythromycin content remained above 90% in the pH 6.3 sample for 12 weeks at 25°C, but losses of 35% in one week and 65% in two weeks occurred in the pH 8.5 sample. At 4°C, 90% of the original activity at both pH values was retained for 12 weeks.

The erythromycin content in the alcoholic solution and in the gel decreased to 90% in three weeks at 25°C.

**STUDY 2:** Stark and Cerise[843] reported that erythromycin was incompatible with the commercial Remederm cream due to the presence of stearate emulsifiers in the formulation. Incorporation of erythromycin powder into the Remederm cream resulted in separation after several days to a week.

**STUDY 3:** Vermeulen et al.[845] reported the stability of erythromycin and benzoyl peroxide in extemporaneously prepared topical gel formulations. The first gel was prepared using carbomer (Carbopol) 940 1 g, propylene glycol 7.5 g, disodium EDTA 25 mg, sodium hydroxide to adjust to pH 7, methyl- and propylparabens, and water. Erythromycin 1.5 g was dissolved in 90% ethanol 4.4 g and mixed into the carbomer 940 gel. Benzoyl peroxide was incorporated into the gel using an ointment mill. The second gel formulation was prepared by dispersing hydroxyethylcellulose 1 g in 17.7 g of water with methyl- and propylparabens as preservatives. Erythromycin 1.5 g was dissolved in 90% ethanol 17.7 g and then propylene glycol 8.8 g was added; the mixture was incorporated into the hydroxyethylcellulose gel, and benzoyl peroxide was incorporated into the gel using an ointment mill. These extemporaneous formulations were compared to commercial Benzamycin (Dermik).

Microbiological assays were performed to determine erythromycin activity; HPLC analysis was used to determine benzoyl peroxide content. The commercial Benzamycin was flocky and unhomogeneous. Microbiological assay found an erythromycin concentration of 140% of the labeled amount at the initial time point, with about 20% loss over three weeks under refrigeration at 6°C. Benzoyl peroxide was not tested. The carbomer 940 gel also became flocky and unhomogeneous due to precipitation of the erythromycin. Analysis found substantial loss of erythromycin; concentrations declined by 20 to 25% over three weeks under refrigeration at 6°C.

However, the hydroxyethylcellulose gel demonstrated much better stability. The much higher ethanol content was sufficient to avoid erythromycin precipitation. Less than 7% loss of erythromycin and no loss of benzoyl peroxide occurred over three weeks at 6°C.

### Rectal

Kassem et al.[427] evaluated the stability and release of erythromycin in several suppository formulations. Suppositories weighing 2 g and containing erythromycin 20 mg were prepared by fusion in bases composed of theobroma oil, Imhausen H-15, Imhausen E-45, macrogol 600 and 4000 (1:3), and Myrj 52. The suppositories were stored at 18 to 27°C for 15 months. A microbiological assay for antibiotic activity found that erythromycin potency in theobroma oil formulation was retained for only six months; in all other formulations erythromycin was stable for 15 months. The drug release properties were similar for all suppositories. The physical properties of most formulations were acceptable, but the H-15 formulation was the best from a hardness and disintegration standpoint. Theobroma oil suppositories were not sufficiently hard. Incorporation of any of several surfactants at 0.5%, including sodium lauryl sulfate, cetrimide, polysorbate 80, Span 80, and Myrj 52, did not affect erythromycin stability in the H-15 suppositories. ■

# ■ Erythromycin Ethylsuccinate

## Properties

Erythromycin ethylsuccinate is a white to slightly yellow odorless and tasteless hygroscopic crystalline powder.[2,3] It contains not less than 765 mcg of erythromycin per milligram calculated on the anhydrous basis.[4] Approximately 1.17 g of the ethylsuccinate is equivalent to 1 g of erythromycin.[3]

### Solubility

Erythromycin ethylsuccinate is very slightly soluble in water but freely soluble in ethanol and soluble in polyethylene glycol.[2,3]

### pH

A 1% aqueous suspension of erythromycin ethylsuccinate has a pH of 6 to 8.5; the oral suspension has a pH of 6.5 to 8.5, while the reconstituted dry oral suspension has a pH of 7 to 9.[4] The pH of reconstituted erythromycin ethylsuccinate for suspension (E.E.S. Granules, Abbott) was 7.6.[19]

### Osmolality

E.E.S. Suspension 200 mg/5 mL (Abbott) had an osmolality of 4475 mOsm/kg determined using a vapor pressure osmometer.[232]

## General Stability Considerations

Erythromycin ethylsuccinate tablets and powder for suspension should be packaged in tight containers and stored at controlled room temperature protected from temperatures exceeding 40°C. The commercial oral suspension should be stored under refrigeration until dispensed. The dispensed suspension of erythromycin ethylsuccinate is stable for 14 days stored at room temperature.[2]

## Stability Reports of Compounded Preparations
### Enteral
**ERYTHROMYCIN ETHYLSUCCINATE COMPATIBILITY SUMMARY**
*Compatible with:* Ensure • Ensure HN • Ensure Plus • Ensure Plus HN • Osmolite • Osmolite HN • Vital
*Incompatible with:* Precitene

**STUDY 1:** Cutie et al.[19] added 5 mL of erythromycin ethylsuccinate suspension (E.E.S. Granules, Abbott) to varying amounts (15 to 240 mL) of Ensure, Ensure Plus, and Osmolite (Ross Laboratories) with vigorous agitation to ensure thorough mixing. The erythromycin ethylsuccinate suspension was physically compatible, distributing uniformly in all three enteral products with no phase separation or granulation.

**STUDY 2:** Altman and Cutie[850] reported the physical compatibility of erythromycin ethylsuccinate granules (E.E.S., Abbott) 5 mL with varying amounts (15 to 240 mL) of Ensure HN, Ensure Plus HN, Osmolite HN, and Vital after vigorous agitation to ensure thorough mixing. The product was physically compatible, distributing uniformly in all four enteral products with no phase separation or granulation.

**STUDY 3:** Ortega de la Cruz et al.[1101] reported the physical incompatibility of an unspecified amount of oral liquid erythromycin ethylsuccinate (Pantomicina, Abbott) with 200 mL of Precitene (Novartis) enteral nutrition diet. Phase separation was observed in 24 hours. ■

# Esomeprazole Magnesium

## Properties

Esomeprazole is the S-isomer of omeprazole. Esomeprazole magnesium trihydrate is a white or nearly white crystalline powder. Commercial esomeprazole magnesium capsules (Nexium) provide 20 or 40 mg of esomeprazole (present as magnesium trihydrate 22.3 or 44.6 mg, respectively) in the form of encapsulated enteric-coated pellets.[7]

### Solubility

Esomeprazole magnesium is slightly soluble in water.[7]

### $pK_a$

Esomeprazole has a basic $pK_a$ of 3.97.[1325]

## General Stability Considerations

Esomeprazole magnesium stability is pH-dependent, rapidly degrading in acidic media, but the drug exhibits better stability at alkaline pH. When buffered to pH 6.8, the half-life ($t_{50}$) is about 19 hours at 25°C and about eight hours at 37°C.[7]

Commercial esomeprazole magnesium capsules (Nexium) should be stored at controlled room temperature near 25°C; the capsules should be dispensed in tight containers.[7]

## Stability Reports of Compounded Preparations

### Oral

Bladh et al.[1245] evaluated the stability of a new packet (sachet) formulation of esomeprazole mixed with water, apple juice, orange juice, and apple sauce at concentrations of 40 mg and 20 mg per 15 mL as well as 10 mg, 5 mg, and 2.5 mg per 5 mL. The stability of esomeprazole was also evaluated in water over a range of pH 3.4 to 5. The samples were stored at temperatures ranging from 5 to 37°C for 60 minutes. HPLC analysis found no difference in esomeprazole stability in any of the mixtures and under any of the storage conditions. The amount of degradation products was less than 0.1% in all samples. Dissolution occurred in about 2 minutes with water and about 9 to 10 minutes in apple or orange juice. Oral dose delivery was equal to or greater than 98%. Pharmacokinetic testing in volunteers after oral dosing found little or no difference in plasma concentration–time profiles.

### Enteral

**STUDY 1:** White et al.[790] evaluated the delivery of esomeprazole 40-mg capsule pellets (Nexium, AstraZeneca) suspended in tap water through a size 8 French polyurethane nasogastric tube, through a size 14 French standard nasogastric tube, and through a size 20 French silicone gastrostomy tube. Two methods were tested. In one method, the contents of one capsule were emptied into a 60-mL syringe with the plunger removed, 50 mL of water was added, and the plunger was replaced, leaving about 5 mL of air. The syringe was shaken for 15 seconds, and the mixture was administered over 30 seconds into each of the tubes. In the second method, only the 14 French nasogastric tube was used. The contents of one capsule were emptied into a 60-mL syringe with the plunger removed, and 25 mL of water was added. Then the plunger was replaced, leaving about 5 mL of air, and the syringe was shaken for about 15 seconds before about 10 mL was administered into the tubes. The syringe was then shaken vigorously for 15 seconds and the remaining 15 mL was administered. Finally, an additional 25 mL of water was added to the syringe and the procedure was repeated.

The first method—using all of the water at once—was much more effective, delivering 98% or more of the esomeprazole pellets through all of the tubes tested. The second two-step method delivered only about 78% of the pellets, leaving many pellets lodged in the syringe.

**STUDY 2:** Sostek et al.[1036] evaluated the bioavailability of esomeprazole pellets removed from capsules, mixed with water in a syringe, and administered using nasogastric tubes. The authors reported that the bioavailability of esomeprazole was similar to that of orally administered esomeprazole.

**STUDY 3:** Shah et al.[1029] reported the delivery of esomeprazole magnesium pellets mixed in various concentrations of Ora-Plus suspending medium through 60 size 14 French nasogastric tubes, narrower size 8 French nasogastric tubes, and shorter size 20 gastrostomy tubes. The esomeprazole pellets were suspended in tap water and Ora-Plus 30, 50, and 70%. The necessary volume of tap water was added to a 60-mL syringe without its plunger. The contents of one esomeprazole capsule were placed into the syringe, followed by the addition of the proper amount of Ora-Plus, bringing the mixture to a final volume of 25 mL, with some residual air to facilitate mixing. The plunger was replaced and the syringe was shaken to ensure dispersion of the pellets. The syringe was connected to the 60 size 14 French nasogastric tubes and the suspension was delivered through it; an additional 25 mL of tap water was added, and the mixture was shaken for 15 seconds and delivered.

There was no difference in the delivery of the esomeprazole pellets among the four mixtures, with the delivery exceeding 99%. Tap water and Ora-Plus 30% were then used for the other tubes. More than 99% of the pellets were delivered through the gastrostomy tubes with both mixtures. However, when water alone was used, about 91% of the pellets were delivered through the size 8 French nasogastric tubes, suggesting the possibility of clogging in tap water alone.

When the pellets were incorporated into the 30% Ora-Plus medium, delivery exceeded 99% with this narrower tube.

**STUDY 4:** Messaouik et al.[1031] reported the delivery of three proton pump inhibitor drugs through polyurethane and also silicone 16 French gastroduodenal tubes. The enteric-coated pellets from capsules of lansoprazole, omeprazole, and esomeprazole were removed from the capsules and mixed with water or apple juice. The mixture was delivered through the tubes and rinsed with 10 or 20 mL of the respective vehicle. No obstruction occurred with any of the esomeprazole samples, but 42% of the lansoprazole and omeprazole samples resulted in clogging. Esomeprazole delivery was 100%, while lansoprazole and omeprazole had reduced drug delivery of about 67 and 61%, respectively. The authors noted that the esomeprazole pellets were substantially smaller in size than those of the other two drugs. The authors concluded that esomeprazole was preferable for nasogastric delivery, compared to omeprazole or lansoprazole.

**STUDY 5:** Messaouik et al.[1089] reported on the compatibility of esomeprazole from microgranule tablets (AstraZeneca) delivered via 16 French silicone Levin type (Vygon) nasogastric tubes each day over 21 days. The delivery of the esomeprazole microgranules from the tablets was performed after dispersion in 50 mL of sterile water. Six individual nasogastric tubes were flushed with 10 mL of sterile water each, and then the syringes with the esomeprazole dispersion were shaken well and the dispersion was delivered into the tubes. After delivery of the drug, the nasogastric tubes were rinsed again with 10 mL of sterile water. The process was repeated each day for 21 days. Delivery of the esomeprazole was found to be complete, with HPLC analysis accounting for about 100% of the dose each day through all of the tubes.

The experiment was repeated with six additional nasogastric tubes (CAIR 9011N) but with the administration of 500 mL of Fresubin Original Fibre (Fresenius) every day by gravity feed. The enteral nutrition was stopped before each administration of the esomeprazole dispersion, and the tubes were rinsed with 10 mL of water. The esomeprazole dispersion was administered, the tube was rinsed again with 10 mL of water, and then the enteral mixture delivery was resumed. HPLC analysis again found esomeprazole delivery to be complete, accounting for about 100% of the dose each day through all six tubes.

The authors concluded that there was no significant difference in esomeprazole availability with concomitant enteral nutrition administration.

**STUDY 6:** Bladh et al.[1245] determined the deliverability of a new packet (sachet) formulation of esomeprazole mixed with water, apple juice, orange juice, and applesauce at

concentrations of 40 mg and 20 mg per 15 mL as well as 10 mg, 5 mg, and 2.5 mg per 5 mL through enteral tubes ranging from 6 to 20 Fr in diameter and from 21 to 127 cm in length. Esomeprazole was determined to be stable for at least 60 minutes after mixing in all of the vehicles. Dose delivery through the enteral tubes was equal to or greater than 96%. Pharmacokinetic testing in volunteers after oral dosing found little or no difference in plasma concentration–time profiles compared to oral dosing.

## Compatibility with Common Beverages and Foods

For patients who have difficulty swallowing esomeprazole magnesium capsules and other proton pump inhibitor drugs, the capsules have been opened and the enteric-coated pellet contents mixed with a slightly acidic food such as yogurt, applesauce, or fruit juice for administration. This approach is believed to help to preserve the drug's integrity until it reaches the alkaline environment of the duodenum.[1029–1035]

**STUDY 1:** Johnson et al.[789] reported the stability of esomeprazole 20-mg capsule pellets (Nexium, AstraZeneca) in 100 mL of the following common beverages and soft foods: apple juice, orange juice, tap water, milk (1.5% butterfat), cultured milk (3% butterfat), and yogurt. After 30 minutes, 500 mL of 0.1 M hydrochloric acid was mixed to simulate the physiologic environment of the stomach. After 20 minutes, the pellets were collected, mixed in pH 11 phosphate buffer, and processed for analysis.

HPLC analysis found that more than 98% of the esomeprazole was recovered from all samples except milk, which only yielded about 14%. This loss was attributed to the pH 6.7 environment of the milk. Apple juice, orange juice, cultured milk, and yogurt all provided acidic environments. Tap water had a pH of 7.9 but also had no buffer capacity. The enteric-coated esomeprazole pellets themselves reduced the pH when suspended in water.

**STUDY 2:** Bladh et al.[1245] evaluated the stability of a new packet (sachet) formulation of esomeprazole mixed with water, apple juice, orange juice, and applesauce at concentrations of 40 mg and 20 mg per 15 mL as well as 10 mg, 5 mg, and 2.5 mg per 5 mL. The stability of esomeprazole was also evaluated in water over a range of pH 3.4 to 5. The samples were stored at temperatures ranging from 5 to 37°C for 60 minutes. HPLC analysis found no difference in esomeprazole stability in any of the mixtures and under any of the storage conditions. The amount of degradation products was less than 0.1% in all samples. Dissolution occurred in about 2 minutes with water and about 9 to 10 minutes in apple or orange juice. Oral dose delivery was equal to or greater than 98%. Pharmacokinetic testing in volunteers after oral dosing found little or no difference in plasma concentration–time profiles. ■

# Estradiol
## (Oestradiol)

## Properties
Estradiol is a natural estrogen that occurs as white or creamy white small crystals or crystalline powder. It is odorless and hydroscopic.[2,3,615] Estradiol hemihydrate 1.03 mg is approximately equivalent to estradiol 1 mg.[3]

### Solubility
Estradiol is practically insoluble in water but has a solubility of 35.7 mg/mL in alcohol at 25°C.[2,3,615] It is sparingly soluble in vegetable oils.[3]

## General Stability Considerations
Estradiol products should be packaged in tight containers and stored at controlled room temperature protected from heat and light.[2-4] Ethinyl estradiol tablets should be stored in well-closed containers between 2 and 30°C.[2]

## Stability Reports of Compounded Preparations
### Oral
Allen[615] reported on an oral formulation of triple hormones in oil-filled capsules. The formula for 100 capsules is shown in Table 38. The powders were weighed accurately and mixed in a mortar. Drops of safflower oil were added to wet the powder thoroughly and form a paste. The remaining safflower oil was added with continued mixing. The mixture was chilled in a refrigerator for about one hour to increase the viscosity and

**TABLE 38.** Formula of Triple Hormones in Oil-Filled Capsules[615]

| Component | | Amount |
|---|---|---|
| Progesterone | | 4 g |
| Estradiol | | 50 mg |
| Testosterone | | 50 mg |
| Safflower oil | qs | 30 mL |

NOTE: Fill 0.3 mL/capsule.

reduce powder settling. A hand-operated capsule machine was loaded with 100 No. 1 empty hard-gelatin capsules, and the caps were removed. Then, 0.3 mL of the triple hormones in oil was transferred by micropipette into the base of each capsule. For clear capsules, additional safflower oil may be added to fill the capsules more completely. The capsule caps were then replaced and locked into place securely. Packaged in tight, light-resistant containers and stored under refrigeration, the capsules were physically stable for at least 60 days.

### Vaginal
**STUDY 1 (SUPPOSITORIES):** Henry et al.[1001] evaluated the stability of estradiol in polyethylene glycol vaginal suppositories. HPLC analysis found good uniformity of distribution of the estradiol in the suppositories. In addition, chemical stability of the estradiol was assessed over 90 days and was found to exhibit less than 10% loss during the storage period.

**STUDY 2 (SOLUTION):** Allen[1278] reported on a compounded formulation of estradiol 0.1% (1-mg/mL) for use as a vaginal solution. The solution had the following formula:

| | | |
|---|---|---|
| Estradiol | | 100 mg |
| Propylene glycol | | 24 mL |
| Pluronic P-105 | | 45 g |
| Purified water | qs | 100 mL |

The recommended method of preparation was to thoroughly mix the estradiol powder and propylene glycol and then add the Pluronic P-105. Finally, purified water sufficient to bring the volume to 100 mL was to be added and the solution mixed well. The solution was to be packaged in tight, light-resistant containers. The author recommended a beyond-use date of six months at room temperature because this formula is a commercial medication in some countries with an expiration date of two years or more. ■

# Ethacrynic Acid
# Sodium Ethacrynate

## Properties
Ethacrynic acid is a white or almost white odorless crystalline powder. It is irritating to eyes, mucous membranes, and skin.[2,3]

### Solubility
Ethacrynic acid is slightly soluble in water but has a solubility of about 625 mg/mL in ethanol. It will dissolve in dilute alkali hydroxide or carbonate in aqueous solution. Sodium ethacrynate has an aqueous solubility of about 70 to 90 mg/mL.[1-3]

### pH
Reconstituted sodium ethacrynate injection has a pH of 6.3 to 7.7.[4]

*pKₐ*

The pK_a of ethacrynic acid is 3.5.[1,2]

## General Stability Considerations

Ethacrynic acid tablets should be packaged in well-closed containers. Both the tablets and the injection should be stored at controlled room temperature and protected from temperatures exceeding 40°C.[2,4,7]

Aqueous solutions of the sodium salt are relatively stable at pH 7 at room temperature but are increasingly unstable as the pH and temperature increase. The reconstituted injection should be discarded after 24 hours.[2,3,7]

## Stability Reports of Compounded Preparations

### Oral

**STUDY 1:** Gupta et al.[62] reported the stability of ethacrynic acid prepared as a pediatric oral liquid dosage form. Ethacrynic acid powder (Merck Sharp & Dohme) was dissolved in ethanol and brought to volume with 50% sorbitol in water to yield a 1-mg/mL concentration with 10%

(vol/vol) ethanol. Methylparaben (0.005%) and propylparaben (0.002%) were added as preservatives. Sodium hydroxide 0.1 N was used to adjust the pH to 7. The product was packaged in amber bottles and stored at 24°C. HPLC analysis found that 96% of the ethacrynic acid remained after 220 days of storage.

**STUDY 2:** Ling and Gupta[706] reported the stability of three oral liquid forms of ethacrynic acid 2.5 mg/mL prepared from crushed tablets. The pulverized tablets were mixed with 0.05 M phosphate buffer; mannitol 10% and sucrose 10% were added to separate samples of the liquid mixture. The samples with and without the sugars were evaluated for stability stored at room temperature of 24 to 26°C and refrigerated at 4 to 6°C. No visible changes occurred in the appearance of the dosage forms, and stability-indicating HPLC analysis found that 4 and 10% loss of drug occurred in three and six days, respectively, at room temperature. The refrigerated samples lost about 8% in 24 days. There was little or no difference in stability among the three liquid forms. ∎

# ▪ Ethambutol Hydrochloride

## Properties

Ethambutol hydrochloride occurs as a white crystalline powder.[3,4,7]

### Solubility

Ethambutol hydrochloride is freely soluble in water, up to as much as 50%, and dimethyl sulfoxide and soluble in ethanol.[1,3,4,1310]

### pH

A 2% aqueous solution of ethambutol hydrochloride has a pH of 3.7 to 4.[3]

### pKₐ

Ethambutol has two apparent dissociation constants; pK_{a1} is 6.35 and pK_{a2} is 9.35.[1310]

## General Stability Considerations

Ethambutol hydrochloride powder and oral tablets should be packaged in tight containers and stored at controlled room temperature.[4,7]

## Compatibility with Common Beverages and Foods

Peloquin et al.[1135] evaluated the stability of several oral antituberculosis drugs, including ethambutol hydrochloride, for compatibility with common foods to facilitate administration to children. Ethambutol hydrochloride tablets were crushed and mixed with 30 g of Hunt's Snackpack no-sugar-added chocolate pudding, Safeway grape jelly, and Safeway creamy peanut butter. The drug–food mixtures were thoroughly mixed and analyzed (no method indicated) for drug content initially and after one, two, and four hours had elapsed. Drug recovery at all time points was ≥93% from the pudding, ≥89% from the jelly, and ≥80% from the peanut butter. The authors recommended the use of the no-sugar-added chocolate pudding and the jelly to aid in administering the drugs to children. ∎

# Ethanol

## Properties

Ethanol is a clear colorless mobile volatile flammable liquid. It has a pleasant characteristic odor and a burning taste.[1]

Dehydrated alcohol contains not less than 99.5% by volume of ethanol.[4]

Alcohol and dextrose injection is provided as a mixture containing ethanol 5% or ethanol 10% and dextrose 5% in water for injection.[4]

### Solubility

Ethanol is freely miscible with water and practically all organic solvents.[1,4]

### pH

Alcohol 5 or 10% and dextrose 5% injection has a pH range of 3.5 to 6.5.[4]

### Osmolality

Alcohol 5% and dextrose 5% injection is hypertonic, having a calculated osmolarity of 1125 mOsm/L. Alcohol 10% and dextrose 5% injection is hypertonic, having a calculated osmolarity of 1995 mOsm/L.[4]

## General Stability Considerations

Ethanol should be packaged in tight containers and stored at controlled room temperature protected from exposure to light.[4] Alcohol and dextrose injections are packaged in tight single-dose containers and stored at controlled room temperature and protected from freezing.[4,7] Dehydrated alcohol injection is packaged in tight single-dose containers and is stored in a cool place away from sources of heat.[7]

## Stability Reports of Compounded Preparations

### Injection

Injections, like other sterile drugs, should be prepared in a suitable clean air environment using appropriate aseptic procedures. When prepared from nonsterile components, an appropriate and effective sterilization method must be employed.

**STUDY 1:** Pomplun et al.[1146] evaluated the stability of ethanol 50% as a heparin-free lock solution for central venous catheters. Ethanol 95% was diluted with sterile water for injection to yield a 50% (vol/vol) concentration. This diluted ethanol was passed through a 0.2-μm Millex GV filter into a sterile container and then refiltered through another 0.2-μm Millex GV filter. The ethanol 50% solution was filled 3 mL into Becton Dickinson Luer Lok polypropylene syringes that were sealed with tip caps.

The filled syringes were stored at room temperature exposed to normal light for 28 days. Chromatography with flame-ionization detection was utilized to determine ethanol concentrations over 28 days. The ethanol concentration was found not to vary substantially during the study period, indicating that ethanol is stable over 28 days under these conditions.

**STUDY 2:** Cober and Johnson[1194] reported the stability of 70% ethanol packaged in 10-mL Becton Dickinson polypropylene syringes and in 12-mL Terumo polypropylene syringes. Dehydrated alcohol injection 98% was diluted with sterile water for injection or with bacteriostatic water for injection preserved with benzyl alcohol to yield the 70% concentration. The sample syringes were stored at 23 to 25°C for 14 days. No visible evidence of precipitation or color or volume change occurred. HPLC analysis found little or no loss of ethanol during the 14-day study period.

### Other

Kucmanic,[1431] Dubowski et al.,[1432] and Gullberg[1433] each conducted stability studies of diluted ethanol solutions in water for use in breath-alcohol testing. Ethanol was diluted to a theoretical concentration of 0.121 g/100 mL. Diluted solutions were packaged in both glass bottles and high-density polyethylene bottles with foam or foil opening seals. Samples were stored at room temperature and under refrigeration and evaluated using breath-alcohol simulators. Although some loss of ethanol had been expected during storage, in actuality little or no loss occurred over periods up to five years. In addition, Chow and Wigmore[1434] found no identifiable by-products of microorganisms when testing aqueous dilutions of ethanol packaged in polyethylene bottles that were stored for 26 years at room temperature. ▪

# ■ Ethionamide

## Properties

Ethionamide occurs as a minute bright yellow crystalline powder with a mild sulfide odor.[1,3,4,7]

### Solubility

Ethionamide is slightly soluble in water and sparingly soluble in ethanol and propylene glycol.[1,3,4,7] Ethionamide is freely soluble in pyridine.[1309]

### pH

Ethionamide 1% slurry in water has a pH between 6 and 7.[4]

## General Stability Considerations

Ethionamide powder and oral tablets should be packaged in tight containers and stored at controlled room temperature.[4,7]

## Compatibility with Common Beverages and Foods

Peloquin et al.[1135] evaluated the stability of several oral anti-tuberculosis drugs, including ethionamide, for compatibility with common foods to facilitate administration to children. Ethionamide tablets were crushed and mixed with 30 g of Hunt's Snackpack no-sugar-added chocolate pudding, Safeway grape jelly, and Safeway creamy peanut butter. The drug–food mixtures were thoroughly mixed and analyzed (no method indicated) for drug content initially and after one, two, and four hours had elapsed. Drug recovery at all time points was ≥93% from the pudding, ≥89% from the jelly, and ≥80% from the peanut butter. The authors recommended the use of no-sugar-added chocolate pudding or grape jelly to aid in administering the drugs to children. ■

# ■ Ethosuximide

## Properties

Ethosuximide occurs as a white to off-white crystalline powder or waxy solid with a characteristic odor.[3,4]

### Solubility

Ethosuximide is freely soluble in water and is very soluble in ethanol.[3,4]

### pH

Ethosuximide oral solution has a pH of 4.5 to 5.8.[4]

## General Stability Considerations

Ethosuximide powder, capsules, and oral solution should be packaged in tight containers and stored at controlled room temperature.[4]

## Stability Reports of Compounded Preparations
### Enteral

Ortega de la Cruz et al.[1101] reported the physical compatibility of an unspecified amount of oral liquid ethosuximide (Zarontin, Parke Davis) with 200 mL of Precitene (Novartis) enteral nutrition diet for a 24-hour observation period. No particle growth or phase separation was observed. ■

# ■ Ethylmorphine Hydrochloride
## (Codethyline)

## Properties

Ethylmorphine hydrochloride is a white to almost white or faint yellow crystalline powder.[1,3]

### Solubility

Ethylmorphine hydrochloride is soluble in water to about 100 mg/mL and in ethanol to about 40 mg/mL.[1,3]

### pH

A 2% aqueous solution of ethylmorphine hydrochloride has a pH of 4.3 to 5.7.[1,3]

### pKa

Ethylmorphine has a pK$_a$ of 7.9.[900]

## General Stability Considerations

Ethylmorphine hydrochloride should be protected from exposure to light.[3]

## Stability Reports of Compounded Preparations
### Oral

Feyel-Dubrokhotov et al.[1312] evaluated the stability of "Saint Lazare potion," an antidiarrheal oral liquid containing codeine and ethylmorphine hydrochloride (codethyline)

6.52 mg/mL in a vehicle containing unspecified amounts of absolute alcohol, simple syrup, and purified water. The preparation was packaged in clear and amber glass bottles and stored at controlled room temperature of 22 to 28°C for 26 weeks. HPLC analysis found little change in drug concentrations in either container during the study period. ■

# Etodolac

## Properties

Etodolac occurs as a white or nearly white crystalline powder.[3,620]

### Solubility

Etodolac is insoluble in water but freely soluble in ethanol and soluble in aqueous polyethylene glycol.[3,620]

### pKa

Etodolac has a pK$_a$ of 4.65.[7]

## General Stability Considerations

Etodolac drug substance, capsules, and oral tablets should be packaged in tight containers and stored at controlled room temperature.[4] The manufacturer states that etodolac tablets and capsules should be stored in the original containers at controlled room temperatures of 20 to 25°C. They should be dispensed in light-resistant containers. The extended-release capsules should also be stored at 20 to 25°C, and they should be protected from excessive heat and humidity.[7] Etodolac in aqueous solution was found to be stable under alkaline conditions but unstable under acidic conditions.

## Stability Reports of Compounded Preparations
### Rectal

**STUDY 1:** Allen[620] reported on an etodolac 200-mg suppository formulation prepared from powder. The suppository

mold should be calibrated for the suppository base to be used. Two bases were cited: (1) 75% polyethylene glycol 1000 with 25% polyethylene glycol 3350 and (2) Polybase, a commercially available polyethylene glycol base. Approximately 10% excess of the suppository mixture should be prepared to allow for loss. The contents of an appropriate number of etodolac 200-mg capsules were emptied into a mortar and pulverized to powder. The weighed base was melted in a suitable beaker at about 60 to 70°C, and the capsule powder was sprinkled on the melted base with stirring until the powder was thoroughly mixed. The mixture was then poured into the molds, leaving a slight excess to allow for contraction upon cooling. The finished suppositories should be packaged in glass containers, plastic bags, or cardboard sleeves. However, unwrapped polyethylene glycol-based suppositories should not be packaged in polystyrene prescription vials.

**STUDY 2:** Molina-Martinez et al.[621] compared the bioavailability of etodolac 200-mg suppositories administered rectally with etodolac 200-mg tablets administered orally to 10 healthy volunteers of both sexes. Plasma levels of etodolac were nearly identical with both routes of administration. The authors indicated that the two routes of administration were bioequivalent and the rectal route was an alternative route for administration of etodolac. ■

# Etoposide
## (VP-16)

## Properties

Etoposide is a white[2,3] to yellowish-brown crystalline powder.[2]

### Solubility

Etoposide has an aqueous solubility of approximately 0.03 mg/mL and a solubility in ethanol of approximately 0.76 mg/mL.[2]

Dilution of the concentrate for injection in aqueous solution results in temporary miscibility. The time to precipitation is related to drug concentration, although it tends to be unpredictable, as is true for many kinds of precipitation phenomena. At the recommended concentrations of 0.2 and 0.4 mg/mL,[7] in 0.9% sodium chloride injection and 5% dextrose injection,

Seargent et al.[359] found the time to precipitation of etoposide to be about 96 and 24 hours, respectively. However, precipitation may occur in shorter periods, especially in differing infusion solutions. Beijnen et al.[360] reported that, at 1 mg/mL in aqueous solution, precipitation occurs in as little as 5 minutes with agitation or in 30 minutes in a static solution. Some 1-mg/mL solutions, however, maintained extended solubility.

### pH

Etoposide concentrate for injection has a pH between 3 and 4 when diluted 1:9 with water.[4,7]

### pK$_a$

Etoposide has a pK$_a$ of 9.8.[1]

## General Stability Considerations

Etoposide concentrate for injection should be stored at 25°C. The capsules should be packaged in tight containers and stored under refrigeration protected from freezing.[2,4]

The maximum stability of etoposide has been reported to occur at pH 5[9] and at pH 5 to 6.15.[490] Decomposition occurs rapidly below pH 3 and above pH 8.[490] In acidic conditions, hydrolysis of the aglycone may occur. In alkaline solutions, epimerization to the *cis*-lactone may occur.[9]

## Stability Reports of Compounded Preparations
### Oral

The stability of etoposide solution for oral use was evaluated by McLeod and Relling.[111] Etoposide injection 20 mg/mL (VePesid, Bristol-Myers Oncology) was diluted to 10 mg/mL with 0.9% sodium chloride injection. One milliliter of the diluted etoposide was drawn into 5-mL plastic oral syringes (Burron Medical) and capped. Samples were stored at 22°C exposed to light and protected from light for 22 days. A stability-indicating HPLC assay was used to assess etoposide content. No loss of etoposide occurred at 22°C either exposed to or protected from light. Furthermore, there was no evidence of precipitation.

# Famotidine

## Properties

Famotidine is a white to yellowish-white or pale yellow odorless crystalline powder having a bitter taste.[2,3]

### Solubility

Famotidine has a solubility of 740 mcg/mL in water and a solubility of 360 mcg/mL in ethanol at 20°C.[2]

### pH

Famotidine injection has a pH of 5 to 5.6. The reconstituted powder for oral suspension has a pH of 6.5 to 7.5.[2]

### pK$_a$

Famotidine has a pK$_a$ of 7.1 at 25°C.[2]

## General Stability Considerations

Famotidine tablets should be packaged in well-closed, light-resistant containers and stored at 40°C or less. The powder for oral suspension should be packaged in tight containers and stored at 40°C or less. When reconstituted, the oral suspension should be stored at less than 30°C; although refrigeration is acceptable, the suspension should be protected from freezing.[2,4,7]

The injection should be stored under refrigeration and protected from freezing. If freezing does occur, the product should be thawed at room temperature or in a warm water bath or under running hot tap water.[2,7] Although refrigerated storage is recommended, Ross[361] reported that Merck indicated the drug was stable for up to 26 weeks at 25°C or less.

Famotidine in aqueous buffer solutions exhibited identical stability at pH 4 and 6, with essentially no loss over the seven-day study period. However, at pH 2 extensive and rapid loss of the drug occurred, with more than 60% loss in 24 hours.[940]

Ibrahim et al.[950] evaluated the compatibility of famotidine with some pharmaceutical excipients. Using differential scanning calorimetry (DSC), infrared spectroscopy, and x-ray diffraction, no interaction between famotidine occurred with Avicel PH 105, cross-linked sodium carboxymethylcellulose (Ac-Di-sol), Emcocel 90, Natrosol HEC, mannitol, starch 1500 (Sta-Rx), and sodium saccharin. However, interactions did occur between famotidine and Emdex and polyethylene glycol 6000.

## Stability Reports of Compounded Preparations

### Injection

Injections, like other sterile drugs, should be prepared in a suitable clean air environment using appropriate aseptic procedures. When prepared from nonsterile components, an appropriate and effective sterilization method must be employed.

**STUDY 1:** Bullock et al.[362] reported that, repackaged in plastic syringes, famotidine injection diluted to 2 mg/mL with sterile water for injection, 0.9% sodium chloride injection, or 5% dextrose injection was stable for 14 days when stored at 4°C.

**STUDY 2:** Shea and Souney[363] found that a 2-mg/mL dilution of famotidine injection in sterile water for injection, 0.9% sodium chloride injection, or 5% dextrose injection, when repackaged into plastic syringes and stored frozen at −20°C, retained 92 to 95% of the initial concentration after eight weeks of storage.

**STUDY 3:** Keyi et al.[651] evaluated the stability of famotidine 0.2 mg/mL in 5% dextrose and 0.9% sodium chloride injections repackaged in polypropylene syringes. The solutions remained clear and colorless, and famotidine levels by HPLC analysis remained within 95% of the initial concentrations after storage of 15 days at 22°C exposed to and protected from light.

### Oral

**STUDY 1:** Quercia et al.[117] reported the stability of an extemporaneously compounded famotidine oral liquid. Twelve famotidine 40-mg tablets (Merck Sharp & Dohme) were crushed and triturated with distilled water to form a paste. The paste was diluted with cherry syrup to 60 mL. The liquid 30 mL was packaged in one-ounce amber glass bottles; the bottles were stored at 4 and 24°C. A stability-indicating HPLC assay was used to determine famotidine content. About 5% famotidine loss occurred in 15 days at 24°C, and about 9% loss occurred in 20 days at 4°C.

**STUDY 2:** Dentinger et al.[715] reported the stability of an extemporaneously compounded famotidine 8-mg/mL oral suspension prepared from tablets at a favorable pH of 5.8. Famotidine 40-mg tablets (Merck) were triturated with sterile water for injection to form a paste and then diluted with equal amounts of Ora-Plus suspending vehicle and Ora-Sweet syrup (Paddock). The suspension was packaged in amber polyethylene terephthalate bottles and stored at room temperature of 23 to 25°C for 95 days. No change in color or odor occurred, and no visible microbial growth was observed. Stability-indicating HPLC analysis found that about 4% famotidine loss occurred in 95 days at room temperature. ▪

#  Fenbendazole
## (Flubendazole)

## Properties
Fenbendazole is a white or almost white powder.[3,4]

*Solubility*
Fenbendazole is practically insoluble in water.[3,4]

## General Stability Considerations
Fenbendazole should be packaged in well-closed light-resistant containers and should be stored at controlled room temperature.[3,4] Ragno et al.[1020] stated that the benzimidazole class of drugs, including fenbendazole, are known to be sensitive to light. The European Pharmacopoeia and the USP require protection from light during storage.[3,4]

## Stability Reports of Compounded Preparations
*Oral*
**STUDY 1:** Ragno et al.[1020] evaluated the stability of fenbendazole in the solid state as well as in liquid form at 20 mg/mL in an ethanol and water medium. Exposure of the solid drug to high temperatures up to 50°C and intense light from a Xenon arc lamp did not result in loss of the drug within 10 hours. Similarly, no loss due to thermal exposure resulted in the liquid form as well. However, exposure of the liquid form to intense light resulted in loss, with about 30% of the fenbendazole decomposing in 10 hours. This loss was similar to mebendazole and less extensive than albendazole. The authors indicated that they believed water must be present for the photodegradation to occur, because no loss occurred in the solid form.

**STUDY 2:** Arias et al.[1341] evaluated the effects of various factors including particle size and morphology, pH, and ionic strength on the sedimentation rate of fenbendazole as an aqueous suspension for veterinary use. The sedimentation of fenbendazole from aqueous suspension is mainly controlled by its hydrophobic nature along with the zeta potential of the drug particles. This also results in poor redispersibility of the drug. The pH of the medium slightly influences the sedimentation ratio and rate with a slightly higher rate as the pH is decreased from pH 9 to 3. Inclusion of aluminum chloride 0.01 M resulted in an improvement with suitable sedimentation and redispersibility of the fenbendazole. ▪

#  Fenoterol Hydrobromide

## Properties
Fenoterol hydrobromide is a white odorless bitter-tasting crystalline powder.[3,7]

*Solubility*
Fenoterol hydrobromide is soluble in water and ethanol.[3,7]

*pH*
A fenoterol hydrobromide 4% aqueous solution has a pH of 4.2 to 5.2.[3]

*Osmolality*
Fenoterol hydrobromide 0.1% inhalation solution (Berotec, Boehringer Ingelheim) has an osmolality of about 275 mOsm/kg, as measured by freezing-point depression.[928]

## General Stability Considerations
Fenoterol hydrobromide powder should be stored at controlled room temperature and protected from exposure to light.[3]

Fenoterol hydrobromide solution packaged in unopened amber glass containers is stored at controlled room temperature and is stable until the expiration date on the package. If the container is opened and the fenoterol hydrobromide solution is undiluted, the drug is stable for 30 days at controlled room temperature. Fenoterol hydrobromide solution diluted with preservative-free sodium chloride 0.9% is stated by the manufacturer to be stable for 24 hours at controlled room temperature. The manufacturer also states that the effects of refrigeration on the drug are not known.[7]

## Compatibility with Other Drugs
### FENOTEROL HYDROBROMIDE COMPATIBILITY SUMMARY

*Compatible with:* Acetylcysteine • Budesonide • Ipratropium hydrobromide

**STUDY 1 (IPRATROPIUM BROMIDE):** The manufacturer indicates that fenoterol hydrobromide 0.1% inhalation solution (Berotec, Boehringer Ingelheim) diluted with ipratropium bromide (Atrovent) in sodium chloride 0.9% preserved with benzalkonium chloride 0.01% is stable for seven days at controlled room temperature.[7]

**STUDY 2 (ACETYLCYSTEINE):** Lee et al.[928] evaluated the stability of acetylcysteine 20% mixed in equal volumes with fenoterol hydrobromide 0.625 mcg/mL in combination inhalation solutions. HPLC analysis of the solutions found acetylcysteine and fenoterol hydrobromide were stable for seven hours at room temperature of 25°C, exhibiting losses of about 7 and 6%, respectively.

**STUDY 3 (MULTIPLE DRUGS):** Gronberg et al.[1027] evaluated the stability and compatibility of budesonide (Pulmicort, concentration unspecified) with several inhalation solutions, including (1) acetylcysteine (Lysomucil) 100 mg/mL 2 parts to 3 parts budesonide, (2) fenoterol hydrobromide (Berotec) 5 mg/mL 8 parts to 1 part budesonide, (3) ipratropium bromide (Atrovent) 0.25 mg/mL 2 parts to 1 part budesonide, and (4) ipratropium bromide 0.125 mg/mL plus fenoterol hydrobromide 0.31 mg/mL (Duovent) in equal parts with budesonide. Samples were stored at room temperature of 22 to 25°C protected from exposure to light. HPLC analysis found all mixtures to be compatible, maintaining greater than 90% of the initial concentrations for at least 18 hours. ▪

# ▪ Fentanyl Citrate

## Properties
Fentanyl citrate is a white crystalline powder or granules having a bitter taste. Fentanyl citrate 157 mcg is approximately equivalent to 100 mcg of fentanyl.[1–3]

### Solubility
Fentanyl citrate has an aqueous solubility of about 25 mg/mL. It is slightly soluble in alcohol.[1,3]

### pH
Fentanyl citrate injection has a pH between 4 and 7.5.[4,7]

### pKₐ

### $pK_a$
The drug is variously reported to have $pK_a$ values at 7.3 and 8.4[7] and at 8.1.[128]

### Osmolality
The osmolality of fentanyl citrate injection was determined to be essentially 0 mOsm/kg.[345]

## General Stability Considerations
Fentanyl citrate injection should be protected from light and stored at controlled room temperature, although brief exposure to temperatures up to 40°C does not affect the product adversely.[2,4,7]

The transmucosal lozenges should be stored in their foil packages at controlled room temperature.[2,7]

Fentanyl citrate exhibits the greatest stability over pH of 3.5 to 7.[364]

## Stability Reports of Compounded Preparations
### Injections
Injections, like other sterile drugs, should be prepared in a suitable clean air environment using appropriate aseptic procedures. When prepared from nonsterile components, an appropriate and effective sterilization method must be employed.

**STUDY 1:** Wilhelm et al.[764] reported the stability of an extemporaneously compounded fentanyl citrate 0.25-mg/mL injection. The injection was prepared using the formula in Table 39 and was autoclaved at 121°C for 15 minutes. HPLC analysis found no degradation after one year of storage.

**STUDY 2 (THREE DRUGS):** Allen[128] reported the compounding of an epidural analgesic formulation. The preparation consisted of fentanyl citrate 1.25 mcg/mL, bupivacaine hydrochloride

**TABLE 39.** Fentanyl Citrate Injection Formula[764]

| Component | Amount |
|---|---|
| Fentanyl citrate | 393 mg |
| Sodium chloride | 9 g |
| Sodium hydroxide 1 mole/L | pH 6 to 6.5 |
| Water for injection | qs 1000 mL |

**TABLE 40.** Epidural Analgesic Formula[128]

| Component | Amount |
|---|---|
| Fentanyl citrate 50 mcg/mL | 2.5 mL |
| Bupivacaine hydrochloride 0.5% | 8.75 mL |
| Epinephrine hydrochloride 1:100,000 | 6.9 mL |
| 0.9% Sodium chloride injection, USP | 81.85 mL |

0.4375 mg/mL, and epinephrine hydrochloride 0.6875 mcg/mL in 0.9% sodium chloride injection, USP (Table 40). Packaged in light-resistant sterile containers, the preparation was physically compatible and chemically stable for 20 days stored under refrigeration at 3°C.

**STUDY 3 (THREE DRUGS):** Kjonniksen et al.[1226] reported the long-term stability of a similar three-drug analgesic mixture for epidural administration. The preparation contained fentanyl citrate 1.93 mcg/mL, bupivacaine hydrochloride 0.935 mg/mL, and epinephrine bitartrate 2.07 mcg/mL with sodium metabisulfite approximately 2 mcg/mL and disodium edetate approximately 0.2 mcg/mL in sodium chloride 0.9%. The formulation was prepared by first compounding a concentrate that was packaged in vials and autoclaved. This concentrate was then diluted 1:11 in 500 mL of sodium chloride 0.9% in polyolefin-lined Excel multilayer bags. The bags were stored refrigerated at 2 to 8°C for 180 days, followed by four days at room temperature of 22°C exposed to ambient laboratory illumination. No visible changes were reported. Stability-indicating HPLC analysis found that none of the drugs exhibited changes in concentration greater than 4%. The authors concluded that this drug combination was stable and could be given a shelf life of six months under refrigeration based on the chemical stability. ■

# ■ Ferrous Sulfate

## Properties
Ferrous sulfate occurs as odorless bluish-green crystals or granules or as a green crystalline powder having a salty astringent taste. Ferrous sulfate contains seven molecules of water of hydration and effloresces in dry air; in moist air it oxidizes to brown or brownish-yellow basic ferric sulfate, which is not used medicinally.[1–3] Ferrous sulfate 300 mg contains about 60 mg of iron.[3]

Dried ferrous sulfate has had part of the water of hydration removed by heating at 40°C. It is composed primarily of the monohydrate and contains 86 to 89% ferrous sulfate on the anhydrous basis. It occurs as a grayish-white to buff powder.[2–4]

### Solubility
Ferrous sulfate is freely soluble in water, having an aqueous solubility of about 1 g/1.5 mL; in boiling water the solubility is about 1 g/0.5 mL. It is practically insoluble in ethanol.[1,3]

Dried ferrous sulfate is slowly but completely soluble in water and practically insoluble in alcohol.[2,3]

### pH
A 5% aqueous solution of ferrous sulfate has a pH of 3 to 4.[3] Ferrous sulfate oral solution has a pH of 1.4 to 5.3.[4] Ferrous sulfate elixir (Feosol, Menley James) had a measured pH of 2.2.[19]

### Osmolality
Ferrous sulfate liquid 60 mg/mL (Roxane) was found to have an osmolality of 4700 mOsm/kg.[233]

## General Stability Considerations
Ferrous sulfate products should be stored in tight containers. Solutions should be protected from light. The oxidation rate is increased by the presence of alkali and exposure to light or heat.[1,3,4]

## Stability Reports of Compounded Preparations
### Enteral
**FERROUS SULFATE COMPATIBILITY SUMMARY**
*Compatible with:* Vivonex T.E.N.
*Incompatible with:* Enrich • Ensure • Ensure HN • Ensure Plus • Ensure Plus HN • Osmolite • Osmolite HN • Precitene • TwoCal HN • Vital

**STUDY 1:** Cutie et al.[19] added 5 mL of ferrous sulfate elixir (Feosol, Menley James) to varying amounts (15 to 240 mL) of Ensure, Ensure Plus, and Osmolite (Ross Laboratories) with vigorous agitation to ensure thorough mixing. The ferrous sulfate elixir was physically incompatible in all three enteral products, forming gels that clogged the feeding tubes. The authors noted that crushed iron tablets also may present problems unless they have been well diluted.

**STUDY 2:** Burns et al.[739] reported the physical compatibility of ferrous sulfate elixir (Feosol) 5 mL with 10 mL of three enteral formulas, including Enrich, TwoCal HN, and Vivonex T.E.N. Visual inspection found no physical incompatibility with Vivonex T.E.N. However, when mixed with the Enrich and TwoCal HN enteral products, thickening occurred and particulates formed.

**STUDY 3:** Altman and Cutie[850] reported the physical incompatibility of ferrous sulfate elixir (Feosol, Menley James) 5 mL with varying amounts (15 to 240 mL) of Ensure HN, Ensure Plus HN, Osmolite HN, and Vital after vigorous agitation to ensure thorough mixing. The ferrous sulfate elixir was physically incompatible with all four enteral products, forming gels that clogged the feeding tubes.

**STUDY 4:** Ortega de la Cruz et al.[1101] reported the physical incompatibility of an unspecified amount of oral liquid ferrous sulfate with vitamin B complex and C (Iberet, Abbott) with 200 mL of Precitene (Novartis) enteral nutrition diet. Particle growth and phase separation were both observed. ■

# Filgrastim

## Properties

Filgrastim, human granulocyte colony-stimulating factor (G-CSF), is a 175 amino acid protein having a molecular weight of 18,800 daltons produced by recombinant DNA technology. Filgrastim has a specific activity of $100 \pm 60$ million units per milligram of active substance.[7]

Filgrastim is formulated as a clear colorless liquid injection. The injections also contain sodium acetate, sorbitol, and polysorbate 80 (Tween 80), in water for injection.[7]

### pH

Filgrastim injections have a pH of 4.[2]

## General Stability Considerations

Filgrastim injections should be stored under refrigeration at 2 to 8°C and protected from freezing. Filgrastim may aggregate if frozen or subjected to temperatures exceeding 30°C. Shaking may result in foaming. The drug may be warmed to room temperature for no more than 24 hours. If left at room temperature for longer than 24 hours, the drug must be discarded.[2,7]

Filgrastim is stable at pH 3.8 to 4.2. At pH values near neutrality, the stability is more limited.[2]

Filgrastim is subject to sorptive loss to container surfaces at low concentrations. For filgrastim diluted 5 to 15 mcg/mL, human albumin 2 mg/mL (0.2%) must be added to the diluent prior to filgrastim. For filgrastim concentrations higher than 15 mcg/mL, the addition of human albumin is unnecessary. The manufacturer indicates that filgrastim should not be diluted to concentrations less than 5 mcg/mL,[7] although one source indicates that the drug is stable in a compatible solution containing 0.2% human albumin down to a concentration of 2 mcg/mL.[2]

Filgrastim should not be diluted with sodium chloride 0.9% (normal saline) due to possible precipitation.[7]

## Stability Reports of Compounded Preparations
### Enteral

Calhoun et al.[881] evaluated the stability of filgrastim 225 nanograms/mL and epoetin alfa 4400 microunits/mL in simulated amniotic fluid for orogastric or nasogastric administration to preterm neonates. The fluid contained sodium chloride 115 mEq/L, sodium acetate 17 mEq/L, potassium chloride 4 mEq/L, and human serum albumin 0.05%. The samples were packaged in polyvinyl chloride (PVC) intravenous infusion bags (Viaflex, Baxter) and were stored at room temperature for 24 hours, refrigerated for 24 hours, and frozen at −80 °C for three weeks. In addition, samples were passed through PVC feeding tubes and the delivered concentrations of the two drugs were measured. Filgrastim and epoetin alfa concentrations were assessed using enzyme-linked immunosorbent assay (Elisa).

No substantial loss of filgrastim was found after 24 hours under refrigeration or frozen for three weeks. However, room temperature storage resulted in a steady loss of filgrastim, totaling about 50% loss in 24 hours. No significant loss of filgrastim occurred as a result of passing the drug through the PVC feeding tubes.

Epoetin alfa was found to be stable for 24 hours at room temperature and frozen for three weeks. Frozen storage for four weeks resulted in a significant loss of about 18%. Approximately 20% loss of epoetin alfa also occurred during priming of the feeding tube, but after the 10-mL priming volume, no further loss occurred during delivery.

The authors recommended storing the filgrastim–epoetin alfa mixture frozen at −80°C for a maximum of three weeks or refrigerated for up to 24 hours. Aliquoted frozen solution should be thawed and then refrigerated prior to use. It may be warmed to room temperature just before administration.

### Injection

Injections, like other sterile drugs, should be prepared in a suitable clean air environment using appropriate aseptic procedures. When prepared from nonsterile components, an appropriate and effective sterilization method must be employed.

The manufacturer of filgrastim indicates that the drug repackaged into polypropylene plastic tuberculin syringes is stable for 24 hours at room temperature and for seven days refrigerated. However, because no preservative is present, use within 24 hours is recommended.[2] ■

# Flecainide Acetate

## Properties
Flecainide acetate is a white to off-white hygroscopic crystalline powder.[2,3,7]

### Solubility
Flecainide acetate has an aqueous solubility of 48.4 mg/mL and a solubility in ethanol of 300 mg/mL at 37°C.[2,7]

### $pK_a$
The $pK_a$ is 9.3.[2,7]

## General Stability Considerations
Flecainide acetate tablets should be packaged in well-closed, light-resistant containers and stored at controlled room temperature.[4,7]

## Stability Reports of Compounded Preparations
### Oral
**STUDY 1:** Flecainide toxicity was reported in an infant being given extemporaneously prepared flecainide acetate syrup 5 mg/mL that was stored at 4°C. Refrigeration caused flecainide crystallization; the crystals appeared as visible floaters. The syrup containing the concentrated flecainide crystals was given to the infant, who then exhibited toxicity.[35]

**STUDY 2:** Wiest et al.[110] evaluated the stability of flecainide acetate in an extemporaneously prepared oral suspension. Forty flecainide acetate 100-mg tablets (Tambocor, 3M Pharmaceuticals) were crushed to fine powder and triturated with a flavored oral diluent (Diluent for Oral Use, Roxane) to form a paste. Additional flavored Diluent for Oral Use containing 33% polyethylene glycol 8000 with 1% ethanol and 0.05% saccharin was added in divided portions to make 800 mL containing flecainide acetate 5 mg/mL. The suspension was packaged as 60 mL in two-ounce amber glass prescription bottles and stored under refrigeration at 5°C and at room temperature for 45 days. A stability-indicating HPLC assay was used to assess flecainide content. Less than 4% flecainide acetate loss occurred in 45 days both under refrigeration and at room temperature. The sediment that formed consisted of the tablet excipients, not precipitated flecainide.

**STUDY 3:** Allen and Erickson[544] evaluated the stability of three flecainide acetate 20-mg/mL oral suspensions extemporaneously compounded from tablets. Vehicles used in this study were (1) an equal parts mixture of Ora-Sweet and Ora-Plus (Paddock), (2) an equal parts mixture of Ora-Sweet SF and Ora-Plus (Paddock), and (3) cherry syrup (Robinson Laboratories) mixed 1:4 with simple syrup. Twenty-four flecainide acetate 100-mg tablets (3M Pharmaceuticals) were crushed and comminuted to fine powder using a mortar and pestle. About 20 mL of the test vehicle was added to the powder and mixed to yield a uniform paste. Additional vehicle was added geometrically, and the solution was brought to the final volume of 120 mL, mixing thoroughly after each addition. The process was repeated for each of the three test suspension vehicles. Samples of each finished suspension were packaged in 120-mL amber polyethylene terephthalate plastic prescription bottles and stored at 5 and 25°C in the dark.

No visual changes or changes in odor were detected during the study. Stability-indicating HPLC analysis found less than 3% flecainide acetate loss in any of the suspensions stored at either temperature after 60 days of storage. ■

# Floxuridine

## Properties
Floxuridine is a white or nearly white odorless powder.[2,7]

### Solubility
Floxuridine is freely soluble in water and soluble in ethanol.[2,7]

### pH
Reconstituted floxuridine injection has a pH of 4 to 5.5.[4]

## General Stability Considerations
The sterile powder should be stored at controlled room temperature and protected from light. The reconstituted solution should be stored under refrigeration and used within two weeks.[2,4,7]

## Stability Reports of Compounded Preparations
### Injection
Injections, like other sterile drugs, should be prepared in a suitable clean air environment using appropriate aseptic procedures. When prepared from nonsterile components, an appropriate and effective sterilization method must be employed.

Stiles et al.[529] evaluated the stability of floxuridine (Roche) 50 and 1 mg/mL in 0.9% sodium chloride injection. The solutions were packaged as 3 mL in 10-mL polypropylene infusion-pump syringes for the CADD-Micro and CADD-LD pumps (Pharmacia Deltec) sealed with Luer tip caps (Becton Dickinson) and stored at 30°C. HPLC analysis found 3% or less loss of the drug after 21 days of storage. ■

# Fluconazole

## Properties

Fluconazole is a white crystalline powder.[2,7]

### Solubility

Fluconazole has an aqueous solubility of 8 mg/mL[2] to 10 mg/mL.[248] In room temperature ethanol, fluconazole has a solubility of 25 mg/mL.[2]

### pH

Fluconazole injection in 0.9% sodium chloride has a pH of 4 to 8. In 5% dextrose injection, the pH range is 3.5 to 6.5.[2,7] Fluconazole oral suspension reconstituted with sterile water for irrigation to a 40-mg/mL concentration is reported to have a pH near 4.2.[1375]

### pKa

Fluconazole has a $pK_a$ of 1.76 at 24°C in 0.1 M sodium chloride.[2]

### Osmolarity

The injections are iso-osmotic, having an osmolarity of around 300 to 315 mOsm/L.[2,7]

## General Stability Considerations

Fluconazole tablets should be packaged in tight containers. Both the tablets and powder for oral suspension should be stored below 30°C. When reconstituted, the oral suspension should be stored between 5 and 30°C and protected from freezing. The reconstituted oral suspension is stable for 14 days and then should be discarded.[2,7] However, Dentinger and Swenson found that the oral suspension was actually stable for at least 70 days at room temperature.[1375] See the *Oral* section below.

Fluconazole injection should be stored between 5 and 30°C (glass bottles) or 5 and 25°C (plastic bags) and protected from freezing. Brief exposure of the injection in plastic bags to temperatures up to 40°C does not affect the product adversely.[2,7]

## Stability Reports of Compounded Preparations

### Oral

**STUDY 1:** Dentinger and Swenson[1375] reported the stability of commercial fluconazole (Diflucan, Pfizer) 40-mg/mL oral suspension in the original bottles and also packaged in amber oral syringes. The oral suspension was reconstituted with sterile water for irrigation yielding the 40-mg/mL concentration. Some of the oral suspension was retained in the original plastic Diflucan bottles while the balance was packaged as 3-mL samples in 3-mL amber polyethylene oral syringes sealed with high-density polyethylene syringe caps.

The samples were stored at room temperature of 22 to 25°C exposed to fluorescent light for 70 days.

No changes in color or visible microbial growth were observed in either of the containers. The pH of the suspension remained near the initial pH of 4.2 in all samples throughout the study. Stability-indicating HPLC analysis found no fluconazole loss occurred in 70 days in either container. The authors stated that reconstituted fluconazole oral suspension is stable in either the original bottles or amber polyethylene oral syringes for at least 70 days.

**STUDY 2:** The stability of fluconazole prepared as an oral liquid was assessed by Yamreudeewong et al.[125] Five fluconazole 500-mg tablets (Diflucan, Pfizer) were crushed to fine powder. The powder was mixed with 500 mL of deionized water to yield a theoretical concentration of 1 mg of fluconazole/mL. The suspension was packaged in 5-mL borosilicate glass vials with Teflon stoppers, and the vials were stored at 4°C in the dark, at 23°C, and at 45°C in a warming oven. A stability-indicating HPLC assay measured fluconazole concentrations over 15 days. No loss of fluconazole occurred at any temperature.

### Topical

Allen[599] described a topical formulation of fluconazole 1.6% in dimethyl sulfoxide (DMSO) for fungal infections of the nails. Fluconazole powder 1.6 g is dissolved in DMSO and brought to a final volume of 100 mL. An equivalent amount of fluconazole tablets crushed to fine powder could also be used, but such use would require filtering the product after the fluconazole goes into solution. The solution is packaged in a tight container with a glass applicator rod. The author indicated that a use period of six months is appropriate for this preparation, although no stability data were provided.

### Ophthalmic

Ophthalmic preparations, like other sterile drugs, should be prepared in a suitable clean air environment using appropriate aseptic procedures. When prepared from nonsterile components, an appropriate and effective sterilization method must be employed.

Allen[603] reported on a 0.2% fluconazole ophthalmic solution. Fluconazole powder 200 mg was weighed and dissolved in 95 mL of sterile 0.9% sodium chloride injection. Additional 0.9% sodium chloride injection was added to bring the total volume to 100 mL, and the solution was thoroughly mixed. The solution was filtered through a 0.2-μm filter into sterile ophthalmic containers. Aqueous solutions of fluconazole are stable for extended periods of time, but an expiration period of 30 days was recommended because no preservative is present. ∎

# Flucytosine

## Properties

Flucytosine is a white to off-white odorless crystalline powder.[1-3,7]

### Solubility

Flucytosine is sparingly soluble in water, having an aqueous solubility of about 1.5 g in 100 mL at 25°C. It is slightly soluble in ethanol.[1-3]

### pKₐ

Flucytosine has cited $pK_a$ values of 2.9[2] to 3.26[1] and 10.71.[2]

## General Stability Considerations

Flucytosine capsules should be stored in tight, light-resistant containers at controlled room temperature and protected from temperatures exceeding 40°C.[2,4]

It is recommended that a solution for intravenous administration be stored at 18 to 25°C; precipitation may occur at lower temperatures, while decomposition to fluorouracil may occur at higher temperatures.[3]

## Stability Reports of Compounded Preparations

### Oral

**USP OFFICIAL FORMULATION:**

> Flucytosine 1 g
> Vehicle for Oral Solution, NF (sugar-containing or sugar-free)
> Vehicle for Oral Suspension, NF (1:1) qs 100 mL

(See the vehicle monographs for information on the individual vehicles.)

Use flucytosine powder or the contents of commercial capsules. Add 20 mL of the vehicle mixture to the powder or capsule contents, and mix to make a uniform paste. Add additional vehicle almost to volume in increments with thorough mixing. Quantitatively transfer to a suitable calibrated tight and light-resistant bottle, bring to final volume with the vehicle mixture, and thoroughly mix yielding flucytosine 10-mg/mL oral suspension. The final liquid preparation should have a pH between 4 and 5. Store the preparation at controlled room temperature or under refrigeration between 2 and 8°C. The beyond-use date is 60 days from the date of compounding at either storage temperature.[4,5]

**STUDY 1:** The extemporaneous formulation of a flucytosine 10-mg/mL oral suspension was described by Nahata and Hipple.[160] The contents of two flucytosine 500-mg capsules were triturated with a small amount of purified water. The suspension was brought to 100 mL with purified water. A stability period of seven days under refrigeration was used, although chemical stability testing was not performed.

**STUDY 2:** Wintermeyer and Nahata[531] evaluated the stability of an extemporaneously compounded flucytosine 10-mg/mL oral solution prepared from capsules. The contents of 10 flucytosine (Roche) 500-mg capsules were triturated in a mortar and pestle with a small amount of distilled water. The mixture was transferred to a 500-mL volumetric flask, and the mortar was rinsed several times with a small amount of distilled water, with the rinses added to the flask. The solution was brought to volume with additional distilled water, yielding a flucytosine concentration of 10 mg/mL. The solution was packaged into 10 two-ounce glass and 10 two-ounce plastic amber prescription bottles. Samples of each were stored at 4 and 25°C. No changes were observed in the physical appearance of the solutions stored at 4°C. HPLC analysis of the refrigerated samples found less than 8% flucytosine loss after storage for 91 days. Although the samples stored at room temperature retained concentration for up to 70 days, they developed a flocculation after storage of 14 days. The source of the flocculation was not evaluated.

**STUDY 3:** Allen and Erickson[543] evaluated the stability of three flucytosine 10-mg/mL oral suspensions extemporaneously compounded from capsules. Vehicles used in this study were (1) an equal parts mixture of Ora-Sweet and Ora-Plus (Paddock), (2) an equal parts mixture of Ora-Sweet SF and Ora-Plus (Paddock), and (3) cherry syrup (Robinson Laboratories) mixed 1:4 with simple syrup. Four flucytosine 250-mg capsules (Roche) were crushed and comminuted to fine powder using a mortar and pestle. About 10 mL of the test vehicle was added to the powder and mixed to yield a uniform paste. Additional vehicle was added geometrically, and the solution was brought to the final volume of 100 mL, mixing thoroughly after each addition. The process was repeated for each of the three test suspension vehicles. Samples of each finished suspension were packaged in 120-mL amber polyethylene terephthalate plastic prescription bottles and stored at 5 and 25°C in the dark.

No visual changes or changes in odor were detected during the study. Stability-indicating HPLC analysis found less than 3% flucytosine loss in any of the suspensions stored at either temperature after 60 days of storage.

**STUDY 4:** VandenBussche et al.[719] reported the stability of two higher concentration flucytosine oral suspensions having a concentration of 50 mg/mL. The contents of flucytosine capsules (ICN Pharmaceuticals) were mixed with (1) an equal parts mixture of Ora-Plus suspending vehicle and strawberry syrup and (2) an equal parts mixture of Ora-Plus and Ora-Sweet SF (sugar-free) syrup (Paddock) to yield the 50-mg/mL suspensions. The suspensions were packaged in amber plastic prescription bottles and stored at room temperature of 23 to 25°C and refrigerated at 3 to 5°C for 90 days. No detectable change in color, odor, or flavor occurred, and no visible microbial growth was observed. Stability-indicating HPLC analysis found no flucytosine loss in either suspension at either temperature in 90 days.

**STUDY 5:** McPhail et al.[752] reported the stability of an extemporaneously prepared flucytosine 113-mg/mL oral suspension. The flucytosine 500-mg tablets (Alcobon, Roche) were pulverized and mixed into a suspending vehicle composed of methylcellulose 0.74%, propylene glycol 2%, ethyl and propyl esters of 4-hydroxybenzoic acid 1%, distilled water 60%, and simple syrup. The suspension was packaged in amber bottles and stored at ambient room temperature near 20°C. The suspension remained physically stable throughout the study. Based on HPLC analysis of flucytosine losses, the time to 10% loss ($t_{90}$) was calculated to be about 4.2 months. Bioavailability of the suspension in volunteers was similar to that of flucytosine tablets.

### Injection

Injections, like other sterile drugs, should be prepared in a suitable clean air environment using appropriate aseptic procedures. When prepared from nonsterile components, an appropriate and effective sterilization method must be employed.

Vermes et al.[1478] evaluated the stability of commercial Ancotil injection (Roche) consisting of flucytosine 10 mg/mL in sodium chloride 0.805% (for isotonicity). The injection for intravenous infusion was stored at elevated temperatures ranging from 40 to 90°C to assess the stability of intact flucytosine, and the formation of the decomposition product fluorouracil, which is toxic and teratogenic. Stability-indicating HPLC analysis found little loss of flucytosine and little formation of fluorouracil after heating at 40°C for 131 days. Decomposition with increased fluorouracil concentration did occur at the higher temperatures. The authors indicated that the drug was proven to be stable for several years at ambient temperature although the increasing amount of fluorouracil in the solution over time should remain a concern. ▪

## ▪ Fludrocortisone Acetate

### Properties

Fludrocortisone acetate is a synthetic glucocorticoid that occurs as white to pale yellow odorless hygroscopic crystals or crystalline powder.[2,3]

### Solubility

Fludrocortisone acetate is essentially insoluble in water, having a solubility of 0.04 mg/mL, and sparingly soluble in alcohol, with a solubility of 20 mg/mL.[1–3,586]

### General Stability Considerations

Fludrocortisone acetate tablets should be stored in well-closed containers at controlled room temperature and protected from excessive heat.[2,4,7]

A difference in fludrocortisone acetate tablet stability packaged in unit dose blister packaging has been reported. Houri et al.[1454] evaluated the stability of 0.55-mcg fludrocortisone acetate tablets (Farmabios) packaged in blister packs composed of polyvinyl chloride, polyvinylidene chloride, and aluminum (PVC/PVDC/AL) and also blister packs composed of polyamide/aluminum and polyvinyl chloride/aluminum (OPA/AL/PVC/AL). In the PVC/PVDC/AL blister packs, drug losses of over 5% and over 40% occurred in 24 months at 25°C and 60% relative humidity and 40°C and 75% relative humidity, respectively. In the OPA/AL/PVC/AL blister packs, no drug loss occurred at 25°C and 60% relative humidity and less than 5% loss occurred at 40°C and 75% relative humidity in 36 months.

### Stability Reports of Compounded Preparations
#### Oral

**STUDY 1:** Woods[586] reported on an extemporaneous oral suspension formulation of fludrocortisone acetate recommended by the manufacturer of Florinef tablets. The formula is shown in Table 41. The liquid suspension vehicle is prepared first. The tablets are crushed to powder, mixed with about 0.5 mL of polysorbate 80, and then incorporated into the suspension vehicle. Although a stability period of 44 days packaged in amber glass bottles was reported, the author suggested an expiration period of two weeks when stored under refrigeration.

**TABLE 41.** Formula of Fludrocortisone Acetate Oral Suspension Prepared from Tablets[586]

| Component | Amount |
|---|---|
| Fludrocortisone acetate 0.1-mg tablets | Number for desired concentration |
| Polysorbate 80 | 0.5 mL |
| Sodium carboxymethylcellulose | 1 g |
| Methylparaben | 150 mg |
| Citric acid monohydrate | 600 mg |
| Lemon spirit (or other flavor) | 0.25 mL |
| Water              qs | 100 mL |

**STUDY 2:** Cisternino et al.[787] evaluated the stability of fludrocortisone acetate oral solution prepared from tablets and from powder. One hundred fludrocortisone acetate 0.05-mg tablets were crushed using a mortar and pestle and triturated with 20 mL of ethanol 90%, making a paste. After 30 minutes, 80 mL of sterile water was added and mixed for 20 minutes, yielding a milky suspension. The suspension was filtered through filter paper, resulting in a clear solution. Drug recovery was about 80%, yielding 40 mcg/mL. A similar 40-mcg/mL solution was prepared from the pure drug powder. Ethanol concentration in both solutions was 17%. The solutions were packaged in amber glass prescription bottles and stored refrigerated at 2 to 6°C, at room temperature of 20 to 26°C, and at elevated temperature of 37 to 43°C.

The drug in solution prepared from tablets was less stable. Fludrocortisone acetate losses of 8% in 14 days under refrigeration and 7% in one day at room temperature were found. In the solution prepared from powder, losses of 3% in 60 days under refrigeration and 8% in seven days at room temperature were found. Losses after those time periods were unacceptable in both solutions. Unacceptable losses occurred rapidly at elevated temperature; 20 to 30% losses occurred in one day.

**STUDY 3:** Severino et al.[1253] evaluated the physical and chemical stability of an extemporaneously prepared oral suspension of fluconazole 10 mg/mL prepared from capsule contents simply suspended in water and compared to a commercial oral liquid preparation stored under refrigeration. The authors reported that the extemporaneously prepared oral suspension was physically "unstable," yielding erratic fluconazole concentrations over 15 days of study. Intestinal absorption proved to be erratic as well, possibly due to variability in drug concentration from inadequate suspension uniformity. ■

# Fluocinolone Acetonide Fluocinonide

## Properties

Fluocinolone acetonide is a fluorinated corticosteroid that occurs as a white or nearly white odorless crystalline powder.[2,3] Fluocinolone acetonide may be anhydrous or occur as the dihydrate.[2]

Fluocinonide is the 21-acetate ester of fluocinolone acetonide. It is a white to cream-colored nearly odorless crystalline powder.[2,3]

### Solubility

Fluocinolone acetonide is practically insoluble in water and has a solubility in ethanol of about 22 mg/mL.[2,3] It is sparingly soluble in propylene glycol.[2]

Fluocinonide is practically insoluble in water and slightly soluble in ethanol.[2,3]

## General Stability Considerations

Fluocinolone acetonide and fluocinonide products should be packaged in tight containers and stored at controlled room temperature protected from excessive heat and from freezing.[2,7]

## Stability Reports of Compounded Preparations
### Topical
**STUDY 1:** Allen[597] described a topical ointment for psoriasis. Each component in the formula shown in Table 42 was accurately weighed. The sulfur and salicylic acid powders were mixed and reduced to fine powder, and a few drops of glycerin were incorporated to make a paste. The paste was mixed with the fluocinonide ointment (Lidex, Medicis), and the mixture was incorporated into the Aquaphor or Aquabase geometrically with thorough mixing. The crude coal tar was added last and mixed well. The final ointment was packaged in tight containers and protected from light. The author indicated that a use period of six months is appropriate for this ointment, although no stability data were provided.

**STUDY 2:** Barnes et al.[665] evaluated the room temperature shelf life periods of fluocinonide 0.05% cream (Stuart) diluted 1:4 and 1:10 by spatulation with Unguentum Merck, Lipobase, and Metosyn diluent. Based on HPLC analysis, a calculated shelf life at 25°C of six months was recommended for the 1:4 dilution in all three bases and for the 1:10 dilution in Unguentum Merck. Due to more rapid loss, only six weeks could be recommended for the 1:10 dilution in Metosyn base.

**TABLE 42.** Psoriasis Topical Ointment Formula[597]

| Component | Amount |
| --- | --- |
| Precipitated sulfur | 3 g |
| Salicylic acid | 1 g |
| Glycerin | Several drops |
| Fluocinonide (Lidex) 0.05% ointment | 24 g |
| Aquaphor or Aquabase | 70 g |
| Coal tar | 2 g |

No recommendation could be made for the 1:10 dilution in Lipobase because of assay variability.

Barnes et al.[665] also reported that fluocinolone acetonide 0.025% diluted 1:4 by spatulation with Unguentum Merck and Lipobase had a calculated shelf life at 25°C of one and two months, respectively.

**STUDY 3:** Barnes et al.[766] reported the instability of fluocinonide 0.05% (Metosyn, Stuart Pharmaceuticals) and fluocinolone acetonide 0.025% (Synalar, ICI Pharmaceuticals) ointments diluted 1 in 10 with compound zinc paste, BP (25% zinc oxide) (Thornton & Ross), stored at controlled room temperature of 24 to 26°C. HPLC analysis found that the fluocinonide 0.05% dilution lost about 34% in seven days and about 66% in 14 days. The fluocinolone acetonide dilution was worse, degrading 66% in seven days.

**STUDY 4:** Ray-Johnson[775] reported the stability of fluocinolone acetonide ointment (Synalar, ICI Pharmaceuticals) diluted to a concentration of 0.013% (1:1 mixture) using Unguentum Merck, an ambiphilic ointment that combines the properties of both an oil-in-water and a water-in-oil emulsion. The diluted ointment was packaged in screw-cap jars and stored refrigerated at 4°C, at room temperature of 18 to 23°C, and at elevated temperature of 32°C. After storage for 32 weeks, the authors reported that HPLC analysis found no unacceptable drug loss at any of the temperatures.

**STUDY 5:** Krochmal et al.[776] reported the stability of fluocinonide 0.05% cream (Lidex) mixed individually with

(1) salicylic acid 2%, (2) liquor carbonis detergens 5%, (3) urea 10%, and (4) a combination of camphor 0.25%, menthol 0.25%, and phenol 0.25%. The topical mixtures were packaged in amber glass jars with polyethylene-lined screw caps and were stored for two months at unspecified ambient temperature.

No physical incompatibilities were observed. HPLC analysis found about 33% loss of fluocinonide in two months in the urea mixture. Acceptable fluocinonide concentrations were found in the other mixtures over two months of storage.

## Compatibility with Other Drugs

**STUDY 1:** Hecker et al.[815] evaluated the compatibility of tazarotene 0.05% gel combined in equal quantity with other topical products, including fluocinonide 0.05% cream and ointment. The mixtures were sealed and stored at 30°C for two weeks. HPLC analysis found less than 10% tazarotene loss in all samples. Fluocinonide also exhibited less than 10% loss in this time frame.

**STUDY 2:** Ohishi et al.[1714] evaluated the stability of fluocinolone acetonide (Flucort) ointment mixed in equal quantity (1:1) with zinc oxide ointment. Samples were stored refrigerated at 5°C. HPLC analysis of the ointment mixtures found that the drugs were stable for 32 weeks at 5°C. ■

# Fluorouracil

## Properties
Fluorouracil is a white to almost white odorless crystalline powder[2,3] that is subject to sublimation.[1]

### Solubility
Fluorouracil is sparingly soluble in water, with 1 g dissolving in about 70 mL. Because of salt formation, the solubility in water increases as the pH increases. The drug is slightly soluble in ethanol, with 1 g dissolving in about 170 mL. Fluorouracil is also slightly soluble in propylene glycol, with 1 g dissolving in 100 mL.[2,3,7]

### pH
A 1% fluorouracil solution has a pH of 4.5 to 5.[3] Fluorouracil injection is adjusted to approximately pH 9.2[7] (range 8.6 to 9.4).[4]

### pK_a
The $pK_a$ is 7.71 at 25°C.[7]

### Osmolality
Fluorouracil injection has an osmolality of around 650 mOsm/kg.[326]

## General Stability Considerations
Fluorouracil topical solution and cream should be packaged in tight containers. Fluorouracil products should be stored at controlled room temperature and protected from freezing and exposure to light.[2–4,7]

Fluorouracil injection is normally colorless to faintly yellow. A slight discoloration does not affect the product adversely.[2,7] However, darker discolorations to dark yellow, dark amber, or brown indicate more extensive decomposition from long-term exposure to high temperatures or intense light. Such products should be discarded.[8,365]

Fluorouracil injection may develop a precipitate if exposed to low temperatures. The precipitate is resolubilized by heating the injection to 60°C and shaking vigorously.[2,7]

Fluorouracil is sensitive to pH. It is subject to slow hydrolysis in alkaline solutions but is quite stable in acidic solutions.[7] However, at pH less than 8, its solubility is reduced and, depending on the drug concentration, precipitation occurs.[9,366] Stiles et al.[367] found fluorouracil precipitation of needlelike crystals forming in two to four hours at about pH 8.6 to 8.7 and immediately at about pH 8.5 or less. At pH values around 8.2 and below, clusters of crystals formed.

## Stability Reports of Compounded Preparations
### Injection
Injections, like other sterile drugs, should be prepared in a suitable clean air environment using appropriate aseptic procedures. When prepared from nonsterile components, an appropriate and effective sterilization method must be employed.

**STUDY 1 (DILUTED):** Stolk and Chandi[368] found no loss of fluorouracil by HPLC analysis when fluorouracil 5 mg/0.5 mL in 0.9% sodium chloride injection was packaged into polypropylene syringes and stored frozen at −20°C for eight weeks. Thawing and refreezing for two more weeks did not result in fluorouracil loss.

**STUDY 2 (REPACKAGED):** Similarly, the undiluted fluorouracil injection has been repackaged in plastic syringes and found to be stable for 28 days at 5 and 25°C.[9]

**STUDY 3 (REPACKAGED):** Stiles et al.[529] evaluated the stability of fluorouracil (Roche) 50 mg/mL undiluted. The solution was packaged as 3 mL in 10-mL polypropylene infusion-pump syringes for the CADD-Micro and CADD-LD pumps (Pharmacia Deltec) sealed with Luer tip caps (Becton Dickinson) and stored at 30°C. HPLC analysis found about 3% loss of the drug after 21 days of storage.

### Oral
Milovanovic and Nairn[70] reported the stability of fluorouracil (Hoffmann-La Roche) 500 mg/10 mL intended for oral administration repackaged in amber glass bottles with screw caps. Spectrophotometric analysis showed little change in absorbance after eight weeks of storage at 30°C. Stored at elevated temperature (50°C), samples became discolored, although the absorbance remained unchanged. Repeated opening and closing of the bottle at weekly intervals had no effect on absorbance. To use an eight-week expiration, the authors recommended that fluorouracil for oral use be protected from light, stored at room temperature, and packaged in tightly sealed containers to avoid absorption of carbon dioxide.

### Ophthalmic
Ophthalmic preparations, like other sterile drugs, should be prepared in a suitable clean air environment using appropriate aseptic procedures. When prepared from nonsterile components, an appropriate and effective sterilization method must be employed.

**STUDY 1:** Fuhrman et al.[732] reported the stability of fluorouracil 10-mg/mL ophthalmic solution prepared from the injection (Pharmacia) diluted in sodium chloride 0.9%. The 10-mg/mL dilution was packaged in polypropylene tuberculin syringes (Becton Dickinson) and was stored at −10, 5, 25, and 40°C for seven days. Visual examination found no precipitation, changes in turbidity, or color change in any sample. Stability-indicating HPLC analysis found about 3 to 5% loss of fluorouracil in all samples. Given the range of storage conditions, it is likely that this is a result of analytical variation rather than decomposition. The authors indicated that the ophthalmic solution was stable for the seven-day test period at all temperatures.

**STUDY 2:** Selva Otaolaurruchi et al.[1543] evaluated the formulation and stability of fluorouracil eyedrops prepared from commercial injection. The eyedrops were prepared by adding the injection into Alcon artificial tears to yield a 0.5% (5 mg/mL) concentration. The eyedrops were stored refrigerated at 8°C without protection from exposure to light. The pH and osmolality of the eyedrops were about pH 9.3 and 310 mOsm/kg and did not change substantially. Ultraviolet spectrographic analysis found the fluorouracil concentration also did not change during six months of storage. ▦

# ▦ Fluoxetine Hydrochloride

## Properties
Fluoxetine hydrochloride is a white to off-white crystalline powder.[1,7]

### Solubility
Fluoxetine hydrochloride has an aqueous solubility of 14 mg/mL.[1,7] In ethanol it has a solubility that is more than 100 mg/mL.[1]

### pH
A 1% fluoxetine hydrochloride solution has a pH of 4.5 to 6.5.[3]

## General Stability Considerations
Fluoxetine hydrochloride capsules and oral solution should be packaged in tight, light-resistant containers. The products should be stored at controlled room temperature.[2,4,7]

## Stability Reports of Compounded Preparations
*Oral*
**STUDY 1:** Peterson et al.[126] reported the stability of fluoxetine hydrochloride solution diluted with five common diluents. Fluoxetine hydrochloride solution (Lilly) was diluted 1:3 and 1:1 to yield concentrations of 1 and 2 mg/mL, respectively, with simple syrup, USP; simple syrup, BP; aromatic elixir, USP; grape-cranberry drink (Cran-Grape, Ocean Spray); and deionized water. The diluted fluoxetine hydrochloride solutions were packaged in 120-mL amber glass bottles and stored at 5 and 30°C for eight weeks. A stability-indicating HPLC assay was used to determine fluoxetine hydrochloride. No test mixture exhibited a concentration less than 98%. Furthermore, degradation product levels did not exceed 0.5%. The diluted fluoxetine hydrochloride solutions were stable for eight weeks at 5 and 30°C.

**STUDY 2:** Geller et al.[242] reported that fluoxetine liquid was visually compatible with simple syrup, purified water, aromatic elixir, and cranberry-grape juice (Ocean Spray).

**STUDY 3:** In a related study, Marshall and Mullen[1178] reported the stability of commercial fluoxetine hydrochloride liquid (Lilly) diluted to concentrations of 5 and 10 mg/mL with simple syrup, USP; simple syrup, BP; aromatic elixir, USP; grape-cranberry drink; and purified water, USP. The diluted fluoxetine hydrochloride solutions were packaged in four-ounce amber polyethylene terephthalate plastic bottles (Owens-Brockway) capped with child-resistant caps and stored at 4 and 30°C for 60 days. No change in odor or visible evidence of color change, cloudiness, or precipitation occurred. A stability-indicating HPLC assay was used to determine fluoxetine hydrochloride. No test mixture exhibited a concentration less than 95%. Furthermore, degradation product levels did not exceed 0.5%. The diluted fluoxetine hydrochloride solutions were stable for 60 days at 4 and 30°C. ■

# ■ Fluphenazine Hydrochloride

## Properties
Fluphenazine hydrochloride is a white or almost white, odorless crystalline powder.[2,3]

### Solubility
Fluphenazine hydrochloride is freely soluble in water, having a solubility of about 1 g/1.4 mL. In ethanol it has a solubility of about 1 g/6.7 mL.[2,3]

### pH
A 5% fluphenazine hydrochloride solution in water has a pH of 1.9 to 2.3. The injection has a pH of 4.8 to 5.2. The oral solution has a pH of 4 to 5, while the pH of the elixir is 5.3 to 5.8.[3,4]

### Osmolality
Fluphenazine hydrochloride elixir 0.5 mg/mL (Squibb) was found to have an osmolality of 1750 mOsm/kg.[233]

## General Stability Considerations
Fluphenazine hydrochloride products should be packaged in tight, light-resistant containers. The products should be stored at controlled room temperature and protected from excessive heat. The liquid products should be protected from freezing.[2,4,7]

The injection may vary in color from colorless to light amber. The drug concentration or efficacy has not been affected when a slight yellowish discoloration exists. However, precipitation, darker solutions, or other discolorations are indications not to use the product.[2,7]

## Compatibility with Other Drugs
**FLUPHENAZINE HYDROCHLORIDE COMPATIBILITY SUMMARY**
*Uncertain or variable compatibility with:* Lithium citrate
• Thioridazine

**STUDY 1:** Theesen et al.[78] evaluated the compatibility of lithium citrate syrup with oral neuroleptic solutions, including fluphenazine hydrochloride concentrate. Lithium citrate syrup (Lithionate-S, Rowell) and lithium citrate syrup (Roxane) at a lithium-ion concentration of 1.6 mEq/mL were combined in volumes of 5 and 10 mL with 10, 20, 40, and 60 mL of fluphenazine hydrochloride oral concentrate 0.5 mg/mL (Squibb) in duplicate, with the order of mixing reversed between samples. Samples were stored at 4 and 25°C for six hours and evaluated for physical compatibility by visual inspection. No incompatibility was observed.

**STUDY 2:** Raleigh[454] evaluated the compatibility of lithium citrate syrup (Philips Roxane) 1.6 mEq lithium/mL with fluphenazine hydrochloride concentrate 5 mg/mL and with fluphenazine hydrochloride elixir 0.5 mg/mL. The fluphenazine hydrochloride products were combined with the lithium citrate syrup in a manner described as "simple random dosage combinations"; the exact amounts tested and test conditions were not specified. When the fluphenazine hydrochloride concentrate was mixed with the lithium citrate syrup, a milky precipitate formed. However, combining the fluphenazine hydrochloride elixir with the lithium citrate syrup did not result in the formation of an observed precipitate.

Raleigh[454] also evaluated the compatibility of thioridazine suspension (Sandoz) (concentration unspecified) with fluphenazine hydrochloride concentrate 5 mg/mL and with fluphenazine hydrochloride elixir 0.5 mg/mL. The fluphenazine hydrochloride products were combined with the lithium citrate elixir and separately with the thioridazine suspension in the manner described as "simple random dosage combinations"; the exact amounts tested and test conditions were not specified. When the fluphenazine hydrochloride concentrate was mixed with the thioridazine suspension, a dense white curdy precipitate formed. However, combining the fluphenazine hydrochloride elixir with the thioridazine suspension did not result in the formation of an observable precipitate.

## Compatibility with Common Beverages and Foods

**FLUPHENAZINE HYDROCHLORIDE COMPATIBILITY SUMMARY**

*Compatible with:* Apricot juice • Grapefruit juice • Milk • Orange juice • Pineapple juice • Prune juice • Saline

solution • Sodas (noncola; noncaffeine) • Tomato juice • V8 vegetable juice • Water

*Incompatible with:* Apple juice • Coffee • Cola sodas • Tea

**STUDY 1:** Geller et al.[242] reported that fluphenazine liquid was visually compatible with water, orange juice, pineapple juice, prune juice, tomato juice, grapefruit juice, apricot juice, vegetable juice (V8), and noncaffeinated noncola sodas, although visual observation in opaque liquids is problematic. Also, fluphenazine was stated[242] to be incompatible with coffee, tea, cola sodas, and apple juice.

**STUDY 2:** In addition to the compatible beverages noted by Geller et al.,[242] Kerr[256] cited the compatibility of fluphenazine with milk and saline solution.

**STUDY 3:** Kulhanek et al.[243] combined equal quantities of fluphenazine 4-mg/mL oral drops with coffee and tea. Flaky precipitation was seen. Analysis found that the precipitate contained fluphenazine. ▪

# ■ Flurandrenolide
## (Flurandrenolone)

## Properties
Flurandrenolide is a white to off-white, fluffy, odorless crystalline powder.[1,3,4,7]

### Solubility
Flurandrenolide is practically insoluble in water and ether, sparingly soluble in ethanol to 1 g in 72 mL, but soluble in methanol to 1 g in 25 mL.[3,4,7,1276]

### pH
Flurandrenolide lotion has a pH between 3.5 and 6.[4,7]

## General Stability Considerations
Flurandrenolide powder should be packaged in tight containers and stored under refrigeration at 2 to 8°C and be protected from exposure to light. Similarly, flurandrenolide cream, lotion, and ointment should be packaged in tight containers and stored under refrigeration at 2 to 8°C and be protected from exposure to light. Flurandrenolide tape should be stored at controlled room temperature.[4,7]

## Stability Reports of Compounded Preparations
*Topical*
Allen[1276] reported on a compounded formulation of flurandrenolide 1-mg/mL for use as a topical film. The preparation had the following formula:

| | | |
|---|---|---|
| Flurandrenolide | | 100 mg |
| Polyvinyl alcohol | | 9 g |
| Povidone | | 11 g |
| Glycerin | | 9 g |
| Ethanol | | 10 g |
| Benzyl alcohol | | 2 g |
| Propylene glycol | | 3 g |
| Disodium edetate | | 20 mg |
| Citric acid | | 100 mg |
| Purified water | qs | 100 mL |

The recommended method of preparation was to mix thoroughly the flurandrenolide powder and propylene glycol, glycerin, and ethanol to form a solution. The polyvinyl alcohol, povidone, benzyl alcohol, disodium edetate, and citric acid were then to be dissolved in 50 mL of purified water, and the flurandrenolide solution was then added with thorough mixing. Finally, purified water sufficient to bring the volume to 100 mL was to be added. The topical film solution was to be packaged in tight, light-resistant containers. The author recommended a beyond-use date of six months at room temperature because this formula is a commercial medication in some countries with an expiration date of two years or more. ■

# Fluticasone Propionate

## Properties

Fluticasone propionate occurs as a white or off-white crystalline powder.[1,3,7] The micronized form is a fine white powder.[4]

### Solubility

Fluticasone propionate is practically insoluble in water, slightly soluble in ethanol and methanol, and freely soluble in dimethyl sulfoxide and dimethylformamide.[1,3,7]

### pH

Fluticasone propionate nasal spray has a pH between 5 and 7.[7]

## General Stability Considerations

Fluticasone propionate should be packaged in tight, light-resistant containers.[4] It should be stored at controlled room temperature not exceeding 30°C[4] and protected from exposure to light.[3] Fluticasone propionate nasal spray should be stored between 4 and 30°C. Fluticasone propionate topical cream and ointment are stored between 2 and 30°C.[7]

Fluticasone propionate appears to be incompatible with polystyrene plastic containers. When fluticasone-17-propionate suspension was stored in polystyrene plastic containers, the drug concentration fell by about 10% and the containers became opaque.[1231]

## Compatibility with Other Drugs

### FLUTICASONE PROPIONATE COMPATIBILITY SUMMARY

*Compatible with:* Albuterol sulfate • Ipratropium bromide • Tobramycin • Tobramycin sulfate
*Incompatible with:* Dornase alfa

**STUDY 1 (ALBUTEROL AND IPRATROPIUM):** Kamin et al.[1231] evaluated the physicochemical compatibility and stability of fluticasone-17-propionate nebulizer suspension (Flutide forte "ready-to-use," GlaxoSmithKline) mixed with ipratropium bromide (Atrovent LS, Boehringer Ingelheim) and albuterol sulfate (Sultanol, GlaxoSmithKline) for combined use in a nebulizer. The test mixtures were prepared in 10-mL glass containers. Fluticasone-17-propionate 2 mg/2 mL was mixed with ipratropium bromide 0.25 mg/2 mL and albuterol 2.5 mg/0.5 mL as the sulfate. No physical changes were observed during the five-hour study period. Stability-indicating HPLC analysis of the drug concentrations found the drug concentrations remained near 100% over five hours as well.

**STUDY 2 (DORNASE ALFA AND TOBRAMYCIN):** Kramer et al.[1383] evaluated the compatibility and stability of a variety of two- and three-drug solution admixtures of inhalation drugs, including fluticasone propionate. The mixtures were evaluated using chemical assays including HPLC, DNase activity assay, and fluorescence immunoassay as well as visual inspection, pH measurement, and osmolality determination. The drug combinations tested and the compatibility results that were reported are shown below.

*Mixture 1*
Dornase alfa (Pulmozyme) 2500 units/2.5 mL
Fluticasone propionate (Flutide forte) 2 mg/2 mL
   Result: Chemically incompatible

*Mixture 2*
Fluticasone propionate (Flutide forte) 2 mg/2 mL
Tobramycin (Tobi) 300 mg/5 mL
   Result: Physically and chemically compatible

*Mixture 3*
Fluticasone propionate (Flutide forte) 2 mg/2 mL
Tobramycin sulfate (Gernebcin) 80 mg/2 mL
   Result: Physically and chemically compatible ■

# Folic Acid

## Properties

Folic acid is a yellow or yellowish-orange to orangish-brown odorless crystalline powder.[2,3]

### Solubility

Folic acid is very slightly soluble in water (about 0.0016 mg/mL) and insoluble in ethanol. It is soluble to about 1% in boiling water. Folic acid is readily soluble in dilute alkali hydroxides and carbonates.[1–3] Folic acid is soluble to about 1 mg/mL in aqueous solutions having a pH of 5.6 or above at room temperature. Below pH 4.5 to 5, folic acid may precipitate from solution in varying times.[369]

### pH

Folic acid injection, prepared with the aid of sodium hydroxide to form the sodium salt, has a pH of 8 to 11.[2–4] An aqueous suspension of 1 g/10 mL has a pH of 4 to 4.8; an aqueous solution in sodium bicarbonate has a pH of 6.5 to 6.8.[1]

### Osmolality

Folic acid 5-mg/mL injection has an osmolality of 186 mOsm/kg.[326]

## General Stability Considerations

Folic acid solutions are sensitive to heat and light and should be stored protected from light and excessive heat. Folic acid tablets should be packaged in well-closed containers and stored protected from light.[2–4] Folic acid solutions exhibit maximum stability at pH 7.6.[793] Solution pH may play a role in the rate of photodegradation of folic acid. The rate of photodegradation proceeds more rapidly in acidic media than in alkaline media.[793,794]

## Stability Reports of Compounded Preparations
### Oral

The stability of an extemporaneous formulation of folic acid 1-mg/mL oral solution was reported by Smith.[170] Folic acid powder was dissolved in distilled water adjusted to pH 8 to 8.5 with sodium hydroxide and preserved with hydroxybenzoates. The solution was evaluated both with and without 0.1% ascorbic acid or 0.1% sodium sulfite. The solutions were packaged as 150 mL in 300-mL brown glass bottles and stored at 4°C and at room temperature. Folic acid content was stable in all formulations at both temperatures for up to two months, although the author recommended a six-week shelf life. However, Woods[428] thought a 30-day shelf life was more reasonable. ■

# Formaldehyde
## (Formalin)

## Properties

Formaldehyde solution is a clear colorless or nearly colorless liquid with a characteristic irritating pungent odor. Formaldehyde solution or formalin is composed of 36.5 to 37% formaldehyde and contains 10 to 15% methanol as a stabilizing agent to delay or prevent polymerization.[1,3,4]

### Solubility

Formaldehyde solution is miscible with water and ethanol.[1]

### pH

Formaldehyde solution has a pH of 2.8 to 4.[1]

## General Stability Considerations

Formaldehyde solution should be packaged in tight containers and stored between 15 and 25°C.[3,4] At low temperatures, it may become cloudy or develop a precipitate of paraformaldehyde.[1,3]

## Stability Reports of Compounded Preparations
### Bladder Irrigation

Sterile bladder irrigations, like other sterile drugs, should be prepared in a suitable clean air environment using appropriate aseptic procedures. When prepared from nonsterile components, an appropriate and effective sterilization method must be employed.

Pesko[419] described the compounding of a sterile 1% formalin solution bladder irrigation solution. The irrigation is prepared by removing 10 mL from a 1-liter container of sterile water for irrigation. Formaldehyde solution, USP, 10 mL is filtered through a 0.22-μm filter into the remaining sterile water for irrigation. [NOTE: Formaldehyde solution, USP, contains 36.5 to 37% formaldehyde. Consequently, a 1% formalin irrigation solution contains 0.37% formaldehyde.] Formaldehyde solutions should be stored at 15 to 25°C in tight containers. The author recommended a 24-hour expiration, although no specific stability information is available. ■

# Formoterol Fumarate

## Properties

Formoterol fumarate occurs as a white, almost white, or yellowish crystalline powder.[3,4,7] Formoterol fumarate 10-mcg/mL inhalation solution (Perforomist, Dey) is a clear, colorless solution with sodium chloride for tonicity and citric acid and sodium citrate as buffers.[7]

### Solubility

Formoterol fumarate is slightly soluble in water and sparingly soluble in ethanol and isopropanol.[3,4,7] It is freely soluble in dimethylsulfoxide.[4]

### pH

Formoterol fumarate as a 0.1% aqueous solution has a pH of 5.5 to 6.5.[3,4] Formoterol fumarate 10-mcg/mL inhalation solution has a pH of about pH 5.[7,1330]

### Osmolality

Formoterol fumarate 10-mcg/mL inhalation solution is isotonic[7] having an osmolality measured at about 299 mOsm/Kg.[1330]

## General Stability Considerations

Formoterol fumarate powder should be packaged in well-closed light-resistant containers and stored at controlled room temperature.[4]

Formoterol fumarate 10-mcg/mL inhalation solution should be stored in its foil pouch under refrigeration. After dispensing, it may be stored between 2 to 25°C for up to three months. Protect from exposure to heat. Formoterol fumarate inhalation solution should only be removed from its foil pouch immediately before use.[7]

## Compatibility with Other Drugs
**FORMOTEROL FUMARATE COMPATIBILITY SUMMARY**

*Compatible with:* Acetylcysteine • Budesonide • Cromolyn sodium • Ipratropium bromide

Akapo et al.[1330] evaluated the compatibility and stability of formoterol fumarate 20-mcg/2 mL inhalation solution (Perforomist, Dey) mixed with several other drugs for administration by inhalation and evaluated over 60 minutes at room temperature of 23 to 27°C. The drugs tested included acetylcysteine 200 mg/2 mL (American Regent), budesonide 0.5 mg/2 mL (AstraZeneca), cromolyn sodium 20 mg/2 mL (Dey), and ipratropium bromide 0.5 mg/2.5 mL (Dey). All of the test samples were physically compatible by visual examination and measurement of osmolality, pH, and turbidity. HPLC analysis of drug concentrations found little or no loss of any of the drugs within the study period.

# Fosfomycin

## Properties

Fosfomycin is available as both calcium and disodium salts. Both are white or almost white powders. Also available is fosfomycin disodium, which is very hygroscopic. Fosfomycin trometamol (tromethamine) is a white or almost white hygroscopic powder.[3]

### Solubility

Fosfomycin calcium is slightly soluble in water and practically insoluble in acetone and methanol. Fosfomycin disodium is very soluble in water, sparingly soluble in methanol, and practically insoluble in dehydrated ethanol. Fosfomycin trometamol is very soluble in water, slightly soluble in ethanol and methanol, and practically insoluble in acetone.[3]

### pH

A 0.1% aqueous solution of fosfomycin calcium has a pH of 8.1 to 9.6. A 5% aqueous solution of fosfomycin disodium has a pH of 9 to 10.5. A 5% aqueous solution of fosfomycin trometamol has a pH of 3.5 to 5.5.[3]

## General Stability Considerations

Fosfomycin calcium and disodium powders should be packaged in tight containers and stored at controlled room temperature protected from exposure to light.[3] Fosfomycin disodium injection reconstituted with dextrose 5% or sodium chloride 0.9% is stable for 24 hours at room temperature.[8]

## Stability Reports of Compounded Preparations
### Injections

Injections, like other sterile drugs, should be prepared in a suitable clean air environment using appropriate aseptic procedures. When prepared from nonsterile components, an appropriate and effective sterilization method must be employed.

Stahlmann and Frey[1102] reported the stability of reconstituted fosfomycin disodium 50 mg/1 mL and 50 mg/3 mL in infusion pump syringes (Braun Melsungen). The drug solutions packaged in the plastic syringes were found to lose about 2 to 3% in six hours at room temperature protected from exposure to light but about 12% in eight hours if exposed to light. When stored under refrigeration, the drug was reported to be stable for 96 hours plus an additional six hours at room temperature protected from exposure to light. When stored frozen at −20°C, the drug was found to be stable for at least 31 days plus 96 hours under refrigeration plus an additional six hours at room temperature protected from exposure to light.

*Enteral*

Ortega de la Cruz et al.[1101] reported the physical compatibility of an unspecified amount of oral liquid fosfomycin (Fosfocina, Em) with 200 mL of Precitene (Novartis) enteral nutrition diet for a 24-hour observation period. No particle growth or phase separation was observed. ∎

# Fumagillin Bicyclohexylammonium

## Properties

Fumagillin is an acyclic antibiotic produced by some strains of Aspergillus fumigatus. It occurs as white to yellow crystalline needles or powder.[1,3,899] Fumagillin bicyclohexylammonium is a water-soluble derivative of fumigillin.[899]

The principal use of fumagillin is in controlling Microsporida in honey bees. It has also been used in the treatment of microsporidial keratoconjunctivitis.[3]

### Solubility

Fumagillin is practically insoluble in water and dilute acids. It is soluble in most organic solvents and in aqueous solutions of bicarbonates and alkali hydroxides.[1] Its solubility in ethanol is 1 mg/mL.

Fumagillin bicyclohexylammonium (Fumidil B) is a water-soluble derivative of fumigillin.[853,899]

## General Stability Considerations

Fumagillin stability is dependent on temperature. Fumigillin also is light-sensitive.[853] The recommended storage is in the dark in evacuated ampules at low temperatures, preferably at −20°C.[1] Fumagillin bicyclohexylammonium should be stored in a dark place and protected from temperatures above 30°C.[899]

## Stability Reports of Compounded Preparations

*Ophthalmic*

Ophthalmic preparations, like other sterile drugs, should be prepared in a suitable clean air environment using appropriate aseptic procedures. When prepared from nonsterile components, an appropriate and effective sterilization method must be employed.

Abdel-Rahman and Nahata[853] reported the stability of fumagillin, an agricultural chemical used to control microsporidiosis in honeybees, prepared as a 70-mcg/mL ophthalmic solution. Fumagillin has been used successfully as a treatment for microsporidial keratoconjunctivitis.[854] The ophthalmic solution was compounded using commercial fumagillin bicyclohexylammonium crystals (Fumidil, Mid-Continent Agrimarketing). Crystals 120 mg were weighed and placed in a vial. Sodium chloride 0.9% 10 mL was added to the vial and swirled. The solution was drawn into a 60-mL syringe. An additional 10 mL of sodium chloride 0.9% was then added as a rinse and drawn into the syringe. Ciba Vision Ophthalmic irrigating solution 20 mL was then added to the vial as a rinse and drawn into the syringe, and the solution was gently shaken to ensure complete dissolution. The solution was sterilized by filtration though a 0.22-μm filter and packaged in plastic dropper bottles.

Sample dropper bottles of the fumagillin 70-mcg/mL ophthalmic solution were stored at room temperature of 25°C exposed to and protected from light and refrigerated at 4°C protected from light for 28 days. No change in color or odor was observed in any of the samples. The room temperature samples exhibited rapid decomposition; losses of about 30 and 17% occurred in seven days in the samples exposed to and protected from light, respectively. However, the refrigerated samples were more stable, exhibiting about 7% loss in 14 days and 13% loss in 28 days. ∎

# Furosemide

## (Frusemide)

## Properties

Furosemide is a white to slightly yellow odorless crystalline powder.[2,3]

### Solubility

Furosemide is practically insoluble in water, has a solubility of about 13 mg/mL in ethanol, and is freely soluble in solutions of alkali hydroxides and other aqueous solutions having a pH above 8.[1–3] It is insoluble in dilute acids[7] and may precipitate from solutions having a pH below 5.5.[2]

### pH

Furosemide injection has a pH of 8 to 9.3.[4]

### pKₐ

The p$K_a$ of furosemide is 3.9.[2,6]

### Osmolality

Furosemide oral solution 10 mg/mL (Lasix, Hoechst-Roussel) was found by Niemiec et al.[232] to have an osmolality of 3938 mOsm/kg, while Dickerson and Melnik[233] reported the osmolality to be 2050 mOsm/kg. Furosemide injection has an osmolality around 290 mOsm/kg.[286,287,345]

## General Stability Considerations

Furosemide injection should be stored at controlled room temperature and protected from light. Exposure to light may result in discoloration; discolored solutions should be discarded.[2,7] Refrigeration may result in precipitation, but the precipitate can be redissolved by rewarming the solution without adversely affecting the product.[370]

Furosemide oral solution should be stored at controlled room temperature protected from light and freezing. The oral tablets should be packaged in well-closed, light-resistant containers and stored at controlled room temperature. Exposure to light may cause discoloration of the tablets; discolored tablets should be discarded.[2,4]

In solution, furosemide is subject to hydrogen-ion catalysis at acidic pH values less than 3.5,[6,697] but at basic pH values of 8 and above hydrolysis is negligible.[6] The addition of alcohol may enhance furosemide stability, but glycerin and sorbitol have no effect.[147] Furosemide solutions are subject to photodegradation;[694–696] photodegradation is minimized at pH 7 but increases as the solution becomes more acidic or basic.[696]

## Stability Reports of Compounded Preparations

### Injection

Injections, like other sterile drugs, should be prepared in a suitable clean air environment using appropriate aseptic procedures. When prepared from nonsterile components, an appropriate and effective sterilization method must be employed.

**STUDY 1 (REPACKAGED):** Neil et al.[371] reported no loss of furosemide by HPLC analysis from a 10-mg/mL solution repackaged into polypropylene syringes and stored at 25°C for 24 hours. Neither exposure to light nor protection from light had a detectable effect on stability in this short study.

**STUDY 2 (EXTEMPORANEOUS INJECTION):** Neil et al.[371] also evaluated autoclaved furosemide 1 mg/mL in sodium chloride 0.9% injection in glass bottles for 34 minutes at 115°C and, using HPLC analysis, found no loss of drug. Storage of the injection at room temperature protected from light resulted in no loss of furosemide in 70 days. However, room temperature storage with exposure to light resulted in a 60% loss in 70 days and formation of a yellow-orange precipitate.

### Oral

**STUDY 1 (LIQUID):** The stability of commercial furosemide oral solution (Lasix, Hoechst-Roussel) 10 mg/mL repackaged as 2 mL in 5-mL amber polypropylene oral syringes (Baxa) and 15-mL amber glass vials (Wheaton Scientific) with suitable closures was investigated by Christensen et al.[86] Samples were stored at 4 and 25°C as well as at several elevated temperatures protected from light and in constant moisture conditions for 180 days. An HPLC assay was used to assess the furosemide content. Furosemide losses of 5% or less occurred at both 4 and 25°C in both containers in 180 days. Substantially greater losses were incurred at higher temperatures. The authors calculated the estimated shelf life at 25°C to reach a furosemide loss of 10%. In polypropylene oral syringes the calculated room temperature shelf life was 475 days, while in the glass vials it was 608 days. About 2% loss of furosemide occurred in the oral solution remaining in the manufacturer's original container after repackaging and storage at room temperature for 180 days.

**STUDY 2 (LIQUID):** Ghanekar et al.[147] evaluated the stability of furosemide in several aqueous vehicles. Furosemide was tested at 1 mg/mL in vehicles containing various percentages of sugar, sorbitol, glycerin, and phosphate buffers and in Syrpalta (Emerson Labs); the samples were stored at elevated temperatures to evaluate the rates of furosemide degradation. The Syrpalta samples and a sorbitol-containing preparation also were tested at 24°C. Furosemide content was determined using a stability-indicating HPLC assay. Furosemide was very unstable in acidic media and very stable in alkaline media. Sugar exerts an adverse effect on furosemide stability. Sorbitol was better than glycerin for furosemide stability; ethanol in place of some water in sorbitol solutions had a mild stabilizing effect. An oral product containing furosemide 1 mg/mL in sorbitol 50% with ethanol 10%, methylparaben 0.005%, propylparaben 0.002%, and sufficient sodium hydroxide to adjust to pH 8.5 exhibited no loss of furosemide in 182 days stored at 24°C. However, in the more acidic Syrpalta, furosemide 1 and 2 mg/mL yielded a mixture with a pH of 7.5; furosemide exhibited about 5 to 6% loss in 30 days and 10 to 12% loss in 60 days stored at 24°C.

**STUDY 3 (POWDER):** Yang et al.[738] reported the stability of furosemide oral powders extemporaneously compounded from commercial tablets. Lasix tablets (Hoechst Marion Roussel) were crushed and pulverized. Furosemide powder equivalent to 40 mg of drug was mixed with 300 mg of lactose, 450 mg of pulverized lactofermin (Biofermin) tablets, or 1.16 g of acetylcysteine granules for use as solid diluents. Powder mixtures were packaged in polyethylene-coated glassine papers and stored for 28 days at ambient temperature (between 15 and 30°C). Stability-indicating HPLC analysis found no loss of furosemide in 28 days when the drug was diluted in either lactose or pulverized lactofermin tablets. In acetylcysteine granules, however, 11% furosemide loss occurred in 14 days.

**STUDY 4 (REPACKAGED TABLETS):** Bowen et al.[1592] evaluated the stability of commercial furosemide tablets repackaged in dose administration aids (Webster-pak) over eight weeks stored at 23 to 27°C and 55 to 65% relative humidity both protected from light and also exposed to tungsten, fluorescent, and indirect daylight. In simulated home use, the tablets were exposed to temperatures ranging from 21.5 to 27.5°C and relative humidity up to 95.5%. In all cases, the tablets were physically stable; weight uniformity, friability, hardness, disintegration, and dissolution were all satisfactory. An unacceptable yellow discoloration was observed in the tablets exposed to light sources, but this discoloration was not indicative of excessive drug decomposition. HPLC analysis found 98% of the furosemide content remaining in these tablets. In tablets under all test conditions, the furosemide content at all time points remained within the British Pharmacopeial range of 95 to 105%. The authors concluded that furosemide tablets repackaged in dose administration aids were chemically and physically stable for at least eight weeks. However, because of the unacceptable tablet discoloration that occurred in tablets exposed to light, they recommended that furosemide tablets be stored in a cool dark place avoiding unnecessary exposure to light, heat, and humidity.

## Enteral
### FUROSEMIDE COMPATIBILITY SUMMARY
*Compatible with:* Enrich • Ensure • Ensure Plus • Osmolite • Osmolite HN • TwoCal HN • Vital • Vivonex T.E.N.

**STUDY 1:** Fagerman and Ballou[52] reported on the compatibility of furosemide (Lasix, Hoechst-Roussel) 10 mg/mL mixed in equal quantities with three enteral feeding formulas. The furosemide was visually compatible with Vital, Osmolite, and Osmolite HN (Ross), with no obvious thickening, granulation, or precipitation.

**STUDY 2:** Holtz et al.[54] evaluated the stability of furosemide oral liquid (Lasix, Hoechst-Roussel) at 40 and 80 mg/L in full-strength Ensure, Ensure Plus, and Osmolite (Ross) over 12 hours at room temperature. Visual inspection revealed no changes such as clumping, gelling, separation, precipitation, or increased viscosity in the mixtures. Osmolalities of the mixtures were largely unaffected by drug addition. Furosemide concentrations were assessed using a fluorometric assay. About 2 to 5% of the furosemide was lost in 12 hours.

**STUDY 3:** Burns et al.[739] reported the physical compatibility of furosemide oral liquid (Lasix) 40 mg/4 mL with 10 mL of three enteral formulas, including Enrich, TwoCal HN, and Vivonex T.E.N. Visual inspection found no physical incompatibility with any of the enteral formulas. ■

# Gabapentin

## Properties

Gabapentin, an amino acid structurally related to the neurotransmitter gamma-aminobutyric acid, is a white to off-white crystalline powder.[7]

Gabapentin is available as oral hard-gelatin capsules with lactose, corn starch, and talc. It is also available as an oral liquid with glycerin, xylitol, purified water, and artificial flavor.[7]

### Solubility

Gabapentin is freely soluble in water at both acidic and alkaline pH.[7] The solubility in water at pH 7.4 exceeds 100 mg/mL.[1]

### $pK_a$

Gabapentin has $pK_a$ values of 3.7 and 10.7.[7]

## General Stability Considerations

Commercial gabapentin capsules (Neurontin, Parke-Davis) should be stored at controlled room temperature. Gabapentin 50-mg/mL oral solution should be stored under refrigeration at 2 to 8°C.[7]

In aqueous solutions, gabapentin exhibits a minimum rate of degradation in the range of pH 5.5 to 6.5.[972]

## Stability Reports of Compounded Preparations
### Oral

**STUDY 1:** Nahata[830] evaluated the stability of two gabapentin oral suspensions prepared from capsules (Neurontin, Parke-Davis). The capsule contents were ground to fine powder using a mortar and pestle. Two vehicles were used to prepare the oral suspensions: (1) an equal parts mixture of simple syrup and 1% methylcellulose and (2) an equal parts mixture of Ora-Plus and Ora-Sweet. The suspensions were packaged in amber plastic prescription bottles and stored at room temperature of 25°C and refrigerated at 4°C.

No changes in appearance or odor of the suspensions occurred. Stability-indicating HPLC analysis found that both formulations exhibited nearly identical stability. About 6 to 7% gabapentin loss occurred in 56 days at room temperature, and 5 to 6% loss occurred in 91 days refrigerated. The author recommended refrigerated storage because of the possibility of microbial growth at room temperature.

**STUDY 2:** Volpe et al.[1240] evaluated the stability differences between intact and split gabapentin 600-mg tablets from three manufacturers. Samples of intact tablets and tablets split using a tablet-splitter from Apothecary Products were stored at 25°C and 60% relative humidity and at 30°C and 60% relative humidity for nine weeks. Using a stability-indicating HPLC analysis, the study found for all three brands of gabapentin tablets the drug content and dissolution remained acceptable throughout the nine-week study period. Analysis did demonstrate a slight increase in a decomposition product, but the increase was only statistically significant when the samples were stored at the elevated temperature. The authors concluded that gabapentin tablets whether whole or split were stable at both storage conditions, and that no significant differences in product quality were observed for the split tablets.

**STUDY 3:** Gupta et al.[1407] evaluated and compared the long-term stability of commercial gabapentin 300-mg capsules (Purepac) in the unopened original containers to the same batch of capsules repackaged into unit dose blister strips by Vanguard Laboratories. Samples were stored for 52 weeks at 25°C and 60% relative humidity and for 13 weeks under accelerated storage at 40°C and 75% relative humidity. Stability-indicating HPLC analysis found no substantial difference in concentration between the gabapentin capsules in original containers and the unit dose blister strips. Little gabapentin loss occurred with over 98% remaining at room temperature in 52 weeks and over 97% remaining under accelerated conditions in 13 weeks. The lactam decomposition product concentrations were between 0.1 and 0.2%, well below the USP acceptability limit of 0.4%. A slight gain in weight in the capsules in unit dose blister strips was ascribed to moisture transmission into the product. ■

# Galactose

## Properties
Galactose is a monosaccharide that exists as crystalline prisms.[1,3]

### Solubility
Galactose as the α-form has an aqueous solubility of about 1 g/0.5 mL. Final solubility at 25°C is about 68%. It is slightly soluble in ethanol. The β-form has an aqueous solubility of about 1 g/1.7 mL.[1]

## Stability Reports of Compounded Preparations
### Aqueous Solutions
The aqueous stability of galactose was investigated by Bhargava et al.[96] They evaluated galactose (Sigma) 5, 10, 20, and 30% in sterile water for injection and 5 and 30% in 0.5, 0.05, and 0.005 M acetate and phosphate buffers, each at pH 4, 5, and 6. The solutions were autoclaved at 121°C for 30 minutes and packaged in type I glass vials. Sample vials were stored at 25°C for six weeks. Galactose concentrations were assessed using a refractive index HPLC analysis. Concentrations of the major degradation product, 5-hydroxymethylfurfural (5-HMF), also were monitored spectrophotometrically. Galactose 5 to 30% in sterile water was stable to autoclaving and maintained stability for at least six weeks at room temperature, exhibiting less than 4% loss. For galactose nonbuffered solutions sterilized by filtration, a shelf life of 4.5 months at 25°C was calculated. Autoclaving of the buffered solutions was unacceptable due to the formation of yellow color (due to excessive amounts of 5-HMF) and/or loss of galactose. Some evacuated flasks contained a residual amount of acetate buffer, rendering them unsuitable containers for autoclaving galactose solutions. ∎

# Ganciclovir
# Ganciclovir Sodium

## Properties
Ganciclovir is a white to off-white crystalline powder. Ganciclovir sodium is a white to off-white lyophilized powder in vials.[7]

### Solubility
Ganciclovir has an aqueous solubility of 2.6 mg/mL at 25°C. Ganciclovir sodium in the intravenous form has an aqueous solubility exceeding 50 mg/mL at 25°C. At physiological pH near pH 7.4, ganciclovir sodium exists as the unionized form, with an aqueous solubility of 6 mg/mL at 37°C.[7] The aqueous solubility at pH 7 and 25°C is 4.3 mg/mL.[1]

### pH
Ganciclovir sodium reconstituted injection is very alkaline, having a pH of about 11.[7]

### pKa
Ganciclovir has $pK_a$ values of 2.2 and 9.4.[7]

### Osmolality
Ganciclovir sodium reconstituted with sterile water for injection to a concentration of 50 mg/mL has an osmolality of 320 mOsm/kg.[900]

## General Stability Considerations
Ganciclovir capsules should be stored between 5 and 25°C. Ganciclovir sodium vials should be stored at controlled room temperature and protected from excessive heat over 40°C.[7]

Ganciclovir sodium reconstituted with sterile water for injection to a concentration of 50 mg/mL is stable for 12 hours at room temperature.[7] Ganciclovir sodium is incompatible with parabens and should not be reconstituted with bacteriostatic water for injection containing paraben preservatives because precipitation will occur.[2]

Although precipitation may occur if the solution is refrigerated,[7] the drug has been reported to be stable for 60 days, with HPLC analysis finding no drug loss. Similarly, the manufacturer of ganciclovir sodium does not recommend freezing the drug in solution. However, in concentrations ranging from 1.4 to 7 mg/mL in sodium chloride 0.9%, no more than 4% loss was found by HPLC analysis after frozen storage for 364 days.[8]

## Stability Reports of Compounded Preparations
### Oral
**USP OFFICIAL FORMULATION:**
> Ganciclovir 10 g
> Vehicle for Oral Solution, NF
> (sugar-containing or sugar-free) qs 100 mL

(See the vehicle monographs for information on the individual vehicles.)

Use ganciclovir powder or commercial capsules. Avoid inhalation and contact of the drug with skin; the use of gloves, mask, and fume hood is recommended. If using capsules, empty the capsule contents providing the correct amount of drug or add the ganciclovir powder to a mortar. Add sufficient vehicle to wet the powder, and mix to make a uniform paste. Add additional vehicle to about half of the final volume and mix. Quantitatively transfer to a suitable calibrated tight and light-resistant bottle. Rinse the mortar with additional vehicle transferring the rinse to the bottle in steps, bring to final volume with the vehicle, and thoroughly mix yielding ganciclovir 100 mg/mL oral suspension. The final liquid preparation should have a pH between 4 and 5. Store the preparation at controlled room temperature. The beyond-use date is 90 days from the date of compounding.[4,5]

Anaizi et al.[688] reported the stability of two ganciclovir 100-mg/mL oral suspensions prepared from 250-mg capsules (Cytovene, Roche). The two suspensions were prepared by emptying the contents of 80 250-mg capsules into glass mortars and wetting the powders with sufficient Ora-Sweet or Ora-Sweet SF. Additional Ora-Sweet or Ora-Sweet SF was added to each suspension in increments with thorough mixing and rinsing of the mortars and pestles. The suspensions were brought to 200 mL with additional vehicle and were packaged in 60-mL amber polyethylene terephthalate plastic prescription bottles. The samples were stored at room temperature of about 23 to 25°C for 123 days.

No visible change or change in odor occurred in the samples of either suspension at any time. Stability-indicating HPLC analysis found that little or no loss of ganciclovir occurred in 123 days of storage at room temperature in either suspension.

### Injection

Injections, like other sterile drugs, should be prepared in a suitable clean air environment using appropriate aseptic procedures. When prepared from nonsterile components, an appropriate and effective sterilization method must be employed.

Ganciclovir sodium in concentrations ranging from 1.4 to 7 mg/mL in sodium chloride 0.9% packaged in polypropylene syringes exhibited 4% or less drug loss by HPLC analysis after storage for seven days at room temperature of 20°C, 80 days refrigerated at 4°C, and 364 days frozen at −20°C.[8] ■

# Gatifloxacin

## Properties

Gatifloxacin is a white to pale yellow prismatic crystalline material. It exists as the sesquihydrate.[1,7]

Gatifloxacin is available as 200- and 400-mg film-coated tablets. The tablets also contain hypromellose, magnesium stearate, methylcellulose, microcrystalline cellulose, polyethylene glycol, polysorbate 80, simethicone, sodium starch glycolate, sorbic acid, and titanium dioxide.[7]

Gatifloxacin injection was a concentrate that was a clear and light yellow to greenish-yellow free-flowing liquid. Gatifloxacin was also provided as premixed infusion solutions in dextrose 5% that were also clear and light yellow to greenish-yellow. The manufacturer indicated that the solution color was not indicative of stability of the injection.[7]

Gatifloxacin is also available as a clear, pale yellow 0.3% sterile ophthalmic solution with benzalkonium chloride 0.005%, edetate disodium, and sodium chloride in purified water. It may contain hydrochloric acid and/or sodium hydroxide to adjust pH.[7]

### Solubility

Gatifloxacin solubility is dependent on pH. Maximum aqueous solubility of 40 to 60 mg/mL occurs in the pH range of 2 to 5.[7]

### pH

Gatifloxacin injection concentrate and the premixed infusion solution had a pH in the range of 3.5 to 5.5. Gatifloxacin ophthalmic solution has the pH adjusted to approximately 6.[7]

### Osmolality

Gatifloxacin ophthalmic solution is near isotonicity, with a range of 260 to 330 mOsm/kg.[7]

## General Stability Considerations

Gatifloxacin dosage forms are stored at controlled room temperature, and the premixed infusion solution in flexible bags should be protected from freezing. Gatifloxacin ophthalmic solution should also be stored at controlled room temperature and protected from freezing.[7]

Admixtures of gatifloxacin concentrate in common infusion solutions (except sodium bicarbonate) are stable for six months frozen at −25 to −20°C. Once thawed, the solutions are stable for an additional 14 days at room temperature or under refrigeration.[7]

## Stability Reports of Compounded Preparations
### Ophthalmic

Ophthalmic preparations, like other sterile drugs, should be prepared in a suitable clean air environment using appropriate

aseptic procedures. When prepared from nonsterile components, an appropriate and effective sterilization method must be employed.

Liu and Huang[965] reported on the stability of a compounded gatifloxacin 0.3% ophthalmic solution. The formulation contained gatifloxacin 0.3%, EDTA sodium 0.01%, sodium chloride 0.83%, benzalkonium chloride 0.005%, and sodium hydroxide or hydrochloric acid to adjust to pH 6 in sterile water. The test solution was stored at room temperature protected from light for one year and at an elevated temperature of 60°C for 10 days with and without exposure to light. The test solutions were physically stable and HPLC analysis found no measurable loss of gatifloxacin during the study periods. A trace amount (0.12%) of degradation product was identified in the elevated temperature samples exposed to light. ▪

# ■ Gentamicin Sulfate

## Properties

Gentamicin sulfate is a white to buff hygroscopic powder consisting of a mixture of three principal components and may also contain one or two minor components.[1–3] It contains not less than 590 units of gentamicin per milligram.[3]

### Solubility

Gentamicin sulfate is soluble in water and insoluble in ethanol and acetone.[1–3]

### pH

A 1 in 25 (4%) aqueous solution has a pH of 3 to 5.5[2–4] as does the intramuscular and intravenous injection.[2,4] The commercial ophthalmic solution is buffered to an approximate pH of 7[132] but may have a pH range of 6.5 to 7.5.[4]

### Osmolality

The 40-mg/mL injection has a reported osmolality of about 160 mOsm/kg,[234] while the 10-mg/mL pediatric injection has been variously reported to have an osmolality of 116 and 212 mOsm/kg.[235]

## General Stability Considerations

Gentamicin sulfate dosage forms are stored in tight containers, usually under controlled room temperature and avoiding excessive heat.[4] Freezing of dilute aqueous solutions of gentamicin sulfate at −20°C for up to 30 days does not affect concentration adversely.[236–238]

Clark et al.[1506] evaluated the in vitro antimicrobial efficacy of gentamicin sulfate with betamethasone otic solution (Garasone, Schering-Plough) after opening. Antimicrobial susceptibility testing against *Pseudomonas aeruginosa* and *Staphylococcus aureus* on agar plates was evaluated over four months of storage. The zones of microbial inhibition remained constant throughout the study period.

## Stability Reports of Compounded Preparations
### Oral
Wamberg et al.[1516] evaluated the stability of an oral preoperative bowel preparation containing three antibiotics. Gentamicin sulfate, nystatin, and vancomycin hydrochloride dry powders were sifted and mixed. The intended oral dose was gentamicin sulfate 240 mg, nystatin 3.5 million units, and vancomycin hydrochloride 250 mg. The mixed antibiotic powder was packaged in unit dose amber glass airtight containers and stored at cool temperatures of 8 to 15°C and protected from exposure to direct sunlight. Using a microbiological assay technique, the authors reported that all of the antibiotics retained activity under these conditions for up to 12 months. If the antibiotics were prepared in a liquid syrup mixture, the activity of vancomycin hydrochloride was lost much more rapidly.

### Injection
Injections, like other sterile drugs, should be prepared in a suitable clean air environment using appropriate aseptic procedures. When prepared from nonsterile components, an appropriate and effective sterilization method must be employed.

**STUDY 1 (REPACKAGED):** Weiner et al.[239] reported that undiluted gentamicin sulfate repackaged from its original containers into plastic syringes may undergo unacceptable (16%) decomposition, with brown precipitate formation in 30 days at 25°C. In glass syringes, concentration loss was 7% in 30 days. However, other studies have not confirmed this loss in plastic syringes.

**STUDY 2 (REPACKAGED):** Kresel et al.[240] found no significant gentamicin loss by enzymatic assay of a 40-mg/mL injection repackaged into polypropylene syringes (Becton Dickinson) and stored for 30 days at 4 and 25°C. Zbrozek et al.[335] reported less than 10% loss of gentamicin in 48 hours at 25°C when a 30-mg/mL solution was repackaged in polypropylene syringes (Becton Dickinson).

**STUDY 3 (REPACKAGED):** Levin et al.[347] reported that gentamicin sulfate 40 mg/mL repackaged in glass cartridges (Tubex) was stable for up to three months at room temperature. Nahata et al.[241] repackaged gentamicin sulfate 10 mg/mL diluted in 0.9% sodium chloride injection in glass syringes (Becton Dickinson).

No loss of gentamicin was found by enzyme-mediated immunoassay after 12 weeks of storage at 4°C.

## Ophthalmic

Ophthalmic preparations, like other sterile drugs, should be prepared in a suitable clean air environment using appropriate aseptic procedures. When prepared from nonsterile components, an appropriate and effective sterilization method must be employed.

**STUDY 1:** McBride et al.[106] reported the stability of a fortified gentamicin sulfate 13.6-mg/mL ophthalmic solution prepared from 165 mL of the commercial ophthalmic solution (Genoptic 3 mg/mL, Allergan) and 66 mL of gentamicin sulfate injection (40 mg/mL, LyphoMed). The solution was prepared in an analogous manner to that used for compounding single-unit admixtures, which usually involves the addition of 2 mL of the 40-mg/mL gentamicin sulfate injection to 5 mL of the 3-mg/mL ophthalmic solution. The final solution was packaged as 7-mL volumes in plastic bottles and stored under refrigeration at 4 to 8°C for 91 days. Gentamicin sulfate stability was evaluated using fluorescence polarization immunoassay and a stability-indicating HPLC assay. No loss of gentamicin sulfate occurred with either technique in 91 days under these storage conditions.

**STUDY 2:** Allen[132] noted that fortified gentamicin ophthalmic solution stored under refrigeration may be chemically stable for 91 days, but the microbiological stability is dependent on the ability of the compounder to assure sterility.

**STUDY 3:** Nahata and Hipple[160] reported the stability of a fortified gentamicin sulfate ophthalmic solution for pediatric use. The solution was prepared from commercial ophthalmic solution and injection, with the final preparation having a concentration of 8 mg/mL. A stability period of 14 days (storage temperature unspecified) was cited.

**STUDY 4:** Arici et al.[796] reported the stability of gentamicin sulfate 13.5-mg/mL stock ophthalmic solutions prepared from injection added to commercial ophthalmic solution and packaged in ophthalmic squeeze bottles at room temperature of 24°C and refrigerated at 4°C. No loss in microbiological activity against *Staphylococcus aureus* and *Pseudomonas aeruginosa* was found over 14 days of storage at both temperatures in both vehicles. However, a substantial decrease in activity against the organisms occurred by 21 days. This is a different result than the study by McBride et al. and may be due to the use of differing methodologies.

**STUDY 5:** El-Shattawy[218] evaluated the stability of 0.3% gentamicin in 36 formulations of ophthalmic ointment bases. Each of 18 different formulations was prepared both with and without 0.1% parabens (methylparaben–propylparaben, 2:1) as preservatives and with 0.1% tocopheryl acetate as an antioxidant. The ointments were stored at 5°C for eight months. Gentamicin activity was assessed microbiologically. Gentamicin antimicrobial activity was unchanged in all formulations for the study period. The preservatives and the antioxidant had no effect on antibiotic activity. The author indicated that the best bases for gentamicin ophthalmic ointments were those containing 75 to 85% castor oil gelled by 15% hydrogenated castor oil or 90 to 95% castor oil gelled by 3 to 5% Aerosil.

**STUDY 6:** Ho et al.[878] evaluated the stability of fortified gentamicin sulfate 8- and 14-mg/mL ophthalmic drops. The ophthalmic drops were prepared by aseptically diluting the commercial injection with sodium chloride 0.9%. The formulation was packaged in sterile amber glass dropper bottles, which were kept refrigerated at 2 to 8°C as stored in a pharmacy. Samples were also stored refrigerated and at room temperature of 26°C as "in-use" containers; patient use was simulated by repeated shaking and opening and closing of the dropper bottles every four hours. HPLC analysis found the sealed bottles stored under refrigeration exhibited about 5 to 8% loss after 42 days. However, under simulated patient-use conditions, the samples were stable for 35 days refrigerated, with 8 to 9% loss, and for seven days at room temperature, with 8 to 10% loss.

**STUDY 7:** Garcia-Valldecabres et al.[880] reported the change in pH of several commercial ophthalmic solutions, including gentamicin sulfate 3 mg/mL (Colircusi Gentamicina), over 30 days after opening. Slight variation in solution pH was found, but no consistent trend and no substantial change occurred.

**STUDY 8 (WITH VANCOMYCIN):** Poveda Andres et al.[1562] evaluated the stability of a solution for use as an irrigation in cataract surgery. The solution was composed of gentamicin sulfate 8 mcg/mL and vancomycin hydrochloride 20 mcg/mL in lactated Ringer's injection and was packaged in glass containers. Using polarized immunofluorescence (TDx) analysis, the concentration of each antibiotic was determined during storage. The vancomycin hydrochloride proved to be the least stable drug and limited the utility time of the solution. The time to 10% loss ($t_{90}$) of vancomycin hydrochloride at room temperature of 20 to 25°C was found to be 136 hours if exposed to light and 190 hours if protected from exposure to light.

## Topical

Sivapunyam et al.[429] evaluated the stability and release characteristics of gentamicin 0.1% in eight ointment bases. The ointments were stored at 4 and 37°C for 40 days. Gentamicin content was assessed microbiologically. The ointment prepared with sodium carboxymethylcellulose base exhibited the best release characteristics and also the longest stability, with 7% loss in 40 days at 4°C and 10% loss in 20 days at 37°C. Second best for release characteristics and stability was the ointment prepared with the polyethylene glycol base. Approximately 9% was lost in 30 days at 4°C, and 10% was lost in 20 days at 37°C.

### Inhalation

Inhalation preparations, like other sterile drugs, should be prepared in a suitable clean air environment using appropriate aseptic procedures. When prepared from nonsterile components, an appropriate and effective sterilization method must be employed.

Tomonaga et al.[1605] evaluated the compatibility and stability of gentamicin sulfate injection mixed with six drugs for inhalation: Alevaire (tyloxapol), Bisolvon (bromhexine hydrochloride), Asthone (chlorprenaline hydrochloride), Asthpul (dl-isoproterenol hydrochloride), Alotec (orciprenaline sulfate), and Ventolin (albuterol sulfate). Using a microbiological activity assay, it was found that in all of these inhalation solutions the gentamicin sulfate content deteriorated substantially in most combinations within one day and within all combinations within three days stored both under refrigeration at 4°C and at room temperature of 20 and 30°C.

### Compatibility with Other Drugs

Gentamicin sulfate is well documented as unstable in the presence of some β-lactam antibiotics such as the penicillins carbenicillin, ticarcillin, piperacillin, and ampicillin. The cephalosporins also may cause gentamicin inactivation, although not usually to the same extent as the penicillins. The extent of gentamicin inactivation is dependent on the specific β-lactam antibiotic, drug concentrations, storage temperature, and duration of storage. In general, compounding gentamicin in the same formulation with β-lactam antibiotics is discouraged. The drugs should be administered separately. ■

# Glyburide
## (Glibenclamide)

### Properties

Glyburide occurs as a white or almost white crystalline powder.[1,3]

### Solubility

Glyburide is practically insoluble in water but soluble in ethanol, methanol, and sparingly soluble in dichloromethane.[1,3] The aqueous solubility of glyburide increases with increasing pH due to salt formation.[7]

### $pK_a$

Glyburide has a $pK_a$ of 5.3.[1]

### General Stability Considerations

Glyburide powder is to be packaged in tight containers. Glyburide oral tablets are to be packaged in well-closed containers and stored at controlled room temperature.[4,7]

### Stability Reports of Compounded Preparations
#### Oral

Bachhav and Patravale[1423] evaluated the stability of a self-emulsifying drug delivery system (SMEDDS) containing glyburide in the self-emulsifying formulation. The drug was dissolved in Capryol 90 as the oil phase and Transcutol P (surfactant) containing hydroxypropylcellulose (to increase viscosity) and Tween-20 (surfactant) were added to the solution with vortexing to prepare the homogenous SMEDDS. Various combinations were explored, but the best formulation contained glyburide 0.75% wt/wt, Capryol 90 15.1% wt/wt, Transcutol P 53% wt/wt, Tween-20 30.3% wt/wt, and hydroxypropylcellulose 0.75% wt/wt. When 0.5 g of this formulation was mixed with 50 mL of distilled water, the average globule size of 5 nm that was formed was satisfactory. However, stability-indicating HPLC analysis found substantial glyburide decomposition within 15 days; losses of 9%, 15%, and 38% occurred at 25°C and 60% relative humidity, 30°C and 65% relative humidity, and 40°C and 75% relative humidity, respectively. The authors then tested the stability of glyburide with each of the components of the formulation, finding glyburide underwent substantial decomposition with all of the components except hydroxypropylcellulose. ■

# Glycerin
## (Glycerol)

## Properties

Glycerin is a clear, colorless, syrupy liquid with a sweet taste, approximately 0.6 time as sweet as sucrose. It is hygroscopic and nearly odorless.[1,3,4,1238] It should contain not more than 5% water according to the USP[4] and not more than 2% water according to the European Pharmacopeia.[3]

### Solubility

Glycerin is miscible with water and with ethanol.[1,4,1238]

### pH

Glycerin solutions are neutral to litmus. Glycerin ophthalmic solution has a pH between 4.5 and 7.5. Glycerin oral solution has a pH between 5.5 and 7.5.[4]

### Osmolality

A 2.6% vol/vol aqueous solution of glycerin is iso-osmotic with serum.[1238]

## General Stability Considerations

Glycerin must not be allowed to contact strong oxidizing agents potassium chlorate or potassium permanganate because an explosion may be produced.[1,3,1238] Glycerin may crystallize if stored at low temperatures. These crystals do not melt until warmed to at least 20°C.[1238]

Glycerin and glycerin oral solution should be packaged in tight containers. Glycerin ophthalmic solution should be packaged in tight 15-mL plastic or glass. Glycerin and glycerin oral solution should be stored at controlled room temperature. Glycerin ophthalmic solution should be stored at controlled room temperature and protected from exposure to light. In addition, the ophthalmic solution should not be used if it contains crystals or precipitation or is cloudy or discolored. Glycerin suppositories should be packaged in well-closed containers and should be stored at controlled room temperature.[4]

## Stability Reports of Compounded Preparations
### Sterile

Like other sterile drugs, sterile glycerin should be prepared in a suitable clean air environment using appropriate aseptic procedures. When prepared from nonsterile components, an appropriate and effective sterilization method must be employed.

McClusky[1237] reported an evaluation of various sterilization techniques for glycerin because sterile glycerin is required for some applications but no commercial form of sterile glycerin is available. Of the techniques that can be used for sterilization, filtration through a 0.22-μm sterilizing filter was deemed by the author to be the best approach for glycerin. However, glycerin's viscosity can make sterilizing filtration a problem.

The technique reported in this article employed a chemically compatible Supor (Baxa) 0.22-μm capsule filter with sterile Baxa fluid transfer set no. 11 and extension set no. 87 tubing attached to the hose barb inlet ends of the filter using a firm twisting motion. The extension set is attached to the outlet of the filter and a sterile Braun fluid dispensing connector is attached to the free end. Glycerin, USP, is placed in a suitable clean depyrogenated glass beaker, and the free end of fluid transfer set is placed in the glycerin. The fluid transfer set is placed in a Baxa repeater peristaltic pump. The tubing is primed at the lowest possible pump setting. As the glycerin reaches the filter, about 100 mL should be permitted to pass by the pump. The pump should then be turned off to allow the filter to become wet, which is stated to take about five minutes. After wetting has occurred, run the pump at the lowest setting because faster pumping may cause a tubing failure to occur. After the air is removed from the tubing sets and filter, aseptically attach a 60-mL sterile syringe to the fluid dispensing connector. Pump about 50 mL of the filtered glycerin into the syringe, and then remove the syringe and fluid dispensing connector. Aseptically seal the syringe using a Luer tip cap. Replace the filter connector with a new sterile fluid connector and attach a new sterile syringe. Repeat this operation until all of the glycerin is packaged.

A filter integrity test (bubble point test) should be performed upon completion of the filling operation. A visual inspection, sterility test, and bacterial endotoxin test should be performed before the syringes are used.

## Compatibility with Other Drug Products

Glycerin is reported to result in a black discoloration if mixtures with bismuth subnitrate or zinc oxide are exposed to light.[3,1238]

# Glycolic Acid

## Properties
Glycolic acid is an organic acid that occurs as odorless hygroscopic crystals.[1,3]

### Solubility
Glycolic acid is soluble in water, ethanol, acetone, and acetic acid.[1]

### pH
Glycolic acid solutions have the following pH values for given concentrations: pH 2.5 for 0.5%, pH 2.3 for 1%, pH 2.2 for 2%, pH 1.9 For 5%, and pH 1.7 For 10%.[1]

### pK$_a$
The pK$_a$ at 25°C is 3.8.[1]

## General Stability Considerations
Crystalline glycolic acid is chemically stable to 50°C. At higher temperatures, polymerization occurs.[903]

Solutions of glycolic acid should be stored above 10°C; at colder temperatures, glycolic acid crystals may form. Such crystals will redissolve with warming and agitation of the solution.[903]

## Stability Reports of Compounded Preparations
### Topical
de Villiers et al.[724] reported the physical and chemical stability of two topical creams of glycolic acid 10%, an alpha-hydroxy acid, subjected to five freeze–thaw cycles of 24 hours at −4°C and then 24 hours at 40°C.

The best results were obtained with the glycolic acid incorporated into hydrophilic ointment, USP. Stability-indicating HPLC analysis found about 6% glycolic acid loss at the end of the five cycles. The physical and chemical stability of the ointment was satisfactory, but the methylparaben and propylparaben preservatives underwent unacceptable losses of about 13 to 16%. In a similar product with the pH adjusted from about pH 1.1 up to about 3.5 using sodium hydroxide solution, the methyl- and propylparabens remained stable, exhibiting less than 5% loss through the freeze–thaw cycles. ∎

# Glycopyrrolate

## Properties
Glycopyrrolate occurs as a white odorless crystalline powder.[1,4] Glycopyrrolate injection is a clear, colorless solution.[7]

### Solubility
Glycopyrrolate is soluble in water and ethanol.[1,4] Aqueous solubility is stated to be about 238 mg/mL, while solubility in ethanol is stated to be about 33 mg/mL.[3]

### pH
Glycopyrrolate injection has a pH of 2 to 3.[4]

### Osmolality
Glycopyrrolate 0.2-mg/mL injection is hypotonic, having an osmolality of 91 mOsm/kg.

## General Stability Considerations
Glycopyrrolate powder and glycopyrrolate tablets should be stored in tight containers at controlled room temperature. Glycopyrrolate injection should also be stored at controlled room temperature.[4,7]

Glycopyrrolate stability is pH-dependent in aqueous solution. The drug is very stable from the pH range of 2 to 3 up to about pH 5. No loss was found in 90 days in a pH 5 phosphate buffer solution. But above pH 6, ester hydrolysis occurs, with increasing rapidity as the solution becomes more alkaline. A 5% loss of glycopyrrolate from solution at 25°C occurred in 48 hours at pH 4 and 5, in 30 hours at pH 6, in four hours at pH 7, and in two hours at pH 8. About 30% loss occurred in a pH 7 buffer solution in 90 days.[8,866]

## Stability Reports of Compounded Preparations
### Oral
**STUDY 1 (FROM TABLETS):** Gupta[722] reported the stability of a glycopyrrolate oral liquid prepared from 1-mg tablets. The tablets were pulverized and the powder was mixed with sufficient water to make a 0.5-mg/mL concentration. The mixture was packaged in an amber plastic prescription bottle and stored at room temperature of 24 to 26°C. Stability-indicating HPLC analysis found about 8% glycopyrrolate loss after 25 days of storage.

**STUDY 2 (FROM TABLETS):** Cober et al.[1664] evaluated the stability of glycopyrrolate 0.5 mg/mL in two oral liquid preparations compounded from commercial 1-mg oral tablets. The tablets were triturated to fine powder using a glass mortar and pestle. The powder was levigated with equal quantity mixtures of Ora-Plus and Ora-Sweet or Ora-Sweet SF using geometric dilution to form a smooth suspension. The suspension was packaged in amber plastic prescription bottles, and the samples were stored at room temperature of 23 to 25°C for 90 days. No change in color, odor, and taste, and no visible microbial growth were reported. No appreciable change in pH was found. Stability-indicating HPLC analysis of drug concentrations found little or no loss of glycopyrrolate occurred in any of the samples during 90 days of storage at room temperature.

**STUDY 3 (FROM POWDER):** Gupta[866] reported improved stability of glycopyrrolate prepared from powder in a phosphate-buffered vehicle. Glycopyrrolate powder was dissolved in 0.05 M phosphate buffer (pH 5.6) with 10% sucrose or sorbitol as sweeteners to yield a final glycopyrrolate concentration of 0.5 mg/mL. The solutions were packaged in amber bottles and stored at room temperature of 24 to 26°C for 129 days. The physical appearance of the solutions remained clear throughout the study. Stability-indicating HPLC analysis found less than 6% glycopyrrolate loss in 129 days in both vehicles. This formulation was considered preferable because of the absence of undissolved tablet components as well as an improved chemical stability.

**STUDY 4 (FROM INJECTION):** Landry et al.[936] evaluated the stability and taste of four glycopyrrolate 0.1-mg/mL oral liquid preparations compounded using glycopyrrolate injection (Sabex). The oral preparations utilized vehicles of (1) an equal parts mixture of Ora-Plus and Ora-Sweet, (2) a simple syrup–methylcellulose 1% mixture (3:7), (3) simple syrup, and (4) sterile water. Equal volumes of glycopyrrolate injection and each of the four vehicles were mixed to yield a glycopyrrolate concentration of 0.1 mg/mL. The stability of the drug was evaluated by stability-indicating HPLC analysis over 35 days stored at room temperature of 25°C and under refrigeration at 2 to 8°C.

No unacceptable drug loss occurred over the 35-day test period at both storage temperatures. The glycopyrrolate concentrations were 93% or greater at all assay points, with most results greater than 95%, indicating that a beyond-use date of 35 days is appropriate for these oral formulations. The authors suggested storage under refrigeration to retard microbial growth. The Ora-Plus–Ora-Sweet mixture was found to have the most acceptable taste, with plain simple syrup being nearly as acceptable. However, sterile water and the methylcellulose mixture failed to adequately mask the drug's bitter taste.

**STUDY 5:** Nahata et al.[1199] evaluated the stability of glycopyrrolate 0.2 mg/mL formulated in two compounded oral suspension formulations. The two vehicles evaluated were 1% methylcellulose–simple syrup (1:10) and Ora-Plus–Ora-Sweet

(1:1). The oral suspensions were packaged in plastic prescription bottles and were stored at room temperature of 25°C and refrigerated at 4°C for 28 days. Stability-indicating HPLC analysis found about 4 to 5% glycopyrrolate loss in 14 days at room temperature and in 28 days refrigerated.

### Topical
Allen[788] reported on the compounding of glycopyrrolate 1% topical solution with benzyl alcohol 1% in purified water adjusted to pH 3 with dilute hydrochloric acid. In addition, glycopyrrolate 1% may be mixed with a small amount of glycerin or propylene glycol to form a paste that is then incorporated into oil-in-water cream and adjusted to approximately pH 3 with dilute hydrochloric acid. The preparations are used in the treatment of hyperhidriosis, especially diabetic gustatory sweating. A beyond-use date of 30 days was suggested as appropriate based on glycopyrrolate stability at pH 3.

### Injection
Injections, like other sterile drugs, should be prepared in a suitable clean air environment using appropriate aseptic procedures. When prepared from nonsterile components, an appropriate and effective sterilization method must be employed.

**STUDY 1:** Stewart et al.[901] evaluated the stability of American Regent glycopyrrolate injection 0.1 mg/mL diluted in sodium chloride 0.9% (normal saline) packaged in Sherwood polypropylene. The drug remained physically stable, and stability-indicating HPLC analysis found that little or no loss occurred in 24 hours at 4 and 23°C.

**STUDY 2:** Storms et al.[902] reported on the stability of undiluted glycopyrrolate injection (Robinul) 0.8 mg/4 mL packaged in 6-mL polypropylene syringes (Becton Dickinson). The syringes were stored refrigerated at 4°C and at room temperature of 25°C exposed to fluorescent light. The injection remained clear and physically stable at both temperatures. HPLC analysis found that little or no loss of glycopyrrolate occurred in 90 days under either storage condition.

## Compatibility with Other Drugs
Fox et al.[1239] evaluated the physical compatibility of hypertonic sodium chloride 7% with several inhalation medications used in treating cystic fibrosis. Glycopyrrolate injection (American Regent) 0.2 mg/mL mixed in equal quantities with extemporaneously compounded hypertonic sodium chloride 7% did not exhibit any visible evidence of physical incompatibility, and the measured turbidity did not increase over the one-hour observation period. ▪

# Granisetron Hydrochloride

## Properties

Granisetron hydrochloride is a white to off-white solid.[7] Granisetron hydrochloride 1.12 mg is equal to granisetron 1 mg.[3,7]

## Solubility

Granisetron hydrochloride is soluble in water and ethanol.[7]

## pH

Granisetron injection in single-dose vials has a pH of 4.7 to 7.3; in multiple-dose vials it has a pH of 4 to 6.[7]

## General Stability Considerations

Granisetron hydrochloride injection products should be stored at controlled room temperature and protected from freezing and light. After initial entry into the multiple-dose vials, the product should be used within 30 days, and any remainder should be discarded. The tablets should be stored at controlled room temperature and protected from light.[7]

## Stability Reports of Compounded Preparations
### Oral

**USP OFFICIAL FORMULATION:**

>   Granisetron (as hydrochloride) 5 mg (5.6 mg of
>      hydrochloride)
>   Vehicle for Oral Solution, NF
>   Vehicle for Oral Suspension, NF (1:1) qs 100 mL

(See the vehicle monographs for information on the individual vehicles.)

Use granisetron hydrochloride powder or commercial tablets. If using tablets, crush or grind to fine powder. Add the vehicle mixture in small portions, and mix to make a uniform paste. Add additional vehicle in increasing volumes to make a pourable suspension. Quantitatively transfer to a suitable calibrated tight and light-resistant bottle, bring to final volume with the vehicle mixture, and thoroughly mix yielding granisetron (as hydrochloride) 50-mcg/mL oral suspension. The final liquid preparation should have a pH between 4 and 4.5. Store the preparation at controlled room temperature or under refrigeration between 2 and 8°C. The beyond-use date is 90 days from the date of compounding at either storage temperature.[4]

**STUDY 1:** Quercia et al.[566] evaluated the stability of a granisetron 0.2-mg/mL oral liquid formulation compounded from tablets. Twelve tablets containing granisetron 1 mg (as the hydrochloride) (Kytril, SmithKline Beecham) were ground to powder and suspended in 30 mL of water. Cherry syrup (Humco) was added, bringing the total volume to 60 mL. The oral liquid formulation was packaged in 30-mL amber plastic prescription bottles. The samples were stored at 5 and 24°C for 14 days. No change in color or consistency was observed in the samples. HPLC analysis found no loss of granisetron at either temperature after 14 days of storage.

**STUDY 2:** Nahata et al.[590] studied the stability of two granisetron hydrochloride 0.05-mg/mL oral suspensions compounded from tablets. Four tablets of granisetron 1 mg (as the hydrochloride) (SmithKline Beecham) were crushed to powder using a mortar and pestle, and 80 mL of 1% methylcellulose in simple syrup or an equal parts mixture of Ora-Plus suspending medium (Paddock) and Ora-Sweet (Paddock) was incorporated with mixing. The suspensions were packaged in two-ounce amber plastic prescription bottles and were stored in a refrigerator at 4°C and in a water bath at 25°C exposed to fluorescent light for 91 days. No change in the physical appearance of any sample was observed. Stability-indicating HPLC analysis found about 4 to 6% granisetron hydrochloride loss in all samples regardless of storage condition.

### Injection

Injections, like other sterile drugs, should be prepared in a suitable clean air environment using appropriate aseptic procedures. When prepared from nonsterile components, an appropriate and effective sterilization method must be employed.

**STUDY 1:** Mayron and Gennaro[477] reported on the stability of granisetron 1 mg/mL (as the hydrochloride) in 5% dextrose injection, 0.9% sodium chloride injection, and bacteriostatic water for injection (containing 0.9% benzyl alcohol) repackaged as 4 mL of solution in 5-mL polypropylene syringes (Becton Dickinson). After storage for 24 hours at room temperature of 20°C exposed to fluorescent light, the solutions remained clear and colorless, and stability-indicating HPLC analysis found no loss of granisetron hydrochloride.

**STUDY 2:** Quercia et al.[478] found granisetron hydrochloride solutions to be stable over 14 days in concentrations of 0.05, 0.07, and 0.1 mg/mL in 0.9% sodium chloride or 5% dextrose injections packaged in polypropylene syringes stored at 5 and 24°C both protected from and exposed to light. No change in color or clarity was observed, and HPLC analysis found no loss of granisetron in any of the samples.

## Compatibility with Common Beverages and Foods

Mayron and Gennaro[477] evaluated the stability of granisetron 1 mg/mL (as the hydrochloride) in four oral liquids: apple juice (Motts), orange juice (Minute Maid), Coca-Cola Classic, and Gatorade lemon-lime electrolyte replacement solution. One milliliter of the granisetron hydrochloride solution was added to 50 mL of each of the oral liquids, yielding a nominal granisetron concentration of 0.02 mg/mL. The samples were stored at room temperature (20°C) exposed to

fluorescent light for one hour. No changes in color or clarity were observed. Stability-indicating HPLC analysis found no loss of granisetron hydrochloride in any sample. No change in color or consistency was observed in the samples. HPLC analysis found no loss of granisetron at either temperature after 14 days of storage. ■

# Griseofulvin

## Properties

Griseofulvin is a white to creamy or yellowish-white odorless powder having a bitter taste.[2,3] Not less than 900 mcg is contained in 1 mg of bulk powder.[4] Microsize griseofulvin is composed of particles predominantly around 4 μm in diameter, while the ultramicrosize griseofulvin particles have a predominant diameter around 1 μm.[2]

### Solubility

Griseofulvin is practically insoluble in water and slightly soluble in ethanol.[1–3]

### pH

Griseofulvin suspension has a pH between 5.5 and 7.5.[4] The suspension (Grifulvin, McNeil) has a measured pH of about 6.4.[19]

## General Stability Considerations

Griseofulvin products should be packaged in tight containers and stored at controlled room temperature protected from temperatures exceeding 40°C.[2,4] The oral suspension should be protected from freezing and light as well.[2]

## Stability Reports of Compounded Products
### Oral

Burgalassi et al.[1537] evaluated the sedimentation rates of several oral liquid suspension formulations of griseofulvin 10 mg/mL using four different suspending mediums: (1) hydroxypropyl

methylcellulose 1%, (2) microcrystalline cellulose 1.5%, (3) carboxymethylcellulose 2.5%, and (4) carrageenan iota 1%. The best suspensions with the slowest sedimentation rates were obtained using microcrystalline cellulose 1.5% and carrageenan iota 1%. The chemical stability of griseofulvin was not tested.

### Enteral
**GRISEOFULVIN COMPATIBILITY SUMMARY**
*Compatible with:* Ensure • Ensure Plus • Osmolite

Cutie et al.[19] added 5 mL of griseofulvin suspension (Grifulvin, McNeil) to varying amounts (15 to 240 mL) of Ensure, Ensure Plus, and Osmolite (Ross Laboratories) with vigorous agitation to ensure thorough mixing. The griseofulvin suspension was physically compatible, distributing uniformly in all three of the enteral products with no phase separation or granulation.

### Topical

Furst et al.[1547] evaluated the stability of griseofulvin 5% incorporated into salicyclic acid ointment and stored at 65°C for 23 days and 55°C for 35 days. The authors noted that these storage periods at elevated temperature corresponded with two years at room temperature. In addition, the formulation was stored for 18 months at room temperature. Ultraviolet spectrophotometric analysis and thin layer chromatography found no loss of griseofulvin occurred in any of the tests. ■

# Guaifenesin

## Properties

Guaifenesin is a white to slightly gray crystalline powder having a bitter aromatic taste. It is odorless or has a faint odor.[1–3]

### Solubility

Guaifenesin is soluble in water;[2] the solubility has been cited as 1 g dissolving in about 20 mL at 25°C.[1] It is more soluble in hot water.[1] Guaifenesin is freely soluble in ethanol, with 1 g dissolving in about 11 mL.[1–3]

### pH

A 1% aqueous solution of guaifenesin has a pH between 5 and 7.[3] Guaifenesin syrup has a pH between 2.3 and 3.[4] Guaifenesin expectorant (Robitussin, Robins) had a measured pH of 2.6.[19]

## General Stability Considerations

Guaifenesin products should be stored in tight containers.[2–4] During storage, guaifenesin powder may become lumpy.[2]

## Stability Reports of Compounded Preparations
### Injection

Injections, like other sterile drugs, should be prepared in a suitable clean air environment using appropriate aseptic procedures. When prepared from nonsterile components, an appropriate and effective sterilization method must be employed.

Grandy and McDonnell[1183] evaluated three concentrations of guaifenesin, 5, 10, and 15% diluted in sterile water for injection, sodium chloride 0.9%, and dextrose 5% as potential veterinary injections for equine use. Guaifenesin concentrations of 33 and 20% in horses have resulted in hemolysis. The hemolytic threshold is believed to be between 16 and 20% in horses. In humans, a 12.5% concentration has resulted in hemolysis.

The 15% concentration in all three diluents was difficult to dissolve and required heating to 37°C. Stored at 21°C, the 15% concentration precipitated within 24 hours, the 10% concentration precipitated in 48 to 72 hours, and the 5% concentration did not precipitate in a week. Warming of precipitated solutions to 37°C redissolved the crystals. Precipitation occurred more rapidly in the sodium chloride 0.9% and dextrose 5% solutions and was slowest in sterile water for injection.

The pH values of all solutions were clinically acceptable in the range of pH 7.03 to 8.53. The solutions in sodium chloride 0.9% and dextrose 5% were hypertonic in the range of 463 to 586 mOsm/kg. Guaifenesin 5% in sterile water for injection was very hypotonic at 184 mOsm/kg, while the 10 and 15% concentrations were slightly hypotonic at 242 and 267 mOsm/kg. The authors concluded that the best option, considering all factors, was to prepare guaifenesin 10% injection in sterile water for injection for equine use.

### Enteral

**GUAIFENESIN COMPATIBILITY SUMMARY**

*Compatible with:* Vital • Vivonex T.E.N.

*Incompatible with:* Enrich • Ensure • Ensure HN • Ensure Plus • Ensure Plus HN • Osmolite • Osmolite HN • TwoCal HN

**STUDY 1:** Cutie et al.[19] added 10 mL of guaifenesin expectorant (Robitussin, Robins) to varying amounts (15 to 240 mL) of Ensure, Ensure Plus, and Osmolite (Ross Laboratories) with vigorous agitation. The guaifenesin expectorant was physically incompatible in all three enteral products, forming a viscous flocculent precipitate that may clog feeding tubes.

**STUDY 2:** Altman and Cutie[850] reported the physical incompatibility of guaifenesin expectorant (Robitussin, Robins) 10 mL with varying amounts (15 to 240 mL) of Ensure HN, Ensure Plus HN, Osmolite HN, and Vital with vigorous agitation to ensure thorough mixing. The guaifenesin expectorant was physically incompatible in Ensure HN, Ensure Plus HN, and Osmolite HN enteral products, forming a viscous flocculent precipitate that may clog feeding tubes. The guaifenesin expectorant was compatible with Vital.

**STUDY 3:** Burns et al.[739] reported the physical compatibility of guaifenesin syrup (Robitussin) 5 mL with 10 mL of three enteral formulas, including Enrich, TwoCal HN, and Vivonex T.E.N. Visual inspection found no physical incompatibility with Vivonex T.E.N. When mixed with Enrich, however, particulates formed and separated. Mixed with TwoCal HN, a thick gelatinous mass formed. ▪

# Halobetasol Propionate

## Properties
Halobetasol propionate is a synthetic corticosteroid that occurs as a white crystalline powder.[7]

### Solubility
Halobetasol propionate is insoluble in water.[7]

## General Stability Considerations
Halobetasol propionate topical products should be stored at controlled room temperature.[7]

## Compatibility with Other Drugs
### HALOBETASOL PROPIONATE COMPATIBILITY SUMMARY
*Compatible with:* Calcipotriene • Tazarotene

**STUDY 1:** Patel et al.[635] evaluated the compatibility of calcipotriene 0.005% ointment mixed by levigation in equal quantities with other topical medications. The mixtures were stored at 5 and 25°C. Drug concentration was assessed by HPLC analysis. Calcipotriene 0.005% ointment mixed with halobetasol propionate 0.05% ointment resulted in no changes in physical appearance and little or no change in either drug concentration in 13 days stored at either temperature. Similarly, mixed with halobetasol propionate 0.05% cream, no changes in physical appearance occurred in 13 days. Both drugs remained stable for 48 hours, but analysis was suspended at that point.

**STUDY 2:** Hecker et al.[815] evaluated the compatibility of tazarotene 0.05% gel combined in equal quantity with other topical products, including halobetasol propionate 0.05% cream and ointment. The mixtures were sealed and stored at 30°C for two weeks. HPLC analysis found that less than 10% tazarotene loss occurred in all samples. Halobetasol propionate in the combinations could not be assayed due to interference with tazarotene. ■

# Haloperidol
# Haloperidol Lactate

## Properties
Haloperidol is a white to yellowish amorphous or microcrystalline powder.[2,3] The injection and the oral solution are prepared with the aid of lactic acid to form haloperidol lactate.[2]

### Solubility
Haloperidol is practically insoluble in water, with an aqueous solubility of about 0.014 mg/mL.[1-3] In ethanol, haloperidol has a solubility of about 16.7 mg/mL.[2,3] The drug is also soluble in dilute acids.[1]

### pH
Haloperidol injection has a pH between 3 and 3.8, while the oral solution has a pH between 2.75 and 3.75.[4] Haloperidol lactate oral drops (Haldol, McNeil) had a measured pH of 3.2.[19]

### $pK_a$
Haloperidol has a $pK_a$ of 8.3.[1,2]

### Osmolality
Haloperidol lactate concentrate 2 mg/mL (McNeil) had an osmolality of 500 mOsm/kg.[233]

## General Stability Considerations

Haloperidol products should be packaged in tight, light-resistant containers and stored at controlled room temperature protected from temperatures of 40°C or more.[2,4] The liquid injection and the oral solution also should be protected from freezing.[2,3] Long-term exposure to light causes discoloration and the formation of a grayish-red precipitate.[8]

Haloperidol was tested for compatibility with three common components of oral dosage forms using differential scanning calorimetry, electron microscopy, infrared spectroscopy, and x-ray diffraction. The drug was found to be incompatible with polyvinyl pyrrolidone (PVP) but compatible with magnesium stearate and lactose.[1593]

## Stability Reports of Compounded Preparations
### Enteral
**HALOPERIDOL LACTATE COMPATIBILITY SUMMARY**
*Compatible with:* Enrich • Ensure • Ensure HN • Ensure Plus • Ensure Plus HN • Osmolite • Osmolite HN • Precitene • TwoCal HN • Vital • Vivonex T.E.N.

**STUDY 1:** Cutie et al.[19] added 1 mL of haloperidol lactate oral drops (Haldol, McNeil) to varying amounts (15 to 240 mL) of Ensure, Ensure Plus, and Osmolite (Ross Laboratories) with vigorous agitation to ensure thorough mixing. The haloperidol lactate drops were physically compatible, distributing uniformly in all three enteral products with no phase separation or granulation.

**STUDY 2:** Burns et al.[739] reported the physical compatibility of haloperidol lactate oral drops (Haldol) 4 mg/2 mL with 10 mL of three enteral formulas, including Enrich, TwoCal HN, and Vivonex T.E.N. Visual inspection found no physical incompatibility with any of the enteral formulas.

**STUDY 3:** Altman and Cutie[850] reported the physical compatibility of haloperidol lactate oral drops (Haldol, McNeil) 1 mL with varying amounts (15 to 240 mL) of Ensure HN, Ensure Plus HN, Osmolite HN, and Vital after vigorous agitation to ensure thorough mixing. Haloperidol lactate oral drops were physically compatible, distributing uniformly in all four enteral products with no phase separation or granulation.

**STUDY 4:** Ortega de la Cruz et al.[1101] reported the physical compatibility of an unspecified amount of oral liquid haloperidol with 200 mL of Precitene (Novartis) enteral nutrition diet for a 24-hour observation period. No particle growth or phase separation was observed.

### Oral
**STUDY 1:** Kleinberg et al.[23] reported the long-term stability of haloperidol oral concentrate 5 mg/2.5 mL (McNeil)

repackaged in 3-mL amber Ped-Pod (Solopak) oral dispensers and stored at 25°C. Haloperidol content was unchanged during 360 days of storage at room temperature when analyzed by HPLC. Furthermore, there were no changes in pH or physical appearance.

**STUDY 2:** Geller et al.[242] reported that haloperidol liquid was visually compatible with water, apple juice, orange juice, tomato juice, and cola sodas, although visual observation in opaque liquids is problematic. Haloperidol was stated to be incompatible with coffee, tea, and saline solutions.

**STUDY 3:** Kulhanek et al.[243] combined equal quantities of haloperidol 2-mg/mL oral drops with coffee and tea. Flaky precipitation was seen. Analysis found that the precipitate contained haloperidol.

## Compatibility with Other Drugs
**HALOPERIDOL LACTATE COMPATIBILITY SUMMARY**
*Incompatible with:* Lithium citrate • Thioridazine

**STUDY 1:** Theesen et al.[78] evaluated the compatibility of lithium citrate syrup with oral neuroleptic solutions, including haloperidol lactate concentrate. Lithium citrate syrups (Lithionate-S, Rowell and the Roxane product) at a lithium-ion concentration of 1.6 mEq/mL were combined in volumes of 5 and 10 mL with 5, 15, 30, and 60 mL of haloperidol lactate oral concentrate 2 mg/mL (McNeil) in duplicate, with the order of mixing reversed between samples. Samples were stored at 4 and 25°C for six hours and evaluated visually for physical compatibility. An opaque turbidity formed immediately in all samples and persisted throughout the study. Thin-layer chromatographic analysis of the sediment layers of incompatible combinations showed high concentrations of the neuroleptic drugs. The authors attributed the incompatibilities to the excessive ionic strength of lithium citrate syrup, which results in salting out of the neuroleptic drug salts in a hydrophobic, viscid layer that adheres tenaciously to container surfaces. The problem persists even when moderate to high doses are diluted 10-fold in water.

**STUDY 2:** Raleigh[454] evaluated the compatibility of lithium citrate syrup (Philips Roxane) 1.6 mEq lithium/mL with haloperidol concentrate 2 mg/mL (as the lactate). The haloperidol concentrate was combined with the lithium citrate syrup in a manner described as "simple random dosage combinations"; the exact amounts tested and test conditions were not specified. No precipitate was observed to form upon mixing.

Raleigh[454] also evaluated the compatibility of thioridazine suspension (Sandoz) (concentration unspecified) with haloperidol concentrate 2 mg/mL (as the lactate). The haloperidol concentrate was combined with the thioridazine suspension in the manner described as "simple random dosage combinations." The exact amounts tested and test conditions were not specified. A dense white curdy precipitate formed.

## Compatibility with Common Beverages and Foods

Theesen et al.[78] noted that fruit and vegetable beverages, which are often used to administer the neuroleptics, also are effective salting-out agents. Unfortunately, visually observing the phenomenon is impaired by the color and opacity of the beverages. ■

# ■ Hemiacidrin
## (Renacidin)

## Properties

Hemiacidrin is a commercial genitourinary irrigant. A 300-g bottle of the powder consists of citric acid (anhydrous) 156 to 171 g, d-gluconic acid (as the lactone) 21 to 30 g, purified magnesium hydroxycarbonate 75 to 87 g, magnesium acid citrate 9 to 15 g, calcium (as the carbonate) 2 to 6 g, and water 17 to 21 g.[7]

As a solution, 100 mL consists of citric acid (anhydrous) 6.602 g, glucono-delta-lactone 0.198 g, magnesium carbonate 3.177 g, and benzoic acid 0.023 g.[7]

### Solubility

Citric acid is very soluble in water, having aqueous solubilities ranging from 540 mg/mL at 10°C to 640 mg/mL at 30°C and 840 mg/mL at 100°C. It is also freely soluble in ethanol. Gluconolactone is freely soluble in water having an aqueous solubility of 59 mg/mL. It is soluble in ethanol at about 10 mg/mL. Magnesium carbonate hydroxide is practically insoluble in water, having a solubility of 1 part in 330 of water. It is insoluble in ethanol.[1,4]

### pH

Hemiacidrin irrigation solution has a pH of 3.5 to 4.2.[7]

### pKₐ

Citric acid has a $pK_1$ of 3.128, a $pK_2$ of 4.761, and a $pK_3$ of 6.396.[1]

## General Stability Considerations

Hemiacidrin should be stored at controlled room temperature, minimizing exposure to excessive heat or cold. Brief exposure to temperatures up to 40°C or down to 5°C does not affect the product adversely. However, it should be protected from freezing.[7]

## Stability Reports of Compounded Preparations
### Irrigation

Irrigations, like other sterile drugs, should be prepared in a suitable clean air environment using appropriate aseptic procedures. When prepared from nonsterile components, an appropriate and effective sterilization method must be employed.

**STUDY 1:** Newton et al.[88] reported on a method of preparation of hemiacidrin (Renacidin, Guardian Chemical) irrigation that is an alternative to the standard autoclaving method recommended by the manufacturer. Sterilization is necessary to avoid microbial growth in the irrigation solution. If flexible polyvinyl chloride bags are used, autoclaving is not possible. However, sealed glass containers pose the potential of excessive pressure buildup from carbon dioxide evolution. As an alternative, sterilization by filtration was recommended.

The contents of a 300-g bottle of hemiacidrin (Renacidin) were added to 3 quarts (2840 mL) of sterile water for irrigation and stirred with a magnetic stirrer for three hours. This period allows completion of effervescence from the evolution of carbon dioxide, which usually subsides in about 30 minutes. The solution was passed through a 2.5-μm depth filter (Millipore, type AP 25) into a clean glass bottle. The depth filter collected the gray-brown siliceous flocculent precipitate invariably present in the initial unfiltered solution. The prefiltered solution was subjected to sterilization filtration. A 0.22-μm filter (Millipore, number GSWP 047 00) with prefilter (Millipore, type AP 15) and a 0.45-μm filter (Millipore, number HAWP 047 00) were connected in series. The 0.45-μm filter served to ensure the physical integrity of the 0.22-μm filter without increasing pressures. The solution was filtered through this filter series and packaged aseptically in sterile, pyrogen-free evacuated liter bottles (950 mL fill) in a Class 100 laminar air flow environment using proper aseptic procedures. No microbial growth was found when a USP sterility test was run on the 0.45-μm filters in either routinely prepared bottles or those with an intentional bacterial contamination. Physicochemical stability of the preparation was monitored by pH, visual examination, vacuum maintenance in the bottles, and absorbance spectrum from 400 to 200 nm. The authors recommended a shelf life of six months stored under refrigeration. Room temperature storage was not recommended due to the potential for excessive pressures from carbon dioxide evolution.

**STUDY 2:** Sewell and Venables,[89] however, strongly recommended autoclaving to sterilize hemiacidrin (Renacidin) irrigation. In their studies, the solution was prepared with pyrogen-free purified water, filtered through a 0.45-μm filter, and filled into rigid 150-mL polypropylene containers that were sealed and autoclaved at 115°C and 34 psi for 30 minutes. The product was stable for 24 months at room temperature by spectrophotometric analysis, pH determination, and visual inspection. Fluid loss did not exceed 0.45% in two years. However, autoclaving at 121°C for one or two hours increased degradation of product content. ■

# Histamine Phosphate

## Properties

Histamine phosphate occurs as colorless odorless prismatic crystals. Both histamine phosphate anhydrous 2.76 mg and histamine phosphate monohydrate 2.93 mg are approximately equivalent to histamine 1 mg.[1,3]

### Solubility

Histamine phosphate has an aqueous solubility of about 250 mg/mL. It is slightly soluble in ethanol.[1,3]

### pH

A 5% histamine phosphate solution has a pH of 3.75 to 3.95.[3] Histamine phosphate injection has a pH between 3 and 6.[4]

### pK$_a$

Histamine has pK$_a$ values of 5.94 and 9.75.[1627]

## General Stability Considerations

Histamine phosphate products should be stored in tight containers protected from light.[3,4]

Histamine salts are subject to photooxidation at high levels of irradiation.[206,207] In a study by Marwaha et al.,[372] exposure to ultraviolet light for four hours caused a 27% loss, considerably more than the 11.5% loss caused by boiling for 4.5 hours. However, normal room light exposure seems to have little or no effect on stability over eight months.[91]

Histamine has a strong tendency to bind to glass.[208] It also can be degraded by microorganisms.[209,210]

## Stability Reports of Compounded Preparations
### Inhalation

Inhalations, like other sterile drugs, should be prepared in a suitable clean air environment using appropriate aseptic procedures. When prepared from nonsterile components, an appropriate and effective sterilization method must be employed.

**STUDY 1:** The stability of histamine (as the diphosphate) in bronchoprovocation solutions was evaluated by Marwaha and Johnson.[91] Histamine diphosphate (ICN Nutritional Biochemicals) was dissolved in a sterile isotonic phosphate buffer, pH 7.4, to yield histamine concentrations of 2 and 8 mg/mL present as the diphosphate salt. The solutions were packaged in sterilized translucent and black plastic bottles (Teflon, FEP Nalgene, Cole-Parmer Instrument Co.) and stored at 12°C exposed to fluorescent light for eight months. Samples also were placed in sterilized polypropylene tubes with snap caps, covered with aluminum foil, and stored frozen at −20°C for a year. Histamine content was assessed by a stability-indicating colorimetric assay. The results were somewhat variable, but at least 90% of the initial histamine amount was retained for eight months at 12°C in both translucent and black bottles. At least 90% was retained for 12 months in polypropylene tubes frozen at −20°C and wrapped in aluminum foil.

**STUDY 2:** McDonald et al.[487] evaluated the stability of histamine phosphate inhalation solution to autoclaving and during storage. Histamine acid phosphate (Sigma) was dissolved in 0.067 M phosphate buffer to yield a concentration of 104.6 mcg/mL, filled into 10-mL brown glass vials, and sealed with rubber closures. The vials were sterilized in an autoclave at 121°C for 15 minutes. HPLC analysis of the solutions found little or no loss after autoclaving and over 12 months of storage at 20°C protected from light. ■

# Hyaluronidase

## Properties

Hyaluronidase is an enzyme that can hydrolyze mucopolysaccharides of the hyaluronic acid type. Hyaluronidase occurs as a white or yellowish-white amorphous powder containing not less than 300 international units of activity per milligram.[3]

## Solubility

Hyaluronidase is soluble in water but is practically insoluble in ethanol and acetone.[3]

## pH

Hyaluronidase 0.3% in water has a pH of 4.5 to 7.5.[3] Hyaluronidase injection, USP, has a pH of 6.4 to 7.4.[4]

## Osmolality

Hyaluronidase injection, USP, 150 units/mL has an osmolality of 300 mOsm/kg.[900]

## General Stability Considerations

Commercial hyaluronidase injection, USP, should be stored under refrigeration. Hyaluronidase for injection, USP, in dry form should be stored at controlled room temperature prior to reconstitution.[4]

Hyaluronidase for injection reconstituted with sodium chloride 0.9% is stable for up to two weeks at temperatures below 25°C.[2] Hyaluronidase in citric acid–sodium citrate buffer (pH 4.5) stored at 4 and 23°C lost about 7 to 8% of its activity over 24 hours and 25 to 33% in 48 hours.[8]

## Stability Reports of Compounded Preparations
### Injection

Injections, like other sterile drugs, should be prepared in a suitable clean air environment using appropriate aseptic procedures. When prepared from nonsterile components, an appropriate and effective sterilization method must be employed.

Allen[742] reported on the formulation and stability of an extemporaneously prepared hyaluronidase 150-units/mL injection. The injection used the formula of the formerly available commercial product shown in Table 43. The author indicated that a beyond-use date of six months stored under refrigeration was appropriate based on the USP and the formerly available commercial product with a three-year expiration date. ▪

**TABLE 43.** Hyaluronidase Injection Formula[743]

| Component | | Amount |
| --- | --- | --- |
| Hyaluronidase | | 15,000 units |
| Sodium chloride | | 850 mg |
| Edetate disodium | | 100 mg |
| Calcium chloride dihydrate | | 53 mg |
| Thimerosal | | 10 mg |
| Sodium phosphate monobasic, anhydrous | | 170 mg |
| Sterile water for injection | qs | 100 mL |
| Sodium hydroxide 1% | qs | pH 6.4 to 7.4 |

# Hydralazine Hydrochloride

## Properties

Hydralazine hydrochloride is a white to off-white odorless crystalline powder.[2,3]

## Solubility

Hydralazine hydrochloride has an aqueous solubility of approximately 40 to 44 mg/mL at 25°C and 30 mg/mL at 15°C. The solubility of hydralazine hydrochloride in ethanol is about 2 mg/mL.[1–3]

## pH

A 2% hydralazine hydrochloride aqueous solution has a pH between 3.5 and 4.2.[3,4] The injection has a pH of 3.4 to 4.4. The oral solution has a pH of 3 to 5.[4]

## pKa

The drug has a pKa of 7.3.[2]

## General Stability Considerations

Hydralazine hydrochloride tablets should be packaged in tight, light-resistant containers and stored at controlled room temperature protected from temperatures of 40°C or more.[2,4] The injection also should be protected from freezing;[2] refrigeration should be avoided because of possible precipitate formation.[370] However, the USP states that the oral solution should be stored under refrigeration.[4]

Hydralazine decomposes in solutions having a pH greater than 7.[862] The pH of maximum stability in dextrose solutions was approximately 3.2 to 4.4,[259] although admixture in dextrose-containing or other infusion solutions is not recommended because of color changes.[2,7] Contact with metal parts and equipment resulted in discolored solutions, often yellow or pink.[374]

Halasi and Nairn[373] reported that exposure of hydralazine hydrochloride products to light increases the decomposition rate during storage. In aqueous solutions, losses of 10% occurred in 9.9 to 12.3 weeks in samples exposed to fluorescent light and in 12.8 to 14.4 weeks in samples stored in the dark.

Lactose, a reducing sugar, is present in the commercial tablets and may accelerate the degradation rate of hydralazine.[373,594]

Lovering et al.[1470] evaluated the formation of hydrazine, a carcinogenic decomposition product of hydralazine, in several dosage forms of the drug evaluated over a two-year storage period at room temperature and elevated temperatures of 37°C and 75% relative humidity. In oral tablets, no significant change in the amount of hydrazine was detected over two years. However, in hydralazine injection, a 20-mg/mL aqueous liquid, hydrazine levels increased from 4.5 to 12 mcg/mL at room temperature and from 4.5 to 76 mcg/mL at 37°C in 16 months.

## Stability Reports of Compounded Preparations
### Oral
#### USP OFFICIAL FORMULATION (ORAL SOLUTION):
Hydralazine hydrochloride

| | | |
|---|---|---|
| for 0.1% (1 mg/mL) | 100 mg | |
| for 1.0% (10 mg/mL) | 1 g | |
| Sorbitol 70% | 40 g | |
| Methylparaben | 65 mg | |
| Propylparaben | 35 mg | |
| Propylene glycol | 10 g | |
| Aspartame | 50 mg | |
| Purified water | qs | 100 mL |

Add hydralazine hydrochloride powder to 30 mL of purified water, add the aspartame, and shake or stir until completely dissolved. Add the sorbitol 70% solution and mix well. In a separate container dissolve methylparaben and propylparaben in propylene glycol. Add this preservative mixture to the hydralazine hydrochloride solution. Add sufficient purified water to bring to volume and mix well, yielding a hydralazine hydrochloride 1-mg/mL or 10-mg/mL solution. Quantitatively transfer the solution to a suitable tight and light-resistant glass or plastic bottle with child-resistant closure. The final liquid preparation should have a pH between 3 and 5. Store the preparation under refrigeration between 2 and 8°C. The beyond-use date is 30 days from the date of compounding.[4,5]

**STUDY 1:** The stability of hydralazine hydrochloride in an extemporaneously compounded syrup and with various potential adjuvants was studied by Alexander et al.[116] The

syrup was prepared by crushing to fine powder 75 hydralazine hydrochloride 50-mg tablets (United Research Laboratories) and adding 250 mL of distilled water to dissolve the drug. A 2250-g quantity of a syrup vehicle composed of 75% (wt/wt) maltitol (Lycasin, Roquette) was added slowly and mixed well. Edetate sodium 3 g and saccharin sodium 3 g were dissolved in 50 mL of distilled water and stirred into the mixture. Thirty milliliters of a preservative stock solution composed of methylparaben 10% (wt/vol) and propylparaben 2% (wt/vol) in propylene glycol also was added. Orange flavoring 3 mL then was added, and the mixture was brought to 3 liters with distilled water. The pH was adjusted with glacial acetic acid to 3.7. The finished syrup was packaged in amber glass bottles and was stored at 5°C and several elevated temperatures. Hydralazine hydrochloride content was assessed using a stability-indicating HPLC assay. Less than 2% loss of hydralazine hydrochloride occurred in two weeks at 5°C. Based on the elevated temperature data, the shelf life at room temperature (25°C) was calculated to be 5.13 days.

In a second series of experiments, Alexander et al.[116] found hydralazine hydrochloride to be incompatible with edetate sodium and sodium bisulfite in aqueous solution. The hydralazine hydrochloride solutions initially were clear and colorless, but they turned yellow immediately after edetate sodium was added and developed a yellow precipitate over time. Sodium bisulfite also caused an immediate yellow discoloration that was even more intense, and eventually a yellow precipitate developed.

**STUDY 2:** Using a stability-indicating HPLC assay, Gupta et al.[259] investigated the stability of 1% hydralazine hydrochloride in aqueous vehicles containing various sweetening agents. Dextrose, fructose, lactose, and maltose exhibited adverse effects on hydralazine hydrochloride stability. Hydralazine losses of 30 to 70% occurred in 24 hours for samples stored in 60-mL amber bottles at 24°C. The use of hydralazine hydrochloride with hydrolyzed sucrose in simple syrup or strawberry syrup was even worse, causing losses of 93 to 95% in one day.

However, Gupta et al.[259] noted that unhydrolyzed 85% sucrose solution was much less problematic with 1% hydralazine hydrochloride. Losses of 10% occurred in about seven days at 24°C. In 0.28 and 0.56 M sorbitol solutions, hydralazine hydrochloride losses were quite low: 4 and 8%, respectively, were lost in 21 days at 24°C. The best hydralazine solution stability occurred in 0.28 M mannitol; no hydralazine hydrochloride loss occurred after 21 days of storage at 24°C.

**STUDY 3:** Allen and Erickson[594] evaluated the stability of three hydralazine hydrochloride 4-mg/mL oral suspensions extemporaneously compounded from tablets. Vehicles used in this study were (1) an equal parts mixture of

Ora-Sweet and Ora-Plus (Paddock), (2) an equal parts mixture of Ora-Sweet SF and Ora-Plus (Paddock), and (3) cherry syrup (Robinson Laboratories) mixed 1:4 with simple syrup. Four hydralazine hydrochloride 100-mg tablets (Rugby) were crushed and comminuted to fine powder using a mortar and pestle. About 15 mL of the test vehicle was added to the powder and mixed to yield a uniform paste. Additional vehicle was added geometrically, and the suspension was brought to the final volume of 100 mL, with thorough mixing after each addition. The process was repeated for each of the three test suspension vehicles. Samples of each finished suspension were packaged in 120-mL amber polyethylene terephthalate plastic prescription bottles and stored at 5 and 25°C.

Cherry syrup was clearly unacceptable for use as the vehicle. Stability-indicating HPLC analysis found nearly 75% hydralazine hydrochloride loss in one day at room temperature in the cherry syrup vehicle. Stability was somewhat better in the other two vehicles but was still not good. At room temperature, losses of 22 and 13%, respectively, occurred in just one day. Under refrigeration, less than 10% loss occurred in one day in the Ora-Sweet vehicle and in two days in the Ora-Sweet SF vehicle. Allen and Erickson suggested that because of the instability of hydralazine hydrochloride, it might be more appropriate to dispense the drug in a manner permitting mixing of the drug and vehicle just before administration. One method suggested was to crush and grind the tablets to powder and fill the correct single doses in individual capsules that could be emptied into a suitable vehicle for administration.

**STUDY 4:** Okeke et al.[862] reported the stability of hydralazine hydrochloride 1- and 10-mg/mL oral solutions having the formula in Table 44 and with or without three or four drops of raspberry flavor added prior to final volume. The preparations were packaged in amber polyethylene terephthalate

**TABLE 44.** Hydralazine Hydrochloride Formulation Evaluated by Okeke et al.[862]

| Component | Amount |
|---|---|
| Hydralazine hydrochloride | 100 mg or 1 g |
| Sorbitol 70% | 40 g |
| Methylparaben | 85 mg |
| Propylparaben | 35 mg |
| Propylene glycol | 10 g |
| Aspartame | 50 mg |
| Raspberry flavor[a] | 3 or 4 drops |
| Purified water | qs 100 mL |

[a]Tested with and without raspberry flavor.

plastic bottles and in amber glass bottles that were stored refrigerated at 5°C and at room temperature of 25°C with 60% relative humidity. Stability-indicating HPLC analysis found that the solution stability was influenced by drug concentration as well as storage temperature, with the higher concentration being more stable.

For the flavored 1-mg/mL concentration at room temperature, hydralazine hydrochloride exhibited 6 to 7% loss in 14 days and 10 to 12% loss in 22 days. The flavored 10-mg/mL concentration was more stable, with about 9 to 10% loss occurring in 92 days. In addition, the solution developed a yellow color, which was more intense in the 10-mg/mL concentration.

Under refrigeration, hydralazine hydrochloride was slightly more stable without the raspberry flavoring. For the flavored 1-mg/mL concentration, 10 to 12% loss occurred in 92 days; without flavor, about 3 to 6% loss occurred in that time period. For the flavored 10-mg/mL concentration, about 5 to 6% loss occurred in 92 days; without flavor, about 2 to 4% loss occurred in that time period. ■

# ■ Hydrochlorothiazide

## Properties

Hydrochlorothiazide is a white or practically white practically odorless crystalline powder.[3,4] Hydrochlorothiazide has a slightly bitter taste.[2]

### Solubility

Hydrochlorothiazide is slightly soluble in water, soluble in ethanol and acetone, and freely soluble in dilute sodium hydroxide solution and dimethylformamide. It is insoluble in dilute acids.[1,3,4]

### $pK_a$

Hydrochlorothiazide has $pK_a$ values of 7.9 and 9.2.[1]

## General Stability Considerations

Hydrochlorothiazide powder and commercial tablets should be stored in well-closed containers at controlled room temperature[4] and protected from light.[7] The commercial oral solution should also be stored at controlled room temperature, and freezing should be avoided.[2,7]

Hydrochlorothiazide is subject to hydrolysis in both very acidic and in basic media.[1363] This hydrolysis is a reversible equilibrium process that is independent of pH in the range of pH 1.5 to 8.2 with the equilibrium favoring hydrochlorothiazide.[1610]

## Stability Reports of Compounded Preparations
### Oral
**STUDY 1:** Totterman et al.[754] reported the stability of hydrochlorothiazide 2-mg/mL oral suspension prepared from powder. Several alternative suspending vehicles were evaluated, and the preferred vehicle was used to test the stability of the drug. The selected test formulation, which had a low osmolality of 20 mOsm/kg, a pH near 3 for drug stability, and a relatively low viscosity for ease of administration through feeding tubes, is shown in Table 45.

The formulations prepared from crushed tablets were deemed to be inferior to those made from the pure hydrochlorothiazide powder. The oral suspension was stored at ambient room temperature protected from exposure to light for 10 weeks. Stability-indicating HPLC analysis found about 8% loss of drug at the end of the 10-week period.

**STUDY 2:** Fajolle et al.[952] prepared an oral suspension of hydrochlorothiazide 5 mg/mL in Ora-Sweet and Ora-Plus. The oral suspension was deemed palatable by patients and easy to administer by nurses. Microbiological assessment met the requirements of the European Pharmacopoeia. Stability of the drug was not evaluated.

**STUDY 3 (WITH SPIRONOLACTONE):** Tagliari et al.[1610] evaluated the physical stability of hydrochlorothiazide compounded as 2.5-mg/mL oral liquid suspensions for pediatric use. In addition to the hydrochlorothiazide 2.5 mg/mL, the two formulas tested contained either carboxymethylcellulose sodium or hydroxypropylmethylcellulose 0.6% as suspending agents, glycerin 2% as a wetting agent, sodium benzoate 0.1% as a preservative, citric acid to adjust to pH 3.3, and water. The hydrochlorothiazide powder was triturated using a mortar and pestle to reduce the particle size, and glycerin was added to wet the powder. The suspending agent was added as a liquid in water with sodium benzoate preservative. Finally, the citric acid was incorporated to adjust pH.

The hydrochlorothiazide content was found to be near 100% of the target concentration in both preparations by HPLC analysis. Carboxymethylcellulose sodium proved to be a superior suspending agent; the oral suspension exhibited acceptable particle size, zeta potential, sedimentation rate, redispersibility, and rheological behavior. However, hydroxypropyl methylcellulose was not acceptable, failing to maintain a good degree of dispersion and exhibiting caking that produced dosing inconsistency.

**STUDY 4:** Allen and Erickson[542] evaluated the stability of three spironolactone 5-mg/mL and hydrochlorothiazide 5-mg/mL oral suspensions extemporaneously compounded from tablets. Vehicles used were (1) an equal parts mixture of Ora-Sweet and Ora-Plus (Paddock), (2) an equal parts mixture of Ora-Sweet SF and Ora-Plus (Paddock), and (3) cherry syrup (Robinson Laboratories) mixed 1:4 with simple syrup. Twenty-four spironolactone and hydrochlorothiazide 25 + 25-mg tablets (Mylan) were crushed and comminuted to fine powder using a mortar and pestle. About 25 mL of the test vehicle was added to the powder and mixed to yield a uniform paste. Additional vehicle was added geometrically and the suspension was brought to the final volume of 120 mL, with thorough mixing after each addition. The process was repeated for each of the three test suspension vehicles. Samples of each finished suspension were packaged in 120-mL amber polyethylene terephthalate plastic prescription bottles and stored at 5 and 25°C in the dark.

No visual changes or changes in odor were detected during the study. Stability-indicating HPLC analysis found less than 4% spironolactone loss in any of the suspensions stored at either temperature after 60 days of storage. Hydrochlorothiazide exhibited less than 3% loss in the Ora-Sweet-containing suspensions but almost 9% loss in the cherry syrup over 60 days at room temperature. ■

**TABLE 45.** Hydrochlorothiazide Formulation[754]

| Component | | Amount |
| --- | --- | --- |
| Hydrochlorothiazide powder | | 2 mg |
| Citric acid | | 0.8 mg |
| Methocel E50 | | 15 mg |
| Distilled water | qs | 1 mL |

# Hydrocortisone

## Properties

Hydrocortisone is a white or almost white odorless bitter-tasting crystalline powder.[1–4]

Hydrocortisone-21-acetate is a white to nearly white odorless crystalline powder.[4]

Hydrocortisone sodium phosphate is a white to light yellow odorless or practically odorless crystalline powder.[4]

Hydrocortisone sodium succinate is a white or nearly white odorless hygroscopic amorphous powder.[4]

Hydrocortisone-17-valerate is a white crystalline solid material.[7]

### Solubility

Hydrocortisone is very slightly soluble in water, having an aqueous solubility of about 0.28 mg/mL.[1–3] The solubility in ethanol is about 15 mg/mL[1] to 25 mg/ml.[3] In propylene glycol, solubility is about 12.7 mg/mL.[1]

Hydrocortisone-21-acetate is insoluble in water with a solubility of only 1 mg/100 mL. It is slightly soluble in ethanol, having a solubility of 450 mg/100 mL and 1.1 mg/g in acetone.[1,4]

Hydrocortisone sodium phosphate is freely soluble in water, with an aqueous solubility of about 667 mg/mL, and is slightly soluble in ethanol.[3,4]

Hydrocortisone sodium succinate is very soluble in water and ethanol and very slightly soluble in acetone.[4]

Hydrocortisone-17-valerate is soluble in ethanol and methanol, sparingly soluble in propylene glycol, and insoluble in water.[7]

### pH

Hydrocortisone sterile suspension has a pH between 5 and 7, while the enema has a pH between 5.5 and 7. The hydrocortisone with acetic acid otic solution has a pH between 2 and 4 when diluted with an equal quantity of water.[4]

Hydrocortisone sodium phosphate injection has a pH of 7.5 to 8.5.[4]

Hydrocortisone sodium succinate injection has a pH of 7 to 8 at 50 mg/mL in aqueous solution.[4]

### pKₐ

The apparent $pK_a$ by titration of hydrocortisone with sodium hydroxide is 11.05 at 25°C.[6]

### Osmolality

Hydrocortisone sodium phosphate 50 mg/mL has an osmolality of 533 mOsm/kg.[900]

Hydrocortisone sodium succinate 50 mg/mL has an osmolality of 260 to 292 mOsm/kg, depending on the method used in the determination.[8]

## General Stability Considerations

Hydrocortisone products are generally stored in tight or well-closed containers at controlled room temperature.[4] Light protection has been recommended.[3]

The pH range of maximum stability of hydrocortisone is 3.5 to 4.5.[6,1126] In phosphate or carbonate buffers at pH 7 to 9, up to 10% degradation occurs in four hours at 26°C. Degradation can be catalyzed by trace metal impurities.[6]

Hydrocortisone does not appear to be susceptible to oxidation from atmospheric oxygen in pH 3.5 buffered solutions. Incorporation of the antioxidants ascorbic acid, cysteine hydrochloride, propyl gallate, and sodium bisulfite had a negligible effect on hydrocortisone stability.[632]

The use of the sodium phosphate or sodium succinate salts to prepare oral products results in reduced bioavailability compared to oral administration of hydrocortisone alcohol.[639]

## Stability Reports of Compounded Preparations
### Oral

**STUDY 1:** Fawcett et al.[474] evaluated the stability and dosage uniformity of two hydrocortisone 2.5-mg/mL oral suspensions. One suspension was prepared from hydrocortisone bulk powder (Sigma Chemical), while the other was prepared from 20-mg tablets (Douglas Pharmaceuticals) crushed to powder. To make 100 mL of the suspension, the vehicle was prepared by dissolving methyl hydroxybenzoate 20 mg, propyl hydroxybenzoate 80 mg, citric acid monohydrate 600 mg, and syrup, BP, 10 mL in hot water. After allowing the solution to cool, it was triturated with carboxymethylcellulose sodium 1 g and allowed to stand overnight. Hydrocortisone powder 250 mg or 12 hydrocortisone 20-mg tablets ground to fine powder using a glass mortar and pestle were triturated with polysorbate 80 0.5 mL, and the previously prepared suspension vehicle was added. The suspension was brought to 100 mL with more water and packaged in amber high-density polyethylene bottles with polypropylene screw caps. Samples were stored at 5, 25, and 40°C in the dark.

The suspensions had an apparent pH of about 3.4. Stability-indicating HPLC analysis found no hydrocortisone degradation within 90 days at 5 and 25°C. At 40°C, the suspension prepared from bulk hydrocortisone powder was stable for 90 days with no loss, but the suspension prepared from crushed tablets was less stable, sustaining about 20% loss. Dosage uniformity was good, with coefficients of variation of 4.5% or less. Microbiological evaluation was not performed, so the authors recommended limiting the shelf life to 30 days.

**STUDY 2:** Gupta[1106] evaluated the stability of a nonaqueous oral suspension of hydrocortisone 2 mg/mL compounded from micronized powder. Hydrocortisone micronized powder 200 mg was mixed with 10 mL of ethanol and stirred for 10 minutes. Glycerin 40 mL was added in portions with trituration. Ora-Sweet syrup was used to bring the oral liquid to a volume of 100 mL. The oral liquid was packaged in amber glass bottles and stored at room temperature for 60 days. No visible changes in physical appearance occurred, and stability-indicating HPLC analysis found little or no loss of hydrocortisone after 60 days of storage.

**STUDY 3:** Gupta[1557] also evaluated the stability of hydrocortisone 0.2 mg/mL compounded from micronized powder. Hydrocortisone micronized powder 20 mg was mixed in 5 mL of ethanol and stirred for three minutes. After this, 40 mL of Humco simple syrup or Ora-Sweet syrup was added and mixed. The liquids were then brought to 100 mL with additional amounts of the respective syrups. Samples were packaged in 15-mL clear glass vials and were stored at a highly elevated temperature of 60°C for 71 days. HPLC analysis found extensive loss of hydrocortisone content after 15 days (the first assay point). Losses were about 12% in Ora-Sweet and 18% in simple syrup. The author recommended using Ora-Sweet as a vehicle for oral hydrocortisone over simple syrup. However, the study did not determine the actual beyond-use period that should be applied.

**STUDY 4:** Rogerson et al.[1643] evaluated the physical stability of oral suspensions of hydrocortisone acetate prepared from tablets. The 20-mg tablets were ground and incorporated into an OraSweet–OraPlus mixture, xanthan gum (% unspecified), and methylcellulose (% unspecified.) Samples of each formulation were stored at ambient conditions for 15 days. Hydrocortisone concentrations declined in the undisturbed upper portion of each suspension; the best suspension was prepared with xanthan gum and did not exhibit hydrocortisone concentrations below therapeutic amounts in the undisturbed upper portion for nine days, while the others were subtherapeutic in three to four days. However, all suspensions easily redispersed with agitation and maintained therapeutic concentrations over the entire 15 days with no significant differences between formulations. The authors stated that the xanthan gum formulation was their preference of these alternatives.

**STUDY 5:** Santovena et al.[1644] evaluated the stability of hydrocortisone 1-mg/mL oral suspension compounded from powder. The oral suspension had the following formula:

| | | |
|---|---|---|
| Hydrocortisone | | 1 g |
| Tween-80 | | 5 mL |
| Sodium carboxymethylcellulose | | 10 g |
| Methyl paraben | | 200 mg |
| Propylparaben | | 800 mg |
| Syrup | | 150 mL |
| Citric acid | | 6 g |
| Purified water | qs | 1000 mL |

To prepare the formulation, 895 mL of purified water was heated to 70°C and the methylparaben was mixed in with constant shaking. Then a propylparaben and citric acid in syrup (previously prepared) mixture was incorporated. The mixture was allowed to cool and the sodium carboxymethylcellulose was added slowly with continuous shaking. This mixture was allowed to stand for 24 hours. The hydrocortisone powder was pulverized using a mortar and pestle, and the Tween 80 was slowly added to it working it in the mortar until a uniform paste was achieved. The previously prepared vehicle was then incorporated gradually. The mixture was brought to volume with purified water, and the mixture was shaken until homogeneous. The mixture was then packaged in opaque light-protected glass bottles and stored at 5 and 40°C for 90 days.

The white opaque appearance of the oral suspension did not change in any of the samples. The pH and viscosity remained constant, and the particle size increased only slightly. Stability-indicating HPLC analysis found that the suspension stored under refrigeration was stable for the entire 90-day test period. However, at elevated temperature of 40°C, 12% loss of hydrocortisone occurred after 10 days of storage.

## Topical
**STUDY 1:** Gupta[212] evaluated the stability of 1% hydrocortisone in various topical bases and with common additives. The bases are cited in Table 46, along with the other drug

**TABLE 46.** Stability of Various Emulsion Formulations of Topical 1% Hydrocortisone[212]

| Hydrocortisone Base | Type | Months | Hydrocortisone Remaining |
|---|---|---|---|
| Polyethylene glycol | Water washable | 54 | 29.6% |
| Petrolatum | Nonpolar | 41 | 96.8% |
| Cold cream | Water-in-oil | 14 | 98.6% |
| Aquaphor | Water-in-oil | 40 | 99.7% |
| HEB | Oil-in-water | 30 | 98.0% |
| Dermovan | Oil-in-water | 12 | 99.1% |
| Zinc oxide shake lotion with water and glycerin 10% | Aqueous | 15 | 45.2% |

additives. The ointments were prepared by trituration, while the zinc oxide shake lotion was prepared by making a smooth paste of the hydrocortisone with the zinc oxide, talc, and glycerin. The paste was then brought to volume with water and continuous stirring. The samples were packaged in glass ointment jars and stored at room temperature. A stability-indicating HPLC method was used to determine hydrocortisone content. Hydrocortisone losses were substantial in the polyethylene glycol base (70% in 54 months) and in the zinc oxide shake lotion (55% in 15 months). Little or no hydrocortisone was lost in the other formulations, as can be seen in Table 46.

**STUDY 2:** Timmons and Gray[430] also found substantial decomposition of hydrocortisone in a zinc oxide lotion. A lotion containing 20% (wt/wt) zinc oxide, 20% (wt/wt) talc, and 30% (wt/wt) glycerin in water was used as the base. Hydrocortisone lotion BPC was combined to yield a final hydrocortisone concentration of 0.1% (wt/wt) with a pH of 8.76. The hydrocortisone content was evaluated using an HPLC assay. When stored at room temperature, hydrocortisone losses were about 7% in one week and 10% in two weeks. Various stabilizing agents also were tried in the lotion formulation, including 1% (wt/wt) citric acid, 0.2% disodium edetate dihydrate, and 0.1% sodium metabisulfite. No method was very effective in stabilizing the hydrocortisone.

**STUDY 3:** Allen[574] described the compounding of hydrocortisone ointment in a nongreasy, nonaqueous, hydrophilic gel. Hydrocortisone powder 1 g was weighed and mixed with about 95 g of propylene glycol. Carbomer 934 (Carbopol 934) 1.5 g was weighed and added to the mixture and mixed thoroughly. Trolamine (triethanolamine) 0.25 to 0.35 g was added slowly until the desired viscosity was achieved. Additional propylene glycol was then added, bringing the total mass to 100 g. The ointment was packaged in a tight, light-resistant container. The nonoleaginous base allows easy removal of the ointment with water and is less messy for the patient to use. Furthermore, the absence of water from the formulation enhances drug stability. The author indicated that this formulation should be stable for at least 30 days.

**STUDY 4:** Krochmal et al.[776] reported the stability of hydrocortisone 17-valerate 0.2% cream (Westcort) mixed individually with (1) salicylic acid 2%, (2) liquor carbonis detergens 5%, (3) urea 10%, and (4) a combination of camphor 0.25%, menthol 0.25%, and phenol 0.25%. The topical mixtures were packaged in amber glass jars with polyethylene-lined screw caps and were stored for two months at unspecified ambient temperature. The salicylic acid mixture underwent physical separation, but uniformity could be obtained with remixing. No other physical incompatibilities were observed. HPLC analysis found acceptable hydrocortisone concentrations in all mixtures over the two-month study period.

**STUDY 5:** Alberg et al.[1060] evaluated the use of a modified hydrophilic ointment base containing ethanol for suitability as a topical vehicle for hydrocortisone-21-acetate 2%. The base was composed of 9% (wt/wt) cetosteryl alcohol, 10.5% (wt/wt) liquid paraffin, 10.5% (wt/wt) white petrolatum, 10% (wt/wt) ethanol 96%, and 60% (wt/wt) water. After incorporation of the hydrocortisone-21-acetate, samples were stored at 21°C for 12 months and at 50°C for three months. Polarized microscopy and scanning electron microscopy showed some drug particle agglomeration, but the authors did not consider this to be problematic. HPLC analysis of drug content found that less than 1% of the drug decomposed under both storage conditions. The authors indicated that this modified base was a good vehicle for topical hydrocortisone-21-acetate.

**STUDY 6 (WITH ZINC OXIDE):** Gander et al.[1225] evaluated the stability of hydrocortisone and hydrocortisone acetate with zinc oxide in hydrophilic and lipophilic topical bases. The hydrophilic paste contained prednisolone or prednisolone acetate 2.5% with zinc oxide 40%, propylene glycol 15%, and aqueous 3.5% hydroxypropylcellulose gel 45%. The lipophilic ointment contained prednisolone or prednisolone acetate 2.5% with zinc oxide 5%, white beeswax 10%, hard paraffin 10%, and almond oil 75%. No metallic utensils were utilized during compounding. Samples were stored at 20 and 37°C protected from exposure to light.

Stability-indicating HPLC analysis found that hydrocortisone in the hydrophilic base underwent 7% degradation in four weeks at 37°C. No loss occurred in a similar formulation that did not contain zinc oxide, indicating that the zinc oxide was responsible for the loss. Also, no statistically significant hydrocortisone loss occurred at 37°C in the lipophilic formulation, even with the zinc oxide present.

Hydrocortisone acetate samples were more stable. In the hydrophilic formulation, hydrocortisone acetate exhibited about 10% loss in 218 days at 37°C. In addition, hydrocortisone acetate in the lipophilic formulation incurred no loss in six months at either 20 or 37°C.

### Iontophoresis

Seth et al.[757] compared the stability of the sodium phosphate and sodium succinate salts of hydrocortisone in solution both passively and during simulated iontophoresis using stability-indicating HPLC analysis. Hydrocortisone sodium succinate was stable in passive solutions for the 24-hour study period. When current was applied to the solutions, the pH increased from about 8 to about 11. For hydrocortisone sodium succinate, short-term iontophoresis up to 30 minutes did not result in measurable decomposition products. However, during long-term iontophoresis of 24 hours, substantial amounts of free hydrocortisone and other decomposition products appeared. Hydrocortisone sodium phosphate solutions also underwent similarly large pH increases during application of current. However, only about 2% decomposition occurred in four hours, the limit of testing for this salt.

### Rectal

Allen[593] reported on a compounded formulation of hydrocortisone suppositories. The formula for each suppository is shown in Table 47. To compound these suppositories, each component was weighed accurately. The polyethylene glycol 3350 and polyethylene glycol 300 were melted using a water bath or low heat. The hydrocortisone powder was then incorporated into the melted glycols and mixed well. The karaya gum was added and mixed well. The mixture was allowed to cool slightly and then poured into suppository molds of suitable size. The suppositories were then packaged in tight, light-resistant containers. A use period of six months was suggested, although no stability data were presented.

### Injection

Injections, like other sterile drugs, should be prepared in a suitable clean air environment using appropriate aseptic procedures. When prepared from nonsterile components, an appropriate and effective sterilization method must be employed.

Gupta and Ling[904] reported that hydrocortisone sodium succinate 10 mg/mL in sodium chloride 0.9% and packaged in polypropylene syringes exhibited losses by HPLC analysis of 5% in three days and 10% in seven days at room temperature of 25°C and 2% in 21 days refrigerated at 5°C.

## Compatibility with Other Drugs

### HYDROCORTISONE COMPATIBILITY SUMMARY

*Compatible with:* Iodochlorhydroxyquin • Menthol • Phenol • Tacrolimus • Zinc oxide

*Incompatible with:* Calcipotriene

*Uncertain or variable compatibility with:* Mupirocin • Zinc oxide

**STUDY 1 (MULTIPLE DRUGS):** Gupta[212] evaluated the stability of 1% hydrocortisone in oil-in-water emulsion bases with added 3% iodochlorhydroxyquin (12 months), or 0.25%

**TABLE 47.** Hydrocortisone Rectal Suppository Formula per Suppository[593]

| Component | Amount |
| --- | --- |
| Hydrocortisone | 100 mg |
| Karaya gum | 500 mg |
| Polyethylene glycol 3350 | 700 mg |
| Polyethylene glycol 300 | 700 mg |

menthol and 0.25% phenol (30 months) at room temperature. Analysis by a stability-indicating HPLC method found no loss of hydrocortisone in either ointment.

**STUDY 2 (CALCIPOTIENE):** Patel et al.[635] evaluated the compatibility of calcipotriene 0.005% ointment mixed by levigation in equal quantities with other topical medications, including hydrocortisone valerate. The mixtures were stored at 5 and 25°C. Drug concentration was assessed by HPLC analysis. The combination was found to be incompatible. Mixed with hydrocortisone valerate 0.2% ointment, no physical changes were observed, but more than 10% calcipotriene loss occurred within 72 hours at both temperatures.

**STUDY 3 (MUROPIROCIN):** Jagota et al.[920] evaluated the compatibility and stability of mupirocin 2% ointment (Bactroban, Beecham) with hydrocortisone 1% cream, ointment, and lotion (Hytone, Dermik) in a 1:1 proportion over periods of up to 60 days at 37°C. The physical compatibility was assessed by visual inspection, while the chemical stability of mupirocin was evaluated by stability-indicating HPLC analysis. The study found that the hydrocortisone products mixed with mupirocin 2% ointment all resulted in separation and layering and may be unacceptable for patient use. However, they could be remixed back to homogeneity, and the mupirocin loss was less than 10% for 60 days (cream), 45 days (ointment), and 15 days (lotion).

**STUDY 4 (TACROLIMUS):** Pappas et al.[1444] evaluated the compatibility and stability of tacrolimus 0.1% ointment (Protopic, Fujisawa) mixed in a 1:1 ratio with hydrocortisone-17-valerate 0.2% (HydroVal, TaroPharma). Samples were packaged in sealed glass vials and stored at 25°C (60% relative humidity), 30°C (60% relative humidity), and 40°C (75% relative humidity) for 28 days. HPLC analysis found no substantial loss of tacrolimus and hydrocortisone-17-valerate during the study period.

**STUDY 5 (ZINC OXIDE):** Nagatani et al.[1526] evaluated the stability of several steroids in steroid ointments when mixed with zinc oxide ointment. The steroid ointments included hydrocortisone butyrate (Locoid). Samples of the mixed ointments were stored at 5 and 30°C until analyzed. Analysis found that the sample of hydrocortisone butyrate ointment mixed with zinc oxide ointment was stable only at 5°C.

**STUDY 6 (ZINC OXIDE):** Ohishi et al.[1714] evaluated the stability of Locoid and Pandel hydrocortisone ointments mixed in equal quantity (1:1) with zinc oxide ointment. Samples were stored refrigerated at 5°C. HPLC analysis of the ointment mixtures found that the drugs were stable for 32 weeks at 5°C. ■

# Hydrogen Peroxide

## Properties

Hydrogen peroxide topical solution, containing 2.5 to 3.5% hydrogen peroxide, is a clear colorless free-flowing liquid that has no odor or a slight odor resembling ozone and a bitter taste.[1–4]

Hydrogen peroxide concentrate contains 29 to 32% hydrogen peroxide and is a strong oxidizing agent. It is caustic and causes burns of skin and mucous membranes.[1–4] Painful oral ulcerations of the mucosa with sloughing and necrosis of the epithelial surface have been reported in one patient upon oral application of hydrogen peroxide 3%.[963]

Hydrogen peroxide solutions may be described in terms of the volume of available oxygen they yield. A 3% (wt/wt) solution is a 10-volume solution corresponding to 10 times its volume in available oxygen. Similarly, a 6% (wt/wt) solution is a 20 volume, 9% (wt/wt) is a 30 volume, 27% (wt/wt) is a 100 volume, and 30% (wt/wt) is a 110 volume solution.[2,3]

## General Stability Considerations

Hydrogen peroxide topical solution should be stored in tight, light-resistant containers at controlled room temperature.[4]

Hydrogen peroxide concentrate is stored in partially filled vented containers at 8 to 15°C.[3,4]

Hydrogen peroxide solutions gradually decompose on standing or upon agitation. Exposure to light or heat or contact with metals, rough surfaces, alkalies, and many oxidizing or reducing agents causes accelerated decomposition.[1–3] It decomposes violently if trace impurities are present.[1] Hydrogen peroxide was found to be most stable when refrigerated, exhibiting a loss of 0.00032 mole/day.[995] At 30°C, decomposition rates in borosilicate glass containers are 0.5 to 1.0% per year.[1693] Hydrogen peroxide is stabilized by phosphate, pyrophosphate, citrate, tartrate, and borate ions.[1702]

## Stability Reports of Compounded Preparations
### Oral

Allen[1374] reported on a compounded formulation of hydrogen peroxide 1.5% with sodium fluoride 0.5-mg/mL for use as an oral rinse. The solution had the following formula:

| | | |
|---|---|---|
| Hydrogen peroxide 3% | 50 mL | |
| Glycerin | 15 mL | |
| Stevioside powder | 40 mg | |
| Sodium fluoride | 50 mg | |
| Polysorbate 20 | 0.15 mL | |
| Flavor concentrate | qs | |
| Paraben preserved water | qs | 100 mL |

The recommended method of preparation was to add the glycerin to the hydrogen peroxide and mix well. Then the stevioside powder, sodium fluoride, flavor, and polysorbate 20 were added and thoroughly mixed until the solution was clear. Finally, paraben preserved water sufficient to bring the volume to 100 mL was added. The solution was to be packaged in tight, light-resistant containers. The author recommended a beyond-use date of six months at room temperature because this formula is a commercial medication in some countries with an expiration date of two years or more.

### Topical

Dannenberg and Peebles[61] reported an explosive reaction from a compounded irrigation solution composed of an equal parts mixture of povidone-iodine solution (Betadine, Purdue-Frederick) and hydrogen peroxide 3%. The mixture at first appeared to be satisfactory, but about 90 minutes after preparation it exploded. The authors suspected that iodine accelerated the evolution of oxygen from the peroxide decomposition, resulting in the explosion in a container with a tightened cap.

## Compatibility with Other Drugs

Hydrogen peroxide 3% is stated to be incompatible with alkalies; ammonia and ammonium carbonates; albumin; balsam of Peru; phenol; charcoal; chlorides; alkali citrates; ferrous, mercurous, or gold salts; hypophosphites; iodides; lime water; permanganates; sulfites; tinctures; and organic compounds in general.[1,3]

# Hydromorphone Hydrochloride

## Properties

Hydromorphone hydrochloride is a white or nearly white odorless crystalline powder.[2,3]

### Solubility

Hydromorphone hydrochloride has an aqueous solubility of about 333 mg/mL. The drug is sparingly soluble in ethanol.[1–3]

### pH

The injection has a pH of 3.5 to 5.5.[4]

## General Stability Considerations

Hydromorphone hydrochloride injection should be stored at controlled room temperature and protected from light and freezing. The tablets should be packaged in tight, light-resistant containers and stored at controlled room temperature. The oral solution should be packaged in light-resistant containers and stored at controlled room temperature. Suppositories should be stored under refrigeration.[2,4]

## Stability Reports of Compounded Preparations
### Injection
Injections, like other sterile drugs, should be prepared in a suitable clean air environment using appropriate aseptic procedures. When prepared from nonsterile components,

an appropriate and effective sterilization method must be employed.

Stiles et al.[529] evaluated the stability of hydromorphone hydrochloride (Knoll) 10 mg/mL undiluted and 0.1 mg/mL in 0.9% sodium chloride injection. The solutions were packaged as 3 mL in 10-mL polypropylene infusion-pump syringes for the CADD-Micro and CADD-LD pumps (Pharmacia Deltec) sealed with Luer tip caps (Becton Dickinson) and stored at 30°C. HPLC analysis found little or no loss of the drug after 30 days of storage. ■

# ■ Hydroquinone

## Properties
Hydroquinone occurs as fine white needles.[4]

### Solubility
Hydroquinone is soluble in water to 58 mg/mL[3] to 71 mg/mL[1] and freely soluble in ethanol to about 250 mg/mL.[1,3,4]

### pH
Hydroquinone topical solution has a pH of 3 to 4.2.[4]

### pKa
Hydroquinone has a pKa of 9.96.[7]

## General Stability Considerations
Hydroquinone should be stored in tight containers and protected from light.[1,4] Contact of the raw material with skin should be avoided.[1]

Hydroquinone in solution turns brown upon exposure to air due to oxidation; it also oxidizes rapidly in alkaline solutions.[1] Formulated products should be stored in tight, light-resistant containers at controlled room temperature.[4,7]

## Stability Reports of Compounded Preparations
### Topical
Matsubayashi et al.[826] evaluated an extemporaneously compounded hydroquinone cream that was intended to mimic commercially available products. Hydroquinone 5 and 10% with glycerin 10%, ascorbic acid 1.6%, and sodium sulfite 0.5% were incorporated into hydrophilic ointment. Substantial chromatographic aberrations occurred during three months of storage at 4 and 25°C. Similar ointments prepared without ascorbic acid and sodium sulfite were slightly better chromatographically, but higher concentrations of decomposition products occurred. Neither formulation was considered acceptable. ■

# ■ Hydroxychloroquine Sulfate

## Properties
Hydroxychloroquine sulfate is a white to almost white odorless crystalline powder having a bitter taste.[1–3] Hydroxychloroquine sulfate 100 mg is approximately equivalent to 77 mg of the base.[3]

### Solubility
Hydroxychloroquine sulfate has a solubility of about 200 mg/mL in water but is practically insoluble in ethanol.[1–3,7]

### pH
A 1% hydroxychloroquine sulfate solution in water has a pH between 3.5 and 5.5.[1,3]

## General Stability Considerations
Hydroxychloroquine sulfate tablets should be packaged in tight, light-resistant containers and stored at controlled room temperature.[2,4]

## Stability Reports of Compounded Preparations
### Oral
Pesko[251] reported on an extemporaneous suspension of hydroxychloroquine. Fifteen hydroxychloroquine sulfate 200-mg tablets were rubbed with a towel moistened with alcohol to remove the coating. The tablets were ground to fine powder and levigated to a paste with 15 mL of Ora-Plus suspending agent. An additional 45 mL of the suspending agent

was added and the mixture was brought to 120 mL with water for irrigation, yielding a suspension containing hydroxychloroquine sulfate 25 mg/mL. The author noted that sugar and artificial flavorings should not be added to the product. A 30-day expiration period was recommended, although stability testing was not performed. ■

# ■ Hydroxyurea

## Properties

Hydroxyurea is a white to off-white hygroscopic powder.[4]

### Solubility

Hydroxyurea is freely soluble in water and hot alcohol,[1,4] but it is practically insoluble in room temperature ethanol.[3]

## General Stability Considerations

Hydroxyurea powder and capsules should be stored in tight containers in a dry atmosphere at controlled room temperature. Hydroxyurea decomposes in the presence of moisture.[4] The manufacturer has stated that hydroxyurea capsules should not be removed from the original packaging and placed in dosing compliance aids.[1622]

## Stability Reports of Compounded Preparations
### Oral

Heeney et al.[890] reported the stability of hydroxyurea oral liquid compounded from commercial capsule contents (Hydrea, Bristol-Myers Squibb). The capsule contents were mixed with room temperature sterile water through vigorous use of a magnetic stirrer to yield a 200-mg/mL concentration. The liquid was then filtered to remove insoluble excipients. Syrpalta (Humco) flavored syrup without color was then added and mixed to result in a final concentration of 100 mg/mL. The oral liquid was packaged in amber plastic bottles and stored at room temperature.

Colorimetric analysis along with functional activity assessment using T lymphocyte proliferation inhibition found that about 5% hydroxyurea loss occurred in up to six months. However, when warming to 41°C was used during preparation to expedite dissolution, an immediate 40% loss of drug and loss of functional activity occurred. ■

# ■ Hydroxyzine Hydrochloride Hydroxyzine Pamoate

## Properties

Hydroxyzine hydrochloride is a white odorless powder having a bitter taste.[1-3]

Hydroxyzine pamoate is a light yellow almost odorless powder. Hydroxyzine pamoate 170 mg is approximately equivalent to hydroxyzine hydrochloride 100 mg.[2,3]

### Solubility

Hydroxyzine hydrochloride is very soluble in water, with an aqueous solubility variously cited as 1 g/mL[3] and less than 700 mg/mL.[1] In ethanol the solubility is about 222 mg/mL.[3]

Hydroxyzine pamoate is practically insoluble in water and ethanol.[2,3]

### pH

Hydroxyzine hydrochloride injection has a pH between 3.5 and 6. Hydroxyzine pamoate suspension has a pH between 4.5 and 7.[2,4]

### $pK_a$

The drug has $pK_a$ values of 2.6 and 7.[2]

### Osmolality

Hydroxyzine hydrochloride syrup 2 mg/mL (Roerig) has an osmolality of 4450 mOsm/kg.[233]

## General Stability Considerations

In general, hydroxyzine products should be packaged in tight, light-resistant containers and stored at controlled room temperature protected from temperatures of 40°C or more. The liquid products also should be protected from freezing.[2]

## Stability Reports of Compounded Preparations
### Injection

Injections, like other sterile drugs, should be prepared in a suitable clean air environment using appropriate aseptic procedures.

When prepared from nonsterile components, an appropriate and effective sterilization method must be employed.

Hydroxyzine hydrochloride 50 mg/mL repackaged into cartridges (Tubex) retained concentration for three months at room temperature.[347]

### Oral

The stability of a methadone pain cocktail formulation was studied by Little et al.[84] The mixture was prepared from methadone hydrochloride injection (Dolophine, Lilly) 10 mg and hydroxyzine pamoate suspension (Vistaril, Pfizer) equivalent to hydroxyzine hydrochloride 25 mg and brought to 15 mL with cherry syrup. Samples were stored at ambient temperature and under refrigeration for two weeks. An HPLC assay was used to determine drug content. The mixtures were stable for two weeks, with about 9% loss of each drug at both temperatures. ■

# ■ Ibuprofen

## Properties
Ibuprofen is a white or nearly white crystalline powder or colorless crystals having a slight characteristic odor.[2,3]

### Solubility
Ibuprofen is practically insoluble in water; with a solubility of less than 0.01 mg/mL, but is very soluble in ethanol.[2,3,7] It will dissolve in dilute aqueous solutions of alkali hydroxides and carbonates.[3]

### pH
Ibuprofen oral suspension has a pH between 3.6 and 4.6.[4]

### $pK_a$
Ibuprofen has a $pK_a$ of 4.43.[7]

## General Stability Considerations
Ibuprofen products should be packaged in well-closed, light-resistant containers and stored at controlled room temperature.[2,4,7]

Using accelerated degradation at 70°C and 75% relative humidity for three weeks, Cory et al.[1621] found that the degradation of ibuprofen in commercial tablets was accelerated when polyethylene glycol or polysorbate 80 was included among the tablet excipients. If either of these excipients was not present in the formulation, ibuprofen degradation did not occur under these conditions.

## Stability Reports of Compounded Preparations
### Oral
STUDY 1: Devi and Rao[537] evaluated various formulation approaches to preparing an oral elixir of ibuprofen containing 100 mg/5 mL. Cosolvent formulations incorporating glycerol, D-sorbitol, propylene glycol, and ethanol in concentrations ranging from 20 to 100% were tested. Ibuprofen solubility increased in propylene glycol and ethanol as the concentrations increased up to about 80%. Glycerol

solutions effected a much more modest improvement in solubility, as did D-sorbitol in the range of 20 to 40%. The authors prepared elixirs of ibuprofen using various combinations of cosolvents. Ibuprofen powder was triturated in a clean dry mortar and pestle with a small amount (0.1%) of polysorbate 80 (Tween 80) to aid in wetting the powder. Saccharin sodium 0.1%, sodium benzoate 0.2%, and tartrazine for color were dissolved in a small amount (unspecified) of water. Banana flavor was dissolved in a mixture of ethanol 15% and propylene glycol 60%, and the aqueous colored preservative solution was incorporated into it. The mixture was added to the mortar contents and triturated until a clear solution resulted. Sorbitol 70% could be substituted for water, also resulting in a clear solution.

Devi and Rao[537] found that other combinations of solvents also resulted in clear solutions. The following solvent concentrations, with the balance being sorbitol 70%, were all clear solutions:

| Ethanol | Propylene Glycol |
| --- | --- |
| 44% | 30% |
| 34% | 40% |
| 28% | 44% |
| 25% | 50% |

However, if chocolate flavor was substituted for banana flavor, a precipitate formed in the elixirs.

STUDY 2: Martin-Viana et al.[1420] evaluated the physical and chemical stability of an ibuprofen 20-mg/mL oral suspension for pediatric use. The oral suspension was prepared using ibuprofen powder with unspecified "auxiliary substances in common use in the pharmaceutical industry." Samples of the oral suspension were packaged in amber glass bottles and stored at room temperature of 28 to 32°C protected from exposure to light for 24 months and at elevated temperature of 40°C and 75% relative humidity for accelerated testing over six months. The oral suspension was a white viscous liquid with a pH near 3.8. At both storage conditions no physical changes were observed and the pH remained essentially unchanged. HPLC analysis of the ibuprofen concentration using the USP HPLC method found no ibuprofen loss in six months at elevated temperature and less than 3% loss in 24 months at room temperature.

**STUDY 3:** Burgalassi et al.[1537] evaluated the sedimentation rates of several oral liquid suspension formulations of ibuprofen 10 mg/mL using four different suspending mediums: (1) hydroxypropyl methylcellulose 1%, (2) microcrystalline cellulose 1.5%, (3) carboxymethylcellulose 2.5%, and (4) carrageenan iota 1%. The best suspensions with the slowest sedimentation rates were obtained using microcrystalline cellulose 1.5% and carrageenan iota 1%. The chemical stability of ibuprofen was not tested.

### Injection

Injections, like other sterile drugs, should be prepared in a suitable clean air environment using appropriate aseptic procedures. When prepared from nonsterile components, an appropriate and effective sterilization method must be employed.

**STUDY 1:** Jain and Jahagirdar[538] evaluated ibuprofen 40-mg/mL injections solubilized with concentrations of sodium benzoate ranging from 5 to 35% (wt/vol). Ibuprofen solubility was improved 68-fold at the 35% sodium benzoate concentration; 30 and 35% concentrations were chosen for further evaluation. The required amount of ibuprofen powder was weighed and added to 150 mL of each concentration of sodium benzoate vehicle in 250-mL flasks along with 0.1% sodium metabisulfite and 0.01% edetic acid (EDTA). The flasks were mechanically shaken for two hours to ensure complete dissolution, and the solutions were aseptically filtered into sterile 10-mL glass

ampules that were pull sealed. Sample ampules were stored at 8, 37, and 45°C. About 7% ibuprofen loss occurred in 28 days at 8°C. At the elevated temperatures, unacceptable losses were found after storage for 15 to 22 days. Physical stability of the solutions was good, with no color change or particulate matter observed. Both formulations produced analgesia in mice after 10-fold dilution in sterile water and an intraperitoneal dose of 100 mg/kg.

**STUDY 2:** Devi and Rao[537] also speculated that an ibuprofen injection might be possible using cosolvents. Ibuprofen solubility was 140 mg/mL in a mixture of ethanol 50%, propylene glycol 20%, and water. Ibuprofen solubility went up to 300 mg/mL in a mixture of ethanol 50%, propylene glycol 30%, and water. However, actual preparation and testing of such injections was not reported.

**STUDY 3:** Cao et al.[996] reported the formulation of ibuprofen injection for veterinary use. The formulation contained polyethylene glycol, ethanol, sodium carbonate, EDTA sodium, and sodium bisulfite in water for injection. No color change and about 2% ibuprofen loss was found after storage for three years at room temperature.

**STUDY 4:** Yeh and Wang[1514] developed a nonaqueous injection formulation of ibuprofen 50 mg/mL. Ibuprofen powder was solubilized in a vehicle composed of an equal parts mixture of propylene glycol and N,N-dimethylacetamide and packaged in type I glass vials. The samples were stored at temperatures of 40, 50, and 60°C for 91 days. Visual examination found no color change or precipitation. HPLC analysis of ibuprofen decomposition at the elevated temperatures was used to calculate the time to 10% decomposition ($t_{90}$). The $t_{90}$ at 25°C was calculated to be 1180 days or over three years. ∎

# Idoxuridine
## (IUdR)

## Properties

Idoxuridine is a white or nearly white crystalline powder or colorless crystals that are practically odorless. Iodine vapor may be liberated upon heating.[1,3]

### Solubility

Idoxuridine is slightly soluble in water, having an aqueous solubility of about 2 mg/mL at 25°C. Its solubility in ethanol is about 2.6 mg/mL. In 0.2 N hydrochloric acid, idoxuridine has a solubility of 2 mg/mL; in 0.2 N sodium hydroxide, the solubility of idoxuridine is 74 mg/mL.[1,3]

### pH

A 0.1% idoxuridine aqueous solution has a pH between 5.5 and 6.5.[1,3] The ophthalmic solution has a pH between 4.5 and 7.[4]

### $pK_a$

The drug has a $pK_a$ of 8.25.[1]

## General Stability Considerations

Idoxuridine products should be packaged in tight, light-resistant containers.[3]

Idoxuridine is hydrolyzed in aqueous alkaline solutions, but it is more stable in acidic environments.[431] Optimum stability is reported to occur near pH 6.[1008] The decomposition products, such as iodouracil, are more toxic than idoxuridine and have less antiviral activity.[2,3]

## Stability Reports of Compounded Products

### Ophthalmic

Ophthalmic preparations, like other sterile drugs, should be prepared in a suitable clean air environment using appropriate aseptic procedures. When prepared from nonsterile components, an appropriate and effective sterilization method must be employed.

Allen[598] reported on a 0.1% idoxuridine ophthalmic solution. Idoxuridine powder 100 mg and thimerosal 2 mg were accurately weighed and dissolved in about 95 mL of sterile water for injection. The mixture was brought to a volume of 100 mL with additional sterile water for injection and was filtered through a sterile 0.2-μm filter into sterile, light-resistant ophthalmic containers. This compounded preparation should be protected from light and refrigerated during storage. The ophthalmic solution has been reported to be stable for at least one year.

### Topical

Tempe et al.[431] evaluated the stability of 0.5% idoxuridine in four topical gel formulations. The composition of the gels is shown in Table 48. The gels were stored for three months at room temperature and at 37°C. Idoxuridine hydrolyzed faster in the alkaline formulations than in the acidic gel. At room temperature, the rate was about 10% greater at pH 8.95 than at pH 5.2. Elevated temperature also increased the hydrolysis rate. Formulations 1, 2, and 3 lost 11, 16, and 14% of the idoxuridine, respectively, in three months at room temperature. Exposure to light increased the loss to about 25% in all three gels. At 37°C, losses were about 55 to 60% in three months. However, after three months, the acidic gel, formulation 4, retained concentration, losing only 5% at room temperature, 7% with light exposure, and 9% at 37°C.

Idoxuridine 5% in dimethyl sulfoxide has been used topically to treat cutaneous lesions of herpes simplex and herpes zoster.[3] ▪

**TABLE 48.** Idoxuridine Gel Formulations Tested for Stability[431]

| Component | Gel Formulation | | | |
|---|---|---|---|---|
| | 1 | 2 | 3 | 4 |
| Idoxuridine | 0.5% | 0.5% | 0.5% | 0.5% |
| Sodium borate | 1.5% | 1.5% | 1.5% | 1.5% |
| Methylparaben | 0.07% | 0.07% | — | 0.07% |
| Propylparaben | 0.03% | 0.03% | — | 0.03% |
| Hydroxyethylcellulose | 2% | 2% | 2% | — |
| Carboxypolymethylene | — | — | — | 2% |
| (pH) | (9.2) | (8.95) | (9.15) | (5.2) |

# ▪ Indinavir Sulfate

## Properties

Indinavir sulfate is a white to off-white hygroscopic crystalline powder.[7]

Indinavir sulfate is formulated as hard-gelatin capsules containing the equivalent of 100, 200, 333, and 400 mg of indinavir as 125, 250, 416, and 500 mg, respectively, of the sulfate salt.[7]

### Solubility

Indinavir sulfate is very soluble in water.[7] Indinavir has an aqueous solubility of 0.015 mg/mL unbuffered and more than 1.5 mg/mL at pH 4.5.[1]

## General Stability Considerations

Indinavir sulfate capsules should be stored in tight containers at controlled room temperature and protected from exposure to moisture. Consequently, the capsules should be dispensed in the original bottles along with the desiccant.[7]

Singh et al.[1257] evaluated the thermal stability of indinavir sulfate using Fourier transform infrared spectroscopy, differential scanning calorimetry, and x-ray diffraction. Indinavir sulfate powder was found to remain stable up to 100°C, degrade slightly at 125°C, and undergo complete degradation at 150°C.

## Stability Reports of Compounded Preparations

### Oral

Hugen et al.[723] reported the stability and bioavailability of an extemporaneously compounded indinavir sulfate (Merck) 10-mg/mL oral liquid prepared from 400-mg capsules. A concentrate was prepared by emptying the contents of indinavir sulfate capsules into purified water and sonicating in a water bath at about 37°C for 60 minutes with repeated stirring. The mixture was passed through filter paper to remove undissolved solids, and the concentrate was brought to a concentration of 100 mg/mL. The concentrate was mixed with a vehicle composed of methylcellulose, saccharin sodium, simple syrup, citric acid, azorubine coloring, sodium hydroxide to adjust to pH 3.0, and lemon oil, resulting in a final indinavir concentration of 10 mg/mL.

During two weeks of storage under refrigeration, the oral liquid formulation remained clear, with no evidence of precipitation, turbidity, or separation of oil droplets. Stability-indicating HPLC analysis found less than 4% indinavir loss in the concentrate at 100 mg/mL and in the final preparation at 10 mg/mL in two weeks of refrigerated storage. In addition, bioequivalence to the commercially available oral capsules was demonstrated. ■

# Indocyanine Green
## (Fox Green)

## Properties
Indocyanine green is a tricarbocyanine dye with infrared absorption properties with peak absorption at about 800 nm.[1,7] The material contains approximately 5% sodium iodide as a contaminant.[1,3,4] Indocyanine green powder is odorless or nearly so and appears as a variety of colors, including olive-brown, dark green, blue-green, dark blue, or black. Aqueous solutions appear deep emerald-green in color.[4]

### Solubility
Indocyanine green is soluble in water and methanol but practically insoluble in most other organic solvents.[4]

### pH
Indocyanine green 1 to 200 in water is about pH 6. Indocyanine green for injection has a pH in the range of 5.5 to 7.5 after reconstitution.[4,7]

## General Stability Considerations
Indocyanine green should be packaged in well-closed containers and stored at controlled room temperature.[4] Indocyanine green for injection should also be stored at controlled room temperature.[4,7] Indocyanine green dissolved in water is stable for about eight hours.[4] The manufacturer of the commercial injection recommends use within six hours after reconstitution.[7]

## Stability Reports of Compounded Preparations
### Injection
Injections, like other sterile drugs, should be prepared in a suitable clean air environment using appropriate aseptic procedures. When prepared from nonsterile components, an appropriate and effective sterilization method must be employed.

Saxena et al.[1660] evaluated the factors that affect the stability of indocyanine green dissolved in water. Using fluorescence spectroscopy, the authors confirmed previous reports[1661,1662] that physicochemical transformations such as aggregation and irreversible degradation occur. These changes result in discoloration, decreased light absorption, and fluorescence. Indocyanine green is relatively unstable in solution; stored protected from light and refrigerated at 4°C the half-life ($t_{1/2}$) is about 20 hours. Factors that increase the rate of degradation include exposure to light and higher temperatures. In room light at 22°C, the $t_{1/2}$ decreases to about 14 hours. More dilute solutions were also found to be less stable.[1660] The pH has also been reported to impact indocyanine green degradation.[1663] ■

# Indomethacin
# Indomethacin Sodium

## Properties
Indomethacin is a pale-yellow to yellowish-tan crystalline powder with a slight odor.[1–3]

Indomethacin sodium (as the trihydrate) in the lyophilized product is a white to yellow cake or powder.[2]

### Solubility
Indomethacin is practically insoluble in water and has a solubility of about 20 mg/mL in ethanol.[1–3] Indomethacin sodium is soluble in both water and ethanol.[1,2]

### pH
The reconstituted injection has a pH between 6 and 7.5.[2,4] Indomethacin oral suspension has a pH between 2.5 and 5.[4] However, Indocin (Merck) oral suspension has a pH of 4 to 5, while the Roxane oral suspension has a pH of 2.9.[2]

### pKₐ
The drug has a p$K_a$ of 4.5.[1,2]

## General Stability Considerations

Indomethacin capsules should be packaged in well-closed containers, stored at controlled room temperature, and protected from temperatures exceeding 40°C. The oral suspension should be packaged in tight, light-resistant containers and stored below 30°C. It should be protected from freezing and from temperatures exceeding 50°C.[2,4] Indomethacin suppositories should be stored at less than 30°C and protected from temperatures exceeding 40°C even for a short time. Indomethacin sodium for injection should be stored below 30°C and protected from exposure to light.[2]

The reconstituted injection is chemically stable for 16 days at room temperature,[2] although the manufacturer recommends discarding any unused solution, because no antibacterial preservative is present.[7] Reconstitution of the injection with solutions having a pH less than 6 may cause the formation of free indomethacin precipitate.[2,7]

Indomethacin degradation in alkaline aqueous solutions was investigated by Hajratwala and Dawson.[216] Indomethacin is stable at neutral or slightly acidic pH but is decomposed by strong alkali. At hydroxide-ion concentrations from 0.001 M up to 0.01 M, the indomethacin decomposition rate got progressively more rapid. The decomposition rate also was accelerated as the temperature was elevated from 20 to 40°C.

Indomethacin as a solid crystalline powder or in solution is sensitive to light. In solution, maximum stability occurs near pH 3.75, with a calculated shelf life of about 8.4 days. Below pH 3 specific acid catalysis is the predominate decomposition mechanism. Specific base catalysis occurs above pH 7.[6]

## Stability Reports of Compounded Preparations
### Oral

**STUDY 1:** Gupta et al.[62] reported the stability of indomethacin prepared as pediatric oral liquid dosage forms. Indomethacin powder and the contents of commercial capsules (Indocin, Merck Sharp & Dohme) were used as sources of the drug. The indomethacin from each source was wetted with ethanol to form a paste or suspension and was brought to volume with simple syrup to yield 2-mg/mL concentrations with 10% (vol/vol) ethanol. Methylparaben (0.005%) and propylparaben (0.002%) were added as preservatives. The products were packaged in amber bottles and stored at 24°C. Analysis by ultraviolet spectrophotometry found that 95% or more of the indomethacin from either source remained after 224 days of storage.

**STUDY 2:** Stewart et al.[414] evaluated the stability of indomethacin (as the bulk powder) in six extemporaneous oral liquid formulations. The indomethacin concentrations ranged from 0.25 mg/mL to 5 mg/mL.

The compositions of the six formulations tested were as follows.

*Formulation 1 (pH 8.15)*

| | | |
|---|---|---|
| Indomethacin | | 25 mg |
| Disodium hydrogen orthophosphate | | 904 mg |
| Potassium dihydrogen orthophosphate | | 46 mg |
| Concentrated chloroform water | | 2.5 mL |
| Purified water | qs | 100 mL |

*Formulation 2*

| | | |
|---|---|---|
| Indomethacin | | 25 mg |
| Alcohol | | 10 mL |
| Methylparaben | | 5 mg |
| Propylparaben | | 2 mg |
| Syrup | qs | 100 mL |

*Formulation 3*

| | | |
|---|---|---|
| Indomethacin | | 25 mg |
| Compound tragacanth powder | | 1.75 g |
| Methylparaben | | 5 mg |
| Propylparaben | | 2 mg |
| Purified water | qs | 100 mL |

*Formulation 4*

| | | |
|---|---|---|
| Indomethacin | | 400 mg |
| Compound tragacanth powder | | 3 g |
| Orange syrup | | 10 mL |
| Sorbitol solution USP | | 10 mL |
| Methylparaben | | 150 mg |
| Propylparaben | | 50 mg |
| Absolute alcohol | | 4 mL |
| Distilled water | qs | 100 mL |

*Formulation 5*

| | | |
|---|---|---|
| Indomethacin | | 200 mg |
| Raspberry syrup | | 20 mL |
| Glycerin | qs | 100 mL |

*Formulation 6*

| | | |
|---|---|---|
| Indomethacin | | 500 mg |
| Imitation vanilla flavor | | 0.1 mL |
| Concentrated chloroform water | | 2.5 mL |
| Syrup | qs | 100 mL |

Indomethacin content was determined using HPLC. Samples of the products were packaged in clear and amber bottles and stored at 2, 20, 35, and 54°C for 12 weeks. The room temperature samples were stored both in the laboratory and exposed to sunlight. The samples also were evaluated for physical stability. Relatively rapid loss of indomethacin occurred in formulations 1 through 4. Formulation 2 also exhibited problems with caking, while both 2 and 4 exhibited color change. Formulations 2, 3, and 4 also had poor uniformity of dosage, with variations of 10 to 12%. Formulations 5 and 6 were much more chemically stable, retaining greater than 90%

concentration for 12 weeks at all but the highest elevated temperature. Unfortunately, both also exhibited problems with caking and difficulty in redispersibility, especially formulation 5.

**STUDY 3:** Burgalassi et al.[1537] evaluated the sedimentation rates of several oral liquid suspension formulations of indomethacin 10 mg/mL using four different suspending mediums: (1) hydroxypropyl methylcellulose 1%, (2) microcrystalline cellulose 1.5%, (3) carboxymethylcellulose 2.5%, and (4) carrageenan iota 1%. The best suspensions with the slowest sedimentation rates were obtained using microcrystalline cellulose 1.5% and carrageenan iota 1%. The chemical stability of indomethacin was not tested.

## Injection

Injections, like other sterile drugs, should be prepared in a suitable clean air environment using appropriate aseptic procedures. When prepared from nonsterile components, an appropriate and effective sterilization method must be employed.

Jain[1262] evaluated the aqueous solubility enhancement of indomethacin by various hydrotropes. Sodium *p*-hydroxybenzoate 1.2 M, sodium benzoate 1.2 M, and nicotinamide 1.2 M all increased indomethacin solubility to about 3, 2, and 1.5 mg/mL, respectively. Samples of indomethacin 1 mg/mL injection were prepared using each of these hydrotropes along with sodium metabisulfite 0.1% (wt/vol) as an antioxidant and stored at 4, 25, and 37°C.

All formulations exhibited physical stability with no color change, turbidity, or precipitation being observed. Stability-indicating HPLC analysis found less than 2% loss at 4°C, 3 to 4% loss at 25°C, and 4 to 7% loss at 37°C in 45 days. The time to 10% drug loss ($t_{90}$) at 25°C was calculated from the degradation rate constants to be 110 days in the sodium *p*-hydroxybenzoate formulation, 185 days in the sodium benzoate formulation, and 227 days in the nicotinamide formulation.

## Ophthalmic

Ophthalmic preparations, like other sterile drugs, should be prepared in a suitable clean air environment using appropriate aseptic procedures. When prepared from nonsterile components, an appropriate and effective sterilization method must be employed.

**STUDY 1:** Vulovic et al.[56] evaluated the stability of extemporaneously prepared indomethacin ophthalmic suspensions. Indomethacin 1% (wt/vol) was compounded with either hydroxypropyl methylcellulose 0.5% (wt/vol) or polyvinyl alcohol 1.4% (wt/vol) as viscolizers and with phenylmercuric nitrate 0.002% (wt/vol) in sodium phosphate (Sorenson's) buffer solution, pH 5.6, and then placed into glass ampules. The suspensions were sterilized at 100°C for 30 minutes and

were stored at 60 to 100°C for an accelerated stability study and for two months at ambient temperature for a particle size distribution study. There was little difference in the stability of indomethacin by spectrophotometry when either viscolizer agent was used. The authors calculated a time to 10% drug loss ($t_{90}$) of approximately 280 days. Furthermore, suspension particle size underwent little change. However, if neither viscolizer was used to protect against crystal growth, particle size increased substantially during heat sterilization. Also, increasing the pH from 6.2 to 8.1 substantially increased the rate of indomethacin decomposition.

**STUDY 2:** The retention of efficacy of phenylmercuric nitrate preservative in indomethacin ophthalmic suspension was investigated by Vulovic et al.[173] An indomethacin 1% (wt/vol) suspension containing hydroxypropyl methylcellulose 0.5% (wt/vol) buffered to pH 5.6 was preserved with 0.002% phenylmercuric nitrate. Antimicrobial activity was retained for at least 28 days despite 90% of the preservative being adsorbed onto the indomethacin powder. In a formulation containing polyvinyl alcohol 1.4% (wt/vol) as the suspending agent, outgrowth of organisms occurred after seven days.

**STUDY 3:** Dimitrova et al.[224] evaluated the stability of 0.5% indomethacin in ophthalmic solutions containing the components shown in Table 49. The formulations were evaluated for indomethacin decomposition at elevated temperature (90°C) using HPLC, thin-layer chromatography, and ultraviolet spectroscopy. The presence of high molecular weight polyoxyethylene as a viscosity additive did not affect the stability of indomethacin. This indomethacin ophthalmic formulation was calculated to be stable for at least one year at room temperature.

**STUDY 4:** Dimitrova et al.[808] also reported on 0.5% indomethacin ophthalmic solutions prepared using Pluronic F-68 (poloxamer 188) 15% and Pluronic F-127 (poloxamer 407) 10% instead of the polysorbate 80, propylene glycol, and polyoxyethylene in the formula in Table 49. The solutions

**TABLE 49.** Indomethacin 0.5% Ophthalmic Solution Composition[224]

| Component | | Amount per 100 mL |
|---|---|---|
| Indomethacin | | 0.5 g |
| Polysorbate 80 | | 1.0 g |
| Propylene glycol | | 10 to 30 g |
| Polyoxyethylene | | 0.4 g |
| Preservative (unspecified) | | 0.004 g |
| Antioxidants (unspecified) | | 0.2 g |
| Phosphate buffer pH 6.8 | qs | 100 mL |

were steam sterilized and evaluated for stability at elevated temperatures from 40°C up to 90°C as well as at ambient temperature protected from light. The poloxamer formulas were judged to be more effective as solubilizers and added appropriate viscosity. However, the rate of indomethacin decomposition at 40°C was about twice the rate of loss of the formulation in Table 49.

**STUDY 5:** Kodym et al.[410] reported the stability of six formulations of indomethacin 1% ophthalmic drops. The best stability was obtained in 0.9% sodium chloride, with a $t_{90}$ of 20 days. Inclusion of an antibacterial preservative reduced the $t_{90}$ to 16 days. In 1.9% boric acid solution, 10% indomethacin loss occurred in only five days.

### Topical
#### USP OFFICIAL FORMULATION (TOPICAL GEL):

| | | |
|---|---|---|
| Indomethacin | | 1.0 g |
| Carbomer 941 | | 2.0 g |
| Purified water | | 10 mL |
| Ethanol 95% | qs | 100 mL |

Place the indomethacin powder in a beaker and dissolve in 55 mL of ethanol 95%. Transfer the solution to a glass mortar, and slowly add the Carbomer 941 so that it is thoroughly distributed. Press out any clumps, forming a smooth gel. Slowly add the purified water with mixing. Bring the gel to volume with additional ethanol 95% and mix well yielding 1% topical gel. Transfer the topical gel to a suitable tight, light-resistant container. Store at controlled room temperature. The beyond-use date is 30 days from the date of compounding.[4,5] Indomethacin has been used topically as a 2.5% solution in ethanol–propylene glycol–dimethylacetamide (19:19:2),[141] as 0.5, 1, and 2.5% solutions in ethanol–propylene glycol–dimethylformamide (1:1:2),[142] as a 2.5% solution in water and mannitol,[143,144] and as a 1% gel in a hydroalcohol base.[145]

**STUDY 1:** Allen and Stiles[140] described a topical formulation of 1% indomethacin (Table 50). Indomethacin was dissolved in ethanol, and carbomer 941 (Carbopol 941) was added and thoroughly dispersed. Water was added while stirring slowly;

**TABLE 50.** Topical Indomethacin Formulation[140]

| Component | Quantity |
|---|---|
| Indomethacin | 1% |
| Carbomer 941 (Carbopol 941) | 2% |
| Distilled water | 10% |
| Ethanol | 87% |

the viscosity slowly increased. Viscosity can be increased further by increasing the ratio of water to ethanol and/or adding a few drops of trolamine (triethanolamine). Stability studies of similar formulations reported a shelf life of 25 months at 25°C in tight, light-resistant containers.

**STUDY 2:** Tomida et al.[219] evaluated the stability of indomethacin in poloxamer (Pluronic F-127, BASF Wyandotte) gels and compared them to aqueous solutions. Pluronic F-127 concentrations ranged from 0 to 30% (wt/vol). At pH 7 and 20°C, the time to 10% indomethacin degradation was 2.7 years in the 20% (wt/vol) gel compared to 48 days in the aqueous solution.

**STUDY 3:** Shawesh et al.[835] evaluated the physical stability of indomethacin 1% in 20 topical formulations of poloxamer (Pluronic F-127) gels. In addition to 20% poloxamer 407 (Pluronic F-127), the topical gel formulations included either hexylene glycol (16, 20, and 24%) or polyethylene glycol 300 (16, 20, and 24%) with or without polyvinyl pyrrolidone 1% or polysorbate 80 (Tween 80) 1% and sterile water qs 100%. The physical stability of all formulations remained acceptable for at least four weeks at room temperature of 18 to 22°C. Consistency and viscosity remained acceptable. However, if stored under refrigeration at 6°C, the gels prepared with hexylene glycol were of lower viscosity than those prepared with polyethylene glycol 300. In addition, indomethacin precipitated from many of the gels prepared with both hexylene glycol and polyethylene glycol 300 but not containing polyvinyl pyrrolidone. If 1% polyvinyl pyrrolidone was present, no indomethacin precipitation occurred in any of the gels. ■

# ■ Infliximab

## Properties
Infliximab is a chimeric human-murine immunoglobulin G1 kappa monoclonal antibody with an approximate molecular weight of 149,100 daltons. It is composed of human constant and murine variable regions. Infliximab binds specifically to human tumor necrosis factor alpha (TN-Falpha) and is used as a biological response modifier.[1,2]

## pH
Infliximab reconstituted injection has a pH of approximately pH 7.2.[7]

## General Stability Considerations
Infliximab in intact commercial single-dose containers should be stored under refrigeration at 2 to 8°C and should

be protected from freezing. The manufacturer states that the reconstituted infliximab solution should be used immediately or be discarded. It should not be stored or re-entered.[7]

In addition, the manufacturer states that infliximab prepared for infusion in a concentration range of 0.4 to 4 mg/mL should begin administration within three hours of preparation. Administration should occur over a period of at least two hours. The total stability period for infliximab diluted for infusion is not stated in the labeling. However, the manufacturer has stated that the drug diluted to a concentration of 0.4 or 4 mg/mL in sodium chloride 0.9% is physically and chemically stable for 24 hours. The manufacturer's recommendation for use within three hours in the labeling results from concern for microbiological contamination during reconstitution and preparation.[7,8]

## Stability Reports of Compounded Formulations
### Ophthalmic
Ophthalmic preparations, like other sterile drugs, should be prepared in a suitable clean air environment using appropriate aseptic procedures. When prepared from nonsterile components, an appropriate and effective sterilization method must be employed.

Beer et al.[1502] evaluated the stability of infliximab reconstituted with sterile water to a high concentration of 50 mg/mL and to 10 mg/mL stored refrigerated at 4°C. Using an infliximab microsphere immunoassay, the authors reported little or no loss of drug occurred over six weeks of storage. The authors stated that infliximab stability was sufficient to permit cost-effective use by compounding pharmacies to prepare intravitreal medication for clinical use. ■

# ■ Interferon Alfa-2b

## Properties
Interferon alfa-2b is one of several cytokine subtypes of interferon alfa produced by leukocytes stimulated by viruses, bacteria, or protozoa as part of the immune response. It exhibits antiviral, antineoplastic, and immunomodulating activities. It inhibits protein synthesis and viral replication and enhances cytotoxic activity. It has a molecular weight of approximately 19 kDa.[1,2]

Interferon alfa-2b injection is a clear, colorless to slightly yellow liquid.[3] The dry powder products are a white to cream color but become a clear, colorless to slightly yellow liquid upon reconstitution.[2,7,9]

### pH
Interferon alfa-2b reconstituted injection has a pH in the range of 6.9 to 7.5.[2]

### Osmolality
The 10-million-unit vials reconstituted with 1 mL of sterile water for injection yield an isotonic solution.[9]

## General Stability Considerations
Interferon alfa-2b liquid and powder injections should be stored under refrigeration at 2 to 8°C for long-term storage.[7] However, intact vials are stable for up to 28 days at room temperature and up to seven days at 45°C.[2,9] Sodium chloride 0.9%, Ringer's injection, and lactated Ringer's injection are recommended for dilution of interferon alfa-2b. However, the drug is incompatible with dextrose injections.[9]

Reconstituted interferon alfa-2b solutions are stable for one month refrigerated and up to two days at room temperature.[7]

Liquid injections of interferon alfa-2b are stated to be stable for 14 days at room temperatures of 30°C and up to seven days at 35°C.[7]

Interferon alfa-2b solutions frozen at −20°C are stable for 56 days, including four freeze–thaw cycles. When frozen at −80°C, interferon alfa-2b is stable for one year.[9]

Interferon alfa-2b is stable over a pH range of 6.5 to 8, exhibiting its greatest stability between pH 6.9 and 7.5.[2,9,1296] It exhibits lower stability at pH 4 and 9.[1296]

## Stability Reports of Compounded Preparations
### Injection
Injections, like other sterile drugs, should be prepared in a suitable clean air environment using appropriate aseptic procedures. When prepared from nonsterile components, an appropriate and effective sterilization method must be employed.

Palmer et al.[1150] evaluated the stability of interferon alfa-2b 3 million units/6 mL of sterile water for injection packaged in 10-mL polypropylene syringes. Test syringes were stored under refrigeration for 14 days and at 37°C for 24 hours. HPLC analysis found changes indicating interconversion between interferon monomers and possible formation of oligomers. The authors stated that prepackaging diluted interferon alfa-2b injection in polypropylene syringes was unacceptable.

### Ophthalmic
Ophthalmic preparations, like other sterile drugs, should be prepared in a suitable clean air environment using appropriate aseptic procedures. When prepared from nonsterile components, an appropriate and effective sterilization method must be employed.

Ruiz et al.[1149] evaluated the stability of interferon alfa-2b in an extemporaneously compounded ophthalmic solution for use in treating acute hemorrhagic conjunctivitis. The interferon alfa-2b (Heberon Alfa R) is an albumin-free formulation of the drug containing in each milliliter interferon alfa-2b 10 million units, polysorbate 80 0.2 mg, sodium chloride 4.89 mg, benzyl alcohol 10 mg, sodium phosphate monobasic dihydrate 3.43 mg, and sodium phosphate dibasic anhydrous 12.68 mg. Using a 1-mL syringe, 0.3 mL of this formulation was added to a 5-mL ophthalmic bottle of eye wash solution containing in each milliliter benzalkonium chloride 0.1 mg, EDTA 1 mg, and sodium chloride 9 mg. Sample bottles were stored refrigerated at 2 to 8°C and at room temperature of 26 to 30°C.

The antiviral activity was assessed using a bioactivity assay and enzyme-linked immunosorbent assay along with visual inspection, sterility testing, and pH testing. The pH remained between 6.7 and 7.3 throughout the study. No visible change in color or clarity was reported, and the test samples remained sterile through the test period. At room temperature, the interferon alfa-2b antiviral activity remained adequate for seven days but declined to about 72% by 15 days. Under refrigeration, the activity was satisfactory for 15 days, but declined to about 73% after 21 days. ■

# ■ Iobenguane Sulfate

## Properties

Iobenguane sulfate occurs as a colorless crystalline material.[1]

Iobenguane sulfate is commercially available labeled with I 131 for diagnostic injection. Each milliliter of the injection contains iobenguane sulfate 0.69 mg and 2.30 mCi of I 131 (as iobenguane sulfate I 131), along with sodium acetate 0.36 mg, acetic acid 0.27 mg, sodium chloride 4.2 mg, methylparaben 0.56 mg, propylparaben 0.056 mg, and benzyl alcohol 0.01 mL in water for injection.[7]

Iobenguane sulfate I 123 is an official USP drug,[4] but it is not commercially available. It is prepared in nuclear pharmacies by compounding.[1220]

### pH

Iobenguane sulfate I 131 has a pH in the range of 4.5 to 7.5. Iobenguane sulfate I 123 has a pH in the range of 6 to 7.5.[4]

## General Stability Considerations

Commercial iobenguane sulfate I 131 injection is packaged in shielded single-dose or multiple-dose containers and stored frozen at −20°C.[4,7] USP requirements for iobenguane sulfate I 123 also require that the prepared injection be packaged in shielded single-dose or multiple-dose containers and stored frozen at −20°C.[4]

## Stability Reports of Compounded Preparations
### Injection

Injections, like other sterile drugs, should be prepared in a suitable clean air environment using appropriate aseptic procedures. When prepared from nonsterile components, an appropriate and effective sterilization method must be employed.

Hinkle et al.[1220] evaluated the extended stability of extemporaneously compounded iobenguane sulfate 2.2 mg/mL as a sterile solution for use by nuclear pharmacies to facilitate preparing the I 123 labeled compound used in diagnostic procedures. Iobenguane sulfate 143 mg was dissolved in 65 mL of sterile water for injection. The solution was sterilized by filtration through a 0.22-μm filter and filled as 0.9 mL of the 2.2-mg/mL solution into sterile Becton Dickinson polycarbonate syringes. After filling the syringes, the residual air was removed, and the syringes were sealed with Luerloc tip caps. The filled syringes were stored refrigerated at 4 to 7°C for 91 days.

The iobenguane sulfate 2.2-mg/mL samples underwent no visible changes in color or clarity during the study. Stability-indicating HPLC analysis found that the drug remained stable, retaining at least 93%, and acceptable for radiolabeling throughout the 91-day study period. ■

# ■ Iodinated Glycerol

## Properties

Iodinated glycerol is an isomeric mixture of iodinated dimers of glycerol and contains 50% organically bound iodine. It is a viscous, essentially tasteless straw-colored liquid having a pungent and bitter aftertaste.[1,3]

### Solubility

Iodinated glycerol is soluble in water, ethanol, and glycerin.[2]

### pH

Iodinate glycerol elixir (Organidin, Wallace) has a measured pH of 4.2.[19]

## General Stability Considerations

Iodinated glycerol should be stored at controlled room temperature and protected from excessive heat and freezing.[2]

## Stability Reports of Compounded Preparations
### Enteral
#### IODINATED GLYCEROL COMPATIBILITY SUMMARY
*Compatible with:* Ensure • Ensure HN • Ensure Plus • Ensure
  Plus HN • Osmolite • Osmolite HN • Vital

**STUDY 1:** Cutie et al.[19] added 10 mL of iodinated glycerol elixir (Organidin, Wallace) to varying amounts (15 to 240 mL) of Ensure, Ensure Plus, and Osmolite (Ross Laboratories) with vigorous agitation to ensure thorough mixing. The iodinated glycerol elixir was physically compatible, distributing uniformly in all three enteral products with no phase separation and only slight granulation.

**STUDY 2:** Altman and Cutie[850] reported the physical compatibility of iodinated glycerol elixir (Organidin, Wallace) 10 mL with varying amounts (15 to 240 mL) of Ensure HN, Ensure Plus HN, Osmolite HN, and Vital with vigorous agitation to ensure thorough mixing. The iodinated glycerol elixir was physically compatible, distributing uniformly in all four enteral products, with no phase separation or granulation. ∎

## ■ Ipratropium Bromide

### Properties
Ipratropium bromide is a quaternary ammonium compound that occurs as white or almost white crystals or crystalline powder having a bitter taste.[1–3]

### Solubility
Ipratropium bromide has solubilities of 90 mg/mL in water and 28 mg/mL in ethanol.[1–3]

### pH
A 1% ipratropium bromide aqueous solution has a pH of 5 to 7.5.[3] The solution for inhalation is adjusted to pH 3 to 4.[7]

### Osmolality
Ipratropium bromide inhalation solution (Atrovent, Boehringer Ingelheim) has an osmolality of about 276 mOsm/kg measured by freezing-point depression.[928]

### General Stability Considerations
Ipratropium bromide inhalation solution should be stored in its foil pouch at controlled room temperature. It should be protected from light; unused vials should be returned to the foil pouch.[7]

  The drug is fairly stable in neutral and acidic solutions but rapidly hydrolyzes in alkaline solutions.[1]

### Stability Reports of Compounded Preparations
#### Nasal Spray
Pesko[252] reported on an extemporaneous nasal spray containing ipratropium bromide 200 mcg/mL (Table 51). The solution was prepared by dissolving ipratropium bromide and ascorbic acid in 10 mL of water and adding it to a mixture of Methocel and polyethylene glycol 400 with stirring. The solution was brought to 25 mL with sterile water. The author recommended refrigerated storage for no longer than 14 days, although no stability data exist.

### Inhalation
Inhalations, like other sterile drugs, should be prepared in a suitable clean air environment using appropriate aseptic procedures. When prepared from nonsterile components, an appropriate and effective sterilization method must be employed.

**STUDY 1:** Blondino and Baker[930] evaluated the physical and chemical stability of four common drugs administered by inhalation, albuterol sulfate 2.5 mg/3 mL, budesonide 0.25 mg/3 mL, cromolyn sodium 20 mg/3 mL, and ipratropium bromide 0.5 mg/3 mL, individually in an acidified (pH 4) 15% ethanol in water solution. The budesonide was dissolved in the ethanol, which was then added to a solution of the other drugs dissolved in the acidified sterile water for irrigation. No visible changes in color or clarity were observed in any of the samples. Ipratropium bromide by itself underwent substantial losses over eight weeks. Losses were 2, about 13 to 15, and 23% at refrigerated, room, and elevated temperature of 40°C. However, ipratropium bromide was more stable in the four-drug combination. See *Compatibility with Other Drugs* below.

**STUDY 2:** Gammon et al.[1303] reported on the stability of a number of drugs used by paramedics when exposed to the temperature range that is found in ambulances as documented by Brown et al.[1304] and Allegra et al.[1318] Undiluted ipratropium

**TABLE 51.** Ipratropium Bromide Nasal Spray Formula[252]

| Component | | Amount |
|---|---|---|
| Ipratropium bromide | | 5 mg |
| Ascorbic acid | | 19.8 mg |
| Methocel | | 19.8 mg |
| Polyethylene glycol 400 | | 1.66 mL |
| Sterile water | qs | 25 mL |

bromide 0.02% solution was stored at temperatures that cycled every 24 hours from −6 to 54°C (2.12 to 129.2°F) for 28 days. The drug was exposed to a total of 336 hours at each of the temperature extremes. The mean kinetic temperature was 33°C. HPLC and ultraviolet spectrophotometry found drug losses of about 15% occurred over the 28-day test period.

## Compatibility with Other Drugs

### IPRATROPIUM BROMIDE COMPATIBILITY SUMMARY

*Compatible with:* Acetylcysteine • Albuterol sulfate (Salbutamol sulfate) • Arformoterol tartrate • Bitolerol • Budesonide • Fenoterol hydrobromide • Fluticasone propionate • Formoterol fumarate • Levalbuterol hydrochloride • Metaproterenol sulfate • Sodium Chloride 7% (hypertonic) • Terbutaline sulfate • Tobramycin sulfate

*Incompatible with:* Dornase alfa

*Uncertain or variable compatibility with:* Cromolyn sodium

**STUDY 1 (ALBUTEROL):** Jacobson and Peterson[652] evaluated the stability of admixtures of ipratropium bromide and albuterol sulfate nebulizer solutions mixed in equal quantities. The drugs were found to retain more than 90% of the initial concentrations after five days stored at 4 and 22°C protected from light or at 22°C exposed to fluorescent light.

**STUDY 2 (CROMOLYN):** Iacono et al.[767] reported the compatibility and stability of 1:1 inhalation solution mixtures of ipratropium bromide 0.25 mg/mL (Atrovent, Boehringer Ingelheim) and cromolyn sodium 10 mg/mL (Intal, Fisons) for use as nebulizer solution combinations. The mixture was incompatible; cloudy turbidity developed immediately upon mixing and persisted for 24 hours. The mixture cleared by 48 hours. Filtration did not isolate a precipitate. HPLC analysis found that no loss of either drug occurred within time periods up to 111 minutes (the study period limit). Addition of albuterol sulfate (salbutamol sulfate) (Respolin, Riker) yielded the same cloudiness in the mixture and similar analytical results. Albuterol was not analyzed. If Intal Spincaps were used as the source of cromolyn sodium, mixing with ipratropium bromide did not result in cloudiness, suggesting that an excipient in the Intal nebulizer solution might be responsible.

**STUDY 3 (CROMOLYN):** Turner[768] reported that the cloudiness of the ipratropium bromide and cromolyn sodium inhalation solution mixture was caused by the formation of a complex of benzalkonium chloride preservative in the ipratropium bromide solution with cromolyn sodium. In an equal parts mixture of the two inhalation solutions, only about 0.8% of the cromolyn sodium was complexed. The complex is an oil and not a crystalline precipitate. Turner indicated that the oily complex would nebulize readily. If unpreserved ipratropium bromide inhalation solution is combined with cromolyn sodium, no cloudiness occurs.

**STUDIES 4 AND 5 (BUDESONIDE):** Smaldone et al.[1026] and McKenzie and Cruz-Rivera[892] evaluated the compatibility and stability of two concentrations of budesonide inhalation suspension with other inhalation medications, including ipratropium bromide. Budesonide (Pulmicort Respules, AstraZeneca) 0.25 mg/2 mL and also 0.5 mg/2 mL was mixed with ipratropium bromide 0.5 mg/2.5 mL. The inhalations were added to new Pari LC Plus nebulizer cups and mixed using a vortex mixer. The samples were examined initially after mixing and over 30 minutes. Visual inspection found no change in color, formation of a precipitate, or non-resuspendability of any of the samples. Stability-indicating HPLC analysis found that ipratropium remained adequately stable throughout the study period with little or no loss over 30 minutes. Budesonide 0.05 mg/2 mL was stable with ipratropium with less than 3% loss. However, budesonide 0.25 mg/2 mL with ipratropium bromide was determined to be about 93% after 30 minutes. Whether this was simply an analytical anomaly is uncertain.

**STUDY 6 (ARFORMOTEROL):** Bonasia et al.[1151] evaluated the physical compatibility and chemical stability of arformoterol 15 mcg/2 mL (as the tartrate) inhalation solution (Brovana, Sepracor) and ipratropium bromide 0.2 mg/2 mL (Atrovent, AstraZeneca). The admixtures were prepared and mixed for homogeneity. Visual inspection found no evidence of precipitation or other physical incompatibility. HPLC analysis found less than 2% change in either drug concentration over the 30-minute test period.

**STUDY 7 (MULTIPLE DRUGS):** Roberts and Rossi[897] reported the compatibility of ipratropium bromide inhalation as both single-use (Atrovent UDV) and multiple-use (Atrovent) forms with budesonide suspension (Pulmicort Respules), albuterol sulfate (salbutamol sodium) (Ventolin) as both single and multiple-use forms, and terbutaline multiple-use (Bricanyl). Visual inspection found that all of the combinations were compatible, with no visible cloudiness.

The combinations with cromolyn sodium (Intal) were more problematic. The unpreserved single-use form of ipratropium bromide was compatible with cromolyn sodium, while the multiple-use form preserved with benzalkonium chloride was not. Marked and immediate cloudiness appeared due to the formation of an oily complex of cromolyn sodium with benzalkonium chloride.

**STUDY 8 (ACETYLCYSTEINE):** Lee et al.[928] evaluated the stability of acetylcysteine 20% mixed in equal volumes with ipratropium bromide 250 mcg/mL in combination inhalation solutions. HPLC analysis of the solutions found that the acetylcysteine mixture with ipratropium bromide was not stable. Ipratropium bromide losses in the inhalation mixture were about 7% in one hour and 11% in two hours at room temperature; the acetylcysteine remained stable for two hours, with a loss of about 2%.

**STUDY 9 (MULTIPLE DRUGS):** Blondino and Baker[930] evaluated the physical and chemical stability of four common drugs administered by inhalation, albuterol sulfate 2.5 mg/3 mL, budesonide 0.25 mg/3 mL, cromolyn sodium 20 mg/3 mL, and ipratropium bromide 0.5 mg/3 mL, individually in an acidified (pH 4) 15% ethanol in water solution and in a four-drug combination. The budesonide was dissolved in the ethanol, which was then added to a solution of the other drugs dissolved in the acidified sterile water for irrigation. No visible changes in color or clarity in any of the samples were observed. Ipratropium bromide by itself underwent substantial losses over eight weeks. Losses were 2, about 13 to 15, and 23% at refrigerated, room, and elevated temperature of 40°C. However, ipratropium bromide was more stable in the four-drug combination, exhibiting little or no loss under refrigeration, 5% loss at room temperature, and about 18% loss at 40°C in eight weeks. The other three drugs remained stable under refrigeration and at room temperature, with no more than 6% loss in eight weeks, but budesonide and albuterol sulfate underwent substantial losses of 27 and 10%, respectively, at 40°C in that time frame.

**STUDY 10 (ADRENERGICS):** Joseph[1024] reported that 1 and 4 mL of preservative-free ipratropium bromide 0.025% inhalation solution is physically and chemically compatible with 5 and 10 mg, respectively, of metaproterenol sulfate 5% solution for up to 30 days at room temperature. Similarly, 2.5 mL of preservative-free ipratropium bromide 0.025% inhalation solution is physically and chemically compatible with 1.25 mL of bitolterol 0.2% solution for at least one hour.

**STUDY 11 (MULTIPLE DRUGS):** White and Hood[1025] reported that albuterol sulfate inhalation solution was compatible with tobramycin sulfate, although no drug concentrations were stated. Adding ipratropium bromide to the inhalation admixture was also stated to be compatible, again without citing drug concentrations.

**STUDY 12 (MULTIPLE DRUGS):** Gronberg et al.[1027] evaluated the stability and compatibility of budesonide (Pulmicort, concentration unspecified) with several inhalation solutions, including (1) acetylcysteine (Lysomucil) 100 mg/mL 2 parts to 3 parts budesonide, (2) fenoterol hydrobromide (Berotec) 5 mg/mL 8 parts to 1 part budesonide, (3) ipratropium bromide (Atrovent) 0.25 mg/mL 2 parts to 1 part budesonide, and (4) ipratropium bromide 0.125 mg/mL plus fenoterol hydrobromide 0.31 mg/mL (Duovent) in equal parts with budesonide. Samples were stored at room temperature of 22 to 25°C protected from exposure to light. HPLC analysis found all mixtures to be compatible, maintaining greater than 90% of the initial concentrations for at least 18 hours.

**STUDY 13 (ALBUTEROL AND CROMOLYN):** Nagtegaal et al.[1046] evaluated the long-term stability of two inhalation solution formulations containing albuterol sulfate 1.5 mg/mL with ipratropium bromide 62.5 mcg/mL, both with and without cromolyn sodium 5 mg/mL. The drugs were compounded in a solution composed of sodium chloride 8.8 mg/mL in water for injection. pH was adjusted to 5.6 for the formulation with cromolyn sodium and to 5.0 for the formulation without the cromolyn. The inhalation solutions were packaged in glass ampules, filtered through a 0.2-μm filter, and autoclaved at 121°C for 15 minutes for sterilization. Samples of the sterilized solutions were stored at room temperature and also at elevated temperatures of 45, 65, and 85°C for accelerated degradation studies.

Ipratropium was the least stable component of the inhalation solutions. Stability-indicating HPLC analysis found the albuterol sulfate with ipratropium bromide inhalation solution (no cromolyn sodium) to be stable, with less than 10% ipratropium loss occurring in 18 months at room temperature and a calculated stability of 3.8 years refrigerated. With cromolyn sodium present in the three-drug inhalation solution, the ipratropium was less stable, exhibiting less than 10% loss for 10 months at room temperature and a calculated stability of 1.9 years refrigerated.

**STUDY 14 (DORNASE ALFA):** Kramer et al.[1216] evaluated the compatibility and stability of mixtures of inhalation solution admixtures with dornase alfa (Pulmozyme, Genentech). The contents of one ampule of dornase alfa (Pulmozyme Respule) were mixed with ipratropium bromide (Atrovent and Atrovent LS, Boehringer Ingelheim) and albuterol sulfate (Sultanol and Sultanol forte FI, GlaxoSmithKline) in the combinations below and were stored at room temperature exposed to room light.

*Mixture 1*
Pulmozyme 2.5 mg/2.5 mL
Atrovent 0.5 mg/2 mL

*Mixture 2*
Pulmozyme 2.5 mg/2.5 mL
Sultanol forte FI 0.5 mL

*Mixture 3*
Pulmozyme 2.5 mg/2.5 mL
Atrovent LS 2 mL
Sultanol 0.5 mL

*Mixture 4*
Pulmozyme 2.5 mL
Atrovent 0.5 mg/2 mL
Sultanol forte FI 0.5 mL

Dornase alfa was physically incompatible with the unpreserved unit-dose forms of Atrovent and Sultanol forte FI, with visible formation of particulates upon mixing. Dornase alfa activity measured using a kinetic colorimetric DNase activity assay remained unchanged over the five-hour test period. Stability-indicating HPLC analysis found that the ipratropium bromide and albuterol sulfate concentrations also remained unchanged.

However, dornase alfa mixed with Atrovent LS and Sultanol inhalation solution, which both contain the preservative benzalkonium chloride in the formulations, resulted in rapid dornase alfa activity losses of 20 to 50% within one to two hours. The ipratropium bromide and albuterol sulfate concentrations remained unchanged. Dornase alfa was found to lose activity due to the preservative benzalkonium chloride and also edetate disodium found in the Atrovent LS formulation. These combinations also resulted in visible particulate formation upon mixing.

Because the unpreserved formulations of ipratropium bromide and albuterol sulfate caused particulate formation when mixed with dornase alfa and because the preserved formulations of these drugs caused extensive and rapid loss of dornase alfa activity along with particulate formation, all of these mixtures of inhalation solutions tested were determined to be incompatible.

**STUDY 15 (ALBUTEROL AND FLUTICASONE):** Kamin et al.[1231] evaluated the physicochemical compatibility and stability of fluticasone-17-propionate nebulizer suspension (Flutide forte "ready-to-use," GlaxoSmithKline) mixed with ipratropium bromide (Atrovent LS, Boehringer Ingelheim) and albuterol sulfate (Sultanol, GlaxoSmithKline) for combined use in a nebulizer. The test mixtures were prepared in 10-mL glass containers. Fluticasone-17-propionate 2 mg/2 mL was mixed with ipratropium bromide 0.25 mg/2 mL and albuterol 2.5 mg/0.5 mL as the sulfate. No physical changes were observed during the five-hour study period. Stability-indicating HPLC analysis of the drug concentrations found that the drug concentrations remained near 100% over five hours as well.

**STUDY 16 (HYPERTONIC SODIUM CHLORIDE):** Fox et al.[1239] evaluated the physical compatibility of hypertonic sodium chloride 7% with several inhalation medications used in treating cystic fibrosis. Ipratropium bromide respiratory solution (Dey) 0.02% mixed in equal quantities with extemporaneously compounded hypertonic sodium chloride 7% did not exhibit any visible evidence of physical incompatibility, and the measured turbidity did not increase over the one-hour observation period.

**STUDY 17 (LEVALBUTEROL):** Yamreudeewong et al.[1328] evaluated the stability of levalbuterol hydrochloride (Xopenex, Sepracor) inhalation solution mixed with ipratropium bromide (Nephron Pharmaceutical Corp.) inhalation solution for combined use as a nebulizer solution. The initial concentrations

were 227 mcg/mL and 90 mcg/mL for the levalbuterol hydrochloride and ipratropium bromide, respectively, in the single sample that was tested. The sample was packaged in a polypropylene screw cap vial and stored at unspecified room temperature for 28 days. Physical stability was not specifically assessed, but no changes were reported. HPLC analysis results found little or no loss of either drug occurred over the 28-day study period.

**STUDY 18 (LEVALBUTEROL):** Bonasia et al.[1329] evaluated the compatibility and stability of levalbuterol hydrochloride inhalation solution (Sepracor) with several drugs including ipratropium bromide (Dey). The drugs were mixed with sodium chloride 0.9% and then with each other to prepare the following solution:

> Levalbuterol 1.25 mg/0.5 mL diluted with 2.5 mL of sodium chloride 0.9%
> Ipratropium bromide 0.5 mg/2.5 mL diluted with 0.5 mL of sodium chloride 0.9%

The drug mixture was prepared in triplicate and evaluated initially and after 30 minutes at an ambient temperature of 21 to 25°C. The authors indicated that a 30-minute study duration was chosen because it represents the approximate time it takes for a patient to administer drugs by nebulization at home. No visible evidence of physical incompatibility was observed with the samples. HPLC analysis found no loss of any of either of the drugs in the samples.

**STUDY 19 (FORMOTEROL):** Akapo et al.[1330] evaluated the compatibility and stability of formoterol fumarate 20 mcg/2 mL inhalation solution (Perforomist, Dey) mixed with ipratropium bromide 0.5 mg/2.5 mL (Dey) for administration by inhalation and evaluated over 60 minutes at room temperature of 23 to 27°C. The test samples were physically compatible by visual examination and measurement of osmolality, pH, and turbidity. HPLC analysis of drug concentrations found little or no loss of either of the drugs within the study period.

**STUDY 20 (ALBUTEROL, DORNASE ALFA, TOBRAMYCIN):** Kramer et al.[1383] evaluated the compatibility and stability of a variety of two- and three-drug solution admixtures of inhalation drugs, including ipratropium bromide. The mixtures were evaluated using chemical assays including HPLC, DNase activity assay, and fluorescence immunoassay as well as visual inspection, pH measurement, and osmolality determination. The drug combinations tested and the compatibility results that were reported are shown below.

*Mixture 1*
Albuterol sulfate (Sultanol) 5 mg/1 mL
Dornase alfa (Pulmozyme) 2500 units/2.5 mL
Ipratropium bromide (Atrovent LS) 0.25 mg/1 mL
   Result: Physically and chemically incompatible

*Mixture 2*
Dornase alfa (Pulmozyme) 2500 units/2.5 mL
Ipratropium bromide (Atrovent) 0.25 mg/2 mL
    Result: Physically and chemically incompatible

*Mixture 3*
Albuterol sulfate (Sultanol) 5 mg/1 mL
Ipratropium bromide (Atrovent LS) 0.25 mg/1 mL

Tobramycin (Tobi) 300 mg/5 mL
    Result: Physically and chemically compatible

*Mixture 4*
Albuterol sulfate (Sultanol) 5 mg/1 mL
Ipratropium bromide (Atrovent LS) 0.25 mg/1 mL
Tobramycin sulfate (Gernebcin) 80 mg/2 mL
    Result: Physically and chemically compatible ▪

# ▪ Isoetharine Hydrochloride

## Properties
Isoetharine hydrochloride is a white to off-white odorless crystalline powder having a bitter, salty taste.[2,3]

## Solubility
Isoetharine hydrochloride is soluble in water and sparingly soluble in ethanol.[2,3]

## pH
A 1% isoetharine hydrochloride solution in water has a pH between 4 and 5.6.[3,4] The commercial solutions have a pH of 2.5 to 5.5.[2,4]

## General Stability Considerations
Isoetharine hydrochloride solutions should be packaged in tight containers and stored at controlled room temperature. The solutions should be protected from exposure to light and should not be used if discoloration occurs or a precipitate appears. Solutions of the drug are stated to be compatible with sterile water, 0.45 and 0.9% sodium chloride, and ethanol 20%; the drug has been used in combination with antibiotics and surfactants such as acetylcysteine and tyloxapol.[2,3]

   Valenzuela et al.[1624] reported the stability of isoetharine hydrochloride exposed to temperatures ranging from 26 to 38°C under simulated summer conditions in paramedic vehicles over four weeks. Gas chromatography coupled with mass spectrometry found no change in the drug over four weeks under these simulated use conditions.

## Stability Reports of Compounded Preparations
### Inhalation
Inhalations, like other sterile drugs, should be prepared in a suitable clean air environment using appropriate aseptic procedures. When prepared from nonsterile components, an appropriate and effective sterilization method must be employed.

**STUDY 1:** Hunke et al.[80] evaluated the stability of isoetharine hydrochloride (Bronkosol 1%, Breon), diluted with distilled water to a final concentration of 0.25% (wt/vol), during ultrasonic nebulization. The nebulizers used were a Mistogen electronic nebulizer (model EN 145), a Monaghan hospital ultrasonic nebulizer (model 675), and a DeVilbiss Pulmosonic (model 25). The drug solution was nebulized for 10 or 15 minutes initially and at 3, 6, 8, and 24 hours. Between runs, the test solutions remained in the nebulizers at ambient temperature but were protected from evaporation. The isoetharine hydrochloride content of the solutions in the nebulizers and in captured nebulized mist was assessed using a stability-indicating chromatographic–ultraviolet spectroscopic method and by thin-layer chromatography. No substantial degradation of isoetharine hydrochloride occurred over 24 hours in either the solution or the mist. Calculated apparent losses averaged less than 2%.

**STUDY 2:** Kleinberg et al.[23] reported the long-term stability of isoetharine hydrochloride solution (Bronkosol, Breon) diluted to 1.67 mg/mL in 0.9% sodium chloride solution and repackaged in 3-mL amber NebuJect (Solopak) nebulizer injectors. Isoetharine hydrochloride loss totaled about 5% in 15 days and 7% in 30 days at 25°C when analyzed by HPLC. Furthermore, there were no changes in pH or physical appearance.

**STUDY 3:** Gupta et al.[1181] evaluated the stability of isoetharine hydrochloride (Barre-National) 1% inhalation solution diluted 1:10 in sodium chloride 0.9% to a concentration of 0.1% and packaged as 3.3 mL of sample solution in 5-mL amber oral syringes sealed with tip caps (Becton Dickinson). At room temperature of 24 to 26°C, diluted isoetharine hydrochloride remained clear and colorless over 60 days, with losses of about 2% by HPLC analysis. However, after 90 days at room temperature, the samples turned pink and the losses increased to about 8%, which was considered unacceptable. The authors indicated that dilution of the antioxidant (acetone sodium bisulfite) in the isoetharine hydrochloride formulation rendered it insufficient to prevent decomposition. However, samples stored refrigerated at 4 to 6°C were stable over 120 days, remaining clear and colorless and exhibiting little or no loss of isoetharine hydrochloride.

## Compatibility with Other Drugs
### ISOETHARINE HYDROCHLORIDE COMPATIBILITY SUMMARY
*Incompatible with:* Acetylcysteine • Atropine sulfate
 • Dexamethasone • Sodium bicarbonate
*Uncertain or variable compatibility with:* Cromolyn sodium

**STUDY 1:** Lesko and Miller[659] evaluated the chemical and physical stability of 1% cromolyn sodium nebulizer solution (Fisons) 2 mL admixed with 0.5 mL of isoetharine hydrochloride 1% inhalation solution. The admixture was visually clear and colorless and remained chemically stable by HPLC analysis for 60 minutes after mixing at 22°C. No loss of cromolyn sodium was found, but about 4% loss of isoetharine hydrochloride occurred.

**STUDY 2:** Owsley and Rusho[517] evaluated the compatibility of several respiratory therapy drug combinations. Isoetharine 1% solution (Bronkosol, Sanofi Winthrop) was combined with the following drug solutions: acetylcysteine 10% solution (Mucosil-10, Dey), atropine sulfate 1 mg/mL (American Regent), cromolyn sodium 10-mg/mL nebulizer solution (Intal, Fisons), dexamethasone (form unspecified) 4 mg/mL (American Regent), and sodium bicarbonate 8.4% injection (Abbott). The test solutions were filtered through 0.22-μm filters into clean vials. The combinations were evaluated over 24 hours (temperature unspecified) using the USP particulate matter test. None of the combinations was compatible. Isoetharine combined with the acetylcysteine, atropine sulfate, cromolyn sodium, dexamethasone, and sodium bicarbonate all formed unacceptable levels of large particulates (≥10 μm). ▪

#  Isoniazid

## Properties
Isoniazid is an odorless white crystalline powder or colorless crystals.[2,3]

### Solubility
Isoniazid is soluble in water to about 125 to 140 mg/mL at 25°C and to about 260 mg/mL at 40°C. In ethanol, the solubility is about 20 mg/mL at 25°C and about 100 mg/mL in boiling alcohol.[1–3]

### pH
A 1% aqueous solution has a pH of 5.5 to 6.5.[1] A 10% aqueous solution has a pH between 6 and 7.5.[4] Isoniazid injection has a pH adjusted to 6 to 7.[4]

## General Stability Considerations
Isoniazid products should be protected from light, air, and excessive heat.[2] Isoniazid tablets should be packaged in well-closed, light-resistant containers, and the syrup should be packaged in tight, light-resistant containers. The injection should be protected from light as well. Isoniazid products should be stored at controlled room temperature protected from temperatures of 40°C or more. The liquid products also should be protected from freezing.[2,4]

Isoniazid injection is subject to crystallization if exposed to low temperatures. If crystallization occurs, the injection should be warmed to room temperature to redissolve the crystals.[2,4]

Solutions of isoniazid may be sterilized by autoclaving at 120°C for 30 minutes.[1,3]

However, studies by Carlin et al.[663] found the formation of potentially carcinogenic levels of hydrazine as a decomposition product in isoniazid syrup stored at room temperature for four months. The formation rate of hydrazine is increased sixfold if stored at an elevated temperature of 40°C. No detectable hydrazine formed in samples stored at 0°C. The authors recommended changing the labeled storage condition for isoniazid syrup to subambient temperature.

Lovering et al.[1470] evaluated the formation of hydrazine, a carcinogenic decomposition product of isoniazid, in several dosage forms of the drug evaluated over a two-year storage period at room temperature and elevated temperatures of 45 and 60°C and 75% relative humidity. In oral tablets, no significant change in the amount of hydrazine was detected over two years. However, in isoniazid elixir, a 10-mg/mL aqueous liquid, hydrazine levels increased from 19.5 to 44 mcg/mL at room temperature and ambient humidity in 23 months.

Isoniazid incompatibility with lactose is well documented. This incompatibility may be of concern if isoniazid tablets containing lactose in the formula are used as the drug source for preparing compounded liquid preparations.[979]

## Stability Reports of Compounded Preparations
### Oral
**STUDY 1:** The extemporaneous formulation of an isoniazid 10-mg/mL oral suspension was described by Nahata and Hipple.[160] Ten isoniazid 100-mg tablets were crushed and mixed thoroughly with 10 mL of purified water. The suspension was brought to 100 mL with 70% sorbitol. The authors stated that a stability period of 21 days under refrigeration was used, although chemical stability testing was not performed.

**STUDY 2:** Carlin et al.[663] reported that unacceptable amounts of carcinogenic hydrazine (a decomposition product of isoniazid) formed in four months when isoniazid syrup 10 mg/mL in sorbitol 70% was stored at room temperature of 23 to 27°C protected from exposure to light. The authors recommended refrigerated storage because virtually no hydrazine formed when the syrup was stored under refrigeration at 0°C.

**STUDY 3:** Seifart et al.[743] reported the stability of an extemporaneously compounded isoniazid 10-mg/mL suspension in a vehicle similar to that of the British Pharmacopoeia. Isoniazid tablets were pulverized using a mortar and pestle. The powder was suspended in a vehicle that included citric acid, sodium sulfate, glycerin, chloroform water, and purified water. Stability-indicating HPLC analysis found that more than 95% of the isoniazid remained after 28 days stored at either 4 or 24°C.

**STUDY 4:** Gupta and Sood[926] evaluated the stability of an isoniazid 10-mg/mL oral solution composed of sorbitol 35% with methylparaben 2 mg/mL and propylparaben 0.2 mg/mL in purified water. The oral solution was packaged in amber glass bottles and stored at room temperature near 25°C. Stability-indicating HPLC analysis found that less than 3% isoniazid loss occurred after 42 days of storage. However, the solution changed from initially being colorless to light brown after 42 days.

**STUDY 5:** Haywood et al.[979] reported that the formula of the tablets used to prepare a compounded oral liquid may be of concern for drug stability. Isoniazid is well documented to be incompatible with lactose and other reducing sugars that cause rapid isoniazid decomposition to a potentially toxic hydrazone compound (1-isonicotinoyl-2-lactosylhydrazone). In this study, commercial compressed tablets of isoniazid BP (Fawns & McAllan) were crushed and incorporated into a liquid medium that contained citric acid, sodium citrate, glycerol, compound hydroxybenzoate solution APF, and purified water to form an oral liquid preparation. The stability was compared to an identical formulation using the pure bulk isoniazid powder as the drug source. Stability-indicating HPLC analysis found isoniazid losses of 22% in one day and 29% in seven days at room temperature and under refrigeration when tablets were used as the drug source. However, the oral liquid preparation compounded from pure isoniazid powder did not exhibit any loss in seven days at either temperature. The authors recommended using only pure isoniazid powder as the drug source and using a sugar-free formulation.

## Rectal

Hudson et al.[1129] evaluated the release of isoniazid from three types of pediatric rectal suppositories, each containing 100 mg of the drug. The suppository bases were the lipophilic bases cocoa butter and Witepsol H15 Base F as well as a water-soluble mixed base composed of polyethylene glycol 3350 600 mg, polyethylene glycol 1000 300 mg, polyethylene glycol 400 100 mg, and silica gel 12 mg. The mixed polyethylene glycol base exhibited a more rapid and complete release of drug; about 70% of the isoniazid was released within six hours. The cocoa butter and Witepsol released about 50 and about 20%, respectively, of the isoniazid in six hours.

## Compatibility with Other Drugs

**STUDY 1 (ASCORBIC ACID):** Seifart et al.[743] reported the stability of an extemporaneously compounded isoniazid 10-mg/mL suspension mixed with ascorbic acid 20 mcg/mL. Although the isoniazid suspension was stable for 28 days alone both at room and refrigeration temperatures, the mixture with ascorbic acid demonstrated much shorter stability. Stability-indicating HPLC analysis found isoniazid losses of 7 and 11% in eight and 12 days, respectively.

**STUDY 2 (RIFAMPIN):** Seifart et al.[743] reported the stability of an extemporaneously compounded isoniazid 10-mg/mL suspension mixed with commercial rifampin 20-mg/mL suspension (Mer National). While both the isoniazid suspension and the rifampin suspension were stable separately, rapid decomposition of both drugs occurred when mixed. Stability-indicating HPLC analysis found about a 10 to 12% loss of both drugs in two days stored at 4°C.

**STUDY 3 (WITH RIFAMPIN AND PYRIDOXINE):** Ved and Deshpande[1628] evaluated the stability of rifampin 20 mg/mL with isoniazid 20 mg/mL and pyridoxine hydrochloride 1 mg/mL in a vehicle composed of sorbitol and propylene glycol (vehicle component concentrations not provided). Samples were stored at 15 and 30°C and at elevated temperatures of 37 and 45°C. The suspension was described as having a good appearance with easy redispersibility and low sedimentation rate. However, microbiological analysis of the rifampin content found 25 and 30% loss of antimicrobial activity after one week of storage at 15 and 30°C, respectively, with higher losses at elevated temperatures. The authors concluded that rifampin was unstable in combination with isoniazid and pyridoxine hydrochloride.

**STUDY 4 (PYRAZINAMIDE):** Seifart et al.[743] reported the stability of an extemporaneously compounded isoniazid 10-mg/mL suspension mixed with an extemporaneously compounded pyrazinamide suspension. To prepare the isoniazid suspension, isoniazid tablets were pulverized using a mortar and pestle. The powder was suspended in a vehicle that included citric acid, sodium sulfate, glycerin, chloroform water, and purified water. To prepare the pyrazinamide suspension, pyrazinamide tablets were pulverized using a mortar and pestle. The powder was suspended in a vehicle that included tragacanth powder, chloroform water, and purified water.

The suspension mixture was stable for 28 days stored at either 4 or 24°C. Stability-indicating HPLC analysis found losses of 8 and 2% for isoniazid and pyrazinamide, respectively, in 28 days. Addition of ascorbic acid 20 mcg/mL to the mixture of suspensions resulted in a 30% loss of isoniazid at 4°C and a 50% loss at 24°C in 28 days.

**STUDY 5 (LACTOSE):** Haywood et al.[979] noted that isoniazid incompatibility with lactose is well documented. This incompatibility may be of concern if isoniazid tablets containing lactose are used as the drug source for preparing compounded liquid preparations. See *Oral Study 5.*

## Compatibility with Common Beverages and Foods

Peloquin et al.[1135] evaluated the stability of several oral antituberculosis drugs, including isoniazid, for compatibility with common foods to facilitate administration to children. Isoniazid tablets were crushed and mixed with 30 g of Hunt's Snackpack no-sugar-added chocolate pudding, Safeway grape jelly, Safeway creamy peanut butter, orange juice, and 7-Up soda. The drug–food mixtures were thoroughly mixed and analyzed (no method indicated) for drug content initially and after one, two, and four hours had elapsed. Drug recovery at all time points was greater than 93% from the pudding, greater than 89% from the jelly, and greater than 80% from the peanut butter. Isoniazid recovery was about 100% from orange juice but as low as 85% in 7-Up. The authors recommended the use of the no-sugar-added chocolate pudding or grape jelly to aid in administering the drugs to children. ■

# Isoproterenol Hydrochloride
## (Isoprenaline Hydrochloride)

## Properties

Isoproterenol hydrochloride is a white or practically white odorless crystalline powder.[2,3]

### Solubility

Isoproterenol hydrochloride is freely soluble in water, with an aqueous solubility of 1 g in 3 mL.[1–3,7] In 95% ethanol, 1 g dissolves in about 50 mL but is less soluble in dehydrated alcohol.[1,3]

### pH

A 1% solution of isoproterenol hydrochloride in water has a pH of about 5.[1–3] Isoproterenol hydrochloride injection has a pH of 2.5 to 4.5[4]; Sanofi Winthrop adjusts the pH of Isuprel injection to 3.5 to 4.5.[7] The isoproterenol inhalation solution has a pH between 2.5 and 5.5.[4] Isoproterenol hydrochloride elixir (Isuprel, Breon) has a pH of 3.7.[19]

### Osmolality

Isoproterenol hydrochloride injection 0.2 mg/mL is reported to have an osmolality of 277 to 293 mOsm/kg, depending on the method of determination.[287]

## General Stability Considerations

Isoproterenol hydrochloride products must be stored in tight, light-resistant containers because air, light, and heat cause discoloration and darkening.[1–3,7,375] Solutions of isoproterenol hydrochloride that have discolored, becoming pink to brownish, or that contain a precipitate should not be used.[2,7]

The stability of isoproterenol hydrochloride in solution is very pH-dependent; pH is a primary determinant of stability.[376] In solutions with a pH above 6, isoproterenol hydrochloride undergoes significant decomposition; 5% loss occurs in about eight hours at pH 6.5 and in about six hours at pH 7.6.[377,378] In 5% dextrose injection at 5 mcg/mL, the drug is stable for at least 24 hours at 25°C from pH 3.7 to 5.7.[377] Products that raise the solution pH above 6 should be avoided.

## Stability Reports of Compounded Preparations
### Oral

Allen[1321] reported on a compounded formulation of isoproterenol compound elixir. The elixir had the following formula:

| | | |
|---|---|---|
| Isoproterenol hydrochloride | | 2.5 mg |
| Phenobarbital | | 6 mg |
| Ephedrine sulfate | | 12 mg |
| Theophylline | | 45 mg |
| Potassium iodide | | 150 mg |
| Ethanol 95% | | 2.85 mL |
| Purified water | qs | 15 mL |

The recommended method of preparation was to dissolve the isoproterenol hydrochloride, ephedrine sulfate, and potassium iodide in 10 mL of purified water. Then the phenobarbital and theophylline were added to the ethanol and thoroughly mixed. The aqueous solution was then added slowly to the ethanol solution and thoroughly mixed. Additional purified water was added to bring the elixir to final volume and was thoroughly mixed. The elixir was to be packaged in tight, light-resistant containers. The author recommended a beyond-use date of six months at room temperature because this formula was formerly a commercial medication in the United States with an expiration date of two years or more.

Also recommended was labeling advising not to use the elixir if crystals, precipitation, or cloudiness appeared.

### Enteral

**ISOPROTERENOL HYDROCHLORIDE COMPATIBILITY SUMMARY**

*Compatible with:* Ensure • Ensure HN • Ensure Plus • Ensure Plus HN • Osmolite • Osmolite HN • Vital

**STUDY 1:** Cutie et al.[19] added 15 mL of isoproterenol hydrochloride elixir (Isuprel, Breon) to varying amounts (15 to 240 mL) of Ensure, Ensure Plus, and Osmolite (Ross Laboratories) with vigorous agitation to ensure thorough mixing. The isoproterenol hydrochloride elixir was physically compatible, distributing uniformly in all three enteral products with no phase separation or granulation.

**STUDY 2:** Altman and Cutie[850] reported the physical compatibility of isoproterenol hydrochloride elixir (Isuprel, Breon) 15 mL with varying amounts (15 to 240 mL) of Ensure HN, Ensure Plus HN, Osmolite HN, and Vital with vigorous agitation to ensure thorough mixing. The elixir was physically compatible, distributing uniformly in all four enteral products with no phase separation or granulation.

### Inhalation

Inhalations, like other sterile drugs, should be prepared in a suitable clean air environment using appropriate aseptic procedures. When prepared from nonsterile components, an appropriate and effective sterilization method must be employed.

Hunke et al.[80] evaluated the stability of isoproterenol hydrochloride (Isuprel 0.5 and 1%, Breon), diluted with distilled water to 0.125% (wt/vol) during ultrasonic nebulization. The nebulizers used were a Mistogen electronic nebulizer (model EN 145), a Monaghan hospital ultrasonic nebulizer (model 675), and a DeVilbiss Pulmosonic (model 25). The drug solution was nebulized for 10 or 15 minutes initially and at 3, 6, 8, and 24 hours. Between runs, the test solutions remained in the nebulizers at ambient temperature but were protected from evaporation. The isoproterenol hydrochloride content of the solutions in the nebulizers and in captured nebulized mist was assessed using a stability-indicating chromatographic–ultraviolet spectroscopic method and by thin-layer chromatography. No substantial degradation of isoproterenol hydrochloride occurred over 24 hours in either the solution or the mist. Calculated apparent losses averaged less than 3%.

## Compatibility with Other Drugs

**STUDY 1:** Lesko and Miller[659] evaluated the chemical and physical stability of 1% cromolyn sodium nebulizer solution (Fisons) 2 mL admixed with 0.5 mL of isoproterenol hydrochloride 0.5% inhalation solution. The admixture was visually clear and colorless and remained chemically stable by HPLC analysis for 60 minutes after mixing at 22°C.

**STUDY 2:** Tomonaga et al.[1605] evaluated the compatibility and stability of gentamicin sulfate injection mixed with six drugs for inhalation including Asthpul (dl-isoproterenol hydrochloride). Using a microbiological activity assay, it was found that in all of these inhalation solutions the gentamicin sulfate content deteriorated substantially in most combinations within one day and within all combinations within three days stored both under refrigeration at 4°C and at room temperature of 20 and 30°C. ■

# ■ Isosorbide Dinitrate

## Properties

Isosorbide dinitrate has been described as hard colorless crystals[1] and as white crystalline rosettes or powder.[3,4]

Diluted isosorbide dinitrate is an ivory-white odorless powder that is a dry mixture of isosorbide dinitrate (at about 25% concentration) and lactose monohydrate, mannitol, or other suitable inert material. The inert material is present to minimize the risk of explosion. Diluted isosorbide dinitrate may also contain a stabilizer such as ammonium phosphate up to 1%.[3,4]

Isosorbide dinitrate injection is available in 0.1 and 0.05% concentrations.[7]

## Solubility

Isosorbide dinitrate is sparingly soluble in water to about 1.09 mg/mL.[1] The drug is very soluble in organic solvents such as acetone and sparingly soluble in ethanol.[1,4]

## General Stability Considerations

NOTE: Undiluted isosorbide dinitrate may explode from exposure to excessive heat or from percussion.[3,4]

Diluted isosorbide dinitrate should be packaged in tight containers and stored at controlled room temperature. Solid oral dosage forms of isosorbide dinitrate are packaged in well-closed containers and stored at controlled room temperature.[4] Isosorbide dinitrate is packaged in ampules, vials, and prefilled syringes and should be stored at controlled room temperature.[7]

Isosorbide dinitrate in the injection has been shown to be subject to loss due to sorption to polyvinyl chloride (PVC) plastic containers and equipment and to various filter media.[8]

## Stability Reports of Compounded Preparations
### Rectal

Hill and Farrands[1114] evaluated several isosorbide dinitrate rectal gel formulations for symptomatic relief of anal fissures, anal fistulae, and hemorrhoids. Sodium benzoate 100 mg was dissolved in isosorbide dinitrate injection 100 mL and the mixture was incorporated into 3 g of each of three gels: Methocel A4C, sodium carboxymethylcellulose, and Bard absorption dressing. Little extraction of the drug was found to occur with the Methocel A4C and the sodium carboxymethylcellulose formulations, rendering them unsuitable. However, use of the Bard absorption dressing resulted in extraction of 72 to 86% of the drug; consequently, this formulation was deemed the best. The formulation was packaged in polypropylene syringes, sealed with tip caps, and stored refrigerated at 6°C for 69 days. HPLC analysis found no formation of decomposition products, no sorption to the plastic syringes, and no substantial change in the amount of isosorbide dinitrate extracted over the 69-day study period. ■

# Isosulfan Blue
## (Sulfan Blue, Patent Blue V, Alphazurine 2G)

## Properties
Isosulfan blue is a blue dye used as a diagnostic agent.[3]

### Solubility
Equilibrium solubility of isosulfan blue in distilled water is around 20 mg/mL at 30°C.[217]

## Stability Reports of Compounded Preparations
### Injection
Injections, like other sterile drugs, should be prepared in a suitable clean air environment using appropriate aseptic procedures. When prepared from nonsterile components, an appropriate and effective sterilization method must be employed.

Newton et al.[217] evaluated extemporaneously prepared isosulfan blue dye injection. Although isosulfan blue has been prepared in lidocaine hydrochloride injection for diagnostic use, the formation of particulate matter following preparation led to an evaluation of the dye formulations. A dye concentration of 1% was prepared in distilled water, in 1% lidocaine hydrochloride injection, and in 1% lidocaine hydrochloride solution in distilled water. The dye solutions were filtered through 0.22-μm filters into sterile 5-mL vials with rubber stoppers and in glass ampules. Representative samples were autoclaved for 20 minutes at 121°C and 15 psi; an equal number of representative samples were not autoclaved.

Combining the dye with lidocaine 1% results in the immediate formation of a sticky dark viscid precipitate. An average of 29.6 mg of precipitate formed from 25 mL of the solution. The greatest amount of precipitate formed immediately; 8.7% of the dye was lost immediately, followed by an additional 2% over the next 120 days. The unsuitability of this combination is obvious. Only the isosulfan blue in distilled water maintained stability, with no loss by spectrophotometric evaluation in 120 days at room temperature. The samples that were not autoclaved remained free of precipitate; the autoclaved samples developed a fine dark precipitate within 285 days of storage at 23°C. The authors recommended using only water as the diluent for isosulfan blue and storing the compounded preparation for no longer than 120 days. ■

# Isradipine

## Properties
Isradipine is a yellow odorless or nearly odorless crystalline powder.[3,7]

### Solubility
Isradipine is essentially insoluble in water, having an aqueous solubility of less than 0.01 mg/mL at 37°C. The drug is soluble in ethanol.[7]

## General Stability Considerations
Isradipine capsules should be packaged in tight, light-resistant containers and stored at less than 30°C.[7]

## Stability Reports of Compounded Preparations
### Oral
MacDonald et al.[199] reported the stability of an extemporaneous oral suspension of isradipine. The isradipine 1-mg/mL suspension was prepared by wetting the powder from 10 isradipine 5-mg capsules (DynaCirc, Sandoz) with a small amount of glycerin and triturating to a fine paste. Simple syrup was used to bring the product to 50 mL. The finished suspension was packaged in two-ounce type III amber glass bottles with child-resistant caps and then stored at 4°C in the dark. A stability-indicating HPLC assay was used to determine the isradipine content. Little or no

drug was lost from the suspension during 35 days of storage at 4°C. Furthermore, there was no detectable change in color or odor or any visible microbiological growth in any sample. ■

## ■ Itraconazole

### Properties

Itraconazole is a white or nearly white powder.[3,7]

### Solubility

Itraconazole is practically insoluble in water and dilute acidic solutions. It is very slightly soluble in ethanol.[1,3]

### pH

Itraconazole oral solution has a target pH of 2.

### pKₐ

Itraconazole has a pKₐ of 3.7.[7,523]

### General Stability Considerations

Itraconazole capsules should be stored at controlled room temperature and protected from light and moisture. The oral solution should be stored at controlled room temperature and protected from freezing.[7]

### Stability Reports of Compounded Preparations
#### Oral

Therapeutic decision-makers must consider potential risks and benefits to specific patients of using extemporaneously compounded itraconazole suspensions when use of the commercial capsules or lower concentration commercial liquid is not feasible. Close monitoring of such patients, including the determination of itraconazole blood levels, may be appropriate.

**STUDY 1:** Jacobson et al.[520] evaluated the stability of an extemporaneously compounded itraconazole 40-mg/mL oral suspension prepared from capsules. The contents of 24 itraconazole 100-mg capsules (Sporanox, Janssen) were emptied into a glass mortar. About 4 to 5 mL of ethanol 95% was added and allowed to stand for 3 or 4 minutes to soften the material; the softened beads were then ground to a paste. Grinding was continued as the ethanol evaporated, resulting in fine powder. About 15 mL of simple syrup was added to the powder, triturated well, and transferred to a two-ounce amber glass bottle with a child-resistant cap. The mortar was rinsed with 15 mL of simple syrup, and this rinse was added to the bottle. Rinsing was repeated with enough syrup to bring the volume of the suspension to 60 mL. The suspension was stored under refrigeration at 4°C for 35 days. No change in color or odor was observed during storage. Stability-indicating HPLC analysis found about 5% itraconazole loss after 35 days of refrigerated storage.

Villarreal and Erush[521] questioned the bioavailability of the suspension just described. The suspension was administered to a patient through a jejunostomy tube. Itraconazole blood levels were not detectable. The authors noted that multiple factors with this patient could have contributed to decreased blood levels. However, they felt that inadequate absorption of itraconazole from the suspension in the gastrointestinal tract was a reasonable conclusion, possibly due to the basic pH of the jejunum environment. Denning et al.[522] and Kintzel et al.[523] also reported unsatisfactory itraconazole blood levels from suspensions.

In response, Jacobson and Johnson[524] agreed that bioavailability of the itraconazole may be uncertain and potentially problematic in some patients. They once again noted that multiple factors in addition to the suspension formulation could influence drug absorption. However, the uncertainty of bioavailability must be weighed against the patient's need when no alternative mode of therapy is possible. In addition, Bhandari et al.[525] and Bhandari and Naranag[526] reported successful treatment with itraconazole suspensions in neonates.

**STUDY 2:** Kintzel et al.[523] reported treatment of two patients by orogastric tubes with two extemporaneously compounded suspensions of itraconazole from capsules that resulted in lower serum levels than equivalent doses of the capsules. One formulation was prepared by pulverizing the capsule beads, adding to sterile water, and incorporating into 20% lipid emulsion. The other formulation was prepared by emptying the capsule contents into 1.5% citric acid in 5% dextrose and allowing the mixture to stand for 20 minutes to permit dissolution. Unfortunately, neither patient achieved serum levels of itraconazole in the desired range after orogastric tube administration of either suspension, although the citric acid formulation yielded higher serum levels than the lipid formulation. Additionally, it was noted that both patients had pathophysiologic factors that decrease absorption.

Kintzel et al.[523] noted that itraconazole oral bioavailability improved markedly in fed patients compared to fasting patients.

**STUDY 3:** Abdel-Rahman and Nahata[1155] evaluated the stability of an oral liquid formulation of itraconazole 20 mg/

mL prepared from commercial capsules. Forty commercial capsules were opened, and the contents were emptied into a mortar and softened with 15 mL of ethanol. The capsule contents were crushed to fine powder using a pestle and diluted to a concentration of 20 mg/mL with an equal parts mixture of Ora-Plus and Ora-Sweet oral vehicles. The itraconazole suspension was packaged in amber plastic prescription bottles, and sample bottles were stored at room temperature of 25°C and refrigerated at 4°C.

No appreciable changes in color, odor, and pH occurred. Stability-indicating HPLC analysis found about 8% loss in 56 days at 25°C and in 70 days refrigerated. The authors stated that the extemporaneous itraconazole formulation was stable for eight weeks at room temperature and refrigerated. ■

# Kanamycin Sulfate

## Properties
Kanamycin sulfate is a white or nearly white odorless crystalline powder.[2,3] Kanamycin sulfate 1.2 g is approximately equivalent to 1 g of kanamycin.[3]

### Solubility
Kanamycin sulfate is freely soluble in water, having an aqueous solubility of about 125 mg/mL. The drug is practically insoluble in ethanol.[1-3]

### pH
A 1% kanamycin sulfate solution in water has a pH of 6.5 to 8.5.[3,4] Kanamycin sulfate injection has a pH of 3.5 to 5.[4]

### Osmolality
Kanamycin sulfate injection has an osmolality of 858 to 952 mOsm/kg, depending on the method used.[287]

## General Stability Considerations
Kanamycin sulfate capsules should be packaged in tight containers. Both the capsules and the injection should be stored at controlled room temperature protected from excessive temperatures of 40°C or more. The injection also should be protected from freezing.[2,4]

## Stability Reports of Compounded Products
### Ophthalmic
Ophthalmic preparations, like other sterile drugs, should be prepared in a suitable clean air environment using appropriate aseptic procedures. When prepared from nonsterile components, an appropriate and effective sterilization method must be employed.

Osborn et al.[169] evaluated the stability of several antibiotics in three artificial tears solutions composed of 0.5% hydroxypropyl methylcellulose. The artificial tears (Lacril, Tearisol, Isopto Tears) were used to dilute kanamycin sulfate to 30 mg/mL. The products were packaged in plastic squeeze bottles and stored at 25°C. A serial dilution bioactivity test was used to estimate the amount of antibiotic activity remaining during seven days of storage. The remaining amounts of antibiotic were 81% in Lacril, 87% in Tearisol, and 82% in Isopto Tears. ■

# Ketamine Hydrochloride

## Properties
Ketamine hydrochloride is a white crystalline powder having a characteristic odor. Ketamine hydrochloride 1.15 mg is approximately equal to 1 mg of ketamine base.[3]

### Solubility
Ketamine hydrochloride has an aqueous solubility of 200 to 250 mg/mL.[1,3] It has solubilities of 71 mg/mL in ethanol and 16.7 mg/mL in dehydrated ethanol.[3]

### pH
A 10% aqueous solution has a pH of 3.5 to 4.1.[3,4] Ketamine hydrochloride injection has a pH of 3.5 to 5.5.[4]

## General Stability Considerations
Ketamine hydrochloride injection should be stored at controlled room temperature and protected from light.[4,5]

## Stability Reports of Compounded Preparations
### Oral
**STUDY 1:** Various approaches have been tried to make ketamine hydrochloride given orally more palatable, particularly to children. Rosen and Rosen[611] tried a number of beverages and flavorings but preferred to add ketamine hydrochloride to a sugar-sweetened gelatin dessert prepared with at least 1.3 mL of gelatin dessert for every milliliter of drug and

allowed to solidify in ice cube trays. The gelatin cubes were made with 100 or 250 mg of ketamine; fractional doses were provided by cutting the cubes. Because the pH of the gelatin preparation was less than 4, the authors indicated the drug should remain stable, but no stability data were provided.

**STUDY 2:** Kaneuchi et al.[958] evaluated several agar-based oral solid gel dosage forms of ketamine hydrochloride used for analgesia to help overcome the extremely bitter taste of the drug that makes oral ingestion of the injection difficult. The oral solid gel dosage units were compounded, incorporating the ketamine hydrochloride into 3% agar bases in water with benzethonium chloride preservative to yield dosage units of 25, 50, and 100 mg of ketamine. Ketamine content and release profiles were unchanged over 12 weeks at room temperature. Addition of sugar or coffee flavoring reduced the release rate. The simpler oral gel form without flavorings was preferable.

**STUDY 3:** Chong et al.[1463] compounded ketamine hydrochloride 25-mg sublingual/oral lozenges for use in treating neuropathic pain and evaluated the stability and bioavailability of the drug in the lozenge formulation. The lozenges weighing 1 g were compounded using the following formula:

| | | |
|---|---|---|
| Ketamine hydrochloride | | 2.5 g |
| Gelatin powder | | 25 g |
| Glycerol | | 40 g |
| Artificial sweetener | | 1 g |
| Amaranth solution | | 1 mL |
| Raspberry essence HC417 | | 1 mL |
| Purified water | qs | 100 g |

The lozenges were formed in a suppository mold and rolled in lactose. The mean sublingual dissolution time was determined to be approximately 10.4 minutes. Samples of the lozenges were stored at room temperature of 25°C and refrigerated at 2 to 8°C. Stability-indicating HPLC analysis of the ketamine hydrochloride concentration found little or no change over 14 weeks of storage at either temperature. Bioavailability of the drug from the lozenges was found to be sufficiently high and reproducible to support the lozenge formulation for use in pain management.

### Injection

Injections, like other sterile drugs, should be prepared in a suitable clean air environment using appropriate aseptic procedures. When prepared from nonsterile components, an appropriate and effective sterilization method must be employed.

**STUDY 1:** Gupta[1292] reported the stability of Abbott ketamine hydrochloride 10 mg/mL in sterile water for injection packaged in Becton-Dickinson 1-mL polypropylene syringes. The samples were stored at 25°C for 30 days. The samples exhibited no visible changes, and HPLC analysis found no drug loss occurred.

**STUDY 2:** Stucki et al.[1291] evaluated the stability of ketamine hydrochloride 1 mg/mL diluted in sodium chloride 0.9% and packaged in 10-mL polypropylene plastic syringes sealed with tamper-evident tip caps. The syringes were stored refrigerated at 2 to 6°C, at room temperature of 23 to 27°C, and at elevated temperature of 38 to 42°C for 12 months. No visible changes occurred in the samples and no increase in subvisible microparticulates was reported. The ketamine concentrations were determined by stability-indicating capillary electrophoresis analysis. The drug concentrations remained above 95% throughout the 12-month test period at all three storage temperatures. ■

# ■ Ketoconazole

## Properties

Ketoconazole is a white to slightly beige odorless crystalline powder.[2,7]

### Solubility

Ketoconazole is practically insoluble in water[2] but is sparingly soluble in ethanol.[3]

### $pK_a$

Ketoconazole has $pK_a$ values of 2.9 and 6.5.[2]

## General Stability Considerations

Ketoconazole powder and tablets should be packaged in well-closed containers and stored at controlled room temperatures protected from temperatures of 40°C or more.[2,4] Ketoconazole oral suspension should be packaged in tight, light-resistant amber containers and stored at controlled room temperature.[4]

Ketoconazole exhibits pH-dependent instability. The drug was found to be stable in the pH range of 5 to 9.[1606] However, ketoconazole was found to be increasingly unstable as the pH dropped from about 5 to about pH 1.[1515,1606] Ketoconazole concentration within the range of 0.25 to 2% had a negligible effect on the degradation rate. However, the concentration of the antioxidant butylated hydroxytoluene (BHT) in an aqueous formulation was found to increase the ketoconazole degradation rate at BHT concentrations higher than 0.1%.[1606]

## Stability Reports of Compounded Preparations
### Oral
**USP OFFICIAL FORMULATION (ORAL SUSPENSION):**

| | |
|---|---|
| Ketoconazole | 2.0 g |
| Cetylpyridium chloride | 10 mg |
| Xanthan gum | 0.15 g |
| Purified water | 30 mL |
| Suspension structured vehicle NF | |
| (sugar-containing or sugar-free) qs | 100 mL |

(See the vehicle monographs for information on the individual vehicles.)

Use ketoconazole powder or commercial tablets. If using tablets, crush or grind to fine powder (40- or 45-mesh sieve) in a glass mortar. Dissolve the cetylpyridium chloride in purified water to yield a 10-mg/10-mL solution. Add the cetylpyridium chloride solution stepwise with mixing to the ketoconazole powder to form a uniform paste. Place 20 mL of purified water in a beaker and using moderate heat and stirring to form a vortex slowly sprinkle the xanthan gum into the vortex to obtain a uniform dispersion. Add this dispersion to the wetted powder paste with mixing to form a smooth mixture. Add sufficient vehicle to bring to volume and thoroughly mix, yielding a ketoconazole 20-mg/mL oral suspension. Quantitatively transfer the solution to a suitable tight and light-resistant amber container. Store the preparation at controlled room temperature. The beyond-use date is 14 days from the date of compounding.[4,5]

**STUDY 1:** Kumer et al.[174] evaluated the stability of two concentrations of ketoconazole ethanolic solution potentially suitable for oral use. Ketoconazole 2.5- and 5-mg/mL solutions in ethanol and water (percentages unspecified) were stored at room temperature and under refrigeration exposed to light and light-protected for 29 days. An HPLC technique was used to assess ketoconazole. No significant effect on drug concentration was found at either ketoconazole concentration under any storage condition.

**STUDY 2:** Allen and Erickson[527] evaluated the stability of three ketoconazole 20-mg/mL oral suspensions extemporaneously compounded from tablets. Vehicles used in this study were (1) an equal parts mixture of Ora-Sweet and Ora-Plus (Paddock), (2) an equal parts mixture of Ora-Sweet SF and Ora-Plus (Paddock), and (3) cherry syrup (Robinson Laboratories) mixed 1:4 with simple syrup. Twelve ketoconazole 200-mg tablets (Janssen) were crushed and comminuted to fine powder using a mortar and pestle. About 20 mL of the test vehicle was added to the powder and mixed to yield a uniform paste. Additional vehicle was added geometrically, and the suspension was brought to the final volume of 120 mL,

with thorough mixing after each addition. The process was repeated for each of the three test suspension vehicles. Samples of each finished suspension were packaged in 120-mL clear polyethylene terephthalate plastic prescription bottles and stored at 5 and 25°C in the dark.

No visual changes or changes in odor were detected during the study. Stability-indicating HPLC analysis found less than 6% ketoconazole loss in all suspensions stored at either temperature after 60 days of storage.

**STUDY 3:** Polnok and Techowanich[910] reported that ketoconazole was stable for at least 60 days at room temperature of 25°C and refrigerated at 5°C when prepared in a suspension medium of sorbitol solution with or without sodium carboxymethylcellulose as a suspending agent.

**STUDY 4:** Skiba et al.[1606] evaluated the stability of a number of possible aqueous formulas of ketoconazole. The authors stated that the best formulation from a stability standpoint for ketoconazole aqueous formulations having a drug concentration in the range of 0.25 to 2% had a pH of 7 and a butylated hydroxytoluene (BHT) concentration of 0.1%.

### Ophthalmic
Ophthalmic preparations, like other sterile drugs, should be prepared in a suitable clean air environment using appropriate aseptic procedures. When prepared from nonsterile components, an appropriate and effective sterilization method must be employed.

Hecq et al.[1221] reported that a ketoconazole 25-mg/mL ophthalmic preparation (formulation unspecified) at pH 3.8 was stable for seven days at ambient room temperature. Ketoconazole is not sufficiently soluble that this unspecified formulation could have been a solution. More likely it was an ophthalmic suspension or possibly an ophthalmic ointment. The analytical method used to determine stability was also not specified.

## Compatibility with Other Drugs
Jagota et al.[920] evaluated the compatibility and stability of mupirocin 2% ointment (Bactroban, Beecham) with ketoconazole 2% cream (Nizoral, Janssen) mixed in a 1:1 proportion over periods of up to 60 days at 37°C. The physical compatibility was assessed by visual inspection, while the chemical stability of mupirocin was evaluated by stability-indicating HPLC analysis. The study found that mupirocin 2% ointment mixed with ketoconazole 2% cream resulted in uncertain compatibility, with separation and layering that may be unacceptable for patient use. However, the mixture could be remixed back to homogeneity, and the mupirocin loss was less than 10% for 45 days, even at 37°C. ∎

# Ketoprofen

## Properties

Ketoprofen is a white to off-white odorless nonhygroscopic granular or crystalline powder.[2,3,7]

### Solubility

Ketoprofen is practically insoluble in water but freely soluble in ethanol.[2,3,7] Ketoprofen was found to have a solubility of about 10 mg/mL in pH 7.4 Sorenson's phosphate buffer at 37°C.[588]

### pK_a

Ketoprofen has a pK_a of 5.9 in methanol–water (3:1).[2,7]

## General Stability Considerations

Ketoprofen capsules should be packaged in tight, light-resistant containers and stored at approximately 25°C.[2] Ketoprofen injection should be stored at controlled room temperature below 30°C protected from exposure to light.[7]

## Stability Reports of Compounded Preparations

### Topical

**STUDY 1:** Allen[153] reported on a topical gel preparation of 1% ketoprofen. The gel was compounded by dissolving the proper amount of ketoprofen in a minimum amount of 95% ethanol. The dissolved ketoprofen was added to a pH 5 buffer composed of 0.1 M citric acid and 0.2 M disodium phosphate (49:51) in purified water. Poloxamer 407 (Pluronic F-127) was added slowly with gentle mixing, avoiding the incorporation of air.[153] In a study of a similar product, the gel was chemically stable for eight months at room temperature and under refrigeration.[154]

**STUDY 2:** Liu et al.[1005] reported the stability of ketoprofen in a self-emulsifying drug delivery system composed of ketoprofen, ethyl oleate, polysorbate 80 (Tween 80), and diethylene glycol monoethyl ether (Transcutol) in a ratio of 5:40:43.5:11.5. Ketoprofen content was found to decline to 88% in five days and 69% in 10 days when stored exposed to illumination. The authors recommended protection from exposure to light during storage.

**STUDY 3:** Allen[1275] reported on a compounded formulation of ketoprofen 20-mg/mL for use as a topical solution. The solution had the following formula:

| | | |
|---|---|---|
| Ketoprofen | | 20 g |
| Pluronic F-127 | | 55 g |
| Ethanol | | 30 mL |
| Purified water | qs | 100 mL |

The recommended method of preparation was to thoroughly mix the ketoprofen powder and ethanol and then add the Pluronic F-127. Finally, purified water sufficient to bring the volume to 100 mL was to be added. The solution was to be packaged in tight, light-resistant containers. The author recommended a beyond-use date of six months at room temperature because this formula is a commercial medication in some countries with an expiration date of two years or more.

### Rectal

Zia et al.[588] determined the release profiles of ketoprofen 50-mg suppositories compounded in eight suppository bases. The test suppository bases were cocoa butter; four Witepsol (HULS America) bases, including H15, W25, W35, and E75; Suppocire AML (Gattefosse); Hydrokote AP5-1 (ABITEC); and an organogel base of Eudragit L100 (Rohm Pharma) 30% in propylene glycol. Most suppositories were prepared by the fusion method; the ketoprofen powder (Barr) was mixed with the melted base and poured into aluminum molds, resulting in approximately 2-g suppositories. For the water-soluble Eudragit suppositories, the drug was mixed with the propylene glycol, and this mixture was added to the Eudragit L100 with stirring for one minute prior to molding.

In vitro drug release testing into pH 7.4 Sorenson's phosphate buffer at 37°C was performed using the USP Rotating Basket Dissolution Test apparatus. The highest rate of drug release was from the cocoa butter base. The absolute bioavailability in rabbits was found to be about 61%, which is quite good for such dosage forms. A closely similar rapid release occurred with Witepsol H15; the authors noted that the ease of handling Witepsol H15 compared to cocoa butter might make this a preferable dosage form. Ketoprofen release was somewhat slower with Witepsol W25 and Suppocire AML. Inadequate release was found from the suppositories made from Witepsol W35, Hydrokote AP5-1, Eudragit L 100, and especially Witepsol E75.

### Injection

Injections, like other sterile drugs, should be prepared in a suitable clean air environment using appropriate aseptic procedures. When prepared from nonsterile components, an appropriate and effective sterilization method must be employed.

Singhai et al.[1560] reported the solubility of ketoprofen in a variety of solubilizing formulations that are different from the commercial injection. The better potential formulations incorporated propylene glycol, polyethylene glycol, sodium benzoate, sodium hydroxybenzoate. The formulation that exhibited the best stability was composed of ketoprofen in polyethylene glycol 600 and water. The calculated shelf life period was 228 days at room temperature.

# Ketorolac Tromethamine
## (Ketorolac Trometamol)

## Properties
Ketorolac tromethamine is a white to off-white crystalline powder.[2,3]

### Solubility
Ketorolac tromethamine is freely soluble in water, with an aqueous solubility of greater than 500 mg/mL. It is slightly soluble in ethanol, with a solubility of about 3 mg/mL.[2,3] Ketorolac tromethamine was found to have a solubility of about 600 mg/mL in pH 7.4 Sorenson's phosphate buffer at 37°C.[588]

### pH
A 1% solution in water has a pH of 5.7 to 6.7.[3,4] Ketorolac tromethamine injection has a pH of 6.9 to 7.9.[4] The ophthalmic solution has a pH adjusted to 7.4.[2]

### $pK_a$
Ketorolac tromethamine has a $pK_a$ of 3.5.[2,7]

## General Stability Considerations
Ketorolac tromethamine products should be stored at controlled room temperature and protected from light. The tablets should also be protected from excessive humidity. Exposure to light for prolonged periods may cause discoloration and precipitate formation in ketorolac tromethamine solutions.[2]

## Stability Reports of Compounded Preparations
### Rectal
Zia et al.[588] determined the release profiles of ketorolac tromethamine 30-mg suppositories compounded in eight suppository bases. The test suppository bases were cocoa butter; four Witepsol (HULS America) bases, including H15, W25, W35, and E75; Suppocire AML (Gattefosse); Hydrokote AP5-1 (ABITEC); and an organogel base of Eudragit L 100 (Rohm Pharma) 30% in propylene glycol. Most suppositories were prepared by the fusion method; the ketorolac tromethamine powder (Lemmon) was mixed with the molten base and poured into aluminum molds, resulting in approximately 2-g suppositories. For the water-soluble Eudragit suppositories, ketorolac tromethamine was mixed with the propylene glycol, and this mixture was added to the Eudragit L 100 with stirring for one minute prior to molding.

In vitro drug release testing into pH 7.4 Sorenson's phosphate buffer at 37°C was performed using the USP Rotating Basket Dissolution Test apparatus. The highest rate of drug release was from the cocoa butter base. The absolute bioavailability in rabbits was about 61% and was stated to be quite good for such dosage forms. A closely similar rapid release occurred with Witepsol H15; the authors indicated that the ease of handling Witepsol H15 compared to cocoa butter may make this a preferable dosage form. Ketorolac release with Witepsol W25 was similar to Witepsol W15, but it was somewhat slower with Suppocire AML. Inadequate release was found from the suppositories made from Witepsol W35, Hydrokote AP5-1, Eudragit L100, and especially Witepsol E75. ■

# Ketotifen Fumarate

## Properties
Ketotifen fumarate is a white to brownish-yellow fine crystalline powder.[3]

### Solubility
Ketotifen fumarate is sparingly soluble in water and slightly soluble in methanol.[3]

### pH
Ketotifen fumarate ophthalmic solution has a pH of 4.4 to 5.8.[7]

### Osmolality
Ketotifen fumarate ophthalmic solution is mildly hypotonic, having an osmolality of 210 to 300 mOsm/kg.[7]

## General Stability Considerations
Ketotifen fumarate powder should be stored at controlled room temperature.[3] Ketotifen fumarate ophthalmic solution is also stored at controlled room temperature.[7]

## Stability Reports of Compounded Preparations
### Enteral
Ortega de la Cruz et al.[1101] reported the physical compatibility of an unspecified amount of oral liquid ketotifen fumarate (Ketasma, Lesvi) with 200 mL of Precitene (Novartis) enteral nutrition diet for the 24-hour observation period. No particle growth or phase separation was observed.

## Ophthalmic

Ophthalmic preparations, like other sterile drugs, should be prepared in a suitable clean air environment using appropriate aseptic procedures. When prepared from nonsterile components, an appropriate and effective sterilization method must be employed.

**STUDY 1:** Abelson et al.[1316] evaluated the safety and efficacy of ketotifen fumarate 0.025% ophthalmic solution in a blinded clinical trial. Based on the success of that study, Abd El-Aleem et al.[1317] developed several ketotifen fumarate ophthalmic preparations—including drops, gel, and ointment—and evaluated their stability.

The preparation procedures are cited below. The authors stated that the formulations were sterilized by autoclaving at 121°C for 20 minutes, although the results of sterility testing were not reported, and the efficacy of autoclaving the formulations, especially the gels and ointments, is uncertain. Samples were evaluated for stability at 25, 35, and 45°C over six months. The pH values of the preparations ranged from about 3.8 to about 6.4 for the various preparations and did not change substantially over the course of the study. No changes in physical attributes occurred. Ultraviolet spectrophotometric analysis of drug concentrations was performed, although the method was not confirmed to be stability indicating.

Solution: Ketotifen fumarate 0.1% wt/vol solution was prepared by mixing the drug in water containing benzalkonium chloride 0.01% wt/vol, sodium metabisulfite 0.03% wt/vol, and 0.1% wt/vol EDTA. Sodium carboxymethyl cellulose 0.25% wt/vol or polyvinylpyrrolidone K30 5% wt/vol was then dispersed on the surface of the solution and stirred until complete dissolution occurred.

The polyvinylpyrrolidone-containing solution exhibited superior stability with less than 2% drug loss in six months at 25 or 35°C and about 5% loss at 45°C. The sodium carboxymethylcellulose-containing solution underwent about 10% drug loss in six months at all temperatures.

Gel: Ketotifen fumarate 0.1% wt/vol solution was prepared by mixing the drug in water containing benzalkonium chloride 0.01% wt/vol, sodium metabisulfite 0.03% wt/vol, and 0.1% wt/vol EDTA. Sodium carboxymethyl cellulose 2.1% wt/vol or polyvinylpyrrolidone K30 50% wt/vol was then dispersed on the surface of the solution, and the preparation was allowed to sit at room temperature for 24 hours until the gel had formed.

The polyvinylpyrrolidone-containing gel exhibited superior stability with no more than 2 to 3% drug loss in six months at all three temperatures. The sodium carboxymethylcellulose-containing solution underwent about 7% drug loss in six months at 25 and 35°C and 8% loss at 45°C.

Ointment: Four bases were evaluated; (1) fatty base (beeswax 10% wt/wt, liquid paraffin 90% wt/wt), (2) absorption base (lanolin 10% wt/wt, petrolatum 90% wt/wt), (3) water soluble base (polyethylene glycol 6000 20% wt/wt, polyethylene glycol 400 80% wt/wt) and (4) polyvinyl alcohol 15% wt/wt. Ketotifen fumarate 0.1% wt/wt was incorporated into the melted bases using geometrical mixing at low temperature. Stirring continued until cooling occurred.

The water-soluble polyethylene glycol base exhibited the best stability with six-month drug losses of 2, 4, and 5% at 25, 35, and 45°C, respectively. All three of the other bases demonstrated losses from 10 to 19% in six months.

**STUDY 2:** Troche Concepcion et al.[1584] evaluated six ketotifen fumarate eyedrop formulations. The formulation that was selected for stability testing contained ketotifen fumarate 0.025% in a vehicle containing glycerin, sodium edetate, benzalkonium chloride, monobasic sodium phosphate, dibasic sodium phosphate, and sodium hydroxide in water. The formulation had a pH near 5.4. Chromatographic analysis of the ketotifen content found little or no loss of ketotifen and no appreciable formation of decomposition products during 12 months of storage at room temperature of 30°C and 65 to 75% relative humidity. ■

# Labetalol Hydrochloride

## Properties

Labetalol hydrochloride is a white or almost white crystalline powder or granules.[2,3,7]

### Solubility

Labetalol hydrochloride has an aqueous solubility of about 20 mg/mL. The drug is freely soluble in ethanol, with a solubility of at least 100 mg/mL.[2,3]

### pH

A 1% solution of labetalol hydrochloride in water has a pH of 4 to 5.[3,4] Labetalol hydrochloride injection has an official pH range of 3 to 4.5.[4] However, both Normodyne (Schering) and Trandate (Glaxo Wellcome) specify a pH range of 3 to 4.[7]

### $pK_a$

The drug has a $pK_a$ of 9.3.[2]

### Osmolality

Labetalol hydrochloride injection has an osmolality of 287 mOsm/kg.[326]

## General Stability Considerations

Labetalol hydrochloride should be packaged in tight, light-resistant containers and stored between 2 and 30°C. The injection also should be stored between 2 and 30°C and should be protected from freezing and exposure to light.[2]

Labetalol hydrochloride exhibits optimal stability in solutions having a pH of 3 to 4.[8,1126] A precipitate may form in alkaline solutions of pH 7.6 to 8.[379]

## Stability Reports of Compounded Preparations
### Oral
#### USP OFFICIAL FORMULATION:

Labetalol hydrochloride 4 g
Vehicle for Oral Solution, NF (sugar-containing or sugar-free)
Vehicle for Oral Suspension, NF (1:1) qs 100 mL

(See the vehicle monographs for information on the individual vehicles.)

Use labetalol hydrochloride powder or commercial tablets. If using tablets, crush or grind to fine powder. Add 20 mL of the vehicle mixture, and mix to make a uniform paste. Add additional vehicle almost to volume in increments with thorough mixing. Quantitatively transfer to a suitable calibrated tight and light-resistant bottle, bring to final volume with the vehicle mixture, and thoroughly mix yielding labetalol hydrochloride 40-mg/mL oral suspension. The final liquid preparation should have a pH between 4 and 5. Store the preparation at controlled room temperature or under refrigeration between 2 and 8°C. The beyond-use date is 60 days from the date of compounding at either storage temperature.[4,5]

**STUDY 1:** Nahata[269] evaluated the stability of labetalol hydrochloride (nominally 10 mg/mL) in several vehicles. One hundred labetalol hydrochloride 100-mg tablets (Trandate, Glaxo) were crushed with a mortar and pestle and transferred to a flask, where they were brought to 1000 mL with each of the following vehicles: distilled water, simple syrup (Humco), apple juice (Cost Cutter), grape juice (Kroger), and orange juice (Cost Cutter). Each mixture was stirred on a magnetic plate for 20 minutes. Although labetalol hydrochloride is soluble in water, many of the other tablet components are not, resulting in formation of a suspension. Each suspension was filtered through a coarse filter and then a 0.45-μm filter; the resulting solution was packaged in amber plastic oval bottles and glass prescription bottles (Owens-Illinois). The samples in plastic bottles were stored at 23 and 4°C for four weeks, except for the distilled water samples, which were stored at 23°C only. The samples in glass bottles also were stored at 23°C only. Labetalol hydrochloride content was determined using a stability-indicating HPLC analysis.

Initial labetalol hydrochloride concentrations ranged from 7.5 to 8.7 mg/mL. No significant loss of labetalol hydrochloride occurred during four weeks. The potential for microbial growth in these formulations was not evaluated. Consequently, refrigerated storage was recommended.

**STUDY 2:** Allen and Erickson[542] evaluated the stability of three labetalol hydrochloride 40-mg/mL oral suspensions extemporaneously compounded from tablets. Vehicles used in this study were (1) an equal parts mixture of Ora-Sweet and Ora-Plus (Paddock), (2) an equal parts mixture of Ora-Sweet SF and Ora-Plus (Paddock), and (3) cherry syrup (Robinson Laboratories) mixed 1:4 with simple syrup. Sixteen labetalol hydrochloride 300-mg tablets (Normodyne, Schering) were crushed and comminuted to fine powder using a mortar and pestle. About 20 mL of the test vehicle was added to the powder and mixed to yield a uniform paste. Additional vehicle was added geometrically, and the suspension was brought to the final volume of 120 mL, with thorough mixing after each addition. The process was repeated for each of the three test suspension vehicles. Samples of each finished suspension were packaged in 120-mL amber polyethylene terephthalate plastic prescription bottles and stored at 5 and 25°C in the dark.

No visual changes or changes in odor were detected during the study. Stability-indicating HPLC analysis found less than 4% labetalol hydrochloride loss in any of the suspensions stored at either temperature after 60 days of storage.

### Injection

Injections, like other sterile drugs, should be prepared in a suitable clean air environment using appropriate aseptic procedures.

When prepared from nonsterile components, an appropriate and effective sterilization method must be employed.

Alffenaar et al.[1100] developed a compounded dosage form of labetalol hydrochloride 5-mg/mL injection in response to a shortage and the unavailability of the commercial dosage form in The Netherlands. The injection was compounded in 40-liter batches. A 200-g quantity of labetalol hydrochloride powder (compliant with Ph Eur, BP, and USP requirements) from Merck was dissolved in 35 liters of water for injection. The solution pH was adjusted to near 4.1 with about 21 mL of 0.1 M hydrochloric acid, and the solution was brought to a volume of 40 liters with additional water for injection. The solution was filtered through a 0.22-μm filter into 1700 ampules containing 20 mL each. The finished ampules were autoclaved at 121°C for 15 minutes. Preliminary testing found that no loss of labetalol hydrochloride occurred upon autoclaving one or even two times.

One lot of the compounded labetalol hydrochloride was evaluated for shelf life over six months, and little or no loss of labetalol hydrochloride was found.[1100] The commercial injection available in the United States includes dextrose, paraben preservatives, and edetate disodium and is nearly isotonic, with an osmolality of 287 mOsm/kg.[326] The compounded labetalol hydrochloride of Alffenaar et al. was very different in formulation, being essentially drug dissolved in water. The compounded injection was reported to be very hypotonic, with a measured osmolality near 28 mOsm/kg.[1100] Care would need to be taken regarding the volume of this hypotonic injection that could be injected undiluted. ■

# ■ Lactic Acid

## Properties

Lactic acid is a colorless to yellowish practically odorless syrupy hygroscopic liquid.[3,4]

### Solubility

Lactic acid is miscible with water, ethanol, and acetone.[1,4]

### pKₐ

Lactic acid has a $pK_a$ of 3.8.[1]

## General Stability Considerations

Lactic acid should be packaged in tight containers and stored at controlled room temperature.[4] Commercial lactic acid topical cream and lotion also should be stored at controlled room temperature.[2]

## Stability Reports of Compounded Preparations

### Topical

de Villiers et al.[724] reported the physical and chemical stability of various topical creams of lactic acid 10%, an alpha-hydroxy acid, subjected to five freeze–thaw cycles of 24 hours at −4°C and then 24 hours at 40°C.

The best results were obtained with the lactic acid incorporated into hydrophilic ointment, USP, with the pH then adjusted up to about 3.5 using sodium hydroxide solution. Stability-indicating HPLC analysis found little or no lactic acid or paraben preservative loss at the end of the five cycles. If the pH was not adjusted and remained at the initial pH near 1.6, the physical and chemical stability of the ointment was satisfactory, but the methylparaben and propylparaben preservatives underwent unacceptable losses of about 15%. ■

# Lactulose

## Properties

Lactulose, a disaccharide sugar containing one molecule each of fructose and galactose, is a white powder or crystals. Lactulose solutions are colorless to yellow sweet viscous liquids.[1–3]

### Solubility

Lactulose is very soluble in water, having an aqueous solubility of 76.4% at 30°C.[1,2] It is slightly soluble in alcohol.[2]

### pH

Lactulose solutions have a pH between 3 and 7.[2] Commercially available solutions have a pH of 2.5 to 6.5.[2,4] Lactulose syrup (Cephulac, Marion Merrell Dow) had a measured pH of 4.8.[19]

### Osmolality

Lactulose syrup 0.67 mg/mL (Roerig) was found to have an osmolality of 3600 mOsm/kg.[233]

## General Stability Considerations

Lactulose solutions should be packaged in tight containers and stored at 2 to 30°C. Freezing should be avoided because increased viscosity makes the solutions difficult or impossible to pour. Warming the solutions to room temperature returns the viscosity to normal. Heat and light exposure cause cloudiness and darkening of the lactulose solutions, although these effects are not indicative of concentration loss.[2–4] Acid hydrolysis yields the two component sugars, fructose and galactose.[1]

## Stability Reports of Compounded Preparations
### Enteral
**LACTULOSE COMPATIBILITY SUMMARY**

*Syrup compatible with:* Ensure • Ensure HN • Ensure Plus • Ensure Plus HN • Osmolite • Osmolite HN • Vital

**STUDY 1:** Cutie et al.[19] added 5 mL of lactulose syrup (Cephulac, Marion Merrell Dow) to varying amounts (15 to 240 mL) of Ensure, Ensure Plus, and Osmolite (Ross Laboratories) with vigorous agitation to ensure thorough mixing. The lactulose syrup was physically compatible, distributing uniformly in all three enteral products with no phase separation or granulation.

**STUDY 2:** Altman and Cutie[850] reported the physical compatibility of lactulose syrup (Cephulac, Merrell-National) 10 mL with varying amounts (15 to 240 mL) of Ensure HN, Ensure Plus HN, Osmolite HN, and Vital after vigorous agitation to ensure thorough mixing. The lactulose syrup was physically compatible, distributing uniformly in all four enteral products with no phase separation or granulation. ∎

# Lamotrigine

## Properties

Lamotrigine is a white to pale-cream powder.[7]

### Solubility

Lamotrigine is very slightly soluble in water, having a solubility of 0.17 mg/mL at 25°C. It is slightly soluble in 0.1 M hydrochloric acid, with a solubility of 4.1 mg/mL at 25°C.[7]

### pKₐ

Lamotrigine has a pK$_a$ of 5.7.[7]

## General Stability Considerations

Lamotrigine tablets should be stored between 15 and 25°C in a dry place and protected from light.[7]

Lamotrigine is highly labile at alkaline pH undergoing rapid hydrolysis compared to acidic environments in which it exhibits a slower rate of hydrolysis.[1529]

## Stability Reports of Compounded Preparations
### Oral

Nahata et al.[674] evaluated the stability of two lamotrigine 1-mg/mL oral suspensions prepared from tablets. One lamotrigine 100-mg tablet was crushed to fine powder using a mortar and pestle. Two vehicles were prepared for use in making the separate oral suspensions for testing; equal parts mixtures of Ora-Plus suspending agent with Ora-Sweet and with Ora-Sweet SF (Paddock) were both prepared. A small amount of the prepared vehicle was incorporated into the tablet powder, making a uniform paste. Additional vehicle was added geometrically with mixing, bringing the volume near 100 mL. The mixture was transferred to a graduate and brought to 100 mL with additional vehicle and mixed. Samples of each of the finished suspensions were packaged in two-ounce amber polyethylene terephthalate bottles and stored at 4 and 25°C for 91 days.

No change in physical appearance or odor occurred in samples of either suspension at either temperature. Stability-indicating HPLC analysis found that all samples retained more than 99% concentration after 91 days. ∎

# Lansoprazole

## Properties

Lansoprazole is a white to brownish-white odorless crystalline powder.[7]

### Solubility

Lansoprazole is practically insoluble in water, sparingly soluble in ethanol, and freely soluble in dimethylformamide.[7]

### pK$_a$

Hellstrom and Vitols[1325] reported that lansoprazole has a pK$_a$ of 4.01. Kristl[1324] found three reported pK$_a$ values of about 8.84, 4.15, and 1.33.

## General Stability Considerations

Lansoprazole is stable exposed to light for up to two months but decomposes in aqueous solutions, with the rate of decomposition increasing with decreasing pH. The half-life in solution at 25°C is about 30 minutes at pH 5 but about 18 hours at pH 7.[7] The presence of citric acid, trisodium citrate, and monosodium citrate in combination with lansoprazole substantially increases the rate of lansoprazole decomposition.[1099]

Lansoprazole is formulated as enteric-coated granules in capsules. Commercial products should be packaged in tight containers and should be stored at controlled room temperature protected from moisture.[7] The manufacturer has stated that lansoprazole orodispersible should not be removed from the original packaging and placed in dosing compliance aids because of its moisture sensitivity.[1622]

## Stability Reports of Compounded Preparations
### Oral

**STUDY 1:** McAndrews and Eastham[672] reported on preparing a lansoprazole suspension from capsule contents. The plungers of 30-mL plastic syringes were removed, and the enteric-coated granule contents of a 15- or 30-mg capsule were emptied into the syringes. Four amounts of 8.4% sodium bicarbonate were used to remove the enteric coating from the granules: 20, 10, 5, and 2.5 mL mixed with 2.5 mL of sterile water for injection. The smaller amounts may be needed for patients with an acid–base imbalance. All of these amounts of sodium bicarbonate were adequate to prepare the suspension in periods ranging from 10 to 20 minutes. Shaking the syringe facilitated suspension formation.

**STUDY 2:** DiGiacinto et al.[782] reported the stability of lansoprazole 3-mg/mL oral suspension extemporaneously compounded from capsules. The contents of lansoprazole 30-mg capsules were mixed with sodium bicarbonate 8.4% injection. The lansoprazole 3-mg/mL suspension was packaged as 5 mL of suspension in 10-mL amber oral syringes and stored at

room temperature of 22°C exposed to fluorescent light and refrigerated at 4°C protected from light. Stability-indicating HPLC analysis found that 10% lansoprazole loss occurred in eight hours at room temperature; furthermore, the oral suspension became a thick paste. Under refrigeration, less than 4% loss occurred in 14 days, but 12% loss occurred by 21 days.

**STUDY 3:** Doan et al.[784] reported that the pharmacokinetics of an extemporaneously compounded lansoprazole 3-mg/mL oral liquid in sodium bicarbonate 8.4% compared favorably to the commercial oral capsules. The contents of one 30-mg lansoprazole capsule were suspended in 10 mL of sodium bicarbonate 8.4%. The simple suspension was administered using a nasogastric tube. Both the oral capsules and the compounded suspension effectively controlled the 24-hour intragastric pH in 36 healthy volunteers. The suspension had some differences in the time to maximum observed concentration and mean plasma concentration, with a lower area under the curve. However, the authors considered the simple oral suspension a suitable option for patients unable to swallow lansoprazole capsules.

**STUDY 4:** Ensom et al.[1564] evaluated the stability of lansoprazole 3 mg/mL oral suspension prepared from commercial capsule contents. The oral suspension was prepared by mixing the contents of 10 lansoprazole 30-mg capsules into 50 mL of sodium bicarbonate 8.4% solution and stirring for 15 minutes at room temperature near 25°C.

An equal parts combination of Ora-Sweet and Ora-Plus was then prepared. Taking a 45-mL portion of this Ora-Sweet/Ora-Plus mixture, the pH was adjusted to about 8.8 using 3.4 mL of 1 N sodium hydroxide solution. The mixture was then brought to 50 mL with additional Ora-Sweet/Ora-Plus mixture. Equal volumes of the initial lansoprazole suspension and the pH-buffered Ora-Sweet/Ora-Plus mixture were then mixed yielding the lansoprazole 3-mg/mL oral suspension. The white, cloudy suspension was packaged in amber glass prescription bottles and stored refrigerated at 4°C and at room temperature of 25°C for 91 days.

The suspension was judged to have an acceptable palatability by 10 tasters. During the storage period, the suspension exhibited no notable changes in color, odor, or pH and was easily resuspended with no caking or clumping at either storage temperature. Stability-indicating HPLC analysis found that the lansoprazole concentration remained at or above 90% of the initial concentration at both temperatures throughout the 91-day study period.

**STUDY 5:** Melkoumov et al.[1712] evaluated the stability of a lansoprazole 3-mg/mL oral liquid formulation in Ora-Blend vehicle (Paddock). Each Prevacid FasTab 30-mg tablet (Abbott Canada) was slightly ground using minimal crushing force to expedite disintegration in 10 mL of Ora-Blend vehicle. The oral liquids were stored protected from exposure to light at room temperature of 21 to 22°C and refrigerated at 4.5 to 5.5°C for seven days. The oral liquid in Ora-Blend was found not to occlude nasogastric catheters of 6, 8, and 10 French. But occlusion did occur in size 5 French nasogastric tubes. Samples stored refrigerated were stable retaining about 99% of the drug for three days and about 96% for seven days and were superior to sodium bicarbonate formulations. However, room temperature samples underwent unacceptable drug loss of 11% in three days and 17% in seven days.

### Enteral

**STUDY 1:** Using water and apple juice as vehicles, Dunn et al.[785] compared the delivery of lansoprazole and omeprazole granules from capsules through 14-French nasogastric tubes. Delivery of the drugs was highly variable, but the values had means of 30% of the lansoprazole granules and 53% of the omeprazole granules. No substantial difference was found with either 15 or 30 mL of either water or apple juice, although the granules in apple juice appeared to be stickier.

**STUDY 2:** Chun et al.[1032] reported that lansoprazole granules from 30-mg capsules each mixed in 40 mL of apple juice in a 60-mL syringe and delivered through 16-French nasogastric tubes to 23 patients resulted in bioavailabilities of lansoprazole that were similar to those achieved using the commercial oral capsules in the same individuals. No statistically significant differences were found in mean time elapsed to peak concentration, mean peak concentration, or area under the curve.

**STUDY 3:** Messaouik et al.[1031] reported the delivery of three proton pump inhibitor drugs through polyurethane and also silicone 16-French gastroduodenal tubes. The enteric-coated pellets from capsules of lansoprazole, omeprazole, and esomeprazole were removed from the capsules and mixed with water or apple juice. The mixtures were delivered through the tubes and rinsed with 10 or 20 mL of the respective vehicle. No obstruction occurred with any of the esomeprazole samples, but 42% of the lansoprazole

and omeprazole samples resulted in clogging. Esomeprazole delivery was 100%, while lansoprazole and omeprazole had reduced drug delivery of about 67 and 61%, respectively. The authors noted that the esomeprazole pellets were substantially smaller in size than those of the other two drugs. The authors concluded that esomeprazole was preferable for nasogastric delivery compared to omeprazole or lansoprazole.

**STUDIES 4, 5, AND 6:** Sharma et al.[1565–1567] in a series of studies evaluated the pharmacodynamics of lansoprazole administered through gastrostomy tubes as intact capsules, nonencapsulated granules, and as a suspension in sodium bicarbonate 8.4%. The authors demonstrated that the degree of acid suppression was similar with all three dosage forms.

**STUDY 7:** Ensom et al.[1564] evaluated the stability of a lansoprazole 3-mg/mL suspension prepared from commercial capsule contents. The suspension was prepared by mixing the contents of 10 lansoprazole 30-mg capsules into 100 mL of sodium bicarbonate 8.4% solution at 37°C. The suspension was packaged in amber glass prescription bottles and stored refrigerated at 4°C and at room temperature of 25°C for 91 days.

This suspension was judged to be unpalatable for oral administration by 10 taste testers but was deemed satisfactory for nasogastric administration. During the storage period, the suspension exhibited no notable changes in color, odor, or pH (near pH 8.7) and was easily resuspended with no caking or clumping at either storage temperature. Stability-indicating HPLC analysis found that the lansoprazole concentration remained at or above 90% of the initial concentration at both temperatures throughout the 91-day study period.

## Compatibility with Common Beverages and Foods

For patients with difficulty in swallowing lansoprazole capsules and other proton pump inhibitor drugs, the capsules have been opened and the enteric-coated pellet contents mixed with a slightly acidic food such as yogurt, applesauce, or fruit juice for administration. This approach is believed to help preserve the drug's integrity until it reaches the alkaline environment of the duodenum.[1029–1035]

# Latanoprost

## Properties

Latanoprost is a colorless to slightly yellow oil.[1,7]

### Solubility

Latanoprost is practically insoluble in water at about 50 mcg/mL but soluble in ethanol to about 200 mg/mL. It is also soluble in acetone, ethyl acetate, and isopropanol.[2,7]

### pH

The commercial latanoprost ophthalmic solution has a pH buffered to about 6.7.[7]

### pKₐ

Latanoprost has an estimated $pK_a$ of 4.88.[2]

### Osmolality

The commercial latanoprost ophthalmic solution is isotonic, having an osmolality of about 267 mOsm/kg.[7]

## General Stability Considerations

Latanoprost commercial sterile ophthalmic solution should be stored under refrigeration and protected from exposure to light. The expiration date stored as directed is 18 months. After opening, the commercial ophthalmic product (Xalatan) may be stored at room temperature up to 25°C for six weeks.[2,7]

## Stability Reports of Compounded Preparations
### Ophthalmic

Ophthalmic preparations, like other sterile drugs, should be prepared in a suitable clean air environment using appropriate aseptic procedures. When prepared from nonsterile components, an appropriate and effective sterilization method must be employed.

**STUDY 1:** Garcia-Valldecabres et al.[880] reported the change in pH of several commercial ophthalmic solutions, including latanoprost 50 mcg/mL (Xalatan), over 30 days after opening. Slight variation in solution pH was found, but no consistent trend and no substantial change occurred.

**STUDY 2:** Varma et al.[1063] evaluated the stability of 0.005% latanoprost ophthalmic solution (Xalatan, Pharmacia & Upjohn) under in-use conditions over four to six weeks stored at uncontrolled room temperatures by patients. HPLC analysis found that the mean drug concentration among the 69 samples tested was 48.3 mcg/mL or about 96.7% of the labeled amount, with 65 samples having concentrations of 90% or greater. However, the remaining five samples had low latanoprost concentrations of 70% for one bottle and 86 to 88% for the other four bottles. The authors concluded that latanoprost ophthalmic solution

was stable for up to six weeks at room temperature under common in-use conditions.

**STUDY 3:** Paolera et al.[1272] evaluated and compared the stability of bimatoprost 0.03% and latanoprost 0.005% ophthalmic solutions during in-use conditions with patients in Sao Paulo, Brazil. Patients were instructed to store and use their ophthalmic solutions as usual and return them between days 28 and 34. The sample solutions were transferred to an analytical laboratory arriving on days 35 through 42. HPLC analysis of the active drugs found no loss of bimatoprost after initiating use of the ophthalmic solution. However, the latanoprost samples averaged only 88% of the active drug remaining. The differing results to the Varma et al.[1063] study were attributed to differences in climate and air-conditioned storage conditions in Los Angeles.

**STUDY 4:** Mochizuki et al.[1284] evaluated the efficacy, tolerability, and safety of latanoprost 0.005% (Xalatan) ophthalmic solution when stored at 30°C compared to refrigerated storage using human volunteers. After storage for four weeks at both 4 and 30°C, no difference in efficacy, tolerability, and safety was observed among the human volunteers.

**STUDY 5 (EMULSION):** Sakai et al.[1517] evaluated the stability of latanoprost 0.005% in an extemporaneously-prepared lipid emulsion ophthalmic liquid. Latanoprost was found to be most stable in medium chain fatty acid triglyceride for use as the lipid phase. The lipid emulsion was composed of latanoprost 0.005% (wt/vol), medium chain fatty acid triglyceride 1% (wt/vol), polyvinyl alcohol 2% (wt/vol), glycerin 2.6% (wt/vol), and sodium acetate 0.2% (wt/vol) to buffer the aqueous phase to pH 5 and 6 or sodium borate 0.2% (wt/vol) to buffer to pH 7. The emulsion was prepared by mixing the polyvinyl alcohol and glycerin with water at 70°C. Separately, latanoprost was dissolved in medium chain fatty acid triglyceride at 70°C. The triglyceride mixture was added to the aqueous mixture and homogenized using a Robomics homogenizer at 8000 RPM for 15 minutes to form a coarse emulsion. The coarse emulsion was then treated using a Microfluidizer M-110EH high-pressure emulsifier for 10 discrete volume cycles. The emulsion was cooled to room temperature and diluted twofold with the sodium acetate or sodium borate buffer. Finally, hydrochloric acid or sodium hydroxide was used to adjust the final pH of the aqueous phase to 5, 6, or 7. The finished emulsion was packaged in glass containers and stored at 25 and 60°C for four weeks. NOTE: No mention of sterilization was made in this report, but ophthalmic preparations must be sterile for clinical use.

Stability-indicating HPLC analysis was used to compare the stability of the latanoprost emulsions with commercial Xalatan ophthalmic solution. No loss of latanoprost occurred within four weeks in any of the emulsion preparations at all tested pH values at both 25 and 60°C. However, latanoprost in commercial Xalatan lost about 2 and 24% in four weeks when stored at 25 and 60°C, respectively. ■

#  Leucovorin Calcium
## (Folinic Acid, Calcium Folinate)

## Properties
Leucovorin calcium is a white to yellowish-white or yellow odorless powder.[2,3]

### Solubility
Leucovorin calcium is very soluble in water, having an aqueous solubility exceeding 500 mg/mL. Leucovorin calcium is practically insoluble in ethanol, with a solubility of less than 1 mg/mL.[2,3]

### pH
A 2.5% aqueous solution has a pH of 6.8 to 8.[3] Leucovorin calcium injection has a pH between 6.5 and 8.5.[4]

### $pK_a$
The drug has $pK_a$ values of 3.1, 4.8, and 10.4.[1]

### Osmolality
Leucovorin calcium 10 mg/mL in sterile water for injection has a pH of 274 mOsm/kg.[326]

## General Stability Considerations
Leucovorin calcium products should be packaged in well-closed containers and stored at controlled room temperature protected from light.[2-4] Reconstituted solutions of leucovorin calcium are chemically stable for seven days both at room temperature and under refrigeration.[2,441] However, if the diluent contains no antibacterial preservative such as benzyl alcohol, immediate use is recommended.[2]

Leucovorin calcium in aqueous solutions is stable over pH 6.5 to 10, with maximum stability occurring at around pH 7.1 to 7.4. An increased rate of decomposition occurs below pH 6.[442]

## Stability Reports of Compounded Preparations
### Ophthalmic
Ophthalmic preparations, like other sterile drugs, should be prepared in a suitable clean air environment using appropriate aseptic procedures. When prepared from nonsterile components, an appropriate and effective sterilization method must be employed.

Pascual and Borrego Dionis[413] reported the stability of an ophthalmic formulation of leucovorin calcium 0.3 mg/mL. The solution contained leucovorin calcium 3 mg, boric acid 200 mg, 1 M sodium hydroxide 0.5 mL, and water to bring to 10 mL. The solution was sterilized by filtering through a 0.22-μm filter. The final preparation had a pH of 7.85 and an osmolarity of 308 mOsm/L. The leucovorin calcium content was determined by using ultraviolet spectrophotometry. Samples were stored under refrigeration and at room temperature for 10 days with virtually no loss of leucovorin calcium.

### Oral
Lauper[786] reported that leucovorin calcium is a stable compound at alkaline pH, and the injection can be mixed in milk, antacids, or other alkaline solutions for oral administration. A 24-hour beyond-use period was recommended. However, because of decomposition of leucovorin calcium at acidic pH, acidic juices should not be used. ■

# ■ Levalbuterol Hydrochloride

## Properties
Levalbuterol hydrochloride is the (R)-enantiomer hydrochloride salt of the racemic albuterol.[1] Levalbuterol hydrochloride is a white to off-white crystalline powder. The commercial inhalation solution (Xopenex, Sepracor) is a clear, colorless sterile preservative-free aqueous solution. Levalbuterol hydrochloride 1.44 mg provides levalbuterol 1.25 mg.[7]

### Solubility
Levalbuterol hydrochloride has an aqueous solubility of about 180 mg/mL.[7]

### pH
Levalbuterol hydrochloride inhalation solution has a pH near 4 with a range of 3.3 to 4.5.[7]

*Osmolality*

Levalbuterol hydrochloride inhalation solution has had its tonicity adjusted with sodium chloride during manufacturing.[7]

## General Stability Considerations

The commercial levalbuterol hydrochloride inhalation solution should be stored at controlled room temperature and retained in its foil packaging. The drug should be protected from light and excessive heat. The drug should be used immediately after opening the foil pouch in which it is supplied. Any drug solution remaining after opening the foil pouch should be discarded. The inhalation solution should be clear and colorless; if it is not clear and colorless, it should not be used and should be discarded. The commercial inhalation solution should be diluted with sterile sodium chloride 0.9% for administration by inhalation.[7]

## Compatibility with Other Drugs

**LEVALBUTEROL HYDROCHLORIDE COMPATIBILITY SUMMARY**

*Compatible with:* Acetylcysteine • Budesonide • Cromolyn sodium • Ipratropium bromide

**STUDY 1 (IPRATROPIUM):** Yamreudeewong et al.[1328] evaluated the stability of levalbuterol hydrochloride (Xopenex, Sepracor) inhalation solution mixed with ipratropium bromide (Nephron Pharmaceuticals) inhalation solution for combined use as a nebulizer solution. The initial concentrations were 227 mcg/mL and 90 mcg/mL for the levalbuterol hydrochloride and ipratropium bromide, respectively, in the single sample that was tested. The sample was packaged in a polypropylene screw cap vial and stored at unspecified room temperature for 28 days. Physical stability was not specifically assessed, but no changes were reported. HPLC analysis results found little or no loss of either drug occurred over the 28-day study period.

**STUDY 2 (MULTIPLE DRUGS):** Bonasia et al.[1329] evaluated the compatibility and stability of levalbuterol hydrochloride inhalation solution (Sepracor) with four drugs: acetylcysteine (Bristol-Meyers Squibb), budesonide (AstraZeneca), cromolyn sodium (Alpharma), and ipratropium bromide (Dey). The two-drug mixtures that were tested are shown below.

1. Levalbuterol 1.25 mg/0.5 mL diluted with 5 mL of sodium chloride 0.9%
   Acetylcysteine 800 mg/4 mL diluted with 0.5 mL of sodium chloride 0.9%
2. Levalbuterol 1.25 mg/0.5 mL diluted with 2 mL of sodium chloride 0.9%
   Budesonide 0.5 mg/2 mL diluted with 2 mL of sodium chloride 0.9%
3. Levalbuterol 1.25 mg/0.5 mL diluted with 2 mL of sodium chloride 0.9%
   Cromolyn sodium 20 mg/2 mL diluted with 0.5 mL of sodium chloride 0.9%
4. Levalbuterol 1.25 mg/0.5 mL diluted with 2.5 mL of sodium chloride 0.9%
   Ipratropium bromide 0.5 mg/2.5 mL diluted with 0.5 mL of sodium chloride 0.9%

The two-drug mixtures were prepared in triplicate and evaluated initially and after 30 minutes at an ambient temperature of 21 to 25°C. The authors indicated that a 30-minute study duration was chosen because it represents the approximate time it takes for a patient to administer drugs by nebulization at home. No visible evidence of physical incompatibility was observed with any of the samples. HPLC analysis found no loss of any of the drugs in the samples. ■

# Levamisole
# Levamisole Hydrochloride

## Properties

Levamisole and levamisole hydrochloride are white or almost white crystalline powders.[3,4]

*Solubility*

Levamisole is slightly soluble in water and freely soluble in ethanol.[3] Levamisole hydrochloride is freely soluble in water (to about 210 mg/mL) and soluble in ethanol.[1,4]

*pH*

A 5% aqueous solution of levamisole hydrochloride has a pH in the range of 3 to 4.5.[3,4]

## General Stability Considerations

Levamisole hydrochloride is stable in acid media.[1] Levamisole hydrochloride powder should be packaged in tight containers and stored at controlled room temperature protected from exposure to light.[3,4] Levamisole hydrochloride oral tablets should be packaged in well-closed containers and stored at controlled room temperature.[4]

## Stability Reports of Compounded Preparations
*Oral*

Chiadmi et al.[1091] evaluated the stability of levamisole 25-mg/mL (as the hydrochloride) oral solutions prepared from powder and from tablets. Using powder, the drug was simply dissolved in sterile water. Using 50-mg tablets, 100 tablets were crushed in a glass mortar and 20 mL of sterile water was added and levigated to create a paste. After 30 minutes, an additional 100 mL of sterile water was added and allowed to stand for 30 minutes, creating a milky-looking suspension. The suspension was filtered into a 200-mL volumetric flask through 7-μm filter paper to remove insoluble tablet components. The solution was brought to volume with additional sterile water to yield the 25-mg/mL (as the hydrochloride) solution. The clear solution was packaged in amber glass prescription bottles and stored refrigerated at about 4°C and at room temperature of about 23°C for 90 days.

All of the samples remained unchanged in appearance, color, and odor throughout the study. Stability-indicating HPLC analysis found that the solutions prepared from powder exhibited better stability than the solutions prepared from tablets, but both were usable. Under refrigeration, the solutions prepared from powder lost little or no levamisole content within 90 days, while the solutions prepared from tablets lost about 6%. At room temperature, the solutions prepared from powder lost about 3% levamisole content within 90 days, while the solutions prepared from tablets lost about 6% in 15 days, 12% in 30 days, and 37% in 90 days. The authors recommended storage of the solutions under refrigeration. They also indicated that a more limited period of use of 30 days was reasonable because no antimicrobial preservative is present in the solutions. ▪

# Levocarnitine
# Levocarnitine Hydrochloride

## Properties
Levocarnitine occurs as hygroscopic white crystals or crystalline powder.[1,3,4]

Levocarnitine hydrochloride also occurs as a crystalline material.[1]

Also see the Carnitine monograph.

### Solubility
Levocarnitine is freely soluble in water and in hot ethanol, but it is practically insoluble in acetone and ether.[4]

Levocarnitine hydrochloride is very soluble in water, soluble in hot ethanol, and slightly soluble in cold ethanol. It is practically insoluble in acetone and ether.[1]

### pH
The United States Pharmacopeia indicates the pH range of a 5% aqueous solution to be 5.5 to 9.5.[4] The European Pharmacopoeia indicates a 5% aqueous solution of levocarnitine has a pH of 6.5 to 8.5.[3]

Levocarnitine injection has a pH range of 6 to 6.5.[4,7] Commercial levocarnitine oral solution has a pH between 4 and 6.[4]

Extemporaneously compounded oral solutions of levocarnitine hydrochloride ranging from 1 to 10% were near pH 2 to 2.5.[1123]

### pK_a
Levocarnitine has a $pK_a$ of 3.8.[7]

## General Stability Considerations
Levocarnitine powder, levocarnitine tablets, and levocarnitine oral solution should be packaged in tight containers and stored at controlled room temperature. Levocarnitine injections in single-dose containers should be stored below 25°C and protected from exposure to light, leaving the vials in their cartons until time of use.[4,7] The injections should also be protected from freezing.[4]

## Stability Reports of Compounded Preparations
### Oral
Tanaka et al.[1123] evaluated the stability of levocarnitine (as the hydrochloride) in simple syrup (pH 2 to 2.5) for use in the oral treatment of infants with levocarnitine deficiency. Concentrations ranging from 1 to 10% were prepared in simple syrup. Little or no drug loss occurred over 90 days when stored refrigerated at 4°C and protected from exposure to light. Samples of these oral solutions were also found not to support the growth of two test organisms, *Rhodotorula* and *Corynebacterium*. The solutions were also autoclaved at 115°C for 30 minutes to sterilize them. No drug loss due to autoclaving was found, and the solutions were found to lose a total of less than 5% over 90 days of storage under refrigeration and protected from exposure to light. ▪

# Levodopa

## Properties

Levodopa is a white to off-white odorless crystalline powder or colorless crystals.[1-3]

### Solubility

Levodopa is slightly soluble in water, having a solubility of about 1.65 mg/mL. It is practically insoluble in ethanol. It is freely soluble in 1 M hydrochloric acid but only sparingly soluble in 0.1 M hydrochloric acid.[1-3]

## pH

A 1% aqueous suspension has a pH of 4.5 to 7.[3]

## pK$_a$

Levodopa has pK$_a$ values of 2.3, 9, 10.2, and 12.5.[1612]

## General Stability Considerations

Levodopa products should be packaged in tight, light-resistant containers and stored at controlled room temperature. In the presence of moisture, levodopa is rapidly oxidized by atmospheric oxygen and darkens in color, indicating loss of concentration. Levodopa products should be protected from light, moisture, and excessive heat.[2-4]

Pappert et al.[1160] evaluated levodopa stability in aqueous solution. A 1-mg/mL solution in water was stable with little or no drug loss for seven days frozen at −4°C and refrigerated at 12°C. However, at 25°C, 10% loss occurred in about 48 hours. Exposure to or protection from light had no effect on drug stability in this study.

## Stability Reports of Compounded Preparations

### Injection

Injections, like other sterile drugs, should be prepared in a suitable clean air environment using appropriate aseptic procedures. When prepared from nonsterile components, an appropriate and effective sterilization method must be employed.

**STUDY 1:** Anderson et al.[481] reported the preparation of levodopa 10-mg/mL injection. Twenty grams of levodopa and 2 g of sodium metabisulfite were added to 1700 mL of sterile water for injection in a clean, sterilized 3-L vessel. The mixture was heated to 60°C with constant stirring, and the pH was adjusted to 2.3 with approximately 85 mL of 1 N hydrochloric acid to solubilize the levodopa. The mixture was brought to a final volume of 2000 mL with sterile water for injection, yielding a nominal levodopa concentration of 10 mg/mL. The bulk solution was filtered through an acid-stable 0.2-μm filter and filled to 30 mL in 30-mL sterile vials. When admixed in 5% dextrose to a concentration between 0.5 and 2 mg/mL,

the pH was near 2.7. Incremental addition of sodium bicarbonate to adjust the pH of a 1-mg/mL admixture resulted in precipitate formation at about pH 5.8. After storage for 18 hours, the solution had become dark brown. When the solution was adjusted to pH 5.25 and less, there were no visible changes. HPLC analysis found no significant loss of levodopa after storage for 96 hours.

**STUDY 2:** In a related study, Stennett et al.[482] evaluated the stability of extemporaneously compounded levodopa injection diluted to a concentration of 1 mg/mL in 5% dextrose injection. The pH of the admixtures was adjusted to pH 5 with sodium acetate injection 2 mEq/mL and to pH 6 with sodium phosphates injection 3 mmol/mL. Samples of the pH 5 admixture were stored at 4°C in the dark, at 25°C under fluorescent light, and at 45°C in an oven. Samples of the pH 6 admixture were stored at 25°C under fluorescent light only.

The pH 5 admixtures stored under refrigeration remained clear and colorless and exhibited no loss by HPLC analysis after 21 days of storage. However, the samples stored at 25°C under fluorescent light, although retaining 97% concentration, became discolored in 14 days. The authors recommended limiting use of this admixture to seven days. At 45°C, the samples became discolored in 12 hours. The pH 6 admixtures began to discolor in 36 hours at 25°C under fluorescent light and were black in 14 days. Concentration losses of about 4% in 36 hours and 13% in 14 days were found.

### Oral

**STUDY 1:** Nahata et al.[829] evaluated the stability of oral suspension formulations of levodopa 5 mg/mL and carbidopa 1.25 mg/mL. Tablets of levodopa and carbidopa were ground to fine powder using a mortar and pestle. A small amount of an equal parts mixture of Ora-Plus and Ora-Sweet was incorporated into the powder to make a paste. Additional vehicle mixture was incorporated geometrically to reach the final volume. The suspensions were packaged in amber plastic prescription bottles and stored at room temperature of 25°C and refrigerated at 4°C.

Stability-indicating HPLC analysis found that the levodopa and carbidopa concentrations were 96 and 92%, respectively, after 28 days of storage at room temperature. Levodopa and carbidopa concentrations were 94 and 93%, respectively, after 42 days of refrigerated storage. The physical appearance of the suspensions did not change during these time periods, but the suspensions became darker yellow during longer storage as more extensive decomposition occurred. Addition of ascorbic acid 2 mg/mL to the formulation resulted in more rapid loss of both drugs.

**STUDY 2:** Lopez Lozano and Moreno Cano[1481] evaluated the stability of an oral liquid formulation of levodopa/carbidopa prepared in an ascorbic-acid-containing aqueous solution. The authors intended this to be preparable by patients at home when needed. The patient would crush the daily number of levodopa/carbidopa tablets that they take and mix them into 500 mL of cold water containing a crushed 1-g tablet of ascorbic acid. If the patient was taking five tablets daily, as an example, that would result in L-dopa/carbidopa/ascorbic acid concentrations of 1/0.25/2 mg/mL, respectively. Any remaining undissolved material was considered unimportant. The test solution was packaged in a glass bottle and stored refrigerated at 0 to 4°C wrapped in aluminum foil for protection from exposure to light. The patient would shake the liquid and withdraw the correct volume for the dose (100 mL in the example) and drink it. Notably, the dose could be changed while keeping the volume of liquid to drink the same.

HPLC analysis found about 3% levodopa loss occurred in 24 hours in the example solution stored under refrigeration and protected from light. Unfortunately, the carbidopa concentration was not tested although it may be similar to the levodopa stability. ■

# ■ Levofloxacin

## Properties
Levofloxacin is pale to light yellowish-white crystals or crystalline powder.[4] Levofloxacin injection is yellow to greenish yellow.[7]

### Solubility
Levofloxacin is slightly soluble in water and ethanol.[4] Levofloxacin solubility is pH dependent. In the pH range of 0.6 to 5.8, the solubility is about 100 mg/mL. Increasing the pH up to 6.7 increases the solubility up to a maximum of 272 mg/mL. Increasing the pH further to 6.9 reduces the solubility to about 50 mg/mL.[2]

### pH
Levofloxacin injection has a pH of 3.8 to 5.8.[7]

## General Stability Considerations
Levofloxacin should be packaged in well-closed containers and stored at controlled room temperature protected from exposure to light.[4] Levofloxacin injection also should be stored at controlled room temperature and protected from light.[7]

## Stability Reports of Compounded Preparations
### Oral
VandenBussche et al.[712] reported the stability of a levofloxacin 50-mg/mL oral suspension prepared from crushed 500-mg tablets (Levaquin, McNeil) mixed with equal amounts of Ora-Plus suspending medium and strawberry syrup, NF. The suspension was packaged in amber plastic prescription bottles and stored at room temperature of 23 to 25°C and refrigerated at 3 to 5°C for 57 days. No observable change in color or odor occurred, and no visible microbial growth was observable. Stability-indicating HPLC analysis found no loss of levofloxacin at either temperature in 57 days. ■

#  Levothyroxine Sodium
## (Thyroxine Sodium)

## Properties
Levothyroxine sodium is a light-yellow to buff-colored or brownish-yellow odorless tasteless hygroscopic powder.[1–3]

### Solubility
Levothyroxine sodium is very slightly soluble in water, having a solubility of about 0.15 mg/mL. It is slightly soluble in alcohol, with a solubility of 3.33 mg/mL. The drug is soluble in dilute mineral acids and alkali hydroxides.[1–3]

Levothyroxine sodium solubility is pH dependent. In the range of pH 3 to 6, solubility was estimated to be only 0.25 mcg/mL. Below pH 2 and above pH 8 aqueous solubility increases.[668]

### pH
Saturated aqueous solutions have a pH of 8.35 to 9.35.[1,3]

### pKa
Levothyroxine has $pK_a$ values of 2.40, 6.87, and 9.96.[668]

## General Stability Considerations
Levothyroxine sodium should be packaged in tight, light-resistant containers and stored at controlled room temperature,[2,4,7] although storage between 8 and 15°C has been required to maintain concentration for some commercial products.[2] Refrigerated storage has also been recommended.[3]

Levothyroxine sodium products should be protected from light, heat, and exposure to air and moisture.[2,4,7] Levothyroxine sodium powder is stable in dry air, but upon exposure to light it may develop a slight pink color.[2,3,502] In aqueous media, exposure to ultraviolet radiation results in deiodination of levothyroxine sodium.[668]

## Stability Reports of Compounded Preparations
### Oral

For infants and children who cannot swallow intact tablets, crushing the tablets and suspending the powder in 5 to 10 mL of water, breast milk, or nonsoybean formula has been recommended. Such suspensions should not be stored for any period of time. The crushed tablet powder may also be sprinkled over applesauce. Food or formulas containing large amounts of soybean, fiber, or iron should not be used for administering the drug.[7]

**STUDY 1:** Alexander et al.[502] determined the projected shelf life of an oral levothyroxine 40-mcg/mL syrup. The syrup formula is shown in Table 52. In a glass mortar and pestle, 200 levothyroxine 0.2-mg tablets (United Research Labs) were crushed to fine powder, and ethanol was added to enhance drug solubility and minimize precipitation potential. The parabens stock solution in propylene glycol and glycerin were then incorporated into the sorbitol 70%. This mixture was added to the mortar contents in small increments, with mixing. Sodium bisulfite, edetate disodium (EDTA disodium), and both sodium phosphates dissolved in 100 mL of distilled water were added to the syrup. If necessary, the pH was adjusted to $7.5 \pm 0.1$ with phosphate buffer. The syrup was flavored with water-soluble banana flavor and brought to a final volume of 1000 mL with water. Samples were stored at four elevated temperatures up to

**TABLE 52.** Levothyroxine Syrup Formula of Alexander et al.[502]

| Component | Amount |
|---|---|
| Levothyroxine 0.2-mg tablets | 200 |
| Ethanol | 50 mL |
| Sorbitol 70% solution | 500 mL |
| Glycerin 99% | 250 mL |
| Parabens stock solution[a] | 10 mL |
| Sodium bisulfite | 1 g |
| EDTA disodium | 1 g |
| Sodium dihydrogen phosphate | 4.768 g |
| Disodium phosphate | 17.719 g |
| Banana flavor | 10 mL |
| Distilled water | qs 1000 mL |

[a]Methylparaben 10% and propylparaben 2% in propylene glycol.

70°C and were analyzed using a stability-indicating HPLC analytical technique. The analysis results were used to project the shelf life of this formulation. The time to 10% concentration loss ($t_{90}$) was projected to be 15 days at 25°C and 47 days at 5°C.

**STUDY 2:** However, Boulton et al.[532] reported a somewhat shorter shelf life based on their evaluation of several levothyroxine sodium 25-mcg/mL oral liquids prepared from tablets and from powder. For the suspensions prepared from tablets, 25 levothyroxine sodium 0.1-mg tablets were crushed to fine powder in a ceramic mortar. For the liquids prepared from the bulk powder, an equivalent amount of powder, 2.5 mg, was weighed and added to a clean ceramic mortar. A total of 40 mL of glycerol was measured. Levothyroxine sodium powder from either source was triturated with a small amount of the glycerol, and the liquid was transferred to a calibrated 100-mL amber high-density polyethylene bottle. The mortar was rinsed with 10-mL portions of glycerol, and the rinses were added to the bottle until all of the glycerol had been used. Then 1 or 2 mL (exact volume not specified) of methylparaben 50 mg/mL in propylene glycol was added to one set of samples but not to another set of samples. The oral liquids were brought to a volume of 100 mL with water. The tablets yielded a yellow suspension, while the powder formed a clear, colorless solution.

For stability testing, the procedure just described was scaled up to provide a sufficient number of samples. Samples of formulations prepared from levothyroxine sodium tablets both with and without methylparaben preservative were stored at about 4°C, at about 25°C, and incubated at elevated temperature around 40°C. The formulations prepared from powder were stored only at 40°C. Stability-indicating HPLC analysis of the samples found significant degradation of the levothyroxine sodium in all samples. The suspension prepared from tablets with no preservative exhibited the best stability, with about 5 to 6% loss in eight days and 10 to 12% loss in 14 days, determined at both 4 and 25°C. The calculated $t_{90}$ periods were 15.7 and 11.2 days at 4 and 25°C, respectively. At 40°C, 11% loss occurred in eight days. The addition of methylparaben reduced pH and stability, but the methylparaben did not reduce viable organisms below the levels found in the unpreserved suspension.

### Injection

Injections, like other sterile drugs, should be prepared in a suitable clean air environment using appropriate aseptic procedures. When prepared from nonsterile components, an appropriate and effective sterilization method must be employed.

Branje and Cremers[1173] evaluated the stability of a compounded levothyroxine sodium injection. The injection had the formula shown in Table 53.

The injection was filtered through a 0.2-μm sterile membrane filter and was heated to 100°C for 30 minutes but was not autoclaved due to drug decomposition. Accelerated stability testing was performed using HPLC analysis. The shelf life under refrigeration was calculated to be one year. In addition to the stability study, four 0.2-μm sterilizing filters were evaluated for compatibility with the levothyroxine sodium injection. Some loss of levothyroxine sodium due to sorption occurred with each filter material tested. The authors indicated that the loss should be considered when compounding the injection. The sorptive loss that occurred was as shown in Table 54. ■

**TABLE 53.** Levothyroxine Sodium Injection Formula of Branje and Cremers[1173]

| Component | | Amount |
|---|---|---|
| Levothyroxine sodium | | 100 mg |
| Ethanol 96% | | 100 mL |
| Propylene glycol | | 400 mL |
| Water for injection | qs | 1000 mL |

**TABLE 54.** Sorptive Loss to Four Sterilizing Filters[1173]

| Filter | Loss |
|---|---|
| Pall N66 polyamide | 35 mcg/cm$^2$ |
| Sartorius SM 116 regenerated cellulose | 17 mcg/cm$^2$ |
| Sartorius SM 113 cellulose nitrate | 32 mcg/cm$^2$ |
| Tuffryn polysulfone | 18 mcg/cm$^2$ |

# ■ Lidocaine Hydrochloride
## (Lignocaine Hydrochloride)

## Properties

Lidocaine hydrochloride is a white crystalline powder having a bitter taste.[2,3] Lidocaine hydrochloride monohydrate 1.23 g and lidocaine hydrochloride 1.16 g are approximately equivalent to lidocaine 1 g.[3]

### Solubility

Lidocaine is practically insoluble in water but very soluble in ethanol, chloroform, and oils.[1,4]

Lidocaine hydrochloride is very soluble in water and ethanol.[1–3]

### pH

A lidocaine hydrochloride 0.5% aqueous solution has a pH of 4 to 5.5.[1,3] Lidocaine hydrochloride injection, topical solution, and oral topical solution have a pH range of 5 to 7.[4] Lidocaine hydrochloride premixed in 5% dextrose injection has a pH range of 3 to 7.[4]

### pK$_a$

The pK$_a$ of lidocaine hydrochloride is 7.86.[2]

## General Stability Considerations

Lidocaine hydrochloride injections and solutions should be stored at controlled room temperature and should be protected from excessive heat and freezing.[2,4] Loss of lidocaine hydrochloride due to sorption to polyvinyl chloride containers from buffered cardioplegia solutions has been reported to occur at room temperature, but refrigeration slows the sorptive process.[642] Precipitation of lidocaine base occurs at a pH around 7.5 to 7.6.[701]

## Stability Reports of Compounded Preparations
### Injection

Injections, like other sterile drugs, should be prepared in a suitable clean air environment using appropriate aseptic procedures. When prepared from nonsterile components, an appropriate and effective sterilization method must be employed.

**STUDY 1:** Hinshaw et al.[701] reported on buffering lidocaine hydrochloride injection with sodium bicarbonate to reduce pain upon injection. The sodium bicarbonate increases the pH, resulting in an increased percentage of the less stable and soluble lidocaine unionized base. Precipitation of lidocaine occurs at pH 7.5 to 7.6.

**STUDY 2:** Meyer and Henneman[702] evaluated the stability of 1% lidocaine hydrochloride injection buffered with sodium bicarbonate using an antibody-binding assay. At pH 6.8, no lidocaine loss occurred in 27 days at room temperature. At pH 7.2, adequate lidocaine stability resulted for 19 days, but 12% loss and crystalline precipitation developed by 27 days. At pH 7.4, 23% lidocaine loss and crystalline precipitation occurred in as little as five days.

**STUDY 3:** Tuleu et al.[1266] buffered lidocaine hydrochloride 2% with epinephrine 5 mcg/mL to around pH 7.5 to 8 with sodium bicarbonate. Stability-indicating HPLC analysis found no loss of lidocaine hydrochloride but nearly 30% loss of epinephrine in six hours at 24°C exposed to light. If kept in the dark, epinephrine losses were about 9% in 20 hours at

24°C. The authors recommended not preparing the buffered injection well in advance of use.

**STUDY 4:** Larson et al.[1386] evaluated the stability of lidocaine hydrochloride 2% with epinephrine hydrochloride 1:100,000 30 mL mixed with sodium bicarbonate 1 mEq/mL 3 mL. Lidocaine hydrochloride concentrations were determined using gas chromatography while epinephrine hydrochloride concentrations were determined by HPLC analysis. At room temperature, lidocaine losses of 11 and 22% occurred in seven days and two weeks, respectively. An epinephrine loss of 28% occurred in seven days. Storage under refrigeration improved stability. A lidocaine loss of 6% occurred in four weeks; epinephrine losses of 2 and 12% occurred in one and three weeks, respectively.

**STUDY 5:** Bartfield et al.[1387] evaluated the stability of lidocaine 0.9% alkalinized by mixing with sodium bicarbonate 0.088 mEq/mL. Fluorescence polarization immunoassay of lidocaine hydrochloride concentrations found 11% loss occurred in seven days at room temperature.

**STUDY 6:** Peterfreund and Datta[1388] evaluated the stability of lidocaine hydrochloride solutions mixed with sodium bicarbonate solutions for alkalinization. Lidocaine hydrochloride solutions of 1 and 1.5% with epinephrine hydrochloride 1:200,000 20 mL were visually compatible for up to five hours at room temperature with sodium bicarbonate solutions of 4% 4 mL and 8.4% 2 mL. However, at a lidocaine hydrochloride concentration of 2% 20 mL mixed with the same sodium bicarbonate solutions, haziness formed. The haze dissipated with gentle shaking.

**STUDY 7:** Bonhomme et al.[1389] evaluated the stability of lidocaine hydrochloride 2% with and without epinephrine hydrochloride 1:100,000 10 mL with sodium bicarbonate 8.4% 1 and 1.5 mL and 1.4% 1.5 mL. All combinations were visually compatible with no increase of particles by microscopic examination and no loss of lidocaine by ion-pair titration or epinephrine by HPLC analysis in six hours.

**STUDY 8:** Murakami et al.[1390] evaluated the compatibility of lidocaine hydrochloride 1% with and without epinephrine hydrochloride 1:100,000 10 mL alkalinized by mixing with sodium bicarbonate 8.4% 1 mL. Cloudiness appeared in some samples. When exposed to light and air, a crystalline precipitate formed.

### Ophthalmic

Ophthalmic preparations, like other sterile drugs, should be prepared in a suitable clean air environment using appropriate aseptic procedures. When prepared from nonsterile components, an appropriate and effective sterilization method must be employed.

Bailey et al.[1525] evaluated the stability of a four-drug combination ophthalmic gel for use as a preoperative medication for cataract surgery. The final drug concentrations in the preoperative gel were cyclopentolate hydrochloride 0.51 mg/mL, phenylephrine hydrochloride 5.1 mg/mL, and tropicamide 0.51 mg/mL in the lidocaine hydrochloride jelly 2%. The gel was prepared by combining 0.3 mL each of cyclopentolate hydrochloride ophthalmic solution 1%, USP, phenylephrine hydrochloride ophthalmic solution 10%, USP, and tropicamide ophthalmic solution 1%, USP in preservative-free lidocaine hydrochloride jelly 2%, USP. The gel was packaged as 0.5 mL in 1-mL polycarbonate plastic syringes with sealed tips. The samples were stored protected from light in an environmental chamber at room temperature of 23 to 27°C and 60% relative humidity and also at refrigerated temperature of 2 to 4°C for 60 days.

The ophthalmic gel exhibited little change in pH and remained sterile and endotoxin-free throughout the 60-day study. Stability-indicating HPLC analysis of drug concentrations over 60 days of storage found no loss of any of the drugs when stored under refrigeration and not more than 4% loss of any drug when stored at room temperature.

### Nasal Solution

Allen[580] reported on the compounding of a lidocaine hydrochloride 40-mg/mL nasal solution for cluster and migraine headaches. Lidocaine hydrochloride 4 g was weighed and transferred to a suitable size graduate. Bacteriostatic water for injection was added, bringing the volume to 100 mL, and the solution was mixed well. The pH of the solution was adjusted to the range of 6 to 7 with sodium hydroxide solution or hydrochloric acid solution. With the addition of methylparaben 200 mg as a preservative, sterile water for injection can be substituted as the diluent. The nasal solution should be packaged in tight containers. Individual doses can be packaged in prefilled syringes, usually in doses from 0.4 to 1 mL (10 to 40 mg). Alternatively, the solution can be packaged in bottles with a calibrated dropper or in calibrated nasal spray bottles. Lidocaine hydrochloride is a stable molecule. However, the author indicated that a relatively short (but unspecified) shelf life would be proper due to the potential for microbial contamination with use.

### Topical

**STUDY 1:** Larson et al.[539] evaluated the stability of a topical anesthetic solution described by Schilling et al.[540] The topical anesthetic was composed of lidocaine hydrochloride 40 mg/L, racemic epinephrine hydrochloride 2.25 mg/mL, tetracaine hydrochloride 5 mg/mL, and sodium metabisulfite 0.63 mg/mL as an aqueous solution. A 500-mL quantity of the solution was prepared by measuring 100 mL of 20% lidocaine hydrochloride injection (Abbott), 50 mL of racemic epinephrine 2.25% as the hydrochloride (Nephron),

125 mL of 2% tetracaine hydrochloride (Winthrop), 315 mg of sodium metabisulfite (Gallipot), and 225 mL of sterile water for irrigation. All of the ingredients were combined and mixed well. The finished solution had a blue tint. The solution was packaged in both clear and amber glass bottles with tight fitting caps, and samples were stored under refrigeration at 4°C in the dark, at room temperatures near 18°C exposed to ambient room light, and at an elevated temperature of 35°C exposed to ambient room light.

The refrigerated samples remained clear with a blue tint, and HPLC analysis of the samples found little or no loss of any of the three drug components after 26 weeks of storage. Epinephrine was the least stable of the drug components at the higher temperatures. Samples in amber containers at room temperature exposed to light were not discolored through four weeks of storage but became discolored in eight weeks, exhibiting about 5% epinephrine loss at that time; the epinephrine loss increased to 9% in 16 weeks. Discoloration was thought to be the result of a highly colored epinephrine decomposition product that was visible when epinephrine decomposition of 2% or more occurred. The lidocaine and tetracaine components both retained at least 95% of the initial concentration throughout 26 weeks of room temperature storage. In clear containers exposed to light, discoloration appeared in as little as one to two weeks. The authors[539] recommended that the solution be packaged only in amber glass bottles and be given expiration dates of 26 weeks under refrigeration and four weeks at room temperature.

The results of a controlled clinical trial of this topical anesthetic solution by Schilling et al.[540] indicated that adequate anesthesia resulted throughout the time period evaluated for stability by Larson et al.[539]

**STUDY 2:** Ohzeki et al.[1247] evaluated a topical eutectic mixture of local anesthetic (EMLA) cream prepared from lidocaine and tetracaine eutectic mixture. Lidocaine and tetracaine form a eutectic mixture at a relatively low temperature of about 18°C. To prepare the cream, lidocaine 2.5 g and tetracaine 2.5 g were mixed, forming a eutectic mixture. Polyoxyethylene hydrogenated castor oil (HCO60) 1.9 g was then incorporated, forming mixture A. Separately, ethyl parahydroxybenzoate (ethyl paraben) 50 mg was dissolved in ultra pure water 100 mL and carboxypolymethylene (Carbopol 934P) 1 g was added and mixed well, forming mixture B. The pH of mixture B was adjusted to 9, but the authors did not state what they used to adjust the pH. Finally mixture A and mixture B were combined and mixed well, forming the topical eutectic mixture of local anesthetic cream.

The cream was stored at room temperature and exposed to room fluorescent light for 90 days. Visual inspection of the cream found no changes in color or appearance, and no separation of a liquid phase occurred within three months of storage. HPLC analysis of the lidocaine content found it

remained unchanged with little or no degradation occurring throughout the entire storage period. The stability of tetracaine was not evaluated.

**STUDY 3:** Abdelmageed et al.[1252] evaluated the stability of morphine sulfate incorporated at a concentration of 1 mg/mL into a combination gel of lidocaine 2% and chlorhexidine gluconate 0.25% (Instillagel, FARCO-Pharma). The compounded gel was packaged in 20-mL Luer-Lok polypropylene syringes sealed with Helapet Combi Caps (Becton Dickinson). Sample syringes were stored at 4°C in the dark, room temperature of 25°C exposed to fluorescent room light, and 37°C in the dark.

Stability-indicating HPLC analysis found little or no loss of any of the three components in 22 days refrigerated and at room temperature. At 37°C over seven months, no loss of morphine sulfate and lidocaine hydrochloride and only about 8% loss of chlorhexidine gluconate occurred.

**STUDY 4:** Vonbach et al.[1336] evaluated the long-term stability of a three-drug solution to be used for bladder iontophoresis. The solution contained lidocaine hydrochloride 20 mg/mL, dexamethasone sodium phosphate 0.2 mg/mL, and epinephrine hydrochloride 0.01 mg/mL in water. The solution was packaged in amber glass vials and clear glass vials. The samples in amber glass vials were stored frozen at −20°C, refrigerated at 4°C, at room temperature of 20°C, and at elevated stress temperature of 40°C. Samples in clear vials were stored at 20°C exposed to daylight.

HPLC analysis found that epinephrine hydrochloride was the least stable component. At 40°C and exposed to daylight, extensive and rapid loss of epinephrine hydrochloride occurred. At 20°C in amber vials, about 10% epinephrine hydrochloride loss occurred in 28 days. Under refrigeration, the loss was about 6% in six months; when frozen, no loss occurred in six months. For lidocaine hydrochloride, little or no loss occurred under any of the test conditions. For dexamethasone sodium phosphate, losses never exceeded 5% in six months at −20, 4, and 20°C. The authors stated that this three-drug iontophoresis solution was stable for three months refrigerated and for six months frozen at −20°C.

### Rectal

Gebauer et al.[733] reported the stability of an extemporaneously prepared rectal gel for the management of tenesmus in colorectal cancer. Oxycodone hydrochloride powder (Macfarlan Smith) was mixed with lidocaine gel (Pharmacia) and hydrogel (Smith and Nephew) to prepare a gel containing oxycodone hydrochloride 0.3% (wt/vol) and lidocaine hydrochloride 1.5% (wt/vol). The gel was packaged in 50-mL syringes and stored refrigerated at 4°C for 12 months. Stability-indicating HPLC analysis found little or no change in concentrations of either of the drugs within 12 months.

## Compatibility with Other Drugs

Gupta[990] evaluated the stability and compatibility of diphenhydramine hydrochloride elixir 2.5 mg/mL (Alpharma) combined in equal volume with lidocaine hydrochloride 2% viscous solution (Roxane) for use in treating stomatitis. This concentrate would be diluted to lower concentrations as needed for the individual patient. After mixing, the concentrated drug combination was packaged in an amber bottle and stored at room temperature of 24 to 26°C for 21 days. The mixture did not undergo any changes in physical appearance during the study. Stability-indicating HPLC analysis found no loss of either drug in 21 days at room temperature. ■

# ■ Liothyronine Sodium

## Properties

Liothyronine sodium is a white to light tan odorless crystalline powder.[3,4]

### Solubility

Liothyronine sodium is very slightly soluble in water and slightly soluble in ethanol[3,4] but practically insoluble in most other organic solvents.[4] The drug dissolves in dilute solutions of alkali hydroxides.[3]

## General Stability Considerations

Liothyronine sodium powder and commercial oral tablets are packaged in tight containers and stored at controlled room temperature.[4] Commercial liothyronine sodium injection is stored under refrigeration at 2 to 8°C.[7]

## Stability Reports of Compounded Preparations

### Injection

Injections, like other sterile drugs, should be prepared in a suitable clean air environment using appropriate aseptic procedures. When prepared from nonsterile components, an appropriate and effective sterilization method must be employed.

Branje and Cremers[1173] evaluated the stability of a compounded liothyronine sodium injection that had a formula that differs from the commercial injection. The compounded injection had the formula shown in Table 55.

The injection was filtered through a 0.2-μm sterile membrane filter and was heated to 100°C for 30 minutes but, due to drug decomposition, was not autoclaved. Accelerated stability testing was performed using HPLC analysis, which found the drug to be very stable in this injection formulation. The shelf life at room temperature was calculated to be 130 years.

In addition to the stability study, three 0.2-μm sterilizing filters were evaluated for compatibility with the liothyronine sodium injection. Some loss of liothyronine sodium due to sorption occurred with each filter material tested. The authors indicated that the loss should be considered when compounding the injection. The sorptive loss that occurred is shown in Table 56. ■

**TABLE 55.** Compounded Liothyronine Sodium Injection[1173]

| Component | | Amount |
|---|---|---|
| Liothyronine sodium | | 100 mg |
| Ethanol 96% | | 100 mL |
| Propylene glycol | | 400 mL |
| Water for injection | qs | 1000 mL |

**TABLE 56.** Sorptive Loss of Liothyronine Sodium Injection[1173]

| Filter | Loss |
|---|---|
| Pall N66 polyamide | 24 mcg/cm$^2$ |
| Sartorius SM 116 regenerated cellulose | 18 mcg/cm$^2$ |
| Sartorius SM 113 cellulose nitrate | 16 mcg/cm$^2$ |

# ■ Lisinopril

## Properties

Lisinopril is a white to off-white odorless crystalline powder.[1,3] Lisinopril 2.72 mg as the dihydrate is approximately equivalent to lisinopril anhydrous 2.5 mg.[3]

### Solubility

Lisinopril has an aqueous solubility of 97 mg/mL but is practically insoluble in ethanol, with a solubility of less than 0.1 mg/mL.[1,3]

## pKₐ

*Lisinopril* has $pK_a$ values at 25°C of 2.5, 4, 6.7, and 10.1.[1]

## General Stability Considerations

Lisinopril products should be packaged in tight containers and stored at controlled room temperature protected from moisture.[4,7]

## Stability Reports of Compounded Preparations

### Oral

**STUDY 1:** Webster et al.[555] evaluated the stability of lisinopril 2-mg/mL oral syrup prepared from powder. One gram of lisinopril powder (Sigma Chemical) was dissolved in distilled water and then incorporated by geometric dilution in syrup to a final volume of 500 mL. The syrup was packaged in amber plastic prescription bottles and stored at 5 and 23°C for 30 days. None of the samples developed any change in color or clarity. HPLC analysis found no loss of lisinopril after 30 days of storage at either temperature. The authors recommended refrigerated storage to inhibit microbial growth.

**STUDY 2:** Rose et al.[735] reported the stability of lisinopril 2-mg/mL oral syrup prepared from crushed tablets. Thirty lisinopril 20-mg tablets (Zestril, Zeneca) were pulverized using a mortar and pestle and mixed with 50 mL of distilled water. The mixture was filtered through filter paper and the filtrate was incorporated into syrup, NF, to yield a 2-mg/mL concentration. The lisinopril syrup was packaged in amber prescription bottles and stored refrigerated at 5 and 23°C for 30 days. No change in color or clarity was visually observed. Stability-indicating HPLC analysis found that less than 2% loss of lisinopril occurred at either temperature in 30 days. The authors recommended storage under refrigeration to inhibit any microbial growth.

**STUDY 3:** Thompson et al.[777] reported the stability of lisinopril 1-mg/mL oral liquid prepared from tablets. Ten 20-mg lisinopril tablets (Prinivil, Merck) were placed in an eight-ounce amber polyethylene terephthalate bottle with 10 mL of purified water and shaken for one minute. Complete dissolution of the lisinopril occurred in 30 seconds, so shaking for one minute was more than adequate. Sodium citrate–citric acid oral solution (Bicitra, Alza) 30 mL to control pH was then added to the bottle, which was reshaken. Ora-Sweet SF (sugar-free) 160 mL was then added to the bottle, which was shaken once again. The entire lisinopril content was fully in solution, but the insoluble tablet components remained. The samples were stored for six weeks at room temperature of 25°C at 35% relative humidity. Some samples were exposed to light in a photostability chamber, while other samples were protected from exposure to light.

The undissolved solids in the suspension settled over time but easily redispersed with shaking; lisinopril was fully dissolved and remained uniform in the liquid even without shaking. Stability-indicating HPLC analysis found that less than 5% loss of lisinopril occurred during six weeks of storage. Exposure to light in the photostability chamber had no adverse effect on lisinopril stability.

**STUDY 4:** Nahata and Morosco[871] evaluated the stability of two formulations of lisinopril 1 mg/mL prepared from commercial tablets. Lisinopril 10-mg tablets (Prinivil, Merck) were crushed to fine powder using a mortar and pestle. The tablet powder was suspended in an equal parts mixture (1:1) of Ora-Sweet and Ora-Plus and also in vehicle composed of methylcellulose 1% (with methyl- and propylparabens) and simple syrup in a 1:13 mixture. The suspensions were packaged in amber plastic prescription bottles and were stored at room temperature of 25°C and under refrigeration at 4°C. Stability-indicating HPLC analysis found that, using the Ora-Sweet–Ora-Plus vehicle, about 5% lisinopril loss occurred in 91 days stored under refrigeration and at room temperature. In the methylcellulose–simple syrup vehicle, lisinopril also exhibited about 5% loss in 91 days when refrigerated. However, the concentration remained acceptable for only 56 days at room temperature, with about 8% loss occurring. ∎

# ▣ Lithium Citrate

## Properties

Lithium citrate is a white odorless deliquescent crystalline powder or granules with a slightly salty or alkaline taste.[1–3] Each gram of lithium citrate contains about 14.3 mEq or 10.6 mmol of lithium.[2,3]

### Solubility

Lithium citrate is very soluble in water, having an aqueous solubility of about 1 g/1.5 mL.[1–3] The drug is only slightly soluble in alcohol.[1,3]

### pH

A 5% lithium citrate solution in water has a pH between 7 and 10.[3,4] Lithium citrate syrup has a pH of 4 to 5.[4] Lithium citrate solution (Cibalith-S, Ciba) had a measured pH of 4.7.[19]

### Osmolality

Lithium citrate syrup 1.6 mEq/mL (Roxane) had an osmolality of 6850 mOsm/kg.[233]

## General Stability Considerations

Lithium citrate powder and solutions should be packaged in tight containers and stored at controlled room temperature.[2,4]

## Stability Reports of Compounded Preparations
### Enteral
#### LITHIUM CITRATE COMPATIBILITY SUMMARY
*Compatible with:* Vital
*Incompatible with:* Ensure • Ensure HN • Ensure HN Plus • Ensure Plus • Osmolite • Osmolite HN

**STUDY 1:** Cutie et al.[19] added 5 mL of lithium citrate oral solution (Cibalith-S, Ciba) to varying amounts (15 to 240 mL) of Ensure, Ensure Plus, and Osmolite (Ross Laboratories) with vigorous agitation to ensure thorough mixing. The lithium citrate oral solution was physically incompatible with the enteral products, which became granular and formed a film.

**STUDY 2:** Altman and Cutie[850] reported the physical compatibility of lithium citrate oral solution (Cibalith-S, Ciba) 5 mL with varying amounts (15 to 240 mL) of Ensure HN, Ensure Plus HN, Osmolite HN, and Vital after vigorous agitation to ensure thorough mixing. The lithium citrate oral solution was physically incompatible with Ensure HN, Ensure Plus HN, and Osmolite HN, becoming granular and forming a film. Lithium citrate oral solution was compatible with Vital, distributing uniformly with no phase separation or granulation.

## Compatibility with Common Beverages and Foods

**STUDY 1:** Geller et al.[242] reported that lithium citrate liquid was visually compatible with water, milk, coffee, tea, fruit juices, and carbonated beverages.

**STUDY 2:** Kerr[256] cited the compatibility of lithium citrate liquid with a number of beverages, including apple juice or cider, apricot juice, cranberry juice, grape juice or drink, grapefruit juice, lemonade, orange juice, pineapple juice, prune juice, Tang, tomato juice, vegetable juice (V8), cola sodas, Mellow-Yellow, orange soda, 7-Up, and Sprite. Kerr also stated that lithium citrate liquid was compatible with soups, puddings, and saline solution.

## Compatibility with Other Drugs
#### LITHIUM CITRATE COMPATIBILITY SUMMARY
*Compatible with:* Doxepin • Loxepine hydrochloride • Mesoridazine besylate • Molindone hydrochloride • Perphenazine • Risperidone • Thiothixene hydrochloride • Trihexyphenidyl hydrochloride
*Incompatible with:* Chlorpromazine hydrochloride • Thioridazine hydrochloride • Trifluoperazine hydrochloride
*Uncertain or variable compatibility with:* Fluphenazine hydrochloride • Haloperidol lactate

**STUDY 1:** McGee et al.[73] reported the formation of a precipitate when lithium citrate syrup (Lithionate-S, Rowell) was added directly to trifluoperazine hydrochloride oral concentrate (Smith Kline & French).

**STUDY 2:** Similarly, Wilson[72] reported precipitate formation when lithium citrate syrup (Lithionate-S, Rowell) and chlorpromazine hydrochloride oral concentrate (Thorazine, Smith Kline & French) were mixed directly together.

**STUDY 3:** Theesen et al.[78] evaluated the compatibility of lithium citrate syrup with 10 neuroleptic solutions. Lithium citrate syrup (Lithionate-S, Rowell) and lithium citrate syrup (Roxane) at a lithium-ion concentration of 1.6 mEq/mL were combined in volumes of 5 and 10 mL with four volumes of the neuroleptic agents in duplicate, reversing the order of mixing between samples. Samples were stored at 4 and 25°C for six hours and evaluated visually for physical compatibility. The specific neuroleptic products, concentrations, and test results are shown in Table 57. Thin-layer chromatography of the sediment layers of incompatible combinations showed high concentrations of the neuroleptic drugs. The authors attributed the incompatibilities to the excessive ionic strength of the lithium citrate syrup. This condition results in salting out of the neuroleptic drug salts in a hydrophobic, viscid layer that adheres tenaciously to container surfaces. The problem persists even when moderate-to-high doses are diluted 10-fold in water.

**STUDY 4:** Raleigh[454] also evaluated the compatibility of lithium citrate syrup (Philips Roxane) 1.6 mEq of lithium/mL with 13 other liquid oral medications. Each of the liquid oral drug products was combined with the lithium citrate syrup in a manner described as "simple random dosage combinations"; the exact amounts tested and test conditions were not specified. Chlorpromazine hydrochloride concentrate (concentration unspecified), fluphenzine concentrate 5 mg/mL, and trifluoperazine concentrate 10 mg/mL each resulted in the formation of a milky precipitate. Thioridazine suspension (Sandoz) (concentration unspecified) formed a dense white curdy precipitate. The other drugs did not form a visible precipitate in combination with lithium citrate syrup. The list of liquid oral drugs tested is presented in Table 58.

**STUDY 5:** Park et al.[1191] evaluated the compatibility of risperidone 1-mg/mL oral liquid (Janssen) mixed with lithium citrate syrup 300 mg/5 mL (Roxane). Various amounts of the two oral liquid drugs were mixed. Lithium citrate volumes ranged from 5 to 20 mL. Risperidone volumes ranged from 1 to 6 mL. Visual inspection found no evidence of haze, precipitation, change in viscosity, or color change over a 30-minute observation period. The authors concluded that these two oral liquids could be mixed together. ∎

**TABLE 57.** Compatibility Observations of Lithium Citrate Syrup 5 and 10 mL Combined with Oral Neuroleptic Drug Solutions by Theesen et al.[78]

| Drug, Concentration | Manufacturer | Volumes (mL) | Results |
|---|---|---|---|
| Chlorpromazine hydrochloride, 100 mg/mL | Lederle | 0.1, 1, 5, 10 | Opaque turbidity forms immediately and separates into two layers |
| Chlorpromazine hydrochloride, 100 mg/mL | Smith Kline & French | 0.1, 1, 5, 10 | Opaque turbidity forms immediately and separates into two layers |
| Fluphenazine hydrochloride, 0.5 mg/mL | Squibb | 10, 20, 40, 60 | No incompatibility visually observed |
| Haloperidol lactate, 2 mg/mL | McNeil | 5, 15, 30, 60 | Opaque turbidity forms immediately and persists |
| Loxapine hydrochloride, 25 mg/mL | Lederle | 3, 6, 9, 15 | No incompatibility visually observed |
| Mesoridazine besylate, 25 mg/mL | Boehringer-Ingelheim | 5, 10, 15, 20 | No incompatibility visually observed |
| Molindone hydrochloride, 20 mg/mL | Endo | 5, 10, 15, 20 | No incompatibility visually observed |
| Perphenazine, 3.2 mg/mL | Schering | 4, 8, 15, 20 | No incompatibility visually observed |
| Thioridazine hydrochloride, 100 mg/mL | Sandoz | 2, 4, 6, 8 | Opaque turbidity forms immediately |
| Thiothixene hydrochloride, 5 mg/mL | Roerig | 2, 6, 12, 24 | No incompatibility visually observed |
| Trifluoperazine hydrochloride, 10 mg/mL | Smith Kline & French | 2, 4, 8, 12 | Opaque turbidity forms immediately |

**TABLE 58.** Compatibility of Selected Oral Drugs with Lithium Citrate Syrup 1.6 mg/mL by Raleigh[454]

| Test Drug | Concentration | Observed Result |
|---|---|---|
| Chlorpromazine hydrochloride concentrate | Unspecified | Milky precipitate |
| Doxepin concentrate | 10 mg/mL (as the hydrochloride) | No precipitate formed |
| Fluphenazine hydrochloride concentrate | 5 mg/mL | Milky precipitate |
| Fluphenazine hydrochloride elixir | 0.5 mg/mL | No precipitate formed |
| Haloperidol concentrate | 2 mg/mL (as the lactate) | No precipitate formed |
| Loxapine concentrate | 25 mg/mL (as the hydrochloride) | No precipitate formed |
| Mesoridazine concentrate | 25 mg/mL (as the besylate) | No precipitate formed |
| Molindone hydrochloride concentrate | 20 mg/mL | No precipitate formed |
| Perphenazine concentrate | 3.2 mg/mL | No precipitate formed |
| Thioridazine suspension | Unspecified | Curdy precipitate |
| Thiothixene concentrate | 5 mg/mL (as the hydrochloride) | No precipitate formed |
| Trifluoperazine concentrate | 10 mg/mL (as the hydrochloride) | Milky precipitate |
| Trihexyphenidyl hydrochloride elixir | 0.4 mg/mL | No precipitate formed |

# Lodoxamide Tromethamine

## Properties

Lodoxamide tromethamine is a white crystalline powder. Lodoxamide tromethamine 1.78 mg is equivalent to 1 mg of lodoxamide.[7]

## Solubility

Lodoxamide tromethamine is water soluble.[7]

## pH

Lodoxamide tromethamine 0.1% commercial ophthalmic solution was found to have a pH of about 5 to 5.4.[880]

## General Stability Considerations

Lodoxamide tromethamine commercial ophthalmic 0.1% solution should be stored at controlled room temperature.[7]

## Stability Reports of Compounded Preparations
### Ophthalmic

Ophthalmic preparations, like other sterile drugs, should be prepared in a suitable clean air environment using appropriate aseptic procedures. When prepared from nonsterile components, an appropriate and effective sterilization method must be employed.

Garcia-Valldecabres et al.[880] reported the change in pH of several commercial ophthalmic solutions, including lodoxamide 1 mg/mL as tromethamine (Alomide, Alcon) over 30 days after opening. Slight variation in solution pH was found, but no consistent trend and no substantial change occurred. ∎

# Loperamide Hydrochloride

## Properties

Loperamide hydrochloride is a white to slightly yellow crystalline or amorphous powder.[2,3]

## Solubility

Loperamide hydrochloride is slightly soluble in water.[1–3] At pH 1.7 it has a solubility of about 1.4 mg/mL. However, the drug is nearly insoluble (0.2 mg/mL) at physiological pH. The drug has solubilities of 53.7 mg/mL in ethanol, 56.4 mg/mL in propylene glycol, and 14 mg/mL in polyethylene glycol 400.[1]

## pH

Loperamide hydrochloride oral solution has a pH of around 5.[2]

## $pK_a$

The drug has a $pK_a$ of 8.66.[1]

## General Stability Considerations

Loperamide hydrochloride products should be packaged in well-closed containers and stored at controlled room temperature.[2,4]

The tablets should be packaged in well-closed, light-resistant containers.[4] The manufacturer has stated that loperamide hydrochloride should not be removed from the original packaging and placed in dosing compliance aids because of its moisture sensitivity.[1622]

Loperamide hydrochloride in aqueous solution is stable over pH 2.1 to 9.7.[2] The pH of maximum stability is approximately 4.5.[716]

## Stability Reports of Compounded Preparations
### Enteral
**LOPERAMIDE HYDROCHLORIDE COMPATIBILITY SUMMARY**
*Compatible with:* Osmolite • Osmolite HN • Vital

Fagerman and Ballou[52] reported on the compatibility of loperamide hydrochloride 1 mg/5 mL (Imodium, McNeil) mixed in equal quantities with three enteral feeding formulas. The loperamide hydrochloride was visually compatible with Vital, Osmolite, and Osmolite HN (Ross), with no obvious thickening, granulation, or precipitation. ∎

# Loratadine

## Properties
Loratadine is a white to off-white powder.[3,4]

### Solubility
Loratadine is relatively insoluble in water but freely soluble in acetone, methanol, and toluene.[3,4]

### pH
Loratadine oral solution has a pH between 2.5 and 3.1.[4]

## General Stability Considerations
Loratadine powder and tablets should be packaged in well-closed containers and stored between 2 and 30°C. Loratadine oral solution should be packaged in tight containers and stored between 2 and 25°C.[4]

## Stability Reports of Compounded Preparations
### Oral
Abounassif et al.[939] reported the photoinstability of loratadine solutions exposed to ultraviolet light of 254 nm and to sunlight in glass vials. The solutions developed a yellow discoloration along with two decomposition products determined by HPLC analysis. The authors noted that the presence of water increases the photolability of loratadine. ■

# Lorazepam

## Properties
Lorazepam is a white or nearly white odorless crystalline powder.[2,3]

### Solubility
Lorazepam is practically insoluble in water, with a reported solubility of 0.08 mg/mL. It is sparingly soluble in ethanol and propylene glycol, having solubilities of 14 and 16 mg/mL, respectively.[1-3] An evaluation of lorazepam solubility in infusion solutions reported the solubilities noted in Table 59.[643]

### pK$_a$
Lorazepam has pK$_a$ values of 1.3 and 11.5.2.

## General Stability Considerations
Lorazepam products should be packaged in tight, light-resistant containers. The tablets should be stored at controlled room temperature, while the injection and oral concentrate should be stored under refrigeration and protected from freezing.[2,4,7] Solutions of lorazepam injection diluted for use should be discarded if they are discolored or contain a precipitate.[2]

**TABLE 59.** Lorazepam Solubilities in Aqueous Solutions[643]

| Solution | Lorazepam Solubility |
| --- | --- |
| 5% Dextrose injection | 0.054 mg/mL |
| Lactated Ringer's injection | 0.062 mg/mL |
| 0.9% Sodium chloride injection | 0.027 mg/mL |
| Water, deionized | 0.054 mg/mL |

## Stability Reports of Compounded Preparations
### Injection
Injections, like other sterile drugs, should be prepared in a suitable clean air environment using appropriate aseptic procedures. When prepared from nonsterile components, an appropriate and effective sterilization method must be employed.

**STUDY 1 (REPACKAGED):** Stiles et al.[529] evaluated the stability of lorazepam (Wyeth) 2 mg/mL undiluted when packaged as 3 mL in 10-mL polypropylene infusion-pump syringes for the CADD-Micro and CADD-LD pumps (Pharmacia Deltec) sealed with Luer tip caps (Becton Dickinson). Samples were stored at 5 and 30°C. HPLC analysis found 12 to 14% loss of the drug in three days and 23 to 26% loss in 10 days at either storage temperature. The authors postulated that the losses could have been due to sorption to container component(s). It was recommended that lorazepam not be stored in these polypropylene infusion-pump syringes.

**STUDY 2 (REPACKAGED):** Share et al.[591] reported the stability of lorazepam (Wyeth) diluted to 1 mg/mL and repackaged in polypropylene syringes. Twenty milliliters of lorazepam 2-mg/mL injection was diluted with an equal quantity of 5% dextrose injection or 0.9% sodium chloride injection and packaged in 60-mL polypropylene syringes (Becton Dickinson). Samples were stored at 22°C exposed to normal room light for 28 hours. No changes in color or clarity were observed, and stability-indicating HPLC analysis found little or no loss of lorazepam in any sample. It was noted that the use of the 2-mg/mL lorazepam injection to prepare the 1-mg/mL sample resulted in higher concentrations of organic solvents that may help to avoid precipitation. ■

# Losartan Potassium

## Properties

Losartan potassium is a white to off-white powder.[4,7]

### Solubility

Losartan potassium is freely soluble in water, soluble in isopropanol and other alcohols, and slightly soluble in some common organic solvents.[4,7]

### pK$_a$

Losartan potassium has a pK$_a$ variously cited as 5 to 6[1] and 3.15.[1370]

## General Stability Considerations

Losartan potassium should be packaged in well-closed containers and stored at controlled room temperature.[4] Losartan potassium tablets are packaged in tight containers and stored at controlled room temperature protected from exposure to light.[7]

## Stability Reports of Compounded Preparations
### Oral

**STUDY 1:** The Merck[7] package insert for Cozaar cites the following compounded formulation of losartan potassium 2.5-mg/mL oral liquid. The oral liquid had the following formula:

| | | |
|---|---|---|
| Losartan potassium 50-mg tablets | | 10 tablets |
| Purified water | | 10 mL |
| Ora-Sweet/Ora-Plus (1:1) | qs | 190 mL |

The recommended preparation procedure was to add 10 mL of purified water to a 240-mL amber polyethylene terephthalate (PET) bottle containing 10 Cozaar tablets and immediately shake the mixture for at least two minutes, allow to stand for one hour, and then shake for at least an additional minute. The mixture is then to be brought to volume with an equal parts mixture of Ora-Sweet and Ora-Plus and shaken for at least an additional minute. The compounded oral liquid suspension should be stored refrigerated at 2 to 8°C. The suspension should be shaken before each use and returned to refrigeration promptly after use. Merck states that the oral liquid preparation can be stored for up to four weeks.

**STUDY 2:** Allen[1369] reported on a similar compounded formulation of losartan potassium 2.5-mg/mL oral liquid. The oral liquid had the following formula:

| | | |
|---|---|---|
| Losartan potassium 50-mg tablets | | 5 tablets |
| Ora-Plus | | 50 mL |
| Ora-Sweet | qs | 100 ml |

The recommended method of preparation was to place five losartan potassium (Cozaar) tablets in a calibrated container and add 50 mL of Ora-Plus. Shake the mixture for at least two minutes, allow to stand for one hour, and then shake for at least an additional minute. The mixture was then to be brought to volume with Ora-Sweet and shaken for at least an additional minute. The oral liquid should be packaged in tight, light-resistant containers. The author referred to the Merck labeling in recommending a beyond-use date of up to four weeks. ∎

# Loxapine Hydrochloride

## Properties

Loxapine hydrochloride injection is a straw-colored solution.[2] Loxapine hydrochloride 28 mg is approximately equivalent to loxapine 25 mg.[3]

### Solubility

Loxapine hydrochloride is slightly soluble in both water and ethanol.[2]

### pH

Loxapine hydrochloride injection and oral solution have a pH of about 5.8.[2]

### pK$_a$

The drug has a pK$_a$ of 6.6.[2]

## General Stability Considerations

Loxapine hydrochloride oral solution should be packaged in well-closed containers. Both the oral solution and the injection should be stored at controlled room temperature and protected from freezing. The injection also should be protected from exposure to light. The injection may darken to a light amber color, but this change does not affect product concentration. However, it should be discarded if it becomes much darker.[2]

## Compatibility with Common Beverages and Foods

Geller et al.[242] reported that loxapine hydrochloride liquid was compatible visually with orange juice, pineapple juice, grapefruit juice, Tang, Kool-Aid, cola sodas, 7-Up, and coffee, although visual observation in opaque liquids is problematic.

## Compatibility with Other Drugs
**LOXAPINE HYDROCHLORIDE COMPATIBILITY SUMMARY**
*Compatible with:* Lithium citrate
*Incompatible with:* Thioridazine

**STUDY 1:** Theesen et al.[78] evaluated the compatibility of lithium citrate syrup with oral neuroleptic solutions, including loxapine hydrochloride concentrate. Lithium citrate syrup (Lithionate-S, Rowell) and lithium citrate syrup (Roxane) at a lithium-ion concentration of 1.6 mEq/mL were combined in volumes of 5 and 10 mL with 3, 6, 9, and 15 mL of loxapine hydrochloride oral concentrate 25 mg/mL (Lederle) in duplicate, with the mixing order reversed between samples. Samples were stored at 4 and 25°C for six hours and evaluated visually for physical compatibility. No incompatibility was observed.

**STUDY 2:** Raleigh[454] evaluated the compatibility of lithium citrate syrup (Philips Roxane) 1.6 mEq lithium/mL with loxapine concentrate 25 mg/mL (as the hydrochloride). The loxapine concentrate was combined with the lithium citrate syrup in a manner described as "simple random dosage combinations"; the exact amounts tested and test conditions were not specified. No precipitate was observed to form upon mixing.

**STUDY 3:** Raleigh[454] also evaluated the compatibility of thioridazine suspension (Sandoz) (concentration unspecified) with loxapine concentrate 25 mg/mL (as the hydrochloride). The loxapine concentrate was combined with the thioridazine suspension in the manner described as "simple random dosage combinations." The exact amounts tested and test conditions were not specified. A dense white curdy precipitate formed.

# Magnesium Sulfate

## Properties

Magnesium sulfate as the heptahydrate is a white crystalline powder or colorless needlelike crystals having a bitter saline taste. The material effloresces in warm dry air.[1-3] Each gram of the heptahydrate contains 8.1 mEq (4.1 mmol) of magnesium. Magnesium sulfate heptahydrate 10.1 g is approximately equivalent to 1 g of magnesium.[2,3]

### Solubility

Magnesium sulfate is very soluble in water, having a solubility variously cited as 0.71 g/mL at 20°C,[1] 0.91 mg/mL at 40°C,[1] and 1.25 g/mL.[3] Aqueous solubility increases to approximately 2 g/mL in boiling water. It is slowly soluble to 1 g/mL in glycerin and slightly soluble in ethanol.[1,3]

### pH

A 5% aqueous solution has a pH of 5 to 9.2. Magnesium sulfate injection has a pH adjusted to 5.5 to 7 when diluted to a 5% concentration. Magnesium sulfate in dextrose injection has a pH of 3.5 to 6.5.[4]

## General Stability Considerations

Magnesium sulfate injections should be stored at controlled room temperature protected from excessive heat and freezing.[2,4] It is incompatible with alkali hydroxides, alkali carbonates, salicylates, tartrates, and phosphates. Some metal ions, including calcium, may react to form insoluble sulfates.[2]

## Stability Reports of Compounded Preparations

### Injection

Injections, like other sterile drugs, should be prepared in a suitable clean air environment using appropriate aseptic procedures. When prepared from nonsterile components, an appropriate and effective sterilization method must be employed.

Sarver et al.[619] reported on the stability of two extemporaneously compounded magnesium sulfate 37-g/L injections for use as dextrose-free alternatives to commercial products. Eighty milliliters of magnesium sulfate injection (American Regent) 0.5 g/mL was added to liter containers of lactated Ringer's injection and to 0.9% sodium chloride injection. Samples were prepared in glass bottles and in polyvinyl chloride (PVC) bags and were stored at room temperature for 91 days.

No signs of gross precipitation were found by visual observation during the three-month storage period. Inductively coupled plasma atomic emission spectroscopy and atomic absorption spectroscopy found little or no change in elemental composition for calcium, magnesium, potassium, sodium, and sulfur and were not different from the control admixture in water. Concentrations all remained within the minimum USP specifications. The authors indicated that magnesium sulfate injection formulations at 37 g/L in lactated Ringer's injection and in 0.9% sodium chloride injection in glass bottles or PVC bags stored at room temperature may be stored for up to three months. ■

# Mebendazole

## Properties

Mebendazole is a white or almost white to slightly yellow powder that is nearly odorless.[3,4]

### Solubility

Mebendazole is soluble in strong acids but practically insoluble in water, dilute acids, and ethanol.[3,4]

### pH

Mebendazole oral suspension has a pH in the range of 6 to 7.[4]

## General Stability Considerations

Mebendazole tablets should be packaged in tight containers, while the oral suspension should be packaged in well-closed containers.[4] The tablets and the oral suspension should be stored at controlled room temperature.[3,4] Ragno et al.[1020] stated that the benzimidazole class of drugs, including mebendazole, are known to be sensitive to light. The European Pharmacopoeia requires protection from light during storage,[3] but the United States Pharmacopeia does not include light protection as a requirement.[4]

## Stability Reports of Compounded Preparations
*Oral*
**STUDY 1:** Ragno et al.[1020] evaluated the stability of mebendazole in the solid state as well as in liquid form at 20 mg/mL in an ethanol and water medium. Exposure of the solid drug to high temperatures up to 50°C and intense light from a Xenon arc lamp did not result in loss of the drug within 10 hours. Similarly, no loss due to thermal exposure resulted in the liquid form. However, exposure of the liquid form to intense light resulted in loss, with about 30% of the mebendazole decomposing in 10 hours. This loss was similar to that for fenbendazole and was less extensive than for albendazole. The authors indicated that they believed water must be present for the photodegradation to occur, because no loss occurred in the solid form.

**STUDY 2:** Agatonnovic-Kustrin et al.[1307] evaluated the purity and stability of mebendazole formulated as a 2% oral suspension. The suspension was formulated in a 10% sucrose solution stabilized with unspecified amounts of cellulose esters and sodium lauryl sulfate and preserved with parabens. In addition, simethicone emulsion was used as a wetting agent along with citric acid and Tutti-Frutti flavoring agents. The emulsion was stored in a well-closed container and stored at controlled room temperature of 23 to 25°C protected from exposure to direct light. Attenuated total reflectance-Fourier transform infrared (ATR-FTIR) spectroscopy was used to assess the purity of mebendazole. Mebendazole was found to exhibit good stability in the suspension for up to 10 days, but more extensive changes occurred after that time. ■

# ■ Mebeverine Hydrochloride

## Properties
Mebeverine hydrochloride exists as a white or almost white crystalline powder.[1,3]

### Solubility
Mebeverine hydrochloride is very soluble in water and freely soluble in ethanol. It is practically insoluble in ether.[3]

### pH
A 2% aqueous solution of mebeverine hydrochloride has a pH from 4.5 to 6.5.[3]

## General Stability Considerations
Mebeverine hydrochloride powder should be packaged in tight containers and stored at controlled room temperature not exceeding 30°C. It should also be protected from exposure to light.[3]

## Stability Reports of Compounded Preparations
*Oral*
Baloglu et al.[1711] evaluated several semisolid oral formulations of mebeverine hydrochloride 20% (200 mg/g) for use as oral mucoadhesive gels in dental procedures. The five oral mucoadhesive gel formulations were as follows:

1. Hydroxypropylmethylcellulose K100M 1.5%
2. Hydroxypropylmethylcellulose K100M 2.0%
3. Hydroxypropylmethylcellulose E50 10%
4. Metolose 2%
5. Hydroxypropylmethylcellulose K100M 2% with Poloxamer 20%

Samples were stored at 25°C and 60% relative humidity, 30°C and 65% relative humidity, and 40°C and 65% relative humidity for three months. No changes in visual appearance, viscosity, texture, or pH occurred in any of the samples under any of the conditions. The concentration of mebeverine hydrochloride was stable at all test conditions throughout the study remaining near 99% of the initial concentration. In vivo testing on rabbits found that formulation 5 had superior mucoadhesive properties and drug release was over a prolonged time period. The authors stated that formulation 5 was the preferred formulation. ■

# ■ Mechlorethamine Hydrochloride
## (Mustine Hydrochloride)

## Properties
Mechlorethamine hydrochloride is a white[3,6] or light yellow-brown[2,7] hygroscopic crystalline powder.[2,3] Contact with this material may cause severe eye and skin irritation and injury. It has been characterized as a carcinogen.[1]

### Solubility
Mechlorethamine hydrochloride is very soluble in water and soluble in ethanol.[1–3,6]

## pH

A 2% aqueous solution has a pH of 3 to 4.[1] A 0.2% solution has a pH of 3 to 5.[4] The injection has a pH of 3 to 5 following reconstitution.[2,4]

## pKa

Mechlorethamine hydrochloride has a $pK_a$ variously cited as 6.1[2] and 6.45.[6]

## General Stability Considerations

Mechlorethamine hydrochloride for injection should be packaged in light-resistant containers and stored at controlled room temperature protected from excessive heat and humidity.[2] The drug is unstable in solution; it should be prepared immediately before use. It is even less stable in neutral and alkaline solutions than in the acidic reconstituted solution.[2,7,8] The drug decomposition rate declines progressively from approximately pH 6 down to pH 1. The maximum decomposition rate occurs at and above the pH range of 7 to 8, corresponding to a half-life of about 29 minutes.[6]

The vials of mechlorethamine hydrochloride injection should not be used if water droplets have formed in the vials before reconstitution. The reconstituted solution should not be used if it is discolored.[2,6–8]

Unused injection solution and empty vials should be neutralized for 45 minutes with an equal volume of a solution containing 5% sodium thiosulfate and 5% sodium bicarbonate before disposal. The same solution should be used to neutralize any spills, such as on gloves, surfaces, and glassware, by soaking for 45 minutes.[2,7]

## Stability Reports of Compounded Preparations
### Topical

Mechlorethamine powder and solution are very irritating to the eyes, skin, and respiratory tract. All preparation steps should be performed in a biological safety cabinet, and personnel should wear adequate protective garb, including gloves and gowns. The drug is unstable in water. Great care should be taken to avoid contact with water during compounding. All equipment should be thoroughly dried before use.[548]

**STUDY 1:** Cummings et al.[495] reported the stability of mechlorethamine hydrochloride in a topical ointment form. Mechlorethamine hydrochloride injection 10 mg (Boots) was dissolved in 1 mL of acetone. This solution was worked into 50 g of white soft paraffin until an even consistency was obtained. The ointment was stored in clear glass jars at 4 and 37°C. HPLC analysis found that the drug was stable for at least 80 days at 4 and one month at 37°C.

**STUDY 2:** Zhang et al.[548] evaluated the stability of mechlorethamine hydrochloride 0.01% in Aquaphor prepared from the injection. Two mechlorethamine hydrochloride 10-mg vials

(Merck) were reconstituted with 1 mL each of dehydrated alcohol (Faulding), yielding 10-mg/mL solutions. Note that the residual unretrievable liquid in the vials necessitates using two vials in order to withdraw one full milliliter of injection. The sodium chloride present in the vials will precipitate. Supernatant liquid 1 mL was withdrawn from the two vials and thoroughly incorporated by spatulation into 100 g of Aquaphor (Beiersdorf) ointment base. The resulting ointment, containing 0.1 mg of mechlorethamine hydrochloride per gram, was packaged in one-ounce plastic ointment jars and stored at 23°C exposed to fluorescent light. Stability-indicating HPLC analysis found that mechlorethamine hydrochloride losses of 10% occurred in as little as seven days.

**STUDY 3:** Ritschel et al.[1311] evaluated the stability of several novel formulations of mechlorethamine hydrochloride 0.02% for topical use. Commercial Mustargen injection (Ovation) 10 mg was used as the source of drug. The vehicles evaluated included Aquaphilic ointment (Medco Labs), Aquaphor ointment (Beiersdorf), Transcutol HP (Gattefosse), and Labrasol (Gattefosse) along with Transcutol 10% in Aquaphor and 10% Transcutol in Labrasol with and without 0.1% butylhydroxytoluene (BHT). NOTE: Labrasol and Transcutol are liquids. The specific formulations are shown in Table 60, and the preparation instructions follow. Stability-indicating HPLC analysis was used to determine drug concentrations.

In Aquaphilic ointment with and without BHT, mechlorethamine hydrochloride losses were extensive and much more rapid than in the other formulations. Over 10% loss occurred in two to three hours, and over 80% loss occurred in 24 hours. The authors stated that Aquaphilic ointment is not a suitable medium for mechlorethamine hydrochloride ointment.

In the Transcutol and Labrasol formulations, drug decomposition proceeded at a higher rate than in Aquaphor. Drug losses of 10% occurred in about two weeks. Adding BHT to the formulations slowed the rate of drug loss with

**TABLE 60.** Mechlorethamine 0.02% Topical Formulations Tested for Stability[1311]

| |
| --- |
| Aquaphilic ointment |
| Aquaphor ointment |
| Aquaphor ointment with BHT |
| Aquaphor ointment with Transcutol 10% |
| Aquaphor ointment with Transcutol 10% and BHT |
| Labrasol |
| Labrasol with BHT |
| Labrasol with Transcutol 10% |
| Labrasol with Transcutol 10% and BHT |
| Transcutol |
| Transcutol with BHT |

10% loss in about four weeks, but it was still much more rapid than in Aquaphor.

In Aquaphor ointment, drug losses were slower. BHT had no effect on drug stability. Drug losses were about 6% in 30 days, 15% in 90 days, and 22% in 288 days. The authors indicated that the use of containers with zero dead space and a higher drug concentration may have resulted in a slower rate of drug loss than was reported in the study by Zhang et al.[548]

The best stability was found in Aquaphor ointment with Transcutol 10% and BHT. Drug losses of only 2% and 7% occurred in 118 days and 288 days, respectively.

To prepare the Aquaphilic ointment samples, mechlorethamine hydrochloride injection 10 mg was reconstituted with 1 mL of absolute alcohol and the liquid withdrawn (leaving the sodium chloride precipitate). This liquid was incorporated into the Aquaphilic ointment in a mortar with mixing. BHT was added to a portion of the formulation. Due to the rapid rate of drug decomposition, the samples were packaged in individual reaction tubes and stirred continuously. The ointment samples were stored at room temperature of about 23°C and protected from exposure to light for up to 49 hours.

To prepare the Aquaphor samples, the ointment base was melted and cooled back to room temperature to facilitate mixing. Mechlorethamine hydrochloride injection 10 mg was reconstituted with 1 mL of absolute alcohol and the liquid withdrawn (leaving the sodium chloride precipitate). This liquid was incorporated into the Aquaphor in a mortar with mixing. BHT was added to a portion of the formulation. The ointment samples were packaged in unguator jars (polyethylene/polypropylene jars with zero dead space) and were stored at room temperature of about 23°C protected from exposure to light for 288 days.

To prepare the Aquaphor with Transcutol 10% samples, the Aquaphor was melted in a beaker. Transcutol was added and mixed thoroughly. The mixed vehicle was cooled to room temperature. Mechlorethamine injection was prepared and incorporated as described previously. BHT was added to a portion of the formulation. The ointment samples were packaged in unguator jars and were stored at room temperature of about 23°C protected from exposure to light for 288 days.

To prepare the Labrasol and Transcutol samples, the respective liquid vehicle was injected into the vial of mechlorethamine hydrochloride with swirling to dissolve the drug. The liquid was transferred to a tared bottle and diluted to volume with additional vehicle. BHT was added to a portion of the formulation. The topical liquid samples were packaged in glass vials with PFTE-lined screw caps and were stored at room temperature of about 23°C protected from exposure to light for 288 days.

To prepare Labrasol with Transcutol 10%, the mechlorethamine hydrochloride was reconstituted with 5 g of Transcutol with swirling to dissolve the drug. The samples were diluted to 50 g with Labrasol with thorough mixing. BHT was added to a portion of the formulation. The topical liquid samples were packaged in glass vials with PFTE-lined screw caps and were stored at room temperature of about 23°C protected from exposure to light for 288 days. ■

# Medium Chain Triglycerides
## (MCT)

## Properties
Medium chain triglycerides are colorless to pale yellow oily liquids from the lipid fraction of coconut oil. The mixture consists largely of triglycerides of $C_8$ (67%) and $C_{10}$ (23%) saturated fatty acids along with small amounts (<5%) of saturated fatty acids that are shorter than $C_8$ and longer than $C_{10}$.[3,5]

### Solubility
Medium chain triglycerides are practically insoluble in water but are miscible with ethanol and fatty oils.[3]

## General Stability Considerations
Medium chain triglycerides oil should be stored in a cool dry place. It may soften or break certain kinds of plastic containers, utensils, and implements. Consequently, nonplastic containers, utensils, and implements are recommended for use with the oil.

Medium chain triglycerides oil may be mixed with fruit juices or sauces and can be used with various food products such as salads, vegetables, and meats.

## Stability Reports of Compounded Preparations
*Enteral*
**MEDIUM CHAIN TRIGLYCERIDES COMPATIBILITY SUMMARY**
*Incompatible with:* Ensure • Ensure HN • Ensure Plus • Ensure Plus HN • Osmolite • Osmolite HN • Precitene • Vital

**STUDY 1:** Cutie et al.[19] added 10 mL of medium chain triglycerides (MCT oil, Mead Johnson) to varying amounts (15 to 240 mL) of Ensure, Ensure Plus, and Osmolite (Ross Laboratories) with vigorous agitation. The oil was physically incompatible, being immiscible with the three enteral feeds studied.

**STUDY 2:** Altman and Cutie[850] reported the physical compatibility of medium chain triglycerides (MCT oil, Mead Johnson)

10 mL with varying amounts (15 to 240 mL) of Ensure HN, Ensure Plus HN, Osmolite HN, and Vital, with vigorous agitation to ensure thorough mixing. The oil was physically incompatible, being immiscible with the four enteral feeds studied.

**STUDY 3:** Ortega de la Cruz et al.[1101] reported the physical incompatibility of an unspecified amount of medium chain triglycerides (MCT, Mead Johnson) with 200 mL of Precitene (Novartis) enteral nutrition diet. Phase separation was observed.

**STUDY 4 (PUMP DAMAGE):** Phelps et al.[258] reported damage to Valleylab model 5000b and 6006 pumps from enteral formulas containing medium chain triglycerides. Visible cracks formed in the cassette within 30 minutes of formula contact. Some cracks became full-thickness and resulted in formula leakage. Administration of the formula was disrupted even in the absence of leakage. The authors noted that MCT oil is a known plasticizer and that polystyrene, foamed polystyrene, and high-impact styrene have been reported to undergo significant deterioration in contact with it. The Valleylab pump cassettes contain a styrene-related plastic. The authors concluded it was likely that the MCT oil produced the deterioration and cracking of the cassettes. ■

# ■ Melatonin

## Properties
Melatonin is pale yellow in color.[1]

### Solubility
Melatonin is slightly soluble in water with a reported solubility of 0.1 mg/mL. It is soluble in ethanol to concentrations variously reported as 8 and 50 mg/mL. If first dissolved in ethanol or propylene glycol, melatonin may remain in solution if diluted with an aqueous solution such as Ringer's solution.[1351]

### pH
The apparent pH of a melatonin 10-mg/mL sublingual solution containing glycerin and ethanol was about pH 3.2.[1353] See *Stability Reports of Compounded Formulations* below.

### pK$_a$
Melatonin has a pK$_a$ of 4.4.[1352]

## General Stability Considerations
Melatonin powder is stated by one supplier to be stored at −20°C. Melatonin in solution is light sensitive and subject to oxidation. Melatonin in a water:isopropanol (70:30) mixture stored at 4°C demonstrated only one spot on TLC evaluation after two weeks of storage.[1351]

## Stability Reports of Compounded Preparations
### Oral
**STUDY 1:** Haywood et al.[1353] evaluated the stability of melatonin extemporaneously compounded as a 10-mg/mL sublingual solution. The sublingual solution had the following composition:

| | | |
|---|---|---|
| Melatonin | | 1000 mg |
| Stevia powder extract | | 250 mg |
| Ethanol 95% | | 15 mL |
| Tutti-Frutti flavor | | 2 mL |
| Glycerin | qs | 100 mL |

The melatonin and stevia powder extract were thoroughly mixed using a mortar and pestle. The powder blend was transferred to a graduated vessel. Ethanol 95% 7.5 mL was used to rinse the mortar and pestle, and the rinse was added to the powder blend. An additional 7.5 mL of ethanol 95% was then added and the mixture was stirred until complete dissolution occurred. Tutti-Frutti flavor 2 mL was added and sufficient glycerin was added to bring to a volume of 100 mL, and the solution was thoroughly mixed. The finished solution was packaged in amber glass prescription bottles and stored refrigerated at 3 to 5°C and at room temperature of 24 to 26°C and 60% relative humidity for 90 days. The samples were protected from exposure to light.

No detectable changes to color, odor, and taste were observed. HPLC analysis of drug concentrations found little or no melatonin loss occurred over 90 days at either storage condition.

**STUDY 2:** Haywood et al.[1353] also evaluated the stability of melatonin extemporaneously compounded as 3-mg oral hard gelatin capsules. The capsules had the following composition:

| | |
|---|---|
| Melatonin | 300 mg |
| Food color powder | 10 mg |
| Methocel E4M | 10 g |
| Lactose anhydrous | 22 g |
| Capsules, size 1 | 100 capsules |

Small amounts of lactose and food color powder were mixed well using a mortar and pestle. The remaining powders were added to the mortar using geometric dilution and thoroughly mixed, ensuring that the particles were reduced to a similar size. The powder mix was filled into size 1 hard gelatin capsules. The capsules were packaged in an amber glass prescription bottle and stored at room temperature of 24 to 26°C and 60% relative humidity and at stress conditions of 39 to 41°C and 75% relative humidity for 90 days. The samples were protected from exposure to light.

No detectable changes to capsule consistency (hardening or softening), clumping of capsules, or other changes were observed. No significant change in weight variation occurred as well. HPLC analysis of drug concentrations found little or no melatonin loss occurred over 90 days at either storage condition.

**STUDY 3:** Johnson et al.[1648] evaluated the stability of four alcohol-free oral liquid formulations of melatonin 1 mg/mL. The oral suspensions were prepared from commercial 3-mg melatonin oral tablets (CVS Pharmacy) triturated to fine powder and suspended in an equal parts mixture (1:1) of Ora-Plus with either Ora-Sweet or Ora-Sweet SF. A second set of test formulations was prepared from commercial oral tablets containing melatonin 3 mg and pyridoxine hydrochloride 10 mg (Natrol) triturated to fine powder and suspended in an equal parts mixture (1:1) of Ora-Plus with either Ora-Sweet or Ora-Sweet SF. The oral suspensions were packaged in amber plastic prescription bottles and stored at room temperature of 23 to 25°C for 90 days.

The oral suspensions exhibited only minor changes in odor, color, or taste during the test period. Stability-indicating HPLC analysis of melatonin concentrations found that not more than 5% loss of melatonin occurred in any of the formulations during 90 days of room temperature storage. ∎

# Meloxicam

## Properties
Meloxicam exists as a pale yellow crystalline powder.[1,3]

### Solubility
Meloxicam is practically insoluble in water, very slightly soluble in ethanol, slightly soluble in acetone, and soluble in dimethylformamide.[3,1504]

Bachhav and Patravale[1588] reported the solubility of meloxicam in several common solvents. The reported meloxicam solubility was highest in n-methyl pyrrolidone at 135 mg/mL and lowest in propylene glycol at 0.5 mg/mL. In Solutol HS 15 (polyoxyethylene esters of 12-hydroxystearic acid), Transcutol P (diethylene glycol monoethyl ether), and Tween 80, the reported solubilities were 9.5, 5.4, and 5.0 mg/mL, respectively.

### $pK_a$
Meloxicam has a $pK_a$ of 4.08 in water and a $pK_a$ of 4.24 in water–ethanol (1:1).[1]

## General Stability Considerations
Meloxicam oral tablets are packaged in blister packs or other tight containers and stored at controlled room temperature protected from moisture. Meloxicam suppositories and liquid suspension for veterinary use are also stored at controlled room temperature.[7]

## Stability Reports of Compounded Preparations
### Oral
Hawkins et al.[1052] evaluated the stability of meloxicam 0.25, 0.5, and 1 mg/mL in extemporaneously compounded oral liquid formulations for use in small exotic pets. Meloxicam (Metacam, Boehringer Ingelheim Vetmedica) was compounded using three diluents individually: deionized water, methylcellulose gel 1%, and a 1:1 mixture of methylcellulose gel and simple syrup. The oral liquids were packaged in amber glass prescription bottles and stored at room temperature of 20 to 24°C and under refrigeration at 3 to 5°C for 28 days.

Meloxicam at all three concentrations in deionized water yielded acceptable suspensions. During storage, the suspensions settled into two layers that were easily redispersed with shaking to apparent uniformity. HPLC analysis found little or no loss of meloxicam in 28 days at both storage temperatures. One assay of the 0.25-mg/mL concentration was low (80%), but that appears to be most likely analytical variation rather than decomposition.

Meloxicam in methylcellulose 1% yielded acceptable suspensions at the 0.25- and 0.5-mg/mL concentrations, but the 1-mg/mL concentration was foamy and had flocculence evident even after shaking. HPLC analysis found little or no loss of drug in 28 days in the 0.25- and 0.5-mg/mL concentrations. However, the 1-mg/mL concentration exhibited erratic drug concentrations as low as 50% of the target concentration.

Meloxicam in the methylcellulose–simple syrup mixture yielded an unacceptable foamy mixture with poor drug content and uniformity.

### Topical
Bachhav and Patravale[1588] developed a topical gel formulation of meloxicam 1% and evaluated its stability. Of several formulations attempted, the best formulation was meloxicam 1% in n-methyl pyrrolidone 10%, Carbopol Ultrez 10 0.75%, triethanolamine 1%, and water for the formula balance. (All percentages are mass/mass.) The topical gel was prepared by dissolving the meloxicam in the n-methyl pyrrolidone. Carbopol Ultrez 10 was dispersed in water using continuous stirring at 1000 rpm until homogeneous. The Carbopol Ultrez 10 dispersion was added to the meloxicam solution with continuous stirring until homogeneous. The mixture was then neutralized by adding the triethanolamine to obtain a transparent gel.

The topical gel demonstrated good viscosity and spreadability. The pH was 7.35 and was physiologically acceptable.

No irritation was found when tested on rabbits. A high degree of skin permeation was demonstrated. The anti-inflammatory activity in rats was found to be twofold higher than commercial piroxicam gel.

Stability of the topical gel was evaluated at 25°C and 60% relative humidity, 30°C and 60% relative humidity, and an accelerated condition of 40°C and 75% relative humidity over six months of storage. HPLC analysis of meloxicam content found the maximum loss over six months was about 4% at the most extreme condition of 40°C and 75% relative humidity. Viscosity, spreadability, and pH remained acceptable throughout. ■

## ■ Melphalan

### Properties
Melphalan is a white or off-white to buff-colored powder that is odorless or has a faint odor.[2,3] Melphalan has been cited as a known carcinogen.[1]

### Solubility
Melphalan is practically insoluble in water and slightly soluble in ethanol and propylene glycol.[1-3]

### pH
Melphalan injection (as the hydrochloride) has a pH of around 7.[2]

### pK$_a$
The drug has a pK$_a$ of 2.5.[2]

### General Stability Considerations
Melphalan tablets should be packaged in well-closed, light-resistant glass containers. The tablets and the injection should be stored at controlled room temperature and protected from exposure to temperatures of 40°C or above and from exposure to light.[2,4]

The reconstituted injection may precipitate if refrigerated.[2,9] Decomposition of melphalan is relatively rapid after reconstitution. The degradation products are less water soluble than melphalan, and a precipitate may form on standing.[9] After dilution in 0.9% sodium chloride injection, melphalan hydrolysis occurs at a rate of almost 1% every 10 minutes.

The manufacturer recommends completing administration within one hour of reconstitution,[2,7,9,380] although others have recommended a maximum of two hours.[9] The presence of chloride ions in the solution reduces the rate of melphalan hydrolysis.[9]

Melphalan exhibits greatest stability in solution at a pH around 3, with a half-life of 5.3 hours; the stability is reduced at pH 5 to 7 and is greatly reduced at pH 9, to a half-life of 3.9 hours.[9]

### Stability Reports of Compounded Preparations
#### Oral
The stability of melphalan 2 mg/mL in an extemporaneous oral suspension was studied by Dressman and Poust.[87] Melphalan 2-mg tablets (Burroughs-Wellcome) were crushed and mixed with Cologel (Lilly) as one-third of the total volume, shaken, and brought to final volume with a 2:1 mixture of simple syrup and cherry syrup, followed by vigorous shaking for 30 seconds and ultrasonication for at least two minutes. The melphalan 2-mg/mL suspension was packaged in amber glass bottles and stored at 5°C and ambient room temperature. An HPLC assay was used to determine the melphalan content of the suspensions. Melphalan was extremely unstable in the suspension, with 15% loss at the initial assay and 80% loss in 24 hours at room temperature. Even under refrigeration, losses of 50% occurred in seven days. The authors concluded that this formulation was unsuitable. ■

## ■ Memantine Hydrochloride

### Properties
Memantine hydrochloride is a white to off-white crystalline powder having a bitter taste.[7]

### Solubility
Memantine hydrochloride is soluble in water to about 35 mg/mL.[7,1012]

### pH
Memantine hydrochloride as an aqueous solution has a pH of 3.5 to 4.5.[1013]

### pK$_a$
Memantine hydrochloride has a pK$_a$ of 10.27.[1012]

## General Stability Considerations

The commercial memantine hydrochloride oral tablets and oral solution are to be stored at controlled room temperature.[7]

## Stability Reports of Compounded Preparations

Yamreudeewong et al.[1014] reported the stability of an extemporaneously compounded oral liquid of memantine hydrochloride. Commercial memantine hydrochloride tablets were triturated to fine powder and incorporated into deionized water to yield a concentration of 0.166 mg/mL. The oral liquid was packaged in glass prescription bottles and stored at 2 and 25°C for 28 days.

Gas chromatography–mass spectrometry was used to evaluate memantine hydrochloride stability. However, the room temperature results exhibited a substantial degree of variability, making interpretation of the drug's room temperature stability difficult. Refrigerated samples were more consistent, indicating little or no drug loss within 28 days. ■

# Menadione
## (Vitamin K3)

## Properties

Menadione occurs as a bright yellow crystalline material that is odorless or has a very faint acrid odor.[1,4]

### Solubility

Menadione is insoluble in water. About 1 g of material will dissolve in 60 mL of ethanol and in 50 mL of vegetable oils.[1,4]

### pH

Alcoholic solutions of menadione are neutral to litmus.[1]

## General Stability Considerations

Menadione is stable in air and may be heated to 120°C without decomposition.[1] Decomposition will occur if exposed to sunlight; menadione should be protected from exposure to light.[1,4] Menadione is also destroyed by alkalies and reducing agents.[1]

Menadione bulk material should be packaged in well-closed light-resistant containers. It may be stored at controlled room temperature.[4]

Huyghebaert et al.[1233] evaluated the stability of menadione bulk powder after the original container had been opened. After the original container was opened, packages were stored, without applying an inert gas layer, for eight months at both room temperature and under refrigeration at 8°C. Stability-indicating HPLC analysis found that the bulk substance remained within the specifications of 95 to 105% for drug content for the entire eight-month period, which is within the margin of error for the analytical method.

## Stability Reports of Compounded Preparations
### Oral

Huyghebaert et al.[1233] evaluated the stability of menadione prepared as oral capsules containing 2 mg of menadione. The vitamin powder was mixed with lactose MH very fine using a mortar and pestle until a uniform mixture was obtained. The powder mixture was filled into size 2 hard-gelatin capsules that were then stored at room temperature and refrigerated at 8°C. Stability-indicating HPLC analysis found that the menadione content in the capsules remained within the specifications of 90 to 110% for drug content for the entire two-month period at both storage temperatures. ■

# Meperidine Hydrochloride
## (Pethidine Hydrochloride)

## Properties

Meperidine hydrochloride is a fine white odorless crystalline powder with a slightly bitter taste.[2,3]

### Solubility

Meperidine hydrochloride is very soluble in water, with a solubility of about 1 g/mL. It is also soluble in ethanol to about 400 mg/mL.[1–3,6]

### pH

A 5% aqueous solution has a pH of about 5.[3] Meperidine hydrochloride injection has a pH of 3.5 to 6. Meperidine hydrochloride syrup has a pH of 3.5 to 4.1.[4]

### pKa

Meperidine hydrochloride has a pKa of 7.7 to 8.15.[6]

## General Stability Considerations

Meperidine hydrochloride products should be packaged in tight (syrup) or well-closed (tablets) light-resistant containers and stored at controlled room temperature protected from excessive heat and light; liquid products should be protected from freezing.[2,4]

Although meperidine hydrochloride is very stable in aqueous solution, it does undergo relatively slow pH-dependent hydrolysis. Maximum stability occurs at about pH 4 to 5, with increasing rates of decomposition as the pH becomes more acidic or more alkaline. The calculated shelf life ($t_{90}$) of meperidine hydrochloride at pH 2 was 246 days; at pH 6.7 it was 763 days.[6]

## Stability Reports of Compounded Preparations
### Injection

Injections, like other sterile drugs, should be prepared in a suitable clean air environment using appropriate aseptic procedures. When prepared from nonsterile components, an appropriate and effective sterilization method must be employed.

**STUDY 1:** Donnelly and Bushfield[589] reported the stability of meperidine hydrochloride in various concentrations repackaged in syringes for patient-controlled analgesia. Preservative-free meperidine hydrochloride injection (Abbott) was diluted to concentrations of 0.25, 1, 10, 20, and 30 mg/mL in 5% dextrose injection (Abbott) and in 0.9% sodium chloride injection (Abbott). The solutions were repackaged in 60-mL polypropylene syringes (Becton Dickinson) and sealed with Luer-lock tip caps (Combi-Caps EPS, The Medi-Dose Group). Samples were stored at 4 and 22°C protected from light for 28 days. Visual inspection showed that the solutions remained clear and free of precipitation. Stability-indicating HPLC analysis found no loss of meperidine hydrochloride in any of the samples.

**STUDY 2:** Strong et al.[644] evaluated the stability of meperidine hydrochloride 5 and 10 mg/mL in 5% dextrose and 0.9% sodium chloride injections in 30-mL plastic syringes sealed with tip caps for use in patient-controlled analgesia. HPLC analysis found that both concentrations were stable for at least 12 weeks at 23°C exposed to and protected from light, at 4°C protected from light, and at −20°C protected from light.

# Mercaptopurine

## Properties

Mercaptopurine is a yellow odorless or nearly odorless crystalline powder.[2,3,6]

### Solubility

Mercaptopurine is insoluble in water, slightly soluble in ethanol (about 1 mg/mL), and soluble in hot ethanol.[1–3,6]

### $pK_a$

Mercaptopurine has $pK_a$ values of 7.6[2] to 7.77[1,6] and 11.17.[1,6]

## General Stability Considerations

Mercaptopurine tablets should be packaged in well-closed containers and stored at controlled room temperature protected from excessive temperatures of 40°C or above.[2,4]

Mercaptopurine is subject to oxidation reactions at room temperature under alkaline conditions.[6] Exposure to light can enhance oxidation.[1223] Maximum stability occurs below pH 8. Stability of the tablet formulation is promoted by minimizing contact with water and light.[6]

## Stability Reports of Compounded Preparations
### Oral

**STUDY 1 (ORAL LIQUID):** The stability of mercaptopurine 50 mg/mL in an extemporaneous oral suspension was studied by Dressman and Poust.[87] Mercaptopurine 50-mg tablets (Burroughs-Wellcome) were crushed and mixed with Cologel (Lilly) as one-third of the total volume, shaken, and brought to final volume with a 2:1 mixture of simple syrup and cherry syrup. Dilution was followed by vigorous shaking for 30 seconds and ultrasonication for at least two minutes. The mercaptopurine 50-mg/mL suspension was packaged in amber glass bottles and stored at 5°C and at ambient room temperature. An HPLC assay was used to determine the mercaptopurine content of the suspensions. Less than 5% loss occurred in 14 days at room temperature. However, the suspension tended to cake, and over 10% loss occurred at both temperatures in 84 days. The authors recommended room temperature storage and a 14-day expiration.

**STUDY 2 (ORAL LIQUID):** Woods and Simonsen[662] recommended an alternative to preparing an oral suspension for dosing children. They suggested removing the plunger from an oral liquid syringe, placing the mercaptopurine dose as 10-mg oral tablets in the barrel, replacing the plunger against the tablets, and capping the syringe. At the time the dose is to be given, about 2 mL of water is drawn into the syringe, and the tablets are allowed to disintegrate over one to three minutes. The dose can be given directly or added to juice.

**STUDY 3 (ORAL LIQUID):** Aliabadi et al.[1223] evaluated the stability of the following four mercaptopurine 50-mg/mL oral suspensions prepared from commercial mercaptopurine 50-mg tablets (Purinethol, Novopharm). (1) The small initial sample suspension formulation was prepared by triturating 10 mercaptopurine tablets and adding 1.7 mL of sterile water for irrigation, 3.3 mL of simple syrup, and cherry syrup to bring the volume of the sample to 10 mL for an initial formulation. The initial formulation was compared to similar formulations that also contained (2) ascorbic acid 10 mg as an antioxidant, (3) sodium phosphate monobasic monohydrate 500 mg as a buffer, and (4) with both the antioxidant and the buffer. The sample formulations were prepared in triplicate and packaged in amber glass bottles stored at room temperature of 19 to 23°C or refrigerated at 4 to 8°C. The samples were evaluated for physical appearance, including color and resuspendability. Stability-indicating HPLC analysis was used to determine chemical stability.

No significant change in appearance was observed in the samples stored at room temperature. However, the refrigerated samples exhibited caking. In addition, a greater rate of drug loss occurred in some of the refrigerated samples. HPLC analysis found that formulation 2, which contained the ascorbic acid antioxidant, exhibited the best stability, reaching the USP limit of 93% in 11 weeks at room temperature and 10 weeks refrigerated. Formulation 1, with no antioxidant or buffer, decomposed to a concentration lower than 93% in five weeks at room temperature and at two weeks refrigerated. Formulation 3, with only the buffer present, exhibited more than 7% loss in four weeks at both temperatures. Formulation 4, with both antioxidant and buffer, exhibited more than 7% loss in nine weeks at both temperatures. The authors recommended the use of formulation 2, with ascorbic acid antioxidant, along with room temperature storage to maximize mercaptopurine stability.

**STUDY 4 (ORAL CAPSULES):** Gambier et al.[974] evaluated the stability of a pediatric oral capsule formulation compounded from commercial oral tablets. Commercial Purinethol tablets were crushed and ground to powder in a mortar and pestle. Lactose was thoroughly mixed in as a diluent. The powder was filled into capsules that contained nominal amounts of drug of 15, 30, and 45 mg. A stability-indicating capillary electrophoresis analytical method was used to determine mercaptopurine concentrations. The capsules were stored at room temperature protected from exposure to light. Less than 10% loss of mercaptopurine occurred, and no detectable decomposition products appeared over 12 months of storage. ■

# ■ Mesalamine
## (Mesalazine)

## Properties

Mesalamine occurs variously as a white, pinkish, grayish, or light tan crystalline powder or needle-shaped crystals. The drug has a slight but characteristic odor.[1–3]

### Solubility

Mesalamine has an aqueous solubility of about 1 mg/mL at 20°C and is also slightly soluble in ethanol.[1–3] It is also soluble in dilute hydrochloric acid and dilute alkali hydroxides.[3]

### pH

A 2.5% mesalamine suspension has a pH of 3.5 to 4.5. Mesalamine rectal suspension has a pH of 3.5 to 5.5 diluted 1 to 10 with water.[4]

### pKa

The drug has pKa values of 3, 6, and 13.9.[2]

## General Stability Considerations

Mesalamine suppositories should be stored between 19 and 26°C.[2]

Mesalamine is unstable exposed to water and light due to oxidation and light-catalyzed degradation.[2] The manufacturer has stated that mesalamine tablets should not be removed from the original packaging and placed in dosing compliance aids because the coating is light and moisture sensitive.[1622] Mesalamine suspension should be stored at controlled room temperature in the unopened foil package. Storage after removal of the foil may result in the off-white to tan suspension darkening. While slight darkening does not affect the product adversely, the manufacturer recommends that dark brown suspensions be discarded.[2,7]

## Stability Reports of Compounded Preparations
### Rectal

**STUDY 1:** The stability of an extemporaneously prepared enema suspension was reported by Montgomery et al.[90] The suspension formula is shown in Table 61. It was prepared by adding methylparaben and propylparaben to 100 mL of propylene glycol and stirring until dissolved. Approximately 3 liters of distilled water was placed in a 4-liter blender and all ingredients, including the parabens solution, were then added. The mixture was blended at low speed for one or two minutes and brought to 4 liters with distilled water. Samples of the suspension were packaged in three-ounce amber glass bottles and then stored at 5 and 25°C for 90 days. The stability of mesalamine was assessed using a stability-indicating HPLC assay. Less than 10% loss occurred over 90 days at both temperatures.

**STUDY 2:** Henderson et al.[197] evaluated the stability of a 1:1 dilution of mesalamine enema suspension with distilled water. The commercial enema suspension (Rowasa 67 mg/mL,

**TABLE 61.** Mesalamine Suspension 100-mg/mL Enema Formulation[90]

| Component | Amount |
|---|---|
| Mesalamine (95%) | 168 g |
| Sodium phosphate dibasic (anhydrous) | 1.6 g |
| Sodium phosphate monobasic (anhydrous) | 17.9 g |
| Sodium chloride | 36 g |
| Sodium ascorbate | 2 g |
| Tragacanth | 16 g |
| Methylparaben | 8 g |
| Propylparaben | 2 g |
| Propylene glycol | 100 mL |
| Distilled water | qs 4000 mL |

Solvay) was diluted with an equal quantity of distilled water to yield a mesalamine concentration of 33 mg/mL. The diluted enema was stored in empty, clean enema containers that had previously contained the undiluted mesalamine enema suspension and was stored at 23°C for 24 hours. A stability-indicating HPLC technique was used to assess mesalamine content. Little loss occurred during the 24-hour study (mesalamine losses of 5% or less). No appreciable change occurred in pH or color.

**STUDY 3:** However, Wiita and Demestihas[504] questioned the therapeutic advisability of using the dilution evaluated by Henderson et al.[197] They noted that dilution of the rectal suspension reduces viscosity and affects the uniformity of the delivered dose. It could also alter flow of the enema to diseased areas and affect the retention time. Dilution potentially could increase systemic absorption and, therefore, adverse effects. In reply,[505] it was noted that such dilutions are often prescribed by gastroenterologists because they have been found to be effective.[506,507] Because the lower concentration is not commercially available, patients are instructed by their physicians to make their own dilutions or have their pharmacists extemporaneously compound them. ■

# Mesna

## Properties
Mesna is a synthetic sulfhydryl compound that occurs as a hygroscopic powder.[2,7]

### Solubility
Mesna is freely soluble in water and sparingly soluble in organic solvents.[1,2]

### pH
Mesna injection has a pH adjusted to 6.5 to 8.5.[7]

### Osmolality
Mesna 100-mg/mL solution has an osmolality of about 1563 mOsm/kg.[900]

## General Stability Considerations
Mesna injection should be stored at controlled room temperature.[2] The solution is clear and colorless and is not light sensitive.[8] When exposed to oxygen, mesna oxidizes to the disulfide form. Solution remaining in opened ampules should be discarded after dose preparation. However, the multiple-dose vials may be used for up to eight days after initial entry. Diluted solutions are stated to be stable for 24 hours at 25°C.[7]

## Stability Reports of Compounded Preparations
### Injection
Injections, like other sterile drugs, should be prepared in a suitable clean air environment using appropriate aseptic procedures. When prepared from nonsterile components, an appropriate and effective sterilization method must be employed.

Goren et al.[609] repackaged 10 mL of mesna 100-mg/mL injection (Mesnex, Asta Pharma) into 20-mL polypropylene syringes (Becton Dickinson). Samples with no air in the syringes were stored at 5, 24, and 35°C for nine days; samples with an equal volume of air drawn into the syringes were stored at 24°C. Colorimetric analysis for thiols and disulfides found little or no change in the mesna concentration throughout the nine-day study period at all three temperatures in the syringes having no air drawn in. The largest decrease occurred in the 35°C samples but was less than 4% after nine days. However, the syringes with air present underwent 10% mesna loss in eight days at 24°C. Consequently, the authors recommended minimizing the exposure of mesna to air in order to slow its conversion to dimesna.

However, storage of mesna injection in syringes has been reported to result in formation of dark or threadlike particles and viscosity changes after 12 hours.[8]

## Compatibility with Common Beverages and Foods
Goren et al.[609] evaluated the stability of mesna injection (Mesnex, Asta Pharma) diluted to concentrations of 50 and 20 mg/mL in orange and grape syrups and of 50, 10, and 1 mg/mL in various beverages used to help mask the sulfur-like taste. Colorimetric analysis for thiols and disulfides found little or no change in the mesna concentration in orange or grape syrup after seven days stored at 24°C. Similarly, little or

no change in mesna concentrations was found in Coca-Cola, Dr. Pepper, Sprite, 7-Up, Pepsi Cola, apple juice, orange juice, ginger ale, whole milk, and chocolate milk in 24 hours stored at 5°C. However, at the lowest mesna concentration of 1 mg/ mL in milk or chocolate milk, losses of 8 to 9% occurred in 24 hours, with commensurate increases in the disulfide form dimesna. The authors of this report recommended that low concentrations of mesna in milk not be stored. ■

# Mesoridazine Besylate

## Properties
Mesoridazine besylate is a white to pale yellow almost odorless powder.[2,3]

### Solubility
Mesoridazine besylate is very soluble in water, having an aqueous solubility of 1 g/mL at 25°C. It has a solubility in ethanol of 90.9 mg/mL at 25°C.[2,3]

### pH
A freshly prepared 1% mesoridazine besylate aqueous solution has a pH between 4.2 and 5.7.[4] The injection has a pH between 4 and 5.[2,4]

## General Stability Considerations
Mesoridazine besylate tablets should be packaged in well-closed, light-resistant containers and stored at controlled room temperatures protected from temperatures of 40°C and above. The oral solution should be packaged in tight, light-resistant glass containers and stored at less than 25°C. The injection should be stored at less than 25°C. Both the injection and the oral solution should be protected from exposure to light and from freezing.[2,4]

## Compatibility with Other Drugs
**MESORIDAZINE BESYLATE COMPATIBILITY SUMMARY**
*Compatible with:* Lithium citrate
*Incompatible with:* Thioridazine

**STUDY 1:** Theesen et al.[78] evaluated the compatibility of lithium citrate syrup with oral neuroleptic solutions, including mesoridazine besylate concentrate. Lithium citrate syrup (Lithionate-S, Rowell) and lithium citrate syrup (Roxane) at a lithium-ion concentration of 1.6 mEq/mL were combined in volumes of 5 and 10 mL with 5, 10, 15, and 20 mL of mesoridazine besylate oral concentrate 25 mg/mL (Boehringer Ingelheim) in duplicate, with the order of mixing reversed between samples. Samples were stored at 4 and 25°C for six hours and visually evaluated for physical compatibility. No incompatibility was observed.

**STUDY 2:** Raleigh[454] evaluated the compatibility of lithium citrate syrup (Philips Roxane) 1.6 mEq lithium/mL with mesoridazine concentrate 25 mg/mL (as the besylate). The mesoridazine concentrate was combined with the lithium citrate syrup in a manner described as "simple random dosage combinations"; the exact amounts tested and test conditions were not specified. No precipitate was observed to form upon mixing.

**STUDY 3:** Raleigh[454] also evaluated the compatibility of thioridazine suspension (Sandoz) (concentration unspecified) with mesoridazine concentrate 25 mg/mL (as the besylate). The mesoridazine concentrate was combined with the thioridazine suspension in the manner described as "simple random dosage combinations." The exact amounts tested and test conditions were not specified. A dense white curdy precipitate formed.

## Compatibility with Common Beverages and Foods
Geller et al.[242] reported that mesoridazine liquid was visually compatible with water, orange juice, grape juice, cranberry juice, and grapefruit juice, although visual observation in opaque liquids is problematic. ■

# Metaproterenol Sulfate
## (Orciprenaline Sulphate)

## Properties
Metaproterenol sulfate is a white to off-white nearly odorless hygroscopic crystalline powder having a bitter taste.[2,3] Metaproterenol sulfate is available in commercial products as a racemic mixture of two isomers.[2]

### Solubility
Metaproterenol sulfate is freely soluble in water and ethanol.[2,3]

### pH
A 10% aqueous solution has a pH of 4 to 5.5. Metaproterenol sulfate inhalation solution has a pH of 2.8 to 4. The syrup has a pH of 2.5 to 4 when mixed 1:4 with water.[4]

## General Stability Considerations
Metaproterenol sulfate products should be packaged in tight (liquids) or well-closed (tablets) light-resistant containers.[4]

They should be stored at controlled room temperature and protected from light. Liquid products should be protected from freezing.[2] The inhalation solution should not be used if it is discolored to a pinkish or darker than slightly yellow color or contains a precipitate.[7]

Metaproterenol sulfate is stable to autoclave sterilization but oxidizes upon exposure to air. This oxidation is more likely in neutral or alkaline solutions, when exposed to light, or in the presence of heavy metal ions.[2]

## Stability Reports of Compounded Preparations
### Inhalation
Inhalations, like other sterile drugs, should be prepared in a suitable clean air environment using appropriate aseptic procedures. When prepared from nonsterile components, an appropriate and effective sterilization method must be employed.

Gupta et al.[1181] evaluated the stability of metaproterenol sulfate (Boehringher Ingelheim) 5% inhalation solution diluted 1:10 in sodium chloride 0.9% to a concentration of 0.5% and packaged as 3.3 mL of sample solution in 5-mL amber oral syringes sealed with tip caps (Becton Dickinson). At both room temperature of 24 to 26°C and refrigerated at 4 to 6°C, diluted metaproterenol sulfate remained clear and colorless over 120 days, with little or no loss of drug content by HPLC analysis.

## Compatibility with Other Drugs
### METAPROTERENOL SULFATE COMPATIBILITY SUMMARY
*Compatible with:* Dexamethasone (unspecified form) • Ipratropium bromide • Sodium chloride 7% (hypertonic)
*Incompatible with:* Acetylcysteine • Atropine sulfate • Sodium bicarbonate
*Uncertain or variable compatibility with:* Cromolyn sodium

**STUDY 1 (CROMOLYN):** Lesko and Miller[659] evaluated the chemical and physical stability of 1% cromolyn sodium nebulizer solution (Fisons) 2 mL admixed with 0.3 mL of metaproterenol sulfate 5% inhalation solution and a dilution of 0.3 mL in 2.5 mL of 0.9% sodium chloride. The admixtures were visually clear and colorless and remained chemically stable by HPLC analysis for 60 minutes after mixing at 22°C.

**STUDY 2 (CROMOLYN):** Emm et al.[628] evaluated the stability and compatibility of cromolyn sodium combinations during short-term preparation and use. Cromolyn sodium 1% nebulizer solution (Fisons) was combined with metaproterenol sulfate 5 and 0.6% (Alupent, Boehringer Ingelheim) and studied over 90 minutes. No visible signs of precipitation or measured changes in pH were found, and HPLC analysis found less than 10% loss of either of the drugs within 90 minutes. Most combinations showed no loss at all. The authors did note that longer term (24-hour) storage might result in discoloration of metaproterenol sulfate solutions; such discolored solutions should not be used.

**STUDY 3 (MULTIPLE DRUGS):** Owsley and Rusho[517] evaluated the compatibility of several respiratory therapy drug combinations. Metaproterenol sulfate 5% solution (Alupent, Boehringer Ingelheim) was combined with the following drug solutions: acetylcysteine 10% solution (Mucosil-10, Dey), atropine sulfate 1 mg/mL (American Regent), cromolyn sodium 10-mg/mL nebulizer solution (Intal, Fisons), dexamethasone (form unspecified) 4 mg/mL (American Regent), and sodium bicarbonate 8.4% injection (Abbott). The test solutions were filtered through 0.22-µm filters into clean vials. The combinations were evaluated over 24 hours (temperature unspecified) using the USP particulate matter test. Only the metaproterenol sulfate–dexamethasone combination was compatible. Metaproterenol sulfate combined with acetylcysteine, atropine sulfate, cromolyn sodium, and sodium bicarbonate formed unacceptable levels of larger particulates (≥10 µm) in all cases.

**STUDY 4 (IPRATROPIUM):** Joseph[1024] reported that 1 and 4 mL of preservative-free ipratropium bromide 0.025% inhalation solution is physically and chemically compatible with 5 and 10 mg, respectively, of metaproterenol sulfate 5% solution for up to 30 days at room temperature.

**STUDY 5 (HYPERTONIC SODIUM CHLORIDE):** Fox et al.[1239] evaluated the physical compatibility of hypertonic sodium chloride 7% with several inhalation medications used in treating cystic fibrosis. Metaproterenol sulfate respiratory solution (Dey) 0.6% mixed in equal quantities with extemporaneously compounded hypertonic sodium chloride 7% did not exhibit any visible evidence of physical incompatibility, and the measured turbidity did not increase over the one-hour observation period. ◼

# Metformin Hydrochloride

## Properties
Metformin hydrochloride occurs as a white or almost white crystalline powder.[1,3,4]

### Solubility
Metformin hydrochloride is freely soluble in water and slightly soluble in ethanol but is practically insoluble in acetone, ether, and chloroform.[1,4,7]

### pH
The pH of a 1% aqueous solution of metformin hydrochloride is 6.68.[7]

### pKₐ
Metformin has a $pK_a$ of 12.4.[7]

## General Stability Considerations
The USP states that metformin hydrochloride should be packaged in well-closed containers, metformin hydrochloride tablets should be packaged in tight containers, and metformin hydrochloride extended-release tablets should be packaged in well-closed light-resistant containers.[4] The manufacturer of the oral dosage forms states that they should be dispensed in light-resistant containers.[7] Metformin hydrochloride drug substance, tablets, and extended-release tablets are stored at controlled room temperature.[4,7]

## Stability Reports of Compounded Preparations
### Oral
Allen[1368] reported on a compounded formulation of metformin hydrochloride 100 mg/mL for use as an oral liquid. The oral liquid had the following formula:

| | | |
|---|---|---|
| Metformin hydrochloride | | 10 g |
| Xylitol | | 40 g |
| Potassium bicarbonate | | 500 mg |
| Potassium sorbate | | 120 mg |
| Saccharin sodium | | 275 mg |
| Hydrochloric acid | | 0.4 mL |
| Wild cherry flavor | | 0.275 mL |
| Purified water | qs | 100 mL |

The recommended method of preparation was to add the potassium bicarbonate and metformin hydrochloride to about 50 mL of purified water and stir until the solids were dissolved. The hydrochloric acid was then diluted with about 5 mL of purified water and added slowly to the metformin hydrochloride solution with stirring. This releases carbon dioxide gas from the potassium bicarbonate. Then the xylitol and potassium sorbate were added and stirred until fully dissolved. Gentle heating may be necessary for dissolution, but the solution should not be heated above 30°C. The solution should then be permitted to cool. The flavor and saccharin sodium were then added and stirred until dissolved. The mixture was then brought to 99 mL with additional purified water. The pH of the solution was measured and adjusted to pH 4.6 to 4.9 using dilute hydrochloric acid if necessary. The solution was brought to 100 mL with additional purified water. The oral liquid was to be packaged in tight, light-resistant containers. The author recommended a beyond-use date of six months at room temperature because this formula is a commercial medication in some countries with an expiration date of two years or more. ∎

# Methacholine Chloride

## Properties
Methacholine chloride occurs as colorless or white hygroscopic crystals or as a white crystalline powder. It is odorless or nearly odorless; the slight odor has been described as one of dead fish.[1,3]

### Solubility
Methacholine chloride is very soluble in water, having an aqueous solubility of 1 g in 1.2 mL. It is also very soluble in ethanol, with a solubility of 1 g in 1.7 mL.[3]

### pH
Aqueous solutions of methacholine chloride are neutral.[1] The reconstituted vials have a pH of 7.[7]

### Osmolality
Methacholine 10 mg/mL diluted in water is hypo-osmotic, having an osmolality less than 120 mOsm/kg. In phosphate-buffered saline, the solution is more nearly iso-osmotic.[728]

## General Stability Considerations
Methacholine chloride vials should be stored at controlled room temperature. The reconstituted solution should be stored under refrigeration at 2 to 8°C.[7]

Methacholine is susceptible to hydrolysis in aqueous solutions, producing acetic acid and methylcholine. The hydrolysis occurs more rapidly at pH values above 6.[728,1473] Methacholine chloride does not appear to be photosensitive.[1097]

## Stability Reports of Compounded Preparations
### Inhalation

Inhalations, like other sterile drugs, should be prepared in a suitable clean air environment using appropriate aseptic procedures. When prepared from nonsterile components, an appropriate and effective sterilization method must be employed.

**STUDY 1:** The stability of methacholine chloride solution for inhalation extemporaneously prepared from pure powder to a concentration of 5 mg/mL in 0.9% sodium chloride for irrigation was reported by MacDonald et al.[75] Samples were stored at 4, 20, 37°C, and higher elevated temperatures. A colorimetric assay was used to determine the methacholine chloride content of the solutions. By using the Arrhenius relationship, a loss of 10% in the solution was calculated to occur in 63 and 156 days at 20 and 4°C, respectively.

**STUDY 2:** Pratter et al.[1579] evaluated the stability of methacholine chloride at concentrations of 1.25, 2.5, 5, 10, and 20 mg/mL in sodium chloride 0.9%. Samples of each concentration were stored at both room temperature and under refrigeration at 4°C. Both HPLC analysis and the colorimetric assay of MacDonald et al.[75] were used and yielded nearly identical results. Little or no loss of methacholine chloride occurred over four months of storage at either room temperature or under refrigeration.

**STUDY 3:** Acar et al.[728] reported the stability of aqueous solutions of methacholine. In deionized water or phosphate-buffered saline (pH 7.4) at a concentration of 50 mg/mL for use as a bulk concentrate, methacholine could be autoclaved at 120°C for 20 minutes with no loss of drug. For longer term storage, methacholine was more stable in deionized water, with no loss occurring in 40 days refrigerated at 2 to 8°C or stored at room temperature of 28 to 32°C. Methacholine 50 mg/mL in phosphate-buffered saline exhibited no loss in 40 days if refrigerated but lost about 6% in 40 days at room temperature. Although more stable in water, the drug is hypo-osmolar. For the best stability, the authors recommended preparing the concentrate in water stored refrigerated and then diluting it to the proper concentration for inhalation challenges with phosphate-buffered saline. The resulting solution has better osmolality and is closer to physiologic pH.

**STUDY 4:** Henn et al.[821] evaluated the stability of methacholine chloride 5 and 10 mg/mL in 0.9% sodium chloride extemporaneously compounded for inhalation. The solutions were packaged in sterile glass vials stored at ambient room temperature, refrigerated at 4°C, and frozen at −20°C protected from light for 365 days. No visible change in color or precipitate formation was observed. Stability-indicating high-performance capillary electrophoresis found that less than 10% drug loss occurred at all temperatures. However, at room temperature an additional peak formed on the chromatogram in 90 days in the 5-mg/mL solution and in 14 days in the 10-mg/mL solution, limiting the recommended use periods to 35 and less than 14 days, respectively.

**STUDY 5:** Asmus et al.[1083] studied the stability of methacholine chloride 0.031 to 32 mg/mL when stored frozen. Methacholine chloride (Provocholine) powder was diluted with a buffered saline solution containing 0.5% sodium chloride, 0.275% sodium bicarbonate, and 0.4% phenol. Eleven methacholine chloride concentrations were tested: 0.031, 0.062, 0.125, 0.25, 0.5, 1, 2, 4, 8, 16, and 32 mg/mL. The solutions were packaged by drawing 3 mL into 5-mL polypropylene plastic syringes and storing the samples frozen at −7 to −20°C. HPLC analysis found that little or no loss of methacholine chloride occurred during six months of frozen storage at concentrations of 0.062 mg/mL and higher. At 0.031 mg/mL, more than 10% loss of drug was observed at five months, leading the authors to recommend storing this dilute solution frozen for no more than four months. After thawing, the methacholine chloride undergoes much more rapid decomposition, with unacceptable losses occurring within 24 hours at room temperature. The authors recommended that unused methacholine chloride solutions, once thawed, should be discarded.

**STUDY 6:** Hayes et al.[1097] found the stability of methacholine chloride in sodium chloride 0.9% or in phosphate-buffered saline to be concentration-dependent, with better stability occurring at high concentrations. HPLC analysis of methacholine chloride 50 mg/mL in either diluent showed that the drug was stable for nine months at room temperature exposed to or protected from light and under refrigeration. Drug loss was about 6.5% in this time period. However, at a concentration of 0.39 mg/mL, losses of about 50% occurred in one to two months in phosphate-buffered saline at room temperature; losses of about 16% occurred in nine months if the solutions were refrigerated or frozen. In sodium chloride 0.9%, the 0.39-mg/mL concentration lost about 11% in nine months both at room temperature and refrigerated.

**STUDY 7:** Watson et al.[1473] evaluated the effects of concentration and varying pH on the stability of methacholine chloride in aqueous solutions. The stabilities of methacholine chloride concentrations ranging from 2.5 to 20 mg/mL prepared in various buffers having pH values of 4 to 9 were tested for stability using the colorimetric method of MacDonald et al.[75] The test solutions were stored at room temperature of 25 to 29°C and refrigerated at 2 to 6°C. Little or no loss of methacholine chloride occurred in buffers of pH 4, 5, and 6 in three weeks at either temperature. However, at pH 7, 8, and especially 9, methacholine hydrolysis was rapid and extensive.

At the worst case of pH 9, drug losses up to 27% and more than 90% occurred under refrigeration and room temperature conditions, respectively, in three weeks. Decomposition was also more extensive at lower concentrations compared to higher concentrations. For comparison, if sodium chloride 0.9% was used as the diluent, less than 10% loss of any of the drug concentrations tested occurred in 15 weeks stored under refrigeration.

**STUDY 8:** Tipton and Ledoux[1658] evaluated the efficacy of acetyl-beta-methacholine 25 mg/mL in sodium chloride 0.9% and packaged in amber bottles and stored at room temperature. The solution was evaluated using an in vitro guinea pig trachea model to assess methacholine-induced contraction. The authors concluded the solution maintained acceptable activity for six months. After 12 months of storage, a partial loss of activity was observed. ■

# Methadone Hydrochloride

## Properties
Methadone hydrochloride is an odorless white crystalline powder or colorless crystals having a bitter taste.[1–3]

### Solubility
Methadone hydrochloride is soluble in water, with an aqueous solubility of 120 mg/mL. It is also soluble in ethanol, with a solubility of 80 mg/mL.[1,2] However, it is practically insoluble in glycerol.[3]

### pH
A 1% aqueous solution of methadone hydrochloride has a pH between 4.5 and 6.5.[4] Methadone hydrochloride oral concentrate and oral solution have pH ranges of 1 to 6 and 1 to 4, respectively. The injection has a pH between 3 and 6.5.[2,4]

## General Stability Considerations
Methadone hydrochloride oral liquid products should be packaged in tight, light-resistant containers. The tablets should be packaged in well-closed containers. The products should be stored at controlled room temperature and protected from light. The oral concentrate should be stored between 15 and 20°C.[2–4]

In solution, methadone hydrochloride is more stable at lower pH values.[989] Free methadone base may precipitate from solutions having a pH above 6. Aqueous solutions may be autoclaved at 120°C for one hour without concentration loss.[1]

## Stability Reports of Compounded Preparations
### Oral
**STUDY 1:** The stability of a methadone pain cocktail formulation was studied by Little et al.[84] The mixture was prepared from methadone hydrochloride injection (Dolophine, Lilly) 10 mg and hydroxyzine pamoate suspension (Vistaril, Pfizer) equivalent to hydroxyzine hydrochloride 25 mg brought to 15 mL with cherry syrup. Samples were stored at ambient temperature and under refrigeration for two weeks. An HPLC assay was used to determine the content of the two drugs. The samples were stable for two weeks, with about 9% loss of each drug at both temperatures.

**STUDY 2:** Eggers[466] evaluated the stability of methadone hydrochloride 0.25 mg/mL in a formulation composed of a commercial elixir of prolintane and vitamins (Catovit, Boehringer Ingelheim) 480 mL, ethanol 192 mL, and distilled water 288 mL. Analysis by both HPLC and gas–liquid chromatography demonstrated little or no methadone hydrochloride or prolintane loss in 111 weeks stored at 20°C. Minimum estimated shelf life to 10% decomposition of the methadone hydrochloride was 4.1 years. Prolintane had even greater stability.

## Compatibility with Common Beverages and Foods
**METHADONE HYDROCHLORIDE COMPATIBILITY SUMMARY**
*Compatible with:* Apple juice • Crystal Light grape • Kool-Aid • Lemonade • Tang

**STUDY 1:** The stability of methadone hydrochloride in several common beverages was tested by Lauriault et al.[109] Methadone hydrochloride 10 mg/mL in distilled water was tested in Tang (General Foods), Kool-Aid grape flavor (General Foods), apple juice, and grape-flavored Crystal Light (General Foods). The dry products were diluted in deionized water according to the manufacturer's instructions. Methadone hydrochloride concentrations of 0.2, 0.8, and 1.5 mg/mL were mixed into the reconstituted beverages and stored at 5°C for 55 days and at 23°C for 17 days. The methadone content was assessed by using a stability-indicating HPLC assay. Less than 5% loss occurred in 17, 11, 9, and 8 days in Kool-Aid, Tang, apple juice, and Crystal Light, respectively. Unfortunately, all solutions stored at room temperature developed visible signs of bacterial growth after the periods indicated, rendering them unacceptable. Methadone hydrochloride 0.2, 0.8, and 1.5 mg/mL in Crystal Light with added 0.1% sodium benzoate showed losses of 2.5% or less in 29 days at 23°C and did not develop any signs of bacterial growth. When samples were stored under refrigeration, less than 5% loss occurred in 55, 49, 47, and 34 days in Kool-Aid, Tang, apple juice, and Crystal Light, respectively, with no obvious bacterial growth.

**STUDY 2:** Allen and Stiles[149] prepared methadone hydrochloride 1-mg/mL solution or suspension in lemonade or other acidic vehicles having a pH of 3 to 6. If tablets are used as the source of the drug, they should be crushed thoroughly and then slowly mixed into the lemonade to prepare a suspension. If a solution is desired, the crushed methadone hydrochloride tablets may be mixed with water to dissolve the drug; the mixture is then filtered to remove the insoluble tablet excipients. The filtered solution is then added to the lemonade. Alternatively, the injection can be used as the drug source to prepare a solution. The chemical stability is at least three months. However, the lemonade may become contaminated with mold or fungal growth and a shorter expiration date may be necessary.

**STUDY 3:** Sochasky et al.[458] evaluated the stability of methadone hydrochloride 1.5 mg/mL in lemonade. The mixture was prepared by dissolving methadone hydrochloride powder and Dutch Mill Flavour Crystals in sterile water for irrigation. Samples were stored at room temperature for 14 weeks, and analysis was performed using liquid chromatography. The methadone hydrochloride concentration remained constant throughout the study period.

**STUDY 4:** Donnelly[908] evaluated the stability of methadone hydrochloride at a high concentration of 5 mg/mL in Tang brand orange-flavor drink. Methadone hydrochloride concentrate 10 mg/mL and methadone hydrochloride powder were added to the Tang. Sodium benzoate was added to the powder mixture to act as a preservative. The commercial concentrate contains sodium benzoate 1.5 mg/mL as a preservative, which the author indicated was sufficient to inhibit bacterial growth. The solutions were stored at room temperature of 22°C and refrigerated at 6°C for 91 days.

The solutions remained clear throughout the study. (Turbidity from bacterial growth occurs at room temperature when powder is used as the drug source if no preservative is present.) Stability-indicating HPLC analysis found little or no methadone hydrochloride loss occurred in 91 days at either storage temperature. ∎

# Methenamine Mandelate
## (Hexamine Mandelate)

## Properties
Methenamine mandelate is composed of 48% methenamine and 52% mandelic acid. It is a white almost odorless crystalline powder or colorless lustrous crystals with a taste variously described as bittersweet or sour. It may develop a slight odor during long-term storage.[1–3,6]

### Solubility
Methenamine mandelate is very soluble in water, with an aqueous solubility of about 1 g/1.5 mL. It has a solubility in ethanol of about 100 to 125 mg/mL at 25°C.[1–3,6]

### pH
Methenamine mandelate aqueous solutions have a pH of about 4.[1] The oral solution when mixed 1 g with 30 mL of water has a pH between 4 and 4.5.[4]

### $pK_a$
Methenamine has a $pK_a$ variously cited as 4.6[6] and 4.8.[2]

## General Stability Considerations
Methenamine mandelate tablets and oral solution should be packaged in well-closed containers, while the oral suspension specifies a tight container.[4] The products should be stored at controlled room temperature and protected from excessive heat.[2]

Methenamine is easily hydrolyzed to formaldehyde and ammonia by acids.[2,6] Because hydrolysis is acid catalyzed, stability increases as the pH approaches neutrality.[6] The drug is darkened by alkalies and ammonium salts. It is incompatible with most alkaloids and some metal salts.[2] Methenamine also undergoes thermal decomposition, yielding ammonia and a carbon residue.[6]

## Stability Reports of Compounded Preparations
### Enteral
#### METHENAMINE MANDELATE COMPATIBILITY SUMMARY
*Uncertain or variable compatibility with:* Ensure • Ensure HN • Ensure Plus • Ensure Plus HN • Osmolite • Osmolite HN • Vital

**STUDY 1:** Cutie et al.[19] added 10 mL of methenamine mandelate suspension (Mandelamine, Parke-Davis) to varying amounts (15 to 240 mL) of Ensure, Ensure Plus, and Osmolite (Ross Laboratories) with vigorous agitation to ensure thorough mixing. The methenamine mandelate suspension was physically incompatible with the enteral products. The mixtures became tacky and gelatinous.

However, when methenamine mandelate granules (Mandelamine, Parke-Davis) were added to varying volumes of the three enteral products, the granules could be dissolved or dispersed without difficulty, forming physically compatible mixtures.[19]

**STUDY 2:** Altman and Cutie[850] reported the physical compatibility of methenamine mandelate suspension (Mandelamine, Parke-Davis) 10 mL with varying amounts (15 to 240 mL) of Ensure HN, Ensure Plus HN, Osmolite HN, and Vital after vigorous agitation to ensure thorough mixing. The methenamine mandelate suspension was physically incompatible with the four enteral products. The mixtures became tacky and gelatinous.

However, when methenamine mandelate granules (Mandelamine, Parke-Davis) were added to varying volumes of Ensure HN, Ensure Plus HN, Osmolite HN, and Vital enteral products, the granules could be dissolved or dispersed without difficulty, forming physically compatible mixtures.[850] ■

# Methimazole
## (Thiamazole)

## Properties
Methimazole occurs as a white to pale buff or pale brown colored crystalline powder and has a faint characteristic odor.[1,3,4]

### Solubility
Methimazole is freely soluble in water and ethanol to about 200 mg/mL for each solvent.[1,3,4]

### pH
Methimazole aqueous solutions are neutral to litmus.[3]

## General Stability Considerations
Methimazole powder and oral tablets should be packaged in well-closed light-resistant containers and stored at controlled room temperature. Protection from exposure to light is required.[3,4]

## Stability Reports of Compounded Preparations
### Topical
Pignato et al.[1645] evaluated the stability of methimazole 5 mg/0.1 mL in poloxamer lecithin organogel (PLO) for veterinary use. Initially, a mixture of soy lecithin 50 g, isopropyl palmitate 50 g, and sorbic acid 200 mg was prepared. Then, separately, poloxamer 407 20 g and sorbic acid 200 mg was mixed in sufficient purified water to make 100 mL of poloxamer 407 20% gel. Methimazole was mixed thoroughly in the lecithin:isopropyl palmitate solution. Then the poloxamer 407 gel was incorporated and the methimazole PLO was thoroughly mixed. The methimazole PLO was packaged in plastic 1-mL syringes, which were placed in plastic storage bags and stored refrigerated at 5°C, at room temperature of 25°C, and elevated temperature of 35°C for 90 days.

The samples stored under refrigeration underwent visible separation into two distinct layers, a clear gel layer and an opaque light yellow gel layer. Samples stored at room temperature did not exhibit any visible changes throughout the study. The samples stored at 35°C had a high thickness and were very difficult to mix. Stability-indicating HPLC analysis found no methimazole loss occurred in 62 days at room temperature and about 3% loss occurred at 35°C. The refrigerated samples exhibited a higher loss of 8%, possibly due to the gel disruption causing the drug to be no longer evenly dispersed. After 90 days of storage, all samples at all temperatures exhibited excessive methimazole losses of about 14 to 17%. The authors recommended room temperature storage and a beyond-use date of 60 days. ■

# Methotrimeprazine
# Methotrimeprazine Hydrochloride
## (Levomepromazine)

## Properties
Methotrimeprazine occurs as a fine practically odorless crystalline powder.[4] Methotrimeprazine hydrochloride occurs as a white or very slightly yellow slightly hygroscopic crystalline powder.[3]

### Solubility
Methotrimeprazine is practically insoluble in water. The drug is sparingly soluble in ethanol at 25°C but is freely soluble in boiling ethanol.[4]

Methotrimeprazine hydrochloride has been stated to be freely soluble in water and in ethanol.[3] The drug's solubility in water and ethanol has also been cited as 0.3 and 0.4% at 20°C.[1]

### pH
A 0.3% aqueous solution of methotrimeprazine hydrochloride has a pH of 4.3.[1] Methotrimeprazine hydrochloride injection has a pH in the range of 3 to 5.[4]

*Osmolality*

Methotrimeprazine hydrochloride injection is isotonic.[7]

## General Stability Considerations

Methotrimeprazine powder should be packaged in well-closed light-resistant containers and stored at controlled room temperature.[4] Methotrimeprazine hydrochloride powder should be packaged in tight containers and stored at controlled room temperature and protected from exposure to light.[3] The injection should be protected from exposure to light during storage as well.[4,8]

Methotrimeprazine hydrochloride is light sensitive[1] and upon exposure to light develops a pink or yellow color; discolored solutions must be discarded.[7]

Methotrimeprazine hydrochloride is incompatible with alkaline pH solutions.[3,8]

## Stability Reports of Compounded Preparations
*Enteral*

Ortega de la Cruz et al.[1101] reported the physical compatibility of an unspecified amount of oral liquid methotrimeprazine hydrochloride (Sinogan) with 200 mL of Precitene (Novartis) enteral nutrition diet for a 24-hour observation period. No particle growth or phase separation was observed. ■

# ■ Methoxsalen

## Properties

Methoxsalen is a psoralen derivative that occurs as white to cream-colored odorless bitter-tasting fluffy needle-like crystals. Methoxsalen topical solution is clear and colorless.[1,4]

*Solubility*

Methoxsalen is practically insoluble in water but is sparingly soluble in boiling water. It is soluble in boiling alcohol, acetone, propylene glycol, and fixed vegetable oils. Methoxsalen dissolves in alkaline solutions, with decomposition, but is restored with neutralization.[1,4]

## General Stability Considerations

Methoxsalen should be stored in well-closed, light-resistant containers. Methoxsalen topical solution and methoxsalen capsules should be packaged in tight, light-resistant containers and stored at controlled room temperature.[4,7]

## Stability Reports of Compounded Preparations
*Topical*

**STUDY 1:** Martens-Lobenhoffer et al.[825] evaluated the stability of methoxsalen 0.05 and 0.005% in two commercial ointment bases (Unguentum Cordes and Cold Cream Naturel) and an extemporaneously prepared gel using carbomer 940 (Carbopol 940). The methoxsalen dissolved in 2-propanol (1.5 or 0.15% for the two concentrations) and water were incorporated into Unguentum Cordes and Cold Cream Naturel. For the carbomer 940 gel, methoxsalen was provided

as a 1% solution in propylene glycol and was mixed with the carbomer 940, sodium hydroxide solution, and water. The ointments were stored at room temperature of 19 to 20°C and refrigerated at a temperature of 5°C protected from light for 88 days.

Unguentum Cordes proved to be the best base for use with methoxsalen. It remained physically intact, and thin-layer chromatography (TLC) found little methoxsalen loss over 88 days at either temperature.

The carbomer 940 gel was unacceptable for this application. Methoxsalen precipitated crystals during preparation, making the gel cloudy. TLC found that only about 40% of the nominal drug concentration was present, and substantial additional drug loss occurred during storage. Methoxsalen was stable in Cold Cream Naturel, but the emulsion broke down after about eight weeks.

**STUDY 2:** Botet Homdedeu and Gamundi Planas[1634] evaluated the stability of methoxsalen 0.5% with salicyclic acid 1% and with and without anthralin 0.2% incorporated into zinc oxide paste (Lassar's paste). Samples were stored refrigerated at 4°C and at ambient room temperature for six months. The samples were white and homogeneous and did not exhibit changes in consistency or appearance during the six-month observation period. Chromatographic evaluation of methoxsalen concentrations found the time to 10% loss ($t_{90}$) was nine months under refrigeration and 6.5 months at room temperature. ■

# Methyl Nicotinate

## Properties

Methyl nicotinate occurs as a white to pale yellow crystalline material.[1,1340]

### Solubility

Methyl nicotinate is stated to be soluble in water, methanol, and ethanol.[1,1340]

## General Stability Considerations

Methyl nicotinate bulk material is packaged in tight containers and stored at controlled room temperature.[1340]

## Stability Reports of Compounded Preparations

### Topical

Ross and Katzman[1339] evaluated the stability of 1 M aqueous solutions of methyl nicotinate used for topical application as a "niacin patch." The solutions had been stored for varying time periods with ages between 5 and 1062 days after preparation. The solutions were packaged in glass containers and were stored refrigerated at 4°C throughout. No visible changes in color or clarity occurred among the solutions of various ages. HPLC analysis of four of the solutions found a very slow rate of decomposition. Hydrolysis of the methyl nicotinate to nicotinic acid occurred at a rate of about 0.5% per year when stored under refrigeration. The oldest solution, which was nearly three years from preparation, exhibited about 1.5% loss. The various solutions, including those that were oldest, did not lose the ability to produce erythema in healthy individuals. ■

# Methyldopa
# Methyldopate Hydrochloride

## Properties

Methyldopa is a white to yellowish odorless powder or colorless or nearly colorless crystals. Methyldopa sesquihydrate 1.13 g is approximately equivalent to 1 g of anhydrous methyldopa.[2,3]

Methyldopate hydrochloride is the ethyl ester of the hydrochloride salt of methyldopa. It is a white or nearly white odorless crystalline powder.[2,3]

### Solubility

Methyldopa is slightly soluble in water, with an aqueous solubility of about 10 mg/mL at 25°C. It also is slightly soluble in ethanol, at about 2.5 mg/mL. In dilute hydrochloric acid, the solubility increases to about 2 g/mL.[1,3,6]

Methyldopate hydrochloride is freely soluble in both water and ethanol. The aqueous solubility of methyldopate hydrochloride is about 1 g/mL, while the solubility in ethanol is about 333 mg/mL.[2,3,6]

### pH

A saturated solution of methyldopa has a pH of around 5.[6] Methyldopa oral suspension has a pH between 3 and 5 or between 3.2 and 3.8 if sucrose is used in the product.[4] Methyldopate hydrochloride injection has a pH between 3 and 4.2.[2]

### pKa

Methyldopa has $pK_a$ values of 2.25, 9.0, 10.35, and 12.6.[6,1609]

### Osmolality

Methyldopa suspension 50 mg/mL (Merck Sharp & Dohme) had an osmolality of 2050 mOsm/kg.[233] Methyldopate hydrochloride injection had an osmolality of 481 mOsm/kg.[326]

## General Stability Considerations

Methyldopa oral suspension should be packaged in tight, light-resistant containers. The tablets should be packaged in well-closed containers. The oral suspension should be stored at less than 26°C and protected from freezing. Methyldopate hydrochloride injection should be stored below 30°C and protected from freezing. The tablets should be stored at controlled room temperature and protected from temperatures of 40°C or more.[2–4]

Oxidizing agents decompose methyldopa.[2,168] Exposure of the injection to air accelerates decomposition.[2] Decomposition of methyldopa in aqueous media occurs primarily through oxidation of the hydroxyl groups. The quinones formed vary from pink to brown. Methyldopa oxidation in aqueous solution is accelerated by ultraviolet light, oxygen, alkalinity, and decreasing drug concentration.[6,168,381]

The drug exhibits the greatest stability at acidic to neutral pH.[2] The injection is stable for at least 24 hours over pH 3.5 to 7.8 in 5% dextrose injection admixtures.[382] However, more than a 5% loss occurred within 24 hours at pH 7.8.[383]

## Stability Reports of Compounded Preparations
### Oral

**STUDY 1:** Gupta et al.[62] reported the stability of methyldopa prepared as pediatric oral liquid dosage forms. Methyldopa from powder (Merck Sharp & Dohme) was dissolved in simple syrup to yield a 25-mg/mL concentration with and without sodium metabisulfite 0.1% and edetate disodium 0.1%. Methyldopate hydrochloride injection (Aldomet, Merck Sharp & Dohme), available with antioxidant, chelating agent, and antimicrobial preservatives, also was used as the source of drug; it was prepared as an equal parts mixture with simple syrup to yield 25 mg/mL. The preparations were packaged in amber bottles and stored at 24°C. By using a colorimetric analytical technique, 99% of the methyldopa was found to remain after 168 days of storage. However, the dosage form made from powder with no antioxidant became discolored, turning from colorless to yellow after 98 days. An antioxidant and a chelating agent are needed to prevent this discoloration.

**STUDY 2:** Newton et al.[168] evaluated the stability of two extemporaneous oral liquid preparations of methyldopa. The first formulation was prepared by crushing 10 methyldopa 250-mg tablets and levigating with simple syrup; the mixture was brought to 50 mL with more simple syrup to yield 250 mg/5 mL. The second formulation also was prepared by crushing 10 methyldopa 250-mg tablets. The crushed tablets were levigated with 25 mL of 0.2 N [0.73% (wt/vol)] hydrochloric acid. This mixture was brought to 50 mL with simple syrup containing 0.5% (wt/vol) citric acid to yield 250 mg/5 mL. The two oral liquid preparations were stored in the dark at 5 and 25°C for 14 days. The methyldopa content was evaluated by spectrophotometry. No significant change in methyldopa concentration occurred in either formulation at either storage temperature over 14 days.

### Enteral
#### METHYLDOPA COMPATIBILITY SUMMARY
*Uncertain or variable compatibility with:* Ensure
- Ensure HN • Ensure Plus • Ensure Plus HN
- Osmolite • Osmolite HN • Vital

Holtz et al.[54] evaluated the stability of methyldopa oral suspension (Aldomet, Merck Sharp & Dohme) at 250 and 500 mg/L in full-strength Ensure, Ensure Plus, and Osmolite (Ross) over 12 hours at room temperature. Visual inspection revealed no changes such as clumping, gelling, separation, precipitation, or increased viscosity. Osmolalities were largely unaffected by drug addition. Methyldopa concentrations were assessed by fluorescent spectroscopy. At the lower concentration, substantial loss of methyldopa occurred. In Ensure, Ensure Plus, and Osmolite, losses in 12 hours were 14, 23, and 10%, respectively. At the 500-mg/L concentration, 11% of the methyldopa was lost in 12 hours in the Ensure Plus mixture, but only 1 to 3% loss occurred in Ensure and Osmolite in this time frame. ■

# ■ Methylene Blue

## Properties
Methylene blue exists as dark green crystals or crystalline powder having a bronzelike luster and is nearly odorless. In water or ethanol, it becomes a deep blue solution.[1,4]

### Solubility
Methylene blue is soluble in water and sparingly soluble in ethanol.[1,4]

### pH
Methylene blue injection has a pH of 3 to 4.5.[4]

### pKₐ
The $pK_a$ of methylene blue is reported to be between 0 and −1.[2]

## General Stability Considerations
Methylene blue is stated to be stable exposed to air. It should be packaged in well-closed containers and stored at room temperature. Methylene blue injection should be packaged in single-dose containers and stored at room temperature.[1,4]

## Stability Reports of Compounded Preparations
### Injection
Injections, like other sterile drugs, should be prepared in a suitable clean air environment using appropriate aseptic procedures. When prepared from nonsterile components, an appropriate and effective sterilization method must be employed.

#### USP OFFICIAL VETERINARY INJECTION FORMULATION:
Methylene blue (as trihydrate) 5 g
Sterile water for injection
or Sodium chloride 0.9% injection qs 500 mL

Accurately weigh the required quantity of methylene blue and dissolve in most of the sterile water for injection or sodium chloride 0.9% injection. Bring to the required volume with additional injection and mix thoroughly, yielding a methylene blue 10-mg/mL solution. Methylene blue injection must be sterilized by filtration or by autoclaving. Package the injection in Type I amber glass single-dose containers. The finished injection should have a pH between 3 and 4.5 and must pass the USP bacterial endotoxin test with NMT 0.17 USP endotoxin unit/mg of methylene blue and the USP sterility test. It must have a methylene blue concentration in the range of 9.5 to 10.5 mg/mL. Store the finished injection at controlled room temperature protected from exposure to light. The beyond-use date is 365 days from the date of compounding.[4]

Hara et al.[1061] reported the long-term stability of extemporaneously compounded methylene blue injection 10 mg/mL in water for injection. The extemporaneously compounded injection conformed to USP requirements for the composition and quality attributes of methylene blue injection, USP. A shelf-life study over 12 months of this compounded injection stored at room temperature found that the drug concentration remained constant.

## Compatibility with Other Drugs

Methylene blue is reported to be incompatible with strong alkalies, iodides, dichromates, and oxidizing and reducing materials.[2] ■

# Methylergonovine Maleate

## Properties
Methylergonovine maleate exists as an odorless white to pinkish-tan microcrystalline powder having a bitter taste. Methylergonovine maleate injection is a clear colorless solution.[1,4,7]

### Solubility
Methylergonovine maleate is slightly soluble in water at about 1 g in 100 mL and in ethanol at about 1 g in 175 mL.[1,3,4]

### pH
Methylergonovine maleate injection has a pH of 2.7 to 3.5.[4]

### Osmolality
In addition to the active drug, methylergonovine maleate injection contains maleic acid and sodium chloride, which bring the injection to near isotonicity.[7]

## General Stability Considerations
Methylergonovine maleate drug substance is packaged in tight, light-resistant containers and stored under refrigeration.

The injection is packaged in single-dose light-resistant containers and is stored under refrigeration. Methylergonovine maleate tablets are packaged in tight, light-resistant containers and stored at room temperature.[4,7]

## Stability Reports of Compounded Preparations
### Oral
Marigny et al.[1051] evaluated the stability of methylergonovine maleate 0.05 mg/mL in a unit-dose oral liquid preparation. In addition to the drug, the formulation contained maleic acid, ethanol, glycerol, purified water, and carbon dioxide. The solution was packaged in polypropylene tubes in 5-mL (0.25 mg) unit doses and stored at room temperature and refrigerated at 5°C protected from exposure to light for 47 days. No precipitation or color change was observed at either temperature. Stability-indicating HPLC analysis found less than 4% loss at either temperature over the 47-day test period. ■

# Methylphenidate Hydrochloride

## Properties
Methylphenidate hydrochloride is a fine white odorless crystalline powder.[2,3]

### Solubility
Methylphenidate hydrochloride is freely soluble in water and soluble in ethanol.[1–3]

### pH
Aqueous solutions of methylphenidate hydrochloride are neutral[1] to acid[3,7] to litmus.

### pKa
Methylphenidate hydrochloride has a $pK_a$ of 8.8[6] or 8.9.[1]

## General Stability Considerations

Methylphenidate hydrochloride tablets should be packaged in tight, light-resistant containers. They should be stored at controlled room temperature and protected from light and, for the extended-release tablets, moisture.[4,7] The manufacturer has stated that methylphenidate hydrochloride tablets should not be removed from the original packaging and placed in dosing compliance aids because of moisture sensitivity.[1622]

Methylphenidate hydrochloride is subject to ester hydrolysis. Methylphenidate hydrochloride in aqueous solutions is stated to be most stable in the pH range of 3 to 4.[6,586]

## Stability Reports of Compounded Preparations
### Oral

Woods[586] reported on an extemporaneous oral liquid formulation of methylphenidate hydrochloride recommended by

**TABLE 62.** Formula of Methylphenidate Hydrochloride Oral Liquid Prepared from Tablets[586]

| Component | Amount |
|---|---|
| Methylphenidate hydrochloride tablets | Sufficient for desired concentration |
| Citric acid monohydrate | 480 mg |
| Sodium citrate | 72 mg |
| Syrup | 50 mL |
| Sorbitol 70% | qs 100 mL |

the manufacturer of Ritalin tablets. The formula is shown in Table 62. An expiration period of 28 days when stored at room temperature was suggested, although no stability data were presented. ■

# Methylprednisolone
# Methylprednisolone Acetate
# Methylprednisolone Sodium Succinate

## Properties

Methylprednisolone free alcohol and methylprednisolone acetate are white or nearly white odorless crystalline powders. Methylprednisolone acetate 44 mg is approximately equivalent to 40 mg of methylprednisolone.[2,3]

Methylprednisolone sodium succinate is a white or almost white odorless hygroscopic amorphous solid. Methylprednisolone sodium succinate 53 mg is approximately equivalent to 40 mg of methylprednisolone.[2,3]

### Solubility

Methylprednisolone is practically insoluble in water and soluble to about 10 mg/mL in ethanol. Methylprednisolone acetate has an aqueous solubility of 0.67 mg/mL and has a solubility in ethanol of 2.5 mg/mL. Methylprednisolone sodium succinate has an aqueous solubility of about 667 mg/mL and has a solubility of about 83 mg/mL in ethanol.[2,3]

### pH

Methylprednisolone acetate sterile suspension has a pH of 3.5 to 7. Methylprednisolone sodium succinate injection has a pH of 7 to 8 after reconstitution.[4]

## General Stability Considerations

Methylprednisolone products should be packaged in tight containers and protected from light.[3,4] They should be stored at controlled room temperature, and the liquid products should be protected from freezing. Reconstituted methylprednisolone sodium succinate injection should be stored at controlled room temperature and used within 48 hours.[2,7] However, freezing of the reconstituted solution at −20°C is reported to result in no loss after four weeks of storage.

Methylprednisolone sodium succinate is subject to ester hydrolysis and acyl migration. The minimum hydrolysis rate occurs at pH 3.5. Between pH 3.4 and 7.4, acyl migration dominates. The solution should not be used if particulate matter is present.[4,6,8]

## Stability Reports of Compounded Preparations
### Injection

Injections, like other sterile drugs, should be prepared in a suitable clean air environment using appropriate aseptic procedures. When prepared from nonsterile components, an appropriate and effective sterilization method must be employed.

**STUDY 1:** Nahata et al.[501] evaluated the stability of methylprednisolone sodium succinate (Upjohn) diluted with sterile water for injection to a concentration of 4 mg/mL. The dilution was packaged in sterile glass vials and stored at 4 and 22°C. HPLC analysis found 10% loss after one day of storage at room temperature. The samples stored at 4°C exhibited 6% loss in seven days and 17% loss in 14 days.

**STUDY 2:** Gupta[729] reported the stability of methylprednisolone sodium succinate (Solu-Medrol, Pharmacia & Upjohn) 10 mg/mL in sodium chloride 0.9% packaged in polypropylene syringes (Monoject and also Becton Dickinson). The sample syringes were stored refrigerated at 4 to 6°C and at room temperature of 24 to 26°C. All sample solutions remained clear throughout the study. Stability-indicating HPLC analysis found about 4% loss in 21 days when refrigerated.

At room temperature, losses of 4, 7, and 11% occurred in two, four, and seven days, respectively. The author recommended restricting the room temperature beyond-use date to four days. ■

# Metoclopramide Hydrochloride

## Properties
Metoclopramide hydrochloride is a white or practically white odorless crystalline powder.[3,4]

### Solubility
Metoclopramide hydrochloride is very soluble in water, having a solubility of 1.43 g/mL. The drug is freely soluble in ethanol, with a solubility of 333 mg/mL.[2,3,4]

### pH
Metoclopramide hydrochloride injection has a pH of 2.5 to 6.5. The oral solution has a pH of 2 to 5.5.[4]

### $pK_a$
Metoclopramide hydrochloride has $pK_a$ values of 0.6 and 9.3.[2]

### Osmolality
Metoclopramide hydrochloride 5-mg/mL injection has an osmolality of 280 mOsm/kg.[8]

## General Stability Considerations
Metoclopramide hydrochloride powder should be packaged in tight containers and stored at controlled room temperature. Metoclopramide hydrochloride tablets and oral solution should be packaged in tight, light-resistant containers and stored at controlled room temperature. The injection also should be stored at controlled room temperature and protected from freezing and exposure to light.[4] Light protection is not needed for dilutions stored up to 24 hours[7] and for the injection if an antioxidant is present.[4] Freezing of the injection has resulted in the formation of large amounts of microparticulates.[905]

Metoclopramide hydrochloride is stable over a pH range of 2 to 9.[2]

## Stability Reports of Compounded Preparations
### Injection
Injections, like other sterile drugs, should be prepared in a suitable clean air environment using appropriate aseptic procedures. When prepared from nonsterile components, an appropriate and effective sterilization method must be employed.

**STUDY 1:** Zhang et al.[905] reported that Robins metoclopramide hydrochloride 5-mg/mL injection repackaged in 3-mL Mini-iMed polypropylene infusion pump syringes was physically and chemically stable for up to seven days at 32°C, for 60 days at room temperature of 23°C, and for 90 days refrigerated at 4°C. HPLC analysis found that little or no loss of drug occurred. The room temperature samples of metoclopramide hydrochloride formed large amounts of microparticulates after 60 days of storage, making the solution unfit for use.

**STUDY 2:** Stewart et al.[901] reported that metoclopramide hydrochloride 2.5 mg/mL diluted with 0.9% sodium chloride injection was physically and chemically stable for at least 24 hours packaged in polypropylene syringes. Stability-indicating HPLC analysis found that little or no loss of drug occurred.

### Enteral
#### METOCLOPRAMIDE HYDROCHLORIDE COMPATIBILITY SUMMARY
> *Compatible with:* TwoCal HN • Vivonex T.E.N.
> *Incompatible with:* Enrich

Burns et al.[739] reported the physical compatibility of metoclopramide hydrochloride oral solution 10 mL with 10 mL of three enteral formulas, including Enrich, TwoCal HN, and Vivonex T.E.N. Visual inspection found no physical incompatibility with the TwoCal HN and Vivonex T.E.N. enteral formulas. However, with Enrich, a thin granular material formed. ■

# Metolazone

## Properties

Metolazone, a quinazoline diuretic, is a white powder.[2]

### Solubility

Metolazone is practically insoluble in water and sparingly soluble in ethanol.[2,7]

### pK_a

The drug has a pK_a of 9.7.[2]

## General Stability Considerations

Metozalone tablets should be packaged in tight, light-resistant containers and should be stored at room temperature.[2,7]

## Stability Reports of Compounded Preparations
### Oral

**USP OFFICIAL FORMULATION:**

> Metolazone 100 mg
> Vehicle for Oral Solution, NF (sugar-containing or
>   sugar-free)
> Vehicle for Oral Suspension, NF (1:1) qs 100 mL

(See the vehicle monographs for information on the individual vehicles.)

Use metolazone powder or commercial tablets. If using tablets, crush or grind to fine powder. Add 20 mL of the vehicle mixture, and mix to make a uniform paste. Add additional vehicle almost to volume in increments with thorough mixing. Quantitatively transfer to a suitable calibrated tight and light-resistant bottle, bring to final volume with the vehicle mixture, and thoroughly mix yielding metolazone 1-mg/mL oral suspension. The final liquid preparation should have a pH between pH 3.6 and 4.6. Store the preparation at controlled room temperature or under refrigeration between 2 and 8°C. The beyond-use date is 60 days from the date of compounding at either storage temperature.[4,5]

**STUDY 1:** The extemporaneous formulation of a metolazone 0.25-mg/mL oral suspension was described by Nahata and Hipple.[160] Eight metolazone 2.5-mg tablets were crushed and mixed thoroughly with a small amount of simple syrup. The suspension was brought to 80 mL with simple syrup. A stability period of 14 days under refrigeration was used, although chemical stability testing was not performed.

**STUDY 2:** Allen and Erickson[527] evaluated the stability of three metolazone 1-mg/mL oral suspensions extemporaneously compounded from tablets. Vehicles used in this study were (1) an equal parts mixture of Ora-Sweet and Ora-Plus (Paddock), (2) an equal parts mixture of Ora-Sweet SF and Ora-Plus (Paddock), and (3) cherry syrup (Robinson Laboratories) mixed 1:4 with simple syrup. Twelve metolazone 10-mg tablets (Fisons) were crushed and comminuted to fine powder using a mortar and pestle. About 20 mL of the test vehicle was added to the powder and mixed to yield a uniform paste. Additional vehicle was added geometrically, and the suspension was brought to the final volume of 120 mL, with thorough mixing after each addition. The process was repeated for each of the three test suspension vehicles. Samples of each finished suspension were packaged in 120-mL clear polyethylene terephthalate plastic prescription bottles and stored at 5 and 25°C protected from light.

No visual changes or changes in odor were detected during the study. Stability-indicating HPLC analysis found that not more than 4% metolazone loss occurred in any of the suspensions stored at either temperature after 60 days of storage.

**STUDY 3:** Nahata et al.[576] evaluated the stability of a metolazone 0.25-mg/mL oral suspension prepared from tablets. A sufficient quantity of metolazone 2.5-mg tablets (Fisons) was crushed and suspended in an equal parts mixture of 1% methylcellulose and simple syrup. The suspension was packaged as 25 mL in two-ounce glass and plastic prescription bottles. Samples were stored at 4 and 25°C for 91 days. No caking or color changes were observed throughout the study. Stability-indicating HPLC analysis found little or no metolazone loss after 91 days stored at 4°C. While little or no loss was found after one week of storage at 25°C, 8% or more had occurred after two weeks of storage. Consequently, refrigerated storage was recommended. ■

# Metoprolol Tartrate

## Properties

Metoprolol tartrate is a white crystalline powder or colorless crystals having a bitter taste.[2,3]

### Solubility

Metoprolol tartrate is very soluble in water, having an aqueous solubility of more than 1 g/mL at 25°C. The drug is freely soluble in ethanol.[2,3]

### pH

A 10% aqueous solution has a pH of 6 to 7.[4] The injection has a pH of 5 to 8.[4] In one report the pH was found to be approximately 7.5.[8]

### pK_a

Metoprolol tartrate has a pK$_a$ of 9.68.[2]

## General Stability Considerations

Metoprolol tartrate should be packaged in tight, light-resistant containers and stored at controlled room temperature. The tablets should be protected from moisture, and the injection should be protected from freezing.[7] Oral tablets stored according to the label requirements have been found to be very stable and exhibit no change in dissolution even many years beyond the labeled expiration date.[1519,1520]

## Stability Reports of Compounded Preparations
### Oral
#### USP OFFICIAL FORMULATION (ORAL SOLUTION):

>    Metoprolol tartrate 1 g
>    Vehicle for Oral Solution, NF (sugar-containing or
>       sugar-free) qs 100 mL

(See the vehicle monograph for information on the Vehicle for Oral Solution, NF.)

Use metoprolol tartrate powder. Add about 20 mL of the vehicle and mix well. Add additional vehicle almost to volume in increments with thorough mixing. Quantitatively transfer to a suitable calibrated tight and light-resistant bottle, bring to final volume with the vehicle, and thoroughly mix yielding metoprolol tartrate 10-mg/mL oral solution. The final liquid preparation should have a pH between 3.6 and 4.6. Store the preparation at controlled room temperature or under refrigeration between 2 and 8°C. The beyond-use date is 60 days from the date of compounding.[4,5]

#### USP OFFICIAL FORMULATION (ORAL SUSPENSION):

>    Metoprolol tartrate 1 g
>    Vehicle for Oral Solution, NF (sugar-containing or
>       sugar-free)
>    Vehicle for Oral Suspension, NF (1:1) qs 100 mL

(See the vehicle monographs for information on the individual vehicles.)

Use metoprolol tartrate powder or commercial tablets. If using tablets, crush or grind to fine powder. Add the vehicle in small increments with thorough mixing yielding metoprolol tartrate 10-mg/mL oral suspension. Quantitatively transfer to a suitable calibrated tight and light-resistant bottle. The final liquid preparation should have a pH between 3.6 and 4.6. Store the preparation at controlled room temperature or under refrigeration between 2 and 8°C. The beyond-use date is 60 days from the date of compounding at either storage temperature.[4,5]

**STUDY 1:** Allen and Erickson[542] evaluated the stability of three metoprolol tartrate 10-mg/mL oral suspensions extemporaneously compounded from tablets. Vehicles used in this study were (1) an equal parts mixture of Ora-Sweet and Ora-Plus (Paddock), (2) an equal parts mixture of Ora-Sweet SF and Ora-Plus (Paddock), and (3) cherry syrup (Robinson Laboratories) mixed 1:4 with simple syrup. Twelve metoprolol tartrate 100-mg tablets (Mylan) were crushed and comminuted to fine powder with a mortar and pestle. About 20 mL of the test vehicle was added to the powder and mixed to yield a uniform paste. Additional vehicle was added geometrically, and the suspension was brought to the final volume of 120 mL, with thorough mixing after each addition. The process was repeated for each of the three test suspension vehicles. Samples of each finished suspension were packaged in 120-mL amber polyethylene terephthalate plastic prescription bottles and stored at 5 and 25°C in the dark.

No visual changes or changes in odor were detected during the study. Stability-indicating HPLC analysis found less than 3% metoprolol tartrate loss in any of the suspensions stored at either temperature after 60 days of storage.

**STUDY 2:** Gupta and Maswoswe[547] evaluated the stability of a metoprolol tartrate 5-mg/mL aqueous mixture prepared from tablets. Ten metoprolol tartrate 50-mg tablets were ground to fine powder, triturated with water, and brought to a final volume of 100 mL with water. The mixture was packaged in a 120-mL amber glass bottle and stored at 25°C for 16 days. Stability-indicating HPLC analysis found little or no loss of metoprolol tartrate.

**STUDY 3:** Modamio et al.[745] reported the stability of metoprolol tartrate (Ciba-Geigy) 0.25 mg/mL in phosphate buffer (pH 7.4) at elevated temperatures. Stability-indicating HPLC analysis found that little or no loss of metoprolol occurred in 84 hours, even at a temperature as high as 90°C.

**TABLE 63.** Metoprolol Tartrate 10-mg/mL Oral Liquid Formulation of Peterson et al.[756]

| Component | | Amount |
|---|---|---|
| Metoprolol tartrate 100-mg tablets | | 10 tablets |
| Compound tragacanth powder | | 3 g |
| Concentrated chloroform water | | 1.25 mL |
| Syrup | | 12.5 mL |
| Distilled water | qs | 100 mL |

**STUDY 4:** Peterson et al.[756] reported the stability of metoprolol tartrate 10-mg/mL oral suspension prepared from 100-mg tablets (Betaloc). The suspension consisted of the formula shown in Table 63. The suspension was packaged in amber glass bottles and stored refrigerated at 5 to 7°C, at room temperature of 21 to 25°C, and at elevated temperature for accelerated decomposition. The physical appearance of the suspension did not change throughout the study. Stability-indicating HPLC analysis found no loss of metoprolol in 60 days refrigerated and less than 10% loss in 28 days at room temperature.

**STUDY 5:** Yamreudeewong et al.[978] evaluated the stability of metoprolol oral liquid compounded from oral tablets. Three metoprolol tartrate 100-mg tablets (Careco Pharmaceutical) were triturated to fine powder with a mortar and pestle, and 20 mL of deionized water was mixed to form a paste. Sorbitol 70% 80 mL was added to the paste and additional deionized water was added to bring to a final volume of 240 mL. The nominal metoprolol concentration was 1.25 mg/mL. The oral liquid was packaged in amber polyethylene terephthalate prescription bottles and stored at ambient room temperature between 22 and 26°C and under refrigeration at 4 to 8°C.

HPLC analysis of the drug concentrations reported highly variable results, indicating the analysis or sampling was apparently not well controlled. Drug concentrations ranged from 89% after one week to 110% after eight weeks at room temperature. Refrigerated samples were nearly as variable. Although the authors reported metoprolol to be stable in the oral liquid, the results of this study do not definitively support that conclusion.

**STUDY 6 (REPACKAGED TABLETS):** Yang et al.[1447] evaluated the stability of commercial metoprolol tartrate 50-mg tablets (Caraco Pharmaceutical Laboratories) repackaged into unit-dose blister strips consisting of two parts, USP Class A certified foil/paper backing and clear plastic cover material. The materials were formed into unit-dose blister packs containing the tablets and were sealed on all four sides. Samples were stored at room temperature of 25°C and 60% relative humidity for 52 weeks and also at elevated temperature of 40°C and 75% relative humidity for 13 weeks. The samples were evaluated for water content, weight change during storage, tablet hardness, drug content, and dissolution.

No significant changes in the metoprolol tartrate tablets occurred in the samples stored at room temperature. However, for the samples stored at elevated temperature and humidity, a significant increase in water content and weight along with a decrease in tablet hardness occurred. The metoprolol tartrate content did not change during the study in either set of samples, but the dissolution of the samples stored at elevated temperature and humidity was significantly altered. These samples showed a faster dissolution (92% in five minutes) compared to the tablets in the original bottles (51% in five minutes). The authors concluded that metoprolol tartrate tablets repackaged into unit-dose blister packs composed of USP Class A materials underwent little or no changes at room temperature and 60% relative humidity over 52 weeks, but underwent significant moisture uptake in hot, humid conditions that adversely affected tablet characteristics and could compromise product quality in such environments. ∎

# ■ Metronidazole
# Metronidazole Benzoate
# Metronidazole Hydrochloride

## Properties

Metronidazole is an odorless white to yellow or cream crystalline powder or crystals.[1-3] Metronidazole has a very bitter taste.[434] Metronidazole benzoate is a white or slightly yellowish crystalline powder or flakes having a bland taste.[434]

Metronidazole hydrochloride is available as an off-white lyophilized powder with mannitol.[2]

## Solubility

Metronidazole is slightly soluble in water, having an aqueous solubility at 20°C of about 10 mg/mL.[1-3,433] Metronidazole solubility in water has been reported to be 8.3, 8.8, and 11.4 mg/mL at 20, 26, and 30°C, respectively.[530] The aqueous solubility in a pH 7.3 phosphate buffer was found to be 12.2 mg/mL at 30°C and calculated to be 9.5 mg/mL at 25°C.

However, solubility in a suspension formulation at 25°C was found to be only 6 mg/mL.[552] In ethanol, its solubility is about 5 mg/mL.[1–3] The benzoate is practically insoluble in water,[3] having a solubility of about 0.2 mg/mL,[960] and is only slightly soluble in ethanol.[3] The hydrochloride is very soluble in water and soluble in ethanol.[1–3]

## pH

The pH of a saturated aqueous solution is 5.8.[552] Metronidazole injection has a pH between 4.5 and 7.[2,4] Reconstituted metronidazole hydrochloride has a pH between 0.5 and 2; after dilution and neutralization, it has a pH between 6 and 7.[7] Metronidazole gel has an apparent pH between 4 and 6.5.[4]

## pK$_a$

Metronidazole has a pK$_a$ of 2.6.[2]

## Osmolarity

Metronidazole injection has an osmolarity of 310 mOsm/L.[7]

## General Stability Considerations

Metronidazole and metronidazole hydrochloride are stable in air but may darken upon exposure to light.[2,3] Products should be packaged in light-resistant containers. Metronidazole tablets and capsules should be packaged in well-closed containers and stored at room temperature of less than 25°C. Metronidazole hydrochloride injection should be stored at less than 30°C. Metronidazole injection should be stored at controlled room temperature and protected from exposure to light and from freezing.[2,7] Refrigeration may result in crystal formation, but the crystals redissolve upon warming to room temperature.[443]

Reconstituted metronidazole hydrochloride is stable for 96 hours stored below 30°C exposed to normal room light. After dilution in infusion solutions, it should be stored at room temperature and discarded after 24 hours. Refrigeration may result in precipitation.[2,7]

The pH rate of decomposition profile of metronidazole shows a pH-independent region in the range of 3.9 to 6.6 with maximum stability reported at both pH 5.1[988] and 5.6.[614] The rate of decomposition increases as the environment becomes more alkaline, particularly above pH 8.[614] In another study, metronidazole in aqueous buffer solutions was reported to be most stable at pH 4, with an increased rate of loss at pH 6.[940]

Reconstituted metronidazole hydrochloride reacts with aluminum due to its low pH, developing a discoloration variously described as orange, rust, and reddish-brown. Consequently, the use of plastic hub needles is recommended. After dilution and neutralization, the reaction does not occur as readily but may still occur if exposure to aluminum lasts six hours or longer.[2,444–446]

## Stability Reports of Compounded Preparations

### Oral

**STUDY 1:** Mathew et al.[432] evaluated the stability of metronidazole 5 and 10 mg/mL in several oral liquid dosage forms. Metronidazole 10-mg/mL formulations were prepared by crushing metronidazole tablets to fine powder and adding either to water or syrup containing citric acid and sodium benzoate. Metronidazole 5 mg/mL was prepared from bulk metronidazole powder added to water or water with methylparaben 0.1%. The metronidazole hydrochloride 10-mg/mL formulations were prepared from the injection by dilution in water or the syrup. The oral liquids were packaged in amber glass bottles and were stored at 25°C. Metronidazole content was assessed using an HPLC assay.

The liquids prepared from tablets sustained substantial decomposition, especially in the syrup. Losses in water were about 16 to 19% in 28 days; in syrup, losses were about 32 to 37% in 28 days. Metronidazole 5 mg/mL prepared from powder sustained no loss after 60 days. Similarly, metronidazole hydrochloride 10 mg/mL prepared from the injection sustained less than 2% loss in water or syrup after 133 days of storage.[432]

**STUDY 2:** Mathew et al.[434] also evaluated the use of Ora-Plus and Ora-Sweet as vehicles for metronidazole oral liquids. Metronidazole tablets were ground to fine powder and mixed with Ora-Plus or a 1:1 mix of Ora-Plus and Ora-Sweet to yield 10-mg/mL concentrations. Metronidazole 5-mg/mL oral liquids were prepared from ground tablets and bulk powder mixed in Ora-Sweet alone. The preparations were packaged in amber glass bottles and stored at 25°C. Metronidazole content was assessed using a stability-indicating HPLC assay. All oral liquid products retained concentration, with no measurable loss after 90 days of storage. This result is in contrast to the substantial loss of drug that occurred when water or syrup was used as the vehicle.[432]

**STUDY 3:** Irwin et al.[461] evaluated the stability and bioavailability of a metronidazole 15-mg/mL oral suspension compounded from tablets. The suspension was prepared by adding about 10 mL of water to 30 metronidazole 250-mg tablets (Apotex) to dissolve the coating and then crushing the tablets with a pestle to form a smooth slurry. A vehicle composed of simple syrup 300 mL with artificial wild cherry flavor 0.6 mL and brought to 500 mL with chocolate syrup (Nestlé) was added slowly to the slurry with thorough mixing. HPLC analysis was used to evaluate samples stored at temperatures ranging from 67.5 to 88°C, and the results were used to predict the preparation's stability at 4 and 25°C. The times to 10% decomposition were estimated to be 204 days at 4°C and 4.3 days at 25°C. The suspension exhibited no evidence of any bacterial or fungal growth over 16 weeks, and it yielded a mean bioavailability of 97.7 ± 7.7% compared to the tablets in eight adult volunteers.

**STUDY 4:** Allen and Erickson[527] evaluated the stability of three metronidazole 50-mg/mL oral suspensions extemporaneously compounded from powder. Vehicles used in this study were (1) an equal parts mixture of Ora-Sweet and Ora-Plus (Paddock), (2) an equal parts mixture of Ora-Sweet SF and Ora-Plus (Paddock), and (3) cherry syrup (Robinson Laboratories) mixed 1:4 with simple syrup. Metronidazole powder (6 g) (Paddock) was weighed and transferred to a mortar and pestle. About 12 mL of the test vehicle was added to the powder and mixed to yield a uniform paste. Additional vehicle was added geometrically, and the suspension was brought to the final volume of 120 mL, with thorough mixing after each addition. The process was repeated for each of the three test suspension vehicles. Samples of each finished suspension were packaged in 120-mL clear polyethylene terephthalate plastic prescription bottles and stored at 5 and 25°C protected from light.

No visual changes or changes in odor were detected during the study. Stability-indicating HPLC analysis found not more than 7% metronidazole loss in any of the suspensions stored at either temperature after 60 days of storage.

**STUDY 5:** Alexander et al.[552] evaluated the physical and chemical stability of an extemporaneously compounded metronidazole 15-mg/mL oral suspension prepared from tablets. The formula for 2000 mL of the suspension is shown in Table 64. The suspending agents Veegum and sodium carboxymethylcellulose were each hydrated in distilled water overnight. Sixty metronidazole 500-mg tablets (Parmed) were crushed and triturated to fine powder in a glass mortar. The hydrated carboxymethylcellulose sodium was slowly added to the tablet powder with constant trituration. The hydrated Veegum and sorbitol 70% were added with constant mixing. The parabens concentrate, saccharin sodium dissolved in 20 mL of distilled water, and pineapple flavor were added to the mixture with constant mixing. The mixture was brought to a final volume of 2000 mL with

**TABLE 64.** Metronidazole 15-mg/mL Oral Suspension Formula[552]

| Component | Amount | |
|---|---|---|
| Metronidazole 500-mg tablets | 60 | |
| Veegum | 0.75% (wt/vol) | |
| Carboxymethylcellulose sodium | 0.75% (wt/vol) | |
| Sorbitol 70% | 60% (wt/vol) | |
| Saccharin sodium | 0.05% (wt/vol) | |
| Parabens concentrate solution[a] | 1% | |
| Artificial pineapple flavor | Amount unspecified | |
| Distilled water | qs | 3000 mL |

[a]Methylparaben 10% and propylparaben 2% in neat propylene glycol.

distilled water. The suspension was packaged in 120-mL amber glass bottles and stored at elevated temperatures (70 to 85°C).

The suspension exhibited sedimentation but was easily resuspended with shaking. Based on stability-indicating HPLC analysis of the samples at elevated temperatures, the time to 10% decomposition ($t_{90}$) at 25°C was calculated to be about 73 years. However, the authors indicated that it would be reasonable to restrict use to 90 days.

**STUDY 6 (BENZOATE):** Mathew et al.[433] evaluated the stability of the benzoate ester of metronidazole prepared as oral suspensions. Although poorly soluble in water, metronidazole benzoate has a bland taste compared to the bitter metronidazole. The suspensions were prepared from metronidazole benzoate powder by mixing with Ora-Plus or mixing with Ora-Plus and Ora-Sweet as a 1:1 mixture. The metronidazole content was assessed using a stability-indicating HPLC assay. After 90 days of storage at room temperature, there was no loss of metronidazole in either formulation. Furthermore, the physical properties and pH did not change.

**STUDY 7 (BENZOATE):** Hoelgaard and Moller[1127] found that crystallization occurs in aqueous metronidazole benzoate suspensions resulting from the conversion of the anhydrous form to the monohydrate form. Consequently, Zietsman et al.[1128] evaluated several suspending agents for use in preparing metronidazole benzoate oral suspensions to determine if these suspending agents would prevent the hydration that leads to crystallization. The formulations included (unnamed) preservatives, sweeteners, flavor, wetting agent, cosolvent, and purified water in addition to one of several suspending agents. The use of povidone and magnesium aluminum silicate formed sediment upon standing. Avicel RC-591 resulted in sedimentation at a concentration of 0.4% mass/vol.

However, Avicel RC-591 at concentrations of 0.8 to 1.4% mass/vol did not exhibit sedimentation. In addition, xanthan gum 0.55 to 1.0% (mass/vol) did not result in sedimentation. The metronidazole benzoate concentrations were between 95 and 105% mass/vol. These suspensions remained physically and chemically stable for three months at 5 and 25°C and 60% relative humidity. Zietsman et al.[1128] concluded that Avicel RC-591 in adequate concentrations showed promise as a suspending agent for metronidazole benzoate.

**STUDY 8 (BENZOATE):** Vu et al.[1295] evaluated the stability of metronidazole benzoate formulated as a 70-mg/mL oral suspension in SyrSpend SF (Gallipot) suspending vehicle. Metronidazole benzoate powder 55.5 g was triturated in a mortar with 1.25 g of propylene glycol to form a smooth paste. SyrSpend SF was added incrementally and mixed well until the mixture was pourable. The liquid was transferred to a graduated container, and the mortar was rinsed three times with small aliquots of SyrSpend SF. Additional vehicle was added to bring the oral suspension to a volume of 750 mL and was mixed well. The oral suspension was packaged as 20-mL portions in 60-mL amber plastic prescription bottles. Samples were stored refrigerated at 2 to 8°C and at room temperature of 25°C and 60% relative humidity for 360 days, and at elevated temperatures of 40 and 55°C for 90 days.

Visual inspection found no changes at any of the storage conditions. Stability-indicating HPLC analysis found little change in drug concentration when refrigerated for 360 days. At room temperature, the drug concentration exhibited little change through 180 days but increased to 112% of the initial concentration after 360 days, possibly indicating loss of water from the formulation over time. Increasing concentration was also observed in the elevated temperature samples, especially the 55°C samples. The times to 10% drug loss ($t_{90}$) were calculated to be two years under refrigeration and one year at room temperature, although shorter periods may be warranted because of potential moisture loss during storage.

### Enteral

Ortega de la Cruz et al.[1101] reported the physical compatibility of an unspecified amount of oral liquid metronidazole (Flagyl, Searle) with 200 mL of Precitene (Novartis) enteral nutrition diet for a 24-hour observation period. No particle growth or phase separation was observed.

### Topical

Allen[530] described the preparation of a topical metronidazole 10-mg/mL solution prepared from bulk powder for the treatment of malodorous decubitus ulcers. The solution was prepared by adding 1 g of metronidazole powder to sufficient purified water to make 100 mL. The mixture was stirred using a magnetic stir bar until the powder had dissolved. The solution was filtered through a 0.22-µm filter and was packaged in a sterile pour bottle or sterile spray bottle. This metronidazole solution is similar to the oral solutions prepared from powder reported by Mathew et al.[432] (See *Oral* section.) Mathew et al. reported no loss of metronidazole at a 5-mg/mL concentration in water in 60 days.

### Injection

Injections, like other sterile drugs, should be prepared in a suitable clean air environment using appropriate aseptic procedures. When prepared from nonsterile components, an appropriate and effective sterilization method must be employed.

Allen[1358] reported on a compounded formulation of metronidazole 2.1-mg/mL in 5.25% dextrose injection. The injection had the following formula:

| | | |
|---|---|---|
| Metronidazole | | 210 mg |
| Dextrose | | 5.25 g |
| Sterile water for injection | qs | 100 mL |

The recommended method of preparation is to dissolve the metronidazole and dextrose in about 90 mL of sterile water for injection that has been heated to about 60°C and mix until the powders are dissolved. The solution is then cooled and brought to 100 mL with additional sterile water for injection. The solution is to be filtered through a suitable 0.2-µm sterilizing filter and packaged in sterile tight, light-resistant containers. If no sterility test is performed, the USP specifies a beyond-use date of 24 hours at room temperature or three days stored under refrigeration because of concern for inadvertent microbiological contamination during preparation. However, if an official USP sterility test for each batch of drug is performed, the author recommended a beyond-use date of six months at room temperature because this formula is similar or the same as a commercial medication in some countries with an expiration date of two years or more.

### Veterinary

Fan et al.[703] reported the stability and dissolution characteristics of metronidazole benzoate in solid lozenge and liquid emulsion dosage forms. Metronidazole benzoate was prepared in chicken-flavored glycerinated gelatin lozenges containing 200 mg of drug in a 1.24-g lozenge. The drug was also prepared as an oil-in-water emulsion by dissolving in olive oil and emulsifying with polysorbate 80 and water. The two dosage forms were stored for 14 days at room temperature and refrigerated at 4°C. Stability-indicating HPLC analysis found little or no loss of drug at either temperature in either dosage form within 14 days. Dissolution testing found 90% release in 12.8 and 83 minutes for the lozenge and emulsion formulations, respectively. ∎

# Mexiletine Hydrochloride

## Properties

Mexiletine hydrochloride is a white to off-white odorless crystalline powder with a slightly bitter taste.[1,3,7]

### Solubility

Mexiletine hydrochloride is freely soluble in water and ethanol.[3,7]

### pH

A 10% aqueous solution of mexiletine hydrochloride has a pH between 3.5 and 5.5.[3,4]

### pKa

The drug has a pKa of 9.2.[7]

## General Stability Considerations

Mexiletine hydrochloride capsules should be packaged in tight containers and stored below 30°C.[4,7]

## Stability Reports of Compounded Preparations

### Oral

**STUDY 1:** The stability of two extemporaneously compounded oral suspensions of mexiletine hydrochloride (Boehringer Ingelheim) 10 mg/mL was evaluated by Nahata et al.[736] The contents of 14 mexiletine hydrochloride 150-mg capsules were pulverized using a mortar and pestle and mixed thoroughly with (1) distilled water and (2) sorbitol solution to form a paste. Additional diluent was added geometrically to reach the 10-mg/mL concentration. The suspensions were packaged in amber plastic prescription bottles and stored refrigerated at 4°C and in a water bath at room temperature of 25°C for up to 91 days.

No change in appearance, color, or odor was observed. Stability-indicating HPLC analysis found the suspension in water to be more stable than the sorbitol suspension. In water, mexiletine was stable for 91 days refrigerated and 70 days at room temperature, with losses of 5 and 6%, respectively, in those time frames. In sorbitol solution, the drug was stable for 28 days refrigerated and 14 days at room temperature, with losses of 5 and 6%, respectively.

**STUDY 2:** Alexander and Kaushik[868] evaluated the stability of a mexiletine hydrochloride 10-mg/mL oral solution prepared from the contents of commercial 150-mg capsules (Novopharm) in the formula shown in Table 65. The contents of the proper quantity of capsules were mixed in simple syrup and sorbitol and stirred with an electric mixer. The rest of the formulation components were then added and brought to volume with reverse osmosis water. The mixture was passed through a Nalgene 0.45-μm nylon filter, packaged in glass bottles, and stored at temperatures of 4, 30, 40, 50, and 60°C over 90 days. Stability-indicating HPLC analysis of the solutions was used to calculate the shelf life at room temperature and under refrigeration. The shelf life was calculated to be 115 days at room temperature and 173 days under refrigeration.

**TABLE 65.** Mexiletine Hydrochloride Oral Solution Formula[868]

| Component | | Amount |
|---|---|---|
| Mexiletine hydrochloride (from commercial capsules) | | 1000 mg |
| Raspberry syrup | | 0.8 mL |
| Sorbitol | | 20 mL |
| Syrup, USP | | 20 mL |
| Paraben concentrate[a] | | 1 mL |
| Purified water | qs | 100 mL |

[a]Methylparaben 10%, propylparaben 2% in propylene glycol.

# Miconazole
# Miconazole Nitrate

## Properties

Miconazole is a white or almost white to pale cream-color powder.[3,4] Miconazole nitrate is a white or almost white crystalline powder having not more than a slight odor.[3,4]

### Solubility

Miconazole is insoluble in water[4] (less than 1.03 mg/L)[1092] but soluble in ethanol (about 1 part in 9.5 parts), isopropanol (1 part in 4 parts), methanol (1 part in 5.3 parts), propylene glycol (1 part in 9 parts), and freely soluble in dimethylformamide.[3,4]

Miconazole nitrate is very slightly soluble in water (about 1 part in 6250 parts) and isopropanol (1 part in 1408 parts), slightly soluble in ethanol (1 part in 312 parts) and propylene glycol (1 part in 119 parts), sparingly soluble in methanol (1 part in 75 parts), soluble in dimethylformamide, and freely soluble in dimethyl sulfoxide.[3,4]

## pH

Miconazole injection has an official pH range from 3.7 to 5.7.[4]

## pKa

Miconazole has a pK$_a$ of about 6.5[1071,1093] to 6.9[1094] and a pK$_b$ of 7.35.[1072]

## General Stability Considerations

Miconazole powder is packaged in well-closed containers and stored at controlled room temperature protected from exposure to light.[3,4] Miconazole injection is packaged in single-dose containers and stored at controlled room temperature.[4]

Miconazole nitrate powder is packaged in well-closed containers and stored at controlled room temperature protected from exposure to light. Miconazole nitrate cream is packaged in collapsible tubes or tight containers and stored at controlled room temperature. Miconazole topical powder is packaged in well-closed containers and stored at controlled room temperature. Miconazole nitrate vaginal suppositories are packaged in tight containers and stored at controlled room temperature.[4]

## Stability Reports of Compounded Preparations

### Ophthalmic Preparations

Ophthalmic preparations, like other sterile drugs, should be prepared in a suitable clean air environment using appropriate aseptic procedures. When prepared from nonsterile components, an appropriate and effective sterilization method must be employed.

Lee and Lai[1196] reported the stability of miconazole prepared in either peanut oil or castor oil for use as ophthalmic preparations. The preparations were sterilized by dry heating at 160°C for 90 minutes. No loss of miconazole from decomposition occurred.

### Oral

Allen[1345] reported on a compounded formulation of miconazole 2% oral gel. The gel had the following formula:

| | | |
|---|---|---|
| Miconazole nitrate | | 2 g |
| Poloxamer F-127 | | 20 g |
| Cremophor RH 40 | | 10 g |
| Propylene glycol | | 10 g |
| Kollidon 90 | | 5 g |
| Saccharin sodium | | 300 mg |
| Orange flavor | qs | |
| Purified water | qs | 100 g |

The recommended method of preparation was to heat the Cremophor RH 40 and poloxamer F-127 and mix well. Incorporate the miconazole nitrate and orange flavor. Then separately heat about 50 mL of purified water to 90°C and add the propylene glycol, saccharin sodium, and finally add the Kollidon 90. Slowly add this aqueous mixture to the miconazole nitrate mixture. Add sufficient purified water to bring to weight, allow the air bubbles to escape, and then cool to room temperature. The gel was to be packaged in tight, light resistant containers. The author recommended a beyond-use date of six months at room temperature because this formula is a commercial medication in some countries with an expiration date of two years or more.

## Compatibility with Other Drugs

Jagota et al.[920] evaluated the compatibility and stability of mupirocin 2% ointment (Bactroban, Beecham) with miconazole nitrate 2% cream and lotion (Monistat Derm, Janssen) mixed in a 1:1 proportion over periods up to 60 days at 37°C. The physical compatibility was assessed by visual inspection, while the chemical stability of mupirocin was evaluated by stability-indicating HPLC analysis. The study found that mupirocin 2% ointment was physically incompatible upon mixing with miconazole nitrate 2% cream and lotion (Monistat Derm, Janssen), resulting in separation and layering along with nearly total loss of mupirocin in 15 days. ■

# ■ Midazolam Hydrochloride

## Properties

Midazolam is a white to yellowish crystalline powder. The hydrochloride salt of midazolam is formed in situ during manufacturing.[2,7]

## Solubility

The aqueous solubility of midazolam hydrochloride is pH-dependent. As the pH decreases and the solution becomes more acidic, the drug's solubility increases (Table 66). A vehicle with a pH of around 4 is required to dissolve midazolam to concentrations of 2 mg/mL.[131,1549]

**TABLE 66.** Midazolam Hydrochloride Solubility–pH Profile at 25°C[2,131,1549]

| pH | Solubility (mg/mL) |
|---|---|
| 6.2 | 0.24 |
| 5.1 | 1.09 |
| 3.8 | 3.67 |
| 3.4 | 10.3 |
| 2.8 | <22 |

## pH

Midazolam hydrochloride injection has a pH adjusted to around 3.[2,7]

## pK_a

The drug has a pK$_a$ of 6.15.[131]

## Osmolality

Midazolam hydrochloride injection 5 mg/mL has an osmolality of 385 mOsm/kg.[2]

## General Stability Considerations

Midazolam injection should be stored at controlled room temperature and protected from light.[2] Midazolam hydrochloride is stable at pH 3 to 3.6.[2] At pH 4 or less, the drug exhibits its best water solubility.[384,1549] At higher pH values, greater lipid solubility occurs.[384] Midazolam hydrochloride exhibited substantive losses when combined in a liquid preparation containing hydroxypropylmethylcellulose 4000.[1570]

## Stability Reports of Compounded Preparations

### Injection

Injections, like other sterile drugs, should be prepared in a suitable clean air environment using appropriate aseptic procedures. When prepared from nonsterile components, an appropriate and effective sterilization method must be employed.

Peterson et al.[653] reported the stability of midazolam hydrochloride 3 mg/mL in 0.9% sodium chloride repackaged in polypropylene syringes (Terumo) and stored for 13 days. No changes to the appearance of the solution occurred. HPLC analysis found 6.5% loss at 20°C and 8.7% loss at 32°C. In glass vials stored at 32°C, losses of 8.9% were found.

Stiles et al.[529] evaluated the stability of midazolam hydrochloride (Roche) 2 mg/mL in 0.9% sodium chloride injection. The solution was packaged as 3 mL in 10-mL polypropylene infusion-pump syringes for the CADD-Micro and CADD-LD pumps (Pharmacia Deltec) sealed with Luer tip caps (Becton Dickinson) and stored at 5 and 30°C. HPLC analysis found less than 6% loss of the drug at 30°C and no loss at 5°C after 10 days of storage.

Trissel and Hassenbusch[1550] evaluated the stability of compounded preservative-free midazolam 2.5- and 5-mg/mL (as hydrochloride) injection intended for investigational intrathecal administration. Midazolam (as hydrochloride) injection was prepared by dissolving midazolam powder in sodium chloride 0.9% to prepare the 2.5-mg/mL concentration and in sodium chloride 0.45% to prepare the 5-mg/mL concentration. The solutions had their pH values adjusted to pH 3.85 and 3.6, respectively, with 1 N hydrochloric acid. These pH values were designed to dissolve the powder and ensure that the solution pH was sufficient to retain the midazolam in solution. The colorless to slightly yellow intrathecal

2.5- and 5-mg/mL injections had osmolalities of 316 and 200 mOsm/kg, respectively. The intrathecal injections were passed through 0.22-μm filters into 20-mL Type I flint glass vials and sealed with rubber closures. The sealed vials were sterilized by autoclaving at 121°C for 30 minutes; there was little or no loss of the midazolam content. Samples of each concentration were stored at 4, 23, and 37°C. Stability-indicating HPLC analysis found 4% or less loss of midazolam after three months of storage at all of the storage temperatures.

### Oral

Various approaches have been tried to make midazolam given orally more palatable, particularly to children. Peterson[610] reported trying several beverages and syrup as vehicles but found that the most palatable form was to mix the midazolam injection in concentrated grape Kool-Aid prepared by reconstituting a two-quart package with two cups of water. Rosen and Rosen[611] also tried a number of beverages and flavorings but preferred to add midazolam to a sugar-sweetened gelatin dessert prepared with at least 1.3 mL of gelatin dessert for every milliliter of drug and allowed to solidify in ice cube trays. The gelatin cubes were made with 5, 10, or 15 mg of midazolam; fractional doses were provided by cutting the cubes. Because the pH of the gelatin preparation was less than 4, the authors indicated the drug should remain stable, but no stability data were provided. (See the stability studies by Bhatt-Mehta et al.[115] and Allen[583] that follow.)

Miethke[1679] reported that midazolam occasionally forms an insoluble crystalline precipitate if combined with saccharin. The precipitate was found to consist of the insoluble saccharinate salt of midazolam.

**STUDY 1 (LIQUID):** The stability of midazolam hydrochloride as an extemporaneously prepared oral liquid was evaluated by Steedman et al.[50] A midazolam concentration of 2.5 mg/mL (as the hydrochloride) was prepared by combining midazolam hydrochloride injection (Versed, Hoffmann-La Roche) in a 1:1 ratio with Syrpalta (Emerson), a flavored dye-free syrup. The oral liquid was packaged in one-ounce amber glass bottles with child-resistant caps and stored at 7, 20, and 40°C for 56 days. The samples were inspected visually for changes in color, odor, or turbidity, but no such changes were observed. A stability-indicating HPLC assay was used to assess intact midazolam content. Little or no loss of midazolam occurred. Although inspected for signs of microbial growth, no formal testing was conducted, because the syrup is not a medium that supports microbial growth and benzyl alcohol (1%) was present in the injection.

**STUDY 2 (LIQUID):** Gregory et al.[693] reported on the stability of an oral solution of midazolam hydrochloride. Midazolam hydrochloride injection was added to simple syrup with 0.23% peppermint oil for flavoring to yield 2.5- and 3-mg/

mL concentrations. The oral liquids were packaged in amber glass prescription bottles. HPLC analysis found midazolam hydrochloride to be stable at room temperature for 14 days in this vehicle. The 2.5-mg/mL concentration exhibited about 10% loss, while the 3-mg/mL concentration exhibited only 3% loss in two weeks. However, after 38 days (the next assay point) unacceptable losses of 25 to 28% occurred.

**STUDY 3 (LIQUID):** Allen[131] reported that several vehicles were tried in an attempt to mask the taste of midazolam above 3 mg/mL; the simple syrup with peppermint oil, cherry syrup, strawberry syrup, and chocolate syrup all failed.

**STUDY 4 (LIQUID):** Ivey et al.[994] evaluated the stability of midazolam 0.5 mg/mL in an oral liquid composed of saccharin 1.2 mg/mL in sterile water. HPLC analysis found that less than 10% drug loss occurred in 84 days stored at ambient room temperature and with room light. No assessment of the taste of the oral solution was reported.

**STUDY 5 (LIQUID):** Soy et al.[406] evaluated the stability of midazolam hydrochloride 1 mg/mL as an oral solution. The oral solution contained 20 mL of midazolam hydrochloride (5 mg/mL), saccharin sodium 240 mg, lemon or strawberry flavor, and purified water 80 mL. The compounded solution was filtered through a 0.5-μm filter. The stability of the midazolam hydrochloride was assessed by HPLC as well as by ultraviolet spectrophotometry. No loss of the drug occurred by either assay technique during 73 days of storage. The strawberry-flavored preparation had greater taste acceptance than the lemon variation.

**STUDY 6 (LIQUID):** Walker et al.[508] evaluated the stability of midazolam at concentrations of 0.35, 0.64, and 1.03 mg/mL prepared from midazolam hydrochloride injection (Versed, Roche) in an orange-flavored syrup. The syrup was prepared by adding 30 mL of distilled water to 50 mL of simple syrup and then adding 0.12 mL of pure orange extract with shaking. One drop each of red and yellow food coloring was added, and additional distilled water was incorporated to bring the volume to 100 mL. The midazolam hydrochloride injection was added to yield the test concentrations. The syrup was packaged in polyethylene containers and stored at 23°C for 102 days. The orange-colored syrup remained clear throughout the study period. Stability-indicating HPLC analysis found midazolam losses of 6.5% or less after storage for 102 days.

**STUDY 7 (LIQUID):** Mehta et al.[763] reported that midazolam 1 mg/mL prepared from the commercial injection (Hypnovel) in syrup, BP, was stable for at least six weeks at room temperature exposed to and protected from light and refrigerated at 4°C when evaluated using stability-indicating HPLC analysis.

**STUDY 8 (LIQUID):** Ruiz Caldes et al.[1563] evaluated the stability of an oral liquid of midazolam hydrochloride prepared from the injection added to simple syrup yielding a concentration of 2.5 mg/mL. The oral liquid was packaged in both glass and plastic containers and was stored at ambient room temperature near 25°C, refrigerated at 2 to 8°C, and frozen at −15°C for 60 days. Visual inspection found no change in odor or appearance during the study including color and clarity. HPLC analysis found little or no loss of midazolam hydrochloride throughout the study.

**STUDY 9 (LIQUID WITH ATROPINE):** Ros and van der Meer[1570] evaluated several compounded formulations of midazolam hydrochloride 1 mg/mL with atropine sulfate 0.1 mg/mL. When hydroxypropyl methylcellulose 4000 (HPMC) was included in the formulations, a substantive loss of midazolam occurred. The formulations without HPMC that were acceptable are as follows:

*Formula 1*

| | | |
|---|---|---|
| Midazolam (as hydrochloride) | | 1500 mg |
| Atropine sulfate | | 150 mg |
| Methyl paraben solution | | 15 mL |
| Syrup | | 200 mL |
| Sodium saccharin | | 7.5 g |
| Fragrance | | 20 mL |
| Red color | | 5 mL |
| Purified water | qs | 150 mL |

*Formula 2*

| | | |
|---|---|---|
| Midazolam (as hydrochloride) | | 1500 mg |
| Atropine sulfate | | 150 mg |
| Sodium chloride | | 1.5 g |
| Sodium saccharin | | 4.5 g |
| Methyl paraben | | 2.25 g |
| Anise oil | | 30 drops |
| Purified water | qs | 1500 mL |

HPLC analysis found that midazolam hydrochloride and atropine sulfate remained stable in both formulas exhibiting little or no loss of either drug over 52 weeks of storage.

**STUDY 10 (GELATIN):** The stability of midazolam hydrochloride in flavored gelatin was evaluated by Bhatt-Mehta et al.[115] Midazolam concentrations of 1 and 2 mg/mL in flavored gelatin (Jello, Kraft) were prepared by adding 30 mL of midazolam injection 5 mg/mL (Versed, Hoffmann-La Roche) to 120 mL of freshly prepared liquid gelatin for a 1-mg/mL concentration and 90 mL of injection to 135 mL of gelatin for a 2-mg/mL concentration. The preparations were packaged in unit-dose cups containing 5 mg/5 mL and 15 mg/7.5 mL, respectively. Samples were stored under refrigeration at 4°C for 14 days or frozen at −20°C for 28 days. A stability-indicating

HPLC assay was used to assess midazolam content. No loss occurred in the refrigerated samples through 14 days and in the frozen samples through 28 days of storage. No change in color or odor occurred, and no evidence of bacterial growth was observed. These preparations are very sweet but produce a bitter aftertaste that is much more intense in the concentrated 2-mg/mL formulation. For this reason, the authors recommended that the midazolam concentration not exceed 1 mg/mL.

**STUDY 11 (GELATIN):** Allen[583] reported on a similar formulation of midazolam in dessert gelatin. The gelatin base (Jello) was prepared by adding six ounces (3/4 cup) of boiling water to a small three-ounce package of the flavored gelatin and allowing it to cool to about 40°C. One milliliter of midazolam 5-mg/mL injection was added to 1.3 mL of the gelatin solution and mixed well. The mixture was transferred to a small ice cube tray or other suitable mold and refrigerated. After the gelatin was set, the dose required was obtained by cutting the gelatin into smaller portions or administering multiple units or by some combination of the two. The gelatin helps to mask the bitter taste of the drug by entrapping it within the gelatin matrix. Although the stability of this specific formulation was not addressed, it should be similarly stable under refrigeration to the gelatin-based preparation of Bhatt-Mehta et al.[115]

### Nasal

Elder et al.[1642] evaluated the stability of a midazolam 50-mg/mL nasal formula for use in treating epilepsy in dogs. The formulation was prepared from midazolam powder. Midazolam 500 mg was wetted with 1 mL of sterile water. Hydrochloric acid 1% 1 mL was added followed by 4 mL of hydroxypropyl methylcellulose 1% (wt/vol) and 0.05 mL of benzalkonium chloride 17% (wt/vol). The apparent pH of the mixture was adjusted to pH 2.8 to 3 with sodium hydroxide 1% (wt/vol). The mixture was then brought to a volume of 10 mL with additional sterile water. The nasal formulation was packaged in clear glass vials and capped. The vials were placed in light-resistant containers with samples stored at room temperature of 22 to 25°C and under refrigeration at 3 to 5°C for up to 90 days.

Visual examination found white precipitate formed in some refrigerated samples as early as three days after preparation and was eventually found in all refrigerated samples. The room temperature samples remained clear, colorless, and free of precipitation for up to 30 days, but white precipitation formed after that time in the room temperature samples as well. Stability-indicating HPLC analysis found less than 10% loss of midazolam after 30 days of storage at room temperature and even under refrigeration. Beyond 30 days drug losses exceeded 25 to 30%. Because of the precipitation in the refrigerated samples, the authors recommended only room temperature storage. ■

# ▨ Milk of Magnesia

## Properties

Milk of magnesia is an opaque white viscous suspension. After milk of magnesia stands for awhile, water will usually separate.[4] Milk of magnesia provides 13.66 mEq of magnesium per 5 mL.[2]

### Solubility

Magnesium hydroxide is practically insoluble in water (about 12.5 mcg/mL).[1]

### pH

Milk of magnesia has a pH near 10.[4]

## General Stability Considerations

Magnesium hydroxide should be packaged in tight containers and stored at controlled room temperature. Milk of magnesia absorbs carbon dioxide and should be packaged in well-closed containers and stored at controlled room temperature.[1,4]

## Stability Reports of Compounded Preparations
### Enteral
**MILK OF MAGNESIA COMPATIBILITY SUMMARY**
*Compatible with:* Enrich • TwoCal HN • Vivonex T.E.N.

Burns et al.[739] reported the physical compatibility of milk of magnesia 20 mL with 10 mL of three enteral formulas, including Enrich, TwoCal HN, and Vivonex T.E.N. Visual inspection found no physical incompatibility with any of the enteral formulas. ■

# Minocycline Hydrochloride

## Properties
Minocycline occurs as a yellow, hygroscopic, crystalline powder.[1,3,4]

### Solubility
Minocycline hydrochloride is sparingly soluble in water, slightly soluble in ethanol, and soluble in alkaline hydroxide and carbonate solutions.[3,4]

### pH
Minocycline 10 mg/mL as the hydrochloride in water is between pH 3.5 and 4.5.[3,4] Minocycline (as hydrochloride) for injection at a minocycline concentration of 10 mg/mL has a pH of 2 to 3.5. Minocycline (as hydrochloride) oral suspension has a pH of 7 to 9.[4]

### Osmolality
Minocycline (as hydrochloride) 20 mg/mL reconstituted with sterile water for injection is hypotonic having a measured osmolality of 107 mOsm/kg.[900]

## General Stability Considerations
Minocycline hydrochloride is slightly hygroscopic, sensitive to exposure to light, and subject to surface oxidation.[1] Minocycline hydrochloride bulk powder should be packaged in tight containers and protected from exposure to light during room temperature storage.[3,4]

Minocycline (as hydrochloride) oral tablets and oral suspension are packaged in tight, light-resistant containers and stored at controlled room temperature.[4]

Minocycline (as hydrochloride) for injection should be stored at controlled room temperature and protected from temperatures above 40°C and from exposure to light. Reconstituted minocycline hydrochloride injection is stated by the manufacturer to be stable for 24 hours at room temperature.[4]

## Stability Reports of Compounded Preparations
### Topical
Chow et al.[1273] evaluated the stability of minocycline hydrochloride 1% wt/wt in various nonaqueous hydrophilic solvents and gels for topical application. HPLC analysis was used to determine minocycline hydrochloride concentration.

Minocycline hydrochloride stability was enhanced by incorporating magnesium chloride 0.82% wt/wt to reduce the rate and extent of minocycline epimerization. Minocycline hydrochloride was found to be most stable in an equal parts glycerin and propylene glycol solvent system with the magnesium chloride present. The magnesium chloride was incorporated into the glycerin portion using heat during the compounding process. When stored at 40°C and protected from exposure to light, the time to 10% loss of minocycline was about 14 days. The authors recommended that the topical gel then be prepared from this mixture, incorporating poly-N-vinylacetamide/sodium acrylate copolymer (PNVA) 3% as the gelling agent. ■

# Minoxidil
# Minoxidil Tartrate

## Properties
Minoxidil is a white to off-white odorless crystalline powder.[2,3,584]

Minoxidil tartrate is a white crystalline powder.[584]

### Solubility
Minoxidil has an aqueous solubility of about 2 mg/mL.[1,584] Its solubility in alcohol is in the range of 15 to 30 mg/mL[1,584] and in propylene glycol is about 75 mg/mL[1] to 90 mg/mL.[584]

Minoxidil tartrate has an aqueous solubility of about 13 mg/mL.[584]

### pKa
Minoxidil has a pKa of 4.6.[2]

## General Stability Considerations
Minoxidil tablets and topical solution should be packaged in tight containers and stored at controlled room temperature.[2,4]

Minoxidil is photolabile, developing a yellow discoloration. Photodegradation is reported to occur at a higher rate at lower concentrations[1174] and at acidic pH.[585] However, minoxidil's overall stability was reported to be greatest at pH 4.5.[998]

## Stability Reports of Compounded Preparations
### Topical
STUDY 1: Allen[584] reported on a topical aqueous solution of minoxidil tartrate for iontophoretic delivery. Minoxidil tartrate 500 mg was dissolved in deionized water and brought to a final volume of 100 mL, resulting in a 5-mg/mL concentration. Packaging in tight, light-resistant containers

was recommended, but no specific expiration period was suggested.

**STUDY 2:** Chinnian and Asker[585] studied factors that influence the photostability of minoxidil solutions. The results indicated that the rate of photodegradation increased with increasing concentration, at lower pH, in the presence of acetate buffer, and in clear glass or polypropylene containers. Photodegradation was reduced if the solutions were packaged in amber or foil-covered glass vials. Solvent composition was also important. Minoxidil was most photostable in water and in water 25% in polyethylene glycol 300; the drug was less stable in water 25% in either ethanol or propylene glycol. The

addition of sodium thiosulfate was the most effective photo-stabilizer tested.

**STUDY 3:** Cao and Ge[1098] evaluated the stability of minoxidil 20 mg/mL in a topical lotion that also contained propylene glycol 10 mL, distilled water 20 mL, and ethanol qs 100 mL. The topical lotion was packaged in glass containers and stored at elevated temperatures ranging from 50 to 80°C to accelerate the rate of decomposition and exposed to ambient room light. The clear yellow lotion underwent a darkening of the yellow color during elevated temperature storage. Using ultraviolet spectrophotometric analysis, the authors calculated the shelf life of the lotion at room temperature of 20°C to be 6.7 months. ■

# ■ Misoprostol

## Properties

Misoprostol is a light yellow viscous oil with a musty odor that is a prostaglandin $E_1$ analog.[1,2,7,946]

### Solubility

Misoprostol is stated to be soluble in water.[1]

## General Stability Considerations

Misoprostol tablets should be packaged in tight containers and stored in a dry place at controlled room temperature.[2,7] The manufacturer has stated that misoprostol tablets should not be removed from the original packaging and placed in dosing compliance aids because of moisture sensitivity.[1622]

## Stability Reports of Compounded Preparations

### Rectal

**STUDY 1:** Hafirassou et al.[946] evaluated the stability of misoprostol rectal suppositories prepared from commercial tablets. Misoprostol 200-mcg tablets (Cytotec, Pharmacia) were crushed using a mortar and pestle. The suppository base (Suppocire, Cooper Industrie) was melted at 37°C using a water bath. The melted suppository base was added to the crushed tablet powder with trituration. The liquid was then transferred to an appropriate size suppository mold and allowed to cool and solidify. Each suppository weighed about 3 g and contained 1 mg of misoprostol or the equivalent of five tablets. Test suppositories were stored for 180 days refrigerated at 3 to 5°C, at room temperature of 23 to 25°C, and at elevated temperature of 55 to 60°C.

Stability-indicating HPLC analysis of the suppositories found that misoprostol losses of 7% occurred in 180 days refrigerated and about 9% occurred in 120 days at room temperature. The samples stored at elevated temperature lost about 11% in 60 days.

**STUDY 2:** Khanderia et al.[986] reported the stability of misoprostol suppositories compounded from misoprostol 200-mcg tablets. The tablets were crushed using a glass mortar and pestle. The tablet powder was added to Polybase, a polyethylene glycol suppository base, melted in a water bath at 60°C with constant stirring. The molded suppositories were stored frozen at −20°C, refrigerated at 2 to 4°C, and at room temperature of 25°C. HPLC analysis found that the suppositories retained more than 90% of the initial drug concentration for 12 months in frozen and refrigerated storage. However, at room temperature, the drug was stable for only one week.

### Injection

Injections, like other sterile drugs, should be prepared in a suitable clean air environment using appropriate aseptic procedures. When prepared from nonsterile components, an appropriate and effective sterilization method must be employed.

Chu et al.[1235] evaluated the preparation and stability of an extemporaneously prepared injection of misoprostol extracted from a commercial tablet formulation. One 200 mcg misoprostol tablet (Cytotec, Pharmacia) was ground to powder, homogenized, and vortexed in 2 mL of water or sodium chloride 0.9% for five minutes. The mixtures were centrifuged at 4500 rpm for three minutes, and the clear supernatant was transferred to amber glass vials. The vials were stored at room temperature and refrigerated at 4°C for six days. Extraction efficiency of misoprostol from the tablet matrix was about 45 to 50% of the labeled amount. The authors indicated that longer extraction periods did not release more drug. This reduced amount of drug should be considered during clinical application.

Using a stability-indicating LC-MS analytical method, misoprostol was stable at both room temperature and at 4°C for six days with no loss of drug occurring. ■

# Mitomycin

## Properties
Mitomycin is a blue-violet crystalline powder.[1–3]

### Solubility
Mitomycin is soluble in water[1–3,7] and freely soluble in organic solvents.[7] Beijnen et al.[1373] reported that mitomycin was soluble to about 0.6 to 0.8 mg/ml in water. However, refrigeration of concentrations above 0.6 mg/mL may result in precipitation.

### pH
An aqueous suspension of mitomycin containing 5 mg/mL has a pH between 6 and 7.5.[4] The reconstituted injection has a pH of 6 to 8.[2,4]

## General Stability Considerations
Mitomycin vials should be stored at controlled room temperature and protected from temperatures of 40°C or more and from exposure to light. The reconstituted mitomycin solution is stable for one week at room temperature and for two weeks when stored under refrigeration.[2,7]

Mitomycin solution stability is extremely pH-sensitive. The drug is stable at neutral pH but rapidly decomposes in solutions having an acidic or basic pH.[266,385–387,463] The decomposition products differ in acidic and basic solutions for this complex, pH-dependent degradation process.[385,388,389] Maximum stability occurs at pH 7[6,366,385,386] to 8,[463] with a time to 10% decomposition of seven days at room temperature.[6] Heating a mitomycin 0.6-mg/mL aqueous solution in 0.9% sodium chloride injection to 100°C resulted in a 24% loss in 30 minutes and a 58% loss in one hour. Freezing the mitomycin solution at −20°C resulted in crystallization; the crystalline material did not redissolve upon thawing and warming.[390]

## Stability Reports of Compounded Preparations
### Ophthalmic
Ophthalmic preparations, like other sterile drugs, should be prepared in a suitable clean air environment using appropriate aseptic procedures. When prepared from nonsterile components, an appropriate and effective sterilization method must be employed.

**STUDY 1:** Allen[152] described the formulation of mitomycin ophthalmic drops. The 5-mg vial of injection (Mutamycin, Bristol-Myers Oncology) was reconstituted with 10 mL of sterile water for injection; the solution was diluted with additional sterile water for injection to yield 25 mL of a mitomycin 0.02% solution. The solution should be packaged in suitable ophthalmic containers. A similar solution (0.5 mg/mL) was chemically stable for one week at room temperature and two weeks under refrigeration.[2] The microbiological stability is dependent on the compounder's ability to ensure the maintenance of sterility initially and during storage.

**STUDY 2:** Fiscella et al.[266] noted that mitomycin injection reconstituted with sterile water for injection to 0.5 mg/mL is stable for seven days at room temperature and 14 days under refrigeration, according to the package insert. However, it also has been shown to be stable frozen at −20°C for up to 52 weeks; the concentration is even retained for up to 17 freeze–thaw cycles. To avoid unnecessary waste, the authors recommended aseptically repackaging the reconstituted mitomycin injection in small single-use aliquots in sealed sterile vials or syringes for subsequent thawing and use.

**STUDY 3:** Kawano et al.[463] studied the stability of mitomycin 0.4% in several ophthalmic solutions. The vehicles tested included solutions of 5% dextrose, 0.9% sodium chloride, water for injection, artificial tears, and pH 8 phosphate buffer. HPLC analysis found the best mitomycin stability to be in the phosphate buffer at pH 8 (Table 67). The use of water for injection as a diluent caused some degree of discomfort in test volunteers.

**STUDY 4:** Segui Gregori et al.[468] studied the stability of mitomycin (Kyowa) 0.02% in 0.9% sodium chloride for use as an ophthalmic solution. Stored at room temperature exposed to or protected from light, mitomycin was stable for seven days. Under refrigeration, the drug exhibited 10% loss in 14 days.

**STUDY 5:** Francoeur et al.[810] evaluated the stability of several formulations of mitomycin ophthalmic solutions used in hospitals in the United States and Canada. Most were found to be stable for the intended period of use, but several were unstable formulations. The authors reported that mitomycin

**TABLE 67.** Stability of Mitomycin in Several Ophthalmic Solution Vehicles[463]

| Vehicle | pH | Results |
|---|---|---|
| 5% Dextrose | 4.8 | 13% loss in 2 days at 25°C; 10% loss in 7 days at 5°C |
| Sterile water for injection | 5.9 | 11% loss in 2 days at 25°C; 12% loss in 14 days at 5°C |
| 0.9% Sodium chloride | 6.1 | 11% loss in 7 days at 25°C and in 14 days at 5°C |
| Artificial tears | 7.3 | 11% loss in 14 days at 25°C; 2% loss in 14 days at 5°C |
| Phosphate buffer | 8.0 | 8% loss in 14 days at 25°C; no loss in 14 days at 5°C |

0.3 mg/mL in 66 mmol/L sodium phosphate with and without dextrose 5% (pH 7.8) and in balanced salts solution was stable. HPLC analysis found little or no loss of mitomycin in 24 hours at 24°C and for one month at 4°C. However, the authors reported 6 and 12% loss of mitomycin in water and 0.9% sodium chloride, respectively, in 24 hours at 24°C, which is not in agreement with other reports.

**STUDY 6:** Georgopoulos et al.[814] evaluated the stability of mitomycin C 0.02 to 0.4 mg/mL in 0.9% sodium chloride with 0.62 mmol of phosphate to buffer the solution to pH 7.2. Samples were stored at room temperature of 22°C, refrigerated at 4°C, and frozen at −20°C and at −196°C in liquid nitrogen. An agar diffusion bioassay was used to assess mitomycin C activity, but the solutions were not analyzed for drug content. The activity determinations were similar over the entire range of concentrations. At room temperature, about 6% loss of activity occurred in one week and about 12%

loss occurred in two weeks. Less than 10% loss of activity occurred in three months under refrigeration, but 11% loss occurred in only one month frozen at −20°C. The authors provided no explanation for this apparent anomaly. No loss occurred in six months in liquid nitrogen. The use of a relatively insensitive and possibly less accurate bioassay in this study should be considered.

**STUDY 7:** Velpandian et al.[945] reported that pH and storage temperature were critical to retaining mitomycin concentration in solutions for ophthalmic administration. Mitomycin in concentrations of 0.15 to 0.6 mg/mL in pH 6, 7, and 8 buffer solutions was found to be much less stable at pH 6, with losses of up to 96 and 41% occurring in 28 days at room temperature and under refrigeration, respectively. The drug was increasingly stable at pH 7 and 8. Drug losses when stored under refrigeration for 28 days were similar at pH 7 and 8, exhibiting about 10 to 20% loss. ■

# ■ Molindone Hydrochloride

## Properties
Molindone hydrochloride is a white to off-white crystalline powder.[2,7]

### Solubility
Molindone hydrochloride is freely soluble in water and ethanol.[2,7]

### pH
A 1% molindone hydrochloride solution in water has a pH between 4 and 5.[4]

### pK$_a$
The drug has a pK$_a$ of 6.94.[2]

## General Stability Considerations
Molindone hydrochloride products should be packaged in tight, light-resistant containers and stored at controlled room temperature protected from light.[2,4]

## Stability Reports of Compounded Preparations
### Oral
Geller et al.[242] reported that molindone liquid was compatible visually with apple juice and orange juice, although visual observation in opaque liquids is problematic.

## Compatibility with Other Drugs
**MOLINDONE HYDROCHLORIDE COMPATIBILITY SUMMARY**
>    *Compatible with:* Lithium citrate
>    *Incompatible with:* Thioridazine

**STUDY 1:** Theesen et al.[78] evaluated the compatibility of lithium citrate syrup with oral neuroleptic solutions including molindone hydrochloride concentrate. Lithium citrate syrup (Lithionate-S, Rowell) and lithium citrate syrup (Roxane) at a lithium-ion concentration of 1.6 mEq/mL were combined in volumes of 5 and 10 mL with 5, 10, 15, and 20 mL of molindone hydrochloride oral concentrate 20 mg/mL (Endo) in duplicate, with the order of mixing reversed between samples. Samples were stored at 4 and 25°C for six hours and evaluated visually for physical compatibility. No incompatibility was observed.

**STUDY 2:** Raleigh[454] evaluated the compatibility of lithium citrate syrup (Philips Roxane) 1.6 mEq lithium/mL with molindone hydrochloride concentrate 20 mg/mL. The molindone hydrochloride concentrate was combined with the lithium citrate syrup in a manner described as "simple random dosage combinations"; the exact amounts tested and test conditions were not specified. No precipitate was observed to form upon mixing.

Raleigh[454] also evaluated the compatibility of thioridazine suspension (Sandoz) (concentration unspecified) with molindone hydrochloride concentrate 20 mg/mL. The molindone hydrochloride concentrate was combined with the thioridazine suspension in the manner described as "simple random dosage combinations." The exact amounts tested and test conditions were not specified. A dense white curdy precipitate formed. ■

# Mometasone Furoate

## Properties
Mometasone furoate is a white to off-white crystalline powder.[4]

### Solubility
Mometasone furoate is insoluble in water, having an aqueous solubility of 70 mcg/mL. It has a solubility in ethanol of 8.3 mg/mL. Mometasone furoate is soluble in acetone. [2,4]

Salgado et al.[1583] reported the solubilities of mometasone furoate in ethanol:aqueous mixed solvents. The solvent mixtures incorporating ethanol and phosphate-buffered saline pH 4.5 (PBS) or water and the drug solubilities found are as shown:

| Solvent Mixture | Solubility (mcg/mL) |
| --- | --- |
| Ethanol:PBS (1:1) | 239.9 |
| Ethanol:PBS (1:3) | 30.4 |
| Ethanol:PBS (2:1) | 5 |
| Ethanol:water (1:1) | 292.7 |
| Ethanol:water (1:3) | 24.0 |
| Ethanol:water (3:1) | 1765.6 |

### pH
Mometasone furoate topical solution has a pH of 4 to 5.[4] Nasonex nasal spray has a pH of 4.3 to 4.9.[7]

## General Stability Considerations
Mometasone furoate powder, cream, ointment, and topical solution should be packaged in well-closed containers and stored at controlled room temperature.[4] The shelf life of the cream and ointment is two years.[2] Nasonex nasal spray should be stored between 2 and 25°C and protected from exposure to light.[7]

Mometasone furoate stability is pH-dependent. Teng et al.[1604] reported that mometasone furoate was most stable below pH 4 while Salgado et al.[1583] stated that mometasone furoate exhibits stability in the range of pH 4 to 4.5. Teng et al.[1604] also reported faster degradation at or above pH 6.7 with very rapid degradation at alkaline pH values.

## Stability Reports of Compounded Preparations
### Topical
Salgado et al.[1583] reported on the stability of mometasone furoate 0.1% (wt/wt) in a topical gel formulation for scalp dermatitis. The gel was prepared using hydroxypropyl methylcellulose 1.5% previously swollen in purified water, isopropyl alcohol, and propylene glycol in a ratio of 40:40:20. The transparent gel was uniform in appearance and exhibited an alcoholic smell. The gel was packaged in aluminum tubes and stored at room temperature of 23 to 27°C and 55 to 65% relative humidity for 365 days. The gel was found to exhibit an increasing viscosity in the first few days after preparation. After this time the viscosity did not change. Microbiological evaluation found the topical gel remained acceptable and kept effectively throughout the study period. HPLC analysis found that no change in drug concentration occurred over 365 days of storage. Permeation studies found that the mometasone furoate gel formulation permeated at a rate of about 0.81% per 48 hours, which the authors stated made it safe for topical application.

## Compatibility with Other Drugs
Hecker et al.[815] evaluated the compatibility of tazarotene 0.05% gel combined in equal quantity with other topical products, including mometasone furoate 0.1% cream. The mixtures were sealed and stored at 30°C for two weeks. HPLC analysis found that less than 10% tazarotene loss occurred in all. Mometasone furoate also exhibited less than 10% loss in two weeks. ■

# Morphine
# Morphine Hydrochloride
# Morphine Sulfate

## Properties
Morphine sulfate occurs as an odorless white crystalline powder, white featherlike crystals, or cubical masses of crystals.[1–3] The drug has five molecules of water of hydration,[2] but it loses some water during standing exposed to air at ordinary room temperature.[1,3]

Morphine hydrochloride occurs as colorless silky crystalline needles or a white or nearly white crystalline powder. The drug effloresces in dry environments.[3]

### Solubility
The solubility of morphine salts is temperature-dependent. Morphine sulfate has an aqueous solubility of 62.5 mg/mL at 25°C;[2,3] the aqueous solubility increases to more than 1 g/mL at 80°C.[1,3] In ethanol, its solubility is 1.75 mg/mL at 25°C.[2]

Morphine hydrochloride is soluble in water and glycerol and slightly soluble in ethanol.[3] The aqueous solubility of morphine hydrochloride is reported to be 49 and 55 mg/mL at 22 and 25°C, but it drops about 30% to 35 mg/mL when refrigerated at 4°C.[977]

## pH

Aqueous solutions of morphine sulfate have a pH around 4.8.[1] Morphine sulfate injection has a pH between 2.5 and 6.5.[4]

Morphine hydrochloride injection diluted in sodium chloride 0.9% is reported to have a pH near 4 to 5.[1045]

## pK$_a$

The drug has pK$_a$ values of 8.31 for the amino group and 9.51 for the phenolic group at 25°C.[6]

## General Stability Considerations

Morphine sulfate loses water of hydration if exposed to air and gradually darkens in color if exposed to light for prolonged periods.[2,3] Morphine sulfate injection should be stored at controlled room temperature and protected from freezing and exposure to light.[2-4] Morphine sulfate oral formulations should be packaged in tight, light-resistant containers and stored at controlled room temperature. The oral solution of morphine sulfate should be protected from freezing.[2]

The solution stability of morphine salts is dependent on pH and the presence of oxygen. They are relatively stable at acidic pH, especially below pH 4. Optimum pH has been stated to be 3.2, but as the pH increases to neutrality or into the alkaline range, degradation increases. The primary mode of degradation is oxidation of the phenolic group[6] with the formation of pseudomorphine and morphine-N-oxide.[977,981]

The degradation of morphine salts is often accompanied by yellow or brownish solution discoloration,[6,981] although the source of this discoloration remains unexplained. The principal decomposition products, pseudomorphine and morphine-N-oxide, are both colorless in solution. Nevertheless, there is a strong correlation between discoloration and some degradation, although even a dark brown solution may exhibit a loss of only a small percentage of morphine concentration.[977,981]

Other factors adversely impacting morphine stability in aqueous solutions include exposure to sunlight, exposure to ultraviolet radiation, presence of iron, and organic impurities that catalyze the drug's degradation. Exposure to higher storage temperatures of 37°C for up to three months has not caused unacceptable stability. Even autoclaving caused only a slight increase in decomposition product formation. The stability of the drug does not appear to be concentration-dependent, at least not when stored at normal temperatures.[977]

## Stability Reports of Compounded Preparations
### Injection

Injections, like other sterile drugs, should be prepared in a suitable clean air environment using appropriate aseptic procedures. When prepared from nonsterile components, an appropriate and effective sterilization method must be employed.

**STUDY 1 (REPACKAGED):** Gove et al.[391] reported that morphine sulfate injection repackaged into plastic syringes with syringe caps (Braun) was stable and within acceptable degradation limits for at least 69 days stored at room temperature.

**STUDY 2 (REPACKAGED):** Hung et al.[392] found less than 3% morphine sulfate loss by HPLC analysis in 12 weeks at 22°C exposed to light when the injection was repackaged into plastic syringes. The loss was even less when the syringes were stored under refrigeration and protected from light.

**STUDY 3 (REPACKAGED):** Similarly, Walker et al.[393] demonstrated that morphine sulfate 10 mg/mL stored in 100-mL glass vials or polyvinyl chloride bags did not sustain any loss within 30 days when stored at 23°C.

**STUDY 4 (REPACKAGED):** Using HPLC analysis, Duafala et al.[654] reported that morphine sulfate 15 mg/mL and 2 mg/mL diluted with sterile water for injection repackaged in 3-mL disposable glass syringes and stored at 4 and 24°C exhibited little or no loss after 12 days of storage.

**STUDY 5 (REPACKAGED):** Strong et al.[644] evaluated the stability of morphine sulfate 1 and 5 mg/mL repackaged into 30-mL plastic syringes for patient-controlled analgesia. HPLC analysis of the samples found that the morphine sulfate was stable for at least six weeks stored at −20, 4, and 23°C protected from light. However, samples stored at 23°C exposed to light were stable for only a week, with some samples developing unacceptable losses after seven days.

**STUDY 6 (REPACKAGED):** Grassby and Hutchings[656] evaluated the stability of morphine sulfate 2 mg/mL in 0.9% sodium chloride injection repackaged into 30- and 50-mL plastic syringes for use in patient-controlled analgesia and in glass vials. Little or no morphine sulfate loss was found by HPLC analysis after six weeks of storage at room temperature protected from light. The maximum loss was about 5% in the 30-mL syringes.

**STUDY 7 (EXTEMPORANEOUS INJECTION):** Deeks et al.[469] evaluated the stability of morphine sulfate 0.2 mg/mL in 0.9% sodium chloride for intrathecal injection using HPLC analysis to determine the formation of its decomposition product, pseudomorphine. Two batches were prepared in glass ampules; one batch was sealed, retaining air in the headspace, while the other was sealed after displacing the air with nitrogen. Ampules from each batch were autoclaved at 115°C for 30 minutes zero, one, two, or three times. Double-autoclaving may result when the autoclaved intrathecal injection container is supplied in an overwrap that is then autoclaved again for sterility. The ampules were then stored protected from light at ambient room temperature (14 to 22°C) and at an elevated temperature of 32°C for 48 weeks. No precipitation or color changes were observed. Pseudomorphine formation was noticeably larger in the samples that had been autoclaved two or three times. Similarly, the samples stored at room temperature contained no more than 15.8 mcg/mL in 48 weeks, while the samples stored at 32°C exhibited higher concentrations, up to 33.5 mcg/mL, after 32 weeks. No difference in pseudomorphine content was found between the nitrogen- and air-filled headspace samples. The authors recommended a shelf life of one year when stored at or below 15°C and autoclaving no more than one time.

**STUDY 8 (EXTEMPORANEOUS INJECTION):** Hernandez Salvador et al.[511] reported the stability of a preservative-free morphine hydrochloride 0.4-mg/mL solution in 0.9% sodium chloride for intrathecal administration. Analysis of the drug concentration using ultraviolet spectroscopy demonstrated little or no loss of morphine hydrochloride after 14 weeks stored at 4 and 25°C.

**STUDY 9 (EXTEMPORANEOUS INJECTION):** Grom and Bander[512] reported on the compounding of morphine sulfate 50-mg/mL injection filled into 10-mL vials. Fifty grams of morphine sulfate powder was wetted with about 40 mL of sterile water for injection to create a slurry, which decreased dissolution time. The slurry was transferred to a glass bottle and diluted further with sterile water for injection to a total volume of 1000 mL and was allowed to dissolve completely. Dissolution was usually completed within 60 minutes; dissolution may be facilitated by warming the solution using warm water running over the bottle. The morphine sulfate 50-mg/mL solution was passed through a 0.8-μm particulate filter and a 0.22-μm sterilizing filter connected in series and filled into 10-mL sterile glass vials by means of a peristaltic pump. The rubber vial stoppers were sealed into the vials using aluminum caps. The 100-vial batch was tested for sterility and bacterial endotoxin content before release. An expiration date of six months stored at room temperature was used for this preparation. Informal analysis of a limited number of long-term stored vials has shown little or no loss in periods up to five years at room temperature.

**STUDY 10 (EXTEMPORANEOUS INJECTION):** Nguyen-Xuan et al.[961] evaluated the long-term stability of morphine sulfate 1 mg/mL in sodium chloride 0.9% packaged in 100-mL polypropylene infusion bags (Polimoon Langeskov). The solutions were autoclaved at 121°C for 20 minutes for sterilization. The solutions were evaluated over 36 months stored at room temperature of 25°C and over six months at temperatures of

30 and 40°C. No visible precipitation appeared in any sample, and electronic particulate evaluation found that subvisible particulates remained within compendial limits throughout the study. Little or no loss of moisture or change in pH occurred in any of the samples. Stability-indicating HPLC analysis found no loss of morphine sulfate in any of the solutions and little formation of degradation products throughout the respective study periods. The authors demonstrated that unlike polyvinyl chloride bags, which cannot be autoclaved and exhibit excessive water loss through evaporation, polypropylene bags can be successfully used for compounding bags of morphine sulfate solutions with long-term (three-year) stability for use in patient-controlled analgesia.

**STUDY 11 (EXTEMPORANEOUS INJECTION):** Steger et al.[983] reported the stability of morphine 40 mg/mL (as the hydrochloride) injection extemporaneously compounded from bulk powder with sodium chloride 3 mg/mL in water for injection adjusted to pH 3.2 with hydrochloric acid and autoclaved at 121°C for periods of 15 and 180 minutes. When a nitrogen purge of the containers was used to eliminate oxygen, little or no loss of morphine occurred during autoclaving, as measured by stability-indicating HPLC analysis.

**STUDY 12 (EXTEMPORANEOUS INJECTION):** Keiner and Kruger[1551] described the compounding of morphine hydrochloride 40 mg/mL injection for use in infusion pumps. The injection was compounded by dissolving 20 g of morphine hydrochloride trihydrate and sodium chloride 1.7 g in 450 g of water for injection. The solution was adjusted to pH 3 to 3.5 using 1 N hydrochloric acid and was brought to a volume of 500 mL with additional water for injection. The injection was sterilized by filtration through a 0.2-μm filter. An expiration date of 12 months when stored at room temperature of 25°C or less protected from light was recommended based on analysis that showed no loss of morphine hydrochloride after that period of storage.

## *Oral*

**STUDY 1 (LIQUID):** Poochikian and Cradock[405] evaluated the stability of the active components in several formulas of Brompton mixtures. The stabilities of morphine 1 mg/mL and cocaine hydrochloride 0.5 mg/mL were assessed using a stability-indicating HPLC assay. The two drugs were incorporated into a 12.5% alcohol–25% syrup vehicle and evaluated at 25°C. The stability of the cocaine was adversely affected compared to a similar mixture utilizing heroin, but the cocaine mixture still had a useful shelf life ($t_{90}$ of the cocaine of 10 days at 25°C).

**STUDY 2 (CAPSULES):** Webster et al.[709] reported the release of drug from extemporaneously compounded morphine sulfate 30-mg slow-release capsules prepared by 15 different pharmacies. Each capsule also contained 100 mg of Methocel E4M and 210 mg of lactose. Dissolution testing found a great deal of consistency, with capsules from 14 of the 15 test pharmacies showing a release of morphine sulfate at a rate near 6.3 mg/hour. The outlier demonstrated a release rate of 4.7 mg/hour. Total (95%) release occurred in six hours with the compounded capsules compared to eight hours for MS Contin.

**STUDY 3 (CAPSULES):** Not unexpectedly, Bogner et al.[710] reported that the release rate of morphine sulfate from extemporaneously prepared 300-mg slow-release capsules was influenced by the amount of Methocel (in this case K-100M) incorporated into the formulation. Various formulas were evaluated, with the Methocel ranging from 53 to 169 mg per capsule. Lactose was present in some of the formulations as well. The most rapid release (44% in one hour and 80% in 2.3 hours) occurred with the lowest Methocel amount (53 mg, or about 11% of the capsule content). In addition, there was a sizable degree of variability of drug release. The highest amount of Methocel (169 mg, or 36% of the capsule content) resulted in a slower and somewhat more consistent release of 24% in the first hour and 80% in 5.2 hours.

**STUDY 4 (LIQUID):** Ng and Yinfoo[980] evaluated the stability of a simple aqueous oral solution of morphine hydrochloride 2 mg/mL with methylparaben 0.65 mg/mL and propylparaben 0.35 mg/mL as preservatives. To prepare the oral liquid, the parabens were first dissolved in a small amount of ethanol 95% and this was incorporated into the morphine hydrochloride solution in purified water, resulting in a final ethanol concentration of 2%. The solution was packaged in colorless glass bottles and in high-density polyethylene containers and stored for 52 weeks at room temperature of 22 to 26°C, with daily exposure to light for one to two hours.

Samples in the high-density polyethylene bottles remained clear and colorless. Stability-indicating HPLC analysis found about a 2% loss of morphine hydrochloride and each of the paraben preservatives. Samples in glass bottles developed a light straw color after six months, and morphine hydrochloride losses of about 5% occurred over 52 weeks.

**STUDY 5 (LIQUID):** Morales et al.[1044] evaluated the stability of a pediatric oral liquid formulation of morphine hydrochloride 4 mg/mL in simple syrup and purified water (3:1). Ultraviolet spectrophotometric analysis found that no significant degradation occurred over 30 days when stored refrigerated at 4°C. However, at room temperature near 25°C, morphine hydrochloride losses of about 7% occurred in 15 days.

**STUDY 6 (LIQUID):** Nahata et al.[1230] evaluated the stability of a diluted oral liquid formulation of morphine elixir for use in infants. Morphine elixir was diluted to a concentration of 0.2 mg/mL with distilled water. The diluted elixir was packaged as 1 mL of solution in oral plastic syringes. Syringes were stored at room temperature of 25°C for 60 days. No changes in physical appearance or pH occurred. Stability-indicating HPLC analysis found that less than 5% morphine loss occurred in the study samples.

**STUDY 7 (LIQUID):** Preechagoon et al.[1647] developed and evaluated the stability of several formulations of morphine sulfate 2-mg/mL oral liquid for pediatric use. All of the formulations that contained sodium metabisulfite were found to be unstable, with morphine losses ranging from 20 to 35% after 35 days of storage at room temperature.

Four oral liquid formulations having the compositions shown in Table 68 all exhibited stability for 13 months at 25°C and 75% relative humidity and also under refrigeration. The oral liquids remained clear, there was no

**TABLE 68.** Morphine Sulfate 2 mg/mL Oral Liquid Formulas Tested[1647]

| Component | | A | B | C | D |
|---|---|---|---|---|---|
| Morphine sulfate | | 360 mg | 360 mg | 360 mg | 360 mg |
| Glycerin | | 67.5 mL | 81 mL | 27 mL | 90 mL |
| Syrup, USP | | 67.5 mL | 54 mL | 90 mL | 40.5 mL |
| Sorbitol | | 22.5 mL | 36 mL | 13.5 mL | 22.5 mL |
| EDTA (wt/vol) | | 0.1% | 0.1% | 0.1% | 0.1% |
| Paraben concentrate (vol/vol) | | 1% | 1% | 1% | 1% |
| Sodium chloride 400 mg/mL[a] | | 2.4 mL | 2.2 mL | 2.2 mL | 2.7 mL |
| Sodium citrate 0.05M buffer | | 9 mL | 13.5 mL | 9 mL | 9 mL |
| Tartaric acid 286 mg/mL[a] | | 2.5 mL | 2.0 mL | 2.2 mL | 1.4 mL |
| Purified water | qs | 180 mL | 180 mL | 180 mL | 180 mL |
| pH adjusted to | | 3.82 | 3.98 | 3.87 | 4.00 |

[a]Solution in purified water.

precipitate formation or substantial discoloration, the pH values remained in the range of 3.8 to 4, and HPLC analysis found the morphine concentration was greater than 97% throughout the study period.

## Enteral

### MORPHINE COMPATIBILITY SUMMARY

*Compatible with:* Isocal • Osmolite • Vital • Vivonex
*Uncertain or variable compatibility with:* Jevity
   • Osmolite HN • Pulmocare

**STUDY 1:** Michelini et al.[22] evaluated the compatibility and stability of oral morphine solution (Roxane) with two commercially available enteral tube feeding products, Isocal and Vivonex reconstituted according to the label instructions. Morphine concentrations of 0.5 and 1 mg/mL were prepared in each enteral product by adding the morphine oral solution to the enteral feed and shaking. Samples were stored at room temperature (25°C) in enteral feeding bags (Travenol) and refrigerated at 4°C in glass containers. The enteral mixtures were analyzed for morphine content using HPLC. After 24 hours at room temperature, 4% or less loss of morphine occurred. When refrigerated for 24 hours first, then followed by 24 hours at room temperature, morphine losses were 7% or less.

**STUDY 2:** Fagerman and Ballou[52] reported on the compatibility of morphine sulfate (Roxane, concentration unspecified) mixed in equal quantities with three enteral feeding formulas. The morphine sulfate was compatible visually with Vital, Osmolite, and Osmolite HN (Ross), with no obvious thickening, granulation, or precipitation.

**STUDY 3:** Udeani et al.[220] evaluated the stability of morphine sulfate at 1 mg/mL in three enteral nutrition products, Jevity, Osmolite HN, and Pulmocare. Morphine sulfate 2 mg/mL (Roxane) was used to prepare the samples. Morphine sulfate stability was evaluated by HPLC analysis after storage at 22 and 37°C for 48 hours. No loss of drug occurred at either temperature in Osmolite HN and Pulmocare. Morphine sulfate losses in Jevity in 48 hours were 2 and 6% at 22 and 37°C, respectively. However, using morphine sulfate 2 mg/mL to prepare the 1-mg/mL concentration in the enteral products caused a pH drop from 6.24 to 4.96, which resulted in phase separation and precipitation. Agitating the mixtures resulted in disappearance of the separation. Using morphine sulfate 20 mg/mL to prepare the 1-mg/mL test mixtures caused little change in the pH and no separation or precipitation. The authors recommended using the 20-mg/mL form of morphine sulfate liquid to add to enteral products to avoid the pH-mediated phase separation and precipitation.

## Rectal

### USP OFFICIAL FORMULATION (FATTY ACID BASE SUPPOSITORIES):

| | |
|---|---|
| Morphine sulfate | 50 mg |
| Silica gel | 25 mg |
| Fatty acid base | qs |

The suppository molds to be used should be calibrated with Fatty Acid Base and the amount of base adjusted according to the calibration. The morphine sulfate powder and silica gel are mixed thoroughly to prepare a uniform powder. The Fatty Acid Base should be melted slowly and evenly. The powder mixture should be added to the melted base and thoroughly mixed. After mixing, the preparation should be poured into the calibrated molds and cooled. After solidifying, the suppositories should be trimmed, wrapped, and packaged in tight containers. Store the finished suppositories under refrigeration between 2 and 8°C. The beyond-use date is 90 days from the date of compounding.[4,5]

### USP OFFICIAL FORMULATION (POLYETHYLENE GLYCOL BASE SUPPOSITORIES):

| | |
|---|---|
| Morphine sulfate | 50 mg |
| Silica gel | 25 mg |
| Polyethylene Glycol Base | qs |

The suppository molds to be used should be calibrated with Polyethylene Glycol Base and the amount of base adjusted according to the calibration. The morphine sulfate powder and silica gel are mixed thoroughly to prepare a uniform powder. The Polyethylene Glycol Base should be melted slowly and evenly. The powder mixture should be added to the melted base and thoroughly mixed. After mixing, the preparation should be poured into the calibrated molds and cooled. After solidifying, the suppositories should be trimmed, wrapped, and packaged in tight containers. Do not use polystyrene containers. Store the finished suppositories under refrigeration between 2 and 8°C. The beyond-use date is 90 days from the date of compounding.[4,5]

Allen[155] reported on the extemporaneous preparation of sustained-release morphine suppositories. Each suppository contained morphine sulfate 25 to 50 mg, alginic acid 25%, with the balance being Witepsol H-15. The alginic acid was passed through a No. 200 sieve and added to melted Witepsol H-15 with stirring. The morphine sulfate powder was added and stirred thoroughly to achieve a uniform dispersion. Sonication of the melted dispersion for 10 minutes was recommended to ensure thorough mixing. The suppositories were formed by pouring the melted mixture into molds and allowing solidification to occur.[155] One study[156] of similar sustained-release suppositories found that they were stable for at least three months under refrigeration at 4°C.

## Topical

**STUDY 1:** Zeppetella et al.[959] reported the stability of morphine sulfate and diamorphine hydrochloride topical gels prepared in Intrasite gel (Smith & Nephew) for use in treating decubitus ulcers. Morphine sulfate (Arum Pharmaceutical) and diamorphine hydrochloride (Arum Pharmaceutical) were incorporated into separate aliquots of the Intrasite gel to form 1.25-mg/mL concentrations. The samples were stored under refrigeration at 4°C in the dark, at room temperature in the dark, and at room temperature exposed to normal mixed fluorescent light and daylight. HPLC analysis of the morphine sulfate samples found no loss of morphine and no evidence of common decomposition products after 28 days of storage under any of the storage conditions.

However, the diamorphine hydrochloride samples were much less stable. The analysis demonstrated rapid loss of intact diamorphine exceeding 10%, along with the formation of 6-monoacetylmorphine within a few days at all temperatures. At room temperature, morphine also began to appear after about two weeks. The authors indicated that this may be acceptable for pain control if the diamorphine hydrochloride is viewed as a prodrug for morphine along with its own analgesic action.

**STUDY 2:** Abdelmageed et al.[1252] evaluated the stability of morphine sulfate incorporated at a concentration of 1 mg/mL into a combination gel of lidocaine 2% and chlorhexidine gluconate 0.25% (Instillagel, FARCO-Pharma). The compounded gel was packaged in 20-mL Luer-lock polypropylene syringes sealed with Helapet Combi Caps (Becton Dickinson). Sample syringes were stored at 4°C in the dark, at room temperature of 25°C exposed to fluorescent room light, and at 37°C in the dark.

Stability-indicating HPLC analysis found little or no loss of any of the three components in 22 days refrigerated and at room temperature. At 37°C over seven months, no loss of morphine sulfate and lidocaine hydrochloride and only about 8% loss of chlorhexidine gluconate occurred. ∎

# ∎ Moxifloxacin Hydrochloride

## Properties

Moxifloxacin hydrochloride is a yellow or slightly yellow, slightly hygroscopic crystalline powder.[1,3,7]

### Solubility

Moxifloxacin hydrochloride is sparingly soluble in water, slightly soluble in ethanol, and practically insoluble in acetone.[3]

### pH

Moxifloxacin hydrochloride 2 mg/mL in water has a pH between 3.9 and 4.6.[3] Moxifloxacin hydrochloride injection (Avelox, Schering Plough) 1.6 mg/mL in sodium chloride 0.8% has a pH adjusted during manufacturing to a range of 4.1 to 4.6.[7] Moxifloxacin hydrochloride 0.5% ophthalmic solution has a pH of about 6.8.[1350]

### $pK_a$

Moxifloxacin is amphoteric, having $pK_a$ values of 6.4 and 9.5.[1460]

### Osmolality

Moxifloxacin hydrochloride 1.6 mg/mL injection has an osmolality near isotonicity. Moxifloxacin hydrochloride ophthalmic solution is an isotonic solution with an osmolality of approximately 290 mOsm/kg. [900,1350]

## General Stability Considerations

Moxifloxacin hydrochloride is slightly hygroscopic. The powder should be packaged in airtight containers and stored at controlled room temperature protected from exposure to light.[3] Moxifloxacin hydrochloride tablets should be stored at controlled room temperature, and high humidity should be avoided. The intravenous injection should also be stored at controlled room temperature and should not be refrigerated so as to avoid drug precipitation. However, the manufacturer of moxifloxacin hydrochloride 0.5% ophthalmic solution states that it may be stored between 2 and 25°C.[7]

## Stability Reports of Compounded Preparations
### Oral

Hutchinson et al.[1349] evaluated the stability of moxifloxacin 20 mg/mL (as hydrochloride) oral liquid prepared extemporaneously from tablets. Three 400-mg moxifloxacin hydrochloride tablets (Avelox, Bayer) were triturated to fine powder using a mortar and pestle. The enteric-coating remnants were removed by sieving. An equal parts mixture of 30-mL of Ora-Sweet and 30-mL of Ora-Plus was prepared and mixed vigorously. About 30 mL of the vehicle mixture was used to levigate the moxifloxacin tablet powder, using geometric dilution to form a smooth suspension. This suspension was transferred to a two-ounce amber plastic prescription bottle, and the suspension was brought to final volume of 60 mL with additional vehicle and closed with a child-resistant cap. Samples were stored at controlled room temperature of 23 to 25°C for 90 days.

No detectable changes in color, odor, and taste and no visible microbial growth were observed. Stability-indicating HPLC analysis found little or no loss of moxifloxacin during 90 days of room temperature storage. ∎

# Mupirocin

## Properties
Mupirocin is a white to off-white crystalline solid.[4]

### Solubility
Mupirocin is very slightly soluble in water and freely soluble in dehydrated ethanol and acetone.[3,4] The aqueous solubility has been stated to be 1 mg/mL, while the solubility in ethanol was reported to be 0.5 mg/mL.[2]

### pH
The pH of a saturated solution of mupirocin in water is 3.5 to 4.5.[4]

## General Stability Considerations
Mupirocin powder should be packaged in tight containers and stored at controlled room temperature. Mupirocin ointment should be packaged in collapsible tubes or well-closed containers and stored at controlled room temperature.[4] Bactroban ointment and cream are stored at controlled room temperature[7] and were reported to be stable for 36 months. However, if stored at an elevated temperature of 37°C, the stability was reduced to nine months.[1705]

Mupirocin is inactivated at pH less than 4 or greater than 7.[2]

## Compatibility with Other Drugs
**MUPIROCIN COMPATIBILITY SUMMARY**
*Compatible with:* Hibiclens liquid soap • Kenalog cream and ointment • Lotrimin solution • Silvadene cream
*Incompatible with:* Keralyt gel • Lotrimin lotion
  • Monistat-Derm cream and lotion • Valisone lotion
*Uncertain or variable compatibility with:* Betamethasone dipropionate • Hytone cream, ointment, and lotion
  • Kenalog lotion • Lotrimin cream • Nizoral cream
  • Valisone cream and ointment • Vytone cream

**STUDY 1 (BETAMETHASONE):** Mourya et al.[859] reported the stability of mupirocin and betamethasone in a combination ointment providing mupirocin 2% (wt/wt) ointment (Glenmark Pharmaceuticals) and betamethasone dipropionate 0.05% (wt/wt) in a base composed of polyethylene glycol 400 and 4000 packaged in aluminum collapsible tubes. Samples were stored at room temperature of 25°C as well as at elevated temperatures of 37 and 40°C with 75% relative humidity. No physical changes in color or ointment consistency were observed. HPLC analysis found little or no loss of either mupirocin or betamethasone in 90 days at any of the storage temperatures.

**STUDY 2 (MULTIPLE DRUGS):** Jagota et al.[920] evaluated the compatibility and stability of mupirocin 2% ointment (Bactroban,

Beecham) with 19 other topical products mixed in a 1:1 proportion over periods up to 60 days at 37°C. The physical compatibility was assessed by visual inspection, while the chemical stability of mupirocin was evaluated by stability-indicating HPLC analysis. The study found that mupirocin 2% ointment was physically compatible and chemically stable for varying periods with triamcinolone 0.1% cream and ointment (Kenalog, Squibb), clotrimazole 1% solution (Lotrimin, Schering), silver sulfadiazine 1% cream (Silvadene, Marion Merrell Dow), and Hibiclens 4% liquid soap (Stuart). See Table 69.

However, betamethasone 0.1% lotion (Valisone, Schering) and clotrimazole 1% lotion (Lotrimin, Schering) were found to be physically incompatible upon mixing with mupirocin ointment. Miconazole nitrate 2% cream and 1% lotion (Monistat Derm, Janssen) as well as salicylic acid 6% gel (Keralyt, Westwood) resulted in separation and layering along with nearly total loss of mupirocin in 15 days. When mixed with triamcinolone 0.1% lotion (Kenalog, Squibb), 12% mupirocin loss occurred within the first assay point of 15 days.

**TABLE 69.** Mupirocin 2% Ointment Compatibility and Stability Results at 37°C[920]

| Drug | Stability (<10% loss) Period | Visual Appearance |
|---|---|---|
| Hibiclens 4% liquid soap | 60 days | Compatible |
| Hytone 1% cream | 60 days | Separation |
| Hytone 1% ointment | 45 days | Separation |
| Hytone 1% lotion | 15 days | Separation |
| Kenalog 0.1% cream | 15 days | Compatible |
| Kenalog 0.1% ointment | 45 days | Compatible |
| Kenalog 0.1% lotion | Unstable | Separation |
| Keralyt 4% gel | Unstable | Incompatible |
| Lotrimin 1% lotion | Not reported | Incompatible |
| Lotrimin 1% cream | 60 days | Separation |
| Lotrimin 1% solution | 60 days | Compatible |
| Monistat-Derm 2% cream | Unstable | Incompatible |
| Monistat-Derm 1% lotion | Unstable | Incompatible |
| Nizoral 2% cream | 45 days | Separation |
| Silvadene 1% cream | 45 days | Compatible |
| Valisone 0.1% cream | 15 days | Separation |
| Valisone 0.1% ointment | 60 days | Separation |
| Valisone 0.1% lotion | Not reported | Incompatible |
| Vytone cream | 60 days | Separation |

Other combinations resulted in uncertain compatibility. Hydrocortisone 1% lotion, cream, and ointment (Hytone, Dermik), ketoconazole 2% cream (Nizoral, Janssen), clotrimazole 1% cream (Lotrimin, Schering), betamethasone 0.1% cream and ointment (Valisone, Schering), and hydrocortisone 1%/ iodoquinol 1% cream (Vytone, Dermik) mixed with mupirocin 2% ointment all resulted in separation and layering and may be unacceptable for patient use. However, they could be remixed back to homogeneity, and the mupirocin loss was less than 10% for varying periods of time. ■

# ■ Mycophenolate Mofetil

## Properties
Mycophenolate mofetil is a white to off-white crystalline powder.[7] This drug has been characterized as a potential teratogen.[503]

### Solubility
Mycophenolate mofetil has an aqueous solubility of 43 mcg/mL at pH 7.4. Aqueous solubility increases as the pH becomes more acidic; the aqueous solubility increases to 4.27 mg/mL at pH 3.6. The drug is also sparingly soluble in ethanol.[7]

### $pK_a$
Mycophenolate mofetil has $pK_a$ values of 5.6 and 8.5.[7]

## General Stability Considerations
Mycophenolate mofetil tablets and capsules should be packaged in tight, light-resistant containers and stored at controlled room temperature.[7]

Mycophenolate mofetil is reported to be stable at acidic pH values. The half-life is 98 days at pH 2 and 118 days at pH 5.1 but only 19 days at pH 7.4.[607]

## Stability Reports of Compounded Preparations
### Oral
**STUDY 1:** Anaizi et al.[503] evaluated the stability of a compounded mycophenolate mofetil 100-mg/mL oral suspension prepared from capsules. This drug is a potential teratogen; unnecessary exposure should be avoided by using an appropriate fume hood or vertical laminar flow hood and wearing gloves during compounding. The contents of 80 mycophenolate mofetil 250-mg capsules (Roche) were emptied into a porcelain mortar. A sufficient amount of sterile water for irrigation was added to wet the powder, and the mixture was levigated to form a smooth paste. Cherry syrup 150 mL was added to the paste in three increments. About one-third was added to the paste in the mortar, mixed well, and transferred to a 250-mL graduate. The mortar was rinsed with another third of the cherry syrup and the rinse was added to the graduate. The mortar was again rinsed with the last third of the cherry syrup, and this rinse also was added to the graduate. Additional cherry syrup was added to bring the volume to 200 mL, and the suspension was mixed. The suspension was packaged in two-ounce amber polyethylene terephthalate G plastic bottles with child-resistant caps and stored at 4 and 24°C for 121 days.

There was no change in pH, color, or odor of the suspension. Stability-indicating HPLC analysis found no loss of drug under refrigeration and 6% loss at room temperature during the 121-day storage period.

**STUDY 2:** Venkataramanan et al.[607] reported on the stability of a mycophenolate mofetil 50-mg/mL oral suspension compounded from capsules. The contents of six 250-mg capsules were emptied into a mortar. The capsule contents were wetted and triturated with 7.5 mL of Ora-Plus (Paddock) to form a smooth paste. About 15 mL of cherry syrup was added to the paste and triturated. The mixture was transferred to an amber bottle and brought to a volume of 30 mL with additional cherry syrup. Samples of this suspension were stored under refrigeration at 5°C, at room temperature near 25°C, and incubated at 37 and 45°C. They were not exposed to direct sunlight and were exposed to room light only during analytical preparation.

No change in color or appearance was observed over 50 days of storage in the samples stored at 5 and 25°C. However, the color darkened and the consistency changed within four weeks in the samples incubated at elevated temperatures. Stability-indicating HPLC analysis found greater than 90% concentration in 11 days at 45°C and in 28 days at 37 and 25°C. Refrigerated samples exhibited no loss in 50 days and retained greater than 90% concentration through 210 days.

**STUDY 3:** Swenson et al.[683] reported the stability of mycophenolate mofetil 100 mg/mL in sugar-free Ora-Plus with aspartame 3-mg/mL artificial cherry flavor (0.4% vol/vol) and FD&C red no. 40 0.05 mg/mL. The suspension was packaged in amber polyethylene terephthalate glycol (PETG) prescription bottles and stored at room temperature of 23 to 25°C and under refrigeration at 2 to 8°C for 120 days.

There was no evidence of microbial growth and no change in the appearance of the suspensions. Stability-indicating HPLC analysis found that no loss of drug occurred within the study period. However, the cherry odor was lost after 28 days at room temperature. The cherry odor was acceptable in the refrigerated samples for 120 days, leading to a recommendation for refrigerated storage. ■

# Nadolol

## Properties
Nadolol is a white to off-white nearly odorless crystalline powder.[4]

## Solubility
Nadolol is soluble in water at pH 2 and slightly soluble at pH 7 to 10. It is freely soluble in ethanol and slightly soluble in isopropanol but insoluble in acetone.[4]

## $pK_a$
Nadolol has a $pK_a$ of 9.67.[2]

## General Stability Considerations
Nadolol powder should be packaged in well-closed containers, and the tablets should be packaged in tight containers and stored at controlled room temperature.[4]

## Stability Reports of Compounded Preparations
### Oral
Yang et al.[738] reported the stability of nadolol oral powders extemporaneously compounded from commercial tablets. Corgard 80-mg tablets (Bristol-Myers Squibb) were crushed and pulverized. Nadolol powder equivalent to 80 mg of drug was mixed with 925 mg of lactose, 750 mg of pulverized lactofermin (Biofermin) tablets, or 1.42 g of acetylcysteine granules for use as solid diluents. Powder mixtures were packaged in polyethylene-coated glassine papers and stored for 28 days at ambient temperature (between 15 and 30°C). Stability-indicating HPLC analysis found no loss of nadolol in 28 days diluted in either lactose or pulverized lactofermin tablets. In acetylcysteine granules about 2% nadolol loss occurred in 28 days. ■

# Naltrexone Hydrochloride

## Properties
Naltrexone hydrochloride is a white crystalline material having a bitter taste.[2,7]

## Solubility
Naltrexone hydrochloride has an aqueous solubility of about 100 mg/mL.[2,7]

## $pK_a$
Naltrexone hydrochloride has a $pK_a$ of 8.13 at 37°C.[2]

## General Stability Considerations
Naltrexone tablets should be packaged in tight containers and stored at controlled room temperature.[7]

## Stability Reports of Compounded Preparations
### Oral
Fawcett et al.[634] evaluated the stability of naltrexone hydrochloride 1-mg/mL oral suspension prepared from tablets and from powder. The formulation that was tested was a modification of the suspension recommended by the National Institute on Drug Abuse, omitting the flavoring and coloring agents and substituting glycerol for sorbitol; the formula is shown in Table 70. Naltrexone hydrochloride (DuPont Merck) tablets were crushed or bulk powder was weighed to provide the correct amount of drug. This material was triturated with the ascorbic acid and sodium benzoate, and the glycerol was added in to form a uniform paste. Distilled water was used to bring the suspension to volume. To remove insoluble components, the suspension

**TABLE 70.** Naltrexone 1-mg/mL Oral Suspension
Formula of Fawcett et al.[634]

| Component | Amount |
|---|---|
| Naltrexone | 1 mg/mL |
| Ascorbic acid | 0.5% |
| Sodium benzoate | 0.1% |
| Glycerol | 20% |
| Distilled water | qs 100% |

prepared from crushed tablets was evaluated both unfiltered and filtered. The suspensions were packaged in 50-mL high-density polyethylene bottles with polypropylene screw caps. The bottles were stored in the dark at 4 and 25°C as well as at an elevated temperature of 70°C.

This formulation appears to have been less than ideal due to its very bitter taste and gritty texture. Filtration improved the appearance (and presumably the texture) of the preparation but did not alter drug content or stability. Stability-indicating HPLC analysis found less than 10% loss within 90 days at either 4 or 25°C; most samples exhibited no loss. However, the suspensions developed a yellow discoloration that the authors indicated limited the preparation's shelf life. They recommended shelf lives of 30 and 60 days stored at 25 and 4°C, respectively. Incorporating the dose into fruit juice or flavored drinks was suggested as a way to overcome the bitter taste.

In addition to this formulation, Fawcett et al.[634] evaluated a 5-mg/mL suspension prepared in a similar manner from powder but stored only at an elevated temperature of 70°C. There was little difference in drug stability compared to the 1-mg/mL concentration under the same conditions. It can be expected to have a similar stability to the lower concentration at 4 and 25°C. The authors of this study suggested using this higher concentration to reduce the required dose volume of an unpleasant tasting preparation.

### Injection

Injections, like other sterile drugs, should be prepared in a suitable clean air environment using appropriate aseptic procedures. When prepared from nonsterile components, an appropriate and effective sterilization method must be employed.

Gupta[1234] evaluated the stability of an extemporaneously prepared naltrexone hydrochloride 1.4-mg/mL injection. The injection was prepared by dissolving 140 mg of naltrexone hydrochloride powder and 878 mg sodium chloride crystals in sterile water for injection, bringing the final volume to 100 mL to yield a 1.4-mg/mL concentration of naltrexone hydrochloride. Although noted as a high-risk preparation, the author did not state a sterilization method. The extemporaneous injection was packaged in 20-mL clear glass vials and stored at 25°C in the dark for 42 days. The visual appearance of the injection did not change throughout the study period. HPLC analysis found about 4% loss of naltrexone hydrochloride occurred after 42 days of storage. ■

# ■ Naproxen

## Properties

Naproxen acid is a white or off-white nearly odorless crystalline powder. Naproxen sodium salt is a white to creamy-white crystalline powder. Each 550 mg of naproxen sodium is equivalent to 500 mg of naproxen.[2,3]

### Solubility

Naproxen acid is practically insoluble in water at low pH but is freely soluble at high pH. It is also soluble in ethanol, with solubility reported to be about 40 mg/mL.[1,7]

Naproxen sodium is freely soluble in water at neutral pH and sparingly soluble in ethanol.[3,7]

### pH

Naproxen oral suspension has a pH of 2.2 to 3.7.[4]

### pK_a

The apparent $pK_a$ of naproxen is 4.15.[2]

## General Stability Considerations

Naproxen tablets should be packaged in well-closed, light-resistant containers and stored at controlled room temperature. The oral suspension should be packaged in tight, light-resistant containers and stored at controlled room temperature protected from excessive heat.[4,7]

## Stability Reports of Compounded Preparations
### Rectal

Santus et al.[1539] evaluated a proposed formulation of naproxen sodium rectal suppositories using calcium and sodium levulinate and Monolena OH-6 fat base for suppositories. Differential scanning calorimetry found that naproxen sodium and calcium levulinate interacted and were incompatible, modifying the technological properties of suppositories.

### Injection

Injections, like other sterile drugs, should be prepared in a suitable clean air environment using appropriate aseptic procedures.

When prepared from nonsterile components, an appropriate and effective sterilization method must be employed.

Peswani and Lalla[488] reported the stability of a compounded injection of naproxen 125 mg/mL present primarily as the sodium salt. Simple aqueous solutions of naproxen sodium were found to have a high pH and to be subject to unacceptable precipitation. Incorporation of propylene glycol 40% as a cosolvent and maintenance of pH near 8 prevented precipitation within six months at 4°C. Extensive evaluation of other potential components led to the selection of the formulation containing the equivalent of naproxen 125 mg/mL primarily as the sodium salt (Table 71).

The pH was adjusted to 7.5 to 8.5 with 0.1 N sodium hydroxide and/or 0.1 N hydrochloric acid. The solution was filtered and filled into 2-mL amber ampules, and a nitrogen purge was used to displace the air prior to sealing. The ampules were sterilized by autoclaving. HPLC analysis was used to determine stability. Autoclaving at 121°C and 15 psi for periods from 15 minutes up to 90 minutes did not result in naproxen

**TABLE 71.** Naproxen Injection Formulation of Peswani and Lalla[488]

| Component | Amount |
| --- | --- |
| Naproxen sodium | 13.626% |
| Naproxen | 0.062% |
| Propylene glycol | 40% |
| Benzyl alcohol | 0.5% |
| Water | qs |

loss. From stability information generated at elevated temperatures up to 90°C, the shelf life of this formulation was predicted to be about four years. Stability was independent of pH in the range of 7.5 to 10. Factors that were found to accelerate decomposition included use of air (not nitrogen) in the headspace, incorporation of chlorobutanol, and exposure to light, which may photocatalyze naproxen oxidation. This preparation of naproxen has not been tested for clinical acceptability. ▪

## ▪ Naratriptan Hydrochloride

### Properties
Naratriptan hydrochloride is a white to pale yellow crystalline powder.[1,7]

#### Solubility
Naratriptan hydrochloride is readily water soluble.[7]

### General Stability Considerations
Naratriptan hydrochloride tablets should be stored at controlled room temperature.[7]

### Stability Reports of Compounded Preparations
#### Oral
**USP OFFICIAL FORMULATION:**

    Naratriptan (as hydrochloride) 50 mg (55 mg of hydrochloride)

    Vehicle for Oral Solution, NF (sugar-containing or sugar-free)

    and Vehicle for Oral Suspension, NF (1:1) qs 100 mL

(See the vehicle monographs for information on the individual vehicles.)

Use naratriptan hydrochloride powder or commercial tablets. If using tablets, crush or grind to fine powder. Add the vehicle mixture in small portions and mix to make a uniform paste. Add additional vehicle in increasing volumes to make a pourable suspension. Quantitatively transfer to a suitable calibrated tight and light-resistant bottle, bring to final volume with the vehicle mixture, and thoroughly mix, yielding naratriptan (as hydrochloride) 0.5-mg/mL oral suspension. The final liquid preparation should have a pH between 4 and 4.5. Store the preparation at controlled room temperature or under refrigeration between 2 and 8°C. The beyond-use date is seven days from the date of compounding when stored at controlled room temperature and 90 days from the date of compounding stored under refrigeration.[4]

Zhang et al.[713] reported the stability of three extemporaneously prepared oral suspensions of naratriptan hydrochloride 0.5 mg/mL prepared from 2.5-mg tablets (Amerge, Glaxo-Wellcome). The tablets were crushed using a mortar and pestle. The tablet powder was incorporated into (1) an equal parts mixture of Ora-Plus and Ora-Sweet, (2) an equal parts mixture of Ora-Plus and Ora-Sweet SF (sugar-free), and (3) Syrpalta. The three suspensions were packaged in amber plastic (polyethylene terephthalate) prescription bottles and stored at ambient room temperature near 23°C for seven days and refrigerated at 4°C for 90 days.

For the Ora-Plus and Ora-Sweet or Ora-Sweet SF formulations, the preparations were easily resuspendable and exhibited no caking and no visible changes at either storage temperature. Stability-indicating HPLC analysis found no loss of naratriptan hydrochloride in seven days at room temperature and 4% or less loss under refrigeration for 30 days. The Syrpalta-containing formulation was inadequate, with a high sedimentation rate, nonuniform dispersion, and unacceptable caking over time. ▪

# Neomycin Sulfate

## Properties

Neomycin sulfate is a mixture of related aminoglycoside antibiotics as the sulfate salts; neomycin B is the predominant component. It occurs as a white to slightly yellow and nearly odorless hygroscopic powder with practically no taste.[1-3] Neomycin sulfate 1 mg contains not less than 600 mcg of neomycin on the dried basis.[4]

### Solubility

Neomycin sulfate is freely soluble in water, having an aqueous solubility of about 333 mg/mL; it may be slowly soluble up to 1 g/mL. In ethanol, the drug is very slightly soluble.[1-3,6]

### pH

A 1% aqueous solution of neomycin sulfate has a pH between 5 and 7.5. The oral solution also has a pH between 5 and 7.5.[3,4]

## General Stability Considerations

Neomycin sulfate and its preparations may discolor if exposed to light.[2] It is also susceptible to atmospheric oxidation. Neomycin sulfate preparations should be packaged in tight, light-resistant containers and stored at controlled room temperature.[2-4,6]

Aqueous solutions of neomycin sulfate are stable over pH 2 to 9. The drug is very stable in highly alkaline solutions. Optimum activity occurs at about pH 7.[1,6]

Neomycin sulfate solutions may be autoclaved at 121°C for 20 minutes without loss of drug in concentrations of 0.5 to 10%.[6]

## Stability Reports of Compounded Preparations

### Ophthalmic

Ophthalmic preparations, like other sterile drugs, should be prepared in a suitable clean air environment using appropriate aseptic procedures. When prepared from nonsterile components, an appropriate and effective sterilization method must be employed.

**STUDY 1:** Osborn et al.[169] evaluated the stability of several antibiotics in three artificial tears solutions composed of 0.5% hydroxypropyl methylcellulose. The three artificial tears solutions (Lacril, Tearisol, Isopto Tears) were used to reconstitute and dilute neomycin to 33 mg/mL. The preparations were packaged in plastic squeeze bottles and stored at 25°C. A serial dilution bioactivity test was used to estimate the amount of antibiotic activity remaining during seven days of storage. The amount after seven days at room temperature was 65% in Lacril, 87% in Tearisol, and 91% in Isopto Tears.

**STUDY 2:** El-Shattawy et al.[436] evaluated the stability and release of neomycin from 10 ophthalmic ointment bases in the presence and absence of benzalkonium chloride 0.35%. The ointments were prepared from neomycin powder in the sterilized ointment bases shown in Table 72. Neomycin content and release were assessed microbiologically. There were significant differences in stability as well as in the release of neomycin from the ointments. The best formulations for release were 7, 8, and 9. However, formulations 7 and 8 lost 16 and 30%, respectively, of the neomycin after 24 months of storage at room temperature. Formulations 2, 3, and 5 were the best from a stability standpoint, losing neomycin 2, 2, and 4%, respectively, after 24 months at room temperature. Considering both release and stability, the best compromise formulations were 9 and 10, which lost 10 and 7%, respectively, after 24 months at room temperature.

### Inhalation

Inhalation solutions, like other sterile drugs, should be prepared in a suitable clean air environment using appropriate

**TABLE 72.** Ophthalmic Ointment Bases Tested for Neomycin Stability[436]

| Component | Formulations | | | | | | | | | |
|---|---|---|---|---|---|---|---|---|---|---|
| | 1 | 2 | 3 | 4 | 5 | 6 | 7 | 8 | 9 | 10 |
| Castor oil | — | — | — | — | — | — | 85% | 84.6% | 80.8% | — |
| Cetyl alcohol | — | — | — | — | 5% | — | — | — | 5% | — |
| Glyceryl monostearate | — | — | — | — | — | 0.5% | — | 0.5% | — | 0.5% |
| Hard paraffin | — | — | — | — | — | — | — | — | — | 19.5% |
| Hydrogenated castor oil | — | — | — | — | — | — | 15% | 14.9% | 14.2% | — |
| Liquid paraffin | — | 10% | 10% | 20% | 19% | 19.9% | — | — | — | 60% |
| Wool fat | 10% | 10% | — | — | — | — | — | — | — | 20% |
| Yellow soft paraffin | 90% | 80% | 90% | 80% | 76% | 79.6% | — | — | — | — |

aseptic procedures. When prepared from nonsterile components, an appropriate and effective sterilization method must be employed.

The potential for microbiological contamination of several extemporaneous aerosol solutions packaged in bulk in 120-mL amber glass bottles or in sterile unit-dose Nebuject syringes was evaluated by Kuhn et al.[81] Solutions composed of neomycin 4%, propylene glycol 10%, and phenylephrine 0.25% were sterilized by filtration through a 0.22-μm filter packaged in the two systems and used by 10 cystic fibrosis patients, each for five days. All Nebuject unit-dose syringes used were found to be sterile. However, the solutions packaged in the amber glass bottles all were found to be heavily contaminated with the bacteria Bacillus cereus after five days of use.

### Topical

Kodym and Bujak[820] evaluated the stability of several formulations of topical ointments of 3% aloe extract with and without added neomycin sulfate. The concentration of the two principal aloe constituents, aloenin and aloin, was determined by thin-layer chromatography and ultraviolet spectroscopy. The antimicrobial activity of neomycin was determined by microbiological assay.

The most successful ointment was prepared in white Vaseline, liquid paraffin, solid paraffin, and cholesterol (amounts not specified). No loss of either principal aloe component or of neomycin activity occurred during two years of storage at 20°C. However, ointment formulations that contained water or propylene glycol underwent about 34 and 42% loss of aloe components and 40 or 19% loss of neomycin activity, respectively. ■

#  Nevirapine

## Properties

Nevirapine is a white to off-white nearly odorless crystalline powder.[1,4]

### Solubility

Nevirapine is practically insoluble in water, having an aqueous solubility at neutral pH of about 0.1 mg/mL.[1,4] At acidic pH values below pH 3, the drug is highly soluble.[1] Nevirapine is slightly soluble in ethanol and slightly insoluble in propylene glycol.[4]

### $pK_a$

Nevirapine has a $pK_a$ of 2.8.[1]

## General Stability Considerations

Nevirapine should be packaged in tight containers and stored at controlled room temperature.[4] Commercial nevirapine (as the hemihydrate) 50-mg/5-mL oral suspension (Viramune, Boehringer Ingelheim) may be stored at room temperature until the labeled expiration date. However, the manufacturer has indicated that the drug should be used within two months after opening.[7]

## Stability Reports of Compounded Preparations

### Oral

Rexroad et al.[1064] evaluated the stability of nevirapine oral suspension 50 mg/5 mL (Boehringer Ingelheim) when packaged in 3-mL amber oral syringes (Exacta-Med Oral Dispensers, Baxa). A 6-mg/0.6-mL quantity of the oral suspension was packaged in each oral syringe. Samples were stored at room temperature of 26°C and high humidity, at elevated temperature of 40°C at both high and low humidity, under refrigeration, and frozen at −30°C for six months.

HPLC analysis of nevirapine concentrations found that the drug exhibited less than 3% loss in six months at room temperature with high humidity, when stored under refrigeration, and when stored frozen. Storage at 40°C in high humidity resulted in a loss of nearly 9% in six months. Storage at 40°C in low humidity (14% relative humidity) resulted in an increase in the nevirapine concentration of about 40 to 50%, presumably due to evaporation of water from the suspension. ■

# ■ Nifedipine

## Properties
Nifedipine is a yellow crystalline powder.[1-3]

### Solubility
Nifedipine is practically insoluble in water but is sparingly soluble in ethanol to about 17 mg/mL.[1-3]

## General Stability Considerations
Nifedipine is light-sensitive.[1,3,999,1651,1652,1709] Nifedipine liquid-filled capsules should be packaged in tight, light-resistant containers and stored between 15 and 25°C.[2,4] The extended-release tablets may be stored at room temperature and protected from exposure to excessive heat of 40°C or more.[2,7] The manufacturer has stated that nifedipine capsules should not be removed from the original packaging and placed in dosing compliance aids because of light sensitivity.[1622,1623]

Because of the light sensitivity of nifedipine,[678] preparation of solutions in the dark or under light with a wavelength of 420 nm has been recommended as a precaution.[3] McCluskey and Brunn[1709] reported that nifedipine topical and oral preparations compounded under fluorescent and red-shaded light, even for brief periods of time, were subpotent due to significant drug degradation. If compounded entirely under gold-shaded light, the preparations were within the intended concentration range.

Grundy et al.[666] reported on the photostability of 10 different brands of nifedipine commercial oral formulations exposed to artificial sunlight for 12 weeks. The products did not undergo appreciable (greater than 10%) decomposition, and all remained within USP requirements.

Valenzuela et al.[1624] reported the stability of nifedipine capsules exposed to temperatures ranging from 26 to 38°C under simulated summer conditions in paramedic vehicles over four weeks. The capsules melted during storage. However, gas chromatography coupled with mass spectrometry found no change in the drug over four weeks under these simulated use conditions.

Liquid capsule contents have been reported as a source of nifedipine for compounding other preparations.[101] However, both Domaratzki and Campbell[667] along with McCluskey and Brunn[1709] reported that inadequate amounts of nifedipine were accessible from the capsules. McCluskey and Brunn[1709] noted that use of nifedipine powder was preferable because of ease of use and drug content accuracy.

## Stability Reports of Compounded Preparations
### Oral
**STUDY 1 (CAPSULE CONTENTS REMOVAL):** Rosen and Johnson[101] evaluated five techniques for removing the contents of nifedipine liquid from the capsule dosage form to facilitate nonstandard dosing. The methods evaluated were:

1. *Scissors method:* Cut off the tip of the capsule with surgical scissors and squeeze out the contents.
2. *Needle puncture and squeeze method:* Puncture the capsule with an 18-gauge needle and squeeze out the contents.
3. *Needle and syringe, one-hole method:* Using a 23-gauge needle on a 1-mL syringe, insert the needle in one end of the capsule and withdraw the liquid while squeezing.
4. *Needle and syringe, two-hole method:* Vent the capsule by puncturing one end with a 23-gauge needle. Using a 23-gauge needle on a 1-mL syringe, insert the needle in the opposite end of the capsule and withdraw the liquid.
5. *Capsule in syringe barrel method:* Place a capsule inside a 5-mL syringe and replace the plunger. Puncture the capsule with a 16-gauge needle through the syringe orifice. Force the liquid from the capsule by pressing on the syringe plunger.

The methods using a needle to withdraw the liquid from the capsule (methods 2, 3, and 4) were the most efficient and reproducible ways of removing the nifedipine liquid. Amounts removed were 97% or more. Furthermore, the one-hole method (method 3) was the most convenient. The other two methods were less efficient, yielding only about 83 to 85% of the liquid.

**STUDY 2 (CAPSULE CONTENTS REMOVAL):** Domaratzki and Campbell[667] noted that nifedipine liquid-filled capsules contain 0.34 mL of liquid drug mixture. They were unable to get much of the drug from the capsules. By using method 3, about 0.23 mL or 6.8 mg (range of 0.16 to 0.27 mL) could be removed. Use of a method similar to method 5 gave quite variable results. Capsule contents of about 0.15 mL or 4.4 mg (range of 0.03 to 0.29 mL) could be recovered.

**STUDY 3 (ORAL POWDER):** Helin et al.[510] evaluated the stability of nifedipine as a powder dilution prepared from crushed timed-release tablets. Because of the drug's light sensitivity, the powders were prepared in a dimly lit room. Five nifedipine 10-mg extended-release tablets (Adalat, Bayer) were crushed to fine powder in a tared metal mortar. By using geometric dilution, the crushed tablets were diluted with lactose to make a total powder mass of 25 g and a resulting nifedipine concentration of 1 mg/500 mg of powder. Powder portions of 500 mg were then weighed and packaged in waxed, sealed powder paper packets. The finished powder papers were stored in black plastic bags for light protection and also without the bag exposed to light. Samples were stored under refrigeration at 6°C and about 64% relative humidity protected from light, at 22°C and about 44% relative humidity protected from light, and at 22°C and about 68% relative humidity exposed to artificial daylight without the light-protective bag. Stability-indicating HPLC analysis found that the nifedipine concentration in the powder dilution was about 92% of the expected concentration, indicating some loss had occurred during compounding. Storage of the powder papers protected from light at either room or refrigeration temperature resulted in little or no loss in 12 months. However, samples exposed to light exhibited a very rapid nifedipine loss. Photodegradation of 20% in three hours, 40% in six hours, and complete loss in three days was found. Consequently, adequate light protection during preparation, storage, and handling for administration is essential.

**STUDY 4 (ORAL POWDER):** Ohkubo et al.[676] evaluated the photodegradation of nifedipine in tablets pulverized for use in pediatric patients unable to swallow the tablets. The tablet powder was packaged in powder papers and exposed to mixed daylight and fluorescent light for 18 hours at 20°C. Stability-indicating HPLC analysis found a nifedipine decrease of 20% in 18 hours, with an increase in the photodegradation product.

**STUDY 5 (SUSPENSION):** Helin-Tanninen et al.[679,680] reported on the preparation and stability of nifedipine from crushed Adalat (Bayer) tablets or pure powder suspended in various concentrations of hypromellose (hydroxypropyl methylcellulose). A hypermellose concentration of 1% was found to be the best for ease of redispersion and homogeneity. The suspension retained physical stability for 28 days. Stability-indicating HPLC analysis yielded variable assay results but indicated that less than 10% nifedipine loss occurred in 28 days at room temperature near 23°C and refrigerated at about 6°C if protected from light exposure at all times. If exposed to light, a 25% loss of nifedipine occurred in three hours.

**STUDY 6 (SUSPENSION):** Nahata et al.[748] reported the stability of nifedipine 4 mg/mL in two oral suspension formulations. Nifedipine 10-mg capsules (Novopharm) were punctured on the top of the capsule to form a vent. The liquid contents were removed using a needle and syringe through the bottom of the capsule. With care and two repetitions per capsule, about 95% of the drug solution was removed. The liquid was mixed with (1) a 13:1 mixture of simple syrup, NF–1% methylcellulose and (2) a 1:1 mixture of Ora-Plus–Ora-Sweet to yield a 4-mg/mL concentration. The finished suspensions were packaged in amber plastic prescription bottles and stored under refrigeration at 4°C and in a water bath at 25°C for 91 days.

No change in color or odor occurred, and the samples were readily resuspended with no caking. Stability-indicating HPLC analysis found no loss of nifedipine in either suspension vehicle in the refrigerated samples in 91 days. At room temperature, nifedipine losses of 1 and 3% occurred in 91 days in the Ora-Plus–Ora-Sweet and simple syrup–1% methylcellulose suspensions, respectively.

**STUDY 7 (SOLUTION):** Dentinger et al.[749] reported the stability of nifedipine 10-mg/mL oral solution prepared from nifedipine powder. Nifedipine 3.2 g was mixed with 190 mL of polyethylene glycol 400 using a stir bar. Glycerin 127 mL was added with stirring, and the cloudy mixture was heated to 95°C until the nifedipine dissolved. The hot clear liquid was filtered through a 1.2-μm glass microfiber filter and then allowed to cool with continued stirring. Peppermint oil 3.2 mL was then added. The finished solution was packaged in amber glass prescription bottles and amber polypropylene oral syringes sealed with tip caps. The prescription bottles were stored at room temperature near 22 to 25°C exposed to fluorescent light. The oral syringes with and without being wrapped in aluminum foil were also stored at room temperature exposed to fluorescent light.

No changes in color or odor and no visible particulates or microbial growth occurred in any sample. Stability-indicating HPLC analysis found that about 6% nifedipine loss occurred in 35 days in the amber glass bottles. In the oral syringes wrapped with foil, no nifedipine loss at all occurred in 14 days; unfortunately the testing was ended at that point. In the oral syringes without foil, 23% nifedipine loss occurred in seven days, indicating that the foil was essential for light protection.

**STUDY 8 (ORAL LIQUIDS):** McCluskey and Brunn[1709] reported the instability of nifedipine powder compounded in oral liquid formulations in fluorescent light and in red-shaded light. The vehicles were a 1:1 mixture of Ora-Plus and Ora-Sweet (Paddock) and also Syrup, USP, as used by Nahata et al.,[748] the polyethylene glycol and glycerin mixture used by Dentinger et al.,[749] and Ora-Blend (Paddock). For the 1:1 mixture of Ora-Plus and Ora-Sweet and also Syrup, USP, nifedipine extracted from liquid-filled capsules was tried with the compounding performed under fluorescent light, but this proved to be inadequate, resulting in subpotent preparations. Nifedipine powder was incorporated into the polyethylene glycol and glycerin mixture under gold-shaded light, resulting in an acceptable drug concentration, but the preparation was unpalatable. The best result occurred using nifedipine powder as the drug source and incorporating it into Ora-Blend while exposed only to gold-shaded light. No loss of nifedipine occurred during compounding. The final preparation proved to have acceptable drug concentration and taste.

**STUDY 9 (SUBLINGUAL ABSORPTION):** van Harten et al.[613] found sublingual absorption from capsule contents to be negligible. Holding capsule contents in the mouth for 20 minutes without swallowing resulted in about 90% of the drug being recovered from the mouth, with no clinical effects observed. For a more rapid onset of action, van Harten et al.[613] and

McAllister[633] recommended that the patient bite the capsule and swallow the contents with water.

### Rectal

Munari[828] reported the stability of nifedipine 0.2% in commercial Excipial cream (used to treat anal fissures) over three months at 25°C and 60% relative humidity, 30°C and 60% relative humidity, and 40°C and 75% relative humidity protected from light. No change in physical appearance was observed, and 4% or less loss of nifedipine occurred in three months under all storage conditions.

### Topical

McCluskey and Brunn[1709] reported the instability of nifedipine powder incorporated into Plastibase 50W (Bristol-Myers Squibb) during compounding in fluorescent light and in red-shaded light. Nifedipine losses of 48% and 36%, respectively, occurred. However, if the compounding was performed under only gold-shaded light with no fluorescent light exposure, drug concentrations near 100% were obtained. Nifedipine extracted from liquid-filled capsules proved to be inadequate, resulting in a subpotent preparation. ■

# ■ Nitrazepam

## Properties

Nitrazepam is a yellow crystalline powder.[1,3]

### Solubility

Nitrazepam is practically insoluble in water but is slightly soluble in ethanol.[1,3]

## General Stability Considerations

Nitrazepam products should be protected from light.[3]

## Stability Reports of Compounded Preparations

### Oral

Amadio et al.[534] evaluated an extemporaneously prepared nitrazepam 1-mg/mL hydroalcoholic oral solution prepared from powder. The formula for the oral solution is shown in Table 73. HPLC analysis of the preparations over nine months under varying storage conditions showed that the

**TABLE 73.** Nitrazepam Oral Solution Formula of Amadio et al.[534]

| Component | | Amount |
|---|---|---|
| Nitrazepam | | 100 mg |
| Ethanol | | 14.4 mL |
| Propylene glycol | | 20 mL |
| Glycerol | | 50 mL |
| Purified water | qs | 100 mL |

formulation was stable and had a potential shelf life of two years stored at 4°C. Attempts to prepare oral suspensions from commercial tablets were unsuccessful because of difficulties in redispersion that might result in inconsistent dosing. The hydroalcoholic solution solved the problem. ■

# ■ Nitrofurantoin

## Properties

Nitrofurantoin is a yellow or orange-yellow odorless crystalline powder having a bitter aftertaste.[1-3]

### Solubility

Nitrofurantoin is slightly soluble in water, having an aqueous solubility of 0.17 to 0.19 mg/mL. In aqueous solution at pH 7.2, the solubility is about 0.37 mg/mL. In ethanol, the drug has a solubility of 0.51 mg/mL. In polyethylene glycol 300, it is freely soluble.[1-3,6]

### pH

Nitrofurantoin oral suspension has a pH between 4.5 and 6.5.[4] Furadantin suspension (Norwich-Eaton) had a measured pH of 6.2.[19]

### pKₐ

The drug has a $pK_a$ of 7[6] or 7.2.[2,6] However, $pK_a$ values of 3.5 and 7.8, with a third $pK_a$ at a higher pH, also have been determined.[6]

## General Stability Considerations

Nitrofurantoin darkens upon prolonged exposure to light or to alkaline materials. Also, it decomposes on contact with metals other than aluminum or stainless steel.[2,3]

Nitrofurantoin products should be packaged in tight, light-resistant containers and stored at controlled room temperature. The products should be protected from excessive temperatures of 40°C and above, and the suspension should be protected from freezing.[2,4,7]

Nitrofurantoin undergoes a reversible hydrolysis reaction. The reaction exhibits specific base catalysis. Nitrofurantoin is stable over a relatively wide pH range, from 5.4 to 9.9. The hydrolysis that occurs at acidic pH is reversed when the pH is increased.[6]

## Stability Reports of Compounded Preparations
### Enteral
#### NITROFURANTOIN COMPATIBILITY SUMMARY
*Compatible with:* Enrich • Ensure • Ensure HN • Ensure Plus • Ensure Plus HN • Osmolite • Osmolite HN • TwoCal HN • Vital • Vivonex T.E.N.

**STUDY 1:** Cutie et al.[19] added 10 mL of nitrofurantoin suspension (Furadantin, Norwich-Eaton) to varying amounts (15 to 240 mL) of Ensure, Ensure Plus, and Osmolite (Ross Laboratories) with vigorous agitation to ensure thorough mixing. The nitrofurantoin suspension was physically compatible,

distributing uniformly in all three enteral products, with no phase separation or granulation.

**STUDY 2:** Burns et al.[739] reported the physical compatibility of nitrofurantoin suspension (Furadantin) 100 mg/20 mL with 10 mL of three enteral formulas, including Enrich, TwoCal HN, and Vivonex T.E.N. Visual inspection found no physical incompatibility with any of the enteral formulas.

**STUDY 3:** Altman and Cutie[850] reported the physical compatibility of nitrofurantoin suspension (Furadantin, Norwich-Eaton) 10 mL with varying amounts (15 to 240 mL) of Ensure HN, Ensure Plus HN, Osmolite HN, and Vital after vigorous agitation to ensure thorough mixing. The elixir was physically compatible, distributing uniformly in all four enteral products, with no phase separation or granulation. ■

# ■ Nitrofurazone

## Properties
Nitrofurazone is a lemon-yellow to brownish-yellow odorless crystalline powder having a bitter aftertaste.[1–3]

### Solubility
Nitrofurazone is only slightly soluble in water, having an aqueous solubility of about 0.23 mg/mL.[1–3] It has solubilities of about 1.7 and 2.9 mg/mL in ethanol and propylene glycol, respectively.[1,3]

### pH
The filtrate from a 1% suspension in water has a pH of 5 to 7.5.[4] The pH of a saturated aqueous solution has been stated to be 6 to 6.5.[1]

## General Stability Considerations
Nitrofurazone darkens upon exposure to light, with the greatest light sensitivity in lower concentrations.[2,3] However, it is stated that the discoloration does not affect concentration.[2] It also discolors on contact with alkalies, turning dark orange.[1,3] The drug is stable over pH 4 to 9.[2]

Nitrofurazone can be autoclaved with little loss of antibacterial efficacy (see *Stability Reports of Compounded Preparations*). Autoclaving nitrofurazone ointment more than once is not recommended due to darkening and deterioration.[2]

Nitrofurazone products should be packaged in tight, light-resistant containers, stored at controlled room temperature, and protected from exposure to direct sunlight, excessive heat, strong fluorescent light, and alkaline materials.[4]

## Stability Reports of Compounded Preparations
### Topical
Phillips and Fisher[533] evaluated the stability of nitrofurazone soluble dressings (Furacin soluble dressing, Roberts) subjected to autoclaving as recommended by the manufacturer. Nitrofurazone in gauze pads was autoclaved at 121°C and 15 to 20 psi for 30 minutes, and nitrofurazone in gauze rolls was autoclaved at 121°C and 20 psi for 45 minutes. The color of the products became much darker yellow after autoclaving. Nitrofurazone losses of 12 to 22% for the pads and 14 to 24% for the rolls were found using the USP chromatographic analytical method. Because the method is not stability indicating, actual decomposition could be greater.

### Nasal
Xu et al.[1681] evaluated the stability of nitrofurazone and ephedrine nasal drops. Samples of the solution were stored at temperatures ranging from 55 to 85°C. The shelf-life of nitrofurazone was calculated to be about 80 days at 25°C. ■

# Nizatidine

## Properties

Nizatidine is an off-white to buff crystalline powder having a bitter taste and a sulfurlike odor.[2,7]

### Solubility

Nizatidine is sparingly soluble in water.[2,7] In aqueous solutions having a pH of 4.5 or 7, nizatidine solubility is about 10 to 33 mg/mL.[1,247]

### pH

A 1% aqueous solution of nizatidine has a pH of 9.[247]

### $pK_a$

The drug has $pK_a$ values of 2.1 and 6.8.[1]

## General Stability Considerations

Nizatidine capsules should be packaged in tight, light-resistant containers and stored at controlled room temperature.[2–4]

## Compatibility with Common Beverages and Foods

**STUDY 1:** The stability of the contents of nizatidine 300-mg capsules (Axid, Lilly) incorporated into several common commercial products to provide extemporaneously prepared oral liquids was evaluated by Lantz and Wozniak.[49] The equivalent of 300 mg of drug in the powder contents of the nizatidine capsules was added to 120 mL of Gatorade Thirst Quencher lemon lime (Stokely-Van Camp), Ocean Spray Cran-Grape cranberry grape drink (Ocean Spray), Speas Farm apple juice (Sundor Brands), and V8 vegetable juice (Campbell Soup). It also was incorporated into 30 mL of aluminum hydroxide and magnesium hydroxide antacid suspension (Maalox, Rorer). Nizatidine in water served as a control. The samples were stored under refrigeration at 5°C and at room temperature for 48 hours. The samples were inspected visually for physical incompatibilities; none of the preparations underwent discoloration or precipitation during the study. However, the capsule contents tended to form clumps in the Maalox suspension and were not adequately dissolved or dispersed. The nizatidine content was assessed using a stability-indicating HPLC assay.

All extemporaneous preparations, including the control solution in water, were stable, exhibiting losses of 2 to 6% in 48 hours. Nizatidine was less stable at room temperature. In Cran-Grape, 9% was lost in eight hours and 26% in 24 hours; in V8, 11% was lost in 48 hours. All other preparations, including the control solution in water, were stable, exhibiting nizatidine losses of 2 to 5% in 48 hours. Although nizatidine was stable in Maalox, a trace decomposition product, nizatidine sulfoxide, was detected within eight hours at both temperatures.

**STUDY 2:** Abdel-Rahman et al.[1156] evaluated the bioequivalence of unopened commercial Axid capsules compared to a commercial nizatidine 15-mg/mL oral liquid (Lyne Laboratories) and nizatidine powder from commercial Axid (Lilly) capsules mixed into apple juice at 1.2 mg/mL and also into Enfamil (Ross Laboratories) infant formula at 15 mg/mL. The commercial oral liquid and the extemporaneous preparation in infant formula exhibited comparable bioavailability to commercial Axid capsules. However, in apple juice there was a marked reduction in bioavailability. The reduced bioavailability in apple juice is consistent with the previous observation of Lantz and Wozniak.[49]

# Norfloxacin

## Properties

Norfloxacin is a white to pale yellow crystalline powder.[1,4,7]

### Solubility

Norfloxacin is slightly soluble in water, ethanol, and acetone. It is very slightly soluble in ethyl acetate. Norfloxacin solubilities in water and ethanol have been reported to be 0.28 mg/mL and 1.9 mg/mL, respectively, at room temperature. Aqueous solubility increases sharply at acidic pH below 5 or basic pH above 10. Norfloxacin solubility in acetone and ethyl acetate is 5.1 and 0.94 mg/mL,[1,4,7] respectively.

### pH

Norfloxacin ophthalmic solution, USP, has a pH of 5 to 5.4.[4]

### $pK_a$

Norfloxacin has $pK_a$ values of 6.34 and 8.75.[1]

## General Stability Considerations

Norfloxacin powder and ophthalmic solution should be packaged in tight, light-resistant containers and stored at controlled room temperature. Norfloxacin tablets should be packaged in well-closed containers and stored at controlled room temperature.[4,7]

Nangia et al.[1468] evaluated the effects of pH, light exposure, heat, and oxygen on the stability of norfloxacin 10 mcg/mL in aqueous solution. The presence of oxygen, exposure to light, and elevated temperature (70°C) all increased the decomposition rate of the drug in aqueous solution at neutral pH; up to

30% loss occurred in 48 hours under these harsh conditions. The drug was most stable at acidic and basic pH (2 and 12, respectively) remaining stable for 48 hours under these conditions.

## Stability Reports of Compounded Preparations
### Oral
**STUDY 1:** Johnson et al.[707] reported the stability of norfloxacin 20 mg/mL extemporaneously prepared as an oral suspension from tablets (Roberts Pharmaceuticals). The 400-mg tablets were crushed using a glass mortar and pestle and the suspension was prepared by adding an equal parts mixture of Ora-Plus suspending medium and strawberry syrup in increments with thorough mixing. The suspension was packaged in amber plastic prescription bottles and stored refrigerated at 3 to 5°C and at room temperature of 23 to 25°C exposed to normal fluorescent room light.

No physical changes to the suspension occurred, and stability-indicating HPLC analysis found that no loss occurred at room temperature and about 7% loss occurred under refrigeration in 56 days. The authors offered no explanation for the better stability at room temperature.

**STUDY 2:** Boonme et al.[801] evaluated the stability of norfloxacin 20-mg/mL oral suspensions having the formula shown in Table 74 prepared from two brands (unspecified) of norfloxacin 400-mg film-coated tablets. The vehicle for the suspension was prepared according to the formula in Table 75. A mortar and pestle were used to mix the tragacanth with the sorbitol solution and glycerin. Paraben concentrate was added, along with saccharin sodium in purified water, peppermint spirit,

**TABLE 74.** Norfloxacin 20-mg/mL Oral Suspension[801]

| Component | | Amount |
|---|---|---|
| Norfloxacin tablets (400 mg) | | 50 tablets |
| Purified water | | 150 mL |
| Suspension vehicle (see Table 75) | qs | 1000 mL |

**TABLE 75.** Suspension Vehicle[801]

| Component | | Amount |
|---|---|---|
| Tragacanth | | 1 g |
| Saccharin sodium | | 1 g |
| Sorbitol solution | | 100 mL |
| Glycerin | | 100 mL |
| Paraben concentrate[a] | | 10 mL |
| Peppermint spirit, BP | | 30 mL |
| Purified water | | 300 mL |
| Syrup, USP | qs | 1000 mL |

[a]Methylparaben 10%, propylparaben 2% in propylene glycol.

and 400 mL of syrup. After thorough mixing, the vehicle was brought to volume with additional syrup. Fifty norfloxacin 400-mg film-coated tablets were triturated to fine powder using a mortar and pestle. Purified water 150 mL was added to dissolve the film coating and form a smooth paste. An 800-mL portion of the suspension vehicle was slowly incorporated, and the suspension was brought to volume with additional vehicle. The finished suspension was packaged in glass containers and stored at ambient temperature for 28 days.

No apparent change in color occurred during storage. HPLC analysis found that no loss of norfloxacin occurred in 28 days using either unspecified brand of tablets as the source of drug.

### Ophthalmic
Ophthalmic preparations, like other sterile drugs, should be prepared in a suitable clean air environment using appropriate aseptic procedures. When prepared from nonsterile components, an appropriate and effective sterilization method must be employed.

Li et al.[1611] evaluated the stability of norfloxacin prepared as eyedrops and stored at room temperature of 25°C. Based on ultraviolet spectrophotometric analysis of drug concentrations, the shelf life was calculated to be 2.5 years. ■

# Noxytiolin
## (Noxythiolin)

## Properties
Noxytiolin exists as a crystalline material.[1]

### Solubility
Noxytiolin is stated to have an aqueous solubility of 100 mg/mL. In ethanol its solubility is 40 mg/mL.[1]

### pH
Noxytiolin 10 mg/mL in water has a pH of 6 to 6.4.[1553] Noxytiolin 25 mg/mL in water has a pH variously cited as 6.3 to 7[1] and pH 6.7 to 7.1.[1553]

## General Stability Considerations
Because of instability in aqueous solution, noxytiolin is supplied in dry form as Noxyflex (with amethocaine hydrochloride) and Noxyflex S (no amethocaine hydrochloride). It is prepared for use as an aqueous solution from the dry material. The action of noxytiolin as a topical antiseptic derives from releasing formaldehyde. In solution, noxytiolin decomposes, releasing N-methylthiourea and formaldehyde until an equilibrium is reached. The extent of the decomposition is dependent on the temperature of the solution, with more rapid decomposition at higher temperatures. The times to

equilibrium were found to be about eight and 40 days at 37 and 20°C, respectively.[1553]

## Stability Reports of Compounded Formulations
### Topical

McCafferty et al.[1553] evaluated the stability of noxytiolin 10- and 25-mg/mL aqueous solutions packaged in glass and polypropylene containers and stored at 37, 20, and 4°C. Concerns had been raised regarding a potential for differing stability between the two container materials. The concentrations of noxytiolin, N-methylthiourea, and formaldehyde were determined using a combination of HPLC analysis and colorimetric determinations.

Noxytiolin decomposition and release of formaldehyde were identical in the solutions packaged in the two types of containers. Equilibrium between the decomposition of noxytiolin and the release of formaldehyde and N-methylthiourea was reached in eight days at 37°C and 40 days at 20°C. Equilibrium was not reached by 40 days at 4°C. The amount of noxytiolin loss after 40 days at 4°C was less than 1% in the 10-mg/mL concentration and about 2.5% in the 25-mg/mL concentration. The presence of amethocaine hydrochloride did not influence noxytiolin stability. ■

#  Nystatin

## Properties

Nystatin is a yellow to light brown hygroscopic powder with an odor similar to cereals. The material contains not less than 4400 nystatin units per milligram of the dried substance. If intended for the preparation of oral suspensions, the substance must have not less than 5000 nystatin units per milligram.[2–4]

### Solubility

Nystatin is very slightly soluble in water and practically insoluble in ethanol. Its aqueous solubility is about 4 mg/mL, while in ethanol its solubility is about 1.2 mg/mL.[1–3]

### pH

A 3% suspension in water has a pH of 6 to 8.[4] Nystatin oral suspension has a pH between 4.5 and 6. Nystatin lotion has a pH of 5.5 to 7.5. A nystatin lozenge dissolved in 100 mL of water at 37°C has a pH of 5 to 7.5 when the solution is allowed to cool to room temperature.[4]

### $pK_a$

The drug has ionization constants of 5.71 and 8.64.[6]

### Osmolality

Nystatin suspension 100,000 units/mL (Squibb) had an osmolality of 3300 mOsm/kg.[233]

### General Stability Considerations

Nystatin decomposes upon exposure to heat and light as well as to moisture and air. Extemporaneous aqueous solutions and suspensions begin losing activity soon after preparation; they should be used soon after preparation and should not be stored.[1–3,6]

In solution, nystatin is most stable at pH 7, but decomposition is accelerated at extremes of pH. At pH 3 and 4, 90% degradation occurs in three and six hours, respectively. Various polyvalent metal ions also enhance degradation.[6]

Nystatin powder should be packaged in tight, light-resistant containers and stored under refrigeration at 2 to 8°C. After opening, the concentration of Nilstat powder can be guaranteed for no longer than 90 days, after which it should be discarded.[2]

Nystatin tablets and oral suspension should be packaged in tight, light-resistant containers and stored at controlled room temperatures. The products should be protected from excessive temperatures of 40°C and more, and the suspension should be protected from freezing.[2] The manufacturer has stated that nystatin should not be removed from the original packaging and placed in dosing compliance aids because of the hygroscopicity of the drug.[1622]

Brown discoloration of nystatin ointment and powder is associated with chemical decomposition. It is not a surface phenomenon but occurs throughout the product.

## Stability Reports of Compounded Preparations
### Oral

**STUDY 1 (LOZENGES):** Allen and Tu[136] reported on nystatin oral lozenges and a frozen Popsicle-type product. The lozenges (Table 76) were prepared by heating Karo syrup, sucrose, and purified water to 120°C and mixing well. The nystatin powder was mixed in quickly, and 4-mL aliquots were placed into molds using a syringe. The frozen form (Table 77) was prepared in the manner described by Dobbins.[661] The nystatin suspension, sorbitol 70%, and banana flavor were mixed, and

**TABLE 76.** Nystatin Lozenges Formulation[136]

| Component | Quantity |
| --- | --- |
| Nystatin powder | 200,000 units |
| Karo syrup | 2.77 g |
| Sucrose | 3.90 g |
| Purified water | 0.33 mL |

**TABLE 77.** Nystatin Frozen Popsicle-Type Formulation[136,661]

| Component | Quantity |
| --- | --- |
| Nystatin suspension 100,000 units/Ml | 75 mL |
| Sorbitol 70% (wt/wt) solution | 60 mL |
| Banana flavoring | 15 mL |
| Distilled water | qs  900 mL |

distilled water was added to make 900 mL. Aliquots of 30 mL were poured into ice cube trays and placed in a freezer for about 1.5 to 2 hours. Junior tongue depressors were inserted, and the preparation then remained in the freezer until frozen firmly. There is no stability information on the lozenges; the frozen form was given an expiration date of two months, as suggested by Squibb.[661]

**STUDY 2 (ORAL RINSE):** Vermerie et al.[917] reported the stability of nystatin in several mouth rinse formulations. Nystatin 14,400 units/mL was evaluated for stability in water, dilute hydrochloric acid (pH 4), and 1.4% sodium bicarbonate (pH 7.9) with and without colloidal silver 0.002% as an antiseptic. The mouth rinses were packaged in tinted glass bottles and stored at room temperature of 22°C and under refrigeration at 4°C. The bottles were opened four times a day to simulate in-use conditions.

Microbiological assays found nystatin stability to be pH-dependent, with the best stability in water and the alkaline suspensions, especially with 0.002% colloidal silver present. In water and the 1.4% sodium bicarbonate solution, nystatin losses of 9 to 10% occurred in four days at room temperature and seven days under refrigeration. With the 0.002% colloidal silver present in the alkaline formulation, 8 to 10% nystatin loss occurred in nine and seven days at 5 and 22°C, respectively. The acidic formulation was less stable, with 10% loss in four days at both temperatures.

**STUDY 3 (ORAL RINSE):** Groeschke et al.[1050] reported the stability of nystatin for oral suspension (Mycostatin) prepared as an antifungal mouth rinse containing nystatin 4580 units/mL in 1.4% sodium bicarbonate. The preparations were packaged in glass and polypropylene containers and stored over 15 days at temperatures ranging from 4 to 37°C exposed to and protected from light. Inspection found no change in color and no precipitation with any test storage condition. HPLC analysis of refrigerated samples found about 8% loss after 15 days. Room temperature samples exhibited 8 to 10% loss in seven days and 10 to 13% loss in 10 days. At elevated temperature of 37°C, losses of 8 and 13% occurred within three and four days, respectively. Exposure to or protection from light made no difference in the result.

**STUDY 4 (POWDER WITH GENTAMICIN AND VANCOMYCIN):** Wamberg et al.[1516] evaluated the stability of an oral preoperative bowel preparation containing three antibiotics. Gentamicin sulfate, nystatin, and vancomycin hydrochloride dry powders were sifted and mixed. The intended oral dose was gentamicin sulfate 240 mg, nystatin 3.5 million units, and vancomycin hydrochloride 250 mg. The mixed antibiotic powder was packaged in unit dose amber glass airtight containers and stored at cool temperatures of 8 to 15°C and protected from exposure to direct sunlight. Using a microbiological assay technique, the authors reported that all of the antibiotics retained adequate concentration under these conditions for up to 12 months. If the antibiotics were prepared in a liquid syrup mixture, the activity of vancomycin hydrochloride was lost much more rapidly.

*Enteral*

Ortega de la Cruz et al.[1101] reported the physical compatibility of an unspecified amount of oral liquid nystatin (Mycostatin, Bristol-Myers Squibb) with 200 mL of Precitene (Novartis) enteral nutrition diet for a 24-hour observation period. No particle growth or phase separation was observed.

*Topical*

**STUDY 1:** Elsner et al.[435] evaluated the stability of a gel formulation for lung mycosis that contained nystatin 50,000 units/g. Several gels were tested, with the best formulation being prepared from 60% hydrogenated palm kernel oil, 16% solid paraffin, and 24% Lipiodol Ultra Fluid, a radiopaque medium. Nystatin content was assessed using both chemical and microbiological techniques. The drug was stable in the gel for at least three months at 37°C. Two samples tested after nine months at 37°C still retained acceptable concentration.

**STUDY 2:** Miligi and Kassem[1009] reported that nystatin was more stable in an ointment absorption base having the composition shown in Table 78.

Miligi and Kassem[1009] also reported that the drug was less stable in an oleaginous base of beeswax 5% and white soft paraffin 95%. Two bases, one containing polyethylene glycols and the other an emulsion base, were found to be unsatisfactory. ∎

**TABLE 78.** Composition of Nystatin in an Ointment Absorption Base[1009]

| Component | Quantity |
| --- | --- |
| Wool alcohol | 6% |
| Hard paraffin | 24% |
| White soft paraffin | 10% |
| Liquid paraffin | 60% |

# Octreotide Acetate

## Properties
Octreotide acetate, an octapeptide analog of somatostatin, is available as a clear sterile buffered solution for injection.[1,7]

## pH
The injection is buffered with lactic acid and has a pH of 3.9 to 4.5.[7]

## General Stability Considerations
Octreotide acetate products should be stored under refrigeration and protected from light. The manufacturer indicates that the injection may be stored at controlled room temperature protected from light for up to 14 days.[7]

## Stability Reports of Compounded Preparations
### Injection
Injections, like other sterile drugs, should be prepared in a suitable clean air environment using appropriate aseptic procedures. When prepared from nonsterile components, an appropriate and effective sterilization method must be employed.

**STUDY 1:** Stiles et al.[514] evaluated the stability of octreotide acetate (Sandoz) 0.2 mg/mL repackaged as 1 mL in 3-mL polypropylene syringes sealed with tip caps (Becton Dickinson). The syringes were stored at 3 and 23°C both protected from and exposed to room light. HPLC analysis found no loss of octreotide in 29 days at 3°C protected from light and about 7 to 9% loss over 15 to 22 days exposed to light. At room temperature, losses in excess of 10% occurred in about two weeks. Consequently, the authors recommended a one-week limitation on room temperature storage whether exposed to light or not.

**STUDY 2:** Ripley et al.[515] also reported on octreotide acetate (Sandoz) stability repackaged in syringes. The injection was repackaged 1 mL (0.2 mg) in 3-mL polypropylene syringes (Terumo) fitted with tip caps and stored under refrigeration at 5°C and frozen at −20°C for 60 days. Light exposure was not addressed. HPLC analysis of the intact drug and its principal decomposition product demonstrated about 6% loss in 60 days under both storage conditions.

# Omeprazole

## Properties
Omeprazole is a white or nearly white powder.[3,7]

### Solubility
Omeprazole is very slightly soluble in water but is freely soluble in ethanol.[3,7] Omeprazole dissolves in dilute solutions of alkali hydroxides.[3]

### pH
A 2% omeprazole solution as the sodium salt has a pH of 10.3 to 11.3.[3]

### $pK_a$
Omeprazole has a $pK_a$ variously cited as 3.97[1325] and 4.2.[1552] The drug is also stated to have a second $pK_a$ of 9.[1552]

## General Stability Considerations
Omeprazole products should be packaged in tight containers and stored at controlled room temperature protected from exposure to light and moisture.[7] Exposure to humidity may result in chemical degradation of omeprazole.[1464] The manufacturer has stated that omeprazole capsules should not be removed from the original packaging and placed in

dosing compliance aids because of the hygroscopicity of the drug.[1622,1623]

The stability of omeprazole is a function of pH. Omeprazole exhibits maximum stability at pH 11 and rapidly decomposes below pH 7.8.[7,562,1467] Increasing temperature accelerates decomposition.[1467] The presence of citric acid, trisodium citrate, and monosodium citrate in combination with omeprazole substantially increases the rate of omeprazole decomposition.[1099] Cyclodextrins have also been shown to accelerate omeprazole decomposition.[1467]

Omeprazole decomposes in gastric acid, leading to poor bioavailability.[586,605] The oral capsules are formulated with the drug in pH-sensitive (enteric-coated) pellets that release the drug at a pH above 6.[561,586,605] Crushing the pellets destroys the protective coating, exposing the drug to gastric pH and degradation.[586,605] Similarly, an omeprazole injection is not active when given orally.[605]

Omeprazole R- and S-isomers have been shown to interact differently with mannitol. Differential scanning calorimetry test results indicated interactions that the authors stated could influence differences in bioavailability.[1271]

As omeprazole decomposes, discoloration appears. A pale yellow discoloration does not necessarily mean unacceptable decomposition. However, suspensions with greater discoloration, including dark yellow, orange, purple, brown, or black, should be discarded.[561,782,1034,1054]

## Stability Reports of Compounded Preparations
### Oral
**STUDY 1 (LIQUID):** Quercia et al.[561] evaluated the stability of an omeprazole 2-mg/mL oral liquid prepared from capsule contents. The oral liquid was prepared by emptying the contents of five omeprazole 20-mg capsules (Astra Merck) into the barrel of a 60-mL polypropylene Luer Lock syringe with an 18-gauge needle attached. The plunger was replaced and 50 mL of 8.4% sodium bicarbonate injection was drawn into the syringe. The needle was removed and one end of a fluid-dispensing connector (Braun) was attached. The other end was attached to a second 60-mL syringe. The omeprazole–sodium bicarbonate mixture was transferred back and forth until the omeprazole had completely dissolved. The mixture was then transferred back into the empty sodium bicarbonate vial. Samples prepared in this manner were stored at −20, 4, and 24°C.

Stability-indicating HPLC analysis found that the omeprazole was stable for 30 days at −20 and 4°C, exhibiting less than 3 and 5% loss, respectively. Furthermore, there was no change in the appearance of the oral liquid stored at these temperatures. However, the room temperature samples exhibited 8% loss in 14 days and 14% loss in 18 days; the color also changed from white to brown during storage. Clinical trials of similar extemporaneously compounded oral liquids have demonstrated the safety and utility of treating patients while maintaining the gastric pH above 5.[563,564]

**STUDY 2 (LIQUID):** McAndrews and Eastham[672] reported on preparing an omeprazole suspension in a similar manner. Shaking the syringe facilitated suspension formation. Such agitation was found to be essential; in static samples, an omeprazole suspension did not form. The authors also evaluated the use of smaller amounts of sodium bicarbonate to remove the enteric coating from the granules for patients with an acid–base imbalance. Four amounts of 8.4% sodium bicarbonate were used: 20, 10, 5, and 2.5 mL mixed with 2.5 mL of sterile water for injection. All of these amounts of sodium bicarbonate were adequate to prepare the suspension of 20 mg of omeprazole.

**STUDY 3 (LIQUID):** DiGiacinto et al.[782] reported the stability of omeprazole 2-mg/mL oral suspension extemporaneously compounded from capsules. The contents of omeprazole 20-mg capsules were mixed with sodium bicarbonate 8.4% injection. The omeprazole 2-mg/mL suspension was packaged as 5 mL of suspension in 10-mL amber oral syringes and stored at room temperature of 22°C exposed to fluorescent light and refrigerated at 4°C protected from light. Stability-indicating HPLC analysis found 10% omeprazole loss occurred in 14 days at room temperature. The suspension became slightly yellow during storage. Under refrigeration, less than 6% loss occurred in 45 days, but 16% loss occurred by 60 days.

**STUDY 4 (LIQUID):** Ogden and Asghar[947] reported on compounding an omeprazole 2-mg/mL oral liquid from commercial Losec (AstraZeneca) capsules. The contents of 10 capsules were emptied into a graduated cylinder. Sodium bicarbonate powder 8.4 g was placed in a separate graduated cylinder and dissolved in 100 mL of sterile water. About 80 mL of sodium bicarbonate 8.4% solution was then added to the capsule contents, and the capsule contents were allowed to dissolve. The omeprazole solution was brought to 100 mL with additional sodium bicarbonate 8.4% solution and mixed well. The oral liquid was packaged in amber glass prescription bottles and assigned a month expiration period when stored in the refrigerator. The authors offered no stability information to support this expiry period.

**STUDY 5 (LIQUID):** Phillips et al.[1034] evaluated the stability of a simplified oral suspension containing omeprazole 2 mg/mL in 8.4% sodium bicarbonate. The suspension was stored in clear glass containers at room temperature, refrigerated, and frozen. The samples were analyzed using stability-indicating HPLC analysis. The refrigerated and frozen samples exhibited little or no omeprazole loss during 24 weeks of storage. However, the room-temperature samples lost about 5% in one week, 13% in two weeks, and 25% in four weeks. A substantial black color change occurred as well, which the authors attributed to light exposure. The use of amber bottles was recommended.

**STUDY 6 (LIQUID):** Graudins et al.[1355] evaluated the stability of an omeprazole 2-mg/mL oral liquid prepared from commercial capsule contents (Probitor, Sandoz) and compared the stability with a similar formulation prepared from omeprazole powder. The oral liquid was prepared by mixing sodium bicarbonate powder 8.4 g with 1 mL of methyl hydroxybenzoate 1% aqueous solution and bringing the mixture to 100 mL with water for irrigation. The contents of ten 20-mg capsules were emptied into the solution, stirred, and allowed to stand under refrigeration for about four to five hours until dispersed. The oral liquid was then mixed with a hand-held blender and packaged in amber glass bottles. A second formulation using this compounding method and using omeprazole powder as the source of drug was also prepared. Samples of each formulation were stored for 45 days refrigerated at 2 to 8°C and at room temperature of 19 to 23°C with periodic shaking of the samples. A separate set of samples was stored for 45 days refrigerated at 2 to 8°C and left undisturbed with no shaking. HPLC analysis of omeprazole concentrations and visual inspection for color changes were performed.

*Capsule Contents Formulation with Periodic Shaking:* Omeprazole concentrations were acceptable for 46 days under refrigeration with about 7% drug loss and for seven days at room temperature with little drug loss. The oral liquid became slightly discolored during the study period with the darkest color in the room temperature samples, but this did not directly correlate with drug concentration.

*Powder Formulation and Both Unshaken Samples:* Unacceptably low drug concentrations were found within one day after preparation. Decreases ranged from 14% to 98%, possibly due to layering and nonuniform distribution of the drug.

The authors stressed the importance of periodic shaking of the omeprazole oral liquid to ensure stability and uniformity.

**STUDY 7 (LIQUID):** Garg et al.[1356] compared the stability of omeprazole 2-mg/mL oral liquid prepared by the shaking method and by the grinding method during eight weeks of storage refrigerated at 2°C and at room temperature of 25°C with 60% relative humidity.

For the shaking method, the contents of ten 20-mg Losec capsules (AstraZeneca) were emptied into a 100-mL bottle, 50 mL of sodium bicarbonate 8.4% aqueous solution was added, and the cap was secured on the bottle. The mixture was vigorously shaken for a minimum of three minutes. If less shaking time is used, not all of the granules are dissolved and the mixture may not be a uniform dispersion. The mixture was transferred to a 100-mL conical flask and the mortar was rinsed with additional sodium bicarbonate 8.4% into the flask. The mixture was brought to 100 mL with additional sodium bicarbonate 8.4% and was transferred to a 100-mL bottle.

For the grinding method, the contents of ten 20-mg Losec capsules (AstraZeneca) were emptied and ground to fine powder using a glass mortar and pestle, and 50 mL of sodium bicarbonate 8.4% aqueous solution was added with gentle stirring. The mixture was transferred to a 100-mL conical flask, and the mortar was rinsed with additional sodium bicarbonate 8.4% into the flask. The mixture was brought to 100 mL with additional sodium bicarbonate 8.4% and was transferred to a 100-mL bottle.

HPLC analysis found that omeprazole 2-mg/mL oral liquid prepared by either method was stable when stored under refrigeration. Drug losses of about 5 to 6% occurred. However, at room temperature the grinding method resulted in enhanced drug loss with less than 10% drug for only one week. The shaking method resulted in retaining 90% of the drug content for at least four weeks. However, room temperature storage by either method resulted in a yellow-brown discoloration forming. The authors recommended using the shaking method of preparation with no crushing of the capsule contents and storing the oral liquid preparation under refrigeration.

**STUDY 8 (LIQUID):** Burnett and Balkin[1054] reported the stability and viscosity of flavored oral suspensions of omeprazole for pediatric administration. The suspensions were prepared using commercial Zegerid (Santarus) omeprazole powder for oral suspension prepared in tap water at concentrations of 0.6, 1.2, 2, 3, and 4 mg/mL. The suspensions were packaged in clear polypropylene tubes and stored at room temperature of 22 to 25°C exposed continuously to light and also refrigerated at 4°C protected from exposure to light.

All of the suspension samples began to develop a light yellow discoloration within 24 hours of preparation, but this did not necessarily indicate an unacceptable degree of decomposition. Stability-indicating HPLC analysis found little or no loss of omeprazole during 28 days of refrigerated storage when protected from exposure to light. Omeprazole 2-, 3-, and 4-mg/mL suspensions stored at room temperature exposed to light also exhibited little or no omeprazole loss in seven days. However, the 1.2-mg/mL suspension lost 3 and 7% after 48 hours and seven days of room temperature storage, respectively. The 0.6-mg/mL suspension lost 4 and 13% after 48 hours and seven days of room temperature storage, respectively. The authors postulated that the increased loss of omeprazole in the more dilute suspensions may result from reduced bicarbonate concentrations. No increase in suspension viscosity was found in any of the samples.

**STUDY 9 (LIQUID):** Johnson et al.[1169] evaluated the stability of reconstituted commercial omeprazole-sodium bicarbonate 2-mg/mL oral suspension. The contents of the commercial powder packets were reconstituted with water to yield a 2-mg/mL oral suspension. The oral suspension was packaged in amber plastic prescription bottles with child-resistant caps and was stored under refrigeration at 3 to 5°C.

Stability-indicating HPLC analysis found that at least 98% of the omeprazole remained over the 45-day stability study. At 45 days, a slight yellowish tint appeared in the suspension. In addition, a partial dose in simulated gastric fluid was stable with at least 93% of the initial concentration over two hours at 37°C. The authors indicated that a bulk supply of the commercial oral suspension can be prepared and stored under refrigeration for at least 45 days.

**STUDY 10 (LIQUID):** Van Der Straeten et al.[1438] evaluated 16 different suggested formulations of omeprazole oral suspensions in an attempt to identify those that are preferable. The authors selected the following formulations as being the best of those tested, taking into account formulation difficulties, physical and chemical stability, acceptability of taste, and bioavailability.

*Formula 1*
Omeprazole 100 mg
Sodium bicarbonate 8.4 g
Guar gum 1 g
Sodium saccharin 750 mg
Simple syrup 20 g
Banana flavor 100 mg
Purified water qs ad 100 gm

Dissolve the sodium bicarbonate in 80 mL of purified water and mix in 17 g of simple syrup. Add the sodium saccharin and ensure complete dissolution. Add the banana flavor and mix well. Separately wet 1 g of guar gum with 2 g of simple syrup and then mix into the aqueous mixture. Separately mix 100 mg of omeprazole with 1 g of simple syrup, and then combine into the aqueous mixture with thorough mixing. Bring the oral suspension to volume with additional purified water and mix well.

*Formula 2*
Omeprazole 100 or 200 mg
Sodium bicarbonate 8.4 g
Xanthan gum 0.5 g
Glycerin 2 g
Banana flavor 100 mg
Sodium saccharin 750 mg
Simple syrup 18 g
Purified water qs ad 100 gm

Dissolve the sodium bicarbonate in 80 mL of purified water and mix in 17 g of simple syrup. Add the sodium saccharin and ensure complete dissolution. Add the banana flavor and mix well. Separately wet 500 mg of xanthan gum with 2 g of glycerin and then mix into the aqueous mixture. Separately mix 100 or 200 mg of omeprazole with 1 g of simple syrup, and then combine into the aqueous mixture with thorough mixing. Bring the oral suspension to volume with additional purified water and mix well.

The authors found these formulas to be preferable. They reported that the oral suspension maintained physical stability for 28 days protected from exposure to light and stored under refrigeration. Furthermore, based on previous stability studies, they stated the alkaline pH near 8.5 would ensure that the omeprazole would remain chemically stable for 28 days under refrigeration as well, although they did not test this assumption themselves.

**STUDY 11 (SOLID):** Woods[586] and Woods and McClintock[605] reported an alternative method to provide reduced doses. A commercial capsule was opened and the pellets were weighed. The amount corresponding to the reduced dose was weighed and repacked into another gelatin capsule. The repacked capsules were packaged in tight containers including a suitable desiccant. Just prior to administration, the repacked capsule contents were added to an acidic beverage (<pH 5.3) and mixed. Drinking the beverage delivers the dose without chewing or crushing the pellets. The author recommended a use period of seven days for the repackaged doses, although no stability data were presented.

*Enteral*
**STUDY 1:** Dunn et al.[785] compared the delivery of lansoprazole and omeprazole granules from capsules through 14 French nasogastric tubes using water and apple juice as vehicles. Delivery of the drugs was highly variable but had means of 30% of the lansoprazole granules and 53% of the omeprazole granules. No substantial difference was found with either 15 or 30 mL of either water or apple juice, although the granules in apple juice appeared to be stickier.

**STUDY 2:** Messaouik et al.[1031] reported the delivery of three proton pump inhibitor drugs through polyurethane and also silicone 16 French gastroduodenal tubes. The enteric-coated pellets from capsules of lansoprazole, omeprazole, and esomeprazole were removed from the capsules and mixed with water or apple juice. The mixture was delivered through the tubes and rinsed with 10 or 20 mL of the respective vehicle. No obstruction occurred with any of the esomeprazole samples, but 42% of the lansoprazole and omeprazole samples resulted in clogging. Esomeprazole delivery was 100%, while lansoprazole and omeprazole had reduced drug delivery of about 67 and 61%, respectively. The authors noted that the esomeprazole pellets were substantially smaller in size than those of the other two drugs. The authors concluded that esomeprazole was preferable for nasogastric delivery, compared to omeprazole or lansoprazole.

## Compatibility with Common Beverages and Foods

For patients with difficulty in swallowing omeprazole capsules and other proton pump inhibitor drugs, the capsules have been opened and the enteric-coated pellet contents mixed with a slightly acidic food such as yogurt, applesauce, or fruit juice for administration. This approach is believed to help preserve the drug's integrity until it reaches the alkaline environment of the duodenum.[1029–1035]

## Compatibility with Other Drugs

Freeman and Trezevant[1421] reported abdominal distention in a patient from the formation of excessive gas (most likely carbon dioxide) when omeprazole oral suspension compounded from capsule contents mixed with sodium bicarbonate 7.5% was administered at the same time with liquid protein solution (Proteinex, Llorens Pharmaceuticals). The same effect could be duplicated in vitro using commercial Zigerid (Santarus), which contains omeprazole 20 mg with sodium bicarbonate 1680 mg, or simply mixing sodium bicarbonate 7.5% solution with the liquid protein solution. The authors recommended separating the administration of omeprazole with sodium bicarbonate and liquid protein solution by at least two hours. ■

# ■ Ondansetron Hydrochloride

## Properties

Ondansetron hydrochloride is a white to off-white crystalline powder.[7] Ondansetron hydrochloride 1.25 mg provides approximately 1 mg of ondansetron base.[3] Ondansetron hydrochloride 1 mg is equivalent to 0.8 mg of ondansetron base.[4,5]

### Solubility

Ondansetron hydrochloride is sparingly soluble in water,[4,7] 0.9% sodium chloride injection,[7] ethanol, and isopropanol.[4]

### pH

Ondansetron hydrochloride injection has a pH between 3.3 and 4.[4,7]

### pK_a

The drug has a $pK_a$ of 7.4.[437]

### Osmolality

The osmolality of ondansetron hydrochloride injection is 281 mOsm/kg.[326]

## General Stability Considerations

Ondansetron hydrochloride injection and tablets should be stored between 2 and 30°C and protected from light.[7] Although the drug is unstable when exposed to intense light, it is stable for about a month exposed to daylight and fluorescent light simultaneously.[394]

The natural pH of ondansetron hydrochloride in aqueous solution is about 4.5.[394,395] If the solution pH increases above the range of 5.7 to 7, a precipitate of free ondansetron base develops.[394,396] The precipitate redissolves if the solution is retitrated with hydrochloric acid.[396] Ondansetron precipitation also has been observed when the drug is combined with alkaline drugs.[396,397]

## Stability Reports of Compounded Preparations
### Injection

Injections, like other sterile drugs, should be prepared in a suitable clean air environment using appropriate aseptic procedures. When prepared from nonsterile components, an appropriate and effective sterilization method must be employed.

**STUDY 1:** Leak and Woodford[394] reported that ondansetron hydrochloride injection diluted with compatible infusion solution was stable for up to seven days at room or refrigerator temperature when repackaged in polypropylene–neoprene plastic syringes (Becton Dickinson) with syringe caps.

**STUDY 2:** Casto[657] evaluated the stability of undiluted ondansetron hydrochloride 2 mg/mL and diluted in 5% dextrose and 0.9% sodium chloride injections to concentrations of 0.25, 0.5, and 1 mg/mL repackaged into polypropylene syringes. HPLC analysis found that the solutions remained stable, with concentrations above 90% for 90 days stored at −20°C, 14 days at 4°C, and 48 hours at 24°C.

### Oral
**USP OFFICIAL FORMULATION (ORAL SUSPENSION):**

    Ondansetron hydrochloride    80 mg
    Vehicle for Oral Solution, NF (sugar-containing or
       sugar-free)
    Vehicle for Oral Suspension, NF (1:1) qs 100 mL

(See the vehicle monographs for information on the individual vehicles.)

Use ondansetron hydrochloride powder or commercial tablets. If using tablets, crush or grind to fine powder. Add about 50 mL of the vehicle mixture in 5-mL portions and mix well after each addition. Quantitatively transfer to a suitable calibrated tight, light-resistant bottle, bring to final volume with the vehicle mixture, and thoroughly mix, yielding ondansetron hydrochloride 1 mg/mL oral suspension. The final liquid preparation should have a pH between 3.6 and 4.6. Store the preparation under refrigeration between 2 and 8°C or at controlled room temperature. The beyond-use date is 42 days from the date of compounding.[4,5]

**STUDY 1:** The stability of ondansetron hydrochloride in cherry syrup, USP, was reported by Graham et al.[112] Ondansetron hydrochloride injection 2 mg/mL (Zofran, Glaxo) was used to prepare solutions for longer-term storage in cherry syrup, USP. Acidic vehicles are preferred because ondansetron precipitates at alkaline pH.[113] The syrup was prepared by adding 84 mL of injection to 231 mL of cherry syrup, USP, to yield a concentration of 0.533 mg/mL and was packaged in 12-ounce amber plastic bottles stored at 4 and 26°C. The ondansetron hydrochloride content of the samples was assessed using a stability-indicating HPLC assay.

In cherry syrup, USP, no ondansetron hydrochloride loss occurred in seven days at either storage temperature. The appearance and color of the solutions did not change.

**STUDY 2:** Williams et al.[267] evaluated the stability of four ondansetron 0.8-mg/mL (as the hydrochloride) syrups compounded from tablets. Ten ondansetron 8-mg tablets were crushed and triturated to fine powder. Flaking of the tablet coating occurred during this process. The fine powder was mixed thoroughly with 50 mL of the suspending vehicle, Ora-Plus (Paddock), in 5-mL increments. The mixture was brought to 100 mL with each of the following syrups and shaken well: cherry syrup, USP; Syrpalta (Humco); Ora-Sweet (Paddock); and Ora-Sweet Sugar-Free (Paddock). The mixtures were packaged in amber plastic vials and stored at 4°C for six weeks. The ondansetron content was analyzed throughout storage using a stability-indicating HPLC assay. Ondansetron concentrations remained above 90% for all but one cherry syrup, USP, sample. The low sample (occurring at day 14) was believed to be the result of dilution and assay variables because subsequent assays out to 42 days were all above 90%. The authors concluded that all four formulations of ondansetron hydrochloride remained stable for 42 days at 4°C. Furthermore, no significant microbial growth occurred for up to 42 days of refrigerated storage.

Williams et al.[267] also measured the osmolalities of the four extemporaneous ondansetron 0.8-mg/mL syrups: cherry syrup, USP, 3328 mOsm/kg; Syrpalta, 2740 mOsm/kg; Ora-Sweet, 2438 mOsm/kg; and Ora-Sweet Sugar-Free, 2444 mOsm/kg.

**STUDY 3:** Vandenbroucke and Robays[460] evaluated the stability of ondansetron hydrochloride 0.8 mg/mL compounded from tablets in a natural raspberry syrup. Nine ondansetron hydrochloride 8-mg tablets were crushed in a mortar and pestle. The powder was transferred to a 100-mL vessel, and the mortar was rinsed with 4.5 mL of water, which was added to the vessel. Natural raspberry syrup 85 mL was mixed with the powder, yielding a preparation that contained ondansetron hydrochloride 8 mg/10 mL. Samples of the syrup were stored at 4, 20, and 30°C for 21 days. An HPLC analytical method was used to determine ondansetron hydrochloride content in the samples. No ondansetron hydrochloride loss was found in

any of the samples at any storage temperature throughout the 21-day evaluation.

**STUDY 4:** Gallardo Lara et al.[1380] developed an ondansetron hydrochloride 8.4-mg/5 mL long-acting oral suspension and evaluated its stability. The formulation consisted of free ondansetron hydrochloride 3 mg/5 mL and an ondansetron hydrochloride-cellulose acetophthalate latex polymer complex to provide ondansetron hydrochloride 5.4 mg/5 mL. The complex was prepared from a solution of ondansetron 30% and cellulose acetophthalate latex 60%. The complex was incubated at 25°C in a thermostatic water bath with 60 rpm shaking for 24 hours. This was then centrifuged for 50 minutes at 14,000 rpm to separate the sediment from the supernatant. Ultraviolet spectrophotometric analysis of the supernatant determined the amount of ondansetron remaining, and the quantity in the sediment was deduced by difference. Stock suspension was prepared by suspending ondansetron hydrochloride 3 mg/5 mL and the ondansetron-cellulose acetophthalate latex polymer complex 135 mg/5 mL (providing 5.4 mg of ondansetron hydrochloride) and allowing the mixture to settle for four hours. The final suspension was prepared by adding the other components to obtain the final composition and volume indicated below.

| | |
|---|---|
| Ondansetron hydrochloride | 3 mg |
| Ondansetron hydrochloride complex | 135 mg |
| Avicel | 1% |
| Kathon CG preservative | 0.1% |
| Glycerin | 10% |
| Xanthan gum CG (E-415) | 0.75% |
| Distilled water qs | 5 mL |

The physical stability of the long-acting suspension was evaluated over 30 days. The suspension was found to have hindered sedimentation and could be easily redispersed to homogeneity by light shaking. The authors concluded that the formulation was acceptable.

An attempt to substitute sodium carboxymethylcellulose for the xanthan gum resulted in a darkening that formed a disagreeable brown color. Additionally, the suspension could not as easily be redispersed. Consequently, this second formulation was considered unacceptable.

**STUDY 5:** Kawamura et al.[1706] prepared an oral syrup formulation of ondansetron hydrochloride in D-sorbitol and used strawberry flavor to mask the unpleasant taste. The syrup was found to be stable for six months when stored at 25°C and 60% relative humidity.

*Rectal*
**STUDY 1:** Tenjarla et al.[565] evaluated the stability, drug release, and flux through rabbit rectal membrane of extemporaneously

compounded ondansetron 8- and 16-mg suppositories in two suppository bases. Ondansetron 8-mg (as the hydrochloride) tablets (Zofran, Glaxo Wellcome) were crushed and ground to fine powder. The displacement factors for Polybase and Fattibase (Paddock) were 0.6 and 1.2, respectively. The tablet mass to be incorporated was 0.25 g/tablet. The amounts of Polybase and Fattibase needed per suppository were computed for the mold to be used, accounting for the tablet mass and base displacement factors. The bases were melted in beakers; for Fattibase, the temperature was kept below 55°C to prevent crystal formation. The tablet powder was incorporated into the melted base with constant stirring. The mixture was poured continuously into suitable suppository molds at a temperature just above congealing. This process permits immediate solidification and prevents the settling of undissolved tablet components. The finished suppositories were stored at 5°C for 28 days.

Stability-indicating HPLC analysis found no loss of ondansetron for both 8- and 16-mg suppositories in both bases and no change in release of the drug after refrigerated storage for 28 days. Rabbit rectal membrane flux studies were inconclusive as to which formulation would be better absorbed.

**STUDY 2:** A study of the pharmacokinetics of ondansetron hydrochloride release from 24-mg extemporaneously compounded suppositories in Polybase (Paddock) found a prolonged absorption period and substantially lower bioavailability compared to 24-mg oral tablets.[714]

## Compatibility with Common Beverages and Foods

### ONDANSETRON HYDROCHLORIDE COMPATIBILITY SUMMARY

*Compatible with:* Apple juice • Coca-Cola • Diet Coke • Hawaiian Punch • Kool Aid • Orange juice • Sprite
*Incompatible with:* Tea

**STUDY 1:** Graham et al.[112] evaluated ondansetron hydrochloride injection 2 mg/mL (Zofran, Glaxo) for use in preparing solutions for short-term administration in orange juice and Coca-Cola. Acidic vehicles such as these are preferred because ondansetron precipitates at alkaline pH.[113] The drug concentrations tested were 0.267 mg/mL (4 mL of the injection in 26 mL of orange juice or Coca-Cola) and 0.067 mg/mL (4 mL of injection in 116 mL of juice or Coca-Cola). The ondansetron hydrochloride content of the samples was assessed using a stability-indicating HPLC assay.

The content in orange juice and Coca-Cola remained at least 97% of the initial amount within 30 and 60 minutes after mixing. The appearance and color of the solutions did not change.

**STUDY 2:** Yamreudeewong et al.[513] evaluated the stability of ondansetron hydrochloride injection (Zofran, Glaxo) added to various beverages for oral administration. The injection was added to apple juice (Tropicana, Sysco), Hawaiian Punch, Rock-a-Dile Red cherry-flavored Kool-Aid powdered drink prepared in water with sugar, Sprite, Diet Coke, and hot Lipton tea. Ondansetron concentrations of 32.8, 64.5, and 95.2 mcg/mL were prepared by adding 2, 4, and 6 mL of the 2-mg/mL injection to 120-mL portions of the beverages. Stability-indicating HPLC analysis found that the ondansetron hydrochloride was chemically stable in the beverages. Little or no loss of ondansetron occurred in apple juice, Hawaiian Punch, and Kool-Aid in 72 hours at room temperature or under refrigeration. Similarly, little or no loss was found in Sprite or Diet Coke in 48 hours under refrigeration or 24 hours at room temperature, the maximum times tested for these beverages. Although the ondansetron was chemically stable in hot tea, a turbid precipitate formed immediately. The precipitate was possibly due to pH or complexing with a component of the tea. Centrifugation and analysis of the supernatant of the tea showed only 40% of the ondansetron concentration present in the supernatant. Treatment of the precipitated tea solution with acid and acetonitrile restored ondansetron concentration to the initial level after one hour. Although chemically stable, ondansetron bioavailability from tea may not be acceptable due to this precipitation. ∎

# ∎ Oseltamivir Phosphate

## Properties

Oseltamivir phosphate is a white crystalline material having a bitter taste.[1,2,7]

### Solubility

Oseltamivir phosphate is very soluble in water, with an aqueous solubility of 588 mg/mL at 25°C.[2]

## General Stability Considerations

Oseltamivir phosphate oral capsules and powder for suspension are stored at controlled room temperature and protected from exposure to light. After reconstitution with water as directed by the manufacturer, the oral suspension provides 12 mg/mL of oseltamivir as the phosphate and should be stored under refrigeration and protected from freezing. The reconstituted oral suspension is stable under these conditions for 10 days, according to the manufacturer.[7]

## Stability Reports of Compounded Preparations
### Oral

**STUDY 1:** The manufacturer of commercial oseltamivir phosphate provides instructions on compounding 15-mg/mL oral liquid suspensions of the drug from commercial capsules using Humco cherry syrup or Paddock Laboratories Ora-Sweet SF (sugar-free) as vehicles when the commercial oral suspension is unavailable.[7] The granular contents of an appropriate number of capsules (see Table 79) are transferred into a clean mortar and triturated to fine powder. About one-third of the appropriate amount of vehicle (see Table 79) is then added and triturated to form a uniform suspension, which is then transferred, using a funnel if necessary, into an amber glass or amber polyethylene terephthalate plastic prescription bottle. The mortar is then rinsed with about one-third of the vehicle, adding the rinse to the prescription bottle. The mortar is rinsed with the final one-third of the vehicle, again adding the rinse to the prescription bottle. A child-resistant cap is then used to close the container, and the mixture is gently shaken to ensure homogeneity. The drug itself will readily dissolve, but the inert ingredients, which are insoluble, make gentle shaking of the suspension necessary.

The manufacturer states that the compounded 15-mg/mL oral suspension prepared in the recommended manner is stable for five days at controlled room temperature and for five weeks under refrigeration. The manufacturer also states that this stability information applies only to the compounded suspension in the vehicles and containers tested. NOTE: The concentration is greater than the commercial suspension, making use of the dosing device packed with the commercial oral suspension inappropriate.

**STUDY 2:** Winiarski et al.[1166] reported the stability of extemporaneously compounded oseltamivir phosphate oral suspensions in a report that appears to provide the foundation for the information in the labeling of the drug. However, this paper[1166] reported somewhat different results. The oseltamivir phosphate 15-mg/mL formulation, preparation, and packaging are the same as in the labeling and as shown in Study 1. Stability in the Humco cherry syrup was found to be as stated

in the labeling: five days at controlled room temperature and five weeks under refrigeration, due to the formation of a glucose-type adduct that exceeded specifications. However, the concentration of oseltamivir phosphate remained at 92 to 95%, even after 35 days of room temperature storage. When Paddock Ora-Sweet SF (sugar-free) was used as the vehicle, the glucose-type adduct did not form, and the oral suspension was found to be stable for at least 35 days, both at controlled room temperature and under refrigeration, with less than 3% drug loss occurring.

**STUDY 3:** Ford et al.[1108] attempted to evaluate the stability of oseltamivir phosphate 12 mg/mL compounded from the contents of capsules in three oral suspension vehicles: PCCA-Plus, PCCA acacia syrup, and methylcellulose 1% oral solution. The gelatin shell of each capsule was removed using a clean razor blade. The capsule contents were ground to fine powder using a mortar and pestle. Each vehicle was then incorporated to yield three individual homogeneous suspensions, each with a 12-mg/mL oseltamivir concentration. The suspensions were packaged in amber vials (composition not noted), and samples were stored at room temperature of 25°C and under refrigeration at 2 to 8°C for 90 days. HPLC analysis results were erratic and variable over the study period, which could be attributed to interference by vehicle components or to lack of control of the analysis, permitting artifactual errors to occur. Although the authors noted no consistent long-term loss indicative of drug decomposition, the variability of the analytical results makes a judgment about the stability of oseltamivir phosphate in these specific vehicles problematic.

**STUDY 4:** Albert and Bockshorn[1208] evaluated the stability of oseltamivir in three simple oral liquid formulations. The test formulations were stored for 84 days at 25°C; Formula 1 was also stored under refrigeration at 6°C. The compositions of the three oral liquids are shown below:

*Formula 1*
Oseltamivir phosphate    1.971 g
Sodium benzoate    0.1 g
Purified water    qs    100 mL

*Formula 2*
Oseltamivir phosphate    1.971 g
Sodium benzoate    0.1 g
Potable water    qs    100 mL

*Formula 3*
Oseltamivir phosphate    1.971 g
Sodium benzoate    0.1 g
Anhydrous citric acid    0.1 g
Potable water    qs    100 mL

**TABLE 79.** Oral Oseltamivir Phosphate 15-mg/mL Liquid Suspensions[7]

| Final Volume | Capsules | Vehicle Volume |
|---|---|---|
| 30 mL | 6 | 29 mL |
| 40 mL | 8 | 38.5 mL |
| 50 mL | 10 | 48 mL |
| 60 mL | 12 | 57 mL |

Stability-indicating HPLC analysis found less than 2% oseltamivir loss at 25°C and no loss at 6°C in Formula 1 using purified water after 84 days. Formula 2 using potable water exhibited a somewhat greater amount of decomposition. About 5% loss occurred in 84 days. In addition, a white precipitate composed of calcium and magnesium phosphates formed. Formula 3 also used potable water, but the incorporation of citric acid reduced the drug loss. Less than 1% loss occurred in 84 days, and no precipitate formed. Sodium benzoate remained stable throughout the study, with no loss found in any of the samples.

**STUDY 5:** Voudrie and Allen[1558] evaluated the stability of oseltamivir phosphate 15 mg/mL prepared from the contents of commercial oral capsules in three oral suspension formulations: cherry syrup, SyrSpend SF, and SyrSpend SF (for reconstitution). The samples were packaged as 25 mL of sample suspension in 60-mL low actinic light plastic prescription bottles. The samples were stored refrigerated at 2 to 8°C for 30 days. Visual inspection found no change in color or homogeneity in any of the samples throughout the study. Stability-indicating HPLC analysis found little or no loss of oseltamivir phosphate in any of the formulations within 30 days. ■

# ■ Oxacillin Sodium

## Properties

Oxacillin sodium is a fine white odorless or nearly odorless crystalline powder.[2,3] Oxacillin sodium contains 815 to 950 mcg of oxacillin per milligram.[4]

### Solubility

Oxacillin sodium is freely soluble in water and slightly soluble in dehydrated ethanol.[2,3]

### pH

A 3% aqueous solution of oxacillin sodium has a pH between 4.5 and 7.5.[3,4] The reconstituted injection has a pH of 6 to 8.5. The reconstituted powder for oral solution has a pH between 5 and 7.5.[4]

### pK$_a$

The drug has a pK$_a$ of 2.8.[2]

### Osmolality

Oxacillin sodium oral solution 250 mg/5 mL (Prostaphlin, Bristol) had an osmolality of 2420 mOsm/kg using a vapor pressure osmometer.[232] Oxacillin sodium reconstituted to 250 mg/1.5 mL with sterile water for injection had an osmolality variously determined as 596 and 657 mOsm/kg, depending on the method of determination.[287]

## General Stability Considerations

Oxacillin sodium products should be packaged in tight containers and stored at controlled room temperature.[2,3]

The reconstituted oxacillin sodium oral solution is stable for three days at room temperature and for 14 days under refrigeration. The reconstituted powder for injection is stable for three days at room temperature and for seven days under refrigeration.[2,7]

## Stability Reports of Compounded Preparations
### Ophthalmic

Ophthalmic preparations, like other sterile drugs, should be prepared in a suitable clean air environment using appropriate aseptic procedures. When prepared from nonsterile components, an appropriate and effective sterilization method must be employed.

Osborn et al.[169] evaluated the stability of several antibiotics in three artificial tears solutions composed of 0.5% hydroxypropyl methylcellulose. The three artificial tears products (Lacril, Tearisol, Isopto Tears) were used to reconstitute and dilute oxacillin sodium to 66 mg/mL. The products were packaged in plastic squeeze bottles and stored at 25°C. A serial dilution bioactivity test was used to estimate the amount of antibiotic activity remaining during seven days of storage. The amount of antibiotic remaining after seven days at room temperature was 74% in Lacril, 94% in Tearisol, and 81% in Isopto Tears. ■

# Oxandrolone

## Properties

Oxandrolone exists as a white to off-white odorless crystalline powder.[1,3,4]

### Solubility

Oxandrolone is practically insoluble in water (1:5200) but freely soluble in chloroform (1:5), and sparingly soluble in ethanol (1:57) and acetone (1:69).[3,4]

## General Stability Considerations

Oxandrolone darkens upon exposure to light. Oxandrolone substance should be packaged in well-closed, light-resistant containers. Oxandrolone oral tablets are to be packaged in tight, light-resistant containers.[4]

## Stability Reports of Compounded Preparations
### Oral Liquid

Johnson et al.[1649] evaluated the stability of an extemporaneously prepared oxandrolone 1-mg/mL oral suspension prepared from commercial oral tablets. The proper amount of tablets was crushed to fine powder in a glass mortar. The powder was incorporated into an equal parts (1:1) mixture of Ora-Plus and Ora-Sweet and also an equal parts mixture of Ora-Plus and Ora-Sweet SF, packaged in amber plastic prescription bottles, and stored at controlled room temperature of 23 to 25°C for 90 days.

No detectable changes in color, odor, and taste occurred, and no visible microbial growth was found. Stability-indicating HPLC analysis found that in both suspending vehicle mixtures, the oxandrolone concentrations demonstrated less than 2% drug loss over 90 days of storage. ■

# Oxybutynin Chloride

## Properties

Oxybutynin chloride occurs as a white to off-white nearly odorless crystalline powder.[2,3]

### Solubility

Oxybutynin chloride is freely soluble in water and ethanol but is relatively insoluble in alkaline solutions.[2,3,7] Oxybutynin is very soluble in acidic solutions. At pH 1, oxybutynin solubility is 77 mg/mL but decreases to 0.8 mg/mL at pH 6 and is down to 0.012 mg/mL at pH over 9.6.[1,1469]

### $pK_a$

The drug has a $pK_a$ of 6.96.[2]

## General Stability Considerations

Oxybutynin chloride oral tablets and syrup should be packaged in tight, light-resistant containers and stored at controlled room temperature.[2,4] The manufacturer has stated that oxybutynin tablets should not be removed from the original packaging and placed in dosing compliance aids because of the hygroscopicity of the drug.[1622]

Oxybutynin stability in aqueous solution is dependent on solution pH. The drug is very stable in acidic and neutral pH aqueous solutions with negligible decomposition in 48 hours. In 0.1 mol/L hydrochloric acid solution, the drug did not degrade during autoclaving at 120°C for 20 minutes.[1469]

## Stability Reports of Compounded Preparations
### Oral

The shelf-life stability of oxybutynin chloride syrup 1 mg/mL (Ditropan, Marion) repackaged in 5-mL unit doses in polypropylene-lined aluminum cups thermosealed with polypropylene-coated closures was evaluated by Van Gansbeke.[31] All samples were stored at 23°C for 18 months. A spectrophotometric assay for which the primary decomposition product of oxybutynin chloride had been shown not to interfere was used to determine the drug content. Oxybutynin chloride concentrations showed little or no loss over 18 months. Furthermore, no loss of weight due to water permeation was found.

### Irrigation

Irrigations, like other sterile drugs, should be prepared in a suitable clean air environment using appropriate aseptic procedures. When prepared from nonsterile components, an appropriate and effective sterilization method must be employed.

**STUDY 1:** Fratta et al.[951] evaluated the stability of an extemporaneously compounded injection of oxybutynin chloride 0.25 and 0.5 mg/mL in sodium chloride 0.9% for intravesical instillation. The solution was sterilized by filtration through a 0.22-μm polyethersulfone filter (Sartorius) and packaged in polypropylene syringes (Becton Dickinson) sealed with a Combi-Lock tip cap (Braun). Sample syringes were stored at ambient room temperature and refrigerated at 4°C. HPLC analysis found little or no change in the drug concentration after 112 days stored under refrigeration. However, 5% loss occurred in six days and 10% loss occurred in about 20 days at ambient room temperature.

**STUDY 2:** Wan and Rickman[1244] evaluated the stability of oxybutynin chloride 0.125-mg/mL solutions for intravesical instillation prepared from commercial tablets. The tablets were either crushed first or simply allowed to dissolve passively. Neither dissolution method made a difference. The tablets were dissolved in tap water (pH 8.6), which simulated homemade solutions, or sodium chloride 0.9% (pH 6.3) each with and without added gentamicin sulfate 0.48 mg/mL. Gas chromatography was used to determine the drug concentration in the solutions over four weeks stored at room temperature. Tap water was unacceptable because of substantial and rapid decline in oxybutynin chloride concentrations with 50% loss occurring in one week. It was also unacceptable because it is not sterile and bacterial contamination is a problem. Oxybutynin chloride dissolved in sodium chloride 0.9%, with or without added gentamicin sulfate was more stable with losses of about 20% in three to four weeks. ▪

## ▪ Oxycodone Hydrochloride

### Properties
Oxycodone hydrochloride occurs as white to off-white hygroscopic crystals or crystalline powder.[1,4]

### Solubility
Oxycodone hydrochloride is soluble at about 100 to 167 mg/mL in water. It is slightly soluble in ethanol.[1,4,7]

### pH
Oxycodone hydrochloride oral solution, USP, has a pH of 1.4 to 4.[4]

### General Stability Considerations
Oxycodone hydrochloride powder, tablets, and oral solution should be packaged in tight containers and stored at controlled room temperature.[4]

### Stability Reports of Compounded Preparations
*Rectal*
Gebauer et al.[733] reported the stability of an extemporaneously prepared rectal gel for the management of tenesmus in colorectal cancer. Oxycodone hydrochloride powder (Macfarlan Smith) was mixed with lidocaine gel (Pharmacia) and hydrogel (Smith and Nephew) to prepare a gel containing oxycodone hydrochloride 0.3% (wt/vol) and lidocaine hydrochloride 1.5% (wt/vol). The gel was packaged in 50-mL syringes and stored refrigerated at 4°C for 12 months. Stability-indicating HPLC analysis found little or no change in the concentrations of either of the drugs within 12 months. ▪

# Pancrelipase
## (Pancreatic Enzymes)

## Properties

Pancrelipase is a mixture of enzymes, primarily lipase along with amylase and protease, obtained from the pancreas of hogs. It is a white, cream, or buff-colored amorphous powder with a characteristic odor. Each milligram of material contains not less than 24 USP units of lipase, not less than 100 USP units of amylase, and not less than 100 USP units of protease activities.[2–4]

## General Stability Considerations

Pancrelipase tablets and capsules should be packaged in tight containers with a desiccant and stored at temperatures not exceeding 25°C. Opened containers should not be refrigerated.[2–4]

Pancrelipase activity is greatest in neutral or slightly alkaline media. It is inactivated by more than traces of acids and by large amounts of alkali hydroxides or carbonates. The drug is inactivated by gastric juice in vivo.[2,3]

## Stability Reports of Compounded Preparations
### Enteral

The results of digestion of 14 enteral diets and changes in osmolality were evaluated by Cha and Randall.[57] The commercial diets (Table 80) were tested by adding to 10 mL of diet formula 5, 10, 20, and 50 mg of pancreatic enzyme concentrate (Viokase, Vioben). Digestion was rapid at ambient temperature; increasing the temperature to 37°C did not change the extent of digestion, as indicated by the insignificant alteration in osmolality. The pH changes observed were small; Isocal, Ensure Plus, Vital, Compleat B, Vipep, and milk exhibited the largest pH changes, around one unit lower. The authors noted that very small amounts of the pancreatic enzyme concentrate produced very rapid increases in osmolality in all diets except Citrotein and milk, which exhibited little change. Increasing the amount of the enzymes did not produce further increases in osmolality. ■

**TABLE 80.** Products Tested for Hydrolysis by Pancreatic Enzymes[57]

| Defined Formula Diets | |
|---|---|
| Flexical | Vivonex |
| Vipep | Vivonex HN |
| Vital | |
| **Low-Residue Diets** | |
| Citrotein | |
| Precision HN | |
| Precision LR | |
| **Liquid Whole Foods** | |
| Compleat B | Isocal |
| Ensure | Milk |
| Ensure Plus | Sustacal |

# Pantoprazole Sodium

## Properties

Pantoprazole sodium is a white to off-white crystalline powder.[1,7] Pantoprazole sodium sesquihydrate 11.28 mg is equivalent to pantoprazole 10 mg.[3]

## Solubility

Pantoprazole sodium is freely soluble in water and slightly soluble in pH 7.4 phosphate buffer.[7]

## pH

Pantoprazole sodium injection reconstituted with sodium chloride 0.9% injection has a pH of 9 to 10.[7]

## pK$_a$

Pantoprazole has pK$_a$ values of 3.92 and 8.19.[1] Hellstrom and Vitols[1325] reported a very similar result with a pK$_a$ of 3.96.

## General Stability Considerations

Pantoprazole sodium tablets should be stored at controlled room temperature, while the injection should be stored under refrigeration at 2 to 8°C and protected from light.[7] Pantoprazole sodium aqueous stability is pH-dependent, with increasing decomposition with decreasing pH.[7] Pantoprazole exhibits maximum stability at pH 9; the drug degrades rapidly in acidic media.[861] The presence of citric acid, trisodium citrate, and monosodium citrate in combination with pantoprazole substantially increases the rate of pantoprazole decomposition.[1099]

## Stability Reports of Compounded Preparations
### Oral

**USP OFFICIAL FORMULATION:**

> Pantoprazole sodium 200 mg
> Sodium bicarbonate 8.4% injection qs 100 mL

Calculate the number of commercial tablets needed for the total amount to be prepared. Gently rub each tablet exterior with a paper towel dampened with ethanol to remove the trademark imprint. If this step is not performed, flecks of dark material will compromise the pharmaceutical elegance of the oral suspension. Crush or grind the tablets to a coarse powder and transfer to a tight, light-resistant calibrated bottle. Add 50 mL of sodium bicarbonate 8.4% injection and agitate until the coating is dissolved. Add additional sodium bicarbonate 8.4% to bring to final volume, and mix thoroughly, yielding pantoprazole sodium 2-mg/mL oral suspension. The final liquid preparation should have a pH between 7.9 and 8.3. Store the preparation under refrigeration between 2 and 8°C. The beyond-use date is 14 days from the date of compounding when stored under refrigeration.[4]

**STUDY 1:** Dentinger et al.[783] reported the stability of a pantoprazole 2-mg/mL oral suspension extemporaneously compounded from 40-mg tablets (Protonix, Wyeth). Twenty pantoprazole 40-mg tablets were rubbed on an ethanol-dampened towel to remove the Protonix imprint. The tablets were triturated and then mixed with 340 mL of sterile water for irrigation. Sodium bicarbonate powder 16.8 g was added and mixed using a magnetic stirrer. Sufficient sterile water for irrigation was added to bring the suspension to 400 mL. The suspension was packaged in amber polyethylene terephthalate bottles and stored at 2 to 8°C for 62 days.

No change in color or odor and no visible microbiological contamination appeared. Stability-indicating HPLC analysis found that the available concentration of pantoprazole was about 95% throughout the 62-day study.

**STUDY 2:** Ferron et al.[861] reported the stability and bioavailability of pantoprazole suspended in sodium bicarbonate solution. A 40-mg pantoprazole tablet was ground to fine powder using a mortar and pestle. The powder was suspended in 20 mL of sodium bicarbonate 4.2% added in increments and used to rinse the mortar and pestle. Pantoprazole concentration was 2 mg/mL, and the sodium bicarbonate concentration was 42 mg/mL. The suspension was packaged in plastic syringes and stored at room temperature of 25°C and 60% relative humidity, refrigerated at 5°C, and frozen at −20°C.

The suspension was homogeneous and remained so throughout the study under all storage conditions. HPLC analysis found little or no pantoprazole loss in one day at room temperature, in two weeks refrigerated, and in three months frozen. Bioavailability was found to be about 25% lower than with the tablet, which is similar to omeprazole and lansoprazole suspensions.

## Compatibility with Common Beverages and Foods

Tammara et al.[1422] evaluated the bioequivalence of three methods of oral administration of pantoprazole delayed-release granules (Protonix, Wyeth) using common beverages and food. Pantoprazole delayed-release granule doses of 40 mg were administered to healthy subjects (1) orally in a teaspoonful of applesauce (Gerber) followed by 240 mL of water, (2) orally in 5 mL of apple juice (Mott's) with additional 5-mL portions of apple juice as necessary to deliver all of the granules followed by water up to a total volume of 240 mL, and (3) using a 60-mL syringe attached to a 16 French nasogastric tube; pantoprazole granules were poured into the barrel of the syringe followed by 10-mL of apple juice and at least two subsequent repetitions of 10-mL of apple juice with water up to a total volume of 240 mL. The plasma drug concentration-time profiles were similar by all three administration methods. These three administration methods were established to be essentially bioequivalent in healthy subjects. ▪

# Papaverine Hydrochloride

## Properties

Papaverine hydrochloride occurs as white crystals or white crystalline powder that is odorless and has a slightly bitter taste.[1,3,4]

### Solubility

Papaverine hydrochloride has an aqueous solubility of about 25 to 33 mg/mL. It is also soluble in chloroform and slightly soluble in ethanol but practically insoluble in ether.[1,3,4]

### pH

A 0.05-M aqueous solution is about pH 3.9, while a 20-mg/mL aqueous solution is about pH 3.3 (range pH 3 to 4.5).[1] Papaverine hydrochloride injection, USP, is not less than pH 3.[4]

## General Stability Considerations

Papaverine hydrochloride powder should be packaged in tight, light-resistant containers, and should be stored at controlled room temperature. Papaverine hydrochloride oral tablets should also be packaged in tight containers and stored at controlled room temperature. Papaverine hydrochloride injection is packaged in single-dose or multiple-dose Type I glass containers and is stored at controlled room temperature.[4]

Piotrowski et al.[1587] evaluated the effects of several photoprotective agents at several concentrations to aid in protecting papaverine hydrochloride 20-mg/mL aqueous injection from ultraviolet radiation-induced decomposition. Of the photoprotective agents tested, methyl-4-hydroxybenzoate 0.1% proved to be the most effective agent followed by propyl-4-hydroxybenzoate 0.1%. The authors found that 4-aminobenzoic acid was less effective, and sodium benzoate was the least effective.

## Stability Reports of Compounded Preparations

### Injection

Injections, like other sterile drugs, should be prepared in a suitable clean air environment using appropriate aseptic technique. When prepared from nonsterile components, an appropriate and effective sterilization method must be employed.

**STUDY 1 (TWO-DRUG MIXTURE):** Tu et al.[1530] evaluated the stability of papaverine hydrochloride (Eli Lilly) and phentolamine mesylate (Ciba-Geigy). Two milliliters were withdrawn from the vial of papaverine hydrochloride injection and injected into the lyophilized phentolamine mesylate vial to reconstitute it. The reconstituted solution was then withdrawn and injected back into the papaverine hydrochloride vial. The final concentrations were papaverine hydrochloride 30 mg/mL and phentolamine mesylate 0.5 mg/mL. Vials were stored refrigerated at 5°C and at controlled room temperature of 25°C for 30 days. No visible color change, haze, or precipitation was observed. Little or no papaverine hydrochloride loss occurred in 30 days at either temperature. After 30 days, phentolamine mesylate losses were about 1 to 3% under refrigeration and 4 to 5% at room temperature.

**STUDY 2 (TWO-DRUG MIXTURE):** Benson and Seifert[1671] evaluated the stability of phentolamine mesylate 0.83 mg/mL in combination with papaverine hydrochloride 25 mg/mL. The mixture was prepared by combining two 5-mg vials of phentolamine mesylate reconstituted with 1 mL of sterile water for injection with one 10-mL vial of papaverine hydrochloride 300 mg. The resulting 12 mL of solution was stored for 40 days at room temperature and under refrigeration. Neither HPLC analysis nor gas chromatography/mass spectrometry analysis found any loss of phentolamine mesylate at either storage temperature within the study period.

**STUDY 3: (THREE-DRUG MIXTURE):** Trissel and Zhang[876] evaluated the stability of a three-drug mixture of alprostadil (prostaglandin $E_1$) 12.5 mcg/mL, papaverine hydrochloride 4.5 mg/mL, and phentolamine mesylate 0.125 mg/mL (commonly called the "Knoxville Formula" or "Trimix") prepared from commercial injections diluted in bacteriostatic sodium chloride 0.9%. The injection mixture was packaged in commercial empty sterile vials and stored at room temperature near 23°C, refrigerated at 4°C, and frozen at −20 and −70°C.

All of the samples remained clear and colorless throughout the study. Stability-indicating HPLC analysis found that alprostadil was the least stable of the three drug components and was the limiting factor in the combination injection. At room temperature, alprostadil losses of 8 and 13% occurred in five and seven days, respectively. Under refrigeration, losses of about 6% in one month and 11% in two months occurred. Frozen at −20 and −70°C, alprostadil losses did not exceed 5% in six months. Subjecting vials frozen at −20°C to four freeze-thaw cycles and warming the vials to room temperature each time resulted in no loss of any drug. The authors recommended a beyond-use date of six months and one month stored frozen at −20°C or under refrigeration, respectively, for batches that have passed a sterility test. They also recommended that room temperature exposure be limited, and vials should be returned to refrigeration as soon as possible after use.

**STUDY 4: (TWO- AND THREE-DRUG MIXTURES):** Soli et al.[1531] evaluated the stability of alprostadil (Upjohn) 0.004 mg/mL with papaverine hydrochloride (Fluka) 6 mg/mL and phentolamine mesylate (Ciba-Geigy) 0.4 mg/mL in two- and three-drug combinations diluted in sodium chloride 0.9%. The injection mixtures were packaged in sterile glass vials and stored refrigerated at 2 to 8°C for 60 days.

No visible precipitation or color change was reported. HPLC analysis found alprostadil losses of about 10 to 13% occurred in five days. No loss of papaverine hydrochloride and about 3 to 4% loss of phentolamine mesylate occurred in 60 days with or without alprostadil present. ▪

# ▪ Parabens
## (Parahydroxybenzoates)

## Properties

Methylparaben exists as white crystalline powder or colorless needles.[1,3,4] Methylparaben sodium is a white hygroscopic crystalline powder.[3,4]

Propylparaben exists as colorless or white crystals or white crystalline powder.[1,3,4] Propylparaben sodium is an odorless hygroscopic white crystalline powder.[3,4]

### Solubility

Methylparaben is slightly soluble in water with an aqueous solubility of about 2.5 mg/mL at 20°C and 3 mg/mL at 25°C. It is freely soluble in ethanol and has solubilities of about 25 mg/mL in warm oil and about 14 mg/mL in warm glycerol.[1,3,4]

Methylparaben sodium is freely soluble in water, sparingly soluble in ethanol, and insoluble in fixed oils.[4]

Propylparaben is very slightly soluble in water, with a solubility cited as one part in 2000 to 2500 parts of water; it is slightly soluble in boiling water at about one part in 400 parts. It is freely soluble in ethanol to about one part in 1.5 parts.[1,3,4]

Propylparaben sodium is freely soluble in water, sparingly soluble in ethanol, and insoluble in fixed oils.[4]

### pH

Methylparaben sodium 0.1% aqueous solution has a pH of 9.5 to 10.5.[3,4]

Propylparaben sodium 0.1% aqueous solution has a pH of 9.5 to 10.5.[3,4]

## General Stability Considerations

Methylparaben and propylparaben should be packaged in well-closed containers and stored at controlled room temperature. Methylparaben sodium and propylparaben sodium should be packaged in tight containers and stored at controlled room temperature.[4]

The antimicrobial effectiveness of parabens may be adversely affected by a variety of excipients and active ingredients. The parabens adsorb to substances like magnesium trisilicate, aluminum calcium trisilicate, talc, polysorbate (Tween) 80, carmellose sodium, and some plastics. Nonionic surfactants and essential oils may reduce the antibacterial activity. Atropine, iron, sorbitol, and weak alkalis or strong acids are stated to be incompatible. Lyophilization may also cause loss of paraben preservatives.[3]

## Stability Reports of Compounded Preparations
### Oral

Ghulam et al.[1254] evaluated the antimicrobial preservative effectiveness of methyl- and propylparabens in methylcellulose 1% and simple syrup, BP, after dilution at ratios of 1:1 and 1:4 to simulate the dilution of preserved oral liquids for pediatric administration. Methylcellulose 1% preserved with methylparaben 0.02% and propylparaben 0.01% and also simple syrup, BP, preserved with methylparaben 0.07% and propylparaben 0.05% were diluted with unpreserved diluents. Dilution greater than 1:1 caused failure of the dilutions to meet the British Pharmacopeia criteria for efficacy of antimicrobial preservative. ▪

# ▪ Paregoric

## Properties

Paregoric, USP, is a formulation containing not less than 35 mg or more than 45 mg of anhydrous morphine per 100 mL. About 950 mL of paregoric is prepared from powdered opium using the formula given in Table 81.[4]

Paregoric may also be prepared using opium or opium tincture to yield anhydrous morphine 40 mg/100 mL and alcohol 45%.[4]

**TABLE 81.** Formula for Preparation of Paregoric[4]

| Compound | Amount |
| --- | --- |
| Powdered opium | 4.3 g |
| Suitable essential oils | |
| Benzoic acid | 3.8 g |
| Diluted alcohol | 900 mL |
| Glycerin | 38 mL |

## General Stability Considerations

Paregoric should be packaged in tight, light-resistant containers and should be stored at controlled room temperature, avoiding exposure to direct sunlight and excessive heat.[4]

## Stability Reports of Compounded Preparations
*Enteral*
**PAREGORIC COMPATIBILITY SUMMARY**
*Compatible with:* Vivonex T.E.N.
*Incompatible with:* Enrich • TwoCal HN

Burns et al.[739] reported the physical compatibility of paregoric 5 mL with 10 mL of three enteral formulas, including Enrich, TwoCal HN, and Vivonex T.E.N. Visual inspection found no physical incompatibility with Vivonex T.E.N. enteral formula. However, separation and globular particle formation occurred when it was mixed with the Enrich and TwoCal HN enteral formulas. ■

# ■ Pefloxacin Mesylate

## Properties

Pefloxacin mesylate is a fine white or nearly white crystalline powder.[1,3]

## Solubility

Pefloxacin mesylate is freely soluble in water and slightly soluble in ethanol.[3] Pefloxacin is soluble in acidic and alkaline solutions.[1]

## pH

Pefloxacin mesylate 10 mg/mL (1%) in water has a pH of 3.5 to 4.5.[3]

## Osmolality

Pefloxacin mesylate 40 mg/mL injection diluted with an equal quantity of dextrose 5% to a concentration of 20 mg/mL was found to have an osmolality near 285 mOsm/kg.[1472]

## General Stability Considerations

Pefloxacin mesylate should be packaged in tight containers and protected from light.[3]

## Stability Reports of Compounded Preparations
*Ophthalmic*

Ophthalmic preparations, like other sterile drugs, should be prepared in a suitable clean air environment using appropriate aseptic procedures. When prepared from nonsterile components, an appropriate and effective sterilization method must be employed.

Gellis et al.[813,1472] evaluated the stability of pefloxacin 40-mg/mL ophthalmic solution compounded from the injection (Bellon). The 80-mg/mL injection was diluted with an equal quantity of dextrose 5% injection, passed through a 0.22-μm filter, and stored frozen at −20°C. The pefloxacin ophthalmic solution remained clear with little or no change in pH and osmolality. Microbiological analysis found no loss of pefloxacin after three months of frozen storage and little or no loss after an additional 48 hours stored under refrigeration. ■

# ■ Penicillamine

## Properties

Penicillamine is a white or nearly white fine crystalline powder with a slight characteristic odor. The D-isomer is used clinically.[2,3]

## Solubility

Penicillamine is freely soluble in water and slightly soluble in ethanol.[2,3]

## pH

A 1% solution in water has a pH of 4.5 to 5.5.[4]

## pKₐ

Penicillamine has pKₐ values of 1.83, 8.03, and 10.83.[2]

## General Stability Considerations

Penicillamine capsules and tablets should be packaged in tight containers and stored at controlled room temperature.[2,4] The pH of maximum stability was determined to be 3 to 4.5.[677]

## Stability Reports of Compounded Preparations
### Oral
**STUDY 1:** deCastro et al.[459] reported on the extemporaneous preparation of D-penicillamine as an oral suspension. The product was compounded as a dry powder mix for reconstitution at the time of dispensing. The contents of 60 D-penicillamine (Cuprimine, Merck Sharp and Dohme) 250-mg capsules were emptied and combined with carboxymethylcellulose 3 g; sucrose 150 g; imitation cherry flavor powder 3 g; citric acid, USP, 300 mg; methylparaben, USP, 360 mg; and propylparaben 60 mg. The contents were then mixed thoroughly. At the time of dispensing, 300 mL of distilled water was added to reconstitute the suspension, yielding a D-penicillamine concentration of 250 mg/5 mL. The authors recommended refrigerated storage for the suspension. A placebo-controlled comparison of the suspension formulation to the intact capsules found comparable reductions in blood lead levels throughout five weeks of evaluation.

**STUDY 2:** Rawlins and Smith[677] reported on the stability of a granular D-penicillamine formulation (Table 82) that then was reconstituted for use.

**TABLE 82.** Granular D-Penicillamine Formulation of Rawlins and Smith[677]

| Component | | Amount |
|---|---|---|
| D-Penicillamine base | | 6.25 g |
| Disodium edetate | | 50 mg |
| Sodium metabisulfite | | 100 mg |
| Methyl- and propylparabens (Nipasept) | | 500 mg |
| Polyvinylpyrrolidone | | 2 g |
| Sucrose | to | 100 g |

The powders were mixed, and a granulation was prepared using the polyvinylpyrrolidone dissolved in isopropanol. The mixture was screened, dried, and rescreened. Contact with metal implements was minimized throughout preparation. Freshly distilled or demineralized water was used to reconstitute the granules.

HPLC analysis found 98 and 95% of the penicillamine remaining after three months stored under refrigeration and at room temperature, respectively. When reconstituted, more than 9% of the D-penicillamine content was retained for 14 days when refrigerated. ▪

# ▪ Penicillin G Potassium
## (Benzylpenicillin Potassium)

## Properties
Penicillin G potassium is a nearly odorless white to slightly yellow crystalline powder or colorless to white crystals.[2,3] Each milligram of penicillin G potassium contains 1440 to 1680 USP units. The powder for injection that also contains sodium citrate as a buffer includes 1355 to 1595 USP units per milligram.[4]

### Solubility
Penicillin G potassium is freely soluble in water and common aqueous infusion solutions but sparingly soluble in ethanol.[2]

### pH
A 6% aqueous solution of penicillin G potassium has a pH between 5 and 7.5.[3,4] The reconstituted powder for injection has a pH of 6 to 8.5, while the frozen injection has a pH of 5.5 to 8.[2,4]

### pK$_a$
The drug has a pK$_a$ of 2.76.[2]

### Osmolality
The osmolality of the powder for injection reconstituted to 250,000 units/mL with sterile water for injection was found to be 776 or 767 mOsm/kg, depending on the method of the determination.[287] Another report cited the osmolality as being 749 mOsm/kg.[286]

## General Stability Considerations
Penicillin G potassium is somewhat hygroscopic, so products should be stored in tight containers. In the dry state the drug is stable when stored at room temperature. After reconstitution, the drug is stable in solution for seven days at temperatures between 2 and 15°C. The frozen injection should be stored at −10 to −20°C.

Penicillin G potassium in aqueous solution is subject to hydrolysis of the β-lactam ring. The process is accelerated by increased temperatures and acidic or alkaline pH. At pH 5.5 and below and above pH 8 inactivation rapidly occurs. The degradation rate is minimized at pH 6.8 to 7. Dilute solutions of penicillin G potassium are more stable than more concentrated solutions.[1–3,398,399]

## Stability Reports of Compounded Preparations
### Ophthalmic
Ophthalmic preparations, like other sterile drugs, should be prepared in a suitable clean air environment using appropriate aseptic procedures.

Osborn et al.[169] evaluated the stability of several antibiotics in three artificial tears solutions composed of 0.5% hydroxypropyl methylcellulose. The three artificial tears products (Lacril, Tearisol, Isopto Tears) were used to reconstitute and dilute penicillin G sodium to 333,000 units/mL. The products were packaged in plastic squeeze bottles and stored at 25°C. A serial dilution bioactivity test was used to estimate the amount of antibiotic activity remaining during seven days of storage. The amount of antibiotic remaining after seven days at room temperature was 24% in Lacril, 23% in Tearisol, and 30% in Isopto Tears. ■

# ■ Penicillin V Potassium
## (Phenoxymethylpenicillin Potassium)

## Properties
Penicillin V potassium is a white odorless crystalline powder.[2,3] Each milligram contains 1525 to 1780 USP units.[4] One gram of penicillin V is contained in 1.1 g of the potassium salt.[3]

### Solubility
Penicillin V potassium is very soluble in water, having an aqueous solubility of about 667 mg/mL. In ethanol, the drug has a solubility of about 6.7 mg/mL at 25°C.[2,4]

### pH
A 3% aqueous solution of penicillin V potassium has a pH between 4 and 7.5. The reconstituted powder for oral solution has a pH of 5 to 7.5.[4] Reconstituted Compocillin VK oral solution (Ross) had a measured pH of 6.5.[19]

### pKa
The drug has a $pK_a$ of 2.73.[2]

### Osmolality
Penicillin V potassium oral solution 250 mg/5 mL (V-Cillin K, Lilly) had an osmolality of 2995 mOsm/kg using a vapor pressure osmometer.[232]

## General Stability Considerations
Penicillin V potassium products should be packaged in tight containers and stored at controlled room temperature. After reconstitution, the oral solution should be stored under refrigeration at 2 to 8°C and should be discarded after 14 days.[2,4]

Jaffe et al.[297] reported the stability of five manufacturers' penicillin V potassium 250-mg/5-mL oral liquids following reconstitution. A spectrophotometric assay specific for the intact drug in the presence of its degradation products was used to assess the penicillin V content of the products when stored at 5°C. Of the products tested, only V-Cillin K (Lilly) retained at least 90% of the labeled amount after 14 days of refrigerated storage.

## Stability Reports of Compounded Preparations
### Oral
**STUDY 1 (REPACKAGED):** Grogan et al.[64] evaluated the stability of penicillin V potassium (Veetids 125, Squibb) oral liquid 125 mg/5 mL repackaged into capped 6-mL plastic oral syringes (Monoject, Sherwood). Samples were stored at 4 and 25°C and at elevated temperatures protected from light to estimate the time to reach 90% of labeled concentration. Spectrophotometric and microbiological assays were used to determine drug content. The drug had an initial concentration of 113.5% of the labeled amount. Although the drug is labeled for an expiration date of 14 days following reconstitution stored under refrigeration in the original container, the repackaged product in oral syringes stored at 4°C dropped from 113.5 to 90% in only 11.5 days. At room temperature, the time to reach 90% was 36 hours. The authors recommended expiration periods of five days at 4°C and 16 hours at 25°C based on products with an initial concentration of 100% of label.

**STUDY 2 (REPACKAGED):** Allen and Lo[65] tested the stability of oral penicillin V potassium solution (V-Cillin K, Lilly) repackaged into unit-dose containers at various temperatures, including frozen storage. The penicillin V potassium was reconstituted to 250 mg/5 mL. Samples of 5 mL were placed in 26-mL amber glass vials with screw-cap closures and stored at −20, −10, 5, and 25°C. Analysis of the penicillin content was performed spectrophotometrically. Frozen at either −20 or −10°C, less than 4% loss occurred in 60 days. However, the effect of freezing on bioavailability, if any, is unknown.[68,69] Under refrigeration, 9% was lost in 20 days, while at room temperature 11.4% was lost in just five days.[65]

### Enteral
**PENICILLIN V POTASSIUM COMPATIBILITY SUMMARY**
*Compatible with:* Ensure • Ensure HN • Ensure Plus • Ensure Plus HN • Osmolite • Osmolite HN • Vital

**STUDY 1:** Cutie et al.[19] added 5 mL of penicillin V potassium oral solution (Compocillin VK, Ross) to varying amounts (15 to 240 mL) of Ensure, Ensure Plus, and Osmolite (Ross Laboratories) with vigorous agitation to ensure thorough mixing. The penicillin V potassium oral solution was physically compatible, distributing uniformly in all three enteral products with no phase separation or granulation.

**STUDY 2:** Altman and Cutie[850] reported the physical compatibility of penicillin V potassium oral solution (Pen-Vee K, Wyeth) 5 mL with varying amounts (15 to 240 mL) of Ensure HN, Ensure Plus HN, Osmolite HN, and Vital after vigorous agitation to ensure thorough mixing. The penicillin V potassium oral solution was physically compatible, distributing uniformly in all four enteral products with no phase separation or granulation. ■

 # Pentoxifylline
## (Oxpentifylline)

### Properties
Pentoxifylline is a white or almost white odorless crystalline powder having a bitter taste.[2,3,572]

### Solubility
Pentoxifylline has an aqueous solubility up to 77 mg/mL at 25°C,[2,572] increasing to 191 mg/mL at 37°C.[1] It has a solubility in ethanol of 63 mg/mL at 22°C.[2]

### pKₐ
Pentoxifylline has a p$K_a$ of 0.28.[2]

### General Stability Considerations
Pentoxifylline products should be packaged in well-closed, light-resistant containers stored at controlled room temperature.[7]

### Stability Reports of Compounded Preparations
#### Oral
**STUDY 1:** Abdel-Rahman and Nahata[572] studied the stability of pentoxifylline 20-mg/mL oral suspension prepared from extended-release tablets. Twelve pentoxifylline 400-mg tablets (Trental, Hoechst-Roussel) were crushed and triturated to fine powder using a mortar and pestle. Distilled water was added to the tablet powder in increments, with trituration. The mixture was transferred to a graduated cylinder and brought to 240 mL with distilled water. The mixture was then transferred to the glass container of a blender. This procedure was repeated three more times, with the mixture added to the blender each time, resulting in 960 mL of the mixture.

The total amount of suspension was thoroughly blended for five minutes. The suspension was packaged as 30 mL in two-ounce amber glass and two-ounce amber plastic prescription bottles, and samples were stored at 4 and 25°C.

No changes in color or odor occurred. Settling of the suspension was observed in all samples after 21 days of storage, but resuspension was readily accomplished by shaking. After 21 days at 4°C, no additional settling occurred. However, considerable additional settling occurred in the room temperature samples, especially in the plastic bottles. Because pentoxifylline is sufficiently soluble to be a solution at the concentration present in this preparation, the settled materials are believed to be less soluble tablet components. HPLC analysis of the samples found little or no loss of pentoxifylline after 91 days of storage in either container at 4 and 25°C.

**STUDY 2:** Cleary et al.[645] reported substantially altered pharmacokinetics of crushed extended-release pentoxifylline tablets compared to intact tablets. The crushed tablets showed a considerable increase in the maximum plasma drug concentration and much shorter mean time to reach that concentration than did intact tablets. The increased plasma concentration was associated with a worse adverse effect profile. No adverse effects from the intact tablets were observed in the test subjects. However, the crushed tablets caused dysgeusia, nausea, vomiting, dizziness, headache, and diaphoresis among the test subjects. The authors indicated that crushed 400-mg tablets could be used in patients unable to take tablets, but the potential benefit should be weighed carefully against the risk of vomiting. ■

 # Pergolide Mesylate
## (Pergolide Mesilate)

### Properties
Pergolide mesylate occurs as a white to off-white crystalline powder.[1,3,4]

### Solubility
Pergolide mesylate is stated to be slightly soluble in water.[1,3,4] Reported aqueous solubilities have ranged from 3.6 to 10 mg/mL.[1348] The drug is also slightly soluble in 0.01 N hydrochloric acid and dehydrated ethanol.[1,3,4] Pergolide mesylate is practically insoluble in 0.1 N sodium hydroxide and 0.1 N hydrochloric acid.[1]

### pKₐ
Pergolide mesylate in dimethylacetamide 66% is reported to have a p$K_a$ of 7.8.[1]

## General Stability Considerations

Pergolide mesylate powder and oral tablets should be packaged in tight, light-resistant containers and stored at controlled room temperature protected from exposure to light.[3,4] Pergolide mesylate is subject to photodecomposition, making protection from exposure to light necessary.[1354] The manufacturer has stated that pergolide mesylate tablets should not be removed from the original packaging and placed in dosing compliance aids because of the hygroscopicity and light sensitivity of the drug.[1622]

## Stability Reports of Compounded Preparations

### Oral

**STUDY 1:** Shank and Ofner[1348] evaluated the stability of pergolide 0.2 mg/mL (as pergolide mesylate 0.262 mg/mL) oral liquid for veterinary use having a formulation shown below.

| | | |
|---|---|---|
| Pergolide mesylate | | 26.2 mg |
| Potassium sorbate | | 200 mg |
| Citric acid | | 100 mg |
| Methylcellulose 4000 cps | | 1 g |
| Purified water | qs | 100 mL |

The preparation was made by first heating about 50 mL of purified water in a beaker almost to boiling and slowly sprinkling methylcellulose onto the liquid while mixing with a stir bar. The beaker was then placed on ice with continued mixing to dissolve the methylcellulose. After overnight hydration, the potassium sorbate and citric acid were added while mixing with a stir bar at a slow setting. Then the pergolide mesylate powder was incorporated. Purified water was added to bring to volume, and the preparation was mixed well. The oral liquid was packaged in vials with Teflon-lined caps. The vials were stored at room temperature of 23 to 26°C protected from exposure to light.[1348] In a subsequent study,[1607] the solubility of pergolide mesylate in this oral liquid at room temperature was determined to be 6.9 mg/mL, meaning the drug is actually in solution.

Stability-indicating HPLC analysis was used to determine pergolide mesylate concentrations during room temperature storage for 16 weeks. Substantial drug loss was found to occur with over 70% of the drug lost in 16 weeks. The time to 10% drug loss ($t_{90}$) was calculated to be 45 days or about 6.5 weeks.[1348]

**STUDY 2:** Davis et al.[1354] also evaluated the stability of pergolide mesylate formulated in an oral liquid having a concentration of 1 mg/mL. The oral liquid was prepared by using Vehicle for Oral Suspension, NF, and Vehicle for Oral Solution, NF. See monographs on these two vehicles for their compositions. (NOTE: These vehicles are similar to Ora-Plus and Ora-Sweet by Paddock Laboratories.)

The oral liquid was prepared by adding 10 mL of Vehicle for Oral Suspension, NF, to the barrel of a 35-mL syringe in which the plunger had been removed and which was connected by a fluid-dispensing connector to a second 35-mL syringe. Pergolide mesylate powder 20 mg was also added into the open barrel of the syringe. The barrel was replaced and depressed to force the mixture through the fluid-dispensing connector into the second syringe. This was repeated for a total of 50 repetitions until a uniform suspension was achieved. Then 10 mL of Vehicle for Oral Solution, NF, was added and mixed in the same manner preparing 20-mL of suspension providing pergolide mesylate 1 mg/mL. Aliquots (0.5 mL) of the suspension were transferred by pipette into 30 plastic sample vials. Triplicate samples were stored at −20, 8, 26, and 37°C protected from exposure to light. Additional samples were stored at 26°C exposed to light.

HPLC analysis found that samples at −20 and 8°C protected from exposure to light remained stable for 35 days, retaining about 93% of the initial pergolide mesylate concentration. The samples at 26 and 37°C protected from exposure to light had acceptable drug concentrations for only about seven days and underwent excessive decomposition within 14 to 21 days. Decomposed samples also developed a brown discoloration. The samples stored at 26°C exposed to light retained only 16% of the drug after storage for 35 days, indicating enhanced decomposition due to the drug's photolability. The authors recommended that compounded oral liquids of pergolide mesylate be stored refrigerated, protected from exposure to light, and that discolored formulations be discarded. ■

# ■ Permethrin

## Properties

Permethrin occurs as colorless crystals to a pale yellow or light orange-brown viscous liquid having a melting point of around 35°C.[1,7] It is a synthetic pyrethroid insecticide.[3] Permethrin is provided as 5% topical cream in a white to off-white vanishing cream base and as a 1% shampoo.[7]

## Solubility

Permethrin is insoluble in water, having an aqueous solubility of less than 1 part per million. It is soluble or miscible with various (unspecified) organic solvents except ethylene glycol.[1]

## General Stability Considerations

Permethrin topical cream and shampoo formulations are stored at controlled room temperature.[7]

## Stability Reports of Compounded Formulations
### Topical
Modamio et al.[1482] reported the stability of 5% permethrin topical cream stored at elevated temperature of about 40°C and relative humidity of about 75% for six months. The other components in the topical formulation prepared by UniPharma S.A. were not cited. Stability-indicating HPLC analysis found less than 5% loss of permethrin occurred after six months while stored at elevated temperature and humidity.

# Perphenazine

## Properties
Perphenazine is a white, creamy-white, or yellowish-white odorless powder with a bitter taste.[2,3]

### Solubility
Perphenazine is practically insoluble in water but is freely soluble in ethanol, having a solubility of 153 mg/mL.[1–3]

### pH
Perphenazine injection has a pH between 4.2 and 5.6.[2,4] The oral solution concentrate has a pH of 4.5 to 4.9.[2]

### Osmolality
Perphenazine injection has an osmolality of 263 mOsm/kg.[326]

## General Stability Considerations
Perphenazine products should be packaged in tight (tablets) or well-closed (syrup, oral solution), light-resistant containers. The tablets should be stored between 2 and 25°C, while the oral solution concentrate should be stored between 2 and 30°C. The injection should be stored at controlled room temperature and protected from temperatures of 40°C or more. The injection and oral solution concentrate also should be protected from freezing.[2,4,7]

Perphenazine is light-sensitive, and products should be protected from light to avoid discoloration. A slightly yellowish discoloration is not indicative of alteration of concentration or efficacy. However, a markedly discolored solution should be discarded.[7]

## Stability Reports of Compounded Preparations
### Oral
**STUDY 1:** Gupta[937] reported the stability of perphenazine 0.5 mg/mL in Ora-Sweet (Paddock Laboratories) using citric acid 50 mg per 50 mg of drug to assist in dissolution. Sodium metabisulfite 0.1% was added as an antioxidant to one portion of the perphenazine oral liquid, while the other portion did not include an antioxidant. The two formulations were packaged in amber glass bottles. Neither version of the oral liquid exhibited a change in appearance at any time during the study. Stability-indicating HPLC analysis found perphenazine losses of about 6 and 8% with and without the antioxidant, respectively, in 30 days at room temperature of 25°C. Losses of perphenazine increased to 13 to 17% in 62 days, indicating that a 30-day beyond-use date is appropriate. Sodium metabisulfite in the formulation offered little stability improvement.

**STUDY 2:** Gupta[1246] also evaluated the stability of perphenazine 0.5 mg/mL in Ora-Sweet (Paddock Laboratories) as well as simple syrup (Humco). The Ora-Sweet had a pH of 4.3 while the simple syrup had a pH of 5.5. Perphenazine powder was mixed with a little glycerin and then incorporated into either the Ora-Sweet or the simple syrup as the vehicle. The oral liquids were packaged in amber glass bottles and stored at room temperature near 25°C. Stability-indicating HPLC analysis found the drug was more stable in the simple syrup with its less acidic pH, exhibiting only 5% loss in 210 days or about seven months. In Ora-Sweet perphenazine losses were 7% in 114 days (less than four months) and 13% in 210 days.

## Compatibility with Common Beverages and Foods
Lever and Hague[1011] and also Geller et al.[242] reported that perphenazine liquid was compatible visually with water, saline solution, milk, orange juice, grape juice, pineapple juice, apricot juice, prune juice, tomato juice, V8 vegetable juice, 7-Up, and orange soda, although visual observation in opaque liquids is problematic. Perphenazine was stated to be incompatible with coffee, tea, cola sodas, and apple juice.

## Compatibility with Other Drugs
### PERPHENAZINE COMPATIBILITY SUMMARY
*Compatible with:* Lithium citrate
*Incompatible with:* Thioridazine

**STUDY 1:** Theesen et al.[78] evaluated the compatibility of lithium citrate syrup with oral neuroleptic solutions, including perphenazine concentrate. Lithium citrate syrup (Lithionate-S, Rowell) and lithium citrate syrup (Roxane) at a lithium-ion concentration of 1.6 mEq/mL were combined in volumes of 5 and 10 mL with 4, 8, 15, and 20 mL of perphenazine oral concentrate 3.2 mg/mL (Schering) in duplicate, with the order of mixing reversed between samples. Samples were stored at 4 and 25°C for six hours and evaluated visually for physical compatibility. No incompatibility was observed.

**STUDY 2:** Raleigh[454] evaluated the compatibility of lithium citrate syrup (Philips Roxane) 1.6 mEq lithium/mL with perphenazine concentrate 3.2 mg/mL. The perphenazine concentrate was combined with the lithium citrate syrup in a manner described as "simple random dosage combinations"; the exact amounts tested and test conditions were not specified. No precipitate was observed to form upon mixing.

Raleigh[454] also evaluated the compatibility of thioridazine suspension (Sandoz) (concentration unspecified) with perphenazine concentrate 3.2 mg/mL. The perphenazine concentrate was combined with the thioridazine suspension in the manner described as "simple random dosage combinations." The exact amounts tested and test conditions were not specified. A dense white curdy precipitate formed. ■

# ■ Phenobarbital
# Phenobarbital Sodium
## (Phenobarbitone)

## Properties
Phenobarbital is an odorless white crystalline powder or colorless crystals having a bitter taste.[1-3]

Phenobarbital sodium is a nearly odorless hygroscopic white powder, white crystalline granules, or flaky crystals having a bitter taste.[1-3]

### Solubility
Phenobarbital is very slightly soluble in water, having an aqueous solubility of about 1 mg/mL. In ethanol it has a solubility of about 100 to 125 mg/mL.[1-3]

Phenobarbital sodium is very soluble in water,[2] with an aqueous solubility cited as 1 g/mL.[1] In ethanol it has a solubility cited as 100 mg/mL.[1] The drug is also freely soluble in propylene glycol.[2]

### pH
A saturated solution of phenobarbital in water has a pH around 5.[3] Phenobarbital elixir (Parke-Davis) had a measured pH of 6.5.[19]

A 10% phenobarbital sodium in water solution has a pH between 9.2 and 10.2.[3,4] Phenobarbital sodium injection also has a pH of 9.2 to 10.2.[4]

### $pK_a$
The drug has a $pK_a$ cited as 7.3,[1] 7.41,[2] and 7.6.[6]

### Osmolality
Phenobarbital sodium injection is very hypertonic, with a measured osmolality of 9285 to 15,570 mOsm/kg, depending on the method of determination.[287,345]

## General Stability Considerations
Phenobarbital is stable to air but subject to hydrolysis to ureide and diamide creating a yellow discoloration. The hydrolysis rate is accelerated in solutions with a pH above 9 and greatly accelerated above pH 11. Hydrolysis is the most important degradation route. In the absence of water, little decomposition occurs.[6,271,1582]

Phenobarbital products should be packaged in tight (elixir) or well-closed (tablets) containers and stored at controlled room temperature. The elixir should be protected from light.

Phenobarbital sodium is not stable in aqueous solution but is stable in polyethylene glycol or propylene glycol. Phenobarbital sodium is stabilized in solution by using a mixed solvent approach with water and organic solvents. Gupta[271] reported that ethanol 20 to 40% exerted the greatest stabilizing effect, followed by propylene glycol 20 to 40%; glycerol 20 to 40% exerted a somewhat lower stabilizing effect.

Free phenobarbital precipitates from solutions of phenobarbital sodium if the pH is too low. The pH at which precipitation occurs is dependent on drug concentration. At 3 mg/mL, free phenobarbital precipitates at pH 7.5 or below; at 20 mg/mL, it precipitates at pH 8.6 or below.[8]

| Concentration | Precipitation pH |
| --- | --- |
| 3 mg/mL | pH 7.5 or below |
| 6 mg/mL | pH 7.9 or below |
| 10 mg/mL | pH 8.3 or below |
| 20 mg/mL | pH 8.6 or below |

## Stability Reports of Compounded Preparations
### Oral
**STUDY 1:** Dietz et al.[480] reported the stability of several oral liquid forms of phenobarbital 4 mg/mL, including emulsions and aqueous solutions with and without propylene glycol. Commercial phenobarbital elixir (Lilly) containing 4 mg/mL of drug was used for comparison. To prepare the emulsions, phenobarbital 400 mg, cholesterol 100 mg, sorbitan trioleate

(Span 85) 1 mL, and corn oil 30 mL were added to a flask, heated to 70°C, and stirred until the phenobarbital was dissolved. Sorenson buffer (2/15 M, pH 5) 65 mL containing 40% (vol/vol) propylene glycol was added to 2 mL of polysorbate 85 (Tween 85). The aqueous and oil phases were heated to 70°C and mixed. The mixture was then cooled to room temperature and adjusted to a final volume of 100 mL with Sorenson buffer, shaken, heated to 70°C, and homogenized three times using a hand homogenizer. A second similar emulsion without the propylene glycol was also prepared. For the solutions, phenobarbital sodium 438 mg (equivalent to phenobarbital 400 mg) was diluted to 100 mL with Sorenson buffer both with and without 40% (vol/vol) propylene glycol. All of the products were packaged in amber glass containers and stored for 56 weeks (temperature unspecified).

HPLC analysis of these oral liquids found that the emulsion and solution containing propylene glycol were stable throughout the 56 weeks, exhibiting no loss of phenobarbital content; this result was also true of the commercial elixir. However, the emulsion and solution without propylene glycol resulted in a decline in phenobarbital content. Phenobarbital losses of 34% in four weeks and 10% in 12 weeks were found for the emulsion and solution without propylene glycol, respectively.

**STUDY 2:** Woods[586] reported on an extemporaneous alcohol-free oral liquid formulation of phenobarbital sodium. The formula is shown in Table 83. An expiration period of 30 days when stored under refrigeration was suggested, although no stability data were presented. The pH should be about 9.4; lower pH values may result in precipitation of free phenobarbital. Consequently, acidic syrups and flavorings should not be added.

**STUDY 3:** Cober and Johnson[1079] evaluated the stability of an alcohol-free phenobarbital 10-mg/mL suspension. Phenobarbital 10-mg tablets were crushed using a glass mortar and pestle. An equal parts mixture of Ora-Plus suspending medium with either Ora-Sweet or Ora-Sweet SF (sugar-free) was added in increments with thorough mixing to bring the suspension to volume, resulting in a phenobarbital 10-mg/mL concentration. The suspension was packaged in amber plastic prescription bottles with child-resistant caps. Samples were stored at room temperature of 23 to 25°C for 115 days.

**TABLE 83.** Formula of Phenobarbital Sodium 30-mg/mL Alcohol-Free Oral Liquid[586]

| Component | Amount |
|---|---|
| Phenobarbital sodium | 1.5 g |
| Glycerol | 35 mL |
| Water | qs 50 mL |

No detectable changes in color, odor, or taste and no visible microbial growth were noted. Stability-indicating HPLC analysis found less than 2% loss of phenobarbital after 115 days of storage at room temperature. The authors did note that mixing this suspension with an equal amount of chocolate syrup immediately before administration improved palatability, masking the bitter aftertaste of the phenobarbital suspension.

**STUDY 4:** Garg et al.[1250] evaluated the physical and chemical stability as well as palatability of an ethanol-free, preservative-free oral liquid formulation of phenobarbital sodium 10 mg/mL. In addition to the phenobarbital sodium, the formulation contained glycerol 10% (vol/vol), sucrose 25% (wt/vol), unspecified suspending agent 0.5% (wt/vol), various flavorings 0.5 to 1% (wt/vol), and distilled water to bring the formulation to volume. Of the 14 flavors tested for palatability, lemon flavor 0.75% (wt/vol) was deemed the best, followed by raspberry, orange, and lime. In addition, the higher viscosity of the formulation tended to decrease the perceived intensity of the bitter taste of the drug compared to a conventional oral liquid formulation.

The test formulation was packaged in glass bottles and stored protected from exposure to light both under refrigeration at 2 to 8°C and at room temperature of 21 to 23°C for four weeks. The appearance, viscosity, pH, and microbial studies showed no change over the 28-day study period. HPLC analysis found no loss of phenobarbital sodium under either storage condition over 28 days, as well. The authors recommended that the pH be maintained at or above 8.6 and that no acidic vehicles or other components be used to help avoid precipitation of free phenobarbital.

**STUDY 5:** Jelveghari and Nokhodchi[1326] evaluated the physical and chemical stability of ethanol-free oral liquid formulations containing phenobarbital 4 mg/mL using a cosolvent approach. The most stable preparations incorporated propylene glycol 10 to 12% (vol/vol) and glycerin 26 to 28% (vol/vol). Also present in the formulations were saccharin sodium 0.35% (wt/wt), butylparaben 0.1% (wt/wt), strawberry oil 5% (wt/wt), sodium acetate trihydrate buffer pH 4.5 25% (vol/vol), and sucrose 32 to 34% (wt/wt). The use of the buffer resulted in a formulation pH of about 3.3 to 3.8. At these cosolvent concentrations no crystal growth was observed and there was no change in color or clarity.

The ethanol-free phenobarbital oral liquid formulations were packaged in bottles under nitrogen atmosphere. Accelerated stability testing at temperatures from 40 to 70°C was performed for 120 days. Ultraviolet spectrophotometry was used to determine the phenobarbital concentration. Based on the determined decomposition rates, the calculated times to 10% drug loss ($t_{90}$) were all over two years at room temperature.

*Enteral*

**PHENOBARBITAL ELIXIR COMPATIBILITY SUMMARY**

*Elixir compatible with:* Enrich • Ensure • Ensure HN • Ensure Plus • Ensure Plus HN • Osmolite • Osmolite HN • TwoCal HN • Vital • Vivonex T.E.N.

**STUDY 1:** Cutie et al.[19] added 5 mL of phenobarbital elixir (Parke-Davis) to varying amounts (15 to 240 mL) of Ensure, Ensure Plus, and Osmolite (Ross Laboratories) with vigorous agitation to ensure thorough mixing. The phenobarbital elixir was physically compatible, distributing uniformly in all three enteral products with no phase separation or granulation.

**STUDY 2:** Burns et al.[739] reported the physical compatibility of phenobarbital elixir 7.5 mL with 10 mL of three enteral formulas, including Enrich, TwoCal HN, and Vivonex T.E.N. Visual inspection found no physical incompatibility with any of these three enteral formulas.

**STUDY 3:** Altman and Cutie[850] reported the physical compatibility of phenobarbital elixir (Parke-Davis) 5 mL with varying amounts (15 to 240 mL) of Ensure HN, Ensure Plus HN, Osmolite HN, and Vital after vigorous agitation to ensure thorough mixing. The phenobarbital elixir was physically compatible, distributing uniformly in all four enteral products with no phase separation or granulation. ■

# ■ Phenoxybenzamine Hydrochloride

## Properties

Phenoxybenzamine hydrochloride is a white or almost white nearly odorless crystalline powder.[2,3] The material has been characterized as a carcinogen.[1]

### Solubility

Phenoxybenzamine hydrochloride is sparingly soluble in water,[3,630] with an aqueous solubility of approximately 40 mg/mL.[2] It is freely soluble in propylene glycol[1] and freely soluble in ethanol,[3] with a solubility of 167 mg/mL.[2]

Phenoxybenzamine hydrochloride stability in aqueous solution is pH-dependent; the drug exhibits better stability at pH values of less than 3 but is unstable in neutral and alkaline solutions.[630]

### $pK_a$

Phenoxybenzamine hydrochloride has a $pK_a$ of about 5 to 5.1.[630]

## General Stability Considerations

Phenoxybenzamine hydrochloride products should be packaged in well-closed containers and stored at controlled room temperature.[4] The drug is unstable at pH greater than 4.5. The presence of sugars also increases the rate of drug loss.[550]

## Stability Reports of Compounded Preparations

### Injection

Injections, like other sterile drugs, should be prepared in a suitable clean air environment using appropriate aseptic procedures. When prepared from nonsterile components, an appropriate and effective sterilization method must be employed.

Nedergaard[765] reported the stability of an extemporaneously compounded phenoxybenzamine hydrochloride nonaqueous injection. The injection was prepared using the formula in Table 84. Ultraviolet light spectrophotometric analysis and thin-layer chromatography found that the drug remained stable for two months at room temperature but underwent unacceptable changes in three months.

### Oral

Lim et al.[550] reported on the stability of phenoxybenzamine hydrochloride 2-mg/mL oral solutions prepared from powder (Tokyo Casei) in 19 vehicle formulations. The phenoxybenzamine hydrochloride powder was triturated with a portion of the vehicle and transferred to a graduate to be brought to volume. The formulations were packaged in 50-mL amber glass prescription bottles and stored at 4°C for 30 days.

None of the vehicles tested exhibited any changes in color, turbidity, or odor. HPLC analysis found extensive phenoxybenzamine hydrochloride losses in most of the vehicles. The best formulations consisted of propylene glycol 1 or 5% with 0.15% citric acid in water; however, less than 10% drug loss occurred only through seven days at 4°C. Preparing these same best formulations using phenoxybenzamine hydrochloride 10-mg capsules (Smith Kline & French) as the source did not alter the stability. Preparation of a 10-mg/mL concentrate in neat propylene glycol resulted in no phenoxybenzamine hydrochloride loss in 30 days at 4°C. It was suggested that the concentrate could be prepared for its more extended stability and then diluted 1:4 with syrup to yield a 2-mg/mL concentration at the time of administration to create a more palatable form. However, the dose must be given within one hour because of relatively rapid drug loss in the syrup. ■

**TABLE 84.** Phenoxybenzamine Hydrochloride Injection Formula[765]

| Component | Amount |
|---|---|
| Phenoxybenzamine hydrochloride | 5 g |
| Ethanol 99% | 197.5 g |
| Propylene glycol | 260 g |
| Hydrochloric acid 1 N | 0.5 g |

# Phenoxyethanol

## Properties

Phenoxyethanol is a colorless, slightly viscous liquid with a faint aromatic odor and a burning taste.[1,3,4]

### Solubility

Phenoxyethanol has a solubility in water of 26.7 mg/mL. It is miscible in ethanol, ether, acetone, and glycerol and slightly soluble in arachis oil and olive oil.[1,3,4]

## General Stability Considerations

Phenoxyethanol should be packaged in tight containers and stored between 8 and 15°C protected from exposure to light.[3,4] Phenoxyethanol antibacterial activity may be reduced by interaction with nonionic surfactants.[3]

## Stability Reports of Compounded Preparations
### Irrigation

Irrigations, like other sterile drugs, should be prepared in a suitable clean air environment using appropriate aseptic technique. When prepared from nonsterile components, an appropriate and effective sterilization method must be employed.

Lee[1632] evaluated the loss upon administration of phenoxyethanol from a 2.2% bladder irrigation solution, which was packaged in glass bottles and administered through polyvinyl chloride (PVC) plastic tubing at a rate of 1 mL/min. Using gas-liquid chromatography, a 20% loss of phenoxyethanol occurred over four hours of delivery through the PVC tubing, and softening of rigid plastic Luer-lock fittings was noted. No loss of phenoxyethanol was found in the solution in glass containers over four hours. ■

# Phentolamine Mesylate

## Properties

Phentolamine mesylate occurs as a white to off-white, odorless, slightly hygroscopic, crystalline powder.[1,3,4]

### Solubility

Phentolamine mesylate has an aqueous solubility of about 500 mg/mL. It is also freely soluble in ethanol. It is only slightly soluble to about 1.5 mg/mL in chloroform.[1,3,4]

### pH

A 10-mg/mL aqueous solution has a pH of about 4.5 to 5.5.[1,4] Phentolamine mesylate injection, USP, has a pH between 4.5 and 6.5.[4]

## General Stability Considerations

Phentolamine mesylate powder should be packaged in tight, light-resistant containers and should be stored at controlled room temperature. Phentolamine mesylate injection is packaged in glass vials and is stored at controlled room temperature.[4]

## Stability Reports of Compounded Preparations
### Injection

Injections, like other sterile drugs, should be prepared in a suitable clean air environment using appropriate aseptic technique. When prepared from nonsterile components, an appropriate and effective sterilization method must be employed.

**STUDY 1 (TWO-DRUG MIXTURE):** Tu et al[1530] evaluated the stability of papaverine hydrochloride (Eli Lilly) and phentolamine mesylate (Ciba-Geigy). Two milliliters were withdrawn from a vial containing papaverine hydrochloride injection and injected into the lyophilized phentolamine mesylate vial to reconstitute it. The reconstituted solution was then withdrawn and injected back into the papaverine hydrochloride vial. The final concentrations were papaverine hydrochloride 30 mg/mL and phentolamine mesylate 0.5 mg/mL. Vials were stored refrigerated at 5°C and at controlled room temperature of 25°C for 30 days. No visible color change, haze, or precipitation was observed. Little or no papaverine hydrochloride loss occurred in 30 days at either temperature. Phentolamine mesylate losses were about 1 to 3% under refrigeration and 4 to 5% at room temperature in 30 days.

**STUDY 2 (TWO-DRUG MIXTURE):** Benson and Seifert[1671] evaluated the stability of phentolamine mesylate 0.83 mg/mL in combination with papaverine hydrochloride 25 mg/mL. The mixture was prepared by combining two 5-mg vials of phentolamine mesylate reconstituted with 1 mL of sterile water for injection with one 10-mL vial of papaverine hydrochloride 300 mg. The resulting 12-mL of solution was stored for 40 days at room temperature and under refrigeration. Neither HPLC analysis nor gas chromatography/mass spectrometry analysis found a loss of phentolamine mesylate at either storage temperature within the study period.

**STUDY 3 (THREE-DRUG MIXTURE):** Trissel and Zhang[876] evaluated the stability of a three-drug mixture of alprostadil (prostaglandin E₁) 12.5 mcg/mL, papaverine hydrochloride 4.5 mg/mL, and phentolamine mesylate 0.125 mg/mL (commonly called the "Knoxville Formula" or "Trimix") prepared from commercial injections diluted in bacteriostatic sodium chloride 0.9%. The injection mixture was packaged in commercial empty sterile vials and stored at room temperature near 23°C, refrigerated at 4°C, and frozen at −20°C and −70°C. All of the samples remained clear and colorless throughout the study. Stability-indicating HPLC analysis found that alprostadil was the least stable of the three drug components and was the limiting factor in the combination injection. At room temperature, alprostadil losses of 8% and 13% occurred in five and seven days, respectively. Under refrigeration, losses of about 6% in one month and 11% in two months occurred. When the injection mixture was frozen at −20°C and −70°C, alprostadil losses did not exceed 5% in six months. Subjecting vials frozen at −20°C to four freeze-thaw cycles and warming the vials to room temperature each time resulted in no loss of any drug. The authors recommended a beyond-use date of six months if stored frozen at −20°C and

one month if stored under refrigeration for batches that have passed a sterility test. They also recommended that room temperature exposure be limited, and vials should be returned to refrigeration as soon as possible after use.

**STUDY 4 (TWO- AND THREE-DRUG MIXTURES):** Soli et al[1531] evaluated the stability of alprostadil (Upjohn) 0.004 mg/mL with papaverine hydrochloride (Fluka) 6 mg/mL and phentolamine mesylate (Ciba-Geigy) 0.4 mg/mL in two- and three-drug combinations diluted in sodium chloride 0.9%. The injection mixtures were packaged in sterile glass vials and stored refrigerated at 2 to 8°C for 60 days.

No visible precipitation or color change was reported. HPLC analysis found alprostadil losses of about 10 to 13% occurred in five days. No loss of papaverine hydrochloride and about 3 to 4% loss of phentolamine mesylate occurred in 60 days, either with or without alprostadil present. ■

# ■ Phenylephrine Hydrochloride

## Properties

Phenylephrine hydrochloride is a white or nearly white odorless crystalline powder or crystals having a bitter taste.[1–3]

### Solubility

Phenylephrine hydrochloride is freely soluble in water, having an aqueous solubility of about 500 mg/mL. It is also soluble in ethanol, having a solubility of 250 mg/mL.[2,3]

### pH

Phenylephrine hydrochloride injection has a pH between 3 and 6.5. The buffered ophthalmic solution has a pH between 4 and 7.5; the unbuffered ophthalmic solution pH is 3 to 4.5.[4]

### Osmolality

Phenylephrine hydrochloride injection 10 mg/mL has an osmolality of 284 mg/mL.[326]

## General Stability Considerations

Phenylephrine hydrochloride products in general should be packaged in tight, light-resistant containers.[4]

Phenylephrine hydrochloride is subject to oxidation and is unstable in aqueous solution, decomposing to colored compounds; the discoloration is accelerated by elevated temperature. Solutions that contain a brown discoloration or precipitate should not be used. However, oxidation and loss of activity occur in the absence of visible discoloration.[2] Contact with ferric ions from amber glass containers also accelerates the discoloration of phenylephrine hydrochloride solutions.[438]

The drug is more stable at acidic pH; at pH 2 no loss occurred in a 2.5-mg/mL solution stored at 97°C for 10 days. However, alkaline pH values, especially above pH 9, increase decomposition.[8]

## Stability Reports of Compounded Preparations
### Ophthalmic

Ophthalmic preparations, like other sterile drugs, should be prepared in a suitable clean air environment using appropriate aseptic procedures. When prepared from nonsterile components, an appropriate and effective sterilization method must be employed.

**STUDY 1:** Ismaiel and Ismaiel[438] evaluated the stabilizing influence of various concentrations of sodium metabisulfite and edetate sodium on phenylephrine hydrochloride 0.25% eyedrops preserved with benzalkonium chloride 0.02%. The eyedrops were packaged in 15-mL amber glass bottles and were stored at room temperature for 12 months. The most stable eyedrops were those containing sodium metabisulfite 0.025% and edetate sodium 0.1%. There was no solution discoloration and no loss of phenylephrine. Furthermore, this formulation was not irritating to the eye.

**STUDY 2:** Perez Maroto et al.[834] reported the stability of extemporaneously compounded phenylephrine hydrochloride 2.5% eyedrops. The eyedrops were prepared by diluting commercial 10% eyedrops with sterile 0.9% sodium chloride. Samples were stored at room temperature of 25°C and refrigerated at 4°C. Spectrophotometric analysis found that little or no phenylephrine loss occurred in 60 days at either temperature. In addition, osmolarity and pH remained within USP limits.

**STUDY 3:** Garcia-Valldecabres et al.[880] reported the change in pH of several commercial ophthalmic solutions, including phenylephrine hydrochloride 1.2 mg/mL (Vistafrin Liquifilm) over 30 days after opening. Slight variation in solution pH was found, but no consistent trend and no substantial change occurred.

**STUDY 4:** Hirowatari et al.[929] reported the stability of a three-component ophthalmic solution for preoperative use. The ophthalmic solution was prepared by mixing commercially available ophthalmic solutions to yield final concentrations of phenylephrine hydrochloride 1.83%, tropicamide 0.17%, and diclofenac sodium 0.03%. The ophthalmic solution admixture was dispensed as 2-mL aliquots packaged in light-resistant bottles. The physical and chemical stability of the mixture was evaluated over one month stored at 10°C protected from light. Visual inspection found that the solution remained unchanged, being clear and colorless; measured pH remained unchanged as well. HPLC analysis found no loss of diclofenac and tropicamide and about 2% loss of phenylephrine in one month.

**STUDY 5 (SOLUTION WITH TROPICAMIDE):** Zuniga Dedorite et al.[1590] evaluated six phenylephrine hydrochloride 10% and tropicamide 1% eyedrop formulations. The formulation that was selected for stability testing contained the two drugs in a vehicle containing polysorbate 80, sodium metabisulfite, benzalkonium chloride, sodium edetate, and hydrochloric acid in water in low-density polyethylene bottles. The formulation had a pH near 4.3. The formulation was physically stable; chromatographic analysis of the phenylephrine and tropicamide concentrations found little or no loss of either drug and no appreciable formation of decomposition products during 12 months of storage at room temperature of 30°C and 65 to 75% relative humidity.

**STUDY 6 (GEL WITH CYCLOPENTOLATE, TROPICAMIDE, LIDOCAINE):** Bailey et al.[1525] evaluated the stability of a four-drug combination ophthalmic gel for use as a preoperative medication for cataract surgery. The final drug concentrations in the preoperative gel were cyclopentolate hydrochloride 0.51 mg/mL, phenylephrine hydrochloride 5.1 mg/mL, and tropicamide 0.51 mg/mL in the lidocaine hydrochloride jelly 2%. The gel was prepared by combining 0.3 mL each of cyclopentolate hydrochloride ophthalmic solution 1%, USP, phenylephrine hydrochloride ophthalmic solution 10%, USP, and tropicamide ophthalmic solution 1%, USP in preservative-free lidocaine hydrochloride jelly 2%, USP. The gel was packaged as 0.5 mL in 1-mL polycarbonate plastic syringes with sealed tips. The samples were stored protected from light in an environmental chamber at room temperature of 23 to 27°C and 60% relative humidity and also at refrigerated temperature of 2 to 4°C for 60 days.

The ophthalmic gel exhibited little change in pH and remained sterile and endotoxin-free throughout the 60-day study. Stability-indicating HPLC analysis of drug concentrations over 60 days found no loss of any of the drugs when stored under refrigeration and not more than 4% loss of any drug when stored at room temperature.

*Nasal*

Gupta and Mosier[1659] evaluated the stability of phenylephrine hydrochloride 0.5% nasal drops in various 0.05 M phosphate buffer solutions ranging from pH 5 to 9. The formulation also contained methylparaben 0.02% and propylparaben 0.01% and sodium bisulfite 0.1%. At pH values of 7 to 9, no loss of phenylephrine hydrochloride was found colorimetrically, but the nasal solution discolored slightly and the parabens decomposed. The best stability occurred at pH 5 to 6. Not only did no loss of phenylephrine hydrochloride occur, but also no discoloration occurred and no fungus growth was observed. A beyond-use date of 18 months at room temperature was suggested. ∎

# ◼ Phenylmercuric Nitrate

## Properties

Phenylmercuric nitrate, USP, is a mixture of phenylmercuric nitrate and phenylmercuric hydroxide. It occurs as a white or pale yellow crystalline powder.[1,4] Phenylmercuric nitrate is used as an antimicrobial preservative.[3,4]

*Solubility*

Phenylmercuric nitrate is very slightly soluble in water to about 1 part in 1250 parts of water. It is slightly soluble in glycerol and ethanol, but practically insoluble in other organic solvents.[1,3,4]

*pH*

Phenylmercuric nitrate as a saturated solution in water is slightly acidic.[4]

## General Stability Considerations

Phenylmercuric nitrate should be packaged in tight, light-resistant containers and stored at controlled room temperature. Phenylmercuric nitrate is sensitive to light.[4] Sorption to plastics and rubber in containers may occur.[3]

Phenylmercuric nitrate is stated to be incompatible with kaolin, magnesium trisilicate, starch, and talc. It is also incompatible with bromides, iodides, chlorides, metals, and ammonium salts. Disodium edetate, sodium thiosulfate, or sodium metabisulfite may produce inactivation.[3] Phenylmercuric nitrate in combination with sodium metabisulfite may exhibit losses of 75 to 100%.[1635]

## Stability Reports of Compounded Preparations
### Ophthalmic
Ophthalmics, like other sterile drugs, should be prepared in a suitable clean air environment using appropriate aseptic procedures. When prepared from nonsterile components, an appropriate and effective sterilization method must be employed.

**STUDY 1:** Parkin[1636] studied the interaction of phenylmercuric nitrate with sodium metabisulfite in eyedrop formulations following heat sterilization. Losses of phenylmercuric nitrate of 75 to 100% have been reported previously. In this study, stability-indicating HPLC analysis determined that phenylmercuric nitrate losses exceeding 75% occurred within 10 hours at 100°C; the losses were true degradation and not simply complexation. The authors conclude that sodium metabisulfite should never be used as an antioxidant together with phenylmercuric nitrate as an antimicrobial preservative in ophthalmic products.

**STUDY 2:** Parkin et al.[1637] evaluated the decomposition of phenylmercuric nitrate 0.002% wt/vol in combination with disodium edetate 0.05% wt/vol in solutions of pH 5 to 8 after heat sterilization at 121°C for 15 min. Stability-indicating HPLC analysis along with atomic absorption spectroscopy found that extensive losses of phenylmercuric nitrate occurred. Losses of 15 and 80% occurred at pH values of 8 and 7, respectively. At pH 5 and 6, total losses of phenylmercuric nitrate occurred. The authors concluded that phenylmercuric nitrate and disodium edetate were chemically incompatible and should never be used in combination.

**STUDY 3:** Parkin[1638] also evaluated the incompatibility of phenylmercuric nitrate 0.002% wt/vol in sulfacetamide 10 and 30% ophthalmic drops containing disodium edetate and sodium metabisulfite. After sterilizing at 121°C for 15 minutes, phenylmercuric nitrate losses of about 25 to 30% occurred.

**STUDY 4:** Barnes[1635] evaluated the compatibility of phenylmercuric nitrate 0.002% wt/vol solution packaged in glass and Steri-Dropper (Helapet) low-density polyethylene dropper bottles. Samples were stored refrigerated at 6 to 8°C for 84 days. No visible changes were observed throughout the study period. HPLC analysis found about 10% loss of phenylmercuric nitrate occurred in glass containers over 84 days of storage. However, in the low-density polyethylene containers, about 25% loss occurred over 84 days. The authors concluded that phenylmercuric nitrate is stable in glass containers for at least 84 days under refrigeration, but in the plastic Steri-Dropper bottles the utility period should be restricted to one month. ■

# ■ Phenytoin
# Phenytoin Sodium

## Properties
Phenytoin is a white or almost white nearly odorless crystalline powder.[1-3]

Phenytoin sodium is a white odorless hygroscopic crystalline powder with a bitter, soapy taste.[1-3] Phenytoin sodium 100 mg is equivalent to 92 mg of phenytoin.[3]

Phenytoin has been characterized as a carcinogen.[1]

## Solubility
Phenytoin is practically insoluble in water. In ethanol, it has a solubility of 14 to 16 mg/mL.[1-3]

Phenytoin sodium is soluble in water, having an aqueous solubility of 15 mg/mL. Due to partial hydrolysis, the solution will be turbid unless the pH is adjusted to greater than 11.7 (the pH of a saturated solution). In ethanol, the solubility is 95 mg/mL.[1-3] It is also freely soluble in propylene glycol.[2]

## pH
A saturated aqueous solution of phenytoin sodium has a pH of 11.7.[1] The pH of phenytoin sodium injection may range from 10 to 12.3[4] but is stated to be adjusted to 12.[2,7] Phenytoin suspension (Dilantin, Parke-Davis) had a measured pH of 4.9.[19]

## pKa
The drug has an apparent $pK_a$ of 8.03 to 8.33.[2]

## Osmolality
Phenytoin sodium suspensions 6 and 25 mg/mL (Parke-Davis) had osmolalities of 2000 and 1500 mOsm/kg, respectively.[233] Phenytoin sodium injection is very hypertonic; measured osmolalities have been erratic, ranging from 3035 to 9740 mOsm/kg, depending on the method of determination.[287,345]

## General Stability Considerations

Phenytoin products should be packaged in tight containers and stored at temperatures less than 30°C. The suspension should be protected from freezing.[2–4,7]

Phenytoin sodium gradually absorbs carbon dioxide during exposure to air, resulting in the formation of free phenytoin.[3] Phenytoin sodium capsules (both extended and prompt) should be packaged in tight containers and stored at less than 30°C. Phenytoin sodium injection should be stored at controlled room temperature and protected from freezing. A precipitate may form in the injection if it is stored under refrigeration or frozen. The precipitate may be redissolved, and the product remains suitable for administration. A slightly yellowish discoloration of the injection does not indicate concentration or efficacy loss, but phenytoin sodium injection with a precipitate should be discarded.[2–4,7]

### Sorption

Also see *Enteral* later in this monograph.

**STUDY 1:** Cacek et al.[102] evaluated the amount of phenytoin from phenytoin oral suspension 25 mg/mL (Dilantin, Parke-Davis) that was delivered through a nasogastric tube (No. 16 French Salem Sump, Argyle). Ten methods of delivery (Table 85) were tested using dilution with 11 mL of various diluents, dilution and flushing with 20 mL after administration, or direct administration of the suspension, as is, without flushing. Phenytoin concentrations were assessed by HPLC.

**TABLE 85.** Phenytoin Suspension Nasogastric Administration Methods and Results[102]

| Diluent[a] | Irrigant[b] | Drug Lost |
|---|---|---|
| 5% Dextrose injection | 5% Dextrose injection | 6% |
| Lactated Ringer's injection | Lactated Ringer's injection | 11% |
| 0.9% Sodium chloride injection | 0.9% Sodium chloride injection | 7% |
| Sterile water for irrigation | Sterile water for irrigation | 20% |
| 5% Dextrose injection | None | 11% |
| Lactated Ringer's injection | None | 14% |
| 0.9% Sodium chloride injection | None | 14% |
| Sterile water for irrigation | None | 16% |
| None (undiluted) | 0.9% Sodium chloride injection | 23% |
| None (undiluted) | None | 89% |

[a]Diluted with 11 mL.
[b]Flushed with 20 mL of irrigant.

Dilution and dilution with flushing yielded the most complete phenytoin delivery. Use of sterile water for irrigation produced significantly lower phenytoin recoveries than use of 5% dextrose injection, 0.9% sodium chloride injection, or lactated Ringer's injection as diluents and irrigants. The most drug undelivered occurred with the undiluted suspension administered without flushing (89% undelivered). The authors indicated that retention of the viscous undiluted suspension in the nasogastric tube and possible binding of the phenytoin to the plastic tube material could occur. The authors suggested dilution of the phenytoin suspension followed by flushing with 20 mL of fluid and air evacuation of the tube with 40 mL of air as the best administration technique.

**STUDY 2:** Ozuna and Friel[182] reported on the loss of phenytoin to a nasogastric tube. Phenytoin suspension 500 mg was diluted with 10 mL of water and delivered through a nasogastric tube (type and size unspecified) sitting in a 37°C water bath. Gas chromatographic analysis showed that 93% of the drug was delivered from the nasogastric tube. The authors did not consider the loss to be significant.

**STUDY 3:** Splinter et al.[103] evaluated the delivery of phenytoin from 12-mL doses of phenytoin oral suspension 125 mg/5 mL (Dilantin-125, Parke-Davis) through latex percutaneous endoscopic gastrostomy Pezzer catheters (20 French 35.5 cm, Bard). Phenytoin content was assessed using an HPLC assay. Several methods of delivery were tested using water dilution, water dilution and irrigation, or neither dilution nor irrigation (Table 86). The greatest amounts of unrecovered phenytoin were in the samples diluted with water, leading to a greater amount of sorption to the catheter. There was little loss of phenytoin when the suspension was undiluted, whether or not it was flushed with water after administration. The authors recommended not diluting the phenytoin suspension with water prior to administration.

**STUDY 4:** Seifert et al.[186] reported the loss of phenytoin when administered through 20 French latex percutaneous endoscopic gastrostomy Pezzer catheters (Bard). Phenytoin

**TABLE 86.** Phenytoin Suspension Percutaneous Endoscopic Gastrostomy Pezzer Catheter Administration Techniques and Results[103]

| Diluent[a] | Irrigant[a] | Drug Lost |
|---|---|---|
| None | 10 mL | 0 to 7% |
| 10 or 12 mL | None | 11.5 to 15% |
| 10 mL | 10 mL | 7.5 to 18.5% |
| None | None | 4% |

[a]Deionized water.

suspension 125 mg/5 mL (Dilantin, Parke-Davis) and phenytoin 100-mg capsules (Dilantin, Parke-Davis) were administered through the catheters using four administration techniques: (1) no dilution or irrigation; (2) irrigation with 10 mL of deionized water after each dose; (3) dilution of the dose with 10 mL of deionized water; and (4) dilution of the dose with 10 mL of deionized water plus irrigation with 10 mL of deionized water after the dose. HPLC analysis was used to determine the amount of phenytoin delivered through the catheters. The authors did not find binding of the drug to the interior of the latex catheter, as has been reported previously. Rather, losses were attributed to physical adherence of the phenytoin to the surface; dislodging of the phenytoin occurs from subsequent dosing. The "mushroom" design of the Pezzer catheter contributes to clogging with medication because it increases the sites of drug collection. Results differing from previous studies were attributed to higher dose dilution, better collection techniques, and differing catheter position. The authors concluded that dilution was better than irrigation for administering phenytoin suspension because it lessened the contact time with the walls of the catheter. Irrigation is superior to dilution for dosing with capsules. Additional dilution and/or irrigation does not add to the amount of phenytoin delivered through the catheter.

## Stability Reports of Compounded Preparations
### Oral
**STUDY 1:** Kleinberg et al.[23] reported the long-term stability of phenytoin oral suspension 100 mg/4 mL (Parke-Davis) repackaged in 5-mL clear Ped-Pod (Solopak) oral dispensers and stored at 25°C. HPLC analysis found no phenytoin loss during 360 days of storage at room temperature. Furthermore, there were no changes in pH or physical appearance.

**STUDY 2:** Jann et al.[626] reported that dietary vanilla pudding was an unacceptable vehicle for administering phenytoin orally. Phenytoin tablets (Dilantin Infatabs, Parke-Davis) were crushed and mixed with vanilla pudding. Low therapeutic serum concentrations of phenytoin resulted. Use of applesauce as the vehicle for the crushed tablets resulted in a significant increase in serum phenytoin levels in most patients.

### Enteral
#### PHENYTOIN COMPATIBILITY SUMMARY
Although phenytoin sodium is physically compatible with many enteral feeds, the drug exhibits possible variable bioavailability. Several reports of interference with oral phenytoin availability in patients also receiving enteral feedings have appeared.[25–28,182,183] Other studies have not found such interference.[184,185] One paper[187] reported variability in the amount of phenytoin protein binding, depending on solution pH and the nature and concentration of the protein. The protein binding of phenytoin increases its solubility to a true supersaturation level.[187] Hennessey[860] demonstrated a lower phenytoin affinity for whey protein compared to casein.

*Physically compatible with:* Enrich • Ensure • Ensure HN • Ensure Plus • Ensure Plus HN • Isocal • Osmolite • Osmolite HN • Sustacal • Sustacal HC • TwoCal HN • Vital • Vivonex T.E.N.

**STUDY 1:** Hooks et al.[55] found a considerable loss of phenytoin availability when it was combined in various concentrations with an enteral feeding product. Phenytoin oral suspension (Dilantin-30 Pediatric Suspension, Parke-Davis) was added to Osmolite enteral feed (Ross) and a control solution of distilled water with hydrochloric acid to adjust the solution to pH 6, the approximate pH of Osmolite. The nominal phenytoin concentration was 10 mcg/mL. In a second phase of the study, phenytoin concentrations were adjusted up and down to determine the concentration effects on phenytoin availability. The samples were filtered using ultrafiltration tubes (Syva) and analyzed for phenytoin content by enzyme-multiplied immunoassay technique (EMIT, Syva).

The amount of phenytoin recovered from the Osmolite samples was significantly smaller (3.7 mcg/mL) than from the pH-adjusted aqueous control (9.9 mcg/mL). Furthermore, phenytoin recovery increased in the samples with a higher phenytoin to Osmolite concentration but decreased when the phenytoin to Osmolite concentration was smaller. There was no substantial loss of phenytoin to the filter membrane. While several potential causes were discussed (i.e., Osmolite's acidic pH, magnesium and calcium content, protein binding), the authors did not identify any specific causes that led to the reduction of the phenytoin availability.

**STUDY 2:** Miller and Strom[29] evaluated the physical compatibility and chemical stability of phenytoin in three enteral formulas. The contents of a phenytoin sodium 100-mg capsule (Dilantin, Parke-Davis) were dissolved in 10 mL of deionized water and then dispersed in 240 mL of Isocal, Sustacal, and Sustacal HC (Mead Johnson) both full-strength and diluted with deionized water to half-strength. In a second group using phenytoin oral suspension (Dilantin, Parke-Davis) 25 mg/mL, 4 mL of the suspension was added to the full- and half-strength enteral formulas. All samples were prepared in duplicate. The samples were stored at 24°C for 24 hours, with visual examination for physical compatibility and HPLC analysis for the determination of intact phenytoin.

None of the samples exhibited any visible changes in 24 hours. Phenytoin content ranged from 95.5 to 103.4% throughout the study, indicating little or no loss of phenytoin.

Ultrafiltration studies for phenytoin–enteral formula interactions found the concentration of free phenytoin to be about 30% in the Isocal and Sustacal but only 18% in the Sustacal HC. Although phenytoin appears to be bound to one or more components in all three enteral formulas, the authors attributed the difference in extent of binding to the much higher fat content of Sustacal HC trapping more of the phenytoin. This binding could account for the effect of enteral nutrient formulas on phenytoin bioavailability. The stability of vitamins or other enteral formula components was not evaluated.

**STUDY 3:** Holtz et al.[54] evaluated the stability of phenytoin oral suspension and injection (Dilantin, Parke-Davis) at 300 mg/L in full-strength Ensure, Ensure Plus, and Osmolite (Ross) over 12 hours at room temperature. Visual inspection revealed no changes such as clumping, gelling, separation, precipitation, or increased viscosity in the mixtures. Osmolalities of the Ensure mixtures were largely unaffected by drug addition.

Phenytoin concentrations were assessed using an HPLC assay. Phenytoin concentrations in the mixtures made with the oral suspension were highly variable, possibly due to pharmaceutical incompatibility rather than drug decomposition. When the injection was used to make the mixtures, phenytoin concentrations showed no loss in Osmolite, 8% loss in Ensure Plus, and 17% loss in Ensure in 12 hours.

**STUDY 4:** Smith et al.[95] evaluated the in vitro recovery of phenytoin from solutions of sodium caseinate 21.8 mg/mL, calcium caseinate 10.7 mg/mL, sodium caseinate–calcium caseinate (2:1) 32.5 mg/mL, calcium chloride 1.5 mg/mL, and distilled water as a control. The phenytoin suspension (Dilantin, Parke-Davis) 125 mg/5 mL was added to each test solution to yield a theoretical phenytoin concentration of 10 mcg/mL. The samples were ultracentrifuged, and phenytoin recovery from the filtrate was determined using an HPLC assay.

Analysis found all samples containing protein and calcium to have reduced phenytoin recovery compared to the distilled water control. Reductions of about 20 to 50% were found. The concentrations of the protein and calcium components tested in this study reflect those found in the enteral product Osmolite. Binding of phenytoin to these components of enteral products, especially sodium caseinate, may contribute to decreased recovery of phenytoin when combined in enteral feeds.

**STUDY 5:** Hennessey[860] reported the degree of phenytoin recovery from suspension and chewable tablets in Ultracal and Replete casein-based enteral feeding formulas. Extraction and HPLC analysis of phenytoin found drug recoveries to be the same for the suspension and the chewable tablets.

Recoveries of about 37 and 32% were found for Ultracal and Replete, respectively. Binding to whey protein was found to be much less, with recoveries of about 82%.

**STUDY 6:** Cutie et al.[19] added 5 mL of phenytoin suspension (Dilantin, Parke-Davis) to varying amounts (15 to 240 mL) of Ensure, Ensure Plus, and Osmolite (Ross Laboratories) with vigorous agitation to ensure thorough mixing. The phenytoin suspension was physically compatible, distributing uniformly in all three enteral products with no phase separation or granulation.

**STUDY 7:** Burns et al.[739] reported the physical compatibility of 16 mL of phenytoin oral suspension (Dilantin) 125 mg/mL with 10 mL of three enteral formulas, including Enrich, TwoCal HN, and Vivonex T.E.N. Visual inspection found no physical incompatibility with any of the enteral formulas.

**STUDY 8:** Altman and Cutie[850] reported the physical compatibility of phenytoin suspension (Dilantin, Parke-Davis) 5 mL with varying amounts (15 to 240 mL) of Ensure HN, Ensure Plus HN, Osmolite HN, and Vital after vigorous agitation to ensure thorough mixing. The phenytoin suspension was physically compatible, distributing uniformly in all four enteral products with no phase separation or granulation.

*Topical*
**STUDY 1:** Allen[592] reported on a topical 5% phenytoin ointment for the treatment of skin wounds and pressure ulcers. The appropriate amount of phenytoin powder was weighed and incorporated into a small amount of appropriate ointment base. Hydrophilic petrolatum, Aquaphor, and Aquabase are suitable for this ointment. The balance of the ointment base was then incorporated geometrically until the mixture was uniform. The ointment was packaged in tight containers. A use period of six months was suggested, although no stability data were presented.

**STUDY 2:** Rhodes et al.[957] reported the stability of simple phenytoin sodium topical suspensions prepared from Parke-Davis and Mylan capsules for use in treating decubitus ulcers topically. The suspensions were compounded by opening the capsules and adding the powder contents to sodium chloride 0.9% to form a 20-mg/mL suspension. The containers used were not cited. HPLC analysis found that no loss of phenytoin occurred during 15 days of storage at both room temperature and under refrigeration at 4°C using phenytoin sodium capsules from either supplier. In addition, no evidence of microbial contamination or growth was found in agar cultures. The authors noted the antibacterial effects of phenytoin, which may have contributed to this result. ■

# Physostigmine
# Physostigmine Salicylate

## Properties

Physostigmine is a white odorless microcrystalline powder. Physostigmine salicylate occurs as white odorless crystals or white powder. The drugs acquire a red tint when exposed to heat, light, air, or traces of metals.[1,4]

### Solubility

Physostigmine is slightly soluble in water and freely soluble in ethanol and fixed oils.[1,4] Physostigmine salicylate is sparingly soluble in water and soluble in ethanol. One gram of physostigmine salicylate dissolves in 75 mL of water at 25°C and in 16 mL of ethanol.[1,4]

### pH

Physostigmine salicylate 5-mg/mL aqueous solution has a pH of 5.8.[1]

### pKₐ

Physostigmine has $pK_a$ values of 6.12 and 12.24.[1]

## General Stability Considerations

Physostigmine and its salts should be packaged in tight, light-resistant containers and stored at controlled room temperature.[4]

Physostigmine is labile in light, oxygen, and alkaline environments.[804] The pH of maximum stability under anaerobic conditions was found to be about 3.4. However, under aerobic conditions, maximum stability was reported as pH 2.2 to 3[805] and as pH 2.8.[804] Physostigmine salicylate is reported to undergo minimum hydrolysis at pH 3.6 at 25°C.[1162]

## Stability Reports of Compounded Preparations

### Injection

Injections, like other sterile drugs, should be prepared in a suitable clean air environment using appropriate aseptic procedures. When prepared from nonsterile components, an appropriate and effective sterilization method must be employed.

**STUDY 1:** Chen et al.[804] reported the shelf-life stability of physostigmine in solution under aerobic and anaerobic conditions. Physostigmine in buffered solutions 60 mcg/mL was sealed in glass ampules with or without nitrogen gas purging. The sealed ampules were stored at various temperatures from 45 to 88°C. Based on HPLC analysis, the shelf life of physostigmine in solution was calculated to be four years at room temperature under anaerobic conditions. However, under aerobic conditions the rate of degradation was up to 33 times higher.

**STUDY 2:** Akkers et al.[1056] evaluated the long-term shelf-life stability of physostigmine salicylate injection 1 mg/mL packaged in ampules. The injection also contained sodium metabisulfite 1 mg/mL, sodium chloride 8 mg/mL, and hydrochloric acid to adjust to pH 3.6 in water for injection. The solution was sterilized by filtration through a 0.22-μm filter and by heating to 100°C for 30 minutes. An accelerated stability study was performed by storing samples at 70, 80, and 90°C and calculating the degradation rate that is expected at room temperature. The authors' calculations indicated that the drug should be stable for three years at room temperature.

**STUDY 3:** Trose and Slowig[1696] evaluated the stability of physostigmine salicylate 0.1% (1 mg/mL) injections and sodium chloride 8.7 mg/mL in water for injection with and without stabilizers. The physostigmine salicylate injection was used to treat intoxications with atropine syndrome, especially from tricyclic antidepressants and phenothiazines. The potential stabilizers tested were sodium pyrosulfite, N-acetylcysteine, and ascorbic acid. In addition, the effect of incorporating a carbon dioxide environment was evaluated.

The N-acetylcysteine under wet sterilization was not a suitable stabilizer. Sodium pyrosulfite prevented injection discoloration during autoclaving, but the injection developed a red discoloration upon storage. Ascorbic acid 1 mg/mL with five minutes of degassing and using a carbon dioxide atmosphere for the injection resulted in usable injections for 12 months stored refrigerated. In addition, the pH only changed from pH 3.2 to 2.9. Without the carbon dioxide environment, no physostigmine salicylate remained after one month of storage.

### Ophthalmic

Ophthalmic preparations, like other sterile drugs, should be prepared in a suitable clean air environment using appropriate aseptic procedures. When prepared from nonsterile components, an appropriate and effective sterilization method must be employed.

**STUDY 1:** Rogers and Smith[1175] evaluated the stability of physostigmine sulfate ophthalmic solutions at concentrations of 0.25, 0.5, and 1%. The formulation was the official ophthalmic solution (BPC) containing sodium metabisulfite 0.2% (wt/vol) and benzalkonium chloride 0.02% (vol/vol) in water. The ophthalmic solutions were sterilized by filtration through a 0.2-μm Millipore GS filter or were filtered and heated at 98 to 100°C for 30 minutes. The solutions were packaged in sterile eyedrop bottles and stored at 25°C for five years.

After five years of storage, the pH declined from the initial pH 3.8 to 3.4, and the initially colorless solution developed a brownish-pink discoloration. However, the physostigmine sulfate content after five years was 98.4% of the initial concentration.

**STUDY 2:** Mair and Miller[1480] evaluated the stability of pilocarpine hydrochloride and physostigmine sulfate in a combination eyedrop formulation after autoclaving. The ophthalmic drops were prepared using the following formula:

| | |
|---|---|
| Pilocarpine hydrochloride | 10 g |
| Physostigmine sulfate | 1.25 g |
| Sodium metabisulfite | 1 g |
| Benzalkonium chloride solution, BP | 0.1 mL |
| Deionized distilled water | qs  500 mL |

The solution was filtered through a 1.2-μm cellulose acetate filter (Sartorius). Some samples were packaged displacing air in the containers with nitrogen. The samples were autoclaved; some samples were autoclaved at 115°C for 30 minutes and other samples at 121°C for 15 minutes.

Stability-indicating HPLC analysis of the samples found that little or no change in pilocarpine hydrochloride concentration occurred with either the autoclaving cycle and with or without nitrogen displacement. However, the physostigmine sulfate proved to be less stable. After autoclaving samples with or without nitrogen displacement, nearly 3% physostigmine sulfate loss occurred. However, the authors noted that this was still within compendial limits. ▪

# Phytonadione
## (Phytomenadione)

## Properties
Phytonadione is a clear yellow to amber viscous liquid that is nearly odorless.[2,3]

### Solubility
Phytonadione is practically insoluble in water. In ethanol it has a solubility of around 14 mg/mL.[2,3] It is freely soluble in fixed oils.[3] The injection is prepared with the aid of solubilizing and dispersing agents.[2]

### pH
Phytonadione injection has a pH between 3.5 and 7.[4]

### Osmolality
Phytonadione injection has a measured osmolality of 325 mOsm/kg.[287]

## General Stability Considerations
Phytonadione is stable to heat and moisture and may be sterilized by autoclaving. However, it is light-sensitive and should be protected from light at all times.[2]

Phytonadione tablets should be packaged in well-closed, light-resistant containers.[4] Phytonadione injection should be protected from exposure to light and also protected from freezing.[2,3] The injection may exhibit a slight opalescence, but this does not affect the concentration or safety of the product.[7]

## Stability Reports of Compounded Preparations
### Injection
Injections, like other sterile drugs, should be prepared in a suitable clean air environment using appropriate aseptic procedures. When prepared from nonsterile components, an appropriate and effective sterilization method must be employed.

Levin et al.[347] reported that phytonadione injection repackaged into Tubex cartridges retained concentration for three weeks at room temperature. However, an insoluble precipitate formed after one month.

### Oral
**STUDY 1:** The extemporaneous formulation of a phytonadione 1-mg/mL oral suspension was described by Nahata and Hipple.[160] Six phytonadione 5-mg tablets were crushed and mixed thoroughly with 5 mL of purified water and 5 mL of 1% methylcellulose. The suspension was brought to 30 mL with 70% sorbitol. A stability period of three days under refrigeration was used, although chemical stability testing was not performed.

**STUDY 2:** Sewell and Palmer[262] reported the stability of vitamin $K_1$ 1 mg/mL in an extemporaneous pediatric liquid formulation. The product was compounded by mixing vitamin $K_1$ 100 mg with 2 g of Cremophor EL. The mixture then was diluted gradually with water for injection to 100 mL, yielding a nominal concentration of 1 mg/mL. The solution was sparged with nitrogen for three minutes and was sterilized by filtration through a 0.2-μm filter. The pediatric liquid was packaged as 1 mL in Baxa amber oral polypropylene syringes that were stored at 5°C for 25 weeks. A stability-indicating HPLC assay was used to assess the vitamin $K_1$ content. The solutions remained clear and pale green; no loss of the vitamin occurred. Sterility tests were conducted immediately after preparation and again after the storage period; no microbial contamination was found. Based on these results, the authors concluded that the extemporaneously compounded formulation could reasonably be given a six-month shelf life.

**STUDY 3:** Wong and Ho[1198] evaluated the stability of phytonadione 2-mg/mL injection packaged in dropper containers for oral administration to neonates. The phytonadione injection was packaged as 0.75 mL in amber soda lime glass dropper bottles with a glass dropper and rubber bulb and also in white polyethylene plastic squeeze dropper bottles with a dropper tip and a detachable cap. The samples were stored for 30 days at room temperature of 25 to 32°C exposed to mixed daylight and fluorescent light for eight hours each day and also stored refrigerated at 4 to 8°C protected from exposure to light. Stability-indicating HPLC analysis found that less than 10% phytonadione loss occurred over 30 days in both containers in the refrigerated samples that were protected from light. At room temperature exposed to light, less than 10% loss occurred in 30 days in the amber containers. However, in the white plastic containers, the room temperature samples exposed to light resulted in rapid loss of phytonadione, with more than 10% loss calculated to occur in about 40 hours. ■

# ■ Pilocarpine Hydrochloride

## Properties
Pilocarpine hydrochloride is a white or nearly white hygroscopic odorless powder or colorless crystals having a bitter taste.[2,3,4]

### Solubility
Pilocarpine hydrochloride has a solubility of 3.33 g/mL in water. In ethanol, the solubility is 333 mg/mL. It is slightly soluble in chloroform and insoluble in ether.[2,3,4] Pilocarpine nitrate 1 g dissolves in 4 mL of water and 75 mL of ethanol but is insoluble in chloroform and ether.[1,3,4]

### pH
A 0.5% pilocarpine hydrochloride solution in carbon dioxide-free water has a pH of 3.5 to 4.5.[3] Pilocarpine hydrochloride ophthalmic solution has a pH of 3.5 to 5.5.[4] The ophthalmic gel has a pH of 4.7 to 4.9.[2] Pilocarpine nitrate 5% aqueous solution has a pH of 3.5 to 4.5.[3] Pilocarpine nitrate ophthalmic solution has a pH of 4 to 5.5.[4]

### pKa
Pilocarpine hydrochloride has apparent $pK_a$ values of 7.15 and 12.57.[2]

## General Stability Considerations
Pilocarpine products should be packaged in tight containers and protected from light.[1,3,4] Solutions of pilocarpine hydrochloride should be dispensed in plastic containers or glass containers that leach very little alkali.[6] The ophthalmic pilocarpine hydrochloride gel should be stored under refrigeration and protected from freezing. After dispensing, the gel may be stored at room temperature for up to eight weeks. Unused portions should be discarded after eight weeks.[2]

Pilocarpine hydrochloride is subject to hydrolysis and epimerization in solution.[6,1158] Hydrolysis of the ester linkage of the lactone ring is the more extensive degradation route and is susceptible to both specific and general acid and base catalysis. Epimerization with subsequent hydrolysis occurs primarily in alkaline solutions.[6]

Pilocarpine hydrochloride is more stable at acidic than at alkaline pH. The pH of maximum stability is around 4 to 5.[6,740,1084,1158,1518,1692] Above pH 6.7, pilocarpine is increasingly present as the free base[1081] and subject to more rapid decomposition.[1197,1518]

Pilocarpine hydrochloride solutions are light-sensitive; exposure to light during long-term storage may result in degradation. Solutions of the drug may be autoclaved for sterilization with little loss of the drug. Filtration also may be used.[6,1158]

Buffered ophthalmic solutions are less stable than unbuffered solutions. Phosphate and acetate buffers increase the rate of pilocarpine hydrochloride degradation as their concentrations increase.[6]

## Stability Reports of Compounded Preparations
### Ophthalmic
Ophthalmic preparations, like other sterile drugs, should be prepared in a suitable clean air environment using appropriate aseptic procedures. When prepared from nonsterile components, an appropriate and effective sterilization method must be employed.

**STUDY 1 (AS HYDROCHLORIDE):** Gibbs and Tuckerman[38] reported on a stable pilocarpine hydrochloride 1% ophthalmic solution. It was formulated without added buffers by dissolving the pilocarpine hydrochloride powder in water and sterilizing through a bacterial-retentive filter. The pH was adjusted initially to about 5, but the solution slowly achieved an equilibrium pH of about 3.8. The solution was stored in glass and polyethylene bottles at elevated temperatures of 45 and 65°C for 118 and 137 days, respectively. The preparation was assayed using a ferric hydroxamate assay for intact pilocarpine. Pilocarpine content reached equilibrium at about 93% of the original content. The solution is self-buffered by a pilocarpine–pilocarpinic acid system at equilibrium. The self-buffered formulation required substantially less alkali for neutralization than the commercial products containing added buffers, which, the authors believed, would result in less irritation upon ocular instillation.

**STUDY 2 (AS HYDROCHLORIDE AND NITRATE):** Kreienbaum and Page[39] evaluated the stability of eight commercial pilocarpine nitrate ophthalmic solution lots from 0.5 to 6% and 242 samples of commercial pilocarpine hydrochloride ophthalmic solutions ranging from 0.25 to 10% using a stability-indicating HPLC assay. The samples were packaged in commercial plastic dropper bottles ranging in size from 1 to 30 mL and had been returned from use sites around the United States. All of the pilocarpine nitrate ophthalmic solution lots were within USP limits for strength and pH. However, eight pilocarpine hydrochloride ophthalmic solution lots exhibited pilocarpine concentrations above the USP limit. All were packaged in small 1- or 2-mL dropper bottles and had evidence of leakage, including the presence of a white residue around the top of the container and a liquid level lower than expected. Presumably some solution had evaporated from the smaller containers, increasing the pilocarpine concentration.

**STUDY 3 (AS HYDROCHLORIDE):** Formulation of a preservative-free pilocarpine hydrochloride ophthalmic solution was described by Allen.[157] Each milliliter of the ophthalmic solution should contain 1.25 mg of pilocarpine hydrochloride in 0.9% sodium chloride injection. If necessary, a dilution of the pilocarpine hydrochloride may be made to obtain the required small amount of drug accurately. The solution should be filtered through a 0.22-μm sterilizing filter into a sterile ophthalmic container. Although pilocarpine hydrochloride ophthalmic solutions are chemically stable for months, the absence of an antibacterial preservative would necessitate preparation in small quantities with short expiration periods.

**STUDY 4 (AS HYDROCHLORIDE WITH TIMOLOL):** Pilatti et al.[740] reported the instability of pilocarpine hydrochloride when prepared with timolol maleate in ophthalmic solutions. For ophthalmic solutions containing pilocarpine hydrochloride 10 and 20 mg/mL, stability-indicating HPLC analysis found about 8% loss of drug after storage for up to 24 months, presumably at room temperature, although the temperature was not specified. However, if timolol maleate 6.8 mg/mL was incorporated into the ophthalmic solution, pilocarpine losses up to 16 and 21% occurred in two and 11 months, respectively.

**STUDY 5 (AS HYDROCHLORIDE):** Garcia-Valldecabres et al.[880] reported the change in pH of several commercial ophthalmic solutions, including pilocarpine hydrochloride 20 mg/mL (Colircusi Pilocarpina) over 30 days after opening. Slight variation in solution pH was found, but no consistent trend and no substantial change occurred.

**STUDY 6 (AS HYDROCHLORIDE WITH PHYSOSTIGMINE):** Mair and Miller[1480] evaluated the stability of pilocarpine hydrochloride and physostigmine sulfate in a combination eyedrop formulation after autoclaving. The ophthalmic drops were prepared using the following formula:

| | |
|---|---|
| Pilocarpine hydrochloride | 10 g |
| Physostigmine sulfate | 1.25 g |
| Sodium metabisulfite | 1 g |
| Benzalkonium chloride solution, BP | 0.1 mL |
| Deionized distilled water: | qs 500 mL |

The solution was filtered through a 1.2-μm cellulose acetate filter (Sartorius). Some samples were packaged displacing air in the containers with nitrogen. The samples were autoclaved; some samples were autoclaved at 115°C for 30 minutes and other samples at 121°C for 15 minutes.

Stability-indicating HPLC analysis of the samples found that little or no change in pilocarpine hydrochloride concentration occurred with either autoclaving cycle and with or without nitrogen displacement. However, the physostigmine sulfate proved to be less stable. After autoclaving samples with or without nitrogen displacement, nearly 3% physostigmine sulfate loss occurred. However, the authors noted that this was still within compendial limits.

**STUDY 7 (AS NITRATE):** Pei and Nan[1691] evaluated the stability of pilocarpine (as nitrate) 10 mg/mL. The formulation contained pilocarpine nitrate 1 g, sodium borate 194 mg, 6% acetic acid 1.42 mL, sodium chloride 500 mg, and phenylmercuric nitrate 1 mg and was brought to 100 mL with distilled water. The ophthalmic solution was autoclaved at 100°C for 30 minutes. Pilocarpine was found to be stable at room temperature for up to six months.

## Oral

**STUDY 1:** Mehta et al.[760] reported the stability of pilocarpine oral solution compounded for use in salivary dysfunction. Pilocarpine 2% eyedrops were used to prepare the formulation shown in Table 87.

**TABLE 87.** Pilocarpine Oral Solution[760]

| Component | Amount |
|---|---|
| Pilocarpine | 0.05% (vol/vol) |
| Benzoic acid solution, BP | 2% (vol/vol) |
| Citric acid | 0.102% (wt/vol) |
| Dihydrogen orthophosphate (12 H$_2$0) | 0.369% (wt/vol) |
| Sorbitol 70% | 80% (vol/vol) |
| Concentrated peppermint emulsion | 2.5% (vol/vol) |
| Sterile water | qs 100% |

The pilocarpine oral solution was packaged in bottles and stored at room temperature exposed to light and light-protected and also refrigerated at 4°C for 12 weeks. Stability-indicating HPLC analysis found about 5 and 6% loss of pilocarpine in 12 weeks when refrigerated and at room temperature, respectively. Exposure to light did not adversely affect the stability in this formulation within the study period.

**STUDY 2:** Fawcett et al.[1197] evaluated the stability of several formulations of pilocarpine hydrochloride oral solutions for use in salivary dysfunction. Four successful formulations with pH values between 4.7 and 5.5 were demonstrated to be stable. The formulas of the more successful formulations are shown in Table 88.

No visible changes occurred in any of the formulations. HPLC analysis found little or no drug loss in formulations 1, 2, and 4 over 90 days at room temperature and under refrigeration. Formulation 3 underwent no loss under refrigeration over 90 days. Although some loss was observed at room temperature, the formulation was still relatively stable, with a calculated time to 10% decomposition of 101 days at room temperature.

One other less successful formulation was buffered to a pH of 7.5. The formulation underwent more rapid decomposition, as was expected from the pH dependency of pilocarpine hydrochloride stability. The time to 10% decomposition at room temperature was about nine days. ▪

**TABLE 88.** Pilocarpine Hydrochloride Oral Solutions for Use in Salivary Dysfunction Formulations[1197]

| Component | 1 | 2 | 3 | 4 |
|---|---|---|---|---|
| Pilocarpine hydrochloride powder | 0.5 g | 7.5 g | | |
| Pilocarpine hydrochloride eyedrops | | | 125 mL | 125 mL |
| Glycerol, BP | 500 mL | 500 mL | 500 mL | 500 mL |
| Citric acid monohydrate | 6.75 g | 6.75 g | 6.75 g | |
| Sodium citrate dihydrate | 28.3 mg | 28.3 mg | 28.3 mg | |
| Hydroxybenzoate solution | 50 mL | 50 mL | 50 mL | |
| Lemon spirit, BPC | 2.5 mL | 2.5 mL | 2.5 mL | 2.5 mL |
| Distilled water qs ad | 2500 mL | 2500 mL | 2500 mL | 2500 mL |
| Pilocarpine hydrochloride concentration | 0.2 mg/mL | 3 mg/mL | 3 mg/mL | 3 mg/mL |
| pH | 5.5 | 5.4 | 5.4 | 4.7 |

# ▪ Piperacillin Sodium

## Properties
Piperacillin sodium is a white to off-white crystalline hygroscopic powder.[3,4] One gram of piperacillin sodium provides 1.85 mEq or 42.5 mg of sodium.[7]

### Solubility
Piperacillin sodium is freely soluble in water and ethanol.[3,4] It is practically insoluble in ethyl acetate.[3]

### pH
A 10% aqueous solution of piperacillin sodium has a pH of 5 to 7.[3] A 40% aqueous solution of piperacillin sodium has a pH of 5.5 to 7.5.[4] Reconstituted piperacillin sodium for injection at a concentration of 200 mg/mL has a pH between 4.8 and 6.8.[4]

### Osmolality
Piperacillin sodium 45 and 70 mg/mL in 5% dextrose had osmolalities of 346 and 389 mOsm/kg, respectively. Piperacillin sodium 45 and 70 mg/mL in 0.9% sodium chloride had osmolalities of 361 and 399 mOsm/kg, respectively.[1147]

### General Stability Considerations
Piperacillin sodium powder should be packaged in tight containers.[4] The bulk powder and commercial freeze-dried powder for injection should be stored at controlled room temperature.[4,7] Exposure to sunlight for one month has resulted in darkening of the powder but no accompanying drug loss. Similarly, darkening of piperacillin sodium solution is not indicative of drug loss. Piperacillin sodium in solution is stable over a pH range of 4.5 to 8.5.[7,1217]

Piperacillin sodium in various infusion solutions has been found to be stable for at least 24 hours at room temperature, 48 hours under refrigeration,[7,1217] and one month frozen,[1217] although frozen storage is not recommended by the manufacturer.[7] Gupta and Davis[1218] reported that piperacillin sodium 10 mg/mL in 5% dextrose and 0.9% sodium chloride was stable for at least 71 days stored frozen at −10°C.

## Stability Reports of Compounded Preparations
### Injection

Injections, like other sterile drugs, should be prepared in a suitable clean air environment using appropriate aseptic procedures. When prepared from nonsterile components, an appropriate and effective sterilization method must be employed.

**STUDY 1 (REPACKAGING):** The manufacturer[1217] has stated that piperacillin sodium 40 mg/mL in sterile water for injection packaged in Glaspak glass and Plastipak polypropylene plastic syringes is chemically stable for at least 32 days frozen.

**STUDY 2 (REPACKAGING):** Borst et al.[313] evaluated the stability of piperacillin 200 and 300 mg/mL in sterile water for injection packaged in Monoject plastic syringes. HPLC analysis found piperacillin sodium was stable, exhibiting not more than 10% loss in two days at 24°C, in 10 days at 4°C, and over three months frozen at −15°C.

**STUDY 3 (REPACKAGING):** Gupta and Ling[730] reported the stability of piperacillin sodium 40 mg/mL in sodium chloride 0.9% packaged in polypropylene syringes (Becton Dickinson) and stored refrigerated at 4 to 6°C and at room temperature of 24 to 26°C. The solutions remained clear throughout the study. Stability-indicating HPLC analysis found that about 2% loss occurred in 28 days under refrigeration. At room temperature, losses of 5 and 9% occurred in three and five days, respectively.

### Ophthalmic

Ophthalmic solutions, like other sterile drugs, should be prepared in a suitable clean air environment using appropriate aseptic procedures. When prepared from nonsterile components, an appropriate and effective sterilization method must be employed.

Lin et al.[1620] evaluated the antibiotic activity of piperacillin sodium 10 and 50% in dextrose 5% extemporaneously prepared for use as eyedrops by microbiological assay. Piperacillin sodium injection was diluted in dextrose 5% to concentrations of 100 and 500 mg/mL. Samples were stored refrigerated at 4°C and stored frozen at −18°C for 28 days. The 10% concentration was for clinical use, while the 50% concentration was for use as a stock solution. The MICs were variable throughout the study but found that the differences between time zero and 28 days were statistically insignificant. ■

# ■ Piroxicam

## Properties

Piroxicam is an off-white, light tan, or light yellow odorless powder. The monohydrate is yellow.[2,3]

### Solubility

Piroxicam is very slightly soluble in water, dilute acids, and most organic solvents. It is only slightly soluble in ethanol and in alkaline aqueous solutions.[2–4]

Shin et al.[1655] found that piroxicam had solubilities of 44.6 mcg/mL in water, 1.9 mg/mL in propylene glycol, and 16.7 mg/mL in polyethylene glycol 400.

Chen et al.[1672] reported on solubilities in cosolvent systems ranging from 10 to 40% of propylene glycol, glycofurol, and N,N-dimethylacetamide (DMA). Piroxicam was most soluble in DMA with solubilities ranging from about 1 mg/mL at 10% to 15 mg/mL at 40%. If nicotinamide 30 mg/mL was added, the solubilities increased to about 5 mg/mL at 10% and 26 mg/mL at 40% DMA. Piroxicam was somewhat less soluble in the other diluents with maxima of 5 mg/mL in propylene glycol 40% and 9 mg/mL in glycofurol 40%. If nicotinamide 30 mg/mL was added, the solubilities increased to about 14 or 15 mg/mL for each solvent at 40%.

### pKₐ

The drug has $pK_a$ values of 1.8 and 5.1.[2,7]

## General Stability Considerations

Piroxicam capsules should be packaged in tight, light-resistant containers and stored at temperatures less than 30°C.[2,4] The manufacturer has stated that piroxicam dispersible tablets should not be removed from the original packaging and placed in dosing compliance aids because of moisture sensitivity.[1622]

Shin et al.[1655] found that piroxicam was unstable in propylene glycol. However, it was much more stable in polyethylene glycol 400, exhibiting no loss when stored at 80°C for 5 days.

Chen et al.[1672] found that piroxicam 20-mg/mL injection was subject to photoinstability when exposed to sunlight for six hours daily over one week. A substantial discoloration was observed. However, the drug was much more stable to elevated temperature, exhibiting less than 10% loss in nine months at elevated temperature of 37 and 45°C and in six months at 60°C.

## Stability Reports of Compounded Preparations
### Oral

Bregni and Iribarren[416] evaluated the stability of 4% piroxicam in the following four suspension systems:

1. Sodium carboxymethylcellulose 0.5% in water
2. Veegum 1% with sodium carboxymethylcellulose 0.5% in water

3. Veegum 2.5% with sodium carboxymethylcellulose 0.5% in water
4. Methylcellulose 2.5% in water

The suspensions were stored at room temperature (20°C) in the dark. The piroxicam content was assessed spectrophotometrically. A loss of 10% occurred in two days in suspensions 1, 2, and 4. Concentration was retained for six days in suspension 3.

## Topical
### USP OFFICIAL FORMULATION (CREAM):

| | |
|---|---|
| White petrolatum | 25 g |
| Stearyl alcohol | 15 g |
| Propylparaben | 60 mg |
| Methylparaben | 150 mg |
| Propylene glycol | 12 g |
| Sodium lauryl sulfate | 1 g |
| Sodium hydroxide 1 N | 2.5 mL |
| Piroxicam | 3 g |
| Purified water | qs   100 g |

In a suitable tared container, mix white petrolatum and stearyl alcohol and heat to 75 to 85°C, forming a clear oil phase. In a separate container, mix propylparaben, methylparaben, propylene glycol, sodium lauryl sulfate, and about 30 mL of purified water. Heat this mixture to 75 to 85°C, forming a clear aqueous phase. Add the aqueous phase into the oil phase with continuous stirring while allowing the mixture to cool to 50°C, forming an emulsion. Using a mortar and pestle, triturate the piroxicam with the sodium hydroxide, forming a suspension. Add the piroxicam suspension quantitatively and stepwise to the emulsion using additional purified water to rinse the mortar. Add additional purified water with stirring to bring the emulsion to the final weight. Package the piroxicam emulsion in tight, light-resistant plastic resealable containers. Store the finished emulsion at controlled room temperature. The beyond-use date is 90 days from the date of compounding.[4,5]

**TABLE 89.** Piroxicam Transdermal Soft Patch Gel Systems[1206]

| Compound | Amount |
|---|---|
| Stearate gel (1000 g) | |
|    Propylene glycol | 500 g |
|    Glycerol | 330 g |
|    Stearic acid | 60 g |
|    Aqueous sodium hydroxide | 10.7% (wt/vol) 100 mL |
|    Piroxicam | 10 g |
| Lipophilic gel (1000 g) | |
|    Adeps solidus | 600 g |
|    Glycerol | 390 g |
|    Piroxicam | 10 g |

Csoka et al.[1206] evaluated the stability of piroxicam formulated as a transdermal "soft-patch" type gel system, which is a semisolid form-retaining gel system consisting of flexible plates containing the drug. The two gels evaluated were a stearate gel and a lipophilic gel. The gels were packaged in sealed plastic containers and stored refrigerated at 4°C, at room temperature of 25°C, and at elevated temperature of 40°C. The gels had the compositions shown in Table 89.

Piroxicam was found to decompose faster in the lipophilic gel. Based on stability-indicating HPLC analysis of drug decomposition, the expiration dates at room temperature were calculated to be 25 months for the stearate gel and 12 months for the lipophilic gel. Under refrigeration, the calculated expiration dates were 37 months for the stearate gel and 22 months for the lipophilic gel. No change in color was observed when the gels were stored at room and refrigeration temperatures. Under refrigeration no color changes were observed. However, at room temperature a slightly yellowish discoloration appeared after 20 months. At 40°C, a brownish discoloration occurred within 20 months. ■

# ■ Polymyxin B Sulfate

## Properties
Polymyxin B sulfate is a white or buff hygroscopic nearly odorless powder. It contains not less than 6000 USP units per milligram of substance.[2,3]

## Solubility
Polymyxin B sulfate is freely soluble in water and slightly soluble in ethanol.[2,3]

## pH
A 0.5% aqueous solution of polymyxin B sulfate has a pH between 5 and 7.5.[4]

## Osmolality
Polymyxin B sulfate reconstituted with sterile water for injection to 50,000 units/mL has a measured osmolality of 10 mOsm/kg.[326]

## General Stability Considerations
In general, polymyxin B sulfate products should be packaged in tight, light-resistant containers. Polymyxin B sulfate sterile powder should be stored at temperatures less than 30°C protected from light.[2–4]

Aqueous solutions are quite stable, retaining concentration for six to 12 months under refrigeration.[2,7] Even so, discarding solutions for parenteral use after 72 hours has been recommended.[7] Polymyxin B in aqueous solutions is stable at acidic pH from 2 to 7, exhibiting maximum stability at pH 3.4.[832] Polymyxin B sulfate is inactivated by strongly acidic solutions and by strongly alkaline solutions. It is incompatible with calcium and magnesium salts.[2]

## Stability Reports of Compounded Preparations
### Inhalation
Inhalations, like other sterile drugs, should be prepared in a suitable clean air environment using appropriate aseptic procedures.

When prepared from nonsterile components, an appropriate and effective sterilization method must be employed.

The potential for microbiological contamination of several extemporaneous aerosol solutions packaged in bulk in 120-mL amber glass bottles or in sterile unit-dose Nebuject syringes was evaluated by Kuhn et al.[81] Solutions composed of polymyxin B sulfate 0.8%, propylene glycol 10%, and phenylephrine 0.25% were sterilized by filtration through a 0.22-μm filter, packaged in the two systems, and used by 10 cystic fibrosis patients, each over five days. All Nebuject unit-dose syringes in the study were found to be sterile. However, the solutions packaged in the amber glass bottles all were heavily contaminated with *Bacillus cereus* after five days of use. ▬

# ▣ Potassium Chloride

## Properties
Potassium chloride is an odorless white crystalline powder or colorless crystals that may be cubical, elongated, or prismatic. Potassium chloride 1.91 g provides approximately 1 g of potassium.[1,3]

### Solubility
Potassium chloride is freely soluble in water, having an aqueous solubility of around 357 mg/mL, increasing to about 555 mg/mL in boiling water. It is practically insoluble in ethanol, with a solubility of about 4 mg/mL.[1,3]

### pH
Potassium chloride injection has a pH of 4 to 8.[4]

Cutie et al.[19] measured the pH of several oral liquid preparations of potassium chloride. Their results are shown in Table 90.

### Osmolality
Potassium chloride elixir 10% (sugar-free) was found by Niemiec et al.[232] to have an osmolality of 3000 mOsm/kg. Dickerson and Melnik[233] found that the osmolalities of various lots of potassium chloride 10% (Adria and Roxane) varied from 3000 to 4350 mOsm/kg. Potassium chloride

injections 1.5 and 2 mEq/mL have osmolalities of 3000 and 4000 mOsm/kg, respectively.[8]

## General Stability Considerations
In general, potassium chloride products should be packaged in tight containers and stored at controlled room temperature.[4]

## Stability Reports of Compounded Preparations
### Enteral
#### POTASSIUM CHLORIDE COMPATIBILITY SUMMARY
Kudsk et al.[58] stated that they believed potassium chloride elixir was the most common drug causing enteral feed clogging. However, variable compatibility results have been reported for various potassium chloride products with many enteral feeds. Some potassium chloride products appear to be less problematic than others. Selection of the appropriate potassium chloride product seems to reduce the chance of enteral feed clogging.

**STUDY 1:** Cutie et al.[19] combined several different potassium chloride products with three enteral feeds with varying results. When 15 mL of Kay Ciel Elixir (Berlex) or one packet of Klorvess Granules (Dorsey) was added to varying amounts (15 to 240 mL) of Ensure, Ensure Plus, and Osmolite (Ross Laboratories) with vigorous agitation to ensure thorough mixing, physically compatible mixtures resulted. The Kay Ciel Elixir and Klorvess Granules were distributed uniformly in all three enteral products with no phase separation or granulation.

However, quite different results were obtained when 15 mL of potassium chloride 10 or 20% liquid (Barre) or Klorvess Syrup (Dorsey) was added to the three enteral feeds. The potassium chloride liquids were incompatible at the mixing interface, while the Klorvess Syrup produced a viscous gelatinous mixture that clogged feeding tubes.

**TABLE 90.** pH Values of Some Potassium Chloride Oral Liquid Products[19]

| Product | pH |
| --- | --- |
| Potassium chloride elixir (Kay Ciel, Berlex) | 6.2 |
| Potassium chloride liquid (Barre) 10% | 4.1 |
| Potassium chloride liquid (Barre) 20% | 3.8 |
| Potassium chloride syrup (Klorvess Syrup, Dorsey) | 2.4 |
| Potassium chloride granules (Klorvess, Dorsey) | 4.7 |

**STUDY 2:** Fagerman and Ballou[52] reported on the compatibility of potassium chloride (K-Lor, Abbott) 20 mEq per packet mixed with three enteral feeding formulas. The K-Lor was compatible visually with Vital, Osmolite, and Osmolite HN (Ross), with no obvious thickening, granulation, or precipitation.

**STUDY 3:** Burns et al.[739] reported the physical compatibility of potassium chloride elixir 15 mL with 10 mL of three enteral formulas, including Enrich, TwoCal HN, and Vivonex T.E.N. Visual inspection found no physical incompatibility with any of the enteral formulas.

**STUDY 4:** Altman and Cutie[850] reported the physical compatibility of several different potassium chloride products. When 15 mL of Kay Ciel Elixir (Berlex) or one packet of Klorvess Granules (Dorsey) was added to varying amounts (15 to 240 mL) of Ensure HN, Ensure Plus HN, Osmolite HN, and Vital with vigorous agitation to ensure thorough mixing, physically compatible mixtures resulted. The Kay Ciel Elixir and Klorvess Granules were distributed uniformly in all three enteral products with no phase separation or granulation.

However, quite different results were obtained when 15 mL of potassium chloride 10 or 20% liquid (Barre) or Klorvess Syrup (Dorsey) was added to Ensure HN, Ensure Plus HN, Osmolite HN, and Vital enteral feeds. The potassium chloride liquids were incompatible at the mixing interface, while the Klorvess Syrup produced a viscous gelatinous mixture that clogged feeding tubes. ■

# ■ Potassium Gluconate

## Properties
Potassium gluconate is a white to yellowish-white odorless crystalline powder or granules having a salty taste. Potassium gluconate (anhydrous) 5.99 g and potassium gluconate monohydrate 6.45 g are approximately equivalent to 1 g of potassium.[1,3]

### Solubility
Potassium gluconate has an aqueous solubility of about 333 mg/mL. It is practically insoluble in ethanol.[1,3]

### pH
Aqueous solutions of potassium gluconate have a pH between 7.5 and 8.5.[1]

## General Stability Considerations
In general, potassium gluconate products should be packaged in tight containers. The elixir should be packaged in tight, light-resistant containers.[3,4]

## Stability Reports of Compounded Preparations
*Enteral*
**POTASSIUM GLUCONATE COMPATIBILITY SUMMARY**
*Compatible with:* Vital
*Incompatible with:* Osmolite • Osmolite HN

Fagerman and Ballou[52] reported on the compatibility of potassium gluconate liquid (Kaon) 20 mEq/15 mL mixed in equal quantities with three enteral feeding formulas. The potassium gluconate was compatible visually with Vital (Ross) but underwent unacceptable thickening when combined with Osmolite and Osmolite HN (Ross). A precipitate formed in the mixtures upon standing. ■

# ■ Potassium Iodide

## Properties
Potassium iodide occurs as an odorless, slightly hygroscopic white granular powder or as transparent, colorless, or opaque, white hexahedral crystals. Each gram of potassium iodide provides about 6 mEq of potassium.[2,3] The oral aqueous solution is clear and colorless. It has a characteristic salty taste and contains 1 g/mL of sodium iodide.[2,4]

Strong iodine solution (Lugol's solution) contains about 5 g of iodine and 10 g of potassium iodide per 100 mL. It is a transparent deep brown, with the odor of iodine.[2,3]

### Solubility
Potassium iodide is very water soluble, having an aqueous solubility of about 1.4 g/mL, increasing to 2 g/mL in boiling water. In ethanol its solubility is about 43 to 45 mg/mL, increasing to 125 mg/mL in boiling alcohol. In glycerol the solubility of potassium iodide is about 500 mg/mL.[1,3]

Iodine readily dissolves in aqueous solutions of potassium iodide, forming potassium triiodide,[3] which is found in strong iodine solution (Lugol's solution).[2]

## pH

Aqueous solutions of potassium iodide are neutral to slightly alkaline, around pH 7 to 9.[1,3] Potassium iodide oral solution (SSKI, Lyne) had a measured pH of 6.6.[19]

## Osmolality

Potassium iodide saturated solution 1 g/mL (Upsher Smith) had an osmolality of 10,950 mOsm/kg.[233]

## General Stability Considerations

Potassium iodide tablets should be packaged in tight containers and stored at controlled room temperature. Potassium iodide oral solution should be packaged in tight, light-resistant containers, stored at controlled room temperature, and protected from temperatures of 40°C and above. Solutions also should be protected from freezing. Crystallization of potassium iodide from the solution may occur under normal conditions, especially if the solution is refrigerated. The crystals redissolve with warming and shaking of the solution.[2]

Oxidation of potassium iodide results in the formation of free iodine, turning solutions brownish-yellow. Discolored solutions should be discarded.[1,2]

## Stability Reports of Compounded Preparations
### Oral

David and Huyck[1161] reported on the ability of various vehicles to disguise the taste of potassium iodide solution. Glycyrrhiza syrup was the most effective, followed by aromatic elixir, cinnamon syrup, and then simple syrup. Glycyrrhiza syrup was also reported to have the best stability.

### Enteral
#### POTASSIUM IODIDE COMPATIBILITY SUMMARY

*Compatible with:* Ensure • Ensure HN • Ensure Plus • Ensure Plus HN • Osmolite • Osmolite HN • Vital

**STUDY 1:** Cutie et al.[19] added 1 and 2 mL of potassium iodide oral solution (SSKI, Lyne) to varying amounts (15 to 240 mL) of Ensure, Ensure Plus, and Osmolite (Ross Laboratories) with vigorous agitation to ensure thorough mixing. The potassium iodide oral solution was physically compatible, distributing uniformly in all three enteral products, with no phase separation and only slight particle growth at the mixture interface.

**STUDY 2:** Altman and Cutie[850] reported the physical compatibility of 1 and 2 mL of potassium iodide oral solution (SSKI, Lyne) with varying amounts (15 to 240 mL) of Ensure HN, Ensure Plus HN, Osmolite HN, and Vital after vigorous agitation to ensure thorough mixing. The potassium iodide oral solution was physically compatible, distributing uniformly in all four enteral products, with no phase separation and only slight particle growth at the mixture interface.

## Compatibility with Common Beverages and Foods

Sadrieh et al.[956] evaluated the stability and palatability of crushed potassium iodide tablets in common foods and beverages in this FDA-sponsored study of compounding counterterrorism drugs for pediatric patients. Potassium iodide proved to be stable in all of the media tested for seven days refrigerated, with at least 90% of the drug remaining. Potassium iodide palatability rankings from best to worst were raspberry syrup, low-fat chocolate milk, orange juice, flat Pepsi, low-fat milk, and water. ∎

# ▪ Potassium Perchlorate

## Properties

Potassium perchlorate is a white crystalline powder or colorless crystals having a slightly salty taste.[1,2]

### Solubility

Potassium perchlorate is sparingly soluble in water, having an aqueous solubility of around 15 mg/mL in cold water and 67 mg/mL in hot water.[1,3] It is practically insoluble in ethanol.[3]

### pH

A 0.1 M solution has a pH of 5 to 6.5.[4]

## General Stability Considerations

Potassium perchlorate decomposes explosively upon exposure to organic matter and oxidizable substances and upon concussion.[3,4] The oral capsules should be packaged in tight, light-resistant containers and stored at controlled room temperature protected from light, heat, and humidity.[3,4] Deformed or discolored capsules should be discarded.[2]

## Stability Reports of Compounded Preparations
### Oral

Williams[171] evaluated the stability of two oral potassium perchlorate solutions. An oral sodium perchlorate solution was prepared by dissolving 5 g of purified sodium perchlorate in 6 mL of water and bringing the mixture to 50 mL with 70% sorbitol, forming a 100-mg/mL final concentration. The solution was filtered through a 0.22-μm filter (Millex) and packaged in butyl rubber-stoppered vials.

Potassium perchlorate oral solution also was prepared by dissolving 6.7 g of dry potassium perchlorate in 500 mL of distilled water. Methylparaben 2 g and propylparaben 200 mg were dissolved in 8.33 mL of ethanol and added to the solution. The mixture was brought to 1000 mL with cherry-flavored syrup (simple syrup flavored with 7.8 mL/1000 mL of imitation cherry flavor). The final concentration of the potassium perchlorate oral solution was 6.7 mg/mL. The oral solution was packaged in ground-glass-stoppered vials.[171]

A gravimetric precipitation assay of tetra-*n*-pentylammonium perchlorate from tetra-*n*-pentylammonium bromide determined perchlorate content. Little or no loss was found after nine months of storage at room temperature (25°C) and under refrigeration (2 to 8°C). However, due to crystal formation, the potassium perchlorate formulations should not be refrigerated. Sodium perchlorate has a much greater aqueous solubility, and refrigeration of the sorbitol formulation is advisable because of the absence of preservatives.[171]

### Injection

Injections, like other sterile drugs, should be prepared in a suitable clean air environment using appropriate aseptic procedures. When prepared from nonsterile components, an appropriate and effective sterilization method must be employed.

Williams[171] also evaluated the stability of potassium perchlorate for intravenous administration prepared by drying the bulk powder at 110°C for one hour and then making a 13.3-mg/mL aqueous solution. The solution was filtered through a 0.22-μm filter (Millex) and packaged in sterile 30-mL glass vials with butyl rubber stoppers. The injection was stable for nine months at room temperature. ■

# ■ Povidone-Iodine

## Properties

Povidone-iodine is a complex of 9 to 12% iodine with povidone (polyvinylpyrrolidone). It is a yellowish-brown or reddish-brown amorphous powder having a characteristic odor.[1,3,4]

### Solubility

Povidone-iodine is soluble in water and in ethanol.[1,3]

### pH

Aqueous solutions of povidone-iodine have a pH near 2 but may be made less acidic by adding sodium bicarbonate.[1] Commercial topical solutions and ointment have a pH between 1.5 and 6.5.[4]

## General Stability Considerations

In general, povidone-iodine products should be packaged in tight containers and stored at controlled room temperature. Adjustment of solution pH with sodium bicarbonate reduces its stability.[1,3,4]

Maloney and O'Neill[1204] noted that povidone-iodine solution has been warmed to body temperature to minimize patient discomfort when applied. Consequently, the stability of commercial povidone-iodine solution stored at 37°C for six months was evaluated. Using the official analytical method, which is a volumetric estimation of available iodine, only about 2% loss occurred in six months when stored at 37°C.

## Stability Reports of Compounded Preparations
### Topical

**STUDY 1:** Dannenberg and Peebles[61] reported an explosive reaction from a compounded irrigation solution composed of an equal parts mixture of povidone-iodine solution (Betadine, Purdue Frederick) and hydrogen peroxide 3%. The mixture at first appeared to be satisfactory, but about 90 minutes after preparation the mixture exploded. The authors suspected that the iodine accelerated the evolution of oxygen from the decomposition of peroxide, resulting in the explosion in a container with a tightened cap.

**STUDY 2:** When 10% povidone-iodine (Betadine, Purdue Frederick) was mixed with denatured 70% alcohol (Lavacol, Parke-Davis), Chatterji and Hood[71] noted a loss of the characteristic yellow color over two or three days stored at room temperature. Assay for free iodine, the source of the antimicrobial action of Betadine, found no free iodine present. Mixing Betadine with 70% ethanol, absolute ethanol, 7% acetone, Lavacol, and isopropanol and testing for free iodine resulted in the loss of 50% free iodine in 72 hours and nearly all free iodine in 96 hours with both the Lavacol and the acetone solution. Small amounts of acetone and methyl isobutyl ketone are components of the Lavacol formulation. Less than 10% free iodine was lost with the other alcohols tested. The authors speculated that the loss of free iodine results from a reaction of iodine with methyl ketones (acetone, methyl isobutyl ketone) to form iodoform.

**STUDY 3:** In vitro and in vivo bactericidal activity of three concentrations of povidone-iodine solution was studied by Ghogawala and Furtado.[163] The three concentrations were commercial 10% povidone-iodine (Betadine, Purdue Frederick) and 2.5 and 1% dilutions. (Dilution of the povidone-iodine concentrate reportedly increased bactericidal activity in vitro due to increased free iodine content.[164,165]) In vitro, the 1% povidone-iodine dilution[163] killed an inoculum of *Staphylococcus aureus* in less than a minute, while the 10% concentrate required four minutes of contact. However, when the inoculum was applied to skin, the bactericidal properties of 10, 2.5, and 1% concentrations were all comparable. In the presence of blood as a competing substrate, the 10% concentration was superior.

**STUDY 4:** Pesko[254] reported on an extemporaneous povidone-iodine paste. The paste was prepared by combining sucrose 800 g with povidone-iodine liquid 80 mL with thorough mixing to avoid clumping. Povidone-iodine ointment 200 g was then incorporated to form the final paste. Although specific stability information does not exist, the author recommended a limited expiration period for the preparation.

### Irrigation

Irrigations, like other sterile drugs, should be prepared in a suitable clean air environment using appropriate aseptic procedures. When prepared from nonsterile components, an appropriate and effective sterilization method must be employed.

Vervloet[1479] evaluated the shelf life of povidone-iodine 2% sterile solution for irrigation packaged in polyvinyl chloride (PVC) plastic bags and also polyethylene (PE) plastic bags. The test irrigation solution was prepared using the following formula:

Povidone-iodine 10% solution: 2,000 mL
Sodium chloride: 90 g
Water for injection: qs 10,000 mL

The solution was passed through a 0.2-μm sterilizing filter (presumably with a large size sufficient for the volume being sterilized) and was packaged in PVC plastic bags or in PE plastic bags. Test samples were stored at unspecified room temperature and refrigerated at 5°C. The authors stated that the povidone-iodine 2% irrigation solutions exhibited over 10% loss in three months when stored at room temperature but adequate stability for a 12-month shelf life when refrigerated. The use of PE bags added no additional stability.

### Ophthalmic

Ophthalmic preparations, like other sterile drugs, should be prepared in a suitable clean air environment using appropriate aseptic procedures. When prepared from nonsterile components, an appropriate and effective sterilization method must be employed.

**STUDY 1:** Takekuma et al.[1695] evaluated the stability of a compounded povidone-iodine 0.625% solution used to disinfect the conjunctival sac in eye surgery. The sample solutions were stored at room temperature of 25°C and refrigerated at 4°C exposed to diffuse light and protected from light for five weeks. No visible changes in any of the samples were observed and the pH changes were minimal. The amount of available iodine after five weeks of storage was 91% at 25°C and 98% at 4°C. The authors concluded that the 0.625% povidone-iodine eye wash was stable for five weeks when stored under refrigeration.

**STUDY 2:** Najafi et al.[1704] reported on a formulation of povidone-iodine 2.5% for use in treating ophthalmia neonatorum. The ophthalmic solution was adjusted to near pH 5 (about the pH of tears) using sodium hydroxide 0.1 N and citric acid 0.5%, which the authors stated also increased stability. The solution was sterilized and packaged in amber bottles. At this concentration, the solution was reported to act effectively against microbial contamination. Treatment of 475 neonates was reported to yield good clinical results against ophthalmia neonatorum. ■

## ■ Pranoprofen

### Properties

Pranoprofen occurs as a white to pale yellowish white crystalline powder.[1136]

### Solubility

Pranoprofen is practically insoluble in water, slightly soluble in ethanol, soluble in acetic acid, and freely soluble in N,N-dimethylformamide.[1136]

### General Stability Considerations

Pranoprofen should be packaged in tight containers and stored at controlled room temperature and protected from exposure to light.[1136]

### Stability Reports of Compounded Preparations
#### Ophthalmic

Ophthalmic preparations, like other sterile drugs, should be prepared in a suitable clean air environment using appropriate aseptic procedures. When prepared from nonsterile components, an appropriate and effective sterilization method must be employed.

Garcia-Valldecabres et al.[880] reported the change in pH of several commercial ophthalmic solutions including pranoprofen 1 mg/mL (Oftalar, Cusi) over 30 days after opening. Slight variation in solution pH was found, but no consistent trend and no substantial change occurred. ■

# Pravastatin Sodium

## Properties
Pravastatin sodium occurs as a white, off-white, or yellowish-white odorless, fine or crystalline hygroscopic powder.[1,3,4,7]

### Solubility
Pravastatin sodium is freely soluble in water to over 300 mg/mL, and soluble in methanol and dehydrated ethanol. It is slightly soluble in isopropanol, and practically insoluble in acetone, ethyl acetate, and ether.[1,3,4,7]

### pH
Pravastatin sodium in a 5% aqueous solution exhibits a pH of 7.2 to 9.[3,4]

### pKₐ
Pravastatin has a p$K_a$ of 4.7.[1305]

## General Stability Considerations
Pravastatin sodium powder should be packaged in tight containers and stored as indicated on the label. According to USP, this may include storage at controlled room temperature or storage under nitrogen atmosphere under refrigeration.[3,4]

Pravastatin sodium oral tablets should be packaged in tight containers and stored at controlled room temperature protected from exposure to moisture and light.[3,4,7]

## Stability Reports of Compounded Preparations
### Enteral
Yano et al.[1306] evaluated pravastatin sodium as a simple aqueous suspension prepared from commercial tablets. Pravastatin sodium tablets were placed in water at 55°C and allowed to disintegrate. The innovator drug tablet formulation disintegrated and could be suspended in the hot water within 10 minutes. However, two generic tablet formulations failed to completely disintegrate unless crushed prior to being added to the hot water. With crushing the tablets, the aqueous suspensions of the two generic formulations were found to result in no problems in tube administration. Both innovator and generic tablet formulations resulted in drug recovery of nearly 100% after tube passage. ■

# Prednisolone
# Prednisolone Acetate
# Prednisolone Sodium Phosphate

## Properties
Prednisolone is a white or nearly white odorless hygroscopic crystalline powder.[2,3] It may be anhydrous or it may contain 1.5 molecules of water of hydration.[2]

Prednisolone acetate occurs as a white to practically white odorless crystalline powder.[1,3,4]

Prednisolone sodium phosphate, the disodium salt of the phosphate ester of prednisolone, is a white or slightly yellow hygroscopic crystalline powder.[1]

### Solubility
Prednisolone is very slightly soluble in water[1-3] (about 1 in 1300).[759] In ethanol the solubility is about 33 mg/mL.[1,3]

Prednisolone acetate is practically insoluble in water and slightly soluble in acetone and ethanol.[3,4]

Prednisolone sodium phosphate is soluble in ethanol[1] and soluble to about 250 mg/mL[4] to 333 mg/mL[691] in water.

### pH
Prednisolone syrup has a pH between 3 and 4.5.[4] A 1% aqueous solution of prednisolone sodium phosphate has a pH of 7.5 to 8.5,[1] and a 5% aqueous solution has a pH of 7.5 to 9.[3] The United States Pharmacopeia indicates that the pH of a 1% aqueous solution is between 7.5 and 10.5.[4]

Prednisolone acetate injectable suspension has a pH of 5 to 7.5. Prednisolone acetate ophthalmic suspension has a pH in the range of 5 to 6.[4]

## General Stability Considerations
Prednisolone is subject to decomposition in alkali or when exposed to ultraviolet light and oxygen.[759] In general, prednisolone products should be stored at controlled room temperature and protected from temperatures of 40°C or more. Liquid dosage forms should be protected from freezing as well. The oral solution should be packaged in tight, light-resistant containers. Solutions and suspensions of prednisolone are heat-labile and should not be autoclaved for sterilization.[2-4] In components of topical formulations, prednisolone acetate may be more stable than prednisolone.[1548] The optimum pH for prednisolone stability in aqueous solutions has been stated to be pH 6.[1684]

## Stability Reports of Compounded Preparations

### Injection

Injections, like other sterile drugs, should be prepared in a suitable clean air environment using appropriate aseptic procedures. When prepared from nonsterile components, an appropriate and effective sterilization method must be employed.

Elema et al.[755] evaluated the stability of prednisolone disodium phosphate injection after autoclave sterilization at 121°C for 20 minutes. Unacceptable decomposition occurred, resulting in free prednisolone and other decomposition products that may result in precipitation. Adjusting the injection pH from 7.3 to 8 resulted in a reduction in free prednisolone to one-third of the previous amount but also resulted in a similar amount of other decomposition products. The authors did not believe that autoclaving of this preparation was an acceptable sterilization technique.

### Oral

**STUDY 1 (PREDNISOLONE):** Gupta[229] reported the stability of prednisolone in various vehicles. Prednisolone powder 500 mg was dissolved in ethanol 100 mL. A 10-mL aliquot then was diluted to 100 mL with water, citrate buffer (pH 4.4), glycerin 50% (vol/vol), sorbitol 50%, or sucrose 50% to yield a 0.5-mg/mL prednisolone concentration. Prednisolone content was determined using a specific HPLC analysis. No crystallization was observed in any vehicle tested. HPLC analysis found little or no loss in 92 days (temperature unspecified) in most vehicles; in water, a 3.5% loss occurred, while the loss in sucrose 50% was 7%. No loss was found in the glycerin 50% and sorbitol 50% solutions. However, in the citrate (pH 4.4) buffer, prednisolone loss was 16% in 30 days and 37% in 92 days.

**STUDY 2 (PREDNISOLONE):** Sullivan[759] reported the stability of prednisolone in a hydroalcoholic oral solution with the formula shown in Table 91. The solution was packaged in amber glass bottles and stored for 52 weeks at room temperature of 22 to 24°C exposed to fluorescent light for 52 weeks. Stability-indicating HPLC analysis found that no loss of prednisolone occurred.

**STUDY 3 (PREDNISOLONE SODIUM PHOSPHATE):** Sullivan and Hobson[691] reported the stability of prednisolone sodium phosphate 10-mg/mL oral solution prepared from bulk

**TABLE 91.** Formula for Hydroalcoholic Prednisolone Oral Solution[759]

| Component | Amount |
| --- | --- |
| Prednisolone | 500 mg |
| Ethanol (100%) | 27 mL |
| Propylene glycol | 20 mL |
| Sucrose | 20 g |
| Orange tincture | 2.4 mL |
| Water for irrigation | qs 100 mL |

**TABLE 92.** Formula for Prednisolone Sodium Phosphate Oral Solution[691]

| Component | Amount |
| --- | --- |
| Prednisolone sodium phosphate | 1000 mg |
| Methylparaben | 65 mg |
| Propylparaben | 35 mg |
| Orange syrup | 40 mL |
| Water for injection | qs 100 mL |

powder (Roussel Uclaf). The oral solution was prepared using the formula shown in Table 92, packaged in 100-mL amber glass bottles, and stored at room temperature of 22 to 24°C for nine months. No change in solution clarity was observed, and stability-indicating HPLC analysis found no loss of drug occurred during the nine-month storage period. It was also reported to have better patient acceptance than more bitter prednisolone oral formulations.

### Ophthalmic

Ophthalmic preparations, like other sterile drugs, should be prepared in a suitable clean air environment using appropriate aseptic procedures. When prepared from nonsterile components, an appropriate and effective sterilization method must be employed.

Akram et al.[1646] developed and tested the stability of a prednisolone acetate 1% ophthalmic suspension prepared from powder. The formula also contained edetate disodium, boric acid and sodium borate buffer, sodium chloride for isotonicity, polysorbate 80, hydroxypropyl methylcellulose (HPMC) 4000 cps, and hydrochloride acid and/or sodium hydroxide to adjust to pH 5 to 6 in purified water.

To reduce the particle size, prednisolone acetate was dissolved in dimethyl sulfoxide (DMSO) and passed through 0.2-μm filters to sterilize the solution. Polyvinyl pyrrolidone (PVP) K-30 1% wt/vol was passed through a 0.2-μm filter into the sterile drug in DMSO solution drop wise. The resulting mixture was passed through another 0.2-μm filter, and the filtrate (the active drug) was washed with sterilized purified water.

The HPMC was slowly dispersed in the appropriate amount of purified water and hydrated. Separately, the edetate disodium, boric acid, benzalkonium chloride, sodium chloride, sodium borate, and polysorbate 80 were dissolved in water for injection. To this solution the hydrated HPMC solution was added, and the pH was adjusted to pH 5 to 6. The solution was passed through a 0.2-μm filter into a calibrated vessel. Under aseptic conditions, the active drug filtrate was transferred into the solution and the mixture was adjusted to volume and thoroughly mixed. The osmolality was near 307 mOsm/kg and had a pH of 5.62. Particle size was between 1 and 3 μm. The prednisolone acetate concentration was 10.4 mg/mL. The final ophthalmic suspension was filled into sterile low-density polyethylene vials under aseptic conditions.

A stability study was conducted on samples stored at 40°C for six months. Based on the decomposition rate of prednisolone acetate, the shelf life of the ophthalmic formulation was calculated to be three years at room temperature.

### Inhalation

Inhalation preparations, like other sterile drugs, should be prepared in a suitable clean air environment using appropriate aseptic procedures. When prepared from nonsterile components, an appropriate and effective sterilization method must be employed.

Mizushima et al.[1677] evaluated the stability of an inhalation solution containing epinephrine and prednisolone succinate. The solution was packaged in 50-mL sterile bottles and was stored for four weeks at room temperature of 23 to 27°C exposed to room light, at room temperature protected from exposure to light using a brown light-resistant bag, and in a cool dark place at 5 to 10°C protected from exposure to light using a brown light-resistant bag.

The solutions stored at room temperature were markedly discolored, whether exposed to or protected from light. HPLC analysis found loss of both drugs and formation of the decomposition products adrenochrome and free prednisolone. Stored in a cool dark place, there was no apparent discoloration, and losses of epinephrine and prednisolone succinate were 14 and 8%, respectively.

### Topical

**STUDY 1:** Gander et al.[1225] evaluated the stability of prednisolone and prednisolone acetate in hydrophilic and lipophilic topical bases with zinc oxide. The hydrophilic paste contained prednisolone or prednisolone acetate 0.5% with zinc oxide 40%, propylene glycol 15%, and aqueous 3.5% hydroxypropylcellulose gel 45%. The lipophilic ointment contained prednisolone or prednisolone acetate 0.5% with zinc oxide 5%, white beeswax 10%, hard paraffin 10%, and almond oil 75%. No metallic utensils were utilized during compounding. Samples were stored at 20 and 37°C protected from exposure to light.

Stability-indicating HPLC analysis found that prednisolone in the hydrophilic base underwent 23% degradation in four weeks at 37°C. No loss occurred in a similar formulation that did not contain zinc oxide, indicating the zinc oxide was responsible for the loss. Also, no statistically significant prednisolone loss occurred at 37°C in the lipophilic formulation, even with the zinc oxide present.

Prednisolone acetate samples were more stable. In the hydrophilic formulation, prednisolone acetate exhibited about 10% loss in 104 days at 37°C and in 136 days at 20°C. In addition, prednisolone acetate in the lipophilic formulation incurred no loss in six months at either 20 or 37°C.

**STUDY 2:** Christen et al.[1548] evaluated the stability of prednisolone and prednisolone acetate in a number of components used in topical formulations stored for 28 days at 37°C protected from exposure to light. HPLC analysis found that prednisolone acetate tended to be more stable than prednisolone. However, prednisolone was very stable in the lipophilic components almond oil and beeswax with no decomposition found after 28 days. Both steroid forms were least stable in the amphiphilic bases Cremophor RH 40 and Lanette. Among the hydrogels, hydroxypropylcellulose 3.5% (Klucel MF) and in Carbopol 934P 0.4% neutralized to pH 7.1 resulted in the best stability while much greater drug loss of about 30% occurred in Bentonite (Veegum) 10%.

### Rectal

Akitoshi et al.[1260] compared the drug release from prednisolone suppositories prepared from pulverized tablets to suppositories compounded from prednisolone powder. The suppositories were prepared using the fusion method with the drug source mixed in Witepsol H-15 and Witepsol E-75 in a ratio of 76:24, respectively. Drug release was evaluated by reciprocating dialysis tube method with tapping and the dialysis tube method using a suppository dissolution apparatus or a paddle apparatus. The drug release was found to be the same by both methods and was similar using both sources of prednisolone.

## Compatibility with Other Drugs

Ohishi et al.[1714] evaluated the stability of Lidomex (prednisolone valeroacetate) ointment mixed in equal quantity (1:1) with zinc oxide ointment. Samples were stored refrigerated at 5°C. HPLC analysis of the ointment mixtures found that the drugs were stable for 32 weeks at 5°C. ▪

# ▪ Prednisone

## Properties

Prednisone is a white or nearly white odorless crystalline powder.[2,3]

### Solubility

Prednisone is very slightly soluble in water. In ethanol, it has a solubility of about 6.7 mg/mL.[1–3]

### pH

Prednisone oral solution has a pH between 2.6 and 4. The syrup has a pH between 3 and 4.5. The injectable suspension has a pH of 3 to 7.[4]

## General Stability Considerations

Prednisone tablets should be packaged in well-closed containers, stored at controlled room temperature, and protected from temperatures of 40°C or above. The oral solutions and syrup should be packaged in tight containers; the solutions should be stored at controlled room temperature.[2,4]

## Stability Reports of Compounded Preparations
### Oral

**STUDY 1:** Gupta et al.[62] reported the stability of prednisone prepared as a pediatric oral liquid dosage form. Prednisone powder (Upjohn) was dissolved in ethanol and brought to volume with simple syrup to yield a 0.5-mg/mL concentration with ethanol 10% (vol/vol). Sodium benzoate 0.1% was added as a preservative. The preparation was packaged in amber bottles and stored at 24°C. Stability-indicating HPLC analysis found that 90% of the prednisone remained after 84 days of storage.

**STUDY 2:** Gupta[229] also reported the stability of prednisone in various vehicles. Prednisone powder 500 mg was dissolved in ethanol 100 mL. A 10-mL aliquot then was diluted to 100 mL with water, citrate buffer (pH 4.4), glycerin 50% (vol/vol), sorbitol 50%, or sucrose 50% to yield a 0.5-mg/mL prednisone concentration. Prednisone content was determined using a specific HPLC analysis. The poor solubility of prednisone resulted in partial crystallization in all vehicles except glycerin 50%, often within two days. Losses due to precipitation were 20 to 40%. In glycerin 50%, prednisone remained in solution; HPLC analysis found no loss in 92 days (temperature unspecified). In addition to precipitation in the presence of sucrose, some prednisone was reduced to prednisolone.

**STUDY 3:** Raitt and Hotaling[226] reported the stability of an oral 10-mg/mL prednisone suspension. The suspension was prepared in a multistep process utilizing the formula in Table 93. The suspending agent was prepared by dissolving 4 g of sodium benzoate in 2000 mL of purified water and then adding 120 g each of tragacanth and acacia, in that order, to prevent clumping; this mixture was shaken vigorously while additional purified water was added until a smooth suspension was formed. Mechanical agitation was suggested because it may take 24 to 48 hours to complete the process. Anise oil 1 mL was added to mask the unpleasant odor, and the suspension was stored in the refrigerator for up to 120 days. The flavoring agent was prepared by combining (2:1) simple syrup preserved with 0.1% sodium benzoate with commercial cherry or wild cherry syrup.

To prepare the prednisone suspension, sodium benzoate 200 mg was dissolved in 200 mL of purified water and 1000 prednisone tablets 5 mg were added. The mixture

**TABLE 93.** Prednisone 10-mg/mL Oral Suspension Formula[226]

| Component | | Amount |
|---|---|---|
| Prednisone (tablets or powder) | | 5000 mg |
| Sodium benzoate solution 0.1% | | 200 mL |
| Tragacanth–acacia suspending mixture | | 100 mL |
|    Tragacanth 3% | | |
|    Acacia 3% | | |
|    Sodium benzoate 0.1% | | |
|    Anise oil 0.025% | | |
| Flavoring mixture | qs | 500 mL |
|    Simple syrup 67% | | |
|    Cherry syrup 33% | | |

was stirred with a magnetic stirrer until complete dissolution occurred; hand stirring worked, but was slow. The tragacanth–acacia suspending agent 100 mL was then added and the mixture was stirred until homogeneous. Finally, the flavoring agent was added to bring the volume to 500 mL, resulting in a prednisone concentration of 10 mg/mL. The suspension was stored in the refrigerator and was free of caking and coalescence, remaining easy to redisperse upon agitation. Analysis by an independent laboratory (method unspecified) found the prednisone content to be unchanged after two months of refrigerated storage.

**STUDY 4:** Dupuis et al.[660] evaluated the palatability and bioavailability of a 5-mg/mL prednisone oral suspension compounded from tablets. In the palatability evaluation, tutti-frutti flavoring was found to be the most palatable compared to chocolate–cherry, banana, wild cherry–anise, and banana–chocolate. For the suspension, 250 prednisone 5-mg tablets (Upjohn) were crushed to fine powder using a mortar and pestle. The powder was levigated with 10 mL of water for irrigation to make a smooth paste. About 200 mL of simple syrup was incorporated and well mixed. Tutti-frutti flavoring 3.5 mL was added and mixed well. The oral suspension was brought to a volume of 250 mL with simple syrup and mixed well. A 60-day shelf life was used, although no stability information was presented. Bioavailability and pharmacokinetic testing in volunteers found no difference between the extemporaneous oral suspension and the oral tablets.

### Enteral

Ortega de la Cruz et al.[1101] reported the physical compatibility of an unspecified amount of oral liquid prednisolone steaglate (Estilsone, Em) with 200 mL of Precitene (Novartis) enteral nutrition diet for a 24-hour observation period. No particle growth or phase separation was observed. ■

# Primaquine Phosphate

## Properties

Primaquine phosphate is an orange-red or yellow odorless crystalline powder or crystals having a bitter taste.[1-3] Primaquine phosphate 13.2 mg is approximately equivalent to primaquine 7.5 mg.[3]

### Solubility

Primaquine phosphate is soluble in water, with an aqueous solubility of around 62.5 mg/mL.[1-3] It is practically insoluble in ethanol.[3]

## General Stability Considerations

Primaquine phosphate tablets should be packaged in well-closed, light-resistant containers and stored at controlled room temperature protected from temperatures of 40°C or more.[2,4]

## Stability Reports of Compounded Preparations
### Oral

The extemporaneous formulation of a primaquine phosphate 6-mg (base)/5-mL oral suspension was described by Nahata and Hipple.[160] Ten primaquine phosphate 15-mg (base) tablets were crushed, mixed thoroughly with 10 mL of 1.5% carboxymethylcellulose, and brought to 125 mL with simple syrup. A stability period of seven days under refrigeration was used, although chemical stability testing was not performed. ■

# Procainamide Hydrochloride

## Properties

Procainamide hydrochloride is a white to tan hygroscopic odorless crystalline powder.[2,3]

### Solubility

Procainamide hydrochloride is very soluble in water, with an aqueous solubility of about 4 g/mL. In ethanol, it has a solubility of about 500 mg/mL.[2,3]

### pH

A 10% procainamide hydrochloride aqueous solution has a pH of 5 to 6.5.[3] Procainamide hydrochloride injection has a pH between 4 and 6.[4]

### pKa

The drug has a $pK_a$ of 9.23.[2]

### Osmolality

Procainamide hydrochloride injection 500 mg/mL has an osmolality of greater than 2000 mOsm/kg.[326]

## General Stability Considerations

In general, procainamide hydrochloride oral products should be packaged in tight containers and stored at controlled room temperature protected from temperatures of 40°C or more. The injection develops a yellow discoloration upon standing; exposure to air causes darkening of the solution. Although discoloration may occur without significant loss of drug concentration, it is recommended that solutions darker than light amber be discarded. Storage of the injection under refrigeration retards oxidation and color development, although storage between 10 and 27°C is considered acceptable.[2,4]

## Stability Reports of Compounded Preparations
### Oral

**STUDY 1:** The stability of procainamide hydrochloride 5, 50, and 100 mg/mL as extemporaneously prepared oral liquids was described by Metras et al.[48] The contents of procainamide hydrochloride 500-mg capsules (Goldline) were triturated to a paste by slowly adding sterile water for irrigation. Cherry syrup or a 1% methylcellulose–cherry syrup mixture (70:30) was added in three increments to bring to volume. The pH values were left unadjusted (pH 6) or were adjusted to 5. The oral liquids were packaged in amber glass bottles and stored at 5 and 25°C for up to 180 days. The 5- and 50-mg/mL concentrations were more stable than the 100-mg/mL concentration. At all three concentrations, procainamide hydrochloride in cherry syrup with the unadjusted pH of 6 was stable for six months stored at 5°C. At room temperature storage, unacceptable losses occurred after 34 days at 5 and 50 mg/mL, but they occurred in as little as seven days in the 100-mg/mL concentration. Procainamide hydrochloride in the other combinations was less stable than in the cherry syrup without pH adjustment.

**STUDY 2:** Swenson[114] reported on the instability of an extemporaneous procainamide hydrochloride syrup. The compounded oral liquid was prepared in imitation wild cherry syrup (Humco) at a theoretical procainamide hydrochloride concentration of 50 mg/mL. Lack of appropriate procainamide blood levels in a pediatric patient led to analysis of the oral liquid, which was found to contain less than 3 mg/mL of procainamide hydrochloride. In evaluating the extensive drug loss, it was found that the syrup has a pH of 3. This low pH resulted in acid-catalyzed hydrolysis of procainamide. To prevent this loss, the pH should be adjusted up to the stability range, pH 4 to 6.

**STUDY 3:** Alexander et al.[118] reported the stability of two oral liquid formulations of procainamide hydrochloride. The first formulation was similar to a standard sucrose-based preparation described in published information.[119] The other formulation was based on a maltitol syrup vehicle.

The sucrose-based preparation was made by stirring the contents of 100 procainamide hydrochloride 500-mg capsules (Danbury Pharmacal) with 330 mL of distilled water. Simple syrup was added geometrically and was levigated until uniform. Paraben concentrate 10 mL (methylparaben 10% and propylparaben 2% in propylene glycol) and cherry flavoring 1 mL were added and mixed well. Simple syrup was used to bring the mixture to volume, yielding a theoretical concentration of 50 mg/mL. The syrup was packaged in 150-mL amber glass bottles and was stored at 5°C and at elevated temperatures from 40 to 70°C.[118]

The maltitol-based preparation was made by mixing the contents of 100 procainamide hydrochloride 500-mg capsules with 200 mL of distilled water. Lycasin (a syrup composed of maltitol 75%) 750 g was added slowly with continuous mixing. Paraben concentrate as in the sucrose-based syrup was added. Sodium bisulfite 1 g, saccharin sodium 1 g, and sodium acetate 2.73 g in 50 mL of distilled water were added to the procainamide mixture and homogenized. Pineapple and apricot flavorings (1 mL each) and 10 mg FD&C yellow no. 6 in 10 mL of distilled water were incorporated, and the mixture was brought to volume with distilled water, yielding a theoretical concentration of 50 mg/mL. The pH was adjusted to 5 with glacial acetic acid. The syrup was packaged in 100-mL portions in amber glass bottles and stored at 5°C and at elevated temperatures from 40 to 70°C.[118]

The procainamide hydrochloride concentration was assessed using a stability-indicating HPLC assay. The sucrose-based formulation had a calculated shelf life at 25°C of 456 days, while the maltitol-based syrup had a calculated shelf life of 127 days. The sucrose formulation stored at 5°C developed crystals in the bottom of the bottle, but the maltitol formulation was satisfactory, retaining 98% of the initial concentration after 180 days at 5°C. Both formulations were initially yellow but turned brown during elevated-temperature storage. Procainamide hydrochloride is extremely bitter; the authors stated that the maltitol syrup with pineapple–apricot flavoring had a much more acceptable flavor than the cherry-flavored simple syrup.[118]

**STUDY 4:** Allen and Erickson[527] evaluated the stability of three procainamide hydrochloride 50-mg/mL oral suspensions extemporaneously compounded from capsules. Vehicles used in this study were (1) an equal parts mixture of Ora-Sweet and Ora-Plus (Paddock), (2) an equal parts mixture of Ora-Sweet SF and Ora-Plus (Paddock), and (3) cherry syrup (Robinson Laboratories) mixed 1:4 with simple syrup. Twenty-four procainamide hydrochloride 250-mg capsules (Rugby) were emptied into a mortar and pestle and comminuted to fine powder. About 20 mL of the test vehicle was added to the powder and mixed to yield a uniform paste. Additional vehicle was added geometrically and brought to the final volume of 120 mL, with thorough mixing after each addition. The process was repeated for each of the three test suspension vehicles. Samples of each finished suspension were packaged in 120-mL clear polyethylene terephthalate plastic prescription bottles and stored at 5 and 25°C protected from light.

No visual changes or changes in odor were detected during the study. Stability-indicating HPLC analysis found not more than 7% procainamide hydrochloride loss in any of the suspensions stored at either temperature after 60 days of storage. ■

# ■ Procarbazine Hydrochloride

## Properties

Procarbazine hydrochloride is a white to pale yellow crystalline powder having a slight odor[2] and a bitter taste.[586] Procarbazine hydrochloride has been characterized as a carcinogen.[1]

### Solubility

Procarbazine hydrochloride is freely soluble in water and sparingly soluble in ethanol.[2]

### $pK_a$

The drug has a $pK_a$ of 6.8.[2]

## General Stability Considerations

Procarbazine hydrochloride capsules should be packaged in tight, light-resistant containers and stored at controlled room temperature protected from temperatures of 40°C and higher. The drug is unstable in aqueous solution.[2]

## Stability Reports of Compounded Preparations
### Oral

**STUDY 1:** The extemporaneous formulation of a procarbazine hydrochloride 5-mg/100-mg oral powder was described by Nahata and Hipple.[160] The contents of one procarbazine hydrochloride 50-mg capsule was triturated and brought to a total mass of 1 g with thorough mixing of small amounts of lactose. A stability period of six months at room temperature was used, although chemical stability testing was not performed.

**STUDY 2:** Sigma-Tau Pharmaceuticals has reported on the stability of an oral suspension formulation of procarbazine prepared from capsules. The contents of 10 procarbazine capsules

were levigated in a mortar with 2 mL of glycerin to make a paste. Strawberry syrup 10 mL was added geometrically and levigated to create a uniform mixture, which was transferred to an amber glass prescription bottle. The mixture was brought to a volume of 50 mL by repeatedly rinsing the mortar and pestle with small amounts of strawberry syrup and adding the rinsings to the prescription bottle. The suspension was stated to be stable for seven days at room temperature protected from exposure to light.[933] ■

# ■ Prochlorperazine
# Prochlorperazine Edisylate
# Prochlorperazine Maleate

## Properties

Prochlorperazine is a clear pale yellow viscous liquid. Prochlorperazine edisylate and maleate are white to light yellow odorless or nearly odorless crystalline powders. Prochlorperazine edisylate 7.5 mg and prochlorperazine maleate 8 mg are approximately equivalent to 5 mg of prochlorperazine.[2,3]

### Solubility

Prochlorperazine is very slightly soluble in water and is freely soluble in ethanol.[2,3]

Prochlorperazine edisylate has an aqueous solubility of about 500 mg/mL and a solubility in ethanol of 0.67 mg/mL at 25°C.[2,3]

Prochlorperazine maleate is practically insoluble in water. In ethanol it has a solubility of about 0.83 mg/mL at 25°C.[1–3]

### pH

Prochlorperazine edisylate injection has a pH between 4.2 and 6.2.[2,4] Prochlorperazine oral solution has a pH of 4.5 to 5.[2]

## General Stability Considerations

In general, prochlorperazine products should be packaged in tight, light-resistant containers, stored at controlled room temperature, and protected from temperatures of 40°C and higher. Liquid products such as the injection and the oral solution should be protected from freezing as well. Prochlorperazine suppositories should be stored at temperatures less than 37°C.[2,4]

## Stability Reports of Compounded Preparations
### Oral

Glass et al.[1594] evaluated the stability of commercial prochlorperazine tablets (Stemetil, Sanofi-Aventis) repackaged in dose administration aids (Webster-pak) over eight weeks

stored at 24 to 26°C and 58.5 to 61.5% relative humidity, both protected from light and also exposed to tungsten, fluorescent, and indirect daylight. The tablets were also tested in simulated home use. In all cases, the tablets were physically stable; weight uniformity, friability, hardness, disintegration, and dissolution were all satisfactory. However, an unacceptable gray to beige discoloration was observed in the tablets exposed to light sources, but this discoloration was not indicative of excessive drug decomposition. In tablets under all test conditions, the prochlorperazine content at all time points remained within the British Pharmacopeial range of 92.5 to 107.5%. The authors concluded that prochlorperazine tablets repackaged in dose administration aids were chemically and physically stable for at least eight weeks. However, because of the unacceptable tablet discoloration that occurred in tablets exposed to light, they recommended that prochlorperazine tablets be stored in a cool dark place and that unnecessary exposure to light, heat, and humidity be avoided.

### Enteral

**PROCHLORPERAZINE COMPATIBILITY SUMMARY**
*Compatible with:* Enrich • TwoCal HN • Vivonex T.E.N.

Burns et al.[739] reported the physical compatibility of prochlorperazine edisylate (Compazine) 10 mg/10 mL with 10 mL of three enteral formulas, including Enrich, TwoCal HN, and Vivonex T.E.N. Visual inspection found no physical incompatibility with any of the enteral formulas.

## Compatibility with Common Beverages and Foods

Hirsch[244] reported that the addition of prochlorperazine liquid products to coffee or tea resulted in the formation of an insoluble precipitate. ■

# Progesterone

## Properties

Progesterone is a natural progestin that is a white, creamy white, or yellowish-white odorless crystalline powder or colorless crystals.[2,3] Progesterone has been characterized as a carcinogen.[1] Progesterone melts at temperatures in the range of 126 to 131°C; the polymorph melts at 121°C.[615]

### Solubility

Progesterone is practically insoluble in water but has a solubility in ethanol of about 125 mg/mL. It also is sparingly soluble in vegetable oils.[2,3]

## General Stability Considerations

Although progesterone is stable upon exposure to air, progesterone commercial products are packaged in tight, light-resistant containers.[2,4] The injection should be stored at controlled room temperature and protected from temperatures of 40°C or more and from freezing.[2]

## Stability Reports of Compounded Preparations
### Oral

Allen[615] reported on an oral formulation of triple hormones in oil-filled capsules. The formula for 100 capsules is shown in Table 94. The powders were accurately weighed and mixed in a mortar. Drops of safflower oil were added to wet the powder thoroughly and form a paste. The remaining safflower oil was added with continued mixing. The mixture was chilled in a refrigerator for about one hour to increase the viscosity and reduce powder settling. A hand-operated capsule machine was loaded with 100 No. 1 empty hard-gelatin capsules, and the caps were removed. Then 0.3 mL of the triple hormones in oil was transferred by micropipette into the base of each capsule. For clear capsules, additional safflower oil may be added to fill the capsules more completely. The capsule caps were then replaced and locked into place securely. Packaged in tight, light-resistant containers and stored under refrigeration, the capsules were physically stable for at least 60 days.

**TABLE 94.** Formula of Triple Hormones in Oil-Filled Capsules[615]

| Component | | Amount |
|---|---|---|
| Progesterone | | 4 g |
| Estradiol | | 50 mg |
| Testosterone | | 50 mg |
| Safflower oil | qs | 30 mL |

Note: Fill 0.3 mL per capsule.

### Vaginal Suppositories
**USP OFFICIAL FORMULATION (FATTY ACID BASE SUPPOSITORIES):**

| | |
|---|---|
| Progesterone (micronized) | 25 to 600 mg |
| Fatty acid base | qs |

The suppository molds should be calibrated with fatty acid base and the amount of base adjusted according to the calibration. Melt the fatty acid base slowly and evenly. Slowly add the progesterone powder to the melted base with stirring. After thorough mixing, the melt should be poured into the calibrated molds and cooled in a refrigerator. After solidifying, the suppositories should be trimmed, wrapped, and packaged in well-closed, light-resistant containers. Store the finished suppositories under refrigeration between 2 and 8°C. The beyond-use date is 90 days from the date of compounding.[4,5]

**USP OFFICIAL FORMULATION (POLYETHYLENE GLYCOL BASE SUPPOSITORIES):**

| | |
|---|---|
| Progesterone (micronized) | 25 to 600 mg |
| Polyethylene glycol base | qs |

The suppository molds should be calibrated with polyethylene glycol base and the amount of base adjusted according to the calibration. Melt the polyethylene glycol base slowly and evenly. Slowly add the progesterone powder to the melted base with stirring. After thorough mixing, the melt should be poured into the calibrated molds and cooled in a refrigerator. After solidifying, the suppositories should be trimmed, wrapped, and packaged in well-closed, light-resistant containers. Do not use polystyrene containers. Store the finished suppositories under refrigeration between 2 and 8°C. The beyond-use date is 90 days from the date of compounding.[4,5]

**STUDY 1:** Fulper at al[104] evaluated the effect of various storage temperatures on the liquefaction times of progesterone 25-mg suppositories compounded from three fatty bases, Fattibase (Paddock), Witepsol H-15 (Riches-Nelson), and cocoa butter (Woltra). The suppositories were prepared by fusion and were packaged in amber plastic prescription vials. The suppositories were cured at 25°C for two days and then stored at 10, 25, and 30°C. Liquefaction time of samples at 37°C was determined visually. There was little difference with or without the progesterone. However, an increase in storage temperature during the 13-week storage greatly increased the liquefaction time. Furthermore, the suppositories prepared with Witepsol H-15 had substantially longer liquefaction times than the other two bases (Table 95). The authors indicated that if liquefaction time is important, the suppositories should be made with Fattibase or cocoa butter and stored under refrigeration.

**TABLE 95.** Liquefaction Times (Minutes) of Progesterone 25-mg Suppositories in Three Bases after Storage for 13 Weeks[104]

| Base | Baseline[a] | 10°C | 25°C | 30°C |
|------|-------------|------|------|------|
| Fattibase | 4.5 | 3.9 | 6.7 | 6.1 |
| Cocoa butter | 3.8 | 4.1 | 6.1 | 5.4 |
| Witepsol H-15 | 5.3 | 8.0 | 16.0 | 16.8 |

[a]Cured for two days at 25°C.

**TABLE 96.** Progesterone Suppository Formulations[439]

| | Quantity | |
|------|---------|------|
| Component | 25 mg[a] | 50 mg |
| Progesterone powder, USP | 44 g | 88 g |
| Polyethylene glycol 400 | 2096 g | 2096 g |
| Polyethylene glycol 6000 | 1392 g | 1392 g |

[a]Yields about 1740 suppositories of 2.03 g.

**STUDY 2:** Roffe et al.[439] reported on the compounding and stability of progesterone 25- and 50-mg suppositories. The suppositories were prepared from the formulas in Table 96 by heating the polyethylene glycols 6000 and 400 to 60°C in a 4-liter Pyrex beaker. The progesterone powder (sieved or triturated to fine powder) then was added. A 2-g suppository mold was lubricated with silicone spray and the hot mass was poured into the cool molds. The filled molds were cooled until the suppositories solidified. The progesterone content was assessed using an ultraviolet spectrophotometric assay. When analyzed after one year of storage, the suppositories were stable. ■

# ■ Promazine Hydrochloride

## Properties
Promazine hydrochloride is a white to yellowish-white nearly odorless and slightly hygroscopic crystalline powder having a bitter taste.[1,3]

### Solubility
Promazine hydrochloride is very soluble in water, having an aqueous solubility stated to be 333 mg/mL.[1,3] It is also very soluble in ethanol, having a solubility of about 500 mg/mL.[3]

### pH
A 5% aqueous solution has a pH of 4.2 to 5.2. The injection has a pH between 4 and 5.5. The oral solution pH is between 5 and 5.5.[3,4]

### Osmolality
Promazine hydrochloride injection 50 mg/mL has an osmolality of 138 mOsm/kg.[326]

## General Stability Considerations
Promazine hydrochloride undergoes oxidation upon exposure to air, developing a pink or blue color.[1] Promazine hydrochloride products are packaged in tight, light-resistant containers. They should be stored at controlled room temperature and protected from temperatures of 40°C and above and exposure to light. The injection also should be protected from freezing.[4]

Promazine hydrochloride injection may develop a slight yellow discoloration, but this change does not affect concentration or efficacy. However, marked discoloration or precipitation necessitates discarding the injection. Promazine hydrochloride is incompatible with alkalies, oxidizing agents, and heavy metals.[1,8] In solution, promazine hydrochloride exhibits the greatest stability at pH 6.5.[327]

## Compatibility with Common Beverages and Foods
### Oral
**PROMAZINE HYDROCHLORIDE COMPATIBILITY SUMMARY**
*Compatible with:* Lemon-lime soda • Prune juice
*Incompatible with:* Apple juice • Coca-Cola • Coffee • Grape juice • Milk • Orange juice • Orange soda • Pineapple juice • Tea • Tomato juice • V8 vegetable juice

**STUDY 1:** Hirsch[244] reported that the addition of promazine hydrochloride liquid products to coffee or tea resulted in the formation of an insoluble precipitate.

**STUDY 2:** Lever and Hague[1011] reported the compatibility of promazine hydrochloride oral concentrate 50 mg mixed with 15 mL of many liquids that might mask the flavor. Compatible liquids included lemon-lime soda and prune juice. Incompatible liquids included apple juice, black coffee, carbonated orange soda, Coca-Cola, grape juice, milk, orange juice, pineapple juice, tea, tomato juice, and V8 vegetable juice. The compatibility with apricot juice was uncertain because the solution's opaqueness may have masked turbidity and/or precipitation. The precipitation occurred within a few minutes of mixing and clung to the sides of the container, indicating unacceptable variable dosage may occur. The authors recommended mixing promazine hydrochloride oral concentrate freshly with only distilled water. ■

# Promethazine Hydrochloride

## Properties

Promethazine hydrochloride occurs as a white to faintly yellow nearly odorless crystalline powder.[2,3]

### Solubility

Promethazine hydrochloride is very soluble in water, with a solubility of 500 mg/mL.[2,3,6] In ethanol, its solubility is 150 mg/mL.[6] In absolute ethanol, the solubility of this drug is 85 mg/mL.[6]

### pH

A 5% aqueous solution of promethazine hydrochloride has a pH between 4 and 5.[4] A 10% aqueous solution has a pH of about 5.3.[1] The injection formulation has a pH between 4 and 5.5.[4] Promethazine hydrochloride syrup (Phenergan, Wyeth) had a measured pH of 5.2.[19]

### pKa

The drug has a $pK_a$ of 9.1.[2]

### Osmolality

Promethazine hydrochloride syrup 1.25 mg/mL (Wyeth) had an osmolality of 3500 mOsm/kg.[233] The osmolality of the 25-mg/mL injection was 291 mOsm/kg.[345]

## General Stability Considerations

Promethazine hydrochloride oxidizes upon exposure to air and moisture, forming a blue discoloration.[1-3] Promethazine hydrochloride products should be protected from exposure to light[2,3] due to potential photolytic decomposition.[6]

Promethazine hydrochloride oral tablets and solution and rectal suppositories should be packaged in tight, light-resistant containers.[2,4] The oral and parenteral products should be stored at room temperatures between 15 and 25°C, while the suppositories should be stored under refrigeration at 2 to 8°C. Liquid promethazine hydrochloride products should be protected from freezing.[2]

Connors et al.[6] indicated that promethazine hydrochloride exhibits greater stability as the pH becomes more acidic. Promethazine hydrochloride is incompatible with alkaline materials; using alkaline materials may result in the precipitation of free promethazine.[2,3]

Thumma and Repka[1378] evaluated the compatibility of promethazine hydrochloride with a variety of potential oral tablet components using differential scanning calorimetry, Fourier transform infrared spectroscopy, and isothermal stress testing. Carbopol 971p, ethylcellulose, Explotab, lactose monohydrate, magnesium stearate, Methocel K100M, microcrystalline cellulose, Pearlitol SD200 (mannitol), talc, and zinc stearate all were found to be compatible with promethazine hydrochloride and acceptable for use in solid oral dosage forms.

## Stability Reports of Compounded Preparations

### Oral

Iqbal and Pasha[1477] evaluated the stability of promethazine hydrochloride 1 mg/mL in several oral liquids prepared with thickening agents and compared the stabilities to that of the drug in sucrose syrup (simple syrup). The thickening agents (CMC sodium 2.5 g, CMC-7MF 5 g, CMC-7HF 5 g, PVP-K90 15 g, and xanthan gum 2.5 g) were hydrated in 200 mL of deionized water to achieve maximum swelling and then were brought to a volume of 400 mL with additional deionized water. In addition, sucrose syrup was prepared for use as a vehicle. Promethazine hydrochloride was added to yield a 1-mg/mL concentration. The liquids were packaged in amber screw cap bottles and were stored at temperatures of 30 to 70°C. The stability of promethazine hydrochloride was determined using ultraviolet spectrophotometry. From the data generated, the times to 10% drug loss ($t_{90}$) were calculated for storage at room temperature. The formulations using CMC sodium, CMC-7HF, and CMC-7MF all resulted in precipitation of some of the drug. Of the other vehicles, the PVP-K90 vehicle generated the shortest time to 10% drug loss, 14.4 days, followed by the xanthan gum at 22.5 days. The longest time to 10% drug loss occurred with promethazine hydrochloride in sucrose syrup; the $t_{90}$ was found to be about 25.7 days at room temperature.

### Rectal

Allen[1392] reported on a compounded formulation of promethazine hydrochloride 2.5-mg/mL for use as a rectal solution. The solution had the following formula:

| | | |
|---|---|---|
| Promethazine hydrochloride | | 250 mg |
| Pluronic P-127 | | 15 g |
| Purified water | qs | 100 mL |

The recommended method of preparation was to dissolve the promethazine hydrochloride powder in about 75 mL of purified water. The Pluronic P-127 was then to be sprinkled onto the solution and thoroughly mixed. Finally, purified water sufficient to bring the volume to 100 mL was to be added. The solution was then to be placed in a refrigerator and allowed to set overnight to effect solution. The solution was to be removed from the refrigerator and be packaged in tight, light-resistant containers. The author recommended a beyond-use date of six months at room temperature because this formula is a commercial medication in some countries with an expiration date of two years or more.

### Enteral

**PROMETHAZINE HYDROCHLORIDE COMPATIBILITY SUMMARY**

*Compatible with:* Ensure • Ensure HN • Ensure Plus • Ensure Plus HN • Osmolite • Osmolite HN • Precitene • Vital

**STUDY 1:** Cutie et al.[19] added 10 mL of promethazine hydrochloride syrup (Phenergan, Wyeth) to varying amounts (15 to 240 mL) of Ensure, Ensure Plus, and Osmolite (Ross Laboratories) with vigorous agitation to ensure thorough mixing. The promethazine hydrochloride syrup was physically compatible, distributing uniformly in all three enteral products with no phase separation or granulation.

**STUDY 2:** Altman and Cutie[850] reported the physical compatibility of promethazine hydrochloride syrup (Phenergan, Wyeth) 10 mL with varying amounts (15 to 240 mL) of Ensure HN, Ensure Plus HN, Osmolite HN, and Vital after vigorous agitation to ensure thorough mixing. The promethazine hydrochloride syrup was physically compatible, distributing uniformly in all four enteral products with no phase separation or granulation.

**STUDY 3:** Ortega de la Cruz et al.[1101] reported the physical compatibility of an unspecified amount of oral liquid promethazine (Frinova, Aventis) with 200 mL of Precitene (Novartis) enteral nutrition diet for a 24-hour observation period. No particle growth or phase separation was observed.

### Compatibility with Common Beverages and Foods

Hirsch[244] reported that the addition of promethazine liquid products to coffee or tea resulted in the formation of an insoluble precipitate.

## Propafenone Hydrochloride

### Properties
Propafenone hydrochloride occurs as a fine white crystalline powder having a bitter taste.[1,4,7]

#### Solubility
Propafenone hydrochloride is soluble in hot water and methanol; slightly soluble in cold water, ethanol, and chloroform; and very slightly soluble in acetone.[1,3,4]

#### pH
Propafenone hydrochloride 0.5% in water has a pH of 5 to 6.2.[3,4]

### General Stability Considerations
Propafenone hydrochloride powder and commercial oral capsules and tablets should be packaged in tight, light-resistant containers and stored at controlled room temperature.[3,4,7]

### Stability Reports of Compounded Preparations
#### Oral
**STUDY 1:** Olguin et al.[1067] reported the stability and bioavailability in animals of propafenone hydrochloride 1.5 mg/mL compounded as an oral suspension prepared from commercial tablets (Norfenon, Abbott). Tablets were crushed, ground to fine powder, and passed through a size 100 mesh for uniformity in particle size. The tablet powder was incorporated into pomegranate syrup (La Madrilena) to yield the 1.5-mg/mL concentration. The oral suspension was packaged in amber plastic screw-cap bottles and stored at room temperature of 10 to 20°C and under refrigeration at 3 to 5°C for 90 days.

No changes in color or odor were observed. HPLC analysis of drug concentrations found that little or no loss of propafenone hydrochloride occurred in 90 days at either storage temperature. In rabbits, the bioavailability was found to be 59.4%, which is consistent with solid dosage forms.

**STUDY 2:** Olguin et al.[1270] reported the stability of propafenone hydrochloride 1.5 mg/mL compounded as an oral suspension prepared from commercial tablets (Norfenon, Abbott) in pomegranate syrup (La Madrilena) in a second article. The formulation was prepared and stored at room temperature and refrigerated for 90 days as in the previous study.

Again, no changes in color or odor were observed. However, in this second report they noted crystallization in the refrigerated samples after 30 days. Such crystallization was not reported in their first study. The reported HPLC test results were virtually identical with the first study and may be a repeat of the previous study data. However, in this work the authors stated a 30-day limitation on the refrigerated samples because of the crystallization but 90-day stability in the room temperature samples.

## Propranolol Hydrochloride

### Properties
Propranolol hydrochloride is a white to off-white odorless crystalline powder having a bitter taste.[2,3]

#### Solubility
Propranolol hydrochloride is soluble in water and ethanol, with a solubility around 50 mg/mL.[2,3]

#### pH
A 1% propranolol hydrochloride aqueous solution has a pH between 5 and 6.[3] The injection has a pH of 2.8 to 4.[4]

### General Stability Considerations
Propranolol hydrochloride products should be packaged in well-closed, light-resistant containers and should be stored

at room temperature (around 25°C). The long-acting formulations should be packaged in tight, light-resistant containers and stored at approximately 25°C protected from light, moisture, freezing, and excessive heat.[2,7] Oral tablets stored according to the label requirements have been found to be very stable and exhibit no change in dissolution even many years beyond the labeled expiration date.[1519, 1520]

The pH of maximum stability was reported to be 2.8 to 4.[20] Brown and Kayes[264] evaluated propranolol hydrochloride stability at pH values between 2.2 and 7.5. They reported maximum stability around pH 3. The drug undergoes rapid decomposition in alkaline solutions.[2,3]

## Stability Reports of Compounded Preparations
### Oral

**STUDY 1 (LIQUID):** Henry et al.[92] reported the stability of an extemporaneous oral suspension of propranolol hydrochloride prepared from commercial 10-mg tablets (Inderal, Ayerst). The tablets were triturated to fine powder and mixed with a commercial suspending vehicle (Diluent Flavored for Oral Use, Roxane) composed of ethanol 1% and saccharin 0.05% in a cherry-flavored 33% polyethylene glycol 8000 base to yield a theoretical propranolol hydrochloride concentration of 1 mg/mL. The suspension was packaged as 12 mL in two-ounce amber glass bottles with safety closures. Samples were stored at 5 and 25°C exposed to daytime fluorescent light for four months. A stability-indicating HPLC assay was used to assess propranolol hydrochloride concentration. No loss occurred at either temperature.

**STUDY 2 (LIQUID):** Gupta and Stewart[20] prepared duplicate propranolol hydrochloride 0.5-mg/mL suspensions from commercial tablets and duplicate 0.5-mg/mL oral solutions from the commercial injection. To prepare the suspension, six 20-mg tablets (Ayerst) were triturated to fine powder; simple syrup containing 600 mg/mL of sucrose, sodium benzoate 0.1%, and an unspecified amount of citric acid was added to bring to 240 mL. The solutions were prepared using 50 mL of the 1-mg/mL commercial injection (Ayerst); again simple syrup was used to bring to 100 mL. The suspensions and solutions were stored in amber glass bottles at 25°C and analyzed by a stability-indicating HPLC assay over 238 days. No loss occurred. Furthermore, no visually observable evidence of fungus growth was noted. The solutions retained their initial clarity throughout the study. In suspensions prepared from tablets, the propranolol hydrochloride was completely dissolved, with only the inactive excipients from the tablets settling to the bottom. A slight greenish-blue color was imparted to the suspensions by the blue color in the tablets.

**STUDY 3 (LIQUID):** Rooney and Creurer[43] noted that this information is not directly applicable when plain simple syrup, USP, is used. The pH of the simple syrup is about 6.8 to 7, much higher than the vehicle of the previous study and too high to maintain propranolol stability. Consequently, to achieve a longer term stability, the pH should be adjusted to approximately 3.

**STUDY 4 (LIQUID):** Brown and Kayes[264] reported on two extemporaneous propranolol suspensions prepared from tablets (Table 97). The unsatisfactory formulation exhibited bacterial growth after five days at 20°C packaged in amber screw-cap bottles. The methyl and propyl hydroxybenzoate preservatives were ineffective in preventing bacterial growth, presumably due to an interaction between the parabens and propranolol. Color increased during storage; in aqueous solution, propranolol hydrochloride produced colored degradation products upon oxidation of its side chain.

However, the satisfactory formulation remained pharmaceutically elegant and retained adequate drug content for up to 14 days at 20°C. The suspension was assayed by ultraviolet spectroscopy and thin-layer chromatography and showed a change in absorbance of 4.5% in 14 days and 6.67% in 28 days. The alcohol content was sufficient to act as a preservative.

**TABLE 97.** Propranolol Hydrochloride Suspension Formulations from Brown and Kayes[264]

| Unsatisfactory Formulation | | Satisfactory Formulation | |
|---|---|---|---|
| Component | % (wt/vol) | Component | % (wt/vol) |
| Inderal | 0.05 | Inderal | 0.05 |
| Sodium carboxymethyl-cellulose [BPC 50 mPa (cps)] | 1.0 | Alcohol 90% | 22.5 |
| Powdered sugar | 20.0 | Compound orange spirit | 0.2 |
| Raspberry flavor | 1.0 | Compound tartrazine solution | 0.2 |
| Methyl hydroxybenzoate | 0.15 | Amaranth solution | 0.05 |
| Propyl hydroxybenzoate | 0.015 | Cetomacrogol 1000 | qs |
| Distilled water | qs 100 | Syrup | 12.5 |
| | | Water | qs 100 |

**STUDY 5 (LIQUID):** Ahmed et al.[440] evaluated the physical and chemical stability of six pediatric formulations (Table 98) of propranolol hydrochloride 1 to 5 mg/mL. The preparations were compounded by dispersing the tablets in distilled water or syrup at 3000 rpm for 10 minutes (Braun Vitamizer). The excipients then were added with continuous mixing using a laboratory mixer (Silverson), and the mixtures were brought to volume with mixing continued for 20 minutes. The oral liquids were packaged in 100-mL amber glass bottles and stored at 4 and 30°C as well as at an elevated temperature for 12 weeks. The stabilities of the propranolol hydrochloride, sodium benzoate, methyl hydroxybenzoate, and propyl hydroxybenzoate were assessed using a stability-indicating HPLC assay. The best extraction of the propranolol hydrochloride from the tablets occurred with formulations E and F; total extraction from the tablets occurred during the compounding process. Formulations B and C were the least efficient, with only 77 to 79% extracted by the compounding procedure. However, after the first week of storage, formulation C had increased to about 97%; formulation B remained unchanged.

No decomposition of propranolol or the preservatives occurred in any formulation after 12 weeks of storage at any storage temperature. However, numerous physical changes were noted. Formulation F developed a precipitate after four weeks at 30°C. Formulation E was difficult to redisperse, especially the 4°C sample. Formulation D developed fungal growth after four weeks. Formulation B became turbid after four weeks at 30°C. Increases in viscosity occurred in formulations A, B, C, and E. None of the formulations was unequivocally recommended, but formulation F, a clear solution, was the best choice as long as no precipitate is present.[440]

**STUDY 6 (LIQUID):** Modamio et al.[745] reported the stability of propranolol hydrochloride (ICI) 0.25 mg/mL in phosphate buffer (pH 7.4) at elevated temperatures. Stability-indicating HPLC analysis found little or no loss of propranolol in 84 hours, even at 90°C.

**STUDY 7 (POWDER):** Yang et al.[738] reported the stability of propranolol oral powders extemporaneously compounded from commercial tablets. Inderal 10-mg tablets (Zeneca) were crushed and pulverized. Propranolol powder equivalent to 10 mg of drug was mixed with 170 mg of lactose, 390 mg of pulverized lactofermin (Biofermin) tablets, or 1.08 g of acetylcysteine granules for use as solid diluents. Powder mixtures were packaged in polyethylene-coated glassine papers and stored for 28 days at ambient temperature (between 15 and 30°C). Stability-indicating HPLC analysis found about 4 to 5% loss of propranolol in 28 days diluted in either lactose or pulverized lactofermin tablets. However, in acetylcysteine granules, 15% propranolol loss occurred in seven days.

### Enteral
**PROPRANOLOL HYDROCHLORIDE COMPATIBILITY SUMMARY**
*Compatible with:* Isocal • Sustacal • Sustacal HC

Strom and Miller[44] evaluated the stability of propranolol hydrochloride (Inderal, Wyeth-Ayerst) from 40-mg tablets and oral liquid (Roxane) 20 mg/5 mL when mixed with three enteral formulas at full- and half-strength. The ground tablets and the oral liquid were dispersed in approximately 10 mL of deionized water and added to 240 mL of Isocal, Sustacal, and Sustacal HC (Mead Johnson) full-strength or half-strength mixed with deionized water. The mixtures were

**TABLE 98.** Propranolol Hydrochloride Pediatric Oral Liquid Formulations[440]

| Component | Formulation | | | | | |
|---|---|---|---|---|---|---|
| | A | B | C | D | E | F |
| Inderal 40-mg tablets | 12.5 | 12.5 | 12.5 | 2.5 | 10 | 12.5 |
| Sodium carboxymethyl-cellulose | — | 1 g | — | — | — | — |
| Compound tragacanth powder | — | — | 2.5 g | — | 0.75 g | — |
| Citric acid | — | — | 0.28 g | 0.5 g | — | 1 g |
| Sodium citrate | — | — | 0.21 g | — | — | — |
| Sodium benzoate | — | — | — | — | — | 0.1 g |
| Concentrated chloroform water | — | — | 2.5 mL | — | — | — |
| Hydroxybenzoate spirit | — | 1 mL | — | — | — | — |
| Alcohol 90% | — | — | — | — | 5 mL | — |
| Raspberry syrup, BP | — | 30 mL | — | — | — | — |
| Cherry syrup, BP | — | — | — | — | — | 40 mL |
| Syrup, BP | — | — | 25 mL | — | qs 100 mL | — |
| Distilled water, mL | qs 100 | qs 100 | qs 100 | qs 100 | 25 | qs 100 |

stirred with a magnetic stirrer for five minutes; duplicate samples of each mixture were stored at 24°C. The mixtures were evaluated visually for precipitation and phase separation, but no physical incompatibilities were observed. A stability-indicating HPLC assay assessed drug content at various times over 24 hours. The propranolol hydrochloride content was between 96 and 104% after 24 hours in each enteral mixture prepared from either drug source. The stabilities of the enteral formula components were not evaluated. ■

# Propylthiouracil

## Properties
Propylthiouracil is a white or nearly white crystalline powder having a bitter taste and a starchlike appearance.[1–3]

### Solubility
Propylthiouracil has an aqueous solubility of about 1.1 mg/mL at 20°C, increasing to about 10 mg/mL in boiling water. In ethanol, it has a solubility of about 16.7 mg/mL.[1,3] It is freely soluble in solutions of alkali hydroxides.[1]

### pH
A saturated aqueous solution of propylthiouracil is neutral to slightly acidic.[1]

## General Stability Considerations
Propylthiouracil tablets should be packaged in well-closed containers and stored at controlled room temperature and protected from temperatures of 40°C and above.[2,4]

## Stability Reports of Compounded Preparations
### Oral
**STUDY 1:** Nahata et al.[682] reported the stability of propylthiouracil 5 mg/mL in two extemporaneously prepared suspensions. Propylthiouracil 50-mg tablets were crushed using a mortar and pestle. One suspension was prepared by incorporating the pulverized tablets into a 1:1 mixture of Ora-Plus and Ora-Sweet. Pulverized tablet powder was incorporated into a 1:1 mixture of methylcellulose 1% and simple syrup to prepare the other suspension.

No visible changes in either suspension occurred over 91 days of storage. Stability-indicating HPLC analysis found that about 4% loss of propylthiouracil occurred in 91 days under refrigeration at 4°C. At room temperature near 25°C, losses of about 6 to 7% occurred in 70 days, with the possibility of excessive losses occurring by 91 days. In addition, the authors warned against extended storage of the suspension at room temperature because microbiological growth might occur.

**STUDY 2:** Alexander and Mitra[914] evaluated the stability of propylthiouracil 5 mg/mL in an extemporaneously compounded oral suspension having the formula noted in Table 99. Carboxymethylcellulose sodium and Veegum were weighed and mixed separately and thoroughly in water and allowed

**TABLE 99.** Propylthiouracil 5-mg/mL Oral Suspension Formula for 3 Liters[914]

| Component | Quantity |
|---|---|
| Propylthiouracil 50-mg tablets | 300 tablets (15 g of drug) |
| Carboxymethylcellulose sodium | 22.5 g |
| Veegum | 15 g |
| Syrup, USP | 750 mL |
| Sorbitol solution 70% wt/vol | 750 mL |
| Saccharin | 4.5 g |
| EDTA disodium | 300 mg |
| Paraben concentrate (Methylparaben 10%, propylparaben 2%, in propylene glycol) | 30 mL |
| Flavor blend (Wild cherry, raspberry, tangerine) | 3 mL |
| Reverse osmosis water | qs 3000 mL |

to hydrate overnight. Propylthiouracil 50-mg tablets were crushed and triturated using a mortar and pestle. The powder was weighed, and hydrated carboxymethylcellulose sodium was added slowly with constant trituration. The hydrated Veegum was then added with trituration. Flavor blend was added and mixed, and the mixture was then transferred into a 4-liter beaker with water washes of the mortar. Electric stirrer and mixer were used, and syrup, USP, was added and mixed. Saccharin dissolved in 40 mL of boiling water, EDTA disodium dissolved in 10 mL of water, and the paraben concentrate were added to the mixture. The mixture was brought to a volume of 3000 mL with reverse osmosis water and mixed thoroughly with an electric mixer. The suspension was packaged in eight-ounce amber glass prescription bottles and stored at several temperatures ranging from 4 to 70°C.

Stability-indicating HPLC analysis performed over 13 weeks found temperature-dependent loss of intact propylthiouracil that was consistent with the work of Nahata et al.[682] Propylthiouracil losses of 4 and 7% occurred in 91 days at 4 and 30°C, respectively. The shelf life of the suspension was calculated to be 248 days refrigerated at 4°C and 127 days at room temperature of 25°C. ■

# Pseudoephedrine Hydrochloride

## Properties

Pseudoephedrine hydrochloride is a white or off-white powder or crystals that are odorless or have a slight characteristic odor.[2,3]

### Solubility

Pseudoephedrine hydrochloride has an aqueous solubility of about 2 g/mL. In ethanol, the solubility is 278 mg/mL.[2]

### pH

A 5% pseudoephedrine hydrochloride aqueous solution has a pH between 4.6 and 6.[4] Pseudoephedrine hydrochloride syrup (Sudafed, Burroughs-Wellcome) has a measured pH of 2.5.[19]

## General Stability Considerations

Pseudoephedrine hydrochloride solid oral dosage forms should be packaged in tight containers, while the syrup should be packaged in tight, light-resistant containers. The products should be stored at controlled room temperature. The syrup should be protected from freezing.[2,4]

## Stability Reports of Compounded Preparations
### Enteral
#### PSEUDOEPHEDRINE HYDROCHLORIDE COMPATIBILITY SUMMARY

*Compatible with:* Vital • Vivonex T.E.N.

*Incompatible with:* Enrich • Ensure • Ensure HN • Ensure Plus • Ensure Plus HN • Osmolite • Osmolite HN • TwoCal HN

**STUDY 1:** Cutie et al.[19] added 10 mL of pseudoephedrine hydrochloride syrup (Sudafed, Burroughs-Wellcome) to varying amounts (15 to 240 mL) of Ensure, Ensure Plus, and Osmolite (Ross Laboratories) with vigorous agitation to ensure thorough mixing. The syrup was incompatible physically with all three enteral products, becoming a viscous gelatinous mass immediately upon mixing.

**STUDY 2:** Burns et al.[739] reported the physical compatibility of pseudoephedrine hydrochloride syrup (Sudafed) 5 mL with 10 mL of three enteral formulas, including Enrich, TwoCal HN, and Vivonex T.E.N. Visual inspection found no physical incompatibility with Vivonex T.E.N. enteral formula. However, when mixed with Enrich, a thick gelatinous mass formed. When mixed with TwoCal HN, a slight thickening occurred.

**STUDY 3:** Altman and Cutie[850] reported the physical incompatibility of pseudoephedrine hydrochloride syrup (Sudafed, Burroughs-Wellcome) 10 mL with varying amounts (15 to 240 mL) of Ensure HN, Ensure Plus HN, Osmolite HN, and Vital after vigorous agitation to ensure thorough mixing. The pseudoephedrine hydrochloride syrup was incompatible physically with Ensure HN, Ensure Plus HN, and Osmolite HN enteral products, becoming a viscous gelatinous mass immediately upon mixing. The pseudoephedrine hydrochloride syrup was compatible with Vital. ■

# Pyrazinamide

## Properties

Pyrazinamide is a white or nearly white almost odorless crystalline powder.[2,3]

### Solubility

Pyrazinamide is sparingly soluble in water, having an aqueous solubility of 15 mg/mL. In dehydrated ethanol it has a solubility of 5.7 mg/mL.[1,3]

### pH

Aqueous solutions of pyrazinamide are pH neutral.[1]

### pK$_a$

Pyrazinamide has a pK$_a$ of 0.5.[1,2]

## General Stability Considerations

Pyrazinamide products should be packaged in well-closed containers and stored at controlled room temperature.[4,7]

## Stability Reports of Compounded Preparations
### Oral

**STUDY 1:** Allen and Erickson[595] evaluated the stability of three pyrazinamide 10-mg/mL oral suspensions extemporaneously compounded from tablets. Vehicles used in this study were (1) an equal parts mixture of Ora-Sweet and Ora-Plus (Paddock), (2) an equal parts mixture of Ora-Sweet SF and Ora-Plus (Paddock), and (3) cherry syrup (Robinson Laboratories) mixed 1:4 with simple syrup. Three pyrazinamide 500-mg tablets (Lederle) were crushed and comminuted to fine powder using a mortar and pestle. About 10 mL of the test vehicle was added to the powder and mixed to yield a uniform paste. Additional vehicle was added geometrically and brought to the final volume of 150 mL, with thorough mixing after each addition. The process was repeated for each of the three test suspension vehicles. Samples (120 mL) of each finished suspension were packaged in 120-mL amber polyethylene terephthalate plastic prescription bottles and stored at 5 and 25°C in the dark. No visual changes or changes in odor were detected. Stability-indicating HPLC analysis found less than 3% pyrazinamide loss in any suspension stored at either temperature after 60 days of storage.

**STUDY 2:** Nahata et al.[601] evaluated the stability of two pyrazinamide 100-mg/mL oral suspensions compounded from tablets. For the first suspension formulation, 140 pyrazinamide 500-mg tablets (Lederle) were crushed in a mortar and pestle and mixed with a small amount of simple syrup. The mixture was transferred to a graduate and additional simple syrup was added, bringing the total to 700 mL of the 100-mg/mL suspension. The process was repeated to yield 1.4 liters of suspension. The suspension was packaged as 60 mL in four-ounce amber plastic and glass prescription bottles and stored at 4 and 25°C.

The second suspension studied by Nahata et al.[601] was compounded by crushing 200 pyrazinamide 500-mg tablets in a mortar and pestle and mixing with 500 mL of 1% methylcellulose (as Citrucel, Marion Merrell Dow) in purified water and 500 mL of simple syrup to make a liter of the 100-mg/mL suspension. The process was repeated to make a second smaller batch. Another 140 pyrazinamide 500-mg tablets were crushed and mixed with 350 mL of 1% methylcellulose and 350 mL of simple syrup. A total of 1.7 liters of pyrazinamide 100-mg/mL suspension in 0.5% methylcellulose was prepared. The suspension was also packaged as 60 mL in four-ounce amber plastic and glass prescription bottles and stored at 4 and 25°C.

Stability-indicating HPLC analysis of both suspensions found less than 10% pyrazinamide loss in 60 days at either temperature. However, the methylcellulose-containing formulation may decrease settling and increase the accuracy of dosing after storage.

**STUDY 3:** Seifart et al.[743] reported the stability of an extemporaneously compounded pyrazinamide 100-mg/mL preparation. Pyrazinamide tablets were pulverized using a mortar and pestle. The powder was suspended in a vehicle that included tragacanth powder, chloroform water, and purified water. Stability-indicating HPLC analysis found that more than 98% of the pyrazinamide remained after 28 days stored at either 4 or 24°C.

## Compatibility with Other Drugs

Seifart et al.[743] reported the stability of an extemporaneously compounded isoniazid 10-mg/mL suspension mixed with an extemporaneously compounded pyrazinamide suspension. To prepare the isoniazid suspension, isoniazid tablets were pulverized using a mortar and pestle. The powder was suspended in a vehicle that included citric acid, sodium sulfate, glycerin, chloroform water, and purified water. To prepare the pyrazinamide suspension, pyrazinamide tablets were pulverized using a mortar and pestle. The powder was suspended in a vehicle that included tragacanth powder, chloroform water, and purified water.

The suspension mixture was stable for 28 days stored at either 4 or 24°C. Stability-indicating HPLC analysis found losses of 8 and 2% for isoniazid and pyrazinamide, respectively, in 28 days. Addition of ascorbic acid 20 mcg/mL to the mixture of suspensions resulted in a 30% loss of isoniazid at 4° and a 50% loss at 24°C in 28 days.

## Compatibility with Common Beverages and Foods

Peloquin et al.[1135] evaluated the stability of several oral antituberculosis drugs, including pyrazinamide, for compatibility with common foods to facilitate administration to children. Pyrazinamide tablets were crushed and mixed with 30 g of Hunt's Snackpack no-sugar-added chocolate pudding, Safeway grape jelly, Safeway creamy peanut butter, orange juice, and 7-Up soda. The drug–food mixtures were thoroughly mixed and analyzed (no method indicated) for drug content initially and after one, two, and four hours had elapsed. Drug recovery at all time points was greater than 93% from the pudding, greater than 89% from the jelly, and greater than 80% from the peanut butter. The authors recommended the use of the no-sugar-added chocolate pudding and the jelly to aid in administering the drugs to children. ■

## ■ Pyridostigmine Bromide

### Properties

Pyridostigmine bromide is a white or nearly white hygroscopic powder having a pleasant odor and a bitter taste.[1,2]

### Solubility

Pyridostigmine bromide is freely soluble in water and ethanol, having a solubility greater than 1 g/mL.[1–3]

### pH

Pyridostigmine bromide injection has a pH between 4.5 and 5.5.[4] Pyridostigmine bromide solution (Mestinon, Roche) has a measured pH of 4.4.[19]

## General Stability Considerations

Pyridostigmine bromide products should be packaged in tight containers. The syrup should be packaged in tight, light-resistant containers, and both the syrup and the injection should be protected from light. The extended-release tablets may develop a mottled appearance because of the hygroscopicity of the drug, but the concentration remains unaffected.[2–4] The manufacturer has stated that pyridostigmine bromide tablets should not be removed from the original packaging and placed in dosing compliance aids.[1622]

Pyridostigmine bromide is unstable in alkaline solutions. Aqueous solutions may be sterilized by steam autoclaving.[1]

## Stability Reports of Compounded Preparations
### Enteral
**PYRIDOSTIGMINE BROMIDE COMPATIBILITY SUMMARY**

*Compatible with:* Ensure • Ensure HN • Ensure Plus • Ensure Plus HN • Osmolite • Osmolite HN • Vital

**STUDY 1:** Cutie et al.[19] added 5 mL of pyridostigmine bromide solution (Mestinon, Roche) to varying amounts (15 to 240 mL) of Ensure, Ensure Plus, and Osmolite (Ross Laboratories) with vigorous agitation to ensure thorough mixing. The pyridostigmine bromide solution was physically compatible, distributing uniformly in all three enteral products with no phase separation or granulation.

**STUDY 2:** Altman and Cutie[850] reported the physical compatibility of pyridostigmine bromide solution (Mestinon, Roche) 5 mL with varying amounts (15 to 240 mL) of Ensure HN, Ensure Plus HN, Osmolite HN, and Vital after vigorous agitation to ensure thorough mixing. The elixir was physically compatible, distributing uniformly in all four enteral products with no phase separation or granulation. ■

# ■ Pyridoxine Hydrochloride

## Properties

Pyridoxine hydrochloride is a white or nearly white almost odorless crystalline powder or crystals with a slightly bitter salty taste.[2,3]

### Solubility

Pyridoxine hydrochloride has an aqueous solubility of about 200 to 222 mg/mL. In ethanol, it has a solubility of about 11 mg/mL[1] or about 8.7 to 10 mg/mL.[3]

### pH

A 5% aqueous solution of pyridoxine hydrochloride has a pH between 2.4 and 3.[3] Pyridoxine hydrochloride injection has a pH between 2 and 3.8.[4]

## General Stability Considerations

Pyridoxine hydrochloride undergoes decomposition slowly when exposed to light. Pyridoxine hydrochloride products should be packaged in well-closed containers and should be stored at controlled room temperature protected from temperatures of 40°C or more and from exposure to light. The injection should be protected from freezing.[2,4]

Pyridoxine hydrochloride is incompatible with oxidizing agents and iron salts and in alkaline solutions.[2]

## Stability Reports of Compounded Preparations
### Oral

The extemporaneous formulation of a pyridoxine hydrochloride 1-mg/mL oral solution was described by Nahata and Hipple.[160] One milliliter of pyridoxine hydrochloride 100-mg/mL injection was mixed thoroughly with 99 mL of syrup and packaged in an amber glass bottle. A stability period of 30 days under refrigeration was used, although chemical stability testing was not performed.

## Compatibility with Other Drugs

Ved and Deshpande[1628] evaluated the stability of rifampin 20 mg/mL with isoniazid 20 mg/mL and pyridoxine hydrochloride 1 mg/mL in a vehicle composed of sorbitol and propylene glycol (vehicle component concentrations not provided). Samples were stored at 15 and 30°C and at elevated temperatures of 37 and 45°C. The suspension was described as having a good appearance with easy redispersibility and low sedimentation rate. However, microbiological analysis of the rifampin content found 25 and 30% loss of antimicrobial activity after one week of storage at 15 and 30°C, respectively, with higher losses at elevated temperatures. The authors concluded that rifampin was unstable in combination with isoniazid and pyridoxine hydrochloride. ■

# Pyrimethamine

## Properties

Pyrimethamine is a white or nearly white almost odorless crystalline powder or crystals.[2,3]

### Solubility

Pyrimethamine is practically insoluble in water and has a solubility of about 5 to 9 mg/mL in ethanol.[1-3]

## General Stability Considerations

Pyrimethamine tablets should be packaged in tight, light-resistant containers and stored between 15 and 25°C.[2,4]

## Stability Reports of Compounded Preparations

### Oral

Pyrimethamine tablets may be crushed for use as a drug source to prepare extemporaneous suspensions. Water, cherry syrup, and sucrose-containing solutions are recommended as vehicles. However, the sucrose tends to affect pyrimethamine stability adversely. If cherry syrup or other sucrose-containing solutions are used, the suspension should be stored at room temperature and used within five to seven days.[2]

**STUDY 1:** The extemporaneous formulation of a pyrimethamine 2-mg/mL oral suspension was described by Nahata and Hipple.[160] Two pyrimethamine 25-mg tablets were crushed and brought to 25 mL with 70% sorbitol, and the solution was thoroughly mixed. A stability period of seven days under refrigeration was used, although chemical stability testing was not performed.

**STUDY 2:** Nahata et al.[509] evaluated the stability of an extemporaneously compounded pyrimethamine 2-mg/mL oral suspension prepared from tablets. Forty pyrimethamine 25-mg tablets were crushed using a mortar and pestle. The powder was mixed completely with 500 mL of an equal parts mixture of simple syrup and methylcellulose 1%. The suspension was packaged in both amber glass and amber plastic prescription bottles and stored under refrigeration at 4°C and at 25°C in a water bath exposed to fluorescent light. No change in the physical appearance of the suspension was observed. Stability-indicating HPLC analysis of the samples found about 7 to 8% drug loss at 25°C and 3 to 4% loss at 4°C after 91 days. No difference in drug concentration occurred between the two different container types.

# Quinapril Hydrochloride

## Properties

Quinapril hydrochloride is a white to off-white amorphous powder or white crystalline solid that may occasionally have a pink cast.[1,4,7]

## Solubility

Quinapril hydrochloride is freely soluble in aqueous solvents.[4,7]

## General Stability Considerations

Quinapril hydrochloride powder and tablets are packaged in well-closed containers and stored at controlled room temperature and protected from exposure to light.[4,7]

The manufacturer states that quinapril hydrochloride is unstable in aqueous solution, degrading 9% in 24 hours at room temperature. The rate of decomposition and the specific degradation products formed are pH-dependent.[1130] The pH of maximum stability for quinapril hydrochloride is in a narrow range of pH 5.5 to 6.5.[1130,1131] The magnesium carbonate excipient in the tablets tends to raise the pH, with dispersal in aqueous solvents well above the most stable range.[1130]

## Stability Reports of Compounded Preparations

### Oral

Freed et al.[1130] evaluated the stability of a number of extemporaneously compounded oral liquid preparations made from quinapril hydrochloride commercial tablets. Because quinapril hydrochloride stability is very pH-dependent,[1131] the most stable formulations were those buffered to be near pH 5.5 to 5.7.

The recommended procedure to prepare these oral liquids is to begin by crushing one K-Phos neutral tablet (Beach Pharmaceuticals) and dissolving the powder in 100 mL of sterile water for irrigation. A 30-mL portion of this solution is transferred to a 200-mL polyethylene terephthalate screw-cap bottle containing 10 quinapril hydrochloride tablets, and the container is then shaken for at least two minutes. The screw cap is removed, and this concentrate is allowed to stand for 15 minutes. After the stand time, the concentrate is shaken again for one minute. A 30-mL portion of Bicitra (Johnson & Johnson) is then added, and the liquid is shaken for another two minutes. A total of 140 mL of Ora-Sweet, Ora-Sweet SF, or syrup, USP, is added, and the suspension is shaken to disperse the ingredients. The suspension is then ready to be used.

The suspensions were reported to be pinkish red with suspended small white particles. No change in appearance was observed in any sample throughout the study. Stability-indicating HPLC analysis found that less than 2% loss of quinapril hydrochloride occurred in 24 hours at room temperature of 25°C and 65% relative humidity, and less than 3% loss occurred in six weeks under refrigeration at 5°C. Of the formulations tested, the suspension prepared with K-Phos solution 15%, Bicitra 15%, and Ora-Sweet SF 70% yielded the best results and was deemed to be the formulation of choice. ■

# Quinidine Sulfate

## Properties

Quinidine sulfate is a white or nearly white odorless crystalline powder or silky needlelike white or colorless crystals having a bitter taste.[2,3] Quinidine sulfate dihydrate 241 mg and quinidine sulfate anhydrous 230 mg are approximately equivalent to quinidine anhydrous 200 mg.[3]

## Solubility

Quinidine sulfate has an aqueous solubility of about 10 mg/mL. In ethanol it has a solubility of about 100 mg/mL.[3]

## pH

A 1% aqueous solution has a pH of 6 to 6.8.[3]

## *pK~a*

Quinidine has pK~a~ values of 4 and 8.6.[2]

## General Stability Considerations

Quinidine sulfate products should be packaged in tight (capsules) or well-closed (tablets) light-resistant containers and stored at controlled room temperature.[4,7]

Quinidine sulfate darkens upon exposure to light.[2,3] Quinidine salts in solution slowly acquire a brownish tint when exposed to light. Discolored solutions should not be used.[2]

## Stability Reports of Compounded Preparations
### *Oral*

**USP OFFICIAL FORMULATION:**

> Quinidine sulfate 1 g
> Vehicle for Oral Solution, NF (sugar-containing or sugar-free)
> Vehicle for Oral Suspension, NF (1:1) qs 100 mL

(See the vehicle monographs for information on the individual vehicles.)

Use quinidine sulfate powder or commercial tablets. If using tablets, crush or grind to fine powder. Add 15 mL of the vehicle mixture and mix to make a uniform paste. Add additional vehicle almost to volume in increments, with thorough mixing. Quantitatively transfer to a suitable calibrated, tight, light-resistant bottle, bring to final volume with the vehicle mixture, and thoroughly mix yielding quinidine sulfate 10-mg/mL oral suspension. The final liquid preparation should have a pH between 3.4 and 4.4. Store the preparation at controlled room temperature or under refrigeration between 2 and 8°C. The beyond-use date is 60 days from the date of compounding at either storage temperature.[4,5]

Allen and Erickson[595] evaluated the stability of three quinidine sulfate 10-mg/mL oral suspensions extemporaneously compounded from tablets. Vehicles used in this study were (1) an equal parts mixture of Ora-Sweet and Ora-Plus (Paddock), (2) an equal parts mixture of Ora-Sweet SF and Ora-Plus (Paddock), and (3) cherry syrup (Robinson Laboratories) mixed 1:4 with simple syrup. Six quinidine sulfate 200-mg tablets (Geneva) were crushed and comminuted to fine powder using a mortar and pestle. About 15 mL of the test vehicle was added to the powder and mixed to yield a uniform paste. Additional vehicle was added geometrically and brought to the final volume of 120 mL, with thorough mixing after each addition. The process was repeated for each of the three test suspension vehicles. Samples of each finished suspension were packaged in 120-mL amber polyethylene terephthalate plastic prescription bottles and stored at 5 and 25°C in the dark.

No visual changes or changes in odor were detected during the study. Stability-indicating HPLC analysis found less than 4% quinidine sulfate loss in any of the suspensions stored at either temperature after 60 days of storage. ▪

# Ramipril

## Properties
Ramipril is a white or almost white crystalline powder.[3]

### Solubility
Ramipril is sparingly soluble in water and soluble in polar organic solvents.[3,7]

## General Stability Considerations
Ramipril products should be packaged in well-closed containers and stored at controlled room temperature.[7]

Hanysova et al.[948] compared the stability of ramipril in aqueous solutions adjusted to pH 3, 5, and 8 and exposed to hydrogen peroxide as an oxidizing agent. Stability-indicating HPLC analysis found ramipril was relatively stable to oxidation and to acidic pH. Little or no change in the drug concentration or increase in decomposition products occurred when ramipril was heated to 90°C for one hour. However, the alkaline pH 8 samples underwent extensive decomposition, with more than half of the ramipril being lost.

## Stability Reports of Compounded Preparations
### Oral
Allen et al.[518] evaluated the stability of ramipril capsule contents incorporated into deionized water, apple juice, and applesauce to be used for patients who cannot swallow solid oral dosage forms. The contents of one ramipril capsule of each of three strengths, 1.25, 2.5, and 5 mg, were incorporated into 120 mL of each of the vehicles with thorough mixing. Samples of each were packaged in polyethylene terephthalate containers and stored at 23°C for 24 hours and 3°C for 48 hours. The ramipril content of the mixtures was determined using a stability-indicating HPLC analysis technique. Little or no loss of ramipril content was found in water or apple juice after storage for 24 hours at 23°C and for 48 hours at 3°C. In the applesauce mixture, losses were slightly higher, in some cases about 5 or 6%. However, the absence of detected decomposition products upon analysis would indicate that incomplete extraction might have been a factor. Nevertheless, the authors indicated that ramipril is stable in water, apple juice, and applesauce under the storage conditions and for the time periods evaluated. ■

# Ranitidine Hydrochloride

## Properties
Ranitidine hydrochloride is a white or off-white to pale yellow granular material or crystalline powder with a bitter taste and a sulfurlike odor.[1–3] Ranitidine hydrochloride 168 mg is equivalent to 150 mg of ranitidine.[2]

### Solubility
Ranitidine hydrochloride has a solubility in water of 660 mg/mL. In ethanol its solubility is 190 mg/mL.[2]

### pH
A 1% aqueous solution of ranitidine hydrochloride has a pH between 4.5 and 6.[4] The injection and oral solution have pH values of 6.7 to 7.3 and 6.7 to 7.5, respectively.[2,4]

### $pK_a$
The drug has $pK_a$ values of 8.2 and 2.7.[2]

## Osmolality

Ranitidine hydrochloride injection 10 mg/mL has an osmolality of 59 mOsm/kg.[345]

## General Stability Considerations

Ranitidine hydrochloride solid dosage forms and solution should be packaged in tight, light-resistant containers. The tablets should be stored in a dry place at controlled room temperature. The capsules should be stored between 2 and 25°C. The effervescent tablets and granules are stored at 2 to 30°C.[2,4] The manufacturer has stated that ranitidine hydrochloride should not be removed from the original packaging and placed in dosing compliance aids because of hygroscopicity of the drug.[1622]

The solution should be stored between 4 and 25° and protected from freezing. The injection should be stored protected from light between 4 and 30°C and protected from freezing. Brief exposure of the injection to temperatures up to 40°C does not affect product stability adversely. A slight darkening of the injection does not indicate concentration loss.[2,4] Ferreira et al.[1116] reported no effect of pH within the range of 5 to 6.6 on the stability of ranitidine hydrochloride.

## Stability Reports of Compounded Preparations

### Injection

Injections, like other sterile drugs, should be prepared in a suitable clean air environment using appropriate aseptic procedures. When prepared from nonsterile components, an appropriate and effective sterilization method must be employed.

Nahata et al.[499] reported on the stability of ranitidine hydrochloride injection (Glaxo) diluted to a ranitidine concentration of 2.5 mg/mL with bacteriostatic water for injection. The diluted solution was repackaged in glass vials and polypropylene syringes (Becton Dickinson). The syringe tip closure was not specified. HPLC analysis was used to quantify the drug concentration.

At 4°C, little or no loss occurred in 28 days, and about 6% ranitidine loss occurred after 91 days of storage in either vials or syringes. Warming refrigerated syringes to 22°C for 72 hours after 91 days of refrigerated storage resulted in another 2% (total 8%) loss. Freshly prepared syringes stored at 22°C exhibited little or no loss in 72 hours.

### Oral

**STUDY 1:** Karnes et al.[98] evaluated the concentration uniformity and stability of a compounded ranitidine hydrochloride suspension. Thirty-six ranitidine (as the hydrochloride) 150-mg tablets (Glaxo) were crushed and suspended in 180 mL of distilled water and diluted with 180 mL of simple syrup, yielding a theoretical concentration of 150 mg/10 mL. The suspension was packaged in two-ounce amber glass bottles. The bottles were sonicated for 15 minutes to ensure

homogeneity and were stored at 25°C. The ranitidine content was assessed using a stability-indicating HPLC assay.

Losses of 9% or less occurred in seven days, but losses up to 18% occurred in 14 days. The authors recommended use within seven days of preparation stored at room temperature. They also noted the rapid sedimentation of the suspension, with about half of the total sediment forming in the first minute after shaking. The dose should be given immediately after shaking to ensure that the proper amount of drug is given.

**STUDY 2:** Schlatter and Saulnier[1109] evaluated the stability of ranitidine hydrochloride 5 mg/mL in several oral solution formulations. Ranitidine hydrochloride powder (Sigma) along with the commercial injection, effervescent 150-mg tablets, and effervescent granules (all from GlaxoWellcome) were used to prepare oral solutions in sterile water for injection in glass volumetric flasks. The sample solutions were stored under refrigeration at 4°C and protected from exposure to light for up to 60 days.

None of the solutions exhibited visible changes associated with microbial growth or crystalline precipitation during the study. Stability-indicating HPLC analysis found that the $t_{90}$ (time to 10% drug loss) under refrigeration was variable, depending on the source of the ranitidine hydrochloride in the formulation. The $t_{90}$ periods were about 28 days for the powder, about 24 days for the injection, about 54 days for the effervescent tablets, and about six days for the effervescent granules.

**STUDY 3:** Ferreira et al.[1116] evaluated the stability of ranitidine hydrochloride in several aqueous solutions, including simple syrup, for various periods at room temperature of 15 to 25°C and one sample refrigerated at 4°C. Test samples included ranitidine hydrochloride (1) 25 mg/mL in water for 153 days at room temperature and 139 days under refrigeration, (2) 25 mg/mL in simple syrup for 153 days at room temperature, (3) 50 mg/mL in water for 139 days at room temperature, and (4) 25 mg/mL in four aqueous buffer solutions of pH 5 to 6.6 for 144 days at room temperature. No changes in color were observed, and HPLC analysis found little or no loss of ranitidine hydrochloride in any of the samples.

**STUDY 4:** Lifshin and Fox[1200] evaluated the stability of an extemporaneously compounded ranitidine hydrochloride oral suspension in Ora-Plus suspending vehicle packaged in plastic containers. The samples were stored at room temperature and under refrigeration for six weeks. No visible change in color was observed. Stability-indicating HPLC analysis found that ranitidine concentrations after six weeks of storage were 84 and 100% at room temperature and under refrigeration, respectively. The authors recommended that ranitidine hydrochloride oral suspension be stored under refrigeration for up to six weeks.

**STUDY 5 (REPACKAGING):** Shah et al.[1222] evaluated the stability of commercial ranitidine 15 mg/mL (as 16.8 mg/mL of the hydrochloride) oral liquid (Zantac syrup, GlaxoSmithKline) repackaged into amber glass unit-dose bottles with aluminum caps. The samples were stored at room temperature of 25°C and 40% relative humidity for 52 weeks and at elevated temperature of 40°C and 25% relative humidity for 13 weeks. No visible changes and no change in pH occurred. Stability-indicating HPLC analysis found no loss of intact ranitidine hydrochloride and no increase in impurities within the study periods.

## Enteral

**RANITIDINE HYDROCHLORIDE COMPATIBILITY SUMMARY**
*Compatible with:* Ensure • Ensure Plus • Jevity • Nutren 1.0
 • Nutren 1.0 with fiber • Nutren 2.0
 • Peptamen • Sustacal

Crowther et al.[472] examined the stability of ranitidine hydrochloride with eight enteral nutrient formulas: Ensure, Ensure Plus, Jevity, Nutren 1.0, Nutren 1.0 with fiber, Nutren

2.0, Peptamen, and Sustacal. Ranitidine hydrochloride syrup 15 mg (base)/mL and 300-mg tablets were used as the sources of drug. To use the tablets, immediately before testing one tablet was crushed to fine powder and mixed with 20 mL of deionized water to yield a nominal ranitidine concentration of 15 mg/mL. The insoluble tablet residue was removed by filtration. The combinations were prepared by adding 0.667 mL (10 mg) of each ranitidine source into 50 mL of each enteral nutrient formula to result in a nominal ranitidine concentration of 0.2 mg/mL. The samples were stored at room temperature (22 to 25°C) under fluorescent light for nine hours followed by dark for 15 hours.

No creaming or flocculation was observed in any sample within 24 hours. HPLC analysis determined that a ranitidine concentration of at least 90% remained in all samples. However, some ranitidine binding to protein or fiber or partitioning into the lipid phase was found after ultrafiltration. The largest bound or partitioned amounts of ranitidine occurred in Nutren 2.0 (28%) and Ensure Plus (29%), the two enteral formulas with the highest protein, carbohydrate, and fat contents, while the lowest amount (8%) occurred with Peptamen. ■

# ■ Ribavirin
## (Tribavirin)

## Properties
Ribavirin is a white crystalline tasteless and odorless powder.[2,4,7]

## Solubility
Ribavirin is freely soluble in water, with a maximum solubility of 142 mg/mL at 25°C, and is slightly soluble in dehydrated ethanol.[4,7]

## pH
Ribavirin for inhalation solution reconstituted as directed in the labeling has a pH of 4 to 6.5.[4]

## Osmolarity
Ribavirin for inhalation solution reconstituted as directed in the labeling has an osmolarity of 82 mOsm/L.[2]

## General Stability Considerations
Ribavirin should be packaged in tight containers.[4] Ribavirin tablets and capsules are stored at controlled room temperature.[7] Ribavirin for inhalation solution is packaged in tight containers and stored in a dry place at controlled room temperature.[4] The reconstituted inhalation solution prepared under properly controlled aseptic conditions is clear and colorless and is stated by the manufacturer to be stable for 24 hours at room temperature.[2,7]

## Stability Reports of Compounded Preparations
### Oral
Chan et al.[909] evaluated the stability of an oral liquid preparation of ribavirin 200 mg/5 mL prepared with the formula in Table 100. Because ribavirin is a potential teratogen, proper safety precautions (goggles, gloves, face mask, protective garb) are required during handling and compounding. In a fume hood, the ribavirin capsule contents were incorporated into tragacanth with alcohol and some water to form a paste. The

**TABLE 100.** Ribavirin 200-mg/5-mL Oral Liquid Preparation Formula[909]

| Component | Amount |
|---|---|
| Ribavirin 200-mg capsule contents | 20 capsules |
| Tragacanth | 1.25 g |
| Alcohol 90% (vol/vol) | 2 mL |
| Benzoic acid solution, BP | 2 mL |
| (Benzoic acid 50 mg/mL in propylene glycol 75% in water) | |
| Compound hydroxybenzoate solution | 1 mL |
| (Methyl hydroxybenzoate 8 mg/mL, propyl hydroxybenzoate 2 mg/mL, in propylene glycol) | |
| Purified water | qs  100 mL |

benzoic acid solution and compound hydroxybenzoate solution were added along with additional water for rinsings. The preparation was brought to volume in the final container with water. Samples of the suspension were stored refrigerated at 4°C for 28 days.

Stability-indicating HPLC analysis found that ribavirin was stable, retaining adequate concentration for at least 28 days under refrigeration. ■

## ■ Rifabutin

### Properties
Rifabutin is an amorphous red-violet powder.[1,3]

### Solubility
Rifabutin is very slightly soluble in water, having an aqueous solubility of about 0.19 mg/mL.[3,7] It is sparingly soluble in ethanol.[3]

### General Stability Considerations
Rifabutin should be protected from light and excessive heat.[3,4] The commercial capsules should be packaged in well-closed containers and kept tightly closed. They should be stored at controlled room temperature.[4,7]

### Stability Reports of Compounded Preparations
#### Oral
Haslam et al.[655] evaluated the stability of two rifabutin 20-mg/mL oral suspensions prepared from capsule contents.

Two vehicles were used for preparing the suspensions; cherry syrup (Humco) was used for one, and an equal parts mixture of Ora-Plus suspending agent and Ora-Sweet (Paddock) was used for the other. Samples for testing were compounded in a manner analogous to the suggested compounding procedure. The suggested method was to empty the contents of eight rifabutin 150-mg capsules into a mortar. About 20 mL of the vehicle was incorporated using a pestle to wet the capsule contents, and this material was transferred to a two-ounce polyethylene terephthalate G bottle. Additional 20-mL aliquots of vehicle were poured into the mortar and mixed, and these rinses were added to the bottle until a total of 60 mL was reached.

Both suspensions were easily resuspended by gentle shaking, even after many weeks of sitting. Stability-indicating HPLC analysis found little or no rifabutin loss in either suspension formulation when stored at 4, 25, and 30°C for 12 weeks. ■

## ■ Rifampin
### (Rifampicin)

### Properties
Rifampin is a nearly odorless red to reddish-brown crystalline powder.[1,3,4]

### Solubility
Rifampin is slightly soluble in water, ethanol, and acetone.[1,3,1302] Aqueous solubility is pH-dependent. At pH 2, the solubility in water is 100 mg/mL. At pH 5.3, the water solubility is 4 mg/mL. At pH 7.5, the solubility drops to 2.8 mg/mL. The aqueous solubility of rifampin can be increased by the addition of ascorbic acid.[6]

### pH
A 1% suspension of rifampin in water has a pH between 4.5 and 6.5.[4] Reconstituted rifampin for injection 60 mg/mL has a pH between 7.8 and 8.8.[4]

### pKa
The drug has pKa values of 1.7 and 7.9.[1,6,1302]

### General Stability Considerations
Rifampin products should be protected from exposure to air, moisture, light, and excessive heat.[2,3,7] The drug is subject to specific acid catalysis at low pH;[6,1559] phosphate buffer also catalyzes the decomposition. Oxidation occurs at alkaline pH in the presence of oxygen, but the oxidation can be slowed by the addition of ascorbic acid. The pH of maximum stability is expected to be near neutrality.[6] Bentonite should not be included in rifampin dosage forms because it adsorbs the drug from solution.[6]

Rifampin capsules should be packaged in tight, light-resistant containers and stored at controlled room temperature and protected from temperatures of 40°C and higher.[2,7]

The powder for injection should be stored at controlled room temperature and protected from light and excessive heat of 40°C and higher.[2,7] The reconstituted injection is stable for 24 hours at room temperature.[2]

## Stability Reports of Compounded Preparations
### Oral
#### USP OFFICIAL FORMULATION:
> Rifampin 1.2 g
> Syrup, NF qs 120 mL

Use rifampin powder or commercial capsules. If using capsules, transfer the capsule contents into a mortar and use a pestle to gently crush the contents into a fine powder. Add about 2 mL of syrup and triturate to make a uniform paste. Add about 10 mL of syrup and triturate to form a suspension. Add additional syrup, up to about 80 mL. Quantitatively transfer to a suitable calibrated, tight, light-resistant 120-mL glass or plastic bottle. Rinse the mortar with successive small portions of syrup and add the rinses to the bottle. Shake vigorously. If necessary, add citric acid or sodium citrate to adjust to pH 5. A flavor may also be added. Bring the oral suspension to a final volume of 120 mL with syrup and shake vigorously, yielding rifampin 10 mg/mL oral suspension. The final liquid preparation should have a pH between pH 4.5 and 5.5. Store the preparation at controlled room temperature. The beyond-use date is 30 days from the date of compounding.[4,5]

**STUDY 1:** The manufacturer's labeling for rifampin capsules[7] cites the following extemporaneously compounded oral suspension containing rifampin 10 mg/mL for use by patients who have difficulty swallowing or who need lower doses. Syrup, NF, simple syrup (Humco), Syrpalta syrup (Emerson), or raspberry syrup (Humco) may be used to prepare the suspension. The contents of four 300-mg oral capsules or eight 150-mg oral capsules are emptied onto weighing paper. If necessary, the capsule contents are crushed with a spatula to produce a fine powder. The rifampin capsule powder is then transferred to a four-ounce amber glass or plastic prescription bottle made of high-density polyethylene, polypropylene, or polycarbonate. Using 20 mL of one of the syrups, the weighing paper and spatula are rinsed into the prescription bottle and the mixture is shaken vigorously. Then 100 mL more syrup is added to bring to volume and the mixture is vigorously shaken to ensure uniformity. The manufacturer notes that this rifampin 10-mg/mL oral suspension is stable for four weeks when stored at room temperature of 22 to 28°C or under refrigeration at 2 to 8°C.

**STUDY 2:** The stability of rifampin as 1% (wt/vol) suspensions in five syrups was evaluated by Krukenberg et al.[93] The compounding procedure cited in the Rifadin package insert was followed. Briefly, the contents of four rifampin 300-mg capsules (Rifadin, Merrell Dow Pharmaceuticals) were gently crushed and then placed in a 120-mL amber glass bottle with a child-resistant closure. The bottle was shaken after the addition of 20 mL of syrup and again after the addition of 100 mL of syrup. Five syrups were used to prepare the suspensions: syrup, NF; simple syrup (Humco); simple syrup (Whiteworth); wild

cherry syrup (Lilly); and Syrpalta (Emerson). The suspensions contained a theoretical rifampin concentration of 10 mg/mL or 1% (wt/vol). Samples were stored at 25 and 4°C for six weeks. Rifampin content was assessed using a stability-indicating HPLC assay and a microbiological assay.

All suspensions exhibited similar stability, with little or no rifampin loss at either temperature after six weeks of storage. Furthermore, the microbiological assay showed no loss of activity in microorganism inhibition.

**STUDY 3:** Allen[138] described 1% rifampin suspensions prepared from the contents of rifampin 300-mg capsules in simple syrup, wild cherry syrup, or Syrpalta. An expiration of four weeks when stored under refrigeration was recommended. Allen also noted that the capsule contents could be mixed with jelly or applesauce as an alternative to suspension compounding for patients unable to take the capsules.

**STUDY 4:** Nahata et al.[246] reported substantially lower rifampin concentrations in suspensions prepared from capsules compared to a solution prepared from the injection. A rifampin 10-mg/mL oral liquid prepared by adding the injection to syrup was compared to suspensions prepared by three different techniques: (1) the manufacturer's recommended approach of transferring capsule contents to a bottle and adding syrup, (2) triturating the capsule contents in syrup to form a paste and then mixing with additional syrup, and (3) triturating the capsule contents with syrup to form a paste, followed by retrituration with syrup to form a slurry and then mixing with additional syrup. The preparations were stored in plastic prescription bottles at 4°C.

By using a stability-indicating HPLC analysis, the oral liquid prepared from the injection was found to retain 97% of the initial concentration through 56 days of storage, but it dropped to 83% after 70 days. In contrast, the three suspensions prepared from capsules had highly variable initial concentrations of 14.5 to 68% of the concentration in the formulation prepared from the injection. Furthermore, the amount of rifampin increased slowly for several weeks. The authors noted that the suspensions prepared from the capsules may not provide the expected dose of rifampin. They could not explain their unusual findings but believed the use of plastic (rather than glass) prescription bottles may have contributed to the variability.

**STUDY 5:** In a parallel study, Nahata et al.[497] reported the stability of rifampin suspensions stored at room temperature. As in the previous study,[246] the rifampin suspension prepared from the injection retained at least 95% of the initial concentration for eight weeks stored at 22°C. Also, as in the previous study, the suspension prepared from capsule contents was again found to have a reduced initial drug concentration (76%), with the concentration changing over time. The authors speculated that this could be due to drug particles that are not well dispersed in spite of vigorous shaking. Once again, the suspension prepared from capsule contents may not deliver the expected dose.

**STUDY 6:** Allen and Erickson[595] evaluated the stability of three rifampin 25-mg/mL oral suspensions extemporaneously compounded from capsules. Vehicles used in this study were (1) an equal parts mixture of Ora-Sweet and Ora-Plus (Paddock), (2) an equal parts mixture of Ora-Sweet SF and Ora-Plus (Paddock), and (3) cherry syrup (Robinson Laboratories) mixed 1:4 with simple syrup. Ten rifampin 300-mg capsules (Ciba-Geigy) were emptied into a mortar. About 20 mL of the test vehicle was added to the powder and mixed to yield a uniform paste. Additional vehicle was added geometrically and brought to the final volume of 120 mL, with thorough mixing after each addition. The process was repeated for each of the three test suspension vehicles. Samples of each finished suspension were packaged in 120-mL amber polyethylene terephthalate plastic prescription bottles and stored at 5 and 25°C in the dark.

No visual changes or changes in odor were detected during the study. Stability-indicating HPLC analysis found 10% or less loss of rifampin in suspensions stored at either temperature after 28 days of storage.

### Enteral

**STUDY 1:** Ortega de la Cruz et al.[1101] reported the physical incompatibility of an unspecified amount of oral liquid rifampin (Rifaldin) with 200 mL of Precitene (Novartis) enteral nutrition diet. Particle growth was observed within three hours.

**STUDY 2:** de Villiers et al.[1569] evaluated the suitability for nasogastric enteral tube administration of five formulas of liquid rifampin prepared from capsule contents using two methods of preparation. The contents of two rifampin commercial capsules (600 mg rifampin) were prepared in 30 mL of the five liquid vehicles noted below.

1. Sterile water for irrigation
2. Syrup, NF
3. Vehicle for Oral Suspension, Sugar Free, NF
4. Ora-Plus:Ora-Sweet (1:1 vol/vol)
5. Ora-Plus

The liquids were prepared by mixing using a glass mortar and pestle and geometric dilution. Each suspension was mixed for five minutes using 25 seconds of continuous mixing followed by five seconds of no mixing. The liquids were also prepared using a prototype INSTA Formulation System device for five minutes using 25 seconds of continuous mixing followed by five seconds of no mixing.

The INSTA device yielded particles of about one-half to one-third the size of those yielded from manual mixing using a mortar and pestle. The sedimentation rate was generally better using the INSTA device compared to manual mixing.

The exception was when syrup, NF, was used. The injectability through an 8 French nasogastric tube was lowest when the preparation was mixed in water and syrup, NF. Using the Vehicle for Oral Suspension, Sugar Free, NF, Ora-Plus, and Ora-Plus/Ora-Sweet mixture, the injectabilities were comparable. When manual mixing was used, the acceptable period was about four days; when the INSTA device was used, the acceptable period was at least seven days.

### Nasal

Rao et al.[858] reported the stability of several potential formulations of rifampin 1% mucoadhesive nasal drops. The formulas that were evaluated are shown in Table 101. The suspensions were all uniform and did not settle rapidly. HPLC analysis found that Formulation 1, with ascorbic acid 0.1%, was the least stable, exhibiting 10% loss in 15 days under refrigeration and 7% loss in 12 days at 36°C. Formulations 2 and 3 were more stable; however, Formulation 3 was preferred because of its less acidic pH (about 6.5) and because it produced no undesirable discomfort. Accelerated stability studies at elevated temperatures project the acceptable stability periods of Formulation 3 to be 17 days at 36°C, 30 days at room temperature of 20°C, and 76 days refrigerated at 8°C.

## Compatibility with Other Drugs

**STUDY 1 (WITH ISONIAZID):** Seifart et al.[743] reported the stability of an extemporaneously compounded isoniazid 10-mg/mL suspension mixed with a commercial rifampin 20-mg/mL suspension (Merrell National). While both the isoniazid suspension and the rifampin suspension were stable separately, rapid decomposition of both drugs occurred when mixed. Stability-indicating HPLC analysis found about 10 to 12% loss of both drugs in two days stored at 4°C.

**STUDY 2 (WITH ISONIAZID AND PYRIDOXINE):** Ved and Deshpande[1628] evaluated the stability of rifampin 20 mg/mL with isoniazid 20 mg/mL and pyridoxine hydrochloride 1 mg/mL in a vehicle composed of sorbitol and propylene glycol (vehicle component concentrations not provided). Samples were stored at 15 and 30°C and at elevated temperatures of 37 and 45°C.

**TABLE 101.** Rifampin Nasal Drops Formulations[858]

| Component | Formulation | | |
| --- | --- | --- | --- |
| | 1 | 2 | 3 |
| Rifampin | 1 g | 1 g | 1 g |
| Hydroxypropyl methylcellulose K4M | 1.25 g | 1.25 g | 1.25 g |
| Tween 80 | 0.3 g | 0.3 g | 0.3 g |
| Ascorbic acid | 0.1 g | 0.5 g | 0.1 g |
| Sodium sulfite | — | — | 0.4 g |
| Water | 100 g | 100 g | 100 g |

The suspension was described as having a good appearance with easy redispersibility and low sedimentation rate. However, microbiological analysis of the rifampin content found 25 and 30% loss of antimicrobial activity after one week of storage at 15 and 30°C, respectively, with higher losses at elevated temperatures. The authors concluded that rifampin was unstable in combination with isoniazid and pyridoxine hydrochloride.

## Compatibility with Common Beverages and Foods

Peloquin et al.[1135] evaluated the stability of several oral antituberculosis drugs, including rifampin, for compatibility with common foods to facilitate administration to children. Rifampin capsules were opened and the contents mixed with 30 g of Hunt's Snackpack no-sugar-added chocolate pudding, Safeway grape jelly, and Safeway creamy peanut butter. The drug–food mixtures were thoroughly mixed and analyzed (no method indicated) for drug content initially and after one, two, and four hours had elapsed. Drug recovery at all time points was greater than 93% from the pudding, greater than 89% from the jelly, and greater than 80% from the peanut butter. To aid in administering the drugs to children, the authors recommended the use of the no-sugar-added chocolate pudding and the jelly. ■

# Rifapentine

## Properties

Rifapentine is a rifamycin derivative antibiotic that occurs as a reddish-brown crystalline material.[1,7]

## General Stability Considerations

Rifapentine oral tablets should be stored at controlled room temperature and protected from exposure to excessive heat and humidity.[7]

## Compatibility with Common Beverages and Foods

Peloquin et al.[1135] evaluated the stability of several oral antituberculosis drugs, including rifapentine, for compatibility with common foods to facilitate administration to children. Rifapentine tablets were crushed and mixed with 30 g of Hunt's Snackpack no-sugar-added chocolate pudding, Safeway grape jelly, Safeway creamy peanut butter, orange juice, and 7-Up soda. The drug–food mixtures were thoroughly mixed and analyzed (no method indicated) for drug content initially and after one, two, and four hours had elapsed. Drug recovery at all time points was greater than 93% from pudding, greater than 89% from jelly, and greater than 80% from peanut butter. Rifapentine recovery was about 100% from orange juice and from 7-Up. The authors recommended the use of the no-sugar-added chocolate pudding and the jelly to aid in administering the drugs to children. However, drug recovery from orange juice and 7-Up make these beverages likely candidates as well. ■

# Rifaximin

## Properties

Rifaximin occurs as a red-orange powder.[1]

### Solubility

Rifaximin is insoluble in water but soluble in ethanol and other alcohols, ethyl acetate, and toluene.[1]

### pH

Compounded oral liquid formulations had pH values near 4.2 to 4.3.[1523]

## General Stability Considerations

Commercial rifaximin oral tablets are to be stored at controlled room temperature.[7]

## Stability Reports of Compounded Preparations
### Oral

Cober et al.[1523] evaluated the stability of extemporaneously compounded oral liquid formulations of rifaximin 20 mg/mL. Commercial rifaximin (Xifaxan) oral tablets were crushed to fine powder and mixed with Ora-Plus suspending medium and an equal volume of either Ora-Sweet or Ora-Sweet SF (sugar-free) making a uniform suspension. The oral liquids were packaged in amber plastic prescription bottles with child-resistant caps and were stored at room temperature of 23 to 25°C for 60 days.

The oral liquid formulations did not exhibit any changes in color, odor, or taste and had no visible microbial growth. Stability-indicating HPLC analysis found that no loss of rifaximin in either formulation occurred in 60 days when stored at room temperature. Although both of the oral liquids were sweet, the drug had a very bitter aftertaste. To help mask the bitter aftertaste, the authors recommended administering chocolate syrup before administering the preparation, or mixing the oral liquid formulation with an equal volume of chocolate syrup immediately before administration, or both. ■

# ■ Risedronate Sodium

## Properties
Risedronate sodium is a white to off-white odorless crystalline powder. The drug exists as the hemipentahydrate.[1,7]

### Solubility
Risedronate sodium is soluble in water and in aqueous solutions but insoluble in organic solvent.[1,7]

### pH
Risedronate sodium 5- and 35-mg tablets dissolved in water have pH values of 6.2 and 5.5, respectively.[941]

## General Stability Considerations
Risedronate sodium (Actonel) tablets should be stored at controlled room temperature.[7]

## Stability Reports of Compounded Preparations
### Oral
**STUDY 1:** Dansereau and Crail[941] evaluated the solubilization and drug delivery of risedronate sodium prepared from tablets as a liquid for oral delivery using common feeding tubes. This study reported that grinding or crushing the hard tablets was unnecessary to prepare a liquid drug formulation because the tablets are designed for rapid dissolution. Adequate dissolution resulted from placing a tablet in two ounces of tap water in a variety of containers (plastic cups, Styrofoam cups, glass cups, Dixie cups, etc.) and allowing it to stand for two minutes, followed by stirring for 30 seconds. The cups were then rinsed with an additional four ounces of water to be compliant with the manufacturer's recommendation for taking this drug with six to eight ounces of water and to ensure removal of the entire amount of drug for analysis. The tablet contents dissolved in two ounces of water were also drawn into polypropylene syringes and delivered down feeding tubes. HPLC assays were performed on the solutions.

Recovered drug from all of the samples was acceptable and ranged from 95.7 to 100.5%. The solutions were also allowed to stand for four hours in all of the containers. No loss of risedronate sodium occurred. The authors recommended not permitting solutions to be stored prior to use.

**STUDY 2:** Dansereau and Crail[1299] evaluated the stability and compatibility of risedronate sodium tablets prepared in food thickeners to facilitate oral administration to patients with dysphagia. One risedronate sodium 35-mg tablet was added to a beaker containing two ounces of purified water at room temperature of 23 to 25°C and allowed to disintegrate. After two minutes, the water were stirred. Then an additional four ounces of water were added and stirred briskly. The appropriate amount of one of the five food thickeners cited in Table 102 was then added to the beaker using the procedure recommended by the manufacturer of the food thickener. Samples were stored at room temperature and tested for risedronate concentration. Stability-indicating HPLC analysis found that little or no loss of risedronate occurred with any of the food thickeners over 24 hours at room temperature. In addition, the drug concentration was no different after 24 hours than with no food thickener present. The authors noted that the safety and effectiveness of administering risedronate sodium in the target patient population have not been studied. ■

**TABLE 102.** Food Thickeners Evaluated for Compatibility with Risedronate Sodium Tablets[1299]

| Brand | Amount/4 fl ounces | Preparation |
|---|---|---|
| Hormel NutraThik | 1 tbsp | Slowly add NutraThik while vigorously hand mixing. Allow to stand for one minute to thicken. |
| Hormel Thick & Easy | 1 tbsp | Not provided |
| Milani Thick-It | 1 to 2 tbsp | Slowly add Thick-It. Stir briskly with a spoon. Allow to stand for 30 seconds to thicken. |
| Resource ThickenUp | 1 tbsp and 1 tsp | Slowly add ThickenUp while stirring briskly for 15 seconds. Allow to stand for one to five minutes to thicken. |
| Thik & Clear | 2 tsp or one scoop | Start stirring and slowly add Thik & Clear. Continue stirring for 30 seconds. Wait for five minutes and stir again. |

# Risperidone

## Properties

Risperidone is a white or almost white to slightly beige crystalline powder.[1,3,7]

### Solubility

Risperidone is practically insoluble in water and sparingly soluble in ethanol, but it dissolves in dilute acids such as 0.1 N HCl.[3,7]

## General Stability Considerations

Risperidone powder should be stored at controlled room temperature and protected from exposure to light.[3] Risperidone tablets are to be stored at controlled room temperature and protected from exposure to light and moisture. Risperidone oral solution should also be stored at controlled room temperature and should be protected from exposure to light and from freezing. Risperidone long-acting injection (Risperdal Consta) packaged as a vial of microencapsulated drug microspheres powder with an accompanying vial of special diluent should be stored under refrigeration and protected from exposure to light. The manufacturer states that the long-acting injection can be kept at temperatures up to 25°C (77°F) for up to seven days. The vials must not be exposed to temperatures exceeding 25°C.[7]

## Compatibility with Other Drugs

Park et al.[1191] evaluated the compatibility of a risperidone 1-mg/mL oral liquid (Janssen) mixed with lithium citrate syrup 300 mg/5 mL (Roxane). Various amounts of the two oral liquid drugs were mixed. Lithium citrate volumes ranged from 5 to 20 mL. Risperidone volumes ranged from 1 to 6 mL. Visual inspection found no evidence of haze, precipitation, change in viscosity, or color change over a 30-minute observation period. The authors concluded that these two oral liquids could be mixed together.

## Compatibility with Common Beverages and Foods

Aki et al.[985,1062] reported the compatibility of risperidone 1-mg/mL oral solution with bottled water and with brewed green tea, black tea, and oolong tea. A commercial risperidone oral solution (Rispadal Liquid, Janssen Pharmaceutica) was mixed with each liquid in a ratio of 3:97 (vol/vol) and allowed to stand at room temperature for 24 hours.

Using isothermal microcalorimetry, the authors found no evidence of interaction when the drug solution was mixed with water. However, when the drug solution was mixed with the tannin-containing beverages green tea, black tea, and oolong tea, cloudiness appeared immediately, with a white precipitate appearing after 24 hours. Using isothermal microcalorimetry, an exothermic reaction was noted, with the risperidone being bound as a 1:1 molar ratio chemical complex to tannin from the teas. The complex was insoluble, leading to an immediate loss of risperidone of nearly 30% from green teas and more than 80% from black tea and oolong tea after 24 hours. The authors indicated that the complexation could reduce the absorption of risperidone. Consequently, this study confirmed the manufacturer's labeling statement not to dilute risperidone with tea. ■

# Rofecoxib

## Properties

Rofecoxib is a white or off-white to light yellow powder.[7]

### Solubility

Rofecoxib is insoluble in water and practically insoluble in octanol, having solubilities of about 9 mcg/mL and 0.12 mg/mL, respectively. It is only slightly soluble in ethanol (0.68 mg/mL), methanol (0.835 mg/mL), and isopropyl acetate and sparingly soluble in acetone. The drug exhibits better solubility in propylene glycol (1.15 mg/mL) and polyethylene glycol 400 (11.2 mg/mL).[7,1504]

### pKa

Rofecoxib has a pKa of 3.5.[7]

## General Stability Considerations

Rofecoxib tablets and oral suspension should be stored below 30°C.[7]

## Stability Reports of Compounded Preparations

### Rectal

Abou-Taleb et al.[1066] evaluated the stability of rofecoxib in several rectal suppository formulations. Rofecoxib 12.5-mg/1-g suppository was prepared using a number of water-soluble suppository bases. Three formulas that gave the highest in vitro drug release were then tested for stability. The selected suppository bases for stability testing were (1) 75% (wt/wt) polyethylene glycol (PEG) 6000 and 25% (wt/wt) PEG 1000, (2) Witepsol E75 with 10% (wt/wt) polysorbate 80, and

(3) cocoa butter with 10% (wt/wt) polysorbate 80. HPLC analysis of the rofecoxib concentration found little or no loss of drug (less than 2%) in 180 days stored at room temperature and under refrigeration. The calculated times to 10% decomposition for the three formulas were (1) 3.5 years at room temperature and 5.4 years under refrigeration, (2) 2.6 years both at room temperature and under refrigeration, and (3) 4.1 years at room temperature and 3.3 years under refrigeration. ▪

## ▪ Rufinamide

### Properties
Rufinamide occurs as a white, odorless, slightly bitter-tasting crystalline powder.[1,7]

### Solubility
Rufinamide is insoluble in water and very slightly soluble in ethanol.[7] Rufinamide exhibits moderate solubility in 0.1 N hydrochloric acid.[1]

### pH
Compounded oral liquid formulations reported by Hutchinson et al.[1483] had pH values near 4.3 to 4.4. See the *Oral* section below.

### General Stability Considerations
Rufinamide oral tablets are to be stored at controlled room temperature and protected from moisture. The manufacturer recommends that the container cap be securely replaced after opening.[7]

### Stability Reports of Compounded Preparations
#### Oral
Hutchinson et al.[1483] evaluated the stability of rufinamide 40-mg/mL oral suspensions compounded from commercial oral tablets (Banzel, Eisai). Rufinamide 400-mg tablets were crushed and triturated to powder using a glass mortar and pestle; Ora-Plus suspending vehicle was then incorporated with thorough mixing, resulting in a homogeneous suspension. In one set of samples, Ora-Sweet in a volume equivalent to the Ora-Plus was incorporated, resulting in a rufinamide concentration of 40 mg/mL. In another set of samples, Ora-Sweet SF (sugar-free) in a volume equivalent to the Ora-Plus was incorporated, also resulting in a 40-mg/mL concentration. The samples were packaged in amber polypropylene plastic bottles with child-resistant caps. Samples of each oral liquid formulation were stored at room temperature of 23 to 25°C for 90 days.

Visual inspection of the two formulations did not reveal any changes, including any microbial growth. The pH values of the two formulations remained at about pH 4.3 to 4.4 throughout the study. Stability-indicating HPLC analysis found no loss of rufinamide in the Ora-Sweet-containing formulation for 56 days and about 8% loss of drug in 90 days. In the Ora-Sweet SF (sugar free) formulation, no loss of rufinamide occurred in 90 days. ▪

# Salicylic Acid

## Properties
Salicylic acid is a white fluffy powder or white or colorless needlelike crystals having a somewhat sweet taste and an acrid aftertaste.[2,3]

### Solubility
Salicylic acid has an aqueous solubility of about 2.2 mg/mL. In boiling water, the solubility increases to about 67 mg/mL. In ethanol, the solubility is about 333 to 370 mg/mL.[1-3]

### pH
A saturated aqueous solution of salicylic acid has a pH of about 2.4.[1]

### pK$_a$
The drug has pK$_a$ values of 2.97 and 13.4.[2]

## General Stability Considerations
Salicylic acid should be packaged in well-closed containers protected from light.[1-3] Exposure to sunlight results in gradual discoloration.[1,2] It is incompatible with iron salts and iodide.[1]

## Stability Reports of Compounded Preparations
### Topical
**STUDY 1:** Pesko[255] reported on an extemporaneous topical formulation for scalp psoriasis. Liquor carbonis detergens (LCD) 20 g was allowed to sit exposed to the environment at room temperature for several hours to permit alcohol evaporation. Salicylic acid 20 g was mixed with a small amount of castor oil and the evaporated LCD was added and mixed thoroughly. A sufficient amount of Unibase then was added to the mixture to make 200 g of a liquidlike cream. The author recommended a 30-day expiration period, although no stability study was reported.

**STUDY 2:** Allen[597] described a topical ointment for psoriasis. Each component in the formula shown in Table 103 was accurately weighed. The sulfur and salicylic acid powders were mixed and reduced to fine powder, and a few drops of glycerin were incorporated to make a paste. The paste was mixed with the fluocinonide ointment (Lidex, Medicis), and the mixture was incorporated into the Aquaphor or Aquabase geometrically with thorough mixing. The crude coal tar was added last and mixed well. The final ointment was packaged in tight containers and protected from light. The author indicated that a use period of six months is appropriate for this ointment, although no stability data were provided.

## Compatibility with Other Drugs
**STUDY 1 (CALCIPOTRIENE):** Patel et al.[635] evaluated the compatibility of calcipotriene 0.005% ointment mixed by levigation with other topical medications. The mixtures were stored at 5 and 25°C. Drug concentration was assessed by HPLC analysis. The combination with salicylic acid was found to be incompatible. When calcipotriene was mixed with salicylic acid to yield a 6% concentration, physical separation occurred in 24 hours, and the calcipotriene content dropped precipitously, losing almost all of the drug in about five hours.

**TABLE 103.** Psoriasis Topical Ointment Formula[597]

| Component | Amount |
| --- | --- |
| Precipitated sulfur | 3 g |
| Salicylic acid | 1 g |
| Glycerin | Several drops |
| Fluocinonide (Lidex) 0.05% ointment | 24 g |
| Aquaphor or Aquabase | 70 g |
| Coal tar | 2 g |

**STUDY 2 (MUPIROCIN):** Jagota et al.[920] evaluated the compatibility and stability of mupirocin 2% ointment (Bactroban, Beecham) with salicylic acid 6% gel (Keralyt, Westwood) mixed in a 1:1 proportion over periods up to 60 days at 37°C. The physical compatibility was assessed by visual inspection, while the chemical stability of mupirocin was evaluated by stability-indicating HPLC analysis. The study found that mupirocin 2% ointment mixed with salicylic acid 6% gel resulted in separation and layering, along with nearly total loss of mupirocin in 15 days.

**STUDY 3 (ANTHRALIN):** Hager and Kaestner[1710] evaluated the stability of anthralin when mixed in an oily formulation with salicyclic acid. The formulation also contained castor oil, peanut oil, and Emulgator MF. Samples were stored at room temperature of 20 to 25°C and refrigerated at 2 to 4°C for six months. The stability of anthralin was found to be dependent on storage temperature and exposure to atmospheric oxygen. The authors recommended that the preparation be stored in completely filled containers for up to two months at room temperature or under refrigeration for up to six months. ■

# ■ Saquinavir

## Properties
Saquinavir is a white to off-white crystalline powder.[1,7]

### Solubility
Saquinavir is insoluble in water.[7]

### $pK_a$
Saquinavir has $pK_a$ values of 1.1 and 7.01 or 7.1. [2,747]

## General Stability Considerations
Saquinavir capsules (Fortovase) should be stored under refrigeration at 2 to 8°C. Once the capsules are brought to room temperature, they should be used within three months.[7]

Tan et al.[747] reported that pH has a significant effect on the stability of saquinavir. The drug is most stable in the pH range of 2 to 4. At higher pH values, oxidative decomposition by atmospheric oxygen is favored.

## Stability Reports of Compounded Preparations
### Oral
Tan et al.[747] reported the stability of saquinavir (Fortovase, Roche) 60 mg/mL in a pH-adjusted alcohol-based oral liquid. The soft-gelatin capsules were slit and the liquid contents squeezed out into a conical graduate. The emptied capsule shell was then slit entirely open and placed in a beaker with an aliquot of absolute alcohol. This was transferred to the graduate, the beaker was rinsed with additional absolute ethanol, and additional absolute ethanol was added to bring the solution to volume. A sufficient amount of citric acid to result in a final concentration of 0.5% (wt/vol) was triturated with purified water using a mortar and pestle, and the solution was transferred to the graduate. The mortar and pestle were rinsed with additional purified water. Syrup was added to result in a concentration of 10% (vol/vol). The prepared oral liquid was packaged in amber glass screw-cap prescription bottles and stored at 5 and 25°C for 30 days. Stability-indicating HPLC analysis found that no loss of saquinavir occurred at either temperature. ■

# ■ Scopolamine Hydrobromide
## (Hyoscine Hydrobromide)

## Properties
Scopolamine hydrobromide occurs as an odorless, colorless, white, or almost white crystalline material or granular powder.[1,3,4]

### Solubility
Scopolamine hydrobromide is freely soluble in water to 667 mg/mL. The drug is soluble in ethanol to about 50 mg/mL. It is only slightly soluble in chloroform and practically insoluble in ether.[1,4]

### pH
Scopolamine hydrobromide at a concentration of 0.05 M in water has a natural pH of 5.85.[1] A 50-mg/mL aqueous solution has a pH between 4 and 5.5.[3] Scopolamine hydrobromide injection has a pH between 3.5 and 6.5. Scopolamine hydrobromide ophthalmic solution has a pH between 4 and 6.[4]

### Osmolality
Scopolamine hydrobromide injection 0.5 mg/mL is isotonic, having an osmolality of 303 mOsm/kg.[345]

## General Stability Considerations
Scopolamine hydrobromide bulk material and oral tablets should be packaged in tight, light-resistant containers and stored at controlled room temperature. The ophthalmic

solution should be packaged in tight containers, and the ophthalmic ointment should be packaged in collapsible ophthalmic ointment tubes. Both ophthalmic forms should be stored at controlled room temperature. Scopolamine hydrobromide injection is to be packaged in light-resistant single- or multiple-dose Type I glass containers and stored at controlled room temperature.[4]

Scopolamine hydrobromide is incompatible with alkaline solutions because of a potential for forming haze or precipitation.[2]

## Stability Reports of Compounded Preparations
### Nasal
Gupta[1408] evaluated the stability of scopolamine hydrobromide 4-mg/mL nasal solution. The solution was prepared by dissolving scopolamine hydrobromide bulk material in 0.9% sodium chloride solution and adjusting the pH to near 5 with a pH 5 buffer solution composed of 0.1 M citric acid and 0.2 M disodium phosphate in purified water. The solution was brought to volume with additional 0.9% sodium chloride solution, and the solution was mixed well. The finished solution was packaged in amber glass prescription bottles and stored at room temperature for 42 days.

The physical appearance of the scopolamine hydrobromide 4-mg/mL nasal solution (which was not described in the article) remained unchanged throughout the study. The initial pH of 4.9 remained unchanged as well. A stability-indicating HPLC analytical method was developed and found no loss of scopolamine hydrobromide over the 42-day study period. ■

# ■ Sildenafil Citrate

## Properties
Sildenafil citrate is a white to off-white crystalline powder.[7]

### Solubility
Sildenafil citrate has an aqueous solubility of 3.5 mg/mL.[7]

### $pK_a$
Sildenafil citrate has $pK_a$ values of 6.5 and 9.2.[2]

## General Stability Considerations
Sildenafil citrate oral tablets (Viagra) should be stored at controlled room temperature.[7]

## Stability Reports of Compounded Preparations
### Oral
**STUDY 1:** Nahata et al.[971] evaluated the extended stability of two oral suspension formulations of sildenafil citrate for use in treating pulmonary hypertension in children. Sildenafil citrate tablets (Viagra, Pfizer) were ground to fine powder and incorporated into two suspension media: (1) methylcellulose 1% and simple syrup mixed in equal quantities and (2) Ora-Sweet and Ora-Plus mixed in equal quantities, yielding a sildenafil citrate concentration of 2.5 mg/mL. The suspensions were packaged in amber polyethylenephthalate plastic prescription bottles and were stored at room temperature of 25°C and also under refrigeration at 4°C for 91 days.

No changes in pH, color, odor, or turbidity occurred with either storage condition. Stability-indicating HPLC analysis found less than a 2% loss of sildenafil citrate at both storage conditions over 91 days of storage.

**STUDY 2:** Nahata et al.[1229] repeated the previous study when oral tablets of sildenafil citrate that are indicated for the treatment of pulmonary hypertension became commercially available. These tablets are formulated with different excipients and dyes. Sildenafil citrate 20-mg tablets (Revatio, Pfizer) were ground to fine powder and incorporated into two suspension media: (1) methylcellulose 1% and simple syrup mixed in equal quantities and (2) Ora-Sweet and Ora-Plus mixed in equal quantities, yielding a sildenafil citrate concentration of 2.5 mg/mL. The suspensions were packaged in amber plastic prescription bottles and were stored at room temperature of 25°C and also under refrigeration at 4°C for 91 days.

As in the previous study, no changes in pH, color, odor, or turbidity occurred with either storage condition. Stability-indicating HPLC analysis found less than a 2% loss of sildenafil citrate under both storage conditions over 91 days of storage. ■

# Silver Lactate

## Properties
Silver lactate monohydrate is a white or slightly gray crystalline powder.[1]

### Solubility
Silver lactate monohydrate is soluble in about 15 parts of water. It is slightly soluble in ethanol.[1]

## General Stability Considerations
Silver lactate monohydrate should be stored in a cool location. The material is light-sensitive and should be protected from exposure to light.[1]

## Stability Reports of Compounded Preparations
### Topical
Mackey et al.[1228] reported the stability of the silver ion content in a topical cream containing silver lactate 1.8% (wt/wt) and allantoin 1.2% (wt/wt) in an unspecified hydrophilic base.

The cream was prepared and packaged in 30-g amber glass ointment jars that were stored at room temperature of 25°C exposed to light and protected from exposure to light, under refrigeration at 4°C protected from exposure to light, and frozen at 0°C protected from exposure to light for 90 days.

The minimum inhibitory concentration was evaluated using *Pseudomonas aeruginosa,* a common pathogen in burn wound sepsis. The testing found substantial and rapid loss of activity in the room temperature and refrigerated samples. Losses were 32% (light) and 23% (dark) in three days at room temperature. Losses were 12 and 16% after three and 10 days of refrigerated storage, respectively. The room temperature and refrigerated samples changed color during storage as well. When the cream was stored frozen at 0°C and not exposed to light, no change in appearance occurred; little or no loss occurred within 15 days and about 5% loss occurred in 90 days. The consistency of the cream after thawing was reported to have no perceptible changes. ■

# Silver Nitrate

## Properties
Silver nitrate occurs as odorless and colorless or white crystals or crystalline powder.[1–3]

### Solubility
Silver nitrate is freely soluble in water, having an aqueous solubility of 1 g/0.4 to 0.5 mL.[1–3,418] In boiling water, the solubility increases to about 10 g/mL.[1,2] In ethanol, it has a solubility of 33 to 37 mg/mL.[1–3] In boiling ethanol, its solubility increases to about 154 mg/mL.[1]

### pH
Aqueous and alcoholic solutions are neutral to litmus.[1] Aqueous solutions of silver nitrate have a pH variously cited as 5.5,[3] approximately 6,[1] and 7.[418] Silver nitrate ophthalmic solution has a pH between 4.5 and 6.[2,4]

## General Stability Considerations
Silver nitrate exposed to light in the presence of organic matter discolors, becoming gray or grayish-black because of reduction to metallic silver. Silver nitrate powder or crystals should be packaged in tight, light-resistant containers. Solutions should be protected from light for long-term storage.[2–4,418]

Silver nitrate ophthalmic solution should be packaged in tight, light-resistant nonmetallic containers, stored at controlled room temperature, and protected from freezing. The solution should not be used when cold.[2–4]

Silver nitrate is stated to be incompatible with various materials, including alkalies, benzalkonium chloride, bromides, carbonates, chlorides, ferrous salts and salts of other metals, halogenated acids and salts, hypophosphites, iodides, morphine and its salts, oils, phosphates, tartrates, thimerosal, thiocyanates, and vegetable extracts.[1,2]

## Stability Reports of Compounded Preparations
### Irrigation
Sterile bladder irrigations, like other sterile drugs, should be prepared in a suitable clean air environment using appropriate aseptic procedures. When prepared from nonsterile components, an appropriate and effective sterilization method must be employed.

Pesko[418] described the compounding of a simple solution of 1% silver nitrate for use as a bladder irrigation to treat bladder hemorrhage. Silver nitrate crystals 10 g were dissolved in 100 mL of sterile water for irrigation, USP, and the solution was sterilized by filtration through a 0.22-μm filter. The sterilized solution was placed in a sterile 1-liter glass bottle and brought to volume with sterile water for injection. The use of glass bottles is required because silver nitrate reacts with polyvinyl chloride bags, forming a black precipitate. Even in glass bottles, silver nitrate may react with alkali in the glass, resulting in a yellow-brown discoloration, especially if the solution is terminally autoclaved. The author recommended use of the solution within 24 hours, although no specific stability information was available. ■

# Silver Sulfadiazine

## Properties
Silver sulfadiazine is a white or creamy-white nearly odorless fluffy crystalline powder.[2,3]

### Solubility
Silver sulfadiazine is practically insoluble in water and in ethanol.[2,3]

## General Stability Considerations
Silver sulfadiazine products should be packaged in tight, light-resistant containers, stored at controlled room temperature, and protected from light.[2-4]

Silver sulfadiazine can develop a yellow discoloration upon exposure to light. It decomposes in moderately strong mineral acids.[3] Free silver may be released from a reaction with most heavy metals, causing a darkening of the topical cream.[2] Discolored products should be discarded.[2]

The silver in silver sulfadiazine may possibly inactivate proteolytic enzymes if used in combination.[2]

## Stability Reports of Compounded Preparations
### Topical
Allen[577] described the preparation of a silver sulfadiazine–hydrocortisone topical nonalcoholic gel. The formula is shown in Table 104. To avoid chemical reactions, only glass, hard rubber, or plastic equipment should be used; stainless steel spatulas should be avoided. The topical gel was prepared by mixing the silver sulfadiazine, hydrocortisone, and carbomer 940 (Carbopol 940) with a small amount of glycerin and mixing until a smooth mixture was formed. Separately, the mineral oil and the polysorbate 80 were combined, mixed, and added to the purified water with thorough mixing. The drug mixture was then incorporated into the oil and surfactant mixture and blended well. Trolamine

**TABLE 104.** Silver Sulfadiazine–Hydrocortisone Gel Formula of Allen[577]

| Component | | Amount |
|---|---|---|
| Silver sulfadiazine | | 1 g |
| Hydrocortisone | | 2 g |
| Carbomer 940 (Carbopol 940) | | 0.5 g |
| Mineral oil | | 8 mL |
| Polysorbate 80 | | 1.5 mL |
| Trolamine (triethanolamine) | | 10 to 15 drops |
| Purified water | qs | 100 g |

(triethanolamine) was added a drop at a time until a gel formed. The gel was mixed until it was uniform and then was packaged in a tight container. Although no specific stability information has been developed on this preparation, silver sulfadiazine is known to react with most heavy metals, resulting in the release of free silver metal and a darkened appearance of the preparation. If this occurs, the preparation must be discarded.

## Compatibility with Other Drugs
Jagota et al.[920] evaluated the compatibility and stability of mupirocin 2% ointment (Bactroban, Beecham) with silver sulfadiazine 1% (Silvadene, Marion Merrell Dow) mixed in a 1:1 proportion over periods of up to 60 days at 37°C. The physical compatibility was assessed by visual inspection, while the chemical stability of mupirocin was evaluated by stability-indicating HPLC analysis. The study found that mupirocin 2% ointment mixed with silver sulfadiazine 1% was physically compatible and chemically stable, with less than 10% mupirocin loss in 45 days. ■

# Simethicone

## Properties
Simethicone is a gray translucent viscous liquid composed of polydimethylsiloxane polymer (molecular weight 14,000 to 21,000) 90.5 to 99% with silicon dioxide 4 to 7% to enhance the defoaming properties of the silicone.[2,3]

### Solubility
Simethicone is practically insoluble and immiscible with water and ethanol.[2,3]

### pH
Simethicone oral suspension has a pH between 3.5 and 4.6.[4] Simethicone drops (Mylicon, Stuart) had a measured pH of 4.5.[19]

## General Stability Considerations
Simethicone chewable and regular tablets should be packaged in well-closed containers and stored at controlled room temperature protected from temperatures of 40°C or higher. The

oral suspension should be packaged in tight, light-resistant containers and stored at controlled room temperatures protected from freezing and temperatures of 40°C or higher.[2,4]

## Stability Reports of Compounded Preparations
### Enteral
#### SIMETHICONE COMPATIBILITY SUMMARY
*Compatible with:* Enrich • Ensure • Ensure HN • Ensure Plus • Ensure Plus HN • Osmolite • Osmolite HN • TwoCal HN • Vital • Vivonex T.E.N.

**STUDY 1:** Cutie et al.[19] added 1.2 mL of simethicone drops (Mylicon, Stuart) to varying amounts (15 to 240 mL) of Ensure, Ensure Plus, and Osmolite (Ross Laboratories) with vigorous agitation to ensure thorough mixing. The simethicone drops were physically compatible, distributing uniformly in all three enteral products with no phase separation or granulation. However, the measured viscosity of the enteral products increased about 20 to 30% in the smallest volume of enteral product, around 15 mL.

**STUDY 2:** Burns et al.[739] reported the physical compatibility of simethicone drops (Mylicon) 0.6 mL with 10 mL of three enteral formulas, including Enrich, TwoCal HN, and Vivonex T.E.N. Visual inspection found no physical incompatibility with any of the enteral formulas.

**STUDY 3:** Altman and Cutie[850] reported the physical compatibility of simethicone drops (Mylicon, Stuart) 1.2 mL with varying amounts (15 to 240 mL) of Ensure HN, Ensure Plus HN, Osmolite HN, and Vital after vigorous agitation to ensure thorough mixing. The simethicone drops were physically compatible, distributing uniformly in all four enteral products with no phase separation or granulation. However, the measured viscosity of the enteral products increased about 20 to 30% in the smallest volume of enteral product, around 15 mL. ■

# ■ Sisomicin Sulfate

## Properties
Sisomicin sulfate occurs as a white to off-white hygroscopic powder material.[1,7] Sisomicin is an aminoglycoside antibiotic chemically related to gentamicin.[3]

### Solubility
Sisomicin sulfate is soluble in water but insoluble in ethanol, acetone, and ether.[7]

### pH
Sisomicin sulfate 40 mg/mL is within the range of pH 3.5 to 5.5. Sisomicin sulfate injection has a pH of 2.5 to 5.5.[4]

## General Stability Considerations
Sisomicin sulfate powder should be packaged in tight containers and should be stored at controlled room temperature. Sisomicin sulfate injection is packaged in single- or multiple-dose Type 1 glass containers.[4,7]

## Stability Reports of Compounded Preparations
### Topical
Yamamura et al.[1631] evaluated the stability of sisomicin sulfate compounded to yield a concentration of 0.3% (wt/wt) in hydrophilic petrolatum for use as a topical ointment to treat burn patients and to treat impetigo. Sisomicin sulfate injection 300 mg in 6 mL was gradually incorporated into 100 g of hydrophilic petrolatum using a mortar and pestle. Tocopherol alcohol solution was added to some of the samples to give a concentration of 0.05% (wt/wt). The sisomicin 0.3% ointment was packaged in porcelain-white plastic jars and was stored refrigerated at 5°C protected from exposure to light; at room temperature near 25°C both protected from and exposed to daylight; and at elevated temperature of 40°C exposed to daylight.

Stability-indicating HPLC analysis found 6 to 7% sisomicin loss on the ointment surface and less than 2% sisomicin loss in the ointment interior over 90 days stored under refrigeration protected from light. However, at room temperature and at 40°C, excessive drug loss was found on the ointment surface in 30 days and on both the surface and the interior after 60 days of storage. The authors stated that this ointment formulation was stable for 90 days when stored under refrigeration. They also noted that sisomicin is more sensitive to surface oxidation than light exposure. ■

# Sodium Bicarbonate

## Properties

Sodium bicarbonate is a white crystalline powder or granules having a salty slightly alkaline taste.[1-3] Each gram of sodium bicarbonate contains about 12 mEq of sodium and bicarbonate ions.[2] Sodium bicarbonate 3.65 g is approximately equivalent to 1 g of sodium.[3]

### Solubility

Sodium bicarbonate is soluble in water, having an aqueous solubility of about 83 mg/mL. It is insoluble in ethanol.[2,3]

### pH

Aqueous solutions of sodium bicarbonate are slightly alkaline to litmus or phenolphthalein.[1] The pH of a freshly prepared 0.1 M solution at 25°C is 8.3.[1] Sodium bicarbonate injection has a pH of 7 to 8.5.[4]

## General Stability Considerations

Sodium bicarbonate tablets and powder should be packaged in well-closed containers and stored at controlled room temperature.[4] Sodium bicarbonate injection should be stored at controlled room temperature and protected from freezing.[2,4]

In moist air, sodium bicarbonate slowly decomposes to sodium carbonate, carbon dioxide, and water. It is stable in dry air.[2]

Heat sterilization may result in the formation of sodium carbonate, a much more alkaline compound. Consequently, the pH of heat-sterilized products should be tested prior to product use to confirm that the pH is appropriate.[2]

Sodium bicarbonate combined with acids in aqueous solutions results in the evolution of carbon dioxide gas bubbles and effervescence.[2]

## Compatibility with Other Drugs

### SODIUM BICARBONATE COMPATIBILITY SUMMARY

*Compatible with:* Sodium chloride 7% (hypertonic)
• Terbutaline sulfate
*Incompatible with:* Albuterol sulfate • Atropine sulfate
• Isoetharine • Liquid protein • Metaproterenol sulfate

**STUDY 1 (MULTIPLE DRUGS):** Owsley and Rusho[517] evaluated the compatibility of several respiratory therapy drug combinations. Sodium bicarbonate 8.4% injection (Abbott) was combined with the following drug solutions: albuterol sulfate 5-mg/mL inhalation solution (Proventil, Schering), atropine sulfate 1 mg/mL (American Regent), isoetharine 1% solution (Bronkosol, Sanofi Winthrop), metaproterenol sulfate 5% solution (Alupent, Boehringer Ingelheim), and terbutaline sulfate 1-mg/mL solution (Brethine, Geigy). The test solutions were filtered through 0.22-μm filters into clean vials. The combinations were evaluated over 24 hours (temperature unspecified) using the USP particulate matter test. Only the terbutaline sulfate–sodium bicarbonate combination was found to be compatible. Sodium bicarbonate combined with the albuterol sulfate, atropine sulfate, isoetharine, and metaproterenol sulfate all formed unacceptable levels of larger particulates (≥10 μm).

**STUDY 2 (HYPERTONIC SODIUM CHLORIDE):** Fox et al.[1239] evaluated the physical compatibility of hypertonic sodium chloride 7% with several inhalation medications used in treating cystic fibrosis. Sodium bicarbonate 8.4% injection (American Regent) mixed in equal quantities with extemporaneously compounded hypertonic sodium chloride 7% did not exhibit any visible evidence of physical incompatibility, and the measured turbidity did not increase over the one-hour observation period.

**STUDY 3 (LIQUID PROTEIN):** Freeman and Trezevant[1421] reported abdominal distention in a patient from the formation of excessive gas (most likely carbon dioxide) when omeprazole oral suspension compounded from capsule contents mixed with sodium bicarbonate 7.5% was administered at the same time with liquid protein solution (Proteinex, Llorens Pharmaceuticals). The same effect could be duplicated in vitro using commercial Zegerid (Santarus), which contains omeprazole 20 mg with sodium bicarbonate 1680 mg or simply mixing sodium bicarbonate 7.5% solution with the liquid protein solution. The authors recommended separating the administration of omeprazole with sodium bicarbonate and liquid protein solution by at least two hours. ∎

# Sodium Gualenate Hydrate
## (Sodium Azulene Sulfonate)

## Properties
Sodium gualenate hydrate (the 3-sulfonate sodium salt of guaiazulene) is a blue semi-solid material that melts at room temperature. It is one of several related molecules that have been termed "Azulene." It is a component found in the plant *Matricaria chamomilla* and exhibits anti-inflammatory properties.[1,3]

## pH
Sodium gualenate hydrate 200:1 aqueous dilution of Azunol 2-mg tablets was stated to have a pH in the range of 6 to 9. Measurement of sodium gualenate 0.006% solution is around pH 8 to 9.[1527]

## General Stability Considerations
Sodium gualenate hydrate tablets (Azunol) are stored at controlled room temperature.[7]

## Stability Reports of Compounded Preparations
### Oral
Ushiyama et al.[1527] evaluated the stability of sodium gualenate gargle solutions compounded from Azunol tablets for use in treating acute oral pain due to oral mucositis from chemotherapy or radiation therapy. Two formulas were tested:

1. Sodium gualenate 0.006% in water
2. Sodium gualenate 0.006% in water with lidocaine hydrochloride 0.08%

The gargle solutions were prepared by dissolving 30 sodium gualenate hydrate 2-mg tablets (Azunol, Nippon Shinaku Co.) in 1000 mL of water. For the gargle solution with lidocaine hydrochloride, 20 mL of lidocaine hydrochloride 4% solution (Xylocaine, AstraZeneca) was then added. Samples of the solutions were stored at 4, 25, and 37°C and protected from light or subjected to light levels of 500 and 1000 lux for seven days.

The loss of sodium gualenate from the solutions was dependent on temperature and light exposure. The best stability occurred in the solutions stored under refrigeration and protected from light with little or no loss of the drug occurring in seven days. At the highest light exposure and at higher temperatures, loss of sodium gualenate was substantial, with residual drug concentrations as low as 18%. ■

# Sodium Hypochlorite

## Properties
Sodium hypochlorite solution, USP, is a clear pale greenish-yellow liquid having the odor of chlorine. It has not less than 4% and not more than 6% by weight of sodium hypochlorite.[4]

Sodium hypochlorite topical solution, USP, contains not less than 0.2% and not more than 0.32% (target 0.25%) of sodium hypochlorite. It is prepared by diluting 5 mL of sodium hypochlorite solution, USP, to 1000 mL in water containing phosphate buffers.[4] See the *Topical* section.

Household bleach contains approximately 5.25% (wt/vol) of sodium hypochlorite[1] and may contain sodium hydroxide to enhance alkalinity.[1442]

### Solubility
Sodium hypochlorite aqueous solubility is 293 mg/mL.[1]

### pH
Sodium hypochlorite topical solution, USP, has a pH of 7.8 to 8.2.[4] Commercial household bleach (sodium hypochlorite 5.25%) has a pH between 11 and 12 and may contain sodium hydroxide to enhance alkalinity.[1442] Acidified sodium hypochlorite solution has a pH between 6 and 6.8.[1362]

### Osmolality
Sodium hypochlorite topical solution, USP, is isotonic, having an osmolality of 300 mOsm/L.[686]

## General Stability Considerations
The stability of sodium hypochlorite solutions is affected by exposure to light[4, 1442, 1443,1703] and heat[1442] and contact with air,[4,1442,1702] metals and metal ions such as $Mn^{2+}$ and $Fe^{2+}$, and organic materials.[1442,1694] Sodium hypochlorite solution, USP, and sodium hypochlorite topical solution, USP, should be packaged in tight, light-resistant containers and should be stored at controlled room temperature.[4]

Sodium hypochlorite topical solution, USP, has a beyond-use date of seven days after the day of compounding.[4]

Sodium hypochlorite solutions should not be mixed with strong acids because of the release of chlorine gas. Sodium hypochlorite solutions also should not be mixed with ammonia because of the release of tosylchloramide sodium gas.[3]

Sodium hypochlorite is most stable at alkaline pH of 10 or higher.[3,1694] Commercial household bleach (sodium hypochlorite 5.25%) is stable, having a pH between 11 and 12, and may contain sodium hydroxide to enhance alkalinity.

Acidified sodium hypochlorite solution is less stable and must be prepared daily. [1362]

## Stability Reports of Compounded Preparations
### Topical

**USP OFFICIAL FORMULATION:**

Sodium hypochlorite solution 5.0 mL
Sodium phosphate monobasic monohydrate 1.02 g
Sodium phosphate dibasic anhydrous 17.61 g
Purified water qs 1000 mL

Dissolve the two sodium phosphate salts in about 500 mL of purified water. Add the sodium hypochlorite solution and bring to 1000 mL with additional purified water. Mix well, making a 0.025% sodium hypochlorite topical solution. Transfer to a tight, light-resistant 1000-mL plastic container. The final liquid preparation should have a pH between 7.8 and 8.2. Store the finished topical solution at controlled room temperature. The beyond-use date is seven days from the date of compounding at either storage temperature. [4,5]

**STUDY 1:** Allen[686] reported the formulation and stability of sodium hypochlorite topical solution, USP. The preparation had the formula shown in Table 105.

The author noted that unscented commercial laundry bleach (approximately 5.25% wt/vol sodium hypochlorite) may be used for this preparation. The final concentration of sodium hypochlorite is 0.025%. The USP has assigned a seven-day beyond-use period to this formulation. [4]

**STUDY 2:** Fabian and Walker[85] reported a longer stability for sodium hypochlorite 1% (Hygeol, Wampole) diluted with distilled water 1:8, 1:12, and 1:20 (0.04 to 0.12% available chlorine) than the official USP beyond-use date of seven days after compounding. [4] The dilutions were packaged in amber glass 2-liter bottles at room temperature of 23°C exposed to sunlight. An iodometric titration was used to determine the amount of available chlorine. At least 98% of the labeled amount of available chlorine was retained over a six-month period. The authors extrapolated from the observed rate of loss that a 23-month beyond-use date was possible.

**TABLE 105.**  Allen's Sodium Hypochlorite Topical Solution Preparation[686]

| Component | Amount |
| --- | --- |
| Sodium hypochlorite solution | 5 mL |
| Monobasic sodium phosphate monohydrate | 1.02 g |
| Dibasic sodium phosphate anhydrous | 17.61 g |
| Purified water | qs  1000 mL |

### Dental

**STUDY 1:** Johnson and Remeikis[1436] evaluated the ability of 1, 2.62, and 5.25% sodium hypochlorite solutions to dissolve tissue for endodontic use in root canal procedures. Sodium hypochlorite 5.25% (Clorox) was evaluated undiluted and also diluted with water to 2.62 and 1% concentrations. The solutions were packaged in polyethylene bottles and were stored at room temperature. The ability of the solutions to dissolve human umbilical cord tissue (a substitute for tooth root pulp) was tested over periods up to 10 weeks. The 5.25% concentration retained its ability to dissolve tissue for at least 10 weeks. The 2.62 and 1% concentrations were able to dissolve the tissue for only one week; the tissue-dissolving ability at these lower concentrations fell off after two weeks and beyond.

**STUDY 2:** Sirtes et al.[964] reported the stability of 1, 2.6, and 5.25% sodium hypochlorite solutions packaged in syringes for endodontic use in root canal procedures upon heating to 45 and 60°C compared to use at room temperature of 20°C. Heating was expected to improve pulp dissolution. Iodometric titration assay results for available chlorine found no loss from heating to 45 or 60°C for one hour compared to the 20°C controls. The heated solutions were also found to be much more efficient in pulp dissolution.

**STUDY 3:** Nicoletti et al.[1412] evaluated the decomposition of sodium hypochlorite 2.5% dental solution at elevated temperatures to predict the shelf life at room temperature using the Arrhenius equation. Samples were kept at 50°C for 23 days and 70°C for seven days in amber glass containers. The decrease in free residual chlorine was determined using the British Pharmacopoeial method of titration with 0.1 N sodium thiosulfate previously standardized with potassium dichromate. From the decrease in amounts of free residual chlorine at elevated temperatures, the shelf life at 20°C was calculated to be 166 days.

**STUDY 4:** Clarkson et al.[1442] evaluated the stability of various sodium hypochlorite concentrations in solutions for endodontic irrigation. The solutions in concentrations of 1 to 4% were packaged in a variety of containers, including plastic bottles with caps that were opened daily, open plastic containers, plastic syringes, and stainless steel bowls. The samples were exposed to light (including sunlight), protected from light, exposed to air or sealed from exposure to air, and heated to 50°C or kept at room temperature. Available chlorine content was determined using an Australian standard titration method with colorimetric determination.

Exposure of solution in plastic bottles and syringes to light was found to increase the rate of chlorine loss. In syringes exposed to indirect sunlight, 25% loss occurred in six days. The samples in syringes protected from light were more stable, but after about three months of storage, damage occurred to the syringe plunger tip, resulting in adhesion to the syringe wall and black discoloration of the solution. Heating the solutions also resulted in loss of chlorine, but not at as fast a rate as caused by exposure to sunlight. Open stainless steel bowls resulted in an increase in chlorine content, probably due to water evaporation. However, some solutions damaged the bowls and caused marked pitting.

The authors concluded that commercial bleach diluted to 1% with demineralized water should be stored tightly closed and protected from exposure to light, and should not be stored for longer than three months. Undiluted bleach should maintain 90% of its initial concentration for at least six months.

**STUDY 5:** Camps et al.[1524] evaluated the stability of sodium hypochlorite 2.5% neutralized with hydrochloric acid to pH 7.5. The neutralized sodium hypochlorite solution rapidly lost chlorine, with a consequent continued decline in pH and loss of effectiveness. The authors recommended a shelf life of such neutralized sodium hypochlorite solutions of two hours after preparation.

## Other

**STUDY 1:** Burger et al.[1167] evaluated the stability of sodium hypochlorite 0.05% used for disinfecting devices. Using the USP analytical method, the authors reported that the solution was stable for at least six months in the refrigerator and that no degradation occurred over 21 days after the container was opened.

**STUDY 2:** Allen[1361] reported on acidified sodium hypochlorite solution used as an antiseptic and disinfectant. The acidified sodium hypochlorite solution was prepared by mixing 1.6 mL of sodium hypochlorite 5% and 1.6 mL of acetic acid 5% with sufficient purified water to bring the volume to 100 mL. Acidified sodium hypochlorite solution should be packaged in tight, light-resistant containers and may be stored at controlled room temperature. Allen stated that the beyond-use period should be limited to 24 hours and that acidified sodium hypochlorite solution should be prepared daily.

The sodium hypochlorite in commercial household bleach with its alkaline pH of 11 to 12 exists nearly totally as chlorite ion (OCl−). In acidified sodium hypochlorite solution that has the pH adjusted down to pH 6 to 6.8, 90% exists as hypochlorous acid (HOCl), which has 80 to 200 times more antimicrobial effect than the chlorite ion. Miner[1362] suggested preparation by adding two ounces of household bleach to a gallon of water and then adding two ounces of white cooking vinegar to make acidified sodium hypochlorite solution. He also noted it should be used in a well-ventilated area, and contact with eyes and mucous membranes should be avoided.

**STUDY 3:** Rutala et al.[1443] evaluated the stability of commercial household bleach diluted with tap water or sterile distilled water in dilutions of 1:100, 1:50, and 1:5 intended for disinfecting purposes. Samples of the test solutions were packaged in five polyethylene container types: translucent open bottle, translucent screw-cap closed bottle, brown opaque bottle closed with a screw cap, translucent wash bottle, and translucent spray bottle. The samples (except for the brown bottle) were stored at ambient laboratory temperature exposed primarily to indirect sunlight and occasional direct sunlight. Available chlorine content was determined using an American Public Health Association standard titration method with colorimetric determination. The bactericidal activity of the test samples was evaluated using *Pseudomonas aeruginosa*, *Staphylococcus aureus*, and *Salmonella choleraesuis*.

The authors found that the sample stored in brown opaque bottles was the most stable, retaining nearly all of the chlorine content for 30 days. However, even the samples in other packaging retained substantial amounts of chlorine. The 1:100 and 1:50 samples retained about 40 to 50% and the 1:5 samples retained over 80% of the initial chlorine concentration after 30 days of storage. The authors found these lower concentrations to still be effective against the microorganisms that were tested. The authors concluded that preparing sodium hypochlorite solutions for disinfecting on a daily basis does not appear to be necessary to ensure biocidal activity and that the disinfecting solutions at the concentrations tested remain sufficiently active for up to 30 days.

**STUDY 4:** Zhu[1476] evaluated the stability of sodium hypochlorite 3% after the addition of 0.1% (vol/vol) of various aqueous flavoring agents. All of the flavoring agents caused an increase in the rate of loss of OCl− over 30 days of storage at 40°C compared to the solution without a flavoring essence. Mint exerted the most damaging effect, causing a 50% loss of OCl− concentration in 30 days. The adverse effects on stability decreased in the following order: mint was greater than lemon; osmanthus flower was greater than pineapple, which was greater than vanilla. ■

# Sodium Phenylacetate and Sodium Benzoate

## Properties

Sodium benzoate is a white, nearly odorless crystalline or granular powder or flakes having a sweetish astringent taste.[1,3,7] Sodium phenylacetate is a crystalline white to off-white powder with a strong offensive odor. The commercial combination product (Ammonul, Ucyclyd Pharma) is a solution containing sodium phenylacetate 100 mg/mL (10%) and sodium benzoate 100 mg/mL (10%) in 50 mL single-use vials.[7]

### Solubility

Sodium benzoate is readily soluble in water, having an aqueous solubility of about 500 to 555 mg/mL, increasing to about 714 mg/mL in boiling water. Its solubility in ethanol is about 13 mg/mL.[1,3,4,7] Sodium phenylacetate is also soluble in water.[7]

### pH

Sodium benzoate plus sodium phenylacetate commercial injection (Ammonul, Ucyclyd Pharma) has a pH of 6 to 8.[7]

## General Stability Considerations

Sodium phenylacetate plus sodium benzoate (Ammonul) solution should be stored at controlled room temperature and protected from excessive heat and from freezing.[7]

## Compatibility with Common Beverages and Foods

Gutteridge and Kuhn[201] reported the compatibility of 10% sodium phenylacetate plus 10% sodium benzoate (Ucephan, McGaw) with several common beverages. The samples were prepared by adding 25 mL of 10% sodium phenylacetate plus 10% sodium benzoate to 60, 120, 180, and 240 mL of Mott's apple juice, Welch's grape juice, Donald Duck orange juice, Hawaiian Punch fruit punch, Coca-Cola Classic, and Robert Corr Natural Soda root beer. The solutions were stored at 26 and 4°C for two hours. With the exception of grape juice stored in the refrigerator, all samples were physically compatible, with no change in pH, turbidity, or color. The grape juice mixtures at 4°C developed large white crystals within 30 minutes that did not redissolve with agitation. ■

# Sodium Phenylbutyrate

## Properties

Sodium phenylbutyrate is an off-white crystalline material. Oral tablets are available containing 500 mg of sodium butyrate with microcrystalline cellulose, magnesium stearate, and colloidal silicon dioxide. Oral granular powder is also available with calcium stearate and colloidal silicon dioxide. One gram of the oral powder provides 0.94 g of sodium butyrate or 3 g of sodium butyrate per 3.2 g of powder.[7] The oral powder provides 125 mg of sodium per gram. The oral tablets provide 62 mg of sodium per 500-mg tablet.[2]

### Solubility

Sodium butyrate is stated by the manufacturer to be very soluble in water to about 500 mg/mL.[7] Caruthers and Johnson[1107] have stated (without reference) that the aqueous solubility of sodium butyrate is only 25 mg/mL. The manufacturer notes that the excipients in the tablets and oral powder are much less soluble in water than is sodium butyrate.[7]

## General Stability Considerations

Sodium butyrate tablets and oral powder are packaged in tight containers and should be stored at room temperature.

After containers of sodium butyrate oral tablets or powder have been opened, they should be tightly reclosed.[7]

## Stability Reports of Compounded Preparations

### Oral

Caruthers and Johnson[1107] evaluated the stability of two extemporaneously compounded oral suspension formulations of sodium butyrate 200 mg/mL prepared from the commercial oral powder. Sodium butyrate powder 12 g was weighed and ground to fine powder in a mortar. Ora-Plus and Ora-Sweet or Ora-Sweet SF were mixed in equal quantities for use as the vehicle. A 30-mL portion of the vehicle was levigated into the ground sodium butyrate powder to form a smooth suspension. The suspension was transferred to a two-ounce amber plastic prescription bottle (bottle composition not cited), and the mortar was rinsed with additional vehicle, bringing the final volume to 60 mL. The sample bottles were stored at controlled room temperature of 23 to 25°C for 90 days.

No visually detectable changes in the color of the suspensions were observed, and no changes in odor or taste were noted. In addition, no microbial growth was observed using either vehicle. Stability-indicating HPLC analysis found that less than a 3% loss of sodium butyrate occurred over 90 days of storage at room temperature.

## Compatibility with Common Beverages and Foods

Sodium butyrate oral powder is intended to be mixed with solid or liquid food for administration. In aqueous liquids, only the drug will dissolve, while the excipients do not. The effects of food on sodium butyrate, if any, have not been determined.[7] ■

# Sodium Polystyrene Sulfonate

## Properties

Sodium polystyrene sulfonate (the sodium salt of sulfonated styrene copolymer) is a cream-colored to golden brown odorless tasteless fine powder.[2,3] Each gram of the cationic exchange resin exchanges between 110 and 135 mg (between 2.81 and 3.45 mEq) of potassium.[2-4]

The commercial suspension is a brown viscous suspension containing 25 g of resin per 100 mL.[2]

### Solubility

Sodium polystyrene sulfonate is insoluble in water.[2,3]

## General Stability Considerations

Sodium polystyrene sulfonate powder and suspension should be packaged in well-closed containers and stored at controlled room temperature protected from temperatures of 40°C or higher.[2,4] To prevent alteration of the potassium exchange characteristics, sodium polystyrene sulfonate resin products should be protected from heat.[137] The suspension also should be protected from freezing.[2,4]

## Stability Reports of Compounded Preparations

### Oral

Allen[137] reported on the formulation of sodium polystyrene sulfonate resin as an oral liquid and as the candy formulation of Johnson et al.[225] The liquid was prepared by suspending 60 g of sodium polystyrene sulfonate powder (Kayexalate, Sanofi Winthrop) in 240 mL of a suitable vehicle such as water, syrup, USP, or sorbitol solution, USP. The vehicle must not contain the significant amounts of potassium that some fruit juices have. The ratio of resin to vehicle may be varied; a 15-g dose of resin is usually in a quantity of vehicle between 20 and 100 mL. The oral liquid should be prepared fresh and used within 24 hours.[137]

The candy (Table 106) was prepared by blending the resin and margarine well. The maple extract was added to the half-and-half, and this addition was thoroughly mixed with the resin mixture until well blended. The powdered sugar was then added, and the mixture was worked into a soft ball. The ball was divided into 40 equal pieces of approximately 23 g containing about 5 g of sodium polystyrene sulfonate resin. The pieces may be formed into a suitable shape and can be covered in powdered sugar if desired. The finished candy pieces should be wrapped separately and stored in a freezer. The candy was stable for at least six weeks frozen at −20°C.[225]

### Rectal Enema

**STUDY 1:** Allen[137] also discussed a sodium polystyrene sulfonate resin as an enema formulation. The enema was prepared by suspending 30 to 50 g of resin in 100 mL of sorbitol solution, USP. Other potential vehicles include 25% sorbitol, 1% methylcellulose, and 10% dextrose. The finished enema should not be too thick because that reduces the effectiveness of the resin. The enema should be prepared fresh and used within 24 hours.

**STUDY 2:** Ohsawa et al.[869] evaluated several sodium polystyrene sulfonate suspensions using dextrose 5% and various suspending agents. The best results were obtained by incorporating the sodium polystyrene sulfonate in dextrose 5% containing Avicel RC 591 NF 1% and mixing at 5000 rpm for three minutes. The suspension was isotonic and stable for at least 90 days at room temperature. ■

**TABLE 106.** Sodium Polystyrene Sulfonate Candy Formulation[225]

| Component | Quantity |
| --- | --- |
| Sodium polystyrene sulfonate powder | 200 g |
| Margarine, softened | ¼ pound |
| Maple extract | 1 tsp |
| Half-and-half (milk and cream) | ¾ cup |
| Powdered sugar, sifted | 1 pound |

# ■ Somatropin

## Properties
Somatropin is a synthetic human growth hormone composed of a polypeptide chain of 191 amino acids. It occurs as a white or nearly white powder having at least 2.5 units per milligram.[3]

## General Stability Considerations
Somatropin products should be stored under refrigeration at 2 to 8°C. After reconstitution of the injection with bacteriostatic water for injection, the somatropin solution is stable for 14 days stored under refrigeration. If sterile water for injection (no preservative) is used to reconstitute the drug, the solution should be stored under refrigeration and used within 24 hours. The solution should be protected from freezing.[7]

## Stability Reports of Compounded Preparations
### Injection
Injections, like other sterile drugs, should be prepared in a suitable clean air environment using appropriate aseptic procedures. When prepared from nonsterile components, an appropriate and effective sterilization method must be employed.

Ray and Chen[617] evaluated the stability of somatropin (Humatrope, Lilly) reconstituted to concentrations of 3.33 and 1 mg/mL with water for injection containing 1.7% glycerin and 0.3% *m*-cresol as a microbiological preservative. One-milliliter aliquots of each concentration were repackaged in 1-mL polypropylene syringes (Plastipak, Becton Dickinson) and 1-mL propylene–ethylene copolymer syringes (Terumo) and capped. Samples were stored under refrigeration at about 4°C for 28 days. Stability-indicating HPLC analysis found somatropin losses of less than 5% in 28 days at either concentration in either syringe. Small losses of *m*-cresol occurred in both syringes, but the concentration remained adequate for microbiological preservation. However, in the propylene–ethylene copolymer syringes (Terumo), the drug solution became cloudy with particulates after 28 days. No visible changes occurred in the polypropylene syringes. The authors recommended a stability period of 28 days in polypropylene syringes and 14 days in propylene–ethylene copolymer syringes. ■

# ■ Sotalol Hydrochloride

## Properties
Sotalol hydrochloride is a white or nearly white crystalline powder.[1,3]

### Solubility
Sotalol hydrochloride is freely soluble in water and soluble in ethanol and propylene glycol.[3,7,457]

### $pK_a$
Sotalol has $pK_a$ values of 8.2 and 9.8.[1]

## General Stability Considerations
Sotalol hydrochloride products should be stored at controlled room temperature.[7] Sotalol hydrochloride is stable in the pH range of 4 to 5.[457]

## Stability Reports of Compounded Preparations
### Oral
**STUDY 1:** Dupuis et al.[457] evaluated the stability of a sotalol hydrochloride 5-mg/mL suspension extemporaneously prepared from tablets. Sotalol hydrochloride (Bristol-Myers) tablets were crushed and levigated with a suspending medium prepared from methylcellulose 1% aqueous solution blended into simple syrup (70:30). The resulting smooth slurry was brought to volume with more of the suspending medium.

Samples were stored at 4, 25, 37, and 45°C for 12 weeks. The samples were evaluated visually for caking, color changes, or any other physical changes. The samples were analyzed using a stability-indicating HPLC assay for sotalol hydrochloride content. The 4 and 25°C samples were evaluated for microbial growth over eight weeks only.

After the 12 weeks of storage, there were no visually observed physical changes, and the sotalol hydrochloride content of all samples at all temperatures remained in the 92 to 96% range. Room temperature samples were found to have microbial growth in as little as four weeks; the refrigerated samples did not develop evidence of microbial growth in eight weeks. Consequently, refrigerated storage was recommended for this preparation of sotalol hydrochloride.

**STUDY 2:** Nahata and Morosco[734] reported the stability of sotalol hydrochloride in two extemporaneously compounded oral suspensions. Sotalol hydrochloride 120-mg tablets (Berlex) were ground to fine powder using a mortar and pestle and mixed into an equal parts mixture of Ora-Plus suspending vehicle and Ora-Sweet syrup and also into a 9:1 mixture of simple syrup and methylcellulose 1%. The suspensions were packaged in amber polyethylenephthalate prescription bottles and stored refrigerated at 4°C and in a water bath at room temperature of 25°C for 91 days.

No changes in physical appearance, color, or odor occurred. Stability-indicating HPLC analysis found sotalol hydrochloride losses of 1 and 4% in 91 days refrigerated and at room temperature, respectively, in the Ora-Plus–Ora-Sweet suspension. In the simple syrup–methylcellulose 1% vehicle, drug losses of 4 and 6% occurred refrigerated and at room temperature, respectively, in 91 days.

**STUDY 3:** Sidhom et al.[935] tested the physical and chemical stability of three sotalol hydrochloride 5-mg/mL oral suspensions prepared from tablets. Sotalol hydrochloride 240-mg tablets (Betapace, Berlex) were comminuted to fine powder using a glass mortar and pestle. The powder was incorporated into three suspending mediums: (1) an equal parts mixture of Ora-Plus suspending vehicle and Ora-Sweet syrup, (2) an equal parts mixture of Ora-Plus suspending vehicle and Ora-Sweet SF syrup, and (3) simple syrup–methylcellulose 1% (1:2.4). The oral suspensions were packaged in amber glass prescription bottles and stored at room temperature of 20 to 25°C and refrigerated at 2 to 8°C for 12 weeks.

No changes in color or odor were noted. The suspensions did not develop caking and were easy to resuspend throughout the study. To resuspend, the methylcellulose-containing suspension required greater shaking for a longer time period than the others. The sugar-containing suspensions stored at room temperature (but not refrigerated) developed crystals around the bottle cap. HPLC analysis results were somewhat erratic, probably due to lack of analytical control. However, there was no indication of consistent loss of drug content in any of the formulations at either storage temperature. After 12 weeks of storage, all analyses remained above 100% of the theoretical concentration of 5 mg/mL. The authors stated that they thought this study supported a beyond-use date of 12 weeks for all formulations at both storage temperatures. ■

# Spironolactone

## Properties

Spironolactone is a white to cream or light tan crystalline powder with a faint mercaptanlike odor.[2,3,545]

### Solubility

Spironolactone is practically insoluble in water, having a solubility of 0.028 mg/mL. Spironolactone solubility at 30°C in 0.1 M acetate buffer having a pH of 4.5 was 35 mcg/mL. The estimated solubility at 5°C was 13.65 mcg/mL.[1–3,545] In ethanol, it has a solubility of about 12.5 mg/mL.[1–3]

## General Stability Considerations

Spironolactone tablets should be packaged in tight, light-resistant containers and stored at controlled room temperature protected from temperatures of 40°C or higher.[2,4] The optimum pH for stability of spironolactone was found to be about 4.5.[612,1126]

## Stability Reports of Compounded Preparations
### Oral

**STUDY 1:** Gupta et al.[62] reported the stability of spironolactone prepared as a pediatric oral liquid dosage form. Spironolactone powder (Searle) was wetted with ethanol to form a suspension and brought to volume with simple syrup to yield a 2-mg/mL concentration with 10% (vol/vol) ethanol. Sodium benzoate 0.1% was added as a preservative. The preparation was packaged in amber bottles and stored at 24°C. HPLC analysis found that 97% of the spironolactone remained after 160 days.

**STUDY 2:** Mathur and Wickman[100] evaluated the stability of three concentrations of spironolactone in extemporaneously prepared suspensions. Two hundred spironolactone tablets (Searle) 25, 50, or 100 mg were used to prepare suspensions containing 2.5, 5, or 10 mg/mL, respectively. The tablets were crushed and ground to fine powder. Purified water 100 mL was added slowly, and the mixture was triturated to make a paste. The film coating was ground with the tablets and dissolved completely in the water. Cherry syrup, NF (Barre), was incorporated gradually to bring the suspensions to volume. The suspensions were homogenized for 10 minutes with a laboratory homogenizer and then packaged in 60-mL type III glass prescription bottles with child-resistant closures. Samples were stored at 5 and 30°C and at 25°C with exposure to intense fluorescent light for four weeks. Spironolactone content was assessed using a stability-indicating HPLC assay.

Suspensions at all three concentrations under all three storage conditions were stable, exhibiting spironolactone losses of less than 5%. Furthermore, there was no noticeable change in color or odor, and microbiological analysis indicated the suspensions remained within acceptable limits.

**STUDY 3:** Nahata et al.[476] evaluated the stability of an extemporaneously compounded spironolactone 1-mg/mL suspension prepared from tablets. Carboxymethylcellulose 1.5% suspension was prepared in 5% ethanol and purified water. Ten spironolactone (Goldline) 25-mg tablets were placed in a blender with a small volume of purified water and allowed to soak for five minutes. Carboxymethylcellulose 1.5% suspension 50 mL and syrup, NF, 100 mL, were added to the blender, and the mixture was blended. The suspension was transferred to a graduated cylinder, and the blender was thoroughly rinsed with purified water. Additional purified water was added to bring the volume to 250 mL, yielding a nominal spironolactone concentration of 1 mg/mL. The suspension was packaged in amber glass prescription bottles and stored at 4 and 22°C. The pH of the suspension was near 6.8 throughout the study. Stability-indicating HPLC analysis determined that no spironolactone loss occurred in 91 days at either 4 or 25°C.

**STUDY 4:** Allen and Erickson[527] evaluated the stability of three spironolactone 25-mg/mL oral suspensions extemporaneously compounded from tablets. Vehicles used in this study were (1) an equal parts mixture of Ora-Sweet and Ora-Plus (Paddock), (2) an equal parts mixture of Ora-Sweet SF and Ora-Plus (Paddock), and (3) cherry syrup (Robinson Laboratories) mixed 1:4 with simple syrup. A total of 120 spironolactone 25-mg tablets (Mylan) was crushed and comminuted to fine powder using a mortar and pestle. About 40 mL of the test vehicle was added to the powder and mixed to yield a uniform paste. Additional vehicle was added geometrically, and the suspension was brought to the final volume of 120 mL, with thorough mixing after each addition. The process was repeated for each of the three test suspension vehicles. Samples of each finished suspension were packaged in 120-mL clear polyethylene terephthalate plastic prescription bottles and stored at 5 and 25°C protected from light.

No visual changes or changes in odor were detected during the study. Stability-indicating HPLC analysis found not more than 7% spironolactone loss in any of the suspensions stored at either temperature after 60 days of storage.

**STUDY 5:** Alexander et al.[545] evaluated the stability of an extemporaneously compounded oral spironolactone 5-mg/mL suspension prepared from tablets. The suspension formula is shown in Table 107. Carboxymethylcellulose sodium and Veegum were individually hydrated in distilled water overnight. A total of 500 spironolactone 25-mg tablets (Mylan) was crushed and triturated to fine powder. The hydrated carboxymethylcellulose sodium was added slowly to the tablet powder with constant trituration. The hydrated Veegum and sorbitol 70% were added separately with constant mixing. Sodium acetate, acetic acid, saccharin sodium, sodium bisulfite, and edetate disodium (disodium EDTA) were dissolved individually in about 20 mL of distilled water and added to the mixture with mixing. The parabens concentrate and pineapple flavor were also added with mixing. The suspension was brought to a final volume of 2500 mL with distilled water and packaged in four-ounce amber bottles. Samples were stored at 5, 30, 50, and 60°C for at least 90 days.

No changes in color, ease of redispersion, or pourability were found in any sample. Less than 10% spironolactone loss was found in 90 days at all temperatures tested by a stability-indicating HPLC analysis.

**STUDY 6:** Pramar et al.[664] evaluated the stability of a spironolactone 2-mg/mL oral liquid prepared using a cosolvent

**TABLE 107.** Spironolactone 5-mg/mL Oral Suspension Formula of Alexander et al.[545]

| Component | Amount |
|---|---|
| Spironolactone 25-mg tablets | 500 |
| Carboxymethylcellulose sodium | 18.75 g |
| Veegum | 18.75 g |
| Parabens concentrate[a] | 25 mL |
| Sorbitol 70% | 1000 mL |
| Sodium acetate | 7.3 g |
| Acetic acid 36% | 14 mL |
| Sodium bisulfite | 2.5 g |
| Edetate disodium (disodium EDTA) | 2.5 g |
| Saccharin sodium | 0.5 g |
| Pineapple flavor | Amount unspecified |
| Distilled water qs | 2500 mL |

[a]Parabens concentrate containing methylparaben 10% and propylparaben 2% in propylene glycol.

blend. The preferred formulation was found to be the one cited in Table 108. Spironolactone, menthol, and benzoic acid were dissolved in the ethanol. The cosolvents propylene glycol, polyethylene glycol 400, and glycerin were then added. The sucrose and saccharin sodium, cherry and sweet flavors, and dye were dissolved in pH 4.5 aqueous citrate or phosphate buffer. The buffer solution was then incorporated, and the mixture was brought to volume with distilled water and

**TABLE 108.** Spironolactone 2-mg/mL Oral Liquid Formulation of Pramar et al.[664]

| Component | Amount |
|---|---|
| Spironolactone | 2 mg/mL |
| Menthol | 0.05% (vol/vol) |
| Benzoic acid | 0.1% (wt/vol) |
| Ethanol | 10% (vol/vol) |
| Polyethylene glycol 400 | 30% (vol/vol) |
| Propylene glycol | 10% (vol/vol) |
| Glycerin | 10% (vol/vol) |
| Sucrose | 10% (wt/vol) |
| Saccharin sodium | 0.01% (wt/vol) |
| Sweet flavor | 0.05% (vol/vol) |
| Cherry flavor | 0.5% (vol/vol) |
| FD&C red no. 40 | 50 ppm |
| Aqueous buffer pH 4.5 | 25% (vol/vol) |

mixed thoroughly. The preparation was packaged in amber glass bottles. HPLC analysis found about 3% loss of spironolactone in the formulation when stored at 40°C for 93 days. The preparation would therefore be expected to be stable for at least two years at room temperature.

**STUDY 7:** Peterson et al.[756] reported the stability of spironolactone 5-mg/mL oral suspension prepared from 25-mg tablets (Aldactone, Searle). The suspension consisted of the formula shown in Table 109. The suspension was packaged in amber glass bottles and stored refrigerated at 5 to 7°C, at room temperature of 21 to 25°C, and at elevated temperature for accelerated decomposition. The physical appearance of the suspension did not change throughout the study. Stability-indicating HPLC analysis found less than 10% loss of spironolactone in 60 days refrigerated and in about 21 days at room temperature.

**STUDY 8:** McKnight[762] reported the stability of spironolactone 2.5 mg/mL in a nonaqueous vehicle for pediatric use. The vehicle contained ethanol 20%, propylene glycol 10%, and glycerol 70% and was packaged in amber glass containers. Stability-indicating HPLC analysis found spironolactone stable for 16 months refrigerated at 4°C and for four months at room temperature of 19°C.

**STUDY 9:** Fajolle et al.[952] prepared an oral suspension of spironolactone 5 mg/mL in Ora-Sweet and Ora-Plus. The oral suspension was deemed palatable by patients and easy to administer by nurses. Microbiological assessment met the requirements of the European Pharmacopoeia. Stability of the drug was not evaluated.

**STUDY 10:** Salgado et al.[970] reported the chemical and microbiological stability of a simple oral suspension of spironolactone 2.5 mg/mL. Spironolactone 25-mg tablets were ground to fine powder in a glass mortar and levigated with 5 mL of sterile water for injection. This mixture was brought to volume in a glass cylinder with simple syrup unpreserved or preserved with 0.2% potassium sorbate, packaged in amber glass bottles, and stored protected from exposure to light at room temperature near 22°C and under refrigeration at about 5°C for 60 days. The suspension was a white, readily dispersible suspension and was unchanged throughout the study. Stability-indicating HPLC analysis found that the suspensions retained spironolactone content, exhibiting less than 5% loss in 60 days. The suspension preserved with 0.2% potassium sorbate was also microbiologically acceptable throughout the study. But the unpreserved suspension failed the specification for microbiological quality after three days at room temperature and after seven days under refrigeration. The preserved formulation did not develop unacceptable microbiological contamination throughout the first 30 days of the study. The unpreserved formulation was deemed unacceptable, while the preserved formulation was acceptable for at least 30 days.

**STUDY 11:** Asiri et al.[1085] evaluated the stability of five spironolactone oral suspension formulations prepared from commercial tablets or powder stored refrigerated at 4°C over 35 days. The five test formulas are shown in Table 110.

**TABLE 109.** Spironolactone 5-mg/mL Oral Liquid Formulation of Peterson et al.[756]

| Component | Amount |
|---|---|
| Spironolactone 25-mg tablets | 20 tablets |
| Compound tragacanth powder | 1.5 g |
| Glycerin | 2.5 mL |
| Benzoic acid solution | 1 mL |
| Concentrated chloroform water | 1.25 mL |
| Syrup | 20 mL |
| Distilled water | qs 100 mL |

**TABLE 110.** Spironolactone Oral Suspensions Tested by Asiri et al.[1085]

| *Formulation 1 (5 mg/mL)* | | |
|---|---|---|
| Spironolactone | | 5 × 100-mg tablets |
| Methylcellulose 2% | | 30 mL |
| Dextrose 70% | qs | 100 mL |
| *Formulation 2 (5 mg/mL)* | | |
| Spironolactone | | 5 × 100-mg tablets |
| Methylcellulose 2% | | 30 mL |
| Syrup, BP | qs | 100 mL |
| *Formulation 3 (10 mg/mL)* | | |
| Spironolactone | | 100 × 100-mg tablets |
| Carboxymethylcellulose 4% | | 30 mL |
| Cherry flavor in glycol and water | | 500 mL |
| Sterile water | qs | 1000 mL |
| *Formulation 4 (5 mg/mL)* | | |
| Spironolactone | | 500 mg |
| Sodium benzoate | | 100 mg |
| Ethanol | | 10% |
| Simple syrup | qs | 100 mL |
| *Formulation 5 (5 mg/mL)* | | |
| Spironolactone | | 5 × 100-mg tablets |
| Distilled water | | 5 mL |
| Methylcellulose 1% with sodium benzoate and syrup | qs | 100 mL |

Stability-indicating HPLC analysis found that spironolactone concentrations remained above 90% in all suspensions for seven days stored under refrigeration. Formulation 1 lost 12% by 14 days and formulations 2 and 5 lost 10% or less by 21 days but exceeded 10% loss within 30 days. Formulation 4 lost 10% by 35 days, while formulation 3 lost only 4% in 35 days. ∎

# Spironolactone and Hydrochlorothiazide

## Properties
Spironolactone is a white to cream or light tan crystalline powder with a faint odor.[2,3]

Hydrochlorothiazide is a white or almost white nearly odorless crystalline powder with a bitter taste.[2,3]

### Solubility
Spironolactone is practically insoluble in water, having a solubility of 0.028 mg/mL. Spironolactone solubility at 30°C in 0.1 M acetate buffer having a pH of 4.5 was 35 mcg/mL. The estimated solubility at 5°C was 13.65 mcg/mL.[1-3,545] In ethanol, it has a solubility of about 12.5 mg/mL.[1-3]

Hydrochlorothiazide is only slightly soluble in water, having an aqueous solubility of 0.6 to 1 mg/mL.[1-3,542] It is soluble in ethanol[1-3] and in solutions of alkali hydroxides.[3]

### $pK_a$
Hydrochlorothiazide has $pK_a$ values of 7.9 and 9.2.[1,2]

## General Stability Considerations
Spironolactone and hydrochlorothiazide tablets should be packaged in tight, light-resistant containers and stored at controlled room temperature.[4,7] The optimum pH for stability of spironolactone is stated to be about 4.5.[542, 545] Phosphate and citrate buffers catalyze spironolactone degradation.[545]

## Stability Reports of Compounded Preparations
### Oral
**STUDY 1:** The extemporaneous formulation of a spironolactone and hydrochlorothiazide 5 + 5-mg/mL oral suspension was described by Nahata and Hipple.[160] Each 25 + 25-mg tablet was added to a blender along with a small amount of purified water. It was soaked for five minutes and blended until uniform. It then was mixed thoroughly with 5 mL per tablet of 1% methylcellulose–simple syrup (70:30). A stability period of 14 days under refrigeration was used, although chemical stability testing was not performed.

**STUDY 2:** Allen and Erickson[542] evaluated the stability of three spironolactone 5-mg/mL and hydrochlorothiazide 5-mg/mL oral suspensions extemporaneously compounded from tablets. Vehicles used in this study were (1) an equal parts mixture of Ora-Sweet and Ora-Plus (Paddock), (2) an equal parts mixture of Ora-Sweet SF and Ora-Plus (Paddock), and (3) cherry syrup (Robinson Laboratories) mixed 1:4 with simple syrup. Twenty-four spironolactone and hydrochlorothiazide 25 + 25-mg tablets (Mylan) were crushed and comminuted to fine powder using a mortar and pestle. About 25 mL of the test vehicle was added to the powder and mixed to yield a uniform paste. Additional vehicle was added geometrically and brought to the final volume of 120 mL, with thorough mixing after each addition. The process was repeated for each of the three test suspension vehicles. Samples of each finished suspension were packaged in 120-mL amber polyethylene terephthalate plastic prescription bottles and stored at 5 and 25°C in the dark.

No visual changes or changes in odor were detected during the study. Stability-indicating HPLC analysis found less than 4% spironolactone loss in any of the suspensions stored at either temperature after 60 days of storage. Hydrochlorothiazide exhibited less than 3% loss in the Ora-Sweet-containing suspensions but almost 9% loss in the cherry syrup over 60 days at room temperature. ∎

# Succinylcholine Chloride
## (Suxamethonium Chloride)

## Properties
Succinylcholine chloride exists as a white or almost white, odorless, hygroscopic crystalline powder having a slightly bitter taste.[1,3,4] It usually contains two molecules of water of hydration, but calculations and labeling are on the anhydrous basis.[4]

### Solubility
Succinylcholine chloride is freely soluble in water, having an aqueous solubility of about 1 g/mL. It is slightly soluble in 95% ethanol, having a solubility of 4.2 mg/mL. It is practically insoluble in ether.[1,3,4]

## pH

The pH of a 2 to 5% aqueous solution of succinylcholine chloride is in the range of pH 3 to 4.5.[1,4] Succinylcholine chloride injection also has a pH in the range of 3 to 4.5.[4] A 0.5% aqueous solution is stated to have a pH of 4 to 5.[3]

## Osmolality

Succinylcholine chloride 50-mg/mL injection has an osmolality of 409 mOsm/kg.[345]

## General Stability Considerations

Succinylcholine chloride powder should be packaged in tight containers and be stored at controlled room temperature.[4] The European Pharmacopeia recommends protection from light during storage.[3] Succinylcholine chloride injection is packaged in single- or multiple-dose containers of Types I or II glass and should be stored under refrigeration.[4]

Succinylcholine chloride exhibits maximum stability in the range of pH 3.75 to 4.5. Succinylcholine chloride is incompatible with alkaline solutions and decomposes at pH greater than 4.5.[1,1489]

Succinylcholine chloride injection is stored under refrigeration for long-term storage to maximize the shelf life of the drug. The manufacturers' labeling indicates the injection is stable for only 14 days stored at room temperature.[7] However, the manufacturers have also made varying statements as to room temperature stability of succinylcholine chloride injection ranging from 30 days to three months.[1491–1493]

Research studies on succinylcholine chloride injection have reported longer periods of stability stored at room temperature. Boehm et al.[1489] reported the loss was about 1% per week when stored at room temperature. Adnet et al.[1494] reported the time to 10% loss analyzed using nuclear magnetic resonance spectroscopy was about five months for the 50-mg/mL concentration and about eight months for the 20-mg/mL concentration. Roy et al.[1495] reported a slightly shorter stability period for the 20-mg/mL concentration with 10% loss in about six months at room temperature using electrospray tandem mass spectrometry. Schmutz and Muhlebach[1490] evaluated the stability of a compounded succinylcholine chloride 10-mg/mL injection using thin layer chromatography and HPLC analysis. They reported 10% loss occurred in about six months at room temperature.

Although refrigerated storage of succinylcholine chloride is required to ensure long-term stability, Kumarvel et al.[1499] and Stone and Fawcett[1500] have reported that syringes prefilled with succinylcholine chloride injection for emergency use froze upon storage in a refrigerator. Refrigerators must be verified to be operating within the compendial temperature range to ensure the availability of stored drugs in emergencies.

## Stability Reports of Compounded Preparations
### Injection

Injections, like other sterile drugs, should be prepared in a suitable clean air environment using appropriate aseptic procedures. When prepared from nonsterile components, an appropriate and effective sterilization method must be employed.

**STUDY 1 (EXTEMPORANEOUS VIALS):** Schmutz and Muhlebach[1490] reported on the compounding of succinylcholine chloride 10-mg/mL injection from powder. The injection was prepared using the following formula, which contained a 4.6% excess of drug to compensate for loss during autoclaving:

| | |
|---|---|
| Succinylcholine chloride, dihydrate (equivalent to 10.46 mg anhydrous) | 11.5 mg |
| Sodium chloride | 6.8 mg |
| Methyl-4-hydroxybenzoate | 1.0 mg |
| Hydrochloric acid | 0.01 mL |
| Water for injection | qs 1 mL |

The methyl-4-hydroxybenzoate preservative was dissolved in hot water, and hydrochloric acid was used to adjust to about pH 3. Sodium chloride was added for isotonicity, and succinylcholine chloride was incorporated. The solution was filtered through a 0.22-μm Gelman HT-Tuffryn filter, filled into 20-mL Type I glass vials, stoppered with ultra clean chlorobutyl rubber closures, and sealed with aluminum covers. The vials were then autoclaved either at 121°C for 20 minutes or at 100°C for 30 minutes. Vials were stored refrigerated at 4 to 6°C protected from light, at room temperature of 20 to 26°C exposed to indirect daylight, and at 20 to 26°C protected from light.

Thin layer chromatography and HPLC analysis found a slight loss due to autoclaving; the formula that was used began with an excess of 4.6% of drug as compensation. The refrigerated samples exhibited little or no drug loss over two years of storage. At room temperature, however, 10% drug loss was found to occur in about six months whether protected from or exposed to light.

**STUDY 2 (REPACKAGED IN SYRINGES):** Fritz et al.[1496] reported the stability of Abbott succinylcholine chloride 20 mg/mL packaged in Becton Dickinson glass and polypropylene syringes with rubber tip caps. Refrigerated at 4°C, little or no loss of drug occurred in 45 days. At room temperature of 22°C and 50% relative humidity, about 5% loss occurred in 45 days. At 37°C and 85% relative humidity, over 10% loss occurred in 30 days.

**STUDY 3 (REPACKAGED IN SYRINGES):** Pramar et al.[1497] reported the stability of Burroughs-Wellcome succinylcholine chloride 20 mg/mL in 5% dextrose and in 0.9% sodium chloride packaged in Monoject polypropylene syringes and wrapped in foil for light protection. Refrigerated at 5°C, HPLC analysis found little or no drug loss in 107 days. At room temperature of 25°C, about 5 or 6% loss occurred in 100 days. At 40°C, about 4% loss occurred in 22 days and 12 to 14% loss in 63 days.

**STUDY 4 (REPACKAGED IN SYRINGES):** Storms et al.[1498] reported the stability of Abbott succinylcholine chloride 20 mg/mL packaged in Becton Dickinson polypropylene syringes. Refrigerated at 4°C, HPLC analysis found little or no loss of drug in 90 days. At room temperature of 25°C exposed to fluorescent light, succinylcholine chloride losses of 6% in 45 days, 10% in 60 days, and 12% in 90 days occurred. ■

# Sucralfate

## Properties

Sucralfate, the hydrous basic aluminum salt of sucrose octasulfate, is a white amorphous powder.[1,2,4]

### Solubility

Sucralfate is practically insoluble in water and ethanol. It is soluble in strong acids and alkalies.[1,2]

### pK$_a$

Sucralfate has a pK$_a$ of 0.43 to 1.19.[1]

## General Stability Considerations

Sucralfate products should be packaged in tight containers and stored at room temperature.[2,4] The suspension should be protected from freezing.[2]

## Stability Reports of Compounded Preparations
### Enteral

Schneider and Ouellette[403] reported a simple way to administer sucralfate tablets through a nasogastric tube to avoid bezoar formation. After removal of the cap and plunger from a 60-mL syringe, a sucralfate tablet was placed inside. The plunger was replaced, and excess air was expelled. Approximately 20 mL of water was drawn into the syringe along with some air to permit shaking, and the cap was replaced. The syringe and its contents were allowed to stand on a flat surface with the tip up for five minutes with occasional shaking, which resulted in tablet disintegration. The resulting suspension was shaken and administered directly from the syringe into the nasogastric tube.

## Compatibility with Other Drugs
### SUCRALFATE COMPATIBILITY SUMMARY

*Incompatible with:* Amphotericin B • Colistin sulfate • Tobramycin sulfate

Feron et al.[467] evaluated the interaction of sucralfate with the antimicrobial drugs colistin sulfate, tobramycin sulfate, and amphotericin B. Sucralfate 500 mg was added to 40 mL of HPLC-grade water and adjusted to pH 3.5 with hydrochloric acid. The antimicrobial drugs were added to separate samples to yield final concentrations of colistin 0.05 mg/mL, tobramycin 0.05 mg/mL, and amphotericin B 0.025 mg/mL. The concentration of each of the drugs was analyzed initially and numerous times over 90 minutes of storage. The colistin concentration was assessed by HPLC, the tobramycin concentration was determined by enzyme immunoassay, and the amphotericin B concentration was found by spectroscopy.

Losses of all three antimicrobials occurred very rapidly and extensively. Colistin, tobramycin, and amphotericin B losses were 88, 99, and 95%, respectively, after five minutes. Furthermore, the loss was not reversible by adjusting pH to near neutrality. The authors noted that the mechanism of interaction is unclear. Similar reductions in bioavailability of quinolones when given with sucralfate have been reported.[608] Concurrent enteral administration of sucralfate and these antimicrobials would result in substantially lower concentrations than would be achieved in the absence of sucralfate. ■

# Sulfacetamide
# Sulfacetamide Sodium

## Properties

Sulfacetamide is a white odorless crystalline powder.[2,3]

Sulfacetamide sodium is a white or yellowish-white odorless crystalline powder having a bitter taste. Sulfacetamide sodium 1.19 g is approximately equivalent to sulfacetamide 1 g.[2,3]

### Solubility

Sulfacetamide is slightly soluble in water, with an aqueous solubility of 6.7 mg/mL.[1,3] It is soluble in ethanol to about 66.7 mg/mL. It is also soluble in dilute mineral acids and in solutions of alkali hydroxides.[3]

Sulfacetamide sodium is very soluble in water, having an aqueous solubility variously cited as approximately 667 mg/mL[1] and 400 mg/mL.[3,6] It is sparingly soluble in ethanol and slightly soluble in dehydrated ethanol.[3]

## pH

Sulfacetamide aqueous solutions are acidic to litmus.[1]

Sulfacetamide sodium 5% aqueous solution has a pH of 8 to 9.5.[4] A 10% aqueous solution is stated to have a pH of 9.[1] Commercial ophthalmic solutions have a pH of 7.4.[2]

## pK$_a$

Sulfacetamide has pK$_a$ values of 2.13 and 5.21.[6]

## General Stability Considerations

Sulfacetamide sodium ophthalmic products should be packaged in tight, light-resistant containers. The solutions should be stored between 2 and 30°C, while the ointment should be stored at controlled room temperature.[2,4]

Exposed to air or heat-sterilized, sulfacetamide sodium solutions may undergo hydrolysis, forming a precipitate of sulfanilamide.[2,6] However, this route of decomposition is not considered a major problem at room temperature.[6] Sterilization by autoclaving solutions at 115°C for 30 minutes has been stated to result in about 1% loss of sulfacetamide. The sulfanilamide produced will not precipitate at room temperature but will if the solution is refrigerated.[6] Oxidation may result in yellowish-brown or reddish-brown discoloration. Discolored products should not be used.[2]

The pH range of minimum rate of decomposition is from pH 5 to 11. Hydrolysis is pH-independent in this broad range.[6]

Sulfacetamide sodium is incompatible with silver-containing products. In combination with zinc sulfate, precipitation may occur, depending on the concentrations involved.[2]

## Stability Reports of Compounded Preparations
### Ophthalmic

Ophthalmic preparations, like other sterile drugs, should be prepared in a suitable clean air environment using appropriate aseptic procedures. When prepared from nonsterile components, an appropriate and effective sterilization method must be employed.

**STUDY 1:** Parkin and Marshall[465] evaluated the stability of phenylmercuric nitrate 0.002% (wt/vol) in sulfacetamide ophthalmic solutions with sodium metabisulfite 0.1% as an antioxidant. Decomposition of the phenylmercuric nitrate during autoclaving occurred primarily due to the presence of the sodium metabisulfite, although additional factors may also play a role. Drug losses ranged from about 30 to 58% after autoclaving at 121°C for 15 minutes.

**STUDY 2 (WITH TRIMETHOPRIM):** Shirwaikar and Rao[1561] evaluated the stability of an ophthalmic drops preparation containing sulfacetamide sodium and trimethoprim. Sulfacetamide sodium 20 g was dissolved in 50 mL of water for injection. Trimethoprim 500 mg was dissolved in 50 mL of propylene glycol. The sulfacetamide sodium solution was slowly added to the trimethoprim solution with stirring. The ophthalmic solution was filtered through a sintered glass filter, filled into vials, and sterilized by autoclaving at 121°C and 30 psig for 30 minutes. The final solution pH was 7.6, and the solution was found to be slightly hypertonic.

Using a rabbit model, the ophthalmic drops were found to be effective against *Staphylococcus aureus*. Using a microbiological assay with several common microorganisms, the formulation was subjected to elevated temperatures of 40, 50, 70, and 90°C. The time to 10% loss (t$_{90}$) at 25°C was calculated to be 4.5 years if the ophthalmic solution was protected from exposure to light. When exposed to light, a discoloration developed in the ophthalmic solution. ■

# ■ Sulfadiazine

## Properties

Sulfadiazine is a white to slightly yellow odorless or nearly odorless powder.[1,4]

### Solubility

Sulfadiazine is practically insoluble in water. It exhibits solubilities of 0.13 mg/mL at pH 5.5 and 2 mg/mL at pH 7.5. It is sparingly soluble in acetone and ethanol.[1,4]

## General Stability Considerations

Sulfadiazine powder slowly darkens upon exposure to light. The drug powder and tablets should be stored in well-closed light-resistant containers.[4]

## Stability Reports of Compounded Preparations
### Oral

Pathmanathan et al.[911] reported the stability of sulfadiazine 200-mg/mL oral suspensions prepared from powder (Sigma) and from crushed oral tablets (Adiazine, Buochara). The tablets were crushed to fine powder using a glass mortar and pestle. Sterile water for irrigation was added to make a paste. The suspension was then brought to volume with additional sterile water for irrigation. A similar approach was utilized to make the suspension from bulk drug powder. The suspensions were packaged in amber glass prescription bottles and stored at room temperature (19 to 25°C) exposed to light and under refrigeration (2 to 6°C).

HPLC analysis of the suspensions found greater losses in the suspension prepared from crushed tablets. Sulfadiazine losses of about 26% occurred in three days at both room and refrigerator temperatures. The suspensions prepared from powder were more stable; losses of 14 and 7% occurred at room temperature and under refrigeration, respectively, in three days. Based on these data, refrigerated storage is needed, and a beyond-use period of three days stored under refrigeration is appropriate. ■

# ■ Sulfisoxazole
# Sulfisoxazole Acetyl
## (Sulphafurazole; Acetyl Sulphafurazole)

## Properties
Sulfisoxazole is a white to yellowish-white odorless crystalline powder or crystals having a bitter taste.[1–3] Sulfisoxazole acetyl is a white to yellowish-white tasteless crystalline powder.[2,3,7] Sulfisoxazole acetyl 1.16 g is approximately equivalent to 1 g of sulfisoxazole.[3]

### Solubility
Sulfisoxazole has an aqueous solubility of about 0.13 mg/mL at 25°C. In ethanol, its solubility is about 20 mg/mL.[2,3]

Sulfisoxazole acetyl is practically insoluble in water and has a solubility of about 5.6 mg/mL in ethanol.[2,3]

### pH
Sulfisoxazole acetyl suspension has a pH between 5 and 5.5.[4] Gantrisin suspension (Roche) has a measured pH of 5.1.[19]

### pKa
Sulfisoxazole has a $pK_a$ of 5.[2]

## General Stability Considerations
Sulfisoxazole and sulfisoxazole acetyl products should be packaged in tight, light-resistant containers protected from light and moisture, including high relative humidity above 60%.[2,4]

## Stability Reports of Compounded Preparations
### Ophthalmic
Ophthalmic preparations, like other sterile drugs, should be prepared in a suitable clean air environment using appropriate aseptic procedures. When prepared from nonsterile components, an appropriate and effective sterilization method must be employed.

Kato et al.[1580] evaluated the stability of an equal volume mixture of epinephrine and sulfisoxazole ophthalmic solution. Samples were stored refrigerated at 5°C and at room temperature of 30°C in the dark. Under refrigeration, the epinephrine content was reduced to 50 and to 10% of the initial concentration in five and 30 days, respectively. The rate of loss increased at room temperature. The color of the mixed solution changed from colorless to dark brown. No change in the concentration of sulfisoxazole occurred in the samples.

### Enteral
**SULFISOXAZOLE COMPATIBILITY SUMMARY**
*Compatible with:* Enrich • Ensure • Ensure HN • Ensure Plus • Ensure Plus HN • Osmolite • Osmolite HN • TwoCal HN • Vital • Vivonex T.E.N.

**STUDY 1:** Cutie et al.[19] added 5 mL of sulfisoxazole acetyl suspension (Gantrisin, Roche) to varying amounts (15 to 240 mL) of Ensure, Ensure Plus, and Osmolite (Ross Laboratories) with vigorous agitation to ensure thorough mixing. The sulfisoxazole acetyl suspension was physically compatible, distributing uniformly in all three enteral products with no phase separation or granulation.

**STUDY 2:** Burns et al.[739] reported the physical compatibility of sulfisoxazole acetyl syrup (Gantrisin) 1 g/10 mL with 10 mL of three enteral formulas, including Enrich, TwoCal HN, and Vivonex T.E.N. Visual inspection found no physical incompatibility with any of the enteral formulas.

**STUDY 3:** Altman and Cutie[850] reported the physical compatibility of sulfisoxazole acetyl suspension (Gantrisin, Roche) 10 mL with varying amounts (15 to 240 mL) of Ensure HN, Ensure Plus HN, Osmolite HN, and Vital after vigorous agitation to ensure thorough mixing. The elixir was physically compatible, distributing uniformly in all four enteral products with no phase separation or granulation. ■

# Sulfur

## Properties

Precipitated elemental sulfur (also known as milk of sulfur) is a very fine pale yellow amorphous or microcrystalline powder. It is odorless and tasteless.[597]

Sublimed elemental sulfur is a fine yellow crystalline powder with a faint odor and taste.[2,3]

### Solubility

Precipitated sulfur is insoluble in water, very slightly soluble in alcohol, and slightly soluble in olive oil to about 10 mg/mL. Wetting can be facilitated by using triethanolamine oleate, alcohol, glycerin, or dilute surfactant.[1,597]

Sublimed sulfur is practically insoluble in water and ethanol and sparingly soluble in olive oil.[2,3]

## General Stability Considerations

Precipitated sulfur and sublimed sulfur should be stored in well-closed containers.[4] Products containing sulfur may react with metals, particularly silver and copper, resulting in discoloration. Contact with topical mercury compounds may result in the formation of hydrogen sulfide, with its foul odor, and may stain the skin.[2,3]

## Stability Reports of Compounded Preparations

### Topical

Allen[597] described a topical ointment for psoriasis. Each component in the formula shown in Table 111 was accurately

**TABLE 111.** Psoriasis Topical Ointment Formula[597]

| Component | Amount |
| --- | --- |
| Precipitated sulfur | 3 g |
| Salicylic acid | 1 g |
| Glycerin | Several drops |
| Fluocinonide (Lidex) 0.05% ointment | 24 g |
| Aquaphor or Aquabase | 70 g |
| Coal tar | 2 g |

weighed. The sulfur and salicylic acid powders were mixed and reduced to fine powder, and a few drops of glycerin were incorporated to make a paste. The paste was mixed with the fluocinonide ointment (Lidex, Medicis), and the mixture was incorporated into the Aquaphor or Aquabase geometrically with thorough mixing. The crude coal tar was added last and mixed well. The final ointment was packaged in tight containers and protected from light. The author indicated that a use period of six months is appropriate for this ointment, although no stability data were provided. ■

# Sumatriptan Succinate

## Properties

Sumatriptan succinate is a white to off-white powder.[7]

### Solubility

Sumatriptan succinate is freely soluble in water and in 0.9% sodium chloride solution.[7]

### pH

Sumatriptan succinate injection has a pH of 4.2 to 5.3. The nasal spray has a pH of approximately 5.5.[2]

### pKa

Sumatriptan succinate has $pK_a$ values of 4.21, 5.67, 9.63, and 12 or greater.[2]

### Osmolality

The injection has an osmolality of 291 mOsm/kg. The nasal sprays have osmolalities of 372 and 742 mOsm/kg for the 5-mg/0.1-mL and 20-mg/0.1-mL concentrations, respectively.[7]

## General Stability Considerations

Sumatriptan succinate products should be stored at a temperature between 2 and 30°C and protected from light.[7]

## Stability Reports of Compounded Preparations

### Injection

Injections, like other sterile drugs, should be prepared in a suitable clean air environment using appropriate aseptic procedures. When prepared from nonsterile components, an appropriate and effective sterilization method must be employed.

Nii et al.[855] reported that undiluted Glaxo Wellcome sumatriptan succinate 6 mg/0.5 mL packaged in polypropylene tuberculin syringes was physically and chemically stable for 24 hours at room temperature and for 72 hours refrigerated exposed to and protected from light. HPLC analysis found no loss of sumatriptan succinate.

### Oral

**USP OFFICIAL FORMULATION:**

> Sumatriptan (as the succinate) 500 mg
> Vehicle for Oral Solution, NF (sugar-containing or sugar-free)
> Vehicle for Oral Suspension, NF (1:1) qs 100 mL

(See the vehicle monographs for information on the individual vehicles.)

Use sumatriptan succinate powder or commercial tablets. If using tablets, crush or grind to fine powder. Add 25 mL of the vehicle mixture in portions and mix thoroughly. Quantitatively transfer to a suitable calibrated tight and light-resistant bottle, bring to final volume with the vehicle mixture, and mix thoroughly, yielding sumatriptan (as succinate) 5-mg/mL oral suspension. The final liquid preparation should have a pH between pH 3.6 and 4.6. Store the preparation under refrigeration between 2 and 8°C. The beyond-use date is 14 days from the date of compounding.[4]

Fish et al.[567] studied the stability of three oral liquid formulations of sumatriptan succinate 5 mg/mL compounded from tablets. Ora-Sweet with Ora-Plus suspension vehicle, Ora-Sweet SF with Ora-Plus, and Syrpalta served as vehicles for the three formulations. Nine sumatriptan succinate 100-mg tablets (Imitrex, Glaxo Wellcome) were crushed to fine powder using a mortar and pestle. For the formulations incorporating Ora-Sweet or Ora-Sweet SF (Paddock), 40 mL of Ora-Plus suspending vehicle (Paddock) was added 5 mL at a time to the tablet powder with thorough mixing. The mixture was transferred to a 180-mL amber glass bottle. The mortar and pestle were rinsed with 10 mL of Ora-Plus five times; rinses were added to the bottle. Ora-Sweet and Ora-Sweet SF were then added, bringing the volume to 180 mL. For Syrpalta, no suspension vehicle was included. The tablet powder was incorporated into the Syrpalta, bringing the volume to 180 mL. The three sumatriptan 5-mg/mL suspensions were stored at 4°C.

No changes in color or appearance of any preparation and no microbial growth was observed. Stability-indicating HPLC analysis found about 5 to 8% sumatriptan loss in 21 days, increasing to 12 to 17% loss in 28 days. ∎

## ■ Sunitinib
## Sunitinib Malate

### Properties
Sunitinib is an orange solid material.[1] Sunitinib malate is a yellow to orange powder.[7]

### Solubility
Sunitinib malate has an aqueous solubility in the pH range of 1.2 to 6.8 exceeding 25 mg/mL.[7] Sunitinib base has a solubility in pH 2 buffer (20 mM KCl/HCl) of only 2.582 mg/mL. Sunitinib base has a solubility in pH 6 buffer (20 mM phosphate) of 0.364 mg/mL.[1] Sunitinib base has a solubility in water of less than 0.1 mg/mL.[1338]

### $pK_a$
Sunitinib malate has a $pK_a$ of 8.95.[7] Sunitinib has a $pK_a$ of 8.5.[1,1338]

### General Stability Considerations
Sunitinib malate commercial capsules are stored at controlled room temperature.[7]

### Stability Reports of Compounded Preparations
### Oral
**STUDY 1 (SUNITINIB MALATE):** Navid et al.[1236] evaluated the stability of a 10-mg/mL oral suspension formulation of

sunitinib malate prepared from commercial 50-mg capsules. The contents of the commercial capsules were incorporated into a 1:1 mixture of Ora-Plus and Ora-Sweet, yielding the 10-mg/mL concentration. The oral suspension was packaged in amber plastic bottles and closed with child-resistant caps. Triplicate samples of the suspension were stored refrigerated at 4°C and also at room temperature. The color, consistency, and odor of the suspension were unchanged throughout study. Liquid chromatography—tandem mass spectrometry assay of drug concentrations—found that greater than 96% of the sunitinib malate was retained for 60 days at both temperatures.

**STUDY 2 (SUNITINIB):** Sistla et al.[1338] evaluated the stability of sunitinib active pharmaceutical ingredient (API) as a powder-in-bottle for use in expediting human clinical trials. Sunitinib powder 50 mg was placed in clear glass vials and stored at 40 to 80°C. Stability-indicating HPLC analysis found no sunitinib loss occurred over 30 days even at the highest temperature.

Sistla et al.[1338] also evaluated the stability of sunitinib after the powder was mixed with apple juice to aid in oral administration. Each 50-mg dose was mixed with 75 mL of apple juice and the stability was determined over two hours at room temperature of about 22 to 25°C exposed to fluorescent light. Once again, stability-indicating HPLC analysis found no loss of sunitinib occurred. The authors stated that the extemporaneous compounding of the doses of sunitinib permitted a more rapid development of the drug. ■

# Suspension Structured Vehicle, NF

## Properties

Suspension Structure Vehicle, NF, is an official vehicle for use in compounding oral liquid preparations. It has the following formula:[4,5]

| | | |
|---|---|---|
| Potassium sorbate | | 150 mg |
| Xanthan gum | | 150 mg |
| Citric acid, anhydrous | | 150 mg |
| Sucrose | | 20 g |
| Purified water | qs | 100 mL |

To prepare, transfer the potassium sorbate to a beaker and dissolve in 50 mL of purified water. Heat the solution on an electric hot plate with stirrer forming a vortex. Slowly add the xanthan gum into the vortex. Apply minimal heat and incorporate the citric acid and the sucrose. Finally the mixture is brought to volume with additional purified water and mixed well.[4,5]

## General Stability Considerations

Suspension Structure Vehicle, NF, should be packaged in tight, light-resistant containers and stored at controlled room temperature. Freezing should be avoided. The beyond-use date is 30 days from the date of compounding.[4,5] ■

# Suspension Structured Vehicle, Sugar-Free, NF

## Properties

Suspension Structure Vehicle, Sugar-Free, NF, is an official vehicle for use in compounding oral liquid preparations. It has the following formula:[4,5]

| | | |
|---|---|---|
| Xanthan gum | | 200 mg |
| Saccharin sodium | | 200 mg |
| Potassium sorbate | | 150 mg |
| Citric acid, anhydrous | | 100 mg |
| Sorbitol | | 2 g |
| Mannitol | | 2 g |
| Glycerin | | 2 mL |
| Purified water | qs | 100 mL |

To prepare, transfer 30 mL of purified water to a beaker and place it on an electric hot plate with stirrer. Using moderate heat and stirring to form a vortex, slowly sprinkle the xanthan gum into the vortex. In a separate beaker dissolve the saccharin sodium, potassium sorbate, and citric acid in 50 mL of purified water. Using moderate heat, incorporate the sorbitol, mannitol, and glycerin. Add the xanthan gum mixture to this latter mixture and mix well. Finally the mixture is brought to volume with additional purified water and mixed well.[4,5]

## General Stability Considerations

Suspension Structure Vehicle, Sugar-Free, NF, should be packaged in tight, light-resistant containers and stored at controlled room temperature. Freezing should be avoided. The beyond-use date is 30 days from the date of compounding.[4,5] ■

# ■ Tacrolimus

## Properties

Tacrolimus occurs as white crystals or crystalline powder.[7]

### Solubility

Tacrolimus is practically insoluble in water (1 to 2 mcg/mL)[1019] but is freely soluble in ethanol.[7]

## General Stability Considerations

Tacrolimus capsules should be stored at controlled room temperature. The injection should be stored between 5 and 25°C.[7] Tacrolimus exhibits a minimum rate of decomposition in the pH range of 2 to 6. Higher pH values result in substantially increased decomposition rates.[646] The tacrolimus formulation contains a large amount of surfactant and will extract diethylhexyl phthalate plasticizer from polyvinyl chloride containers and delivery equipment.[2]

## Stability Reports of Compounded Preparations

### Injection

Injections, like other sterile drugs, should be prepared in a suitable clean air environment using appropriate aseptic procedures. When prepared from nonsterile components, an appropriate and effective sterilization method must be employed.

Taormina et al.[647] reported the stability of tacrolimus (Fujisawa) 0.1 mg/mL in 0.9% sodium chloride injection repackaged as 20 mL in 30-mL plastic syringes (Becton Dickinson). HPLC analysis found no decrease in tacrolimus concentration in 24 hours stored at 24°C exposed to normal room light and protected from light.

### Oral

**STUDY 1:** Jacobson et al.[541] evaluated the stability of a tacrolimus 0.5-mg/mL oral suspension prepared from capsules. The contents of six 5-mg tacrolimus capsules were placed in a glass mortar and pestle. The vehicle for the suspension was prepared by thoroughly mixing equal parts of Ora-Plus (Paddock)

suspending medium and simple syrup. A small amount of the prepared vehicle was triturated with the powder to wet it and yield a paste. About 15 mL of the vehicle was added to the paste, triturated well, and transferred to a two-ounce amber prescription bottle with a childproof cap. Both glass and plastic bottles were used in this testing. The mortar was rinsed with about 15 mL of the prepared vehicle, and the rinse was transferred to the bottle. The rinsing was repeated with the prepared vehicle until a final volume of 60 mL was reached. Samples in each bottle type were stored at room temperature of 25°C for 56 days.

The suspension, which was sweet with a slightly bitter aftertaste, had no detectable change in color, odor, or pH. Stability-indicating HPLC analysis found little or no loss of tacrolimus throughout the study period in either container type.

**STUDY 2:** Han et al.[1018] reported the physical and microbiological stability of a tacrolimus 0.5-mg/mL oral suspension similar to the formulation of Jacobson et al.[541] (see Study 1). Tacrolimus capsule contents (Prograf, Fujisawa) were combined with Ora-Plus suspending medium (Paddock) using a mortar and pestle. Syrup, BP, was added in increments to this mixture and mixed well. The suspension was homogenized using a Silverson high-shear mixer. The finished suspension was packaged in screw-cap amber glass bottles and stored at 22 to 26°C for eight weeks. Laser light diffraction and microscopic examination determined that no change in particle size occurred during the storage period. Furthermore, microbial contamination of the suspension by bacteria or fungi was not detected using the BP sterility test. The authors concluded the formulation was not only chemically stable (see Study 1) but also physically and microbiologically stable for eight weeks.

**STUDY 3:** Elefante et al.[1017] evaluated the stability of an oral suspension formulation of tacrolimus at a concentration of 1 mg/mL. This concentration was believed to be less prone to dosage errors than other concentrations. The contents of six tacrolimus 5-mg capsules (Prograf, Fujisawa) were transferred to an amber plastic bottle followed by the addition of 5 mL of sterile water to wet the powder and form a slurry upon shaking. An equal parts mixture of Ora-Plus suspending medium and Ora-Sweet syrup (Paddock) was then added to bring the suspension to final volume, resulting in the 1-mg/mL tacrolimus concentration. The suspension was stored in amber plastic-cap bottles at room temperature of 23 to 26°C for up to 132 days. Stability-indicating HPLC analysis on days 78 and 132 after preparation found about 2 to 3% tacrolimus loss over this time period, which exceeded four months.

## Compatibility with Other Drugs

### TACROLIMUS OINTMENT COMPATIBILITY SUMMARY

*Compatible with:* Clobetasol propionate • Desoximetasone • Hydrocortisone-17-valerate

**STUDY 1:** Levitt et al.[727] evaluated the stability of desoximetasone and tacrolimus in an equal parts mixture of their commercial ointments. A 1:1 (wt/wt) mixture of Topicort (Taro Pharmaceuticals) containing desoximetasone 0.25% and Protopic (Fujisawa Healthcare) containing tacrolimus 0.1% was prepared and packaged in glass vials. The samples were stored

for 28 days at the following conditions of temperature and relative humidity: 25°C and 60% relative humidity, 30°C and 60% relative humidity, and 40°C and 75% relative humidity.

For all storage conditions, no change in appearance occurred, with the ointment remaining white until a slight yellow discoloration was observed after 28 days. HPLC analysis found that desoximetasone remained stable throughout the study with little or no loss occurring. Tacrolimus also was stable. A wider range in concentration occurred (90 to 107%), but this appears to have been more likely due to analytical variability. After 28 days, all samples were near their starting concentrations.

**STUDY 2:** Pappas et al.[1444] evaluated the compatibility and stability of tacrolimus 0.1% ointment (Protopic, Fujisawa) mixed in a 1:1 ratio with three other topical ointments: clobetasol propionate 0.05% (Dermovate, TaroPharma), desoximetasone 0.25% (TopiCort, TaroPharma), and hydrocortisone valerate 0.2% (HydroVal, TaroPharma). Samples were packaged in sealed glass vials and stored at 25°C (60% relative humidity), 30°C (60% relative humidity), and 40°C (75% relative humidity) for 28 days. HPLC analysis found no substantial loss of tacrolimus and the other drugs during the study period. ■

 # Talc

## Properties

Purified talc is a hydrated magnesium silicate approximating the formula $Mg_6(Si_2O_5)_4(OH)_4$. It also may contain varying amounts of aluminum silicate and iron. It occurs as a very fine, light, and white or nearly white to grayish-white powder having no odor. It has an oily, nongritty feel and readily adheres to skin. Purified talc should be free of asbestos fibers.[1,3]

### Solubility

Talc is practically insoluble in water, ethanol, dilute mineral acids, and dilute alkali hydroxides.[3]

## General Stability Considerations

Talc should be packaged in well-closed containers and stored at controlled room temperature.[4]

## Stability Reports of Compounded Preparations

### Sterile Powder

Sterile preparations should be prepared in suitable clean air environments using appropriate aseptic procedures. When prepared from nonsterile components, an appropriate and effective sterilization method must be employed.

Vaughan and Bishop[473] evaluated the effectiveness of three sterilization methods on talc powder from J. T. Baker. Samples of 2.2 g were weighed and packaged individually

in 3.5 × 9-inch ViewPack (SBW Medical Products) sterilization pouches. Each of these pouches was then enclosed in a Tower DualPeel 5 × 15-inch sterilization pouch (Baxter). Samples were sterilized with exposure to dry heat of 118°C for six hours, steam heat of 118°C for five hours at 24 psi, and gamma irradiation of 30.7 to 34.6 kGy over 1000 minutes. The unsterilized talc showed no growth with anaerobic and mycobacterial cultures. However, there was slight growth of Aspergillus species (not fumigatus) and a gram-positive Bacillus species that was not identified. No growth occurred in any of the sterilized samples through 30 days of storage; for the irradiated and steam-heat sterilized samples, no growth occurred through 90 days. The dry heat-sterilized samples developed a slight growth of coagulase-negative *Staphylococcus* when tested in thioglycolate broth but not in blood agar plates.

### Topical

Allen[575] reported on the preparation of a dusting powder formulation for treating prickly heat. Zinc oxide 10 g and cornstarch 45 g were blended using geometric dilution. Talc 45 g was added to the mixture and thoroughly blended. The dusting powder was packaged in shaker containers. The author indicated that this preparation should have a stability period of at least one year. ■

# Tazarotene

## Properties

Tazarotene is a white solid material.[1] Tazarotene topical creams are white, whereas the topical gels are colorless to light yellow and translucent.[7]

### Solubility

Tazarotene is soluble in water.[907]

## General Stability Considerations

Tazarotene topical creams and gels should be stored at controlled room temperature.[7]

Hecker et al.[816] reported that tazarotene 0.1% gel was stable during exposure to ultraviolet light (UVA and UVB) during phototherapy. HPLC analysis found that no loss of tazarotene occurred.

## Compatibility with Other Drugs

Hecker et al.[815] evaluated the compatibility of tazarotene 0.05% gel combined in equal quantity with 17 other topical products, as shown in Table 112. The mixtures were sealed and stored at 30°C for two weeks. HPLC analysis found that less

**TABLE 112.** Topical Products Tested for Compatibility with Tazarotene[815]

| Drug | Forms |
|------|-------|
| Betamethasone dipropionate 0.05% | Cream, gel, ointment, lotion |
| Calcipotriene 0.005% | Cream, ointment |
| Clobetasol propionate 0.05% | Cream, gel, ointment, scalp solution |
| Diflorasone diacetate 0.05% | Cream, ointment |
| Fluocinonide 0.05% | Cream, ointment |
| Halobetasol propionate 0.05% | Cream, ointment |
| Mometasone furoate 0.1% | Cream |

than 10% tazarotene loss occurred in all combinations except betamethasone dipropionate gel and clobetasol propionate gel, which had losses of 13.1 and 18.8%, respectively. Calcipotriene and halobetasol propionate in the combinations could not be assayed due to interference with tazarotene. All of the other drugs exhibited less than 10% loss in two weeks. ∎

# Tegaserod Maleate

## Properties

Tegaserod maleate is a white to off-white crystalline powder. Tegaserod maleate 1.385 mg is equivalent to 1 mg of tegaserod.[7]

### Solubility

Tegaserod maleate is very slightly soluble in water and slightly soluble in ethanol.[7]

## General Stability Considerations

Tegaserod maleate tablets should be stored at controlled room temperature and protected from exposure to moisture.[7]

## Compatibility with Common Beverages and Foods

Carrier et al.[872] evaluated the stability, compatibility, and dissolution of crushed tegaserod mesylate 6-mg tablets

(Zelnorm, Novartis) in common foods and beverages. The tablets were crushed using a metal teaspoon. The tablet powder equivalent to 6 mg of drug was mixed with tap water, apple juice, orange juice, milk, applesauce, yogurt, and chocolate–hazelnut spread.

The drug was stable and compatible in apple juice, orange juice, and applesauce, and the mixtures were homogeneous. The dissolution profile of tegaserod mesylate in apple juice was comparable to the intact tablets; it was not comparable in orange juice and applesauce. The test results in yogurt and chocolate–hazelnut spread were inconclusive. The mixture with milk was not homogeneous and the delivered dose was incomplete. In addition, apple juice masked the drug taste and was considered to be the preferred vehicle for tegaserod mesylate. ∎

# Temozolomide

## Properties

Temozolomide is a white to light tan or light pink crystalline powder. Temozolomide is a prodrug that must be hydrolyzed to the alkylating active moiety at neutral and alkaline pH values.[7]

NOTE: This antineoplastic drug should be handled as a hazardous substance. The drug causes the rapid appearance of tumors in laboratory animals. The manufacturer recommends that the capsules not be opened. However, for patients who are unable to swallow the capsules, no other choice is available. If the capsules are opened or damaged, great care should be taken to avoid inhalation or contact with the capsule contents.[7]

### Solubility

Temozolomide has an aqueous solubility of 3 to 4 mg/mL.[1040] Sampson et al.[1041] reported the saturation solubility to be 3.1 mg/ml in sodium chloride 0.9%. However, a proprietary microcrystalline formulation increased this by about 10-fold.

### pH

Temozolomide 3.1 mg/mL in sodium chloride 0.9% was reported to have a neutral pH.[1041] Extemporaneously compounded oral liquid formulations were acidified to a pH near 4.[1039] See the *Oral* section below.

## General Stability Considerations

Temozolomide capsules are packaged in amber glass bottles with polypropylene caps and are stored at controlled room temperature. In aqueous solution, the drug is stable at mildly acidic pH values but hydrolyzes at neutral pH and rapidly hydrolyzes at alkaline pH.[7]

Temozolomide stability in aqueous media is pH sensitive. Kodawara et al.[1382] reported that maximum stability occurs in the pH range of pH 2 to 6. Above pH 6, temozolomide decomposition rapidly increases, especially in alkaline solutions.

## Stability Reports of Compounded Preparations

### Oral

Trissel et al.[1039] developed two extemporaneously compounded oral liquid formulations of temozolomide and evaluated their stability over 60 days. Using proper containment and personnel protection for handling hazardous substances, the temozolomide 10-mg/mL oral suspensions were prepared from capsule contents for use in brain tumor patients who are unable to swallow the capsules. Compounding the oral suspensions in a suitable protective environment could reduce the potential drug exposure to healthcare workers or a patient's family members who would otherwise be required to open the hazardous temozolomide capsules for dosing.

To make 100 mL of the suspensions, the capsule contents from ten 100-mg capsules were mixed thoroughly in a mortar with 500 mg of povidone K-30 powder and triturated. NOTE: The povidone K-30 powder is required to keep early crystallization from occurring. Citric acid 25 mg in 1.5 mL of purified water was added to acidify the mixture and wet the powder, forming a paste. Ora-Plus suspension vehicle 50 mL was added in two increments: a small amount of Ora-Plus suspending vehicle was triturated to uniformity and then the balance was incorporated. The resulting mixture was transferred to a graduated cylinder and brought to a volume of 100 mL with Ora-Sweet or Ora-Sweet SF (sugar-free). The finished oral liquid formulations had pH values near 4. The finished suspensions were packaged in amber plastic prescription bottles. They were stored at ambient room temperature near 23°C for 21 days and refrigerated at 2 to 8°C for 60 days.

Both the Ora-Sweet and Ora-Sweet SF formulations were found to be pharmaceutically acceptable as long as the povidone K-30 was also incorporated. (Preliminary attempts to prepare the suspensions without the povidone K-30 resulted in rapid crystal growth within a few days, rendering the mixtures unacceptable.) The room temperature samples darkened over 14 days and were distinctly brownish by 21 days. The refrigerated samples remained unchanged over 60 days of storage. Stability-indicating HPLC analysis found that the temozolomide decomposed at room temperature near 23°C, with losses of about 6% in seven days and about 14% in 14 days in the sugar-containing Ora-Sweet formulation. In the sugar-free Ora-Sweet SF formulation, losses were a bit less, with about 5% loss in seven days and more than 10% loss in 21 days at room temperature. Samples stored under refrigeration exhibited no loss of temozolomide in 60 days. Consequently, the authors recommended refrigerated storage. The suspensions should not be kept at room temperature for longer than seven days using Ora-Sweet and for 14 days using Ora-Sweet SF.

## Compatibility with Common Beverages and Foods

Kodawara et al.[1382] reported that the stability of temozolomide was pH-sensitive, and the drug's stability in various beverages was directly related to the pH of the beverage. In beverages in the pH range of 2 to 6, temozolomide exhibited better stability than in beverages with higher pH. In no case was the stability very good, but in milk and various water products having pH values from 6.68 to 8.95, 10% loss occurred in as little as five minutes. In beverages including Coca-Cola (pH 2.6), Aquarius soft drink (pH 3.6), Pocari Sweat (pH 3.6), and Tropicana apple juice (pH 3.8), temozolomide retained 90% of the original concentration for up to 60 minutes. ∎

# Terbinafine Hydrochloride

## Properties
Terbinafine hydrochloride is a white to off-white fine crystalline powder.[7]

### Solubility
Terbinafine hydrochloride is slightly soluble in water and soluble in ethanol.[7]

### $pK_a$
Terbinafine hydrochloride has a pK$_a$ of 7.13.[1210]

## General Stability Considerations
Terbinafine hydrochloride tablets should be packaged in tight containers and stored below 25°C protected from light. The topical cream should be stored between 5 and 30°C. The topical solution has conflicting storage information. It is labeled for storage between 5 and 25°C, but the label also states that the solution should not be refrigerated.[7]

## Stability Reports of Compounded Preparations
### Oral
**USP OFFICIAL FORMULATION:**

> Terbinafine (as hydrochloride) 2.5 g (2.81 g of hydrochloride)
> Vehicle for Oral Solution, NF
> Vehicle for Oral Suspension, NF (1:1) qs 100 mL

(See the vehicle monographs for information on the individual vehicles.)

Use terbinafine hydrochloride powder or commercial tablets. If using tablets, crush or grind to fine powder. Add the vehicle mixture in small portions and mix to make a uniform paste. Add additional vehicle in increasing volumes to make a pourable suspension. Quantitatively transfer to a suitable calibrated tight and light-resistant bottle, bring to final volume with the vehicle mixture, and mix thoroughly, yielding terbinafine (as hydrochloride) 25-mg/mL oral suspension. The final liquid preparation should have a pH between 5.3 and 5.7. Store the preparation at controlled room temperature or under refrigeration between 2 and 8°C. The beyond-use date is 30 days from the date of compounding at either storage temperature.[4]

Abdel-Rahman and Nahata[673] reported on the stability of a terbinafine 25-mg/mL (as the hydrochloride) oral suspension prepared from tablets. Twenty terbinafine 250-mg (as the hydrochloride) tablets were crushed to fine powder using a mortar and pestle. The suspension vehicle was prepared by mixing equal amounts of Ora-Plus suspending agent and Ora-Sweet (Paddock). A small amount of this vehicle was added to the tablet powder and triturated to a smooth paste. Additional vehicle was incorporated in increments and, when pourable, was transferred to a graduate. The volume was brought to 200 mL with additional vehicle. Samples were packaged in two-ounce amber polyethylene prescription bottles and stored at 4 and 25°C. No changes in color or odor were observed. Stability-indicating HPLC analysis found 4% loss at 4°C and 7% loss at 25°C in 42 days. Unacceptable losses (>10%) occurred after that time. ∎

# Terbutaline Sulfate

## Properties
Terbutaline sulfate is a white to grayish-white crystalline powder. It may be odorless or have a faint odor of acetic acid.[2,3]

### Solubility
Terbutaline sulfate is freely soluble in water, having an aqueous solubility of 250 mg/mL. It is slightly soluble in ethanol,[2,3] with a solubility of 1.2 mg/mL.[1]

### pH
Terbutaline sulfate injection has a pH between 3 and 5.[4]

### $pK_a$
Terbutaline sulfate has pK$_a$ values of 8.8, 10.1, and 11.2.[1]

### Osmolality
Terbutaline sulfate injection 1 mg/mL has an osmolality of 283 mOsm/kg.[326]

## General Stability Considerations
Terbutaline sulfate is sensitive to light and excessive heat; drug solutions should not be used if discolored. Tablets should be packaged in tight containers and stored at controlled room temperature. Similarly, the injection should be stored at controlled room temperature and protected from light. Terbutaline sulfate is stable in aqueous solution from pH 1 to 7.[2]

## Stability Reports of Compounded Preparations
### Injection
Injections, like other sterile drugs, should be prepared in a suitable clean air environment using appropriate aseptic procedures. When prepared from nonsterile components, an appropriate and effective sterilization method must be employed.

**STUDY 1:** Glascock et al.[400] reported the stability of terbutaline sulfate 1 mg/mL repackaged into plastic syringes (Becton

Dickinson) composed of polypropylene and fitted with tip caps. Samples stored in the dark at 4 and 25°C were stable, with about 5 or 6% loss after 60 days. Samples stored at 25°C and exposed to light had an increased rate of drug loss. Losses of 5 and 11% occurred after storage for 28 and 60 days, respectively. Furthermore, a yellow discoloration formed after 60 days.

**STUDY 2:** Raymond[401] also reported the stability of terbutaline sulfate 1 mg/mL with 0.25 mL of solution repackaged in tuberculin syringes (Becton Dickinson) and sealed with tip caps. Losses of about 5% in seven weeks occurred in samples stored at 4 and 24°C whether protected from or exposed to light.

### Oral
**USP OFFICIAL FORMULATION:**
> Terbutaline sulfate 100 mg
> Syrup, NF qs 100 mL

Use terbutaline sulfate powder or commercial tablets. If using tablets, crush or grind to fine powder. Add syrup, NF, in increasing volumes to make a pourable suspension. Quantitatively transfer to a suitable calibrated tight and light-resistant bottle, bring to final volume with syrup, NF, and mix thoroughly, yielding terbutaline sulfate 1 mg/mL oral suspension. Store the preparation under refrigeration between 2 and 8°C. The beyond-use date is 30 days from the date of compounding when stored under refrigeration.[4]

The stability of terbutaline sulfate extemporaneously compounded as a 1-mg/mL oral liquid was assessed by Horner and Johnson.[45] Oral liquids were prepared by crushing 5-mg tablets (Brethine, Ciba-Geigy), adding 5 mL of purified water to form a paste, and then mixing with simple syrup, NF (Humco), to yield the 1-mg/mL concentration. The final preparation was a suspension because of insoluble tablet excipients. An oral solution also was prepared from analytical-grade terbutaline sulfate powder dissolved in 5 mL of purified water and added to simple syrup, NF. The oral suspension and the solution were stored in amber type III glass bottles with child-resistant caps refrigerated at 4°C in the dark for 55 days.

No visually observable change in color and no change in odor or pH was noted. A stability-indicating HPLC assay was used to determine terbutaline sulfate. The suspension remained stable, with no loss of drug; the solution exhibited a small amount of loss, retaining about 92% concentration after 55 days.

### Inhalation
Inhalations, like other sterile drugs, should be prepared in a suitable clean air environment using appropriate aseptic procedures. When prepared from nonsterile components, an appropriate and effective sterilization method must be employed.

Gupta et al.[1181] evaluated the stability of terbutaline sulfate (Ciba-Geigy) 0.1% injection diluted 1:10 in 0.9% sodium chloride to a concentration of 0.01% (0.1 mg/mL) and packaged as 3.3 mL of sample solution in 5-mL amber oral syringes sealed with tip caps (Becton Dickinson). At both room temperature of 24 to 26°C and refrigerated at 4 to 6°C, diluted terbutaline sulfate remained clear and colorless over 120 days, with little or no loss of drug content as determined by ultraviolet spectrophotometric analysis.

### Enteral
Ortega de la Cruz et al.[1101] reported the physical compatibility of an unspecified amount of oral liquid terbutaline sulfate (Terbasmin, Em) with 200 mL of Precitene (Novartis) enteral nutrition diet for a 24-hour observation period. No particle growth or phase separation was observed.

## Compatibility with Other Drugs
### TERBUTALINE SULFATE COMPATIBILITY SUMMARY
*Compatible with:* Budesonide • Ipratropium bromide
  • Sodium bicarbonate 8.4% • Sodium chloride
  7% (hypertonic)
*Incompatible with:* Acetylcysteine • Atropine sulfate
  • Dexamethasone
*Uncertain or variable compatibility with:* Cromolyn sodium

**STUDY 1 (CROMOLYN):** Lesko and Miller[659] evaluated the chemical and physical stability of 1% cromolyn sodium nebulizer solution (Fisons) 2 mL admixed with 0.5 mL of terbutaline sulfate 0.1% inhalation solution. The admixture was visually clear and colorless and remained chemically stable by HPLC analysis for 60 minutes after mixing at 22°C.

**STUDY 2 (MULTIPLE DRUGS):** Owsley and Rusho[517] evaluated the compatibility of several respiratory therapy drug combinations. Terbutaline sulfate 1-mg/mL solution (Brethine, Geigy) was combined with the following drug solutions: acetylcysteine 10% solution (Mucosil-10, Dey), atropine sulfate 1 mg/mL (American Regent), cromolyn sodium 10-mg/mL nebulizer solution (Intal, Fisons), dexamethasone (form unspecified) 4 mg/mL (American Regent), and sodium bicarbonate 8.4% injection (Abbott). The test solutions were filtered through 0.22-µm filters into clean vials. The combinations were evaluated over 24 hours (temperature unspecified) using the USP particulate matter test. Only the terbutaline sulfate–sodium bicarbonate combination was found to be compatible. Terbutaline sulfate combined with the acetylcysteine, atropine sulfate, cromolyn sodium, and dexamethasone all formed unacceptable levels of larger particulates (≥10 µm).

**STUDY 3 (MULTIPLE DRUGS):** Roberts and Rossi[897] reported the compatibility of terbutaline sulfate inhalation solution (Bricanyl) with ipratropium bromide (Atrovent) as both single and multiple-use forms as well as with cromolyn sodium (Intal) and budesonide (Pulmicort Respules). Visual inspection found that all of the combinations were compatible, with no visible cloudiness.

**STUDY 4 (HYPERTONIC SODIUM CHLORIDE):** Fox et al.[1239] evaluated the physical compatibility of hypertonic sodium chloride 7% with several inhalation medications used in treating cystic fibrosis. Terbutaline sulfate injection (Sicor) 1 mg/mL mixed in equal quantities with extemporaneously compounded hypertonic sodium chloride 7% did not exhibit any visible evidence of physical incompatibility, and the measured turbidity did not increase over the one-hour observation period. ■

# ■ Terpin Hydrate

## Properties

Terpin hydrate is a white powder or colorless crystals having a slight odor and a bitter taste. It effloresces in dry air.[1,3] The elixir contains between 1.53 and 1.87 g of terpin hydrate per 100 mL.[4]

### Solubility

Terpin hydrate has an aqueous solubility of about 4 to 5 mg/mL at 20°C. The solubility in boiling water is much higher, about 29 mg/mL. In ethanol, its solubility is about 77 mg/mL, increasing to 333 mg/mL in boiling alcohol.[1,3]

### pH

A 1% solution of terpin hydrate in hot water is neutral.[3] Terpin hydrate elixir (Barre) had a measured pH of 6.5.[19]

## General Stability Considerations

Terpin hydrate products should be packaged in tight containers.[4] If crystals form in terpin hydrate elixir, they may be redissolved with warming and gentle shaking.[3]

## Stability Reports of Compounded Preparations
*Enteral*

**TERPIN HYDRATE COMPATIBILITY SUMMARY**

*Compatible with:* Ensure • Ensure HN • Ensure Plus • Ensure Plus HN • Osmolite • Osmolite HN • Vital

**STUDY 1:** Cutie et al.[19] added 10 mL of terpin hydrate elixir (Barre) to varying amounts (15 to 240 mL) of Ensure, Ensure Plus, and Osmolite (Ross Laboratories) with vigorous agitation to ensure thorough mixing. The terpin hydrate elixir was physically compatible, distributing uniformly in all three enteral products with no phase separation or granulation.

**STUDY 2:** Altman and Cutie[850] reported the physical compatibility of terpin hydrate elixir (Barre) 10 mL with varying amounts (15 to 240 mL) of Ensure HN, Ensure Plus HN, Osmolite HN, and Vital after vigorous agitation to ensure thorough mixing. The terpin hydrate elixir was physically compatible, distributing uniformly in all four enteral products with no phase separation or granulation. ■

# ■ Testosterone

## Properties

Testosterone occurs as white or creamy-white odorless crystals or crystalline powder.[2,3,615]

### Solubility

Testosterone is practically insoluble in water, freely soluble in ethanol, slightly soluble in ethyl oleate, and soluble in vegetable oils.[2,3,615] Testosterone has a solubility in dehydrated ethanol of about 167 mg/mL.[3]

### pH

Testosterone suspension has a pH of 4 to 7.5.[4]

## General Stability Considerations

Testosterone products should be stored at controlled room temperature and protected from excessive heat and from freezing.[2,4] The manufacturer has stated that testosterone should not be removed from the original packaging and placed in dosing compliance aids because the drug is sensitive to moisture.[1622] A precipitate may form in injections stored at low temperatures, but the precipitate should redissolve after warming to room temperature and shaking. Use of a syringe or needle that is wet may result in cloudiness. This cloudiness is stated not to affect concentration.[2]

## Stability Reports of Compounded Preparations
*Oral*

**STUDY 1 (TRIPLE HORMONE CAPSULES):** Allen[615] reported on an oral formulation of triple hormones in oil-filled capsules. The formula for 100 capsules is shown in Table 113. The powders were accurately weighed and mixed in a mortar.

**TABLE 113.** Formula of Triple Hormones
in Oil-Filled Capsules[615]

| Component | Amount |
|---|---|
| Progesterone | 4 g |
| Estradiol | 50 mg |
| Testosterone | 50 mg |
| Safflower oil | qs 30 mL |

Note: Fill 0.3 mL per capsule.

Drops of safflower oil were added to wet the powder thoroughly and form a paste. The remaining safflower oil was added with continued mixing. The mixture was chilled in a refrigerator for about one hour to increase the viscosity and reduce powder settling. A hand-operated capsule machine was loaded with 100 No. 1 empty hard-gelatin capsules, and the caps were removed. Then 0.3 mL of the triple hormones in oil was transferred by micropipette into the base of each capsule. For clear capsules, additional safflower oil may be added to fill the capsules more completely. The capsule caps were then replaced and locked into place securely. These capsules, packaged in tight, light-resistant containers and stored under refrigeration, are physically stable for at least 60 days.

**STUDY 2 (SUSPENSION):** Gupta[921] reported that testosterone formulated as a 50-mg/mL suspension in Ora-Plus and assayed by a stability-indicating HPLC method underwent no loss of drug concentration in 45 days when stored at an elevated temperature of 40°C. This result indicates that the drug is expected to be stable for at least one year at 25°C, although this was not actually tested. In addition, the suspension underwent no visible changes.

### Injection

Injections, like other sterile drugs, should be prepared in a suitable clean air environment using appropriate aseptic procedures. When prepared from nonsterile components, an appropriate and effective sterilization method must be employed.

White and Quercia[698] reported the instability of testosterone prepared extemporaneously for intravenous infusion. Testosterone powder (Paddock) was dissolved in 95% ethanol, yielding a 1-mg/mL solution that was sterilized by filtration using a Millex-FG filter. The concentrate was used to prepare a testosterone 1-mcg/mL large-volume infusion in sodium chloride 0.9% (Abbott) in polyvinyl chloride bags. No visible changes occurred, but radioimmunoassay found rapid loss of testosterone. At room temperature near 24°C protected from exposure to light, 15% loss occurred in one hour and 23% loss occurred in three hours. Refrigerated at 4°C protected from exposure to light, the drug remained stable for nine hours, but 25% loss occurred in 24 hours. ∎

# ∎ Tetracaine
# Tetracaine Hydrochloride
## (Amethocaine Hydrochloride)

## Properties
Tetracaine is a white or light yellow waxy solid.[2,3]

Tetracaine hydrochloride is a white odorless hygroscopic crystalline powder having a bitter taste.[1–3]

### Solubility
Tetracaine is slightly soluble in water and soluble in ethanol to about 200 mg/mL.[2,3]

Tetracaine hydrochloride has an aqueous solubility of about 133 to 143 mg/mL. The drug also is soluble in ethanol to about 25 mg/mL.[1–3]

### pH
Aqueous solutions of tetracaine hydrochloride are neutral to litmus.[1] Tetracaine hydrochloride sterile powder reconstituted to a 1% concentration has a pH of 5 to 6. Tetracaine hydrochloride injection has a pH between 3.2 and 6. Tetracaine hydrochloride in dextrose injection has a pH of 3.5 to 6. The ophthalmic solution has a pH range of 3.7 to 6. The topical solution has a pH of 4.5 to 6.[4]

### pKa
The drug has a pKa of 8.39.[1,2]

## General Stability Considerations
In general, tetracaine hydrochloride products should be packaged in tight, light-resistant containers and stored under refrigeration. They should be protected from freezing.[2–4]

In solution, tetracaine hydrochloride is subject to hydrolysis and the formation of a crystalline precipitate of *p*-butylaminobenzoic acid. Cloudy or discolored solutions or those with a precipitate should be discarded. The injection packaged in ampules may be autoclaved once at 121°C at 15 psi for 15 minutes for sterilization. However, autoclaving increases the chance of crystals forming; unused autoclaved ampules should be discarded. The sterile powder in ampules also may be autoclaved similarly. Autoclaving changes the drug's appearance and increases its dissolution rate, but it does not affect its concentration.[2,3] The presence of sodium chloride 0.9% has been stated to aid the stability of tetracaine hydrochloride during autoclaving as well as prevent hemolysis upon injection.[1209]

Tetracaine hydrochloride stability in aqueous solution is pH-dependent, being less stable in alkaline solutions.[1043,1047,1048] The drug exhibits good stability over a pH range of about 3 to 5,[1047,1193] with a maximum stability at pH 3.8[1043,1176] to 4.[1163] Tetracaine hydrochloride is incompatible with alkali hydroxides or carbonates; oily liquid tetracaine base is generated.[2,3]

## Stability Reports of Compounded Preparations
### Ophthalmic

Ophthalmic preparations, like other sterile drugs, should be prepared in a suitable clean air environment using appropriate aseptic procedures. When prepared from nonsterile components, an appropriate and effective sterilization method must be employed.

Parkin and Marshall[465] evaluated the stability of phenylmercuric nitrate 0.002% (wt/vol) in tetracaine hydrochloride ophthalmic solution with sodium metabisulfite 0.1% as an antioxidant. Decomposition of the phenylmercuric nitrate during autoclaving occurred primarily due to the presence of the sodium metabisulfite, although other factors may also play a role. Losses ranged from about 50 to 63% after autoclaving at 121°C for 15 minutes.

### Injection

Injections, like other sterile drugs, should be prepared in a suitable clean air environment using appropriate aseptic procedures. When prepared from nonsterile components, an appropriate and effective sterilization method must be employed.

He and Yu[1424] evaluated the stability of compounded tetracaine hydrochloride injection stored at room temperature of 25°C and at elevated temperature of 40°C with 75% relative humidity. HPLC analysis found drug loss of 6% after three months at elevated temperature. Stored at room temperature, tetracaine hydrochloride losses were about 4% in one year and 11% in two years. The authors stated that an expiration date of one year at room temperature for the tetracaine hydrochloride injection was appropriate.

### Topical
#### USP OFFICIAL FORMULATION (T-A-C TOPICAL SOLUTION):

| | | |
|---|---|---|
| Cocaine hydrochloride | | 4 g |
| Tetracaine hydrochloride | | 1 g |
| Epinephrine hydrochloride injection (1:1000) | | 25 mL |
| Benzalkonium chloride | | 10 mg |
| Edetate disodium | | 6.4 mg |
| Sodium chloride injection 0.9% | | 35 mL |
| Purified water | qs | 100 mL |

**TABLE 114.** T-A-C Topical Solution Formulations[150]

| Component | Formula 1 | Formula 2 |
|---|---|---|
| Tetracaine hydrochloride 2% | 2 mL | 1 mL |
| Epinephrine hydrochloride 0.1% | 1 mL | 2 mL |
| Cocaine hydrochloride 4% solution | 1 mL | — |
| Cocaine hydrochloride powder | — | 0.472 g[a] |
| 0.9% Sodium chloride injection | — | qs 4 mL |

[a]Cocaine hydrochloride 11.8%

The cocaine hydrochloride and tetracaine hydrochloride powders should be dissolved in 25 mL of purified water. Add the epinephrine hydrochloride (1:1000) injection and mix. Dissolve the edetate disodium separately in sodium chloride 0.9% and bring to a final volume of 35 mL with additional sodium chloride 0.9%. Dissolve the benzalkonium chloride separately in a quantity of purified water and bring to a final volume of 10 mL with additional purified water. Combine the three solutions, add additional purified water to bring the solution to volume, and mix well. Package the solution in a suitable sterile, tight, light-resistant container. The final liquid preparation should have a pH between pH 4 and 6. Store the preparation under refrigeration between 2 and 8°C. The beyond-use date is 30 days from the date of compounding.[4,5]

**STUDY 1:** Stiles and Allen[150] reported on two formulations for topical anesthesia. The T-A-C solutions (Table 114) include tetracaine hydrochloride (Pontocaine, Winthrop), epinephrine hydrochloride (Adrenalin, Parke-Davis), and cocaine hydrochloride. The first formula is a simple combination of three commercial products. The second formula is a cocaine-fortified solution prepared from cocaine hydrochloride powder. The T-A-C solutions should be terminally filtered to remove any contaminants. The solutions are unstable in light and air and should be freshly prepared for use.

**STUDY 2:** Pesko[253] also reported on a T-A-C topical formulation (Table 115). The author recommended mixing the components aseptically, with filtration through a 0.22-μm filter; an expiration period of 14 days protected from light was recommended, although stability studies have not been conducted.

**TABLE 115.** T-A-C Topical Anesthetic Solution Formula[253]

| Component | Amount |
|---|---|
| Tetracaine hydrochloride 2% | 50 mL |
| Epinephrine hydrochloride 0.1% | 2.5 mL |
| Cocaine hydrochloride 10% | 40 mL |
| Sterile water for irrigation | 7.5 mL |

**STUDY 3:** Jeter and Mueller[404] described the preparation of a sterile T-A-C gel formulation. The formulation was prepared from T-A-C sterile solution composed of tetracaine hydrochloride 0.5%, epinephrine hydrochloride 1:2000, and cocaine hydrochloride 11.8% that had been sterilized using a 0.2-μm filter (Millex-GS). T-A-C solution 12 mL was added to a 1-g jar of sterile Gelfoam powder and thoroughly mixed. The mixture was divided into six 2.5-g portions using a gas-sterilized measuring spoon and was packaged into heat-sterilized and depyrogenated amber open-mouth vials and sealed with gas-sterilized tamper-evident aluminum lids. The preparation should be protected from light to prevent oxidation of the epinephrine turning the mixture brown. The authors use a three-month date on this T-A-C gel preparation, although specific stability was not assessed.

**STUDY 4:** Larson et al.[539] evaluated the stability of a topical anesthetic solution described by Schilling et al.[540] The topical anesthetic was composed of lidocaine hydrochloride 40 mg/mL, racemic epinephrine hydrochloride 2.25 mg/mL, tetracaine hydrochloride 5 mg/mL, and sodium metabisulfite 0.63 mg/mL as an aqueous solution. A 500-mL quantity of the solution was prepared by measuring 100 mL of 20% lidocaine hydrochloride injection (Abbott), 50 mL of racemic epinephrine 2.25% as the hydrochloride (Nephron), 125 mL of 2% tetracaine hydrochloride (Winthrop), 315 mg of sodium metabisulfite (Gallipot), and 225 mL of sterile water for irrigation. All of the ingredients were combined and mixed well. The finished solution had a blue tint. The solution was packaged in both clear and amber glass bottles with tight-fitting caps, and samples were stored under refrigeration at 4°C in the dark, at room temperatures near 18°C exposed to ambient room light, and at an elevated temperature of 35°C exposed to ambient room light.

The refrigerated samples remained clear with a blue tint, and HPLC analysis of the samples found little or no loss of any of the three drug components after 26 weeks of storage. The epinephrine was the least stable of the drug components at the higher temperatures. Samples in amber containers at room temperature exposed to light were not discolored through four weeks of storage but became discolored in eight

weeks, exhibiting about 5% epinephrine loss at that time; the epinephrine loss increased to 9% in 16 weeks. Discoloration was thought to be the result of a highly colored epinephrine decomposition product that was visible when epinephrine decomposition of 2% or more occurred. The lidocaine and tetracaine components both retained at least 95% of the initial concentration throughout 26 weeks of room temperature storage. In clear containers exposed to light, discoloration appeared in as little as one to two weeks. The authors[539] recommended that the solution be packaged only in amber glass bottles and be given expiration dating of 26 weeks under refrigeration and four weeks at room temperature.

The results of a controlled clinical trial of this topical anesthetic solution by Schilling et al.[540] indicated that adequate anesthesia resulted throughout the time period evaluated for stability by Larson et al.[539]

**STUDY 5:** Ohzeki et al.[1247] evaluated a topical eutectic mixture of local anesthetic (EMLA) cream prepared from lidocaine and tetracaine eutectic mixture. Lidocaine and tetracaine form a eutectic mixture at a relatively low temperature of about 18°C. To prepare the cream, lidocaine 2.5 g and tetracaine 2.5 g were mixed forming a eutectic mixture. Polyoxyethylene hydrogenated castor oil (HCO-60) 1.9 g was then incorporated, forming mixture A. Separately, ethyl parahydroxybenzoate (ethyl paraben) 50 mg was dissolved in ultra pure water 100 mL and carboxypolymethylene (Carbopol 934P) 1 g was added and mixed well, forming mixture B. The pH of mixture B was adjusted to 9, but the authors did not state what they used to adjust the pH. Finally mixture A and mixture B were combined and mixed well, forming the topical eutectic mixture of local anesthetic cream.

The cream was stored at room temperature and exposed to room fluorescent light for 90 days. Visual inspection of the cream found no changes in color or appearance, and no separation of a liquid phase occurred within three months of storage. HPLC analysis of the lidocaine content found it remained unchanged with little or no degradation occurring throughout the entire storage period. The stability of tetracaine was not evaluated. ■

# Tetracycline
# Tetracycline Hydrochloride

## Properties

Tetracycline and tetracycline hydrochloride are yellow odorless moderately hygroscopic crystalline powders.[2,3,578] Tetracycline hydrochloride 1.08 g is approximately equivalent to tetracycline 1 g.[3]

## *Solubility*

Tetracycline has an aqueous solubility of about 0.4 mg/mL. Its solubility in ethanol is about 20 mg/mL at 25°C.[2,3]

Tetracycline hydrochloride has an aqueous solubility of about 100 mg/mL. Its solubility in ethanol is about 10 mg/mL at 25°C.[2,3]

## pH

A 10-mg/mL aqueous tetracycline suspension has a pH between 3 and 7. The oral suspension has a pH of 3.5 to 6.[4] Tetracycline syrup (Sumycin, Squibb) has a measured pH of 4.3.[19]

A 10-mg/mL aqueous solution of tetracycline hydrochloride has a pH between 1.8 and 2.8. The injection has a pH of 2 to 3. Tetracycline topical solution has a pH of 1.9 to 3.5 when reconstituted as directed.[4]

## General Stability Considerations

Tetracycline and tetracycline hydrochloride darken upon exposure to strong sunlight and moist air.[2,3] Fluorescent light exposure had little or no effect on tetracycline stability.[940] Tetracycline oral products should be packaged in tight, light-resistant containers and stored at controlled room temperature. The suspension should be protected from freezing.[2–4]

Tetracycline hydrochloride solutions are less stable at a pH lower than 2.[3] Tetracycline hydrochloride in aqueous buffer solutions was reported to exhibit similar stability over the pH range of 2 to 6, but the solutions were most stable at pH 4.[940] Tetracycline hydrochloride is rapidly decomposed in alkaline hydroxide solutions.[578] Aqueous solutions become turbid on standing due to hydrolysis and precipitation of free tetracycline.[3]

Tetracycline hydrochloride was reported to be more stable in topical ointments based on petrolatum, lanolin, and paraffin oil than in polyethylene glycol-based ointments.[1697]

## Stability Reports of Compounded Preparations
### Oral

**USP OFFICIAL FORMULATION:**

| | | |
|---|---|---|
| Tetracycline hydrochloride | | 2.5 g |
| Cetylpyridium chloride 1 | | 0 mg |
| Xanthan gum | | 150 mg |
| Dibasic sodium phosphate | | 60 mg |
| Monobasic sodium phosphate | | 650 mg |
| Sodium hydroxide | | 300 mg |
| Purified water | | 35 mL |
| Suspension Structured Vehicle, NF (sugar-containing or sugar-free) | qs | 100 mL |

(See the vehicle monographs for information on the individual vehicles.)

Dissolve the two sodium phosphate salts in 25 mL of purified water. Separately dissolve the cetylpyridium chloride in purified water, diluting to obtain 5 mL of the solution containing 10 mg of the cetylpyridium chloride, and then mix with 5 mL of the phosphates solution. Add this solution in divided portions by mixing with the tetracycline hydrochloride powder in a glass mortar. Make certain to wet the powder completely, resulting in a smooth paste.

Transfer the remaining 20 mL of the phosphates solution to a beaker. Apply moderate heat and stir to form a vortex. Slowly sprinkle the xanthan gum into the vortex to produce a uniform dispersion. Transfer this dispersion to the tetracycline hydrochloride paste in the glass mortar and mix until uniform and smooth. Add 20 mL of Suspension Structured Vehicle, NF (sugar-containing or sugar-free). Dissolve the sodium hydroxide in 5 mL of purified water and slowly add the solution to the drug mixture while mixing. Bring to volume with additional Suspension Structured Vehicle, NF (sugar-containing or sugar-free), yielding a tetracycline hydrochloride 25 mg/mL oral suspension. This final mixture should be passed through a hand homogenizer before transferring to a tight, light-resistant container. The final liquid preparation should have a pH between pH 3.5 and 6. Store the preparation at controlled room temperature and protect from freezing. The beyond-use date is 30 days from the date of compounding.[4,5]

**STUDY 1:** Allen[578] described the compounding of an oral tetracycline 25-mg/mL suspension prepared from tetracycline or tetracycline hydrochloride powder or from the commercial capsules. Tetracycline hydrochloride 2.5 g or tetracycline base 2.31 g (equivalent to 2.5 g of the hydrochloride) was weighed or the contents of 10 tetracycline hydrochloride 250-mg capsules were placed in a mortar. About 10 mL of Ora-Plus suspending vehicle (Paddock) was mixed with the powder to form a smooth paste. An additional 40 mL of Ora-Plus was added geometrically. Sufficient Ora-Sweet (Paddock) was incorporated to bring the final volume to 100 mL, and this suspension was thoroughly mixed, resulting in a concentration of 125 mg/5 mL of tetracycline hydrochloride, or the equivalent of the base. The prepared suspension was packaged in tight, light-resistant containers and stored under refrigeration or at room temperature. The author indicated that these suspensions are stable for at least 30 days.

**STUDY 2:** Allen and Erickson[595] evaluated the stability of three tetracycline hydrochloride 25-mg/mL oral suspensions extemporaneously compounded from capsules. Vehicles used in this study were (1) an equal parts mixture of Ora-Sweet and Ora-Plus (Paddock), (2) an equal parts mixture of Ora-Sweet SF and Ora-Plus (Paddock), and (3) cherry syrup (Robinson Laboratories) mixed 1:4 with simple syrup. Six tetracycline hydrochloride 500-mg capsules (Mylan) were emptied into a mortar. About 20 mL of the test vehicle was added to the powder and mixed to yield a uniform paste. Additional vehicle was added geometrically, and the suspension was brought to the final volume of 120 mL, with thorough mixing after each addition. The process was repeated for each of the three test suspension vehicles. Samples of each finished suspension were packaged in 120-mL amber polyethylene terephthalate plastic prescription bottles and stored at 5 and 25°C in the dark.

No visual changes or changes in odor were detected during the study. Cherry syrup was unacceptable for use as the vehicle. Stability-indicating HPLC analysis found nearly 10% tetracycline hydrochloride loss in two days at room temperature and in seven days under refrigeration in the cherry syrup vehicle. Stability was somewhat better in the other two vehicles. In the Ora-Sweet SF-containing vehicle, 10% loss occurred in seven days at 25°C and in 10 days refrigerated at 5°C. Stability was much better in the Ora-Sweet-containing vehicle. About 10% loss occurred in 28 days and 35 days at 25 and 5°C, respectively. Allen and Erickson suggested that because of the instability of the tetracycline hydrochloride, a formulation prepared with tetracycline base powder would have better stability because it is insoluble and would be simply suspended in the vehicle rather than being dissolved in it as tetracycline hydrochloride is. The use of tetracycline base was described by Allen[578] and was discussed previously. (See Study 1.)

### Enteral

**TETRACYCLINE COMPATIBILITY SUMMARY**
*Compatible with:* Enrich • Ensure • Ensure HN • Ensure Plus • Ensure Plus HN • Osmolite • TwoCal HN • Vital • Vivonex T.E.N.
*Uncertain or variable compatibility with:* Osmolite HN

**STUDY 1:** Cutie et al.[19] added 5 mL of tetracycline syrup (Sumycin, Squibb) to varying amounts (15 to 240 mL) of Ensure, Ensure Plus, and Osmolite (Ross Laboratories) with vigorous agitation to ensure thorough mixing. The tetracycline syrup was physically compatible, distributing uniformly in these products with no phase separation or granulation.

**STUDY 2:** Fagerman et al.[51] reported a case of a febrile patient who was given tetracycline suspension via a feeding tube coadministered along with Osmolite HN. The patient did not improve, remaining febrile for nine days, at which time the tetracycline and Osmolite HN coadministration was discovered.

The schedule was changed to administration of tetracycline one hour before or two hours after a feeding. After three days, the patient was no longer febrile. The authors believed that the enteral feed contained sufficient divalent and trivalent cations to inactivate tetracycline, although serum tetracycline was not measured.

**STUDY 3:** Burns et al.[739] reported the physical compatibility of tetracycline syrup (Sumycin) 250 mg/10 mL with 10 mL of three enteral formulas, including Enrich, TwoCal HN, and Vivonex T.E.N. Visual inspection found no physical incompatibility with any enteral product.

**STUDY 4:** Altman and Cutie[850] reported the physical compatibility of tetracycline syrup (Achromycin, Lederle) 5 mL with varying amounts (15 to 240 mL) of Ensure HN, Ensure Plus HN, Osmolite HN, and Vital after vigorous agitation to ensure thorough mixing. The tetracycline syrup was physically compatible, distributing uniformly in all four enteral products with no phase separation or granulation.

### Topical

Kawamoto et al.[1288] evaluated the stability of tetracycline and nadifloxacin when their respective topical ointments were mixed. Mixing the two ointments together resulted in tetracycline turning from yellow to brown. HPLC analysis found about 40% tetracycline loss occurred, declining over the 12-day test period. However, the nadifloxacin content did not change over the test period. The authors stated that the loss of tetracycline was attributable to the alkaline pH of the nadifloxacin ointment, which contains sodium hydroxide in the formulation. Furthermore, the antibacterial activity was inhibited by the admixture of the two ointments. The authors recommended that tetracycline and nadifloxacin ointments not be mixed for use in treating impetigo contagiosa. ■

## ■ Thalidomide

### Properties

Thalidomide occurs as a white to off-white crystalline powder.[1,4] Severe life-threatening birth defects may occur if thalidomide is administered during pregnancy.[7]

### Solubility

Thalidomide is sparingly soluble in water;[1,4] its aqueous solubility has been variously cited as about 45 to 60 mcg/mL,[1] about 50 mcg/mL,[1314] and about 80 mcg/mL.[1315] The drug is also sparingly soluble in ethanol, acetone, ethyl acetate, butyl acetate, and glacial acetic acid. Thalidomide is very soluble in dimethylformamide and pyridine.[1,4] Thalidomide is soluble in dimethylsulfoxide.[7]

The incorporation of hydroxypropyl-beta-cyclodextrin into aqueous solutions increases thalidomide solubility; concentrations of about 1.5 to 1.7 mg/mL have been reached.[1314,1315]

### General Stability Considerations

Thalidomide powder and oral capsules should be packaged in tight containers, stored at controlled room temperature, and protected from exposure to light. The oral capsules should not be repackaged from the original packages.[4,7]

## Stability Reports of Compounded Preparations
### Oral
Alvarez et al.[1313] reported on the solubility and stability effects of the inclusion of hydroxypropyl-beta-cyclodextrin in aqueous solutions of thalidomide for use as an oral solution in HIV-infected children. The solubility of thalidomide increased with increasing concentrations of hydroxypropyl-beta-cyclodextrin. With hydroxypropyl-beta-cyclodextrin concentrations ranging from 0 to 25%, the aqueous solubility of thalidomide increased from less than 0.1 mg/mL to about 1.5 mg/mL. HPLC analysis of drug concentrations found that relatively rapid drug decomposition occurred at room temperature of 25°C, both with and without hydroxypropyl-beta-cyclodextrin present in the solution. Drug losses of about 10% occurred in as little as two days with losses of 30 to 40% in nine days. However, drug solutions stored refrigerated at 4 to 8°C exhibited much better stability. With hydroxypropyl-beta-cyclodextrin present in the solution, no thalidomide decomposition occurred over nine days of refrigerated storage.

This is a very different result than the relatively slower aqueous thalidomide decomposition found and reported by Eriksson et al.[1315] for aqueous solutions at room temperature.

### Injection
Injections, like other sterile drugs, should be prepared in a suitable clean air environment using appropriate aseptic procedures. When prepared from nonsterile components, an appropriate and effective sterilization method must be employed.

Eriksson et al.[1315] evaluated the stability of thalidomide at a concentration of 0.2 mg/mL in dextrose 5% as a potential intravenous infusion solution. Thalidomide powder was added to dextrose 5% in glass containers and was vigorously shaken and ultrasonicated at 33°C for 15 minutes. The procedure was repeated four times. The solution was then passed through a 0.22-μm Millex-GS sterilization filter. Although an HPLC method was used to determine drug concentrations, the stability-indicating capability of the method was not addressed. The authors reported that they found no loss of thalidomide over nine days stored at 23°C.

This is a very different result than the relatively rapid aqueous thalidomide decomposition found and reported by Alvarez et al.[1313] for aqueous solutions at room temperature. ■

# ■ Theophylline

## Properties
Theophylline is a white odorless crystalline powder with a bitter taste. It may be anhydrous or contain a molecule of water of hydration.[2–4]

### Solubility
Theophylline has an aqueous solubility of about 8.3 mg/mL.[1] The solubility in water increases with increasing pH,[2] and it is soluble in alkali hydroxide solutions.[1,3] Its solubility in ethanol is about 12.5 mg/mL.[3]

### pH
Theophylline in dextrose injection has a pH between 3.5 and 6.5.[4] Cutie et al.[19] measured the pH of some theophylline oral products: Theolair Liquid (Riker), 3.7; Elixophyllin (Berlex), 3.4; Elixophyllin-KI (Berlex), 3.3; and Asbron Elixir (Dorsey), 6.2.

### pK_a
Theophylline has a $pK_a$ of 8.77 and $pK_b$ values of 11.5 and 13.5.[1]

### Osmolality
Theophylline solution 5.33 mg/mL made by Berlex had an osmolality of 800 mOsm/kg, while the product made by Pharmaceutical Associates had an osmolality of 600 mOsm/kg.[233]

## General Stability Considerations
Theophylline products should be packaged in well-closed containers and stored at controlled room temperature.[4,7]

Theophylline 1 mg/mL in 5% dextrose is stable during autoclaving at 120°C for 20 minutes.[402]

## Stability Reports of Compounded Preparations
### Oral
**STUDY 1 (REPACKAGING):** The stability of commercial theophylline elixir (Elixophyllin, Berlex) 5.33 mg/mL repackaged as 2 mL in 5-mL amber polypropylene oral syringes (Baxa) and 15-mL amber glass vials (Wheaton Scientific) with suitable closures was investigated by Christensen et al.[86] Samples were stored at 4 and 25°C as well as at several elevated temperatures protected from light and in constant moisture conditions for 180 days. An HPLC assay was used to assess theophylline content.

Approximately 2% theophylline loss occurred at both 4 and 25°C in both containers in 180 days. At higher temperatures, evaporation of the alcohol component led to substantially increased theophylline losses due to precipitation. No loss of theophylline content of the elixir remaining in the manufacturer's original container after repackaging was found after 180 days at room temperature.

**STUDY 2 (REPACKAGING):** The shelf-life stability of a commercial nonalcoholic liquid theophylline preparation (Aquaphyllin, Ferndale) was evaluated by Johnson and Drabik.[30] The product containing theophylline 80 mg/15 mL was repackaged in four types of unit-dose oral syringes (Exacta-Med Liquid Dispensers, Baxa); 50 syringes of each type were prepared. Samples of 2 mL were packaged in 3-mL clear and amber oral syringes. Samples of 7 mL were packaged in 10-mL clear and amber oral syringes. Excess air was expelled from the syringes; a syringe cap was used to seal each syringe. The syringes were stored at 24°C exposed to fluorescent light for 180 days. A stability-indicating HPLC assay determined theophylline content.

Little or no loss occurred in any sample over 180 days; at least 96% of the initial theophylline content remained in the samples. Similarly, the mean volumes of the samples did not vary substantially. At least 96% of the initial volume remained in each sample. Although no loss of theophylline was observed in the samples exposed to light, the authors recommended protection from light because of the potential for discoloration of liquid theophylline preparations.

**STUDY 3 (EXTEMPORANEOUS COMPOUNDING):** Johnson et al.[984] reported the stability of four alcohol-free oral liquid formulations of theophylline 5 mg/mL compounded from bulk theophylline, USP, powder and also from crushed theophylline extended-release tablets (Major Pharmaceuticals). The vehicle consisted of an equal parts mixture of Ora-Plus with Ora-Sweet or Ora-Sweet SF. The suspensions were packaged in 60-mL amber plastic prescription bottles and stored at room temperature of 23 to 25°C for 90 days. No changes in color, odor, or taste were noted. The formula prepared from theophylline bulk powder in Ora-Plus–Ora-Sweet SF was the most palatable, while the formulation prepared from crushed tablets in Ora-Plus–Ora-Sweet was the least palatable, though it was still acceptable. Stability-indicating HPLC analysis found no theophylline loss occurred in any of the formulations in 90 days stored at room temperature.

## Enteral

### THEOPHYLLINE COMPATIBILITY SUMMARY

*Compatible with:* Enrich • Ensure • Ensure HN • Ensure Plus • Ensure Plus HN • Osmolite HN • Precitene • TwoCal HN • Vital • Vivonex T.E.N.
*Uncertain or variable compatibility with:* Osmolite

**STUDY 1:** Cutie et al.[19] added 15 mL of several theophylline-containing oral solutions (Theolair Liquid, Riker; Elixophyllin and Elixophyllin-KI, Berlex; Asbron Elixir, Dorsey) to varying amounts (15 to 240 mL) of Ensure, Ensure Plus, and Osmolite (Ross Laboratories) with vigorous agitation to ensure thorough mixing. The theophylline-containing solutions were compatible physically, distributing uniformly in all three enteral products with no phase separation or granulation.

**STUDY 2:** Holtz et al.[54] evaluated the stability of theophylline oral liquid (Slophyllin, Rorer) at 300 mg/L in full-strength Ensure, Ensure Plus, and Osmolite (Ross) over 12 hours at room temperature. Visual inspection revealed no changes such as clumping, gelling, separation, precipitation, or increased viscosity in the mixtures. Osmolality of the mixtures was increased substantially to around 900 mOsm/kg, approximately doubling the osmolality of the enteral products. Theophylline concentrations were assessed using a fluorescence polarization immunoassay (TDx, Abbott). About 6% loss occurred in the two Ensure products, while a 3% loss occurred in Osmolite in 12 hours.

**STUDY 3:** Fagerman and Ballou[52] reported on the compatibility of theophylline (Theostat 80) 80 mg/15 mL mixed in equal quantities with three enteral feeding formulas. The theophylline was compatible visually with Vital, Osmolite, and Osmolite HN (Ross), with no obvious thickening, granulation, or precipitation.

**STUDY 4:** Gal and Layson[213] reported a case of impaired theophylline absorption in a patient receiving continuous nasogastric feedings at 100 mL/hr with Osmolite. Theophylline 300-mg tablets (Theo-Dur) given every 12 hours and theophylline liquid 200 mg (Slo-Phyllin) given along with the continuous Osmolite failed to maintain adequate theophylline blood levels. Adequate blood levels were established using the liquid theophylline 200 mg every six hours when the Osmolite was interrupted for one hour before and after theophylline dosing.

**STUDY 5:** Burns et al.[739] reported the physical compatibility of theophylline oral liquid (Theolair) 30 mL with 10 mL of three enteral formulas, including Enrich, TwoCal HN, and Vivonex T.E.N. Visual inspection found no physical incompatibility with any of the enteral formulas.

**STUDY 6:** Altman and Cutie[850] reported the physical compatibility of several theophylline-containing oral solutions (Elixophyllin and Elixophyllin-KI, Berlex; Asbron Elixir, Dorsey) 15 mL with varying amounts (15 to 240 mL) of Ensure HN, Ensure Plus HN, Osmolite HN, and Vital after vigorous agitation to ensure thorough mixing. The theophylline-containing solutions were physically compatible, distributing uniformly in all four enteral products with no phase separation or granulation.

**STUDY 7:** Ortega de la Cruz et al.[1101] reported the physical compatibility of an unspecified amount of oral liquid theophylline (Elixifilin, Yamanouchi and Eufilina, Byk) with 200 mL of Precitene (Novartis) enteral nutrition diet for a 24-hour observation period. No particle growth or phase separation was observed.

### Topical

Kizu et al.[840] evaluated the skin penetration and stability of a variety of topical preparations of theophylline. An oleaginous ointment, a glycerin-based gel, and an isopropanol–water-based gel yielded less transdermal penetration ability than did the washable isopropyl myristate-based ointment formulation shown in Table 116, which proved to be the best. Using a mortar heated to 70°C with a water jacket, the rapeseed oil and glyceryl monostearate were dissolved and then cooled to room temperature. Isopropyl myristate was added gradually with grinding. Theophylline was then added with grinding to yield a 2% concentration having a uniform consistency; the 2% concentration demonstrated slightly better

**TABLE 116.** Washable Theophylline Ointment Formula[840]

| Component | Amount |
|---|---|
| Theophylline | 2 g |
| N-Methyl-2-pyrrolidone | 10 g |
| Hydrogenated rape seed oil | 26.5 g |
| Glyceryl monostearate | 4.5 g |
| Isopropyl myristate | qs   100 g |

transdermal penetration than did the 1 and 5% theophylline concentrations.

Samples of the washable gel formulation were stored at 5, 20, and 30°C protected from light and at ambient room temperature exposed to normal room light. HPLC analysis found that theophylline was stable, with little or no change in concentration throughout the 56-day test period stored under all of the tested conditions. ◼

## ◼ Thiamine Hydrochloride

### Properties

Thiamine hydrochloride is a white crystalline material having a slight yeastlike odor and a bitter taste.[1,2,4]

### Solubility

Thiamine hydrochloride is freely soluble in water, with an aqueous solubility of about 1 g/mL. It is soluble in glycerin (55 mg/mL), ethanol (10 mg/mL), and propylene glycol.[1,4]

### pH

Thiamine hydrochloride 1 and 0.1% solutions in water have pH values of 3.13 and 3.58, respectively.[1] Thiamine hydrochloride injection has a pH in the range of 2.5 to 4.5.[4]

### pK_a

Thiamine has $pK_a$ values of 4.8 and 9.0.[2]

### General Stability Considerations

Thiamine hydrochloride as the anhydrous form rapidly absorbs water to about 4% when exposed to air. It should be packaged in tight, light-resistant containers and stored at controlled room temperature.[4]

Thiamine hydrochloride tablets, elixir, and oral solution should be stored in tight, light-resistant containers at controlled room temperature. Thiamine hydrochloride injection should also be stored at controlled room temperature and protected from freezing and light.[2,4]

Thiamine hydrochloride is stable in acidic solutions at pH 4 or lower but unstable at alkaline and neutral pH values. Maximum stability in aqueous solution has been variously reported to be at pH 2[6] and at pH 3.5.[913]

### Stability Reports of Compounded Preparations
#### Oral

**STUDY 1:** El-Khawas and Boraie[913] evaluated the effect on the stability of thiamine hydrochloride 5 mg/mL of several potential oral vehicles, including high-concentration sugar solutions and a sugar-free vehicle of 30% sorbitol. The sugar-containing formulations included 40% sucrose, 21% dextrose, 21% fructose, and a 21% dextrose–21% fructose mixture. The test formulations also contained a citric phosphate pH 3.5 buffer (McIlvain's) and 0.1% methylparaben preservative. The oral liquids were packaged in screw-cap amber glass bottles and were stored at room temperature of 25°C and at elevated temperatures.

Stability-indicating HPLC analysis found that decomposition of the thiamine hydrochloride proceeded more rapidly in the sugar-containing vehicles than in 30% sorbitol. Projecting from the elevated temperature decomposition rates, the room temperature shelf life in the 30% sorbitol vehicle was calculated to be 41.4 months but only 6.9 months in 40% sucrose.

**STUDY 2:** Ensom and Decarie[1344] evaluated the stability of thiamine hydrochloride 100 mg/mL extemporaneously compounded from powder as an oral suspension in an equal parts mixture of Ora-Sweet and Ora-Plus. The oral suspension was packaged in amber plastic prescription bottles and stored refrigerated at 4°C protected from exposure to light and at room temperature of 25°C exposed to fluorescent light for 91 days. No substantial changes in physical appearance,

odor, color, or taste of the suspensions occurred during the study period at either temperature. The suspensions were easily resuspended with no caking or clumping throughout the study. Stability-indicating HPLC analysis found less than 5% thiamine hydrochloride loss occurred in 91 days under either storage condition. ■

# Thioguanine

## Properties
Thioguanine is a pale yellow nearly odorless crystalline powder.[2,3]

### Solubility
Thioguanine is insoluble in both water and ethanol but dissolves in aqueous solutions of alkali hydroxides.[2,3]

### pKa
Thioguanine has a $pK_a$ of 8.1.[2]

## General Stability Considerations
Thioguanine tablets should be packaged in tight containers, stored at controlled room temperature, and protected from temperatures of 40°C or more.[2,3]

## Stability Reports of Compounded Preparations
### Oral
**STUDY 1:** The stability of thioguanine 40 mg/mL in an extemporaneous oral suspension was studied by Dressman and Poust.[87] Thioguanine 40-mg tablets (Burroughs-Wellcome) were crushed and mixed with Cologel (Lilly) as one-third of the total volume, shaken, and brought to final volume with a 2:1 mixture of simple syrup and cherry syrup followed by vigorous shaking for 30 seconds and ultrasonication for at least two minutes. The thioguanine 40-mg/mL suspension was packaged in amber glass bottles and stored at 5°C and at ambient room temperature. An HPLC assay was used to determine thioguanine levels. Less than 10% loss occurred in 84 days at either temperature.

**STUDY 2:** Aliabadi et al.[1665] evaluated the stability of thioguanine 20-mg/mL oral liquid preparations compounded from commercial oral tablets. The thioguanine oral tablets were triturated to fine powder that was then used to prepare three formulations. The first formulation was similar to the formulation of Dressman and Poust[87] and served as a reference formulation. The powder was combined with 3.33 mL of 1% methylcellulose and brought to a 10 mL volume with simple syrup. The test formulations were prepared by combining the tablet powder with 5 mL of Ora-Plus suspending vehicle and bringing to a 10-mL volume with Ora-Sweet. A third formulation that included ascorbic acid 1 mg/mL was added to some of the Ora-Plus/Ora-Sweet formulation to determine if the presence of an antioxidant improved thioguanine stability. The oral liquids were packaged in amber glass bottles and stored at room temperature of 19 to 23°C for up to 91 days.

No caking occurred with any of the samples. A slight yellow discoloration appeared in the formulation containing ascorbic acid after 70 days of storage. After 35 days, the pH rose steadily through 70 days in all samples. Using a stability-indicating LCMS analytical method, the authors found that the thioguanine concentration remained above the minimum USP concentration of 93% for at least 63 days in the reference formulation and in the Ora-Plus/Ora-Sweet formulation without ascorbic acid. The inclusion of ascorbic acid 0.1% appeared to keep the thioguanine concentration above 93% for 91 days in most samples. However, one failed after about 70 days, and the authors determined that ascorbic acid was not effective in consistently increasing the shelf life of the thioguanine suspension. ■

# Thioridazine
# Thioridazine Hydrochloride

## Properties
Thioridazine is a white to slightly yellow crystalline powder that is odorless or has a slight odor.[1,4]

Thioridazine hydrochloride is a white to slightly yellow crystalline or granular powder having a slight odor and a very bitter taste.[3,4] Thioridazine hydrochloride 110 mg is equivalent to thioridazine 100 mg.[2,3]

### Solubility
Thioridazine base is practically insoluble in water[1,4,455] with a solubility of about 0.0336 mcg/mL.[1297] Thioridazine base is soluble in ethanol to about 167 mg/mL and is freely soluble in dehydrated ethanol.[4,1297]

Thioridazine hydrochloride is soluble in water, having an aqueous solubility of 111 mg/mL. It is also soluble in ethanol, having a solubility of 100 mg/mL.[2,3]

## pH

A 1% thioridazine hydrochloride aqueous solution has a pH of 4.2 to 5.2.[3,4] Thioridazine hydrochloride oral solution (Mellaril, Sandoz) had a measured pH of 3.8.[19] Thioridazine oral suspension has a pH between 8 and 10.[4]

## pK$_a$

Thioridazine has a pK$_a$ of 9.5.[1, 1297]

## Osmolality

Thioridazine suspension 20 mg/mL (Sandoz) had an osmolality of 2050 mOsm/kg.[233]

## General Stability Considerations

Thioridazine and thioridazine hydrochloride products should be packaged in tight, light-resistant containers. The oral solution should be stored at controlled room temperature and protected from temperatures above 30°C and from freezing. The tablets should be stored at controlled room temperature and protected from temperatures of 40°C or more.[2,4]

Thioridazine hydrochloride is the soluble acid form of the drug. Thioridazine base is insoluble in water. The soluble acidic form can be converted to the insoluble base by raising the solution above pH 6.[455]

Exposure of thioridazine products to light may result in small color changes due to colored free radical formation. Storage in the dark causes the product to revert to its colorless form. The slightly discolored product has been shown to exhibit no change in toxicity in animal experiments.[455]

## Compatibility with Common Beverages and Foods

**STUDY 1:** Geller et al.[242] reported that thioridazine liquid was visually compatible with acidic juices such as orange juice and grapefruit juice, although visual observation in opaque liquids is problematic. Thioridazine was stated to be incompatible with water, milk, coffee, tea, nonacidic juices such as apple juice, grape juice, pineapple juice, and prune juice, and with cola sodas.

**STUDY 2:** In addition to results with these beverages, Kerr[256] cited the compatibility of thioridazine with cranberry juice, lemonade, Mellow-Yellow, 7-Up, and Sprite. Thioridazine was stated to be incompatible with apricot juice, tomato juice, and V8. Although compatible with canned orange juice, thioridazine in higher concentrations may be incompatible with orange juice from frozen concentrate, Kerr noted.

**STUDY 3:** Lever and Hague[1011] reported the precipitation and/or color change of thioridazine oral concentrate 50 mg mixed with 15 mL of most liquids tried to mask the flavor including

Coca-Cola, Bubble Up, carbonated orange, milk, black coffee, tea, pineapple juice, prune juice, orange juice, V8, grape juice, apple juice, and tomato juice. Apricot juice was uncertain and may have masked turbidity and/or precipitation. The precipitation occurred within a few minutes of mixing and clung to the sides of the container, indicating that unacceptable variable dosage may occur. The authors recommended mixing thioridazine oral concentrate freshly with only distilled water.

**STUDY 4:** The drug manufacturer has evaluated a number of juices and other diluents and flavored vehicles for compatibility with thioridazine hydrochloride concentrate. The drug–diluent combinations were evaluated for precipitation,

**TABLE 117.** Compatibility Results of Thioridazine Liquid Concentrates with Various Diluents[455]

| Compatible | Incompatible |
|---|---|
| **Thioridazine Concentrate 30 mg/mL** | |
| Lemon juice from frozen concentrate | Milk |
| Grapefruit juice from frozen concentrate | Orange drink |
| Lemonade from frozen concentrate | Pineapple juice |
| Orange juice from frozen concentrate | Tomato juice |
| Fresh orange juice | Apricot juice |
| | Apple juice |
| | V8 vegetable juice |
| | Cider (type unspecified) |
| | Grape juice and grape drink |
| | Coca-Cola and Tab |
| | Coffee, tea, tap water |
| **Thioridazine Concentrate 100 mg/mL** | |
| Canned grapefruit juice | Apple cider and apple juice |
| Hi C orange/pineapple | Canned grape drink |
| Canned orange juice | Carrot juice |
| Cranberry juice cocktail | V8 juice |
| 7-Up | Hawaiian Punch |
| Canada Dry ginger ale | Tomato juice |
| Lemonade from frozen concentrate | Coca-Cola |
| Kool-Aid, lemon | Coffee and tea |
| Sprite | Grapefruit juice from frozen concentrate |
| Mellow-Yellow | Orange juice from frozen concentrate |
| Tap water (acidified) | Milk and tap water |

flocculation, and/or color changes. The test results appear in Table 117.[455]

Thioridazine hydrochloride is soluble in acidic media but insoluble above pH 6. Consequently, it remains dissolved in acidified or distilled water but precipitates in the more alkaline tap water containing dissolved alkali carbonates. The drug also precipitates in contact with tannic acid; tannates are found in coffee, tea, and Coca-Cola. Although the drug is soluble in acidic media, it does interact with pectin, gums, and carboxymethylcellulose sodium, forming insoluble associations. Such materials are present in some natural products and are added to some synthetic fruit drinks. The manufacturer does not consider the precipitates to seriously alter therapeutic efficacy because exposure to the gastrointestinal juices should reverse the reactions.[455]

The problematic nature of visual observation of incompatibilities in opaque liquids may be the reason for differences in the compatibility designations among the studies. Because the compatibility determinations require making judgments, it is possible that different investigators come to differing compatibility conclusions.

### Enteral

#### THIORIDAZINE HYDROCHLORIDE COMPATIBILITY SUMMARY

*Compatible with:* Vivonex T.E.N.

*Incompatible with:* Enrich • Ensure • Ensure HN • Ensure Plus • Ensure Plus HN • Osmolite • Osmolite HN • TwoCal HN • Vital

**STUDY 1:** Cutie et al.[19] added 2 mL (200 mg) of thioridazine hydrochloride oral solution (Mellaril, Sandoz) to varying amounts (15 to 240 mL) of Ensure, Ensure Plus, and Osmolite (Ross Laboratories) with vigorous agitation to ensure thorough mixing. The thioridazine hydrochloride oral solution was physically incompatible with the enteral products, which developed granulation. The granules clogged the feeding tubes.

**STUDY 2:** Burns et al.[739] reported the physical compatibility of thioridazine hydrochloride oral concentrate (Mellaril) 150 mg/5 mL with 10 mL of three enteral formulas including Enrich, TwoCal HN, and Vivonex T.E.N. Visual inspection found no physical incompatibility with the Vivonex T.E.N. However, when mixed with Enrich and TwoCal HN, a thinner consistency with particulates formed.

**STUDY 3:** Altman and Cutie[850] reported the physical incompatibility of 5 mL of thioridazine hydrochloride oral solution (Mellaril, Sandoz) with varying amounts (15 to 240 mL) of Ensure HN, Ensure Plus HN, Osmolite HN, and Vital with vigorous agitation to ensure thorough mixing. The thioridazine hydrochloride oral solution was physically incompatible with all four enteral products, which developed granulation. The granules clogged the feeding tubes.

## Compatibility with Other Drugs

#### THIORIDAZINE HYDROCHLORIDE COMPATIBILITY SUMMARY

*Compatible with:* Trifluoperazine hydrochloride

*Incompatible with:* Chlorpromazine hydrochloride
• Doxepin hydrochloride • Haloperidol lactate • Lithium citrate • Loxapine hydrochloride • Mesoridazine besylate • Molindone hydrochloride • Perphenazine • Thiothixene hydrochloride • Trihexyphenidyl hydrochloride

*Uncertain or variable compatibility with:* Fluphenazine hydrochloride

**STUDY 1:** Theesen et al.[78] evaluated the compatibility of lithium citrate syrup with oral neuroleptic solutions, including thioridazine hydrochloride concentrate. Lithium citrate syrup (Lithionate-S, Rowell) and lithium citrate syrup (Roxane) at a lithium-ion concentration of 1.6 mEq/mL were combined in volumes of 5 and 10 mL with 2, 4, 6, and 8 mL of thioridazine hydrochloride oral concentrate 100 mg/mL (Sandoz) in duplicate, with the order of mixing reversed between samples. Samples were stored at 4 and 25°C for six hours and evaluated visually for physical compatibility.

An opaque turbidity formed immediately in all samples. Thin-layer chromatographic analysis of the sediment layers of incompatible combinations showed high concentrations of the neuroleptic drugs. The authors attributed the incompatibilities to the excessive ionic strength of the lithium citrate syrup, which results in salting out of the neuroleptic drug salts in a hydrophobic, viscid layer that adheres tenaciously to container surfaces. The problem persists even when moderate-to-high doses are diluted 10-fold in water.

Theesen et al.[78] also noted that fruit and vegetable beverages, which are often used to administer the neuroleptics, are effective salting-out agents. Unfortunately, visually observing the phenomenon is impaired by the color and opacity of the beverages.

**STUDY 2:** Raleigh[454] evaluated the compatibility of thioridazine suspension (Mellaril-S, Sandoz) (concentration unspecified) with 13 other liquid oral medications. Each liquid oral drug product was combined with the thioridazine suspension in a manner described as "simple random dosage combinations"; the exact amounts tested and test conditions were not specified. All but two of the oral liquid dosage forms tested resulted in the formation of a dense white curdy precipitate. Only fluphenazine hydrochloride elixir 0.5 mg/mL (but not concentrate) and trifluoperazine 10-mg/mL concentrate (as hydrochloride) did not result in the curdy precipitate. The list of liquid oral drugs tested is presented in Table 118. ■

**TABLE 118.** Compatibility of Selected Oral Drugs with Thioridazine Suspension[454]

| Test Drug | Concentration | Observed Result |
|---|---|---|
| Chlorpromazine hydrochloride concentrate | Unspecified | Curdy precipitate |
| Doxepin concentrate (as the hydrochloride) | 10 mg/mL | Curdy precipitate |
| Fluphenazine hydrochloride concentrate | 5 mg/mL | Curdy precipitate |
| Fluphenazine hydrochloride elixir | 0.5 mg/mL | No precipitate formed |
| Haloperidol concentrate | 2 mg/mL (as the lactate) | Curdy precipitate |
| Lithium citrate elixir | 1.6 mEq lithium/mL | Curdy precipitate |
| Loxapine concentrate | 25 mg/mL (as the hydrochloride) | Curdy precipitate |
| Mesoridazine concentrate | 25 mg/mL (as the besylate) | Curdy precipitate |
| Molindone hydrochloride concentrate | 20 mg/mL | Curdy precipitate |
| Perphenazine concentrate | 3.2 mg/mL | Curdy precipitate |
| Thiothixene concentrate | 5 mg/mL (as the hydrochloride) | Curdy precipitate |
| Trifluoperazine concentrate | 10 mg/mL (as the hydrochloride) | No precipitate formed |
| Trihexyphenidyl hydrochloride Elixir | 0.4 mg/mL | Curdy precipitate |

# ◼ Thiothixene
# Thiothixene Hydrochloride

## Properties

Thiothixene is a white to tan nearly odorless crystalline powder. Thiothixene 1 mg is approximately equivalent to thiothixene hydrochloride dihydrate 1.25 mg or thiothixene hydrochloride anhydrous 1.16 mg.[2,3]

Thiothixene hydrochloride is a white or almost white crystalline powder having a slight odor.[2,3]

### Solubility

Thiothixene is practically insoluble in water, having an aqueous solubility of 0.1 mg/mL at 25°C. Its solubility in dehydrated ethanol is about 9 mg/mL.[2,3]

Thiothixene hydrochloride is very soluble in water, having an aqueous solubility of about 125 mg/mL at 25°C. In dehydrated ethanol its solubility is about 3.7 mg/mL.[2,3]

### pH

Thiothixene injection has a pH of 2.5 to 3.5. Reconstituted thiothixene hydrochloride for injection has a pH between 2.3 and 3.7. The oral solution has a pH between 2 and 3.[4] Thiothixene hydrochloride oral solution concentrate (Navane, Roerig) had a measured pH of 2.6.[19]

## General Stability Considerations

Thiothixene and thiothixene hydrochloride products are adversely affected by light.[2,3] In general, the products should be packaged in tight, light-resistant containers.[4] The reconstituted injection is stable for 48 hours at room temperature.[2,7] Thiothixene hydrochloride is stable in aqueous solution over a narrow pH range of 2 to 4.[2]

## Stability Reports of Compounded Preparations

### Injection

Injections, like other sterile drugs, should be prepared in a suitable clean air environment using appropriate aseptic procedures. When prepared from nonsterile components, an appropriate and effective sterilization method must be employed.

Allen[1359] reported on a compounded formulation of thiothixene hydrochloride 2-mg/mL injection. The injection had the following formula:

| | |
|---|---|
| Thiothixene hydrochloride | 200 mg |
| Benzyl alcohol | 900 mg |
| Propyl gallate | 20 mg |
| Dextrose | 5 g |
| Sterile water for injection | qs 100 mL |

The recommended method of preparation is to add the thiothixene hydrochloride, benzyl alcohol, propyl gallate, and dextrose to about 95 mL of sterile water for injection and stir until dissolved. Sterile water for injection sufficient to bring the volume to 100 mL is added, and the solution is mixed well. The solution is to be filtered through a suitable 0.2-μm sterilizing filter and packaged in sterile tight, light-resistant containers. If no sterility test is performed, the USP specifies a beyond-use date of 24 hours at room temperature or three days stored under refrigeration because of concern for inadvertent microbiological contamination during preparation. However, if an official USP sterility test for each batch of drug is performed, the author recommended a beyond-use date of six months at room temperature because this formula is similar or the same as a commercial medication in some countries with an expiration date of two years or more.

### Enteral
**THIOTHIXENE HYDROCHLORIDE COMPATIBILITY SUMMARY**
*Compatible with:* Enrich • Ensure • Ensure HN • Ensure Plus • Ensure Plus HN • Osmolite • Osmolite HN • TwoCal HN • Vital • Vivonex T.E.N.

**STUDY 1:** Cutie et al.[19] added 2 mL of thiothixene hydrochloride concentrate (Navane, Roerig) to varying amounts (15 to 240 mL) of Ensure, Ensure Plus, and Osmolite (Ross Laboratories) with vigorous agitation to ensure thorough mixing. The thiothixene hydrochloride concentrate was physically compatible, distributing uniformly in all three enteral products with no phase separation or granulation. The concentrate should be added very slowly with vigorous agitation to ensure that no incompatibility occurs.

**STUDY 2:** Burns et al.[739] reported the physical compatibility of thiothixene hydrochloride concentrate (Navane) 25 mg/5 mL with 10 mL of three enteral formulas, including Enrich, TwoCal HN, and Vivonex T.E.N. Visual inspection found no physical incompatibility with any of the enteral formulas.

**STUDY 3:** Altman and Cutie[850] reported the physical compatibility of 2 mL of thiothixene hydrochloride concentrate (Navane, Roerig) with varying amounts (15 to 240 mL) of Ensure HN, Ensure Plus HN, Osmolite HN, and Vital after vigorous agitation to ensure thorough mixing. The thiothixene hydrochloride concentrate was physically compatible, distributing uniformly in all four enteral products with no phase separation or granulation.

## Compatibility with Other Drugs
**THIOTHIXENE HYDROCHLORIDE COMPATIBILITY SUMMARY**
*Compatible with:* Lithium citrate
*Incompatible with:* Thioridazine hydrochloride

**STUDY 1:** Theesen et al.[78] evaluated the compatibility of lithium citrate syrup with oral neuroleptic solutions, including thiothixene hydrochloride concentrate. Lithium citrate syrup (Lithionate-S, Rowell) and lithium citrate syrup (Roxane) at a lithium-ion concentration of 1.6 mEq/mL were combined in volumes of 5 and 10 mL with 2, 6, 12, and 24 mL of thiothixene hydrochloride oral concentrate 5 mg/mL (Roerig) in duplicate, with the order of mixing reversed between samples. Samples were stored at 4 and 25°C for six hours and evaluated visually for physical compatibility. No incompatibility was observed.

**STUDY 2:** Raleigh[454] evaluated the compatibility of lithium citrate syrup (Philips Roxane) 1.6 mEq lithium/mL with thiothixene concentrate 5 mg/mL (as the hydrochloride). The thiothixene concentrate was combined with the lithium citrate syrup in a manner described as "simple random dosage combinations"; the exact amounts tested and test conditions were not specified. No precipitate was observed to form upon mixing.

Raleigh[454] also evaluated the compatibility of thioridazine suspension (Sandoz) (concentration unspecified) with thiothixene concentrate 5 mg/mL (as the hydrochloride). The thiothixene concentrate was combined with the thioridazine suspension in the manner described as "simple random dosage combinations." The exact amounts tested and test conditions were not specified. A dense white curdy precipitate formed.

## Compatibility with Common Beverages and Foods
Geller et al.[242] reported that thiothixene liquid was compatible visually with water, milk, prune juice, apricot juice, orange juice, tomato juice, grapefruit juice, and cranberry juice, although visual observation in opaque liquids is problematic. Thiothixene was incompatible with cola sodas and apple juice.

In addition to these beverages, Kerr[256] cited the compatibility of thiothixene with V8 vegetable juice. ■

# ■ Tiagabine Hydrochloride

## Properties
Tiagabine hydrochloride is a white to off-white odorless crystalline powder.[1,4,7] Tiagabine hydrochloride 55 mg is equivalent to tiagabine 50 mg.[3]

### Solubility
Tiagabine hydrochloride is sparingly soluble in water to about 30 mg/mL (3%).[1,7]

### pKa
Tiagabine hydrochloride has $pK_a$ values of 3.3 and 9.4.[1]

## General Stability Considerations
Tiagabine hydrochloride powder should be packaged in tight, light-resistant containers and stored at temperatures not exceeding 30°C.[4] Tiagabine hydrochloride tablets should be stored at controlled room temperature and protected from exposure to light and moisture.[7]

## Stability Reports of Compounded Preparations
### Oral
**USP OFFICIAL FORMULATION:**

>Tiagabine hydrochloride 100 mg
>Vehicle for Oral Solution, NF
>and Vehicle for Oral Suspension, NF (1:1) qs 100 mL

(See the vehicle monographs for information on the individual vehicles.)

Use tiagabine hydrochloride powder or commercial tablets. If using tablets, crush or grind to fine powder. Add the vehicle mixture in small portions and mix to make a uniform paste. Add additional vehicle in increasing volumes to make a pourable suspension. Quantitatively transfer to a suitable calibrated, tight, light-resistant bottle, bring to final volume with the vehicle mixture, and thoroughly mix, yielding tiagabine hydrochloride 1mg/mL oral suspension. The final liquid preparation should have a pH between 4 and 4.5. Store the preparation at controlled room temperature or under refrigeration between 2 and 8°C. The beyond-use date is 60 days from the date of compounding at controlled room temperature and 90 days from the date of compounding when stored under refrigeration.[4]

**STUDY 1:** Nahata and Morosco[778] reported the stability of tiagabine hydrochloride 1 mg/mL in two oral liquid formulations. Tiagabine hydrochloride 12-mg tablets (Gabitril, Cephalon) were pulverized using a mortar and pestle and mixed with two suspending vehicles composed of (1) simple syrup, NF–methylcellulose 1% in a 6:1 ratio and (2) Ora-Sweet–Ora-Plus in an equal parts mixture. The suspensions were packaged in amber plastic prescription bottles and stored refrigerated at 4 and at 25°C in a water bath for 91 days.

No change in the physical appearance or odor of either tiagabine hydrochloride formulation occurred. Stability-indicating HPLC analysis found tiagabine hydrochloride losses of 3% in 91 days refrigerated and 8% in 42 days at 25°C in the simple syrup–methylcellulose 1% vehicle. The stability of tiagabine hydrochloride was better with the Ora-Sweet–Ora-Plus vehicle, exhibiting a loss of 2% in 91 days at 4°C and an 8% loss in 70 days at 25°C.

**STUDY 2:** Haase et al.[863] evaluated the stability of tiagabine hydrochloride 2- and 6-mg/mL oral suspensions prepared from tablets (Gabitril, Cephalon). The appropriate quantity of tiagabine hydrochloride tablets was triturated to fine powder using a mortar and pestle. A smooth paste was then prepared using a small amount of sterile water for injection, and the paste was then mixed with an equal parts mixture of Ora-Sweet and Ora-Plus to prepare the final concentrations. The suspensions were packaged in amber plastic prescription bottles and were stored at room temperature of 23 to 25°C in the dark and under refrigeration at 3 to 5°C for up to 61 days.

No change in odor was detected, but the solutions stored at room temperature developed a blue discoloration after 30 days of storage. Stability-indicating HPLC analysis found the 6-mg/mL concentration stable throughout the 61-day storage period at both room temperature and under refrigeration. However, the 2-mg/mL concentration was stable at room temperature for only 61 days. The refrigerated samples exhibited 10% loss in 42 days and 12% loss in 61 days. Whether this result is real or anomalous is unknown. Because of less desirable palatability for the 6-mg/mL concentration and the 2-mg/mL concentration at room temperature, the authors recommended using the 2-mg/mL concentration and storing it under refrigeration for up to 42 days. ■

# ■ Ticarcillin Disodium

## Properties
Ticarcillin disodium is a white, creamy-white, or pale yellow hygroscopic noncrystalline powder.[1,4]

### Solubility
Ticarcillin disodium is freely soluble in water, having an aqueous solubility exceeding 1 g/mL.[1,4]

### pH
Reconstituted ticarcillin disodium injection has a pH in the range of 6 to 8.[7]

### Osmolality
Ticarcillin disodium 30 mg/mL in 5% dextrose and in 0.9% sodium chloride is reported to have osmolalities of 420 and 442 mOsm/kg, respectively.[8]

## General Stability Considerations
Intact vials of ticarcillin disodium and ticarcillin disodium–clavulanate potassium are stored at room temperature. The manufacturer recommends use of solutions of ticarcillin 200 and 300 mg/mL as the disodium salt in bulk packages within 24 hours at room temperature or 72 hours under refrigeration.[7]

Ticarcillin disodium in solution forms polymer conjugation products upon standing, which can result in hypersensitivity reactions. A greater amount of polymer conjugates form at high concentrations and at room temperature. Use of reconstituted solutions or storage under refrigeration has been suggested.[8]

Diluted in 5% dextrose or 0.9% sodium chloride, ticarcillin disodium is reported to be stable for up to 72 hours at room temperature, 14 days under refrigeration, and 270 days frozen at −20°C.[7,8]

## Stability Reports of Compounded Preparations
### Ophthalmic

Ophthalmic preparations, like other sterile drugs, should be prepared in a suitable clean air environment using appropriate aseptic procedures. When prepared from nonsterile components, an appropriate and effective sterilization method must be employed.

Blondeel et al.[973] evaluated the stability of ticarcillin disodium ophthalmic solution stored under a variety of conditions for extended periods. Commercial ticarcillin disodium injection was reconstituted with sterile water for injection and mixed into sufficient sodium chloride 0.9% to prepare a 5-mg/mL solution. The solution was passed through a 0.22-μm filter into 10-mL sterile glass bottles and sealed with sterile polyvinyl chloride plastic caps with plastic droppers.

Samples were stored at room temperature of 25°C exposed to and protected from light and also refrigerated at 4°C for 15 days. A second set of samples was stored frozen at −20°C for nine weeks before being thawed and stored at room temperature and refrigerated as the first set of samples was.

No visually observable changes such as color changes or precipitation were noted. Stability-indicating HPLC analysis found that ticarcillin disodium retained acceptable stability, with less than 10% loss after three days at room temperature (light exposure did not affect the results) and after seven days under refrigeration. Frozen storage for nine weeks did not substantially alter the stability of the solutions after thawing. Less than 10% ticarcillin loss occurred in three days at room temperature and in six days refrigerated. The authors concluded that the ticarcillin disodium ophthalmic solution could be prepared and stored frozen for up to nine weeks before thawing for use without adversely affecting the drug's stability. ■

# ■ Timolol Maleate

## Properties

Timolol maleate is a white or practically white odorless or practically odorless powder.[1,3,4,7]

### Solubility

Timolol maleate is soluble in water and ethanol and is sparingly soluble in propylene glycol.[1,4]

### pH

Timolol maleate 20 mg/mL in water has a pH of 3.8 to 4.3.[3,4] Timolol maleate ophthalmic solution has a pH of 6.5 to 7.5.[4] Timolol 5 mg/mL (as the maleate) ophthalmic solution was found to have a pH near 6.9.[880]

### pKₐ

Timolol maleate has a $pK_a$ of 9.[2]

### Osmolality

Timolol maleate ophthalmic solution is isotonic.[7]

## General Stability Considerations

Timolol maleate powder and tablets should be packaged in well-closed containers and stored at controlled room temperature and protected from exposure to light.[4,7] Timolol maleate ophthalmic solution should be packaged in tight, light-resistant containers and stored at controlled room temperature protected from light and from freezing.[4,7]

Timolol maleate is stable in solution up to pH 12,[1] but the pH of maximum stability is 6.8.[740]

## Stability Reports of Compounded Preparations
### Ophthalmic

Ophthalmic preparations, like other sterile drugs, should be prepared in a suitable clean air environment using appropriate aseptic procedures. When prepared from nonsterile components, an appropriate and effective sterilization method must be employed.

**STUDY 1:** Pilatti et al.[740] reported the instability of pilocarpine hydrochloride when prepared with timolol maleate in ophthalmic solutions. For ophthalmic solutions containing pilocarpine hydrochloride 10 and 20 mg/mL, stability-indicating HPLC analysis found about 8% loss of drug after storage of up to 24 months, presumably at room temperature, although the temperature was not specified. However, if timolol maleate 6.8 mg/mL was incorporated into the ophthalmic solution, pilocarpine losses up to 16 and 21% occurred in two and 11 months, respectively.

**STUDY 2:** Garcia-Valldecabres et al.[880] reported the change in pH of several commercial ophthalmic solutions, including timolol 5 mg/mL (as the maleate) (Cusimolol) over 30 days after opening. Slight variation in solution pH was found, but no consistent trend and no substantial change occurred. ■

# ■ Tinidazole

## Properties

Tinidazole occurs as colorless crystals[1] or as a nearly white or pale yellow crystalline powder.[3,4]

### Solubility

Tinidazole is practically insoluble in water,[3,4] having an aqueous solubility of about 6 mg/mL.[1132] Tinidazole is sparingly soluble in methanol and soluble in acetone.[3,4]

### pKa

Tinidazole has a pKa of 2.11.[1132]

## General Stability Considerations

Tinidazole powder is packaged in tight containers and stored at controlled room temperature protected from exposure to light.[3,4] Tinidazole tablets are also packaged in tight containers and stored at controlled room temperature protected from exposure to light.[7]

Tu et al.[1133] evaluated the stability of tinidazole 2 mg/mL in aqueous solution. Tinidazole was most stable in the range of pH 4 to 5. Above pH 6, decomposition proceeds much more rapidly. When tinidazole is exposed to light, the time to 10% loss ($t_{90}$) was found to be about 17.5 days, indicating protection from exposure to light is necessary. From elevated temperature degradation studies, the $t_{90}$ at 25°C was calculated to be 890 days.

Salo et al.[1134] confirmed pH 4 to 5 as the pH range of maximum stability. They too noted the more rapid loss of tinidazole at pH greater than 6 to 7.

## Stability Reports of Compounded Preparations
### Oral

The manufacturer of tinidazole tablets provides the following information on an extemporaneously compounded oral suspension prepared from commercial tablets. Four 500-mg oral tablets were pulverized using a mortar and pestle. Approximately 10 mL of artificial cherry syrup was added to this powder and mixed until smooth. The suspension was then transferred into a graduate amber container. Several small rinses of cherry syrup were used to transfer any remaining drug in the mortar to the final suspension for a final volume of 30 mL. This suspension was stated to be stable for seven days at room temperature. The suspension should be shaken well before administration.[7] ■

# ■ Tobramycin
# Tobramycin Sulfate

## Properties

Tobramycin is a white or nearly white hygroscopic powder. The sulfate is formed in situ during manufacturing.[2,3] Each milligram of tobramycin sulfate contains tobramycin 634 to 739 mcg.[4]

### Solubility

Tobramycin has an aqueous solubility of about 667 mg/mL. It is very slightly soluble in ethanol.[2,3]

### pH

A 4% tobramycin sulfate aqueous solution has a pH between 6 and 8. Tobramycin injection has a pH between 3 and 6.5.[4]

## General Stability Considerations

Tobramycin sulfate products should be stored at controlled room temperature; they should not be used if discolored. The commercially available injections in 0.9% sodium chloride should be stored at controlled room temperature and protected from excessive temperatures of 40°C or more and from freezing.[2] The reconstituted injection from powder is stable for 24 hours at room temperature and for 96 hours under refrigeration.[2,7] Frozen in the original container at −20°C, reconstituted tobramycin sulfate solution is stable for up to 12 weeks.[2]

Tobramycin inhalation solution (Tobi) should be stored under refrigeration at 2 to 8°C and protected from exposure to intense light. If refrigeration is unavailable, the inhalation solution may be stored at room temperature up to 25°C for a maximum of 28 days. The inhalation solution is slightly yellow but may darken if not stored in a refrigerator. The color change does not indicate a change in product quality as long as it has been stored within the recommended storage conditions.[7]

Several studies showed good solution stability of tobramycin sulfate diluted in 5% dextrose injection and 0.9% sodium chloride injection over 28 to 35 days frozen at −20°C.[448–450]

Tobramycin sulfate is stable in solution over pH 1 to 11 at 5 to 27°C for up to several weeks. Furthermore, it can be autoclaved without concentration loss.[447]

## Stability Reports of Compounded Preparations
### Injection
Injections, like other sterile drugs, should be prepared in a suitable clean air environment using appropriate aseptic procedures. When prepared from nonsterile components, an appropriate and effective sterilization method must be employed.

**STUDY 1:** Seitz et al.[451] packaged a 40-mg/mL tobramycin sulfate solution in Monoject plastic syringes and stored them at 4 and 25°C. No substantial concentration change was found by radioimmunoassay after two months at either temperature.

**STUDY 2:** Zbrozek et al.[335] diluted tobramycin sulfate to 30 mg/mL in 0.9% sodium chloride injection and packaged the solution in polypropylene syringes (Becton Dickinson). The syringes were stored at 25°C exposed to fluorescent light for 48 hours. No loss of tobramycin sulfate was observed using fluorescence polarization immunoassay (TDx).

### Ophthalmic
Ophthalmic preparations, like other sterile drugs, should be prepared in a suitable clean air environment using appropriate aseptic procedures. When prepared from nonsterile components, an appropriate and effective sterilization method must be employed.

**STUDY 1:** McBride et al.[106] reported the stability of a fortified tobramycin sulfate 13.6-mg/mL ophthalmic solution prepared from 165 mL of the commercial ophthalmic solution (Tobrex 3 mg/mL, Alcon) and 66 mL of tobramycin sulfate injection (Nebcin 40 mg/mL, Lilly). The solution was prepared in an analogous manner to that used for compounding single-unit admixtures, which usually involves the addition of 2 mL of the 40-mg/mL tobramycin sulfate injection to 5 mL of the 3-mg/mL ophthalmic solution. The final preparation was packaged as 7-mL volumes in plastic bottles and stored under refrigeration at 4 to 8°C for 91 days. Tobramycin sulfate stability was evaluated using a fluorescence polarization immunoassay and a stability-indicating HPLC assay. Little or no drug loss was shown by either method.

**STUDY 2:** Charlton et al.[582] studied the stability of tobramycin 15 mg/mL as the sulfate salt in an ophthalmic solution. The solution was prepared by aseptically adding 3.75 mL of tobramycin sulfate 40-mg/mL injection (Elkins-Sinn) to 6.25 mL of artificial tears (Liquifilm Tears, Allergan). The samples were stored in the original artificial tears bottles at 4 and 25°C for 28 days. The osmolality remained unchanged; the pH changed somewhat, from about 6 to about 5.3 to 5.6, but remained within the acceptable range. Antimicrobial activity, estimated by Kirby-Bauer microbial growth inhibition, exceeded the minimum and remained unchanged for 28 days at both storage temperatures.

**STUDY 3:** Bowe et al.[627] evaluated the stability of tobramycin 15 mg/mL as the sulfate salt in a fortified ophthalmic preparation. A vial of the injection was diluted with methylcellulose artificial tears. Samples were stored at 24 and 4°C in the dark for four weeks. Antibiotic activity was determined by measuring minimum inhibitory concentration against common ophthalmic pathogens. No loss of antibiotic activity was found at either storage temperature over four weeks.

**STUDY 4:** Arici et al.[796] reported the stability of tobramycin sulfate 13.5-mg/mL stock ophthalmic solutions prepared from injection added to commercial ophthalmic solution and packaged in ophthalmic squeeze bottles at room temperature of 24°C and refrigerated at 4°C. No loss in microbiological activity against *Staphylococcus aureus* and *Pseudomonas aeruginosa* was found over 28 days of storage at both temperatures in both vehicles. Small increases in pH of 0.4 pH unit or less occurred.

Also see *Compatibility with Other Drugs.*

### Inhalation
Inhalation preparations, like other sterile drugs, should be prepared in a suitable clean air environment using appropriate aseptic procedures. When prepared from nonsterile components, an appropriate and effective sterilization method must be employed.

Walker and Madden[975] reported the stability of tobramycin sulfate 32 mg/mL in an inhalation solution with phenylephrine 0.125% in propylene glycol and distilled water. The inhalation solution was packaged in plastic bottles and stored at room temperature of 23°C and refrigerated at 4°C. HPLC analysis found that the tobramycin sulfate remained stable for eight weeks, exhibiting less than 10% loss.

Also see *Compatibility with Other Drugs.*

## Compatibility with Other Drugs
### TOBRAMYCIN AND TOBRAMYCIN SULFATE
### COMPATIBILITY SUMMARY
*Compatible with:* Albuterol sulfate • Budesonide • Fluticasone propionate • Iprotropium bromide • Sodium chloride 7% (hypertonic) • Vancomycin hydrochloride
*Incompatible with:* β-Lactam antibiotics • Dornase alfa • Sucralfate

**STUDY 1 (β-LACTAMS):** Tobramycin sulfate is well documented to be unstable in the presence of some β-lactam antibiotics, such as the penicillins carbenicillin, ticarcillin, piperacillin, and ampicillin. The cephalosporins also may cause inactivation, although not usually to the same extent as the pencillins. The extent of tobramycin inactivation is dependent on the specific β-lactam antibiotic, drug concentrations, storage temperature, and duration of storage. In general, compounding tobramycin in the same formulation with β-lactam antibiotics is discouraged. The drugs should be administered separately.

**STUDY 2 (SUCRALFATE):** Feron et al.[467] evaluated the interaction of sucralfate with the antimicrobial drugs colistin sulfate, tobramycin sulfate, and amphotericin B. Sucralfate 500 mg was added to 40 mL of HPLC-grade water and adjusted to pH 3.5 with hydrochloric acid. The antimicrobial drugs were added to separate samples to yield final concentrations of colistin 0.05 mg/mL, tobramycin 0.05 mg/mL, and amphotericin B 0.025 mg/mL. The concentration of each drug was analyzed initially and numerous times over 90 minutes of storage. The colistin concentration was assessed by HPLC, tobramycin concentration was determined by enzyme immunoassay, and amphotericin B concentration was found by spectroscopy.

Losses of all three antimicrobials occurred very rapidly and extensively. Colistin, tobramycin, and amphotericin B losses were 88, 99, and 95%, respectively, after five minutes. Furthermore, the loss was not reversible by adjusting pH to near neutrality. The authors noted that the mechanism of interaction is unclear. However, concurrent enteral administration of sucralfate and these antimicrobials would result in substantially lower concentrations than would be achieved in the absence of sucralfate.

**STUDY 3 (ALBUTEROL):** Gooch[1023] evaluated the compatibility and stability of albuterol sulfate (Ventolin) 1.4 mg/mL and tobramycin sulfate (Nebcin) 11.4 mg/mL in sodium chloride 0.9% for inhalation in treating infections in cystic fibrosis patients. Samples were prepared in screw-cap glass containers and stored refrigerated at 4°C for seven days. No precipitation or color change was observed. HPLC analysis found no loss of albuterol in seven days. Fluorescence immunoassay of tobramycin found no loss in seven days as well.

**STUDY 4 (MULTIPLE DRUGS):** White and Hood[1025] reported that albuterol sulfate inhalation solution was compatible with tobramycin sulfate, although no drug concentrations were stated. Adding ipratropium bromide to the inhalation admixture was also stated to be compatible, again without citing drug concentrations.

**STUDY 5 (HYPERTONIC SODIUM CHLORIDE):** Fox et al.[1239] evaluated the physical compatibility of hypertonic sodium chloride 7% with several inhalation medications used in treating cystic fibrosis. Tobramycin injection (Hospira) 40 mg/mL (as sulfate) mixed in equal quantities with extemporaneously compounded hypertonic sodium chloride 7% did not exhibit any visible evidence of physical incompatibility, and the measured turbidity did not increase over the one-hour observation period.

**STUDY 6 (DORNASE ALFA):** Kramer et al.[1301] also evaluated the compatibility and stability of inhalation solution admixtures of tobramycin inhalation solutions with dornase alfa (Pulmozyme, Roche). Dornase alfa (Pulmozyme Respule 2500 units/2.5 mL) was mixed with 300 mg/5 mL of tobramycin inhalation solution (Tobi, Chiron) or with tobramycin sulfate inhalation solution

(Gernebcin, Infectopharm) 80 mg/2 mL and was stored at room temperature exposed to room light.

The admixtures were incompatible. Visible brown discoloration was observed within an hour after mixing, and the odor of the mixtures worsened. Fluorescence immunoassay (TDx/TDxFLx) assay of gentamicin concentrations found no loss over 24 hours in either combination. However, kinetic colorimetric DNase activity assays of dornase alfa found decreased activity. Dornase alfa mixed with Tobi resulted in a decrease of dornase alfa of about 23% in 24 hours. With Gernebcin, dornase alfa activity decreased 24% in as little as one hour, and loss of nearly 40% occurred 24 hours after mixing. The more rapid and extensive loss with Gernebcin was attributed to the presence of sodium metabisulfite in the Gernebcin formulation. When dornase alfa was mixed with sodium metabisulfite 0.05% solution, dornase alfa loss of 40% occurred in about one hour.

**STUDY 7 (VANCOMYCIN):** Anton Torres et al.[1629] evaluated the stability of tobramycin sulfate 0.01 mg/mL (10 mcg/mL) with vancomycin hydrochloride 0.05 mg/mL (50 mcg/mL) in BSS-Plus (Alcon) used for intraocular irrigation during cataract surgery. Vials of the two antibiotics were reconstituted and appropriate amounts of the solutions were passed through a 0.22-μm filter as they were added unto the BSS-Plus solution. Samples were stored refrigerated at 4 to 8°C and at room temperature for 14 days. The samples remained visually clear with no precipitation. TDx/FLx polarizing immunofluorescence analysis of tobramycin content found less than 10% loss occurred over 14 days of storage under refrigeration and over 12 days at room temperature. HPLC analysis of vancomycin concentration found less than 10% loss occurred over 14 days of storage under refrigeration and over eight days at room temperature. The authors concluded that the mixed antibiotic irrigation solution could be prepared and stored for 14 days if kept refrigerated.

**STUDY 8 (MULTIPLE DRUGS):** Kramer et al.[1383] evaluated the compatibility and stability of a variety of two- and three-drug solution admixtures of inhalation drugs, including tobramycin and tobramycin sulfate. The mixtures were evaluated using chemical assays including HPLC, DNase activity assay, and fluorescence immunoassay as well as visual inspection, pH measurement, and osmolality determination. The drug combinations tested and the compatibility results that were reported are shown below.

*Mixture 1*
Dornase alfa (Pulmozyme) 2500 units/2.5 mL
Tobramycin (Tobi) 300 mg/5 mL
   Result: Physically and chemically incompatible

*Mixture 2*
Dornase alfa (Pulmozyme) 2500 units/2.5 mL
Tobramycin sulfate (Gernebcin) 80 mg/2 mL
   Result: Physically and chemically incompatible

*Mixture 3*
Albuterol sulfate (Sultanol) 5 mg/1 mL
Ipratropium bromide (Atrovent LS) 0.25 mg/1 mL
Tobramycin sulfate (Tobi) 300 mg/5 mL
    Result: Physically and chemically compatible

*Mixture 4*
Albuterol sulfate (Sultanol) 5 mg/1 mL
Ipratropium bromide (Atrovent LS) 0.25 mg/1 mL
Tobramycin sulfate (Gernebcin) 80 mg/2 mL
    Result: Physically and chemically compatible

*Mixture 5*
Budesonide (Pulmicort) 1 mg/2 mL
Tobramycin sulfate (Tobi) 300 mg/5 mL
    Result: Physically and chemically compatible

*Mixture 6*
Budesonide (Pulmicort) 1 mg/2 mL
Tobramycin sulfate (Gernebcin) 80 mg/2 mL
    Result: Physically and chemically compatible

*Mixture 7*
Fluticasone propionate (Flutide forte) 2 mg/2 mL
Tobramycin (Tobi) 300 mg/5 mL
    Result: Physically and chemically compatible

*Mixture 8*
Fluticasone propionate (Flutide forte) 2 mg/2 mL
Tobramycin sulfate (Gernebcin) 80 mg/2 mL
    Result: Physically and chemically compatible ■

# ■ Tramadol Hydrochloride

## Properties
Tramadol hydrochloride is a white, bitter, odorless, crystalline powder.[1,7]

## Solubility
Tramadol hydrochloride is readily soluble in both water and ethanol.[7]

## pH
Tramadol hydrochloride injection has a pH near 6.7.[7]

## pK$_a$
Tramadol hydrochloride has a pK$_a$ of 9.41.[7]

## Osmolarity
Tramadol hydrochloride injection has an osmolarity of 320 to 380 mOsm/L.[7]

## General Stability Considerations
Tramadol hydrochloride tablets should be packaged in a tight container and stored at controlled room temperature. The injection also should be stored at controlled room temperature.[7]

## Stability Reports of Compounded Preparations
### Oral
**STUDY 1:** Wagner et al.[779] reported the stability of tramadol hydrochloride 5-mg/mL prepared from tablets in strawberry syrup and a sugar-free vehicle. Tramadol hydrochloride 50-mg tablets (Ortho-McNeil) were pulverized using a mortar and pestle and mixed with two vehicles consisting of equal parts mixtures of Ora-Plus with strawberry syrup or Ora-Sweet SF. The oral liquid formulations were packaged in amber plastic prescription bottles and stored refrigerated at 3 to 5°C and at room temperature of 23 to 25°C for 91 days.

No change in color, odor, or taste occurred. Stability-indicating HPLC analysis found no loss of tramadol hydrochloride in either formulation in 91 days at either room temperature or under refrigeration.

**STUDY 2:** Johnson et al.[867] evaluated the stability of two oral suspension formulations providing tramadol hydrochloride 7.5 mg/mL and acetaminophen 65 mg/mL prepared from commercial Ultracet tablets (Ortho-McNeil). Ultracet tablets were crushed using a mortar and pestle. The tablet powder was suspended in an equal parts mixture of Ora-Plus and strawberry syrup or in an equal parts mixture of Ora-Plus and Ora-Sweet SF for a sugar-free preparation. The suspensions were packaged in amber plastic prescription bottles and stored refrigerated at 3 to 5°C and at room temperature of 23 to 25°C for 90 days.

No change in color, odor, or taste occurred. Stability-indicating HPLC analysis found no loss of tramadol hydrochloride or acetaminophen in either formulation in 90 days at either room temperature or under refrigeration.

### Injection
Injections, like other sterile products, should be prepared in a suitable clean air environment using appropriate aseptic procedures. When prepared from nonsterile components, an appropriate and effective sterilization method must be employed.

Gupta[1224] evaluated the stability of an extemporaneously compounded tramadol hydrochloride 50-mg/mL injection. Tramadol hydrochloride powder 5 g and sodium acetate 1.66 g were dissolved in sterile water for injection, bringing the final volume to 100 mL. The injection was filled into 20-mL clear glass vials. The sterilization method was not noted. The sample vials were stored at room temperature and protected from exposure to light for the 42-day test period.

No changes in the appearance of the injection samples were noted throughout the study. Stability-indicating HPLC analysis found no loss of tramadol hydrochloride over the 42-day test period. ■

# Trazodone Hydrochloride

## Properties

Trazodone hydrochloride is a white to off-white odorless crystalline powder with a bitter taste.[2,3,421]

### Solubility

Trazodone hydrochloride is sparingly soluble in water and ethanol.[1-3]

### pKa

The drug has a pK$_a$ of 6.7.[2,134,421]

## General Stability Considerations

Trazodone hydrochloride tablets should be packaged in tight, light-resistant containers and stored at controlled room temperature protected from temperatures of 40°C or more.[2,4]

## Stability Reports of Compounded Preparations
### Oral

**STUDY 1:** Allen[134] reported on extemporaneous oral suspensions of trazodone hydrochloride. Trazodone hydrochloride tablets (Desyrel, Mead Johnson) were pulverized, and a suitable vehicle was added in increments. Then the powder was mixed well to form a suspension. Cherry syrup, simple syrup, Syrpalta, and Cologel were recommended as suitable vehicles. When stored in a tight, light-resistant container under refrigeration, the suspension was stable for 30 days.

**STUDY 2:** Pesko[421] described a similar oral liquid formulation containing trazodone hydrochloride 10 mg/mL. Fifteen trazodone hydrochloride 300-mg tablets were crushed to fine powder in a glass mortar. Approximately 20 mL of Syrpalta was mixed with the powder. The mortar was washed with additional Syrpalta into an amber glass bottle and brought to 450 mL with Syrpalta. When stored in tightly closed amber bottles under refrigeration, the suspension was stable for 30 days. ■

# Treprostinil Sodium

## Properties

Treprostinil sodium is a synthetic analog of the prostaglandin epoprostenol (also known as prostacyclin) that is used as a vasodilating agent.[3]

### pH

Treprostinil sodium injection is adjusted to a pH in the range of 6 to 7.2.[7]

## General Stability Considerations

Treprostinil sodium injection in intact containers stored at controlled room temperature is stable until the labeled expiration date. The vials of treprostinil sodium injection are preserved with metacresol; the manufacturer indicates that each vial may be used for up to 30 days after the initial penetration of the septum.[7]

Undiluted treprostinil sodium injection in a reservoir syringe is stated to be stable at 37°C for 72 hours. Treprostinil sodium injection diluted in sterile water for injection or sodium chloride 0.9% to a concentration as low as 0.004 mg/mL (4 mcg/mL) is stated to be stable for 48 hours at 37°C.[7]

## Stability Reports of Compounded Preparations
### Injection

Injections, like other sterile products, should be prepared in a suitable clean air environment using appropriate aseptic procedures. When prepared from nonsterile components, an appropriate and effective sterilization method must be employed.

Xu et al.[875] reported the stability of treprostinil (as sodium) injection (Remodulin, United Therapeutics) 1, 2.5, 5, and 10 mg/mL repackaged in 3-mL MiniMed plastic syringe pump reservoirs. Three milliliters of each concentration was packaged in the syringe pump reservoirs and sealed with tip caps. The samples were stored for 60 days at 37, 23, 4, and −20°C. All of the samples remained clear and colorless throughout the study; measured turbidity and microparticulate content did not increase. Stability-indicating HPLC analysis found that treprostinil was stable for 60 days at all temperatures, with all concentrations being 95% or greater. ∎

# ■ Tretinoin

## Properties

Tretinoin (all *trans*-retinoic acid) is a derivative of vitamin A that occurs as a yellow to light orange crystalline powder.[2,3]

### Solubility

Tretinoin is practically insoluble in water but is slightly soluble in ethanol.[2,3] The solubility of tretinoin in oils was reported to be as follows: castor oil, 6.04 mg/mL; isopropyl myristate, 3.45 mg/mL; maize oil, 3.0 mg/mL; Miglyol 812, 2.91 mg/mL; and olive oil, 2.2 mg/mL.[1111]

## General Stability Considerations

Tretinoin products should be packaged in tight, light-resistant containers and stored at controlled room temperature protected from light and freezing.[3,4,7]

Tretinoin is very sensitive to exposure to light, heat, and air.[3] Consequently, bulk tretinoin powder should be kept stored under an atmosphere of an inert gas. Any tretinoin powder remaining in an opened container should be used as soon as possible.[3,4]

## Stability Reports of Compounded Preparations

### Oral

Caviglioli et al.[803] evaluated the shelf-life stability of tretinoin 10 mg with lactose (Tablettose Meggle) in hard-gelatin capsules packaged in amber glass screw-cap bottles and stored protected from light at temperatures from 30 to 60°C and in the freezer. Based on stability-indicating HPLC analysis, the time to 10% loss of tretinoin ($t_{90}$) was calculated to be 453 days at 25°C, 678 days at 19°C, and 1289 days (3.5 years) frozen.

### Topical

**STUDY 1:** Gatti et al.[744] reported the stability of a mixed retinoid solution containing tretinoin, beta-carotene, and vitamin A palmitate (vehicle not specified). Stability-indicating HPLC analysis found about 20% loss at ambient temperature and refrigerated in 90 days if protected from exposure to light. If stored at ambient temperature exposed to sunlight and to air, tretinoin losses of 60 to 70% and vitamin A palmitate loss of 30% resulted.

**STUDY 2:** Brisaert and Plaizier-Vercammen[802] evaluated the photostability of tretinoin 0.05% (wt/vol) in propylene glycol and ethanol (1:2) for use as a topical lotion. The lotion was prepared under dim red light before deliberate photoirradiation using a high-pressure xenon lamp. HPLC analysis found that extensive and rapid photodegradation occurred. Only about 20% of the tretinoin remained after 30 minutes of photoirradiation. The most harmful wavelength was in the region of 420 nm. Surfactants failed to improve the photostability. Incorporation of chrysoin 0.025% yellow dye provided some protection by absorbing far higher in the 420-nm region than the tretinoin but did not color the skin at that concentration.

**STUDY 3:** In a subsequent study, Brisaert and Plaizier-Vercammen[1111] evaluated the photostability of tretinoin 0.05% (wt/vol) in several oils and topical creams prepared from those oils. For this testing, tretinoin was dissolved in castor oil, maize oil, olive oil, isopropyl myristate, and Miglyol 812. In addition, topical creams were prepared by incorporating the tretinoin in the oil phase using ultrasonication for 10 minutes. This was followed by the addition of the emulsifier Pemulen TR-1 (acrylic acid and alkyl methacrylate), Tween 80, and water to bring to volume. The pH was adjusted to pH 5 using Neutrol TE (tetrahydroxypropyl ethylenediamine) and the cream was homogenized for two minutes in a low shear mixer, followed by passing the cream through an ointment mill. Samples of the creams were photoirradiated using a xenon lamp, and the samples were evaluated for tretinoin loss using HPLC analysis.

In the oils, tretinoin was most stable in the castor oil, followed by olive oil, compared to the other oils tested. In the cream formulations, tretinoin was most stable in the olive oil-based cream, followed by the castor oil-based cream, compared to the others. This surprising difference was discovered to be related to microscopic crystals of tretinoin that had formed in the olive oil formulation due to tretinoin's lower solubility in that oil. The drug degrades less rapidly in the solid state, resulting in the stability difference. However, considering solubility and stability together, the authors concluded that castor oil was the most suitable choice for the oil phase of a topical cream.

**STUDY 4:** Brisaert et al.[1172] evaluated the stability of tretinoin 0.05% in one lotion and four hydrogel formulations to determine the impact of formulation components on tretinoin stability and the impact of exposure to light. HPLC analysis found that the incorporation of Brij 35 S into a Carbopol gel greatly reduced the stability of tretinoin compared to the other formulations. Time to 10% drug decomposition was eight and 17 days at room temperature and under refrigeration, respectively. In the other formulations, tretinoin remained stable for 100 days if stored protected from light. If exposed to light, the time to 10% loss was less than one hour in the lotion and 2.5 and 10 hours in two of the gels. The most stable formulation to light exposure was found to be a Carbopol gel, but, under microscopic examination, this formulation was also found to have many large tretinoin crystals. None of these test formulas appeared to be particularly successful. The stability of tretinoin in these topical preparations was adversely affected by formulation components, exposure to light, and higher temperatures.

**STUDY 5:** Tashtoush et al.[1255] evaluated the photodegradation of tretinoin 0.025% and isotretinoin 0.025% in ethanol and dermatological cream preparations exposed to solar simulated light, UVA light, and visible light. HPLC analysis found that tretinoin was more sensitive to photodegradation than isotretinoin. Both compounds undergo photolysis and photoisomerization. The UVA component of solar simulated light was found to be the major contributor to photodegradation of both compounds. The degradation by photolysis was more extensive in cream preparations than in ethanol. Because UVA penetrates deeply into skin, the authors postulated that tretinoin degradation may contribute to the photosensitivity that is associated with tretinoin therapy.

**STUDY 6 (WITH ALPHA-BISABOLOL):** Allen[1488] reported on a compounded formulation of tretinoin 0.05% with (-)alpha-bisabolol as a penetration enhancer for use as a topical gel. The gel had the following formula to make 100 g:

| | |
|---|---|
| Tretinoin | 50 mg |
| (-)Alpha-bisabolol | 100 mg |
| Polyethylene glycol 400 | 5 g |
| Cremophor RH400 | 6 g |
| Butylated hydroxytoluene | 40 mg |
| Methylparaben | 200 mg |
| Propylparaben | 50 mg |
| Pluronic F-127 | 18.5 g |
| Purified water | 70.3 g |

The recommended method of preparation was to heat the purified water to about 75°C, dissolve the parabens in the water, then allow to cool to about 40°C. Separately mix the tretinoin, polyethylene glycol 400, Cremophor RH400, butylated hydroxytoluene, and (-)alpha-bisabolol and heat to about 40°C. Add the aqueous phase to the oil phase, and heat the mixture to about 50°C. Dissolve about 14 g of Pluronic F-127, and mix well. Cool the mixture to about 6°C, add the remainder of the components with sufficient purified water to bring to 100 g, and mix thoroughly. Keep the mixture in cool temperature until the air bubbles have escaped. Package the gel in tight, light-resistant containers. The author recommended a beyond-use date of six months at room temperature.

## Compatibility with Other Drugs

Martin et al.[636] evaluated the stability of retinoid preparations, including tretinoin 0.025% gel, combined with an equal quantity of benzoyl peroxide 10% lotion. The mixture was packaged in 10-mL plastic syringes and stored exposed to normal room fluorescent light over 24 hours. Retinoid content was assessed by HPLC analysis. The tretinoin mixture underwent substantial and rapid decomposition of about 50% in two hours and 95% in 24 hours. Even without exposure to fluorescent light, the strongly oxidizing benzoyl peroxide reduced the tretinoin concentration to about 80% of the initial amount in 24 hours. ■

# ■ Triamcinolone Acetonide

## Properties

Triamcinolone acetonide is a white to cream-colored crystalline powder that has no more than a slight odor.[4] Triamcinolone acetonide 11 mg is equivalent to triamcinolone 10 mg.[3]

### Solubility

Triamcinolone acetonide is practically insoluble in water and is sparingly soluble in dehydrated ethanol.[4] Triamcinolone acetonide is also sparingly soluble in acetone and ethyl acetate.[1]

### pH

Triamcinolone acetonide injectable suspension has a pH of 5 to 7.5.[4] Triamcinolone acetonide nasal spray has a pH of 4.5 to 6.[7]

## General Stability Considerations

Triamcinolone acetonide powder should be packaged in well-closed containers, stored at controlled room temperature, and protected from exposure to light.[3,4] Triamcinolone acetonide cream, lotion, ointment, nasal spray, and dental paste should be packaged in tight containers and stored at controlled room temperature. Triamcinolone acetonide injectable suspension should be stored at controlled room temperature and protected from exposure to light.[4,7]

Triamcinolone acetonide in solutions exhibits a minimum rate of drug loss at pH 3.4. Above pH 5.5, the rate of decomposition increases.[922]

## Stability Reports of Compounded Preparations
### Oral
Ungphaiboon et al.[923] evaluated the stability of a triamcinolone acetonide 1-mg/mL oral mouthwash formulation prepared with the formula in Table 119. The mouthwash was packaged in amber glass bottles and stored at room temperature of 30°C and at elevated temperatures ranging from 45 to 80°C and 75% relative humidity. No precipitation or color change was observed in the 30 and 45°C samples; however, a yellowish brown discoloration appeared at the higher temperatures. Stability-indicating HPLC analysis found that about 6% drug loss occurred in 1064 days. Projected shelf life was calculated to be at least 3.9 years at room temperature.

### Injection
Injections, like other sterile products, should be prepared in a suitable clean air environment using appropriate aseptic procedures. When prepared from nonsterile components, an appropriate and effective sterilization method must be employed.

Bitter et al.[1241] evaluated several techniques for preparing preservative-free triamcinolone acetonide injection suitable for intravitreal injection. No commercial product without retinal toxic components such as benzyl alcohol or solubilizers was available for this use. Two methods using commercial dosage forms as starting materials were found to be inadequate. The membrane filtration method of Nishimura et al.[1242] was an ad hoc preparation unsuitable for standardized production. The centrifugation method of Hernaez-Ortega and Soto-Pedre[1243] was found to result in a larger median particle size, with some particles as large as 0.1 mm. In addition, residual benzyl alcohol was still present even after two washings.

Instead, Bitter et al.[1241] prepared a preservative-free triamcinolone acetonide 40 mg/mL injectable suspension using direct suspending of micronized (less than 15 μm)

**TABLE 119.** Triamcinolone Acetonide Mouthwash Formula[923]

| Component | Amount |
|---|---|
| Triamcinolone acetonide | 100 mg |
| Menthol | 50 mg |
| Ethanol | 10 mL |
| Propylene glycol | 30 mL |
| Glycerin | 20 mL |
| Sorbitol 70% | 20 mL |
| Sodium saccharin | 100 mg |
| Sodium metabisulfite | 20 mg |
| Disodium EDTA | 100 mg |
| in water | 5 mL |
| Sorbitol 70%           qs | 100 mL |

triamcinolone acetonide particles. Micronized triamcinolone acetonide 5.2 g was suspended in 120.25 mL of balanced salt solution (BSS, Cytosol Ophthalmics) in a 250-mL injection vial containing a magnetic stir bar. The vial was autoclaved at 121°C, 15 psi for 20 minutes. Vitrax II (AMO Switzerland)—a hyaluronic acid product—9.75 mL was added aseptically to adjust viscosity. The bulk injection was aliquoted as 1-mL units containing 40 mg of the drug suitable for intravitreal injection.

The direct suspension technique of micronized triamcinolone acetonide produced a relatively uniform suspension with mean particle size of 11 μm and no particle larger than 38 μm. Autoclaving did not change the particle size. The suspension remained homogeneously resuspendable throughout the 12-month study period with color and particle size remaining unchanged. HPLC analysis found little or no change in drug concentration when stored for 12 months at room temperature and at 40°C. The authors noted that this compounding technique could be used for alternative concentrations as well as the one studied.

Triamcinolone acetonide extemporaneously compounded as a preservative-free injection may undergo particle size aggregation that may contribute to acute and severe elevations in intraocular pressure after intravenous administration.[1599,1600]

## Compatibility with Other Drugs
### Topical
**STUDY 1 (MULTIPLE DRUGS):** Krochmal et al.[776] reported the stability of triamcinolone acetonide 0.1% cream (Kenalog) mixed individually with (1) salicylic acid 2%, (2) liquor carbonis detergens 5%, (3) urea 10%, and (4) a combination of camphor 0.25%, menthol 0.25%, and phenol 0.25%. The topical mixtures were packaged in amber glass jars with polyethylene-lined screw caps and were stored for two months at unspecified ambient temperature.

No physical incompatibilities were observed. HPLC analysis found about 45% loss of triamcinolone acetonide in two months in the urea mixture. Acceptable triamcinolone acetonide concentrations were found in the other mixtures over two months of storage.

**STUDY 2 (MUPIROCIN):** Jagota et al.[920] evaluated the compatibility and stability of mupirocin 2% ointment (Bactroban, Beecham) with triamcinolone 0.1% cream, ointment, and lotion (Kenalog, Squibb) mixed in a 1:1 proportion over periods of up to 60 days at 37°C. The physical compatibility was assessed by visual inspection, whereas the chemical stability of mupirocin was evaluated by stability-indicating HPLC analysis. The study found that mupirocin 2% ointment was physically compatible and chemically stable, with less than 10% mupirocin loss in 15 days in triamcinolone 0.1% cream and in 45 days in triamcinolone 0.1% ointment.

However, when mixed with triamcinolone 0.1% lotion (Kenalog, Squibb), 12% mupirocin loss occurred within the first assay point of 15 days.

**STUDY 3 (ZINC OXIDE):** Nagatani et al.[1526] evaluated the stability of several steroids in steroid ointments when mixed with zinc oxide ointment. The steroid ointments included triamcinolone (Kenacort AG). Samples of the mixed ointments were stored at 5 and 30°C until analyzed. Analysis found no loss of triamcinolone when stored at 5 and 30°C.

**STUDY 4 (ZINC OXIDE):** Ohishi et al.[1714] evaluated the stability of Kenacort ointment mixed in equal quantity (1:1) with zinc oxide ointment. Samples were stored refrigerated at 5°C. HPLC analysis of the ointment mixtures found that the drugs were stable for 32 weeks at 5°C. ■

# ■ Trifluoperazine Hydrochloride

## Properties

Trifluoperazine hydrochloride is a white to cream-colored or pale yellow nearly odorless hygroscopic crystalline powder with a bitter taste. Trifluoperazine 1 mg is equivalent to 1.2 mg of the hydrochloride.[2,3]

### Solubility

Trifluoperazine hydrochloride has an aqueous solubility of about 286 mg/mL. Its solubility in ethanol is about 90.9 mg/mL.[2,3]

### pH

A 5% aqueous solution of trifluoperazine hydrochloride has a pH between 1.7 and 2.6. The injection has a pH of 4 to 5, while the syrup has a pH between 2 and 3.2.[4]

## General Stability Considerations

Trifluoperazine hydrochloride in aqueous solution is readily oxidized by atmospheric oxygen.[3] Trifluoperazine hydrochloride syrup is packaged in tight, light-resistant containers, while the tablets are packaged in well-closed, light-resistant containers.[4] Trifluoperazine hydrochloride products should be stored at controlled room temperature and protected from excessive temperatures of 40°C or more and from exposure to light. Liquid products such as the injection and syrup also should be protected from freezing. Exposure of the injection to light results in discoloration. Concentration has not been affected if a slight yellowish discoloration appears. However, a markedly discolored solution should not be used.[2,4]

## Compatibility with Common Beverages and Foods

**TRIFLUOPERAZINE HYDROCHLORIDE**
**COMPATIBILITY SUMMARY**

*Compatible with:* Apricot juice • Grapefruit juice • Lemon-lime soda • Milk • Orange juice • Pineapple juice • Prune juice • Puddings • Soups • Tomato juice • V8 vegetable juice • Water
*Incompatible with:* Apple juice • Grape juice
*Uncertain or variable compatibility with:* Coffee • Cola sodas • Orange soda • Tea

**STUDY 1:** Geller et al.[242] reported that trifluoperazine liquid was compatible visually with water, milk, coffee, tea, grapefruit juice, orange juice, pineapple juice, prune juice, tomato juice, sodas, and orange and simple syrups, although visual observation in opaque liquids is problematic.

**STUDY 2:** In addition to these beverages, Kerr[256] cited the compatibility of trifluoperazine with V8 vegetable juice and the cola sodas, Mellow-Yellow, 7-Up, Sprite, orange soda, and soups and puddings.

**STUDY 3:** Lever and Hague[1011] reported the compatibility results of trifluoperazine hydrochloride oral concentrate 5 mg mixed with 15 mL of a number of task-masking liquids. Those liquids determined to be compatible included apricot juice, prune juice, lemon-lime soda, milk, pineapple juice, orange juice, V8 vegetable juice, and tomato juice. Those that were incompatible demonstrated turbidity, precipitation, and/or color changes and included Coca-Cola, carbonated orange soda, coffee, tea, grape juice, and apple juice. The precipitation occurred within a few minutes of mixing and the precipitate clung to the sides of the container, indicating that unacceptable variable dosage may occur. The authors recommended mixing trifluoperazine hydrochloride oral concentrate freshly with only distilled water.

The problematic nature of visual observation of incompatibilities in opaque liquids may be the reason for differences in the compatibility designations among the studies. Because the compatibility determinations require making judgments, it is possible that different investigators come to differing compatibility conclusions.

## Compatibility with Other Drugs

**TRIFLUOPERAZINE HYDROCHLORIDE**
**COMPATIBILITY SUMMARY**

*Compatible with:* Thioridazine suspension
*Incompatible with:* Lithium citrate syrup

**STUDY 1:** McGee et al.[73] reported the formation of a precipitate when lithium citrate syrup (Lithionate-S, Rowell) was directly added to trifluoperazine hydrochloride oral concentrate (Smith Kline & French).

**STUDY 2:** Theesen et al.[78] evaluated the compatibility of lithium citrate syrup with oral neuroleptic solutions, including trifluoperazine hydrochloride concentrate. Lithium citrate syrup (Lithionate-S, Rowell) and lithium citrate syrup (Roxane) at a lithium-ion concentration of 1.6 mEq/mL were combined in volumes of 5 and 10 mL with 2, 4, 8, and 12 mL of trifluoperazine hydrochloride oral concentrate 10 mg/mL (Smith Kline & French) in duplicate, with the order of mixing reversed between samples. Samples were stored at 4 and 25°C for six hours and evaluated visually for physical compatibility. An opaque turbidity formed immediately in all samples. Thin-layer chromatographic analysis of the sediment layers showed high concentrations of the neuroleptic drugs. The authors attributed the incompatibilities to the excessive ionic strength of the lithium citrate syrup, which results in salting out of the neuroleptic drug salts in a hydrophobic, viscid layer that adheres tenaciously to container surfaces. The problem persists even when moderate-to-high doses are diluted 10-fold in water.

Theeson et al.[78] also commented that fruit and vegetable beverages, which are often used to administer the neuroleptics, are effective salting-out agents. Unfortunately, visually observing the phenomenon is impaired by the color and opacity of these beverages.

**STUDY 3:** Raleigh[454] evaluated the compatibility of lithium citrate syrup (Philips Roxane) 1.6 mEq lithium/mL with trifluoperazine concentrate 10 mg/mL (as the hydrochloride). The trifluoperazine concentrate was combined with the lithium citrate syrup in a manner described as "simple random dosage combinations"; the exact amounts tested and test conditions were not specified. A milky precipitate formed upon mixing.

Raleigh[454] also evaluated the compatibility of thioridazine suspension (Sandoz) (concentration unspecified) with trifluoperazine concentrate 10 mg/mL (as the hydrochloride). The trifluoperazine concentrate was combined with the thioridazine suspension in the manner described as "simple random dosage combinations." The exact amounts tested and test conditions were not specified. No precipitate was observed to form upon mixing. ■

# ■ Trihexyphenidyl Hydrochloride
## (Benzhexol Hydrochloride)

## Properties
Trihexyphenidyl hydrochloride is a white or creamy-white nearly odorless crystalline powder[2,3] that is reported to have a bitter taste and to induce local tingling and numbness.[586]

### Solubility
Trihexyphenidyl hydrochloride has solubilities of 10 mg/mL in water and 59 mg/mL in ethanol.[1,2,586]

### pH
A saturated aqueous solution of trihexyphenidyl hydrochloride has a pH of 5.2 to 6.2.[3] A 1% aqueous solution has a pH of 5.5 to 6.[1] Trihexyphenidyl hydrochloride elixir has a pH of 2 to 3.[4]

## General Stability Considerations
Trihexyphenidyl hydrochloride products should be packaged in tight containers and stored at controlled room temperature. The elixir should be protected from freezing.[4,7] Trihexyphenidyl hydrochloride is stated to be incompatible with oxidizing agents.[2]

## Stability Reports of Compounded Preparations
### Oral
Woods[586] reported on an extemporaneous alcohol-free oral liquid formulation of trihexyphenidyl hydrochloride. The formula is shown in Table 120. An expiration period of two

**TABLE 120.** Formula of Alcohol-Free Trihexyphenidyl Hydrochloride 2-mg/5-mL Oral Liquid Prepared from Tablets[586]

| Component | | Amount |
|---|---|---|
| Trihexyphenidyl hydrochloride 2-mg tablets | | 20 |
| Citric acid monohydrate | | 200 mg |
| Parabens | | 0.1% |
| Syrup | | 50 mL |
| Water | qs | 100 mL |

weeks when stored at room temperature or under refrigeration was suggested, although no stability data were presented. While this formulation is alcohol-free, the author noted that the addition of 5% ethanol may result in a better formulation.

## Compatibility with Other Drugs
**TRIHEXYPHENIDYL HYDROCHLORIDE**
**COMPATIBILITY SUMMARY**
*Compatible with:* Lithium citrate
*Incompatible with:* Thioridazine hydrochloride

Raleigh[454] evaluated the compatibility of lithium citrate syrup (Philips Roxane) 1.6 mEq lithium/mL with trihexyphenidyl hydrochloride elixir 0.4 mg/mL. The trihexyphenidyl hydrochloride elixir was combined with the lithium citrate syrup in a manner described as "simple random dosage combinations"; the exact amounts tested and test conditions were not specified. No precipitate was observed to form upon mixing.

Raleigh[454] also evaluated the compatibility of thioridazine suspension (Sandoz) (concentration unspecified) with trihexyphenidyl hydrochloride elixir 0.4 mg/mL. The trihexyphenidyl hydrochloride elixir was combined with the thioridazine suspension in the manner described as "simple random dosage combinations." The exact amounts tested and test conditions were not specified. A dense white curdy precipitate formed. ■

# ■ Trimethobenzamide Hydrochloride

## Properties
Trimethobenzamide hydrochloride is a white crystalline powder with a slightly phenolic odor.[1,4]

### Solubility
Trimethobenzamide hydrochloride is freely soluble in water with a solubility at 25°C greater than 500 mg/mL.[1,3,4] The drug is also soluble in 1 in 59 in ethanol.[3,4]

### pH
Trimethobenzamide hydrochloride injection has a pH in the 4.5 to 5.5 range.[4] However, the manufacturer states that the actual pH is adjusted to approximately 5 during manufacturing.[7]

### $pK_a$
Trimethobenzamide has a $pK_a$ of 8.3.[1377]

## General Stability Considerations
Trimethobenzamide hydrochloride powder and oral capsules should be packaged in well-closed containers and stored at controlled room temperature. The injection should be packaged in single-dose or multiple-dose type I glass containers and stored at controlled room temperature.[4]

Trimethobenzamide hydrochloride 50 mg/mL in water is stable to autoclaving at 120°C for 20 minutes over a pH range of 3 to 7.[1]

## Stability Reports of Compounded Preparations
### Oral
Guneri and Ozyazici[1376] evaluated the stability of several oral liquid preparations of trimethobenzamide hydrochloride 20 mg/mL varying in pH and type of sweetener. The various formulas were evaluated for drug decomposition at elevated temperatures of 40, 60, and 80°C using thin layer chromatography, and the decomposition rate at room temperature was calculated. A vehicle composed of propylene glycol 20%, ethanol 11%, and glycerin 5% in distilled water was used. The drug in either formula sweetened with aspartame was found to be much less stable than in those sweetened with sorbitol. The formulas tested and the calculated estimates of times for 10% decomposition ($t_{90}$) at room temperature for the four formulas are shown below. ■

| Formula | | 1 | 2 | 3 | 4 |
|---|---|---|---|---|---|
| Trimethobenzamide hydrochloride | | 1 g | 1 g | 1 g | 1 g |
| Aspartame | | 5 g | 5 g | | |
| Sorbitol | | | | 5 g | 5 g |
| Citric acid/phosphate buffer pH 4 | | 25 mL | | 25 mL | |
| Phosphate buffer pH 7 | | | 25 mL | | 25 mL |
| Vehicle | qs | 50 mL | 50 mL | 50 mL | 50 mL |
| $t_{90}$ at 25°C | | 10.4 days | 18.9 hours | 134 days | 243 days |

# ■ Trimethoprim
# Trimethoprim Sulfate

## Properties
Trimethoprim occurs as white to cream-colored or yellowish odorless bitter-tasting crystals or crystalline powder.[1,3,4]

Trimethoprim sulfate is a white to off-white crystalline powder.[4]

### Solubility
Trimethoprim is very slightly soluble in water, having an aqueous solubility of about 0.4 mg/mL. The drug is slightly soluble in ethanol and has a solubility in propylene glycol of 25.7 mg/mL.[1,3,4]

Trimethoprim sulfate is soluble in water, ethanol, dilute mineral acids, and in fixed alkalies.[4]

## pH

The pH of the trimethoprim and sulfamethoxazole injection is in the range of 9.5 to 10.5.[7]

Trimethoprim sulfate 0.5 mg/mL in aqueous solution has a pH of 7.5 to 8.5.[4]

## $pK_a$

Trimethoprim has a $pK_a$ of 6.6.[1]

## Osmolality

Santiero and Sagraves[1147] reported the osmolalities of several concentrations of trimethoprim and sulfamethoxazole injection diluted in dextrose 5%. See Table 121.

Santiero and Sagraves[1147] also reported that trimethoprim and sulfamethoxazole injection diluted in sodium chloride 0.9% to a concentration of 1.6 + 8 mg/mL had an osmolality of 833 mOsm/Kg.

## General Stability Considerations

Trimethoprim and sulfamethoxazole oral tablets should be packaged in tight, light-resistant containers and stored at controlled room temperature and protected from exposure to moisture and light. Trimethoprim and sulfamethoxazole oral suspension should be stored at controlled room temperature and protected from light. Trimethoprim and sulfamethoxazole injection should be stored at controlled room temperature and not refrigerated.[4,7]

Trimethoprim sulfate powder should be packaged in well-closed containers and stored at controlled room temperature.[4]

## Stability Reports of Compounded Preparations

### Oral

Nahata[1148] evaluated the stability of trimethoprim 10 mg/mL in an extemporaneous formulation. Trimethoprim oral tablets were crushed to powder using a mortar and pestle. The vehicle, composed of an equal parts mixture of methylcellulose 1% and syrup, was then added with mixing. Samples of

**TABLE 121.** Osmolalities of Several Trimethoprim and Sulfamethoxazole Injection Concentrations[1147]

| Trimethoprim Concentration (mg/mL) | | Sulfamethoxazole Concentration (mg/mL) | Osmolalities (mOsm/kg) |
|---|---|---|---|
| 0.8 | + | 4 | 541 |
| 1.1 | + | 5.5 | 669 |
| 1.6 | + | 8 | 798 |

the suspension were packaged in glass and plastic prescription bottles and were stored at both room temperature of 25°C and under refrigeration at 4°C.

Stability-indicating HPLC analysis found about 8% loss in 91 days refrigerated, but 8% loss occurred in only 42 days at room temperature. There was no difference in the stability in glass or plastic prescription bottles.

### Ophthalmic

Ophthalmic preparations, like other sterile drugs, should be prepared in a suitable clean air environment using appropriate aseptic procedures. When prepared from nonsterile components, an appropriate and effective sterilization method must be employed.

Shirwaikar and Rao[1561] evaluated the stability of an ophthalmic drops preparation containing sulfacetamide sodium and trimethoprim. Sulfacetamide sodium 20 g was dissolved in 50 mL of water for injection. Trimethoprim 500 mg was dissolved in 50 mL of propylene glycol. The sulfacetamide sodium solution was slowly added to the trimethoprim solution with stirring. The ophthalmic solution was filtered through a sintered glass filter, filled into vials, and sterilized by autoclaving at 121°C and 30 psig for 30 minutes. The final solution pH was 7.6, and the solution was found to be slightly hypertonic.

The ophthalmic drops were found to be effective against *Staphylococcus aureus* using a rabbit model. Using a microbiological assay with several common microorganisms, the formulation was subjected to elevated temperatures of 40, 50, 70, and 90°C. The time to 10% loss ($t_{90}$) at 25°C was calculated to be 4.5 years if the ophthalmic solution was protected from exposure to light. When exposed to light, a discoloration developed in the ophthalmic solution.

### Injection

Injections, like other sterile drugs, should be prepared in a suitable clean air environment using appropriate aseptic procedures. When prepared from nonsterile components, an appropriate and effective sterilization method must be employed.

Tu et al.[1657] developed a nonaqueous formulation and trimethoprim for intravenous injection. Trimethoprim was formulated in a cosolvent system composed of 52% N,N-dimethylacetamide and 48% propylene glycol. The injection was packaged in type I glass ampules. Samples were stored at elevated temperatures ranging up to 140°C. Visual inspection found no precipitation occurred. However, the solution was initially colorless but developed a yellow then brown discoloration. The discoloration was found to be associated with the degree of trimethoprim degradation. The time to 10% trimethoprim degradation ($t_{90}$) at 25°C was calculated from elevated temperature results to be 885 days. ∎

# Triprolidine Hydrochloride

## Properties

Triprolidine hydrochloride is a white crystalline powder having a slight unpleasant odor.[2,3]

### Solubility

Aqueous solutions of triprolidine hydrochloride are alkaline to litmus.[1] Triprolidine hydrochloride has an aqueous solubility of 476 mg/mL. Its solubility in ethanol is 556 mg/mL.[2,3]

### pH

Triprolidine hydrochloride syrup has a pH between 5.6 and 6.6.[4] Triprolidine hydrochloride with pseudoephedrine hydrochloride syrup (Actifed, Burroughs-Wellcome) had a measured pH of 5.6.[19]

### pK$_a$

The drug has pK$_a$ values of 3.6 and 9.3.[2]

## General Stability Considerations

Triprolidine hydrochloride products should be packaged in tight, light-resistant containers and stored at controlled room temperature in a dry place. Freezing of liquid products should be avoided.[2,4]

## Stability Reports of Compounded Preparations

### Oral

Allen[1320] reported on a compounded formulation of triprolidine hydrochloride 0.25-mg/mL oral syrup. The syrup had the following formula:

| | | |
|---|---|---|
| Triprolidine hydrochloride | 25 mg | |
| Sucrose | 60 g | |
| Glycerin | 10 mL | |
| Sorbitol 70% solution | 10 mL | |
| Propylene glycol | 1.5 mL | |
| Methylparaben | 100 mg | |
| Propylparaben | 30 mg | |
| Saccharin sodium | 50 mg | |
| Menthol | 5 mg | |
| Apricot flavor | qs | |
| Sodium citrate | 115 mg | |
| Purified water | qs | 100 mL |

The recommended method of preparation was to heat 40 mL of purified water to 90 to 95°C, add the parabens, and mix until dissolved. The sucrose was then added and mixed at high speed for one hour while maintaining the temperature at 90 to 95°C. The solution was cooled to about 50°C with slow-speed mixing. The saccharin sodium and sodium citrate were added and mixed well. This was followed by the addition of glycerin and sorbitol solution, and the preparation was again mixed well. The solution was removed from the heat and allowed to cool. The triprolidine hydrochloride was then added, and the solution was mixed well. The menthol was dissolved in the apricot flavor and propylene glycol; then the mixture was added to drug solution and mixed well. Finally, purified water was added to bring the syrup to volume.

The syrup should be packaged in tight, light-resistant containers. The author recommended a beyond-use date of six months at room temperature because this formula is a commercial medication in some countries with an expiration date of two years or more.

### Enteral

**TRIPROLIDINE HYDROCHLORIDE COMPATIBILITY SUMMARY**

*Compatible with:* Ensure • Ensure HN • Ensure Plus • Ensure Plus HN • Osmolite • Osmolite HN • Vital

**STUDY 1:** Cutie et al.[19] added 5 mL of triprolidine hydrochloride with pseudoephedrine hydrochloride syrup (Actifed, Burroughs-Wellcome) to varying amounts (15 to 240 mL) of Ensure, Ensure Plus, and Osmolite (Ross Laboratories) with vigorous agitation to ensure thorough mixing. The syrup was compatible physically, distributing uniformly in all three enteral products with no phase separation or granulation.

**STUDY 2:** Altman and Cutie[850] reported the physical compatibility of triprolidine hydrochloride with pseudoephedrine hydrochloride syrup (Actifed, Burroughs-Wellcome) 5 mL with varying amounts (15 to 240 mL) of Ensure HN, Ensure Plus HN, Osmolite HN, and Vital after vigorous agitation to ensure thorough mixing. The syrup was compatible physically, distributing uniformly in all products, with no phase separation or granulation.

Also see the *Pseudoephedrine Hydrochloride* monograph. ∎

# Tropicamide

## Properties

Tropicamide occurs as white or practically white crystals or crystalline powder that is odorless or nearly odorless.[1,4]

### Solubility

Tropicamide is slightly soluble in water and freely soluble in chloroform, dimethyl sulfoxide, and solutions of strong acids.[3,4]

### pH

Tropicamide ophthalmic solution has a pH between 4 and 5.8.[4,7]

### pK$_a$

Tropicamide has a pK$_a$ of 5.4.[1068]

## General Stability Considerations

Tropicamide should be packaged in tight, light-resistant containers and stored at controlled room temperature. Tropicamide ophthalmic solution should also be packaged in tight containers and stored at controlled room temperature. Refrigerated storage or freezing of the solution should be avoided.[4,7] Tropicamide decomposes gradually in neutral and alkaline solutions.[1212, 1687, 1690]

## Stability Reports of Compounded Preparations
### Ophthalmic

Ophthalmic preparations, like other sterile drugs, should be prepared in a suitable clean air environment using appropriate aseptic procedures. When prepared from nonsterile components, an appropriate and effective sterilization method must be employed.

**STUDY 1 (SOLUTION):** Pohloudek-Fabini et al.[1690] evaluated the stability of tropicamide 0.5% ophthalmic solution. Tropicamide was highly stable, exhibiting less than 5% decomposition in 70 days stored at 70°C and less than 2% loss in 1825 days (five years) at 20°C.

**STUDY 2 (SOLUTION WITH DICLOFENAC AND PHENYLEPHRINE):** Hirowatari et al.[929] reported the stability of a three-component ophthalmic solution for preoperative use. The ophthalmic solution was prepared by mixing commercially available ophthalmic solutions to yield final concentrations of phenylephrine hydrochloride 1.83%, tropicamide 0.17%, and diclofenac sodium 0.03%. The ophthalmic solution admixture was dispensed as 2-mL aliquots packaged in light-resistant bottles. The physical and chemical stability of the mixture was evaluated over one month stored at 10°C protected from exposure to light. Visual inspection found the solution remained unchanged, being clear and colorless; measured pH remained unchanged as well. HPLC analysis found no loss of diclofenac and tropicamide and about 2% loss of phenylephrine in one month.

**STUDY 3 (SOLUTION WITH PHENYLEPHRINE):** Zuniga Dedorite et al.[1590] evaluated six phenylephrine hydrochloride 10% and tropicamide 1% eyedrop formulations. The formulation that was selected for stability testing contained the two drugs in a vehicle containing polysorbate 80, sodium metabisulfite, benzalkonium chloride, sodium edetate, and hydrochloric acid in water in low-density polyethylene bottles. The formulation had a pH near 4.3. The formulation was physically stable; chromatographic analysis of the phenylephrine and tropicamide concentrations found little or no loss of either drug and no appreciable formation of decomposition products during 12 months of storage at room temperature of 30°C and 65 to 75% relative humidity.

**STUDY 4 (GEL WITH CYCLOPENTOLATE, LIDOCAINE, AND PHENYLEPHRINE):** Bailey et al.[1525] evaluated the stability of a four-drug combination ophthalmic gel for use as a preoperative medication for cataract surgery. The final drug concentrations in the preoperative gel were cyclopentolate hydrochloride 0.51 mg/mL, phenylephrine hydrochloride 5.1 mg/mL, and tropicamide 0.51 mg/mL in the lidocaine hydrochloride jelly 2%. The gel was prepared by combining 0.3 mL each of cyclopentolate hydrochloride ophthalmic solution 1%, USP, phenylephrine hydrochloride ophthalmic solution 10%, USP, and tropicamide ophthalmic solution 1%, USP in preservative-free lidocaine hydrochloride jelly 2%, USP. The gel was packaged as 0.5 mL in 1-mL polycarbonate plastic syringes with sealed tips. The samples were stored protected from light in an environmental chamber at room temperature of 23 to 27°C and 60% relative humidity and also at refrigerated temperature of 2 to 4°C for 60 days.

The ophthalmic gel exhibited little change in pH and remained sterile and endotoxin-free throughout the 60-day study. Stability-indicating HPLC analysis of drug concentrations over 60 days' storage found no loss of any of the drugs when stored under refrigeration and not more than 4% loss of any drug when stored at room temperature. ■

# Trovafloxacin Mesylate

## Properties

Trovafloxacin mesylate is a white to off-white powder.[7]

### Solubility

Trovafloxacin is practically insoluble in water, having a solubility of 14 mcg/mL at pH 7. The drug has increased solubility at pH values below 2 and above 11.[906]

### pH

Alatrofloxacin mesylate injection has a pH of 3.5 to 4.3.[7]

## General Stability Considerations

Trovafloxacin mesylate tablets should be packaged in tight containers and stored at controlled room temperature.[7]

## Stability Reports of Compounded Preparations

### Oral

Gupta[689] reported the stability of a trovafloxacin 10-mg/mL oral liquid dosage form prepared from 200-mg tablets (Trovan, Pfizer). Five 200-mg Trovan tablets were pulverized using a mortar and pestle and mixed with 4 mL of 0.1 N hydrochloric acid to fully dissolve the drug. The mixture was brought to 100 mL with 10% mannitol solution. The preparation had a final pH of 3.4. The oral suspension was packaged in four-ounce amber glass bottles and stored at room temperature of 24 to 26°C for 14 days.

No visible changes or change in pH occurred. Stability-indicating HPLC analysis found no loss of trovafloxacin within the 14-day study period. ■

# Ubidecarenone
## (Coenzyme Q₁₀)

## Properties
Ubidecarenone, also known as coenzyme $Q_{10}$, occurs as a yellow to orange crystalline powder.[3,4]

### Solubility
Ubidecarenone is practically insoluble in water, very slightly soluble in dehydrated ethanol, and soluble in ether.[3,4]

## General Stability Considerations
Ubidecarenone gradually decomposes and darkens upon exposure to light. Ubidecarenone powder, capsules, and tablets should be packaged in tight, light-resistant containers and stored at controlled room temperature protected from exposure to light.[3,4]

## Stability Reports of Compounded Preparations
### Oral
Allen[1281] reported on a compounded formulation of carnitine 400-mg/mL with ubidecarenone (coenzyme $Q_{10}$) 10-mg/mL oral solution. The solution had the following formula:

| | | |
|---|---|---|
| Carnitine | | 40 g |
| Ubidecarenone (Coenzyme $Q_{10}$) | | 1 g |
| Polyethylene glycol 400 | | 1 g |
| Cremophor RH40 | | 4 g |
| Methylparaben | | 200 mg |
| Purified water | qs | 100 mL |

The recommended method of preparation was to place the ubidecarenone (coenzyme $Q_{10}$), polyethylene glycol 400, Cremophor RH40, methylparaben, and 50 mL of purified water powder in an appropriate container, heat the mixture to about 60°C, and stir thoroughly. The mixture was then cooled, the carnitine was added along with sufficient purified water to bring the volume to 100 mL, and the preparation was thoroughly mixed. The solution was to be packaged in tight, light-resistant containers. The author recommended a beyond-use date of six months at room temperature because this formula is a commercial medication in some countries with an expiration date of two years or more. ■

# Urea

## Properties
Urea is a white crystalline powder or colorless to white crystals that are nearly odorless. An ammonia odor develops upon standing.[2,3] Urea has an unpleasant saline taste.[2]

### Solubility
Urea is freely soluble in water, having an aqueous solubility of about 1 g/mL.[1–3] In 95% ethanol it has a solubility of about 100 mg/mL, increasing to about 1 g/mL in boiling 95% ethanol. In dehydrated ethanol its solubility is about 50 mg/mL.[1]

### pH
Aqueous solutions have a neutral pH.[3] Reconstitution of the injection results in a pH of about 5.5 to 7.[2]

### Osmolality
A 1.63% urea aqueous solution is isotonic.[2]

## General Stability Considerations
Urea bulk material should be packaged in well-closed containers.[4] Reconstituted solutions are stable for 48 hours under refrigeration.[2]

Urea solutions cannot be sterilized by heat due to instability. Upon standing or heating or in the presence of acids or alkalies, urea decomposes to ammonia and carbon dioxide.[2]

## Stability Reports of Compounded Preparations
### Topical

**STUDY 1:** Two topical formulations of 40% urea were reported by Allen.[135] The urea particle size was first reduced to fine powder using an electric mill. A dispersion of 2% Carbopol 941 (carbomer 941) in water was prepared, and the viscosity was adjusted by adding 0.5 mL of trolamine (triethanolamine) for each 30 g of the Carbopol 941 dispersion. The finely powdered urea then was dissolved into the Carbopol 941 dispersion, forming a 40% urea clear gel. Alternatively, the Carbopol 941 dispersion was mixed geometrically in equal quantity into Velvachol, forming a uniform mixture. The finely powdered urea then was mixed geometrically with the combination mixture to form the 40% urea formulation. Both formulations were physically stable for at least 30 days at room temperature.

**STUDY 2:** Allen and Li[546] evaluated the physical stability of 42 topical urea formulations having urea concentrations ranging from 10 to 40%. Dermatological bases evaluated included Aquaphor (Beiersdorf), Carbopol 941 (B. F. Goodrich), Lanaphilic (Medco), polyethylene glycol ointment, NF, Velvachol (Texas Pharmacal), and white petrolatum (Fisher Scientific), as well as combinations of Carbopol 941 with Velvachol and with white petrolatum. For formulations having no water, urea powder (Baker) was weighed and triturated in a mortar to reduce particle size. The powder was mixed with a small amount of the base followed by successive additions with thorough mixing until the total amount of base was incorporated. For formulations that contained water, the urea powder was thoroughly mixed with the water and then incorporated into the Aquaphor or Velvachol. Carbopol gels 2 and 3% were prepared by sprinkling the Carbopol 941 powder onto rapidly mixing water and stirring with a magnetic stirrer until thoroughly mixed. Trolamine (triethanolamine) at a concentration of 2% was added to the Carbopol (carbomer) gels to thicken them. For two-component mixed bases, the two bases were combined first, and then the urea powder was incorporated. The preparations were packaged in one-ounce plastic jars and stored at 23°C for 60 days. The topical preparations were observed for physical changes in uniformity, color, and texture.

The formulations described in Table 122 were all found to be acceptable throughout the 60-day storage period. However, the bases that worked best across the entire urea concentration range were Carbopol 3% and Carbopol 2 and 3% with Velvachol.

All of the Aquaphor and polyethylene glycol ointment samples were grainy, as were the urea 10 and 20%

**TABLE 122.** Topical Formulations of Urea Having Acceptable Physical Stability[546]

| Carbopol 3% | Carbopol 2% | Velvachol | White Petrolatum | Water | Urea |
|---|---|---|---|---|---|
| 90% | — | — | — | — | 10% |
| 80% | — | — | — | — | 20% |
| 70% | — | — | — | — | 30% |
| 60% | — | — | — | — | 40% |
| — | 45% | 45% | — | — | 10% |
| — | 40% | 40% | — | — | 20% |
| — | 35% | 35% | — | — | 30% |
| — | 30% | 30% | — | — | 40% |
| 45% | — | 45% | — | — | 10% |
| 40% | — | 40% | — | — | 20% |
| 35% | — | 35% | — | — | 30% |
| 30% | — | 30% | — | — | 40% |
| — | — | 70% | — | — | 30% |
| — | — | 60% | — | — | 40% |
| — | — | 50% | — | 30% | 20% |
| — | — | 40% | — | 20% | 40% |
| — | — | — | 90% | — | 10% |
| — | — | — | 70% | — | 30% |

combinations in Velvachol. In Lanaphilic base, urea 10 to 30% separated and at 40% the formulation was grainy. All of the Carbopol 941 2% only preparations were found to be thin. Carbopol 2% combined with white petrolatum separated, and these preparations were unacceptable. Strangely, urea 10 and 30% in white petrolatum were acceptable but the 20 and 40% concentrations were not.

## Compatibility with Other Drugs

Tezuka[1708] evaluated the stability of betamethasone in a series of compounded topical mixtures with urea-containing products. The betamethasone-17-valerate topical products evaluated in the testing included Rinderon V ointment, cream, and lotion and Rinderon VG lotion. The urea-containing products included Urepeal, Urepeal L, Keratinamin Kowa ointment, Pastaron, Pastaron 20, Pastaron soft, and Pastaron 10 lotion. The products were mixed in equal quantity by weight, and samples were stored at room temperature and elevated temperature of 40°C and 75% relative humidity for four weeks. There were no visible changes to the samples. All samples except the Urepeal mixtures resulted in substantial decomposition of the betamethasone when stored at room temperature; results were much worse at elevated temperature, with losses up to 84%. The authors recommended a short expiration period and storage at room temperature. ■

# ■ Ursodiol
## (Ursodeoxycholic Acid)

## Properties

Ursodiol is a naturally occurring bile acid that is a white or nearly white crystalline powder having a bitter taste.[1,3,7]

### Solubility

Ursodiol is practically insoluble in water but is freely soluble in ethanol and glacial acetic acid.[1,3,7]

## General Stability Considerations

Ursodiol products should be packaged in tight containers and stored at controlled room temperature.[7]

## Stability Reports of Compounded Preparations
### Oral

**STUDY 1:** Johnson and Nesbitt[471] evaluated the stability of an extemporaneously prepared oral suspension of ursodiol (Summit Pharmaceuticals) 60 mg/mL prepared from 300-mg capsules. To make 60 mL of suspension, the contents of 12 powder-filled capsules were placed in a glass mortar. A small amount of glycerin, USP, was added to wet the powder. It was then triturated to a fine paste. Simple syrup was added in increments; 15 mL was triturated well with the paste and then was transferred to a two-ounce amber glass bottle with a child-resistant cap. The mortar was rinsed with 10 mL of simple syrup several times, with the rinses being added to the bottle. The suspension was brought to a final volume of 60 mL with simple syrup. A similar suspension was prepared from ursodiol bulk powder for comparison. Triplicate samples of the suspensions were stored under refrigeration at 4°C and protected from light for 35 days.

There was no change in color or odor and no visible microbiological growth in any sample. HPLC analysis found little or no loss of ursodiol in either formulation after 35 days. The suspension prepared from the capsule contents retained 96.5% of the initial ursodiol concentrations, while the suspension prepared from bulk powder was nearly 100%.

**STUDY 2:** Mallett et al.[568] studied the stability of an oral 25-mg/mL suspension of ursodiol compounded from capsules. An initial attempt to suspend the capsule contents in water was unsuccessful due to rapid settling, and an alternative formulation was developed. The contents of 10 ursodiol 300-mg capsules (Summit Pharmaceuticals) were emptied into a glass mortar, and any clumps were broken up. The powder was levigated with 10 mL of glycerin to obtain a smooth mixture. A total of 60 mL of Ora-Plus (Paddock) was added in geometric proportions and thoroughly incorporated to achieve a smooth mixture. The mixture was transferred to a bottle large enough to permit shaking. The mortar was rinsed with a small amount of orange syrup, NF, and the rinse was added to the bottle. Additional orange syrup, NF, was added to bring the final volume to 120 mL. The suspension was shaken thoroughly and packaged in a four-ounce amber plastic (polyethylene) prescription bottle. Samples were stored at 4 and 23°C protected from light for 60 days. HPLC analysis of the suspension found no loss of ursodiol at either storage temperature throughout the study period.

**STUDY 3:** Nahata et al.[681] reported the stability of two extemporaneously compounded oral suspensions of ursodiol (Actigall). Ursodiol 20 mg/mL was prepared from capsules by grinding the capsule contents and incorporating them into (1) a 1:1 mixture of Ora-Sweet and Ora-Plus and (2) a 1:1 mixture of simple syrup and methylcellulose 1%.

Stability-indicating HPLC analysis found that 3% or less change in ursodiol concentration occurred in 91 days stored at room temperature of 25°C and refrigerated at 4°C.

**STUDY 4:** Johnson and Streetman[711] reported the stability of ursodiol 50-mg/mL oral suspensions extemporaneously compounded from crushed 250-mg tablets (Urso, Axcan Pharma) combined with equal parts mixtures of Ora-Plus and strawberry syrup or Ora-Plus with Ora-Sweet SF (sugar-free). The suspensions were packaged in amber plastic prescription bottles and stored for 90 days at room temperature of 23 to 25°C or refrigerated at 3 to 5°C. No detectable changes in color, odor, or taste occurred, and no microbial growth was observed. Stability-indicating HPLC analysis found little or no loss of ursodiol in either suspension formulation during the 90-day study period. ■

# Valacyclovir Hydrochloride

## Properties

Valacyclovir hydrochloride is a white to off-white crystalline powder.[1,7] Valacyclovir hydrochloride 1.11 g is equivalent to valacyclovir 1 g.[3]

### Solubility

Valacyclovir hydrochloride has an aqueous solubility of 174 mg/mL.[1,7]

### $pK_a$

Valacyclovir hydrochloride has $pK_a$ values of 1.90, 7.47, and 9.43.[7]

## General Stability Considerations

Valacyclovir hydrochloride capsules should be stored at controlled room temperature.[7]

Valacyclovir exhibits pH-dependent stability. The drug is most stable at acidic pH, with maximum stability occurring at pH 4 and lower. At pH 1.8, valacyclovir exhibited only 2% hydrolysis in 24 hours. However, as the pH increased above 4, especially at neutral and alkaline pH values, hydrolysis occurred much more rapidly.[1042]

## Stability Reports of Compounded Preparations

### Oral

Fish et al.[687] reported the stability of three oral suspensions of valacyclovir 50 mg/mL (as the hydrochloride) prepared from crushed 500-mg caplets (Glaxo Wellcome). The 18 caplets were pulverized using a mortar and pestle for each suspension. For suspensions 1 and 2, the powder was mixed with 40 mL of Ora-Plus suspending vehicle added in 5-mL increments with thorough mixing. For suspension 3, 40 mL of Syrpalta was used for this step. The mixtures were transferred into 180-mL amber glass bottles with five thorough rinsings of the mortar and pestle with 10 mL of Ora-Plus (suspensions 1 and 2) or Syrpalta (suspension 3), for a total of 50 mL. Suspension 1 was brought to a volume of 180 mL with Ora-Sweet. Suspension 2 was brought to 180 mL with Ora-Sweet SF. Suspension 3 was brought to 180 mL with additional Syrpalta. All suspensions were stored refrigerated at 4°C for 60 days.

No change in appearance or evidence of microbial growth occurred over 60 days. Stability-indicating HPLC analysis found that no decomposition products formed in 60 days. However, 10% loss of valacyclovir concentration occurred in 21 days in suspensions 1 and 2 in Ora-Sweet and Ora-Sweet SF. In Syrpalta, 10% loss occurred in 35 days. The authors indicated that some observed variations in drug concentrations may have been due to inadequate mixing.

# Valganciclovir Hydrochloride

## Properties

Valganciclovir hydrochloride is a white to off-white crystalline powder.[1,7]

### Solubility

Valganciclovir hydrochloride has an aqueous solubility of 70 mg/mL at pH 7 and 25°C.[7]

### pKₐ

Valganciclovir hydrochloride has a pKₐ of 7.6.[7]

## General Stability Considerations

Valganciclovir hydrochloride capsules should be stored at controlled room temperature.[7]

Valganciclovir was found to be most stable at pH values of 3.5 and lower. The drug was found to be relatively unstable at pH 5.[851]

## Stability Reports of Compounded Preparations

### Oral

**STUDY 1:** Anaizi et al.[851] reported the stability of valganciclovir 90-mg/mL oral suspension prepared from oral tablets. Commercial Valcyte (Roche) 450-mg tablets were triturated to a paste with sterile water for irrigation containing sodium benzoate. The paste was mixed with chocolate–cherry syrup to yield the 90-mg/mL concentration. The suspension contained sodium benzoate 1 mg/mL as a preservative. The suspension was packaged in amber polyethylene terephthalate prescription bottles, and the bottles were stored refrigerated at 2 to 8°C for 125 days.

No detectable changes in color or odor and no microbial growth occurred. The suspension remained easily resuspended throughout the study period. Stability-indicating HPLC analysis found no loss of valganciclovir at any time throughout the study. The authors indicated that the valganciclovir suspension was stable for at least 125 days.

**STUDY 2:** Henkin et al.[781] reported the stability of valganciclovir 30- and 60-mg/mL oral suspensions prepared from tablets. Valganciclovir 450-mg tablets were crushed and triturated to fine powder using a mortar and pestle. The tablet powder was incorporated into an equal parts mixture of Ora-Plus and Ora-Sweet by triturating the powder with 1-mL increments of Ora-Plus to form a paste and then gradually adding the balance of the Ora-Plus. This mixture was transferred to a container and the mortar was scraped and then rinsed with Ora-Sweet using four repetitions. The suspension was brought to volume with Ora-Sweet and shaken well. The suspension was packaged in amber glass bottles and stored refrigerated at 4°C and at room temperature of 20 to 24°C for 45 days.

No changes in appearance of the suspension occurred. Stability-indicating HPLC analysis found losses of 7 and 4% in the 30- and 60-mg/mL formulations stored under refrigeration after 35 days. By 45 days both concentrations were below 90%. At room temperature, variability in the results occurred. The average of the samples at both concentrations remained above 90% through 35 days, but several individual samples fell below 90% in 20 days (30 mg/mL) and 25 days (both 30 and 60 mg/mL). Whether these occurrences resulted from assay variability or not, storage under refrigeration may be appropriate.

**STUDY 3:** Zhang et al.[1015] evaluated the stability of valganciclovir hydrochloride 20-mg/mL oral suspension compounded from commercial Valcyte (Roche) oral tablets in Ora-Plus and Ora-Sweet intended for pediatric administration. Commercial valganciclovir hydrochloride tablets were pulverized and triturated to fine powder using a mortar and pestle. A small amount of purified water was used to wet the powder. The wet powder was thoroughly mixed with a portion of the Ora-Sweet syrup and then with an equal quantity of Ora-Plus suspending medium and mixed thoroughly until homogeneous. The mixture was brought to volume with an equal parts mixture of Ora-Sweet and Ora-Plus and was thoroughly mixed. The suspension was packaged in amber plastic screw-cap prescription bottles and stored for 60 days at ambient room temperature near 23°C exposed to normal fluorescent light and also at 4°C in the dark.

The suspension remained pharmaceutically acceptable, not exhibiting visually obvious settling or layering. Furthermore, no caking of the solids was observed throughout the study. The suspension was easily resuspended, poured very easily, and remained homogeneous throughout the 60-day study period. The pH of the suspension was near pH 4 and remained nearly unchanged. Stability-indicating HPLC analysis found little or no change in valganciclovir hydrochloride over 60 days when stored under refrigeration. Stored at room temperature, about 8% loss of valganciclovir hydrochloride from the initial concentration occurred in 60 days. Refrigerated storage was recommended to minimize loss of the drug during storage. ■

# Valproic Acid
# Valproate Sodium

## Properties

Valproic acid is a clear colorless to pale yellow slightly viscous liquid having a characteristic odor.[2,3]

Valproate sodium is a white or nearly white almost odorless crystalline hygroscopic powder with a saline taste.[2,3]

### Solubility

Valproic acid is slightly soluble in water but freely soluble in ethanol. It also dissolves in alkaline hydroxide solutions. Valproate sodium is very soluble in both water and ethanol, having a solubility of about 200 mg/mL.[2,3]

### pH

Valproate sodium syrup has a pH between 7 and 8.[4]

### pK$_a$

Valproate sodium has a pK$_a$ of 4.8.[1,2] Valproic acid has a pK$_a$ of 4.6.[1]

## General Stability Considerations

Valproic acid capsules should be packaged in tight containers and stored at controlled room temperature.[4] The manufacturer has stated that valproate sodium tablets should not be removed from the original packaging and placed in dosing compliance aids because the drug is hygroscopic.[1622] Valproate sodium syrup should be packaged in tight containers, stored at controlled room temperature, and protected from excessive temperatures of 40°C or more and from freezing.[2,4]

## Stability Reports of Compounded Preparations
### Oral

**STUDY 1 (REPACKAGED TABLETS):** Llewelyn et al.[1585] reported on the stability of sodium valproate 100-mg immediate-release tablets (Eplim, Sanofi-Aventis) repackaged from the original containers into polyethylene blisters with foil backing as dose administration aids. Samples were stored refrigerated at 2 to 8°C, in an incubator at room temperature of 23 to 27°C, and in a temperature and humidity cabinet at 43 to 47°C and 70 to 80% relative humidity for 56 days. At all storage conditions, substantial weight gain of the tablets occurred due to the hygroscopicity of sodium valproate. Dissolution was also affected under all storage conditions, resulting in failure to comply with British Pharmacopeial requirements. However,

HPLC analysis of drug content found no significant loss of the drug at any storage condition.

**STUDY 2 (REPACKAGED SYRUP):** The stability of valproate sodium syrup in several types of unit-dose containers was studied by Sartnurak and Christensen.[83] Two-milliliter samples of valproate sodium syrup (Depakene, Abbott) 250 mg/5 mL were repackaged into polypropylene oral syringes (Baxa), glass oral syringes (M.P.L.), and glass vials (Wheaton Scientific) and fitted with suitable caps. Samples were stored at 4, 25, and 60°C protected from light and from variation in moisture conditions. A gas chromatographic assay determined valproic acid content. Less than 5% loss occurred at both 4 and 25°C in glass syringes and vials in 180 days. However, in the polypropylene syringes, about 12% was lost in 20 days at room temperature and 10% was lost in 90 days at 4°C, primarily due to sorption to the syringe materials. Most lost drug was recovered through desorption of the syringes with chloroform. At 60°C, losses were much greater, ranging from around 10% in the glass containers to 20% in the polypropylene syringes in 30 days.

**STUDY 3 (EMULSION):** Veerman et al.[205] evaluated the bioavailability of an extemporaneously prepared valproic acid emulsion and compared it to the bioavailability of the commercial syrup. Ninety-six valproic acid 250-mg liquid-filled capsules (Depakene, Abbott) were warmed under a heat lamp and punctured with a 16-gauge needle; the contents (0.3–0.4 mL) were squeezed into a mortar and pestle. Acacia 20.6 g was triturated with the capsule contents to form a uniform mixture. Water 41 mL was added and stirred vigorously to make the primary emulsion. Additional water was used to bring to 240 mL, making a theoretical valproic acid concentration of 100 mg/mL. The emulsion was stored at 25°C throughout the study. The extemporaneous emulsion demonstrated bioavailability similar to the commercial syrup both initially and after 30 days, making it a suitable alternative for patients unable to tolerate the syrup.

### Enteral

Ortega de la Cruz et al.[1101] reported the physical compatibility of an unspecified amount of oral liquid valproate sodium (Depakine, Sanofi) with 200 mL of Precitene (Novartis) enteral nutrition diet for a 24-hour observation period. No particle growth or phase separation was observed. ∎

# Valsartan

## Properties

Valsartan is a white to practically white fine crystalline powder.[1,7]

### Solubility

Valsartan is slightly soluble in water and soluble in ethanol.[1,7] Valsartan sodium salt is soluble in water to about 5 mg/mL.[1385]

### pH

Valsartan sodium aqueous solutions have a pH near 5.5.[1385]

## General Stability Considerations

Valsartan powder and commercial oral tablets should be packaged in airtight containers, stored at controlled room temperature, and protected from exposure to moisture.[3,4,7]

## Stability Reports of Compounded Preparations

### Oral

**STUDY 1:** The commercial oral tablet labeling[7] for valsartan (Diovan) tablets cites a method for preparing a valsartan 4-mg/mL oral suspension from the commercial tablets. The labeling states that eight valsartan 80-mg oral tablets should be placed in an amber glass bottle and 80 mL of Ora-Plus oral suspending vehicle should be added. The bottle should be shaken for a minimum of two minutes and then allowed to stand for a minimum of one hour. After the standing time, the suspension should again be shaken, this time for a minimum of one additional minute. After shaking is completed, 80 mL of Ora-Sweet SF sugar-free oral sweetening vehicle should

be added to the bottle, and again the suspension should be shaken for at least 10 seconds to disperse the ingredients. Once the suspension has become homogeneous, it can be stored in the amber glass bottle with a child-resistant screw-cap closure for 30 days at controlled room temperature or up to 75 days if stored refrigerated at 2 to 8°C. Of course, the oral suspension should be shaken for at least 10 seconds prior to use.

**STUDY 2:** Allen[1384] reported a similar oral suspension formulation in a standardized size of 100 mL for valsartan 4-mg/mL oral suspension. For this formulation, 400 mg of valsartan as five 80-mg commercial oral tablets is placed in the amber glass bottle and 50 mL of Ora-Plus is added. The shaking and standing times for the preparation procedure that Allen recommended are identical to those in the oral tablet labeling for valsartan (Diovan) tablets.[7] The bottle should be shaken for a minimum of two minutes and then allowed to stand for a minimum of one hour. After the standing time, the suspension should again be shaken, this time for a minimum of one additional minute. After shaking is completed, the oral suspension is brought to 100 mL with Ora-Sweet SF, and the suspension is shaken for an additional 10 seconds to disperse the ingredients. Once the suspension has become homogeneous, Allen follows the commercial tablet labeling recommendation, stating that the oral suspension can be stored in the amber glass bottle for 30 days at controlled room temperature or up to 75 days if stored refrigerated at 2 to 8°C. Again, the oral suspension should be shaken for at least 10 seconds prior to use.

# Vancomycin Hydrochloride

## Properties

Vancomycin hydrochloride, a tricyclic glycopeptide antibiotic, is an amphoteric, freely flowing powder having a bitter taste. It is reported to range in color from white or almost white to tan or brown. It contains not less than 900 mcg of vancomycin per milligram. Vancomycin hydrochloride 1.03 mg is approximately equivalent to vancomycin 1 mg. Each milligram of vancomycin is equivalent to 1 million units.[2–4]

### Solubility

Vancomycin hydrochloride is freely soluble in water, having an aqueous solubility greater than 100 mg/mL.[1–3] The drug is slightly soluble in ethanol.[3]

### pH

A 5% aqueous solution of vancomycin hydrochloride has a pH between 2.5 and 4.5.[3,4] Vancomycin injection has a pH of 3

to 5.[4] Reconstituted vancomycin hydrochloride injection has a pH of about 3.9.[8] Vancomycin hydrochloride for oral solution has a pH of 2.5 to 4.5 when reconstituted as directed.[4]

### Osmolality

Vancomycin hydrochloride 50 mg/mL in sterile water for injection has an osmolality of 57 mOsm/kg.[286]

## General Stability Considerations

Vancomycin hydrochloride capsules are packaged in tight containers. Vancomycin hydrochloride products should be stored at controlled room temperature.[2,7]

Reconstituted vancomycin hydrochloride injection is stable for up to 14 days at room temperature or under refrigeration.[2,8] At 5 mg/mL in 5% dextrose injection and 0.9% sodium chloride injection, no loss occurred during 63 days of storage frozen at −10°C or under refrigeration at 5°C.[452]

Vancomycin hydrochloride is most stable in solutions, having a pH between 3 and 5.[453] A 1-mg/mL vancomycin hydrochloride solution at pH 1.4 lost 19% concentration after five days at 24°C; at pH 5.6 it lost 10% in 17 days at 24°C; and at pH 7.1 it lost 11% in five days at 24°C.[452]

## Stability Reports of Compounded Preparations
### Injection
Injections, like other sterile drugs, should be prepared in a suitable clean air environment using appropriate aseptic procedures. When prepared from nonsterile components, an appropriate and effective sterilization method must be employed.

Wood et al.[658] evaluated the stability of vancomycin hydrochloride 10 mg/mL in sterile water for injection, 5% dextrose injection, and 0.9% sodium chloride injection repackaged into plastic syringes. Stored under refrigeration, the solutions were physically and chemically stable for 84 days, with 4% or less vancomycin hydrochloride loss by HPLC analysis. At 25°C, losses were much greater. Losses of 10% occurred in varying periods from 29 to 62 days, depending on diluent and brand of syringe. In addition, a white flocculent precipitate appeared in all room temperature samples after eight weeks of storage.

### Oral
**STUDY 1 (ORAL SOLUTION):** Mallet et al.[479] evaluated the stability of reconstituted vancomycin hydrochloride oral solution at 0, 4, and 25°C over 90 days of storage. HPLC analysis found no loss of drug in samples stored frozen at 0°C or refrigerated at 4°C. However, room temperature storage was not acceptable due to the formation of a precipitate on the sixth day of storage. In addition, the room temperature samples exhibited about 17% loss after 90 days of storage.

**STUDY 2 (ORAL SOLUTION):** Ensom et al.[1609] evaluated the stability of vancomycin hydrochloride 25 mg/mL extemporaneously compounded from vancomycin hydrochloride injection (Pharmaceutical Partners of Canada). The oral liquid was prepared by diluting the reconstituted injection with a vehicle composed of an equal parts mixture of Ora-Sweet and distilled water. The oral liquid formulation was packaged in 15-mL opaque blue polyethylene unit-dose cups and also in 50-mL amber polyvinyl chloride prescription bottles. The samples were stored at room temperature of 25°C and under refrigeration at 4°C for 75 days.

The physical characteristics of the oral liquid in the unit-dose cups did not change over 75 days of storage at either temperature. In the plastic bottles, no changes in physical characteristics were observed over 75 days stored under refrigeration, but at room temperature a white precipitate was observed beginning on day 63. Stability-indicating HPLC analysis found no loss of vancomycin hydrochloride in the oral liquid formulation in 75 days when stored under refrigeration. However, at room temperature vancomycin loss

neared or reached 10% within 63 days. The authors recommended limiting room temperature storage to 30 days for the oral liquid in unit-dose cups and 26 days in plastic prescription bottles, which were the calculated periods to 90% of initial concentration with 95% confidence.

**STUDY 3 (BUCCAL GEL):** Benyahya et al.[1086] evaluated the stability of vancomycin extemporaneously compounded in a buccal gel for oropharyngeal staphylococcal infections in immunocompromised children. Vancomycin 1 g was incorporated into a gel composed of Carbopol 934P 1 g, demineralized water 99 g, dilute ammonia five drops, tartrazine 0.5% aqueous solution 10 drops, and methyl- and propylparabens 0.15 g each. A fluorescence immunoassay technique was used to determine vancomycin concentrations. During nine days of refrigerated storage, no bacterial or fungal contamination was detected. The gel's adhesive properties were judged to be acceptable. However, reported vancomycin concentrations were erratic, ranging from about 75 to 95%, indicating that the method was inadequate to determine the drug concentration accurately.

**STUDY 4 (POWDER WITH GENTAMICIN AND NYSTATIN):** Wamberg et al.[1516] evaluated the stability of an oral preoperative bowel preparation containing three antibiotics. Gentamicin sulfate, nystatin, and vancomycin hydrochloride dry powders were sifted and mixed. The intended oral dose was gentamicin sulfate 240 mg, nystatin 3.5 million units, and vancomycin hydrochloride 250 mg. The mixed antibiotic powder was packaged in unit dose amber glass airtight containers and stored at cool temperatures of 8 to 15°C and protected from exposure to direct sunlight. The authors reported that, when using a microbiological assay technique, all of the antibiotics retained concentration under these conditions for up to 12 months. If the antibiotics were prepared in a liquid syrup mixture, the activity of vancomycin hydrochloride was lost much more rapidly.

### Ophthalmic
Ophthalmic preparations, like other sterile drugs, should be prepared in a suitable clean air environment using appropriate aseptic procedures. When prepared from nonsterile components, an appropriate and effective sterilization method must be employed.

**STUDY 1:** Osborn et al.[169] evaluated the stability of several antibiotics in three artificial tears solutions composed of 0.5% hydroxypropyl methylcellulose. Vancomycin was insoluble in the artificial tears as reconstituting agents, so the drug was first dissolved in water before dilution with the tears. The artificial tears (Lacril, Tearisol, Isopto Tears) then were used to dilute vancomycin to 31 mg/mL. The formulations were packaged in plastic squeeze bottles and stored at 25°C. A serial dilution bioactivity test used to estimate the amount of antibiotic activity remaining after seven days of storage found

89% in Lacril, 107% in Tearisol, and 112% in Isopto Tears, compared to the activity of vancomycin reference solution.

**STUDY 2:** Ross and Abate[204] reported the formulation and use of an extemporaneously prepared vancomycin hydrochloride ophthalmic solution. In a laminar airflow hood, 9 mL of artificial tears (Tearisol) was removed from a 15-mL squeeze bottle. A vial containing vancomycin hydrochloride 500 mg was reconstituted with 10 mL of sterile water for injection; 10.2 mL of the solution was added to the squeeze bottle, resulting in a vancomycin concentration of about 31 mg/mL. The preparation was stored under refrigeration and used within five days. The solutions remained visually clear throughout their use.

**STUDY 3:** Fleischer et al.[228] evaluated the physical properties and antimicrobial activity of several extemporaneous vancomycin ophthalmic solutions. Vancomycin hydrochloride injection was used as the drug source; it was diluted with sterile water, sterile normal saline, or phosphate-buffered artificial tears solution (Goldline Teargen) to vancomycin concentrations of 5, 30, and 50 mg/mL. The solutions were stored at room temperature and 4°C for 14 days.

The antimicrobial activity was assessed microbiologically; all solutions retained full antimicrobial activity for 14 days. However, the solutions prepared with sterile water and saline were much more irritating to patients' eyes, probably related to the preparations' low osmolalities and pH values. The pH values and osmolalities were more physiologic with the phosphate-buffered saline (Table 123). A precipitate formed in two solutions; the 5 mg/mL in sterile water and the 50 mg/mL in phosphate-buffered saline both developed gross visible precipitation after one week. The precipitate redissolved with high-speed vortex mixing.

**STUDY 4:** Charlton et al.[582] studied the stability of vancomycin 50 mg/mL (as the hydrochloride) in an ophthalmic solution. The solution was prepared by aseptically adding 10 mL of artificial tears (Liquifilm Tears, Allergan) to a 500-mg vial of vancomycin hydrochloride (Elkins-Sinn) for injection. The ophthalmic solution samples were stored in the original artificial tears bottles at 4 and 25°C for 28 days. The osmolality remained unchanged. However, the pH dropped substantially within as little as seven days to pH 3.5 or below, a level not well tolerated by the eye. Antimicrobial activity, estimated by Kirby-Bauer microbial growth inhibition, exceeded the minimum and remained unchanged for 28 days at either storage temperature. Because of the substantial change of pH to an unacceptably low level, the authors could not recommend a use period for this preparation.

**STUDY 5:** Fuhrman and Stroman[596] evaluated the stability of vancomycin 31 mg/mL (as the hydrochloride) in an extemporaneously prepared ophthalmic solution. A 500-mg vial of vancomycin (as the hydrochloride) (Lilly) was reconstituted with 5 mL of sterile water for injection and shaken, and the contents were allowed to dissolve, yielding a 100-mg/mL solution. By using a syringe and needle, a total of 4.6 mL was removed from a 15-mL ophthalmic dropper bottle of artificial tears (Tears Naturale, Alcon) and replaced with 4.6 mL of the reconstituted vancomycin hydrochloride injections, yielding a concentration near 31 mg/mL. The cap was replaced on the dropper bottle and the contents mixed well. The ophthalmic solution was stored at −10, 4, and 25°C under fluorescent light and at an elevated temperature of 40°C not exposed to light.

The samples stored at 40°C developed crystalline particles in three days; the samples at lower temperatures remained clear and colorless throughout the study. The pH remained

**TABLE 123.** Physical Properties of Ophthalmic Solutions of Vancomycin Hydrochloride[228]

| Vancomycin | Initial | | Day 14 | |
|---|---|---|---|---|
| Concentration | pH | mOsm/kg | pH | mOsm/kg |
| *Sterile Water* | | | | |
| 5 mg/mL | 3.2 | 16 | 3.3 | 60 |
| 30 mg/mL | 2.8 | 39 | 2.7 | 65 |
| 50 mg/mL | 2.7 | 65 | 2.8 | 95 |
| *0.9% Sodium Chloride* | | | | |
| 5 mg/mL | 3.4 | 316 | 3.4 | 325 |
| 30 mg/mL | 3.2 | 340 | 3.3 | 347 |
| 50 mg/mL | 3.1 | 357 | 3.1 | 364 |
| *Phosphate-Buffered Saline*[a] | | | | |
| 5 mg/mL | 6.2 | 323 | 6.2 | 327 |
| 30 mg/mL | 5.9 | 348 | 5.8 | 364 |
| 50 mg/mL | 5.8 | 381 | 5.7 | 390 |

[a]Artificial tears solution.

substantially unchanged throughout, at about pH 5, a pH value that the authors noted could be irritating to the eye. Stability-indicating HPLC analysis found 7% loss of vancomycin in seven days and 15% loss in 10 days at 25°C. At −10 and 4°C, 10% losses were found in 60 and 21 days, respectively. However, relatively large deviations occurred at these two temperatures, and the authors indicated that the results were not conclusive. After three days at 40°C, nearly 17% loss occurred.

**STUDY 6:** Arici et al.[796] reported the stability of vancomycin hydrochloride 31.1-mg/mL stock ophthalmic solutions prepared from injection in sodium chloride 0.9% and in artificial tears (Liquifilm Tears) and packaged in ophthalmic squeeze bottles at room temperature of 24°C and refrigerated at 4°C. No loss in microbiological activity against *Staphylococcus aureus* and *Pseudomonas aeruginosa* was found over 28 days of storage at both temperatures in both vehicles. Small increases in pH of 0.25 pH unit or less occurred.

**STUDY 7:** Barbault et al.[797] evaluated the stability of vancomycin hydrochloride 50-mg/mL ophthalmic solution prepared from injection in sodium chloride 0.9% and packaged in glass ophthalmic bottles at room temperature of 25°C and refrigerated at 4°C. Stability-indicating HPLC analysis found adequate drug stability, with about 93% remaining in 15 days at room temperature and 97% remaining in 32 days refrigerated. By 20 days at room temperature, the vancomycin hydrochloride concentration had declined to about 85%. The refrigerated samples became unclear by 22 days.

**STUDY 8:** Peyron et al.[870] reported the stability of vancomycin hydrochloride 50-mg/mL ophthalmic solution compounded from injection in sodium chloride 0.9%. HPLC analysis found about 9% loss in 45 days under refrigeration at 4°C and in 15 days at room temperature of 25°C.

**STUDY 9:** Achach and Peroux[1058] evaluated the stability of several anti-infective drugs in ophthalmic solutions, including vancomycin hydrochloride. Compounded vancomycin hydrochloride 50 mg/mL in sodium chloride 0.9% was stored under refrigeration at 4°C for 12 days. No changes in color or turbidity were observed, and no change in osmolality was found. The analytical method was not specified, but the authors reported that the vancomycin hydrochloride retained 100% through four days, only 84% at eight days, and back to 98% in 12 days. The eight-day analysis appears to have been in error. The authors recommended a four-day use period, but the data seem to support a longer use period. Also see other previously discussed studies documenting longer use periods.

**STUDY 10:** Moreno et al.[1188] evaluated the stability of 5% vancomycin hydrochloride ophthalmic solution prepared in isotonic sodium chloride 0.9%. Based on HPLC analysis at

elevated temperatures, the time to 10% drug loss ($t_{90}$) was calculated to be about 45 days when the solution was stored at 25°C. However, this seems to be a much longer time period than other studies have found for stability at room temperature, so caution in applying this study result is warranted.

**STUDY 11:** Dobrinas et al.[1195] evaluated the stability of ophthalmic injections, including vancomycin hydrochloride (Lilly) 22.5 mg/mL in balanced salt solution simulating aqueous humor solution. The solutions were passed through 0.22-μm filters for sterilization. Samples were packaged as 1 mL in syringes (composition not cited) and were stored frozen at less than −18°C for six months. HPLC analysis found less than 5% drug concentration change and no formation of decomposition products. In addition, the drug remained stable after thawing and storing at room temperature for six hours.

**STUDY 12:** Chedru-Legros et al.[1219] evaluated the physical and chemical stability of vancomycin hydrochloride fortified ophthalmic solution prepared from the commercial injection. Vancomycin hydrochloride (Dakota Pharm) was diluted in dextrose 5%, yielding a 50-mg/mL concentration. The solution was passed through a Millipore 0.22-μm filter, packaged in clear glass containers, and stored frozen at −20°C over a 75-day test period. The vancomycin hydrochloride solutions had a pH of 3.8 and an osmolality of 351 mOsm/kg, neither of which changed substantially throughout the study. HPLC analysis found little or no change in drug concentration over the 75-day study period. The authors stated that the vancomycin hydrochloride fortified ophthalmic solution could be stored frozen for 75 days. However, after thawing, refrigerated storage and discarding after three days was recommended.

**STUDY 13:** Chedru-Legros et al.[1591] also evaluated the physical and chemical stability of several fortified ophthalmic solutions. Vancomycin hydrochloride injection powder (Dakota Pharm) 50 mg/mL in balanced salt solution (Alcon), dextrose 5% (B. Braun), and sodium chloride 0.9% (B. Braun) packaged in 20-mL glass bottles fitted with pipettes mounted on screw caps were stored frozen at −20°C protected from exposure to light for 6 months. The solution was filtered through 0.22-μm Millipore filters as it was added to each bottle. No visible instability was observed, and osmolality and pH were acceptable. HPLC analysis of vancomycin hydrochloride in sodium chloride 0.9% (but not the other diluents) found little or no change in vancomycin hydrochloride concentration at any time point during six months of frozen storage. The authors concluded that such fortified antibiotic solutions could be stored frozen at −20°C for six months.

**STUDY 14:** McLellan et al.[1282] tried to evaluate the long-term stability of vancomycin hydrochloride 14-mg/mL ophthalmic solution containing benzalkonium chloride 0.005%. The ophthalmic solution was prepared by reconstituting a 500-mg vial of vancomycin hydrochloride injection with 10 mL of sodium chloride 0.9%. The reconstituted solution was mixed with 1.35 mL of benzalkonium chloride 1.33-mg/mL solution and then brought to a final volume of 35.5 mL with additional sodium chloride 0.9%. This solution was passed through a 0.22-μm filter and packaged in sterile ophthalmic bottles. Samples were stored at controlled room temperature of 20 to 25°C for 60 days and frozen at −10 to −20°C for six months.

HPLC analysis was attempted, but the authors acknowledged that the method used was *not* stability indicating. Analytical results were wildly variable and inconsistent throughout the study. Some assays showed that the amount of vancomycin present had increased to as much as 134% of the starting concentration. While the authors stated that this paper supported the use of the vancomycin hydrochloride ophthalmic solution for 60 days at room temperature and six months frozen, in actuality the analytical results are much too variable to provide any support for such a conclusion. This is especially true since the method was *not* stability indicating, an attribute that is generally regarded as necessary for publication in all reliable journals. Because many more reliable studies have not supported a 60-day stability period at room temperature, caution should be observed. This study should *not* be relied upon as the basis for an extended stability period.

**STUDY 15:** Lindquist et al.[1581] reported the stability and activity of vancomycin when added at a concentration of 0.2 mg/mL to Dexsol (Chiron Ophthalmics) corneal storage solution. Vancomycin levels were found to decline at a rate of about 7% per month using an agar diffusion bioassay. However, it was also found that corneal tissue tends to concentrate vancomycin over time, although no differences in endothelial cell count or cell death were seen.

**STUDY 16:** Karampatakis et al.[1597] evaluated the stability of vancomycin hydrochloride 50 mg/mL ophthalmic drops in balanced salt solution (BSS). Vancomycin hydrochloride injection (Lilly) 500-mg vials were reconstituted using BSS (Alcon), resulting in a 50-mg/mL solution. The solution was packaged in polypropylene containers and stored at room temperature of 24°C and under refrigeration at 4°C, both exposed to and protected from light. Organoleptic evaluations were performed, and microbiological assays against *Pseudomonas aeruginosa* and *Staphylococcus aureus* were used to determine antibiotic activity.

The solutions became turbid within 24 hours at both storage temperatures. The antibacterial activity was found to be maintained throughout the four-week study period. Drug stability was unaffected by exposure to or protection from exposure to light.

**STUDY 17:** Sautou-Miranda et al.[1614] evaluated the stability of vancomycin hydrochloride 25-mg/mL ophthalmic

solution stored frozen for 90 days. Lyophilized vancomycin hydrochloride (Dakota Pharm) was dissolved in dextrose 5% (Lavoisier), passed through 0.22-μm filters, and packaged in 5-mL glass bottles with rubber stoppers and crimped aluminum seals. The test samples were stored frozen at −20°C for up to 90 days and were evaluated by physical and chemical analyses. Samples were thawed for testing by exposure to ambient temperature or warm (38°C) running water before testing. Other samples were thawed and placed under refrigeration at 2 to 8°C for 48 hours before evaluation.

The thawing procedures did not affect the stability of the vancomycin hydrochloride. Similarly, storage of thawed solutions for 48 hours under refrigeration did not affect stability. None of the samples showed visible particulates at any point. The pH values were unchanged, remaining between pH 3.6 to 3.9. The osmolality of all test samples at each time point was near 319 mOsm/kg. HPLC analysis of vancomycin content found no significant difference in drug concentration in any of the samples at any time; no sample varied by more than 5% from the concentration at the initial analysis. The authors concluded that vancomycin hydrochloride 25-mg/mL ophthalmic solution in dextrose 5% could be successfully stored frozen at −20°C for up to three months.

**STUDY 18:** Lin et al.[1620] evaluated the antibiotic activity of vancomycin hydrochloride 5 and 25% in dextrose 5% extemporaneously prepared for use as eyedrops by microbiological assay. Vancomycin hydrochloride injection was diluted in dextrose 5% to concentrations of 50 and 250 mg/mL. Samples were stored refrigerated at 4°C and stored frozen at −18°C for 28 days. The 5% concentration was for clinical use, while the 25% concentration was for use as a stock solution. The minimum inhibitory concentrations were variable throughout the study, but the differences between those at time zero and 28 days were statistically insignificant.

Also see *Compatibility with Other Drugs*.

## Compatibility with Other Drugs

**STUDY 1 (AMIKACIN):** Lin et al.[955] reported the activity retention and physiological characteristics of a mixed ophthalmic solution of amikacin (Bristol-Myers Squibb) 20 mg/mL and vancomycin (Lilly) 50 mg/mL prepared from reconstituted injections in sterile water for injection. The mixed antibiotic solution packaged in standard ophthalmic dispensing bottles remained clear and colorless throughout 14 days of storage refrigerated at 4°C. The admixed ophthalmic solution had a pH of 5 to 5.2 and an osmolarity of about 200 mOsm/L. Antimicrobial activity evaluated by the disk diffusion method found no significant differences in the zones of inhibition.

**STUDY 2 (AMIKACIN):** Hui et al.[1153] reported the compatibility of amikacin sulfate 0.4 mg/0.1 mL and vancomycin hydrochloride 1 mg/0.1 mL for ophthalmic use prepared in sodium chloride 0.9% and in balanced salt solution plus (BSS Plus), both with and without dexamethasone sodium phosphate 0.4 mg/0.1 mL. The drugs were mixed together in 4 mL of sodium chloride 0.9%, BSS Plus, and vitreous obtained from cadaver eyes. The samples were incubated at 37°C. Amikacin sulfate and vancomycin hydrochloride concentrations were evaluated using TDx analysis, while dexamethasone sodium phosphate was assayed using HPLC.

No precipitation was observed for amikacin sulfate mixed with vancomycin hydrochloride in all three media and in the human vitreous, either with or without dexamethasone sodium phosphate. In addition, no substantial loss of either antibiotic occurred within 48 hours. However, about 13% dexamethasone sodium phosphate loss occurred. The authors indicated that the amikacin–vancomycin two-drug combination was preferred for use in the treatment of infective endophthalmitis.

**STUDY 3 (CEFTAZIDIME):** Lifshitz et al.[915] reported the incompatibility of vancomycin hydrochloride with ceftazidime in intravitreal and subconjunctival injections. Two cases of immediate precipitation, described as yellowish-white, occurred upon sequential administration of vancomycin hydrochloride and ceftazidime. The intravitreal doses of vancomycin hydrochloride and ceftazidime were 1 mg/0.1 mL and 2.2 mg/0.1 mL, respectively. The subconjunctival doses of vancomycin hydrochloride and ceftazidime were 25 mg/0.25 mL and 100 mg/0.5 mL, respectively. The intravitreal opacities created by the precipitation cleared gradually over two months.

**STUDY 4 (CEFTAZIDIME):** McLellan and Papadopoulos[916] reported another case of immediate precipitation upon sequential subconjunctival administration of ceftazidime 100 mg/0.5 mL and vancomycin hydrochloride 25 mg/0.5 mL. The white precipitation appeared immediately and dissipated over time. Intravitreal administration of ceftazidime 2 mg/0.1 mL and vancomycin hydrochloride 1 mg/0.1 mL was reported not to result in precipitate formation in this one case. However, the reports of precipitation with the combined ophthalmic use of these drugs and the reports of possible concentration-dependent precipitation in injections indicate the need for caution.

**STUDY 5 (CEFTAZIDIME):** Kwok et al.[1154] reported the precipitation of ceftazidime 2.2 mg/0.1 mL and vancomycin hydrochloride 1 mg/0.1 mL for ophthalmic use in sodium chloride 0.9% and balanced salt solution plus glutathione (BSS Plus). The drugs were mixed together in 4 mL of sodium chloride 0.9%, BSS Plus, and vitreous obtained from cadaver eyes. The samples were incubated at 37°C. Ceftazidime and vancomycin hydrochloride were analyzed by HPLC.

Ceftazidime precipitated by itself and in combination with vancomycin hydrochloride in all three media. The extent of precipitation was about 54% if prepared in sodium chloride 0.9% and about 88% if prepared in BSS Plus. Vancomycin precipitation was negligible. After precipitation, the concentration of ceftazidime may not have been sufficiently high for antibacterial activity against common microorganisms.

**STUDY 6 (CIPROFLOXACIN):** Hui et al.[873] reported the compatibility of ciprofloxacin 0.2 mg/0.1 mL and vancomycin hydrochloride 1 mg/0.1 mL for ophthalmic use in sodium chloride 0.9% and balanced salt solution plus glutathione (BSS Plus). The drugs were mixed together in 4 mL of sodium chloride 0.9%, BSS Plus, and vitreous obtained from cadaver eyes. The samples were incubated at 37°C. Ciprofloxacin was analyzed by HPLC, while vancomycin was evaluated using TDx analysis.

Ciprofloxacin precipitated by itself and in combination with vancomycin hydrochloride in all three media. The extent of precipitation (up to 45%) was similar in both sodium chloride 0.9% and BSS Plus. Vancomycin precipitation was negligible. Even after precipitation, the ciprofloxacin concentration was greater than the minimum inhibitory concentration. The authors indicated that the two drugs could be used together in the treatment of infective endophthalmitis.

**STUDY 7 (WITH GENTAMICIN):** Poveda Andres et al.[1562] evaluated the stability of a solution for use as an irrigation in cataract surgery. The solution was composed of gentamicin sulfate 8 mcg/mL and vancomycin hydrochloride 20 mcg/mL in lactated Ringer's injection and was packaged in glass containers. Using polarized immunofluorescence (TDx) analysis, the concentration of each antibiotic was determined during storage. The vancomycin hydrochloride proved to be the least stable drug and limited the utility time of the solution. The time to 10% loss ($t_{90}$) of vancomycin hydrochloride at room temperature of 20 to 25°C was found to be 136 hours if exposed to light and 190 hours if protected from exposure to light.

**STUDY 8 (WITH TOBRAMYCIN):** Anton Torres et al.[1629] evaluated the stability of tobramycin sulfate 0.01 mg/mL (10 mcg/mL) with vancomycin hydrochloride 0.05 mg/mL (50 mcg/mL) in BSS-Plus (Alcon) used for intraocular irrigation during cataract surgery. Vials of the two antibiotics were reconstituted, and appropriate amounts of the solutions were passed through a 0.22-μm filter as they were added unto the BSS-Plus solution. Samples were stored refrigerated at 4 to 8°C and at room temperature for 14 days. The samples remained visually clear with no precipitation. TDx/FLx polarizing immunofluorescence analysis of tobramycin content found less than 10% loss occurred over 14 days of storage under refrigeration and for 12 days at room temperature. HPLC analysis of vancomycin concentration found less than 10% loss occurred over 14 days of storage under refrigeration and for eight days at room temperature. The authors concluded that the mixed antibiotic irrigation solution could be prepared and stored for 14 days if kept refrigerated. ◼

# Vasopressin
# Vasopressin Tannate

## Properties

Vasopressin is a polypeptide hormone causing the contraction of vascular and other smooth muscles and of diuresis.[1,3,4] Vasopressin injection contains 8-L-arginine vasopressin and 8-L-lysine vasopressin as a clear and colorless or practically colorless injection with a faint characteristic odor.[4]

Vasopressin tannate is 8-L-arginine vasopressin and 8-L-lysine vasopressin as the tannate.[1,3]

### Solubility

Vasopressin is water soluble.[1]

### pH

Vasopressin injection has a pH from 2.4 to 4.5.[4,7]

### Osmolality

Vasopressin injection is nearly isotonic due to the presence of sodium chloride 0.9% in the formulation.[7]

## General Stability Considerations

Vasopressin is packaged in tight containers of Type I glass and stored under refrigeration. Vasopressin injection is packaged in single- or multiple-dose containers of Type I glass and stored at controlled room temperature. The injection should be protected from freezing.[4]

## Stability Reports of Compounded Preparations
### Injection

Injections, like other sterile drugs, should be prepared in a suitable clean air environment using appropriate aseptic procedures. When prepared from nonsterile components, an appropriate and effective sterilization method must be employed.

Allen[1360] reported on a compounded formulation of vasopressin tannate 5-units/mL in oil injection. The injection had the following formula:

| | | |
|---|---|---|
| Vasopressin tannate | | 500 pressor units |
| Peanut oil | qs | 100 mL |

The recommended method of preparation is to dissolve the vasopressin tannates in the peanut oil. The solution is to be filtered through a suitable 0.2-μm sterilizing filter and packaged in sterile tight, light-resistant containers. If no sterility test is performed, the USP specifies a beyond-use date of 24 hours when stored at room temperature or three days when stored under refrigeration because of concern for inadvertent microbiological contamination during preparation. However, if an official USP sterility test for each batch of drug is performed, the author recommended a beyond-use date of six months at room temperature because in some countries this formula is similar or the same as a commercial medication that has an expiration date of two years or more. ■

# Vehicle for Oral Solution, NF

## Properties

Vehicle for Oral Solution, NF, is an official vehicle for use in compounding oral liquid preparations.[4,5] It is similar to commercial vehicles used for compounding oral liquids such as Ora-Sweet (Paddock). Vehicle for Oral Solution, NF, has the following formula:[4,5]

| | | |
|---|---|---|
| Sucrose | | 80 g |
| Glycerin | | 5 g |
| Sorbitol | | 5 g |
| Sodium phosphate, dibasic | | 120 mg |
| Citric acid | | 200 mg |
| Potassium sorbate | | 100 mg |
| Methylparaben | | 100 mg |
| Purified water | qs | 100 mL |

To prepare, heat about 30 mL of purified water to 70 to 75°C and add the glycerin and methylparaben, stirring until the methylparaben is dissolved. The sodium phosphate, dibasic, citric acid, potassium sorbate, and sorbitol should then be added and mixed well. Then the sucrose should be added and mixed until dissolution has occurred. The mixture should then be removed from the heat and cooled. Finally the mixture should be brought to volume with additional purified water and mixed well. The pH should be between 4 and 5 and adjusted if necessary.[4,5]

## General Stability Considerations

Vehicle for Oral Solution, NF, should be packaged in tight, light-resistant containers and stored at controlled room temperature. In the absence of specific supporting stability information for longer storage, a beyond-use date of six months from the date of preparation may be used.[4,5] ■

# Vehicle for Oral Solution, Sugar Free, NF

## Properties

Vehicle for Oral Solution, Sugar Free, NF, is an official vehicle for use in compounding oral liquid preparations.[4,5] It is similar to commercial vehicles used for compounding oral liquids such as Ora-Sweet SF (Paddock). Vehicle for Oral Solution, Sugar Free, NF, has the following formula:[4,5]

| | | |
|---|---|---|
| Xanthan gum | | 50 mg |
| Glycerin | | 10 mL |
| Sorbitol solution | | 25 mL |
| Saccharin sodium | | 100 mg |
| Citric acid monohydrate | | 1.5 g |
| Sodium citrate | | 2 g |
| Potassium sorbate | | 100 mg |
| Methylparaben | | 100 mg |
| Purified water | qs | 100 mL |

To prepare, heat about 60 mL of purified water to 70 to 75°C and add the methylparaben, stirring until the methylparaben is dissolved. Remove the mixture from the heat. The glycerin, sorbitol solution, saccharin sodium, citric acid monohydrate, sodium citrate, potassium sorbate, and xanthan gum should then be added and mixed well. Finally the mixture should be brought to volume with additional purified water and mixed well. The pH should be between 4 and 5 and adjusted if necessary.[4,5]

## General Stability Considerations

Vehicle for Oral Solution, Sugar Free, NF, should be packaged in tight, light-resistant containers and stored at controlled room temperature. In the absence of specific supporting stability information for longer storage, a beyond-use date of six months from the date of preparation may be used.[4,5] ■

# Vehicle for Oral Suspension, NF

## Properties

Vehicle for Oral Suspension, NF, is an official vehicle for use in compounding oral liquid preparations.[4,5] It is similar to commercial vehicles used for compounding oral liquids such as Ora-Plus (Paddock). Vehicle for Oral Suspension, NF, has the following formula:[4,5]

| | | |
|---|---|---|
| Cellulose, microcrystalline | | 800 mg |
| Xanthan gum | | 200 mg |
| Carrageenan | | 150 mg |
| Carboxymethylcellulose sodium (High Viscosity) | | 25 mg |
| Citric acid | | 250 mg |
| Sodium phosphate, dibasic | | 120 mg |
| Simethicone | | 0.1 mL |
| Potassium sorbate | | 100 mg |
| Methylparaben | | 100 mg |
| Purified water | qs | 100 mL |

To prepare, heat about 90 mL of purified water to 70 to 75°C and add the methylparaben, followed by the citric acid, sodium phosphate, dibasic, and potassium sorbate, and mix. Remove the mixture from the heat. With constant mixing, slowly sprinkle the microcrystalline cellulose, xanthan gum, carrageenan, and carboxymethylcellulose. Continue to stir the mixture until the suspending agents are fully hydrated. Add the simethicone, and mix well. Finally, the mixture should be brought to volume with additional purified water and mixed well. The pH should be between 4 and 5 and adjusted if necessary.[4,5]

## General Stability Considerations

Vehicle for Oral Suspension, NF, should be packaged in tight, light-resistant containers and stored at controlled room temperature. In the absence of specific supporting stability information for longer storage, a beyond-use date of six months from the date of preparation may be used.[4,5] ■

# Venlafaxine Hydrochloride

## Properties
Venlafaxine hydrochloride occurs as a white to off-white crystalline powder.[1,3,7]

### Solubility
Venlafaxine hydrochloride is freely soluble in water, having an aqueous solubility of 572 mg/mL.[1,3,7] It is also freely soluble in methanol and soluble in dehydrated ethanol.[3]

### pH
De Rosa and Sharley reported that venlafaxine hydrochloride 7.5 mg/mL in their oral suspension and solution had pH values of 3.2 and 3.4, respectively.[1337] Kervela et al. reported that venlafaxine hydrochloride 5 mg/mL in a compounded oral liquid formulation in Ora-Sweet and Ora-Plus had a pH near 4.[1439] See *Oral* section below.

### pK$_a$
Venlafaxine has a pK$_a$ of 9.4.[7,1458]

## General Stability Considerations
Venlafaxine hydrochloride oral tablets and extended-release capsules are to be dispensed in a well-closed container and should be stored at controlled room temperature in a dry place.[7]

## Stability Reports of Compounded Preparations
### Oral
**STUDY 1:** De Rosa and Sharley[1337] developed and evaluated the stability of two liquid preparations, a suspension and a solution of venlafaxine hydrochloride 75 mg/10 mL, prepared from extended-release capsules. The suspension had the following formula:

| | | |
|---|---|---|
| Venlafaxine hydrochloride extended-release capsules | | 15 |
| Tragacanth mucilage | | 60 mL |
| Sorbitol compound syrup | | 75 mL |
| Benzoic acid 5% solution | | 6 mL |
| Freshly distilled water | qs | 300 mL |

The suspension was prepared by emptying the contents of Effexor-XR capsules into a mortar and pulverizing them to a fine powder using a pestle. A small amount of freshly distilled water was added to produce a smooth magma. The tragacanth mucilage suspending agent, sorbitol compound syrup sweetener, and benzoic acid solution preservative were then added in that order with thorough mixing. The suspension was brought to volume with distilled water.

The solution had the following formula:

| | | |
|---|---|---|
| Venlafaxine hydrochloride extended-release capsules | | 15 |
| Sorbitol compound syrup | | 75 mL |
| Benzoic acid 5% solution | | 6 mL |
| Freshly distilled water | qs | 300 mL |

The solution was prepared by emptying the contents of Effexor-XR capsules into a mortar and pulverizing them to a fine powder using a pestle. Water was added to dissolve the drug, and the solution was filtered through Whatman 40 filter paper using a Büchner funnel under pressure. The mortar was rinsed with 25 mL of water and the resultant solution was run through the filter paper. This procedure was repeated twice. The sorbitol compound sweetener and benzoic acid solution preservative were added to the solution and mixed thoroughly.

The samples of suspension and solution were packaged in 100-mL amber glass bottles and stored at controlled room temperature of 20 to 24°C for 30 days. HPLC analysis found that the solution samples exhibited no loss over the 30-day test period. The suspension samples lost 3 to 6% over the same period. The authors stated that either formulation was suitable for administration via an enteral feeding tube (and presumably orally as well) and was stable for 30 days at room temperature, protected from exposure to light.

**STUDY 2:** Kervela et al.[1439] evaluated the stability of venlafaxine hydrochloride 5 mg/mL in an extemporaneously compounded oral liquid formulation. Venlafaxine hydrochloride commercial oral tablets (Effexor, Wyeth) were crushed to fine powder using a mortar and pestle. The powder was incorporated into an equal parts mixture of Ora-Sweet and Ora-Plus (Paddock) in a volumetric flask and shaken to ensure thorough mixing. The oral liquid was packaged in 50-mL amber glass bottles, and samples were stored refrigerated at 4°C and at room temperature of 22 to 28°C. Stability-indicating HPLC analysis found that drug decomposition exceeded 10% after 15 days refrigerated and after 10 days at room temperature.

**STUDY 3:** Donnelly et al.[1640] evaluated the stability of venlafaxine hydrochloride compounded as an oral liquid suspension from extended-release capsule contents (Novopharm). The capsule contents were reduced to powder using a high-speed blender and incorporated into an equal parts (50:50) mixture of Ora-Plus and Ora-Sweet and also into simple syrup with a target concentration of 15 mg/mL. The oral liquids were packaged in amber plastic bottles, and samples were stored at 5 and 23°C for 28 days.

Visual inspection found that there was no color change during the observation period and all samples remained easy to resuspend. The Ora-Plus/Ora-Sweet mixtures exhibited a slower rate of settling than the simple syrup mixtures. Stability-indicating HPLC analysis determined that the actual initial concentrations of the samples were about 12 mg/mL,

somewhat below the target concentration. Little or no loss of venlafaxine hydrochloride occurred in the Ora-Plus/Ora-Sweet samples over 28 days at either storage temperature. In simple syrup, venlafaxine hydrochloride losses of about 2 and 7% occurred in 28 days under refrigeration and at room temperature, respectively. ■

# ■ Verapamil Hydrochloride

## Properties

Verapamil hydrochloride is a white or almost white nearly odorless crystalline powder having a bitter taste.[2,3]

### Solubility

Verapamil hydrochloride is soluble in water,[2,3] having an aqueous solubility variously cited as 70 mg/mL[1] at 21°C and pH 4.24 and as 83 mg/mL.[1] The drug is sparingly soluble in ethanol, having a solubility in dehydrated ethanol of 26 mg/mL[1–3] but increasing to more than 100 mg/mL in 95% ethanol. In propylene glycol, verapamil hydrochloride has a solubility of 93 mg/mL.[1]

### pH

A 5% aqueous solution of verapamil hydrochloride has a pH of 4.5 to 6.5.[4] A 0.1% aqueous solution has a pH of 5.25.[1] Verapamil hydrochloride injection has a pH of 4 to 6.5.[4]

### pK$_a$

Verapamil hydrochloride has a pK$_a$ of 8.6.[1]

### Osmolality

Verapamil hydrochloride 2.5-mg/mL injection has an osmolality of 290 mOsm/kg.[345]

## General Stability Considerations

Verapamil hydrochloride products should be packaged in tight, light-resistant containers and stored at controlled room temperature protected from light.[4] The injection should be stored at controlled room temperature and protected from light and freezing.[2,7]

Aqueous solutions of verapamil hydrochloride 0.5 mg/mL are reported to be stable; storage at 50°C for 105 days resulted in no decomposition. Maximum stability occurs in the pH range of 3.2 to 5.6.[671] Verapamil hydrochloride injection is stated to be physically compatible in infusion solutions, having a pH in the range of 3 to 6. It is both physically and chemically stable in most infusion solutions, although dilution in 1/6 M sodium lactate injection is not recommended. In solutions having a pH above 6, however, precipitation may occur.[2] Intentional decomposition of verapamil

hydrochloride in aqueous solution using sodium hydroxide resulted in the formation of an insoluble white precipitate.[671]

Verapamil hydrochloride has been found not to interact with several common solid oral dosage form excipients including microcrystalline cellulose, magnesium stearate, hydroxypropyl methylcellulose, polyvinylpyrrolidone, and talc.[1462]

## Stability Reports of Compounded Preparations
### Oral
#### USP OFFICIAL FORMULATION (ORAL SOLUTION):

Verapamil hydrochloride 5 g
Vehicle for Oral Solution, NF (sugar-containing or sugar-free), qs 100 mL

(See the vehicle monograph for information on the Vehicle for Oral Solution, NF.)

Use verapamil hydrochloride powder. Add about 40 mL of the vehicle and mix well. Add additional vehicle almost to volume in increments with thorough mixing. Quantitatively transfer to a suitable calibrated, tight, light-resistant bottle, bring to final volume with the vehicle, and thoroughly mix, yielding verapamil hydrochloride 50-mg/mL oral solution. The final liquid preparation should have a pH between 3.8 and 4.8. Store the preparation at controlled room temperature or under refrigeration between 2 and 8°C. The beyond-use date is 60 days from the date of compounding.[4,5]

#### USP OFFICIAL FORMULATION (ORAL SUSPENSION):

Verapamil hydrochloride 5 g
Vehicle for Oral Solution, NF (sugar-containing or sugar-free) and Vehicle for Oral Suspension, NF (1:1) qs 100 mL

(See the vehicle monographs for information on the individual vehicles.)

Use verapamil hydrochloride powder or commercial tablets. If using tablets, crush or grind to fine powder. Add about 40 mL of the vehicle mixture and mix to make a uniform paste. Add additional vehicle almost to volume in increments with thorough mixing. Quantitatively transfer to a suitable calibrated, tight, light-resistant bottle, bring to final volume with the vehicle mixture, and thoroughly mix, yielding verapamil hydrochloride 50-mg/mL oral suspension. The final liquid preparation should have a pH between 3.8 and 4.8. Store the preparation at controlled room temperature or under refrigeration between 2 and 8°C. The beyond-use date is 60 days from the date of compounding.[4,5]

**STUDY 1:** Nahata[684] reported the stability of a verapamil hydrochloride 50-mg/mL extemporaneously prepared oral suspension. Verapamil hydrochloride 80-mg tablets were crushed using a mortar and pestle. The powder was triturated with purified water to wet it and then incorporated into a 1:1 mixture of methylcellulose 1% and simple syrup. The suspension was packaged in equal numbers of glass and plastic prescription bottles. Samples were stored at room temperature of 25°C and refrigerated at 4°C.

Stability-indicating HPLC analysis found verapamil hydrochloride losses of about 6 to 7% in 91 days at room temperature. Refrigerated samples exhibited less loss, near 4% in 91 days.

**STUDY 2:** Allen and Erickson[542] evaluated the stability of three verapamil hydrochloride 50-mg/mL oral suspensions extemporaneously compounded from tablets. Vehicles used in this study were: (1) an equal parts mixture of Ora-Sweet and Ora-Plus (Paddock), (2) an equal parts mixture of Ora-Sweet SF and Ora-Plus (Paddock), and (3) cherry syrup (Robinson Laboratories) mixed 1:4 with simple syrup. Seventy-five verapamil hydrochloride 80-mg tablets (Schein) were crushed and comminuted to fine powder using a mortar and pestle. About 40 mL of the test vehicle was added to the powder and mixed to yield a uniform paste. Additional vehicle was added geometrically and brought to the final volume of 120 mL, with thorough mixing after each addition. The process was repeated for each of the three test suspension vehicles. Samples of each finished suspension were packaged in 120-mL amber polyethylene terephthalate plastic prescription bottles and stored at 5 and 25°C in the dark.

No visual changes or changes in odor were detected during the study. Stability-indicating HPLC analysis found less

**TABLE 124.** Formula of Verapamil Hydrochloride 10-mg/mL Oral Liquid Prepared from Standard Tablets[586]

| Component | | Amount |
|---|---|---|
| Verapamil hydrochloride 80-mg tablets | | 10 |
| Syrup | | 40 mL |
| Glycerin | | 10 mL |
| Water | qs | 80 mL |

than 6% verapamil hydrochloride loss in any suspension stored at either temperature after 60 days of storage.

**STUDY 3:** Woods[586] reported on an extemporaneous oral liquid formulation of verapamil hydrochloride at a lower concentration, 10 mg/mL, prepared from standard tablets. The formula is shown in Table 124. An expiration period of seven days when stored under refrigeration protected from light was suggested, although no stability data were presented. The oral liquid should be adjusted to pH 4 to 5 with hydrochloric acid if necessary. Precipitation may occur above pH 6.

**STUDY 4:** Voudrie et al.[1686] evaluated the stability of verapamil hydrochloride 50 mg/mL compounded from powder in two different oral liquid vehicles. Formula 1: Verapamil hydrochloride powder was incorporated in an equal parts mixture of Ora-Sweet SF and Ora-Plus (Paddock). Verapamil hydrochloride powder was added to a glass flask and a portion of the Ora-Sweet SF was mixed using a stir plate. Additional Ora-Sweet SF was added; then an equal amount of Ora-Plus was incorporated and thoroughly mixed. The oral liquid formulation was packaged in a 60-mL amber prescription bottle and stored refrigerated at 2 to 8°C for 60 days. Formula 2: Verapamil hydrochloride powder was incorporated into SyrSpend SF (Gallipot) using a stir plate until homogeneous. The oral liquid was again packaged in a 60-mL amber prescription bottle and stored refrigerated at 2 to 8°C for 60 days.

Stability-indicating HPLC analysis found no loss of verapamil hydrochloride in either oral liquid formulation during 60 days of refrigerated storage. However, an unidentified solid precipitate was found in the Ora-Sweet SF-OraPlus formulation just after preparation. ■

# Vidarabine

## Properties

Vidarabine monohydrate, a purine nucleoside, is a white to off-white crystalline powder.[2,3]

### Solubility

Vidarabine is very slightly soluble in water, having an aqueous solubility of 0.45 mg/mL at 25°C.[2,3]

### $pK_a$

Vidarabine-5′-phosphate has $pK_a$ values at 4.04 and 6.6.[1512]

## General Stability Considerations

Vidarabine ophthalmic ointment should be stored at controlled room temperature and protected from excessive temperatures of 40°C or more and from freezing.[2]

Vidarabine as the 5′-phosphate exhibits maximum stability in aqueous solutions at pH 9 to 9.5. The rate of decomposition due to hydrolysis increases rapidly at pH values below 8.[1512]

## Stability Reports of Compounded Preparations

### Topical

Morita et al.[412] reported the stability of several formulations of vidarabine 3% ointment. Several bases were used, including Plastibase, Plastibase with liquid paraffin, Plastibase with white petrolatum, and white petrolatum. The vidarabine content was determined using an HPLC assay. Vidarabine content remained above 90% in all ointments tested for at least 21 days stored at both 20 and 30°C. The consistency of all ointments also remained good.

### Injection

Injections, like other sterile drugs, should be prepared in a suitable clean air environment using appropriate aseptic procedures. When prepared from nonsterile components, an appropriate and effective sterilization method must be employed.

Stolk et al.[1511] reported on the stability of a solution of vidarabine 450 mg/L (450 mcg/mL) in dextrose 5% when subjected to autoclaving and to long-term storage. Using HPLC and TLC analyses, the authors found that the drug solution could be autoclaved for one hour at 100°C or 20 minutes at 120°C with no decomposition occurring. However, autoclaving for one hour at 120°C resulted in about 2% vidarabine loss. After autoclaving the solution for 20 minutes at 120°C, storage for eight months resulted in about 4% drug loss. ■

# Vitamin A
## (Retinol, Vitamin A Alcohol)

## Properties

Vitamin A is a light yellow to red oil with a slight odor. It may solidify upon refrigeration. It may be odorless or have a slightly fishy odor.[2,3]

### Solubility

Vitamin A is insoluble in water but is soluble in ethanol and vegetable oils. As a solid it may be dispersible in water.[2,3]

### pH

Vitamin A injection has a pH of 6.5 to 7.1.[2]

## General Stability Considerations

Vitamin A is unstable in air and light[2,744] and very susceptible to oxidation in most dosage forms.[6,744] Oral products should be packaged in tight, light-resistant containers. Vitamin A products should be stored at controlled room temperature and protected from excessive temperatures of 40°C or more and from freezing.[2,4]

Semenazato et al.[1603] compared the stability of various forms of vitamin A in solution. Retinol was the least stable, followed by retinyl acetate, with retinyl palmitate being the most stable. Even so, retinyl palmitate underwent about 30% degradation in 60 days at room temperature.

Huyghebaert et al.[1233] evaluated the stability of retinol acetate cold-water-soluble form bulk powder. The cold-water-soluble form of vitamin A acetate is embedded in a coarse gelatin matrix. After the original container had been opened, packages were stored without applying an inert gas layer for eight months at both room temperature and under refrigeration at 8°C. Stability-indicating HPLC analysis found that the bulk substance remained within the specifications of 95 to 105% for drug content for the entire eight-month period.

## Stability Reports of Compounded Preparations

### Ophthalmic

Ophthalmic preparations, like other sterile drugs, should be prepared in a suitable clean air environment using appropriate aseptic procedures. When prepared from nonsterile components, an appropriate and effective sterilization method must be employed.

Mehta and Calvert[1187] evaluated the stability of extemporaneously compounded retinoic acid ophthalmic drops. Retinoic acid was dissolved in absolute ethanol and then aseptically incorporated into sterile arachis oil to yield 0.05% retinoic acid ophthalmic drops. The finished formulation was packaged in 8-mL amber bottles. Samples were stored at −20°C, refrigerated at 4°C, and stored at room temperature exposed to and protected from light. HPLC analysis found no more than 4% loss of retinoic acid over eight weeks in the samples protected from exposure to light at all three temperatures. However, the room temperature sample exposed to daylight underwent about 50% loss by the second day.

## Enteral

**STUDY 1:** Bryant and Neufeld[230] evaluated the potential for loss of vitamin A from the enteral feeding products Osmolite (Ross) and Vivonex High Nitrogen (Norwich-Eaton) in polyvinyl chloride (PVC) feeding bags. A 360-mL volume of each enteral feed was placed in PVC enteral delivery systems and delivered at 60 mL/hour over six hours at room temperature exposed to light. The simulated enteral delivery was repeated over three days. The amount of vitamin A delivered was determined spectrophotometrically; no loss of vitamin A to the PVC containers was found. The authors speculated that the other substances in the enteral feeding products prevented the vitamin A from undergoing the sorption to PVC that occurs in parenteral solutions containing vitamin A.

**STUDY 2:** Davis et al.[231] reported on the stability of vitamin A in two enteral feeding containers stored frozen and thawed for use. Travasorb STD (Travenol) was placed in PVC bags (Corpak) and in low-density polyethylene coextruded with 5% ethylvinyl acetate (Polar Bag, Corpak). The bags of enteral feeds were stored at −20°C for 90 days; the bags were then thawed in water at 25°C and were left at 25°C for another 12 hours. The vitamin A content was assessed by HPLC analysis. No significant loss of vitamin A occurred due to sorption or the freeze–thaw process.

## Oral

Huyghebaert et al.[1233] evaluated the stability of retinol acetate cold-water-soluble form prepared as oral capsules containing 10,000 units of vitamin A. The cold-water-soluble form of vitamin A acetate is embedded in a coarse gelatin matrix. The vitamin powder was mixed with lactose 80 mesh by gently shaking the two powders together in a glass bottle. More vigorous mixing using a mortar and pestle was avoided in order to prevent destruction of the vitamin's gelatin matrix. The powder mixture was filled into size 2 hard-gelatin capsules that were then stored at room temperature and refrigerated at 8°C. Stability-indicating HPLC analysis found that the vitamin A content in the capsules remained within the specifications of 90 to 110% for drug content for the entire two-month period at both storage temperatures.

## Topical

**STUDY 1:** Gatti et al.[744] reported the stability of a mixed retinoid solution containing tretinoin, beta-carotene, and vitamin A palmitate (vehicle not specified). Stability-indicating HPLC analysis found about 20% loss at ambient temperature and refrigerated in 90 days if protected from exposure to light. If stored at ambient temperature exposed to sunlight and to air, tretinoin losses of 60 to 70% and vitamin A palmitate loss of 30% resulted.

**STUDY 2:** Ji and Seo[839] reported the stability of 5% retinyl palmitate in four topical formulations with tocopheryl acetate. The preparations included creams prepared as water-in-oil (w/o), oil-in-water (o/w), water-in-silicone (w/s), and multilamellar vesicles (liposomes). The stability of retinyl palmitate was best in the liposome formulation. HPLC analysis found about 10% loss in two months and 20% loss in 12 months at 25°C. The drug was much less stable in the other cream formulations, losing about 30% (w/s and o/w) to 40% (w/o) in two months.

**STUDY 3:** Carlotti et al.[912] reported some factors influencing the stability of retinyl palmitate in topical formulations. The authors reported that emulsions are better topical vehicles than hydroxyethyl cellulose gels in protecting retinyl palmitate from photodecomposition because the drug is protected by the oil phase. The stability is not affected by pH in the range of 5.6 to 7; however, the drug is less stable at pH 4 and 8. Incorporation of a high concentration of sunscreen into the topical product improves the photostability of retinyl palmitate. Similarly, the presence of an antioxidant such as butylated hydroxytoluene (BHT) increases retinyl palmitate stability.

**STUDY 4:** Bonhomme et al.[1164] evaluated the stability of retinoic acid in a topical gel formulation. The gel was prepared by mixing 1 g of retinoic acid with 93 g of sweet almond oil and 6 g of Cab-O-Sil. HPLC analysis found no loss of retinoic acid over six months when the gel was stored under refrigeration.

**STUDY 5:** Tashtoush et al.[1255] evaluated the photodegradation of tretinoin 0.025% and isotretinoin 0.025% in ethanol and dermatological cream preparations exposed to solar simulated light, UVA light, and visible light. HPLC analysis found that tretinoin was more sensitive to photodegradation than isotretinoin. Both compounds undergo photolysis and photoisomerization. The UVA component of solar simulated light was found to be the major contributor to photodegradation of both compounds. The degradation by photolysis was more extensive in cream preparations than in ethanol. Because UVA penetrates deeply into skin, the authors postulated that tretinoin degradation may contribute to the photosensitivity that is associated with tretinoin therapy.

**STUDY 6:** Akhavan and Levitt[1269] evaluated the stability of retinol 0.3% in a commercial topical cream (Lustra-Ultra, Taro-Pharma) that also contains hydroquinone 4%, the sunscreens avobenzene 3% and octinoxate 7.5%, along with vitamins C and E. The stability of the retinol component was evaluated at about 37°C and exposed to varying combinations of full spectrum light (visible and ultraviolet) and air or nitrogen. HPLC analysis found that the retinol in samples exposed to air or nitrogen and light decomposed about 5% in one hour and 9% in four hours. If protected from light, retinol losses were about 3% or less in four hours. The authors concluded that the retinol in this topical formulation was stable, exhibiting less than 10% loss under simulated-use conditions.

**STUDY 7:** Semenazato et al.[1603] evaluated the stability of retinyl palmitate incorporated into a variety of topical formulations. For monophasic commercial products such as oils, lipsticks, and coated tablets, little loss of retinyl palmitate was found. However, for commercial products that were emulsified biphasic formulations, the retinyl palmitate content was highly varied; in one case the drug degraded to 7.7% of the labeled amount. The authors found that the surfactant system used in emulsified formulations is the most important factor influencing the degradation rate of retinyl palmitate. ▪

# Vitamin E
## (Tocopherols)

## Properties
Vitamin E is a group of related clear yellow or greenish-yellow nearly odorless viscous oils that may solidify when cold.[2,3]

### Solubility
Vitamin E is insoluble in water but is soluble in ethanol and miscible with vegetable oils.[2,3]

## General Stability Considerations
Tocopherols are unstable in air and light, but their esters are stable in both.[2] Vitamin E products should be packaged in tight, light-resistant containers.[2–4]

Huyghebaert et al.[1233] evaluated the stability of DL-alpha-tocopherol acetate cold-water-soluble form bulk powder after opening the original container. The cold-water-soluble form of vitamin E is embedded in a coarse gelatin matrix. After opening the original container, packages were stored without applying an inert gas layer for eight months at both room temperature and under refrigeration at 8°C. Stability-indicating HPLC analysis found the bulk substance remained within the specifications of 95 to 105% for drug content for the entire eight-month period.

## Stability Reports of Compounded Preparations
### Enteral
Davis et al.[231] reported on the stability of vitamin E in two enteral feeding containers stored frozen and thawed for use. Travasorb STD (Travenol) was placed in polyvinyl chloride bags (Corpak) and in low-density polyethylene coextruded with 5% ethylvinyl acetate (Polar Bag, Corpak). The bags of enteral feeds were stored at −20°C for 90 days; the bags were then thawed in water at 25°C and were left at 25°C for another 12 hours. The vitamin E content was assessed by HPLC analysis. No significant loss of vitamin E occurred due to sorption or the freeze–thaw process.

### Oral
Huyghebaert et al.[1233] evaluated the stability of DL-alpha-tocopherol acetate cold-water-soluble form prepared as oral capsules containing 100 mg of vitamin E. The cold-water-soluble form of vitamin A acetate is embedded in a coarse gelatin matrix. The vitamin powder was mixed with lactose 90 mesh by gently shaking the two powders together in a glass bottle. More vigorous mixing using a mortar and pestle was avoided in order to prevent destruction of the vitamin's gelatin matrix. The powder mixture was filled into size 0 hard-gelatin capsules that were then stored at room temperature and refrigerated at 8°C. Stability-indicating HPLC analysis found that the vitamin E content in the capsules remained within the specifications of 90 to 110% for drug content for the entire two-month period at both storage temperatures.

### Topical
Rozman and Gasperlin[1267] evaluated the stability of ascorbic acid 0.4% (wt/wt) and vitamin E 1% (wt/wt) in several topical microemulsion formulations. Oil/water, water/oil, and gel-like microemulsions were evaluated. The microemulsions had the components shown in Table 125. The surfactant and cosurfactant were blended in a 1:1 mass ratio. The isopropyl myristate and distilled water were then added and mixed using a magnetic stirrer for 30 minutes.

**TABLE 125.** Percentage (wt/wt) Composition of Microemulsion Formulations[1267]

| Component | Water/Oil | Oil/Water | Gel-like |
|---|---|---|---|
| Distilled water | 10 | 45 | 60 |
| Isopropyl myristate | 60 | 25 | 10 |
| Tween 40[a] | 15 | 15 | 15 |
| Imwitor 308[b] | 15 | 15 | 15 |

[a]Polyoxyethylene (20) sorbitan monopalmitate
[b]Glyceryl caprylate

The finished topical preparations were packaged in glass containers and stored at controlled room temperature of 21 to 23°C. Stability indicating HPLC analysis found that the vitamin E was stable with no difference between any of the formulations, with light exposure, and with the presence of ascorbic acid. Indeed, the presence of ascorbic acid protected the vitamin E from decomposition over 56 days of storage. Vitamin C decomposed during storage about 30 to 40% over 28 days. ■

# ■ Voriconazole

## Properties

Voriconazole is a white or lightly colored crystalline powder. The commercially available injection is provided as a white lyophilized powder in vials containing voriconazole 200 mg and sulfobutyl ether β-cyclodextrin sodium 3200 mg. A vacuum is present in the intact vials.[7]

### Solubility

Voriconazole exhibits low aqueous solubility, but the commercial injection incorporates a β-cyclodextrin excipient to enhance aqueous solubility. The reconstituted injection is fully dissolved, with a voriconazole concentration of 10 mg/mL and a β-cyclodextrin concentration of 160 mg/mL.[7]

### pH

The pH of the commercial voriconazole injection is not provided by the manufacturer[7]; however, Dupuis et al.[1367] found the pH of reconstituted voriconazole injection 10 mg/mL to be near pH 7. Al-Badriyeh et al.[1415] reported that voriconazole 10 mg/mL had a pH of 6.3 while voriconazole 20-mg/mL solution had a pH of 6.1. Isla Tejera et al.[1049] also reported the pH of an extemporaneously compounded ophthalmic solution to be near pH 7 when compounded in a buffered saline solution (BSS). See the *Ophthalmic* stability report for the formulation.

### Osmolality

The osmolality of the commercial voriconazole injection is not provided by the manufacturer[7]; however, Dupuis et al.[1367] found the osmolarity of reconstituted voriconazole injection to be near 562 mOsm/L. Isla Tejera et al.[1049] reported the osmolarity of an extemporaneously compounded ophthalmic solution to be near isotonicity in the range of 265 to 285 mOsm/L. See the *Ophthalmic* stability report for the formulation.

## General Stability Considerations

Commercial voriconazole tablets and injection in intact vials should be stored at controlled room temperature. Intact containers of the commercial powder for oral suspension should be stored refrigerated.[7]

After reconstitution with sterile water for injection, voriconazole injection should be stored refrigerated; according to the manufacturer, chemical and physical stability of the reconstituted solution has been demonstrated for 24 hours.[7]

After reconstitution of the oral suspension, the manufacturer states it may be stored at controlled room temperature and protected from refrigeration and freezing. The manufacturer indicates that the shelf life of the reconstituted suspension is 14 days at room temperature, after which any remaining suspension should be discarded.[7]

Exposure of voriconazole powder and solution to ultraviolet light resulted in decomposition. This did not occur with the commercial tablets, indicating some protection is afforded by the tablet excipients. In aqueous solution, voriconazole is stable at acidic and neutral pH but is extensively degraded at alkaline pH. Voriconazole decomposition products were found not to have antifungal activity.[1364] However, Dupuis et al.[1367] found that a reconstituted voriconazole 10-mg/mL solution in sterile water for injection exhibited less than 10% drug loss over 30 days of storage at room temperature of 21 to 27°C, both exposed to and protected from exposure to light.

## Stability Reports of Compounded Preparations
### Ophthalmic

Ophthalmic preparations, like other sterile drugs, should be prepared in a suitable clean air environment using appropriate aseptic procedures. When prepared from nonsterile components, an appropriate and effective sterilization method must be employed.

**STUDY 1:** Isla Tejera et al.[1049] reported the stability of an extemporaneously compounded voriconazole 3-mcg/mL ophthalmic solution prepared from the commercial injection. The injection was reconstituted with sterile water for injection according to the manufacturer's recommendation, yielding a voriconazole concentration of 10 mg/mL. A 0.5-mL (5-mg) quantity of the reconstituted solution was diluted to 50 mL in a syringe with sterile sodium chloride 0.9%. A 0.3-mL (0.1-mg) quantity of this solution was then diluted in 9.7 mL of sterile buffered saline solution (BSS), yielding a final concentration of 3 mcg/mL. The solution was evaluated for stability over 30 days stored at ambient room temperature of 22 to 24°C and refrigerated at 2 to 8°C.

Ultraviolet spectrophotometric analysis found little or no change in drug concentration over 21 days at either storage temperature. The osmolarity ranged from about 265 to 285 mOsm/L throughout the study, and pH exhibited little change. Antifungal activity against five species remained at 100% through 21 days, but after 30 days the activity against *Candida albicans* and *Candida parapsilosis* had diminished to 56 and 88%, respectively. The authors concluded that the ophthalmic preparation was stable and active for 21 days at both room temperature and under refrigeration.

**STUDY 2:** Lau et al.[1323] evaluated the long-term stability of voriconazole 1% (10 mg/mL) eyedrops compounded from the commercial injection (Vfend IV, Pfizer). The eyedrops were prepared by reconstituting the injection with sterile water for injection and were preserved with benzalkonium chloride 0.01%. The drops were packaged in amber high-density polyethylene eyedropper bottles. Samples were stored refrigerated at 2 to 8°C for 75 days. Stability-indicating HPLC analysis found less than 2% drug loss occurred over the 75-day study period. Based on the calculated rate of drug decomposition, the authors suggested that less than 10% voriconazole loss would occur over 120 days of refrigerated storage.

**STUDY 3:** Dupuis et al.[1367] evaluated the stability of voriconazole 10 mg/mL (1%) eyedrops prepared from commercial injection (Vfend, Pfizer). Voriconazole powder for injection 200 mg was reconstituted with 19 mL of sterile water for injection resulting in 20 mL of a 10-mg/mL voriconazole solution. This solution was passed through a 0.2-μm filter and packaged in sterile containers. Samples were stored for 30 days (1) at room temperature of 21 to 27°C not protected from light in colorless glass containers, (2) at room temperature of 21 to 27°C protected from light in amber glass containers, and (3) refrigerated at 2 to 6°C protected from light in amber glass containers. The osmolarity of the solution was found to be near 562 mOsm/L and remained unchanged throughout the study period in all samples. The pH was about 7, which remained unchanged in all samples. HPLC analysis found less than 10% loss of voriconazole occurred in 30 days under any of the storage conditions.

**STUDY 4:** Al-Badriyeh et al.[1415] evaluated the stability of voriconazole 20 mg/mL (2%) and 10 mg/mL (1%) ophthalmic solutions prepared from the commercial injection (Vfend, Pfizer). For the 2% solution, voriconazole powder for injection 200 mg was reconstituted with 9 mL of water for injection containing benzalkonium chloride 0.01%, resulting in a 20-mg/mL solution. For the 1% solution, voriconazole

powder for injection 200 mg was reconstituted with 19 mL of water for injection containing benzalkonium chloride 0.01%, resulting in a 10-mg/mL solution. The solutions were filtered through a 0.22-μm filter and packaged in light-resistant high-density polyethylene plastic eyedrop bottles. Samples of both concentrations were stored refrigerated at 2 to 8°C. Samples of the 2% solution were also stored at room temperature of 25°C in an environmental chamber and at 40°C in a dry oven. The solutions were clear and colorless and remained so throughout the study. The 1% and 2% solutions had pH values near 6.3 and 6.1, respectively, which remained unchanged throughout the study. Stability-indicating HPLC analysis found little or no loss of voriconazole from the 2% solution in 16 weeks (112 days) stored under refrigeration and at room temperature. About 2% loss occurred in eight weeks stored at 40°C. Approximately 6% loss occurred after 32 weeks (224 days) stored refrigerated, which the authors stated was representative of an unstable solution. However, they acknowledged that in practice a loss less than 10% is an accepted stability standard. The voriconazole 1% solution also exhibited little or no loss in 14 weeks (98 days) when stored refrigerated.

**STUDY 5:** Senthilkumari et al.[1650] evaluated the stability of voriconazole for ophthalmic use prepared from voriconazole injection. The injection was reconstituted and diluted to concentrations of 1 and 10 mg/mL using sterile water for injection. The solutions were packaged in sterile eyedroppers and stored at 4 and 37°C for 30 days. Using tandem mass spectroscopy, the authors found 3% or less loss of voriconazole occurred in 30 days even at 37°C.

## Oral

Nguyen et al.[1456] evaluated the stability of two voriconazole 40-mg/mL oral liquid formulations for veterinary use. Commercial voriconazole oral tablets were pulverized using a mortar and pestle and incorporated into (1) an equal parts mixture of Ora-Plus and Ora-Sweet, and (2) a 3:1 mixture of Ora-Plus and deionized water resulting in a 40-mg/mL concentration. Forty-mL samples of the oral liquids were packaged in oversized (200 mL) amber glass bottles and in 60-mL amber plastic bottles and were stored at room temperature of 20 to 22°C and refrigerated at 3 to 7°C for 30 days.

No visible changes in odor or color were observed in any of the samples. The 3:1 Ora-Plus and deionized water formulation was substantially more viscous than the equal parts mixture of Ora-Plus and Ora-Sweet. HPLC analysis of drug concentrations found lower initial concentrations in the samples packaged in oversized bottles. Over the 30-day study period, variability in the measured drug concentrations was reported, but no sample demonstrated a loss exceeding 10% at either storage condition at any time point. The authors concluded that voriconazole was stable in the two formulations for 30 days at both room temperature and refrigerated. ∎

# Warfarin Sodium

## Properties

Warfarin sodium is a white odorless amorphous or crystalline hygroscopic powder with a slightly bitter taste.[2,3]

### Solubility

Warfarin sodium is very soluble in water and ethanol, having solubilities greater than 1 g/mL.[2,3]

### pH

A 1% aqueous solution has a pH variously cited as between 7.2 and 8.3[4] and between 7.6 and 8.6.[3] The injection has a pH of 8.1 to 8.3.[7]

### pK$_a$

The pK$_a$ of warfarin is 5.[6]

## General Stability Considerations

Warfarin sodium is discolored by light. The oral tablets should be packaged in tight, light-resistant containers, stored at controlled room temperature, and protected from excessive temperatures of 40°C or more.[2-4]

Warfarin sodium in solutions having a pH below 8 is subject to precipitation because of the very low solubility of warfarin (about 0.004 mg/mL).[6,322] Clear solutions are formed at pH values above 8.[322]

Warfarin sodium also is subject to pH-dependent sorption to polyvinyl chloride plastic. Sorptive losses increase as the pH is lowered from 7.[323,324,331] However, over pH 2 to 7, little loss of drug due to sorption to polypropylene containers occurred.[324]

## Stability Reports of Compounded Preparations

### Oral

Sharley et al.[1207] evaluated the stability of a warfarin sodium 1-mg/mL oral liquid formulation compounded from tablets (Coumadin and Marevan) and warfarin sodium clathrate powder. The oral liquid formulations were preserved with Compound Hydroxybenzoate Solution APF (8% methyl- and 2% propyl hydroxybenzoate in propylene glycol) and had the following compositions:

**From Tablets**

| | | |
|---|---|---|
| Warfarin 5-mg tablets | | 60 tablets |
| Sodium phosphate | | 3 g |
| Tragacanth mucilage | | 60 mL |
| Sorbitol compound syrup | | 75 mL |
| Compound Hydroxybenzoate Solution APF | | 3 mL |
| Distilled water | qs | 300 mL |

**From Powder**

| | | |
|---|---|---|
| Warfarin sodium clathrate | | 324.3 mg |
| Sodium phosphate | | 3 g |
| Compound Hydroxybenzoate Solution APF | | 3 mL |
| Distilled water | qs | 300 mL |

For the tablet-derived formulations, the warfarin sodium 5-mg tablets from each source were ground to fine powder using a mortar and pestle, and sufficient water was added to create a smooth magma. Sodium phosphate was dissolved in 50 mL of distilled water and added to the magma. Tragacanth mucilage, sorbitol compound syrup, and Compound Hydroxybenzoate Solution APF were then added and mixed well. The suspension was brought to volume with distilled water.

For the powder-derived formulation, the warfarin sodium clathrate powder was dissolved in half of the distilled water. The sodium phosphate was dissolved in 50 mL of distilled water and was added to warfarin solution. Compound Hydroxybenzoate Solution APF was added and mixed well. The solution was brought to volume with distilled water.

The oral liquid formulations were packaged in amber glass bottles and were stored at 20 to 24°C for 28 days. Stability-indicating HPLC analysis found no loss of warfarin in any of the oral liquids over the test period.

### Enteral

Kuhn et al.[99] evaluated the recovery of warfarin when mixed with the enteral nutrient formula Osmolite (Ross). Warfarin sodium 16.7, 33.3, and 66.7 mcg/mL in full-strength Osmolite and warfarin sodium 5 mg in 150 mL of Osmolite full-strength and mixed 2:1 and 1:1 with water were tested by ultrafiltration to remove the protein-bound fraction. A stability-indicating HPLC assay determined warfarin concentrations. Significant reductions in the amount of recoverable warfarin occurred, although varying the warfarin sodium concentration from 16.7 to 67.7 mcg/mL in full-strength Osmolite did not have a significant effect; reductions of 35 to 38% occurred in all three concentrations. Sufficient dilution of Osmolite had a greater effect on the amount of recoverable warfarin; reductions of 37, 35, and 13% occurred in the full-strength, 2:1, and 1:1 dilutions, respectively. Although the amount of recoverable warfarin was decreased considerably when mixed with Osmolite in this in vitro study, the authors indicated that further research was necessary to extrapolate any results to the clinical setting.

Several reports noted that the vitamin K content of enteral feeding formulations antagonizes the effectiveness of warfarin sodium on anticoagulation.[188–195,222] Because the antagonism is particularly problematic when initiating or discontinuing enteral feeding, close monitoring is required at those times.[189] ■

# White Lotion

## Properties

White lotion, USP, is prepared by dissolving zinc sulfate 40 g and sulfurated potash 40 g in separate 450-mL portions of purified water and filtering each solution. The sulfurated potash solution is added slowly to the zinc sulfate solution with constant stirring. Additional purified water is then added to bring the final volume to 1000 mL, and the solution is mixed thoroughly.[4]

Sulfurated potash (liver of sulfur) is a mixture of potassium polysulfides (mostly trisulfide) and occurs as yellowish-brown lumps with a slight hydrogen sulfide odor.[1,4]

Zinc sulfate is available as the monohydrate and the heptahydrate. The monohydrate occurs as white crystalline powder or granules. The heptahydrate occurs as odorless powder, crystals, or granules and may cake.[1,4] Zinc sulfate 220 mg is equivalent to zinc 50 mg. One gram of zinc sulfate provides 3.5 mmol of zinc.[3]

### Solubility

Sulfurated potash is freely soluble in water to about 500 mg/mL, usually leaving a slight residue. It is partially soluble in ethanol.[1,4]

Zinc sulfate heptahydrate has an aqueous solubility of 1 g in 0.6 mL and 400 mg/mL in glycerol but is insoluble in ethanol. Zinc sulfate monohydrate is also soluble in water but is practically insoluble in ethanol.[1]

### pH

Sulfurated potash aqueous solutions are alkaline.[3] Zinc sulfate aqueous solutions have a pH of about 4.5.[1] A 50-mg/mL solution has a pH in the range of 4.4 to 5.6.[3]

## General Stability Considerations

White lotion, USP, should be packaged in tight containers and stored at controlled room temperature.[4]

Sulfurated potash is incompatible with acids and acid salts, ethanol, and water containing too much carbon dioxide. Sulfurated potash may discolor metals and metal implements.[1]

Zinc sulfate should be packaged in tight containers and stored at controlled room temperature.[4]

## Stability Reports of Compounded Preparations

### Topical

Boonme[737] reported the ability of three antioxidants to curtail yellow discoloration resulting from oxidation of sulfides in white lotion, USP. White lotion containing zinc sulfate and sulfurated potash 40 mg/mL, each in purified water with 0.1% sodium sulfite, was prepared. No yellow discoloration appeared in the lotion in 20 days at room temperature near 28°C; without the sodium sulfite, yellow discoloration appeared in as little as seven days. Attempts to use ascorbic acid and sodium metabisulfite as antioxidants failed. ■

# Zidovudine

## Properties
Zidovudine exists as a white to yellowish to beige or brownish powder or crystalline material.[1,3,4,7]

### Solubility
Zidovudine is sparingly soluble in water,[3,4] having an aqueous solubility at 25°C of 20 mg/mL[7] to 25 mg/mL.[1] Zidovudine is soluble in ethanol.[4]

### pH
Zidovudine injection has a pH of approximately 5.5,[7] with a range of 3.5 to 7 at a concentration of 30 mg/mL in 0.12 M potassium chloride.[4] Zidovudine oral solution has a pH in the range of 3 to 4 when a volume of the oral solution equivalent to 150 mg of zidovudine is mixed 3:1 with 0.12 M potassium chloride.[4]

## General Stability Considerations
Zidovudine bulk material, oral capsules, oral tablets, injection, and oral solution should be packaged in tight, light-resistant containers and should be stored at controlled room temperature.[4]

## Stability Reports of Compounded Preparations
### Oral
Szof et al.[1201] evaluated the stability of an extemporaneously compounded oral liquid formulation of zidovudine. Zidovudine capsules were opened using a small grinding mill and transferred to a beaker, and sterile water was added. The powder and capsule fragments were filtered out. The resulting filtered solution yielded an initial zidovudine concentration of 10 mg/mL. The solution was packaged in amber glass bottles. Samples were stored under refrigeration. Analysis found that the drug remained stable for at least 22 weeks. ▩

# Zinc Oxide

## Properties
Zinc oxide is a very fine, odorless, amorphous, white to yellowish-white powder free from gritty particles.[1,3,4] Zinc oxide is a component of calamine lotion.[4]

### Solubility
Zinc oxide is practically insoluble in water and ethanol but is soluble in dilute acetic or mineral acids.[1,3,4]

## General Stability Considerations
Zinc oxide sublimes at normal atmospheric pressure[1] and gradually absorbs carbon dioxide from the air.[4] Zinc oxide powder should be packaged in well-closed containers and stored at controlled room temperature. Zinc oxide paste and ointment should be packaged in well-closed containers and stored at controlled room temperature, avoiding prolonged exposure to temperatures exceeding 30°C.[4]

## Compatibility with Other Drugs
STUDY 1: Nagatani et al.[1526] evaluated the stability of several steroids in steroid ointments when mixed with zinc oxide ointment. The steroid ointments included betamethasone valerate (Rinderon VG), hydrocortisone butyrate (Locoid), triamcinolone (Kenacort AG), and diflucortolone valerate (Nerisona). Samples of the mixed ointments were stored at 5 and 30°C until analyzed. Analysis found no loss of most of the steroids when stored at 5 and 30°C with the exception of hydrocortisone butyrate. The sample of hydrocortisone

**TABLE 126.** Corticosteroid Ointments Found to Be Stable Mixed 1:1 with Zinc Oxide Ointment[1714]

| | |
|---|---|
| Adcortin (halcinonide) | Locorten (flumetasone) |
| Decaderm (dexamethasone) | Methaderm (dexamethasone propionate) |
| Dermovate (clobetasol propionate) | Nerisona (diflucortolone valerate) |
| Flucort (fluocinolone acetonide) | Pandel (hydrocortisone) |
| Kenacort (triamcinolone) | Propaderm (beclomethasone dipropionate) |
| Kindavate (clobetasol propionate) | Rinderon (betamethasone) |
| Lidomex (prednisolone valeroacetate) | Topsym (flucinolone polivalente) |
| Locoid (hydrocortisone) | Visderm (amcinonide) |

butyrate ointment mixed with zinc oxide ointment was stable only at 5°C.

**STUDY 2:** Ohishi et al.[1714] evaluated the stability of corticosteroid ointments (Table 126), each mixed in equal quantity (1:1) with zinc oxide ointment. Samples were stored refrigerated at 5°C. HPLC analysis of the ointment mixtures found that all of the drugs were stable for 32 weeks at 5°C. ■

#  Zinc Sulfate

## Properties
Zinc sulfate is available as the monohydrate and as the heptahydrate. The monohydrate occurs as white crystalline powder or granules. The heptahydrate occurs as odorless powder, crystals, or granules and may cake.[1,4] Zinc sulfate 220 mg is equivalent to zinc 50 mg. One gram of zinc sulfate provides 3.5 mmol of zinc.[3]

### Solubility
Zinc sulfate heptahydrate has an aqueous solubility of 1 g in 0.6 mL and 400 mg/mL in glycerol but is insoluble in ethanol. Zinc sulfate monohydrate is also soluble in water but is practically insoluble in ethanol.[1]

### pH
Zinc sulfate aqueous solutions have a pH of about 4.5.[1] A 50-mg/mL solution has a pH in the range of 4.4 to 5.6.[3]

## General Stability Considerations
Zinc sulfate should be packaged in tight containers and stored at controlled room temperature.[4]

## Stability Reports of Compounded Preparations
*Enteral*
**ZINC SULFATE COMPATIBILITY SUMMARY**
Compatible with: Vivonex T.E.N.
Incompatible with: Enrich • TwoCal HN

Burns et al.[739] reported the physical compatibility of zinc sulfate capsules 220 mg with 10 mL of three enteral formulas, including Enrich, TwoCal HN, and Vivonex T.E.N. Visual inspection found no physical incompatibility with the Vivonex T.E.N. enteral formula. However, when mixed with Enrich, a soft gelatinous mass formed. When mixed with TwoCal HN, a hard mass formed. ■

#  Zonisamide

## Properties
Zonisamide occurs as a white crystalline powder.[1,7]

### Solubility
Zonisamide has an aqueous solubility of 0.8 mg/mL. The solubility of zonisamide in 0.1 N hydrochloric acid is 0.5 mg/mL. The drug is soluble in ethanol, ethyl acetate, and acetic acid.[1]

### pKa
Zonisamide has a pKa of 10.2.[1]

## General Stability Considerations
Zonisamide capsules are to be stored at controlled room temperature in a dry place and protected from light.[7]

## Stability Reports of Compounded Preparations
### Oral
Abobo et al.[1372] reported on the stability of two extemporaneously compounded oral liquid formulations of zonisamide 10 mg/mL prepared from commercial capsules. The oral liquids were prepared using the contents of zonisamide capsules incorporated into simple syrup or methylcellulose 0.5% (wt/vol) sugar-free vehicle using a mortar and pestle. The two suspensions were packaged in amber plastic prescription bottles, and samples were stored for 28 days at room temperature of 23 to 25°C exposed to normal fluorescent light and refrigerated at 3 to 5°C.

The authors reported no visible changes in color or odor for either formulation in any of the samples. Similarly, no microbial growth was observed in the refrigerated samples of either formulation and in the room temperature samples of the simple syrup formulation. However, the methylcellulose formulation remained acceptable for only seven days, developing substantial microbial growth within 14 days. Stability-indicating HPLC analysis found little or no loss within 28 days in any of the samples. The authors stated that these zonisamide oral liquids were stable for 28 days under refrigeration, but only the simple syrup formulation was acceptable for 28 days at room temperature. ■

# Appendix I: Beyond-Use Dating for Compounded Nonsterile Preparations*

The United States Pharmacopeia[4] (USP) Chapter <795> (entitled Pharmaceutical Compounding—Nonsterile Preparations) has established a quality standard for compounding that includes the assignment of beyond-use dates to compounded nonsterile preparations. The following overview contains the opinions of the author regarding beyond-use dating for compounded nonsterile preparations, including consideration of the standards in USP Chapter <795>.

## Beyond-Use Date

"Beyond-use date" is the correct term for the date after which an extemporaneously compounded preparation should not be used. (The term "expiration date" applies to manufactured drugs.) Usually the beyond-use date for a preparation is assigned by the person compounding the preparation and should be for a conservative period from the date (or time) of compounding that reasonably ensures that the preparation's pharmaceutical quality—chemical, physical, and microbiological integrity—is maintained. The intended duration of therapy should also be included in a beyond-use date assignment.

## Assigning a Beyond-Use Date to a Compounded Nonsterile Preparation Based on Stability Information

Both general and drug-specific stability information from reputable sources should be considered when assigning a beyond-use date. The expected type of decomposition mechanism and the labeled storage conditions need to be factored in. All available information, especially published articles from peer-reviewed journals, should be considered. Drug manufacturers may also be sources of stability information, particularly general information on the known stability behavior of a drug. Using education and experience, the person compounding the preparation should consider carefully the available information in relation to the specific preparation composition and should determine a suitable period of use based on the preparation's chemical and physical integrity.

## Assigning a Beyond-Use Date to a Nonsterile Preparation in the Absence of Stability Information

Many times, no drug-specific stability information exists that can be applied to a compounded preparation. When no published studies have appeared and no reference work provides the needed basis for determining a beyond-use date, the USP offers general guidance in assigning a beyond-use date.[4] It can be reasonably expected that preparations compounded following USP guidelines will maintain adequate pharmaceutical integrity and quality. Pharmacists who use these USP benchmarks have a professional resource to rely on and to support the beyond-use dates on their compounded preparations. In the absence of specific drug stability information that documents the drug's stability, beyond-use dates exceeding these recommended periods have little or no scientific and professional support.

In the absence of stability information, the following beyond-use dates can be assigned to the types of compounded

*Disclaimer

Although Lawrence Trissel was a member of the USP Nonsterile Compounding Committee, this appendix was prepared in his individual capacity and not as a member of that committee or as a USP representative. The views and opinions presented are entirely those of Mr. Trissel. They do not necessarily reflect the views of USP, nor should they be construed as an official explanation or interpretation of USP Chapter <795>.

nonsterile products noted when packaged in tight, light-resistant containers but shall not be later than the earliest expiration date on the container of any component of the preparation:

- **Nonaqueous formulations**—Not later than the remaining time until the earliest expiration date of any active pharmaceutical ingredient or six months, whichever is earlier, stored at controlled room temperature.

- **Water-containing oral formulations**—Not more than 14 days stored at controlled refrigeration temperature.

- **Water-containing topical/dermal and mucosal liquid and semisolid preparations**—Not more than 30 days at controlled room temperature.

# Appendix II: Beyond-Use Dating for Compounded Sterile Preparations*

The United States Pharmacopeia[4] (USP) Chapter <797> (entitled Pharmaceutical Compounding—Sterile Preparations) has established a quality standard for compounding that includes the assignment of beyond-use dates to compounded sterile preparations. The following overview contains the opinions of the author regarding beyond-use dating for compounded sterile preparations, including consideration of the standards in USP Chapter <797>.

## Beyond-Use Date

"Beyond-use date" is the correct term for the date after which an extemporaneously compounded preparation should not be used. (The term "expiration date" applies to manufactured drugs.) Usually the beyond-use date for a preparation is assigned by the person compounding the preparation and should be for a conservative period from the date (or time) of compounding that reasonably ensures that the preparation's pharmaceutical quality—chemical and physical integrity, sterility, and nonpyrogenicity—is maintained. The intended duration of the therapy also should be included in a beyond-use date assignment.

## Factors That Must Be Considered in Arriving at a Suitable Beyond-Use Date for a Compounded Sterile Preparation

Chemical and physical integrity are among the principal pharmaceutical quality concerns when assigning a beyond-use date to a compounded sterile preparation, just as they are for nonparenteral products. These factors are related to the specific drug and formulation. However, additional important factors include the sterility and nonpyrogenicity of the final preparations—attributes that are not related to a specific drug. Instead they are dependent on the components, preparation method, adequacy of the compounding environment, and capabilities and performance of compounding personnel. Like all humans, compounding personnel are covered in microbes that are shed constantly. In addition, no one is perfect; we all are subject to unintentional work performance errors that can directly transfer microbes to compounded sterile preparations. It must always be kept in mind that the single most important factor in ensuring that manually compounded preparations are, in fact, sterile is the adequacy of personnel work practices.

Sterility, the absence of living microorganisms, is an essential quality for all preparations that are administered by routes that bypass the patient's natural protective systems. Such preparations include injections by the intravascular, intramuscular, subcutaneous, intrathecal, and epidural routes, as well as ophthalmic preparations, inhalation solutions, and topical preparations to be used on broken or abraded skin. Failure to ensure sterility has been a cause of much patient morbidity and mortality. For injections to be administered by the intravascular, intrathecal, and epidural routes, the absence of bacterial endotoxins or pyrogens—fever-producing lipopolysaccharides from microorganisms—is also essential. The qualities of sterility and nonpyrogenicity can limit the beyond-use date that should be assigned to a compounded sterile preparation as much as chemical and physical concerns can.

Most beyond-use date decisions are made by the person compounding the preparation. The decision for each specific

*Disclaimer

Although Lawrence Trissel was a member of the USP Sterile Compounding Committee, this appendix was prepared in his individual capacity and not as a member of that committee or as a USP representative. The views and opinions presented are entirely those of Mr. Trissel. They do not necessarily reflect the views of USP, nor should they be construed as an official explanation or interpretation of <797>.

compounded sterile preparation is generally a two-step process based on:

1. The compounded sterile preparation's chemical and physical stability, and
2. The risk of the preparation being contaminated with microbes during compounding.

Using the considerations above, the beyond-use date applied should be the shorter of the two. There may not always be a single answer in all cases. Knowledgeable compounding personnel with good intent can and do address such decisions differently.

In Chapter <797> the USP has provided examples of compounding that would fall into various risk categories. However, that chapter cannot provide all the answers for the myriad of compounding preparations, practices, and situations that exist and how each would fall under the risk-level categories. The person performing the compounding is responsible for deciding what risk category applies in each specific case. These important decisions have consequence not only for the beyond-use date (and therefore patient safety); they also may have consequence should a legal or regulatory question arise about the nature of the compounding.

Always remember that risk categories and beyond-use dating are not about the convenience or profit of compounding personnel. They are and should always be about patient safety. A good way of thinking about these compounding decisions is as if one's own loved one—a child, spouse, or parent—were to be the recipient. This is known as the "Loved-One Test." What would we want for people for whom we care deeply? If compounding decisions (including beyond-use dating) are made as if the preparation were going to be administered to a loved one, then we have a better chance of our moral compass pointing correctly. It is also important to think about how one would explain and defend compounding decisions to a legal or regulatory body should a need arise. If compounding personnel make these decisions with such moral considerations in mind, in addition to applying the appropriate scientific and technical information, they will be on firmer moral and professional ground and can feel more confident about their beyond-use date decisions.

To arrive at a suitable beyond-use date, consider all the factors discussed below and assign the most conservative beyond-use date based on whichever factor results in the shortest time frame.

## Assigning a Beyond-Use Date to a Compounded Sterile Preparation Using Stability Information

When assigning a beyond-use date, compounding personnel should consider both general and drug-specific stability information from reputable sources. The expected type of decomposition mechanism and the storage conditions need to be factored in. The available stability information, especially published articles from peer-reviewed journals, should be considered. Drug manufacturers may also be sources of stability information, particularly general information on the known stability behavior of a drug. Using education and experience, the person compounding the preparation should consider carefully the available information in relation to the specific preparation composition and should determine a suitable period of use based on the chemical and physical integrity of the preparation. The microbiological integrity of the preparation (addressed below) should then be considered. The beyond-use date that is assigned should be based on the chemical and physical integrity of the preparation, especially the least stable component, or the microbiological limit based on contamination risk category for the appropriate type of preparation (low-risk level, low-risk level with 12-hour beyond-use date, medium-risk level, high-risk level, immediate use), whichever is shorter.

## Assigning a Beyond-Use Date to a Compounded Sterile Preparation in the Absence of Stability Information

In general, more stability information is available for compounded sterile preparations than for nonparenteral preparations. Even so, drug-specific stability information and data that can be applied to a compounded sterile preparation sometimes do not exist. When no published studies have appeared and no reference work provides the needed basis for determining a beyond-use date, USP Chapter <797> refers to the general beyond-use date guidance in USP Chapter <795> for assigning a beyond-use date.[4] Compounding personnel need to be familiar with USP Chapter <795> (entitled Pharmaceutical Compounding—Nonsterile Preparations) as well as Chapter <797>. (See Appendix I: Beyond-Use Dating for Compounded Nonsterile Preparations.) This general guidance from the USP for physical and chemical stability factors is the same for sterile and nonsterile preparations.

In the absence of applicable stability information for water-containing preparations, a chemical stability beyond-use date of not more than 14 days when stored at refrigeration temperatures may be applied. Refer to USP Chapter <795> or to Appendix I in this book for other types of preparations. For compounded sterile preparations, the microbiological contamination risk of the preparation must then be considered. (See below.) The beyond-use date that is assigned should be based on the maintenance of the chemical and physical integrity of the preparation from the general guidance or the microbiological limit based on contamination risk category for the type of preparation (low-risk level, low-risk level with 12-hour beyond-use date, medium-risk level, high-risk level, immediate use), whichever is shorter.

## Assigning a Beyond-Use Date to a Compounded Sterile Preparation on the Basis of Microbiological Contamination Risk Category

The risk of inadvertent microbiological contamination in a compounded preparation intended to be sterile results from the sterility or contamination of the starting drugs, components, and equipment used during compounding; the complexity of the preparation procedure; adequacy of the compounding environment; and the competency of the personnel who compound the preparation. USP Chapter <797> offers general guidance in assigning a microbiological beyond-use date based on the contamination risk category and storage condition of the compounded sterile preparation when the chemical and physical stability of the compounded sterile preparation permit.

### *Low-Risk Level Compounded Sterile Preparations*

Low-risk level compounded sterile preparations are those that are prepared by simple aseptic transfer of sterile drugs from one container to another using sterile equipment and devices in a suitable controlled air environment meeting ISO Class 5 (Class 100). To be low-risk level, the drugs and components must always be in sealed containers and not exposed to the open compounding environment at any time. Examples would include simple injections, infusions, admixtures, and ophthalmic products that are compounded using no more than three commercially manufactured sterile drug packages, using commercially manufactured sterile equipment such as needles and syringes, and involving only simple aseptic transfers.

If the low-risk level compounded sterile preparation passes a USP sterility test, then the chemical and physical stability of the preparation can be used to assign the beyond-use date. In the absence of passing a sterility test, the maximum microbiological beyond-use date periods (if chemical and physical stability permit) by storage condition are:

- Room temperature—48 hours
- Refrigeration—14 days
- Frozen—45 days

### *Low-Risk Level with 12-Hour or Less Beyond-Use Date Compounded Sterile Preparations*

Low-risk level compounded sterile preparations with a 12-hour or less beyond-use date is a risk category that applies to situations in which it may not be physically possible for a compounding site to have an ISO Class 7 buffer area for the placement of each and every ISO Class 5 primary engineering control, such as a laminar airflow workbench. In such cases, the primary engineering control may have to be located in a room that does not qualify as an ISO Class 7 buffer area—what is termed an uncontrolled environment. However, to qualify for this risk category, the space must be a segregated compounding area dedicated solely to compounding sterile preparations. The instance that can serve as a model is low-volume short-term sterile compounding performed in an ISO Class 5 hood located in a segregated space within a satellite pharmacy. However, compounding personnel must recognize that aseptic compounding in an uncontrolled environment has a higher potential of producing contaminated doses than if a real ISO Class 7 buffer area cleanroom with full cleanroom work practices is utilized.

Because of the higher risk of microbial contamination in doses prepared in an uncontrolled environment, this risk category includes a short time limitation on the acceptable beyond-use period. If a trace contamination from airborne microorganisms should occur, the microbiological beyond-use period limitation not exceeding 12 hours should be short enough to avoid an overgrowth of the microorganisms to hazardous or fatal levels. Even so, using this kind of compounding environment for a patient population with compromised immune systems is probably a bad idea.

Several limitations are placed on the sterile compounding that can be performed under this risk category:

1. The ISO Class 5 primary engineering control must be located in a segregated compounding area, not in a high-traffic area, with no unsealed windows or doors that connect outdoors and that is restricted to preparing low-risk level compounded sterile preparations.
2. The segregated compounding area cannot be adjacent to food preparation areas, warehouses, construction sites, or other locations that may increase the risk of contamination.
3. The segregated compounding area must meet all the cleaning and disinfecting requirements in Chapter <797> for a real ISO Class 7 buffer area.
4. All personnel cleansing and garbing requirements for low-risk level compounding still apply.
5. All quality assurance, personnel testing, environmental sampling, and other requirements for low-risk level compounding still apply.
6. No hazardous drugs (such as chemotherapy doses) are permitted to be compounded because little or no containment for personnel safety is present.
7. Administration must begin within 12 hours of preparation or as stated in the package insert for the drug, whichever is shorter, so as not to permit an overgrowth of the microorganisms to hazardous or fatal levels.

### *Medium-Risk Level Compounded Sterile Preparations*

Medium-risk level compounded sterile preparations are those that:

- Are prepared in a suitable controlled air environment meeting ISO Class 5 (Class 100) by multiple or complex aseptic transfers of sterile commercial drugs from sealed containers using sterile commercial equipment and devices, or
- Are prepared for the use of multiple patients or one patient on numerous occasions over time, or
- Require a compounding process of an unusually long duration for dissolution or mixing.

The commercial drugs and components must always be in sealed containers and not exposed to the open compounding environment at any time. Although only aseptic transfers of commercial drugs are performed, the number and/or complexity of the aseptic manipulations make this category of aseptic compounded preparations at greater risk for contamination. Examples would include parenteral nutrition admixtures, solutions prepared for infusion device reservoirs with multiple components, or multiple individual dosage units subdivided from a pooled reservoir.

If the medium-risk level compounded sterile preparation passes a USP sterility test, then the chemical and physical stability of the preparation can be used to assign the beyond-use date. In the absence of passing a sterility test, the maximum microbiological beyond-use date periods (if chemical and physical stability permit) by storage condition are:

- Room temperature—30 hours
- Refrigeration—nine (9) days
- Frozen—45 days

### High-Risk Level Compounded Sterile Preparations
High-risk level compounded sterile preparations are at higher risk of being inadvertently contaminated or not adequately sterilized during compounding, resulting in patient injury or death. High-risk level sterile compounding should be performed only by personnel with sufficient specialized education, training, and experience in high-risk level compounding to be truly knowledgeable and competent. In addition, high-risk level sterile compounding should be performed only in adequate sterile compounding operations with outstanding facilities, work practices, and quality assurance. Even then, high-risk level sterile compounding should be reserved only for occasions when the patient's medical need can be met in no safer way, such as by using commercially manufactured drugs. Manual compounding of sterile preparations can never hope to achieve the quality assurance level of a commercial drug manufactured under the Food and Drug Administration's Good Manufacturing Practices (GMPs).

Examples of high-risk level compounded sterile preparations include those compounded from nonsterile components and/or with nonsterile equipment, or those compounded from sterile components but in an environment inferior to ISO Class 5 (Class 100), including inadvertent touch contamination. All such preparations require sterilization. Autoclaving is the preferred sterilization method when drug stability permits. Filtration through a 0.2-µm filter is a common alternative sterilization technique but is less certain than autoclaving to result in sterility.

If the high-risk level compounded sterile preparation passes a USP sterility test, the chemical and physical stability of the preparation can be used to assign the beyond-use date. In the absence of passing a sterility test, the maximum microbiological beyond-use date periods (if chemical and physical stability permit) by storage condition are:

- Room temperature—24 hours
- Refrigeration—three (3) days
- Frozen—45 days

### Immediate-Use Compounded Sterile Preparations
The immediate-use category of compounded sterile preparations is intended for occasions when no delay occurs between sterile drug preparation and administration. Examples include aseptic drug preparation in emergency rooms or in the back of ambulances during "code" situations, a dose prepared by a nurse at the patient's bedside, and preparation in any other location where there is no delay between sterile drug preparation and administration.

The immediate-use category states that the doses are exempted from all requirements in USP Chapter <797> only if all the following requirements are met:

1. Only simple aseptic transfers involving three or fewer drug packages are involved, including an infusion solution.
2. No delays or interruptions occur during dose preparation.
3. The dose is prepared using good aseptic technique and care is taken to prevent direct contact contamination of the ingredients and critical sites during preparation.
4. No hazardous drugs are involved, because no protection from exposure exists for the compounding personnel.
5. The maximum amount of time from the beginning of compounding to the start of administration is not more than one hour. If more than one hour has elapsed, the dose must be destroyed.
6. No dose storage beyond one hour from the start of preparation and no recycling of doses is permitted.
7. The dose must be fully and completely labeled if not administered by the preparer, or the administration must be witnessed by the preparer.

If these requirements cannot be or are not met, then all requirements of USP Chapter <797> must be followed.

## Conclusion

USP Chapter <797> is first and foremost about patient safety. Far too many patients have been injured or died as a result of unacceptable, poor quality sterile compounding practices. Chapter <797> has established a meaningful and achievable quality standard for compounding sterile preparations. The standard is being adopted and implemented in the better quality sterile compounding facilities across the United States and around the world. Appropriate beyond-use dating is one critical part of achieving the patient safety goals of Chapter <797>.

# References

1. O'Neil MJ. The Merck index, 14th ed. Rahway, NJ: Merck & Co., Inc. 2006.
2. McEvoy J, ed. AHFS drug information 2011. Bethesda, MD: American Society of Health System Pharmacists; 2011. Also selected information from prior editions.
3. Sweetman SC, ed. Martindale: the complete drug reference, 37th ed. London: The Pharmaceutical Press; 2011. Also selected information from prior editions.
4. The United States Pharmacopeia, 34th revision. Rockville, MD: The United States Pharmacopeial Convention; 2011.
5. USP Pharmacists' Pharmacopeia, 2nd ed. Rockville, MD: The United States Pharmacopeial Convention; 2008.
6. Connors KA, Amidon GL, Stella VJ. Chemical stability of pharmaceuticals, 2nd ed. New York: John Wiley and Sons; 1986.
7. Manufacturer's official product labeling (package insert) and other product labeling.
8. Trissel LA. Handbook on injectable drugs, 16th ed. Bethesda, MD: American Society of Health System Pharmacists; 2010.
9. Allwood M, Stanley A, Wright P, eds., The cytotoxics handbook, 3rd ed. Oxford: Radcliffe Medical Press; 1997.
10. Mirkin BL, Newman TJ. Efficacy and safety of captopril in the treatment of severe childhood hypertension: report of the international collaborative study group. Pediatrics. 1985;75:1091–1100.
11. O'Dea RF, Mirkin BL, Alward CT, et al. Treatment of neonatal hypertension with captopril. J Pediatr. 1988;113:403–6.
12. Pereira CM, Tam YK. Stability of captopril in tap water. Am J Hosp Pharm. 1992;49:612–5.
13. Lee T, Notari RE. Kinetics and mechanism of captopril oxidation in aqueous solution under controlled oxygen partial pressure. Pharm Res. 1987;4:98–103.
14. Nahata MC, Morosco RS, Hipple TF. Stability of captopril in three liquid dosage forms. Am J Hosp Pharm. 1994;51:95–6.
15. Anaizi NH, Swenson C. Instability of aqueous captopril solutions. Am J Hosp Pharm. 1993;50:486–8.
16. Chan DS, Sato AK, Claybaugh JR. Degradation of captopril in solutions compounded from tablets and standard powder. Am J Hosp Pharm. 1994;51:1205–7.
17. Levin VA, Zackheim HS, Liu J. Stability of carmustine for topical application. Arch Dermatol. 1982;118:450–1.
18. Chan KK, Zackheim HS. Stability of nitrosourea solutions. Arch Dermatol. 1973;107:298, 782.
19. Cutie AJ, Altman E, Lenkel L. Compatibility of enteral products with commonly employed drug additives. J Parenter Enter Nutr. 1983;7:186–91.
20. Gupta VD, Stewart KR. Stability of propranolol hydrochloride suspension and solution compounded from injection or tablets. Am J Hosp Pharm. 1987;44:360–1.
21. de Vries H, Beijersbergen van Henegiouwen GMJ, Huf PA. Photochemical decomposition of chloramphenicol in a 0.25% eyedrop and in a therapeutic intraocular concentration. Int J Pharm. 1984;20:265–71.
22. Michelini TJ, Bhargova VO, DuBe JE. Stability of oral morphine sulfate solution in two enteral tube feeding products. Am J Hosp Pharm. 1988;45:628–30.
23. Kleinberg ML, Stauffer GL, Latiolais CJ. Stability of five liquid drug products after unit dose repackaging. Am J Hosp Pharm. 1980;37:680–2.
24. Eisenberg MG, Kang N. Stability of citrated caffeine solutions for injectable and enteral use. Am J Hosp Pharm. 1984;41:2405–6.
25. Bauer LA. Interference of oral phenytoin absorption by continuous nasogastric feedings. Neurology. 1982;32:570–2.
26. Worden JP, Wood CA, Workman CH. Phenytoin and nasogastric feedings. Neurology. 1984;34:132.
27. Hatton RC. Dietary interaction with phenytoin. Clin Pharm. 1984;3:110–1.
28. Saklad JJ, Graves RH, Sharp WP. Interaction of oral phenytoin with enteral feedings. J Parenter Enter Nutr. 1986;10:322–3.
29. Miller SW, Strom JG Jr. Stability of phenytoin in three enteral nutrient formulas. Am J Hosp Pharm. 1988; 45:2529–32.
30. Johnson CE, Drabik BT. Stability of alcohol-free theophylline liquid repackaged in plastic oral syringes. Am J Hosp Pharm. 1989;46:980–1.
31. Van Gansbeke B. Stability of oxybutynin chloride syrup after repackaging in unit doses. Am J Hosp Pharm. 1991;48:1265–6.

32. Ptachcinski RJ, Walker S, Burkart GJ, et al. Stability and availability of cyclosporine stored in plastic syringes. Am J Hosp Pharm. 1986;43:692–4.

33. Sylvestri MF, Makoid MC. Stability of ampicillin trihydrate suspension in amber plastic oral syringes. Am J Hosp Pharm. 1986;43:1496–8.

34. Sylvestri MF, Makoid MC, Adams SC. Stability of dicloxacillin sodium oral suspension stored in polypropylene syringes. Am J Hosp Pharm. 1987;44:1401–5.

35. Stuart A, Wren C, Bain H. Is there a genetic factor in flecainide toxicity? Br Med J. 1989;298:117–8.

36. Tu YH, Stiles ML, Allen LV Jr, et al. Stability of amoxicillin trihydrate–potassium clavulanate in original containers and unit dose oral syringes. Am J Hosp Pharm. 1988;45:1092–9.

37. Sylvestri MF, Makoid MC, Cox BE. Stability of cephalexin monohydrate suspension in polypropylene syringes. Am J Hosp Pharm. 1988;45:1353–6.

38. Gibbs IS, Tuckerman MM. Formulation of a stable pilocarpine hydrochloride solution. J Pharm Sci. 1974;63:276–9.

39. Kreienbaum MA, Page DP. Stability of pilocarpine hydrochloride and pilocarpine nitrate ophthalmic solutions submitted by U.S. hospitals. Am J Hosp Pharm. 1986;43:109–17.

40. Ahmed I, Day P. Stability of cefazolin sodium in various artificial tear solutions and aqueous vehicles. Am J Hosp Pharm. 1987;44:2287–90.

41. Closson RG. Liquid dosage form of chloroquine. Drug Intell Clin Pharm. 1988;22:347.

42. Migton JM, Kennon L, Sideman M, et al. A stability study of clindamycin hydrochloride and phosphate salts in topical formulations. Drug Dev Ind Pharm. 1984;10:563–73.

43. Rooney M, Creurer I. Stability of propranolol hydrochloride suspension and solution. Am J Hosp Pharm. 1988;45:530–1.

44. Strom JG Jr, Miller SW. Stability of drugs with enteral nutrient formulas. DICP Ann Pharmacother. 1990;24:130–4.

45. Horner RK, Johnson CE. Stability of an extemporaneously compounded terbutaline sulfate oral liquid. Am J Hosp Pharm. 1991;48:293–5.

46. Attia MA, El-Sourdy HA, Shanawany SM. Stability of chlortetracycline hydrochloride and chloramphenicol in some ophthalmic ointment bases. Pharmazie. 1985;40:629–31.

47. Levinson ML, Johnson CE. Stability of an extemporaneously compounded clonidine hydrochloride oral liquid. Am J Hosp Pharm. 1992;49:122–5.

48. Metras JI, Swenson CF, McDermott MP. Stability of procainamide hydrochloride in an extemporaneously compounded oral liquid. Am J Hosp Pharm. 1992;49:1720–4.

49. Lantz MD, Wozniak TJ. Stability of nizatidine in extemporaneous oral liquid preparations. Am J Hosp Pharm. 1990; 47:2716–9.

50. Steedman SL, Koonce JR, Wynn JE, et al. Stability of midazolam hydrochloride in a flavored, dye-free oral solution. Am J Hosp Pharm. 1992;49:615–8.

51. Fagerman KE, McGuigan S, Pixley B. Potential interaction between enteral feeding solutions and oral tetracycline. Nutr Clin Pract. 1986;1:257–8.

52. Fagerman KE, Ballou AE. Drug compatibilities with enteral feeding solutions coadministered by tube. Nutr Supp Serv. 1988;8:31–2.

53. Rochard EB, Rogeron B. Stability of cimetidine hydrochloride in Sondalis Iso enteral nutrition formula. Am J Hosp Pharm. 1991;48:1681–2.

54. Holtz L, Milton J, Sturek JK. Compatibility of medications with enteral feedings. J Parenter Enter Nutr. 1987; 11:183–6.

55. Hooks MA, Longe RL, Taylor, AT, et al. Recovery of phenytoin from an enteral nutrient formula. Am J Hosp Pharm. 1986;43:685–8.

56. Vulovic N, Primorac M, Stupar M, et al. Some studies into the properties of indomethacin suspension intended for ophthalmic use. Int J Pharm. 1989; 55:123–8.

57. Cha CJM, Randall HT. Osmolality of liquid and defined formula diets: the effect of hydrolysis by pancreatic enzymes. J Parenter Enter Nutr. 1981;5:7–10.

58. Kudsk KA, Campbell SM, O'Brien TO. Postoperative jejunal feedings following complicated pancreatitis. Nutr Clin Pract. 1990;5:14–7.

59. Clark-Schmidt AL, Garnett WR, Lowe DR, et al. Loss of carbamazepine suspension through nasogastric feeding tubes. Am J Hosp Pharm. 1990;47:2034–7.

60. Lowe DR, Fuller SH, Pesko LJ, et al. Stability of carbamazepine suspension after repackaging into four types of single-dose containers. Am J Hosp Pharm. 1989;46:982–4.

61. Dannenberg E, Peebles J. Betadine–hydrogen peroxide irrigation solution incompatibility. Am J Hosp Pharm. 1978;35:525.

62. Gupta VD, Gibbs CW Jr, Ghanekar AG. Stability of pediatric liquid dosage forms of ethacrynic acid, indomethacin, methyldopate hydrochloride, prednisone, and spironolactone. Am J Hosp Pharm. 1978;35:1382–5.

63. Tortorici MP. Formulation of a cimetidine oral suspension. Am J Hosp Pharm. 1979;36:22.

64. Grogan LJ, Jensen BK, Makoid MC, et al. Stability of penicillin V potassium in unit dose oral syringes. Am J Hosp Pharm. 1979;36:205–8.

65. Allen LV Jr, Lo P. Stability of oral liquid penicillins in unit dose containers at various temperatures. Am J Hosp Pharm. 1979;36;209–11.

66. Sparkman HE. Experience with cimetidine oral suspension formulation. Am J Hosp Pharm. 1979;36:600.

67. Tempio JS. Extemporaneous preparation of cimetidine oral suspension. Am J Hosp Pharm. 1979;36:1039.

68. Perrin JH. Freezing of suspensions. Am J Hosp Pharm. 1979;36:1157, 1160.

69. Allen LV Jr, Lo P. Freezing of suspensions. Am J Hosp Pharm. 1979;36:1160, 1163.

70. Milovanovic D, Nairn JG. Stability of fluorouracil in amber glass bottles. Am J Hosp Pharm. 1980;37:164–5.

71. Chatterji DC, Hood JC. Incompatibility of povidone-iodine with Lavacol. Am J Hosp Pharm. 1980;37:464, 466.

72. Wilson JE. Lithium-drug interaction alert. ASHP Newsl. 1979;12:6.

73. McGee JL, Alexander B, Perry PJ. Lithium citrate syrup and trifluoperazine hydrochloride concentrate incompatibility: effect on serum lithium levels. Am J Hosp Pharm. 1980;37:1052.

74. Gupta VD. Stability of aqueous suspensions of disulfiram. Am J Hosp Pharm. 1981;38:363–4.

75. MacDonald NC, Whitmore CK, Makoid MC, et al. Stability of methacholine chloride in bronchial provocation test solutions. Am J Hosp Pharm. 1981;38:868–71.

76. Dugas JE. Stability of chlorpromazine liquid in unit dose packaging. Am J Hosp Pharm. 1981;38:1276.

77. Swerling R. Dilution of oral and intravenous aminophylline preparations. Am J Hosp Pharm. 1981;38:1359–60.

78. Theesen KA, Wilson JE, Newton DW, et al. Compatibility of lithium citrate syrup with 10 neuroleptic solutions. Am J Hosp Pharm. 1981;38:1750–3.

79. Bertino JS Jr, Reed MD, Lambert PW, et al. Stability of extemporaneous formulation of injectable cholecalciferol. Am J Hosp Pharm. 1981;38:1932–3.

80. Hunke WA, Katz E, Gorman WG. Stability of some bronchodilator solutions during ultrasonic nebulization. Am J Hosp Pharm. 1982;39:297–300.

81. Kuhn RJ, Lubin AH, Jones PR, et al. Bacterial contamination of aerosol solutions used to treat cystic fibrosis. Am J Hosp Pharm. 1982;39:308–9.

82. Mathur LK, Lai PK, Shively CD. Stability of disopyramide phosphate in cherry syrup. Am J Hosp Pharm. 1982;39:309–10.

83. Sartnurak S, Christensen JM. Stability of valproate sodium syrup in various unit dose containers. Am J Hosp Pharm. 1982; 39:627–9.

84. Little TL, Tielke VM, Carlson RK. Stability of methadone pain cocktails. Am J Hosp Pharm. 1982;39:646–7.

85. Fabian TM, Walker SE. Stability of sodium hypochlorite solutions. Am J Hosp Pharm. 1982;39:1016–7.

86. Christensen JM, Lee RY, Parrott KA. Stability of three oral drug products repackaged in unit dose containers. Am J Hosp Pharm. 1983;40:612–5.

87. Dressman JB, Poust RI. Stability of allopurinol and of five antineoplastics in suspension. Am J Hosp Pharm. 1983;40:616–8.

88. Newton DW, Pollock GR, Narducci WA, et al. Preparation and sterilization by filtration of Renacidin irrigation. Am J Hosp Pharm. 1984;41:121–4.

89. Sewell GJ, Venables B. Preparation of Renacidin irrigation. Am J Hosp Pharm. 1985;42:537.

90. Montgomery HA, Smith FM, Scott BE, et al. Stability of 5-aminosalicylic acid suspension. Am J Hosp Pharm. 1986;43:118–20.

91. Marwaha RK, Johnson BF. Long-term stability study of histamine in sterile bronchoprovocation solutions. Am J Hosp Pharm. 1986;43:380–3.

92. Henry DW, Repta AJ, Smith FM, et al. Stability of propranolol hydrochloride suspension compounded from tablets. Am J Hosp Pharm. 1986;43:1492–5.

93. Krukenberg CC, Mischler PG, Massad EN, et al. Stability of 1% rifampin suspensions prepared in five syrups. Am J Hosp Pharm. 1986;43:2225–8.

94. Souney PF, Braun L, Steele L, et al. Stability of bacitracin solution frozen in glass vials or plastic syringes. Am J Hosp Pharm. 1987;44:1125–6.

95. Smith OB, Longe RL, Altman RE, et al. Recovery of phenytoin from solutions of caseinate salts and calcium chloride. Am J Hosp Pharm. 1988;45:365–8.

96. Bhargava VO, Rahman S, Newton DW. Stability of galactose in aqueous solutions. Am J Hosp Pharm. 1989;46:104–8.

97. St. Claire R L III, Caudill WL. Stability of cefuroxime axetil in beverages. Am J Hosp Pharm. 1989;46:256.

98. Karnes HT, Harris SR, Garnett WR, et al. Concentration uniformity of extemporaneously prepared ranitidine suspension. Am J Hosp Pharm. 1989;46:304–7.

99. Kuhn TA, Garnett WR, Wells BK, et al. Recovery of warfarin from an enteral nutrient formula. Am J Hosp Pharm. 1989;46:1395–9.

100. Mathur LK, Wickman A. Stability of extemporaneously compounded spironolactone suspensions. Am J Hosp Pharm. 1989;46:2040–2.

101. Rosen WJ, Johnson CE. Evaluation of five procedures for measuring nonstandard doses of nifedipine liquid. Am J Hosp Pharm. 1989;46:2313–7.

102. Cacek AT, DeVito JM, Koonce JR. In vitro evaluation of nasogastric administration methods for phenytoin. Am J Hosp Pharm. 1986;43:689–92.

103. Splinter MY, Seifert CF, Bradberry JC, et al. Recovery of phenytoin suspension after in vitro administration through percutaneous endoscopic gastrostomy Pezzer catheters. Am J Hosp Pharm. 1990; 47:373–7.

104. Fulper LD, Cleary RW, Harland EC, et al. Liquefaction times of fatty-type suppositories with and without progesterone. Am J Hosp Pharm. 1990;47:602–3.

105. Taketomo CK, Chu SA, Cheng MH, et al. Stability of captopril in powder papers under three storage conditions. Am J Hosp Pharm. 1990;47:1799–1801.

106. McBride HA, Martinez DR, Trang JM, et al. Stability of gentamicin sulfate and tobramycin sulfate in extemporaneously prepared ophthalmic solutions at 8°C. Am J Hosp Pharm. 1991;48:507–9.

107. Nahata MC, Jackson DS. Stability of cefadroxil in reconstituted suspension under refrigeration and at room temperature. Am J Hosp Pharm. 1991;48:992–3.

108. Alexander KS, Haribhakti RP, Parker GA. Stability of acetazolamide in suspension compounded from tablets. Am J Hosp Pharm. 1991;48:1241-4.

109. Lauriault G, LeBelle MJ, Lodge BA, et al. Stability of methadone in four vehicles for oral administration. Am J Hosp Pharm. 1991;48:1252-6.

110. Wiest DB, Garner SS, Pagacz LR, et al. Stability of flecainide acetate in an extemporaneously compounded oral suspension. Am J Hosp Pharm. 1992; 49:1467–70.

111. McLeod HL, Relling MV. Stability of etoposide solution for oral use. Am J Hosp Pharm. 1992;49:2784–5.

112. Graham CL, Dukes GE, Fox JL, et al. Stability of ondansetron hydrochloride injection in extemporaneously prepared oral solutions. Am J Hosp Pharm. 1993;50:106–8.

113. Fox JL, Ikeda M. Precipitation of ondansetron in alkaline solutions. N Engl J Med. 1991;325:1315–6.

114. Swenson CF. Importance of following instructions when compounding. Am J Hosp Pharm. 1993;50:261.

115. Bhatt-Mehta V, Johnson CE, Kostoff L, et al. Stability of midazolam hydrochloride in extemporaneously prepared flavored gelatin. Am J Hosp Pharm. 1993;50:472–5.

116. Alexander KS, Pudipeddi M, Parker GA. Stability of hydralazine hydrochloride syrup compounded from tablets. Am J Hosp Pharm. 1993;50:683–6.

117. Quercia RA, Jay GT, Fan C, et al. Stability of famotidine in an extemporaneously prepared oral liquid. Am J Hosp Pharm. 1993;50:691-3.

118. Alexander KS, Pudipeddi M, Parker GA. Stability of procainamide hydrochloride syrups compounded from capsules. Am J Hosp Pharm. 1993;50:693–8.

119. Handbook on extemporaneous formulations. Bethesda, MD: American Society of Hospital Pharmacists; 1987.

120. Crawford SY. Tap water should not be used. Am J Hosp Pharm. 1993;50:1579.

121. Anaizi NH, Swenson C. Tap water should not be used. Am J Hosp Pharm. 1993;50:1579.

122. Tam YK, Pereira CM. Tap water should not be used. Am J Hosp Pharm. 1993;50:1579.

123. Fiscella RG, Lam J, Schell J, et al. Cocaine hydrochloride topical solution prepared for ophthalmic use. Am J Hosp Pharm. 1993;50:1572, 1574.

124. Johnson CE, Hart SM. Stability of extemporaneously compounded baclofen oral liquid. Am J Hosp Pharm. 1993; 50:2353–5.

125. Yamreudeewong W, Lopez-Anaya A, Rappaport H. Stability of fluconazole in an extemporaneously prepared oral liquid. Am J Hosp Pharm. 1993;50:2366–7.

126. Peterson JA, Risley DS, Anderson PN, et al. Stability of fluoxetine hydrochloride in fluoxetine solution diluted with common pharmaceutical diluents. Am J Hosp Pharm. 1994;51:1342–5.

127. Garner SS, Wiest DB, Reynolds ER. Stability of atenolol in an extemporaneously compounded oral liquid. Am J Hosp Pharm. 1994;51:508–11.

128. Allen LV. Formula for epidural analgesia. U.S. Pharmacist. 1993;18:88-9, 105–6.

129. Allen LV. Plasminogen activator. U.S. Pharmacist. 1992;17:64–5, 70–1.

130. Allen LV. Cyclosporine ophthalmic drops. U.S. Pharmacist. 1992;17:78–9, 88.

131. Allen LV. Midazolam oral solution. U.S. Pharmacist. 1991;16:66–7.

132. Allen LV. Fortified garamycin ophthalmic solution. U.S. Pharmacist. 1991;16:75–6.

133. Allen LV. Busulfan oral suspension. U.S. Pharmacist. 1990;15:94–5.

134. Allen LV. Preparing a trazodone HCl suspension. U.S. Pharmacist. 1990;15:54, 57.

135. Allen LV. Preparation of urea topical products. U.S. Pharmacist. 1990;15:70, 72.

136. Allen LV, Tu YH. Nystatin lozenges and popsicles. U.S. Pharmacist. 1990;15:64–6.

137. Allen LV. Kayexalate preparation. U.S. Pharmacist. 1990;15:60, 62–3.

138. Allen LV. Rifampin suspension. U.S. Pharmacist. 1989;14:102–3.

139. Allen LV. Diazepam oral suspension. U.S. Pharmacist. 1989;14:64–5.

140. Allen LV, Stiles ML. Topical indomethacin. U.S. Pharmacist. 1988;13:52–3.

141. Snyder DS. Topical indomethacin and sunburn. Br J Dermatol. 1974;90:91–3.

142. Kaidbey KH, Kurban AK. The influence of corticosteroids and topical indomethacin on sunburn erythema. J Invest Dermatol. 1976;66:153–6.

143. Gschnait F, Schwartz T, Seiser A. Topical indomethacin protects from UVB and UVA irradiation. Arch Dermatol Res. 1984;276:131–2.

144. Schwartz T, Gschnait F, Greiter F. Photoprotective effect of topical indomethacin, an experimental study. Dermatologica. 1985;171:450–8.

145. Farr PM, Diffey BL. A quantitative study of the effect of topical indomethacin on cutaneous erythema induced by UVB and UVC radiation. Br J Dermatol. 1986;115:453–66.

146. Stiles ML, Allen LV. Allopurinol mouthwash. U.S. Pharmacist. 1988;137–8.

147. Ghanekar AG, Gupta VD, Gibbs CW. Stability of furosemide in aqueous systems. J Pharm Sci. 1978;67:808–11.

148. Clark PI, Sievin ML. Allopurinol mouthwash and 5-fluorouracil induced oral toxicity. Eur J Surg Oncol. 1985;11:267.

149. Allen LV, Stiles ML. Methadone lemonade. U.S. Pharmacist. 1988;132–3.

150. Stiles ML, Allen LV. T-A-C solution. U.S. Pharmacist. 1988;13:86–8.

151. Allen LV. Preservative-free albuterol solution. U.S. Pharmacist. 1993;18:90–2, 100.

152. Allen LV. Preparing mitomycin ophthalmic solution. U.S. Pharmacist. 1993;18:84–5.

153. Allen LV. Ketoprofen gel. U.S. Pharmacist. 1993;18:98–100.

154. Chi SC, Jun HW. Anti-inflammatory activity of ketoprofen gel in carrageenan-induced paw edema in rats. J Pharm Sci. 1990;79:974–7.

155. Allen LV. Morphine sustained release suppositories. U.S. Pharmacist. 1994; 19:88–90.

156. Kawashima S, Inoue Y, Shimeno T, et al. Studies on sustained-release suppositories III. Rectal absorption of morphine in rabbits and prolongation of its absorption by alginic acid addition. Chem Pharm Bull. 1990;38:498–505.

157. Allen LV. Preservative-free pilocarpine solution. U.S. Pharmacist. 1994;19:104–5.

158. Burckart GJ, Hammond RW, Akers MJ. Stability of extemporaneous suspensions of carbamazepine. Am J Hosp Pharm. 1981;38:1929–31.

159. St. Claire RL, Wilbourne DK, Caudill WL. Stability of cefuroxime axetil in apple juice. Pediatr Infect Dis J. 1988;7:744.

160. Nahata MC, Hipple TF. Pediatric drug formulations. Cincinnati, Ohio: Harvey Whitney Books; 1992.

161. Marino MT, Kucera RF, Almquist AF, et al. Foaming of nebulized albuterol. Am J Hosp Pharm. 1989;46:921.

162. Sojka DA. Foaming of nebulized albuterol. Am J Hosp Pharm. 1989;46:921.

163. Ghogawala Z, Furtado D. In vitro and in vivo bactericidal activities of 10%, 2.5%, and 1% povidone-iodine solution. Am J Hosp Pharm. 1990;47:1562–6.

164. Berkelman RL, Holland BW, Anderson RL. Increased bactericidal activity of dilute preparations of povidone-iodine solutions. J Clin Microbiol. 1982; 15:632–9.

165. Berkelman RL, Lewin S, Allen JR, et al. Pseudobacteremia attributed to contamination of povidone-iodine with *Pseudomonas cepacia*. Ann Intern Med. 1981; 95:32–6.

166. Mutch RS, Hutson PR. Stability of antipyrine plus caffeine in intravenous solution. Am J Hosp Pharm. 1991;48: 1267–70.

167. Gupta VD. Stability of cocaine hydrochloride solutions at various pH values as determined by high-pressure liquid chromatography. Int J Pharm. 1982;10:249–57.

168. Newton DW, Rogers AG, Becker CH, et al. Extemporaneous preparation of methyldopa in two syrup vehicles. Am J Hosp Pharm. 1975;32:817–21.

169. Osborn E, Baum JL, Ernst C, et al. The stability of ten antibiotics in artificial tear solutions. Am J Ophthamol. 1976;82:775–80.

170. Smith SG. A folic acid solution for oral use. Pharm J. 1976;216:109.

171. Williams CC. Stability of aqueous perchlorate formulations. Am J Hosp Pharm. 1977;34:93–5.

172. Newton DW, Schulman SG, Becker CH. Limitations of compounding diazepam suspensions from tablets. Am J Hosp Pharm. 1976;33:450–2.

173. Vulovic N, Primorac M, Stupar M, et al. Some studies on the preservation of indomethacin suspensions intended for ophthalmic use. Pharmazie. 1990; 45:678–9.

174. Kumer KP, Okonomah AD, Bradshaw WG, et al. Stability of ketoconazole in ethanolic solutions. Drug Dev Ind Pharm. 1991;17:577–80.

175. Lacina NC, Orr RJ, Peters LS, et al. Topical clindamycin for acne. Part 1: current prescribing practices. Am Pharm. 1978;NS18:30–3.

176. Orr RJ, Lacina NC, Peters LS, et al. Topical clindamycin for acne. Part 2: guidelines for extemporaneous compounding. Am Pharm. 1978;NS18:23–6.

177. Algra RJ, Rosen T, Waisman M. Topical clindamycin in acne vulgaris: safety and stability. Arch Dermatol. 1977; 113:1390–1.

178. Horn JR, Anderson GD. Stability of an extemporaneously compounded cisapride suspension. Clin Ther. 1994;16:169–72.

179. Horn JR. Caution with preparing suspensions. Am Pharm. 1994;NS34:4–5.

180. Parasrampuria J, Gupta VD. Development of oral liquid dosage forms of acetazolamide. J Pharm Sci. 1990; 79:835–6.

181. Parasrampuria J, Gupta VD. Preformulation studies of acetazolamide: effect of pH, two buffer species, ionic strength, and temperature on its stability. J Pharm Sci. 1989;78:855–7.

182. Ozuna J, Friel P. Effect of enteral tube feeding on serum phenytoin levels. J Neurosurg Nursing. 1984;16:289–91.

183. Maynard GA, Jones KM, Guidry JR. Phenytoin absorption from tube feedings. Arch Intern Med. 1987;147:1821.

184. Nishimura LY, Armstrong EP, Plezia PM, et al. Influence of enteral feedings on phenytoin sodium absorption from capsules. Drug Intell Clin Pharm. 1988; 22:130–3.

185. Krueger KA, Garnett WR, Comstock TJ, et al. Effect of two administration schedules of an enteral nutrient formula on phenytoin bioavailability. Epilepsia. 1987; 28:706–12.

186. Seifert CF, McGoodwin PL, Allen LV. Phenytoin recovery from percutaneous endoscopic gastrostomy Pezzer catheters after long-term in vitro administration. J Parenter Enter Nutr. 1993;17:370–4.

187. Rosen A, Machera P. The effect of protein on the dissolution of phenytoin. J Pharm Pharmacol. 1984;36:723–7.

188. Lader E, Yang L, Clarke A. Warfarin dosage and vitamin K in Osmolite. Ann Intern Med. 1980;93:373–4.

189. Watson AJM, Pegg M, Green JRB. Enteral feeds may antagonize warfarin. Br Med J. 1984;288:557.

190. O'Reilly RA, Rytand DA. "Resistance" to warfarin due to unrecognized vitamin K supplementation. N Engl J Med. 1980; 303:160–1.

191. Lee M, Schwartz RN, Sharifi R. Warfarin resistance and vitamin K. Ann Intern Med. 1981;94:140–1.

192. Landau J, Moulds RFW. Warfarin resistance caused by vitamin K in intestinal feeds. Med J Aust. 1982;ii: 263–4.

193. Parr MD, Record KE, Griffith GL, et al. Effect of enteral nutrition on warfarin therapy. Clin Pharm. 1982;1:274–6.

194. Westfall LK. An unrecognized cause of warfarin resistance. Drug Intell Clin Pharm. 1981;15:131.

195. Martin JE, Lutomski DM. Warfarin resistance and enteral feedings. J Parenter Enter Nutr. 1989;13:206–8.

196. Mueller DW. Improved extemporaneous formulation of cyclosporine ophthalmic drops. Am J Hosp Pharm. 1994;51:3080–1.

197. Henderson LM, Johnson CE, Berardi RR. Stability of mesalamine in rectal suspension diluted with distilled water. Am J Hosp Pharm. 1994;51:2955–7.

198. Piccolo ML, Toossi Z, Goldman M. Effect of coadministration of a nutritional supplement on ciprofloxacin absorption. Am J Hosp Pharm. 1994;51:2697–9.

199. MacDonald JL, Johnson CE, Jacobson P. Stability of isradipine in an extemporaneously compounded oral liquid. Am J Hosp Pharm. 1994;51:2409–11.

200. Nahata MC, Morosco RS, Hipple TF. Stability of captopril in liquid containing ascorbic acid or sodium ascorbate. Am J Hosp Pharm. 1994;51:1707–8.

201. Gutteridge C, Kuhn RJ. Compatibility of 10% sodium benzoate plus 10% sodium phenylacetate with various flavored vehicles. Am J Hosp Pharm. 1994; 51:2508, 2510.

202. Donnelly RF, Tirona RG. Stability of citrated caffeine injectable solution in glass vials. Am J Hosp Pharm. 1994;51:512–4.

203. Vandenbossche GMR, Vanhaecke E, De Muynck C, et al. Stability of topical erythromycin formulations. Int J Pharm. 1991;67:195–9.

204. Ross J, Abate MA. Topical vancomycin for the treatment of Staphylococcus epidermidis and methicillin-resistant Staphylococcus aureus conjunctivitis. DICP Ann Pharmacother. 1990;24:1050, 1053.

205. Veerman MW, Hatton RC, Knight ME. Relative bioavailability and stability of an extemporaneously prepared valproic acid emulsion. Clin Pharm. 1991;10:382–4.

206. Lukton A, Weisbrod R, Schlesinger J. The effect of imidazole ring substitution on the rate of its photooxidation. Photochem Photobiol. 1965;4:277–9.

207. Straight R, Spikes JD. Sensitized photooxidation of amino acids: effects on the reactivity of their primary amine groups with fluorescamine and ophthalaldehyde. Photochem Photobiol. 1978;27:565–9.

208. Verburg KM, Henry DP. Binding of histamine by glass surfaces. Agents Actions. 1984;14:633–6.

209. Satake K, Ando S, Fujita H. Bacterial oxidation of some primary amines. J Biochem. 1953;40:299–315.

210. Rothberg S, Hayaishi O. Studies on oxygenases enzymatic oxidation of imadazoleacetic acid. J Biol Chem. 1957; 229:897–903.

211. Ehrsson H, Eksborg S, Wallin I, et al. Degradation of chlorambucil in aqueous solution. J Pharm Sci. 1980;69:1091–4.

212. Gupta VD. Effect of vehicles and other active ingredients on stability of hydrocortisone. J Pharm Sci. 1978; 67:299–302.

213. Gal P, Layson R. Interference with oral theophylline absorption by continuous nasogastric feedings. Ther Drug Monitor. 1986;8:421–3.

214. Bass J, Miles MV, Tennison MB, et al. Effects of enteral tube feeding on the absorption and pharmacokinetic profile of carbamazepine suspension. Epilepsia. 1989;30:364–9.

215. Timmins P, Jackson IM, Wang YJ. Factors affecting captopril stability in aqueous solution. Int J Pharm. 1982;11:329–36.

216. Hajratwala BR, Dawson JE. Kinetics of indomethacin degradation I: presence of alkali. J Pharm Sci. 1977;66:27–9.

217. Newton DW, Rogers AG, Becker CH, et al. Evaluation of preparations of patent blue (Alphazurine 2G) dye for parenteral use. Am J Hosp Pharm. 1975;32:912–7.

218. El-Shattawy HH. Effect of various ophthalmic ointment bases on carbenicillin and gentamicin stability. Drug Dev Ind Pharm. 1982;8:487–96.

219. Tomida H, Kuwada N, Kiryu S. Hydrolysis of indomethacin in Pluronic F-127 gels. Acta Pharm Suec. 1988;25:87–96.

220. Udeani GO, Bass J, Johnston TP. Compatibility of oral morphine sulfate solution with enteral feeding products. Ann Pharmacother. 1994;28:451–5.

221. Nahata MC, Zingarelli JR, Durrell DE. Stability of caffeine injection in intravenous admixtures and parenteral nutrition solutions. DICP Ann Pharmacother. 1989;23:466–7.

222. Zallman JA, Lee DP, Jeffrey PL. Liquid nutrition as a cause of warfarin resistance. Am J Hosp Pharm. 1981;38:1174.

223. Brooke D, Davis RE, Bequette RJ. Chemical stability of cyclophosphamide in aromatic elixir USP. Am J Hosp Pharm. 1973;30:618–20.

224. Dimitrova E, Bogdanova S, Minkov E, et al. High-molecular weight polyoxyethylene as an additive in ophthalmic solutions. Int J Pharm. 1993;93:21–6.

225. Johnson K, Cazee C, Gutch C, et al. Sodium polystyrene sulfonate resin candy for control of potassium in chronic dialysis patients. Clin Nephrol. 1976;5:266–8.

226. Raitt JR, Hotaling WH. Preparation of stable prednisone suspension. Am J Hosp Pharm. 1973;30:923–4.

227. Bialer MG, Baron EJ, Harper RG. Erythromycin bioactivity is stable in ophthalmic ointment used for prophylaxis of neonatal gonococcal conjunctivitis. Antimicrob Agents Chemother. 1987;31:954–5.

228. Fleischer AB, Hoover DL, Khan JA, et al. Topical vancomycin formulation for methicillin-resistant Staphylococcus epidermidis blepharoconjunctivitis. Am J Ophthalmol. 1986;101:283–7.

229. Gupta VD. High-pressure liquid chromatographic evaluation of aqueous vehicles for preparation of prednisolone and prednisone liquid dosage forms. J Pharm Sci. 1979;68:908–10.

230. Bryant CA, Neufeld NJ. Differences in vitamin A content of enteral feeding solutions following exposure to a polyvinyl chloride enteral feeding system. J Parenter Enter Nutr. 1982;6:403–5.

231. Davis AT, Fagerman KE, Downer FD, et al. Effect of enteral feeding bag composition and freezing and thawing upon vitamin stability in an enteral feeding solution. J Parenter Enter Nutr. 1986;10:245–6.

232. Niemiec PW, Vanderveen TW, Morrison JI, et al. Gastrointestinal disorders caused by medication and electrolyte solution osmolality during enteral nutrition. J Parenter Enter Nutr. 1983;7:387–9.

233. Dickerson RN, Melnik G. Osmolality of oral drug solutions and suspensions. Am J Hosp Pharm. 1988;45:832–4.

234. Leff RD, Roberts RJ. Effect of intravenous fluid and drug solution coadmistration on final-infusate osmolality, specific gravity, and pH. Am J Hosp Pharm. 1982;39:468–71.

235. Ernst JA, Williams JM, Glick MR, et al. Osmolality of substances used in the intensive care nursery. Pediatrics. 1983;72:347–52.

236. Dinel BA, Ayotte DL, Behme RJ, et al. Stability of antibiotic admixtures frozen in minibags. Drug Intell Clin Pharm. 1977; 11:542–8.

237. Holmes CJ, Ausman RK, Walter CW, et al. Activity of antibiotic admixtures subjected to different freeze-thaw treatments. Drug Intell Clin Pharm. 1980;14:353–7.

238. Marble DA, Bosso JA, Townsend RJ. Compatibility of clindamycin phosphate with amikacin sulfate at room temperature and with gentamicin sulfate and tobramycin sulfate under frozen conditions. Drug Intell Clin Pharm. 1986;20:960–3.

239. Weiner B, McNeely DJ, Kluge RM, et al. Stability of gentamicin sulfate following unit dose packaging, Am J Hosp Pharm. 1976;33:1254–9.

240. Kresel JJ, Smith AL, Siber GR. Stability of gentamicin in plastic syringes. Am J Hosp Pharm. 1997;34:570.

241. Nahata MC, Hipple TF, Strausbaugh SD. Stability of gentamicin sulfate diluted in 0.9% sodium chloride injection in glass syringes. Hosp Pharm. 1987;22:1131–2.

242. Geller JL, Gaulin BD, Barreira PJ. A practitioner's guide to use of psychotropic medication in liquid form. Hosp Comm Psychiatry. 1992;43:969–71.

243. Kulhanek F, Linde OK, Meisenberg G., et al. Precipitation of antipsychotic drugs in interactions with coffee or tea. Lancet. 1979;2:1130–1.

244. Hirsch SR. Precipitation of antipsychotic drugs in interaction with coffee or tea. Lancet. 1979;2:1131.

245. Andrews CD, Essex A. Captopril suspension. Pharm J. 1986;237:734–5.

246. Nahata MC, Morosco RS, Hipple TF. Effect of preparation method and storage on rifampin concentration in suspensions. Ann Pharmacother. 1994;28:182–5.

247. Pesko LJ. Compounding: nizatidine oral liquid. Am Druggist. 1994;209:53–4.

248. Pesko LJ. Compounding: fluconazole oral. Am Druggist. 1993;209:53–4.

249. Pesko LJ. Compounding: baclofen oral liquid. Am Druggist. 1993;207:56.

250. Pesko LJ. Compounding: mitomycin ophthalmic. Am Druggist. 1992;207:63.

251. Pesko LJ. Compounding: hydroxychloroquine. Am Druggist. 1993;207:57.

252. Pesko LJ. Paroxysmal rhinorrhea. Am Druggist. 1991;203:35.

253. Pesko LJ. TAC solution. Am Druggist. 1991;203:110.

254. Pesko LJ. Povidone-iodine/sugar paste. Am Druggist. 1991;203:62.

255. Pesko LJ. Compounding: scalp psoriasis. Am Druggist. 1992;205:55.

256. Kerr LE. Oral liquid neuroleptics. J Psychosocial Nurs. 1986;24:33–5.

257. Murray JB, Al-Shora HI. Stability of cocaine in aqueous solution. J Clin Pharm. 1978;3:1–6.

258. Phelps SJ, Dorf A, Catarau EM, et al. An interaction between enteral formulas containing medium-chain triglycerides and the Valleylab volumetric infusion pump set. Nutr Supp Serv. 1986;6:16–9.

259. Gupta VD, Stewart KR, Bethea C. Stability of hydralazine hydrochloride in aqueous vehicles. J Clin Hosp Pharm. 1986;11:215–23.

260. Hebron B, Scott H. Shelf life of cefuroxime eye-drops when dispensed in artificial tear preparations. Int J Pharm Pract. 1993;2:163–7.

261. Suleiman MS, Najib NM, Abdelhameed ME. Stability of diltiazem hydrochloride in aqueous sugar solutions. J Clin Pharm Ther. 1988;13:417–22.

262. Sewell GJ, Palmer AJ. The formulation and stability of a unit-dose oral vitamin K1 preparation. J Clin Pharm Ther. 1988;13:73–6.

263. Lee TY, Notari RE. Kinetics and mechanism of captopril oxidation in aqueous solution under controlled oxygen partial pressure. Pharm Res. 1987;4:98–103.

264. Brown GC, Kayes JB. The stability of suspensions prepared extemporaneously from solid oral dosage forms. J Clin Pharm. 1976;1:29–37.

265. Strom JG, Kalu AU. Formulation and stability of diazepam suspension compounded from tablets. Am J Hosp Pharm. 1986;43:1489–91.

266. Fiscella RG, Proffitt DF, Weisbecker CA. Stability of mitomycin for ophthalmic use. Am J Hosp Pharm. 1992;49:2440.

267. Williams CL, Sanders PL, Laizure SC, et al. Stability of ondansetron hydrochloride in syrups compounded from tablets. Am J Hosp Pharm. 1994;51:806–9.

268. Kinzel PE, Trausch DE, Copfer AL. Otic administration of amphotericin B 0.25% in sterile water. Ann Pharmacother. 1994;28:333–5.

269. Nahata MC. Stability of labetalol hydrochloride in distilled water, simple syrup, and three fruit juices. DICP Ann Pharmacother. 1991;25:465–9.

270. Brower JF, Juenge EC, Page DP, et al. Decomposition of aminophylline in suppository formulations. J Pharm Sci. 1980;69:942–5.

271. Gupta VD. Effect of ethanol, glycerol, and propylene glycol on the stability of phenobarbital sodium. J Pharm Sci. 1984;73:1661–2.

272. Gupta VD, Parasrampuria J. Quantitation of acetazolamide in pharmaceutical dosage forms using high-performance liquid chromatography. Drug Dev Ind Pharm. 1987;13:147–57.

273. Parasrampuria J, Gupta VD, Stewart KR. Stability of acetazolamide sodium in 5% dextrose or 0.9% sodium chloride injection. Am J Hosp Pharm. 1987;44:358–60.

274. Trissel LA, Martinez JF. Compatibility of allopurinol sodium with selected drugs during simulated Y-site administration. Am J Hosp Pharm. 1994;51:1792–9.

275. Parker EA. Compatibility digest. Am J Hosp Pharm. 1970;27:67–9.

276. Edward M. pH—an important factor in the compatibility of additives in intravenous therapy. Am J Hosp Pharm. 1967;24:440–9.

277. Swerling R. Dilution of oral and intravenous aminophylline preparations. Am J Hosp Pharm. 1981;38:1359–60.

278. Boak LR. Aminophylline stability. Can J Hosp Pharm. 1987;40:155.

279. Nahata MC, Morosco RS, Hipple TF. Stability of aminophylline in bacteriostatic water for injection stored in plastic syringes at two temperatures. Am J Hosp Pharm. 1992;49:2962–3.

280. Dalton-Bunnow MF, Halvacks FJ. Update on room temperature stability of drug products labeled for refrigerated storage. Am J Hosp Pharm. 1990; 47:2522–4.

281. Bonner DP, Mechlinski W, Schaffner CP. Stability studies with amphotericin B and amphotericin B methyl ester. J Antibiotics. 1975;28:132–5.

282. Shadomy S, Brummer DL, Ingroff AV. Light sensitivity of prepared solutions of amphotericin B. Am Rev Resp Dis. 1973;107:303–4.

283. Block ER, Bennett JE. Stability of amphotericin B in infusion bottles. Antimicrob Agents Chemother. 1973; 4:648–9.

284. Riffkin C. Incompatibilities of manufactured parenteral products. Am J Hosp Pharm. 1963;20:19–22.

285. Huber RC, Riffkin C. Inline final filters for removing particles from amphotericin B infusions. Am J Hosp Pharm. 1975;32:173–6.

286. Leff RD, Roberts RJ. Effect of intravenous fluid and drug solution coadministration on final-infusate osmolality, specific gravity, and pH. Am J Hosp Pharm. 1982;39:468–71.

287. Ernst JA, Williams JM, Glick MR, et al. Osmolality of substances used in the intensive care nursery. Pediatrics. 1983;72:347–52.

288. Lynn B. Pharmaceutics of semi-synthetic penicillins. Chem Drug. 1967;187:134–6.

289. Savello DR, Shangraw RF. Stability of sodium ampicillin solutions in the frozen and liquid states. Am J Hosp Pharm. 1971;28:754–9.

290. Stratton M, Sandmann BJ. Stability studies of ampicillin sodium intravenous fluids using optical activity. Bull Parenter Drug Assoc. 1975;29:286–95.

291. Warren E, Snyder RJ, Thompson CO, et al. Stability of ampicillin in intravenous solutions. Mayo Clin Proc. 1972;47:34–5.

292. Hou JP, Poole JW. Kinetics and mechanism of degradation of ampicillin in solution. J Pharm Sci. 1969;58:447–54.

293. Riffanti EF, King JC. Effect of pH on the stability of sodium ampicillin solutions. Am J Hosp Pharm. 1974;31:745–51.

294. Stjernstrom G, Olson OT, Nyqvist H, et al. Studies on the stability and compatibility of drugs in infusion fluids. VI. Factors affecting the stability of ampicillin. Acta Pharm Suec. 1978;15:33–50.

295. Mitra AK, Narurkar MM. Kinetics of azathioprine degradation in aqueous solution. Int J Pharm. 1986;35:165–71.

296. Johnson CA, Porter WA. Compatibility of azathioprine sodium with intravenous fluids. Am J Hosp Pharm. 1981;38:871–5.

297. Jaffe JM, Certo NM, Pirakitikulr P, et al. Stability of several brands of ampicillin and penicillin V potassium oral liquids following reconstitution. Am J Hosp Pharm. 1976;33:1005–10.

298. Lund W, Cowe HJ. Stability of dry powder formulations. Pharm J. 1986; 237:179–80.

299. Anon. Moisture hardens carbamazepine tablets, FDA finds. Am J Hosp Pharm. 1990;47:958.

300. Lowe MMJ. More information on hardening of carbamazepine tablets. Am J Hosp Pharm. 1991;48:2130–1.

301. Zia H, Tehrani M, Zargarbashi R. Kinetics of carbenicillin degradation in aqueous solution. Can J Pharm Sci. 1974;9:112–7.

302. Acred P, Brown DM, Knudsen ET, et al. New semi-synthetic penicillin active against pseudomonas pyocyanea. Nature (London). 1967;215:25–30.

303. Sterchele JA. Update on stability guidelines for routinely refrigerated drug products. Am J Hosp Pharm. 1987;44:2698, 2701.

304. Arbus MH. Room temperature stability guidelines for carmustine. Am J Hosp Pharm. 1988;45:531.

305. Chan KK, Zackheim HS. Stability of nitrosourea solutions. Arch Dermatol. 1973;107:298.

306. Kleinman LM, Davignon JP, Cradock JC, et al. Investigational drug information. Drug Intell Clin Pharm. 1976;10:48–9.

307. Davignon JP, Yang KW, Wood HB, et al. Formulation of three nitrosoureas for intravenous use. Cancer Chemother Rep. 1973;4:7–11.

308. Laskar PA, Ayres JW. Degradation of carmustine in aqueous media. J Pharm Sci. 1977;66:1073–6.

309. Frederiksson K, Lundgren P, Landersjo L. Stability of carmustine—kinetics and compatibility during administration. Acta Pharm Suec. 1986;23:115–24.

310. Benvenuto JA, Anderson RW, Kerkof K, et al. Stability and compatibility of antitumor agents in glass and plastic containers. Am J Hosp Pharm. 1981;38:1914–8.

311. Benvenuto JA, Adams SC, Vyas HM, et al. Pharmaceutical issues in infusion chemotherapy stability and compatiblity, in Lokich JJ (ed). Cancer chemotherapy by infusion. Chicago:Precept Press. 1987;100–13.

312. Bornstein M, Thomas PN, Coleman DL, et al. Stability of parenteral solutions of cefazolin sodium. Am J Hosp Pharm. 1974;31:296–8.

313. Borst DL, Sesin GP, Cersosimo RJ. Stability of selected beta-lactam antibiotics stored in plastic syringes. NITA J. 1987;10:369–72.

314. Senholzi CS, Kerus MP. Crystal formation after reconstituting cefazolin sodium with 0.9% sodium chloride injection. Am J Hosp Pharm. 1985;42:129–30.

315. Bornstein M. Templeton RJ. Crystal formation after reconstituting cefazolin sodium with 0.9% sodium chloride injection. Am J Hosp Pharm. 1985;42:2436.

316. Yamana T, Tsuji A. Comparative stability of cephalosporins in aqueous solutions: kinetics and mechanisms of degradation. J Pharm Sci. 1976;65:1563–74.

317. Parker EA. Compatibility digest. Am J Hosp Pharm. 1970;27:492–3.

318. Boylan JC, Simmons JL, Winely CL. Stability of frozen solutions of sodium cephalothin and cephaloridine. Am J Hosp Pharm. 1972;29:687–9.

319. Dinel BA, Ayotte DL, Behme RJ, et al. Stability of antibiotic admixtures frozen in minibags. Drug Intell Clin Pharm. 1977;11:542–8.

320. Holmes CJ, Ausman RK, Walter CW, et al. Activity of antibiotic admixtures subjected to different freeze-thaw treatments. Drug Intell Clin Pharm. 1980;14:353–7.

321. Kleinberg ML, Stauffer GL, Prior RB, et al. Stability of antibiotics frozen and stored in disposable hypodermic syringes. Am J Hosp Pharm. 1980;37:1087–8.

322. Rusmin S, Welton S, DeLuca P, et al. Effect of inline filtration on the potency of drugs administered intravenously. Am J Hosp Pharm. 1977;34:1071–4.

323. Ennis CE, Merritt RJ, Neff ON. In vitro study of inline filtration of medications commonly administered to pediatric cancer patients. J Parenter Enter Nutr. 1983;7:156–8.

324. Kane M, Jay M, DeLuca PP. Binding of insulin to a continuous ambulatory peritoneal dialysis system. Am J Hosp Pharm. 1986;43:81–8.

325. Kirschenbaum BE, Latiolais CJ. Injectable medications — a guide to stability and reconstitution. New York:McMahon Group. 1993.

326. Trissel LA, Martinez JF, Xu QA. Data on file. Pharmaceutical Analysis Laboratory, University of Texas, M. D. Anderson Cancer Center. Jan. 24, 1994.

327. Ammar HO, Salama HA, El-Nimr AE. Studies on the stability of injectable solutions of some phenothiazines, part 1:effect of pH and buffer systems. Pharmazie. 1975;30:368–9.

328. D'Arcy PF, Thompson KM. Stability of chlorpromazine hydrochloride added to intravenous infusion fluids. Pharm J. 1973;210:28.

329. DeVane CL, Wailand LA. Stability of chlorpromazine in five milliliter vials. Can J Hosp Pharm. 1984;37:9.

330. Kowaluk EA, Roberts MS, Blackburn HD, et al. Interactions between drugs and polyvinyl chloride infusion bags. Am J Hosp Pharm. 1981;38:1308–14.

331. Kowaluk EA, Roberts MS, Polack AE. Interactions between drugs and intravenous delivery systems. Am J Hosp Pharm. 1982;39:460–7.

332. Kowaluk EA, Roberts MS, Polack AE. Drug loss in polyolefin infusion systems. Am J Hosp Pharm. 1983;40:118–9.

333. Niemiec PW, Vanderveen TW. Compatibility considerations in parenteral nutrient solutions, Am J Hosp Pharm. 1984;41:893–911.

334. Riebe KW, Oesterling TO. Parenteral development of clindamycin-2-phosphate. Bull Parenter Drug Assoc. 1972; 26:139–45.

335. Zbrozek AS, Marbel DA, Bosso JA, et al. Compatibility and stability of clindamycin phosphate-aminoglycoside combinations within polypropylene syringes. Drug Intell Clin Pharm. 1987;21:806–10.

336. Marble DA, Bosso JA, Townsend RJ. Compatibility of clindamycin phosphate with aztreonam in polypropylene syringes and with cefoperazone sodium, cefonicid sodium,

and cefuroxime sodium in partial-fill glass bottles. Drug Intell Clin Pharm. 1988;22:54–7.

337. Lesko LJ, Marion A, Ericson J, et al. Stability of trimethoprim–sulfamethoxazole injection in two infusion fluids. Am J Hosp Pharm. 1981;38:1004–6.

338. Kaufman MB, Scavone JM, Foley JJ. Stability of undiluted trimethoprim–sulfamethoxazole for injection in plastic syringes. Am J Hosp Pharm. 1992; 49:2782–3.

339. Brooke D, Scott JA, Bequette RJ. Effect of briefly heating cyclophosphamide solutions. Am J Hosp Pharm. 1975; 32:44–5.

340. Kirk B, Melia CD, Wilson JV, et al. Chemical stability of cyclophosphamide injection. Br J Parenter Ther. 1984;5:90–7.

341. Ptachcinski RJ, Logue LW, Burkhart GJ, et al. Stability and availability of cyclosporine in 5% dextrose injection or 0.9% sodium chloride injection. Am J Hosp Pharm. 1986;43:94–7.

342. Parr MD, Barton SD, Haver VM, et al. Cyclosporine binding to components in medication administration sets. Drug Intell Clin Pharm. 1988;22:173–4.

343. Pearson SD, Trissel LA. Leaching of diethylhexyl phthalate from polyvinyl chloride containers by selected drugs and formulation components. Am J Hosp Pharm. 1993;50:1405–9.

344. Venkataramanan R, Burkhart GJ, Ptachcinski RJ, et al. Leaching of diethylhexyl phthalate from polyvinyl chloride bags into intravenous cyclosporine solution. Am J Hosp Pharm. 1986;43:2800–2.

345. Bretschneider H. Osmolalities of commercially supplied drugs often used in anesthesia. Anesthes Analg. 1987; 66:361–2.

346. Mayer W, Erbe S, Voigt R. Analysis and stability of various pharmaceutically interesting benzodiazepines. Pharmazie. 1972;27:32–42.

347. Levin HJ, Fieber RS, Levi RS. Stability data for Tubex filled by hospital pharmacists. Hosp Pharm. 1973;8:310–1.

348. Newton DW, Driscoll DF, Goudreau JL, et al. Solubility characteristics of diazepam in aqueous admixture solutions: theory and practice. Am J Hosp Pharm. 1981;38:179–82.

349. Wisnes M, Jeppsson R, Sjoberg B. Diazepam absorption to infusion sets and plastic syringes. Acta Anaesthesiol Scand. 1981;25:93–6.

350. Smith FM, Nuessle NO. Stability of diazepam injection repackaged in glass unit-dose syringes. Am J Hosp Pharm. 1982;39:1687–90.

351. Speaker TJ, Turco SJ, Nardone DA, et al. A study of the interaction of selected drugs and plastic syringes. J Parenter Sci Technol. 1991;45:212–7.

352. DeMuynck C, De Vroe C, Remon JP, et al. Binding of drugs to end-line filters: a study of four commonly administered drugs in intensive care units. J Clin Pharm Ther. 1988;13:335–40.

353. Kuhlman J, Abshagen U, Rietbrock N. Cleavage of glycosidic bonds of digoxin and derivatives as function of pH and time. Naunyn Schmiedebergs Arch Pharmacol. 1973;276:149–56.

354. Gault MH, Charles JD, Sugden DL, et al. Hydrolysis of digoxin by acid. J Pharm Pharmacol. 1977;29:27–32.

355. Sternson LA, Shaffer RD. Kinetics of digoxin stability in aqueous solution. J Pharm Sci. 1978;67:327–30.

356. Khalil SA, El-Masury S. Instability of digoxin in acid medium using a nonisotopic method. J Pharm Sci. 1978; 67:1358–60.

357. Caille G, Dube LM, Theoret Y, et al. Stability study of diltiazem and two of its metabolites using a high performance liquid chromatographic method. Biopharma Drug Dispos. 1989;10:107–14.

358. Ray JB, Newton DW, Nye MT, et al. Droperidol stability in intravenous admixtures. Am J Hosp Pharm. 1983;40:94–7.

359. Seargent LE, Kobrinsky NL, Sus CJ, et al. In vitro stability and compatibility of daunorubicin, cytarabine, and etoposide. Cancer Treat Rep. 1987;71:1189–92.

360. Beijnen JH, Beijnen-Bandhoe AU, Dubbelman AC, et al. Chemical and physical stability of etoposide and teniposide in commonly used infusion fluids. J Parenter Sci Technol. 1991;45:108–12.

361. Ross MB. Additional stability guidelines for routinely refrigerated drug products. Am J Hosp Pharm. 1988;45:1498–9.

362. Bullock LS, Fitzgerald JF, Mazur HI. Stability of intravenous famotidine stored in polyvinyl-chloride syringes. DICP Ann Pharmacother. 1989;23:588–90.

363. Shea BF, Souney PF. Stability of famotidine frozen in polypropylene syringes. Am J Hosp Pharm. 1990;47:2073–4.

364. Theuer H, Scherbel G, Windsheimer U. Stabilitatsuntersuchungen von fentanylcitrat i.v. Krankenhaus pharmazie. 1991; 12:233–45.

365. Sesin GP, Millette LA, Weiner B. Stability study of 5-fluorouracil following repackaging in plastic disposable syringes and multidose vials. Am J IV Ther Clin Nutr. 1982;9:23–5, 29, 30.

366. Williams DA. Stability and compatibility of admixtures of antineoplastic drugs. In Lokich JJ (ed). Cancer chemotherapy by infusion, 2nd ed. Chicago:Precept Press; 1990.

367. Stiles ML, Allen LV Jr, Tu YH. Stability of fluorouracil administered through four portable infusion pumps, Am J Hosp Pharm. 1989;46:2036–40.

368. Stolk LML, Chandi LS. Stabiliteit van fluorouracil (0,5-5 mg) in polypropyleen spuiten bij −20 °C. Ziekenhuisfarmacie. 1991;7:12–3.

369. Barker A, Hebron BS, Beck PR, et al. Folic acid and total parenteral nutrition. J Parenter Enter Nutr. 1984;8:3–7.

370. Romankiewicz JA, McManus J, Gotz VP, et al. Medications not to be refrigerated. Am J Hosp Pharm. 1979;36:1541–5.

371. Neil JM, Fell AF, Smith G. Evaluation of the stability of frusemide in intravenous infusions by reversed-phase high-performance liquid chromatography. Int J Pharm. 1984;22:105–26.

372. Marwaha RK, Johnson BF, Wright GE. Simple stability-indicating assay for histamine solutions. Am J Hosp Pharm. 1985;42:1568–71.

373. Halasi S, Nairn JG. Stability of hydralazine hydrochloride in parenteral solutions. Can J Hosp Pharm. 1990;43:237–41.

374. Enderlin G. Discoloration of hydralazine injection. Am J Hosp Pharm. 1984;41:634.

375. Smith G, Hasson K, Clements JA. Effects of ascorbic acid and disodium edetate on the stability of isoprenaline hydrochloride injection. J Clin Hosp Pharm. 1984;9:209–15.

376. Newton DW, Fung EYY, Williams DA. Stability of five catecholamines and terbutaline sulfate in 5% dextrose injection in the absence and presence of aminophylline. Am J Hosp Pharm. 1981;38:1314–9.

377. Parker EA. Compatibility digest. Am J Hosp Pharm. 1974;31:775.

378. Leach JK, Strickland RD, Millis DL, et al. Biological activity of dilute isoproterenol solution stored for long periods in plastic bags. Am J Hosp Pharm. 1977;34:709–12.

379. Yuen PH, Taddei CR, Wyka BE, et al. Compatibility and stability of labetalol hydrochloride in commonly used intravenous solutions. Am J Hosp Pharm. 1983;40:1007–9.

380. NCI investigational drugs pharmaceutical data. Bethesda, MD: National Cancer Institute, 1990.

381. Hartshorn EA: Oxidation of methyldopa hydrochloride in alkaline media. Am J Hosp Pharm. 1975;32:244.

382. Parker EA. Compatibility digest. Am J Hosp Pharm. 1974;31:1076.

383. Parker EA. Oxidation of methyldopate hydrochloride in alkaline media. Am J Hosp Pharm. 1975;32:244.

384. Forman JK, Souney PF. Visual compatibility of midazolam hydrochloride with common preoperative injectable medications. Am J Hosp Pharm. 1987;44:2298–9.

385. Edwards D, Selkirk AB, Taylor RB. Determination of the stability of mitomycin C by high-performance liquid chromatography. Int J Pharm. 1979;4:21–6.

386. Beijnen JH, Underberg WJM. Degradation of mitomycin C in acidic solution. Int J Pharm. 1985;24:219–29.

387. Beijnen JH, den Hartigh J, Underberg WJM. Quantitative aspects of the degradation of mitomycin C in alkaline solution. J Pharm Biomed Anal. 1985;3:59–69.

388. Beijnen JH, Fokkens RH, Rosing H, et al. Degradation of mitomycin C in acid phosphate and acetate buffer. Int J Pharm. 1986;32:111–21.

389. Beijnen JH, Lingeman H, Van Munster HA, et al. Mitomycin antitumor agents: a review of their physicochemical and analytical properties and stability. J Pharm Biomed Anal. 1986;4:275–95.

390. Stolk LML, Fruijter A, Umans R. Stability after freezing and thawing of solutions of mitomycin C in plastic minibags for intravesical use. Pharm Weekbl Sci Ed. 1986;8:286–8.

391. Gove LF, Gordon NH, Miller J, et al. Prefilled syringes for self-administration of epidural opiates. Pharm J. 1985;234:378–9.

392. Hung CT, Young M, Gupta PK. Stability of morphine solutions in plastic syringes determined by reversed-phase ion-pair liquid chromatography. J Pharm Sci. 1988;77:719–23.

393. Walker SE, Coons C, Matte D, et al. Hydromorphone and morphine stability in portable infusion pump cassettes and minibags. Can J Hosp Pharm. 1988; 41:177–82.

394. Leak RE, Woodford JD. Pharmaceutical development of ondansetron injection, Eur J Cancer Clin Oncol. 1989;25(Suppl 1):S67–S69.

395. MacKinnon JWM, Collin DT. The chemistry of ondansetron. Eur J Cancer Clin Oncol. 1989;25(Suppl 1):S61.

396. Jarosinski PF, Hirschfield S. Precipitation of ondansetron in alkaline solutions. N Engl J Med. 1991;325:1315–6.

397. Trissel LA, Tramonte SM, Grilley BJ. Visual compatibility of ondansetron hydrochloride with other selected drugs during simulated Y-site injection. Am J Hosp Pharm. 1991;48:988–92.

398. Ong JTH, Kostenbauder HB. Effect of self-association on rate of penicillin G degradation in concentrated aqueous solutions. J Pharm Sci. 1975;64:1378–80.

399. Parker EA. Compatibility digest. Am J Hosp Pharm. 1969;26:543–4.

400. Glascock JC, DiPiro JT, Cadwallader DE, et al. Stability of terbutaline sulfate repackaged in disposable plastic syringes. Am J Hosp Pharm. 1987;44:2291–3.

401. Raymond GG. Stability of terbutaline sulfate injection stored in plastic tuberculin syringes, Drug Intell Clin Pharm. 1988;22:303–5.

402. Askerud L, Finholt P, Karlsen J. Intravenous infusion of theophylline in 5% dextrose solution—formulation and stability. Medd Nor Farm Selsk. 1981;43:17–24.

403. Schneider JS, Ouellette SM. Sucralfate administration via nasogastric tube. N Engl J Med. 1984;310:990.

404. Jeter JH, Mueller D. TAC Gel—a sterile formulation. Ann Emerg Med. 1994; 23:600.

405. Poochikian GK, Cradock JC. Stability of Brompton mixtures: determination of heroin (diacetylmorphine) and cocaine in the presence of their hydrolysis products. J Pharm Sci. 1980;69:637–9.

406. Soy D, Lopez MC, Salvador L, et al. Stability of an oral midazolam solution for premedication in paediatric patients. Pharm World Sci. 1994;16:260–4.

407. Accordino A, Chambers R, Thompson B. A short-term stability study of an oral solution of dexamethasone. Aust J Hosp Pharm. 1994;24:312–6.

408. Boulton DW, Woods DJ, Fawcett JP, et al. The stability of an enalapril maleate oral solution prepared from tablets. Aust J Hosp Pharm. 1994;24:151–6.

409. Fawcett JP, Woods DJ, Hayes P, et al. The stability of acetylcysteine eyedrops. Aust J Hosp Pharm. 1993;23:18–21.

410. Kodym A, Paczkowska B, Szczepanski J. Technologia kropli do oczu z indometacyna (in) z ocena tolerancji i skutecznosci laczniczej. Farm Pol. 1990;46:397–403.

411. Jain NK, Agrawal RK, Singhai AK. Formulation of aqueous injection of carbamazepine. Pharmazie. 1990;45:221–2.

412. Morita K, Tsuchiya M, Watanabe S, et al. Pharmaceutical evaluation of 3% vidarabine ointment. Jpn J Hosp Pharm. 1988;14:241–5.

413. Pascual I, Borrego Dionis A. Coliro de folinato calcico. Formulacion y elaboracion. Farm Clin. 1987;4:564, 565, 568, 569, 571.

414. Stewart PJ, Doherty PG, Bostock JM, et al. The stability of extemporaneously prepared paediatric formulations of indomethacin. Aust J Hosp Pharm. 1985;15:55–60.

415. Hirota Y, Ohue K, Haga I, et al. Stability of clindamycin phosphate lotion prepared in hospital pharmacy. Jpn J Hosp Pharm. 1985;11:439–43.

416. Bregni C, Iribarren NA. Rheological study and stability of piroxicam aqueous suspensions. Boll Chim Farm. 1984;123:183–7.

417. Loprinzi CL, Burnham NL, O'Connell MJ, et al. Allopurinol mouthwash kinetic and stability information based on normal volunteers. Hosp Pharm. 1989;24:353, 354, 373.

418. Pesko LJ. Silver nitrate irrigation. Am Druggist. 1991;204:56.

419. Pesko LJ. Formalin irrigation. Am Druggist. 1991;204:64.

420. Pesko LJ. Amphotericin B nasal spray. Am Druggist. 1991;204:72.

421. Pesko LJ. Compounding: trazodone oral liquid. Am Druggist. 1992;205:58.

422. Odusote MO, Nasipuri RN. Effect of pH and storage conditions on the stability of a novel chloroquine phosphate syrup formulation. Pharm Ind. 1988;50:367–9.

423. Van Doorne H, Wieringa NF, Bosch EH, et al. The suitability of some preservatives in chloroquine phosphate syrup. Pharm Weekbl Sci Ed. 1988;10:170–2.

424. Fujii T, Kubota A, Togawa K, et al. Evaluation of aminophylline suppositories prepared in a hospital pharmacy. Tokai J Exp Clin Med. 1982;7:371–83.

425. Barnes AR, Nash S. Stability of bendrofluazide in a low-dose extemporaneously prepared capsule. J Clin Pharm Ther. 1994;19:89–93.

426. Fawcett JP, Stark G, Tucker IG, et al. Stability of dantrolene oral suspension prepared from capsules. J Clin Pharm Ther. 1994;19:349–53.

427. Kassem A, Said S, Shalaby S. In-vitro release and stability of erythromycin suppository. J Drug Res Egypt. 1977; 9:161–6.

428. Woods D. Extemporaneous formulation in pharmacy practice. Part 1. Folic acid oral solution. NZ Pharm. 1993;13:34.

429. Sivapunyam R, Ramakrishnan, Palanichamy S. A scanning study of ointment bases for the stability and release of gentamycin. Ind J Hosp Pharm. 1983;20:238–41.

430. Timmons P, Gray EA. Degradation of hydrocortisone in a zinc oxide lotion. J Clin Hosp Pharm. 1983;8:79–85.

431. Tempe M, Jonvel P, Spitter J, et al. Stability study of idoxuridine in gels by very high speed liquid chromatography. J Pharm Belg. 1985;40:222–8.

432. Mathew M, Gupta VD, Bethea C. The development of oral liquid dosage forms of metronidazole. J Clin Pharm Ther. 1993;18:291–4.

433. Mathew M, Gupta VD, Bethea C. Stability of metronidazole benzoate in suspensions. J Clin Pharm Ther. 1994;19:31–4.

434. Mathew M, Gupta VD, Bethea C. Stability of metronidazole in solutions and suspensions. J Clin Pharm Ther. 1994;19:27–9.

435. Elsner Z, Leszczynska-Bakal H, Pawlak E, et al. Gel with nystatin for treatment of lung mycosis. Pol J Pharmacol Pharm. 1976;28:349–52.

436. El-Shattawy HH, Kassem AA, Bayomi MA, et al. Release studies of neomycin from different ophthalmic ointment bases. Drug Dev Ind Pharm. 1994;20:1599–1604.

437. Pesko LJ. Compounding: ondansetron oral liquid. Am Druggist. 1994;209:49–50.

438. Ismaiel SA, Ismaiel EEA. A study on eye drops containing phenylephrine. Pharmazie. 1975;30:59.

439. Roffe BD, Zimmer RA, Derewicz HJ. Preparation of progesterone suppositories. Am J Hosp Pharm. 1977;34:1344–6.

440. Ahmed GH, Stewart PJ, Tucker IG. The stability of extemporaneous paediatric formulations of propranolol hydrochloride. Aust J Hosp Pharm. 1988;18:312–8.

441. Carstens G. Calcium-folinat uberlegungen zur stabilitat und zum einsatz verschieder zubereitungen. Krankenhauspharmazie. 1989;10:478–82.

442. Mozzi G, Conegliani B, Lomi R, et al. Stabilita del calcio folinato in soluzioni acquose in funzione del pH e delta temperatura. Boll Chim Farm. 1986; 125:424–8.

443. Cano SB, Golgiewicz FL. Storage requirements for metronidazole injection. Am J Hosp Pharm. 1986;43:2983, 2985.

444. Little GB, Boylan JC. I.V. Flagyl reacts with aluminum. Hosp Pharm. 1981;16:627.

445. Schell KH, Copland JR. Metronidazole hydrochloride–aluminum interaction. Am J Hosp Pharm. 1985;42:1040, 1042.

446. Struthers BJ, Parr RJ. Clarifying the metronidazole hydrochloride–aluminum interaction. Am J Hosp Pharm. 1985;42:2660.

447. Dienstag JL, Neu HC. Tobramycin: new aminoglycoside antibiotic. Clin Med. 1975;82:13–9.

448. Holmes CJ, Ausman RK, Kundsin RB, et al. Effect of freezing and microwave thawing on the stability of six antibiotic admixtures in plastic bags. Am J Hosp Pharm. 1982;39:104–8.

449. Marble DA, Bosso JA, Townsend RJ. Compatibility of clindamycin phosphate and amikacin sulfate at room temperature and with gentamicin sulfate and tobramycin sulfate under frozen conditions. Drug Intell Clin Pharm. 1986;20:960–3.

450. Awang DVC, Graham KC. Microwave thawing of frozen drug solutions. Am J Hosp Pharm. 1987;44:2256.

451. Seitz DJ, Archambault JR, Kresel JJ, et al. Stability of tobramycin sulfate in plastic syringes. Am J Hosp Pharm. 1980;37:1614–5.

452. Gupta VD, Stewart KR, Nohria S. Stability of vancomycin hydrochloride in 5% dextrose and 0.9% sodium chloride injections, Am J Hosp Pharm. 1986;43:1729–31.

453. Mann JM, Coleman DL, Boylan JC. Stability of parenteral solutions of sodium cephalothin, cephaloridine, potassium penicillin G, and vancomycin HCl. Am J Hosp Pharm. 1971;28:760–3.

454. Raleigh F. Incompatibilities with orally administered liquid neuroleptic medication combinations. Hosp Pharm. 1981; 16:486–7.

455. Anon. Information Bulletin, Sandoz Pharmaceuticals. Hosp Pharm. 1982;17:168.

456. Allen LV. Clonazepam suspension. A liquid oral dosage form of clonazepam may be prepared for administration to geriatric and pediatric patients. US Pharm. 1995;20:84–5.

457. Dupuis LL, James G, Bacola G. Stability of a sotalol hydrochloride oral liquid formulation. Can J Hosp Pharm. 1988;41:121–3.

458. Sochasky C, Johannesson B, Isaacs E, et al. Methadone stability in lemonade. Can J Hosp Pharm. 1987;40:188.

459. deCastro FJ, Jaeger RW, Rolfe UT. An extemporaneously prepared penicillamine suspension used to treat lead intoxication. Hosp Pharm. 1977;12:446–7.

460. Vandenbroucke J, Robays H. Stabiliteitstudie van ondansetron HCL (Zofran®) in frambozensiroop. Farm Tijdschrift Belg. 1993;6:20–2.

461. Irwin DB, Dupuis LL, Prober CG, et al. The acceptability, stability and relative bioavailability of an extemporaneous metronidazole suspension. Can J Hosp Pharm. 1987;40:42–6.

462. Pramar Y, Gupta VD, Bethea C. Stability of captopril in some aqueous systems. J Clin Pharm Ther. 1992;17:185–9.

463. Kawano K, Arai C, Anzai K, et al. Pharmaceutical study of mitomycin C ophthalmic solutions. Jpn J Hosp Pharm. 1993;19:230–3.

464. Zia H, Shalchian N, Borhanian F. Kinetics of amoxycillin degradation in aqueous solutions. Can J Pharm Sci. 1977;12:80–3.

465. Parkin JE, Marshall CA. The instability of phenylmercuric nitrate in APF ophthalmic products containing sodium metabisulphite. Aust J Hosp Pharm. 1990;20:434–6.

466. Eggers NJ. Stability of a methadone and prolintane mixture. Aust J Hosp Pharm. 1978;8:91–2.

467. Feron B, Adair CG, Gorman SP, et al. Interaction of sucralfate with antibiotics used for selective decontamination of the gastrointestinal tract. Am J Hosp Pharm. 1993;50:2550–3.

468. Segui Gregori MI, Sanchez Pinero J, Lobato Ballesteros M. Formulacion y estudio de la estabilidad de una solucion de mitomicina-C para uso oftalmico en el tratamiento postoperatorio del pterigion. Farm Clin. 1993;10:526–33.

469. Deeks T, Davis S, Nash S. Stability of an intrathecal morphine injection formulation. Pharm J. 1983;230:495–7.

470. Wong CY, Wang DP, Chang LC. Stability of cefazolin in pluronic F-127 gels. Drug Dev Ind Pharm. 1997;23:603–5.

471. Johnson CE, Nesbitt J. Stability of ursodiol in an extemporaneously compounded oral liquid. Am J Health Syst Pharm. 1995;52:1798–800.

472. Crowther RS, Bellanger R, Szauter KEM. In vitro stability of ranitidine hydrochloride in enteral nutrient formulas. Ann Pharmacother. 1995;29:859–62.

473. Vaughan LM, Bishop TD. Sterilization of talc USP for intrapleural use. Ann Pharmacother. 1994;28:1309–10.

474. Fawcett JP, Boulton DW, Jiang R, et al. Stability of hydrocortisone oral suspensions prepared from tablets and powder. Ann Pharmacother. 1995;29:987–90.

475. Nahata MC, Morosco RS, Hipple TF. Stability of cisapride in a liquid dosage form at two temperatures. Ann Pharmacother. 1995;29:125–6.

476. Nahata MC, Morosco RS, Hipple TF. Stability of spironolactone in an extemporaneously prepared suspension at two temperatures. Ann Pharmacother. 1993;27:1198–99.

477. Mayron D, Gennaro AR. Stability and compatibility of granisetron hydrochloride in i.v. solutions and oral liquids and during simulated Y-site injection with selected drugs. Am J Health Syst Pharm. 1996;53:294–304.

478. Quercia RA, Zhang J, Fan C, et al. Stability of granisetron hydrochloride in polypropylene syringes. Am J Health Syst Pharm. 1996;53:2744–6.

479. Mallet L, Sesin GP, Ericson J, et al. Storage of vancomycin oral solution. N Engl J Med. 1982;307:445.

480. Dietz NJ, Cascella PJ, Houglum JE, et al. Phenobarbital stability in different dosage forms: alternatives for elixirs. Pharm Res. 1988;5:803–5.

481. Anderson JL, Stennett DJ, Stewart J, et al. IV levodopa preparation and sterilization by filtration. Hosp Form. 1985;20;926, 930.

482. Stennett DJ, Christensen JM, Anderson JL, et al. Stability of levodopa in 5% dextrose injection at pH 5 or 6. Am J Hosp Pharm. 1986;43:1726–8.

483. Nahata MC, Zingarelli J, Durrell DE. Stability of caffeine citrate injection in intravenous admixtures and parenteral nutrition solutions. J Clin Pharm Ther. 1989;14:53–5.

484. Nahata MC, Zingarelli JR, Hipple TF. Stability of caffeine injection stored in plastic and glass syringes. DICP Ann Pharmacother. 1989;23:1035.

485. Bosanquet AG, Clarke HE. Chlorambucil: stability of solutions during preparation and storage. Cancer Chemother Pharmacol. 1986;18:176–9.

486. Stewart PJ, Owen WR. Prediction of the stability of chlorambucil in pharmaceutical solutions. Aust J Pharm Sci. 1980; 9:15–8.

487. McDonald C, Parkin JE, Richardson CA, et al. Stability of solutions of histamine acid phosphate after sterilization by heating in an autoclave. J Clin Pharm Ther. 1990;15:41–4.

488. Peswani KS, Lalla JK. Naproxen parenteral formulation studies. J Parenter Sci Technol. 1990;44:336–42.

489. Phillips M, Agarwal RP, Brodeur RJ, et al. Stability of an injectable disulfiram formulation sterilized by gamma irradiation. Am J Hosp Pharm. 1985;42:343–5.

490. Shah JC, Chen JR, Chow D. Preformulation study of etoposide: identification of physiochemical characteristics responsible for the low and erratic oral bioavailability of etoposide. Pharm Res. 1989;6:408–12.

491. Miethke T. Stabilitat von cocain-atropinmischinjektion. Pharm Ztg. 1991;136:39–41.

492. Maloney TJ. Stability of apomorphine hydrochloride solutions. Aust J Hosp Pharm. 1985;15:34.

493. Nahata MC, Morosco RS, Hipple TF. Stability of cimetidine hydrochloride and of clindamycin phosphate in water for injection stored in glass vials at two temperatures. Am J Hosp Pharm. 1993; 50:2559–61.

494. Nahata MC, Roberts DL, Hipple TF. Formulation of caffeine injection for i.v. administration. Am J Hosp Pharm. 1987; 44:1308–12.

495. Cummings J, Maclellan A, Langdon SJ, et al. The long term stability of mechlorethamine hydrochloride (nitrogen mustard) ointment measured by HPLC. J Pharm Pharmacol. 1993;45:6–9.

496. Schuster F, Baum S, Guntner S. Allopurinol zur intravenosen anwendung. Krankenhauspharmazie. 1995; 16:244–51.

497. Nahata MC, Morosco RS, Hipple TF. Stability of rifampin in two suspensions at room temperature. J Clin Pharm Ther. 1994;19:263–5.

498. Schlatter JL, Saulnier JL. Bethanechol chloride oral solutions: stability and use in infants. Ann Pharmacother. 1997;31: 294–6.

499. Nahata MC, Morosco RS, Fox J. Stability of ranitidine hydrochloride in water for injection in glass vials and plastic syringes. Am J Health Syst Pharm. 1996;53:1588–90.

500. Lugo RA, Nahata MC. Stability of diluted dexamethasone sodium phosphate injection at two temperatures. Ann Pharmacother. 1994;28:1018–9.

501. Nahata MC, Morosco RS, Hipple TF. Stability of diluted methylprednisolone sodium succinate injection at two temperatures. Am J Hosp Pharm. 1994;51:2157–9.

502. Alexander KS, Kothapalli MR, Dollimor D. Stability of an extemporaneously formulated levothyroxine sodium syrup compounded from commercial tablets. Int J Pharm Compound. 1997;1:60–4.

503. Anaizi NH, Swenson CF, Dentinger PJ. Stability of mycophenolate mofetil in an extemporaneously compounded oral liquid. Am J Health Syst Pharm. 1998;55:926–9.

504. Wiita B, Demestihas E. Diluted effect from diluted mesalamine rectal suspension? Am J Health Syst Pharm. 1995;52:2136–7.

505. Berardi RR, Johnson CE, Henderson LM. Diluted effect from diluted mesalamine rectal suspension? Am J Health Syst Pharm. 1995;52:2136–7.

506. Biddle WL, Greenberger NJ, Swan JT, et al. 5-Aminosalicylic acid enemas: effective agents in maintaining remission in left-sided ulcerative colitis. Gastroenterology. 1988;94:1075–9.

507. Campieri M, Gionchetti P, Belluzzi A, et al. Optimum dosage of 5-aminosalicylic acid as rectal enemas in patients with active ulcerative colitis. Gut. 1991;32:929–31.

508. Walker SE, Grad HA, Haas DA, et al. Stability of parenteral midazolam in an oral formulation. Anesth Prog. 1997;44:17–22.

509. Nahata MC, Morosco RS, Hipple TF. Stability of pyrimethamine in a liquid dosage formulation stored for three months. Am J Health Syst Pharm. 1997;54:2714–6.

510. Helin MM, Kontra KM, Naaranlahti TJ, et al. Content uniformity and stability of nifedipine in extemporaneously compounded oral powders. Am J Health Syst Pharm. 1998;55:1299–301.

511. Hernandez Salvador M, Merida Rodriguez C, Gomez-Huarez Molina F. Estudio de la estabilidad de una solucion de clorhidrato de morfina en solucion salina 0.9% sin conservantes para uso epidural. Farm Clin. 1992;9:514–7.

512. Grom JA, Bander LC. Compounding of preservative-free high-concentration morphine sulfate injection. Am J Health Syst Pharm. 1995;52:2125–7.

513. Yamreudeewong W, Danthi SN, Hill RA, et al. Stability of ondansetron hydrochloride injection in various beverages. Am J Health Syst Pharm. 1995;52:2011–4.

514. Stiles ML, Allen LV, Resztak KE, et al. Stability of octreotide acetate in polypropylene syringes. Am J Hosp Pharm. 1993;50:2356–8.

515. Ripley RG, Ritchie DJ, Holstad SG. Stability of octreotide acetate in polypropylene syringes at 5 and −20°C. Am J Health Syst Pharm. 1995;52:1910–1.

516. Ghnassia LT, Yau DF, Kaye KI, et al. Stability of cyclosporine in an extemporaneously compounded paste. Am J Health Syst Pharm. 1995;52:2204–7.

517. Owsley HD, Rusho WF. Compatibility of common respiratory therapy drug combinations. Int J Pharm Compound. 1997;1:121–2.

518. Allen LV, Stiles ML, Prince SJ, et al. Stability of ramipril in water, apple juice, and applesauce. Am J Health Syst Pharm. 1995;52:2433–6.

519. Gannon PM, Charest CA, Kasparian SS. Prostaglandin E2 vaginal suppositories for cervical ripening: a formulation and stability study. Hosp Pharm. 1995;30:791–3, 796–8, 801–2, 805.

520. Jacobson PA, Johnson CE, Walters JR. Stability of itraconazole in an extemporaneously compounded oral liquid. Am J Health Syst Pharm. 1995;52:189–91.

521. Villarreal JD, Erush SC. Bioavailability of itraconazole from oral liquids in question. Am J Health Syst Pharm. 1995;52:1707–8.

522. Denning D, Tucker R, Hanson L, et al. Treatment of invasive aspergillosis with itraconazole. Am J Med. 1989;86:791–800.

523. Kintzel PE, Rollins CJ, Yee WJ, et al. Low itraconazole serum concentrations following administration of itraconazole suspension to critically ill allogeneic bone marrow transplant recipients. Ann Pharmacother. 1995;29:140–3.

524. Jacobson PA, Johnson CE. Bioavailability of itraconazole from oral liquids in question. Am J Health Syst Pharm. 1995;52:1708.

525. Bhandari V, Narang A, Kumar B, et al. Itraconazole therapy for disseminated candidiasis in a very low birth-weight neonate. J Pediatr Child Health. 1992;28:323–4.

526. Bhandari V, Narang A. Oral itraconazole therapy for disseminated candidiasis in low birth weight infants. J Pediatr. 1992;120:330.

527. Allen LV, Erickson MA. Stability of ketoconazole, metolazone, metronidazole, procainamide hydrochloride, and spironolactone in extemporaneously compounded oral liquids. Am J Health Syst Pharm. 1996;53:2073–8.

528. Young D, Fadiran EO, Change KU, et al. Stability of ticarcillin disodium in polypropylene syringes. Am J Health Syst Pharm. 1995;52:890, 892.

529. Stiles ML, Allen LV, Prince SJ. Stability of deferoxamine mesylate, floxuridine, fluorouracil, hydromorphone hydrochloride, lorazepam, and midazolam hydrochloride in polypropylene infusion-pump syringes. Am J Health Syst Pharm. 1996;53:1583–8.

530. Allen LV. Metronidazole solution. US Pharm. 1995;20:92–3.

531. Wintermeyer SM, Nahata MC. Stability of flucytosine in an extemporaneously compounded oral liquid. Am J Health Syst Pharm. 1996;53:407–9.

532. Boulton DW, Fawcett JP, Woods DJ. Stability of an extemporaneously compounded levothyroxine sodium oral liquid. Am J Health Syst Pharm. 1996;53:1157–61.

533. Phillips C, Fisher E. Effect of autoclaving on stability of nitrofurazone soluble dressing. Am J Health Syst Pharm. 1996;53:1169–71.

534. Amadio M, Chambers R, McDonald C. Formulation of nitrazepam oral solution. Aust J Hosp Pharm. 1996;26:371.

535. Sabol BJ, Kerr TM, Schroeder LA. Stability of captopril solution. US Pharm. 1996;21:HS-28, HS-31, HS-32, HS-36.

536. Mathew M, Gupta VD. The stability of captopril in aqueous systems. Drug Stability. 1996;1:161–5.

537. Devi PN, Rao YM. Solubilization of ibuprofen by using cosolvents, surfactants and formulation of an elixir and injectable preparation. East Pharm. 1995;38:137–40.

538. Jain NK, Jahagirdar A. Formulation and evaluation of ibuprofen injections. Pharmazie. 1989;44:727–8.

539. Larson JA, Uden DL, Schilling CG. Stability of epinephrine hydrochloride in an extemporaneously compounded topical anesthetic solution of lidocaine, racepinephrine, and tetracaine. Am J Health Syst Pharm. 1996;53:659–62.

540. Schilling CG, Bank DE, Borchert BA, et al. Tetracaine, epinephrine (Adrenalin), and cocaine (TAC) versus lidocaine, epinephrine, and tetracaine (LET) for anesthesia of lacerations in children. Ann Emerg Med. 1995;25:203–8.

541. Jacobson PA, Johnson CE, West NJ, et al. Stability of tacrolimus in an extemporaneously compounded oral liquid. Am J Health Syst Pharm. 1997;54:178–80.

542. Allen LV, Erickson MA. Stability of labetalol hydrochloride, metoprolol tartrate, verapamil hydrochloride, and spironolactone with hydrochlorothiazide in extemporaneously compounded oral liquids. Am J Health Syst Pharm. 1996;53:2304–9.

543. Allen LV, Erickson MA. Stability of acetazolamide, allopurinol, azathioprine, clonazepam, and flucytosine in extempo-

raneously compounded oral liquids. Am J Health Syst Pharm. 1996;53:1944–9.

544. Allen LV, Erickson MA. Stability of baclofen, captopril, diltiazem hydrochloride, dipyridamole, and flecainide acetate in extemporaneously compounded oral liquids. Am J Health Syst Pharm. 1996; 53:2179–84.

545. Alexander KS, Vangala SSKS, White DB, et al. The formulation development and stability of spironolactone suspension. Int J Pharm Compound. 1997;1:195–9.

546. Allen LV, Li XH. Physical stability of urea topical formulations. Int J Pharm Compound. 1997;1:168–71.

547. Gupta VD, Maswoswe J. Quantitation of metoprolol tartrate and propranolol hydrochloride in pharmaceutical dosage forms: stability of metoprolol in aqueous mixture. Int J Pharm Compound. 1997;1:125–7.

548. Zhang Y, Trissel LA, Johansen JF, et al. Stability of mechlorethamine hydrochloride 0.01% ointment in Aquaphor® base. Int J Pharm Compound. 1998;2:89–91.

549. Lye MYF, Yow KL, Lim LY, et al. Effects of ingredients on stability of captopril in extemporaneously prepared oral liquids. Am J Health Syst Pharm. 1997;54:2483–7.

550. Lim LY, Tan LL, Chan EWY, et al. Stability of phenoxybenzamine hydrochloride in various vehicles. Am J Health Syst Pharm. 1997;54:2073–8.

551. Alexander KS, Davar N, Parker GA. Stability of allopurinol suspension compounded from tablets. Int J Pharm Compound. 1997;1:128–31.

552. Alexander KS, Vangala SSKS, Dollimore, D. The formulation development and stability of metronidazole suspension. Int J Pharm Compound. 1997;1:200–5.

553. Trissel LA, Xu QA, Hassenbusch SJ. Development of clonidine hydrochloride injections for epidural and intrathecal administration. Int J Pharm Compound. 1997;1:274–7.

554. Gupta DV, Maswoswe J. Stability of bethanechol chloride in oral liquid dosage forms. Int J Pharm Compound. 1997;1:278–9.

555. Webster AA, English BA, Rose DJ. The stability of lisinopril as an extemporaneous syrup. Int J Pharm Compound. 1997;1:352–3.

556. Patel D, Doshi DH, Desai A. Short-term stability of atenolol in oral liquid formulations. Int J Pharm Compound. 1997;1:437–9.

557. Roy JJ, Besner JG. Stability of clonazepam suspension in HSC vehicle. Int J Pharm Compound. 1997;1:440–1.

558. Nation RL, Hackett LP, Dusci LJ. Uptake of clonazepam by plastic intravenous infusion bags and administration sets. Am J Hosp Pharm. 1983;40:1692–3.

559. Hooymans PM, Janknegt R, Lohman JJHM. Comparison of clonazepam sorption to polyvinyl chloride-coated and polyethylene-coated tubings. Pharm Weekbl Sci Ed. 1990;12:188–9.

560. Bureau A, Lahet JJ, D'Athis P, et al. Compatibilite PVC-psychotropes au cours d'une perfusion. J Pharm Clin. 1995;14:26–30.

561. Quercia RA, Fan C, Liu X, et al. Stability of omeprazole in an extemporaneously prepared oral liquid. Am J Health Syst Pharm. 1997;54:1833–6.

562. Mathew M, Gupta VD, Bailey RE. Stability of omeprazole solutions at various pH values as determined by high-performance liquid chromatography. Drug Dev Ind Pharm. 1995;21:965–71.

563. Phillips JD, Metzler MH, Palmieri TL, et al. A prospective study of simplified omeprazole suspension for the prophylaxis of stress-related mucosal damage. Crit Care Med. 1996;24:1793–800.

564. Carrol M, Trudeau W. Nasogastric administration of omeprazole for control of gastric pH. Proc. 10th World Congress Gastroenterol. Los Angeles, CA; Oct 3, 1994; Abstract 22:P.

565. Tenjarla SN, Ward ES, Fox JL. Ondansetron suppositories: extemporaneous preparation, drug release, stability and flux through rabbit rectal membrane. Int J Pharm Compound. 1998;2:83–8.

566. Quercia RA, Zhang J, Fan C, et al. Stability of granisetron hydrochloride in an extemporaneously prepared oral liquid. Am J Health Syst Pharm. 1997;54:1404–6.

567. Fish DN, Beall HD, Goodwin SD, et al. Stability of sumatriptan succinate in extemporaneously prepared oral liquids. Am J Health Syst Pharm. 1997;54:1619–22.

568. Mallett MS, Hagan RL, Peters DA. Stability of ursodiol 25 mg/mL in an extemporaneously prepared oral liquid. Am J Health Syst Pharm. 1997;54:1401–4.

569. Al-Achi A, Greenwood R, Koo J. Need for quality-control testing of extemporaneously prepared oral solids. Am J Health Syst Pharm. 1996;53:1194–5.

570. Allen LV. Chewable lozenges for peds/geriatric use. US Pharm. 1993;18:102–4.

571. Anaizi NH, Swenson CF, Dentinger PJ. Stability of acetylcysteine in an extemporaneously compounded ophthalmic solution. Am J Health Syst Pharm. 1997;54:549–53.

572. Abdel-Rahman SM, Nahata MC. Stability of pentoxifylline in an extemporaneously prepared oral suspension. Am J Health Syst Pharm. 1997;54:1301–3.

573. Islam MS, Asker AF. Photoprotection of daunorubicin hydrochloride with sodium sulfite. PDA J Pharm Sci Technol. 1995;49:122–6.

574. Allen LV. Hydrocortisone ointment. US Pharm. 1997;22:96–8.

575. Allen LV. Prickly heat dusting powder. US Pharm. 1997; 22:88–90.

576. Nahata MC, Morosco RS, Hipple TF. Stability of metolazone in a liquid dosage form. Hosp Pharm. 1997;32:691–3.

577. Allen LV. Silver sulfadiazine/hydrocortisone gel. US Pharm. 1996;21:122–3.

578. Allen LV. Tetracycline oral liquid. US Pharm. 1996;21:102–4.

579. Gatti R, Gotti R, Cavrini V, et al. Stability study of prostaglandin E1 (PGE1) in physiological solutions by liquid chromatography (HPLC). Int J Pharm. 1995;115:113–7.

580. Allen LV. Lidocaine HCl nasal solution. US Pharm. 1996;21: 98–100.

581. Nahata MC, Morosco RS, Hipple TF. Stability of enalapril maleate in three extemporaneously prepared oral liquids. Am J Health Syst Pharm. 1998;55:1155–7.

582. Charlton JF, Dalla KP, Kniska A. Storage of extemporaneously prepared ophthalmic antimicrobial solutions. Am J Health Syst Pharm. 1998;55:463–6.

583. Allen LV. Midazolam gelatin cubes for children. US Pharm. 1996;21:134, 136–7.

584. Allen LV. Minoxidil solution for iontophoresis. US Pharm. 1996;21:116–8.

585. Chinnian D, Asker AF. Photostability profiles of minoxidil solutions. PDA J Pharm Sci Technol. 1996;50:94–8.

586. Woods DJ. Formulation in pharmacy practice. Dunedin, New Zealand: Health-Care Otago;1993.

587. Hunter DA, Yoo SD, Fox JL, et al. Stability of albuterol in continuous nebulization. Int J Pharm Compound. 1998;2:394–6.

588. Zia H, Rashed SF, Quadir M, et al. Ketorolac tromethamine and ketoprofen suppositories: release profiles and bioavailability of a cocoa butter base formula in rabbits. Int J Pharm Compound. 1998;2:390–3.

589. Donnelly RF, Bushfield TL. Chemical stability of meperidine hydrochloride in polypropylene syringes. Int J Pharm Compound. 1998;2:463–5.

590. Nahata MC, Morosco RS, Hipple TF. Stability of granisetron hydrochloride in two oral suspensions. Am J Health Syst Pharm. 1998;55:2511–3.

591. Share MJ, Harrison RD, Folstad J, et al. Stability of lorazepam 1 and 2 mg/mL in glass bottles and polypropylene syringes. Am J Health Syst Pharm. 1998;55:2013–5.

592. Allen LV. Phenytoin ointment for skin disorders. US Pharm. 1998;23:92–3.

593. Allen LV. HC suppositories for hemorrhoid relief. US Pharm. 1998;23:110–1.

594. Allen LV, Erickson MA. Stability of alprazolam, chloroquine phosphate, cisapride, enalapril maleate, and hydralazine hydrochloride in extemporaneously compounded oral liquids. Am J Health Syst Pharm. 1998;55:1915–20.

595. Allen LV, Erickson MA. Stability of bethanechol chloride, pyrazinamide, quinidine sulfate, rifampin, and tetracycline hydrochloride in extemporaneously compounded oral liquids. Am J Health Syst Pharm. 1998;55:1804–9.

596. Fuhrman LC, Stroman RT. Stability of vancomycin in an extemporaneously compounded ophthalmic solution. Am J Health Syst Pharm. 1998;55:1386–8.

597. Allen LV. Psoriasis ointment. Int J Pharm Compound. 1998;2:305.

598. Allen LV. Idoxuridine 0.1% ophthalmic solution. Int J Pharm Compound. 1998;2:232.

599. Allen LV. Fluconazole in dimethyl-sulfoxide. Int J Pharm Compound. 1998;2:299.

600. Allen LV. Benzocaine 2% anesthetic gel. Int J Pharm Compound. 1998;2:296.

601. Nahata MC, Morosco RS, Peritore SP. Stability of pyrazinamide in two suspensions. Am J Health Syst Pharm. 1995;52:1558–60.

602. Allen LV. Ascorbic acid 10% ophthalmic solution. Int J Pharm Compound. 1998;2:224.

603. Allen LV. Fluconazole 0.2% ophthalmic solution. Int J Pharm Compound. 1998;2:228.

604. Allen LV. Amphotericin B 2-mg/mL ophthalmic solution. Int J Pharm Compound. 1998;2:223.

605. Woods DJ, McClintock AD. Omeprazole administration. Ann Pharmacother. 1993;27:651.

606. Nahata MC. Stability of amiodarone in an oral suspension stored under refrigeration and at room temperature. Ann Pharmacother. 1997;31:851–2.

607. Venkataramanan R, McCombs JR, Zuckerman S, et al. Stability of mycophenolate mofetil as an extemporaneous suspension. Ann Pharmacother. 1998;32:755–7.

608. Lomaestro BM, Bailie GR. Quinolone-cation interactions: a review. DICP Ann Pharmacother. 1991;25:1249–58.

609. Goren MP, Lyman BA, Li JT. The stability of mesna in beverages and syrup for oral administration. Cancer Chemother Pharmacol. 1991;28:298–301.

610. Peterson MD. Making oral midazolam palatable for children. Anesthesiology. 1990;73:1053.

611. Rosen DA, Rosen KR. A palatable gelatin vehicle for midazolam and ketamine. Anesthesiology. 1991;75:914–5.

612. Pramar Y, Gupta VD. Preformulation studies of spironolactone: effect of pH, two buffer species, ionic strength, and temperature on stability. J Pharm Sci. 1991;80:551–3.

613. van Harten J, Danhof M, Burggraaf K, et al. Negligible sublingual absorption of nifedipine. Lancet. 1987;2:1363–4.

614. Wang DP, Yeh MK. Degradation kinetics of metronidazole in solution. J Pharm Sci. 1993;82:95–8.

615. Allen LV. Progesterone-estradiol-testosterone in oil capsules. Int J Pharm Compound. 1998;2:54.

616. Johnson CE, Wong DV, Hoppe HL, et al. Stability of ciprofloxacin in an extemporaneous oral liquid dosage form. Int J Pharm Compound. 1998;2:314–7.

617. Ray LR, Chen DA. Stability of somatropin stored in plastic syringes for 28 days. Am J Health Syst Pharm. 1998;55:1508–11.

618. Kumar R, Reznik DA, Gupta RB, et al. The stability of amphotericin B in a mixture of Fungizone® and Optimoist®. Int J Pharm Compound. 1998;2:311–3.

619. Sarver JG, Pryka R, Alexander KS, et al. Stability of magnesium sulfate in 0.9% sodium chloride and lactated Ringer's solutions. Int J Pharm Compound. 1998;2:385–8.

620. Allen LV. Etodolac suppositories. US Pharm. 1996;21:90–2.

621. Molina-Martinez IT, Herrero R, Gutierrez JA, et al. Bioavailability and bioequivalence of two formulations of etodolac (tablets and suppositories). J Pharm Sci. 1993;82:211–3.

622. Jaffe GJ, Green GDJ, Abrams GW. Stability of recombinant tissue plasminogen activator. Am J Ophthalmol. 1989;108:90–1.

623. Ward C, Weck S. Dilution and storage of recombinant tissue plasminogen activator (Activase) in balanced salt solutions. Am J Ophthalmol. 1990;109:98–9.

624. Grewing R, Mester U, Low M. Clinical experience with tissue plasminogen activator stored at −20°C. Ophthalmol Surg. 1992;23:780–1.

625. Mirochnick M, Barnett E, Clarke DF, et al. Stability of chloroquine in an extemporaneously prepared suspension stored at three temperatures. Pediatr Infect Dis J. 1994;13:827–8.

626. Jann MW, Bean J, Fidone GS. Interaction of dietary pudding with phenytoin. Pediatrics. 1986;78:952–3.

627. Bowe BE, Snyder JW, Eiferman RA. An in vitro study of the potency and stability of fortified ophthalmic antibiotic preparations. Am J Ophthalmol. 1991;111:686–9.

628. Emm T, Metcalf JE, Lesko LJ, et al. Update on the physical-chemical compatibility of cromolyn sodium nebulizer solution: bronchodilator inhalant solution admixtures. Ann Allergy. 1991;66:185–9.

629. de Castro FJ, Jaeger RW, Peters A, et al. Apomorphine: clinical trial of stable solution. Clin Toxicol. 1978;12:65–8.

630. Adams WP, Kostenbauder HB. Phenoxybenzamine stability in aqueous ethanolic solution. I. application of potentiometric pH stat analysis to determine kinetics. Int J Pharm. 1985;25:293–312.

631. Algra RJ, Rosen T, Waisman M. Topical clindamycin in acne vulgaris. Arch Dermatol. 1977;113:1390–1.

632. Gupta VD. The effect of some formulation adjuncts on the stability of hydrocortisone. Drug Dev Ind Pharm. 1985;11:2083–97.

633. McAllister RG. Kinetics and dynamics of nifedipine after oral and sublingual doses. Am J Med. 1986;81(Suppl 6A):2–5.

634. Fawcett JP, Morgan NC, Woods DJ. Formulation and stability of naltrexone oral liquid for rapid withdrawal from methadone. Ann Pharmacother. 1997;31:1291–5.

635. Patel B, Siskin S, Krazmien BA, et al. Compatibility of calcipotriene with other topical medications. J Am Acad Dermatol. 1998;38:1010–1.

636. Martin B, Meunier C, Montels D, et al. Chemical stability of adapalene and tretinoin when combined with benzoyl peroxide in presence and in absence of visible light and ultraviolet radiation. Br Assoc Dermatol. 1998;139(Suppl 52):8–1.

637. Deeks T. Oral atropine sulfate mixtures. Pharm J. 1983;230:481.

638. Wood MJ, Irwin WJ, Scott DK. Stability of doxorubicin, daunorubicin, epirubicin in plastic syringes and minibags. J Clin Pharm Ther. 1990;15:279–89.

639. Fleisher D, Johnson KC, Stewart BH, et al. Oral absorption of 21-corticosteroid esters: a function of aqueous solubility and enzyme activity and distribution. J Pharm Sci. 1986;75:934–9.

640. Lau DWC, Law S, Walker SE, et al. Dexamethasone phosphate stability and contamination of solutions stored in syringes. PDA J Pharm Sci Technol. 1996;50:261–7.

641. Donnelly RF, Yen M. Epinephrine stability in plastic syringes and glass vials. Can J Hosp Pharm. 1996;49:62–5.

642. Lackner TE, Baldus D, Butter CD, et al. Lidocaine stability in cardioplegic solution stored in glass bottles and polyvinyl chloride bags. Am J Hosp Pharm. 1983;40:97–101.

643. Newton DW, Narducci WA, Leet WA, et al. Lorazepam solubility in sorption from intravenous admixture solutions. Am J Hosp Pharm. 1983;40:424–7.

644. Strong ML, Schaaf LJ, Pankaskie MC, et al. Shelf-lives and factors affecting the stability of morphine sulphate and meperidine (pethidine) hydrochloride in plastic syringes for use in patient-controlled analgesic devices. J Clin Pharm Ther. 1994;19:361–9.

645. Cleary JD, Evans PC, Hikal AH, et al. Administration of crushed extended-release pentoxifylline tablets: bioavailability and adverse effects. Am J Health Syst Pharm. 1999;56:1529–34.

646. Namika Y, Fujiwara A, Kihara N, et al. Factors affecting tautomeric phenomenon of a novel potent immunosuppressant (FK506) on the design for injectable formulation. Drug Dev Ind Pharm. 1995;21:809–22.

647. Taormina D, Abdallah HY, Venkataramanan R, et al. Stability and sorption of FK 506 in 5% dextrose injection and 0.9% sodium chloride injection in glass, polyvinyl chloride, and polyolefin containers. Am J Hosp Pharm. 1992;49:119–22.

648. Teraoka K, Minakuchi K, Tsuchiya K, et al. Compatibility of ciprofloxacin infusion with other injections. Jpn J Hosp Pharm. 1995;21:541–50.

649. Kaijser GP, Aalbers T, Beijnen JH, et al. Chemical stability of cyclophosphamide, trofosfamide, and 2- and 3-dechloroethyl-fosfamide in aqueous solutions. J Oncol Pharm Pract. 1996;2:15–21.

650. Kawano K, Matsunaga A, Terade K, et al. Loss of diltiazem hydrochloride in solutions in polyvinyl chloride containers or intravenous administration set—hydrolysis and sorption. Jpn J Hosp Pharm. 1994;20:537–41.

651. Keyi X, Gagnon N, Bisson C, et al. Stability of famotidine in polyvinyl chloride minibags and polypropylene syringes and compatibility of famotidine with selected drugs. Ann Pharmacother. 1993;27:422–6.

652. Jacobson GA, Peterson GM. Stability of iprotropium bromide and salbutamol nebuliser admixtures. Int J Pharm Pract. 1995;3:169–73.

653. Peterson GM, Khoo BHC, Galloway JG, et al. A preliminary study of the stability of midazolam in polypropylene syringes. Aust J Hosp Pharm. 1991;21:115–8.

654. Duafala ME, Kleinberg ML, Nacov C, et al. Stability of morphine sulfate in infusion devices and containers for intravenous administration. Am J Hosp Pharm. 1990;47:143–6.

655. Haslam JL, Egodage KL, Chen Y, et al. Stability of rifabutin in two extemporaneously compounded oral liquids. Am J Health Syst Pharm. 1999;56:333–6.

656. Grassby PF, Hutchings L. Factors affecting the physical and chemical stability of morphine sulphate solutions stored in syringes. Int J Pharm Pract. 1993;2:39–43.

657. Casto DT. Stability of ondansetron stored in polypropylene syringes. Ann Pharmacother. 1994;28:712–4.

658. Wood MJ, Lund R, Beavan M. Stability of vancomycin in plastic syringes measured by high-performance liquid chromatography. J Clin Pharm Ther. 1995;20:319–25.

659. Lesko LJ, Miller AK. Physical-chemical compatibility of cromolyn sodium nebulizer solution-bronchodilator inhalant solution admixtures. Ann Allerg. 1984;53:236–8.

660. Dupuis LL, Zahn DA, Silverman ED, et al. Palatability and relative bioavailability of an extemporaneous prednisone suspension. Can J Hosp Pharm. 1990;43:101–5.

661. Dobbins JC. A frozen nystatin preparation. Hosp Pharm. 1983;18:452–3.

662. Woods DJ, Simonsen K. Administration of liquid dose forms of 6-mercaptopurine. J Paediatr Child Health. 1995;31:62–3.

663. Carlin A, Gregory N, Simmons J. Stability of isoniazid in isoniazid syrup; formation of hydrazine. J Pharm Biomed Anal. 1998;17:885–90.

664. Pramar Y, Gupta VD, Bethea C. Development of a stable oral liquid dosage form of spironolactone. J Clin Pharm Ther. 1992;17:245–8.

665. Barnes AR, Chapman SB, Irwin WJ. Chemical stability of fluocinolone acetonide ointment and fluocinonide cream diluted in emollient bases. J Clin Pharm Ther. 1995;20:265–9.

666. Grundy JS, Kherani R, Foster RT. Photostability determination of commercially available nifedipine oral dosage formulations. J Pharm Biomed Anal. 1994;12:1529–35.

667. Domaratzki J, Campbell S. Nifedipine administration in the critically ill. Can J Hosp Pharm. 1988;41:34–5.

668. Won CM. Kinetics of degradation of levothyroxine in aqueous solution and in solid state. Pharm Res. 1992;9:131–7.

669. Barnes AR, Hebron BS, Smith J. Stability of caffeine oral formulations for neonatal use. J Clin Pharm Ther. 1994;19:391–6.

670. How TH, Loo WY, Yow KL, et al. Stability of cefazolin sodium eye drops. J Clin Pharm Ther. 1998;23:41–7.

671. Gupta VD. Quantitation and stability of verapamil hydrochloride using high-performance liquid chromatography. Drug Dev Ind Pharm. 1985;11:1497–506.

672. McAndrews KL, Eastham JH. Omeprazole and lansoprazole suspensions for nasogastric administration. Am J Health Syst Pharm. 1999;56:81.

673. Abdel-Rahman SM, Nahata MC. Stability of terbinafine hydrochloride in an extemporaneously prepared oral suspension at 25 and 4°C. Am J Health Syst Pharm. 1999;56:243–5.

674. Nahata MC, Morosco RS, Hipple TF. Stability of lamotrigine in two extemporaneously prepared oral suspensions at 4 and 25°C. Am J Health Syst Pharm. 1999;56:240–2.

675. Barnes AR. Determination of ceftazidime and pyridine by HPLC: application to a viscous eye drop formulation. J Liq Chromatogr. 1995;18:3117–28.

676. Ohkubo T, Noro H, Sugawara K. High-performance liquid chromatographic determination of nifedipine and a trace photodegradation product in hospital prescriptions. J Pharm Biomed Anal. 1992;10:67–70.

677. Rawlins DA, Smith JM. Formulation of a readily-prepared D-penicilliamine oral liquid. J Clin Hosp Pharm. 1982;7:141–3.

678. Matsuda Y, Teraoka R, Sigimoto I. Comparative evaluation of photostability of solid-state nifedipine under ordinary and intensive light irradiation conditions. Int J Pharm. 1989;54:211–21.

679. Helin-Tanninen M, Naaranlahti T, Kontra K, et al. Enteral suspension of nifedipine for neonates. Part 1. Formulation of nifedipine suspension for hospital use. J Clin Pharm Ther. 2001;26:49–57.

680. Helin-Tanninen M, Naaranlahti T, Kontra K, et al. Enteral suspension of nifedipine for neonates. Part 2. Stability of an extemporaneously compounded nifedipine suspension. J Clin Pharm Ther. 2001;26:59–66.

681. Nahata MC, Morosco RS, Hipple TF. Stability of ursodiol in two extemporaneously prepared oral suspensions. J App Ther Res. 1999;2:221–4.

682. Nahata MC, Morosco RS, Trowbridge JM. Stability of propylthiouracil in extemporaneously prepared oral suspensions at 4 and 25°C. Am J Health Syst Pharm. 2000;57:1141–3.

683. Swenson CF, Dentinger PJ, Anaizi NH. Stability of mycophenolate mofetil in an extemporaneously compounded sugar-free oral liquid. Am J Health Syst Pharm. 1999;56:2224–6.

684. Nahata MC. Stability of verapamil in an extemporaneous liquid dosage form. J App Ther. 1997;1:271–3.

685. Gupta VD. Stability of cocaine hydrochloride solutions at various pH values as determined by high-pressure liquid chromatography. Int J Pharm. 1982;10:249–57.

686. Allen LV. Modified solution for wounds and burns. US Pharm. 1999;24:95–8.

687. Fish DN, Vidaurri VA, Deiter RG. Stability of valacyclovir hydrochloride in extemporaneously prepared oral liquids. Am J Health Syst Pharm. 1999;56:1957–60.

688. Anaizi NH, Swenson CF, Dentinger PJ. Am J Health Syst Pharm. 1999;56:1738–41.

689. Gupta VD. Stability of an oral liquid dosage form of trovafloxacin mesylate and its quantitation in tablets using high-performance liquid chromatography. Int J Pharm Compound. 2000;4:233–5.

690. Barnes AR. Determination of caffeine and potassium sorbate in an neonatal oral solution by HPLC. Int J Pharm. 1992;80:267–70.

691. Sullivan JA, Hobson LT. Prednisolone disodium phosphate stability in a prednisolone oral solution. Aust J Hosp Pharm. 1994;24:397–8.

692. Tiefenbacher EM, Haen E, Pryzbilla B, et al. Photodegradation of some quinolones used as antimicrobial therapeutics. J Pharm Sci. 1994;83:463–7.

693. Gregory DF, Koestner JA, Tobias JD. Stability of midazolam prepared for oral administration. South Med J. 1993;86:771–2.

694. Moore DE, Sithipitaks V. Photolytic degradation of furosemide. J Pharm Pharmacol. 1983;35:489–93.

695. Bundgaard H, Norgaard T, Nielson NM. Photodegradation and hydrolysis of furosemide and furosemide esters in aqueous solutions. Int J Pharm. 1988;42:217–24.

696. Asker AF, Ferdous AJ. Photodegradation of furosemide solutions. PDA J Pharm Sci Technol. 1996;50:158–62.

697. Cruz JE, Maness DD, Yakatan GJ. Kinetics and mechanism or hydrolysis of furosemide. Int J Pharm. 1979;2:275–81.

698. White CM, Quercia R. Stability of extemporaneously prepared sterile testosterone solution in 0.9% sodium chloride solution large-volume parenterals in plastic bags. Int J Pharm Compound. 1999;3:156–7.

699. Proot P, Schepdael AV, Raymakers AA, et al. Stability of adenosine in infusion. J Pharm Biomed Anal. 1998;17:415–8.

700. Lee DKT, Wang DP. Formulation development of allopurinol suppositories and injectables. Drug Dev Ind Pharm. 1999;25:1205–8.

701. Hinshaw KD, Fiscella R, Sugar J. Preparation of pH-adjusted local anesthetics. Ophthalmol Surg. 1995;26:194–9.

702. Meyer G, Henneman PL. Buffered lidocaine. Ann Emerg Med. 1991;20:218–9.

703. Fan B, Stewart JT, White CA. Stability and dissolution of lozenge and emulsion formulations of metronidazole benzoate. Int J Pharm Compound. 2001;5:153–6.

704. Allen LV. Anthralin 1% in lipid crystals cream. Int J Pharm Compound. 2001;5:205.

705. Lindahl A. Embedding of dithranol in lipid crystals. Acta Derm Venereol Suppl (Stockh). 1992;172:13–6.

706. Ling J, Gupta VD. Stability of oral liquid dosage forms of ethacrynic acid. Int J Pharm Compound. 2001;5:232–3.

707. Johnson CE, Price J, Hession JM. Stability of norfloxacin in an extemporaneously prepared oral liquid. Am J Health Syst Pharm. 2001;58:577–9.

708. Justice J, Kupiec TC, Matthews P, et al. Stability of Adderall in extemporaneously compounded oral liquids. Am J Health Syst Pharm. 2001;58:1418–21.

709. Webster KD, Al-Achi A, Greenwood R. In vitro studies on the release of morphine sulfate from compounded slow-

release morphine-sulfate capsules. Int J Pharm Compound. 1999;3:409–11.

710. Bogner RH, Szwejkowski J, Houston A. Release of morphine sulfate from compounded slow-release capsules: the effect of formulation on release. Int J Pharm Compound. 2001;5:401–5.

711. Johnson CE, Streetman DD. Stability of oral suspensions of ursodiol made from tablets. Am J Health Syst Pharm. 2002;59:361–3.

712. VandenBussche HL, Johnson CE, Fontana EM, et al. Stability of levofloxacin in an extemporaneously compounded oral liquid. Am J Health Syst Pharm. 1999;56:2316–8.

713. Zhang YP, Trissel LA, Fox JL. Naratriptan hydrochloride in extemporaneously compounded oral suspensions. Int J Pharm Compound. 2000;4:69–71.

714. VanDenBerg CM, Kazmi Y, Stewart J, et al. Pharmacokinetics of three formulations of ondansetron hydrochloride in healthy volunteers: 24-mg oral tablet, rectal suppository, and i.v. infusion. Am J Health Syst Pharm. 2000;57:1046–50.

715. Dentinger PJ, Swenson CF, Anaizi NH. Stability of famotidine in an extemporaneously compounded oral liquid. Am J Health Syst Pharm. 2000;57:1340–2.

716. Nahata MC, Morosco RS, Trowbridge JM. Stability of dapsone in two oral liquid dosage forms. Ann Pharmacother. 2000;34:848–50.

717. Tu YH, Allen LV, Wang DP. Stability of loperamide hydrochloride in aqueous solutions as determined by high performance liquid chromatography. Int J Pharm. 1998;51:157–60.

718. Nahata MC, Morosco RS, Hipple TF. Stability of amlodipine besylate in two liquid dosage forms. J Am Pharm Assoc. 1999;39:375–7.

719. VandenBussche HL, Johnson CE, Yun J, et al. Stability of flucytosine 50 mg/mL in extemporaneous oral liquid formulations. Am J Health Syst Pharm. 2002;59:1853–5.

720. Cook TJ, Shenoy SS. Stability of calcitonin salmon in nasal spray at elevated temperatures. Am J Health Syst Pharm. 2002;59:713–5.

721. Trissel LA, Zhang Y, Xu QA. Stability of 4-aminopyridine and 3,4-diaminopyridine oral capsules. Int J Pharm Compound. 2002;6:155–7.

722. Gupta VD. Stability of an oral liquid dosage form of glycopyrrolate prepared from tablets. Int J Pharm Compound. 2001;5:480–1.

723. Hugen PWH, Burger DM, ter Hofstede HJM, et al. Development of an indinavir oral liquid for children. Am J Health Syst Pharm. 2000;57:1332–9.

724. de Villiers MM, Narsai K, van der Watt JG. Physicochemical stability of compounded creams containing alphahydroxy acids. Int J Pharm Compound. 2000;4:72–5.

725. Taha EI, Zaghloul AAA, Kassem AA, et al. Salbutamol sulfate suppositories: influence of formulation on physical parameters and stability. Pharm Dev Technol. 2003;8:21–30.

726. Pramar Y, Gupta VD, Bethea C, et al. Stability of cefuroxime axetil in suspensions. J Clin Pharm Ther. 1991;16:341–4.

727. Levitt J, Feldman T, Riss I, et al. Compatibility of desoximetasone and tacrolimus. J Drugs Dermatol. 2003;2:640–2.

728. Acar V, Houri JJ, Le Hoang MD, et al. Stability of stored methacholine solutions: study of hydrolysis kinetic by IP-LC. J Pharm Biomed Anal. 2001;25:861–9.

729. Gupta VD. Chemical stability of methylprednisolone sodium succinate after reconstitution in 0.9% sodium chloride injection and storage in polypropylene syringes. Int J Pharm Compound. 2001;5:148–50.

730. Gupta VD, Ling J. Stability of piperacillin sodium after reconstitution in 0.9% sodium chloride injection and storage in polypropylene syringes for pediatric use. Int J Pharm Compound. 2001;5:230–1.

731. Kommanaboyina B, Lindauer RF, Rhodes CT, et al. Some studies of the stability of compounded cefazolin ophthalmic solution. Int J Pharm Compound. 2000;4:146–9.

732. Fuhrman LC, Godwin DA, Davis RA. Stability of 5-fluorouracil in an extemporaneously compounded ophthalmic solution. Int J Pharm Compound. 2000;320–3.

733. Gebauer MG, McClure AF, Vlhakis TL. Stability indicating HPLC method for the estimation of oxycodone and lidocaine in rectal gel. Int J Pharm. 2001;223:49–54.

734. Nahata MC, Morosco RS. Stability of sotalol in two liquid formulations at two temperatures. Ann Pharmacother. 2003;37:506–9.

735. Rose DJ, Webster AA, English BA, et al. Stability of lisinopril syrup (2 mg/mL) extemporaneously compounded from tablets. Int J Pharm Compound. 2000;4(5):398–9.

736. Nahata MC, Morosco RS, Hipple TF. Stability of mexiletine in two extemporaneous liquid formulations stored under refrigeration and at room temperature. J Am Pharm Assoc. 2000;40:257–9.

737. Boonme P. Effect of some antioxidants on the color change of white lotion. Int J Pharm Compound. 2000;4(2):154–5.

738. Yang YHK, Lin TR, Huang YF, et al. Stability of furosemide, nadalol and propranolol hydrochloride in extemporaneously compounded powder packets from tablets. Chin Pharm J. 2000;52:51–8.

739. Burns PE, McCall L, Wirsching R. Physical compatibility of enteral formulas with various common medications. J Am Diet Assoc. 1988;88:1094–6.

740. Pilatti C, Torre MC, Chiale C, et al. Stability of pilocarpine ophthalmic solutions. Drug Dev Ind Pharm. 1999;25:801–5.

741. Gairard AC, Demirdjian S, Vasseur R, et al. Stabilite de l'adrenaline dans des solutions pour irrigations intra-oculaires. J Fr Ophtalmol. 1996;19:743–7.

742. Allen LV. Hyaluronidase 150 u/mL injection. US Pharm. 2001;26:68–71.

743. Seifart HI, Parkin DP, Donald PR. Stability of isoniazid, rifampin and pyrazinamide in suspensions used for the treatment of tuberculosis in children. Pediatr Infect Dis J. 1991;10:827–31.

744. Gatti R, Gioia MG, Cavrini V. Analysis and stability study of retinoids in pharmaceuticals by LC and fluorescence detection. J Pharm Biomed Anal. 2000;23:147–59.

745. Modamio P, Lastra CF, Montejo O, et al. Development and validation of liquid chromatography methods for the quantitation of propranolol, metoprolol, atenolol and bisoprolol: application in solution stability studies. Int J Pharm. 1996;130:137–40.

746. Mehta AC, Calvert RT, Ryatt KS. Betamethasone 17-valerate: an investigation into its stability in Benovate after dilution with emulsifying ointment: quantitation of degradation products. Br J Pharm Pract. 1982;4:10–3

747. Tan LK, Thenmozhiyal JC, Ho PC. Stability of extemporaneously prepared saquinavir formulations. J Clin Pharm Ther. 2003;28:457–63.

748. Nahata MC, Morosco RS, Willhite EA. Stability of nifedipine in two oral suspensions stored at two temperatures. J Am Pharm Assoc. 2002;42:865–7.

749. Dentinger PJ, Swenson CF, Anaizi NH. Stability of nifedipine in an extemporaneously compounded oral solution. Am J Health Syst Pharm. 2003;60:1019–22.

750. Cornarakis-Lentzos M, Cowin PR. Dilutions of corticosteroid creams and ointments—a stability study. J Pharm Biomed Anal. 1987;5:707–16.

751. Fiscella RG, Le H, Lam TT, et al. Stability of cyclosporine 1% in artificial tears. J Ocular Pharmacol Ther. 1996;12:1–4.

752. McPhail DC, Johnson JR, Pryce-Jones RH. Stability and bioavailability of a flucytosine suspension. Int J Pharm Pract. 1994;235:235–9.

753. Fawcett JP, Woods DJ, Ferry DG, et al. Stability of amiloride hydrochloride oral liquids prepared from tablets and powder. Aust J Hosp Pharm. 1995;25:19–23.

754. Totterman AM, Luukkonen P, Riukka L, et al. Formulation of enteral hydrochlorothiazide suspension for premature infants. Eur J Hosp Pharm. 1994;4:65–9.

755. Elema MO, Nobus CMC, Stolk LKL, et al. Prednisoloninjectievloeistof FNA nader bekeken. Ziekenhuisfarmacie. 1997;13: 17–20.

756. Peterson GM, Meaney MF, Reid CA. Stability of extemporaneously prepared mixtures of metoprolol and spirono-lactone. Aust J Hosp Pharm. 1989;19:344–6.

757. Seth SC, Allen LV, Pinnamaraju P. Stability of hydrocortisone salts during iontophoresis. Int J Pharm. 1994;106:7–14.

758. Tucker IG, Geddes JA, Stewart PJ. Dose variations from an extemporaneously prepared chlorothiazide suspension. Aust J Hosp Pharm. 1982;12:59–63.

759. Sullivan JA. Prednisolone stability in a hydroalcoholic prednisolone mixture. Aust J Hosp Pharm. 1991;21:239–41.

760. Mehta AC, Hart-Davies S, Bedford C. Stability of pilocarpine oral solution. Hosp Pharm Pract. 1992;2:726–7.

761. Oldham GB. Formulation and stability of cefuroxime eyedrops. Int J Pharm Pract. 1991;1:19–22.

762. McKnight DL. The formulation of spironolactone solution for paediatric use. Aust J Hosp Pharm. 1993;23:83.

763. Mehta AC, Hart-Davies S, Bedford C. Chemical stability of midazolam syrup. Hosp Pharm Pract. 1993,3:224–5.

764. Wilhelm AJ, Dieleman HG, de Vogel EM. Formulation and shelf-life of fentanyl 1 mL=0.25 mg injection. Ziekenhuisfarmacie. 1993;9:102–4.

765. Nedergaard M. Phenoxybenzamine in solutions. J Hosp Pharm. 1969;27:174–6.

766. Barnes AR, Nash S, Watkiss SB. Stability of steroid ointments diluted with compound zinc paste B.P. J Clin Pharm Ther. 1991;16:103–9.

767. Iacono M, Johnson GJ, Bury RW. An investigation of the compatibility of ipratropium and sodium cromoglycate nebulizer solutions. Aust J Hosp Pharm. 1987;17:158–61.

768. Turner M. Compatibility of nebulizer solutions. Aust J Hosp Pharm. 1987;17:240–1.

769. Fraser BD. Stability of caffeine citrate injection in polypropylene syringes at room temperature. Am J Health Syst Pharm. 1997;54:1106, 1108.

770. Boonsaner P, Remon JP, De Rudder D. The stability and blanching efficiency of some Betnelan-V cream dilutions. J Clin Hosp Pharm. 1986;11:101–6.

771. Yip YW, Po ALW. The stability of betamethasone-17-valerate in semi-solid bases. J Pharm Pharmacol. 1979; 31:400–2.

772. Boersma HH, Groothuijsen HJG, Stolk LML, et al. Goed houdbaar onder normale omstandigheden. Pharm Weekblad. 1999;134:1444–8.

773. Hatem A, Marton S, Csoka G, et al. Preformulation studies of atenolol in oral liquid dosage form. I. Effect of pH and temperature. Acta Pharm Hung. 1996;66:177–80.

774. Wiernikowski JT, Crowther M, Clase CM, et al. Stability and sterility of recombinant tissue plasminogen activator at −30°C Lancet. 2000;2221–2.

775. Ray-Johnson ML. Effect of diluents on corticosteroid stability. Br J Pharm Pract. 1981;3:24–7.

776. Krochmal L, Wang JCT, Patel B, et al. Topical corticosteroid compounding: effects on physicochemical stability and skin penetration rate. J Am Acad Dermatol. 1989;21:979–84.

777. Thompson KC, Zhao Z, Mazakas JM, et al. Characterization of an extemporaneous liquid formulation of lisinopril. Am J Health Syst Pharm. 2003;60:69–74.

778. Nahata MC, Morosco RS. Stability of tiagabine in two oral liquid vehicles. Am J Health Syst Pharm. 2003;60:75–7.

779. Wagner DS, Johnson CE, Cichon-Hensley BK, et al. Stability of oral liquid preparations of tramadol in strawberry syrup and sugar-free vehicle. Am J Health Syst Pharm. 2003;60:1268–70.

780. Kaila N, El-Ries M, Riga A, et al. Formulation development and stability testing of extemporaneous suspension prepared from dapsone tablets. Int J Pharm Compound. 2003;7:233–9.

781. Henkin CC, Griener JC, Ten Eick AP. Stability of valganciclovir in extemporaneously compounded liquid formulations. Am J Health Syst Pharm. 2003;60:687–90.

782. DiGiacinto JL, Olsen KM, Bergman KL, et al. Stability of suspension formulations of lansoprazole and omeprazole stored in amber-colored plastic oral syringes. Ann Pharmacol. 2000; 34:600–5.

783. Dentinger PJ, Swenson CF, Anaizi NH. Stability of pantoprazole in an extemporaneously compounded oral liquid. Am J Health Syst Pharm. 2002;59:953–6.

784. Doan TT, Wang Q, Griffin JS, et al. Comparative pharmacokinetics and pharmacodynamics of lansoprazole oral capsules and suspension in healthy subjects. Am J Health Syst Pharm. 2001;58:1512–9.

785. Dunn A, White CM, Reddy P, et al. Delivery of omeprazole and lansoprazole granules through a nasogastric tube in vitro. Am J Health Syst Pharm. 1999;56:2327–30.

786. Lauper RD. Leucovorin calcium administration and preparation. Am J Hosp Pharm. 1978;35:377–8.

787. Cisternino S, Schlatter J, Saulnier JL. Stability of fludrocortisone acetate solutions prepared from tablets and powder. Eur J Pharm Biopharm. 2003;55:209–13.

788. Allen LV. Glycopyrrolate 1% topical soln/cream. US Pharm. 2000;25:80–5.

789. Johnson DA, Roach AC, Carlsson AS, et al. Stability of esomeprazole capsule contents after in vitro suspension in common soft foods and beverages. Pharmacotherapy. 2003;23:731–4.

790. White CM, Kalus JS, Quercia R, et al. Delivery of esomeprazole magnesium enteric-coated pellets through small caliber and standard nasogastric tubes and gastrostomy tubes in vitro. Am J Health Syst Pharm. 2003;59:2085–8.

791. Andrisano V, Hrelia P, Gotti R, et al. Photostability and phototoxicity studies on diltiazem. J Pharm Biomed Anal. 2001;25:589–97.

792. Andrisano V, Gotti R, Leoni A, et al. Photodegradation studies on atenolol by liquid chromatography. J Pharm Biomed Anal. 1999;21:851–7.

793. Akhtar MJ, Khan MA, Ahmad I. Photodegradation of folic acid in aqueous solution. J Pharm Biomed Anal. 1999;25:269–75.

794. Akhtar MJ, Khan MA, Almad I. Identification of photoproducts of folic acid and its degradation pathways in aqueous solution. J Pharm Biomed Anal. 2003;31:579–88.

795. Alldredge BK, Venteicher R, Calderwood TS. Stability of diazepam rectal gel in ambulance-like environments. Am J Emerg Med. 2002;20:88–91.

796. Arici MK, Sumer Z, Guler C, et al. In vitro potency and stability of fortified ophthalmic antibiotics. Aust NZ J Ophthamol. 1999;27:426–30.

797. Barbault S, Aymard G, Feldman D, et al. Stability of vancomycin eye drops. J Pharm Clin. 1999;18:183–9.

798. Bonhomme-Faivre L, Mathieu MC, Depraetere P, et al. Formulation of a charcoal suspension for intratumoral injection. Study of galenical excipients. Drug Dev Ind Pharm. 1999;25:175–86.

799. Bonhomme L, Mathieu MC, Seiller M, et al. Charcoal labeling of mammary tumors in mice prior to chemotherapy and surgery. Eur J Pharm Sci. 1993;1:103–8.

800. Bonhomme-Faivre L, Mathieu MC, Orbach-Arbouys S, et al. Studies on toxicity of charcoal used in tattooing of tumors. Eur J Pharm Sci. 1996;4:95–100.

801. Boonme P, Phadoongsombut N, Phoomborplub P, et al. Stability of extemporaneous norfloxacin suspension. Drug Dev Ind Pharm. 2000;16:777–9.

802. Brisaert M, Plaizier-Vercammen J. Investigation on the photostability of a tretinoin lotion and stabilization with additives. Int J Pharm. 2000;199:49–57.

803. Caviglioli G, Parodi B, Posocco V, et al. Stability of hard gelatin capsules containing retinoic acid. Drug Dev Ind Pharm. 2000;26:995–1001.

804. Chen JL, Yeh MK, Chiang CH. Anaerobic stability of aqueous physostigmine solution. Drug Dev Ind Pharm. 2000;26:1007–11.

805. Yang ST, Wilken LO. Effects of autoclaving on the stability of physostigmine salicylate in buffer solutions. J Parenter Sci Technol. 1988;42:62–7.

806. Chen QM, Li W, Liu ZS. The stability of carboplatin suppository. Chin Pharm J. 2000;25:168–71.

807. de Blois AW, Grouls RJE, Ackerman EW, et al. Development of a stable solution of 5-aminolaevulinic acid for intracutaneous injection in photodynamic therapy. Lasers Med Sci. 2002;17:208–15.

808. Dimitrova E, Bogdanova S, Mitcheva M, et al. Development of model aqueous ophthalmic solution of indomethacin. Drug Dev Ind Pharm. 2000;26:1297–301.

809. Elfsson B, Wallin I, Eksborg S, et al. Stability of 5-aminolevulinic acid in aqueous solution. Eur J Pharm Sci. 1998;7:87–91.

810. Francoeur AM, Assalian A, Lesk MR, et al. A comparative study of the chemical stability of various mitomycin C solutions used in glaucoma filtering surgery. J Glaucoma. 1999;8:242–6.

811. Frøyland K, Klem W, Tennesen HH, et al. Formulation and stability of cyclosporin eye drops. Eur Hosp Pharm. 1999;5:159–61.

812. Gallarate M, Carlotti ME, Trotta M, et al. On the stability of ascorbic acid in emulsified systems for topical and cosmetic use. Int J Pharm. 1999;188:233–41.

813. Gellis C, Sautou-Miranda V, Arvouet A, et al. Impact of deep-freezing on pefloxacin eye drops stability. J Pharm Clin. 1999;18:179–82.

814. Georgopoulos M, Vass C, Vatanparast Z, et al. Activity of dissolved mitomycin C after different methods of long-term storage. J Glaucoma. 2002;11:17–20.

815. Hecker D, Worsley J, Yueh G, et al. In vitro compatibility of tazarotene with other topical treatments of psoriasis. J Am Acad Dermatol. 2000;42:1008–11.

816. Hecker D, Worsley J, Yueh G, et al. Interactions between tazarotene and ultraviolet light. J Am Acad Dermatol. 1999;41:927–30.

817. Huang YB, Tsai YH, Chang JS, et al. Effect of antioxidants and anti-irritants on the stability, skin irritation and penetration capacity of captopril gel. Int J Pharm. 2002;241:345–51.

818. Kakehi K, Nakano M, Nishiura S, et al. Examination of the stability of chloral hydrate and its preparation by capillary electrophoresis. Yakugaku Zasshi. 1999;119:410–6.

819. Kizu J, Tsuchiya M, Watanabe S, et al. Preparation and clinical application of 2% diflunisal oral ointment for painful lesions of the oral mucosa. Yakugaku Zasshi. 2001;121:819–35.

820. Kodym A, Bujak T. Physicochemical and microbiological properties as well as stability of ointments containing aloe extract (Aloe arborescens Mill.) or aloe extract associated to neomycin sulphate. Pharmazie. 2002;57:834–7.

821. Henn S, Monfort P, Vigneron JH, et al. Stability of methacholine chloride in isotonic sodium chloride using capillary electrophoresis assay. J Clin Pharm Ther. 1999;24:365–8.

822. Gad Kariem EA, Abounassif MA, Hagga ME, et al. Photodegradation kinetic study and stability-indicating assay of danazol using high-performance liquid chromatography. J Pharm Biomed Anal. 2000; 23:413–20.

823. Barnes AR, Nash S. Stability of ceftazidime in a viscous eye drop formulation. J Clin Pharm Ther. 1999;24:299–302.

824. Li YNB, Moore DE, Tattam BN. Photodegradation of amiloride in aqueous solution. Int J Pharm 1999;183:109–16.

825. Martens-Lobenhoffer J, Rinke M, Losche D, et al. Long-term stability of 8-methoxypsoralen in ointments for topical PUVA therapy ('cream-PUVA'). Skin Pharmacol Appl Skin Physiol. 1999;12:266–70.

826. Matsubayashi T, Sakaeda T, Kita T, et al. Pharmaceutical and clinical assessment of hydroquinone ointment prepared by extemporaneous nonsterile compounding. Biol Pharm Bull. 2002;25:92–6.

827. Midhat V, Sabira H, Aida D, et al. Stability study for sulfamethoxazole + trimethoprim suspension. Farm Vestnik. 1999;50:357–8.

828. Munari L. Fissures anales: traitment topique par le diltiazem et la nifedipine (suite). J Suisse Pharm. 2002;18:622–3.

829. Nahata MC, Morosco RS, Leguire LE. Development of two stable oral suspensions of levodopa-carbidopa for children with amblyopia. J Pediatr Ophthamol Stabismus. 2000;37:333–7.

830. Nahata MC. Development of two stable oral suspensions for gabapentin. Pediatr Neurol. 1999;20:195–7.

831. Ohtani M, Nakai T, Ohsawa K, et al. Effect of admixture of betamethasone butyrate propionate ointment on preservative efficacy. Yakugaku Zasshi. 2002;122:1153–8.

832. Orwa JA, Govaerts C, Gevers K, et al. Study of the stability of polymyxins B1, E1, and E2 in aqueous solution using liquid chromatography and mass spectrometry. J Pharm Biomed Anal. 2002;19:203–12.

833. Pasto L, Vuelta MF, Marti T, et al. Oily solutions for ophthalmic use: microbial stability. Eur Hosp Pharm. 1999;5:79–82.

834. Perez Maroto MT, Luque Infantes R, Santolaya Perrin R, et al. Estabilidad y esterilidad de un colirio de fenilefrina al 2,5%. Farm Hosp. 1999;23:48–52.

835. Shawesh A, Kallioinen S, Antikainen O, et al. Influence of storage time and temperature on the stability of indomethacin Pluronic F-127 gels. Pharmazie. 2002;57:690–4.

836. Liu J, Chan SY, Ho PC. Effects of sucrose, citric buffer and glucose oxidase on the stability of captopril in liquid formulations. J Clin Pharm Ther. 1999;24:145–50.

837. Chun AHC, Shi HH, Achari R, et al. Lansoprazole: administration of the contents of a capsule dosage formulation through a nasogastric tube. Clin Ther. 1996;18:833–42.

838. Li SF, Zhang Y, Wang Y, et al. Stability study of norfloxacin eye drops. Chin J Hosp Pharm. 2000;347–8.

839. Ji HG, Seo BS. Retinyl palmitate at 5% in a cream: its stability, efficacy and effect. Cosmet Toiletries. 1999;114:61–4, 66–8.

840. Kizu J, Yazawa M, Yasuno N, et al. A new theophylline ointment and rapid and high transdermal delivery. J Nippon Hosp Pharm. 1999;25:239–48.

841. Rieutord A, Arnaud P, Dauphin JF, et al. Stability and compatibility of an aerosol mixture including N-acetylcysteine, netilimicin and betamethasone. Int J Pharm. 1999;190:103–7.

842. Spiclin P, Gasperlin M, Kmetec V. Stability of ascorbyl palmitate in topical microemulsions. Int J Pharm. 2001;222:271–9.

843. Stark FS, Cerise C. Incomaptibilite de l'erythromycin avec la crème Remederm. Schweizer Apothekerzeitung. 2002;12:399–400.

844. Valenta C. Stabilitat: Cyproteron-acetat in magistralen Zubereitungen. Osterreich Apotheker-Zeitung. 2002;56:676–8.

845. Vermeulen B, Remon JP, Nelis H. The formulation and stability of erythromycin-benzoyl peroxide in a topical gel. Int J Pharm. 1999;178:137–41.

846. Wu FLL, Shen LJ, Yang CC, et al. Stability of extemporaneous suspensions of cotrimoxazole. Chin Pharm J. 1999;51:93–102.

847. Wuis EW, Burger DM, Beelen M, et al. Stability of dithranol in creams. Pharm World Sci. 1999;21:275–7.

848. Ros JJW, van der Meer YG. Preparation, analysis and stability of oil-in-water creams containing dithranol. Eur J Hosp Pharm. 1991;1:77–84.

849. Ros JJW, Boer Y. Preparation of dithranol containing compounds. Pharm Weekbl. 1998;133:1504–7.

850. Altman E, Cutie AJ. Compatibility of enteral products with commonly employed drug additives. Nutr Supp Serv. 1984;4:8–17.

851. Anaizi NH, Dentinger PJ, Swenson CF. Stability of valganciclovir in an extemporaneously compounded oral liquid. Am J Health Syst Pharm. 2002;59:1267–70.

852. Peyron F, Elias R, Ibrahim E, et al. Stability of amphotericin B in 5% dextrose ophthalmic solution. Int J Pharm Compound. 1999;3:316–20.

853. Abdel-Rahman SM, Nahata MC. Stability of fumagillin in an extemporaneously prepared ophthalmic solution. Am J Health Syst Pharm. 1999;56:547–50.

854. Diesenhouse MC, Wilson LA, Corrent GF, et al. Treatment of microsporidial conjunctivitis with topical fumagillin. Am J Ophthamol. 1993;115:293–8.

855. Nii LJ, Chin A, Cao TM, et al. Stability of sumatriptan succinate in polypropylene syringes. Am J Health Syst Pharm. 1999;56:983–5.

856. Austria R, Semenzato A, Bettero A. Stability of vitamin C derivatives in solution and topical formulations. J Pharm Biomed Anal. 1997;795–801.

857. Nahata MC, Morosco RS, Hipple TF. Stability of amiodarone in extemporaneous oral suspensions prepared from commercially available vehicles. J Pediatr Pharm Pract. 1999;4:186–9.

858. Rao YM, Devi KM, Chary RBR. Stability studies of rifampicin mucoadhesive nasal drops. Indian J Pharm Sci. 1999;61:366–70.

859. Mourya VK, Shete JS, Chaudhari GN, et al. Evaluation of mupirocin in mupirocin ointment and mupirocin with betamethasone dipropionate ointment. East Pharm. 2000;43:143–5.

860. Hennessey DD. Recovery of phenytoin from feeding formulas and protein mixtures. Am J Health Syst Pharm. 2003;60:1850–2.

861. Ferron GM, Ku S, Abell M, et al. Oral bioavailability of pantoprazole suspended in sodium bicarbonate solution. Am J Health Syst Pharm. 2003;60:1324–9.

862. Okeke CC, Medwick T, Nairn G, et al. Stability of hydralazine hydrochloride in both flavored and nonflavored extemporaneous preparations. Int J Pharm Compound. 2003;7:313–9.

863. Haase MR, Khan MA, Bonilla J. Stability of two concentrations of tiagabine in an extemporaneously compounded suspension. Int J Pharm Compound. 2003;7:485–9.

864. Alexander KS, Thyagarajapuram N. Formulation and accelerated stability studies for an extemporaneous suspension of amiodarone hydrochloride. Int J Pharm Compound. 2003;7:389–93.

865. Johnson CE, Wagner DS, Bussard WE. Stability of dolasetron in two oral liquid vehicles. Am J Health Syst Pharm. 2003;60:2242–4.

866. Gupta VD. Stability of oral liquid dosage forms of glycopyrrolate prepared with the use of powder. Int J Pharm Compound. 2003;7:386–8.

867. Johnson CE, Wagner DS, DeLoach SL, et al. Stability of tramadol hydrochloride– acetaminophen (Ultracet) in strawberry syrup and in a sugar-free vehicle. Am J Health Syst Pharm. 2004;61:54–7.

868. Alexander KS, Kaushik S. Extemporaneous formulation and stability testing of mexiletine hydrochloride solution. Int J Pharm Compound. 2004;8:147–52.

869. Ohsawa K, Ohtani M, Kariya S, et al. Preparation of sodium polystyrene sulfonate suspension enema for treatment of hyperkalemia. Jpn J Hosp Pharm. 2000;26:380–7.

870. Peyron F, Ibrahim E, Elias R, et al. Stability of fortified ophthalmic solutions of amphotericin B, ceftazidime, and vancomycin. J Pharm Clin. 1999; 18:48–52.

871. Nahata MC, Morosco RS. Stability of lisinopril in two liquid dosage forms. Ann Pharmacother. 2004;38:396–9.

872. Carrier MN, Garinot O, Vitzling C. Stability and compatibility of tegaserod from crushed tablets mixed in beverages and foods. Am J Health Syst Pharm. 2004;61:1135–42.

873. Hui M, Kwok AKH, Pang CP, et al. An in vitro study on the compatibility and precipitation of a combination of ciprofloxacin and vancomycin in human vitreous. Br J Ophthamol. 2004;88:218–2.

874. Dix J, Weber RJ, Frye RF, et al. Stability of atropine sulfate prepared for mass chemical terrorism. Clin Toxicol. 2003;41:771–5.

875. Xu QA, Trissel LA, Pham L. Physical and chemical stability of treprostinil sodium injection packaged in plastic syringe pump reservoirs. Int J Pharm Compound. 2004;8:228–30.

876. Trissel LA, Zhang Y. Long-term stability of Trimix: a three-drug injection used to treat erectile dysfunction. Int J Pharm Compound. 2004;8:231–3.

877. Yoshida N, Arai K, Kohda Y. Stability of chloral hydrate in aqueous solution for rectal use. Jpn J Hosp Pharm. 2000;26:198–201.

878. Ho PC, Soh H, Lim SW. Stability of extemporaneously prepared gentamicin ophthalmic solutions. Ann Pharmacother. 2001;35:1293–4.

879. Benziane H, Simon L, Carde A, et al. Formulation and stability of hospital preparation of 3,4-diaminopyridine capsules. Ann Pharm Fr. 2003;61:418–24.

880. Garcia-Valldecabres M, Lopez-Alemany A, Refojo MF. pH stability of ophthalmic solutions. Optometry. 2004;75:161–8.

881. Calhoun DA, Juul SE, McBryde EV, et al. Stability of filgrastim and epoetin alfa in a system designed for enteral administration in neonates. Ann Pharmacother. 2000;34:1257–61.

882. Anon. 3,4-Diaminopyridine. Chemicalland, Seoul, South Korea. Data on file.

883. Anon. 4-Aminopyridine. Extoxnet, Extension Toxicology Network. Cornell University, Ithaca, NY. Data on file.

884. Ketkar VA, Kolling WM, Nardviriyakul N, et al. Stability of undiluted and diluted adenosine at three temperatures in syringes and bags. Am J Health Syst Pharm. 1998;55:466–70.

885. Naud C, Marti B, Fernandez C, et al. Stability of adenosine 6 mcg/mL in 0.9% sodium chloride solution. Am J Health Syst Pharm. 1998;55:1161–4.

886. Calis KA, Cullinane AM, Horne MK. Bioactivity of cryopreserved alteplase solutions. Am J Health Syst Pharm. 1999;56:2056–7.

887. Casasin Edo T, Roca Massa M, Soy Munne D. Drug distribution system for anesthesia use of preloaded syringes. Stability study. Farm Hosp. 1996;20:55–9.

888. Betnesol injection [package insert]. Berkshire, UK: Celltech Pharmaceuticals Limited; 2003.

889. Dribben WH, Porto SM, Jeffords BK. Stability and microbiology of inhalant N-acetylcysteine used as an intravenous solution for the treatment of acetaminophen poisoning. Ann Emerg Med. 2003;40:9–13.

890. Heeney MM, Whorton MR, Howard TA, et al. Chemical and functional analysis of hydroxyurea oral solutions. J Pediatr Hematol Oncol. 2004;26:179–84.

891. Green AE, Banks S, Jay M, et al. Stability of nimodipine solution in oral syringes. Am J Health Syst Pharm. 2004;61:1493–6.

892. McKenzie JE, Cruz-Rivera M. Compatibility of budesonide inhalation suspension with four nebulizing solutions. Ann Pharmacother. 2004;38:967–72.

893. Trissel LA, Flora KP, Vishnuvajjala R, et al. NCI investigational drugs pharmaceutical data. Bethesda, MD: National Cancer Institute. 1994.

894. Williams DA. Stability and compatibility of admixtures of antineoplastic drugs, in Lokich JJ (ed), Cancer chemotherapy by infusion, 2nd ed. Chicago: Precept Press. 1990.

895. Sewell GJ, Riley CM, Rowland CG. The stability of carboplatin in ambulatory continuous infusion regimes. J Clin Pharm Ther. 1987;12:427–32.

896. Stewart JT, Warren FW, Johnson SM, et al. Stability of ceftazidime in plastic syringes and glass vials under various storage conditions. Am J Hosp Pharm. 1992;49:2765–8.

897. Roberts GW, Rossi SOP. Compatibility of nebulizer solutions. Aust J Hosp Pharm. 1993;23:35–7.

898. Naughton CA, Duppong LM, Forbes KD, et al. Stability of multidose, preserved formulation epoetin alfa in syringes for three and six weeks. Am J Health Syst Pharm. 2003;60:464–8.

899. Anon. Fumidil B. Mid-Continent Agrimarketing, Inc. Data on file.

900. Trissel LA. Drug information—osmolalities. Data on file.

901. Stewart JT, Warren FW, King DT, et al. Stability of ondansetron hydrochloride and 12 medications in plastic syringes. Am J Health Syst Pharm. 1998;55:2630–4.

902. Storms ML, Stewart JT, Warren FW. Stability of glycopyrrolate injection at ambient temperature and 4°C in polypropylene syringes. Int J Pharm Compound. 2003;7:65–7.

903. Anon. Glycolic acid. E. I. du Pont de Nemours and Company. 1997. Data on file.

904. Gupta VD, Ling J. Stability of hydrocortisone sodium succinate after reconstitution in 0.9% sodium chloride injection and storage in polypropylene syringes for pediatric use. Int J Pharm Compound. 2000;4:396–7.

905. Zhang Y, Trissel LA, Martinez JF, et al. Stability of metoclopramide hydrochloride in plastic syringes. Am J Health Syst Pharm. 1996;53:1300–2.

906. Anon. pSol-3 intrinsic solubility. Woburn, MA: Pion Inc. Data on file.

907. Anon. Allergan Material Safety Data Sheet—Zorac. Irvine, CA: Allergan, Inc. June 20, 2001.

908. Donnelly R. Chemical stability of methadone concentrate and powder diluted in orange-flavored drink. Int J Pharm Compound. 2004;8:489–91.

909. Chan JP, Tong HHY, Chow AHL, et al. Stability of extemporaneous oral ribavirin liquid preparation. Int J Pharm Compound. 2004;8:486–8.

910. Polnok A, Techowanich S. Stability of extemporaneous keto-conazole suspensions. Thai J Hosp Pharm. 2004;14:27–34.

911. Pathmanathan U, Halgrain D, Chiadmi F, et al. Stability of sulfadiazine oral liquids prepared from tablets and powder. J Pharm Pharm Sci. 2004;7:84–7.

912. Carlotti ME, Rossatto V, Gallarte M, et al. Vitamin A palmitate photostability and stability over time. J Cosmet Sci. 2004;55:233–52.

913. El-Khawas M, Boraie NA. Stability and compatibility of thiamine hydrochloride in liquid dosage forms at various temperatures. Acta Pharm. 2000;50:219–28.

914. Alexander KS, Mitra P. Stability of an extemporaneously compounded propylthiouracil suspension. Int J Pharm Compound. 2005;9:82–6.

915. Lifshitz T, Lapid-Gortzak R, Finkelman Y, et al. Vancomycin and ceftazidime incompatibility upon intravitreal injection. Br J Ophthalmol. 2000;84:117–8.

916. McLellan C, Papadopoulos A. Precipitate formation after subconjunctival administration of ceftazidime and vancomycin. Hosp Pharm. 2005;40:154–5.

917. Vermerie N, Malbrunot C, Azar M, et al. Stability of nystatin in mouthrinses: effect of pH, temperature, concentration and colloidal silver addition, studied using an in vitro antifungal activity. Pharm World Sci. 1997;19:197–201.

918. Trissel LA, Xu QA. Shelf-life study of clonidine HCl 1.5 mg/mL in original vials and repackaged in 30-mL syringes. Data on file. 2000.

919. Uebel RA, Wium CA, Schmidt AC. Stability evaluation of a prostaglandin E1 saline solution packed in insulin syringes. Int J Impot Res. 2001;13:16–7.

920. Jagota NK, Stewart JT, Warren FW, et al. Stability of mupirocin ointment (Bactroban) admixed with other proprietary dermatological products. J Clin Pharm Ther. 1992;17:181–4.

921. Gupta VD. Quantitation of testosterone in a suspension using high-performance liquid chromatography. Int J Pharm Compound. 1999;3:239–40.

922. Gupta VD. Stability of triamcinolone acetonide solutions as determined by high-performance liquid chromatography. J Pharm Sci. 1983;72:1453–6.

923. Ungphaiboon S, Nittayananta W, Vuddhakul V, et al. Formulation and efficacy of triamcinolone acetonide mouthwash for treating oral lichen planus. Am J Health Syst Pharm. 2005;62:485–91.

924. Fraccaro A, Grion AM, Gaion RM, et al. Prostaglandin E1 diluted solution: preparation technique and stability evaluation. Int J Impot Res. 1993;5:43–5.

925. Shulman NH, Fyfe RK. Shelf-life determination of prostaglandin E1 injections. J Clin Pharm Ther. 1995;20:41–4.

926. Gupta VD, Sood A. Chemical stability of isoniazid in an oral liquid dosage form. Int J Pharm Compound. 2005;9:165–6.

927. McHugh RC, Sangha ND, McCarty MA, et al. A topical azithromycin preparation for the treatment of acne vulgaris and rosacea. J Dermatolog Treat. 2004;15:295–302.

928. Lee TY, Chen CM, Lee CN, et al. Compatibility and osmolality of inhaled N-acetylcysteine nebulizing solution with fenoterol and ipratropium. Am J Health Syst Pharm. 2005;62:828–33.

929. Hirowatari T, Tokuda K, Kamei Y, et al. Evaluation of a new preoperative ophthalmic solution. Can J Ophthalmol. 2005;40:58–62.

930. Blondino FE, Baker M. Stability of nebulizer admixtures. Int J Pharm Compound. 2005;9:323–6.

931. Klous MG, Nuijen B, van den Brink W, et al. Pharmaceutical development of an intravenous dosage form of diacetylmorphine hydrochloride. PDA J Pharm Sci Technol. 2004;58:287–95.

932. Supattanakul P. The stability of an enalapril maleate oral syrup prepared from tablets. Thai J Hosp Pharm. 2004;14:122–32.

933. Dana W. Data on file. Procarbazine suspension 10 mg/mL. Aug 18, 2005.

934. Kumana CR, Au WY, Lee NSL, et al. Systemic availability of arsenic from oral arsenictrioxide used to treat patients with hematological malignancies. Eur J Clin Pharmacol. 2002;58:521–6.

935. Sidhom MB, Rivera N, Almoazen N, et al. Stability of sotalol hydrochloride in extemporaneously prepared oral suspension formulations. Int J Pharm Compound. 2005;9:402–6.

936. Landry C, Forest JM, Hildgen P. Stability and subjective taste acceptability of four glycopyrrolate solutions for oral administration. Int J Pharm Compound. 2005;9:396–8.

937. Gupta VD. Chemical stability of perphenazine in oral liquid dosage forms. Int J Pharm Compound. 2005;9:484–6.

938. Chong E, Dumont RJ, Hamilton DP, et al. Stability of aminophylline in extemporaneously prepared oral suspensions. J Informed Pharmacother. 2000;2:100–6.

939. Abounassif MA, El-Obeid HA, Hadkariem EA. Stability studies on some benzocycloheptane antihistaminic agents. J Pharm Biomed Anal. 2005;36:1001–8.

940. Wu YQ, Fasshi R. Stability of metronidazole, tetracycline HCl and famotidine alone and in combination. Int J Pharm. 2005;290:1–13.

941. Dansereau RJ, Crail DJ. Extemporaneous procedures for dissolving risedronate tablets for oral administration and for feeding tubes. Ann Pharmacother. 2005;39:63–7.

942. Smith D. Data on file. Arsenic trioxide injection, 1 mg/mL, 2001.

943. Valenta C, Auner BG, Loibl I. Skin permeation and stability studies of 5-aminolevulinic acid in a new gel and patch preparation. J Control Release. 2005;107:495–501.

944. Valenta C, Schultz K. Influence of carrageenan on the rheology and skin permeation of microemulsion formulations. J Control Release. 2004;95:257–65.

945. Velpandian T, Saluja V, Ravi AK, et al. Evaluation of the stability of extemporaneously prepared ophthalmic formulation of mitomycin C. J Ocular Pharmacol Ther. 2005;21:217–22.

946. Hafirassou H, Chiadmi F, Schlatter J, et al. Stability of misoprostol in suppositories. Am J Health Syst Pharm. 2005;62:1192–4.

947. Ogden S, Asghar J. Compounding notes—preparing omeprazole liquid from Losec capsules. NZ Pharm. 2005;25:20.

948. Hanysova L, Vaclavkova M, Dohnal J, et al. Stability of ramipril in solvents of different pH. J Pharm Biomed Anal. 2005;37:1179–83.

949. Bouligand J, Storme T, Laville I, et al. Quality control and stability study using HPTLC: applications to cyclophosphamide

in various pharmaceutical products. J Pharm Biomed Anal. 2005;38:180–5.

950. Ibrahim AS, Hassan MA, El-Mohsen MG, et al. Compatibility study of famotidine with some pharmaceutical excipients using different techniques. Bull Pharm Sci Assiut Univ. 2004;27:281–91.

951. Fratta A, Bordenave J, Boissinot C, et al. Development of an oxybutynin chloride solution: from formulation to quality control. Ann Pharm Fr. 2005;63:162–6.

952. Fajolle V, Dujois C, Darbord J-C, et al. Oral suspensions of spironolactone, hydrochlorothiazide and captopril: microbiological stability study and clinical use review. J Pharm Clin. 2005; 24:23–9.

953. Ensom MHH, Decarie D, Hamilton DP. Stability of domperidone in extemporaneously compounded suspensions. J Informed Pharmacother. 2002;8:100–4.

954. Escribano Garcia MJ, Torrado Duran S, Torrado Duran JJ. Stability study of an aqueous formulation of captopril at 1 mg/mL. Farm Hosp. 2005;29:30–6.

955. Lin JM, Tsai YY, Fu YL. The fixed combination of fortified vancomycin and amikacin ophthalmic solution—VA solution: in vitro study of the potency and stability. Cornea. 2005;24:717–21.

956. Sadrieh N, Brower J, Yu L, et al. Stability, dose uniformity, and palatability of three counterterrorism drugs—human subject and electronic tongue studies. Pharm Res. 2005;22:1747–56.

957. Rhodes RS, Kuykendall JR, Heyneman CA, et al. Stability of phenytoin sodium suspensions for the treatment of open wounds. Int J Pharm Compound. 2006;10:74–7.

958. Kaneuchi MK, Kohri N, Senbongi K, et al. Preparing and evaluation of oral dosage form of ketamine considering simplicity for preparation in hospital and ease for patients to take. (1) Preparations using agar. Yakugaki Zasshi. 2005;125:187–96.

959. Zeppetella G, Joel SP, Ribiero MDC. Stability of morphine sulphate and diamorphine hydrochloride in Intrasite gel.™ Palliat Med. 2005;19:131–6.

960. Bempong DK, Manning RG, Mirza T, et al. A stability-indicating HPLC assay for metronidazole benzoate. J Pharm Biomed Anal. 2005;38:776–80.

961. Nguyen-Xuan T, Griffiths W, Kern C, et al. Stability of morphine sulfate in polypropylene infusion bags for use in patient-controlled analgesia pumps for postoperative pain management. Int J Pharm Compound. 2006;10:69–73.

962. Chandibhamar V, Yadav MR, Murthy RSR. Studies on the development of taste-masked suspension for chloroquine. Boll Chim Famac. 2004;143:377–82.

963. Shetty K. Hydrogen peroxide burn of the oral mucosa. Ann Pharmacother. 2006;40:351.

964. Sirtes G, Waltimo T, Schaetzle M, et al. The effects of temperature on sodium hypochlorite short-term stability, pulp dissolution capacity, and antimicrobial efficacy. J Endod. 2005;31:669–71.

965. Liu BL, Huang JL. Study on the stability of gatifloxacin eye drops. Chin J Antibiotics. 2004;29:539–41.

966. McCarron PA, Donnelly RF, Andrews GP, et al. Stability of 5-aminolevulinic acid in novel nonaqueous gel and patch-type systems intended for topical application. J Pharm Sci. 2005;94:1756–71.

967. Novo M, Huttman G, Diddens H. Chemical instability of 5-aminolevulinic acid used in the fluorescence diagnosis of bladder tumors. J Photochem Photobiol B. 1996;34:143–8.

968. Bunke A, Zerbe O, Schmid H, et al. Degradation mechanism and stability of 5-aminolevulinic acid. J Pharm Sci. 2000;89:1335–41.

969. Gadmar OB, Moan J, Scheie E, et al. The stability of 5-aminolevulinic acid in solution. J Photochem Photobiol B. 2002;67:187–93.

970. Salgado AC, Rosa ML, Duarte MA, et al. Stability of spironolactone in an extemporaneously prepared aqueous suspension: the importance of microbiological quality of compounded paediatric formulations. Eur J Hosp Pharm Sci. 2005;11:68–73.

971. Nahata MC, Morosco RS, Brady MT. Extemporaneous sildenafil citrate oral suspensions for the treatment of pulmonary hypertension in children. Am J Health Syst Pharm. 2006;63:254–7.

972. Zour E, Lodhi SA, Nesbitt, RU, et al. Stability studies of gabapentin in aqueous solutions. Pharm Res. 1992;9:595–600.

973. Blondeel S, Pelloquin A, Pointereau-Bellanger A, et al. Effect of freezing on stability of a fortified 5 mg/mL ticarcillin ophthalmic solution. Can J Hosp Pharm. 2005;58:65–70.

974. Gambier N, Vigneron J, Menetre S, et al. Stability of 6-mercaptopurine in capsules for paediatric patients using a capillary electrophoresis assay. Eur J Hosp Pharm Sci. 2006;12:13–5.

975. Walker SE, Madden MJ. Thermal stability of tobramycin in a tobramycin/phenylephrine inhalation solution. Can J Hosp Pharm. 1986;39:92–5.

976. Gupta VD. Chemical stability of diclofenac sodium injection. Int J Pharm Compound. 2006;10:154–5.

977. Vermeire A, Remon JP. Stability and compatibility of morphine. Int J Pharm. 1999;187:17–51.

978. Yamreudeewong W, Dolence EK, Pahl D. Stability of two extemporaneously prepared oral metoprolol and carvediol liquids. Hosp Pharm. 2006;41:254–9.

979. Haywood A, Mangan M, Grant G, et al. Extemporaneous isoniazid mixture: stability implications. J Pharm Pract Res. 2005;35:181–2.

980. Ng K, Yinfoo IR. Stability of morphine mixture with hydroxybenzoates as preservatives. Aust J Hosp Pharm. 1988;18:168–70.

981. Roksvaag PO, Fredrikson JB, Waaler T. High-performance liquid chromatographic assay of morphine and the main degradation product pseudomorphine. A study of pH, discoloration and degradation in 1- to 43-year-old morphine injections. Pharm Acta Helv. 1980;55:198–202.

982. Sletten DM, Nickander KK, Low PA. Stability of acetylcholine chloride solution in autonomic testing. J Neurol Sci. 2005;234:1–3.

983. Steger PJK, Martinelli EF, Muhlebach SF. Stability of high-dose morphine chloride injection upon heat sterilization: comparison of UV spectroscopy and HPLC. J Clin Pharm Ther. 1996;21:73–8.

984. Johnson CE, vanDeKoppel S, Myers E. Stability of anhydrous theophylline in extemporaneously prepared alcohol-free oral suspensions. Am J Health Syst Pharm. 2005;62:2518–20.

985. Aki H, Ohta M, Fukusumi K, et al. Evaluation of compatibility of risperidone with soft drinks and interactions of risperidone with tea tannin using isothermal titration microcalimetry. Iryo Yakugaku. 2006;32:190–8.

986. Khanderia S, Patel J, Pearlman MD, et al. Stability of misoprostol in rectal suppositories. ASHP Midyear Clinical Meeting. 2005:40:p–7D.

987. Siden R, Johnson CE. Stability of a flavored formulation of acetylcysteine for oral administration. Am J Health Syst Pharm. 2008;65:558–61.

988. Suresh P, Gupta BK. pH and stability of metronidazole benzoate suspensions. J Instit Chem (India). 1993;65:22–3.

989. Bamio-Nuez A, Artalejo-Orgega B, Fauli Trillo C, et al. Influence of pH on the stability of methadone hydrochloride in oral solutions. Farm Clin. 1989;6:618–22.

990. Gupta VD. Chemical stability of diphenhydramine hydrochloride from an elixir and lidocaine hydrochloride from a viscous solution when mixed together. Int J Pharm Compound. 2006;10:237–9.

991. Whittet TD, Robinson AE. Stability of sodium p-aminohippurate. Pharm J. 1964;193:39–40.

992. Donskaya NG, Lapidus VL, Libinson GS, et al. Stability of dicloxacillin in aqueous solutions. Khimiko-Farmatsevticheskii Zhurnal. 1976;10:108–11.

993. Tuchel N, Antonescu V, Gheorghe V, et al. Drug stability. II. Stability of chloral hydrate solutions. Farmacia (Bucharest). 1961;9:205–12.

994. Ivey MM, Kuhn RJ, Pedigo N. Stability of an oral midazolam solution for pediatric preanesthesia. ASHP Midyear Clinical Meeting. 1991;26:p–583D.

995. Komorsky-Lovric S, Markicevic B. Farm Glasnik. 1990;46:3–8.

996. Cao Y, Cao H, Cao H, et al. Preparation of stable ibuprofen injection for veterinary use. Faming Zhuanlii Shenqing Gongkai Shuomingshu. 1990;1–8.

997. Farrington EA, Bawdon RA, Fox JL. Stability of cefuroxime axetil suspensions. ASHP Midyear Clinical Meeting. 1991;26:p–580E.

998. Kim KS. Stability of minoxidil in aqueous solution. Yakhak Hoechi. 1986;30:228–31.

999. Gambaro V, Caligara M, Pesce E. Study of the light stability of nifedipine in a drop formulation under therapeutic use conditions. Boll Chim Farm. 1985;124:13–8.

1000. Meijer E, van Loenen AC. Stability study of adenosine injection. Ziekenhaus-farmacie. 1993;9:10–3.

1001. Henry DW, Kuestermeyer GL, Oszko MA, et al. Chemical stability and content uniformity of 17-beta-estradiol in polyethylene glycol base vaginal suppositories. ASHP Annual Meeting. 1993;50:p–144E.

1002. Tuchel N, Lenhardt E. Stability of chloral hydrate solutions. II. Farmacia (Bucharest). 1961;9:351–6.

1003. Taguchi M, Horiuchi T, Mimura Y, et al. Studies on hospital preparation "chloral hydrate syrup." Byoin Yakugaki. 1999;25:546–51.

1004. Narang PK, Bird G, Crouthamel WG. High-performance liquid chromatography assay for benzocaine and p-aminobenzoic acid including preliminary stability data. J Pharm Sci. 1980;69:1384–7.

1005. Liu H, Li S, Yuan Y, et al. Self-emulsifying preparation and chemical stability study of ketoprofen. Shenyang Yaoke Daxue Xuebao. 2005;22:5–7, 14.

1006. Remon JP, Gyselinck P, De Vos G, et al. Stability studies of Carbopol gels with benzoyl peroxide. Farm Tijdschrift Belg. 1979;56:279–90.

1007. Puigdellivol E, Carral ME, Dalmau JM. Comparative study of cefadroxil and cephalexin in solid form and in solution. Afinidad. 1981;38:337–42.

1008. Racz I, Mezei J, Farkas M, et al. Stability testing of the virostatic drug 5-iodo-2-deoxyuridine in solution. 1. Testing of the pH dependence of hydrolytic decomposition and interpretation of the decomposition mechanism. Acta Pharm Technol. 1980;26:29–37.

1009. Miligi MF, Kassem AA. Stability of nystatin ointment. Bull Faculty Pharm (Cairo). 1978;15:67–80.

1010. Aye RD, Koch H. Development of mydriatic eye drops. 1. Stability conditions of cyclopentolate. Deutsche Apotheker Zeitung. 1974;114:661–4.

1011. Lever PG, Hague JR. Observations on phenothiazine concentrates and diluting agents. Am J Psychiatry. 1964;120:1000–2.

1012. Freudenthaler S, Meineke I, Schreeb KH, et al. Influence of urine pH and urinary flow on the renal excretion of memantine. Br J Clin Pharmacol. 1998;46:541–6.

1013. Anon. Memantine hydrochloride. Accessed at www.chemicalland21.com/life-science/phar/ MEMANTINE.htm. July 9, 2006.

1014. Yamreudeewong W, Teixeira MG, Mayer GE. Stability of memantine in an extemporaneously prepared oral liquid. Int J Pharm Compound. 2006;10:316–7.

1015. Zhang Y, Trissel LA, Koontz SE. Valganciclovir hydrochloride stability in an extemporaneously compounded oral suspension. Data on file. 2002.

1016. Lv FF, Li N, Zheng LQ, et al. Studies on the stability of the chloramphenicol in the microemulsion free of alcohols. Eur J Pharm Biopharm. 2006;62:288–94.

1017. Elefante A, Muindi J, West K, et al. Long-term stability of a patient-convenient 1 mg/mL suspension of tacrolimus for accurate maintenance of stable therapeutic levels. Bone Marrow Transplant. 2006;37:781–4.

1018. Han J, Beeton A, Long PF, et al. Physical and microbiological stability of an extemporaneous tacrolimus suspension for paediatric use. J Clin Pharm Ther. 2006;31:167–72.

1019. Yamashita K, Tochiomi N, Okimoto K, et al. Establishment of new preparation method for solid dispersion formulation of tacrolimus. Int J Pharm. 2003;267:79–91.

1020. Ragno G, Risoli A, Ioele G, et al. Photo- and thermal-stability studies on benzimidazole anthelmintics by HPLC and GC-MS. Chem Pharm Bull. 2006;54:802–6.

1021. Kamin W, Schwabe A, Kramer I. Inhalation solutions: which ones are allowed to be mixed? Physico-chemical compatibility of drug solutions in nebulizers. J Cyst Fibros. 2006;5:205–13.

1022. Roberts GW, Badcock NR, Jarvinen AO. Cystic fibrosis inhalation therapy: stability of a combined salbutamol/colistin solution. Aust J Hosp Pharm. 1992;22:378–80.

1023. Gooch MD. Stability of albuterol and tobramycin when mixed for aerosol administration. Respir Care. 1991;36:1387–90.

1024. Joseph JC. Compatibility of nebulizer solution admixtures. Ann Pharmacother. 1997;31:487–9.

1025. White TS, Hood JC. Stability of selected antibiotic solutions with albuterol and ipratropium inhalation solutions in the treatment of pulmonary infections. ASHP Midyear Clinical Meeting. 1998;33:p43E.

1026. Smaldone GC, McKenzie J, Cruz Rivera M, et al. Budesonide inhalation suspension is chemically compatible with other nebulizing formulations. Chest (Abstract). 2000;118:98.

1027. Gronberg S, Magnusson P, Bladh N. Chemical compatibility of budesonide inhalation suspension (Pulmicort) with other nebulizing products. American Thoracic Society (Abstract). 2001.

1028. Amer MM, Takla KF. Studies on the stability of some pharmaceutical formulations. V. Stability of erythromycin. Bull Faculty Pharmacy (Cairo). 1978;15:325–9.

1029. Shah SA, Sander S, Coleman CI, et al. Delivery of esomeprazole magnesium through nasogastric and gastrostomy tubes using an oral liquid vehicle as a suspending agent in vitro. Am J Health Syst Pharm. 2006;63:1882–7.

1030. Welage LS. Overview of pharmacologic agents for acid suppression in critically ill patients. Am J Health Syst Pharm (Suppl). 2005;62:S4–10.

1031. Messaouik D, Sauto-Miranda V, Bagel-Boithias S, et al. Comparative study and optimisation of the administration mode of three proton pump inhibitors by nasogastric tube. Int J Pharm. 2005;299:65–72.

1032. Chun AHC, Shi HH, Achai R, et al. Lansoprazole: administration of the contents of a capsule dosage formulation through a nasogastric tube. Clin Ther. 1996;18:833–42.

1033. Freston J, Chiu Y-L, Pan WJ, et al. Effects on 24-hour intragastric pH: a comparison of lansoprazole administered nasogastrically in apple juice and pantoprazole administered intravenously. Am J Gastroenterol. 2001;96:2058–65.

1034. Phillips JO, Metzler M, Johnson K. The stability of simplified omeprazole suspension (SOS). Crit Care Med (Suppl). 1998;26:101A.

1035. Tsai WL, Poon SK, Yu HK, et al. Nasogastric lansoprazole is effective in suppressing gastric acid secretion in critically ill patients. Aliment Pharmacol Ther. 2000;14:123–7.

1036. Sostek MB, Chen Y, Skammer W, et al. Esomeprazole administered through a nasogastric tube provides bioavailability similar to oral dosing. Aliment Pharmacol Ther. 2003;18:581–6.

1037. Lv FF, Zheng LQ, Tung CH. Phase behavior of the microemulsions and the stability of the chloramphenicol in microemulsion-based ocular drug delivery system. Int J Pharm. 2005;301:237–46.

1038. Chen LM, Wang B, Zhou Y, et al. Chemical stability of the aqueous clobetasol propionate solution. J Chin Pharm Univ. 2006;37:226–9.

1039. Trissel LA, Zhang Y, Koontz SE. Temozolomide stability in extemporaneously compounded oral suspensions. Int J Pharm Compound. 2006;10:396–9.

1040. Anon. Temozolomide NSC 362856 Pharm Data, National Cancer Institute. Accessed at: http://dtpws4.ncicrf.gov/PHARM_DATA/362856.html. July 17, 2006.

1041. Sampson JH, Villavicencio AT, McLendon RE, et al. Treatment of neoplastic meningitis with intrathecal temozolomide. Clin Cancer Res. 1999;5:1183–8.

1042. Granero GE, Amidon GL. Stability of valacyclovir: implications for its oral bioavailability. Int J Pharm. 2006; 317:14–8.

1043. Wang DP. Stability of tetracaine in aqueous systems. J Taiwan Pharm Assoc. 1983;35:132–41.

1044. Morales ME, Gallardo V, Lopez G, et al. Stability of an oral liquid formulation of morphine for pediatric use. Farm Hosp. 2006;30:29–32.

1045. Roos PJ, Glerum JH, Meilink JW. Stability of morphine hydrochloride in a portable pump reservoir. Pharm Week Sci. 1992;14:23–6.

1046. Nagtegaal JE, de Jong A, de Waard WJ, et al. Formulation and shelf life of two solutions for inhalation. Ziekenhuisfarmacie. 1997;13:23–9.

1047. Kersten D, Gober B, Ulriketimm. The stability of aqueous tetracaine solutions in nonisothermal short-time test with linear increase in temperature. Pharmazie. 1981;36:341–4.

1048. Harb N. The stability of amethocaine hydrochloride solutions. J Hosp Pharm. 1969;26:44–5.

1049. Isla Tejera B, Garzas Martin de Almagro C, Cardenas Aransuna M, et al. Stability and in vitro activity of voriconazole eyedrops at a concentration of 3 mcg/mL. Farm Hosp. 2005;29:331–4.

1050. Groeschke J, Solassol I, Bressolle F, et al. Stability of amphotericin B and nystatin in antifungal mouthrinses containing sodium hydrogen carbonate. J Pharm Biomed Anal. 2006;42:362–6.

1051. Marigny K, Lohezic-Ledevehat F, Aubin F, et al. Stability of oral liquid preparations of methylergometrine. Pharmazie. 2006;61:701–5.

1052. Hawkins MG, Karriker MJ, Wiebe V, et al. Drug distribution and stability in extemporaneous preparations of meloxicam and carpofen after dilution and suspension at two storage temperatures. J Am Vet Med Assoc. 2006;229:968–74.

1053. Bakri SJ, Snyder MR, Pulido JS, et al. Six-month stability of bevacizumab (Avastin) binding to vascular endothelial growth factor after withdrawal into a syringe and refrigeration or freezing. Retina. 2006;26:519–22.

1054. Burnett JE, Balkin ER. Stability and viscosity of a flavored omeprazole oral suspension for pediatric use. Am J Health Syst Pharm. 2006;63:2240–7.

1055. Ameer B, Callahan RJ, Dragotakes SC, et al. Preparation and stability of an oral suspension of dipyridamole. J Pharm Technol. 1989;5:202–5.

1056. Akkers MRJ, Holtmaat I, van Niel JCC. Analysis, stability and a (ring) survey of physostigmine salicylate injection 1 ml = 1 mg FNA. Ziekenhuisfarmacie. 1994;10:44–9.

1057. Accordino A, Chambers RA, Thompson BC. The stability of a topical solution of cocaine hydrochloride. Aust J Hosp Pharm. 1996;26:629–33.

1058. Achach K, Peroux E. Antibiotic ophthalmic solutions: stability study. J Pharm Clin. 1999;65–6.

1059. Ammar AA, Marzouk MA. Carbamazepine suppositories: influence of physical parameters and stability. Egypt J Biomed Sci. 2004;16:84–96.

1060. Alberg U, Folger M, Mueller-Goymann CC. Investigations on the long-term stability of corticosteroids in modified water containing hydrophilic ointment. Pharm Pharmacol Lett. 1998;8:53–6.

1061. Hara C, Hashizume T, Tanaka J, et al. Problems in hospital preparation of acute poisoning antagonist (methylene blue injection). Japan J Toxicol. 2006;19:257–63.

1062. Aki H, Okamoto Y, Kimura T. Compatibility and stability tests of risperidone with soft-drinks by isothermal titration microcalorimetry. J Therm Anal Calorimetry. 2006;85:681–4.

1063. Varma R, Winarko J, Kiat-Winarko T, et al. Concentration of latanoprost ophthalmic solution after 4 to 6 weeks' use in an eye clinic setting. Invest Ophthalmol Vis Sci. 2006;47:222–5.

1064. Rexroad VE, Parsons TL, Hamzeh FM, et al. Stability of nevirapine suspension in prefilled oral syringes used for reduction of mother-to-child HIV transmission. J Acquir Immune Defic Syndr. 2006;43:373–5.

1065. Kodym A, Zawisza T, Buzka K, et al. Influence of additives and storage temperature on physicochemical and microbiological properties of eye drops containing cefazolin. Acta Pol Pharm. 2006;63:225–34.

1066. Abou-Taleb AE, Abdel-Rhman AA, Samy EM, et al. Formulation and stability of rofecoxib suppositories. J Drug Deliv Sci Technol. 2006;16:389–96.

1067. Olguin HJ, Perez CF, Perez JF, et al. Bioavailability of an extemporaneous suspension of propafenone made from tablets. Biopharm Drug Disp. 2006;27:241–5.

1068. Titcomb LC. Mydriatic-cycloplegic drugs and corticosteroids. Pharm J. 1999;263:900–5.

1069. Allen LV. Desonide 0.05% otic liquid. Int J Pharm Compound. 2003;7:298.

1070. Gupta VD. Chemical stability of desonide in ear drops. Int J Pharm Compound. 2007;11:79–81.

1071. Beggs WH. Influence of alkaline pH on the direct lethal action of miconazole against Candida albicans. Biomed Life Sci. 1992;120:11–3.

1072. Xu QA, Trissel LA. Stability-indicating HPLC methods for drug analysis, third edition. Washington, DC: American Pharmacists Association; 2008.

1073. Wen-Lin Chow J, Decarie D, Dumont RJ, et al. Stability of dexamethasone in extemporaneously prepared oral suspensions. Can J Hosp Pharm. 2001;54:96–101.

1074. Anon. Specifications L-4-Boronophenyl-alanine. Interpharma Praha. Accessed at www. interpharmapraha.com/html/ pecif_bpa. shtml. March 8, 2007.

1075. van Rij CM, Sinjewel A, van Loenen AC, et al. Stability of 10B-L-boronophenyl-alanine-fructose injection. Am J Health Syst Pharm. 2005;62:2608–10.

1076. Mori Y, Suzuki A, Yoshino K, et al. Complex formation of p-borono-phenylalanine with some monosaccharides. Pigment Cell Res. 1989;2:273–7.

1077. Hatanaka H, Komada F, Shiono M, et al. Tissue distribution of para-borono-phenylalanine administered orally as a cyclodextrin inclusion complex to melanoma-bearing hamsters. Pigment Cell Res. 1992;5:38–40.

1078. Pitois A, Aldave de las Heras L, Zampolli A, et al. Capillary electrophoresis-electrospray mass spectrometry and Hr-ICP-MS for the detection and quantification of 10B-boronophenylalanine (10B-BPA) used in boron neutron capture therapy. Anal Bioanal Chem. 2006;384:751–60.

1079. Cober MP, Johnson CE. Stability of an extemporaneously prepared alcohol-free phenobarbital suspension. Am J Health Syst Pharm. 2007;64:644–6.

1080. Schmutz A, Thormann W. Assessment of impact of physicochemical drug properties on monitoring drug levels by micellar electrokinetic capillary chromatography with direct serum injection. Electrophoresis. 1994;15:1295–303.

1081. Anderson RA. Stability of pilocarpine solutions. Can J Pharm Sci. 1967;2:25–6.

1082. Dentinger PJ, Swenson CF, Anaizi NH. Stability of amphotericin B in an extemporaneously compounded oral suspension. Am J Health Syst Pharm. 2001;58:1021–4.

1083. Asmus MJ, Vaughan LM, Malcolm R, et al. Stability of frozen methacholine solutions in unit-dose syringes for bronchoprovocation. Chest. 2002;121:1634–7.

1084. Baeschlin K, Etter JC. Stability of pilocarpine in an aqueous medium. II. Kinetic study of pilocarpine hydrolysis. Pharm Acta Helv. 1969;44:339–47.

1085. Asiri YA, Bawazir SA, Al-Hadiya BM, et al. Stability of extemporaneously prepared spironolactone suspensions in Saudi hospitals. Saudi Pharm J. 2001;9:106–12.

1086. Benyahya N, Bou P, Hary L. Chemical stability and microbial contamination of a 1% vancomycin buccal gel. J Pharm Clin. 1994;13:102–5.

1087. Maia AM, Baby AR, Pinto CASO, et al. Influence of sodium metabisulfite and glutathione on the stability of vitamin C in O/W emulsion and extemporaneous aqueous gel. Int J Pharm. 2006;322:130–5.

1088. Kiser TH, Oldland AR, Fish DN. Stability of acetylcyteine solution repackaged in oral syringes and associated cost savings. Am J Health Syst Pharm. 2007;64:762–6.

1089. Messaouik D, Sautou-Miranda V, Balayssac D, et al. Is the administration of esomeprazole through a nasogastric tube modified by concomitant delivery of a nutrition mixture? Eur J Hosp Pharm Sci. 2006;12:100–4.

1090. Mueller BA, Brierton DG, Abel SR, et al. Effect of enteral feeding with Ensure on oral bioavailabilities of ofloxacin and ciprofloxacin. Antimicrob Agents Chemother. 1994;38:2101–5.

1091. Chiadmi F, Iyer A, Cisternino S, et al. Stability of levamisole oral solutions prepared from tablets and powder. J Pharm Pharm Sci. 2005;8:322–5.

1092. Piel G, Hayette MP, Pavoni E, et al. In vitro comparison of the antimycotic activity of a miconazole-HP-beta-cyclodextrin solution with miconazole surfactant solution. J Antimicrob Chemother. 2001;48:83–7.

1093. Beggs WH. Fungistatic activity of miconazole against Candida albicans in relation to pH and concentration of nonprotonated drug. Mycoses. 1989;32:239–44.

1094. Dash AK, Cudworth GC. Evaluation of an acetic acid ester of monoglyceride as a suppository base with unique properties. AAPS PharmSciTech. 2001;2(13):1–7.

1095. Raust JA, Goulay-Dufay S, Le Hoang MD, et al. Stability studies of ionized and nonionized 3,4-diaminopyridine: hypothesis of degradation pathways and chemical structure of degradation products. J Pharm Biomed Anal. 2007;43:83–8.

1096. Hongo T, Hikage S, Sato A. Stability of benzoyl peroxide in methyl alcohol. Dental Materials J. 2006;25:298–302.

1097. Hayes RD, Beach JR, Rutherford DM, et al. Stability of methacholine chloride solutions under different storage conditions over a 9 month period. Eur Respir J. 1998;11:946–8.

1098. Cao J, Ge Y. Content determination of minoxidil in lotion and its stability. Zhongguo Yiyuan Yaoxue Zazhi. 1999; 19:287–9.

1099. Ekpe A, Jacobsen T. Effect of various salts on the stability of lansoprazole, omeprazole, and pantoprazole as determined by high-performance liquid chromatography. Drug Dev Ind Pharm. 1999;25:1057–65.

1100. Alffenaar JWC, van der Heiden J, Herder RE, et al. Rapid development of pharmacy prepared labetalol injection as the solution for Trandate drug discontinuity. Eur J Hosp Pharm Sci. 2006;12:123–8.

1101. Ortega de la Cruz C, Fernandez Gallardo LC, Damas Fernandez-Figares M, et al. Physico-chemical compatibility of drugs with enteral nutrition. Nutr Hosp. 1993;8:105–8.

1102. Stahlmann SA, Frey OR. Stability study of pentoxifylline PVC infusion containers. Krankenhausfarmazie. 1998;19:553–7.

1103. Landersjo L, Kallstrand G. Studies on the stability and compatibility of drugs in infusion solutions. III. Factors affecting the stability of cloxacillin. Acta Pharm Suec. 1974;11:563–80.

1104. Bundgaard H, Ilver K. Kinetics of degradation of cloxacillin sodium in aqueous solution. Dansk Tidsskr Farm. 1970; 44:365–80.

1105. Lynn B. Pharmaceutical aspects of semi-synthetic penicillins. J Hosp Pharm. 1970;28:71–86.

1106. Gupta VD. Chemical stability of hydrocortisone in an oral liquid dosage form without suspending agents. Int J Pharm Compound. 2007;11:259–61.

1107. Caruthers RL, Johnson CE. Stability of extemporaneously prepared sodium phenylbutyrate oral suspensions. Am J Health Syst Pharm. 2007;64:1513–5.

1108. Ford SM, Kloessel LG, Grabenstein JD. Stability of oseltamivir in various extemporaneous liquid preparations. Int J Pharm Compound. 2007;11:162–74.

1109. Schlatter J, Saulnier J-L. Stability of ranitidine oral solutions prepared from commercial forms. Eur Hosp Pharm. 1998;4:23–5.

1110. Juarez JC, Lopez RM, Hueto M, et al. Stability of amphotericin B in water solutions for inhalation. Eur Hosp Pharm. 1997;3:59–60.

1111. Brisaert M, Plaizier-Vercammen JA. Investigation on the photostability of tretinoin in creams. Int J Pharm. 2007;334:56–61.

1112. Nahata MC, Morosco RS, Hipple TF, et al. Development of stable oral suspensions of ciprofloxacin. J App Ther Res. 2000;3:61–5.

1113. Rotheli-Simmen B, Martinelli E, Muhlebach S. Formulation of a stable calcium gluconate gel for topical treatment of hydrofluoric acid burns. Eur Hosp Pharm. 1996;2:176–80.

1114. Hill SS, Farrands P. Formulation and stability of isosorbide dinitrate gel. Eur Hosp Pharm. 2000;6:1–5.

1115. Chow DSL, Bhagwatwar HP, Phadung-pojna S, et al. Stability-indicating high-performance liquid chromatographic assay of busulfan in aqueous and plasma samples. J Chromatogr B. 1997;704:277–88.

1116. Ferreira MO, Bahia MF, Costa P. Stability of ranitidine hydrochloride in different aqueous solutions. Eur J Hosp Pharm Sci. 2004;10:60–3.

1117. Griffiths W, Ing H, Fleury-Souverain S, et al. The stability of ready-to-use (RTU) ephedrine hydrochloride in polypropylene syringes for use in maternal hypotension. Eur J Hosp Pharm Sci. 2005;11:107–10.

1118. Casasin Edo T, Roca Massa M. Sistema de distribucion de medicamentos utiliz-ados en anesthesia mediante jeringas precargadas. Estudio de establilidad. Farm Hosp. 1996;20:55–9.

1119. Storms ML, Stewart JT, Warren FW. Stability of ephedrine sulfate at ambient temperature and 4°C in polypropylene syringes. Int J Pharm Compound. 2001;5:394–6.

1120. Lewis B, Jarvi E, Cady P. Atropine and ephedrine adsorption to syringe. J Am Assoc Nurse Anesth. 1994;62:257–60.

1121. Kodym A, Zawaisza T, Napierala B, et al. Influence of additives and storage temperature of physicochemical and microbiological properties of eye drops containing ceftazidime. Acta Pol Pharm. 2006;63:507–13.

1122. Partin JM, Poust RI, Cox FO, et al. Stability of busulfan in suspension. Pharm Res. 1988;5:845–50.

1123. Tanaka H, Asakura M, Doi C, et al. Optimum preparation of levocarnitine chloride solution in the hospital pharmacy. Yakugaku Zasshi. 2006;126:805–9.

1124. Gupta VD. Chemical stability of cyproheptadine hydrochloride in an oral liquid dosage form. Int J Pharm Compound. 2007;11:347–8.

1125. Buur JL, Baynes RE, Yeatts JL, et al. Analysis of diltiazem in Lipoderm transdermal gel using reversed-phase high-performance liquid chromatography applied to homogenization and stability studies. J Pharm Biomed Anal. 2005;38:60–5.

1126. Glass BD, Haywood A. Stability considerations in liquid dosage forms extemporaneously prepared from commercially available products. J Pharm Pharm Sci. 2006;9:398–426.

1127. Hoelgaard A, Moller N. Hydrate formation of metronidazole benzoate in aqueous suspensions. Int J Pharm. 1983;15:213–221.

1128. Zietsman S, Kilian G, Worthington M, et al. Formulation development and stability studies of aqueous metronidazole benzoate suspensions containing various suspending agents. Drug Dev Ind Pharm. 2007;33:191–7.

1129. Hudson KC, Asbill CS, Webster AA. Isoniazid release from suppositories compounded with selected bases. Int J Pharm Compound. 2007;11:433–7.

1130. Freed AL, Silbering SB, Kolodsick KJ, et al. The development and stability assessment of extemporaneous pediatric formulations of Accupril. Int J Pharm. 2005;304:135–44.

1131. Freed AL, Kolodsick K, Silbering S, et al. The development and stability assessment of extemporaneous pediatric formulations of Accupril. Paper presented at the AAPS Annual Meeting and Exposition, 2003.

1132. Caplar V, Sunjic V, Kajfez F, et al. Physicochemical properties and identification methods of tinidazole. Acta Pharm Jugoslav. 1974;24:147–51.

1133. Tu J, Wu Z, Zhao A. Study on stability of tinidazole solution. Zhongguo Yaoke Daxue. 1996;27:525–7.

1134. Salo JP, Yli-Kauuhaluoma J, Salomies H. On the hydrolytic behavior of tinidazole, metronidazole, and ornidazole. J Pharm Sci. 2003;92:739–46.

1135. Peloquin CA, Durbin D, Childs J, et al. Stability of antituberculosis drugs mixed in food. Clin Infect Dis. 2007;45:521.

1136. Anon. Pranoprofen. Japanese Pharmacopeia, 14th ed. 2002; 699–700.

1137. Kaplan MA, Coppola WP, Nunning BC, et al. Pharmaceutical properties and stability of amikacin, part i. Curr Ther Res Clin Exp. 1976;20:352–8.

1138. Nunning BC, Granatek AP. Physical compatibility and chemical stability of amikacin sulfate in large-volume parenteral solutions, part ii. Curr Ther Res Clin Exp. 1976;20:359–68.

1139. Holmes CJ, Ausman RK, Kundsin RB, et al. Effect of freezing and microwave radiation thawing on the stability of six antibiotic admixtures in plastic bags. Am J Hosp Pharm. 1982;39:104–8.

1140. Zbrozek AS, Marble DA, Bosso JA, et al. Compatibility and stability of clindamycin phosphate–aminoglycoside combinations within polypropylene syringes. Drug Intell Clin Pharm. 1987;21:806–10.

1141. Poochikian GK, Cradock JC. Heroin: stability and formulation approaches. Int J Pharm. 1983;13:219–26.

1142. Beaumont IM. Stability study of aqueous solutions of diamorphine and morphine using HPLC. Pharm J. 1982;229:39–41.

1143. Kirk B, Hain WR. Diamorphine injection BP incompatibility. Pharm J. 1985;235:426.

1144. Allwood MC. The stability of diamorphine alone and in combination with antiemetics in plastic syringes. Palliat Med. 1991;5:330–3.

1145. Kleinberg ML, Duafala ME, Nacov C, et al. Stability of heroin hydrochloride in infusion devices and containers for intravenous administration. Am J Hosp Pharm. 1990;47:377–81.

1146. Pomplun M, Johnson JJ, Johnston S, et al. Stability of a heparin-free 50% ethanol lock solution for central venous catheters. J Oncol Pharm Pract. 2007;13:33–7.

1147. Santiero ML, Sagraves R. Osmolality of small-volume i.v. admixtures for pediatric patients. Am J Hosp Pharm. 1990;47:1359–64.

1148. Nahata MC. Stability of trimethoprim in an extemporaneous liquid dosage form. J Pediatr Pharm Pract. 1997;2:82–4.

1149. Ruiz L, Rodriguez I, Baez R, et al. Stability of an extemporaneously prepared recombinant human interferon alfa-2b eye drop formulation. Am J Health Syst Pharm. 2007;64:1716–9.

1150. Palmer AJ, Sewell GJ, Rowland CG. Qualitative studies on a-interferon-2b in prolonged continuous infusion regimes using gradient elution high-performance liquid chromatography. J Clin Pharm Ther. 1988;13:225–31.

1151. Bonasia P, Cook C, Cheng Y, et al. Compatibility of arformoterol tartrate inhalation solution with three nebulized drugs. Curr Med Res Opin. 2007;23:2477–83.

1152. Foppa T, Murakami FS, Silva MAS. Development, validation and stability study of pediatric atenolol syrup. Pharmazie. 2007;62:519–21.

1153. Hui M, Kwok AKH, Pang CP, et al. An in vitro study on the compatibility and concentrations of combinations of vancomycin, amikacin, and dexamethasone in human vitreous. Eye. 2007;21:643–8.

1154. Kwok AKH, Hui M, Pang CP, et al. An in vitro study of ceftazidime and vancomycin concentrations in various fluid media: implications for use in treating endophthalmitis. Invest Ophthalmol Vis Sci. 2002;43:1182–8.

1155. Abdel-Rahman SM, Nahata MC. Stability of itraconazole in an extemporaneous suspension. J Pediatr Pharm Pract. 1998;3:115–8.

1156. Abdel-Rahman SM, Johnson FK, Gouthier-Dubois G, et al. The bioequivalence of nizatidine (Axid) in two extemporaneously and one commercially prepared oral liquid formulations compared with capsule. J Clin Pharmacol. 2003;43:148–53.

1157. Mehta AC, Hart-Davies S, Payne J, et al. Stability of amoxicillin and potassium clavulanate in co-amoxiclav oral suspension. J Clin Pharm Ther. 1994;19:313–5.

1158. Plaizier-Vercammen JA. Parameter-effecten op de stabiliteit en activiteit van pilocarpine collyria, een overzicht. Farm Tijdschrift Belg. 1993;70:2–9

1159. Pappert EJ, Lipton JW, Goetz CG, et al. The stability of carbidopa in solution. Movement Disorders. 1997;12:608–10.

1160. Pappert EJ, Buhrfiend C, Lipton JW, et al. Levodopa stability in solution: time course, environmental effects, and practical recommendations for clinical use. Movement Disorders. 1996;11:24–6.

1161. David M, Huyck CL. Stability of potassium iodide in various disguising agents. Am J Hosp Pharm. 1958;15:586–7.

1162. Christenson I. Kinetics of hydrolysis of physostigmine salicylate. 1969;6:287–98.

1163. Breinlich J. Stability of sterilized tetracaine hydrochloride solutions. Pharm Zeit. 1965;110:579–82.

1164. Bonhomme L, Duleba B, Beugre T, et al. HPLC determination of the stability of retinoic acid in gel formulation. Int J Pharm. 1990;65:R9–R10.

1165. Morand K, Bartoletti AC, Bochot A, et al. Liposomal amphotericin B eye drops to treat fungal keratitis: physico-chemical and formulation stability. Int J Pharm. 2007;344:150–3.

1166. Winiarski AP, Infeld MH, Tscherne R, et al. Preparation and stability of extemporaneous oral liquid formulations of oseltamivir using commercially available capsules. J Am Pharm Assoc. 2007;47:747–55.

1167. Burger DM, Hendriks E, Wuis EW. Stability of sodium hypochlorite solution 0.05%. Ziekenhuisfarmacie. 1996;12:133–4.

1168. Reddy GT, Kumar TM, Veena. Formulation and evaluation of alendronate sodium gel for the treatment of bone resorptive lesions in periodontitis. Drug Deliv. 2005;12:217–22.

1169. Johnson CE, Cober MP, Ludwig JL. Stability of partial doses of omeprazole–sodium bicarbonate oral suspension. Ann Pharmacother. 2007;41:1954–61.

1170. Anon. Cysteamine hydrochloride. Sigma-Aldrich. Available at: http://www.sigmaaldrich.com/catalog/search/ProductDetail/FLUKA/30078. Accessed 12/24/2007.

1171. Brodrick A, Broughton HM, Oakley RM. The stability of an oral liquid formulation of cysteamine. J Clin Hosp Pharm. 1981;6:67–70.

1172. Brisaert MG, Everaerts I, Plaizier-Vercammen JA. Chemical stability of tretinoin in dermatological preparations. Pharm Acta Helv. 1995;70:161–6.

1173. Branje JA, Cremers HMHG. Preparation and stability of levothyroxine and liothyronine injections. Ziekenhuisfarmacie. 1988;4:1–5.

1174. Ekiz-Gucer N, Reisch J. Photostability of minoxidil in the liquid and solid state. Acta Pharm Turc. 1990;32:103–6.

1175. Rogers AR, Smith G. Stability of physostigmine eye-drops BPC. Pharm J. 1973;211:353–5.

1176. Huang L. Effect of pH and temperature on the stability of 1% tetracaine hydrochloride injection. Yaoxue Tongbao. 1982;17:208–9.

1177. Munoz CA, Odalys Fernandez V, Leopoldo Nunez dela F, et al. Chemical and physical parameters influence on the formulation design of budesonide 0.025% creams. Rev Mexicana Ciencias Farm. 2005;36:25–30.

1178. Marshall TM, Mullen MV. The stability of extemporaneously prepared solutions of fluoxetine HCl. Pharm Sci Comm. 1994;4:143–5.

1179. Dentinger PJ, Swenson CF. Stability of codeine phosphate in an extemporaneously compounded syrup. Am J Health Syst Pharm. 2007;64:2569–73.

1180. Gupta VD. Effect of some formulation adjuncts on the stability of benzoyl peroxide. J Pharm Sci. 1982;71:585–7.

1181. Gupta VD, Parasrampuria J, Gardner SN. Chemical stabilities of isoetharine hydrochloride, metaproterenol sulphate and terbutaline sulphate after mixing with normal saline for respiratory therapy. J. Clin Pharm Ther. 1988;13:165–9.

1182. Green PG, Kennedy CT, Forbes DR. Anthralin stability in various vehicles. J Am Acad Dermatol. 1987;16:984–8.

1183. Grandy JL, McDonnell WN. Evaluation of concentrated solutions of guaifenesin for equine anesthesia. J Am Vet Med Assoc. 1980;176:619–22.

1184. Goodwin BL, Weg MW, Moore FG. Stability of dinoprostone in Witepsol-based pessaries. Pharm J. 1986;237:801.

1185. Gauger LJ. Extemporaneous preparation of a dinoprostone gel for cervical ripening. Am J Hosp Pharm. 1983;40:2195–6.

1186. Gauger LJ. Hydroxyethylcellulose gel as a dinoprostone vehicle. Am J Hosp Pharm. 1984;41:1761–2.

1187. Mehta AC, Calvert RT. Stability of retinoic acid eye-drops. Pharm J. 1987;238:214–5.

1188. Moreno S, Roca M, del Pozo A. Stability of vancomycin eye-drops measured by high-resolution liquid chromatography. Cien Pharm. 1996;6:77–81.

1189. Chellquist EM, Gorman WG. Benzoyl peroxide solubility and stability in hydric solvents. Pharm Res. 1992;9:1341–6.

1190. Fuentes de Frutos MJ, Mataix Sanjuan A, Garcia Sanchez F, et al. Stability of apomorphine in dosifiers for intranasal administration. Farm Hosp. 1995;19:222–4.

1191. Park SH, Gill MA, Dopheide JA. Visual compatibility of risperidone solution and lithium citrate syrup. Am J Health Syst Pharm. 2003;60:612–3.

1192. Hudson SJ, Jones MF, Nolan S, et al. Stability of premixed syringes of diamorphine and hyperbaric bupivacaine. Int J Obstet Anesth. 2005;14:284–7.

1193. Goeber B, Timm U, Pfeifer S. Comparative studies on the stability of aqueous drug solutions in the isothermal and nonisothermal short-time test as well as in the long-time test. Part 1. Stability of aqueous tetracaine solutions in the isothermal short-time test and long-time test. Pharmazie. 1979;34:161–4.

1194. Cober MP, Johnson CE. Stability of 70% alcohol solutions in polypropylene syringes for use in ethanol-lock therapy. Am J Health Syst Pharm. 2007;64:2480–2.

1195. Dobrinas M, Fleury-Souverain S, Bonnabry P, et al. Off-label antibiotic preparation. Ophthamology. 2007;114:2095.

1196. Lee R, Lai H. Stability of miconazole on dry heating in vegetable oils. Aust J Hosp Pharm. 1985;15:233–4.

1197. Fawcett JP, Tucker IG, Davies NM, et al. Formulation and stability of pilocarpine oral solution. Int J Pharm Pract. 1994;3:14–8.

1198. Wong VK, Ho PC. Stability of Konkion repacked in dropper bottles for oral administration. Aust J Hosp Pharm. 1996;26:641–4.

1199. Nahata MC, Morosco RS, Zuacha JA. Stability of glycopyrrolate in two extemporaneously prepared oral suspensions stored at two temperatures. ASHP Midyear Clinical Meeting. 2004;p490E.

1200. Lifshin LS, Fox JL. Stability of extemporaneously prepared ranitidine hydrochloride suspension. Paper presented at the ASHP Annual Meeting. 1992;120E.

1201. Szof CA, Garrett SD, Kauffman RE. Study to determine the stability of extemporaneously compounded zidovudine solution. Paper presented at the ASHP Midyear Clinical Meeting. 1989;212E.

1202. McCoy KS. Compounded colistimethate as possible cause of fatal acute respiratory distress syndrome. N Engl J Med. 2007;357:2310–1.

1203. Li J, Milne RW, Nation RL, et al. Stability of colistin and colistin methanesulfonate in aqueous media and plasma as determined by high-performance liquid chromatography. Antimicrob Agents Chemother. 2003;47:1364–70.

1204. Maloney T, O'Neill B. Stability of povidone-iodine antiseptic solution stored at 37°C. Med J Aust. 1986;144:389.

1205. Berger-Gryllaki M, Podilsky G, Widmer N, et al. The development of a stable oral solution of captopril for paediatric patients. Eur J Hosp Pharm Sci. 2007;13:67–72.

1206. Csoka G, Marton S, Zelko R, et al. Stability of diclofenac sodium and piroxicam in new transderm "soft-patch" type gel systems. STP Pharm Sci. 2000;10:415–8.

1207. Sharley NA, Yu AMC, Williams DB. Stability of mixtures formulated from warfarin tablets or powder. J Pharm Pract Res. 2007;37:95–7.

1208. Albert K, Bockshorn J. Chemical stability of oseltamivir in oral solutions. Pharmazie. 2007;62:678–82.

1209. Zhou J, Liu R, Li Y, et al. Effect of sodium chloride on the stability of 1% tetracaine hydrochloride injection. Yaoxue Tongbao. 1986;21:147–8.

1210. Murakawa Y, Takasugi M. Stability and physicochemical properties of Lamisil 125 tablet. Kagaku Ryoho No Ryoiki. 1998;14:293–6.

1211. Lee KC, Lee YJ, Hyun M, et al. Degradation of synthetic salmon calcitonin in aqueous solution. Pharm Res. 1992;9:1521–3.

1212. Gober B, Timm U, Wendlandt S, et al. Stability of tropicamide. Pharmazie. 1975;30:610–1.

1213. Prins AMA, Vos K, Franssen EJF. Instability of topical ciclosporin emulsion for nail psoriasis. Dermatology. 2007;215:362–3.

1214. Fittler A, Mayer A, Kocsis B, et al. Stability testing of amphotericin B nasal spray solutions with chemical and biological analysis. Acta Pharm Hung. 2007;77:159–64.

1215. Kodym A, Zawisza T, Taberska J, et al. Physicochemical and microbiological properties of eye drops containing cefuroxime. Acta Pol Pharm. 2006;63:293–9.

1216. Kramer I, Schwabe A, Lichtinghagen R, et al. Physicochemical compatibility of nebulizable drug mixtures containing dornase alfa and ipratropium and/or albuterol. Pharmazie. 2007;62:760–6.

1217. Pipracil [hospital formulary monograph]. Wayne, NJ; Lederle Laboratories; 1981.

1218. Gupta VD, Davis DD. Stability of piperacillin sodium in dextrose 5% and sodium chloride 0.9% injections. Am J IV Ther Clin Nutr. 1984;11:14–5, 18–9.

1219. Chedru-Legros V, Fines-Guyon M, Cherel A, et al. Fortified antibiotic (amikacin, ceftazidime, vancomycin) eye drop stability assessment at −20°C. J Fr Ophthamol. 2007;30:807–13.

1220. Hinkle GH, Dura JV, Morosco RS, et al. Extended stability of iobenguane under simulated clinical conditions. Am J Health Syst Pharm. 2008;65:142–4.

1221. Hecq JD, Jacquet S, Cavrenne P, et al. Stabilite physicochemique de solutions ophthalmiques renforcees pretes a l'emploi: une revue de la literature. Pharmakon. 2006;38:3–11.

1222. Shah RB, Prasanna HR, Rothman B, et al. Stability of ranitidine syrup repackaged in unit-dose containers. Am J Health Syst Pharm. 2008;65:325–9.

1223. Aliabadi HM, Romanick M, Desai S, et al. Effect of buffer and antioxidant on stability of a mercaptopurine suspension. Am J Health Syst Pharm. 2008;65:441–7.

1224. Gupta VD. Chemical stability of tramadol hydrochloride injection. Int J Pharm Compound. 2008;12:161–2.

1225. Gander B, Kloeti F, Christen P, et al. Stability of corticosteroids in zinc oxide-containing hydrophilic paste and lipophilic ointment. Eur J Pharm Biopharm. 1991;37:64–8.

1226. Kjonniksen I, Brustugun J, Niemi G, et al. Stability of epidural analgesic solution containing adrenaline, bupivacaine and fentanyl. Acta Anaesthiol Scand. 2000;44:864–7.

1227. Tamura T, Takayama K, Satoh H, et al. Evaluation of oil/water-type cyclosporine gel ointment with commercially available oral solution. Drug Dev Ind Pharm. 1997;23:285–91.

1228. Mackey C, Tanja JJ, Zaenz RV. Stability of a pharmacy-prepared silver lactate burn cream. Hosp Pharm. 1975;10:59–60.

1229. Nahata MC, Morosco RS, Zuacha J. Stability of sildenafil citrate in two extemporaneously prepared oral dosage forms stored under refrigeration and at room temperature. Paper presented at the ASHP Midyear Clinical Meeting, 2007.

1230. Nahata MC, Morosco RS, Zuacha J. Stability of diluted morphine liquid in amber oral syringes extemporaneously prepared and stored at room temperature. Paper presented at the ASHP Midyear Clinical Meeting, 2007.

1231. Kamin W, Schwabe A, Kramer I. Physicochemical compatibility of fluticasone-17-propionate nebulizer suspension with ipratropium and albuterol nebulizer solutions. Int J Chron Obstruct Pulmon Dis. 2007;2:599–607.

1232. Lee MG. A study on the stability of prostaglandin E2 in methyl hydroxyethyl cellulose gel by gas chromatography. J Clin Hosp Pharm. 1982;7:67–70.

1233. Huyghebaert N, De Beer J, Vervaet C, et al. Compounding of vitamin A, D3, E, and K3 supplements for cystic fibrosis patients: formulation and stability study. J Clin Pharm Ther. 2007;32:489–96.

1234. Gupta VD. Chemical stability of naltrexone hydrochloride injection. Int J Pharm Compound. 2008;12:274–5.

1235. Chu KO, Wang CC, Pang CP, et al. Method to determine stability and recovery of carboprost and misoprotol in infusion preparations. J Chromatogr B. 2007;857:83–91.

1236. Navid F, Christensen R, Minkin P, et al. Stability of sunitinib in oral suspension. Ann Pharmacother. 2008;42:962–6.

1237. McCluskey SV. Sterilization of glycerin. Am J Health Syst Pharm. 2008;65:1173–6.

1238. Price JC. Glycerin. In: Rowe RC, Shesky PJ, Owen SC, eds. Handbook of pharmaceutical excipients. 5th ed. Washington, DC: American Pharmacist Association: 2006;301–3.

1239. Fox LM, Foushee JA, Jackson DJ, et al. Visual compatibility of common nebulizer medications with 7% sodium chloride solution. Am J Health Syst Pharm. 2011;68:1032–5.

1240. Volpe DA, Gupta A, Ciavarella AB, et al. Comparison of the stability of split and intact gabapentin tablets. Int J Pharm. 2008;350:65–9.

1241. Bitter C, Suter K, Figueiro V, et al. Preservative-free triamcinolone acetonide suspension developed for intravitreal injection. J Ocul Pharmacol Ther. 2008;24:62–9.

1242. Nishimura A, Kobayashi A, Segawa Y, et al. Isolating triamcinolone acetonide particles for intravitreal use with a porous membrane filter. Retina. 2003;23:777–9.

1243. Hernaez-Ortega MC, Soto-Pedre E. A simple and rapid method for purification of triamcinolone acetonide suspension for intravitreal injection. Ophthalmic Surg Lasers Imaging. 2004;35:350–1.

1244. Wan J, Rickman C. The durability of intravesical oxybutynin solutions over time. J Urol. 2007;178:1768–70.

1245. Bladh N, Blychert E, Johansson K, et al. A new esomeprazole packet (sachet) formulation for suspension: in vitro characteristics and comparative pharmacokinetics versus intact capsules/tablets in healthy volunteers. Clin Ther. 2007;29:640–9.

1246. Gupta VD. Chemical stability of perphenazine in two commercially available vehicles for oral liquid dosage forms. Int J Pharm Compound. 2008;12:372–4.

1247. Ohzeki K, Kitahara M, Suzuki N, et al. Local anesthetic cream prepared from lidocaine-tetracaine eutectic mixture. Yakugaku Zasshi. 2008;128:611–6.

1248. Kwiecien A, Krzek J, Zylewski M. Stability of chosen beta-adreonolytic drugs of different polarity in basic environment. J AOAC Int. 2008;91:322–31.

1249. Uzunovic A, Vranic E. Stability of cefuroxime axetil oral suspension at different temperature storage conditions. Bosn J Basic Med Sci. 2008;8:93–7.

1250. Garg S, Svirskis D, Myftiu J, et al. Properties of a formulated paediatric phenobarbitone oral liquid. J Pharm Pract Res. 2008;38:28–31.

1251. Hames H, Seabrook JA, Matsui D, et al. A palatability study of a flavored dexamethasone preparation versus prednisolone liquid in children. Can J Clin Pharmacol. 2008;15:e95–8.

1252. Abdelmageed R, Labyad N, Watson DG, et al. Evaluation of the stability of morphine sulphate in combination with Instillagel. J Clin Pharm Ther. 2008;33:263–71.

1253. Severino P, Zanchetra B, Franco LM, et al. Intestinal absorption and physical chemical stability in fluconazole extemporaneous preparations. Lat Am J Pharm. 2007;26:744–7.

1254. Ghulam A, Keen K, Tuleu C, et al. Poor preservative efficacy versus quality and safety of pediatric extemporaneous liquids. Ann Pharmacother. 2007;41:857–60.

1255. Tashtoush BM, Jacobson EL, Jacobson MK. UVA is the major contributor to the photodegradation of tretinoin and isotretinoin: Implications for development of improved pharmaceutical formulations. Int J Pharm. 2008;352:123–8.

1256. Stulzer HK, Rodrigues PO, Cardoso TM, et al. Studies between captopril and pharmaceutical excipients used in tablets formulations. J Therm Anal Calorim. 2008;91:323–8.

1257. Singh P, Premkumar L, Mehrotra R, et al. Evaluation of thermal stability of indinavir sulphate using diffuse reflectance infrared spectroscopy. J Pharm Biomed Anal. 2008;47:248–54.

1258. Desai D, Rao V, Guo H, et al. Stability of low concentrations of guanine-based antivirals in sucrose or maltitol solutions. Int J Pharm. 2007;342:87–94.

1259. Ling J, Gupta VD. Stability of acyclovir sodium after reconstitution in 0.9% sodium chloride injection and storage in polypropylene syringes for pediatric use. Int J Pharm Compound. 2001;5:75–7.

1260. Akitoshi T, Shoko O, Tomoko N, et al. Controlled release of prednisolone from suppository prepared using powder of pulverized tablet. Yakugaku Zasshi. 2008;128:641–8.

1261. Majekodunmi BD, Lau-Cam CA, Nash RS. Stability of benzoyl peroxide in aromatic ester-containing formulations. Pharm Dev Technol. 2007;12:609–20.

1262. Jain AK. Solubilization of indomethacin using hydrotropes for aqueous injection. Eur J Pharm Biopharm. 2008;68:701–14.

1263. Vahdat L, Sunderland VB. Kinetics of amoxicillin and clavulanate degradation alone and in combination in aqueous solution under frozen conditions. Int J Pharm. 2007;342:95–104.

1264. Haynes RK. From artemisinin to new artemisinin antimalarials: biosynthesis, extraction, old and new derivatives, stereochemistry and medicinal chemistry requirements. Curr Top Med Chem. 2006;6:509–37.

1265. Gaudin K, Barbaud A, Boyer C, et al. In vitro release and stability of an artesunate rectal gel suitable for pediatric use. Int J Pharm. 2008;353:1–7.

1266. Tuleu C, Allam J, Gill H, et al. Short term stability of pH-adjusted lidocaine-adrenaline epidural solution used for emergency caesarean section. Int J Obstet Anesth. 2008;17:118–22.

1267. Rozman B, Gasperlin M. Stability of vitamins C and E in topical microemulsions for combined antioxidant therapy. Drug Deliv. 2007;14:235–45.

1268. Soltani A, Prokop AF, Vaezy S. Stability of alteplase in presence of cavitation. Ultrasonics. 2008;48:109–16.

1269. Akhavan A, Levitt J. Assessing retinol stability in a hydroquinone 4%/retinol 0.3% cream in the presence of antioxidants and sunscreen under simulated-use conditions: a pilot study. Clin Ther. 2008;30:543–7.

1270. Juarez Olguin H, Flores Perez C, Ramirez Mendiola B, et al. Extemporaneous suspension of propafenone: attending lack of pediatric formulations in Mexico. Pediatr Cardiol. 2008;29:1077–81.

1271. Agatonovic-Kustrin S, Markovic N, Ginic-Markovic M, et al. Compatibility studies between mannitol and omeprazole sodium isomers. J Pharm Biomed Anal. 2008;48:356–60.

1272. Paolera MD, Kasahara N, Umbelino CC, et al. Comparative study of the stability of bimatoprost 0.03% and latanoprost 0.005%: a patient-use study. BMC Ophthalmol. 2008;8:11.

1273. Chow KT, Chan LW, Heng PWS. Formulation of hydrophilic non-aqueous gel: drug stability in different solvents and rheological behavior of gel matrices. Pharm Res. 2008;25:207–17.

1274. Allen LV Jr. Acetaminophen 10-mg/mL rectal solution. Int J Pharm Compound. 2008;12:447.

1275. Allen LV Jr. Ketoprofen 20% topical solution. Int J Pharm Compound. 2008;12:452.

1276. Allen LV Jr. Flurandrenolide topical film. Int J Pharm Compound. 2008;12:451.

1277. Allen LV Jr. Piroxicam 5% and dexpanthenol 5% gel. Int J Pharm Compound. 2008;12:454.

1278. Allen LV Jr. Estradiol 0.1% vaginal solution. Int J Pharm Compound. 2008;12:450.

1279. Allen LV Jr. Dichlorobenzyl alcohol tooth gel. Int J Pharm Compound. 2008;12:449.

1280. Anon. Opinion of the scientific committee on cosmetic products and non-food products intended for consumers concerning 2,4-dichlorobenzyl alcohol (DCBA) dated 10 January, 2003. Available at: http://ec.europa.eu/health/ph_risk/committees/sccp/documents/out189_en.pdf.

1281. Allen LV Jr. Carnitine 400-mg/mL and coenzyme Q10 10-mg/mL solution. Int J Pharm Compound. 2008;12:448.

1282. McLellan C, Pasedis S, Dohlman CH. Testing the long-term stability of vancomycin ophthalmic solution. Int J Pharm Compound. 2008;12:456–9.

1283. Jain SP, Shah SP, Rajadhyaksha NS, et al. In situ ophthalmic gel of ciprofloxacin hydrochloride for once a day sustained delivery. Drug Dev Ind Pharm. 2008;34:445–52.

1284. Mochizuki H, Itakara H, Takamatsu M, et al. Efficacy of latanoprost when stored at room temperature. Hiroshima J Med Sci. 2008;57:69–72.

1285. Urmi C, Bijaya G, Ashish J. The effect of some common excipients on stability of salbutamol sulphate. J Pharm Res. 2008;7:97–100.

1286. Echezarreta-Lopez MM, Otaro-Mazoy I, Ramirez HL, et al. Solubilization and stabilization of sodium dicloxacillin by cyclodextrin inclusion. Curr Drug Discov Technol. 2008;5:140–5.

1287. Patel NA, Patel NJ, Patel RP. Design and evaluation of transdermal drug delivery system for curcumin as an anti-inflammatory drug. Drug Dev Ind Pharm. 2009;35(2):234–42.

1288. Kawamoto T, Tayama Y, Sawa A, et al. Effect of the admixture of tetracycline and nadifloxacin ointments on their stability and their antibacterial activity. Yakugaku Zasshi. 2008;128:1221–6.

1289. Aqil M, Bhavna, Chowdry I, et al. Transdermal therapeutic system of enalapril maleate using piperidine as penetration enhancer. Curr Drug Deliv. 2008;5:148–52.

1290. Williams NT. Medication administration through enteral feeding tubes. Am J Health Syst Pharm. 2008;65:2347–57.

1291. Stucki MC, Fleury-Souverain S, Sautter AM, et al. Development of ready-to-use ketamine hydrochloride syringes for safe use in post-operative pain. Eur J Hosp Pharm Sci. 2008;14:14–8.

1292. Gupta VD. Stability of ketamine hydrochloride injection after reconstitution in water for injection and storage in 1-mL tuberculin polypropylene syringes for pediatric use. Int J Pharm Compound. 2002;6:316–7.

1293. Allen LV Jr. Emetine hydrochloride 30-mg/mL injection. Int J Pharm Compound. 2008;12:68.

1294. Donnelly RF, Corman C. Physical compatibility and chemical stability of a concentrated solution of atropine sulfate (2 mg/mL) for use as an antidote in nerve agent casualties. Int J Pharm Compound. 2008;12:550–2.

1295. Vu NT, Aloumanis V, Ben MJ, et al. Stability of metronidazole benzoate in SyrSpend SF one-step suspension system. Int J Pharm Compound. 2008;12:558–64.

1296. Chen WQ, Liu YD. The influences of additives on the stability of interferon alpha in liquid state. Chinese J New Drugs. 2008;17:1425–8.

1297. Anon. Thioridazine. Tuberculosis. 2008;88:164–7.

1298. Anon. Cycloserine. Tuberculosis. 2008;88:100–1.

1299. Dansereau RJ, Crail DJ. Compatibility of risedronate sodium tablets with food thickeners. Am J Health Syst Pharm. 2008;65:2133–6.

1300. Fasani E, Albini A, Gemme S. Mechanism of the photochemical degradation of amlodipine. Int J Pharm. 2008;352:197–201.

1301. Kramer I, Schwabe A, Lichtinghagen R, et al. Physicochemical compatibility of mixtures of dornase alfa and tobramycin containing nebulizer solutions. Pediatr Pulmonol. 2009;44:134–41.

1302. Anon. Rifampin. Tuberculosis. 2008;88:151–4.

1303. Gammon DL, Su S, Huckfeldt R, et al. Alteration in prehospital drug concentration after thermal exposure. Am J Emerg Med. 2008;26:566–73.

1304. Brown LH, Bailey LC, Medwick T, et al. Medication storage on US ambulances: a prospective multi-center observational study. Pharm Forum. 2003;29:540–7.

1305. Watson H, Junker-Buchheit A. Fractionation of acidic, neutral, and basic drugs from plasma with polymeric SPE cation exchange, Bond Elut Plexa PCX. Varian Application Note. 2007; SI-01013. Available at: http://www.chem.agilent.com/Library/applications/SI-01013.pdf.

1306. Yano K, Ikarashi N, Ito K, et al. Comparison of the pravastatin original and generic drugs in the simple suspension method. Iryo Yakugaku. 2008;34:699–704.

1307. Agatonovic-Kustrin S, Glass BD, Mangan M, et al. Analysing the crystal purity of mebendazole raw material and its stability in a suspension. Int J Pharm. 2008;361:245–50.

1308. Anon. Gatifloxacin. Tuberculosis. 2008;88:109–11.

1309. Anon. Ethionamide. Tuberculosis. 2008;88:106–8.

1310. Anon. Ethambutol. Tuberculosis. 2008;88:102–5.

1311. Ritschel WA, Ye W, Buhse L, et al. Stability of the nitrogen mustard mechlorethamine in novel formulations for dermatological use. Int J Pharm. 2008;362:67–73.

1312. Feyel-Dobrokhotov AC, Baylatry MT, Guntz C, et al. Stability of codeine and codethyline hydrochloride potion by liquid chromatography in real conservation conditions at St Antoine University Hospital in Paris. J Pharm Clin. 2008;27:93–9.

1313. Alvarez C, Calero J, Menendez JC, et al. Effects of hydroxypropyl-beta-cyclodextrin on the chemical stability and the aqueous solubility of thalidomide enantiomers. Pharmazie. 2008;63:511–3.

1314. Krenn M, Gamcsik MP, Vogelsang GB. Improvements in solubility and stability of thalidomide upon complexation with hydroxypropyl-beta-cyclodextrin. J Pharm Sci. 1992;81:685–9.

1315. Eriksson T, Bjorkman S, Roth B, et al. Intravenous formulations of the enantiomers of thalidomide: pharmacokinetic and initial pharmacodynamic characterization in man. J Pharm Pharmacol. 2000;52:807–17.

1316. Abelson MB, Chapin MJ, Kapik BM, et al. Efficacy of ketotifen fumarate 0.025% ophthalmic solution compared with placebo in the conjunctival allergen challenge model. Arch Ophthalmol. 2003;121:626–30.

1317. Abd El-Aleem HM, Sakr FM, Soliman OA, et al. Formulation and in-vivo evaluation of ketotifen fumarate ophthalmic preparations in rabbit eye. Alex J Pharm Sci. 2008; 22:13–21.

1318. Allegra JR, Brennan J, Lanier V, et al. Storage temperatures of out-of-hospital medications. Acad Emerg Med. 1996;6:1098–103.

1319. Geller RJ, Lopez GP, Cutler S, et al. Atropine availability as an antidote for nerve agent casualties: validated rapid reformulation of high-concentration atropine from bulk powder. Ann Emerg Med. 2003;41:453–6.

1320. Allen LV Jr. Triprolidine hydrochloride 0.25-mg/mL syrup. Int J Pharm Compound. 2007;11:161.

1321. Allen LV Jr. Isoproterenol compound elixir. Int J Pharm Compound. 2007;11:246.

1322. Aboshiha J, Weir R, Singh P, et al. To what extent does lack of refrigeration of generic chloramphenicol eye-drops used in India decrease their purity and what are the implications for Europe? Br J Ophthalmol. 2008;92:609–11.

1323. Lau D, Leung L, Fullinfaw R, et al. Chemical stability of voriconazole 1% eye drops. J Pharm Pract Res. 2008;38:179–82.

1324. Kristl A. Acido-basic properties of proton pump inhibitors in aqueous solution. Drug Dev Ind Pharm. 2009;35:114–7.

1325. Hellstrom PM, Vitols S. The choice of proton pump inhibitor: does it matter? Basic Clin Pharmacol Toxicol. 2004;94:106–11.

1326. Jelveghari M, Nokhodchi A. Development and chemical stability of alcohol-free phenobarbital solution for use in pediatrics: a technical note. AAPS PharmSciTech. 2008;9:939–43.

1327. Adepoju-Bello AA, Coker HA, Eboka CJ, et al. The physicochemical and antibacterial properties of ciprofloxacin-Mg++ complex. Nig Q J Hosp Med. 2008;18:133–6.

1328. Yamreudeewong W, Teixeira MG, Mayer GE. Stability of levalbuterol in a mixture of levalbuterol and ipratropium nebulizer solution. Hosp Pharm. 2008;43:303–6.

1329. Bonasia PJ, McVicar WK, Bill W, et al. Chemical and physical compatibility of levalbuterol inhalation solution concentrate mixed with budesonide, ipratropium bromide, cromolyn sodium, or acetylcysteine sodium. Respir Care. 2008;12:1716–22.

1330. Akapo S, Gupta J, Matinez E, et al. Compatibility and aerosol characteristics of formoterol fumarate mixed with other nebulizing solutions. Ann Pharmacother. 2008;42:1416–24.

1331. Kristensen S, Lao YE, Braenden JU. Influence of formulation properties on chemical stability of captopril in aqueous preparations. Pharmazie. 2008;63:872–7.

1332. Hiatt AN, Ferruzzi MG, Taylor LS, et al. Impact of deliquescence on the chemical stability of vitamins B1, B6, and C in powder blends. J Agric Food Chem. 2008;56:6471–9.

1333. Brouwers J, Vermeire K, Schols D, et al. Development and in vitro evaluation of chloroquine gels as microbicides against HIV-1 infection. Virology. 2008;378:306–10.

1334. Wallace SJ, Li J, Rayner CR, et al. Stability of colistin methanesulfonate in pharmaceutical products and solutions for administration to patients. Antimicrob Agents Chemother. 2008;52:3047–51.

1335. Hollenwaeger M, Challandes IB, Fallab CL, et al. Development and stability testing of a concentrated injectable clonidine solution for intrathecal analgesia. Eur J Hosp Pharm Sci. 2009;15:3–5.

1336. Vonbach P, Angelika H, Dubied A. Long-term stability of an aqueous solution containing lidocaine hydrochloride, dexamethasone sodium phosphate and adrenaline hydrochloride for electromotive administration. Eur J Hosp Pharm Sci. 2009;15:6–10.

1337. De Rosa NF, Sharley NA. Stability of venlafaxine hydrochloride liquid formulations suitable for administration via enteral feeding tubes. J Pharm Pract Res. 2008;38:212–5.

1338. Sistla A, Sunga A, Phung K, et al. Powder-in-bottle formulation of SU011248. Enabling rapid progression into human clinical trials. Drug Dev Ind Pharm. 2004;30:19–25.

1339. Ross BM, Katzman M. Stability of methylnicotinate in aqueous solution as utilized in the "niacin patch test". BMC Res Notes. 2008;1:89.

1340. Anon. Methyl nicotinate. MP Biomedicals. 2009. Available at: http://www.mpbio.com. Accessed April 14, 2009.

1341. Arias JL, Lopez-Viota M, Clares B, et al. Stability of fenbendazole suspensions for veterinary use correlation between zeta potential and sedimentation. Eur J Pharm Sci. 2008;34:257–62.

1342. Patel NA, Patel NJ, Patel RP. Formulation and evaluation of curcumin gel for topical application. Pharm Dev Technol. 2009;14:80–9.

1343. Vidal NLG, Zubata PD, Simionato LD, et al. Dissolution stability study of cefadroxil extemporaneous suspensions. Dissolut Technol. 2008;15:29–36.

1344. Ensom MHH, Decarie D. Stability of thiamine in extemporaneously compounded suspensions. Can J Hosp Pharm. 2005;58:26–30.

1345. Allen LV Jr. Miconazole 2% oral gel. Int J Pharm Compound. 2007;11:512.

1346. Allen LV Jr. Diclofenac sodium 75-mg/mL injection. Int J Pharm Compound. 2007;11:421.

1347. Xu M, Warren FW, Bartlett MG. Stability of low-concentration ceftazidime in 0.9% sodium chloride injection and balanced salt solutions in plastic syringes under various storage conditions. Int J Pharm Compound. 2009;13:166–9.

1348. Shank BR, Ofner CM. Stability of pergolide mesylate oral liquid at room temperature. Int J Pharm Compound. 2009;13:254–8.

1349. Hutchinson DJ, Johnson CE, Klein KC. Stability of extemporaneously prepared moxifloxacin oral suspensions. Am J Health Syst Pharm. 2009;66:665–7.

1350. Shen Z, Shen T, Wientjes MG, et al. Intravesical treatments of bladder cancer: review. Pharm Res. 2008;25:1500–10.

1351. Anon. Melatonin. St. Louis, MO: Sigma-Aldrich; 2009

1352. He H, Lin M, Han Z, et al. The formation and properties of the melatonin radical: a photolysis study of melatonin with 248 nm laser light. Org Biomol Chem. 2005;3:1568–74.

1353. Haywood A, Burrell A, van Breda K, et al. Stability of melatonin in an extemporaneously compounded sublingual solution and hard gelatin capsule. Int J Pharm Compound. 2009;13:170–4.

1354. Davis JL, Kirk LM, Davidson GS, et al. Effects of compounding and storage conditions on stability of pergolide mesylate. J Am Vet Med Assoc. 2009;234:385–9.

1355. Graudins LV, Kivi NJ, Tattam BN. Stability of omeprazole mixtures: implications for paediatric prescribing. J Pharm Pract Res. 2008;38:276–9.

1356. Garg S, Svirskis D, Al-Kabban M, et al. Chemical stability of extemporaneously compounded omeprazole formulations: a comparison of two methods of compounding. Int J Pharm Compound. 2009;13:250–3.

1357. Allen LV Jr. Chloroquine phosphate 64.5-mg/mL injection. Int J Pharm Compound. 2009;13:154.

1358. Allen LV Jr. Metronidazole 2.1-mg/mL in 5.25% dextrose injection. Int J Pharm Compound. 2009;13:157.

1359. Allen LV Jr. Thiothixene hydrochloride 2-mg/mL injection. Int J Pharm Compound. 2009;13:159.

1360. Allen LV Jr. Vasopressin tannate 5 pressor units per mL in oil injection. Int J Pharm Compound. 2009;13:161.

1361. Allen LV Jr. Acidified sodium hypochlorite for antisepsis and disinfection. Int J Pharm Compound. 2009;13:152.

1362. Miner N. Effective use of bleach as an antiseptic and disinfectant. Microbe. 2006;June letters.

1363. Deventer K, Baele G, Van Eenoo P, et al. Stability of selected chlorinated thiazide diuretics. J Pharm Biomed Anal. 2009;49:519–24.

1364. Adams AIH, Gosmann G, Schneider PH, et al. LC studies of voriconazole and structural elucidation of its major degradation product. Chromatographia. 2009;69:S115–22.

1365. Dupuis LL, Lingertat-Walsh K, Walker SE. Stability of an extemporaneous oral liquid aprepitant formulation. Support Care Cancer. 2009;17:701–6.

1366. Takano R, Furumoto K, Shiraki K, et al. Rate-limiting steps of oral absorption for poorly water-soluble drugs in dogs; prediction from a miniscale dissolution test and a physiologically-based computer simulation. Pharm Res. 2008;25:2334–44.

1367. Dupuis A, Tournier N, Le Moal G, et al. Preparation and stability of voriconazole eye drop solution. Antimicrob Agent Chemother. 2009;53:798–9.

1368. Allen LV Jr. Metformin hydrochloride 100-mg/mL oral liquid. Int J Pharm Compound. 2007;11:155.

1369. Allen LV Jr. Losartan potassium 2.5-mg/mL oral liquid. Int J Pharm Compound. 2007;11:248.

1370. Cagigal E, Gonzalez L, Alonso RM, et al. pKa determination of angiotensin II receptor antagonists (ARA II) by spectrofluorimetry. J Pharm Biomed Anal. 2001;26:477–86.

1371. Allen LV Jr. Benazepril 2-mg/mL oral liquid. Int J Pharm Compound. 2007;11:247.

1372. Abobo C, Wei B, Liang D. Stability of zonisamide in extemporaneously compounded oral suspensions. Am J Health Syst Pharm. 2009;66:1105–9.

1373. Beijnen JH, Van Gijn R, Underberg WJM. Chemical stability of the antitumor drug mitomycin C in solutions for intravesical instillation. J Parenter Sci Technol. 1990;44:332–5.

1374. Allen LV Jr. Hydrogen peroxide 1.5% and sodium fluoride 0.5-mg/mL oral rinse. Int J Pharm Compound. 2009;13:71.

1375. Dentinger PJ, Swenson CF. Stability of reconstituted fluconazole oral suspension in plastic bottles and oral syringes. Ann Pharmacother. 2009;43:485–9.

1376. Guneri T, Ozyazici M. Effect of temperature, pH, and sweetening agents on stability of trimethobenzamide hydrochloride syrups: evaluation by factorial design. Acta Pharmaceutica Sciencia. 2008;50:11–22.

1377. Blessel KW, Rudy BC, Senkowski, BZ. Trimethobenzamide hydrochloride. Anal Profiles Drug Subs. 1973;2:551–70.

1378. Thumma S, Repka MA. Compatibility studies of promethazine hydrochloride with tablet excipients by means of thermal and non-thermal methods. Pharmazie. 2009;64:183–9.

1379. Vahdat L, Sunderland B. The influence of potassium clavulanate on the rate of amoxicillin sodium degradation in phosphate and acetate buffers in the liquid state. Drug Dev Ind Pharm. 2009;35:471–9.

1380. Gallardo Lara V, Lopez-Viota Gallardo M, Morales Hernandez ME, et al. Ondansetron: design and development of oral pharmaceutical suspensions. Pharmazie. 2009;64:90–3.

1381. Yillar DO, Akcasu A, Akkan G, et al. The stability of choline ascorbate. J Basic Clin Physiol Pharmacol. 2008;19:177–83.

1382. Kodawara T, Mizuno Y, Taue H, et al. Evaluation of stability of temozolomide in solutions after opening the capsule. Yakugaku Zasshi. 2009;129:353–7.

1383. Kramer I, Schwabe A, Kamin W. Physikalisch-chemische kompatibilittat von mischinhalationslolungen und -suspension zur simultanen feuchtinhalationstherapie bei mukoviszidose-patienten. Krankenhauspharmazie. 2007;28:170–8.

1384. Allen LV Jr. Valsartan 4-mg/mL oral liquid. Int J Pharm Compound. 2008;12:269.

1385. Wang J, Chen L, Zhao Z, et al. Valsartan lowers brain beta-amyloid protein levels and improves spatial learning in a mouse model of Alzheimer disease. J Clin Invest. 2007;117:3393–402.

1386. Larson PO, Ragi G, Swandby M, et al. Stability of buffered lidocaine and epinephrine used for local anesthesia. J Dermatol Surg Oncol. 1991;17:411–4.

1387. Bartfield JM, Homer PJ, Ford DT, et al. Buffered lidocaine as a local anesthetic: an investigation of shelf life. Ann Emerg Med. 1992;21:16–9.

1388. Peterfreund RA, Datta S. pH adjustment of local anesthetic solutions with sodium bicarbonate: laboratory evaluation of alkalinization and precipitation. Reg Anesth. 1989;14:265–70.

1389. Bonhomme L, Postaire E, Touratier S, et al. Chemical stability of lignocaine (lidocaine) and adrenaline (epinephrine) in pH-adjusted parenteral solutions. J Clin Pharm Ther. 1988;13:257–61.

1390. Murakami CS, Odland PB, Ross BK. Buffered local anesthetics and epinephrine degradation. J Dermatol Surg Oncol. 1994;20:192–5.

1391. Allen LV Jr. Coal tar 2% and allantoin 2.5% cream. Int J Pharm Compound. 2007;11:508.

1392. Allen LV Jr. Promethazine hydrochloride 2.5-mg/mL rectal solution. Int J Pharm Compound. 2008;12:267.

1393. Weerasinghe CA, Lewis DO, Mathews JM, et al. Aquatic photodegradation of albendazole and its major metabolites. 1. Photolysis rate and half-life for reactions in a tube. J Agric Food Chem. 1992;40:1413–8.

1394. Weerasinghe CA, Mathews JM, Wright RS, et al. Aquatic photodegradation of albendazole and its major metabolites. 2. Reaction quantum yield, photlysis rate, and half-life in the environment. J Agric Food Chem. 1992;40:1419–21.

1395. Hernandez-Luis F, Hernandez-Campos A, Yepez-Mulia L, et al. Synthesis and hydrolytic stability studies of albendazole carrier prodrugs. Bioorg Med Chem Lett. 2001;11:1359–62.

1396. Wu Z, Razzak M, Tucker IG, et al. Physicochemical characterization of ricobendazole: 1. Solubility, lipophilicity, and ionization characteristics. J Pharm Sci. 2005;94:983–93.

1397. Wu Z, Tucker IG, Razzak M, et al. Stability of ricobendazole in aqueous solutions. J Pharm Biomed Anal. 2009;49:1282–6.

1398. To TP, Ellis AG, Ching MS, et al. Stability of a formulated N-acetylcysteine capsule for prevention of contrast-induced nephropathy. J Pharm Pract Res. 2008;38:219–22.

1399. Ahuja M, Drake AS, Sharma SK, et al. Stability studies on aqueous and oily ophthalmic solutions of diclofenac. Yakugaku Zasshi. 2009;129:495–502.

1400. Carvallo Santos V, Brandao Periera JF, Brandao Haga R, et al. Stability of clavulanic acid under variable pH, ionic strength and temperature conditions. A new kinetic approach. Biochem Eng J. 2009;45:89–93.

1401. Haginaka J, Nakagawa T, Uno T. Stability of clavulanic acid in aqueous solutions. Chem Pharm Bull. 1981;29:3334–41.

1402. Dhawan S, Medhi B, Chopra S. Formulation and evaluation of diltiazem hydrochloride gels for the treatment of anal fissures. Sci Pharm. 2009;77:465–82.

1403. Allen LV Jr. Benzyl benzoate lotion. Int J Pharm Compound. 2008;12:261.

1404. Allen LV Jr. African scalp lotion. Int J Pharm Compound. 2008;12:260.

1405. Braun C, Wiegrebe W. Note concerning the stability of dithranol ointments. Pharm Zeit. 1988;133:691–2.

1406. Decker WJ, Corby DG, Combs HF. A stable parenteral solution of apomorphine. Clin Toxicol. 1981;18:763–72.

1407. Gupta A, Ciavarella AB, Rothman B, et al. Stability of gabapentin 300-mg capsules repackaged in unit dose containers. Am J Health Syst Pharm. 2009;66:1376–80.

1408. Gupta VD. Chemical stability of scopolamine hydrobromide nasal solution. Int J Pharm Compound. 2009;13:438–9.

1409. Gupta VD. Chemical stability of amitriptyline hydrochloride in oral liquid dosage forms. Int J Pharm Compound. 2009;13:445–6.

1410. Enever RP, Po ALW, Millard BJ, et al. Decomposition of amitriptyline hydrochloride in aqueous solution: identification of decomposition products. J Pharm Sci. 1975;64:1497–9.

1411. Enever RP, Po ALW, Shotton E. Factors influencing decomposition rate of amitriptyline hydrochloride in aqueous solution. J Pharm Sci. 1977;66:1087–9.

1412. Nicoletti MA, Siqueira EL, Bombana AC, et al. Shelf-life of a 2.5% sodium hypochlorite solution as determined by Arrhenius equation. Braz Dent J. 2009;20:27–31.

1413. Allen LV Jr. Chlorhexidine 2% gel. Int J Pharm Compound. 2007;11:507.

1414. Allen LV Jr. Chlorhexidine gel. Int J Pharm Compound. 2008;12:151.

1415. Al-Badriyeh D, Li J, Stewart K, et al. Stability of extemporaneously prepared voriconazole ophthalmic solution. Am J Health Syst Pharm. 2009;66:1478–83.

1416. Mathews KG, Linder KE, Davidson GS, et al. Assessment of clotrimazole gels for in vitro stability and in vivo retention in the frontal sinus of dogs. Am J Vet Res. 2009;70:640–7.

1417. Zhang N, Kannan R, Okamoto CT, et al. Characterization of brimodipine transport in retinal pigment epithelium. Invest Ophthalmol Vis Sci. 2006;47:287–94.

1418. Ali MS, Khatri AR, Munir MI, et al. A stability-indicating assay of brimonidine tartrate ophthalmic solution and stress testing using HPLC. Chromatographia. 2009;70:539–44.

1419. Kumar V, Shah RP, Malik S, et al. Compatibility of atenolol with excipients: LC-MS/TOF characterization of degradation/interaction products, and mechanisms of their formation. J Pharm Biomed Anal. 2009;49:880–8.

1420. Martin-Viana NDLP, Lacarrere IGM, Apan JMG, et al. Development of formulation for oral suspension ibuprofen 100 mg/5mL for pediatric use. Rev Cubana Farm. 2009;43:1–11.

1421. Freeman KL, Trezevant MS. Interaction between liquid protein solution and omeprazole suspension. Am J Health Syst Pharm. 2009;66:1901–2.

1422. Tammara B, Weisel K, Katz A, et al. Bioequivalence among three methods of administering pantoprazole granules in healthy subjects. Am J Health Syst Pharm. 2009;66:1923–8.

1423. Bachhav Y, Patravale V. SMEDDS of glyburide: Formulation, in vitro evaluation, and stability studies. AAPS PharmSciTech. 2009;10:482–7.

1424. He Y, Yu YC. Study on the stability of tetracaine hydrochloride injection. Chin Pharm J. 2001;36:33–5.

1425. Moreno AD, da Silva MFC, Salgado HRN. Stability studies of azithromycin in ophthalmic preparations. Braz J Pharm Sci. 2009;46:219–26.

1426. Yuhas LM, Fuerst JK, Timpano JM, et al. Personal communication of $pK_a$ values of CP-62,993, azithromycin, assigned using 1H NMR spectroscopy. New York: Pfizer Laboratories; November 24, 2009.

1427. Brustugun J, Lao YE, Fagernaes C, et al. Long-term stability of extemporaneously prepared captopril oral liquids in glass bottles. Am J Health Syst Pharm. 2009;66:1722–5.

1428. Lanzanova FA, Argenta D, Arend MZ, et al. LC and LC-MS evaluation of stress degradation behavior of carvediol. J Liq Chromatogr Relat Technol. 2009;32:526–43.

1429. Shi S, Chen H, Tang X. Formulation, stability and degradation kinetics of intravenous cinnarizine lipid emulsion. Int J Pharm. 2009;373:147–55.

1430. Tokumura T, Ichikawa T, Sugawara N, et al. Kinetics of degradation of cinnarizine in aqueous solution. Chem Pharm Bull. 1985;33:2069–72.

1431. Kucmanic J. Long-term stability of ethanol solutions for breath-alcohol tests. J Anal Toxicol. 2009;33:328–31.

1432. Dubowski KM, Goodson EE, Sample Jr M. Storage stability of simulator ethanol solutions for vapor-alcohol control tests in breath-alcohol analysis. J Anal Toxicol. 2000;26:406–10.

1433. Gullberg RG. Common legal challenges and responses in forensic breath-alcohol determination. Forensic Sci Rev. 2004;16:91–101.

1434. Chow BLC, Wigmore JG. Technical note: the stability of aqueous alcohol standard used in breath-alcohol testing after twenty-six years storage. Can Soc Forensic Sci J. 2005;38:21–4.

1435. Kerc J, Srcic S, Urleb U, et al. Compatibility study between acetylcysteine and some commonly used tablet excipients. J Pharm Pharmacol. 1992;44:515–8.

1436. Johnson BR, Remeikis NA. Effective shelf-life of prepared sodium hypochlorite solution. J Endod. 1993;19:40–3.

1437. de la Paz N, Martinez L, Munoz A, et al. Desarrollo de una formulacion de alprazolam solucion oral at 0.01%. Acta Farm Bonaerense. 2009;28:55–61.

1438. Van Der Straeten F, Al-Fandi A, De Peepe K, et al. Preparation magistrale d'une suspension d'omeprazole. J Pharm Belg. 2009;2:54–63.

1439. Kervela JG, Castagnet S, Chiadmi F, et al. Assessment of stability in extemporaneously prepared venlafaxine solutions. Eur J Hosp Pharm Pract. 2009;15:30–2.

1440. Castenada B, Ortiz-Cala W, Gallardo-Cabrera C, et al. Stability studies of alprazolam tablets: effects of chemical interactions with some excipients in pharmaceutical solid preparations. J Phys Org Chem. 2009;22:807–14.

1441. Belgamwar VS, Chauk DS, Mahajan HS, et al. Formulation and evaluation of in situ gelling system of dimenhydrinate for nasal administration. Pharm Dev Technol. 2009;14:240–8.

1442. Clarkson RM, Moule AJ, Podlich HM. The shelf-life of sodium hypochlorite irrigating solutions. Aust Dent J. 2001;46:269–76.

1443. Rutala WA, Cole EC, Thomann CA, et al. Sterility and bactericidal activity of chlorine solutions. Infect Control Hosp Epidemiol. 1998;19:323–7.

1444. Pappas L, Kiss A, Levitt J. Compatibility of tacrolimus ointment with corticosteroid ointments of varying potencies. J Cutan Med Surg. 2009;13:140–5.

1445. Rizwan M, Aqil M, Azeem A, et al. Study of the degradation kinetics of carvediol by use of a validated stability-indicating LC method. Chromatographia. 2009;70:1283–6.

1446. Barboza F, Vecchia DD, Tagliari MP, et al. Differential scanning calorimetry as a screening technique in compatibility studies of acyclovir extended release formulations. Pharm Chem J. 2009;43:363–8.

1447. Yang Y, Gupta A, Carlin AS, et al. Comparative stability of repackaged metoprolol tartrate tablets. Int J Pharm. 2009;385:92–7.

1448. Moberg-Wolff E. Potential clinical impact of compounded versus noncompounded intrathecal baclofen. Arch Phys Med Rehabil. 2009;90:1815–20.

1449. Shivare UD, Jain VB, Mathur KB, et al. Formulation development and evaluation of diclofenac sodium gel using water soluble polyacrylamide polymer. Dig J Nanomater Bios. 2009;4:285–90.

1450. Monajjemzadeh F, Hassanzadeh D, Valizadeh H, et al. Compatibility studies of acyclovir and lactose in physical mixtures and commercial tablets. Eur J Pharm Biopharm. 2009;73:404–13.

1451. Sosnowska K, Winnicka K, Czajkowska-Kosnik A. Stability of extemporaneous enalapril maleate suspensions for pediatric use prepared from commercially available tablets. Acta Pol Pharm. 2009; 66:321–6.

1452. Song HJ, Zhang YM, Gao LH. Study on light stability of curcumin in absolute alcohol. Chin Pharm J. 2009;44:468–70.

1453. Le Quan P, Dubouch S, Clement R, et al. Stability study of azathioprine solution for pediatric use. Pharm World Sci. 2009;31:70.

1454. Houri JJ, Le Hoang MD, Berleur MP, et al. Influence of primary packaging on stability studies of fludrocortisone acetate tablets. Eur J Hosp Pharm Sci. 2009;15:71–7.

1455. Kennedy R, Groepper D, Tagen M, et al. Stability of cyclophosphamide in extemporaneous oral suspensions. Ann Pharmacother. 2010;44:295–301.

1456. Nguyen KQ, Hawkins MG, Taylor IT, et al. Stability and uniformity of extemporaneous preparations of voriconazole in two liquid suspension vehicles at two storage temperatures. Am J Vet Res. 2009;70:908–14.

1457. Martinez-Garcia MA, Perpina-Torderon M, Vila V, et al. Analysis of the stability of stored adenosine 5'-monophosphate for bronchoprovocation. Pulm Pharmacol Ther 2002;15:157–60.

1458. Ellingros VL, Perry PJ. Venlafaxine: a heterocyclic antidepressant. Am J Hosp Pharm. 1994;51:3003–46.

1459. Demir-bas M, Sautou V, Montager A, et al. Stability study of 20 mg/mL amikacin eye drops. Poster presented at: ASHP Summer Meeting; June 16, 2009; Rosemont, IL.

1460. Carceles CM, Villamayor L, Escuderon E, et al. Pharmacokinetics and milk penetration of moxifloxacin after intramuscular administration to lactating goats. Vet J. 2007;173:452–5.

1461. Martin TP, Hayes P, Collins DM. Tablet dispersion as an alternative to formulation of oral liquid dosage forms. Aust J Hosp Pharm. 1993; 23: 378–86.

1462. Nunes RS, Semaan FS, Riga AT, et al. Thermal behavior of verapamil hydrochloride and its association with excipients. J Therm Anal Calorim. 2009;97:349–53.

1463. Chong C, Schug SA, Page-Sharp M, et al. Development of a sublingual/oral formulation of ketamine for use in neuropathic pain. Clin Drug Investig. 2009;29:317–24.

1464. Murakami FS, Mendes C, Pereira RN, et al. Stability study of 20 mg omeprazole gastro-resistant tablets. Lat Am J Pharm. 2009;28:645–52.

1465. Rawas-Qalaji M, Simons FER, Collins D, et al. Long-term stability of epinephrine dispensed in unsealed syringes for the first-aid treatment of anaphylaxis. Ann Allergy Asthma Immunol. 2009;102:500–3.

1466. Sosnowska K, Winnicka K, Czajkowska-Kosnik A. Comparison of the stability of pediatric enalapril maleate suspensions prepared from various commercially available tablets. Farmacja Polska. 2009;65:243–6.

1467. El-Badry M, Taha E, Alanazi FK, et al. Study of omeprazole stability in aqueous solution: influence of cyclodextrins. J Drug Del Sci Tech. 2009;19:347–51.

1468. Nangia A. A stability study of aqueous solution of norfloxacin. Drug Dev Ind Pharm. 1991;17:681–94.

1469. Miyamoto E, Kawashima S, Murata Y, et al. Physicochemical properties of oxybutynin. Analyst. 1994;119:1489–92.

1470. Lovering EG, Matsui F, Curran NM, et al. Hydrazine levels in formulations of hydralazine, isoniazid, and phenelzine over a 2-year period. J Pharm Sci. 1983;72:965–7.

1471. Schlatter J, Saulnier JL. Stability of enalapril solutions prepared from tablets in sterile water. Aust J Hosp Pharm. 1997;27:395–7.

1472. Gellis C, Sautou-Miranda V, Pinon V, et al. Impact of deep-freezing on stability of 40 mg/mL pefloxacin eye drops. J Pharm Clin. 1999;18:88–9.

1473. Watson BL, Cormier RA, Harbeck RJ. Effect of pH on the stability of methacholine chloride in solution. Respir Med. 1998;92:588–92.

1474. Lee RLH. Stability of dithranol (Anthralin) in various vehicles. Aust J Hosp Pharm. 1987;17:254–8.

1475. Weller PJ, Newman CM, Middleton KR, et al. Stability of a novel dithranol ointment formulation, containing ascorbyl palmitate as an antioxidant. J Clin Pharm Ther. 1990;15:419–23.

1476. Zhu ZL. Stability of sodium hypochlorite in the presence of aqueous flavoring essence. J Surfactants Deterg. 1998;1:251–2.

1477. Iqbal Z, Pasha M. Stability of promethazine-HCl in non-sucrose syrup vehicles and sucrose syrup. Sci Int. 1999;11:215–6.

1478. Vermes A, van Der Sijs H, Guchelaar HJ. An accelerated stability study of 5-flucytosine in intravenous solution. Pharm World Sci. 1999;21:35–9.

1479. Vervloet E. Stability of povidone iodine 2% I polyethylene irrigation bags. Ziekenhuisfarmacie. 1989;5:90–2.

1480. Mair AE, Miller JHMcB. Effect of sterilization by autoclaving on the stability of eye drops of pilocarpine hydrochloride and physostigmine sulphate. J Clin Hosp Pharm. 1984;9:217–24.

1481. Lopez Lozano JJ, Moreno Cano R. Preparation of levodopa/carbidopa solution in ascorbic acid (citridopa) and chromatographic and electrochemical assessment of its stability over 24 hours. Neurologia. 1995;10:155–8.

1482. Modamio P, Lastra CF, Sebarroja J, et al. Stability of 5% permethrin cream used for scabies treatment. Pediatr Infect Dis J. 2009;28:668.

1483. Hutchinson DJ, Liou Y, Best R, et al. Stability of extemporaneously prepared rufinamide oral suspensions. Ann Pharmacother. 2010;44:462–5.

1484. Burchett DK, Darko W, Zhra J, et al. Mixing and compatibility guide for commonly used aerosolized medications. Am J Health Syst Pharm. 2010;67:227–30.

1485. Schier JG, Ravikumar PR, Nelson LS, et al. Preparing for chemical terrorism: stability of injectable atropine sulfate. Acad Emerg Med. 2004;11:329–34.

1486. Kondritzer AA, Zvirblis P. Stability of atropine in aqueous solution. J Am Pharm Assoc. 1957;46:531–5.

1487. Allen LV, Jr. Alum 6.5% topical cream. Int J Pharm Compound. 2008;12:147.

1488. Allen LV, Jr. Tretinoin and (-)alpha-bisabolol gel. Int J Pharm Compound. 2008;12:360.

1489. Boehm JJ, Dutton DM, Poust RI. Shelf life of unrefrigerated succinylcholine chloride injection. Am J Hosp Pharm. 1984;41:300–2.

1490. Schmutz CW, Muhlebach SF. Stability of succinylcholine chloride injection. Am J Hosp Pharm. 1991;48:501–6.

1491. Ross MB, Additional stability guidelines for routinely refrigerated drug products. Am J Hosp Pharm. 1988;45:1498–9.

1492. Cohen V, Jellinek SP, Teperikis L, et al. Room-temperature storage of medications labeled for refrigeration. Am J Health Syst Pharm. 2007;64:1711–5.

1493. Taiwo T. Personal communication from Medical Communications, Hospira Laboratories. Feb. 23, 2009.

1494. Adnet F, LeMoyec L, Smith CE, et al. Stability of succinylcholine chloride solutions stored at room temperature studied by nuclear magnetic resonance spectroscopy. Emerg Med J. 2007;24:168–9.

1495. Roy JJ, Boismenu D, Mamer OA, et al. Room temperature stability of injectable succinylcholine chloride. Int J Pharm Compound. 2008;12:83–5.

1496. Fritz BL, Lockhart HE, Giacin JR. Chemical stability of selected pharmaceuticals repackaged in glass and plastic. Pharm Tech. 1988;12:44–52.

1497. Pramar YV, Moniz D, Hobbs D. Chemical stability and adsorption of succinylcholine chloride injections in disposable plastic syringes. J Clin Pharm Ther. 1994;19:195–8.

1498. Storms ML, Stewart JT, Warren FW. Stability of succinylcholine chloride injection at ambient temperature and 4°C in polypropylene syringes. Int J Pharm Compound. 2003;7:68–70.

1499. Kumarvel V, Gandhimani P, Cundill G. Frozen succinylcholine. Anaesthesia. 2006;61:2002.

1500. Stone J, Fawcett W. A case of frozen succinylcholine encountered during emergency cesarean delivery. Anesth Analg. 2002;95:1465.

1501. Donnelly RF, Pascuet E, Ma C, et al. Stability of diclofenac sodium oral suspensions packaged in amber polyvinyl chloride bottles. Can J Hosp Pharm. 2010;63:25–30.

1502. Beer PM, Wong SJ, Schartman JP, et al. Infliximab stability after reconstitution, dilution, and storage under refrigeration. Retina. 2010;30:81–4.

1503. Donnelly RF, Pascuet E, Ma C, et al. Stability of celecoxib oral suspension. Can J Hosp Pharm. 2009;62:464–8.

1504. Seedher N, Bhatia S. Solubility enhancement of Cox-2 inhibitors using various solvent systems. AAPS PharmSciTech. 2003;4(3):E33.

1505. Celebrex [product monograph]. Kirkland, Quebec: Pfizer Canada; 2009.

1506. Clark MP, Pangilinan L, Wang A, et al. The shelf life of antimicrobial ear drops. Laryngoscope. 2010;120:565–9.

1507. Plavix [product monograph]. Laval, Quebec: Sanofi-Aventis Canada; 2009.

1508. Skillman KL, Caruthers RL, Johnson CE. Stability of an extemporaneously prepared clopidogrel oral suspension. Am J Health Syst Pharm. 2010;67:559–61.

1509. Ng Ying Kin KMK, Lal S, Thavundayil JX. Stability of apomorphine hydrochloride in aqueous sodium bisulphate solutions. Prog Neuropsychopharmachol Biol Pschiatry. 2001;25:1461–8.

1510. Loeb AJ. Preparation of caffeine and sodium benzoate injection. Am J Hosp Pharm. 1983;40:2120.

1511. Stolk LML, Huisman W, Nordemann HD, et al. Formulation of a stable vidarabine fluid. Pharm Weekbl Sci Ed. 1981;5:57–60.

1512. Hong WH, Szulczewski DH. Stability of vidarabine-5'-phosphate in aqueous solutions. J Parenter Sci Technol. 1984;38:60–4.

1513. Hald JG. Stability of isoniazid in aqueous solutions containing sucrose or sorbitol. Dansk Tidsskr Farm. 1969;43:156–9.

1514. Yeh MK, Wang DP. Stability of nonaqueous ibuprofen preparation. Chin Pharm J. 1992;44:141–7.

1515. Junyaprasert VB, Techowanich S, Vajragupta O, et al. Effect of pH, buffer concentration, ionic strength and temperature on ketoconazole stability in acidic solutions. Mahidol J Pharm Sci. 2003;30:17–26.

1516. Wamberg T, Neilsen ML, Scheibel JH. The stability of a preoperative oral bowel preparation containing gentamicin, vancomycin and nystatin. Arch Pharm Chem Sci Ed. 1980;8:1–4.

1517. Sakai Y, Yasueda SI, Ohtori A. Stability of latanoprost in an ophthalmic lipid emulsion using polyvinyl alcohol. Int J Pharm. 2005;305:176–9.

1518. Jacob M, Duru M, Duru C. Stability and determination of pilocarpine in collyrium. Pharm Acta Helv. 1981;56:59–63.

1519. Jasinka M, Karwowski B, Orzulak-Michalak D, et al. Stability studies of expired tablets of metoprolol tartrate and propranolol hydrochloride. Part 1. Content determination. Acta Pol Pharm. 2009;66:697–701.

1520. Jasinka M, Karwowski B, Orzulak-Michalak D, et al. Stability studies of expired tablets of metoprolol tartrate and propranolol hydrochloride. Part 2. Dissolution study. Acta Pol Pharm. 2009;66:703–7.

1521. Hernando EP, Alvarez LAD. Formulacion de soluciones oftalmicas de ciclosporina en colirio al 2 por 100 para la practica clinica. Anal Real Acad Nac F. 2009;75:911–22.

1522. Ohtani M, Yamaoka Y, Matsumoto M, et al. Compatibility of adapalene gel (Differin gel) with other kinds of ointments or creams. J Pharm Sci Technol. 2009;69:470–6.

1523. Cober MP, Johnson CE, Lee J, et al. Stability of extemporaneously prepared rifaximin oral suspensions. Am J Health Syst Pharm. 2010;67:287–9.

1524. Camps J, Pommel L, Aubut V, et al. Shelf life, dissolving action, and antibacterial activity of a neutralized 2.5% sodium hypochlorite solution. Oral Surg, Oral Med, Oral Pathol, Oral Radiol, Endod. 2009;108: e66–73.

1525. Bailey C, Aloumanis V, Walker B, et al. Stability of preoperative cataract surgery gel in polycarbonate syringes. Int J Pharm Compound. 2009;13:564–8.

1526. Nagatani K, Oishi T, Iwasaki S, et al. Ointment mixture 3. Concentration change of steroids in the mixture. Hiroshimaken Byoin Yakuzaishikai. 1986;20:16–8.

1527. Ushiyama M, Ikeda R, Nitta T, et al. Stability of hospital preparations of Azunol water gargles for pain relief in oral cancer patients with oral mucositis. Cancer Ther. 2009;7:277–81.

1528. Sikora A, Oszczapopwicz I, Tejchman B, et al. A method for the obtaining of increased viscosity eye drops containing amikacin. Acta Pol Pharm. 2005;62:31–7.

1529. Shrivastava PK, Shrivastava SK. Stress studies and the estimation of lamotrigine in pharmaceutical formulation by validated RP-HPLC method. Indian J Pharm Educ Res. 2009;43:156–61.

1530. Tu YH, Allen Jr LV, Wang DP. Stability of papaverine hydrochloride and phentolamine mesylate in injectable mixtures. Am J Hosp Pharm. 1987;44:2524–7.

1531. Soli M, Bertaccini A, Caparelli A, et al. Vasoactive cocktails for erectile dysfunction: chemical stability of PGE1, papaverine and phentolamine. J Urol. 1998;160:551–5.

1532. Lebwohl M, Quijije J, Gilliard J, et al. Topical calcitriol is degraded by ultraviolet light. J Invest Dermatol. 2003;121:594–5.

1533. Pecosky DA, Parasrampuria J, Li LC, et al. Stability and sorption of calcitriol in plastic tuberculin syringes. Am J Hosp Pharm. 1992;49:1463–6.

1534. Vargas-Ruiz M, Bushman LR, Hillegas MS, et al. Stability of parenteral calcitriol in a tuberculin syringe. Poster presented at: ASHP Midyear Clinical Meeting; December 1994; Miami Beach, FL.

1535. Hamada C, Hayashi K, Shou I, et al. Pharmacokinetics of calcitriol and maxacalcitol administered into peritoneal dialysate bags in peritoneal dialysis patients. Perit Dial Int. 2005;25:570–5.

1536. Vieth R, Ledermann SE, Kooh SW, et al. Losses of calcitriol to peritoneal dialysis bags and tubing. Perit Dial Int. 1989;9:277–80.

1537. Burgalassi S, Lodi A, Giannaccini B, et al. Formulation and stability of pharmaceutical suspensions. Boll Chim Farm. 1993;132:60–1.

1538. Sankar V, Chandrasekaran AK, Durga S, et al. Formulation and stability evaluation of diclofenac sodium ophthalmic gels. Indian J Pharm Sci. 2005;67:473–6.

1539. Santus G, Giordano F, Gazzaniga A, et al. Preformulation studies on naproxen sodium suppositories. Eur J Pharm Biopharm. 1994;40:243–5.

1540. Cedarbaum JM. Stability of levodopa/carbidopa solutions. Mov Disord. 1997;12:625.

1541. Malkki-Laine L, Purra K, Kahkonen K, et al. Decomposition of salbutamol in aqeuous solutions. II. The effect of buffer species, pH, buffer concentration and antioxidants. Int J Pharm. 1995;117:189–95.

1542. Wall BP, Sunderland VB. A preliminary study on the stability of salbutamol in aqueous solution. Aust J Hosp Pharm. 1976;6:156–60.

1543. Selva Otaolaurruchi J, Marco Garbayo JL, Luque Infantes R, et al. Preparacion y estabilidad del colirio de 5-fluorouracilo. Rev Asoc Esp Farm Hosp. 1987;11:189–92.

1544. Church WH, Hu SS, Henry AJ. Thermal degradation of injectable epinephrine. Am J Emerg Med. 1994;12:306–9.

1545. Grant TA, Carroll RG, Church WH, et al. Environmental temperature variations cause degradations in epinephrine concentration and biological activity. Am J Emerg Med. 1994;12:319–22.

1546. Rudland SV, Annus T, Dickinson J, et al. Adrenaline degradation in general practice. Br J Gen Pract. 1997;47:827–8.

1547. Furst W, Roth B, Albrecht G. Zur stabilitat von griseofulvin in standardrezepturen. Pharmazie. 1985;40:65.

1548. Christen P, Kloeti F, Gander B. Stability of predinisolone and prednisolone acetate in various vehicles used in semi-solid topical preparations. J Clin Pharm Ther. 1990;15:325–9.

1549. Andersin R. Solubility and acid-base behaviour of midazolam in media of different pH, studied by ultraviolet spectrophotometry with multicomponent software. J Pharm Biomed Anal. 1991;9:451–5.

1550. Trissel LA, Hassenbusch SJ. Extended stability of compounded preservative-free midazolam (as hydrochloride) injection. Nov 13, 2001. Data on file.

1551. Keiner D, Kruger L. Morphin-Losung (40 mg/ml) fur Dosierpumpen. Krankenhauspharmazie. 2010;31:158–9.

1552. Iuga C, Bojita M. Stability study of omeprazole. Farmacia 2010;58:203–10.

1553. McCafferty DF, Furness K, Anderson L. Stability of noxythiolin solutions stored in plastic and glass containers. J Clin Hosp Pharm. 1984;9:241–7.

1554. Tiphine T, Armand A, Navas D, et al. Physicochemical stability of a 1 mg/mL dexamethasone acetate oral suspension. Eur J Hosp Pharm Sci. 2010;16:17–22.

1555. Ramuth S, Flanagan RJ, Taylor DM. A liquid clozapine preparation for oral administration in hospital. Pharm J. 1996;257:190–1.

1556. Walker SE, Baker D, Law S. Stability of clozapine stored in oral suspension vehicles at room temperature. Can J Hosp Pharm. 2005;58:279–84.

1557. Gupta VD. Chemical stability of hydrocortisone in Humco simple syrup and Ora-Sweet vehicle. Int J Pharm Compound. 2010;14:76–7.

1558. Voudrie MA, Allen DB. Stability of oseltamivir phosphate in SyrSpend SF, cherry syrup, and SyrSpend SF (for reconstitution). Int J Pharm Compound. 2010;14:82–5.

1559. Shishoo CJ, Shah SA, Rathod IS, et al. Stability of rifampicin in dissolution medium in the presence of isoniazid. Int J Pharm. 1999;190:109–23.

1560. Singhai AK, Jain S, Jain NK. Stability studies on aqueous injections of ketoprofen. Pharmazie. 1997;52:226–8.

1561. Shirwaikar A, Rao PG. Formulation and evaluation of ophthalmic drops containing sulphacetamide sodium and trimethoprim. Indian J Pharm Sci. 1995;57:143–7.

1562. Poveda Andres JL, Hermenegildo Caudevilla M, Sanchez Alcaraz A, et al. Stability of gentamicin and vancomycin in cataract surgery irrigation solutions. Farm Clin. 1997; 14:54–9.

1563. Ruiz Caldes MJ, San Martin Ciges E, Ezquer Borras J, et al. Jarabe de midazolam: estudio de estabilidad de una solucion oral para uso hospitalario. Farm Hosp. 1995;19:41–4.

1564. Ensom MHH, Decarie D, Sheppard I. Stability of lansoprazole in extemporaneously compounded suspensions for nasogastric or oral administration. Can J Hosp Pharm. 2007;60:184–91.

1565. Sharma VK, Ugheoke EA, Vasudeva R, et al. The pharmacodynamics of lansoprazole administered via gastrostomy as intact, non-encapsulated granules. Aliment Pharmacol Ther. 1998;12:1171–4.

1566. Sharma VK, Vasudeva R, Howden CW. Simplified lansoprazole suspension—a liquid formulation of lansoprazole—effectively suppresses intragastric acidity when administered through a gastrostomy. Am J Gastroenterol. 1999;94:1813–7.

1567. Sharma VK, Peyton B, Spears T, et al. Oral pharmacokinetics of omeprazole and lansoprazole after single and repeated doses as intact capsules or as suspensions in sodium bicarbonate. Aliment Pharmacol Ther. 2000;14:887–92.

1568. Freire FD, Aragao CFS, De Lima E Maura TFA, et al. Compatibility study between chlorpropamide and excipients in their physical mixtures. J Therm Anal Calorim. 2009;97:355–7.

1569. de Villiers M, Vogel L, Bogenwschutz MC, et al. Compounding rifampin suspensions with improved injectability for nasogastric enteral feeding tube administration. Int J Pharm Compound. 2010;14:250–6.

1570. Ros JJW, Van der Meer YG. Formulering en haudbaarheid van een midazolam-atropine drank. Ziekenhuisfarmacie. 1989;5:87–90.

1571. Plunkett G. Stability of allergen extracts used in skin testing and immunotherapy. Curr Opin Otolaryngol Head Neck Surg. 2008;16:285–91.

1572. Nelson HS, Ikle D, Buchmeier A. Studies of allergen extract stability: the effects of dilution and mixing. J Allergy Clin Immunol. 1996;98:382–8.

1573. Bousquet J, Djjoukadar F, Hewitt B, et al. Comparison of the stability of a mite and pollen extract stored in normal conditions of use. Clin Allergy. 1985;15:29–35.

1574. Anderson MC, Baer H. Antigenic and allergenic changes during storage of a pollen extract. J Allergy Clin Immunol. 1981;69:3–10.

1575. Cadot P, Lejoy M, Stevens EAM. The effect of sucrose on the quality of ryegrass (Lolium perenne) pollen extracts. Allergy. 1995;50:941–51.

1576. Center JG, Shuller N, Zeleznick LD. Stability of antigen E content of commercially prepared ragweed pollen extracts. J Allergy Clin Immunol. 1974;54:305–10.

1577. Vijay HM, Young NM, Bernstein H. Studies on Alternaria allergens, VI: stability of the allergen components of Alternaria tenuis extracts under a variety of storage conditions. Int Arch Allergy Appl Immunol. 1987;83:325–8.

1578. Cadot P, van Hoeyveld EM, Ceuppens JL, et al. Composition and stability of allergenic extracts from gamma-irradiated rye grass (Lolium perenne) pollen. Clin Exp Allergy. 1999;29:1248–55.

1579. Pratter MR, Woodman TF, Irwin RS, et al. Stability of stored methacholine chloride solutions. Clinically useful information. Am Rev Respr Dis. 1982;126:717–9.

1580. Kato Y, Koizumi H, Yokoyama T. Stability on the mixing of epinephrine and sulfisoxazole ophthalmic solution. J Jpn Soc Hosp Pharm. 1988;24:17–20.

1581. Lindquist TD, Roth BP, Frische TR. Stability and activity of vancomycin in corneal storage media. Cornea. 1993;12:222–7.

1582. Cadorniga R, Lastres JL, Ballesteros MP. Effects of pH and solubilizing agents on phenobarbital stability. Boll Chim Farm. 1980;119:405–16.

1583. Salgado ACGB, da Silva AMNN, Machado MCJC, et al. Development, stability and in vitro permeation studies of gels containing mometasone furoate for the treatment of dermatitis of the scalp. Brazilian J Pharm Sci. 2010;46:109–14.

1584. Troche Concepcion Y, Garcia Pena CM, Romero Diaz JA, et al. Diseno de una formulacion de ketotifeno 0,025% colirio. Rev Cubana Farm. 2010;44:168–77.

1585. Llewelyn VK, Mangan MF, Glass BD. Stability of sodium valproate tablets repackaged into dose administration aids. J Pharm Pharmacol. 2010;62:833–43.

1586. Chan K, Swindon J, Donyai P, et al. Pilot study of the short-term physico-chemical stability of atenolol tablets stored in a multi-compartment compliance aid. Eur J Hosp Pharm Sci. 2007;13:60–6.

1587. Piotrowski K, Hermann TW, Pawelska A. Photostabilization of papaverine hydrochloride solutions. Acta Pol Pharm. 2010;67:321–6.

1588. Bachav YG, Patravale VB. Formulation of meloxicam gel for topical application: in vitro and in vivo evaluation. Acta Pharm. 2010;60:153–63.

1589. Yildirim N, Topbas S, Usluer G, et al. Stability of cefazolin sodium as eyedrops in various solvents. Mikrobiyol Bul. 1991;25:272–6.

1590. Zuniga Dedorite GA, Garcia Pena CM, Botet Garcia M, et al. Diseno de una formulacion de fenilefrina 10% y tropicamida 1% colirio. Rev Cubana Farm. 2010;44:178–88.

1591. Chedru-Legros V, Fines-Guyon M, Cherel A, et al. In vitro stability of fortified ophthalmic antibiotics stored at −20 °C for 6 months. Cornea. 2010;29:807–11.

1592. Bowen L, Mangan M, Haywood A, et al. Stability of frusemide tablets repackaged in dose administration aids. J Pharm Pract Res. 2007;37:178–81.

1593. Bruni G, Berbenni V, Milanese C, et al. Drug-excipient compatibility studies in binary and ternary mixtures by physico-chemical techniques. J Therm Anal Calorim. 2010;102:193–201.

1594. Glass B, Mangan M, Haywood A. Prochlorperazine tablets repackaged into dose administration aids: can the patient be assured of quality? J Clin Pharm Ther. 2009;34:161–9.

1595. Haywood A, Mangan M, Glass B. Stability implications of repackaged paracetamol tablets into dose administration aids. J Pharm Pract Res. 2006;36:25–8.

1596. Kerddonfak S, Manuyakorn W, Kamchaisatian W, et al. The stability and sterility of epinephrine prefilled syringe. Asian Pac J Allergy Immunol. 2010;28:53–7.

1597. Karampatakis V, Papanikolaou T, Giannousis M, et al. Stability and antibacterial potency of ceftazidime and vancomycin eyedrops reconstituted in BSS against *Pseudomonas aeruginosa* and *Staphylococcus aureus*. Acta Ophthamol. 2009;87:555–8.

1598. Cope M, Bautista-Parra F. The degradation of salbutamol in ethanolic solutions. J Pharm Biomed Anal. 2010;52:210–5.

1599. Kleinman ME, Westhouse SJ, Ambati J, et al. Triamcinolone crystal size. Ophthalmology. 2010;117:1654.

1600. Moshfeghi AA, Nugent AK, Nomoto H, et al. Triamcinolone acetonide preparations: impact of crystal size on in vitro behavior. Retina. 2009;29:689–98.

1601. Soliman II, Soliman NA, Abdou EM. Formulation and stability study of chlorpheniramine maleate nasal gel. Pharm Dev Technol. 2010;15:484–91.

1602. Ryatt KS, Feather JW, Mehta A, et al. The stability and blanching efficacy of betamethasone-17-valerate in emulsifying ointment. Br J Dermatol. 1982;107;71–6.

1603. Semenazato A, Secchieri M, Bau A. Stability of vitamin A palmitate in topical formulations. Farmaco. 1992;47:1407–17.

1604. Teng XW, Cutler DC, Davies NM. Degradation kinetics of mometasone furoate in aqueous systems. Int J Pharm. 2003;259:129–41.

1605. Tomonaga F, Murase S, Kagaya M, et al. Compatibility of commercial gentamicin injection with six drugs for inhalation. Jpn J Hosp Pharm. 1982;7:351–9.

1606. Skiba M, Skiba-Lahiani M, Marchais H, et al. Stability assessment of ketoconazole in aqueous formulations. Int J Pharm. 2000;198:1–6.

1607. Shank BR, Ofner CM. Multitemperature stability and degradation characteristics of pergolide mesylate oral liquid. J Pharm Pract. 2010;23:570–4.

1608. Fohl AL, Johnson CE, Cober MP. Stability of extemporaneously prepared acetylcysteine 1% and 10% solutions for treatment of meconium ileus. Am J Health Syst Pharm. 2011;68:69–72.

1609. Ensom MHH, Decarie D, Lakhani A. Stability of vancomycin 25 mg/mL in Ora-Sweet and water in unit-dose cups and plastic bottles at 4°C and 25°C. Can J Hosp Pharm. 2010;63:366–72.

1610. Tagliari MP, Stulzer HK, Assreuy J, et al. Evaluation of physicochemical characteristics of suspensions containing hydrochlorothiazide developed for pediatric use. Lat Am J Pharm. 2009;28:734–40.

1611. Li S, Zhang Y, Yang Y, et al. Stability study of norfloxacin in eye drops. Zhongguo Yiyuan Yaoxue Zazhi. 2000;20:347–8.

1612. Lippold BC, Jaeger I. Stability and dissociation constants of L-dopa and alpha-L-methyldopa. Arch Pharm. 1973;306:106–17.

1613. Marton S, Hatem A, Csoka G, et al. Preformulation study of atenolol containing solutions. I. The pH dependence of thermostability. Acta Pharm Hung. 2001;71:192–5.

1614. Sautou-Miranda V, Libert F, Grand-Boyer A, et al. Impact of deep freezing on the stability of 25 mg/mL vancomycin ophthalmic solutions. Int J Pharm. 2002;234:205–12.

1615. Yuan LC, Samuels GJ, Visor GC. Stability of cidofovir in 0.9% sodium chloride injection and in 5% dextrose injection. Am J Health Syst Pharm. 1996;53:1939–43.

1616. Ennis RD, Dahl TC. Stability of cidofovir in 0.9% sodium chloride injections for five days. Am J Health Syst Pharm. 1997;54:2204–6.

1617. Stiles J, Gwin W, Pogranichniy R. Stability of 0.5% cidofovir stored under various conditions for up to 6 months. Vet Ophthalmol. 2010;13:275–7.

1618. Hennere G, Havard L, Bonan B, et al. Stability of cidofovir in extemporaneously prepared syringes. Am J Health Syst Pharm. 2005;62:508–9.

1619. Rojanarata T, Tankul J, Woranaipinich C, et al. Stability of fortified cefazolin ophthalmic solutions prepared in artificial tears containing surfactant-based versus oxidant-based preservatives. J Ocul Pharmacol Th. 2010;26:485–90.

1620. Lin CP, Tsai MC, Sun CY, et al. Stability of self-prepared fortified antibiotic eyedrops. Kaohsiung J Med Sci. 1999;15:80–6.

1621. Cory WC, Harris C, Martinez S. Accelerated degradation of ibuprofen in tablets. Pharm Dev Technol. 2010;15:636–43.

1622. Church C, Smith J. How stable are medicines moved from original packs into compliance aids? Pharm J. 2006;276:75–81.

1623. Walker R. Stability of medicinal products in compliance aids. Pharm J. 1992;248:124–6.

1624. Valenzuela TD, Criss EA, Hammargen WM, et al. Thermal stability of prehospital medications. Ann Emerg Med. 1989;18:173–6.

1625. Donyai P. Quality of medicines stored together in multicompartment compliance aids. J Clin Pharm Ther. 2010;35:533–43.

1626. Garcia MJE, Duran ST, Duran JJT. Studies of the stability of aqueous solutions of captopril at a 1 mg/mL concentration. Farm Hosp. 2005;29:30–6.

1627. van der Schans MJ, Reijenga JC, Everaerts FM. Quality control of histamine and methacholine in diagnostic solutions with capillary electrophoresis. J Chromatogr A. 1996;735:387–93.

1628. Ved S, Deshpande SG. Stability of rifampicin in the presence of isoniazid and pyridoxine hydrochloride in liquid dosage form. East Pharm. 1990;33:155–6.

1629. Anton Torres R, Borras Blasco J, Esteban Rodriguez A, et al. Estabilidad una mezcla de tobramycina 0,01 mg/ml y vancomicina 0,05 mg/ml en BSS-PLUS para irrigacion-lavado intraocular. Farm Hosp. 2001;25:327–31.

1630. Blanchard C, Donnelly R. Clomipramine suspensions formulations. Can J Hosp Pharm. 1989;42:248.

1631. Yamamura K, Nakao M, Yamada J, et al. High-performance liquid chromatography determination of the stability of sisomicin in hydrophilic petrolatum ointment. Chem Pharm Bull. 1983;31:3632–6.

1632. Lee MG. Phenoxyethanol absorption by polyvinyl chloride. J Clin Hosp Pharm. 1984;9: 353–5.

1633. Stolk LML, Gerrits M, Wiltink EHH, et al. Formulation and stability of a beclomethasone diproprionate enema. Pharm Weekbl Sci Ed. 1989;11:20–2.

1634. Botet Homdedeu F, Gamundi Planas C. Estabilidad del 8-Methoxipsoralen en Pasta Lassar. Cir Far. 1986;44:39–52.

1635. Barnes AR. Compatibility of a commercially available low-density polyethylene eye-drop container with antimicrobial preservatives and potassium ascorbate. J Clin Pharm Ther. 1995;20:341–4.

1636. Parkin JE. High-performance liquid chromatographic investigation of the interaction of phenylmercuric nitrate and sodium metabisulphite in eye drop formulations. J Chromatogr. 1990;511:233–42.

1637. Parkin JE, Duffy MB, Loo CN. The chemical degradation of phenylmercuric nitrate by disodium edetate during heat sterilization at pH values commonly encountered in ophthalmic products. J Clin Pharm Ther. 1992;17:307–14.

1638. Parkin JE. The decomposition of phenylmercuric nitrate in sulphacetamide drops during heat sterilization. J Pharm Pharmacol. 1993;45:1024–7.

1639. Hill DG, Barnes AR. Compatibility of phenylmercuric acetate with cefuroxime and ceftazidime eye drops. Int J Pharm. 1997;147:127–9.

1640. Donnelly RF, Wong K, Goddard R, et al. Stability of venlafaxine immediate-release suspensions. Int J Pharm Compound. 2011;15:81–4.

1641. Gupta VD, Gupta VS. Chemical stability of brompheniramine maleate in an oral liquid dosage form. Int J Pharm Compound. 2011;15:78–80.

1642. Elder DL, Zheng B, White CA, et al. Stability of midazolam intranasal formula for the treatment of status epilepticus in dogs. Int J Pharm Compound. 2011;15:74–7.

1643. Rogerson A, Hiom S, Smith JC, et al. Physical stability of extemporaneously prepared oral hydrocortisone suspensions. J Pharm Pharmacol. 2010;62:1460.

1644. Santovena A, Llabres M, Farina JB. Quality control and physical and chemical stability of hydrocortisone oral suspension: an interlaboratory study. Int J Pharm Compound. 2010;14:430–5.

1645. Pignato A, Pankaskie M, Birnie C. Stability of methimazole in poloxamer lecithin organogel to determine beyond-use date. Int J Pharm Compound. 2010;14:522–5.

1646. Akram M, Naqvi SBS, Guahar S. Development of new ophthalmic suspension prednisolone acetate 1%. Pak J Pharm Sci. 2010;23:149–54.

1647. Preechagoon D, Sumyai V, Tontisirin K, et al. Formulation development and stability testing of oral morphine solution utilizing preformulation approach. J Pharm Pharm Sci. 2005;18:362–9.

1648. Johnson CE, Cober MP, Thome T, et al. Stability of an extemporaneous alcohol-free melatonin suspension. Am J Health Syst Pharm. 2011;68:420–3.

1649. Johnson CE, Cober MP, Hawkins KA, et al. Stability of extemporaneously prepared oxandrolone oral suspensions. Am J Health Syst Pharm. 2011;68:519–21.

1650. Senthilkumari S, Lalitha P, Prajna NV, et al. Single and multi-dose ocular kinetics and stability analysis of extemporaneous formulation of topical voriconazole in humans. Curr Eye Res. 2010;35:953–60.

1651. Sousa LA, Beezer A, Clapham D, et al. Assessment of drug photostability using a photocalorimeter. J Pharm Pharmacol. 2010;62:1202.

1652. Sousa LA, Beezer A, Clapham D, et al. The use of photocalorimetry to assess the photostability of nifedipine solutions. J Pharm Pharmacol. 2010;62:1214–5.

1653. Maheshwari RK, Indurkhya A. Formulation and evaluation of aceclofenac injection made by mixed hydrotropic solubilization technique. Iran J Pharm Res. 2010;9:233–42.

1654. Davis SN, Vermeulen L, Banton J, et al. Activity and dosage of alteplase dilution for clearing occlusions of venous-access devices. Am J Health Syst Pharm. 2000;57:1039–45.

1655. Shin YS, Young H, Lee CH. The stability of piroxicam in propylene glycol. J Kor Pharm Sci. 1988;18:203–8.

1656. Wiseman EH, Change YH, Lombardino JG. Piroxicam, a novel anti-inflammatory agent. Arzneimittelforschung. 1976;26:1300.

1657. Tu YH, Wang DP, Allen LV. Stability of a nonaqueous trimethoprim preparation. Am J Hosp Pharm. 1989;46:301–4.

1658. Tipton WR, Ledoux RA. Functional shelf life of methacholine and atropine methylnitrate solutions. Ann Allergy. 1986;556:117–9.

1659. Gupta VD, Mosier RL. Stability of phenylephrine hydrochloride nasal drops. Am J Hosp Pharm. 1972;29:870–3.

1660. Saxena V, Sadoqi M, Shao J. Degradation kinetics of indocyanine green in aqueous solution. J Pharm Sci. 2003;92:2090–7.

1661. Philip R, Penzkofer A, Baumler W, et al. Absorption and fluorescence spectroscopic investigation of indocyanine green. J Photoch Photobio A. 1996;96:137–48.

1662. Holzer W, Mauerer M, Penzkofer A, et al. Photostability and thermal stability of indocyanine green. J Photoch Photobio B. 1998;47:155–64.

1663. Bjornsson OG, Murphy R, Chadwick VS. Physicochemical studies of indocyanine green (ICG): Absorbance/concentration relationship, pH tolerance and assay precision in various solvents. Experientia. 1982;38:1441–2.

1664. Cober MP, Johnson CE, Sudekum D, et al. Stability of extemporaneously prepared glycopyrrolate oral suspensions. Am J Health Syst Pharm. 2011;68:843–5.

1665. Aliabadi HM, Romanick M, Somayaji V, et al. Stability of compounded thioguanine oral suspensions. Am J Health Syst Pharm. 2011;68:900–8.

1666. Charlton A, Kniska GM, Chao JV, et al. Stability of topical fortified antibiotic solutions. Invest Ophthalmol Vis Sci. 1992;33:937.

1667. Thoma K, Holzmann C. Dithranol formulations. Stability in pharmaceuticals. Deutsche Apotheker Zeit. 1997;137:3280–6.

1668. Regdon E, Regdon G, Kedvessy G. Chemical stability of acetylsalicyclic acid (ASA) in suppositories with a lipophilic base. Pharm Indust. 1981;43:388–90.

1669. Thielens I. Stability of N-acetyl-L-cysteine in pharmaceutical syrups. Farm Tijdschr Belg. 1984;61:445–8.

1670. Van Loenen AC, De Jong A, Van der Meer YG, et al. Formulation and shelf-life of acetylcysteine injections. Pharm Weekbl. 1985;120:313–7.

1671. Benson GS, Seifert WE. Is phentolamine stable in solution with papaverine? J Urol. 1988;140:970–1.

1672. Chen SJ, Hsu MY, Chen CY, et al. Solubility profiles, $pK_a$ values and stability of piroxicam in parenteral solutions. Chin Pharm J. 1991;43:27–43.

1673. Hubicka U, Krzek J, Walczak M. Stability of ciprofloxacin and norfloxacin in the presence and absence of metal ions in acidic solutions. Pharm Dev Technol. 2010;15:532–44.

1674. Allen LV Jr. Carvediol 1-mg/mL oral suspension. Int J Pharm Compound. 2010;14:423.

1675. Buontempo F, Bernabeu E, Glisoni RJ, et al. Carvediol stability in paediatric oral liquid formulations. Farm Hosp. 2010;34:293–7.

1676. Loftsson T, Vogensen SB, Desbos VC, et al. Carvediol: solubilization and cyclodextrin complexation. A technical note. AAPS PharmSciTech. 2008;24:187–92.

1677. Mizushima N, Tanaka Y, Fujii T, et al. Stability of a mixed solution of epinephrine and prednisolone succinate for inhalation prepared in our hospital. Byoin Yakugaku. 1992;18:458–65.

1678. Oda M, Kiyama Y, Oka M, et al. Physicochemical properties and stability of etodolac. Iyakuhin Kenkyu. 1991;22:152–64.

1679. Miethke T. Incompatibility of midazolam and saccharin. Pharm Ztg. 1991;136:43–4.

1680. Xiao Z, Sun DH, Jin J, et al. Study on preparation technique of doxycycline gel. Yaoxue Fuwu Yu Yanjiu. 2005;5:160–1.

1681. Xu L, Chen E, Cai F. Estimation of the stability of furacillin and ephedrine nasal drops. Zhongguo Yaoxue Zazhi. 1997;32:150–1.

1682. Anon. Stability of dithranol in Nutra-D cream. Aust J Hosp Pharm. 1979;9:135–6.

1683. Werchan D, Meffert H. Stability of dithranol preparations. Pharmazie. 1981;36:445–6.

1684. Umer S. Stability of aqueous prednisolone solutions. Ann Univ Mariae Curie-Sklodowska. 1975;30:151–7.

1685. McCluskey SV, Lovely JK. Stability of droperidol 0.625 mg/mL diluted with sodium chloride injection and stored in polypropylene syringes. Int J Pharm Compound. 2011; 15:170–3.

1686. Voudrie MA, Alexander B, Allen DB. Stability of verapamil hydrochloride in SyrSpend SF compared to sorbitol containing syrup and suspending vehicles. Int J Pharm Compound. 2011;15:255–8.

1687. Timm U, Goeber B, Doehnert H, et al. Stability of drugs. Part 5. Analysis and stability of tropicamide. Pharmazie. 1977;32:331–5.

1688. Wong CY, Wang DP, Wang CN. Stability of ceftizoxime in pluronic F-127 gels. Chin Pharm J. 2001;53:333–7.

1689. Hirikawa M, Yoshikawa M, Otsubo K, et al. Compatibility of tolperisone hydrochloride and dantrolene sodium in ground mixture. Byoin Yakugaku. 1996;22:521–6.

1690. Pohloudek-Fabini R, Martin E, Gallasch V. Stability of tropicamide solutions. Pharmazie. 1982;37:184–7.

1691. Pei YJ, Nan GZ. Buffering of pilocarpine ophthalmic solution. Yaoxue Tongbao. 1980;15:3–5.

1692. Porst H, Kny L. The stability of pilocarpine hydrochloride in eye drops. Pharmazie. 1985;40:23–9.

1693. Roth EM Jr., Shanley ES. Stability of pure hydrogen peroxide. J Ind Eng Chem. 1953;45:2343–9.

1694. Sheng M, Ma F, Yang WW. Study on stability of aqueous sodium hypochlorite solution. Huagong Jishu Yu Kaifa. 2005;34:8–10.

1695. Takekuma Y, Shiga H, Yamashita Y, et al. Studies on the stability of 0.625% povidone-iodine for eye washing. Iryo Yakugaku. 2003;29:62–5.

1696. Trose D, Slowig P. Stabilization of physostigmine salicylate injection solutions. Pharmazie. 1985;40:124–6.

1697. Pastia A, Gaita G, Verbuta A, et al. Stability of tetracycline hydrochloride in some ointments and syrups. Rev Med Chir Soc Med Nat Iasi. 1975;79:455–9.

1698. Kodym A, Pawlowska M, Ruminski JK, et al. Stability of cefepime in aqueous eyedrops. Pharmazie. 2011;66:17–23.

1699. Stewart JT, Warren FW. Stability of cefepime hydrochloride injection in polypropylene syringes at −20°C, 4°C, and 22–24°C. Am J Health Syst Pharm. 1999;56:457–9.

1700. Stewart JT, Maddox FC. Stability of cefepime hydrochloride in polypropylene syringes. Am J Health Syst Pharm 1999;56:1134.

1701. Ling J, Gupta VD. Stability of cefepime hydrochloride after reconstitution in 0.9% sodium chloride injection and storage in polypropylene syringes for pediatric use. Int J Pharm Compound. 2001;5:151–2.

1702. Nicoli WD, Smith AF. Stability of dilute alkaline solutions of hydrogen peroxide. J Ind Eng Chem. 1955;47:2548–54.

1703. Nicoletti MA, Magalhaes JF. Stability study of commercial sodium hypochlorite solution (chlorine water). Rev Farm Bioquim Univ Sao Paulo. 1995;31:53–60.

1704. Najafi RB, Samani SM Pishva N, et al. Formulation and clinical evaluation of povidone-iodine ophthalmic drop. Iran J Pharm Res. 2003;2:157–60.

1705. Nakagawa K, Matsuda S. Stability and physicochemical properties of Bactroban nasal ointment. Kagaku Ryoho no Royiki. 1996;12:1293–6.

1706. Kawamura K, Sugibayashi N, Otsuka K, et al. Formulation and stability of zofran syrup. Kagaku Ryoho no Royiki. 1999;15:1603–6.

1707. Allen LV Jr. Aminacrine hydrochloride 0.1% cream. Int J Pharm Compound. 2008;12:148.

1708. Tezuka T. The stability of mixtures of betamethasone-17-valerate ointments and urea containing preparations. Nippon Hifuka Gakkai Zasshi. 1996;106:1307–12.

1709. McCluskey SV, Brunn GJ. Nifedipine in compounded oral and topical preparations. Int J Pharm Compound. 2011;15:166–9.

1710. Hager K, Kaestner AM Stability of dithranol-salicylic acid oily formulation. Pharmazie. 1983;38:70–1.

1711. Baloglu E, Karavana SY, Hyusein IY, et al. Design and formulation of mebeverine HCl semisolid formulations for intra-orally administration. AAPS PharmSciTech. 2010;11:181–8.

1712. Melkoumov A, Soukrati A, Elkin I, et al. Quality evaluation of extemporaneous delayed-release liquid formulations of lansoprazole. Am J Health Syst Pharm. 2011;68:2069–74.

1713. Hergert LY, Sperando NR, Allemandi DA, et al. Preparation and characterization of a new alternative formulation of sodium p-aminosalicyclate. Int J Pharm Compound. 2011;15:344–9.

1714. Ohishi T, Shinagawa R, Harada Y, et al. The stability of a mixture corticosteroid ointment and zinc oxide ointment. Hiroshimaken Buoin Yakuzaishikai. 1990;25:16–8.

# Index

Monograph title drugs are in ALL CAPS; brand name drugs are in *italics*.

## A

*Abactrim,* 139
ACECLOFENAC, 1
ACETAMINOPHEN, 2
ACETAZOLAMIDE, 3
ACETAZOLAMIDE SODIUM, 3
ACETYLCHOLINE CHLORIDE, 5
ACETYLCYSTEINE, 5
Acetylsalicylic acid, 50
Acetyl sulphafurazole, 452
*Achromycin,* 467
Aciclovir, 9
Aciclovir sodium, 9
*Actifed,* 490
*Actigall,* 495
*Actonel,* 429
ACYCLOVIR, 9
ACYCLOVIR SODIUM, 9
*Adalat,* 351, 352
ADAPALENE, 10
ADDERALL, 10
*Adenocard,* 11
*Adenoscan,* 11
ADENOSINE, 11
*Adiazine,* 451
Adrenaline, 186
African Scalp Lotion, 43
ALA, 29
ALBENDAZOLE, 12
ALBENDAZOLE SULFOXIDE, 12
ALBUTEROL SULFATE, 12
*Alcobon,* 209
Alcohol, dehydrated, 197
*Aldactone,* 447
*Aldomet,* 319
ALENDRONATE SODIUM, 17

*Alka-Seltzer,* 50
ALLERGEN EXTRACTS, 17
ALLOPURINOL, 19
ALOE, 21
*Alomide,* 295
*Alphagan P,* 70
Alphazurine 2G, 267
ALPRAZOLAM, 21
ALPROSTADIL, 22
ALTEPLASE, 23
ALUM, 24
Aluminum potassium sulfate, 24
*Alupent,* 311
*AmBisome,* 39
AMBROXOL HYDROCHLORIDE, 25
*Amerge,* 348
Amethocaine hydrochloride, 463
AMIKACIN SULFATE, 25
AMILORIDE HYDROCHLORIDE, 27
AMINACRINE, 27
AMINACRINE HYDROCHLORIDE, 27
Aminoacridine hydrochloride, 27
AMINOHIPPURATE SODIUM, 28
AMINOLEVULINIC ACID, 29
AMINOPHYLLINE, 30
4-AMINOPYRIDINE, 31
AMINOSALICYLATE SODIUM, 32
AMIODARONE HYDROCHLORIDE, 32
AMITRIPTYLINE HYDROCHLORIDE, 33
AMLODIPINE BESYLATE, 34
AMMONIUM LACTATE, 34
*Ammonul,* 442
AMOXICILLIN, 35
AMOXICILLIN TRIHYDRATE–CLAVULANATE POTASSIUM, 36
*Amoxil,* 35
*Amphojel,* 42

AMPHOTERICIN B, 38
AMPICILLIN, 40
*Ancotil,* 209
*Antabuse,* 175
ANTACIDS, 42
*Antebate,* 65
ANTHRALIN, 42
ANTIPYRINE, 44
*Anzemet,* 177
4-AP, 31
*Apiretal,* 2
APOMORPHINE HYDROCHLORIDE, 45
APREPITANT, 46
*Aquaphyllin,* 469
*Aralen HCl,* 111
ARFORMOTEROL TARTRATE, 46
ARSENIC TRIOXIDE, 47
Artemether, 47
ARTESUNATE, 47
*Asbron Elixir,* 468, 469
ASCORBIC ACID, 48
ASPIRIN, 50
*Asthpul,* 266
ATENOLOL, 50
ATROPINE, 52
ATROPINE SULFATE, 52
ATROPINE SULFATE WITH DIPHENOXYLATE
        HYDROCHLORIDE, 173
*Atrovent,* 259, 260, 261, 262
*Augmentin,* 36
*Avastin,* 68
*Avelox,* 343
*Axid,* 355
AZATHIOPRINE, 54
AZITHROMYCIN, 56
*Azunol,* 439

**B**
BACITRACIN, 57
BACLOFEN, 58
*Bactrim,* 139
*Bactroban,* 330, 344
*Banzel,* 431
Beclometasone dipropionate, 59
BECLOMETHASONE DIPROPIONATE, 59
*Beconase,* 59
BELLADONNA WITH PHENOBARBITAL, 60
*Benadryl,* 172
BENAZEPRIL HYDROCHLORIDE, 60
Bendrofluazide, 61
BENDROFLUMETHIAZIDE, 61
*Bentyl,* 165
*Benylin DM,* 172
*Benzamycin,* 191
Benzhexol hydrochloride, 487
BENZOCAINE, 62
BENZOYL PEROXIDE, 62

BENZYL BENZOATE, 63
Benzylpenicillin potassium, 375
*Berotec,* 202, 203
*Betadine,* 245, 400
*Betaloc,* 325
BETAMETHASONE, 64
*Betapace,* 445
BETHANECHOL CHLORIDE, 66
*Betnelan-V,* 65
*Betnovate,* 64, 65
BEVACIZUMAB, 68
BIMATOPROST, 68
*Bisolvon,* 71
BISOPROLOL FUMARATE, 69
BITOLTEROL MESYLATE, 69
BORONOPHENYLALANINE–FRUCTOSE COMPLEX, 69
*Brethine,* 461
*Bricanyl,* 461
BRIMONIDINE TARTRATE, 70
BROMHEXINE HYDROCHLORIDE, 71
BROMPHENIRAMINE MALEATE, 71
Brompton mixtures, 134, 340
*Bronkosol,* 262, 263
*Brovana,* 46
BUDESONIDE, 72
BUPIVACAINE HYDROCHLORIDE, 74
BUSULFAN, 75

**C**
CAFFEINE, 76
CALCIPOTRIENE, 78
CALCITONIN-SALMON, 78
CALCITRIOL, 79
Calcium folinate, 282
CALCIUM GLUBIONATE, 80
CALCIUM GLUCONATE, 80
*Capoten,* 82, 83, 84, 85
CAPTOPRIL, 81
CARBAMAZEPINE, 86
CARBENICILLIN DISODIUM, 88
CARBENICILLIN INDANYL SODIUM, 88
CARBIDOPA, 88
CARBOPLATIN, 89
CARBOPROST TROMETHAMINE, 90
CARMUSTINE, 90
CARNITINE, 91
CARPROFEN, 91
CARTEOLOL HYDROCHLORIDE, 92
CARVEDILOL, 93
*Catapres,* 130
*Ceclor,* 94
CEFACLOR, 94
CEFADROXIL, 94
*Cefalexgobens,* 105
CEFAZOLIN SODIUM, 95
CEFEPIME HYDROCHLORIDE, 97
CEFTAZIDIME, 98

*Ceftin,* 101, 102
CEFTIZOXIME SODIUM, 101
CEFUROXIME AXETIL, 101
CEFUROXIME SODIUM, 102
CELECOXIB, 104
CEPHALEXIN, 104
CEPHALOTHIN SODIUM, 105
*Cephulac,* 278
CHARCOAL, 106
CHLORAL HYDRATE, 106
CHLORAMBUCIL, 107
CHLORAMPHENICOL, 108
CHLORAMPHENICOL PALMITATE, 108
CHLORAMPHENICOL SODIUM SUCCINATE, 108
CHLORHEXIDINE DIACETATE, 110
*Chloromycetin,* 110
CHLOROQUINE HYDROCHLORIDE, 110
CHLOROQUINE PHOSPHATE, 110
CHLOROTHIAZIDE, 113
CHLOROTHIAZIDE SODIUM, 113
CHLORPHENIRAMINE MALEATE, 114
CHLORPROMAZINE HYDROCHLORIDE, 114
CHLORPROTHIXENE, 117
CHLORTETRACYCLINE HYDROCHLORIDE, 117
CHOLECALCIFEROL, 118
*Cibalith-S,* 292, 293
CIDOFOVIR, 119
CIMETIDINE, 119
CIMETIDINE HYDROCHLORIDE, 119
CINNARIZINE, 121
*Ciprodex,* 122
CIPROFLOXACIN, 122
CIPROFLOXACIN HYDROCHLORIDE, 122
CISAPRIDE, 124
*Clamoxyl,* 35
Clavulanate potassium, 36
CLINDAMYCIN HYDROCHLORIDE, 125
CLINDAMYCIN PALMITATE HYDROCHLORIDE, 125
CLINDAMYCIN PHOSPHATE, 125
CLOBETASOL PROPIONATE, 127
CLOMIPRAMINE HYDROCHLORIDE, 128
CLONAZEPAM, 128
CLONIDINE HYDROCHLORIDE, 129
CLOPIDOGREL BISULFATE, 130
*Clorox,* 440
CLOTRIMAZOLE, 131
CLOXACILLIN SODIUM, 131
CLOZAPINE, 132
COAL TAR, 132
COAL TAR SOLUTION, 132
Co-Amoxiclav, 36
COCAINE HYDROCHLORIDE, 134
CODEINE, 136
CODEINE PHOSPHATE, 136
CODEINE SULFATE, 136
Codethyline, 198
Coenzyme Q$_{10}$, 493

*Colircusi Atropina,* 53
*Colircusi Cicloplejico,* 143
*Colircusi Gentamicina,* 225
*Colircusi Pilocarpina,* 393
COLISTIMETHATE SODIUM, 137
COLISTIN SULFATE, 137
*Coly-Mycin,* 138
*Compazine,* 408
*Compocillin VK,* 376
*Cordarone,* 33
*Coreg,* 93
*Corgard,* 346
*Cortoftal,* 127
CO-TRIMOXAZOLE, 138
*Coumadin,* 516
*Cozaar,* 297
CROMOLYN SODIUM, 139
*Cuprimine,* 375
CURCUMIN, 142
*Cusimolol,* 477
CYCLOPENTOLATE HYDROCHLORIDE, 143
CYCLOPHOSPHAMIDE, 143
CYCLOSERINE, 145
CYCLOSPORINE, 145
CYPROHEPTADINE HYDROCHLORIDE, 148
CYPROTERONE ACETATE, 148
CYSTEAMINE BITARTRATE, 149
CYSTEAMINE HYDROCHLORIDE, 149
*Cytotec,* 335
*Cytovene,* 223
*Cytoxan,* 144

**D**

DANAZOL, 150
DANTROLENE SODIUM, 150
3,4-DAP, 158
DAPSONE, 151
DAUNORUBICIN HYDROCHLORIDE, 152
*Decadron,* 156
DEFEROXAMINE MESYLATE, 153
Dehydrated alcohol, 197
*Depakene,* 499
*Depakine,* 499
*Dermovate,* 127
Desferrioxamine mesylate, 153
DESONIDE, 153
DESOXIMETASONE, 153
*Desyrel,* 482
DEXAMETHASONE, 154
DEXAMETHASONE ACETATE, 154
DEXAMETHASONE SODIUM PHOSPHATE, 154
DEXCHLORPHENIRAMINE MALEATE, 157
DEXPANTHENOL, 157
Dextroamphetamine salts, 10
DEXTROMETHORPHAN HYDROBROMIDE, 158
Dextrose, 197
Diacetylmorphine hydrochloride, 159

3,4-DIAMINOPYRIDINE, 158
DIAMORPHINE HYDROCHLORIDE, 159
*Diamox,* 3
DIAZEPAM, 160
DICHLOROBENZYL ALCOHOL, 162
DICLOFENAC SODIUM, 162
DICLOXACILLIN SODIUM, 164
DICYCLOMINE HYDROCHLORIDE, 165
*Differin,* 10
DIFLORASONE DIACETATE, 165
*Diflucan,* 207
DIFLUCORTOLONE VALERATE, 166
DIFLUNISAL, 166
Difluorophenylsalicylic acid, 166
DIGOXIN, 167
*Dilantin,* 386, 387, 388, 389
DILTIAZEM HYDROCHLORIDE, 168
DIMENHYDRINATE, 170
*Dimetane,* 71, 72
*Dimetapp,* 71, 72
DINOPROSTONE, 170
Dioctyl sodium sulfosuccinate, 176
*Diovan,* 500
DIPHENHYDRAMINE HYDROCHLORIDE, 172
DIPHENOXYLATE HYDROCHLORIDE WITH ATROPINE
     SULFATE, 173
*Diprosone,* 65
DIPYRIDAMOLE, 173
DISOPYRAMIDE PHOSPHATE, 174
DISULFIRAM, 175
Dithranol, 42
*Ditropan,* 368
DOCUSATE SODIUM, 176
DOLASETRON MESYLATE, 176
*Dolophine,* 314
DOMPERIDONE, 177
DOMPERIDONE MALEATE, 177
*Donnatal,* 60
DORNASE ALFA, 178
DOXEPIN HYDROCHLORIDE, 179
DOXYCYCLINE, 180
DOXYCYCLINE HYCLATE, 180
DROPERIDOL, 181
*Duricef,* 94
*DynaCirc,* 267
*Dynapen,* 164

**E**
*Edex,* 22
*E.E.S.,* 192
*Effexor,* 508
*Eleblock,* 92
*Elixifilin,* 470
*Elixophyllin,* 468, 469
*Elixophyllin-KI,* 468, 469
EMETINE HYDROCHLORIDE, 182

ENALAPRIL MALEATE, 183
*Enarenal,* 184
ENTECAVIR, 185
*Entocort EC,* 72
EPHEDRINE HYDROCHLORIDE, 185
EPHEDRINE SULFATE, 185
EPINEPHRINE, 186
*Eplim,* 499
EPOETIN ALFA, 189
ERYTHROMYCIN, 190
ERYTHROMYCIN ETHYLSUCCINATE, 192
ESOMEPRAZOLE MAGNESIUM, 193
*Estilsone,* 405
ESTRADIOL, 195
ETHACRYNIC ACID, 195
ETHAMBUTOL HYDROCHLORIDE, 196
ETHANOL, 197
ETHIONAMIDE, 198
ETHOSUXIMIDE, 198
ETHYLMORPHINE HYDROCHLORIDE, 198
ETODOLAC, 199
ETOPOSIDE, 199
*Eufilina,* 470

**F**
FAMOTIDINE, 201
FENBENDAZOLE, 202
FENOTEROL HYDROBROMIDE, 202
FENTANYL CITRATE, 203
*Feosol,* 204, 205
FERROUS SULFATE, 204
FILGRASTIM, 205
*Flagyl,* 328
FLECAINIDE ACETATE, 206
*Florinef,* 209
FLOXURIDINE, 206
Flubendazole, 202
FLUCONAZOLE, 207
*Flucort,* 211
FLUCYTOSINE, 208
FLUDROCORTISONE ACETATE, 209
FLUOCINOLONE ACETONIDE, 210
FLUOCINONIDE, 210
FLUOROURACIL, 211
FLUOXETINE HYDROCHLORIDE, 212
FLUPHENAZINE HYDROCHLORIDE, 213
FLURANDRENOLIDE, 214
Flurandrenolone, 214
FLUTICASONE PROPIONATE, 215
*Flutide,* 215
FOLIC ACID, 216
Folinic acid, 282
FORMALDEHYDE, 216
Formalin, 216
FORMOTEROL FUMARATE, 217
*Fortaz,* 100

*Fortical,* 79
*Fortovase,* 433
*Fosfocina,* 218
FOSFOMYCIN, 217
Fox green, 252
*Frinova,* 412
Fructose, 69
Frusemide, 219
FUMAGILLIN BICYCLOHEXYLAMMONIUM, 218
*Fumidil,* 218
*Fungizone,* 38, 39
*Furacin,* 354
*Furadantin,* 353, 354
FUROSEMIDE, 219

## G

GABAPENTIN, 221
*Gabitril,* 476
GALACTOSE, 222
GANCICLOVIR, 222
GANCICLOVIR SODIUM, 222
*Gantrisin,* 452
*Garasone,* 224
GATIFLOXACIN, 223
G-CSF, 205
*Genoptic,* 225
GENTAMICIN SULFATE, 224
*Gernebcin,* 480, 481
Glibenclamide, 226
GLYBURIDE, 226
GLYCERIN, 227
Glycerol, 227
GLYCOLIC ACID, 228
GLYCOPYRROLATE, 228
GRANISETRON HYDROCHLORIDE, 230
*Grifulvin,* 231
GRISEOFULVIN, 231
GUAIFENESIN, 231

## H

*Haldol,* 233, 234
HALOBETASOL PROPIONATE, 233
HALOPERIDOL, 233
HALOPERIDOL LACTATE, 233
*Heberon Alfa R,* 256
*Hemabate,* 90
HEMIACIDRIN, 235
Heroin, 159
Hexamine mandelate, 315
HISTAMINE PHOSPHATE, 236
human granulocyte colony-stimulating factor, 205
*Humatrope,* 444
HYALURONIDASE, 237
HYDRALAZINE HYDROCHLORIDE, 237
*Hydrea,* 247
HYDROCHLOROTHIAZIDE, 239

HYDROCHLOROTHIAZIDE AND SPIRONOLACTONE, 448
HYDROCORTISONE, 241
HYDROGEN PEROXIDE, 245
HYDROMORPHONE HYDROCHLORIDE, 245
HYDROQUINONE, 246
*HydroVal,* 244
HYDROXYCHLOROQUINE SULFATE, 246
HYDROXYUREA, 247
HYDROXYZINE HYDROCHLORIDE, 247
HYDROXYZINE PAMOATE, 247
*Hygeol,* 440
Hyoscine hydrobromide, 433
*Hypnovel,* 332

## I

*Iberet,* 205
IBUPROFEN, 249
IDOXURIDINE, 250
*Imitrex,* 454
*Imodium,* 295
*Inderal,* 413, 414
INDINAVIR SULFATE, 251
*Indocin,* 252, 253
INDOCYANINE GREEN, 252
INDOMETHACIN, 252
INDOMETHACIN SODIUM, 252
INFLIXIMAB, 255
*Instillagel,* 290
*Intal,* 140, 141
INTERFERON ALFA-2B, 256
IOBENGUANE SULFATE, 257
IODINATED GLYCEROL, 257
Iodine, 398, 400
IPRATROPIUM BROMIDE, 258
ISOETHARINE HYDROCHLORIDE, 262
ISONIAZID, 263
Isoprenaline hydrochloride, 265
ISOPROTERENOL HYDROCHLORIDE, 265
ISOSORBIDE DINITRATE, 266
ISOSULFAN BLUE, 267
ISRADIPINE, 267
*Isuprel,* 265, 266
ITRACONAZOLE, 268
IUdR, 250

## K

KANAMYCIN SULFATE, 270
*Kaon,* 398
*Kay Ciel Elixir,* 397, 398
*Kayexalate,* 443
*Keflex,* 104, 105
*Kefzol,* 96
*Kenacort,* 486
*Kenalog,* 485, 486
*Keralyt,* 433
KETAMINE HYDROCHLORIDE, 270

*Ketasma,* 274
KETOCONAZOLE, 271
KETOPROFEN, 273
Ketorolac trometamol, 274
KETOROLAC TROMETHAMINE, 274
KETOTIFEN FUMARATE, 274
*Kindavate,* 127
*K-Lor,* 398
*Klorvess,* 397, 398
Knoxville Formula, 23
*Kytril,* 230

## L

LABETALOL HYDROCHLORIDE, 276
LACTIC ACID, 277
LACTULOSE, 278
LAMOTRIGINE, 278
*Lanacordin,* 167
*Lanoxin,* 167
LANSOPRAZOLE, 279
*Larotid,* 35
*Lasix,* 219, 220
LATANOPROST, 281
LEUCOVORIN CALCIUM, 282
LEVALBUTEROL HYDROCHLORIDE, 282
LEVAMISOLE, 283
LEVAMISOLE HYDROCHLORIDE, 283
*Levaquin,* 286
Levoamphetamine salts, 10
LEVOCARNITINE, 284
LEVOCARNITINE HYDROCHLORIDE, 284
LEVODOPA, 285
LEVOFLOXACIN, 286
Levomepromazine, 316
LEVOTHYROXINE SODIUM, 286
*Levulan Kerastick,* 29
*Lidex,* 210, 211
LIDOCAINE HYDROCHLORIDE, 288
*Lidomex,* 404
Lignocaine hydrochloride, 288
*Lioresal,* 58
LIOTHYRONINE SODIUM, 291
Liquor carbonis detergens, 132
LISINOPRIL, 291
*Lithionate-S,* 293
LITHIUM CITRATE, 292
*Locoid,* 244
LODOXAMIDE TROMETHAMINE, 295
*Lomotil,* 173
LOPERAMIDE HYDROCHLORIDE, 295
LORATIDINE, 296
LORAZEPAM, 296
LOSARTAN POTASSIUM, 297
*Losec,* 360, 361
*Lotensin,* 61
*Lotrimin,* 131
LOXAPINE HYDROCHLORIDE, 297

*Lugol's solution,* 398
*Lysomucil,* 8

## M

MAGNESIUM SULFATE, 299
*Mandelamine,* 315, 316
*Marevan,* 516
MCT, 302
MEBENDAZOLE, 299
MEBEVERINE HYDROCHLORIDE, 300
MECHLORETHAMINE HYDROCHLORIDE, 300
MEDIUM CHAIN TRIGLYCERIDES, 302
MELATONIN, 303
*Mellaril,* 472, 473
MELOXICAM, 304
MELPHALAN, 305
MEMANTINE HYDROCHLORIDE, 305
MENADIONE, 306
MEPERIDINE HYDROCHLORIDE, 306
Mercaptamine bitartrate, 149
Mercaptamine hydrochloride, 149
MERCAPTOPURINE, 307
MESALAMINE, 308
Mesalazine, 308
MESNA, 309
*Mesnex,* 309
MESORIDAZINE BESYLATE, 310
*Mestinon,* 417, 418
*Metacam,* 304
METAPROTERENOL SULFATE, 310
METFORMIN HYDROCHLORIDE, 312
METHACHOLINE CHLORIDE, 312
*Methaderm,* 156
METHADONE HYDROCHLORIDE, 314
METHENAMINE MANDELATE, 315
METHIMAZOLE, 316
METHOTRIMEPRAZINE, 316
METHOTRIMEPRAZINE HYDROCHLORIDE, 316
METHOXSALEN, 317
METHYL NICOTINATE, 318
METHYLDOPA, 318
METHYLDOPATE HYDROCHLORIDE, 318
METHYLENE BLUE, 319
METHYLERGONOVINE MALEATE, 320
METHYLPHENIDATE HYDROCHLORIDE, 320
METHYLPREDNISOLONE, 321
METHYLPREDNISOLONE ACETATE, 321
METHYLPREDNISOLONE SODIUM SUCCINATE, 321
METOCLOPRAMIDE HYDROCHLORIDE, 322
METOLAZONE, 323
METOPROLOL TARTRATE, 324
*Metosyn,* 211
METRONIDAZOLE, 325
METRONIDAZOLE BENZOATE, 325
METRONIDAZOLE HYDROCHLORIDE, 325
MEXILETINE HYDROCHLORIDE, 329
*Miacalcin,* 78, 79

MICONAZOLE, 329
MICONAZOLE NITRATE, 329
*Midamor,* 27
MIDAZOLAM HYDROCHLORIDE, 330
MILK OF MAGNESIA, 333
MINOCYCLINE HYDROCHLORIDE, 334
MINOXIDIL, 334
MINOXIDIL TARTRATE, 334
MISOPROSTOL, 335
MITOMYCIN, 336
MOLINDONE HYDROCHLORIDE, 337
MOMETASONE FUROATE, 338
*Monistat Derm,* 330
MORPHINE, 338
MORPHINE HYDROCHLORIDE, 338
MORPHINE SULFATE, 338
*Motilium,* 177
MOXIFLOXACIN HYDROCHLORIDE, 343
*Mucomyst,* 7, 8
*Mucosan,* 25
*Mucosil-10,* 8
*Mucosil-20,* 7
MUPIROCIN, 344
*Mustargen,* 301
Mustine hydrochloride, 300
*Mutamycin,* 336
MYCOPHENOLATE MOFETIL, 345
*Mycostatin,* 358
*Mylanta,* 42
*Myleran,* 75
*Mylicon,* 437

## N

NADOLOL, 346
NALTREXONE HYDROCHLORIDE, 346
NAPROXEN, 347
NARATRIPTAN HYDROCHLORIDE, 348
*Nasonex,* 338
*Navane,* 475
*Nebcin,* 479, 480
*Neo-Calglucon,* 80
NEOMYCIN SULFATE, 349
*Nerisone,* 166
*Neurontin,* 221
NEVIRAPINE, 350
*Nexium,* 193, 194
NIFEDIPINE, 351
NITRAZEPAM, 353
NITROFURANTOIN, 353
NITROFURAZONE, 354
NIZATIDINE, 355
*Nizoral,* 272
*Norfenon,* 412
NORFLOXACIN, 355
*Normodyne,* 276
*Norpace,* 175
*Norvasc,* 34

*Noxyflex,* 356
*Noxyflex S,* 356
Noxythiolin, 356
NOXYTIOLIN, 356
NYSTATIN, 357

## O

Obetrol, 10
*Occupress,* 92
OCTREOTIDE ACETATE, 359
Oestradiol, 195
*Oftalar,* 401
OMEPRAZOLE, 359
*Omnipen,* 40, 41
ONDANSETRON HYDROCHLORIDE, 363
*Orbenin,* 132
Orciprenaline sulphate, 310
*Organidin,* 257, 258
OSELTAMIVIR PHOSPHATE, 365
OXACILLIN SODIUM, 367
OXANDROLONE, 368
Oxpentifylline, 377
OXYBUTYNIN CHLORIDE, 368
OXYCODONE HYDROCHLORIDE, 369

## P

*Paceron,* 33
PAH, 28
*Panamax,* 2
Pancreatic enzymes, 370
PANCRELIPASE, 370
*Pandel,* 244
*Pantomicina,* 192
PANTOPRAZOLE SODIUM, 370
PAPAVERINE HYDROCHLORIDE, 372
para-Aminohippuric acid, 28
PARABENS, 373
Paracetamol, 2
Parahydroxybenzoates, 373
PAREGORIC, 373
*Parvolex,* 6
PAS sodium, 32
Patent blue V, 267
*Pathocil,* 164
PEFLOXACIN MESYLATE, 374
*Penbritin,* 40, 41
PENICILLAMINE, 374
PENICILLIN G POTASSIUM, 375
PENICILLIN V POTASSIUM, 376
PENTOXIFYLLINE, 377
*Pen-Vee K,* 377
*Perforomist,* 217
Pergolide mesilate, 377
PERGOLIDE MESYLATE, 377
*Periactin,* 148
PERMETHRIN, 378
PERPHENAZINE, 379

*Persantine,* 174
Pethidine hydrochloride, 306
Phenazone, 44
*Phenergan,* 412
PHENOBARBITAL, 380
PHENOBARBITAL SODIUM, 380
PHENOBARBITAL WITH BELLADONNA, 60
Phenobarbitone, 380
PHENOXYBENZAMINE HYDROCHLORIDE, 382
PHENOXYETHANOL, 383
Phenoxymethylpenicillin potassium, 376
PHENTOLAMINE MESYLATE, 383
PHENYLEPHRINE HYDROCHLORIDE, 384
PHENYLMERCURIC NITRATE, 385
PHENYTOIN, 386
PHENYTOIN SODIUM, 386
PHYSOSTIGMINE, 390
PHYSOSTIGMINE SALICYLATE, 390
Phytomenadione, 391
PHYTONADIONE, 391
PILOCARPINE HYDROCHLORIDE, 392
PIPERACILLIN SODIUM, 394
PIROXICAM, 395
*Polaramine,* 157
*Polycillin,* 41
POLYMYXIN B SULFATE, 396
*Pontocaine,* 464
Potassium alum, 24
POTASSIUM CHLORIDE, 397
POTASSIUM GLUCONATE, 398
POTASSIUM IODIDE, 398
POTASSIUM PERCHLORATE, 399
POVIDONE-IODINE, 400
PRANOPROFEN, 401
PRAVASTATIN SODIUM, 402
PREDNISOLONE, 402
PREDNISOLONE ACETATE, 402
PREDNISOLONE SODIUM PHOSPHATE, 402
PREDNISONE, 404
PRIMAQUINE PHOSPHATE, 406
*Principen,* 40, 41
*Prinivil,* 292
*Probitor,* 361
PROCAINAMIDE HYDROCHLORIDE, 406
PROCARBAZINE HYDROCHLORIDE, 407
PROCHLORPERAZINE, 408
PROCHLORPERAZINE EDISYLATE, 408
PROCHLORPERAZINE MALEATE, 408
PROGESTERONE, 409
*Prograf,* 456, 457
PROMAZINE HYDROCHLORIDE, 410
PROMETHAZINE HYDROCHLORIDE, 411
*Propaderm,* 59, 60
PROPAFENONE HYDROCHLORIDE, 412
PROPRANOLOL HYDROCHLORIDE, 412
*Propulsid,* 124

PROPYLTHIOURACIL, 415
Prostaglandin E$_1$, 22
Prostaglandin E$_2$, 170
*Prostaphlin,* 367
*Prostin E2,* 170, 171
*Prostin VR,* 22
*Protonix,* 371
*Protopic,* 154, 457
*Proventil,* 13, 14
*Provocholine,* 313
PSEUDOEPHEDRINE HYDROCHLORIDE, 416
*Psoriatec,* 42
*Pulmicort Respules,* 72, 73, 74
*Pulmozyme,* 178, 179
*Purinethol,* 307
PYRAZINAMIDE, 416
PYRIDOSTIGMINE BROMIDE, 417
PYRIDOXINE HYDROCHLORIDE, 418
PYRIMETHAMINE, 419

**Q**
QUINAPRIL HYDROCHLORIDE, 420
QUINIDINE SULFATE, 420

**R**
RAMIPRIL, 422
RANITIDINE HYDROCHLORIDE, 422
*Redoxon,* 48
*Remodulin,* 482
Renacidin, 235
Retinol, 511
*Revatio,* 434
*Rhinocort Aqua,* 72
RIBAVIRIN, 424
Ricobendazole, 12
RIFABUTIN, 425
*Rifadin,* 426
*Rifaldin,* 427
Rifampicin, 425
RIFAMPIN, 425
RIFAPENTINE, 428
RIFAXIMIN, 428
*Rimadyl,* 92
*Rinderon VG,* 66
*Riopan,* 42
RISEDRONATE SODIUM, 429
*Rispadal Liquid,* 430
*Risperdal Consta,* 430
RISPERIDONE, 430
*Ritalin,* 321
*Robinul,* 229
*Robitussin,* 232
ROFECOXIB, 430
*Romilar,* 158
*Rowasa,* 308
RUFINAMIDE, 431

## S

Salbutamol sulphate, 12
SALICYLIC ACID, 432
*Sandimmune,* 145, 146
SAQUINAVIR, 433
SCOPOLAMINE HYDROBROMIDE, 433
*Septra,* 139
SILDENAFIL CITRATE, 434
*Silvadene,* 436
SILVER LACTATE, 435
SILVER NITRATE, 435
SILVER SULFADIAZINE, 436
SIMETHICONE, 436
*Sinequan,* 179, 180
*Sinogan,* 317
SISOMICIN SULFATE, 437
*SK-Ampicillin,* 41
*Slo-Phyllin,* 469
*Slophylline,* 469
Sodium azulene sulfonate, 439
SODIUM BENZOATE AND SODIUM PHENYLACETATE, 442
SODIUM BICARBONATE, 438
Sodium chromoglycate, 139
SODIUM ETHACRYNATE, 195
SODIUM GUALENATE HYDRATE, 439
SODIUM HYPOCHLORITE, 439
SODIUM PHENYLACETATE AND SODIUM BENZOATE, 442
SODIUM PHENYLBUTYRATE, 442
SODIUM POLYSTYRENE SULFONATE, 443
*Solu-Medrol,* 321
SOMATROPIN, 444
*Somophylline,* 30
SOTALOL HYDROCHLORIDE, 444
SPIRONOLACTONE, 445
SPIRONOLACTONE AND HYDROCHLOROTHIAZIDE, 448
*Sporanox,* 268
SSKI, 399
*Stemetil,* 408
SUCCINYLCHOLINE CHLORIDE, 448
SUCRALFATE, 450
*Sudafed,* 416
SULFACETAMIDE, 450
SULFACETAMIDE SODIUM, 450
SULFADIAZINE, 451
Sulfamethoxazole, 138
Sulfan blue, 267
SULFISOXAZOLE, 452
SULFISOXAZOLE ACETYL, 452
SULFUR, 453
Sulphafurazole, 452
*Sultanol,* 15, 16
*Sultanol forte FI,* 15, 16
SUMATRIPTAN SUCCINATE, 453
*Sumycin,* 466, 467
SUNITINIB, 454
SUNITINIB MALATE, 454

SUSPENSION STRUCTURED VEHICLE, NF, 455
SUSPENSION STRUCTURED VEHICLE, SUGAR-FREE, NF, 455
Suxamethonium chloride, 448
*Synalar,* 211

## T

TACROLIMUS, 456
T-A-C solutions, 188
*Tagamet,* 119, 120
TALC, 457
*Tambocor,* 206
TAZAROTENE, 458
TEGASEROD MALEATE, 458
*Tegretol,* 86, 87
TEMOZOLOMIDE, 458
*Tenormin,* 51
*Terbasmin,* 461
TERBINAFINE HYDROCHLORIDE, 460
TERBUTALINE SULFATE, 460
TERPIN HYDRATE, 462
TESTOSTERONE, 462
TETRACAINE, 463
TETRACAINE HYDROCHLORIDE, 463
TETRACYCLINE, 465
TETRACYCLINE HYDROCHLORIDE, 465
THALIDOMIDE, 467
*Theo-Dur,* 469
*Theolair Liquid,* 468, 469
THEOPHYLLINE, 468
*Theostat 80,* 469
Thiamazole, 316
THIAMINE HYDROCHLORIDE, 470
THIOGUANINE, 471
THIORIDAZINE, 471
THIORIDAZINE HYDROCHLORIDE, 471
THIOTHIXENE, 474
THIOTHIXENE HYDROCHLORIDE, 474
*Thorazine,* 115
Thyroxine sodium, 286
TIAGABINE HYDROCHLORIDE, 475
TICARCILLIN DISODIUM, 476
TIMOLOL MALEATE, 477
TINIDAZOLE, 478
*Tobi,* 480, 481
TOBRAMYCIN, 478
TOBRAMYCIN SULFATE, 478
*Tobrex,* 479
Tocopherols, 513
*Topicort,* 154
*Tornalate,* 69
t-PA, 23
TRAMADOL HYDROCHLORIDE, 481
*Trandate,* 276
TRAZODONE HYDROCHLORIDE, 482
*Trental,* 377
TREPROSTINIL SODIUM, 482

TRETINOIN, 483
TRIAMCINOLONE ACETONIDE, 484
*Triaminic,* 114
*Tribavirin,* 424
*Tridesilon,* 153
TRIFLUOPERAZINE HYDROCHLORIDE, 486
TRIHEXYPHENIDYL HYDROCHLORIDE, 487
TRIMETHOBENZAMIDE HYDROCHLORIDE, 488
TRIMETHOPRIM, 488
Trimethoprim–sulfamethoxazole, 138
TRIMETHOPRIM SULFATE, 488
Triple hormones, 195
TRIPROLIDINE HYDROCHLORIDE, 490
*Trisenox,* 47
TROPICAMIDE, 491
TROVAFLOXACIN MESYLATE, 492
*Trovan,* 492
*Tylenol,* 2

## U

UBIDECARENONE, 493
*Ucephan,* 442
*Ultracet,* 2
UREA, 493
*Urso,* 496
Ursodeoxycholic acid, 495
URSODIOL, 495
*Utimox,* 35

## V

VALACYCLOVIR HYDROCHLORIDE, 497
*Valcyte,* 498
VALGANCICLOVIR HYDROCHLORIDE, 498
*Valisone,* 66
*Valium,* 161
VALPROATE SODIUM, 499
VALPROIC ACID, 499
VALSARTAN, 500
*Vancenase,* 59
VANCOMYCIN HYDROCHLORIDE, 500
VASOPRESSIN, 506
VASOPRESSIN TANNATE, 506
*Vasotec,* 183
*V-Cillin K,* 376
*Veetids,* 376

VEHICLE FOR ORAL SOLUTION, NF, 506
VEHICLE FOR ORAL SOLUTION, SUGAR FREE, NF, 507
VEHICLE FOR ORAL SUSPENSION, NF, 507
VENLAFAXINE HYDROCHLORIDE, 508
*Ventolin,* 13, 14, 15, 16
*VePesid,* 200
VERAPAMIL HYDROCHLORIDE, 509
*Versed,* 331, 332
*Vfend,* 515
*Viagra,* 434
VIDARABINE, 511
*Viokase,* 370
*Viramune,* 350
*Vistaril,* 248
VITAMIN A, 511
Vitamin A alcohol, 511
Vitamin D$_3$, 118
VITAMIN E, 513
Vitamin K$_3$, 306
VORICONAZOLE, 514
VP-16, 199

## W

WARFARIN SODIUM, 516
WHITE LOTION, 517

## X

*Xalatan,* 281, 282
*Xifaxan,* 428
*Xopenex,* 282, 283

## Z

*Zantac,* 424
*Zarontin,* 198
*Zebeta,* 69
*Zegerid,* 361, 363
*Zelnorm,* 458
*Zestril,* 292
ZIDOVUDINE, 518
*Zinacef,* 102, 103
ZINC OXIDE, 518
ZINC SULFATE, 519
*Zofran,* 364, 365
ZONISAMIDE, 519